Rational (Reciprocal) Function

$f(x) = \dfrac{1}{x}$

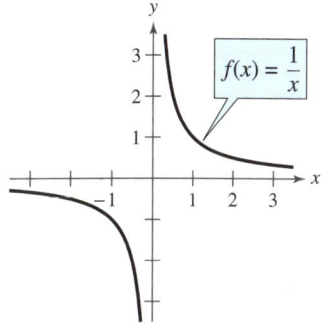

Domain: $(-\infty, 0) \cup (0, \infty)$
Range: $(-\infty, 0) \cup (0, \infty)$
No intercepts
Decreasing on $(-\infty, 0)$ and $(0, \infty)$
Odd function
Origin symmetry
Vertical asymptote: y-axis
Horizontal asymptote: x-axis

Exponential Function

$f(x) = a^x, \; a > 0, \; a \neq 1$

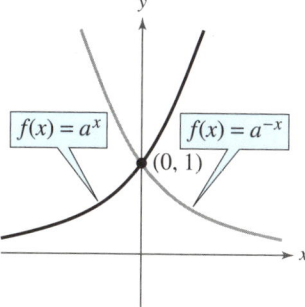

$f(x) = a^x$ $f(x) = a^{-x}$
$(0, 1)$

Domain: $(-\infty, \infty)$
Range: $(0, \infty)$
Intercept: $(0, 1)$
Increasing on $(-\infty, \infty)$
 for $f(x) = a^x$
Decreasing on $(-\infty, \infty)$
 for $f(x) = a^{-x}$
Horizontal asymptote: x-axis
Continuous

Logarithmic Function

$f(x) = \log_a x, \; a > 0, \; a \neq 1$

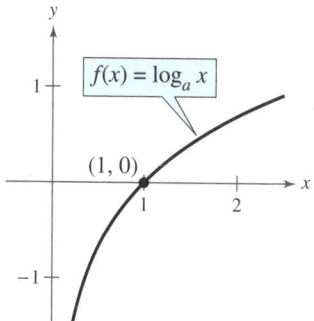

$f(x) = \log_a x$
$(1, 0)$

Domain: $(0, \infty)$
Range: $(-\infty, \infty)$
Intercept: $(1, 0)$
Increasing on $(0, \infty)$
Vertical asymptote: y-axis
Continuous
Reflection of graph of $f(x) = a^x$
 in the line $y = x$

SYMMETRY

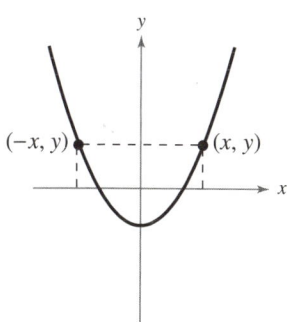

$(-x, y)$ (x, y)

y-Axis Symmetry

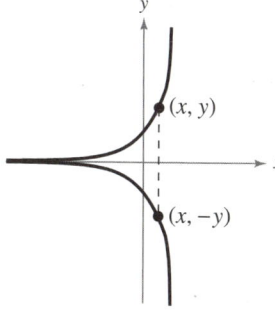

(x, y)
$(x, -y)$

x-Axis Symmetry

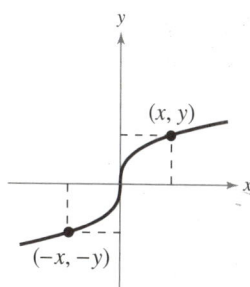

(x, y)
$(-x, -y)$

Origin Symmetry

College Algebra

Seventh Edition

Ron Larson

The Pennsylvania State University
The Behrend College

Robert Hostetler

The Pennsylvania State University
The Behrend College

With the assistance of David C. Falvo

The Pennsylvania State University
The Behrend College

Houghton Mifflin Company   Boston   New York

Publisher : Richard Stratton
Sponsoring Editor: Cathy Cantin
Development Manager: Maureen Ross
Development Editor: Lisa Collette
Editorial Associate: Elizabeth Kassab
Supervising Editor: Karen Carter
Senior Project Editor: Patty Bergin
Editorial Assistant: Julia Keller
Art and Design Manager: Gary Crespo
Executive Marketing Manager: Brenda Bravener-Greville
Director of Manufacturing: Priscilla Manchester
Cover Design Manager: Tony Saizon

Cover Image: Ryuichi Okano/Photonica

Printed in the U.S.A.

Library of Congress Catalog Card Number: 2005929677

Instructor's exam copy;

ISBN 13: 978-0-618-64311-0
ISBN 10: 0-618-64311-7

For orders, use student text ISBNs:

ISBN 13: 978-0-618-64310-3
ISBN 10: 0-618-64310-9

3456789–DOW–10 09 08 07 06

Contents

CONTENTS

A Word from the Authors

Welcome to *College Algebra:* Seventh Edition. We are pleased to present this new edition of our textbook in which we focus on making the mathematics accessible, supporting student success, and offering instructors flexible teaching options.

Accessible to Students

Over the years we have taken care to write this text with the student in mind. Paying careful attention to the presentation, we use precise mathematical language and a clear writing style to develop an effective learning tool. We believe that every student can learn mathematics, and we are committed to providing a text that makes the mathematics of the college algebra course accessible to all students. For the Seventh Edition, we have revised and improved many text features designed for this purpose.

Throughout the text, we now present solutions to many examples from multiple perspectives—algebraically, graphically, and numerically. The side-by-side format of this pedagogical feature helps students to see that a problem can be solved in more than one way and to see that different methods yield the same result. The side-by-side format also addresses many different learning styles.

We have found that many college algebra students grasp mathematical concepts more easily when they work with them in the context of real-life situations. Students have numerous opportunities to do this throughout the Seventh Edition. The new *Make a Decision* feature has been added to the text in order to further connect real-life data and applications and motivate students. They also offer students the opportunity to generate and analyze mathematical models from large data sets. To reinforce the concept of functions, each function is introduced at the first point of use in the text with a definition and description of basic characteristics. Also, all elementary functions are presented in a summary on the endpapers of the text for convenient reference.

We have carefully written and designed each page to make the book more readable and accessible to students. For example, to avoid unnecessary page turning and disruptions to students' thought processes, each example and corresponding solution begins and ends on the same page.

Supports Student Success

During more than 30 years of teaching and writing, we have learned many things about the teaching and learning of mathematics. We have found that students are most successful when they know what they are expected to learn and why it is important to learn the concepts. With that in mind, we have enhanced the thematic study thread throughout the Seventh Edition.

Each chapter begins with a list of applications that are covered in the chapter and that serve as a motivational tool by connecting section content to real-life situations. Using the same pedagogical theme, each section begins with a set of section learning objectives—*What You Should Learn*. These are followed by an engaging real-life application—*Why You Should Learn It*—that motivates students and illustrates an area where the mathematical concepts will be applied in an example or exercise in the section. The *Chapter Summary—What Did You Learn?*—at the end of each chapter is a section-by-section overview that ties the learning objectives from the chapter to sets of *Review Exercises* at the end of each chapter.

Throughout the text, other features further improve accessibility. *Study Tips* are provided throughout the text at point-of-use to reinforce concepts and to help students learn how to study mathematics. *Technology, Writing About Mathematics, Historical Notes,* and *Explorations* have been expanded in order to reinforce mathematical concepts. Each example with worked-out solution is now followed by a *Checkpoint*, which directs the student to work a similar exercise from the exercise set. The *Section Exercises* now begin with a *Vocabulary Check*, which gives the students an opportunity to test their understanding of the important terms in the section. A new *Prerequisite Skills Review* is offered at the beginning of each exercise set. *Synthesis Exercises* check students' conceptual understanding of the topics in each section. The new *Make a Decision* exercises further connect real-life data and applications and motivate students. *Skills Review Exercises* provide additional practice with the concepts in the chapter or previous chapters. *Chapter Tests*, at the end of each chapter, and periodic *Cumulative Tests* offer students frequent opportunities for self-assessment and to develop strong study- and test-taking skills.

The use of technology also supports students with different learning styles. *Technology* notes are provided throughout the text at point-of-use. These notes call attention to the strengths and weaknesses of graphing technology, as well as offer alternative methods for solving or checking a problem using technology. These notes also direct students to the *Graphing Technology Guide*, on the textbook website, for keystroke support that is available for numerous calculator models. The use of technology is optional. This feature and related exercises can be omitted without the loss of continuity in coverage of topics.

Numerous additional text-specific resources are available to help students succeed in the college algebra course. These include "live" online tutoring, instructional DVDs, and a variety of other resources, such as tutorial support and self-assessment, which are available on the HM mathSpace® CD-ROM, the Web, and in Eduspace®. In addition, the *Online Notetaking Guide* is a notetaking guide that helps students organize their class notes and create an effective study and review tool.

Flexible Options for Instructors

From the time we first began writing textbooks in the early 1970s, we have always considered it a critical part of our role as authors to provide instructors with flexible programs. In addition to addressing a variety of learning styles, the optional features within the text allow instructors to design their courses to meet their instructional needs and the needs of their students. For example, the

Explorations throughout the text can be used as a quick introduction to concepts or as a way to reinforce student understanding.

Our goal when developing the exercise sets was to address a wide variety of learning styles and teaching preferences. New to this edition are the *Vocabulary Check* questions, which are provided at the beginning of every exercise set to help students learn proper mathematical terminology. In each exercise set we have included a variety of exercise types, including questions requiring writing and critical thinking, as well as real-data applications. The problems are carefully graded in difficulty from mastery of basic skills to more challenging exercises. Some of the more challenging exercises include the *Synthesis Exercises* that combine skills and are used to check for conceptual understanding and the new *Make a Decision* exercises that further connect real-life data and applications and motivate students. *Skills Review Exercises*, placed at the end of each exercise set, reinforce previously learned skills. In addition, Houghton Mifflin's Eduspace® website offers instructors the option to assign homework and tests online—and also includes the ability to grade these assignments automatically.

Several other print and media resources are also available to support instructors. The *Online Instructor Success Organizer* includes suggested lesson plans and is an especially useful tool for larger departments that want all sections of a course to follow the same outline. The *Instructor's Edition* of the *Student Notetaking Guide* can be used as a lecture outline for every section of the text and includes additional examples for classroom discussion and important definitions. This is another valuable resource for schools trying to have consistent instruction and it can be used as a resource to support less experienced instructors. When used in conjunction with the *Student Notetaking Guide* these resources can save instructors preparation time and help students concentrate on important concepts.

Instructors who stress applications and problem solving, or exploration and technology, coupled with more traditional methods will be able to use this text successfully.

We hope you enjoy the Seventh Edition.

Ron Larson
Robert Hostetler

Acknowledgments

We would like to thank the many people who have helped us prepare the text and the supplements package. Their encouragement, criticisms, and suggestions have been invaluable to us.

Seventh Edition Reviewers

Rick Adkins, *Indiana University of Pennsylvania*; Shane Brewer, *College of Eastern Utah—San Juan Campus*; Susan Brown, *Gadsden State Community College*; Connie Buller, *Metropolitan Community College*; Matt Calhoun, *Southwest Mississippi Community College*; Alex Clark, *University of North Texas*; Tampa Clark, *Jacksonville College*; Jia Feng, *Cape Fear Community College*; Sudhir Goel, *Valdosta State University*; Ruth Henley, *J. Sargeant Reynolds Community College*; Cheryl Kane, *University of Nebraska Lincoln*; Lynette King, *Gadsden State Community College*; Gloria Kittel, *State University of West Georgia*; Barbara Kurt, *Saint Louis Community College—Meramec*; Andreas Lazari, *Valdosta State University*; Shinemin Lin, *Savannah State University*; Kelly McCoun, *McMurry University (TX)*; Carol Poos, *Southwestern Illinois College*; David Ray, *University of Tennessee at Martin*; Michael Sakowski, *Lake Superior College*; Steve Shaff, *Sauk Valley Community College*; Bryan Stewart, *Tarrant County College*; Mark Tom, *College of the Sequoias*; Kathy Wagner, *Big Sandy Community and Technical College, Pikeville Campus*; Ernest Whitmore, *John Brown University*; Kary Williams, *Eastern Maine Community College*; Professor Wilson, *St. Cloud Technical College*; Fred Worthing, *Henderson State University*; Tom Worthing, *Hutchinson Community College*

We would like to thank the staff of Larson Texts, Inc. who assisted in preparing the manuscript, rendering the art package, and typesetting and proofreading the pages and supplements.

On a personal level, we are grateful to our wives, Deanna Gilbert Larson and Eloise Hostetler for their love, patience, and support. Also, a special thanks goes to R. Scott O'Neil.

If you have suggestions for improving this text, please feel free to write us. Over the past three decades we have received many useful comments from both instructors and students, and we value these very much.

Ron Larson
Robert Hostetler

Textbook Features and Highlights

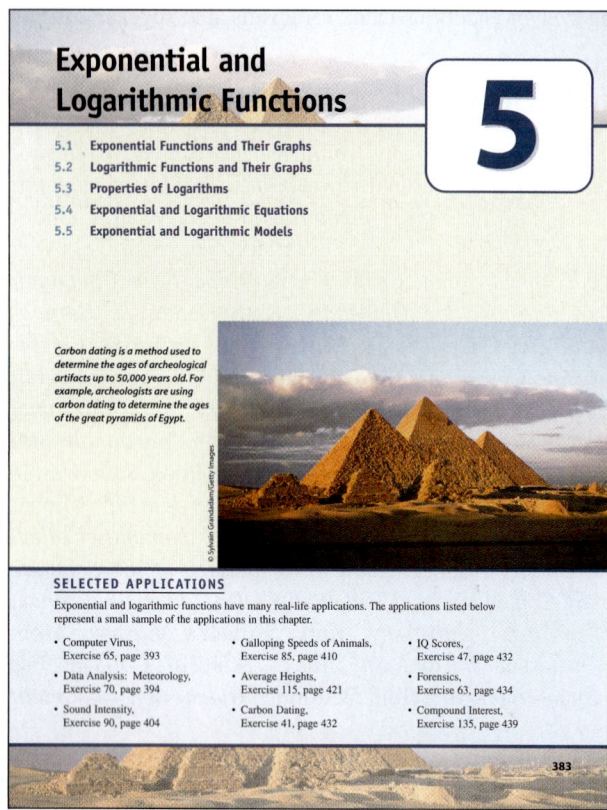

Exponential and Logarithmic Functions

5

5.1 Exponential Functions and Their Graphs
5.2 Logarithmic Functions and Their Graphs
5.3 Properties of Logarithms
5.4 Exponential and Logarithmic Equations
5.5 Exponential and Logarithmic Models

Carbon dating is a method used to determine the ages of archeological artifacts up to 50,000 years old. For example, archeologists are using carbon dating to determine the ages of the great pyramids of Egypt.

SELECTED APPLICATIONS

Exponential and logarithmic functions have many real-life applications. The applications listed below represent a small sample of the applications in this chapter.

- Computer Virus, Exercise 65, page 393
- Data Analysis: Meteorology, Exercise 70, page 394
- Sound Intensity, Exercise 90, page 404
- Galloping Speeds of Animals, Exercise 85, page 410
- Average Heights, Exercise 115, page 421
- Carbon Dating, Exercise 41, page 432
- IQ Scores, Exercise 47, page 432
- Forensics, Exercise 63, page 434
- Compound Interest, Exercise 135, page 439

383

• Chapter Opener

Each chapter begins with a comprehensive overview of the chapter concepts. The photograph and caption illustrate a real-life application of a key concept. Section references help students prepare for the chapter.

• Applications List

An abridged list of applications, covered in the chapter, serve as a motivational tool by connecting section content to real-life situations.

• "What You Should Learn" and "Why You Should Learn It"

Sections begin with *What You Should Learn*, an outline of the main concepts covered in the section, and *Why You Should Learn It*, a real-life application or mathematical reference that illustrates the relevance of the section content.

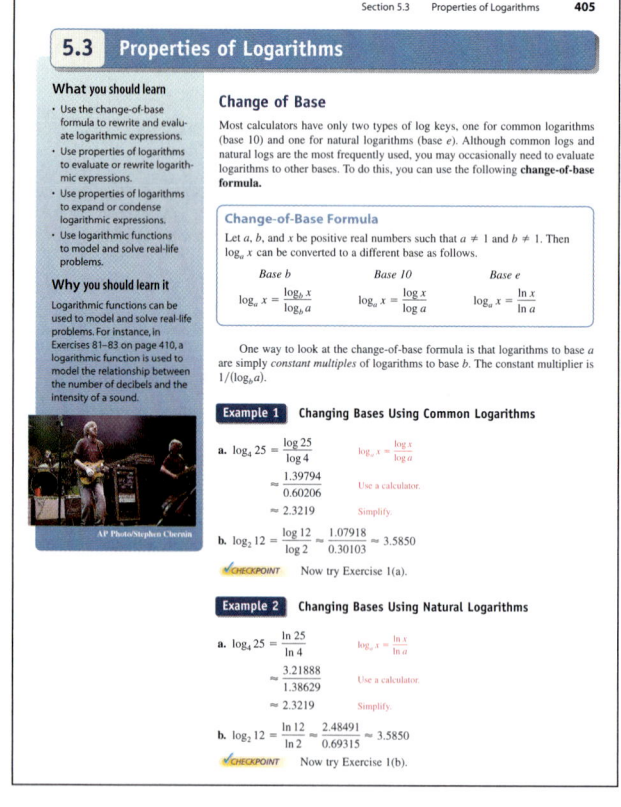

5.3   Properties of Logarithms

What you should learn
- Use the change-of-base formula to rewrite and evaluate logarithmic expressions.
- Use properties of logarithms to evaluate or rewrite logarithmic expressions.
- Use properties of logarithms to expand or condense logarithmic expressions.
- Use logarithmic functions to model and solve real-life problems.

Why you should learn it
Logarithmic functions can be used to model and solve real-life problems. For instance, in Exercises 81–83 on page 410, a logarithmic function is used to model the relationship between the number of decibels and the intensity of a sound.

AP Photo/Stephen Chernin

Change of Base

Most calculators have only two types of log keys, one for common logarithms (base 10) and one for natural logarithms (base e). Although common logs and natural logs are the most frequently used, you may occasionally need to evaluate logarithms to other bases. To do this, you can use the following **change-of-base formula.**

Change-of-Base Formula

Let a, b, and x be positive real numbers such that $a \neq 1$ and $b \neq 1$. Then $\log_a x$ can be converted to a different base as follows.

Base b	Base 10	Base e
$\log_a x = \dfrac{\log_b x}{\log_b a}$	$\log_a x = \dfrac{\log x}{\log a}$	$\log_a x = \dfrac{\ln x}{\ln a}$

One way to look at the change-of-base formula is that logarithms to base a are simply *constant multiples* of logarithms to base b. The constant multiplier is $1/(\log_b a)$.

Example 1   Changing Bases Using Common Logarithms

a. $\log_4 25 = \dfrac{\log 25}{\log 4}$ $\log_a x = \dfrac{\log x}{\log a}$

$\approx \dfrac{1.39794}{0.60206}$ Use a calculator.

≈ 2.3219 Simplify.

b. $\log_2 12 = \dfrac{\log 12}{\log 2} \approx \dfrac{1.07918}{0.30103} \approx 3.5850$

✓**CHECKPOINT**   Now try Exercise 1(a).

Example 2   Changing Bases Using Natural Logarithms

a. $\log_4 25 = \dfrac{\ln 25}{\ln 4}$ $\log_a x = \dfrac{\ln x}{\ln a}$

$\approx \dfrac{3.21888}{1.38629}$ Use a calculator.

≈ 2.3219 Simplify.

b. $\log_2 12 = \dfrac{\ln 12}{\ln 2} \approx \dfrac{2.48491}{0.69315} \approx 3.5850$

✓**CHECKPOINT**   Now try Exercise 1(b).

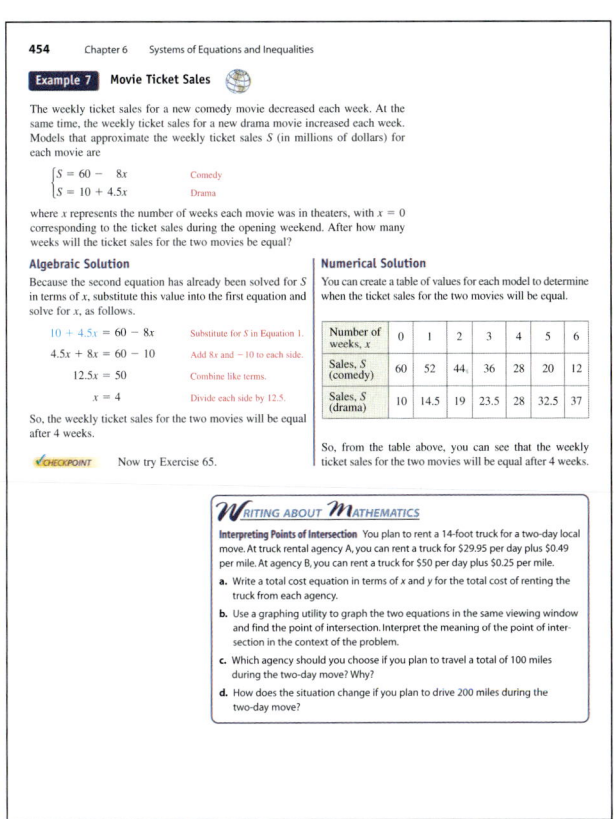

• Examples

Many examples present side-by-side solutions with multiple approaches—algebraic, graphical, and numerical. This format addresses a variety of learning styles and shows students that different solution methods yield the same result.

• Checkpoint

The *Checkpoint* directs students to work a similar problem in the exercise set for extra practice.

• Explorations

The *Exploration* engages students in active discovery of mathematical concepts, strengthens critical thinking skills, and helps them to develop an intuitive understanding of theoretical concepts.

• Study Tips

Study Tips reinforce concepts and help students learn how to study mathematics.

• Technology

The *Technology* feature gives instructions for graphing utilities at point of use.

• Additional Features

Additional carefully crafted learning tools, designed to connect concepts, are placed throughout the text. These learning tools include *Writing About Mathematics*, *Historical Notes,* and an extensive art program.

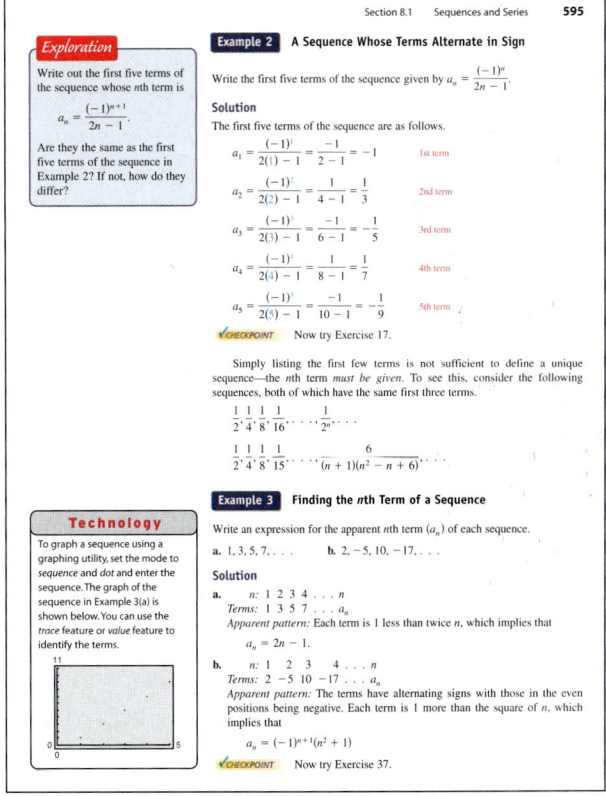

FEATURES

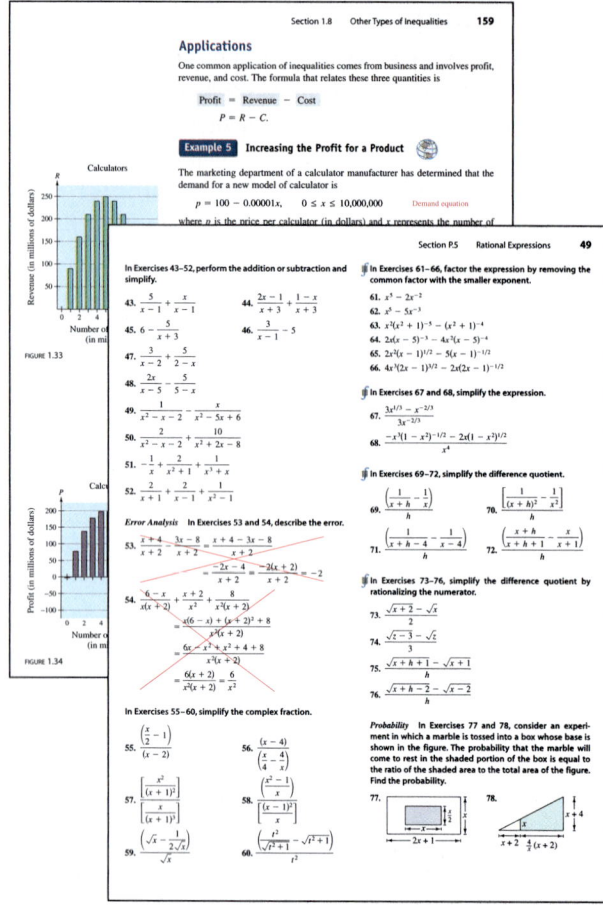

• Real-Life Applications

A wide variety of real-life applications, many using current real data, are integrated throughout the examples and exercises. The 🌐 indicates an example that involves a real-life application.

• Algebra of Calculus

Throughout the text, special emphasis is given to the algebraic techniques used in calculus. Algebra of Calculus examples and exercises are integrated throughout the text and are identified by the symbol ∫.

• Section Exercises

The section exercise sets consist of a variety of computational, conceptual, and applied problems.

• Vocabulary Check

Section exercises begin with a *Vocabulary Check* that serves as a review of the important mathematical terms in each section.

• Prerequisite Skills Review

Extra practice and a review of algebra skills, needed to complete the section exercise sets, are offered to the students and available in Eduspace®.

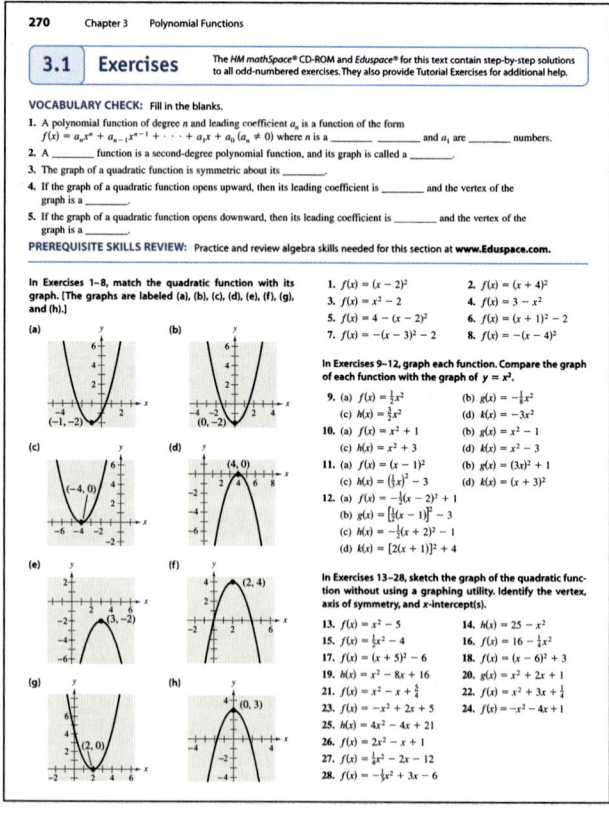

• Model It

These multi-part applications that involve real data offer students the opportunity to generate and analyze mathematical models.

First panel (Section 7.1):

Section 7.1   Matrices and Systems of Equations   **537**

82. *Electrical Network*   The currents in an electrical network are given by the solution of the system

$$\begin{cases} I_1 - I_2 + I_3 = 0 \\ 3I_1 + 4I_2 \quad\ = 18 \\ I_2 + 3I_3 = 6 \end{cases}$$

where I_1, I_2, and I_3 are measured in amperes. Solve the system of equations using matrices.

83. *Partial Fractions*   Use a system of equations to write the partial fraction decomposition of the rational expression. Solve the system using matrices.

$$\frac{4x^2}{(x+1)^2(x-1)} = \frac{A}{x-1} + \frac{B}{x+1} + \frac{C}{(x+1)^2}$$

84. *Partial Fractions*   Use a system of equations to write the partial fraction decomposition of the rational expression. Solve the system using matrices.

$$\frac{8x^2}{(x-1)^2(x+1)} = \frac{A}{x+1} + \frac{B}{x-1} + \frac{C}{(x-1)^2}$$

85. *Finance*   A small shoe corporation borrowed $1,500,000 to expand its line of shoes. Some of the money was borrowed at 7%, some at 8%, and some at 10%. Use a system of equations to determine how much was borrowed at each rate if the annual interest was $130,500 and the amount borrowed at 10% was 4 times the amount borrowed at 7%. Solve the system using matrices.

86. *Finance*   A small software corporation borrowed $500,000 to expand its software line. Some of the money was borrowed at 9%, some at 10%, and some at 12%. Use a system of equations to determine how much was borrowed at each rate if the annual interest was $52,000 and the amount borrowed at 10% was $2\frac{1}{2}$ times the amount borrowed at 9%. Solve the system using matrices.

In Exercises 87 and 88, use a system of equations to find the specified equation that passes through the points. Solve the system using matrices. Use a graphing utility to verify your results.

87. Parabola: **88.** Parabola:
$y = ax^2 + bx + c$ $y = ax^2 + bx + c$

89. *Mathematical Modeling*   A videotape of the path of a ball thrown by a baseball player was analyzed with a grid covering the TV screen. The tape was paused three times, and the position of the ball was measured each time. The coordinates obtained are shown in the table. (x and y are measured in feet.)

Horizontal distance, x	Height, y
0	5.0
15	9.6
30	12.4

(a) Use a system of equations to find the equation of the parabola $y = ax^2 + bx + c$ that passes through the three points. Solve the system using matrices.

(b) Use a graphing utility to graph the parabola.

(c) Graphically approximate the maximum height of the ball and the point at which the ball struck the ground.

(d) Analytically find the maximum height of the ball and the point at which the ball struck the ground.

(e) Compare your results from parts (c) and (d).

Model It

90. *Data Analysis: Snowboarders*   The table shows the numbers of people y (in millions) in the United States who participated in snowboarding for selected years from 1997 to 2001. (Source: National Sporting Goods Association)

Year	Number, y
1997	2.8
1999	3.3
2001	5.3

(a) Use a system of equations to find the equation of the parabola $y = at^2 + bt + c$ that passes through the points. Let t represent the year, with $t = 7$ corresponding to 1997. Solve the system using matrices.

(b) Use a graphing utility to graph the parabola.

(c) Use the equation in part (a) to estimate the number of people who participated in snowboarding in 2003. How does this value compare with the actual 2003 value of 6.3 million?

(d) Use the equation in part (a) to estimate y in the year 2008. Is the estimate reasonable? Explain.

• Synthesis and Skills Review Exercises

Each exercise set concludes with the two types of exercises.

Synthesis exercises promote further exploration of mathematical concepts, critical thinking skills, and writing about mathematics. The exercises require students to show their understanding of the relationships between many concepts in the section.

Skills Review Exercises reinforce previously learned skills and concepts.

Make a Decision exercises, found in selected sections, further connect real-life data and applications and motivate students. They also offer students the opportunity to generate and analyze mathematical models from large data sets.

Second panel (Chapter 5):

394   Chapter 5   Exponential and Logarithmic Functions

Model It

69. *Data Analysis: Biology*   To estimate the amount of defoliation caused by the gypsy moth during a given year, a forester counts the number x of egg masses on $\frac{1}{40}$ of an acre (circle of radius 18.6 feet) in the fall. The percent of defoliation y the next spring is shown in the table. (Source: USDA, Forest Service)

Egg masses, x	Percent of defoliation, y
0	12
25	44
50	81
75	96
100	99

A model for the data is given by

$$y = \frac{100}{1 + 7e^{-0.069x}}.$$

(a) Use a graphing utility to create a scatter plot of the data and graph the model in the same viewing window.

(b) Create a table that compares the model with the sample data.

(c) Estimate the percent of defoliation if 36 egg masses are counted on $\frac{1}{40}$ acre.

(d) You observe that $\frac{2}{3}$ of a forest is defoliated the following spring. Use the graph in part (a) to estimate the number of egg masses per $\frac{1}{40}$ acre.

70. *Data Analysis: Meteorology*   A meteorologist measures the atmospheric pressure P (in pascals) at altitude h (in kilometers). The data are shown in the table.

Altitude, h	Pressure, P
0	101,293
5	54,735
10	23,294
15	12,157
20	5,069

A model for the data is given by $P = 107,428e^{-0.150h}$.

(a) Sketch a scatter plot of the data and graph the model on the same set of axes.

(b) Estimate the atmospheric pressure at a height of 8 kilometers.

Synthesis

True or False?   In Exercises 71 and 72, determine whether the statement is true or false. Justify your answer.

71. The line $y = -2$ is an asymptote for the graph of $f(x) = 10^x - 2$.

72. $e = \dfrac{271,801}{99,990}$

Think About It   In Exercises 73–76, use properties of exponents to determine which functions (if any) are the same.

73. $f(x) = 3^{x-2}$ **74.** $f(x) = 4^x + 12$
$\quad g(x) = 3^x - 9$ $\quad g(x) = 2^{2x+6}$
$\quad h(x) = \frac{1}{9}(3^x)$ $\quad h(x) = 64(4^x)$

75. $f(x) = 16(4^{-x})$ **76.** $f(x) = e^{-x} + 3$
$\quad g(x) = \left(\frac{1}{4}\right)^{x-2}$ $\quad g(x) = e^{3-x}$
$\quad h(x) = 16(2^{-2x})$ $\quad h(x) = -e^{x-3}$

77. Graph the functions given by $y = 3^x$ and $y = 4^x$ and use the graphs to solve each inequality.
(a) $4^x < 3^x$ (b) $4^x > 3^x$

78. Use a graphing utility to graph each function. Use the graph to find where the function is increasing and decreasing, and approximate any relative maximum or minimum values.
(a) $f(x) = x^2 e^{-x}$ (b) $g(x) = x2^{1-x}$

79. *Graphical Analysis*   Use a graphing utility to graph
$$f(x) = \left(1 + \frac{0.5}{x}\right)^x \quad \text{and} \quad g(x) = e^{0.5}$$
in the same viewing window. What is the relationship between f and g as x increases and decreases without bound?

80. *Think About It*   Which functions are exponential?
(a) $3x$ (b) $3x^2$ (c) 3^x (d) 2^{-x}

Skills Review

In Exercises 81 and 82, solve for y.

81. $x^2 + y^2 = 25$ **82.** $x - |y| = 2$

In Exercises 83 and 84, sketch the graph of the function.

83. $f(x) = \dfrac{2}{9+x}$ **84.** $f(x) = \sqrt{7-x}$

85. *Make a Decision*   To work an extended application analyzing the population per square mile of the United States, visit this text's website at college.hmco.com. (Data Source: U.S. Census Bureau)

FEATURES

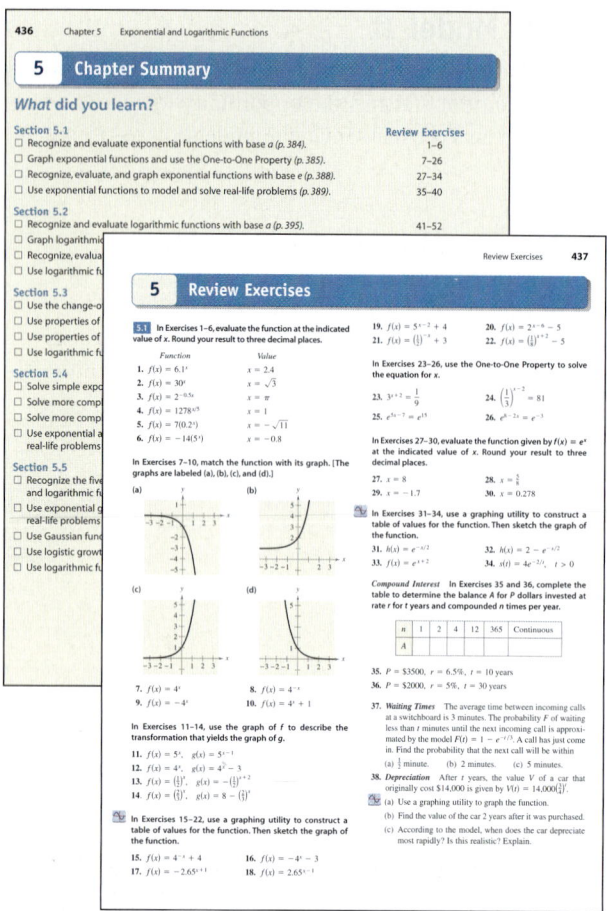

• Chapter Summary

The *Chapter Summary* "*What Did You Learn?*" is a section-by-section overview that ties the learning objectives from the chapter to sets of Review Exercises for extra practice.

• Review Exercises

The chapter *Review Exercises* provide additional practice with the concepts covered in the chapter.

• Chapter Tests and Cumulative Tests

Chapter Tests, at the end of each chapter, and periodic *Cumulative Tests* offer students frequent opportunities for self-assessment and to develop strong study and test-taking skills.

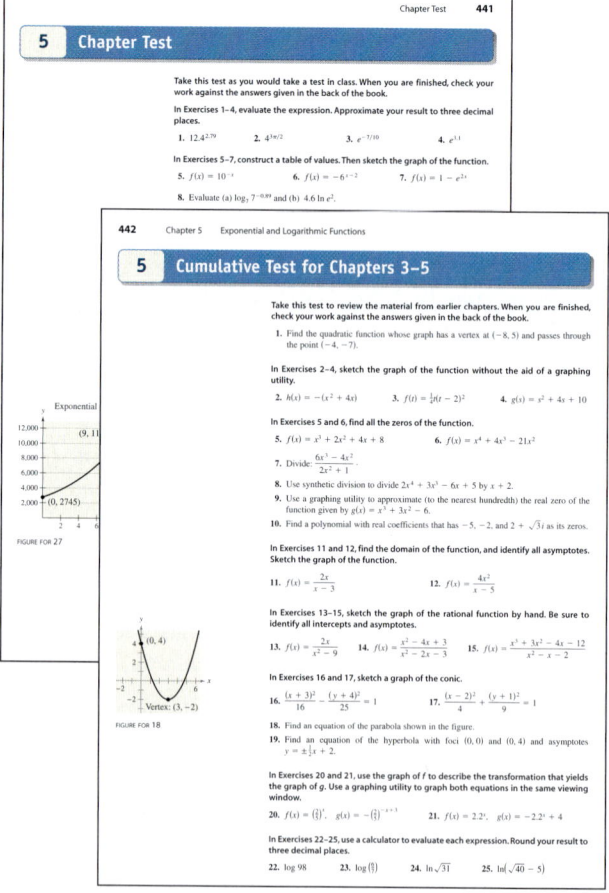

• Proofs in Mathematics

At the end of every chapter, proofs of important mathematical properties and theorems are presented as well as discussions of various proof techniques.

• P.S. Problem Solving

Each chapter concludes with a collection of thought-provoking and challenging exercises that further explore and expand upon the chapter concepts. These exercises have unusual characteristics that set them apart from traditional text exercises.

Proofs in Mathematics

What does the word *proof* mean to you? In mathematics, the word *proof* is used to mean simply a valid argument. When you are proving a statement or theorem, you must use facts, definitions, and accepted properties in a logical order. You can also use previously proved theorems in your proof. For instance, the Distance Formula is used in the proof of the Midpoint Formula below. There are several different proof methods, which you will see in later chapters.

The Midpoint Formula   (p. 61)

The midpoint of the line segment joining the points (x_1, y_1) and (x_2, y_2) is given by the Midpoint Formula

$$\text{Midpoint} = \left(\frac{x_1 + x_2}{2}, \frac{y_1 + y_2}{2}\right).$$

The Cartesian Plane

The Cartesian plane was named after the French mathematician René Descartes (1596–1650). While Descartes was lying in bed, he noticed a fly buzzing around on the square ceiling tiles. He discovered that the position of the fly could be described by which ceiling tile the fly landed on. This led to the development of the Cartesian plane. Descartes felt that a coordinate plane could be used to facilitate description of the positions of objects.

Proof

Using the figure, you must show that $d_1 = d_2$ and $d_1 + d_2 = d_3$.

By the Distance Formula, you obtain

$$d_1 = \sqrt{\left(\frac{x_1 + x_2}{2} - x_1\right)^2 + \left(\frac{y_1 + y_2}{2} - y_1\right)^2}$$

$$= \frac{1}{2}\sqrt{(x_2 - x_1)^2 + (y_2 - y_1)^2}$$

$$d_2 = \sqrt{\left(x_2 - \frac{x_1 + x_2}{2}\right)^2 + \left(y_2 - \frac{y_1 + y_2}{2}\right)^2}$$

$$= \frac{1}{2}\sqrt{(x_2 - x_1)^2 + (y_2 - y_1)^2}$$

$$d_3 = \sqrt{(x_2 - x_1)^2 + (y_2 - y_1)^2}$$

So, it follows that $d_1 = d_2$ and $d_1 + d_2 = d_3$.

74

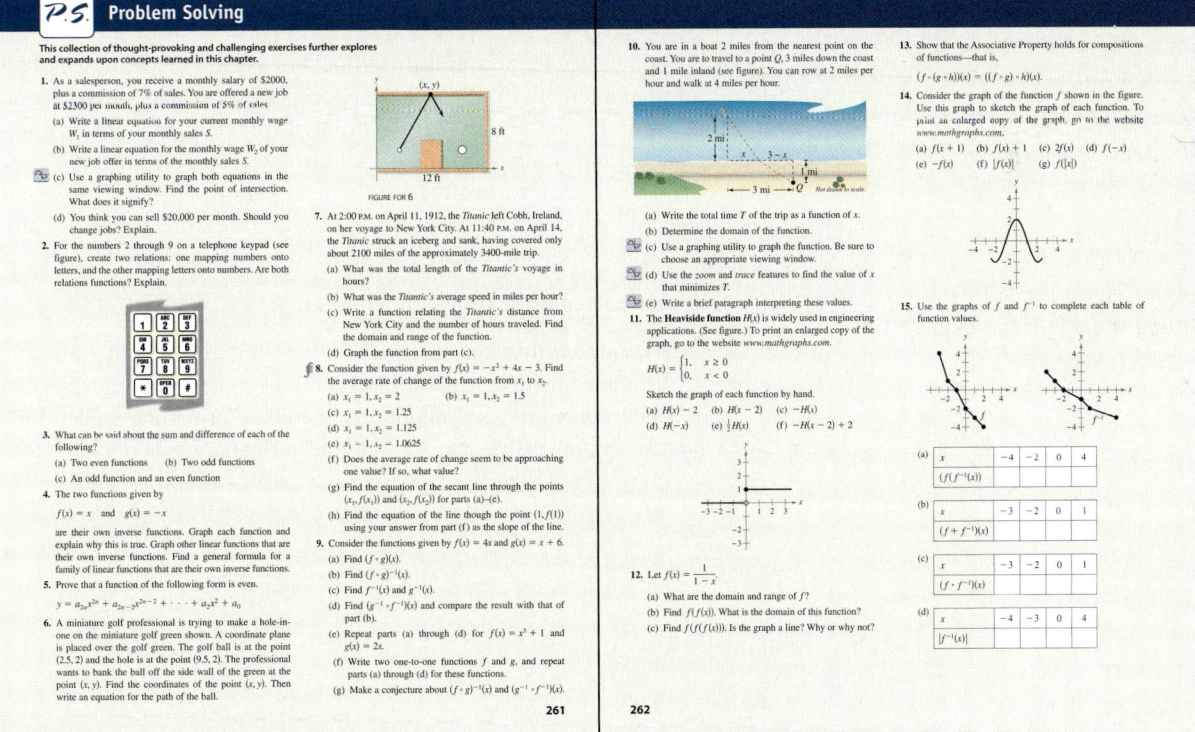

FEATURES

Supplements

Supplements for the Instructor

College Algebra, Seventh Edition, has an extensive support package for the instructor that includes:

Instructor's Annotated Edition (IAE)

Online Complete Solutions Guide

Online Instructor Success Organizer

Online Teaching Center: This free companion website contains an abundance of instructor resources.

HM ClassPrep™ with HM Testing (powered by Diploma™): This CD-ROM is a combination of two course management tools.

- *HM Testing* (powered by *Diploma*™) offers instructors a flexible and powerful tool for test generation and test management. Now supported by the Brownstone Research Group's market-leading *Diploma*™ software, this new version of *HM Testing* significantly improves on functionality and ease of use by offering all the tools needed to create, author, deliver, and customize multiple types of tests—including authoring and editing algorithmic questions. *Diploma*™ is currently in use at thousands of college and university campuses throughout the United States and Canada.

- HM ClassPrep™ also features supplements and text-specific resources for the instructor.

Eduspace®: Eduspace®, powered by Blackboard®, is Houghton Mifflin's customizable and interactive online learning tool. Eduspace® provides instructors with online courses and content. By pairing the widely recognized tools of Blackboard with quality, text-specific content from Houghton Mifflin Company, Eduspace® makes it easy for instructors to create all or part of a course online. This online learning tool also contains ready-to-use homework exercises, quizzes, tests, tutorials, and supplemental study materials.

Visit *www.eduspace.com* for more information.

Eduspace® with eSolutions: Eduspace® with eSolutions combines all the features of Eduspace® with an electronic version of the textbook exercises and the complete solutions to the odd-numbered exercises, providing students with a convenient and comprehensive way to do homework and view course materials.

Supplements for the Student

College Algebra, Seventh Edition, has an extensive support package for the student that includes:

Study and Solutions Guide

Online Student Notetaking Guide

Instructional DVDs

Online Study Center: This free companion website contains an abundance of student resources.

HM mathSpace® CD-ROM: This tutorial CD-ROM provides opportunities for self-paced review and practice with algorithmically generated exercises and step-by-step solutions.

Eduspace®: Eduspace®, powered by Blackboard®, is Houghton Mifflin's customizable and interactive online learning tool for instructors and students. Eduspace® is a text-specific, web-based learning environment that your instructor can use to offer students a combination of practice exercises, multimedia tutorials, video explanations, online algorithmic homework and more. Specific content is available 24 hours a day to help you succeed in your course.

Eduspace® with eSolutions: Eduspace® with eSolutions combines all the features of Eduspace® with an electronic version of the textbook exercises and the complete solutions to the odd-numbered exercises. The result is a convenient and comprehensive way to do homework and view your course materials.

Smarthinking®: Houghton Mifflin has partnered with Smarthinking® to provide an easy-to-use, effective, online tutorial service. Through state-of-the-art tools and whiteboard technology, students communicate in real-time with qualified e-instructors who can help the students understand difficult concepts and guide them through the problem-solving process while studying or completing homework.

Three levels of service are offered to the students.

Live Tutorial Help provides real-time, one-on-one instruction.

Question Submission allows students to submit questions to the tutor outside the scheduled hours and receive a reply usually within 24 hours.

Independent Study Resources connects students around-the-clock to additional educational resources, ranging from interactive websites to Frequently Asked Questions.

Visit *smarthinking.com* for more information.

*Limits apply; terms and hours of SMARTHINKING® service are subject to change.

SUPPLEMENTS

Prerequisites

P

An expanded version of Sections P.1–P.4 is available on the text-specific website at *college.hmco.com* and on the *Interactive* version of this text. This expanded version contains the following sections: Operations with Real Numbers; Properties of Real Numbers; Algebraic Expressions; Operations with Polynomials; Factoring Polynomials; Factoring Trinomials.

You can use the Cartesian plane to organize real-life data such as the number of pieces of mail handled by the U.S. Postal Service each year.

Mary Kate Denny/PhotoEdit

SELECTED APPLICATIONS

Prealgebra concepts have many real-life applications. The applications listed below represent a small sample of the applications in this chapter.

- Budget Variance,
  Exercises 61–64, page 10

- Period of a Pendulum,
  Exercise 105, page 23

- Stopping Distance,
  Exercise 108, page 32

- Interactive Money Management,
  Exercise 84, page 50

- Meteorology,
  Exercise 22, page 65

- Data Analysis: Mail,
  Exercise 65, page 67

- Engineering,
  Exercise 42, page 71

- Astronomy,
  Exercise 8, page 76

P.1   Review of Real Numbers and Their Properties

Real Numbers

Real numbers are used in everyday life to describe quantities such as age, miles per gallon, and population. Real numbers are represented by symbols such as

$$-5, 9, 0, \frac{4}{3}, 0.666 \ldots, 28.21, \sqrt{2}, \pi, \text{ and } \sqrt[3]{-32}.$$

Here are some important **subsets** (each member of subset B is also a member of set A) of the real numbers. The three dots, called *ellipsis points*, indicate that the pattern continues indefinitely.

$$\{1, 2, 3, 4, \ldots\} \qquad \text{Set of natural numbers}$$

$$\{0, 1, 2, 3, 4, \ldots\} \qquad \text{Set of whole numbers}$$

$$\{\ldots, -3, -2, -1, 0, 1, 2, 3, \ldots\} \qquad \text{Set of integers}$$

A real number is **rational** if it can be written as the ratio p/q of two integers, where $q \neq 0$. For instance, the numbers

$$\frac{1}{3} = 0.3333 \ldots = 0.\overline{3}, \frac{1}{8} = 0.125, \text{ and } \frac{125}{111} = 1.126126 \ldots = 1.\overline{126}$$

are rational. The decimal representation of a rational number either repeats $\left(\text{as in } \frac{173}{55} = 3.1\overline{45}\right)$ or terminates $\left(\text{as in } \frac{1}{2} = 0.5\right)$. A real number that cannot be written as the ratio of two integers is called **irrational.** Irrational numbers have infinite nonrepeating decimal representations. For instance, the numbers

$$\sqrt{2} = 1.4142135 \ldots \approx 1.41 \quad \text{and} \quad \pi = 3.1415926 \ldots \approx 3.14$$

are irrational. (The symbol $\approx$ means "is approximately equal to.") Figure P.1 shows subsets of real numbers and their relationships to each other.

Real numbers are represented graphically by a **real number line.** The point 0 on the real number line is the **origin.** Numbers to the right of 0 are positive, and numbers to the left of 0 are negative, as shown in Figure P.2. The term **nonnegative** describes a number that is either positive or zero.

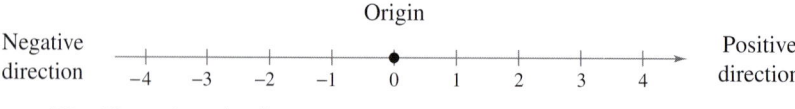

FIGURE P.2   *The real number line*

As illustrated in Figure P.3, there is a *one-to-one correspondence* between real numbers and points on the real number line.

Every real number corresponds to exactly one point on the real number line.

FIGURE P.3   *One-to-one*

Every point on the real number line corresponds to exactly one real number.

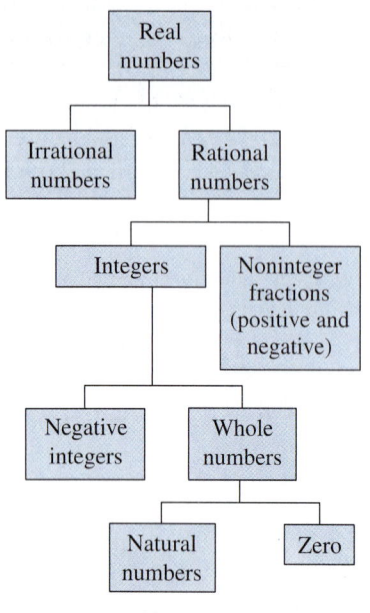

FIGURE P.1   *Subsets of real numbers*

Ordering Real Numbers

One important property of real numbers is that they are *ordered*.

> ### Definition of Order on the Real Number Line
>
> If a and b are real numbers, a is less than b if $b - a$ is positive. The **order** of a and b is denoted by the **inequality** $a < b$. This relationship can also be described by saying that b is *greater than a* and writing $b > a$. The inequality $a \le b$ means that a is *less than or equal to b*, and the inequality $b \ge a$ means that b is *greater than or equal to a*. The symbols $<$, $>$, $\le$, and $\ge$ are *inequality symbols*.

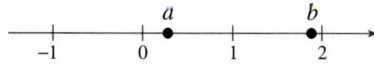

FIGURE **P.4**   *a* < *b if and only if a lies to the left of b.*

Geometrically, this definition implies that $a < b$ if and only if a lies to the *left* of b on the real number line, as shown in Figure P.4.

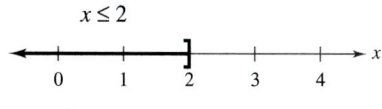

FIGURE **P.5**

Example 1   Interpreting Inequalities

Describe the subset of real numbers represented by each inequality.

a. $x \le 2$ **b.** $-2 \le x < 3$

Solution

a. The inequality $x \le 2$ denotes all real numbers less than or equal to 2, as shown in Figure P.5.

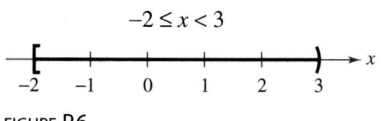

FIGURE **P.6**

b. The inequality $-2 \le x < 3$ means that $x \ge -2$ *and* $x < 3$. This "double inequality" denotes all real numbers between -2 and 3, including -2 but not including 3, as shown in Figure P.6.

✓**CHECKPOINT**   Now try Exercise 19.

Inequalities can be used to describe subsets of real numbers called **intervals.** In the bounded intervals below, the real numbers a and b are the **endpoints** of each interval. The endpoints of a closed interval are included in the interval, whereas the endpoints of an open interval are not included in the interval.

> ### Bounded Intervals on the Real Number Line
>
> | *Notation* | *Interval Type* | *Inequality* | *Graph* |
> |---|---|---|---|
> | $[a, b]$ | Closed | $a \le x \le b$ | |
> | (a, b) | Open | $a < x < b$ | |
> | $[a, b)$ | | $a \le x < b$ | |
> | $(a, b]$ | | $a < x \le b$ | |

STUDY TIP

The reason that the four types of intervals at the right are called *bounded* is that each has a finite length. An interval that does not have a finite length is *unbounded* (see page 4).

The symbols ∞, **positive infinity,** and $-\infty$, **negative infinity,** do not represent real numbers. They are simply convenient symbols used to describe the unboundedness of an interval such as $(1, \infty)$ or $(-\infty, 3]$.

Unbounded Intervals on the Real Number Line

Notation	Interval Type	Inequality	Graph
$[a, \infty)$		$x \geq a$	
(a, ∞)	Open	$x > a$	
$(-\infty, b]$		$x \leq b$	
$(-\infty, b)$	Open	$x < b$	
$(-\infty, \infty)$	Entire real line	$-\infty < x < \infty$	

Example 2   **Using Inequalities to Represent Intervals**

Use inequality notation to describe each of the following.

a. c is at most 2.   **b.** m is at least -3.

c. All x in the interval $(-3, 5]$

Solution

a. The statement "c is at most 2" can be represented by $c \leq 2$.

b. The statement "m is at least -3" can be represented by $m \geq -3$.

c. "All x in the interval $(-3, 5]$" can be represented by $-3 < x \leq 5$.

✔CHECKPOINT   Now try Exercise 31.

Example 3   **Interpreting Intervals**

Give a verbal description of each interval.

a. $(-1, 0)$ **b.** $[2, \infty)$ **c.** $(-\infty, 0)$

Solution

a. This interval consists of all real numbers that are greater than -1 and less than 0.

b. This interval consists of all real numbers that are greater than or equal to 2.

c. This interval consists of all negative real numbers.

✔CHECKPOINT   Now try Exercise 29.

The **Law of Trichotomy** states that for any two real numbers a and b, *precisely* one of three relationships is possible:

$$a = b, \quad a < b, \quad \text{or} \quad a > b. \qquad \text{Law of Trichotomy}$$

Exploration

Absolute value expressions can be evaluated on a graphing utility. When an expression such as $|3 - 8|$ is evaluated, parentheses should surround the expression, as shown below.

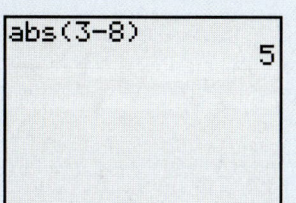

Evaluate each expression. What can you conclude?

a. $|6|$ **b.** $|-1|$

c. $|5 - 2|$ **d.** $|2 - 5|$

Absolute Value and Distance

The **absolute value** of a real number is its *magnitude*, or the distance between the origin and the point representing the real number on the real number line.

> ### Definition of Absolute Value
>
> If a is a real number, then the absolute value of a is
>
> $$|a| = \begin{cases} a, & \text{if } a \geq 0 \\ -a, & \text{if } a < 0 \end{cases}.$$

Notice in this definition that the absolute value of a real number is never negative. For instance, if $a = -5$, then $|-5| = -(-5) = 5$. The absolute value of a real number is either positive or zero. Moreover, 0 is the only real number whose absolute value is 0. So, $|0| = 0$.

Example 4   Evaluating the Absolute Value of a Number

Evaluate $\dfrac{|x|}{x}$ for (a) $x > 0$ and (b) $x < 0$.

Solution

a. If $x > 0$, then $|x| = x$ and $\dfrac{|x|}{x} = \dfrac{x}{x} = 1$.

b. If $x < 0$, then $|x| = -x$ and $\dfrac{|x|}{x} = \dfrac{-x}{x} = -1$.

✔CHECKPOINT   Now try Exercise 47.

> ### Properties of Absolute Values
>
> **1.** $|a| \geq 0$ **2.** $|-a| = |a|$
>
> **3.** $|ab| = |a||b|$ **4.** $\left|\dfrac{a}{b}\right| = \dfrac{|a|}{|b|}, \quad b \neq 0$

Absolute value can be used to define the distance between two points on the real number line. For instance, the distance between -3 and 4 is

$$|-3 - 4| = |-7|$$
$$= 7$$

as shown in Figure P.7.

FIGURE P.7   *The distance between -3 and 4 is 7.*

> ### Distance Between Two Points on the Real Number Line
>
> Let a and b be real numbers. The **distance between a and b** is
>
> $$d(a, b) = |b - a| = |a - b|.$$

Algebraic Expressions

One characteristic of algebra is the use of letters to represent numbers. The letters are **variables,** and combinations of letters and numbers are **algebraic expressions.** Here are a few examples of algebraic expressions.

$$5x, \qquad 2x - 3, \qquad \frac{4}{x^2 + 2}, \qquad 7x + y$$

> ### Definition of an Algebraic Expression
>
> An **algebraic expression** is a collection of letters **(variables)** and real numbers **(constants)** combined using the operations of addition, subtraction, multiplication, division, and exponentiation.

The **terms** of an algebraic expression are those parts that are separated by *addition.* For example,

$$x^2 - 5x + 8 = x^2 + (-5x) + 8$$

has three terms: x^2 and $-5x$ are the **variable terms** and 8 is the **constant term.** The numerical factor of a variable term is the **coefficient** of the variable term. For instance, the coefficient of $-5x$ is -5, and the coefficient of x^2 is 1.

To **evaluate** an algebraic expression, substitute numerical values for each of the variables in the expression. Here are two examples.

Expression	Value of Variable	Substitute	Value of Expression
$-3x + 5$	$x = 3$	$-3(3) + 5$	$-9 + 5 = -4$
$3x^2 + 2x - 1$	$x = -1$	$3(-1)^2 + 2(-1) - 1$	$3 - 2 - 1 = 0$

When an algebraic expression is evaluated, the **Substitution Principle** is used. It states that "If $a = b$, then a can be replaced by b in any expression involving a." In the first evaluation shown above, for instance, 3 is *substituted* for x in the expression $-3x + 5$.

Basic Rules of Algebra

There are four arithmetic operations with real numbers: *addition, multiplication, subtraction,* and *division,* denoted by the symbols $+$, $\times$ or $\cdot$, $-$, and $\div$ or $/$. Of these, addition and multiplication are the two primary operations. Subtraction and division are the inverse operations of addition and multiplication, respectively.

> ### Definitions of Subtraction and Division
>
> **Subtraction:** Add the opposite.   **Division:** Multiply by the reciprocal.
>
> $$a - b = a + (-b) \qquad\qquad \text{If } b \neq 0, \text{ then } a/b = a\left(\frac{1}{b}\right) = \frac{a}{b}.$$
>
> In these definitions, $-b$ is the **additive inverse** (or opposite) of b, and $1/b$ is the **multiplicative inverse** (or reciprocal) of b. In the fractional form a/b, a is the **numerator** of the fraction and b is the **denominator.**

Because the properties of real numbers below are true for variables and algebraic expressions as well as for real numbers, they are often called the **Basic Rules of Algebra.** Try to formulate a verbal description of each property. For instance, the first property states that *the order in which two real numbers are added does not affect their sum.*

Basic Rules of Algebra

Let a, b, and c be real numbers, variables, or algebraic expressions.

Property		*Example*
Commutative Property of Addition:	$a + b = b + a$	$4x + x^2 = x^2 + 4x$
Commutative Property of Multiplication:	$ab = ba$	$(4 - x)x^2 = x^2(4 - x)$
Associative Property of Addition:	$(a + b) + c = a + (b + c)$	$(x + 5) + x^2 = x + (5 + x^2)$
Associative Property of Multiplication:	$(ab)c = a(bc)$	$(2x \cdot 3y)(8) = (2x)(3y \cdot 8)$
Distributive Properties:	$a(b + c) = ab + ac$	$3x(5 + 2x) = 3x \cdot 5 + 3x \cdot 2x$
	$(a + b)c = ac + bc$	$(y + 8)y = y \cdot y + 8 \cdot y$
Additive Identity Property:	$a + 0 = a$	$5y^2 + 0 = 5y^2$
Multiplicative Identity Property:	$a \cdot 1 = a$	$(4x^2)(1) = 4x^2$
Additive Inverse Property:	$a + (-a) = 0$	$5x^3 + (-5x^3) = 0$
Multiplicative Inverse Property:	$a \cdot \dfrac{1}{a} = 1, \quad a \neq 0$	$(x^2 + 4)\left(\dfrac{1}{x^2 + 4}\right) = 1$

Because subtraction is defined as "adding the opposite," the Distributive Properties are also true for subtraction. For instance, the "subtraction form" of $a(b + c) = ab + ac$ is $a(b - c) = ab - ac$.

Properties of Negation and Equality

Let a and b be real numbers, variables, or algebraic expressions.

Property	*Example*
1. $(-1)a = -a$	$(-1)7 = -7$
2. $-(-a) = a$	$-(-6) = 6$
3. $(-a)b = -(ab) = a(-b)$	$(-5)3 = -(5 \cdot 3) = 5(-3)$
4. $(-a)(-b) = ab$	$(-2)(-x) = 2x$
5. $-(a + b) = (-a) + (-b)$	$-(x + 8) = (-x) + (-8)$
	$\qquad\qquad = -x - 8$
6. If $a = b$, then $a \pm c = b \pm c$.	$\frac{1}{2} + 3 = 0.5 + 3$
7. If $a = b$, then $ac = bc$.	$4^2 \cdot 2 = 16 \cdot 2$
8. If $a \pm c = b \pm c$, then $a = b$.	$1.4 - 1 = \frac{7}{5} - 1 \Longrightarrow 1.4 = \frac{7}{5}$
9. If $ac = bc$ and $c \neq 0$, then $a = b$.	$3x = 3 \cdot 4 \Longrightarrow x = 4$

STUDY TIP

The "or" in the Zero-Factor Property includes the possibility that either or both factors may be zero. This is an **inclusive or,** and it is the way the word "or" is generally used in mathematics.

Properties of Zero

Let a and b be real numbers, variables, or algebraic expressions.

1. $a + 0 = a$ and $a - 0 = a$ **2.** $a \cdot 0 = 0$

3. $\dfrac{0}{a} = 0,$ $a \neq 0$ **4.** $\dfrac{a}{0}$ is undefined.

5. Zero-Factor Property: If $ab = 0$, then $a = 0$ or $b = 0$.

Properties and Operations of Fractions

Let a, b, c, and d be real numbers, variables, or algebraic expressions such that $b \neq 0$ and $d \neq 0$.

1. Equivalent Fractions: $\dfrac{a}{b} = \dfrac{c}{d}$ if and only if $ad = bc$.

2. Rules of Signs: $-\dfrac{a}{b} = \dfrac{-a}{b} = \dfrac{a}{-b}$ and $\dfrac{-a}{-b} = \dfrac{a}{b}$

3. Generate Equivalent Fractions: $\dfrac{a}{b} = \dfrac{ac}{bc},$ $c \neq 0$

4. Add or Subtract with Like Denominators: $\dfrac{a}{b} \pm \dfrac{c}{b} = \dfrac{a \pm c}{b}$

5. Add or Subtract with Unlike Denominators: $\dfrac{a}{b} \pm \dfrac{c}{d} = \dfrac{ad \pm bc}{bd}$

6. Multiply Fractions: $\dfrac{a}{b} \cdot \dfrac{c}{d} = \dfrac{ac}{bd}$

7. Divide Fractions: $\dfrac{a}{b} \div \dfrac{c}{d} = \dfrac{a}{b} \cdot \dfrac{d}{c} = \dfrac{ad}{bc},$ $c \neq 0$

STUDY TIP

In Property 1 of fractions, the phrase "if and only if" implies two statements. One statement is: If $a/b = c/d$, then $ad = bc$. The other statement is: If $ad = bc$, where $b \neq 0$ and $d \neq 0$, then $a/b = c/d$.

Example 5   **Properties and Operations of Fractions**

a. Equivalent fractions: $\dfrac{x}{5} = \dfrac{3 \cdot x}{3 \cdot 5} = \dfrac{3x}{15}$ **b.** Divide fractions: $\dfrac{7}{x} \div \dfrac{3}{2} = \dfrac{7}{x} \cdot \dfrac{2}{3} = \dfrac{14}{3x}$

c. Add fractions with unlike denominators: $\dfrac{x}{3} + \dfrac{2x}{5} = \dfrac{5 \cdot x + 3 \cdot 2x}{3 \cdot 5} = \dfrac{11x}{15}$

✓**CHECKPOINT**   Now try Exercise 103.

If a, b, and c are integers such that $ab = c$, then a and b are **factors** or **divisors** of c. A **prime number** is an integer that has exactly two positive factors—itself and 1—such as 2, 3, 5, 7, and 11. The numbers 4, 6, 8, 9, and 10 are **composite** because each can be written as the product of two or more prime numbers. The number 1 is neither prime nor composite. The **Fundamental Theorem of Arithmetic** states that every positive integer greater than 1 can be written as the product of prime numbers in precisely one way (disregarding order). For instance, the *prime factorization* of 24 is $24 = 2 \cdot 2 \cdot 2 \cdot 3$.

P.1   Exercises

The *HM mathSpace®* CD-ROM and *Eduspace®* for this text contain step-by-step solutions to all odd-numbered exercises. They also provide Tutorial Exercises for additional help.

VOCABULARY CHECK: Fill in the blanks.

1. A real number is _____ if it can be written as the ratio $\dfrac{p}{q}$ of two integers, where $q \neq 0$.

2. _____ numbers have infinite nonrepeating decimal representations.

3. The distance between a point on the real number line and the origin is the _____ _____ of the real number.

4. A number that can be written as the product of two or more prime numbers is called a _____ number.

5. An integer that has exactly two positive factors, the integer itself and 1, is called a _____ number.

6. An algebraic expression is a collection of letters called _____ and real numbers called _____.

7. The _____ of an algebraic expression are those parts separated by addition.

8. The numerical factor of a variable term is the _____ of the variable term.

9. The _____ _____ states that if $ab = 0$, then $a = 0$ or $b = 0$.

In Exercises 1–6, determine which numbers in the set are (a) natural numbers, (b) whole numbers, (c) integers, (d) rational numbers, and (e) irrational numbers.

1. $-9, -\frac{7}{2}, 5, \frac{2}{3}, \sqrt{2}, 0, 1, -4, 2, -11$

2. $\sqrt{5}, -7, -\frac{7}{3}, 0, 3.12, \frac{5}{4}, -3, 12, 5$

3. $2.01, 0.666 \ldots, -13, 0.010110111 \ldots, 1, -6$

4. $2.3030030003 \ldots, 0.7575, -4.63, \sqrt{10}, -75, 4$

5. $-\pi, -\frac{1}{3}, \frac{6}{3}, \frac{1}{2}\sqrt{2}, -7.5, -1, 8, -22$

6. $25, -17, -\frac{12}{5}, \sqrt{9}, 3.12, \frac{1}{2}\pi, 7, -11.1, 13$

In Exercises 7–10, use a calculator to find the decimal form of the rational number. If it is a nonterminating decimal, write the repeating pattern.

7. $\frac{5}{8}$

8. $\frac{1}{3}$

9. $\frac{41}{333}$

10. $\frac{6}{11}$

In Exercises 11 and 12, approximate the numbers and place the correct symbol (< or >) between them.

11.
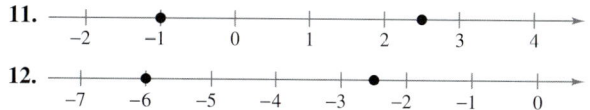

12.

In Exercises 13–18, plot the two real numbers on the real number line. Then place the appropriate inequality symbol (< or >) between them.

13. $-4, -8$

14. $-3.5, 1$

15. $\frac{3}{2}, 7$

16. $1, \frac{16}{3}$

17. $\frac{5}{6}, \frac{2}{3}$

18. $-\frac{8}{7}, -\frac{3}{7}$

In Exercises 19–30, (a) give a verbal description of the subset of real numbers represented by the inequality or the interval, (b) sketch the subset on the real number line, and (c) state whether the interval is bounded or unbounded.

19. $x \leq 5$

20. $x \geq -2$

21. $x < 0$

22. $x > 3$

23. $[4, \infty)$

24. $(-\infty, 2)$

25. $-2 < x < 2$

26. $0 \leq x \leq 5$

27. $-1 \leq x < 0$

28. $0 < x \leq 6$

29. $[-2, 5)$

30. $(-1, 2]$

In Exercises 31–38, use inequality notation to describe the set.

31. All x in the interval $(-2, 4]$

32. All y in the interval $[-6, 0)$

33. y is nonnegative.

34. y is no more than 25.

35. t is at least 10 and at most 22.

36. k is less than 5 but no less than -3.

37. The dog's weight W is more than 65 pounds.

38. The annual rate of inflation r is expected to be at least 2.5% but no more than 5%.

In Exercises 39–48, evaluate the expression.

39. $|-10|$

40. $|0|$

41. $|3 - 8|$

42. $|4 - 1|$

43. $|-1| - |-2|$

44. $-3 - |-3|$

45. $\dfrac{-5}{|-5|}$

46. $-3|-3|$

47. $\dfrac{|x + 2|}{x + 2}, \quad x < -2$

48. $\dfrac{|x - 1|}{x - 1}, \quad x > 1$

In Exercises 49–54, place the correct symbol (<, >, or =) between the pair of real numbers.

49. $|-3|$ [ ] $-|-3|$

50. $|-4|$ [ ] $|4|$

51. -5 [ ] $-|5|$

52. $-|-6|$ [ ] $|-6|$

53. $-|-2|$ [ ] $-|2|$

54. $-(-2)$ [ ] -2

In Exercises 55–60, find the distance between a and b.

55. $a = 126, b = 75$

56. $a = -126, b = -75$

57. $a = -\frac{5}{2}, b = 0$

58. $a = \frac{1}{4}, b = \frac{11}{4}$

59. $a = \frac{16}{5}, b = \frac{112}{75}$

60. $a = 9.34, b = -5.65$

Budget Variance In Exercises 61–64, the accounting department of a sports drink bottling company is checking to see whether the actual expenses of a department differ from the budgeted expenses by more than $500 or by more than 5%. Fill in the missing parts of the table, and determine whether each actual expense passes the "budget variance test."

	Budgeted Expense, b	Actual Expense, a	$|a-b|$	$0.05b$
61. Wages	$112,700	$113,356		
62. Utilities	$9,400	$9,772		
63. Taxes	$37,640	$37,335		
64. Insurance	$2,575	$2,613		

Model It

65. *Federal Deficit* The bar graph shows the federal government receipts (in billions of dollars) for selected years from 1960 through 2000. (Source: U.S. Office of Management and Budget)

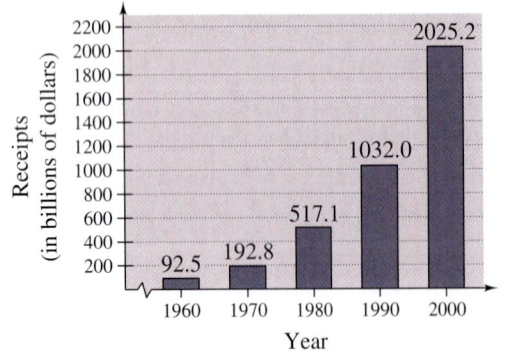

(a) Complete the table. (*Hint:* Find |Receipts − Expenditures|.)

Model It (continued)

Year	Expenditures (in billions)	Surplus or deficit (in billions)
1960	$92.2	
1970	$195.6	
1980	$590.9	
1990	$1253.2	
2000	$1788.8	

(b) Use the table in part (a) to construct a bar graph showing the magnitude of the surplus or deficit for each year.

66. *Veterans* The table shows the number of living veterans (in thousands) in the United States in 2002 by age group. Construct a circle graph showing the percent of living veterans by age group as a fraction of the total number of living veterans. (Source: Department of Veteran Affairs)

Age group	Number of veterans
Under 35	2213
35–44	3290
45–54	4666
55–64	5665
65 and older	9784

In Exercises 67–72, use absolute value notation to describe the situation.

67. The distance between x and 5 is no more than 3.

68. The distance between x and -10 is at least 6.

69. y is at least six units from 0.

70. y is at most two units from a.

71. While traveling on the Pennsylvania Turnpike, you pass milepost 326 near Valley Forge, then milepost 351 near Philadelphia. How many miles do you travel during that time period?

72. The temperature in Chicago, Illinois was 48° last night at midnight, then 82° at noon today. What was the change in temperature over the 12-hour period?

In Exercises 73–78, identify the terms. Then identify the coefficients of the variable terms of the expression.

73. $7x + 4$

74. $6x^3 - 5x$

75. $\sqrt{3}x^2 - 8x - 11$

76. $3\sqrt{3}x^2 + 1$

77. $4x^3 + \dfrac{x}{2} - 5$

78. $3x^4 - \dfrac{x^2}{4}$

In Exercises 79–84, evaluate the expression for each value of x. (If not possible, state the reason.)

Expression   *Values*

79. $4x - 6$ (a) $x = -1$ (b) $x = 0$

80. $9 - 7x$ (a) $x = -3$ (b) $x = 3$

81. $x^2 - 3x + 4$ (a) $x = -2$ (b) $x = 2$

82. $-x^2 + 5x - 4$ (a) $x = -1$ (b) $x = 1$

83. $\dfrac{x + 1}{x - 1}$ (a) $x = 1$ (b) $x = -1$

84. $\dfrac{x}{x + 2}$ (a) $x = 2$ (b) $x = -2$

In Exercises 85–96, identify the rule(s) of algebra illustrated by the statement.

85. $x + 9 = 9 + x$

86. $2\left(\frac{1}{2}\right) = 1$

87. $\dfrac{1}{h + 6}(h + 6) = 1, \quad h \neq -6$

88. $(x + 3) - (x + 3) = 0$

89. $2(x + 3) = 2 \cdot x + 2 \cdot 3$

90. $(z - 2) + 0 = z - 2$

91. $1 \cdot (1 + x) = 1 + x$

92. $(z + 5)x = z \cdot x + 5 \cdot x$

93. $x + (y + 10) = (x + y) + 10$

94. $x(3y) = (x \cdot 3)y = (3x)y$

95. $3(t - 4) = 3 \cdot t - 3 \cdot 4$

96. $\frac{1}{7}(7 \cdot 12) = \left(\frac{1}{7} \cdot 7\right)12 = 1 \cdot 12 = 12$

In Exercises 97–104, perform the operation(s). (Write fractional answers in simplest form.)

97. $\frac{3}{16} + \frac{5}{16}$

98. $\frac{6}{7} - \frac{4}{7}$

99. $\frac{5}{8} - \frac{5}{12} + \frac{1}{6}$

100. $\frac{10}{11} + \frac{6}{33} - \frac{13}{66}$

101. $12 \div \frac{1}{4}$

102. $-\left(6 \cdot \frac{4}{8}\right)$

103. $\dfrac{2x}{3} - \dfrac{x}{4}$

104. $\dfrac{5x}{6} \cdot \dfrac{2}{9}$

105. (a) Use a calculator to complete the table.

n	1	0.5	0.01	0.0001	0.000001
$5/n$					

(b) Use the result from part (a) to make a conjecture about the value of $5/n$ as n approaches 0.

106. (a) Use a calculator to complete the table.

n	1	10	100	10,000	100,000
$5/n$					

(b) Use the result from part (a) to make a conjecture about the value of $5/n$ as n increases without bound.

Synthesis

True or False? In Exercises 107 and 108, determine whether the statement is true or false. Justify your answer.

107. If $a < b$, then $\dfrac{1}{a} < \dfrac{1}{b}$, where $a \neq b \neq 0$.

108. Because $\dfrac{a + b}{c} = \dfrac{a}{c} + \dfrac{b}{c}$, then $\dfrac{c}{a + b} = \dfrac{c}{a} + \dfrac{c}{b}$.

109. *Exploration* Consider $|u + v|$ and $|u| + |v|$, where $u \neq v \neq 0$.

(a) Are the values of the expressions always equal? If not, under what conditions are they unequal?

(b) If the two expressions are not equal for certain values of u and v, is one of the expressions always greater than the other? Explain.

110. *Think About It* Is there a difference between saying that a real number is positive and saying that a real number is nonnegative? Explain.

111. *Think About It* Because every even number is divisible by 2, is it possible that there exist any even prime numbers? Explain.

112. *Writing* Describe the differences among the sets of natural numbers, whole numbers, integers, rational numbers, and irrational numbers.

In Exercises 113 and 114, use the real numbers A, B, and C shown on the number line. Determine the sign of each expression.

113. (a) $-A$

(b) $B - A$

114. (a) $-C$

(b) $A - C$

115. *Writing* Can it ever be true that $|a| = -a$ for a real number a? Explain.

P.2 Exponents and Radicals

What you should learn

- Use properties of exponents.
- Use scientific notation to represent real numbers.
- Use properties of radicals.
- Simplify and combine radicals.
- Rationalize denominators and numerators.
- Use properties of rational exponents.

Why you should learn it

Real numbers and algebraic expressions are often written with exponents and radicals. For instance, in Exercise 107 on page 23, you will use an expression involving rational exponents to find the time required for a funnel to empty for different water heights.

Integer Exponents

Repeated *multiplication* can be written in **exponential form.**

Repeated Multiplication	Exponential Form
$a \cdot a \cdot a \cdot a \cdot a$	a^5
$(-4)(-4)(-4)$	$(-4)^3$
$(2x)(2x)(2x)(2x)$	$(2x)^4$

Exponential Notation

If a is a real number and n is a positive integer, then

$$a^n = \underbrace{a \cdot a \cdot a \cdots a}_{n \text{ factors}}$$

where n is the **exponent** and a is the **base.** The expression a^n is read "a to the nth **power.**"

An exponent can also be negative. In Property 3 below, be sure you see how to use a negative exponent.

Properties of Exponents

Let a and b be real numbers, variables, or algebraic expressions, and let m and n be integers. (All denominators and bases are nonzero.)

Property	Example
1. $a^m a^n = a^{m+n}$	$3^2 \cdot 3^4 = 3^{2+4} = 3^6 = 729$
2. $\dfrac{a^m}{a^n} = a^{m-n}$	$\dfrac{x^7}{x^4} = x^{7-4} = x^3$
3. $a^{-n} = \dfrac{1}{a^n} = \left(\dfrac{1}{a}\right)^n$	$y^{-4} = \dfrac{1}{y^4} = \left(\dfrac{1}{y}\right)^4$
4. $a^0 = 1, \quad a \neq 0$	$(x^2 + 1)^0 = 1$
5. $(ab)^m = a^m b^m$	$(5x)^3 = 5^3 x^3 = 125 x^3$
6. $(a^m)^n = a^{mn}$	$(y^3)^{-4} = y^{3(-4)} = y^{-12} = \dfrac{1}{y^{12}}$
7. $\left(\dfrac{a}{b}\right)^m = \dfrac{a^m}{b^m}$	$\left(\dfrac{2}{x}\right)^3 = \dfrac{2^3}{x^3} = \dfrac{8}{x^3}$
8. $\lvert a^2 \rvert = \lvert a \rvert^2 = a^2$	$\lvert (-2)^2 \rvert = \lvert -2 \rvert^2 = (2)^2 = 4$

Technology

You can use a calculator to evaluate exponential expressions. When doing so, it is important to know when to use parentheses because the calculator follows the order of operations. For instance, evaluate $(-2)^4$ as follows

Scientific:

(2 +/−) y^x 4 =

Graphing:

((−) 2) ^ 4 ENTER

The display will be 16. If you omit the parentheses, the display will be −16.

It is important to recognize the difference between expressions such as $(-2)^4$ and -2^4. In $(-2)^4$, the parentheses indicate that the exponent applies to the negative sign as well as to the 2, but in $-2^4 = -(2^4)$, the exponent applies only to the 2. So, $(-2)^4 = 16$ and $-2^4 = -16$.

The properties of exponents listed on the preceding page apply to *all* integers m and n, not just to positive integers as shown in the examples in this section.

Example 1   Using Properties of Exponents

Use the properties of exponents to simplify each expression.

a. $(-3ab^4)(4ab^{-3})$ **b.** $(2xy^2)^3$ **c.** $3a(-4a^2)^0$ **d.** $\left(\dfrac{5x^3}{y}\right)^2$

Solution

a. $(-3ab^4)(4ab^{-3}) = (-3)(4)(a)(a)(b^4)(b^{-3}) = -12a^2b$

b. $(2xy^2)^3 = 2^3(x)^3(y^2)^3 = 8x^3y^6$

c. $3a(-4a^2)^0 = 3a(1) = 3a, \quad a \neq 0$

d. $\left(\dfrac{5x^3}{x}\right)^2 = \dfrac{5^2(x^3)^2}{y^2} = \dfrac{25x^6}{y^2}$

✔CHECKPOINT   Now try Exercise 25.

Example 2   Rewriting with Positive Exponents

Rewrite each expression with positive exponents.

a. x^{-1} **b.** $\dfrac{1}{3x^{-2}}$ **c.** $\dfrac{12a^3b^{-4}}{4a^{-2}b}$ **d.** $\left(\dfrac{3x^2}{y}\right)^{-2}$

Solution

a. $x^{-1} = \dfrac{1}{x}$ Property 3

b. $\dfrac{1}{3x^{-2}} = \dfrac{1(x^2)}{3} = \dfrac{x^2}{3}$ The exponent -2 does not apply to 3.

c. $\dfrac{12a^3b^{-4}}{4a^{-2}b} = \dfrac{12a^3 \cdot a^2}{4b \cdot b^4} = \dfrac{3a^5}{b^5}$ Properties 3 and 1

d. $\left(\dfrac{3x^2}{y}\right)^{-2} = \dfrac{3^{-2}(x^2)^{-2}}{y^{-2}}$ Properties 5 and 7

$\qquad\qquad = \dfrac{3^{-2}x^{-4}}{y^{-2}}$ Property 6

$\qquad\qquad = \dfrac{y^2}{3^2x^4}$ Property 3

$\qquad\qquad = \dfrac{y^2}{9x^4}$ Simplify.

✔CHECKPOINT   Now try Exercise 33.

STUDY TIP

Rarely in algebra is there only one way to solve a problem. Don't be concerned if the steps you use to solve a problem are not exactly the same as the steps presented in this text. The important thing is to use steps that you understand and, of course, steps that are justified by the rules of algebra. For instance, you might prefer the following steps for Example 2(d).

$$\left(\frac{3x^2}{y}\right)^{-2} = \left(\frac{y}{3x^2}\right)^2 = \frac{y^2}{9x^4}$$

Note how Property 3 is used in the first step of this solution. The fractional form of this property is

$$\left(\frac{a}{b}\right)^{-m} = \left(\frac{b}{a}\right)^m.$$

Historical Note

The French mathematician Nicolas Chuquet (ca. 1500) wrote *Triparty en la science des nombres*, in which a form of exponent notation was used. Our expressions $6x^3$ and $10x^2$ were written as $.6.^3$ and $.10.^2$. Zero and negative exponents were also represented, so x^0 would be written as $.1.^0$ and $3x^{-2}$ as $.3.^{2m}$. Chuquet wrote that $.72.^1$ divided by $.8.^3$ is $.9.^{2m}$. That is, $72x \div 8x^3 = 9x^{-2}$.

Scientific Notation

Exponents provide an efficient way of writing and computing with very large (or very small) numbers. For instance, there are about 359 billion billion gallons of water on Earth—that is, 359 followed by 18 zeros.

$$359,000,000,000,000,000,000$$

It is convenient to write such numbers in **scientific notation.** This notation has the form $\pm c \times 10^n$, where $1 \le c < 10$ and n is an integer. So, the number of gallons of water on Earth can be written in scientific notation as

$$3.59 \times 100,000,000,000,000,000,000 = 3.59 \times 10^{20}.$$

The *positive* exponent 20 indicates that the number is *large* (10 or more) and that the decimal point has been moved 20 places. A *negative* exponent indicates that the number is *small* (less than 1). For instance, the mass (in grams) of one electron is approximately

$$9.0 \times 10^{-28} = 0.0000000000000000000000000009.$$

28 decimal places

Example 3   **Scientific Notation**

Write each number in scientific notation.

a. 0.0000782   **b.** 836,100,000

Solution

a. $0.0000782 = 7.82 \times 10^{-5}$ **b.** $836,100,000 = 8.361 \times 10^8$

✔**CHECKPOINT**   Now try Exercise 37.

Example 4   **Decimal Notation**

Write each number in decimal notation.

a. 9.36×10^{-6} **b.** 1.345×10^2

Solution

a. $9.36 \times 10^{-6} = 0.00000936$ **b.** $1.345 \times 10^2 = 134.5$

✔**CHECKPOINT**   Now try Exercise 41.

Technology

Most calculators automatically switch to scientific notation when they are showing large (or small) numbers that exceed the display range.

To *enter* numbers in scientific notation, your calculator should have an exponential entry key labeled

   or   EXP.

Consult the user's guide for your calculator for instructions on keystrokes and how numbers in scientific notation are displayed.

Radicals and Their Properties

A **square root** of a number is one of its two equal factors. For example, 5 is a square root of 25 because 5 is one of the two equal factors of 25. In a similar way, a **cube root** of a number is one of its three equal factors, as in $125 = 5^3$.

Definition of *n*th Root of a Number

Let a and b be real numbers and let $n \geq 2$ be a positive integer. If

$$a = b^n$$

then b is an ***n*th root of *a*.** If $n = 2$, the root is a **square root.** If $n = 3$, the root is a **cube root.**

Some numbers have more than one *n*th root. For example, both 5 and -5 are square roots of 25. The *principal square root* of 25, written as $\sqrt{25}$, is the positive root, 5. The **principal *n*th root** of a number is defined as follows.

Principal *n*th Root of a Number

Let a be a real number that has at least one *n*th root. The **principal *n*th root of *a*** is the *n*th root that has the same sign as a. It is denoted by a **radical symbol**

$$\sqrt[n]{a}. \qquad \text{Principal } n\text{th root}$$

The positive integer n is the **index** of the radical, and the number a is the **radicand.** If $n = 2$, omit the index and write $\sqrt{a}$ rather than $\sqrt[2]{a}$. (The plural of index is *indices*.)

A common misunderstanding is that the square root sign implies both negative and positive roots. This is not correct. The square root sign implies only a positive root. When a negative root is needed, you must use the negative sign with the square root sign.

Incorrect: $\sqrt{4} = \pm 2$ *Correct:* $-\sqrt{4} = -2$ *and* $\sqrt{4} = 2$

Example 5   Evaluating Expressions Involving Radicals

a. $\sqrt{36} = 6$ because $6^2 = 36$.

b. $-\sqrt{36} = -6$ because $-\left(\sqrt{36}\right) = -\left(\sqrt{6^2}\right) = -(6) = -6$.

c. $\sqrt[3]{\dfrac{125}{64}} = \dfrac{5}{4}$ because $\left(\dfrac{5}{4}\right)^3 = \dfrac{5^3}{4^3} = \dfrac{125}{64}$.

d. $\sqrt[5]{-32} = -2$ because $(-2)^5 = -32$.

e. $\sqrt[4]{-81}$ is not a real number because there is no real number that can be raised to the fourth power to produce -81.

✔CHECKPOINT   Now try Exercise 51.

Here are some generalizations about the *n*th roots of real numbers.

Generalizations About *n*th Roots of Real Numbers			
Real Number a	Integer n	Root(s) of a	Example
$a > 0$	$n > 0$, is even.	$\sqrt[n]{a}$, $-\sqrt[n]{a}$	$\sqrt[4]{81} = 3$, $-\sqrt[4]{81} = -3$
$a > 0$ or $a < 0$	n is odd.	$\sqrt[n]{a}$	$\sqrt[3]{-8} = -2$
$a < 0$	n is even.	No real roots	$\sqrt{-4}$ is not a real number.
$a = 0$	n is even or odd.	$\sqrt[n]{0} = 0$	$\sqrt[5]{0} = 0$

Integers such as 1, 4, 9, 16, 25, and 36 are called **perfect squares** because they have integer square roots. Similarly, integers such as 1, 8, 27, 64, and 125 are called **perfect cubes** because they have integer cube roots.

Properties of Radicals

Let a and b be real numbers, variables, or algebraic expressions such that the indicated roots are real numbers, and let m and n be positive integers.

Property	*Example*
1. $\sqrt[n]{a^m} = \left(\sqrt[n]{a}\right)^m$	$\sqrt[3]{8^2} = \left(\sqrt[3]{8}\right)^2 = (2)^2 = 4$
2. $\sqrt[n]{a} \cdot \sqrt[n]{b} = \sqrt[n]{ab}$	$\sqrt{5} \cdot \sqrt{7} = \sqrt{5 \cdot 7} = \sqrt{35}$
3. $\dfrac{\sqrt[n]{a}}{\sqrt[n]{b}} = \sqrt[n]{\dfrac{a}{b}}, \quad b \neq 0$	$\dfrac{\sqrt[4]{27}}{\sqrt[4]{9}} = \sqrt[4]{\dfrac{27}{9}} = \sqrt[4]{3}$
4. $\sqrt[m]{\sqrt[n]{a}} = \sqrt[mn]{a}$	$\sqrt[3]{\sqrt{10}} = \sqrt[6]{10}$
5. $\left(\sqrt[n]{a}\right)^n = a$	$\left(\sqrt{3}\right)^2 = 3$
6. For n even, $\sqrt[n]{a^n} = |a|$.	$\sqrt{(-12)^2} = |-12| = 12$
For n odd, $\sqrt[n]{a^n} = a$.	$\sqrt[3]{(-12)^3} = -12$

A common special case of Property 6 is $\sqrt{a^2} = |a|$.

Example 6 Using Properties of Radicals

Use the properties of radicals to simplify each expression.

a. $\sqrt{8} \cdot \sqrt{2}$ **b.** $\left(\sqrt[3]{5}\right)^3$ **c.** $\sqrt[3]{x^3}$ **d.** $\sqrt[6]{y^6}$

Solution

a. $\sqrt{8} \cdot \sqrt{2} = \sqrt{8 \cdot 2} = \sqrt{16} = 4$

b. $\left(\sqrt[3]{5}\right)^3 = 5$

c. $\sqrt[3]{x^3} = x$

d. $\sqrt[6]{y^6} = |y|$

✓CHECKPOINT   Now try Exercise 61.

Simplifying Radicals

An expression involving radicals is in **simplest form** when the following conditions are satisfied.

1. All possible factors have been removed from the radical.

2. All fractions have radical-free denominators (accomplished by a process called *rationalizing the denominator*).

3. The index of the radical is reduced.

To simplify a radical, factor the radicand into factors whose exponents are multiples of the index. The roots of these factors are written outside the radical, and the "leftover" factors make up the new radicand.

Example 7　Simplifying Even Roots

Perfect 4th power　　Leftover factor

a. $\sqrt[4]{48} = \sqrt[4]{16 \cdot 3} = \sqrt[4]{2^4 \cdot 3} = 2\sqrt[4]{3}$

Perfect square　　Leftover factor

b. $\sqrt{75x^3} = \sqrt{25x^2 \cdot 3x}$　　　Find largest square factor.

$\phantom{\sqrt{75x^3}} = \sqrt{(5x)^2 \cdot 3x}$

$\phantom{\sqrt{75x^3}} = 5x\sqrt{3x}$　　　Find root of perfect square.

c. $\sqrt[4]{(5x)^4} = |5x| = 5|x|$

✔CHECKPOINT　　Now try Exercise 63(a).

Example 8　Simplifying Odd Roots

Perfect cube　　Leftover factor

a. $\sqrt[3]{24} = \sqrt[3]{8 \cdot 3} = \sqrt[3]{2^3 \cdot 3} = 2\sqrt[3]{3}$

Perfect cube　　Leftover factor

b. $\sqrt[3]{24a^4} = \sqrt[3]{8a^3 \cdot 3a}$　　　Find largest cube factor.

$\phantom{\sqrt[3]{24a^4}} = \sqrt[3]{(2a)^3 \cdot 3a}$

$\phantom{\sqrt[3]{24a^4}} = 2a\sqrt[3]{3a}$　　　Find root of perfect cube.

c. $\sqrt[3]{-40x^6} = \sqrt[3]{(-8x^6) \cdot 5}$　　　Find largest cube factor.

$\phantom{\sqrt[3]{-40x^6}} = \sqrt[3]{(-2x^2)^3 \cdot 5}$

$\phantom{\sqrt[3]{-40x^6}} = -2x^2\sqrt[3]{5}$　　　Find root of perfect cube.

✔CHECKPOINT　　Now try Exercise 63(b).

Radical expressions can be combined (added or subtracted) if they are **like radicals**—that is, if they have the same index and radicand. For instance, $\sqrt{2}$, $3\sqrt{2}$, and $\frac{1}{2}\sqrt{2}$ are like radicals, but $\sqrt{3}$ and $\sqrt{2}$ are unlike radicals. To determine whether two radicals can be combined, you should first simplify each radical.

Example 9 Combining Radicals

a. $2\sqrt{48} - 3\sqrt{27} = 2\sqrt{16 \cdot 3} - 3\sqrt{9 \cdot 3}$ Find square factors.

$= 8\sqrt{3} - 9\sqrt{3}$ Find square roots and multiply by coefficients.

$= (8 - 9)\sqrt{3}$ Combine like terms.

$= -\sqrt{3}$ Simplify.

b. $\sqrt[3]{16x} - \sqrt[3]{54x^4} = \sqrt[3]{8 \cdot 2x} - \sqrt[3]{27 \cdot x^3 \cdot 2x}$ Find cube factors.

$= 2\sqrt[3]{2x} - 3x\sqrt[3]{2x}$ Find cube roots.

$= (2 - 3x)\sqrt[3]{2x}$ Combine like terms.

✓**CHECKPOINT** Now try Exercise 71.

Rationalizing Denominators and Numerators

To rationalize a denominator or numerator of the form $a - b\sqrt{m}$ or $a + b\sqrt{m}$, multiply both numerator and denominator by a **conjugate**: $a + b\sqrt{m}$ and $a - b\sqrt{m}$ are conjugates of each other. If $a = 0$, then the rationalizing factor for $\sqrt{m}$ is itself, $\sqrt{m}$. For cube roots, choose a rationalizing factor that generates a perfect cube.

Example 10 Rationalizing Single-Term Denominators

Rationalize the denominator of each expression.

a. $\dfrac{5}{2\sqrt{3}}$ **b.** $\dfrac{2}{\sqrt[3]{5}}$

Solution

a. $\dfrac{5}{2\sqrt{3}} = \dfrac{5}{2\sqrt{3}} \cdot \dfrac{\sqrt{3}}{\sqrt{3}}$ $\sqrt{3}$ is rationalizing factor.

$= \dfrac{5\sqrt{3}}{2(3)}$ Multiply.

$= \dfrac{5\sqrt{3}}{6}$ Simplify.

b. $\dfrac{2}{\sqrt[3]{5}} = \dfrac{2}{\sqrt[3]{5}} \cdot \dfrac{\sqrt[3]{5^2}}{\sqrt[3]{5^2}}$ $\sqrt[3]{5^2}$ is rationalizing factor.

$= \dfrac{2\sqrt[3]{5^2}}{\sqrt[3]{5^3}}$ Multiply.

$= \dfrac{2\sqrt[3]{25}}{5}$ Simplify.

✓**CHECKPOINT** Now try Exercise 79.

Example 11   **Rationalizing a Denominator with Two Terms**

$$\frac{2}{3 + \sqrt{7}} = \frac{2}{3 + \sqrt{7}} \cdot \frac{3 - \sqrt{7}}{3 - \sqrt{7}}$$

Multiply numerator and denominator by conjugate of denominator.

$$= \frac{2(3 - \sqrt{7})}{3(3) + 3(-\sqrt{7}) + \sqrt{7}(3) - (\sqrt{7})(\sqrt{7})}$$

Use Distributive Property.

$$= \frac{2(3 - \sqrt{7})}{(3)^2 - (\sqrt{7})^2}$$

Simplify.

$$= \frac{2(3 - \sqrt{7})}{9 - 7}$$

Square terms of denominator.

$$= \frac{2(3 - \sqrt{7})}{2} = 3 - \sqrt{7}$$

Simplify.

✓CHECKPOINT   Now try Exercise 81.

Sometimes it is necessary to rationalize the numerator of an expression. For instance, in Section P.5 you will use the technique shown in the next example to rationalize the numerator of an expression from calculus.

Example 12   **Rationalizing a Numerator**

$$\frac{\sqrt{5} - \sqrt{7}}{2} = \frac{\sqrt{5} - \sqrt{7}}{2} \cdot \frac{\sqrt{5} + \sqrt{7}}{\sqrt{5} + \sqrt{7}}$$

Multiply numerator and denominator by conjugate of numerator.

$$= \frac{(\sqrt{5})^2 - (\sqrt{7})^2}{2(\sqrt{5} + \sqrt{7})}$$

Simplify.

$$= \frac{5 - 7}{2(\sqrt{5} + \sqrt{7})}$$

Square terms of numerator.

$$= \frac{-2}{2(\sqrt{5} + \sqrt{7})} = \frac{-1}{\sqrt{5} + \sqrt{7}}$$

Simplify.

✓CHECKPOINT   Now try Exercise 85.

Rational Exponents

Definition of Rational Exponents

If a is a real number and n is a positive integer such that the principal nth root of a exists, then $a^{1/n}$ is defined as

$a^{1/n} = \sqrt[n]{a}$, where $1/n$ is the **rational exponent** of a.

Moreover, if m is a positive integer that has no common factor with n, then

$$a^{m/n} = (a^{1/n})^m = (\sqrt[n]{a})^m \quad \text{and} \quad a^{m/n} = (a^m)^{1/n} = \sqrt[n]{a^m}.$$

The symbol ∫ indicates an example or exercise that highlights algebraic techniques specifically used in calculus.

STUDY TIP

Rational exponents can be tricky, and you must remember that the expression $b^{m/n}$ is not defined unless $\sqrt[n]{b}$ is a real number. This restriction produces some unusual-looking results. For instance, the number $(-8)^{1/3}$ is defined because $\sqrt[3]{-8} = -2$, but the number $(-8)^{2/6}$ is undefined because $\sqrt[6]{-8}$ is not a real number.

The numerator of a rational exponent denotes the *power* to which the base is raised, and the denominator denotes the *index* or the *root* to be taken.

$$b^{m/n} = \left(\sqrt[n]{b}\right)^m = \sqrt[n]{b^m}$$

When you are working with rational exponents, the properties of integer exponents still apply. For instance,

$$2^{1/2}2^{1/3} = 2^{(1/2)+(1/3)} = 2^{5/6}.$$

Example 13   Changing from Radical to Exponential Form

a. $\sqrt{3} = 3^{1/2}$

b. $\sqrt{(3xy)^5} = \sqrt[2]{(3xy)^5} = (3xy)^{(5/2)}$

c. $2x\sqrt[4]{x^3} = (2x)(x^{3/4}) = 2x^{1+(3/4)} = 2x^{7/4}$

✓CHECKPOINT   Now try Exercise 87.

Example 14   Changing from Exponential to Radical Form

a. $(x^2 + y^2)^{3/2} = \left(\sqrt{x^2+y^2}\right)^3 = \sqrt{(x^2+y^2)^3}$

b. $2y^{3/4}z^{1/4} = 2(y^3z)^{1/4} = 2\sqrt[4]{y^3z}$

c. $a^{-3/2} = \dfrac{1}{a^{3/2}} = \dfrac{1}{\sqrt{a^3}}$ d. $x^{0.2} = x^{1/5} = \sqrt[5]{x}$

✓CHECKPOINT   Now try Exercise 89.

Technology

There are four methods of evaluating radicals on most graphing calculators. For square roots, you can use the *square root key* ⎷. For cube roots, you can use the *cube root key* ∛. For other roots, you can first convert the radical to exponential form and then use the *exponential key* ⌃, or you can use the *xth root key* ⎷. Consult the user's guide for your calculator for specific keystrokes.

Rational exponents are useful for evaluating roots of numbers on a calculator, for reducing the index of a radical, and for simplifying expressions in calculus.

Example 15   Simplifying with Rational Exponents

a. $(-32)^{-4/5} = \left(\sqrt[5]{-32}\right)^{-4} = (-2)^{-4} = \dfrac{1}{(-2)^4} = \dfrac{1}{16}$

b. $(-5x^{5/3})(3x^{-3/4}) = -15x^{(5/3)-(3/4)} = -15x^{11/12}, \quad x \neq 0$

c. $\sqrt[9]{a^3} = a^{3/9} = a^{1/3} = \sqrt[3]{a}$ Reduce index.

d. $\sqrt[3]{\sqrt{125}} = \sqrt[6]{125} = \sqrt[6]{(5)^3} = 5^{3/6} = 5^{1/2} = \sqrt{5}$

e. $(2x-1)^{4/3}(2x-1)^{-1/3} = (2x-1)^{(4/3)-(1/3)}$

$$= 2x - 1, \quad x \neq \frac{1}{2}$$

f. $\dfrac{x-1}{(x-1)^{-1/2}} = \dfrac{x-1}{(x-1)^{-1/2}} \cdot \dfrac{(x-1)^{1/2}}{(x-1)^{1/2}}$

$$= \frac{(x-1)^{3/2}}{(x-1)^0}$$

$$= (x-1)^{3/2}, \quad x \neq 1$$

✓CHECKPOINT   Now try Exercise 99.

P.2 | Exercises

VOCABULARY CHECK: Fill in the blanks.

1. In the exponential form a^n, n is the _____ and a is the _____.

2. A convenient way of writing very large or very small numbers is called _____ _____.

3. One of the two equal factors of a number is called a _____ _____ of the number.

4. The _____ _____ _____ of a number is the nth root that has the same sign as a, and is denoted by $\sqrt[n]{a}$.

5. In the radical form, $\sqrt[n]{a}$ the positive integer n is called the _____ of the radical and the number a is called the _____.

6. When an expression involving radicals has all possible factors removed, radical-free denominators, and a reduced index, it is in _____ _____.

7. The expressions $a + b\sqrt{m}$ and $a - b\sqrt{m}$ are _____ of each other.

8. The process used to create a radical-free denominator is know as _____ the denominator.

9. In the expression $b^{m/n}$, m denotes the _____ to which the base is raised and n denotes the _____ or root to be taken.

In Exercises 1 and 2, write the expression as a repeated multiplication problem.

1. 8^5

2. $(-2)^7$

In Exercises 3 and 4, write the expression using exponential notation.

3. $(4.9)(4.9)(4.9)(4.9)(4.9)(4.9)$

4. $(-10)(-10)(-10)(-10)(-10)$

In Exercises 5–12, evaluate each expression.

5. (a) $3^2 \cdot 3$ (b) $3 \cdot 3^3$

6. (a) $\dfrac{5^5}{5^2}$ (b) $\dfrac{3^2}{3^4}$

7. (a) $(3^3)^0$ (b) -3^2

8. (a) $(2^3 \cdot 3^2)^2$ (b) $\left(-\frac{3}{5}\right)^3\left(\frac{5}{3}\right)^2$

9. (a) $\dfrac{3 \cdot 4^{-4}}{3^{-4} \cdot 4^{-1}}$ (b) $32(-2)^{-5}$

10. (a) $\dfrac{4 \cdot 3^{-2}}{2^{-2} \cdot 3^{-1}}$ (b) $(-2)^0$

11. (a) $2^{-1} + 3^{-1}$ (b) $(2^{-1})^{-2}$

12. (a) $3^{-1} + 2^{-2}$ (b) $(3^{-2})^2$

In Exercises 13–16, use a calculator to evaluate the expression. (If necessary, round your answer to three decimal places.)

13. $(-4)^3(5^2)$

14. $(8^{-4})(10^3)$

15. $\dfrac{3^6}{7^3}$

16. $\dfrac{4^3}{3^{-4}}$

In Exercises 17–24, evaluate the expression for the given value of x.

Expression	Value
17. $-3x^3$	$x = 2$
18. $7x^{-2}$	$x = 4$
19. $6x^0$	$x = 10$
20. $5(-x)^3$	$x = 3$
21. $2x^3$	$x = -3$
22. $-3x^4$	$x = -2$
23. $4x^2$	$x = -\frac{1}{2}$
24. $5(-x)^3$	$x = -\frac{1}{3}$

In Exercises 25–30, simplify each expression.

25. (a) $(-5z)^3$ (b) $5x^4(x^2)$

26. (a) $(3x)^2$ (b) $(4x^3)^0$

27. (a) $6y^2(2y^0)^2$ (b) $\dfrac{3x^5}{x^3}$

28. (a) $(-z)^3(3z^4)$ (b) $\dfrac{25y^8}{10y^4}$

29. (a) $\dfrac{7x^2}{x^3}$ (b) $\dfrac{12(x + y)^3}{9(x + y)}$

30. (a) $\dfrac{r^4}{r^6}$ (b) $\left(\dfrac{4}{y}\right)^3\left(\dfrac{3}{y}\right)^4$

In Exercises 31–36, rewrite each expression with positive exponents and simplify.

31. (a) $(x + 5)^0, \quad x \neq -5$ (b) $(2x^2)^{-2}$

32. (a) $(2x^5)^0, \quad x \neq 0$ (b) $(z + 2)^{-3}(z + 2)^{-1}$

33. (a) $(-2x^2)^3(4x^3)^{-1}$ (b) $\left(\dfrac{x}{10}\right)^{-1}$

34. (a) $(4y^{-2})(8y^4)$ (b) $\left(\dfrac{x^{-3}y^4}{5}\right)^{-3}$

35. (a) $3^n \cdot 3^{2n}$ (b) $\left(\dfrac{a^{-2}}{b^{-2}}\right)\left(\dfrac{b}{a}\right)^3$

36. (a) $\dfrac{x^2 \cdot x^n}{x^3 \cdot x^n}$ (b) $\left(\dfrac{a^{-3}}{b^{-3}}\right)\left(\dfrac{a}{b}\right)^3$

In Exercises 37–40, write the number in scientific notation.

37. Land area of Earth: 57,300,000 square miles

38. Light year: 9,460,000,000,000 kilometers

39. Relative density of hydrogen: 0.0000899 gram per cubic centimeter

40. One micron (millionth of a meter): 0.00003937 inch

In Exercises 41–44, write the number in decimal notation.

41. Worldwide daily consumption of Coca-Cola: 4.568×10^9 ounces (Source: The Coca-Cola Company)

42. Interior temperature of the sun: 1.5×10^7 degrees Celsius

43. Charge of an electron: 1.6022×10^{-19} coulomb

44. Width of a human hair: 9.0×10^{-5} meter

In Exercises 45 and 46, evaluate each expression without using a calculator.

45. (a) $\sqrt{25 \times 10^8}$ (b) $\sqrt[3]{8 \times 10^{15}}$

46. (a) $(1.2 \times 10^7)(5 \times 10^{-3})$ (b) $\dfrac{(6.0 \times 10^8)}{(3.0 \times 10^{-3})}$

In Exercises 47–50, use a calculator to evaluate each expression. (Round your answer to three decimal places.)

47. (a) $750\left(1 + \dfrac{0.11}{365}\right)^{800}$

(b) $\dfrac{67,000,000 + 93,000,000}{0.0052}$

48. (a) $(9.3 \times 10^6)^3(6.1 \times 10^{-4})$

(b) $\dfrac{(2.414 \times 10^4)^6}{(1.68 \times 10^5)^5}$

49. (a) $\sqrt{4.5 \times 10^9}$ (b) $\sqrt[3]{6.3 \times 10^4}$

50. (a) $(2.65 \times 10^{-4})^{1/3}$ (b) $\sqrt{9 \times 10^{-4}}$

In Exercises 51–56, evaluate each expression without using a calculator.

51. (a) $\sqrt{9}$ (b) $\sqrt[3]{\dfrac{27}{8}}$

52. (a) $27^{1/3}$ (b) $36^{3/2}$

53. (a) $32^{-3/5}$ (b) $\left(\dfrac{16}{81}\right)^{-3/4}$

54. (a) $100^{-3/2}$ (b) $\left(\dfrac{9}{4}\right)^{-1/2}$

55. (a) $\left(-\dfrac{1}{64}\right)^{-1/3}$ (b) $\left(\dfrac{1}{\sqrt{32}}\right)^{-2/5}$

56. (a) $\left(-\dfrac{125}{27}\right)^{-1/3}$ (b) $-\left(\dfrac{1}{125}\right)^{-4/3}$

In Exercises 57–60, use a calculator to approximate the number. (Round your answer to three decimal places.)

57. (a) $\sqrt{57}$ (b) $\sqrt[5]{-27^3}$

58. (a) $\sqrt[3]{45^2}$ (b) $\sqrt[6]{125}$

59. (a) $(-12.4)^{-1.8}$ (b) $\left(5\sqrt{3}\right)^{-2.5}$

60. (a) $\dfrac{7 - (4.1)^{-3.2}}{2}$ (b) $\left(\dfrac{13}{3}\right)^{-3/2} - \left(\dfrac{3}{2}\right)^{13/3}$

In Exercises 61 and 62, use the properties of radicals to simplify each expression.

61. (a) $\left(\sqrt[3]{4}\right)^3$ (b) $\sqrt[5]{96x^5}$

62. (a) $\sqrt{12} \cdot \sqrt{3}$ (b) $\sqrt[4]{(3x^2)^4}$

In Exercises 63–74, simplify each radical expression.

63. (a) $\sqrt{8}$ (b) $\sqrt[3]{54}$

64. (a) $\sqrt[3]{\dfrac{16}{27}}$ (b) $\sqrt{\dfrac{75}{4}}$

65. (a) $\sqrt{72x^3}$ (b) $\sqrt{\dfrac{18^2}{z^3}}$

66. (a) $\sqrt{54xy^4}$ (b) $\sqrt{\dfrac{32a^4}{b^2}}$

67. (a) $\sqrt[3]{16x^5}$ (b) $\sqrt{75x^2y^{-4}}$

68. (a) $\sqrt[4]{3x^4y^2}$ (b) $\sqrt[5]{160x^8z^4}$

69. (a) $2\sqrt{50} + 12\sqrt{8}$ (b) $10\sqrt{32} - 6\sqrt{18}$

70. (a) $4\sqrt{27} - \sqrt{75}$ (b) $\sqrt[3]{16} + 3\sqrt[3]{54}$

71. (a) $5\sqrt{x} - 3\sqrt{x}$ (b) $-2\sqrt{9y} + 10\sqrt{y}$

72. (a) $8\sqrt{49x} - 14\sqrt{100x}$ (b) $-3\sqrt{48x^2} + 7\sqrt{75x^2}$

73. (a) $3\sqrt{x+1} + 10\sqrt{x+1}$ (b) $7\sqrt{80x} - 2\sqrt{125x}$

74. (a) $-\sqrt{x^3 - 7} + 5\sqrt{x^3 - 7}$ (b) $11\sqrt{245x^3} - 9\sqrt{45x^3}$

In Exercises 75–78, complete the statement with <, =, or >.

75. $\sqrt{5} + \sqrt{3}$ ▢ $\sqrt{5 + 3}$

76. $\sqrt{\dfrac{3}{11}}$ ▢ $\dfrac{\sqrt{3}}{\sqrt{11}}$

77. 5 ▢ $\sqrt{3^2 + 2^2}$

78. 5 ▢ $\sqrt{3^2 + 4^2}$

In Exercises 79–82, rationalize the denominator of the expression. Then simplify your answer.

79. $\dfrac{1}{\sqrt{3}}$

80. $\dfrac{5}{\sqrt{10}}$

81. $\dfrac{2}{5 - \sqrt{3}}$

82. $\dfrac{3}{\sqrt{5} + \sqrt{6}}$

In Exercises 83–86, rationalize the numerator of the expression. Then simplify your answer.

83. $\dfrac{\sqrt{8}}{2}$

84. $\dfrac{\sqrt{2}}{3}$

85. $\dfrac{\sqrt{5} + \sqrt{3}}{3}$

86. $\dfrac{\sqrt{7} - 3}{4}$

In Exercises 87–94, fill in the missing form of the expression.

Radical Form	Rational Exponent Form
87. $\sqrt{9}$	
88. $\sqrt[3]{64}$	
89.	$32^{1/5}$
90.	$-(144^{1/2})$
91. $\sqrt[3]{-216}$	
92.	$(-243)^{1/5}$
93. $\left(\sqrt[4]{81}\right)^{3}$	
94.	$16^{5/4}$

In Exercises 95–98, perform the operations and simplify.

95. $\dfrac{(2x^2)^{3/2}}{2^{1/2}x^4}$

96. $\dfrac{x^{4/3}y^{2/3}}{(xy)^{1/3}}$

97. $\dfrac{x^{-3} \cdot x^{1/2}}{x^{3/2} \cdot x^{-1}}$

98. $\dfrac{5^{-1/2} \cdot 5x^{5/2}}{(5x)^{3/2}}$

In Exercises 99 and 100, reduce the index of each radical.

99. (a) $\sqrt[4]{3^2}$ (b) $\sqrt[6]{(x + 1)^4}$

100. (a) $\sqrt[6]{x^3}$ (b) $\sqrt[4]{(3x^2)^4}$

In Exercises 101 and 102, write each expression as a single radical. Then simplify your answer.

101. (a) $\sqrt{\sqrt{32}}$ (b) $\sqrt{\sqrt[4]{2x}}$

102. (a) $\sqrt{\sqrt{243(x + 1)}}$ (b) $\sqrt{\sqrt[3]{10a^7b}}$

103. *Period of a Pendulum*   The period T (in seconds) of a pendulum is

$$T = 2\pi\sqrt{\dfrac{L}{32}}$$

where L is the length of the pendulum (in feet). Find the period of a pendulum whose length is 2 feet.

The symbol *f* indicates an example or exercise that highlights algebraic techniques specifically used in calculus.

The symbol indicates an exercise or a part of an exercise in which you are instructed to use a graphing utility.

104. *Erosion*   A stream of water moving at the rate of v feet per second can carry particles of size $0.03\sqrt{v}$ inches. Find the size of the largest particle that can be carried by a stream flowing at the rate of $\frac{3}{4}$ foot per second.

Model It

105. *Mathematical Modeling*   A funnel is filled with water to a height of h centimeters. The formula

$$t = 0.03[12^{5/2} - (12 - h)^{5/2}], \quad 0 \le h \le 12$$

represents the amount of time t (in seconds) that it will take for the funnel to empty.

(a) Use the *table* feature of a graphing utility to find the times required for the funnel to empty for water heights of $h = 0$, $h = 1$, $h = 2$, . . . $h = 12$ centimeters.

(b) What value does t appear to be approaching as the height of the water becomes closer and closer to 12 centimeters?

106. *Speed of Light*   The speed of light is approximately 11,180,000 miles per minute. The distance from the sun to Earth is approximately 93,000,000 miles. Find the time for light to travel from the sun to Earth.

Synthesis

True or False?   **In Exercises 107 and 108, determine whether the statement is true or false. Justify your answer.**

107. $\dfrac{x^{k+1}}{x} = x^{k}$ **108.** $(a^n)^k = a^{n^k}$

109. Verify that $a^0 = 1$, $a \ne 0$. (*Hint:* Use the property of exponents $a^m/a^n = a^{m-n}$.)

110. Explain why each of the following pairs is not equal.

(a) $(3x)^{-1} \ne \dfrac{3}{x}$ (b) $y^3 \cdot y^2 \ne y^6$

(c) $(a^2b^3)^4 \ne a^6b^7$ (d) $(a + b)^2 \ne a^2 + b^2$

(e) $\sqrt{4x^2} \ne 2x$ (f) $\sqrt{2} + \sqrt{3} \ne \sqrt{5}$

111. *Exploration*   List all possible digits that occur in the units place of the square of a positive integer. Use that list to determine whether $\sqrt{5233}$ is an integer.

112. *Think About It*   Square the real number $2/\sqrt{5}$ and note that the radical is eliminated from the denominator. Is this equivalent to rationalizing the denominator? Why or why not?

P.3 Polynomials and Special Products

What you should learn

- Write polynomials in standard form.
- Add, subtract, and multiply polynomials.
- Use special products to multiply polynomials.
- Use polynomials to solve real-life problems.

Why you should learn it

Polynomials can be used to model and solve real-life problems. For instance, in Exercise 98 on page 30, polynomials are used to model the cost, revenue, and profit for producing and selling hats.

David Noton/Masterfile

Polynomials

The most common type of algebraic expression is the **polynomial.** Some examples are $2x + 5$, $3x^4 - 7x^2 + 2x + 4$, and $5x^2y^2 - xy + 3$. The first two are *polynomials in x* and the third is a *polynomial in x and y*. The terms of a polynomial in x have the form ax^k, where a is the **coefficient** and k is the **degree** of the term. For instance, the polynomial

$$2x^3 - 5x^2 + 1 = 2x^3 + (-5)x^2 + (0)x + 1$$

has coefficients 2, -5, 0, and 1.

Definition of a Polynomial in x

Let $a_0, a_1, a_2, \ldots, a_n$ be real numbers and let n be a nonnegative integer. A polynomial in x is an expression of the form

$$a_n x^n + a_{n-1} x^{n-1} + \cdots + a_1 x + a_0$$

where $a_n \neq 0$. The polynomial is of **degree** n, a_n is the **leading coefficient,** and a_0 is the **constant term.**

Polynomials with one, two, and three terms are called **monomials, binomials,** and **trinomials,** respectively. In **standard form,** a polynomial is written with descending powers of x.

Example 1 Writing Polynomials in Standard Form

	Polynomial	*Standard Form*	*Degree*
a.	$4x^2 - 5x^7 - 2 + 3x$	$-5x^7 + 4x^2 + 3x - 2$	7
b.	$4 - 9x^2$	$-9x^2 + 4$	2
c.	8	$8 \ (8 = 8x^0)$	0

✓**CHECKPOINT** Now try Exercise 11.

A polynomial that has all zero coefficients is called the zero polynomial, denoted by 0. No degree is assigned to this particular polynomial. For polynomials in more than one variable, the degree of a *term* is the sum of the exponents of the variables in the term. The degree of the *polynomial* is the highest degree of its terms. For instance, the degree of the polynomial $-2x^3y^6 + 4xy - x^7y^4$ is 11 because the sum of the exponents in the last term is the greatest. The leading coefficient of the polynomial is the coefficient of the highest-degree term. Expressions are not polynomials if a variable is underneath a radical or if a polynomial expression (with degree greater than 0) is in the denominator of a term. The following expressions are not polynomials.

$$x^3 - \sqrt{3x} = x^3 - (3x)^{1/2} \qquad \text{The exponent "1/2" is not an integer.}$$

$$x^2 + \frac{5}{x} = x^2 + 5x^{-1} \qquad \text{The exponent "−1" is not a nonnegative integer.}$$

Operations with Polynomials

You can add and subtract polynomials in much the same way you add and subtract real numbers. Simply add or subtract the *like terms* (terms having the same variables to the same powers) by adding their coefficients. For instance, $-3xy^2$ and $5xy^2$ are like terms and their sum is

$$-3xy^2 + 5xy^2 = (-3 + 5)xy^2$$
$$= 2xy^2.$$

Example 2   **Sums and Differences of Polynomials**

a. $(5x^3 - 7x^2 - 3) + (x^3 + 2x^2 - x + 8)$

$= (5x^3 + x^3) + (-7x^2 + 2x^2) - x + (-3 + 8)$ Group like terms.

$= 6x^3 - 5x^2 - x + 5$ Combine like terms.

b. $(7x^4 - x^2 - 4x + 2) - (3x^4 - 4x^2 + 3x)$

$= 7x^4 - x^2 - 4x + 2 - 3x^4 + 4x^2 - 3x$ Distributive Property

$= (7x^4 - 3x^4) + (-x^2 + 4x^2) + (-4x - 3x) + 2$ Group like terms.

$= 4x^4 + 3x^2 - 7x + 2$ Combine like terms.

✓**CHECKPOINT**   Now try Exercise 33.

To find the *product* of two polynomials, use the left and right Distributive Properties. For example, if you treat $5x + 7$ as a single quantity, you can multiply $3x - 2$ by $5x + 7$ as follows.

$$(3x - 2)(5x + 7) = 3x(5x + 7) - 2(5x + 7)$$

$$= (3x)(5x) + (3x)(7) - (2)(5x) - (2)(7)$$

$$= 15x^2 + 21x - 10x - 14$$

Product of	Product of	Product of	Product of
First terms	**Outer terms**	**Inner terms**	**Last terms**

$$= 15x^2 + 11x - 14$$

Note in this **FOIL Method** (which can only be used to multiply two binomials) that the outer (O) and inner (I) terms are like terms and can be combined.

Example 3   **Finding a Product by the FOIL Method**

Use the FOIL Method to find the product of $2x - 4$ and $x + 5$.

Solution

F   O   I   L

$$(2x - 4)(x + 5) = 2x^2 + 10x - 4x - 20$$

$$= 2x^2 + 6x - 20$$

✓**CHECKPOINT**   Now try Exercise 55.

When multiplying two polynomials, be sure to multiply each term of one polynomial by *each* term of the other. A vertical arrangement is helpful.

Example 4 **A Vertical Arrangement for Multiplication**

Multiply $x^2 - 2x + 2$ by $x^2 + 2x + 2$ using a vertical arrangement.

Solution

$$
\begin{array}{r}
x^2 - 2x + 2 \\
\times\ x^2 + 2x + 2 \\
\hline
x^4 - 2x^3 + 2x^2 \\
2x^3 - 4x^2 + 4x \\
2x^2 - 4x + 4 \\
\hline
x^4 + 0x^3 + 0x^2 + 0x + 4 = x^4 + 4
\end{array}
$$

Write in standard form.
Write in standard form.
$x^2(x^2 - 2x + 2)$
$2x(x^2 - 2x + 2)$
$2(x^2 - 2x + 2)$
Combine like terms.

So, $(x^2 - 2x + 2)(x^2 + 2x + 2) = x^4 + 4$.

✓**CHECKPOINT** Now try Exercise 59.

Special Products

Some binomial products have special forms that occur frequently in algebra. You do not need to memorize these formulas because you can use the Distributive Property to multiply. However, becoming familiar with these formulas will enable you to manipulate the algebra more quickly.

Special Products

Let u and v be real numbers, variables, or algebraic expressions.

Special Product	*Example*

Sum and Difference of Same Terms

$(u + v)(u - v) = u^2 - v^2$ 　　　　 $(x + 4)(x - 4) = x^2 - 4^2$
　　　　　　　　　　　　　　　　　　　　　　$= x^2 - 16$

Square of a Binomial

$(u + v)^2 = u^2 + 2uv + v^2$ 　　　　 $(x + 3)^2 = x^2 + 2(x)(3) + 3^2$
　　　　　　　　　　　　　　　　　　　　　　$= x^2 + 6x + 9$

$(u - v)^2 = u^2 - 2uv + v^2$ 　　　　 $(3x - 2)^2 = (3x)^2 - 2(3x)(2) + 2^2$
　　　　　　　　　　　　　　　　　　　　　　$= 9x^2 - 12x + 4$

Cube of a Binomial

$(u + v)^3 = u^3 + 3u^2v + 3uv^2 + v^3$ 　　 $(x + 2)^3 = x^3 + 3x^2(2) + 3x(2^2) + 2^3$
　　　　　　　　　　　　　　　　　　　　　　$= x^3 + 6x^2 + 12x + 8$

$(u - v)^3 = u^3 - 3u^2v + 3uv^2 - v^3$ 　　 $(x - 1)^3 = x^3 - 3x^2(1) + 3x(1^2) - 1^3$
　　　　　　　　　　　　　　　　　　　　　　$= x^3 - 3x^2 + 3x - 1$

| **Example 5** | **Sum and Difference of Same Terms** |

Find the product of $5x + 9$ and $5x - 9$.

Solution

The product of a sum and a difference of the *same* two terms has no middle term and takes the form $(u + v)(u - v) = u^2 - v^2$.

$$(5x + 9)(5x - 9) = (5x)^2 - 9^2 = 25x^2 - 81$$

✔CHECKPOINT   Now try Exercise 61.

STUDY TIP

When squaring a binomial, note that the resulting middle term is always *twice* the product of the two terms.

| **Example 6** | **Square of a Binomial** |

Find $(6x - 5)^2$.

Solution

The square of a binomial has the form $(u - v)^2 = u^2 - 2uv + v^2$.

$$(6x - 5)^2 = (6x)^2 - 2(6x)(5) + 5^2 = 36x^2 - 60x + 25$$

✔CHECKPOINT   Now try Exercise 67.

| **Example 7** | **Cube of a Binomial** |

Find $(3x + 2)^3$.

Solution

The cube of a binomial has the form

$$(u + v)^3 = u^3 + 3u^2v + 3uv^2 + v^3.$$

Note the *decrease* of powers of $u = 3x$ and the *increase* of powers of $v = 2$.

$$(3x + 2)^3 = (3x)^3 + 3(3x)^2(2) + 3(3x)(2^2) + 2^3$$
$$= 27x^3 + 54x^2 + 36x + 8$$

✔CHECKPOINT   Now try Exercise 69.

| **Example 8** | **The Product of Two Trinomials** |

Find the product of $x + y - 2$ and $x + y + 2$.

Solution

By grouping $x + y$ in parentheses, you can write the product of the trinomials as a special product.

$$(x + y - 2)(x + y + 2) = [(x + y) - 2][(x + y) + 2]$$

(with labels: Difference over -2, Sum over $+2$)

$$= (x + y)^2 - 2^2 \qquad \text{Sum and difference of same terms}$$
$$= x^2 + 2xy + y^2 - 4$$

✔CHECKPOINT   Now try Exercise 59.

Application

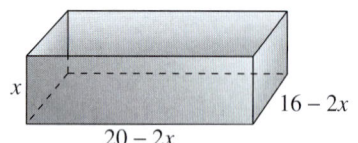

FIGURE P.8

Example 9 Volume of a Box

An open box is made by cutting squares from the corners of a piece of metal that is 16 inches by 20 inches, as shown in Figure P.8. The edge of each cut-out square is x inches. Find the volume of the box when $x = 1$, $x = 2$, and $x = 3$.

Solution

The volume of a rectangular box is equal to the product of its length, width, and height. From the figure, the length is $20 - 2x$, the width is $16 - 2x$, and the height is x. So, the volume of the box is

$$\text{Volume} = (20 - 2x)(16 - 2x)(x)$$
$$= (320 - 72x + 4x^2)(x)$$
$$= 320x - 72x^2 + 4x^3.$$

When $x = 1$ inch, the volume of the box is

$$\text{Volume} = 320(1) - 72(1)^2 + 4(1)^3$$
$$= 252 \text{ cubic inches.}$$

When $x = 2$ inches, the volume of the box is

$$\text{Volume} = 320(2) - 72(2)^2 + 4(2)^3$$
$$= 384 \text{ cubic inches.}$$

When $x = 3$ inches, the volume of the box is

$$\text{Volume} = 320(3) - 72(3)^2 + 4(3)^3$$
$$= 420 \text{ cubic inches.}$$

✓**CHECKPOINT** Now try Exercise 101.

WRITING ABOUT **M**ATHEMATICS

Mathematical Experiment In Example 9, the volume of the open box is given by

$$\text{Volume} = 320x - 72x^2 + 4x^3.$$

You want to create a box that has as much volume as possible. From Example 9, you know that by cutting one-, two-, and three-inch squares from the corners, you can create boxes whose volumes are 252, 384, and 420 cubic inches, respectively. What are the possible values of x that make sense in the problem? Write your answer as an interval. Try several other values of x to find the size of the squares that should be cut from the corners to produce a box that has maximum volume. Write a summary of your findings.

P.3   Exercises

VOCABULARY CHECK

In Exercises 1–5, fill in the blanks.

1. For the polynomial $a_n x^n + a_{n-1} x^{n-1} + \cdots + a_1 x + a_0$, the degree is _____, the leading coefficient is _____, and the constant term is _____.

2. A polynomial in x in standard form is written with _____ powers of x.

3. A polynomial with one term is called a _____, while a polynomial with two terms is called a _____, and a polynomial with three terms is called a _____.

4. To add or subtract polynomials, add or subtract the _____ _____ by adding their coefficients.

5. The letters in "FOIL" stand for the following.

    F _____   O _____   I _____   L _____

In Exercises 6–8, match the special product form with its name.

6. $(u + v)(u - v) = u^2 - v^2$ (a) A binomial sum squared

7. $(u + v)^2 = u^2 + 2uv + v^2$ (b) A binomial difference squared

8. $(u - v)^2 = u^2 - 2uv + v^2$ (c) The sum and difference of same terms

---

In Exercises 1–6, match the polynomial with its description. [The polynomials are labeled (a), (b), (c), (d), (e), and (f).]

(a) $3x^2$ (b) $1 - 2x^3$

(c) $x^3 + 3x^2 + 3x + 1$ (d) 12

(e) $-3x^5 + 2x^3 + x$ (f) $\frac{2}{3}x^4 + x^2 + 10$

1. A polynomial of degree 0

2. A trinomial of degree 5

3. A binomial with leading coefficient -2

4. A monomial of positive degree

5. A trinomial with leading coefficient $\frac{2}{3}$

6. A third-degree polynomial with leading coefficient 1

In Exercises 7–10, write a polynomial that fits the description. (There are many correct answers.)

7. A third-degree polynomial with leading coefficient -2

8. A fifth-degree polynomial with leading coefficient 6

9. A fourth-degree binomial with a negative leading coefficient

10. A third-degree binomial with an even leading coefficient

In Exercises 11–22, (a) write the polynomial in standard form, (b) identify the degree and leading coefficient of the polynomial, and (c) state whether the polynomial is a monomial, a binomial, or a trinomial.

11. $14x - \frac{1}{2}x^5$ 12. $2x^2 - x + 1$

13. $-3x^4 + 2x^2 - 5$ 14. $7x$

15. $x^5 - 1$ 16. $-y + 25y^2 + 1$

17. 3 18. $t^2 + 9$

19. $1 + 6x^4 - 4x^5$ 20. $3 + 2x$

21. $4x^3 y$ 22. $-x^5 y + 2x^2 y^2 + xy^4$

In Exercises 23–28, determine whether the expression is a polynomial. If so, write the polynomial in standard form.

23. $2x - 3x^3 + 8$ 24. $2x^3 + x - 3x^{-1}$

25. $\dfrac{3x + 4}{x}$ 26. $\dfrac{x^2 + 2x - 3}{2}$

27. $y^2 - y^4 + y^3$ 28. $\sqrt{y^2 - y^4}$

In Exercises 29–46, perform the operation and write the result in standard form.

29. $(6x + 5) - (8x + 15)$

30. $(2x^2 + 1) - (x^2 - 2x + 1)$

31. $-(x^3 - 2) + (4x^3 - 2x)$

32. $-(5x^2 - 1) - (-3x^2 + 5)$

33. $(15x^2 - 6) - (-8.3x^3 - 14.7x^2 - 17)$

34. $(15.2x^4 - 18x - 19.1) - (13.9x^4 - 9.6x + 15)$

35. $5z - [3z - (10z + 8)]$

36. $(y^3 + 1) - [(y^2 + 1) + (3y - 7)]$

37. $3x(x^2 - 2x + 1)$ 38. $y^2(4y^2 + 2y - 3)$

39. $-5z(3z - 1)$ 40. $(-3x)(5x + 2)$

41. $(1 - x^3)(4x)$ 42. $-4x(3 - x^3)$

43. $(2.5x^2 + 3)(3x)$ 44. $(2 - 3.5y)(2y^3)$

45. $-4x\left(\frac{1}{8}x + 3\right)$ 46. $2y\left(4 - \frac{7}{8}y\right)$

In Exercises 47–54, perform the operation.

47. Add $7x^3 - 2x^2 + 8$ and $-3x^3 - 4$.

48. Add $2x^5 - 3x^3 + 2x + 3$ and $4x^3 + x - 6$.

49. Subtract $x - 3$ from $5x^2 - 3x + 8$.

50. Subtract $-t^4 + 0.5t^2 - 5.6$ from $0.6t^4 - 2t^2$.

51. Multiply $-6x^2 + 15x - 4$ and $5x + 3$.

52. Multiply $4x^4 + x^3 - 6x^2 + 9$ and $x^2 + 2x + 3$.

53. $(x^2 + 9)(x^2 - x - 4)$ **54.** $(x - 2)(x^2 + 2x + 4)$

In Exercises 55–92, multiply or find the special product.

55. $(x + 3)(x + 4)$ **56.** $(x - 5)(x + 10)$

57. $(3x - 5)(2x + 1)$ **58.** $(7x - 2)(4x - 3)$

59. $(x^2 - x + 1)(x^2 + x + 1)$

60. $(x^2 + 3x - 2)(x^2 - 3x - 2)$

61. $(x + 10)(x - 10)$ **62.** $(2x + 3)(2x - 3)$

63. $(x + 2y)(x - 2y)$ **64.** $(2x + 3y)(2x - 3y)$

65. $(2x + 3)^2$ **66.** $(4x + 5)^2$

67. $(2x - 5y)^2$ **68.** $(5 - 8x)^2$

69. $(x + 1)^3$ **70.** $(x - 2)^3$

71. $(2x - y)^3$ **72.** $(3x + 2y)^3$

73. $(4x^3 - 3)^2$ **74.** $(8x + 3)^2$

75. $[(m - 3) + n][(m - 3) - n]$

76. $[(x + y) + 1][(x + y) - 1]$

77. $[(x - 3) + y]^2$ **78.** $[(x + 1) - y]^2$

79. $(2r^2 - 5)(2r^2 + 5)$ **80.** $(3a^3 - 4b^2)(3a^3 + 4b^2)$

81. $\left(\frac{1}{2}x - 3\right)^2$ **82.** $\left(\frac{2}{3}t + 5\right)^2$

83. $\left(\frac{1}{3}x - 2\right)\left(\frac{1}{3}x + 2\right)$ **84.** $\left(2x + \frac{1}{5}\right)\left(2x - \frac{1}{5}\right)$

85. $(1.2x + 3)^2$ **86.** $(1.5y - 3)^2$

87. $(1.5x - 4)(1.5x + 4)$ **88.** $(2.5y + 3)(2.5y - 3)$

89. $5x(x + 1) - 3x(x + 1)$

90. $(2x - 1)(x + 3) + 3(x + 3)$

91. $(u + 2)(u - 2)(u^2 + 4)$ **92.** $(x + y)(x - y)(x^2 + y^2)$

In Exercises 93–96, find the product. (The expressions are not polynomials, but the formulas can still be used.)

93. $(\sqrt{x} + \sqrt{y})(\sqrt{x} - \sqrt{y})$ **94.** $(5 + \sqrt{x})(5 - \sqrt{x})$

95. $(x - \sqrt{5})^2$ **96.** $(x + \sqrt{3})^2$

97. Cost, Revenue, and Profit An electronics manufacturer can produce and sell x radios per week. The total cost C (in dollars) for producing x radios is $C = 73x + 25{,}000$ and the total revenue R (in dollars) is $R = 95x$.

(a) Find the profit P in terms of x.

(b) Find the profit obtained by selling 5000 radios per week.

98. Cost, Revenue, and Profit An artisan can produce and sell x hats per month. The total cost C (in dollars) for producing x hats is $C = 460 + 12x$ and the total revenue R (in dollars) is $R = 36x$.

(a) Find the profit P in terms of x.

(b) Find the profit obtained by selling 42 hats per month.

99. Compound Interest After 2 years, an investment of $500 compounded annually at an interest rate r will yield an amount of $500(1 + r)^2$.

(a) Write this polynomial in standard form.

(b) Use a calculator to evaluate the polynomial for the values of r shown in the table.

r	$2\frac{1}{2}\%$	3%	4%	$4\frac{1}{2}\%$	5%
$500(1 + r)^2$					

(c) What conclusion can you make from the table?

100. Compound Interest After 3 years, an investment of $1200 compounded annually at an interest rate r will yield an amount of $1200(1 + r)^3$.

(a) Write this polynomial in standard form.

(b) Use a calculator to evaluate the polynomial for the values of r shown in the table.

r	2%	3%	$3\frac{1}{2}\%$	4%	$4\frac{1}{2}\%$
$1200(1 + r)^3$					

(c) What conclusion can you make from the table?

101. Volume of a Box A take-out fast-food restaurant is constructing an open box by cutting squares from the corners of a piece of cardboard that is 18 centimeters by 26 centimeters (see figure). The edge of each cut-out square is x centimeters.

(a) Find the volume of the box in terms of x.

(b) Find the volume when $x = 1$, $x = 2$, and $x = 3$.

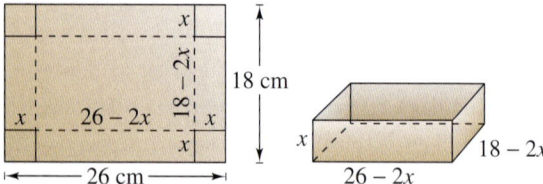

102. Volume of a Box An overnight shipping company is designing a closed box by cutting along the solid lines and folding along the broken lines on the rectangular piece of corrugated cardboard shown in the figure. The length and width of the rectangle are 45 centimeters and 15 centimeters, respectively.

(a) Find the volume of the shipping box in terms of x.

(b) Find the volume when $x = 3$, $x = 5$, and $x = 7$.

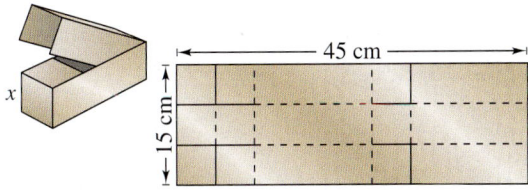

FIGURE FOR 102

103. *Geometry* Find the area of the shaded region in each figure. Write your result as a polynomial in standard form.

(a)
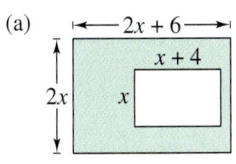

(b)
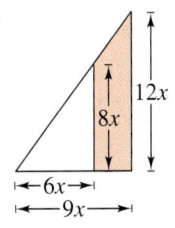

(c)
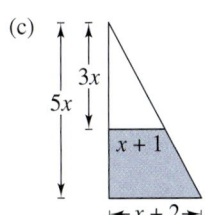

(d)
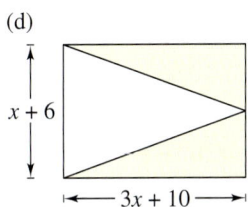

104. *Geometry* Find the area of the shaded region in each figure. Write your result as a polynomial in standard form.

(a)
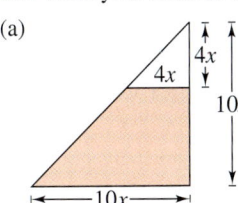

(b)
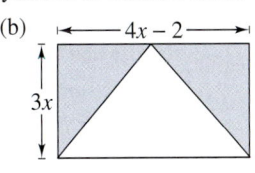

(c)
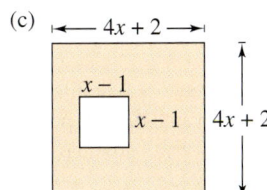

(d)
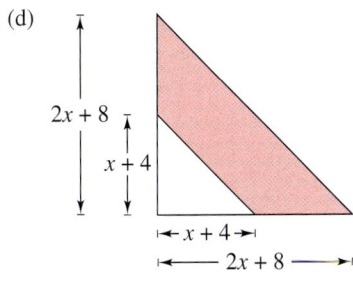

Geometry **In Exercises 105 and 106, find a polynomial that represents the total number of square feet for the floor plan shown in the figure.**

105.
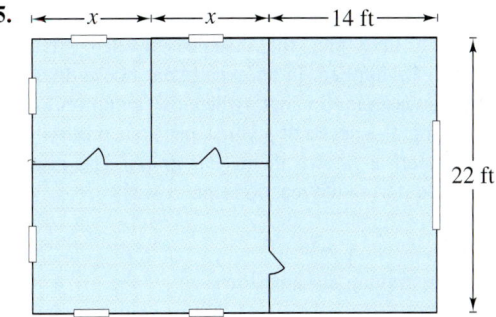

106.
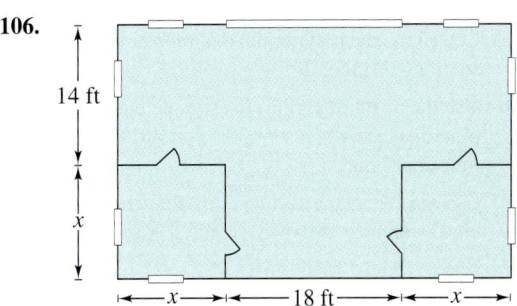

107. *Engineering* A uniformly distributed load is placed on a one-inch-wide steel beam. When the span of the beam is x feet and its depth is 6 inches, the safe load S (in pounds) is approximated by

$$S_6 = (0.06x^2 - 2.42x + 38.71)^2.$$

When the depth is 8 inches, the safe load is approximated by

$$S_8 = (0.08x^2 - 3.30x + 51.93)^2.$$

(a) Use the bar graph to estimate the difference in the safe loads for these two beams when the span is 12 feet.

(b) How does the difference in safe load change as the span increases?

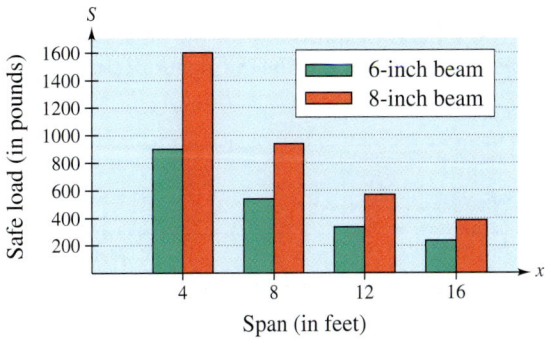

Model It

108. *Stopping Distance* The stopping distance of an automobile is the distance traveled during the driver's reaction time plus the distance traveled after the brakes are applied. In an experiment, these distances were measured (in feet) when the automobile was traveling at a speed of x miles per hour on dry, level pavement, as shown in the bar graph. The distance traveled during the reaction time R was

$$R = 1.1x$$

and the braking distance B was

$$B = 0.0475x^2 - 0.001x + 0.23.$$

(a) Determine the polynomial that represents the total stopping distance T.

(b) Use the result of part (a) to estimate the total stopping distance when $x = 30$, $x = 40$, and $x = 55$ miles per hour.

(c) Use the bar graph to make a statement about the total stopping distance required for increasing speeds.

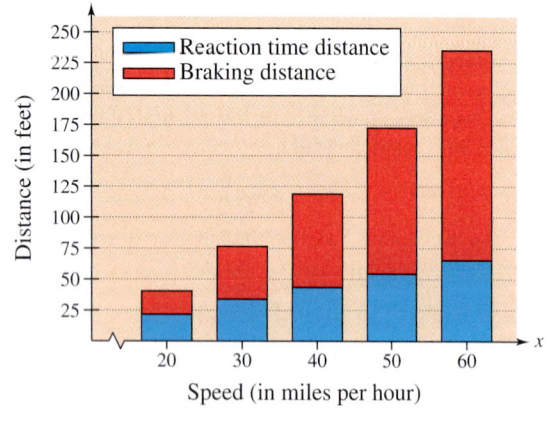

Geometry In Exercises 109 and 110, use the area model to write two different expressions for the area. Then equate the two expressions and name the algebraic property that is illustrated.

109.

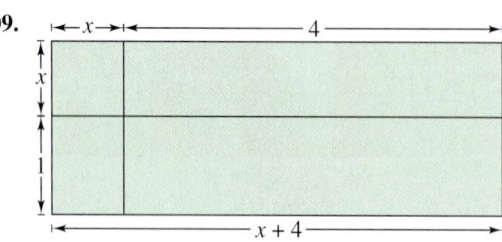

110.

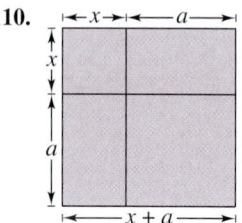

Synthesis

True or False? In Exercises 111 and 112, determine whether the statement is true or false. Justify your answer.

111. The product of two binomials is always a second-degree polynomial.

112. The sum of two binomials is always a binomial.

113. Find the degree of the product of two polynomials of degrees m and n.

114. Find the degree of the sum of two polynomials of degrees m and n if $m < n$.

115. *Writing* A student's homework paper included the following.

$$(x - 3)^2 = x^2 + 9$$

Write a paragraph fully explaining the error and give the correct method for squaring a binomial.

116. A third-degree polynomial and a fourth-degree polynomial are added.

(a) Can the sum be a fourth-degree polynomial? Explain or give an example.

(b) Can the sum be a second-degree polynomial? Explain or give an example.

(c) Can the sum be a seventh-degree polynomial? Explain or give an example.

117. *Think About It* Must the sum of two second-degree polynomials be a second-degree polynomial? If not, give an example.

118. *Think About It* When the polynomial $-x^3 + 3x^2 + 2x - 1$ is subtracted from an unknown polynomial, the difference is $5x^2 + 8$. If it is possible, find the unknown polynomial.

119. *Logical Reasoning* Verify that $(x + y)^2$ is not equal to $x^2 + y^2$ by letting $x = 3$ and $y = 4$ and evaluating both expressions. Are there any values of x and y for which $(x + y)^2 = x^2 + y^2$? Explain.

P.4   **Factoring Polynomials**

What you should learn

- Remove common factors from polynomials.
- Factor special polynomial forms.
- Factor trinomials as the product of two binomials.
- Factor polynomials by grouping.

Why you should learn it

Polynomial factoring can be used to solve real-life problems. For instance, in Exercise 135 on page 40, factoring is used to develop an alternative model for the rate of change of an autocatalytic chemical reaction.

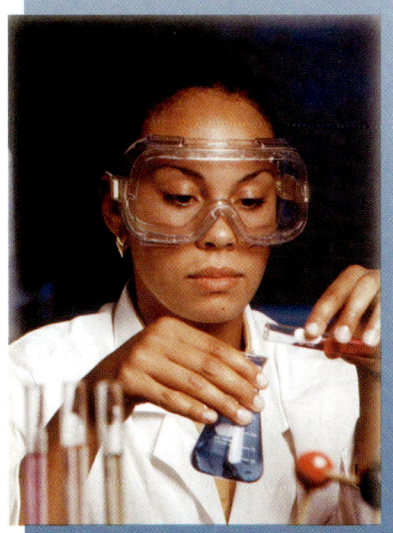

Mitch Wejnarowicz/The Image Works

Polynomials with Common Factors

The process of writing a polynomial as a product is called **factoring.** It is an important tool for solving equations and for simplifying rational expressions.

Unless noted otherwise, when you are asked to factor a polynomial, you can assume that you are looking for factors with integer coefficients. If a polynomial cannot be factored using integer coefficients, then it is **prime** or **irreducible over the integers.** For instance, the polynomial $x^2 - 3$ is irreducible over the integers. Over the *real numbers*, this polynomial can be factored as

$$x^2 - 3 = \left(x + \sqrt{3}\right)\left(x - \sqrt{3}\right).$$

A polynomial is **completely factored** when each of its factors is prime. For instance

$$x^3 - x^2 + 4x - 4 = (x - 1)(x^2 + 4) \qquad \text{Completely factored}$$

is completely factored, but

$$x^3 - x^2 - 4x + 4 = (x - 1)(x^2 - 4) \qquad \text{Not completely factored}$$

is not completely factored. Its complete factorization is

$$x^3 - x^2 - 4x + 4 = (x - 1)(x + 2)(x - 2).$$

The simplest type of factoring involves a polynomial that can be written as the product of a monomial and another polynomial. The technique used here is the Distributive Property, $a(b + c) = ab + ac$, in the *reverse* direction.

$$ab + ac = a(b + c) \qquad \text{a is a common factor.}$$

Removing (factoring out) any common factors is the first step in completely factoring a polynomial.

Example 1   Removing Common Factors

Factor each expression.

a. $6x^3 - 4x$

b. $-4x^2 + 12x - 16$

c. $(x - 2)(2x) + (x - 2)(3)$

Solution

a. $6x^3 - 4x = 2x(3x^2) - 2x(2)$ $\qquad$ $2x$ is a common factor.

$\qquad\qquad\quad = 2x(3x^2 - 2)$

b. $-4x^2 + 12x - 16 = -4(x^2) + (-4)(-3x) + (-4)4$ $\quad$ -4 is a common factor.

$\qquad\qquad\qquad\qquad = -4(x^2 - 3x + 4)$

c. $(x - 2)(2x) + (x - 2)(3) = (x - 2)(2x + 3)$ $\qquad$ $(x - 2)$ is a common factor.

✓**CHECKPOINT**   Now try Exercise 7.

Factoring Special Polynomial Forms

Some polynomials have special forms that arise from the special product forms on page 26. You should learn to recognize these forms so that you can factor such polynomials easily.

Factoring Special Polynomial Forms

Factored Form	*Example*

Difference of Two Squares

$$u^2 - v^2 = (u + v)(u - v)$$ $$9x^2 - 4 = (3x)^2 - 2^2 = (3x + 2)(3x - 2)$$

Perfect Square Trinomial

$$u^2 + 2uv + v^2 = (u + v)^2$$ $$x^2 + 6x + 9 = x^2 + 2(x)(3) + 3^2 = (x + 3)^2$$

$$u^2 - 2uv + v^2 = (u - v)^2$$ $$x^2 - 6x + 9 = x^2 - 2(x)(3) + 3^2 = (x - 3)^2$$

Sum or Difference of Two Cubes

$$u^3 + v^3 = (u + v)(u^2 - uv + v^2)$$ $$x^3 + 8 = x^3 + 2^3 = (x + 2)(x^2 - 2x + 4)$$

$$u^3 - v^3 = (u - v)(u^2 + uv + v^2)$$ $$27x^3 - 1 = (3x)^3 - 1^3 = (3x - 1)(9x^2 + 3x + 1)$$

One of the easiest special polynomial forms to factor is the difference of two squares. The factored form is always a set of *conjugate pairs*.

$$u^2 - v^2 = (u + v)(u - v)$$ Conjugate pairs

Difference   Opposite signs

To recognize perfect square terms, look for coefficients that are squares of integers and variables raised to *even powers*.

STUDY TIP

In Example 2, note that the first step in factoring a polynomial is to check for any common factors. Once the common factors are removed, it is often possible to recognize patterns that were not immediately obvious.

Example 2   **Removing a Common Factor First**

$$3 - 12x^2 = 3(1 - 4x^2)$$ 3 is a common factor.

$$= 3[1^2 - (2x)^2]$$

$$= 3(1 + 2x)(1 - 2x)$$ Difference of two squares

✔CHECKPOINT   Now try Exercise 21.

Example 3   **Factoring the Difference of Two Squares**

a. $(x + 2)^2 - y^2 = [(x + 2) + y][(x + 2) - y]$

$$= (x + 2 + y)(x + 2 - y)$$

b. $16x^4 - 81 = (4x^2)^2 - 9^2$

$$= (4x^2 + 9)(4x^2 - 9)$$ Difference of two squares

$$= (4x^2 + 9)[(2x)^2 - 3^2]$$

$$= (4x^2 + 9)(2x + 3)(2x - 3)$$ Difference of two squares

✔CHECKPOINT   Now try Exercise 25.

A **perfect square trinomial** is the square of a binomial, and it has the following form.

$$u^2 + 2uv + v^2 = (u + v)^2 \qquad \text{or} \qquad u^2 - 2uv + v^2 = (u - v)^2$$

Like signs   Like signs

Note that the first and last terms are squares and the middle term is twice the product of u and v.

Example 4   Factoring Perfect Square Trinomials

Factor each trinomial.

a. $x^2 - 10x + 25$

b. $16x^2 + 24x + 9$

Solution

a. $x^2 - 10x + 25 = x^2 - 2(x)(5) + 5^2 = (x - 5)^2$

b. $16x^2 + 24x + 9 = (4x)^2 + 2(4x)(3) + 3^2 = (4x + 3)^2$

✓**CHECKPOINT**   Now try Exercise 31.

The next two formulas show the sums and differences of cubes. Pay special attention to the signs of the terms.

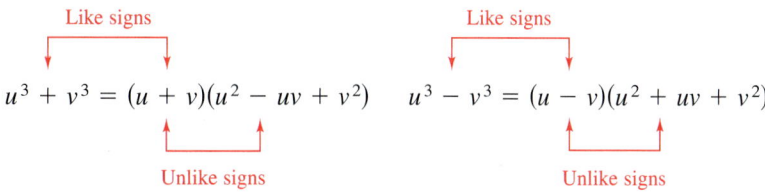

Like signs   Like signs

$$u^3 + v^3 = (u + v)(u^2 - uv + v^2) \qquad u^3 - v^3 = (u - v)(u^2 + uv + v^2)$$

Unlike signs   Unlike signs

Example 5   Factoring the Difference of Cubes

Factor $x^3 - 27$.

Solution

$$x^3 - 27 = x^3 - 3^3 \qquad\qquad \text{Rewrite 27 as } 3^3.$$

$$= (x - 3)(x^2 + 3x + 9) \qquad \text{Factor.}$$

✓**CHECKPOINT**   Now try Exercise 39.

Example 6   Factoring the Sum of Cubes

a. $y^3 + 8 = y^3 + 2^3$ Rewrite 8 as 2^3.

$$= (y + 2)(y^2 - 2y + 4) \qquad \text{Factor.}$$

b. $3(x^3 + 64) = 3(x^3 + 4^3)$ Rewrite 64 as 4^3.

$$= 3(x + 4)(x^2 - 4x + 16) \qquad \text{Factor.}$$

✓**CHECKPOINT**   Now try Exercise 41.

Exploration

Rewrite $u^6 - v^6$ as the difference of two squares. Then find a formula for completely factoring $u^6 - v^6$. Use your formula to factor $x^6 - 1$ and $x^6 - 64$ completely.

Trinomials with Binomial Factors

To factor a trinomial of the form $ax^2 + bx + c$, use the following pattern.

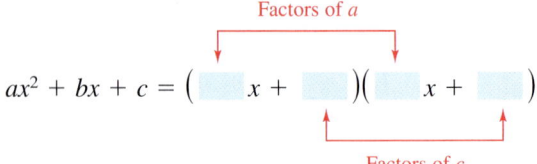

Factors of a

$$ax^2 + bx + c = (\quad x + \quad)(\quad x + \quad)$$

Factors of c

The goal is to find a combination of factors of a and c such that the outer and inner products add up to the middle term bx. For instance, in the trinomial $6x^2 + 17x + 5$, you can write all possible factorizations and determine which one has outer and inner products that add up to $17x$.

$$(6x + 5)(x + 1), \ (6x + 1)(x + 5), \ (2x + 1)(3x + 5), \ (2x + 5)(3x + 1)$$

You can see that $(2x + 5)(3x + 1)$ is the correct factorization because the outer (O) and inner (I) products add up to $17x$.

$$\text{F} \quad \text{O} \quad \text{I} \quad \text{L} \qquad \text{O} + \text{I}$$

$$(2x + 5)(3x + 1) = 6x^2 + 2x + 15x + 5 = 6x^2 + 17x + 5.$$

Example 7 Factoring a Trinomial: Leading Coefficient Is 1

Factor $x^2 - 7x + 12$.

Solution

The possible factorizations are

$$(x - 2)(x - 6), \quad (x - 1)(x - 12), \quad \text{and} \quad (x - 3)(x - 4).$$

Testing the middle term, you will find the correct factorization to be

$$x^2 - 7x + 12 = (x - 3)(x - 4).$$

✓CHECKPOINT   Now try Exercise 51.

Example 8 Factoring a Trinomial: Leading Coefficient Is Not 1

Factor $2x^2 + x - 15$.

Solution

The eight possible factorizations are as follows.

$$(2x - 1)(x + 15) \qquad (2x + 1)(x - 15)$$
$$(2x - 3)(x + 5) \qquad (2x + 3)(x - 5)$$
$$(2x - 5)(x + 3) \qquad (2x + 5)(x - 3)$$
$$(2x - 15)(x + 1) \qquad (2x + 15)(x - 1)$$

Testing the middle term, you will find the correct factorization to be

$$2x^2 + x - 15 = (2x - 5)(x + 3). \qquad \text{O} + \text{I} = 6x - 5x = x$$

✓CHECKPOINT   Now try Exercise 59.

STUDY TIP

Factoring a trinomial can involve trial and error. However, once you have produced the factored form, it is an easy matter to check your answer. For instance, you can verify the factorization in Example 7 by multiplying out the expression $(x - 3)(x - 4)$ to see that you obtain the original trinomial, $x^2 - 7x + 12$.

Factoring by Grouping

Sometimes polynomials with more than three terms can be factored by a method called **factoring by grouping.** It is not always obvious which terms to group, and sometimes several different groupings will work.

Example 9   Factoring by Grouping

Use factoring by grouping to factor $x^3 - 2x^2 - 3x + 6$.

Solution

$$x^3 - 2x^2 - 3x + 6 = (x^3 - 2x^2) - (3x - 6) \qquad \text{Group terms.}$$

$$= x^2(x - 2) - 3(x - 2) \qquad \text{Factor each group.}$$

$$= (x - 2)(x^2 - 3) \qquad \text{Distributive Property}$$

✔CHECKPOINT   Now try Exercise 67.

Factoring a trinomial can involve quite a bit of trial and error. Some of this trial and error can be lessened by using factoring by grouping. The key to this method of factoring is knowing how to rewrite the middle term. In general, to factor a trinomial $ax^2 + bx + c$ by grouping, choose factors of the product ac that add up to b and use these factors to rewrite the middle term. This technique is illustrated in Example 10.

Example 10   Factoring a Trinomial by Grouping

Use factoring by grouping to factor $2x^2 + 5x - 3$.

Solution

In the trinomial $2x^2 + 5x - 3$, $a = 2$ and $c = -3$, which implies that the product ac is -6. Now, -6 factors as $(6)(-1)$ and $6 - 1 = 5 = b$. So, you can rewrite the middle term as $5x = 6x - x$. This produces the following.

$$2x^2 + 5x - 3 = 2x^2 + 6x - x - 3 \qquad \text{Rewrite middle term.}$$

$$= (2x^2 + 6x) - (x + 3) \qquad \text{Group terms.}$$

$$= 2x(x + 3) - (x + 3) \qquad \text{Factor groups.}$$

$$= (x + 3)(2x - 1) \qquad \text{Distributive Property}$$

So, the trinomial factors as $2x^2 + 5x - 3 = (x + 3)(2x - 1)$.

✔CHECKPOINT   Now try Exercise 73.

> **STUDY TIP**
>
> Another way to factor the polynomial in Example 9 is to group the terms as follows.
>
> $x^3 - 2x^2 - 3x + 6$
>
> $= (x^3 - 3x) - (2x^2 - 6)$
>
> $= x(x^2 - 3) - 2(x^2 - 3)$
>
> $= (x^2 - 3)(x - 2)$
>
> As you can see, you obtain the same result as in Example 9.

Guidelines for Factoring Polynomials

1. Factor out any common factors using the Distributive Property.

2. Factor according to one of the special polynomial forms.

3. Factor as $ax^2 + bx + c = (mx + r)(nx + s)$.

4. Factor by grouping.

P.4 | Exercises

VOCABULARY CHECK

In Exercises 1–4, fill in the blanks.

1. The process of writing a polynomial as a product is called _____.

2. If a polynomial cannot be factored with integer coefficients it is called prime or _____ over the integers.

3. A polynomial is _____ _____ when each of its factors is prime.

4. If a polynomial has more than three terms, a method of factoring called _____ _____ _____ may be used.

5. Match the factored form of the polynomial with its name.
   (a) $u^2 - v^2 = (u + v)(u - v)$ (i) Perfect square trinomial
   (b) $u^3 - v^3 = (u - v)(u^2 + uv + v^2)$ (ii) Difference of two squares
   (c) $u^2 - 2uv + v^2 = (u - v)^2$ (iii) Difference of two cubes

---

In Exercises 1–4, find the greatest common factor of the expressions.

1. 90, 300
2. 36, 84, 294
3. $12x^2y^3$, $18x^2y$, $24x^3y^2$
4. $15(x + 2)^3$, $42x(x + 2)^2$

In Exercises 5–12, factor out the common factor.

5. $3x + 6$
6. $5y - 30$
7. $2x^3 - 6x$
8. $4x^3 - 6x^2 + 12x$
9. $x(x - 1) + 6(x - 1)$
10. $3x(x + 2) - 4(x + 2)$
11. $(x + 3)^2 - 4(x + 3)$
12. $(3x - 1)^2 + (3x - 1)$

In Exercises 13–18, find the greatest common factor such that the remaining factors have only integer coefficients.

13. $\frac{1}{2}x + 4$
14. $\frac{1}{3}y + 5$
15. $\frac{1}{2}x^3 + 2x^2 - 5x$
16. $\frac{1}{3}y^4 - 5y^2 + 2y$
17. $\frac{2}{3}x(x - 3) - 4(x - 3)$
18. $\frac{4}{5}y(y + 1) - 2(y + 1)$

In Exercises 19–28, completely factor the difference of two squares.

19. $x^2 - 81$
20. $x^2 - 49$
21. $32y^2 - 18$
22. $4 - 36y^2$
23. $16x^2 - \frac{1}{9}$
24. $\frac{4}{25}y^2 - 64$
25. $(x - 1)^2 - 4$
26. $25 - (z + 5)^2$
27. $9u^2 - 4v^2$
28. $25x^2 - 16y^2$

In Exercises 29–38, factor the perfect square trinomial.

29. $x^2 - 4x + 4$
30. $x^2 + 10x + 25$
31. $4t^2 + 4t + 1$
32. $9x^2 - 12x + 4$
33. $25y^2 - 10y + 1$
34. $36y^2 - 108y + 81$

35. $9u^2 + 24uv + 16v^2$
36. $4x^2 - 4xy + y^2$
37. $x^2 - \frac{4}{3}x + \frac{4}{9}$
38. $z^2 + z + \frac{1}{4}$

In Exercises 39–46, factor the sum or difference of cubes.

39. $x^3 - 8$
40. $x^3 - 27$
41. $y^3 + 64$
42. $z^3 + 125$
43. $8t^3 - 1$
44. $27x^3 + 8$
45. $u^3 + 27v^3$
46. $64x^3 - y^3$

In Exercises 47–50, factor a negative real number from the polynomial and then write the polynomial factor in standard form.

47. $25 - 5x^2$
48. $5 + 3y^2 - y^3$
49. $-2t^3 + 4t + 6$
50. $-3x^5 - 3x^2 + 6x + 9$

In Exercises 51–64, factor the trinomial.

51. $x^2 + x - 2$
52. $x^2 + 5x + 6$
53. $s^2 - 5s + 6$
54. $t^2 - t - 6$
55. $20 - y - y^2$
56. $24 + 5z - z^2$
57. $x^2 - 30x + 200$
58. $x^2 - 13x + 42$
59. $3x^2 - 5x + 2$
60. $2x^2 - x - 1$
61. $5x^2 + 26x + 5$
62. $12x^2 + 7x + 1$
63. $-9z^2 + 3z + 2$
64. $-5u^2 - 13u + 6$

In Exercises 65–72, factor by grouping.

65. $x^3 - x^2 + 2x - 2$
66. $x^3 + 5x^2 - 5x - 25$
67. $2x^3 - x^2 - 6x + 3$
68. $5x^3 - 10x^2 + 3x - 6$
69. $6 + 2x - 3x^3 - x^4$
70. $x^5 + 2x^3 + x^2 + 2$
71. $6x^3 - 2x + 3x^2 - 1$
72. $8x^5 - 6x^2 + 12x^3 - 9$

In Exercises 73–78, factor the trinomial by grouping.

73. $3x^2 + 10x + 8$

74. $2x^2 + 9x + 9$

75. $6x^2 + x - 2$

76. $6x^2 - x - 15$

77. $15x^2 - 11x + 2$

78. $12x^2 - 13x + 1$

In Exercises 79–112, completely factor the expression.

79. $6x^2 - 54$

80. $12x^2 - 48$

81. $x^3 - 4x^2$

82. $x^3 - 9x$

83. $x^2 - 2x + 1$

84. $16 + 6x - x^2$

85. $1 - 4x + 4x^2$

86. $-9x^2 + 6x - 1$

87. $2x^2 + 4x - 2x^3$

88. $2y^3 - 7y^2 - 15y$

89. $9x^2 + 10x + 1$

90. $13x + 6 + 5x^2$

91. $\frac{1}{81}x^2 + \frac{2}{9}x - 8$

92. $\frac{1}{8}x^2 - \frac{1}{96}x - \frac{1}{16}$

93. $3x^3 + x^2 + 15x + 5$

94. $5 - x + 5x^2 - x^3$

95. $x^4 - 4x^3 + x^2 - 4x$

96. $3u - 2u^2 + 6 - u^3$

97. $\frac{1}{4}x^3 + 3x^2 + \frac{3}{4}x + 9$

98. $\frac{1}{5}x^3 + x^2 - x - 5$

99. $(t - 1)^2 - 49$

100. $(x^2 + 1)^2 - 4x^2$

101. $(x^2 + 8)^2 - 36x^2$

102. $2t^3 - 16$

103. $5x^3 + 40$

104. $4x(2x - 1) + (2x - 1)^2$

105. $5(3 - 4x)^2 - 8(3 - 4x)(5x - 1)$

106. $2(x + 1)(x - 3)^2 - 3(x + 1)^2(x - 3)$

107. $7(3x + 2)^2(1 - x)^2 + (3x + 2)(1 - x)^3$

108. $7x(2)(x^2 + 1)(2x) - (x^2 + 1)^2(7)$

109. $3(x - 2)^2(x + 1)^4 + (x - 2)^3(4)(x + 1)^3$

110. $2x(x - 5)^4 - x^2(4)(x - 5)^3$

111. $5(x^6 + 1)^4(6x^5)(3x + 2)^3 + 3(3x + 2)^2(3)(x^6 + 1)^5$

112. $\dfrac{x^2}{2}(x^2 + 1)^4 - (x^2 + 1)^5$

Geometric Modeling **In Exercises 113–116, match the factoring formula with the correct "geometric factoring model." [The models are labeled (a), (b), (c), and (d).] For instance, a factoring model for**

$$2x^2 + 3x + 1 = (2x + 1)(x + 1)$$

is shown in the figure.

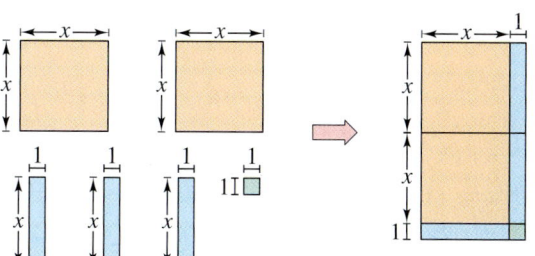

(a)

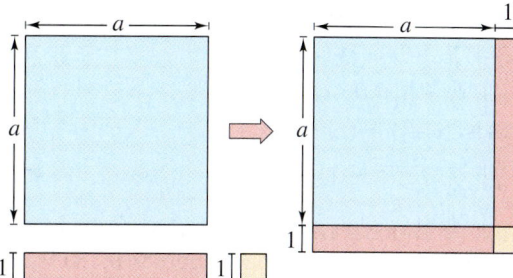

(b)

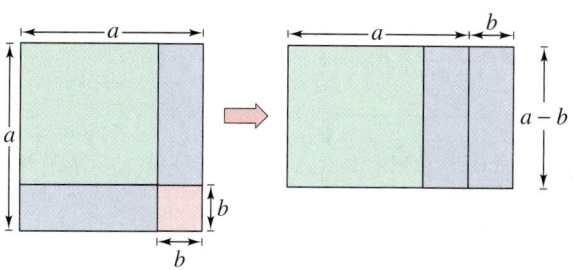

(c)

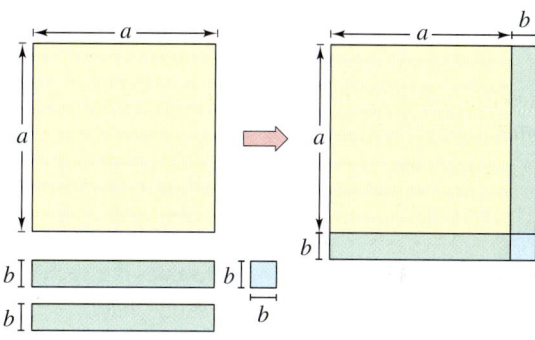

(d)

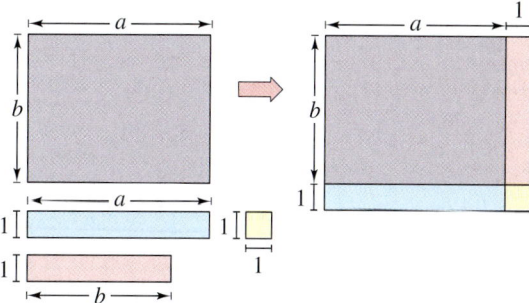

113. $a^2 - b^2 = (a + b)(a - b)$

114. $a^2 + 2ab + b^2 = (a + b)^2$

115. $a^2 + 2a + 1 = (a + 1)^2$

116. $ab + a + b + 1 = (a + 1)(b + 1)$

Geometric Modeling In Exercises 117–120, draw a "geometric factoring model" to represent the factorization.

117. $3x^2 + 7x + 2 = (3x + 1)(x + 2)$

118. $x^2 + 4x + 3 = (x + 3)(x + 1)$

119. $2x^2 + 7x + 3 = (2x + 1)(x + 3)$

120. $x^2 + 3x + 2 = (x + 2)(x + 1)$

Geometry In Exercises 121–124, write an expression in factored form for the area of the shaded portion of the figure.

121.

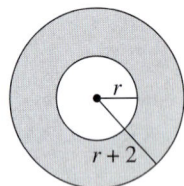

122.

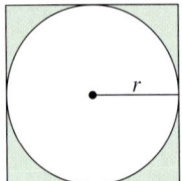

123.

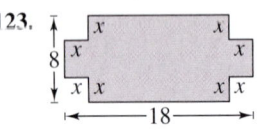

124.

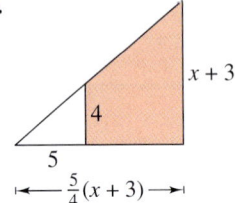

In Exercises 125–128, find all values of b for which the trinomial can be factored.

125. $x^2 + bx - 15$ **126.** $x^2 + bx + 50$

127. $x^2 + bx - 12$ **128.** $x^2 + bx + 24$

In Exercises 129–132, find two integer values of c such that the trinomial can be factored. (There are many correct answers.)

129. $2x^2 + 5x + c$ **130.** $3x^2 - 10x + c$

131. $3x^2 - x + c$ **132.** $2x^2 + 9x + c$

133. *Error Analysis* Describe the error.

$$9x^2 - 9x - 54 = (3x + 6)(3x - 9)$$
$$= 3(x + 2)(x - 3)$$

134. *Think About It* Is $(3x - 6)(x + 1)$ completely factored? Explain.

135. *Chemistry* The rate of change of an autocatalytic chemical reaction is $kQx - kx^2$, where Q is the amount of the original substance, x is the amount of substance formed, and k is a constant of proportionality. Factor the expression.

Model It

136. *Geometry* The volume V of concrete used to make the cylindrical concrete storage tank shown in the figure is $V = \pi R^2 h - \pi r^2 h$ where R is the outside radius, r is the inside radius, and h is the height of the storage tank.

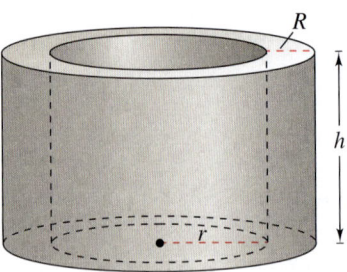

(a) Factor the expression for the volume.

(b) From the result of part (a), show that the volume of concrete is

2π(average radius)(thickness of the tank)h.

(c) An 80-pound bag of concrete mix yields $\frac{3}{5}$ cubic foot of concrete. Find the number of bags required to construct a concrete storage tank having the following dimensions.

Outside radius, $R = 4$ feet

Inside radius, $r = 3\frac{2}{3}$ feet

Height, h feet

 (d) Use the *table* feature of a graphing utility to create a table showing the number of bags of concrete required to construct the storage tank in part (c) with heights of $h = \frac{1}{2}$, $h = 1$, $h = \frac{3}{2}$, $h = 2, \ldots ,$ $h = 6$ feet.

Synthesis

True or False? In Exercises 137 and 138, determine whether the statement is true or false. Justify your answer.

137. The difference of two perfect squares can be factored as the product of conjugate pairs.

138. The sum of two perfect squares can be factored as the binomial sum squared.

139. Factor $x^{2n} - y^{2n}$ completely.

140. Factor $x^{3n} + y^{3n}$ completely.

141. Factor $x^{3n} - y^{2n}$ completely.

142. *Writing* Explain what is meant when it is said that a polynomial is in factored form.

143. Give an example of a polynomial that is prime with respect to the integers.

P.5   Rational Expressions

What you should learn

- Find domains of algebraic expressions.
- Simplify rational expressions.
- Add, subtract, multiply, and divide rational expressions.
- Simplify complex fractions and rewrite difference quotients.

Why you should learn it

Rational expressions can be used to solve real-life problems. For instance, in Exercise 84 on page 50, a rational expression is used to model the projected number of households banking and paying bills online from 2002 through 2007.

© Dex Images, Inc./Corbis

Domain of an Algebraic Expression

The set of real numbers for which an algebraic expression is defined is the **domain** of the expression. Two algebraic expressions are **equivalent** if they have the same domain and yield the same values for all numbers in their domain. For instance, $(x + 1) + (x + 2)$ and $2x + 3$ are equivalent because

$$(x + 1) + (x + 2) = x + 1 + x + 2$$
$$= x + x + 1 + 2$$
$$= 2x + 3.$$

Example 1   Finding the Domain of an Algebraic Expression

a. The domain of the polynomial

$$2x^3 + 3x + 4$$

is the set of all real numbers. In fact, the domain of any polynomial is the set of all real numbers, unless the domain is specifically restricted.

b. The domain of the radical expression

$$\sqrt{x - 2}$$

is the set of real numbers greater than or equal to 2, because the square root of a negative number is not a real number.

c. The domain of the expression

$$\frac{x + 2}{x - 3}$$

is the set of all real numbers except $x = 3$, which would result in division by zero, which is undefined.

✓CHECKPOINT   Now try Exercise 1.

The quotient of two algebraic expressions is a *fractional expression*. Moreover, the quotient of two *polynomials* such as

$$\frac{1}{x}, \qquad \frac{2x - 1}{x + 1}, \qquad \text{or} \qquad \frac{x^2 - 1}{x^2 + 1}$$

is a **rational expression.** Recall that a fraction is in simplest form if its numerator and denominator have no factors in common aside from ±1. To write a fraction in simplest form, divide out common factors.

$$\frac{a \cdot \cancel{c}}{b \cdot \cancel{c}} = \frac{a}{b}, \quad c \neq 0$$

The key to success in simplifying rational expressions lies in your ability to *factor* polynomials.

Simplifying Rational Expressions

When simplifying rational expressions, be sure to factor each polynomial completely before concluding that the numerator and denominator have no factors in common.

In this text, when a rational expression is written, the domain is usually not listed with the expression. It is *implied* that the real numbers that make the denominator zero are excluded from the expression. Also, when performing operations with rational expressions, this text follows the convention of listing *by the simplified expression* all values of x that must be specifically excluded from the domain in order to make the domains of the simplified and original expressions agree.

Example 2   Simplifying a Rational Expression

Write $\dfrac{x^2 + 4x - 12}{3x - 6}$ in simplest form.

Solution

$$\frac{x^2 + 4x - 12}{3x - 6} = \frac{(x + 6)(x - 2)}{3(x - 2)} \qquad \text{Factor completely.}$$

$$= \frac{x + 6}{3}, \qquad x \neq 2 \qquad \text{Divide out common factors.}$$

Note that the original expression is undefined when $x = 2$ (because division by zero is undefined). To make sure that the simplified expression is *equivalent* to the original expression, you must restrict the domain of the simplified expression by excluding the value $x = 2$.

✔**CHECKPOINT**   Now try Exercise 19.

Sometimes it may be necessary to change the sign of a factor to simplify a rational expression, as shown in Example 3.

Example 3   Simplifying Rational Expressions

Write $\dfrac{12 + x - x^2}{2x^2 - 9x + 4}$ in simplest form.

Solution

$$\frac{12 + x - x^2}{2x^2 - 9x + 4} = \frac{(4 - x)(3 + x)}{(2x - 1)(x - 4)} \qquad \text{Factor completely.}$$

$$= \frac{-(x - 4)(3 + x)}{(2x - 1)(x - 4)} \qquad (4 - x) = -(x - 4)$$

$$= -\frac{3 + x}{2x - 1}, \qquad x \neq 4 \qquad \text{Divide out common factors.}$$

✔**CHECKPOINT**   Now try Exercise 25.

STUDY TIP

In Example 2, do not make the mistake of trying to simplify further by dividing out terms.

$$\frac{x + 6}{3} = \frac{x + 6}{3} = x + 2$$

Remember that to simplify fractions, divide out common *factors*, not terms.

Operations with Rational Expressions

To multiply or divide rational expressions, use the properties of fractions discussed in Section P.1. Recall that to divide fractions, you invert the divisor and multiply.

Example 4   **Multiplying Rational Expressions**

$$\frac{2x^2 + x - 6}{x^2 + 4x - 5} \cdot \frac{x^3 - 3x^2 + 2x}{4x^2 - 6x} = \frac{(2x - 3)(x + 2)}{(x + 5)(x - 1)} \cdot \frac{x(x - 2)(x - 1)}{2x(2x - 3)}$$

$$= \frac{(x + 2)(x - 2)}{2(x + 5)}, \qquad x \neq 0, x \neq 1, x \neq \tfrac{3}{2}$$

✓**CHECKPOINT**   Now try Exercise 39.

In Example 4 the restrictions $x \neq 0$, $x \neq 1$, and $x \neq \tfrac{3}{2}$ are listed with the simplified expression in order to make the two domains agree. Note that the value $x = -5$ is excluded from both domains, so it is not necessary to list this value.

Example 5   **Dividing Rational Expressions**

$$\frac{x^3 - 8}{x^2 - 4} \div \frac{x^2 + 2x + 4}{x^3 + 8} = \frac{x^3 - 8}{x^2 - 4} \cdot \frac{x^3 + 8}{x^2 + 2x + 4} \qquad \text{\color{red}Invert and multiply.}$$

$$= \frac{(x - 2)(x^2 + 2x + 4)}{(x + 2)(x - 2)} \cdot \frac{(x + 2)(x^2 - 2x + 4)}{(x^2 + 2x + 4)}$$

$$= x^2 - 2x + 4, \qquad x \neq \pm 2 \qquad \text{\color{red}Divide out common factors.}$$

✓**CHECKPOINT**   Now try Exercise 41.

To add or subtract rational expressions, you can use the LCD (least common denominator) method or the *basic definition*

$$\frac{a}{b} \pm \frac{c}{d} = \frac{ad \pm bc}{bd}, \qquad b \neq 0, d \neq 0. \qquad \text{\color{red}Basic definition}$$

This definition provides an efficient way of adding or subtracting *two* fractions that have no common factors in their denominators.

Example 6   **Subtracting Rational Expressions**

$$\frac{x}{x - 3} - \frac{2}{3x + 4} = \frac{x(3x + 4) - 2(x - 3)}{(x - 3)(3x + 4)} \qquad \text{\color{red}Basic definition}$$

$$= \frac{3x^2 + 4x - 2x + 6}{(x - 3)(3x + 4)} \qquad \text{\color{red}Distributive Property}$$

$$= \frac{3x^2 + 2x + 6}{(x - 3)(3x + 4)} \qquad \text{\color{red}Combine like terms.}$$

✓**CHECKPOINT**   Now try Exercise 49.

STUDY TIP

When subtracting rational expressions, remember to distribute the negative sign to all the terms in the quantity that is being subtracted.

For three or more fractions, or for fractions with a repeated factor in the denominators, the LCD method works well. Recall that the least common denominator of several fractions consists of the product of all prime factors in the denominators, with each factor given the highest power of its occurrence in any denominator. Here is a numerical example.

$$\frac{1}{6} + \frac{3}{4} - \frac{2}{3} = \frac{1 \cdot 2}{6 \cdot 2} + \frac{3 \cdot 3}{4 \cdot 3} - \frac{2 \cdot 4}{3 \cdot 4}$$ The LCD is 12.

$$= \frac{2}{12} + \frac{9}{12} - \frac{8}{12}$$

$$= \frac{3}{12}$$

$$= \frac{1}{4}$$

Sometimes the numerator of the answer has a factor in common with the denominator. In such cases the answer should be simplified. For instance, in the example above, $\frac{3}{12}$ was simplified to $\frac{1}{4}$.

Example 7   Combining Rational Expressions: The LCD Method

Perform the operations and simplify.

$$\frac{3}{x - 1} - \frac{2}{x} + \frac{x + 3}{x^2 - 1}$$

Solution

Using the factored denominators $(x - 1)$, x, and $(x + 1)(x - 1)$, you can see that the LCD is $x(x + 1)(x - 1)$.

$$\frac{3}{x - 1} - \frac{2}{x} + \frac{x + 3}{(x + 1)(x - 1)}$$

$$= \frac{3(x)(x + 1)}{x(x + 1)(x - 1)} - \frac{2(x + 1)(x - 1)}{x(x + 1)(x - 1)} + \frac{(x + 3)(x)}{x(x + 1)(x - 1)}$$

$$= \frac{3(x)(x + 1) - 2(x + 1)(x - 1) + (x + 3)(x)}{x(x + 1)(x - 1)}$$

$$= \frac{3x^2 + 3x - 2x^2 + 2 + x^2 + 3x}{x(x + 1)(x - 1)}$$ Distributive Property

$$= \frac{3x^2 - 2x^2 + x^2 + 3x + 3x + 2}{x(x + 1)(x - 1)}$$ Group like terms.

$$= \frac{2x^2 + 6x + 2}{x(x + 1)(x - 1)}$$ Combine like terms.

$$= \frac{2(x^2 + 3x + 1)}{x(x + 1)(x - 1)}$$ Factor.

✔CHECKPOINT   Now try Exercise 51.

Complex Fractions and the Difference Quotient

Fractional expressions with separate fractions in the numerator, denominator, or both are called **complex fractions.** Here are two examples.

$$\frac{\left(\dfrac{1}{x}\right)}{x^2 + 1} \quad \text{and} \quad \frac{\left(\dfrac{1}{x}\right)}{\left(\dfrac{1}{x^2 + 1}\right)}$$

A complex fraction can be simplified by combining the fractions in its numerator into a single fraction and then combining the fractions in its denominator into a single fraction. Then invert the denominator and multiply.

Example 8   Simplifying a Complex Fraction

$$\frac{\left(\dfrac{2}{x} - 3\right)}{\left(1 - \dfrac{1}{x - 1}\right)} = \frac{\left[\dfrac{2 - 3(x)}{x}\right]}{\left[\dfrac{1(x - 1) - 1}{x - 1}\right]} \qquad \text{Combine fractions.}$$

$$= \frac{\left(\dfrac{2 - 3x}{x}\right)}{\left(\dfrac{x - 2}{x - 1}\right)} \qquad \text{Simplify.}$$

$$= \frac{2 - 3x}{x} \cdot \frac{x - 1}{x - 2} \qquad \text{Invert and multiply.}$$

$$= \frac{(2 - 3x)(x - 1)}{x(x - 2)}, \qquad x \neq 1$$

✔CHECKPOINT   Now try Exercise 57.

Another way to simplify a complex fraction is to multiply its numerator and denominator by the LCD of all fractions in its numerator and denominator. This method is applied to the fraction in Example 8 as follows.

$$\frac{\left(\dfrac{2}{x} - 3\right)}{\left(1 - \dfrac{1}{x - 1}\right)} = \frac{\left(\dfrac{2}{x} - 3\right)}{\left(1 - \dfrac{1}{x - 1}\right)} \cdot \frac{x(x - 1)}{x(x - 1)} \qquad \text{LCD is } x(x - 1).$$

$$= \frac{\left(\dfrac{2 - 3x}{x}\right) \cdot x(x - 1)}{\left(\dfrac{x - 2}{x - 1}\right) \cdot x(x - 1)}$$

$$= \frac{(2 - 3x)(x - 1)}{x(x - 2)}, \qquad x \neq 1$$

The next three examples illustrate some methods for simplifying rational expressions involving negative exponents and radicals. These types of expressions occur frequently in calculus.

To simplify an expression with negative exponents, one method is to begin by factoring out the common factor with the *smaller* exponent. Remember that when factoring, you *subtract* exponents. For instance, in $3x^{-5/2} + 2x^{-3/2}$ the smaller exponent is $-\frac{5}{2}$ and the common factor is $x^{-5/2}$.

$$3x^{-5/2} + 2x^{-3/2} = x^{-5/2}[3(1) + 2x^{-3/2-(-5/2)}]$$

$$= x^{-5/2}(3 + 2x^1)$$

$$= \frac{3 + 2x}{x^{5/2}}$$

Example 9   **Simplifying an Expression**

Simplify the following expression containing negative exponents.

$$x(1 - 2x)^{-3/2} + (1 - 2x)^{-1/2}$$

Solution

Begin by factoring out the common factor with the *smaller exponent*.

$$x(1 - 2x)^{-3/2} + (1 - 2x)^{-1/2} = (1 - 2x)^{-3/2}[x + (1 - 2x)^{(-1/2)-(-3/2)}]$$

$$= (1 - 2x)^{-3/2}[x + (1 - 2x)^1]$$

$$= \frac{1 - x}{(1 - 2x)^{3/2}}$$

✔CHECKPOINT   Now try Exercise 65.

A second method for simplifying an expression with negative exponents is shown in the next example.

Example 10   **Simplifying an Expression with Negative Exponents**

$$\frac{(4 - x^2)^{1/2} + x^2(4 - x^2)^{-1/2}}{4 - x^2}$$

$$= \frac{(4 - x^2)^{1/2} + x^2(4 - x^2)^{-1/2}}{4 - x^2} \cdot \frac{(4 - x^2)^{1/2}}{(4 - x^2)^{1/2}}$$

$$= \frac{(4 - x^2)^1 + x^2(4 - x^2)^0}{(4 - x^2)^{3/2}}$$

$$= \frac{4 - x^2 + x^2}{(4 - x^2)^{3/2}}$$

$$= \frac{4}{(4 - x^2)^{3/2}}$$

✔CHECKPOINT   Now try Exercise 67.

Example 11   **Rewriting a Difference Quotient**

The following expression from calculus is an example of a *difference quotient*.

$$\frac{\sqrt{x + h} - \sqrt{x}}{h}$$

Rewrite this expression by rationalizing its numerator.

Solution

$$\frac{\sqrt{x + h} - \sqrt{x}}{h} = \frac{\sqrt{x + h} - \sqrt{x}}{h} \cdot \frac{\sqrt{x + h} + \sqrt{x}}{\sqrt{x + h} + \sqrt{x}}$$

$$= \frac{\left(\sqrt{x + h}\right)^2 - \left(\sqrt{x}\right)^2}{h\left(\sqrt{x + h} + \sqrt{x}\right)}$$

$$= \frac{h}{h\left(\sqrt{x + h} + \sqrt{x}\right)}$$

$$= \frac{1}{\sqrt{x + h} + \sqrt{x}}, \quad h \neq 0$$

Notice that the original expression is undefined when $h = 0$. So, you must exclude $h = 0$ from the domain of the simplified expression so that the expressions are equivalent.

✓**CHECKPOINT**   Now try Exercise 73.

Difference quotients, such as that in Example 11, occur frequently in calculus. Often, they need to be rewritten in an equivalent form that can be evaluated when $h = 0$. Note that the equivalent form is not simpler than the original form, but it has the advantage that it is defined when $h = 0$.

P.5   Exercises

VOCABULARY CHECK: Fill in the blanks.

1. The set of real numbers for which an algebraic expression is defined is the _____ of the expression.

2. The quotient of two algebraic expressions is a fractional expression and the quotient of two polynomials is a _____ _____.

3. Fractional expressions with separate fractions in the numerator, denominator, or both are called _____ fractions.

4. To simplify an expression with negative exponents, it is possible to begin by factoring out the common factor with the _____ exponent.

5. Two algebraic expressions that have the same domain and yield the same values for all numbers in their domains are called _____.

6. An important rational expression, such as $\dfrac{(x + h)^2 - x^2}{h}$, that occurs in calculus is called a _____ _____.

In Exercises 1–8, find the domain of the expression.

1. $3x^2 - 4x + 7$

2. $2x^2 + 5x - 2$

3. $4x^3 + 3, \quad x \geq 0$

4. $6x^2 - 9, \quad x > 0$

5. $\dfrac{1}{x - 2}$

6. $\dfrac{x + 1}{2x + 1}$

7. $\sqrt{x + 1}$

8. $\sqrt{6 - x}$

In Exercises 9 and 10, find the missing factor in the numerator such that the two fractions are equivalent.

9. $\dfrac{5}{2x} = \dfrac{5()}{6x^2}$

10. $\dfrac{3}{4} = \dfrac{3()}{4(x + 1)}$

In Exercises 11–28, write the rational expression in simplest form.

11. $\dfrac{15x^2}{10x}$

12. $\dfrac{18y^2}{60y^5}$

13. $\dfrac{3xy}{xy + x}$

14. $\dfrac{2x^2y}{xy - y}$

15. $\dfrac{4y - 8y^2}{10y - 5}$

16. $\dfrac{9x^2 + 9x}{2x + 2}$

17. $\dfrac{x - 5}{10 - 2x}$

18. $\dfrac{12 - 4x}{x - 3}$

19. $\dfrac{y^2 - 16}{y + 4}$

20. $\dfrac{x^2 - 25}{5 - x}$

21. $\dfrac{x^3 + 5x^2 + 6x}{x^2 - 4}$

22. $\dfrac{x^2 + 8x - 20}{x^2 + 11x + 10}$

23. $\dfrac{y^2 - 7y + 12}{y^2 + 3y - 18}$

24. $\dfrac{x^2 - 7x + 6}{x^2 + 11x + 10}$

25. $\dfrac{2 - x + 2x^2 - x^3}{x^2 - 4}$

26. $\dfrac{x^2 - 9}{x^3 + x^2 - 9x - 9}$

27. $\dfrac{z^3 - 8}{z^2 + 2z + 4}$

28. $\dfrac{y^3 - 2y^2 - 3y}{y^3 + 1}$

In Exercises 29 and 30, complete the table. What can you conclude?

29.

x	0	1	2	3	4	5	6
$\dfrac{x^2 - 2x - 3}{x - 3}$							
$x + 1$							

30.

x	0	1	2	3	4	5	6
$\dfrac{x - 3}{x^2 - x - 6}$							
$\dfrac{1}{x + 2}$							

31. *Error Analysis* Describe the error.

$$\dfrac{5x^3}{2x^3 + 4} = \dfrac{5x^3}{2x^3 + 4} = \dfrac{5}{2 + 4} = \dfrac{5}{6}$$

32. *Error Analysis* Describe the error.

$$\dfrac{x^3 + 25x}{x^2 - 2x - 15} = \dfrac{x(x^2 + 25)}{(x - 5)(x + 3)}$$
$$= \dfrac{x(x + 5)(x - 5)}{(x - 5)(x + 3)} = \dfrac{x(x + 5)}{x + 3}$$

Geometry **In Exercises 33 and 34, find the ratio of the area of the shaded portion of the figure to the total area of the figure.**

33.

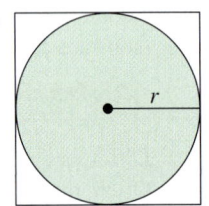

34.

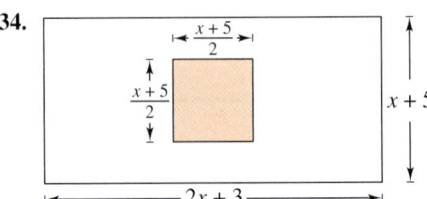

In Exercises 35–42, perform the multiplication or division and simplify.

35. $\dfrac{5}{x - 1} \cdot \dfrac{x - 1}{25(x - 2)}$

36. $\dfrac{x + 13}{x^3(3 - x)} \cdot \dfrac{x(x - 3)}{5}$

37. $\dfrac{r}{r - 1} \cdot \dfrac{r^2 - 1}{r^2}$

38. $\dfrac{4y - 16}{5y + 15} \cdot \dfrac{2y + 6}{4 - y}$

39. $\dfrac{t^2 - t - 6}{t^2 + 6t + 9} \cdot \dfrac{t + 3}{t^2 - 4}$

40. $\dfrac{x^2 + xy - 2y^2}{x^3 + x^2y} \cdot \dfrac{x}{x^2 + 3xy + 2y^2}$

41. $\dfrac{x^2 - 36}{x} \div \dfrac{x^3 - 6x^2}{x^2 + x}$

42. $\dfrac{x^2 - 14x + 49}{x^2 - 49} \div \dfrac{3x - 21}{x + 7}$

In Exercises 43–52, perform the addition or subtraction and simplify.

43. $\dfrac{5}{x-1} + \dfrac{x}{x-1}$

44. $\dfrac{2x-1}{x+3} + \dfrac{1-x}{x+3}$

45. $6 - \dfrac{5}{x+3}$

46. $\dfrac{3}{x-1} - 5$

47. $\dfrac{3}{x-2} + \dfrac{5}{2-x}$

48. $\dfrac{2x}{x-5} - \dfrac{5}{5-x}$

49. $\dfrac{1}{x^2-x-2} - \dfrac{x}{x^2-5x+6}$

50. $\dfrac{2}{x^2-x-2} + \dfrac{10}{x^2+2x-8}$

51. $-\dfrac{1}{x} + \dfrac{2}{x^2+1} + \dfrac{1}{x^3+x}$

52. $\dfrac{2}{x+1} + \dfrac{2}{x-1} + \dfrac{1}{x^2-1}$

Error Analysis   In Exercises 53 and 54, describe the error.

53. $\dfrac{x+4}{x+2} - \dfrac{3x-8}{x+2} = \dfrac{x+4-3x-8}{x+2}$

$$= \dfrac{-2x-4}{x+2} = \dfrac{-2(x+2)}{x+2} = -2$$

54. $\dfrac{6-x}{x(x+2)} + \dfrac{x+2}{x^2} + \dfrac{8}{x^2(x+2)}$

$$= \dfrac{x(6-x)+(x+2)^2+8}{x^2(x+2)}$$

$$= \dfrac{6x-x^2+x^2+4+8}{x^2(x+2)}$$

$$= \dfrac{6(x+2)}{x^2(x+2)} = \dfrac{6}{x^2}$$

In Exercises 55–60, simplify the complex fraction.

55. $\dfrac{\left(\dfrac{x}{2}-1\right)}{(x-2)}$

56. $\dfrac{(x-4)}{\left(\dfrac{x}{4}-\dfrac{4}{x}\right)}$

57. $\dfrac{\left[\dfrac{x^2}{(x+1)^2}\right]}{\left[\dfrac{x}{(x+1)^3}\right]}$

58. $\dfrac{\left(\dfrac{x^2-1}{x}\right)}{\left[\dfrac{(x-1)^2}{x}\right]}$

59. $\dfrac{\left(\sqrt{x}-\dfrac{1}{2\sqrt{x}}\right)}{\sqrt{x}}$

60. $\dfrac{\left(\dfrac{t^2}{\sqrt{t^2+1}}-\sqrt{t^2+1}\right)}{t^2}$

In Exercises 61–66, factor the expression by removing the common factor with the smaller exponent.

61. $x^5 - 2x^{-2}$

62. $x^5 - 5x^{-3}$

63. $x^2(x^2+1)^{-5} - (x^2+1)^{-4}$

64. $2x(x-5)^{-3} - 4x^2(x-5)^{-4}$

65. $2x^2(x-1)^{1/2} - 5(x-1)^{-1/2}$

66. $4x^3(2x-1)^{3/2} - 2x(2x-1)^{-1/2}$

In Exercises 67 and 68, simplify the expression.

67. $\dfrac{3x^{1/3} - x^{-2/3}}{3x^{-2/3}}$

68. $\dfrac{-x^3(1-x^2)^{-1/2} - 2x(1-x^2)^{1/2}}{x^4}$

In Exercises 69–72, simplify the difference quotient.

69. $\dfrac{\left(\dfrac{1}{x+h} - \dfrac{1}{x}\right)}{h}$

70. $\dfrac{\left[\dfrac{1}{(x+h)^2} - \dfrac{1}{x^2}\right]}{h}$

71. $\dfrac{\left(\dfrac{1}{x+h-4} - \dfrac{1}{x-4}\right)}{h}$

72. $\dfrac{\left(\dfrac{x+h}{x+h+1} - \dfrac{x}{x+1}\right)}{h}$

In Exercises 73–76, simplify the difference quotient by rationalizing the numerator.

73. $\dfrac{\sqrt{x+2} - \sqrt{x}}{2}$

74. $\dfrac{\sqrt{z-3} - \sqrt{z}}{3}$

75. $\dfrac{\sqrt{x+h+1} - \sqrt{x+1}}{h}$

76. $\dfrac{\sqrt{x+h-2} - \sqrt{x-2}}{h}$

Probability   In Exercises 77 and 78, consider an experiment in which a marble is tossed into a box whose base is shown in the figure. The probability that the marble will come to rest in the shaded portion of the box is equal to the ratio of the shaded area to the total area of the figure. Find the probability.

77.

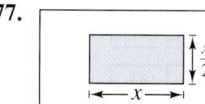

78.

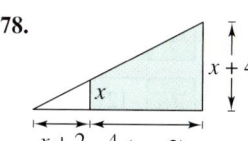

Errors Involving Exponents

Potential Error	Correct Form	Comment
$(x^2)^3 = x^5$	$(x^2)^3 = x^{2 \cdot 3} = x^6$	Multiply exponents when raising a power to a power.
$x^2 \cdot x^3 = x^6$	$x^2 \cdot x^3 = x^{2+3} = x^5$	Add exponents when multiplying powers with like bases.
$2x^3 = (2x)^3$	$2x^3 = 2(x^3)$	Exponents have priority over coefficients.
$\dfrac{1}{x^2 - x^3} = x^{-2} - x^{-3}$	Leave as $\dfrac{1}{x^2 - x^3}$.	Do not move term-by-term from denominator to numerator.

Errors Involving Radicals

Potential Error	Correct Form	Comment
$\sqrt{5x} = 5\sqrt{x}$	$\sqrt{5x} = \sqrt{5}\sqrt{x}$	Radicals apply to every factor inside the radical.
$\sqrt{x^2 + a^2} = x + a$	Leave as $\sqrt{x^2 + a^2}$.	Do not apply radicals term-by-term.
$\sqrt{-x + a} = -\sqrt{x - a}$	Leave as $\sqrt{-x + a}$.	Do not factor minus signs out of square roots.

Errors Involving Dividing Out

Potential Error	Correct Form	Comment
$\dfrac{a + bx}{a} = 1 + bx$	$\dfrac{a + bx}{a} = \dfrac{a}{a} + \dfrac{bx}{a} = 1 + \dfrac{b}{a}x$	Divide out common factors, not common terms.
$\dfrac{a + ax}{a} = a + x$	$\dfrac{a + ax}{a} = \dfrac{a(1 + x)}{a} = 1 + x$	Factor before dividing out.
$1 + \dfrac{x}{2x} = 1 + \dfrac{1}{x}$	$1 + \dfrac{x}{2x} = 1 + \dfrac{1}{2} = \dfrac{3}{2}$	Divide out common factors.

A good way to avoid errors is to *work slowly*, *write neatly*, and *talk to yourself*. Each time you write a step, ask yourself why the step is algebraically legitimate. You can justify the step below because *dividing the numerator and denominator by the same nonzero number produces an equivalent fraction*.

$$\frac{2x}{6} = \frac{2 \cdot x}{2 \cdot 3} = \frac{x}{3}$$

Example 1   Using the Property for Adding Fractions

Describe and correct the error. $\dfrac{1}{2x} + \dfrac{1}{3x} = \dfrac{1}{5x}$

Solution

When adding fractions, use the property for adding fractions: $\dfrac{1}{a} + \dfrac{1}{b} = \dfrac{b + a}{ab}$.

$$\frac{1}{2x} + \frac{1}{3x} = \frac{3x + 2x}{6x^2} = \frac{5x}{6x^2} = \frac{5}{6x}$$

✔**CHECKPOINT**   Now try Exercise 17.

Some Algebra of Calculus

In calculus it is often necessary to take a simplified algebraic expression and "unsimplify" it. See the following lists, taken from a standard calculus text.

Unusual Factoring

Expression	*Useful Calculus Form*	*Comment*
$\dfrac{5x^4}{8}$	$\dfrac{5}{8}x^4$	Write with fractional coefficient.
$\dfrac{x^2 + 3x}{-6}$	$-\dfrac{1}{6}(x^2 + 3x)$	Write with fractional coefficient.
$2x^2 - x - 3$	$2\left(x^2 - \dfrac{x}{2} - \dfrac{3}{2}\right)$	Factor out the leading coefficient.
$\dfrac{x}{2}(x + 1)^{-1/2} + (x + 1)^{1/2}$	$\dfrac{(x + 1)^{-1/2}}{2}[x + 2(x + 1)]$	Factor out factor with lowest power.

Writing with Negative Exponents

Expression	*Useful Calculus Form*	*Comment*
$\dfrac{9}{5x^3}$	$\dfrac{9}{5}x^{-3}$	Move the factor to the numerator and change the sign of the exponent.
$\dfrac{7}{\sqrt{2x - 3}}$	$7(2x - 3)^{-1/2}$	Move the factor to the numerator and change the sign of the exponent.

Writing a Fraction as a Sum

Expression	*Useful Calculus Form*	*Comment*
$\dfrac{x + 2x^2 + 1}{\sqrt{x}}$	$x^{1/2} + 2x^{3/2} + x^{-1/2}$	Divide each term by $x^{1/2}$.
$\dfrac{1 + x}{x^2 + 1}$	$\dfrac{1}{x^2 + 1} + \dfrac{x}{x^2 + 1}$	Rewrite the fraction as the sum of fractions.
$\dfrac{2x}{x^2 + 2x + 1}$	$\dfrac{2x + 2 - 2}{x^2 + 2x + 1}$	Add and subtract the same term.
	$= \dfrac{2x + 2}{x^2 + 2x + 1} - \dfrac{2}{(x + 1)^2}$	Rewrite the fraction as the difference of fractions.
$\dfrac{x^2 - 2}{x + 1}$	$x - 1 - \dfrac{1}{x + 1}$	Use long division. (See Section 3.3.)
$\dfrac{x + 7}{x^2 - x - 6}$	$\dfrac{2}{x - 3} - \dfrac{1}{x + 2}$	Use the method of partial fractions. (See Section 6.4.)

Inserting Factors and Terms

Expression	Useful Calculus Form	Comment
$(2x - 1)^3$	$\dfrac{1}{2}(2x - 1)^3(2)$	Multiply and divide by 2.
$7x^2(4x^3 - 5)^{1/2}$	$\dfrac{7}{12}(4x^3 - 5)^{1/2}(12x^2)$	Multiply and divide by 12.
$\dfrac{4x^2}{9} - 4y^2 = 1$	$\dfrac{x^2}{9/4} - \dfrac{y^2}{1/4} = 1$	Write with fractional denominators.
$\dfrac{x}{x + 1}$	$\dfrac{x + 1 - 1}{x + 1} = 1 - \dfrac{1}{x + 1}$	Add and subtract the same term.

The next five examples demonstrate many of the steps in the preceding lists.

Example 2 **Factors Involving Negative Exponents**

Factor $x(x + 1)^{-1/2} + (x + 1)^{1/2}$.

Solution

When multiplying factors with like bases, you add exponents. When factoring, you are undoing multiplication, and so you *subtract* exponents.

$$x(x + 1)^{-1/2} + (x + 1)^{1/2} = (x + 1)^{-1/2}[x(x + 1)^0 + (x + 1)^1]$$
$$= (x + 1)^{-1/2}[x + (x + 1)]$$
$$= (x + 1)^{-1/2}(2x + 1)$$

✓CHECKPOINT Now try Exercise 23.

Another way to simplify the expression in Example 2 is to multiply the expression by a fractional form of 1 and then use the Distributive Property.

$$x(x + 1)^{-1/2} + (x + 1)^{1/2} = [x(x + 1)^{-1/2} + (x + 1)^{1/2}] \cdot \frac{(x + 1)^{1/2}}{(x + 1)^{1/2}}$$

$$= \frac{x(x + 1)^0 + (x + 1)^1}{(x + 1)^{1/2}} = \frac{2x + 1}{\sqrt{x + 1}}$$

Example 3 **Inserting Factors in an Expression**

Insert the required factor: $\dfrac{x + 2}{(x^2 + 4x - 3)^2} = ()\dfrac{1}{(x^2 + 4x - 3)^2}(2x + 4)$.

Solution

The expression on the right side of the equation is twice the expression on the left side. To make both sides equal, insert a factor of $\frac{1}{2}$.

$$\frac{x + 2}{(x^2 + 4x - 3)^2} = \left(\frac{1}{2}\right)\frac{1}{(x^2 + 4x - 3)^2}(2x + 4)$$

Right side is multiplied and divided by 2.

✓CHECKPOINT Now try Exercise 25.

Example 4   **Rewriting Fractions**

Explain why the two expressions are equivalent.

$$\frac{4x^2}{9} - 4y^2 = \frac{x^2}{\frac{9}{4}} - \frac{y^2}{\frac{1}{4}}$$

Solution

To write the expression on the left side of the equation in the form given on the right side, multiply the numerators and denominators of both terms by $\frac{1}{4}$.

$$\frac{4x^2}{9} - 4y^2 = \frac{4x^2\left(\frac{1}{4}\right)}{9\left(\frac{1}{4}\right)} - 4y^2\frac{\left(\frac{1}{4}\right)}{\left(\frac{1}{4}\right)} = \frac{x^2}{\frac{9}{4}} - \frac{y^2}{\frac{1}{4}}$$

✓CHECKPOINT   Now try Exercise 29.

Example 5   **Rewriting with Negative Exponents**

Rewrite each expression using negative exponents.

a. $\dfrac{-4x}{(1 - 2x^2)^2}$ **b.** $\dfrac{2}{5x^3} - \dfrac{1}{\sqrt{x}} + \dfrac{3}{5(4x)^2}$

Solution

a. $\dfrac{-4x}{(1 - 2x^2)^2} = -4x(1 - 2x^2)^{-2}$

b. Begin by writing the second term in exponential form.

$$\frac{2}{5x^3} - \frac{1}{\sqrt{x}} + \frac{3}{5(4x)^2} = \frac{2}{5x^3} - \frac{1}{x^{1/2}} + \frac{3}{5(4x)^2}$$

$$= \frac{2}{5}x^{-3} - x^{-1/2} + \frac{3}{5}(4x)^{-2}$$

✓CHECKPOINT   Now try Exercise 39.

Example 6   **Writing a Fraction as a Sum of Terms**

Rewrite each fraction as the sum of three terms.

a. $\dfrac{x^2 - 4x + 8}{2x}$ **b.** $\dfrac{x + 2x^2 + 1}{\sqrt{x}}$

Solution

a. $\dfrac{x^2 - 4x + 8}{2x} = \dfrac{x^2}{2x} - \dfrac{4x}{2x} + \dfrac{8}{2x}$

$$= \frac{x}{2} - 2 + \frac{4}{x}$$

b. $\dfrac{x + 2x^2 + 1}{\sqrt{x}} = \dfrac{x}{x^{1/2}} + \dfrac{2x^2}{x^{1/2}} + \dfrac{1}{x^{1/2}}$

$$= x^{1/2} + 2x^{3/2} + x^{-1/2}$$

✓CHECKPOINT   Now try Exercise 43.

P.6 Exercises

VOCABULARY CHECK: Fill in the blanks.

1. To write the expression $\dfrac{2}{\sqrt{x}}$ with negative exponents, move $\sqrt{x}$ to the _____ and change the sign of the exponent.

2. When dividing fractions, multiply by the _____.

In Exercises 1–18, describe and correct the error.

1. $2x - (3y + 4) = 2x - 3y + 4$

2. $5z + 3(x - 2) = 5z + 3x - 2$

3. $\dfrac{4}{16x - (2x + 1)} = \dfrac{4}{14x + 1}$

4. $\dfrac{1 - x}{(5 - x)(-x)} = \dfrac{x - 1}{x(x - 5)}$

5. $(5z)(6z) = 30z$

6. $x(yz) = (xy)(xz)$

7. $a\left(\dfrac{x}{y}\right) = \dfrac{ax}{ay}$

8. $(4x)^2 = 4x^2$

9. $\sqrt{x + 9} = \sqrt{x} + 3$

10. $\sqrt{25 - x^2} = 5 - x$

11. $\dfrac{2x^2 + 1}{5x} = \dfrac{2x + 1}{5}$

12. $\dfrac{6x + y}{6x - y} = \dfrac{x + y}{x - y}$

13. $\dfrac{1}{a^{-1} + b^{-1}} = \left(\dfrac{1}{a + b}\right)^{-1}$

14. $\dfrac{1}{x + y^{-1}} = \dfrac{y}{x + 1}$

15. $(x^2 + 5x)^{1/2} = x(x + 5)^{1/2}$

16. $x(2x - 1)^2 = (2x^2 - x)^2$

17. $\dfrac{3}{x} + \dfrac{4}{y} = \dfrac{7}{x + y}$

18. $\dfrac{1}{2y} = (1/2)y$

In Exercises 19–38, insert the required factor in the parentheses.

19. $\dfrac{3x + 2}{5} = \dfrac{1}{5}(\quad)$

20. $\dfrac{7x^2}{10} = \dfrac{7}{10}(\quad)$

21. $\tfrac{2}{3}x^2 + \tfrac{1}{3}x + 5 = \tfrac{1}{3}(\quad)$

22. $\tfrac{3}{4}x + \tfrac{1}{2} = \tfrac{1}{4}(\quad)$

23. $x^2(x^3 - 1)^4 = (\quad)(x^3 - 1)^4(3x^2)$

24. $x(1 - 2x^2)^3 = (\quad)(1 - 2x^2)^3(-4x)$

25. $\dfrac{4x + 6}{(x^2 + 3x + 7)^3} = (\quad)\dfrac{1}{(x^2 + 3x + 7)^3}(2x + 3)$

26. $\dfrac{x + 1}{(x^2 + 2x - 3)^2} = (\quad)\dfrac{1}{(x^2 + 2x - 3)^2}(2x + 2)$

27. $\dfrac{3}{x} + \dfrac{5}{2x^2} - \dfrac{3}{2}x = (\quad)(6x + 5 - 3x^3)$

28. $\dfrac{(x - 1)^2}{169} + (y + 5)^2 = \dfrac{(x - 1)^3}{169(\quad)} + (y + 5)^2$

29. $\dfrac{9x^2}{25} + \dfrac{16y^2}{49} = \dfrac{x^2}{(\quad)} + \dfrac{y^2}{(\quad)}$

30. $\dfrac{3x^2}{4} - \dfrac{9y^2}{16} = \dfrac{x^2}{(\quad)} - \dfrac{y^2}{(\quad)}$

31. $\dfrac{x^2}{1/12} - \dfrac{y^2}{2/3} = \dfrac{12x^2}{(\quad)} - \dfrac{3y^2}{(\quad)}$

32. $\dfrac{x^2}{4/9} + \dfrac{y^2}{7/8} = \dfrac{9x^2}{(\quad)} + \dfrac{8y^2}{(\quad)}$

33. $x^{1/3} - 5x^{4/3} = x^{1/3}(\quad)$

34. $3(2x + 1)x^{1/2} + 4x^{3/2} = x^{1/2}(\quad)$

35. $(1 - 3x)^{4/3} - 4x(1 - 3x)^{1/3} = (1 - 3x)^{1/3}(\quad)$

36. $\dfrac{1}{2\sqrt{x}} + 5x^{3/2} - 10x^{5/2} = \dfrac{1}{2\sqrt{x}}(\quad)$

37. $\dfrac{1}{10}(2x + 1)^{5/2} - \dfrac{1}{6}(2x + 1)^{3/2} = \dfrac{(2x + 1)^{3/2}}{15}(\quad)$

38. $\dfrac{3}{7}(t + 1)^{7/3} - \dfrac{3}{4}(t + 1)^{4/3} = \dfrac{3(t + 1)^{4/3}}{28}(\quad)$

In Exercises 39–42, write the expression using negative exponents.

39. $\dfrac{3x^2}{(2x - 1)^3}$

40. $\dfrac{x + 1}{x(6 - x)^{1/2}}$

41. $\dfrac{4}{3x} + \dfrac{4}{x^4} - \dfrac{7x}{\sqrt[3]{2x}}$

42. $\dfrac{x}{x - 2} + \dfrac{1}{x^2} + \dfrac{8}{3(9x)^3}$

In Exercises 43–48, write the fraction as a sum of two or more terms.

43. $\dfrac{16 - 5x - x^2}{x}$

44. $\dfrac{x^3 - 5x^2 + 4}{x^2}$

45. $\dfrac{4x^3 - 7x^2 + 1}{x^{1/3}}$

46. $\dfrac{2x^5 - 3x^3 + 5x - 1}{x^{3/2}}$

47. $\dfrac{3 - 5x^2 - x^4}{\sqrt{x}}$

48. $\dfrac{x^3 - 5x^4}{3x^2}$

In Exercises 49–60, simplify the expression.

49. $\dfrac{-2(x^2 - 3)^{-3}(2x)(x + 1)^3 - 3(x + 1)^2(x^2 - 3)^{-2}}{[(x + 1)^3]^2}$

50. $\dfrac{x^5(-3)(x^2 + 1)^{-4}(2x) - (x^2 + 1)^{-3}(5)x^4}{(x^5)^2}$

51. $\dfrac{(6x + 1)^3(27x^2 + 2) - (9x^3 + 2x)(3)(6x + 1)^2(6)}{[(6x + 1)^3]^2}$

52. $\dfrac{(4x^2 + 9)^{1/2}(2) - (2x + 3)(\frac{1}{2})(4x^2 + 9)^{-1/2}(8x)}{[(4x^2 + 9)^{1/2}]^2}$

53. $\dfrac{(x + 2)^{3/4}(x + 3)^{-2/3} - (x + 3)^{1/3}(x + 2)^{-1/4}}{[(x + 2)^{3/4}]^2}$

54. $(2x - 1)^{1/2} - (x + 2)(2x - 1)^{-1/2}$

55. $\dfrac{2(3x - 1)^{1/3} - (2x + 1)(\frac{1}{3})(3x - 1)^{-2/3}(3)}{(3x - 1)^{2/3}}$

56. $\dfrac{(x + 1)(\frac{1}{2})(2x - 3x^2)^{-1/2}(2 - 6x) - (2x - 3x^2)^{1/2}}{(x + 1)^2}$

57. $\dfrac{1}{(x^2 + 4)^{1/2}} \cdot \dfrac{1}{2}(x^2 + 4)^{-1/2}(2x)$

58. $\dfrac{1}{x^2 - 6}(2x) + \dfrac{1}{2x + 5}(2)$

59. $(x^2 + 5)^{1/2}(\frac{3}{2})(3x - 2)^{1/2}(3)$
$\qquad + (3x - 2)^{3/2}(\frac{1}{2})(x^2 + 5)^{-1/2}(2x)$

60. $(3x + 2)^{-1/2}(3)(x - 6)^{1/2}(1)$
$\qquad + (x - 6)^3(-\frac{1}{2})(3x + 2)^{-3/2}(3)$

Model It

61. *Athletics*   An athlete has set up a course for training as part of her regimen in preparation for an upcoming triathlon. She is dropped off by a boat 2 miles from the nearest point on shore. The finish line is 4 miles down the coast and 2 miles inland (see figure). She can swim 2 miles per hour and run 6 miles per hour. The time t (in hours) required for her to reach the finish line can be approximated by the model

$$t = \frac{\sqrt{x^2 + 4}}{2} + \frac{\sqrt{(4 - x)^2 + 4}}{6}$$

where x is the distance down the coast (in miles) to which she swims and then leaves the water to start her run.

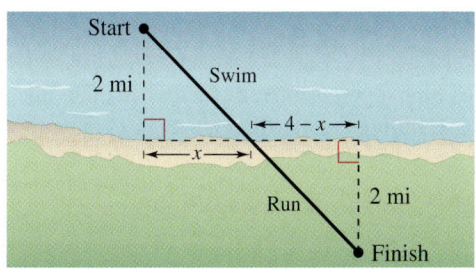

(a) Find the times required for the triathlete to finish when she swims to the points $x = 0.5$, $x = 1.0, \ldots, x = 3.5$, and $x = 4.0$ miles down the coast.

(b) Use your results from part (a) to determine the distance down the coast that will yield the minimum amount of time required for the triathlete to reach the finish line.

Model It (continued)

(c) The expression below was obtained using calculus. It can be used to find the minimum amount of time required for the triathlete to reach the finish line. Simplify the expression.

$$\tfrac{1}{2}x(x^2 + 4)^{-1/2} + \tfrac{1}{6}(x - 4)(x^2 - 8x + 20)^{-1/2}$$

62. (a) Verify that $y_1 = y_2$ analytically.

$$y_1 = x^2\left(\frac{1}{3}\right)(x^2 + 1)^{-2/3}(2x) + (x^2 + 1)^{1/3}(2x)$$

$$y_2 = \frac{2x(4x^2 + 3)}{3(x^2 + 1)^{2/3}}$$

(b) Complete the table and demonstrate the equality in part (a) numerically.

x	-2	-1	$-\frac{1}{2}$	0	1	2	$\frac{5}{2}$
y_1							
y_2							

Synthesis

True or False?   **In Exercises 63–66, determine whether the statement is true or false. Justify your answer.**

63. $x^{-1} + y^{-2} = \dfrac{y^2 + x}{xy^2}$

64. $\dfrac{1}{x^{-2} + y^{-1}} = x^2 + y$

65. $\dfrac{1}{\sqrt{x} + 4} = \dfrac{\sqrt{x} - 4}{x - 16}$

66. $\dfrac{x^2 - 9}{\sqrt{x} - 3} = \sqrt{x} + 3$

In Exercises 67–70, find and correct any errors. If the problem is correct, state that it is correct.

67. $x^n \cdot x^{3n} = x^{3n^2}$

68. $(x^n)^{2n} + (x^{2n})^n = 2x^{2n^2}$

69. $x^{2n} + y^{2n} = (x^n + y^n)^2$

70. $\dfrac{x^{2n} \cdot x^{3n}}{x^{3n} + x^2} = \dfrac{x^{5n}}{x^{3n} + x^2}$

71. *Think About It*   You are taking a course in calculus, and for one of the homework problems you obtain the following answer.

$$\tfrac{1}{10}(2x - 1)^{5/2} + \tfrac{1}{6}(2x - 1)^{3/2}$$

The answer in the back of the book is $\tfrac{1}{15}(2x - 1)^{3/2}(3x + 1)$. Show how the second answer can be obtained from the first. Then use the same technique to simplify each of the following expressions.

(a) $\tfrac{2}{3}x(2x - 3)^{3/2} - \tfrac{2}{15}(2x - 3)^{5/2}$

(b) $\tfrac{2}{3}x(4 + x)^{3/2} - \tfrac{2}{15}(4 + x)^{5/2}$

P.7   The Rectangular Coordinate System and Graphs

What you should learn

- Plot points in the Cartesian plane.
- Use the Distance Formula to find the distance between two points.
- Use the Midpoint Formula to find the midpoint of a line segment.
- Use a coordinate plane to model and solve real-life problems.

Why you should learn it

The Cartesian plane can be used to represent relationships between two variables. For instance, in Exercise 61 on page 67, a graph represents the minimum wage in the United States from 1950 to 2004.

© Ariel Skelly/Corbis

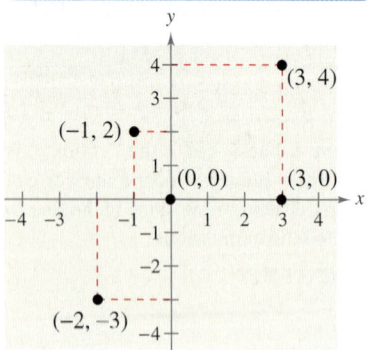

FIGURE P.11

The Cartesian Plane

Just as you can represent real numbers by points on a real number line, you can represent ordered pairs of real numbers by points in a plane called the **rectangular coordinate system,** or the **Cartesian plane,** named after the French mathematician René Descartes (1596–1650).

The Cartesian plane is formed by using two real number lines intersecting at right angles, as shown in Figure P.9. The horizontal real number line is usually called the **x-axis,** and the vertical real number line is usually called the **y-axis.** The point of intersection of these two axes is the **origin,** and the two axes divide the plane into four parts called **quadrants.**

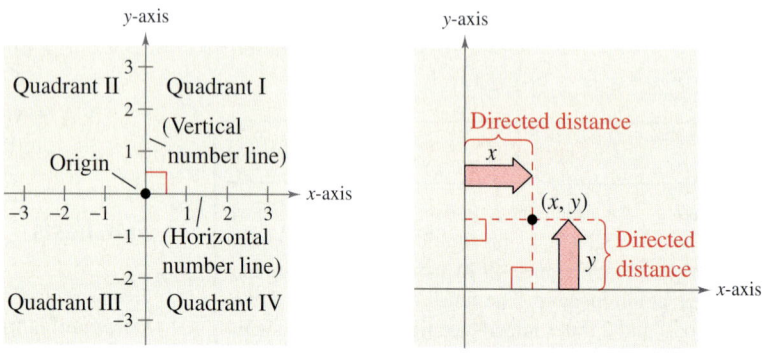

FIGURE P.9   FIGURE P.10

Each point in the plane corresponds to an **ordered pair** (x, y) of real numbers x and y, called **coordinates** of the point. The **x-coordinate** represents the directed distance from the y-axis to the point, and the **y-coordinate** represents the directed distance from the x-axis to the point, as shown in Figure P.10.

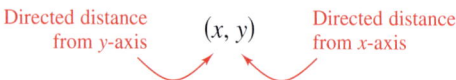

The notation (x, y) denotes both a point in the plane and an open interval on the real number line. The context will tell you which meaning is intended.

Example 1   Plotting Points in the Cartesian Plane

Plot the points $(-1, 2)$, $(3, 4)$, $(0, 0)$, $(3, 0)$, and $(-2, -3)$.

Solution

To plot the point $(-1, 2)$, imagine a vertical line through -1 on the x-axis and a horizontal line through 2 on the y-axis. The intersection of these two lines is the point $(-1, 2)$. The other four points can be plotted in a similar way, as shown in Figure P.11.

✔CHECKPOINT   Now try Exercise 3.

The beauty of a rectangular coordinate system is that it allows you to *see* relationships between two variables. It would be difficult to overestimate the importance of Descartes's introduction of coordinates *in* the plane. Today, his ideas are in common use in virtually every scientific and business-related field.

Example 2   **Sketching a Scatter Plot**

From 1990 through 2003, the amounts A (in millions of dollars) spent on skiing equipment in the United States are shown in the table, where t represents the year. Sketch a scatter plot of the data.   (Source: National Sporting Goods Association)

Solution

To sketch a *scatter plot* of the data shown in the table, you simply represent each pair of values by an ordered pair (t, A) and plot the resulting points, as shown in Figure P.12. For instance, the first pair of values is represented by the ordered pair $(1990, 475)$. Note that the break in the t-axis indicates that the numbers between 0 and 1990 have been omitted.

Year, t	Amount, A
1990	475
1991	577
1992	521
1993	569
1994	609
1995	562
1996	707
1997	723
1998	718
1999	648
2000	495
2001	476
2002	527
2003	464

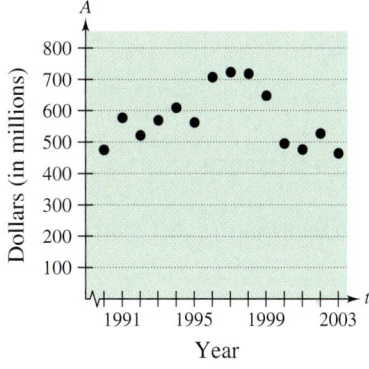

FIGURE P.12

✓CHECKPOINT   Now try Exercise 21.

In Example 2, you could have let $t = 1$ represent the year 1990. In that case, the horizontal axis would not have been broken, and the tick marks would have been labeled 1 through 14 (instead of 1990 through 2003).

Technology

The scatter plot in Example 2 is only one way to represent the data graphically. You could also represent the data using a bar graph and a line graph. If you have access to a graphing utility, try using it to represent graphically the data given in Example 2.

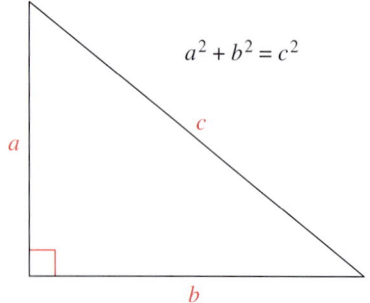

$$a^2 + b^2 = c^2$$

a

c

b

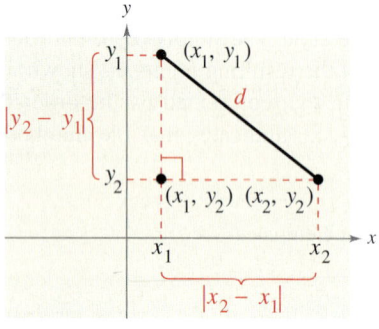

The Distance Formula

Recall from the Pythagorean Theorem that, for a right triangle with hypotenuse of length c and sides of lengths a and b, you have

$$a^2 + b^2 = c^2 \qquad \text{Pythagorean Theorem}$$

as shown in Figure P.13. (The converse is also true. That is, if $a^2 + b^2 = c^2$, then the triangle is a right triangle.)

Suppose you want to determine the distance d between two points (x_1, y_1) and (x_2, y_2) in the plane. With these two points, a right triangle can be formed, as shown in Figure P.14. The length of the vertical side of the triangle is $|y_2 - y_1|$, and the length of the horizontal side is $|x_2 - x_1|$. By the Pythagorean Theorem, you can write

$$d^2 = |x_2 - x_1|^2 + |y_2 - y_1|^2$$
$$d = \sqrt{|x_2 - x_1|^2 + |y_2 - y_1|^2} = \sqrt{(x_2 - x_1)^2 + (y_2 - y_1)^2}.$$

This result is the **Distance Formula.**

The Distance Formula

The distance d between the points (x_1, y_1) and (x_2, y_2) in the plane is

$$d = \sqrt{(x_2 - x_1)^2 + (y_2 - y_1)^2}.$$

Example 3 **Finding a Distance**

Find the distance between the points $(-2, 1)$ and $(3, 4)$.

Algebraic Solution

Let $(x_1, y_1) = (-2, 1)$ and $(x_2, y_2) = (3, 4)$. Then apply the Distance Formula.

$$d = \sqrt{(x_2 - x_1)^2 + (y_2 - y_1)^2} \qquad \text{Distance Formula}$$
$$= \sqrt{[3 - (-2)]^2 + (4 - 1)^2} \qquad \begin{array}{l}\text{Substitute for}\\ x_1, y_1, x_2, \text{and } y_2.\end{array}$$
$$= \sqrt{(5)^2 + (3)^2} \qquad \text{Simplify.}$$
$$= \sqrt{34} \qquad \text{Simplify.}$$
$$\approx 5.83 \qquad \text{Use a calculator.}$$

So, the distance between the points is about 5.83 units. You can use the Pythagorean Theorem to check that the distance is correct.

$$d^2 \overset{?}{=} 3^2 + 5^2 \qquad \text{Pythagorean Theorem}$$
$$\left(\sqrt{34}\right)^2 \overset{?}{=} 3^2 + 5^2 \qquad \text{Substitute for } d.$$
$$34 = 34 \qquad \text{Distance checks.} \checkmark$$

Graphical Solution

Use centimeter graph paper to plot the points $A(-2, 1)$ and $B(3, 4)$. Carefully sketch the line segment from A to B. Then use a centimeter ruler to measure the length of the segment.

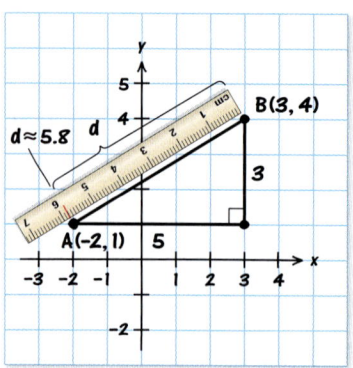

FIGURE P.15

The line segment measures about 5.8 centimeters, as shown in Figure P.15. So, the distance between the points is about 5.8 units.

✓**CHECKPOINT** Now try Exercises 31(a) and (b).

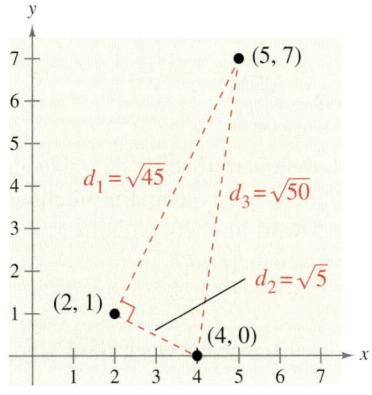

FIGURE **P.16**

Example 4   **Verifying a Right Triangle**

Show that the points $(2, 1)$, $(4, 0)$, and $(5, 7)$ are vertices of a right triangle.

Solution

The three points are plotted in Figure P.16. Using the Distance Formula, you can find the lengths of the three sides as follows.

$$d_1 = \sqrt{(5 - 2)^2 + (7 - 1)^2} = \sqrt{9 + 36} = \sqrt{45}$$

$$d_2 = \sqrt{(4 - 2)^2 + (0 - 1)^2} = \sqrt{4 + 1} = \sqrt{5}$$

$$d_3 = \sqrt{(5 - 4)^2 + (7 - 0)^2} = \sqrt{1 + 49} = \sqrt{50}$$

Because

$$(d_1)^2 + (d_2)^2 = 45 + 5 = 50 = (d_3)^2$$

you can conclude by the Pythagorean Theorem that the triangle must be a right triangle.

✓CHECKPOINT   Now try Exercise 41.

The Midpoint Formula

To find the **midpoint** of the line segment that joins two points in a coordinate plane, you can simply find the average values of the respective coordinates of the two endpoints using the **Midpoint Formula.**

> ### The Midpoint Formula
>
> The midpoint of the line segment joining the points (x_1, y_1) and (x_2, y_2) is given by the Midpoint Formula
>
> $$\text{Midpoint} = \left(\frac{x_1 + x_2}{2}, \frac{y_1 + y_2}{2} \right).$$

For a proof of the Midpoint Formula, see Proofs in Mathematics on page 74.

Example 5   **Finding a Line Segment's Midpoint**

Find the midpoint of the line segment joining the points $(-5, -3)$ and $(9, 3)$.

Solution

Let $(x_1, y_1) = (-5, -3)$ and $(x_2, y_2) = (9, 3)$.

$$\text{Midpoint} = \left(\frac{x_1 + x_2}{2}, \frac{y_1 + y_2}{2} \right) \qquad \text{Midpoint Formula}$$

$$= \left(\frac{-5 + 9}{2}, \frac{-3 + 3}{2} \right) \qquad \text{Substitute for } x_1, y_1, x_2, \text{ and } y_2.$$

$$= (2, 0) \qquad \text{Simplify.}$$

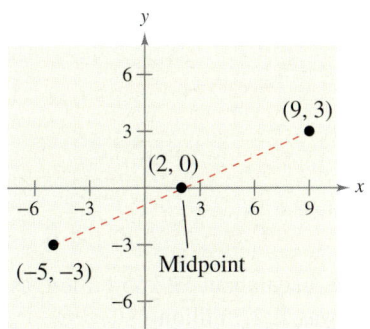

FIGURE **P.17**

The midpoint of the line segment is $(2, 0)$, as shown in Figure P.17.

✓CHECKPOINT   Now try Exercise 31(c).

Applications

Example 6   Finding the Length of a Pass

During the third quarter of the 2004 Sugar Bowl, the quarterback for Louisiana State University threw a pass from the 28-yard line, 40 yards from the sideline. The pass was caught by the wide receiver on the 5-yard line, 20 yards from the same sideline, as shown in Figure P.18. How long was the pass?

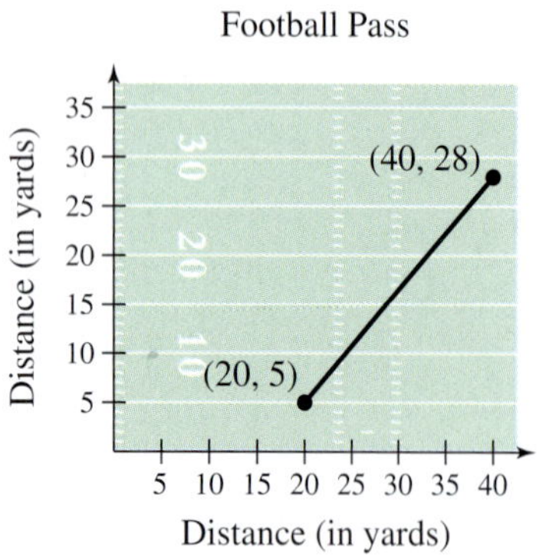

Solution

You can find the length of the pass by finding the distance between the points (40, 28) and (20, 5).

$$d = \sqrt{(x_2 - x_1)^2 + (y_2 - y_1)^2}$$ Distance Formula

$$= \sqrt{(40 - 20)^2 + (28 - 5)^2}$$ Substitute for x_1, y_1, x_2, and y_2.

$$= \sqrt{400 + 529}$$ Simplify.

$$= \sqrt{929}$$ Simplify.

$$\approx 30$$ Use a calculator.

So, the pass was about 30 yards long.

✔CHECKPOINT   Now try Exercise 47.

In Example 6, the scale along the goal line does not normally appear on a football field. However, when you use coordinate geometry to solve real-life problems, you are free to place the coordinate system in any way that is convenient for the solution of the problem.

Example 7   Estimating Annual Revenue

FedEx Corporation had annual revenues of $20.6 billion in 2002 and $24.7 billion in 2004. Without knowing any additional information, what would you estimate the 2003 revenue to have been?   (Source: FedEx Corp.)

Solution

One solution to the problem is to assume that revenue followed a linear pattern. With this assumption, you can estimate the 2003 revenue by finding the midpoint of the line segment connecting the points (2002, 20.6) and (2004, 24.7).

$$\text{Midpoint} = \left(\frac{x_1 + x_2}{2}, \frac{y_1 + y_2}{2} \right)$$ Midpoint Formula

$$= \left(\frac{2002 + 2004}{2}, \frac{20.6 + 24.7}{2} \right)$$ Substitute for x_1, y_1, x_2, and y_2.

$$= (2003, 22.65)$$ Simplify.

So, you would estimate the 2003 revenue to have been about $22.65 billion, as shown in Figure P.19. (The actual 2003 revenue was $22.5 billion.)

✔CHECKPOINT   Now try Exercise 49.

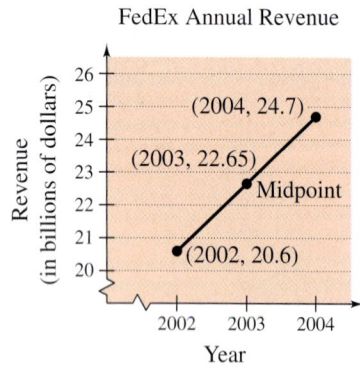

Football Pass

FIGURE P.18

FedEx Annual Revenue

FIGURE P.19

Paul Morrell

Much of computer graphics, including this computer-generated goldfish tessellation, consists of transformations of points in a coordinate plane. One type of transformation, a translation, is illustrated in Example 8. Other types include reflections, rotations, and stretches.

Example 8   Translating Points in the Plane

The triangle in Figure P.20 has vertices at the points $(-1, 2)$, $(1, -4)$, and $(2, 3)$. Shift the triangle three units to the right and two units upward and find the vertices of the shifted triangle, as shown in Figure P.21.

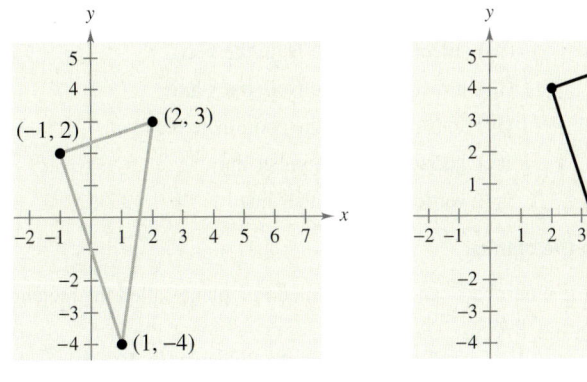

FIGURE P.20   FIGURE P.21

Solution

To shift the vertices three units to the right, add 3 to each of the *x*-coordinates. To shift the vertices two units upward, add 2 to each of the *y*-coordinates.

Original Point	Translated Point
$(-1, 2)$	$(-1 + 3, 2 + 2) = (2, 4)$
$(1, -4)$	$(1 + 3, -4 + 2) = (4, -2)$
$(2, 3)$	$(2 + 3, 3 + 2) = (5, 5)$

✓CHECKPOINT   Now try Exercise 51.

The figures provided with Example 8 were not really essential to the solution. Nevertheless, it is strongly recommended that you develop the habit of including sketches with your solutions—even if they are not required.

𝒲RITING ABOUT 𝑀ATHEMATICS

Extending the Example Example 8 shows how to translate points in a coordinate plane. Write a short paragraph describing how each of the following transformed points is related to the original point.

Original Point	Transformed Point
(x, y)	$(-x, y)$
(x, y)	$(x, -y)$
(x, y)	$(-x, -y)$

P.7   Exercises

VOCABULARY CHECK

1. Match each term with its definition.

(a) *x*-axis   (i) point of intersection of vertical axis and horizontal axis

(b) *y*-axis   (ii) directed distance from the *x*-axis

(c) origin   (iii) directed distance from the *y*-axis

(d) quadrants   (iv) four regions of the coordinate plane

(e) *x*-coordinate   (v) horizontal real number line

(f) *y*-coordinate   (vi) vertical real number line

In Exercises 2–4, fill in the blanks.

2. An ordered pair of real numbers can be represented in a plane called the rectangular coordinate system or the _____ plane.

3. The _____ _____ is a result derived from the Pythagorean Theorem.

4. Finding the average values of the representative coordinates of the two endpoints of a line segment in a coordinate plane is also known as using the _____ _____.

In Exercises 1 and 2, approximate the coordinates of the points.

1.
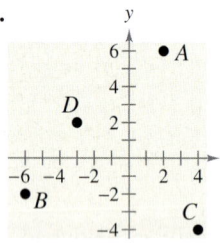

2.
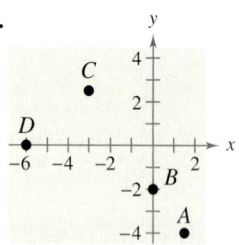

In Exercises 3–6, plot the points in the Cartesian plane.

3. $(-4, 2), (-3, -6), (0, 5), (1, -4)$

4. $(0, 0), (3, 1), (-2, 4), (1, -1)$

5. $(3, 8), (0.5, -1), (5, -6), (-2, 2.5)$

6. $\left(1, -\frac{1}{3}\right), \left(\frac{3}{4}, 3\right), (-3, 4), \left(-\frac{4}{3}, -\frac{3}{2}\right)$

In Exercises 7–10, find the coordinates of the point.

7. The point is located three units to the left of the *y*-axis and four units above the *x*-axis.

8. The point is located eight units below the *x*-axis and four units to the right of the *y*-axis.

9. The point is located five units below the *x*-axis and the coordinates of the point are equal.

10. The point is on the *x*-axis and 12 units to the left of the *y*-axis.

In Exercises 11–20, determine the quadrant(s) in which (x, y) is located so that the condition(s) is (are) satisfied.

11. $x > 0$ and $y < 0$ **12.** $x < 0$ and $y < 0$

13. $x = -4$ and $y > 0$ **14.** $x > 2$ and $y = 3$

15. $y < -5$ **16.** $x > 4$

17. $x < 0$ and $-y > 0$ **18.** $-x > 0$ and $y < 0$

19. $xy > 0$ **20.** $xy < 0$

In Exercises 21 and 22, sketch a scatter plot of the data shown in the table.

21. *Number of Stores*   The table shows the number y of Wal-Mart stores for each year x from 1996 through 2003. (Source: Wal-Mart Stores, Inc.)

Year, *x*	Number of stores, *y*
1996	3054
1997	3406
1998	3599
1999	3985
2000	4189
2001	4414
2002	4688
2003	4906

22. *Meteorology*   The table shows the lowest temperature on record y (in degrees Fahrenheit) in Duluth, Minnesota, for each month x, where $x = 1$ represents January.   (Source: NOAA)

Month, x	Temperature, y
1	−39
2	−39
3	−29
4	−5
5	17
6	27
7	35
8	32
9	22
10	8
11	−23
12	−34

In Exercises 23–26, find the distance between the points. (*Note:* In each case, the two points lie on the same horizontal or vertical line.)

23. $(6, -3), (6, 5)$

24. $(1, 4), (8, 4)$

25. $(-3, -1), (2, -1)$

26. $(-3, -4), (-3, 6)$

In Exercises 27–30, (a) find the length of each side of the right triangle, and (b) show that these lengths satisfy the Pythagorean Theorem.

27.
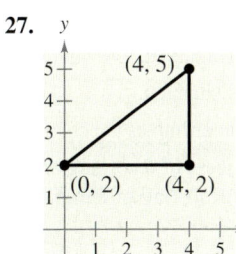

28.
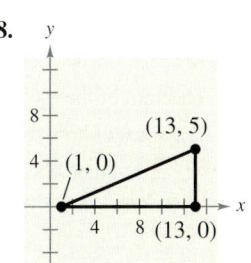

29.
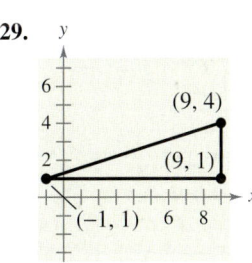

30.
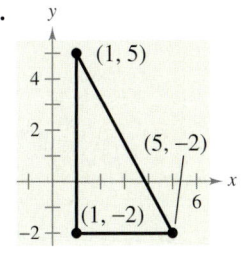

In Exercises 31–40, (a) plot the points, (b) find the distance between the points, and (c) find the midpoint of the line segment joining the points.

31. $(1, 1), (9, 7)$ **32.** $(1, 12), (6, 0)$

33. $(-4, 10), (4, -5)$ **34.** $(-7, -4), (2, 8)$

35. $(-1, 2), (5, 4)$ **36.** $(2, 10), (10, 2)$

37. $\left(\frac{1}{2}, 1\right), \left(-\frac{5}{2}, \frac{4}{3}\right)$

38. $\left(-\frac{1}{3}, -\frac{1}{3}\right), \left(-\frac{1}{6}, -\frac{1}{2}\right)$

39. $(6.2, 5.4), (-3.7, 1.8)$

40. $(-16.8, 12.3), (5.6, 4.9)$

In Exercises 41 and 42, show that the points form the vertices of the indicated polygon.

41. Right triangle: $(4, 0), (2, 1), (-1, -5)$

42. Isosceles triangle: $(1, -3), (3, 2), (-2, 4)$

43. A line segment has (x_1, y_1) as one endpoint and (x_m, y_m) as its midpoint. Find the other endpoint (x_2, y_2) of the line segment in terms of $x_1, y_1, x_m,$ and y_m.

44. Use the result of Exercise 43 to find the coordinates of the endpoint of a line segment if the coordinates of the other endpoint and midpoint are, respectively,

(a) $(1, -2), (4, -1)$ and (b) $(-5, 11), (2, 4)$.

45. Use the Midpoint Formula three times to find the three points that divide the line segment joining (x_1, y_1) and (x_2, y_2) into four parts.

46. Use the result of Exercise 45 to find the points that divide the line segment joining the given points into four equal parts.

(a) $(1, -2), (4, -1)$ (b) $(-2, -3), (0, 0)$

47. *Sports*   A soccer player passes the ball from a point that is 18 yards from the endline and 12 yards from the sideline. The pass is received by a teammate who is 42 yards from the same endline and 50 yards from the same sideline, as shown in the figure. How long is the pass?

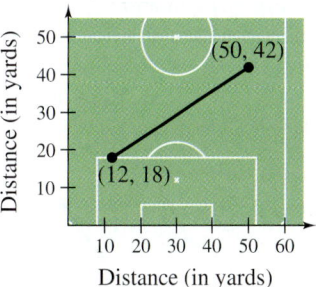

48. *Flying Distance*   An airplane flies from Naples, Italy in a straight line to Rome, Italy, which is 120 kilometers north and 150 kilometers west of Naples. How far does the plane fly?

Sales   In Exercises 49 and 50, use the Midpoint Formula to estimate the sales of Big Lots, Inc. and Dollar Tree Stores, Inc. in 2002, given the sales in 2001 and 2003. Assume that the sales followed a linear pattern. (Source: Big Lots, Inc.; Dollar Tree Stores, Inc.)

49. Big Lots

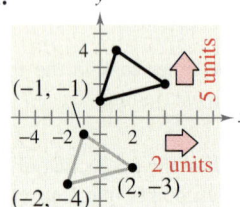

Year	Sales (in millions)
2001	$3433
2003	$4174

50. Dollar Tree

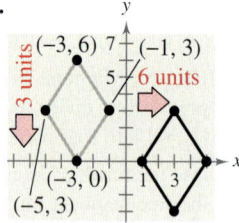

Year	Sales (in millions)
2001	$1987
2003	$2800

In Exercises 51–54, the polygon is shifted to a new position in the plane. Find the coordinates of the vertices of the polygon in its new position.

51.

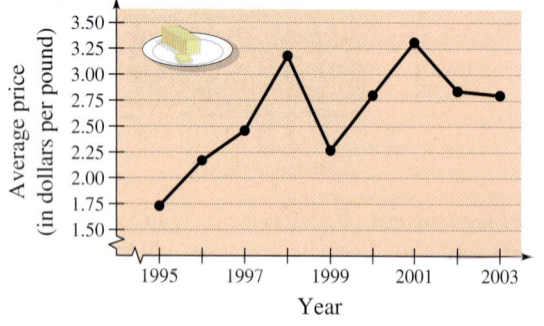

52.

53. Original coordinates of vertices: $(-7, -2), (-2, 2),$
$(-2, -4), (-7, -4)$

Shift: eight units upward, four units to the right

54. Original coordinates of vertices: $(5, 8), (3, 6), (7, 6), (5, 2)$

Shift: 6 units downward, 10 units to the left

Retail Price   In Exercises 55 and 56, use the graph below, which shows the average retail price of 1 pound of butter from 1995 to 2003. (Source: U.S. Bureau of Labor Statistics)

55. Approximate the highest price of a pound of butter shown in the graph. When did this occur?

56. Approximate the percent change in the price of butter from the price in 1995 to the highest price shown in the graph.

Advertising   In Exercises 57 and 58, use the graph below, which shows the cost of a 30-second television spot (in thousands of dollars) during the Super Bowl from 1989 to 2003. (Source: USA Today Research and CNN)

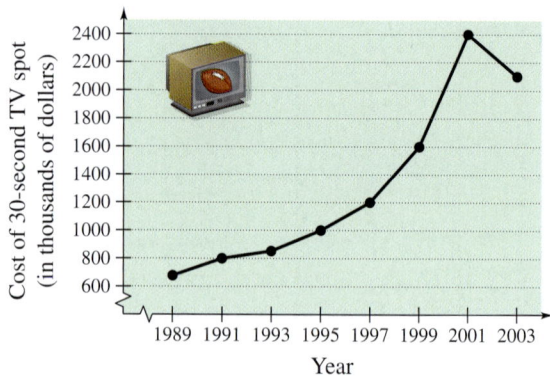

57. Approximate the percent increase in the cost of a 30-second spot from Super Bowl XXIII in 1989 to Super Bowl XXXV in 2001.

58. Estimate the percent increase in the cost of a 30-second spot (a) from Super Bowl XXIII in 1989 to Super Bowl XXVII in 1993 and (b) from Super Bowl XXVII in 1993 to Super Bowl XXXVII in 2003.

59. *Make a Conjecture*   Plot the points $(2, 1), (-3, 5),$ and $(7, -3)$ on a rectangular coordinate system. Then change the sign of the *x*-coordinate of each point and plot the three new points on the same rectangular coordinate system. Make a conjecture about the location of a point when each of the following occurs.

(a) The sign of the *x*-coordinate is changed.

(b) The sign of the *y*-coordinate is changed.

(c) The signs of both the *x*- and *y*-coordinates are changed.

60. *Music*   The graph shows the numbers of recording artists who were elected to the Rock and Roll Hall of Fame from 1986 to 2004.

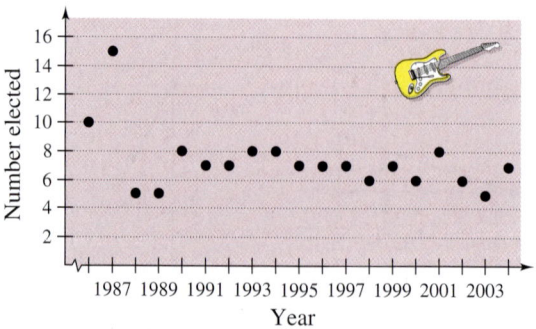

(a) Describe any trends in the data. From these trends, predict the number of artists elected in 2008.

(b) Why do you think the numbers elected in 1986 and 1987 were greater than in other years?

Model It

61. Labor Force   Use the graph below, which shows the minimum wage in the United States (in dollars) from 1950 to 2004.   (Source: U.S. Department of Labor)

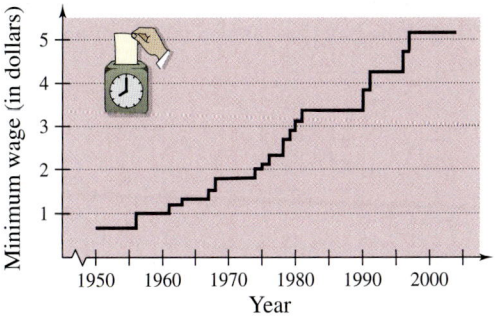

(a) Which decade shows the greatest increase in minimum wage?

(b) Approximate the percent increases in the minimum wage from 1990 to 1995 and from 1995 to 2004.

(c) Use the percent increase from 1995 to 2004 to predict the minimum wage in 2008.

(d) Do you believe that your prediction in part (c) is reasonable? Explain.

62. Sales   Pepsi Bottling Group, Inc. had sales of $6603 million in 1996 and $10,800 million in 2004. Use the Midpoint Formula to estimate the sales in 1998, 2000, and 2002. Assume that the sales followed a linear pattern. (Source: Pepsi Bottling Group, Inc.)

63. Sales   The Coca-Cola Company had sales of $18,546 million in 1996 and $21,900 million in 2004. Use the Midpoint Formula to estimate the sales in 1998, 2000, and 2002. Assume that the sales followed a linear pattern. (Source: The Coca-Cola Company)

64. Data Analysis: Exam Scores   The table shows the mathematics entrance test scores x and the final examination scores y in an algebra course for a sample of 10 students.

x	22	29	35	40	44
y	53	74	57	66	79

x	48	53	58	65	76
y	90	76	93	83	99

(a) Sketch a scatter plot of the data.

(b) Find the entrance exam score of any student with a final exam score in the 80s.

(c) Does a higher entrance exam score imply a higher final exam score? Explain.

65. Data Analysis: Mail   The table shows the number y of pieces of mail handled (in billions) by the U.S. Postal Service for each year x from 1996 through 2003.   (Source U.S. Postal Service)

Year, x	Pieces of mail, y
1996	183
1997	191
1998	197
1999	202
2000	208
2001	207
2002	203
2003	202

(a) Sketch a scatter plot of the data.

(b) Approximate the year in which there was the greatest decrease in the number of pieces of mail handled.

(c) Why do you think the number of pieces of mail handled decreased?

66. Data Analysis: Oil Production   The table shows the daily crude oil production y (in thousands of barrels) in Saudi Arabia for each year x from 1995 through 2001.   (Source: U.S. Energy Information Administration)

Year, x	Oil production, y
1995	8231
1996	8218
1997	8362
1998	8389
1999	7833
2000	8404
2001	8031

(a) Sketch a scatter plot of the data.

(b) In which year was the daily oil production the greatest? In which year was it the least?

(c) Determine the percent increase in daily oil production between the years you found in part (b).

(d) Use the scatter plot to describe any trends in the data.

67. *Data Analysis: Athletics*   The table shows the numbers of men's M and women's W college basketball teams for each year x from 1994 through 2003.   (Source: National Collegiate Athletic Association)

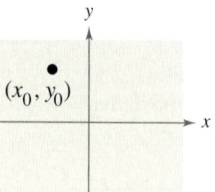

Year, x	Men's teams, M	Women's teams, W
1994	858	859
1995	868	864
1996	866	874
1997	865	879
1998	895	911
1999	926	940
2000	932	956
2001	937	958
2002	936	975
2003	967	1009

(a) Sketch a scatter plot of both sets of data on the same set of coordinate axes.

(b) Find the year in which the numbers of men's and women's teams were nearly equal.

(c) Find the year in which the difference between the numbers of men's and women's teams was the greatest. What is this difference?

68. *Collinear Points*   Three or more points are *collinear* if they all lie on the same line. Use the steps below to determine if the set of points $\{A(2, 3), B(2, 6), C(6, 3)\}$ and the set of points $\{A(8, 3), B(5, 2), C(2, 1)\}$ are collinear.

(a) For each set of points, use the Distance Formula to find the distances from A to B, from B to C, and from A to C. What relationship exists among these distances for each set of points?

(b) Plot each set of points in the Cartesian plane. Do all the points of either set appear to lie on the same line?

(c) Compare your conclusions from part (a) with the conclusions you made from the graphs in part (b). Make a general statement about how to use the Distance Formula to determine collinearity.

Synthesis

True or False?   **In Exercises 69 and 70, determine whether the statement is true or false. Justify your answer.**

69. In order to divide a line segment into 16 equal parts, you would have to use the Midpoint Formula 16 times.

70. The points $(-8, 4)$, $(2, 11)$, and $(-5, 1)$ represent the vertices of an isosceles triangle.

71. *Think About It*   What is the y-coordinate of any point on the x-axis? What is the x-coordinate of any point on the y-axis?

72. *Think About It*   When plotting points on the rectangular coordinate system, is it true that the scales on the x- and y-axes must be the same? Explain.

In Exercises 73–76, use the plot of the point (x_0, y_0) in the figure. Match the transformation of the point with the correct plot. [The plots are labeled (a), (b), (c), and (d).]

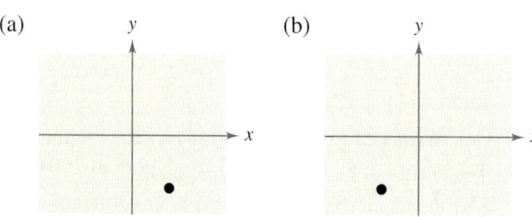

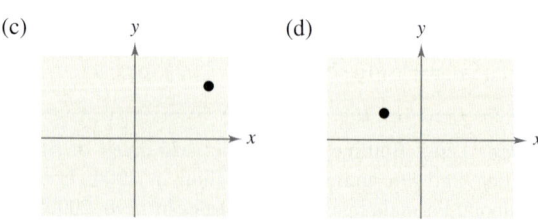

73. $(x_0, -y_0)$ **74.** $(-2x_0, y_0)$

75. $\left(x_0, \tfrac{1}{2}y_0\right)$ **76.** $(-x_0, -y_0)$

77. *Proof*   Prove that the diagonals of the parallelogram in the figure intersect at their midpoints.

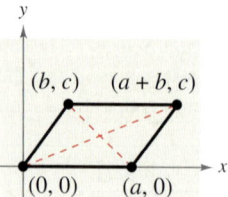

P	**Chapter Summary**

What did you learn?

P Review Exercises

P.1 In Exercises 1 and 2, determine which numbers in the set are (a) natural numbers, (b) whole numbers, (c) integers, (d) rational numbers, and (e) irrational numbers.

1. $\left\{ 11, -14, -\frac{8}{9}, \frac{5}{2}, \sqrt{6}, 0.4 \right\}$

2. $\left\{ \sqrt{15}, -22, -\frac{10}{3}, 0, 5.2, \frac{3}{7} \right\}$

In Exercises 3 and 4, use a calculator to find the decimal form of each rational number. If it is a nonterminating decimal, write the repeating pattern. Then plot the numbers on the real number line and place the appropriate inequality sign ($<$ or $>$) between them.

3. (a) $\frac{5}{6}$ (b) $\frac{7}{8}$ 4. (a) $\frac{9}{25}$ (b) $\frac{5}{7}$

In Exercises 5 and 6, give a verbal description of the subset of real numbers represented by the inequality, and sketch the subset on the real number line.

5. $x \le 7$ 6. $x > 1$

In Exercises 7 and 8, find the distance between a and b.

7. $a = -92, b = 63$ 8. $a = -112, b = -6$

In Exercises 9–12, use absolute value notation to describe the situation.

9. The distance between x and 7 is at least 4.

10. The distance between x and 25 is no more than 10.

11. The distance between y and -30 is less than 5.

12. The distance between z and -16 is greater than 8.

In Exercises 13–16, evaluate the expression for each value of x. (If not possible, state the reason.)

Expression	Values	
13. $12x - 7$	(a) $x = 0$	(b) $x = -1$
14. $x^2 - 6x + 5$	(a) $x = -2$	(b) $x = 2$
15. $-x^2 + x - 1$	(a) $x = 1$	(b) $x = -1$
16. $\dfrac{x}{x-3}$	(a) $x = -3$	(b) $x = 3$

In Exercises 17–20, identify the rule of algebra illustrated by the statement.

17. $2x + (3x - 10) = (2x + 3x) - 10$

18. $4(t + 2) = 4 \cdot t + 4 \cdot 2$

19. $0 + (a - 5) = a - 5$

20. $\dfrac{2}{y+4} \cdot \dfrac{y+4}{2} = 1, \quad y \ne -4$

In Exercises 21–26, perform the operation without using a calculator.

21. $|-3| + 4(-2) - 6$ 22. $\dfrac{|-10|}{-10}$

23. $\frac{5}{18} \div \frac{10}{3}$ 24. $(16 - 8) \div 4$

25. $6[4 - 2(6 + 8)]$ 26. $-4[16 - 3(7 - 10)]$

P.2 In Exercises 27 and 28, simplify each expression.

27. (a) $3x^2(4x^3)^3$ (b) $\dfrac{5y^6}{10y}$

28. (a) $(-2z)^2(8z^3)$ (b) $\dfrac{36x^5}{9x^{10}}$

In Exercises 29 and 30, rewrite each expression with positive exponents and simplify.

29. (a) $\dfrac{6^2 u^3 v^{-3}}{12u^{-2}v}$ (b) $\dfrac{3^{-4}m^{-1}n^{-3}}{9^{-2}mn^{-3}}$

30. (a) $(x + y^{-1})^{-1}$ (b) $\left(\dfrac{x^{-3}}{y}\right)\left(\dfrac{x}{y}\right)^{-1}$

In Exercises 31 and 32, write the number in scientific notation.

31. *Sales for Pep Boys in 2003:*
    $2,134,300,000   (Source: Pep Boys)

32. *Number of meters in 1 foot:* 0.3048

In Exercises 33 and 34, write the number in decimal notation.

33. *Distance between the sun and Jupiter:* 4.837×10^8 miles

34. *Ratio of day to year:* 2.74×10^{-3}

In Exercises 35–38, simplify each expression.

35. (a) $\sqrt[3]{27^2}$ (b) $\sqrt{49^3}$

36. (a) $\sqrt[3]{\frac{64}{125}}$ (b) $\sqrt{\frac{81}{100}}$

37. (a) $\left(\sqrt[3]{216}\right)^3$ (b) $\sqrt[4]{32^4}$

38. (a) $\sqrt[3]{\dfrac{2x^3}{27}}$ (b) $\sqrt[5]{64x^6}$

In Exercises 39 and 40, simplify each expression.

39. (a) $\sqrt{50} - \sqrt{18}$ (b) $2\sqrt{32} + 3\sqrt{72}$

40. (a) $\sqrt{8x^3} + \sqrt{2x}$ (b) $\sqrt{18x^5} - \sqrt{8x^3}$

41. Writing Explain why $\sqrt{5u} + \sqrt{3u} \neq 2\sqrt{2u}$.

42. Engineering The rectangular cross section of a wooden beam cut from a log of diameter 24 inches (see figure) will have a maximum strength if its width w and height h are $w = 8\sqrt{3}$ and $h = \sqrt{24^2 - (8\sqrt{3})^2}$. Find the area of the rectangular cross section and write the answer in simplest form.

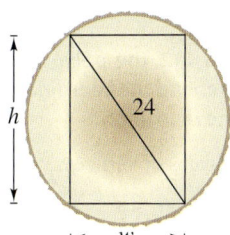

In Exercises 43 and 44, rationalize the denominator of the expression. Then, simplify your answer.

43. $\dfrac{1}{2 - \sqrt{3}}$

44. $\dfrac{1}{\sqrt{5} + 1}$

In Exercises 45 and 46, rationalize the numerator of the expression. Then, simplify your answer.

45. $\dfrac{\sqrt{7} + 1}{2}$

46. $\dfrac{\sqrt{2} - \sqrt{11}}{3}$

In Exercises 47–50, simplify the expression.

47. $(16)^{3/2}$

48. $(64)^{-2/3}$

49. $(3x^{2/5})(2x^{1/2})$

50. $(x - 1)^{1/3}(x - 1)^{-1/4}$

P.3 **In Exercises 51–54, write the polynomial in standard form. Identify the degree and leading coefficient.**

51. $3 - 11x^2$

52. $3x^3 - 5x^5 + x - 4$

53. $-4 - 12x^2$

54. $12x - 7x^2 + 6$

In Exercises 55–58, perform the operation and write the result in standard form.

55. $-(3x^2 + 2x) + (1 - 5x)$

56. $8y - [2y^2 - (3y - 8)]$

57. $2x(x^2 - 5x + 6)$

58. $(3x^3 - 1.5x^2 + 4)(-3x)$

In Exercises 59–64, find the product.

59. $(3x - 6)(5x + 1)$

60. $\left(x - \dfrac{1}{x}\right)(x + 2)$

61. $(2x - 3)^2$

62. $(6x + 5)(6x - 5)$

63. $(3\sqrt{5} + 2)(3\sqrt{5} - 2)$

64. $(x - 4)^3$

65. Surface Area The surface area S of a right circular cylinder is $S = 2\pi r^2 + 2\pi rh$.

(a) Draw a right circular cylinder of radius r and height h. Use the figure to explain how the surface area formula was obtained.

(b) Find the surface area when the radius is 6 inches and the height is 8 inches.

66. Geometry Find a polynomial that represents the total number of square feet for the floor plan shown in the figure.

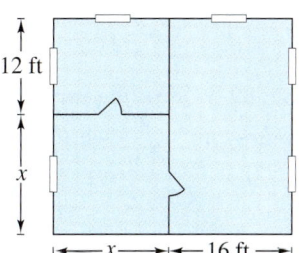

P.4 **In Exercises 67–76, completely factor the expression.**

67. $x^3 - x$

68. $x(x - 3) + 4(x - 3)$

69. $25x^2 - 49$

70. $x^2 - 12x + 36$

71. $x^3 - 64$

72. $8x^3 + 27$

73. $2x^2 + 21x + 10$

74. $3x^2 + 14x + 8$

75. $x^3 - x^2 + 2x - 2$

76. $x^3 - 4x^2 + 2x - 8$

P.5 **In Exercises 77 and 78, find the domain of the expression.**

77. $\dfrac{1}{x + 6}$

78. $\sqrt{x + 4}$

In Exercises 79 and 80, write the rational expression in simplest form.

79. $\dfrac{x^2 - 64}{5(3x + 24)}$

80. $\dfrac{x^3 + 27}{x^2 + x - 6}$

In Exercises 81–84, perform the indicated operation and simplify.

81. $\dfrac{x^2 - 4}{x^4 - 2x^2 - 8} \cdot \dfrac{x^2 + 2}{x^2}$

82. $\dfrac{4x - 6}{(x - 1)^2} \div \dfrac{2x^2 - 3x}{x^2 + 2x - 3}$

83. $\dfrac{1}{x - 1} + \dfrac{1 - x}{x^2 + x + 1}$

84. $\dfrac{3x}{x + 2} - \dfrac{4x^2 - 5}{2x^2 + 3x - 2}$

In Exercises 85 and 86, simplify the complex fraction.

85. $\dfrac{\left[\dfrac{3a}{(a^2/x) - 1}\right]}{\left(\dfrac{a}{x} - 1\right)}$

86. $\dfrac{\left(\dfrac{1}{2x - 3} - \dfrac{1}{2x + 3}\right)}{\left(\dfrac{1}{2x} - \dfrac{1}{2x + 3}\right)}$

In Exercises 87 and 88, simplify the difference quotient.

87. $\dfrac{\left[\dfrac{1}{2(x + h)} - \dfrac{1}{2x}\right]}{h}$

88. $\dfrac{\dfrac{1}{x + h - 3} - \dfrac{1}{x - 3}}{h}$

P.6 **In Exercises 89–94, describe and correct the error.**

89. $10(4 \cdot 7) = 40 \cdot 70$

90. $\left(\frac{1}{3}x\right)\left(\frac{1}{3}y\right) = \frac{1}{3}xy$

91. $(3^4)^4 = 3^8$

92. $\sqrt{3^2 + 4^2} = 3 + 4$

93. $(5 + 8)^2 = 5^2 + 8^2$

94. $\dfrac{7 + 5(x + 3)}{x + 3} = 12$

In Exercises 95–98, insert the required factor in the parentheses.

95. $\frac{2}{3}x^4 - \frac{3}{8}x^3 + \frac{5}{6}x^2 = \frac{1}{24}x^2(\quad\quad)$

96. $\dfrac{t}{\sqrt{t + 1}} - \sqrt{t + 1} = \dfrac{1}{\sqrt{t + 1}}(\quad\quad)$

97. $2x(x^2 - 3)^{1/3} - 5(x^2 - 3)^{4/3} = (x^2 - 3)^{1/3}(\quad\quad)$

98. $y(y - 1)^{5/4} - y^2(y - 1)^{1/4} = y(y - 1)^{1/4}(\quad\quad)$

In Exercises 99 and 100, write the fraction as a sum of two or more terms.

99. $\dfrac{x^3 + 5x^2 + 7}{x}$

100. $\dfrac{2x^3 - x^2 + 4}{x^{1/2}}$

In Exercises 101 and 102, simplify the expression.

101. $\dfrac{x(x + 2)^{-1/2} + (x + 2)^{1/2}}{(x + 2)^{3/2}}$

102. $\dfrac{\frac{2}{3}x(4 + x)^{-1/2} - \frac{2}{15}(4 + x)^{1/2}}{(4 + x)^{5/2}}$

P.7 **In Exercises 103 and 104, plot the points in the Cartesian plane.**

103. $(2, 2), (0, -4), (-3, 6), (-1, -7)$

104. $(5, 0), (8, 1), (4, -2), (-3, -3)$

In Exercises 105 and 106, determine the quadrant(s) in which (x, y) is located so that the condition(s) is (are) satisfied.

105. $x > 0$ and $y = -2$

106. $y > 0$

In Exercises 107–110, (a) plot the points, (b) find the distance between the points, and (c) find the midpoint of the line segment joining the points.

107. $(-3, 8), (1, 5)$

108. $(-2, 6), (4, -3)$

109. $(5.6, 0), (0, 8.2)$

110. $(0, -1.2), (-3.6, 0)$

In Exercises 111 and 112, the polygon is shifted to a new position in the plane. Find the coordinates of the vertices of the polygon in its new position.

111. Original coordinates of vertices:

$(4, 8), (6, 8), (4, 3), (6, 3)$

Shift: three units downward, two units to the left

112. Original coordinates of vertices:

$(0, 1), (3, 3), (0, 5), (-3, 3)$

Shift: five units upward, four units to the left

113. *Sales* The Cheesecake Factory had annual sales of $539.1 million in 2001 and $773.8 million in 2003. Use the Midpoint Formula to estimate the sales in 2002. (Source: The Cheesecake Factory, Inc.)

114. *Meteorology* The apparent temperature is a measure of relative discomfort to a person from heat and high humidity. The table shows the actual temperatures x (in degrees Fahrenheit) versus the apparent temperatures y (in degrees Fahrenheit) for a relative humidity of 75%.

x	70	75	80	85	90	95	100
y	70	77	85	95	109	130	150

(a) Sketch a scatter plot of the data shown in the table.

(b) Find the change in the apparent temperature when the actual temperature changes from 70°F to 100°F.

Synthesis

True or False? In Exercises 115 and 116, determine whether the statement is true or false. Justify your answer.

115. A binomial sum squared is equal to the sum of the terms squared.

116. $x^n - y^n$ factors as conjugates for all values of n.

117. *Think About It* Is the following statement true for all nonzero real numbers a and b? Explain.

$$\dfrac{ax - b}{b - ax} = -1$$

P Chapter Test

Take this test as you would take a test in class. When you are finished, check your work against the answers given in the back of the book.

1. Place < or > between the real numbers $-\frac{10}{3}$ and $-|-4|$.

2. Find the distance between the real numbers -5.4 and $3\frac{3}{4}$.

3. Identify the rule of algebra illustrated by $(5 - x) + 0 = 5 - x$.

In Exercises 4 and 5, evaluate each expression without using a calculator.

4. (a) $27\left(-\frac{2}{3}\right)$ (b) $\dfrac{5}{18} \div \dfrac{5}{8}$ (c) $\left(-\dfrac{3}{5}\right)^3$ (d) $\left(\dfrac{3^2}{2}\right)^{-3}$

5. (a) $\sqrt{5} \cdot \sqrt{125}$ (b) $\dfrac{\sqrt{27}}{\sqrt{2}}$ (c) $\dfrac{5.4 \times 10^8}{3 \times 10^3}$ (d) $(3 \times 10^4)^3$

In Exercises 6 and 7, simplify each expression.

6. (a) $3z^2(2z^3)^2$ (b) $(u - 2)^{-4}(u - 2)^{-3}$ (c) $\left(\dfrac{x^{-2}y^2}{3}\right)^{-1}$

7. (a) $9z\sqrt{8z} - 3\sqrt{2z^3}$ (b) $(4x^{3/5})(x^{1/3})$ (c) $\sqrt[3]{\dfrac{16}{v^5}}$

8. Write the polynomial $x + 4x^4 - 5 - 3x^2$ in standard form. Identify the degree and leading coefficient.

In Exercises 9–12, perform the operation and simplify.

9. $(x^2 + 3) - [3x + (8 - x^2)]$

10. $\left(x + \sqrt{5}\right)\left(x - \sqrt{5}\right)$

11. $\dfrac{8x}{x - 3} + \dfrac{24}{3 - x}$

12. $\dfrac{\left(\dfrac{2}{x} - \dfrac{2}{x + 1}\right)}{\left(\dfrac{4}{x^2 - 1}\right)}$

13. Factor (a) $2x^4 - 3x^3 - 2x^2$ and (b) $x^3 + 2x^2 - 4x - 8$ completely.

14. Rationalize each denominator. (a) $\dfrac{16}{\sqrt[3]{16}}$ (b) $\dfrac{6}{1 - \sqrt{3}}$

15. Find the domain of $\dfrac{1 - x}{4 - x}$.

16. Multiply: $\dfrac{y^2 + 8y + 16}{2y - 4} \cdot \dfrac{8y - 16}{(y + 4)^3}$.

17. A T-shirt company can produce and sell x T-shirts per day. The total cost C (in dollars) for producing x T-shirts is $C = 1480 + 6x$, and the total revenue R (in dollars) is $R = 15x$. Find the profit obtained by selling 225 T-shirts per day.

18. Plot the points $(-2, 5)$ and $(6, 0)$. Find the coordinates of the midpoint of the line segment joining the points and the distance between the points.

19. Write an expression for the area of the shaded region in the figure at the left, and simplify the result.

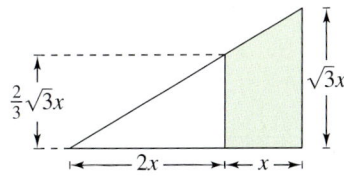

FIGURE FOR 19

Proofs in Mathematics

What does the word *proof* mean to you? In mathematics, the word *proof* is used to mean simply a valid argument. When you are proving a statement or theorem, you must use facts, definitions, and accepted properties in a logical order. You can also use previously proved theorems in your proof. For instance, the Distance Formula is used in the proof of the Midpoint Formula below. There are several different proof methods, which you will see in later chapters.

The Midpoint Formula *(p. 61)*

The midpoint of the line segment joining the points (x_1, y_1) and (x_2, y_2) is given by the Midpoint Formula

$$\text{Midpoint} = \left(\frac{x_1 + x_2}{2}, \frac{y_1 + y_2}{2} \right).$$

The Cartesian Plane

The Cartesian plane was named after the French mathematician René Descartes (1596–1650). While Descartes was lying in bed, he noticed a fly buzzing around on the square ceiling tiles. He discovered that the position of the fly could be described by which ceiling tile the fly landed on. This led to the development of the Cartesian plane. Descartes felt that a coordinate plane could be used to facilitate description of the positions of objects.

Proof

Using the figure, you must show that $d_1 = d_2$ and $d_1 + d_2 = d_3$.

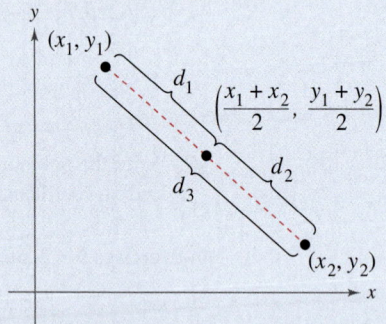

By the Distance Formula, you obtain

$$d_1 = \sqrt{\left(\frac{x_1 + x_2}{2} - x_1 \right)^2 + \left(\frac{y_1 + y_2}{2} - y_1 \right)^2}$$

$$= \frac{1}{2} \sqrt{(x_2 - x_1)^2 + (y_2 - y_1)^2}$$

$$d_2 = \sqrt{\left(x_2 - \frac{x_1 + x_2}{2} \right)^2 + \left(y_2 - \frac{y_1 + y_2}{2} \right)^2}$$

$$= \frac{1}{2} \sqrt{(x_2 - x_1)^2 + (y_2 - y_1)^2}$$

$$d_3 = \sqrt{(x_2 - x_1)^2 + (y_2 - y_1)^2}$$

So, it follows that $d_1 = d_2$ and $d_1 + d_2 = d_3$.

This collection of thought-provoking and challenging exercises further explores and expands upon concepts learned in this chapter.

1. The NCAA states that the men's and women's shots for track and field competition must comply with the following specifications. (Source: NCAA)

	Men's	Women's
Weight (minimum)	7.26 kg	4.0 kg
Diameter (minimum)	110 mm	95 mm
Diameter (maximum)	130 mm	110 mm

   (a) Find the maximum and minimum volumes of both the men's and women's shots.

   (b) The *density* of an object is an indication of how heavy the object is. To find the density of an object, divide its mass (weight) by its volume. Find the maximum and minimum densities of both the men's and women's shots.

   (c) A shot is usually made out of iron. If a ball of cork has the same volume as an iron shot, do you think they would have the same density? Explain your reasoning.

2. Find an example for which

   $$|a - b| > |a| - |b|,$$

   and an example for which

   $$|a - b| = |a| - |b|.$$

   Then prove that

   $$|a - b| \geq |a| - |b|$$

   for all a, b.

3. A major feature of Epcot Center at Disney World is called Spaceship Earth. The building is shaped as a sphere and weighs 1.6×10^7 pounds, which is equal in weight to 1.58×10^8 golf balls. Use these values to find the approximate weight (in pounds) of one golf ball. Then convert the weight to ounces. (Source: Disney.com)

4. The average life expectancies at birth in 2001 for men and women were 74.4 years and 79.8 years, respectively. Assuming an average healthy heart rate of 70 beats per minute, find the numbers of beats in a lifetime for a man and for a woman. (Source: National Center for Health Statistics)

5. The accuracy of an approximation to a number is related to how many significant digits there are in the approximation. Write a definition of significant digits and illustrate the concept with examples.

6. The table shows the census population y (in millions) of the United States for each census year x from 1950 to 2000. (Source: U.S. Census Bureau)

Year, x	Population, y
1950	151.33
1960	179.32
1970	203.30
1980	226.54
1990	248.72
2000	281.42

   (a) Sketch a scatter plot of the data. Describe any trends in the data.

   (b) Find the increase in population from each census year to the next.

   (c) Over which decade did the population increase the most? the least?

   (d) Find the percent increase in population from each census year to the next.

   (e) Over which decade was the percent increase the greatest? the least?

7. Find the annual depreciation rate r from the bar graph below. To find r by the declining balances method, use the formula

   $$r = 1 - \left(\frac{S}{C}\right)^{1/n}$$

   where n is the useful life of the item (in years), S is the salvage value (in dollars), and C is the original cost (in dollars).

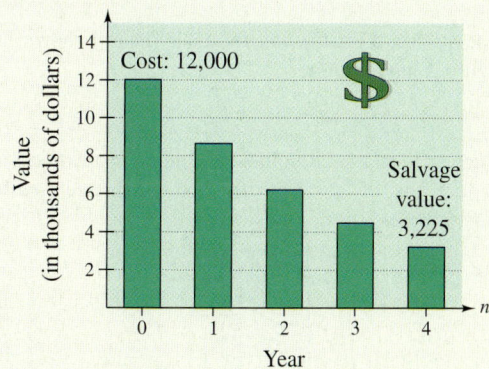

8. Johannes Kepler (1571–1630), a well-known German astronomer, discovered a relationship between the average distance of a planet from the sun and the time (or period) it takes the planet to orbit the sun. People then knew that planets that are closer to the sun take less time to complete an orbit than planets that are farther from the sun. Kepler discovered that the distance and period are related by an exact mathematical formula.

The table shows the average distances x (in astronomical units) and periods y (in years) for the five planets that are closest to the sun. By completing the table, can you rediscover Kepler's relationship? Write a paragraph that summarizes your conclusions.

Planet	Mercury	Venus	Earth	Mars	Jupiter
x	0.387	0.723	1.000	1.524	5.203
$\sqrt{x}$					
y	0.241	0.615	1.000	1.881	11.863
$\sqrt[3]{y}$					

9. A stained glass window is designed in the shape of a rectangle with a semicircular arch (see figure). The width of the window is 2 feet and the perimeter is approximately 13.14 feet. Find the smallest amount of glass required to construct the window.

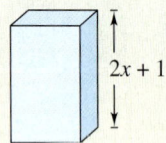

— 2 ft —

10. The volume V of the box (in cubic inches) shown in the figure is modeled by

$$V = 2x^3 + x^2 - 8x - 4$$

where x is measured in inches. Find an expression for the surface area of the box. Then find the surface area when $x = 6$ inches.

$2x + 1$

11. Verify that $y_1 \neq y_2$ by letting $x = 0$ and evaluating y_1 and y_2.

$$y_1 = 2x\sqrt{1 - x^2} - \frac{x^3}{\sqrt{1 - x^2}}$$

$$y_2 = \frac{2 - 3x^2}{\sqrt{1 - x^2}}$$

Change y_2 so that $y_1 = y_2$.

12. Prove that

$$\left(\frac{2x_1 + x_2}{3}, \frac{2y_1 + y_2}{3} \right)$$

is one of the points of trisection of the line segment joining (x_1, y_1) and (x_2, y_2). Find the midpoint of the line segment joining

$$\left(\frac{2x_1 + x_2}{3}, \frac{2y_1 + y_2}{3} \right)$$

and (x_2, y_2) to find the second point of trisection.

13. Use the results of Exercise 12 to find the points of trisection of the line segment joining each pair of points.

(a) $(1, -2), (4, 1)$ (b) $(-2, -3), (0, 0)$

14. Although graphs can help visualize relationships between two variables, they can also be used to mislead people. The graphs shown below represent the same data points.

(a) Which of the two graphs is misleading, and why? Discuss other ways in which graphs can be misleading.

(b) Why would it be beneficial for someone to use a misleading graph?

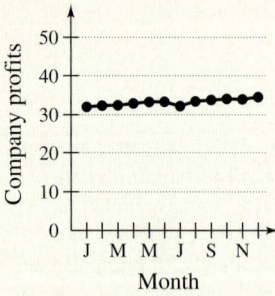

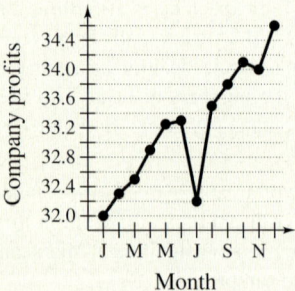

Equations, Inequalities, and Mathematical Modeling

Real-life data can be modeled by several types of equations. A rational equation can be used to model the voting-age population of the United States.

© Marc Serota/Reuters/Corbis

SELECTED APPLICATIONS

Equations and inequalities have many real-life applications. The applications listed below represent a small sample of the applications in this chapter.

1.1 Graphs of Equations

What you should learn

- Sketch graphs of equations.
- Find x- and y-intercepts of graphs of equations.
- Use symmetry to sketch graphs of equations.
- Find equations of and sketch graphs of circles.
- Use graphs of equations in solving real-life problems.

Why you should learn it

The graph of an equation can help you see relationships between real-life quantities. For example, in Exercise 67 on page 88, a graph can be used to estimate the life expectancies of children who are born in the years 2005 and 2010.

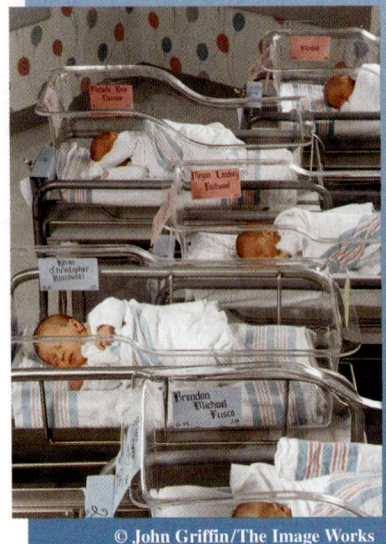

© John Griffin/The Image Works

The Graph of an Equation

In Section P.7, you used a coordinate system to represent graphically the relationship between two quantities. There, the graphical picture consisted of a collection of points in a coordinate plane.

Frequently, a relationship between two quantities is expressed as an **equation in two variables.** For instance, $y = 7 - 3x$ is an equation in x and y. An ordered pair (a, b) is a **solution** or **solution point** of an equation in x and y if the equation is true when a is substituted for x and b is substituted for y. For instance, $(1, 4)$ is a solution of $y = 7 - 3x$ because $4 = 7 - 3(1)$ is a true statement.

In this section you will review some basic procedures for sketching the graph of an equation in two variables. The **graph of an equation** is the set of all points that are solutions of the equation.

Example 1   Determining Solutions

Determine whether (a) $(2, 13)$ and (b) $(-1, -3)$ are solutions of the equation $y = 10x - 7$.

Solution

a. $y = 10x - 7$ Write original equation.

   $13 \stackrel{?}{=} 10(2) - 7$ Substitute 2 for x and 13 for y.

   $13 = 13$ $(2, 13)$ is a solution. ✔

Because the substitution does satisfy the original equation, you can conclude that the ordered pair $(2, 13)$ *is* a solution of the original equation.

b. $y = 10x - 7$ Write original equation.

   $-3 \stackrel{?}{=} 10(-1) - 7$ Substitute -1 for x and -3 for y.

   $-3 \neq -17$ $(-1, -3)$ is not a solution.

Because the substitution does not satisfy the original equation, you can conclude that the ordered pair $(-1, -3)$ *is not* a solution of the original equation.

✔**CHECKPOINT**   Now try Exercise 1.

The basic technique used for sketching the graph of an equation is the **point-plotting method.**

Sketching the Graph of an Equation by Point Plotting

1. If possible, rewrite the equation so that one of the variables is isolated on one side of the equation.

2. Make a table of values showing several solution points.

3. Plot these points on a rectangular coordinate system.

4. Connect the points with a smooth curve or line.

When making a table of solution points, be sure to use positive, zero, and negative values of x.

Example 2   Sketching the Graph of an Equation

Sketch the graph of $y = 7 - 3x$.

Solution

Because the equation is already solved for y, construct a table of values that consists of several solution points of the equation. For instance, when $x = -1$,

$$y = 7 - 3(-1)$$

$$= 10$$

which implies that $(-1, 10)$ is a solution point of the graph.

x	$y = 7 - 3x$	(x, y)
-1	10	$(-1, 10)$
0	7	$(0, 7)$
1	4	$(1, 4)$
2	1	$(2, 1)$
3	-2	$(3, -2)$
4	-5	$(4, -5)$

From the table, it follows that

$$(-1, 10), (0, 7), (1, 4), (2, 1), (3, -2), \text{ and } (4, -5)$$

are solution points of the equation. After plotting these points, you can see that they appear to lie on a line, as shown in Figure 1.1. The graph of the equation is the line that passes through the six plotted points.

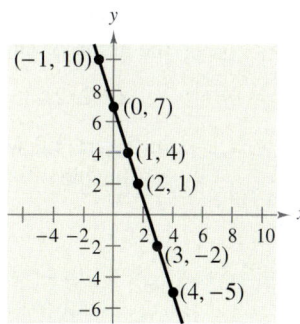

FIGURE 1.1

✓CHECKPOINT   Now try Exercise 5.

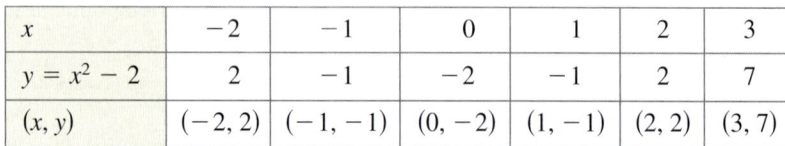

Example 3   **Sketching the Graph of an Equation**

Sketch the graph of

$$y = x^2 - 2.$$

Solution

Because the equation is already solved for y, begin by constructing a table of values.

x	-2	-1	0	1	2	3
$y = x^2 - 2$	2	-1	-2	-1	2	7
(x, y)	$(-2, 2)$	$(-1, -1)$	$(0, -2)$	$(1, -1)$	$(2, 2)$	$(3, 7)$

Next, plot the points given in the table, as shown in Figure 1.2. Finally, connect the points with a smooth curve, as shown in Figure 1.3.

STUDY TIP

One of your goals in this course is to learn to classify the basic shape of a graph from its equation. For instance, you will learn that the *linear equation* in Example 2 has the form

$$y = mx + b$$

and its graph is a line. Similarly, the *quadratic equation* in Example 3 has the form

$$y = ax^2 + bx + c$$

and its graph is a parabola.

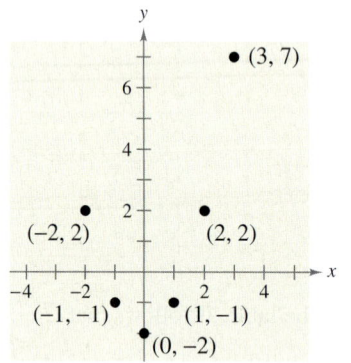

FIGURE 1.2

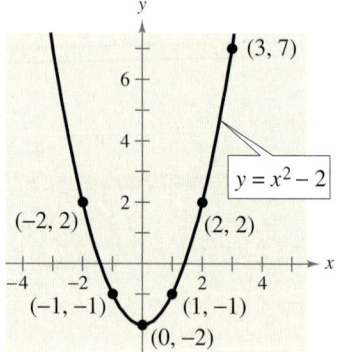

FIGURE 1.3

✔**CHECKPOINT**   Now try Exercise 7.

The point-plotting method demonstrated in Examples 2 and 3 is easy to use, but it has some shortcomings. With too few solution points, you can misrepresent the graph of an equation. For instance, if only the four points

$$(-2, 2), (-1, -1), (1, -1), \text{ and } (2, 2)$$

in Figure 1.2 were plotted, any one of the three graphs in Figure 1.4 would be reasonable.

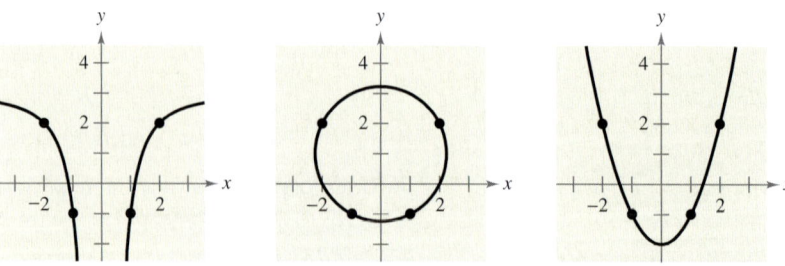

FIGURE 1.4

> ## Technology
>
> To graph an equation involving *x* and *y* on a graphing utility, use the following procedure.
>
> 1. Rewrite the equation so that *y* is isolated on the left side.
> 2. Enter the equation into the graphing utility.
> 3. Determine a *viewing window* that shows all important features of the graph.
> 4. Graph the equation.
>
> For more extensive instructions on how to use a graphing utility to graph an equation, see the *Graphing Technology Guide* on the text website at *college.hmco.com*.

Intercepts of a Graph

It is often easy to determine the solution points that have zero as either the *x*-coordinate or the *y*-coordinate. These points are called **intercepts** because they are the points at which the graph intersects or touches the *x*- or *y*-axis. It is possible for a graph to have no intercepts, one intercept, or several intercepts, as shown in Figure 1.5.

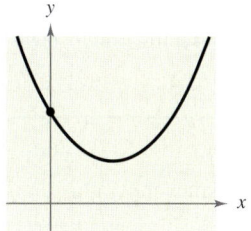

No x-intercepts
One y-intercept

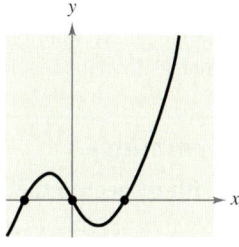

Three x-intercepts
One y-intercept

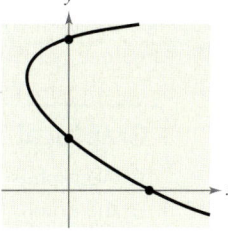

One x-intercept
Two y-intercepts

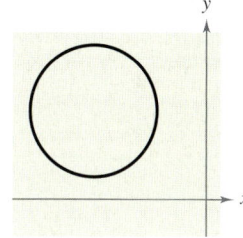

No intercepts

FIGURE 1.5

Note that an *x*-intercept can be written as the ordered pair $(x, 0)$ and a *y*-intercept can be written as the ordered pair $(0, y)$. Some texts denote the *x*-intercept as the *x*-coordinate of the point $(a, 0)$ [and the *y*-intercept as the *y*-coordinate of the point $(0, b)$] rather than the point itself. Unless it is necessary to make a distinction, we will use the term *intercept* to mean either the point or the coordinate.

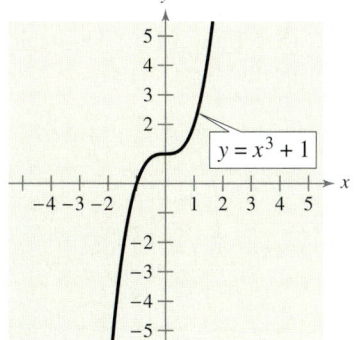

FIGURE 1.6

Example 4   Identifying *x*- and *y*-Intercepts

Identify the *x*- and *y*-intercepts of the graph of

$$y = x^3 + 1$$

shown in Figure 1.6.

Solution

From the figure, you can see that the graph of the equation $y = x^3 + 1$ has an *x*- intercept (where *y* is zero) at $(-1, 0)$ and a *y*-intercept (where *x* is zero) at $(0, 1)$.

 CHECKPOINT   Now try Exercise 9.

Symmetry

Graphs of equations can have **symmetry** with respect to one of the coordinate axes or with respect to the origin. Symmetry with respect to the x-axis means that if the Cartesian plane were folded along the x-axis, the portion of the graph above the x-axis would coincide with the portion below the x-axis. Symmetry with respect to the y-axis or the origin can be described in a similar manner, as shown in Figure 1.7.

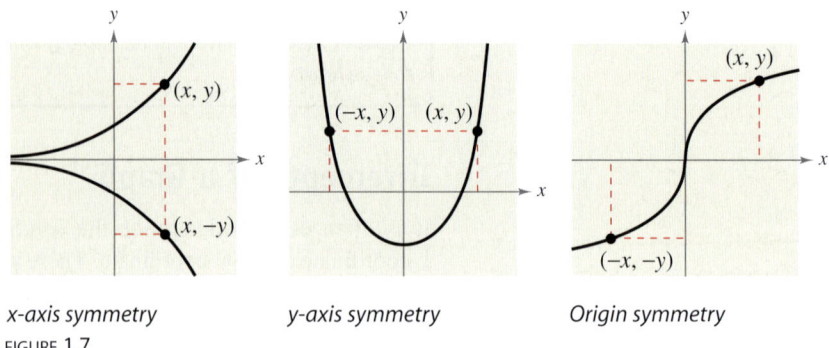

x-axis symmetry   *y-axis symmetry*   *Origin symmetry*

FIGURE 1.7

Knowing the symmetry of a graph *before* attempting to sketch it is helpful, because then you need only half as many solution points to sketch the graph. There are three basic types of symmetry, described as follows.

Graphical Tests for Symmetry

1. A graph is **symmetric with respect to the x-axis** if, whenever (x, y) is on the graph, $(x, -y)$ is also on the graph.

2. A graph is **symmetric with respect to the y-axis** if, whenever (x, y) is on the graph, $(-x, y)$ is also on the graph.

3. A graph is **symmetric with respect to the origin** if, whenever (x, y) is on the graph, $(-x, -y)$ is also on the graph.

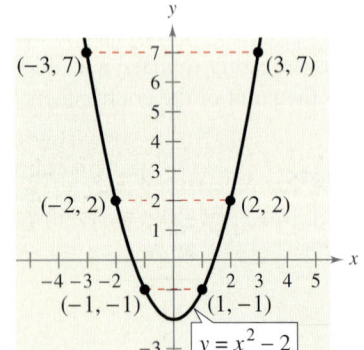

FIGURE 1.8 *y-axis symmetry*

Example 5   Testing for Symmetry

The graph of $y = x^2 - 2$ is symmetric with respect to the y-axis because the point $(-x, y)$ is also on the graph of $y = x^2 - 2$. (See Figure 1.8.) The table below confirms that the graph is symmetric with respect to the y-axis.

x	-3	-2	-1	1	2	3
y	7	2	-1	-1	2	7
(x, y)	$(-3, 7)$	$(-2, 2)$	$(-1, -1)$	$(1, -1)$	$(2, 2)$	$(3, 7)$

✓**CHECKPOINT**   Now try Exercise 15.

> ## Algebraic Tests for Symmetry
>
> **1.** The graph of an equation is symmetric with respect to the x-axis if replacing y with $-y$ yields an equivalent equation.
>
> **2.** The graph of an equation is symmetric with respect to the y-axis if replacing x with $-x$ yields an equivalent equation.
>
> **3.** The graph of an equation is symmetric with respect to the origin if replacing x with $-x$ and y with $-y$ yields an equivalent equation.

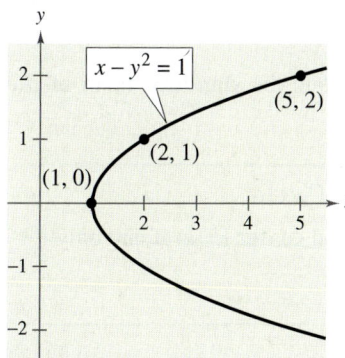

FIGURE **1.9**

STUDY TIP

Notice that when creating the table in Example 6, it is easier to choose y-values and then find the corresponding x-values of the ordered pairs.

Example 6   **Using Symmetry as a Sketching Aid**

Use symmetry to sketch the graph of

$$x - y^2 = 1.$$

Solution

Of the three tests for symmetry, the only one that is satisfied is the test for x-axis symmetry because $x - (-y)^2 = 1$ is equivalent to $x - y^2 = 1$. So, the graph is symmetric with respect to the x-axis. Using symmetry, you only need to find the solution points above the x-axis and then reflect them to obtain the graph, as shown in Figure 1.9.

y	$x = y^2 + 1$	(x, y)
0	1	$(1, 0)$
1	2	$(2, 1)$
2	5	$(5, 2)$

✔**CHECKPOINT**   Now try Exercise 29.

Example 7   **Sketching the Graph of an Equation**

Sketch the graph of

$$y = |x - 1|.$$

Solution

This equation fails all three tests for symmetry and consequently its graph is not symmetric with respect to either axis or to the origin. The absolute value sign indicates that y is always nonnegative. Create a table of values and plot the points as shown in Figure 1.10. From the table, you can see that $x = 0$ when $y = 1$. So, the y-intercept is $(0, 1)$. Similarly, $y = 0$ when $x = 1$. So, the x-intercept is $(1, 0)$.

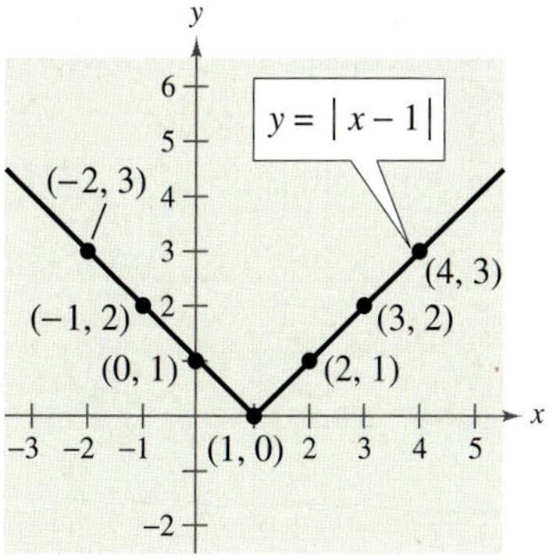

x	-2	-1	0	1	2	3	4
$y = |x - 1|$	3	2	1	0	1	2	3
(x, y)	$(-2, 3)$	$(-1, 2)$	$(0, 1)$	$(1, 0)$	$(2, 1)$	$(3, 2)$	$(4, 3)$

FIGURE **1.10**

✔**CHECKPOINT**   Now try Exercise 33.

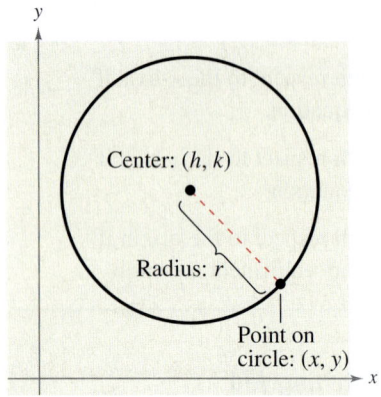

FIGURE 1.11

Throughout this course, you will learn to recognize several types of graphs from their equations. For instance, you will learn to recognize that the graph of a second-degree equation of the form

$$y = ax^2 + bx + c$$

is a parabola (see Example 3). The graph of a **circle** is also easy to recognize.

Circles

Consider the circle shown in Figure 1.11. A point (x, y) is on the circle if and only if its distance from the center (h, k) is r. By the Distance Formula,

$$\sqrt{(x - h)^2 + (y - k)^2} = r.$$

By squaring each side of this equation, you obtain the **standard form of the equation of a circle.**

Standard Form of the Equation of a Circle

The point (x, y) lies on the circle of **radius** r and **center** (h, k) if and only if

$$(x - h)^2 + (y - k)^2 = r^2.$$

STUDY TIP

To find the correct h and k, from the equation of the circle in Example 8, it may be helpful to rewrite the quantities $(x + 1)^2$ and $(y - 2)^2$, using subtraction.

$$(x + 1)^2 = [x - (-1)]^2,$$

$$(y - 2)^2 = [y - (2)]^2$$

So, $h = -1$ and $k = 2$.

From this result, you can see that the standard form of the equation of a circle *with its center at the origin*, $(h, k) = (0, 0)$, is simply

$$x^2 + y^2 = r^2. \qquad \text{Circle with center at origin}$$

Example 8   **Finding the Equation of a Circle**

The point $(3, 4)$ lies on a circle whose center is at $(-1, 2)$, as shown in Figure 1.12. Write the standard form of the equation of this circle.

Solution

The radius of the circle is the distance between $(-1, 2)$ and $(3, 4)$.

$$r = \sqrt{(x - h)^2 + (y - k)^2} \qquad \text{Distance Formula}$$

$$= \sqrt{[3 - (-1)]^2 + (4 - 2)^2} \qquad \text{Substitute for } x, y, h, \text{ and } k.$$

$$= \sqrt{4^2 + 2^2} \qquad \text{Simplify.}$$

$$= \sqrt{16 + 4} \qquad \text{Simplify.}$$

$$= \sqrt{20} \qquad \text{Radius}$$

Using $(h, k) = (-1, 2)$ and $r = \sqrt{20}$, the equation of the circle is

$$(x - h)^2 + (y - k)^2 = r^2 \qquad \text{Equation of circle}$$

$$[x - (-1)]^2 + (y - 2)^2 = \left(\sqrt{20}\right)^2 \qquad \text{Substitute for } h, k, \text{ and } r.$$

$$(x + 1)^2 + (y - 2)^2 = 20. \qquad \text{Standard form}$$

✓**CHECKPOINT**   Now try Exercise 53.

You will learn more about writing equations of circles in Section 4.4.

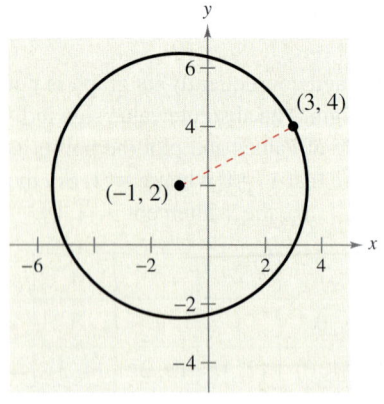

FIGURE 1.12

Application

In this course, you will learn that there are many ways to approach a problem. Three common approaches are illustrated in Example 9.

A *Numerical Approach:* Construct and use a table.
A *Graphical Approach:* Draw and use a graph.
An *Algebraic Approach:* Use the rules of algebra.

Example 9   **Recommended Weight**

The median recommended weight y (in pounds) for men of medium frame who are 25 to 59 years old can be approximated by the mathematical model

$$y = 0.073x^2 - 6.99x + 289.0, \quad 62 \leq x \leq 76$$

where x is the man's height (in inches).   (Source: Metropolitan Life Insurance Company)

a. Construct a table of values that shows the median recommended weights for men with heights of 62, 64, 66, 68, 70, 72, 74, and 76 inches.

b. Use the table of values to sketch a graph of the model. Then use the graph to estimate *graphically* the median recommended weight for a man whose height is 71 inches.

c. Use the model to confirm *algebraically* the estimate you found in part (b).

Solution

a. You can use a calculator to complete the table, as shown at the left.

b. The table of values can be used to sketch the graph of the equation, as shown in Figure 1.13. From the graph, you can estimate that a height of 71 inches corresponds to a weight of about 161 pounds.

Height, x	Weight, y
62	136.2
64	140.6
66	145.6
68	151.2
70	157.4
72	164.2
74	171.5
76	179.4

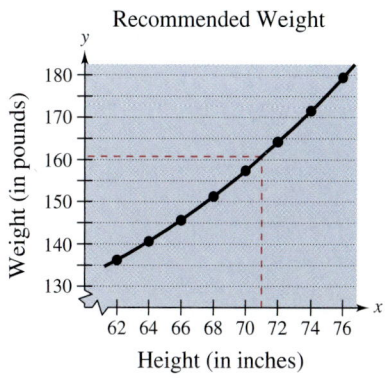

FIGURE 1.13

c. To confirm algebraically the estimate found in part (b), you can substitute 71 for x in the model.

$$y = 0.073(71)^2 - 6.99(71) + 289.0 \approx 160.70$$

So, the graphical estimate of 161 pounds is fairly good.

✔CHECKPOINT   Now try Exercise 67.

1.1   Exercises

The *HM mathSpace®* CD-ROM and *Eduspace®* for this text contain step-by-step solutions to all odd-numbered exercises. They also provide Tutorial Exercises for additional help.

VOCABULARY CHECK: Fill in the blanks.

1. An ordered pair (a, b) is a _____ of an equation in x and y if the equation is true when a is substituted for x and b is substituted for y.

2. The set of all solution points of an equation is the _____ of the equation.

3. The points at which a graph intersects or touches an axis are called the _____ of the graph.

4. A graph is symmetric with respect to the _____ if, whenever (x, y) is on the graph, $(-x, y)$ is also on the graph.

5. The equation $(x - h)^2 + (y - k)^2 = r^2$ is the standard form of the equation of a _____ with center _____ and radius _____.

6. When you construct and use a table to solve a problem, you are using a _____ approach.

PREREQUISITE SKILLS REVIEW: Practice and review algebra skills needed for this section at **www.Eduspace.com.**

In Exercises 1–4, determine whether each point lies on the graph of the equation.

	Equation		Points
1.	$y = \sqrt{x + 4}$	(a) $(0, 2)$	(b) $(5, 3)$
2.	$y = x^2 - 3x + 2$	(a) $(2, 0)$	(b) $(-2, 8)$
3.	$y = 4 - |x - 2|$	(a) $(1, 5)$	(b) $(6, 0)$
4.	$y = \frac{1}{3}x^3 - 2x^2$	(a) $\left(2, -\frac{16}{3}\right)$	(b) $(-3, 9)$

In Exercises 5–8, complete the table. Use the resulting solution points to sketch the graph of the equation.

5. $y = -2x + 5$

x	-1	0	1	2	$\frac{5}{2}$
y					
(x, y)					

6. $y = \frac{3}{4}x - 1$

x	-2	0	1	$\frac{4}{3}$	2
y					
(x, y)					

7. $y = x^2 - 3x$

x	-1	0	1	2	3
y					
(x, y)					

8. $y = 5 - x^2$

x	-2	-1	0	1	2
y					
(x, y)					

In Exercises 9–12, find the x- and y-intercepts of the graph of the equation.

9. $y = 16 - 4x^2$

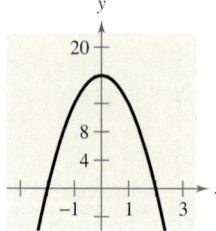

10. $y = (x + 3)^2$

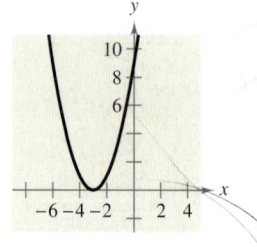

11. $y = 2x^3 - 4x^2$

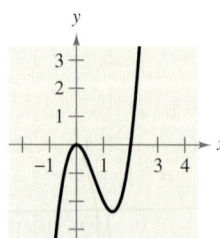

12. $y^2 = x + 1$

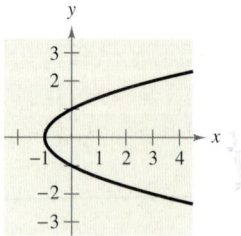

In Exercises 13–16, assume that the graph has the indicated type of symmetry. Sketch the complete graph of the equation. To print an enlarged copy of the graph, go to the website *www.mathgraphs.com*.

13.

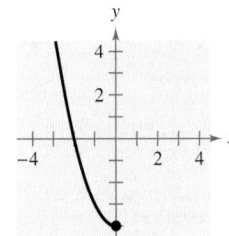

y-Axis symmetry

14.

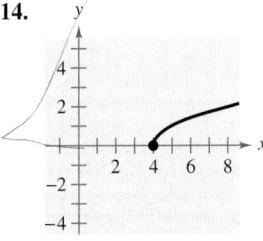

x-Axis symmetry

15.

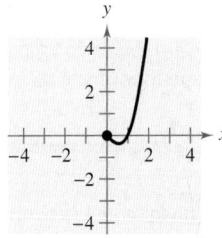

Origin symmetry

16.

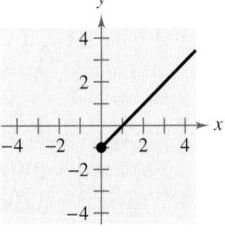

y-Axis symmetry

In Exercises 17–24, use the algebraic tests to check for symmetry with respect to both axes and the origin.

17. $x^2 - y = 0$

18. $x - y^2 = 0$

19. $y = x^3$

20. $y = x^4 - x^2 + 3$

21. $y = \dfrac{x}{x^2 + 1}$

22. $y = \dfrac{1}{x^2 + 1}$

23. $xy^2 + 10 = 0$

24. $xy = 4$

In Exercises 25–36, use symmetry to sketch the graph of the equation.

25. $y = -3x + 1$

26. $y = 2x - 3$

27. $y = x^2 - 2x$

28. $y = -x^2 - 2x$

29. $y = x^3 + 3$

30. $y = x^3 - 1$

31. $y = \sqrt{x - 3}$

32. $y = \sqrt{1 - x}$

33. $y = |x - 6|$

34. $y = 1 - |x|$

35. $x = y^2 - 1$

36. $x = y^2 - 5$

 In Exercises 37–48, use a graphing utility to graph the equation. Use a standard setting. Approximate any intercepts.

37. $y = 3 - \frac{1}{2}x$

38. $y = \frac{2}{3}x - 1$

39. $y = x^2 - 4x + 3$

40. $y = x^2 + x - 2$

41. $y = \dfrac{2x}{x - 1}$

42. $y = \dfrac{4}{x^2 + 1}$

43. $y = \sqrt[3]{x}$

44. $y = \sqrt[3]{x + 1}$

45. $y = x\sqrt{x + 6}$

46. $y = (6 - x)\sqrt{x}$

47. $y = |x + 3|$

48. $y = 2 - |x|$

In Exercises 49–56, write the standard form of the equation of the circle with the given characteristics.

49. Center: $(0, 0)$; radius: 4

50. Center: $(0, 0)$; radius: 5

51. Center: $(2, -1)$; radius: 4

52. Center: $(-7, -4)$; radius: 7

53. Center: $(-1, 2)$; solution point: $(0, 0)$

54. Center: $(3, -2)$; solution point: $(-1, 1)$

55. Endpoints of a diameter: $(0, 0)$, $(6, 8)$

56. Endpoints of a diameter: $(-4, -1)$, $(4, 1)$

In Exercises 57–62, find the center and radius of the circle, and sketch its graph.

57. $x^2 + y^2 = 25$

58. $x^2 + y^2 = 16$

59. $(x - 1)^2 + (y + 3)^2 = 9$

60. $x^2 + (y - 1)^2 = 1$

61. $\left(x - \frac{1}{2}\right)^2 + \left(y - \frac{1}{2}\right)^2 = \frac{9}{4}$

62. $(x - 2)^2 + (y + 3)^2 = \frac{16}{9}$

63. *Depreciation*   A manufacturing plant purchases a new molding machine for \$225,000. The depreciated value y (drop in value) after t years is given by $y = 225,000 - 20,000t, \ 0 \le t \le 8$. Sketch the graph of the equation.

64. *Consumerism*   You purchase a jet ski for \$8100. The depreciated value y after t years is given by $y = 8100 - 929t, \ 0 \le t \le 6$. Sketch the graph of the equation.

65. *Geometry*   A regulation NFL playing field (including the end zones) of length x and width y has a perimeter of $346\frac{2}{3}$ or $\frac{1040}{3}$ yards.

(a) Draw a rectangle that gives a visual representation of the problem. Use the specified variables to label the sides of the rectangle.

(b) Show that the width of the rectangle is $y = \dfrac{520}{3} - x$ and its area is $A = x\left(\dfrac{520}{3} - x\right)$.

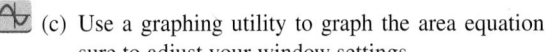 (c) Use a graphing utility to graph the area equation. Be sure to adjust your window settings.

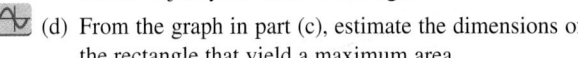 (d) From the graph in part (c), estimate the dimensions of the rectangle that yield a maximum area.

(e) Use your school's library, the Internet, or some other reference source to find the actual dimensions and area of a regulation NFL playing field and compare your findings with the results of part (d).

The symbol indicates an exercise or a part of an exercise in which you are instructed to use a graphing utility.

66. *Geometry*   A soccer playing field of length x and width y has a perimeter of 360 meters.

(a) Draw a rectangle that gives a visual representation of the problem. Use the specified variables to label the sides of the rectangle.

(b) Show that the width of the rectangle is $w = 180 - x$ and its area is $A = x(180 - x)$.

 (c) Use a graphing utility to graph the area equation. Be sure to adjust your window settings.

 (d) From the graph in part (c), estimate the dimensions of the rectangle that yield a maximum area.

(e) Use your school's library, the Internet, or some other reference source to find the actual dimensions and area of a regulation Major League Soccer field and compare your findings with the results of part(d).

Model It

67. *Population Statistics*   The table shows the life expectancies of a child (at birth) in the United States for selected years from 1920 to 2000.   (Source: U.S. National Center for Health Statistics)

Year	Life expectancy, y
1920	54.1
1930	59.7
1940	62.9
1950	68.2
1960	69.7
1970	70.8
1980	73.7
1990	75.4
2000	77.0

A model for the life expectancy during this period is

$$y = -0.0025t^2 + 0.574t + 44.25, \quad 20 \le t \le 100$$

where y represents the life expectancy and t is the time in years, with $t = 20$ corresponding to 1920.

(a) Sketch a scatter plot of the data.

(b) Graph the model for the data and compare the scatter plot and the graph.

(c) Determine the life expectancy in 1948 both graphically and algebraically.

(d) Use the graph of the model to estimate the life expectancies of a child for the years 2005 and 2010.

(e) Do you think this model can be used to predict the life expectancy of a child 50 years from now? Explain.

68. *Electronics*   The resistance y (in ohms) of 1000 feet of solid copper wire at 68 degrees Fahrenheit can be approximated by the model $y = \dfrac{10{,}770}{x^2} - 0.37, \ 5 \le x \le 100$ where x is the diameter of the wire in mils (0.001 inch).

(Source: American Wire Gage)

(a) Complete the table.

x	5	10	20	30	40	50
y						

x	60	70	80	90	100
y					

(b) Use the table of values in part (a) to sketch a graph of the model. Then use your graph to estimate the resistance when $x = 85.5$.

(c) Use the model to confirm algebraically the estimate you found in part (b).

(d) What can you conclude in general about the relationship between the diameter of the copper wire and the resistance?

Synthesis

True or False?   **In Exercises 69 and 70, determine whether the statement is true or false. Justify your answer.**

69. A graph is symmetric with respect to the x-axis if, whenever (x, y) is on the graph, $(-x, y)$ is also on the graph.

70. A graph of an equation can have more than one y-intercept.

 71. *Think About It*   Suppose you correctly enter an expression for the variable y on a graphing utility. However, no graph appears on the display when you graph the equation. Give a possible explanation and the steps you could take to remedy the problem. Illustrate your explanation with an example.

72. *Think About It*   Find a and b if the graph of $y = ax^2 + bx^3$ is symmetric with respect to (a) the y-axis and (b) the origin. (There are many correct answers.)

Skills Review

73. Identify the terms: $9x^5 + 4x^3 - 7$.

74. Rewrite the expression using exponential notation.

$$-(7 \times 7 \times 7 \times 7)$$

In Exercises 75–78, simplify the expression.

75. $\sqrt{18x} - \sqrt{2x}$

76. $\dfrac{55}{\sqrt{20} - 3}$

77. $\sqrt[6]{t^2}$

78. $\sqrt[3]{\sqrt{y}}$

What you should learn

- Identify different types of equations.
- Solve linear equations in one variable.
- Solve equations that lead to linear equations.
- Find x- and y-intercepts of graphs of equations algebraically.
- Use linear equations to model and solve real-life problems.

Why you should learn it

Linear equations are used in many real-life applications. For example, in Exercise 98 on page 97, linear equations can be used to model the relationship of the number of married women in the civilian workforce over time.

© Andrew Douglas/Masterfile

Equations and Solutions of Equations

An **equation** in x is a statement that two algebraic expressions are equal. For example

$$3x - 5 = 7, \; x^2 - x - 6 = 0, \text{ and } \sqrt{2x} = 4$$

are equations. To **solve** an equation in x means to find all values of x for which the equation is true. Such values are **solutions.** For instance, $x = 4$ is a solution of the equation

$$3x - 5 = 7$$

because $3(4) - 5 = 7$ is a true statement.

The solutions of an equation depend on the kinds of numbers being considered. For instance, in the set of rational numbers, $x^2 = 10$ has no solution because there is no rational number whose square is 10. However, in the set of real numbers, the equation has the two solutions $x = \sqrt{10}$ and $x = -\sqrt{10}$.

An equation that is true for *every* real number in the domain of the variable is called an **identity.** For example

$$x^2 - 9 = (x + 3)(x - 3) \qquad \text{Identity}$$

is an identity because it is a true statement for any real value of x. The equation

$$\frac{x}{3x^2} = \frac{1}{3x} \qquad \text{Identity}$$

where $x \neq 0$, is an identity because it is true for any nonzero real value of x.

An equation that is true for just *some* (or even none) of the real numbers in the domain of the variable is called a **conditional equation.** For example, the equation

$$x^2 - 9 = 0 \qquad \text{Conditional equation}$$

is conditional because $x = 3$ and $x = -3$ are the only values in the domain that satisfy the equation. The equation $2x - 4 = 2x + 1$ is conditional because there are no real values of x for which the equation is true. Learning to solve conditional equations is the primary focus of this chapter.

Linear Equations in One Variable

Definition of a Linear Equation

A **linear equation in one variable** x is an equation that can be written in the standard form

$$ax + b = 0$$

where a and b are real numbers with $a \neq 0$.

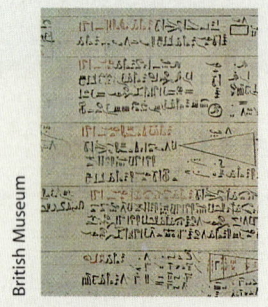

British Museum

Historical Note

This ancient Egyptian papyrus, discovered in 1858, contains one of the earliest examples of mathematical writing in existence. The papyrus itself dates back to around 1650 B.C., but it is actually a copy of writings from two centuries earlier. The algebraic equations on the papyrus were written in words. Diophantus, a Greek who lived around A.D. 250, is often called the Father of Algebra. He was the first to use abbreviated word forms in equations.

A linear equation has exactly one solution. To see this, consider the following steps. (Remember that $a \neq 0$.)

$$ax + b = 0 \qquad \text{Write original equation.}$$

$$ax = -b \qquad \text{Subtract } b \text{ from each side.}$$

$$x = -\frac{b}{a} \qquad \text{Divide each side by } a.$$

To solve a conditional equation in x, isolate x on one side of the equation by a sequence of **equivalent** (and usually simpler) **equations,** each having the same solution(s) as the original equation. The operations that yield equivalent equations come from the Substitution Principle and the Properties of Equality studied in Chapter P.

Generating Equivalent Equations

An equation can be transformed into an *equivalent equation* by one or more of the following steps.

	Given Equation	*Equivalent Equation*
1. Remove symbols of grouping, combine like terms, or simplify fractions on one or both sides of the equation.	$2x - x = 4$	$x = 4$
2. Add (or subtract) the same quantity to (from) *each* side of the equation.	$x + 1 = 6$	$x = 5$
3. Multiply (or divide) *each* side of the equation by the same *nonzero* quantity.	$2x = 6$	$x = 3$
4. Interchange the two sides of the equation.	$2 = x$	$x = 2$

Exploration

Use a graphing utility to graph the equation $y = 3x - 6$. Use the result to estimate the x-intercept of the graph. Explain how the x-intercept is related to the solution of the equation $3x - 6 = 0$, as shown in Example 1(a).

Example 1   Solving a Linear Equation

a. $3x - 6 = 0$ Original equation

$\quad\quad 3x = 6$ Add 6 to each side.

$\quad\quad\; x = 2$ Divide each side by 3.

b. $5x + 4 = 3x - 8$ Original equation

$\quad 2x + 4 = -8$ Subtract $3x$ from each side.

$\quad\quad 2x = -12$ Subtract 4 from each side.

$\quad\quad\; x = -6$ Divide each side by 2.

✓**CHECKPOINT**   Now try Exercise 27.

After solving an equation, you should check each solution in the original equation. For instance, you can check the solution to Example 1(a) as follows.

$$3x - 6 = 0 \qquad \text{Write original equation.}$$

$$3(2) - 6 \overset{?}{=} 0 \qquad \text{Substitute 2 for } x.$$

$$0 = 0 \qquad \text{Solution checks. } \checkmark$$

Try checking the solution to Example 1(b).

Some equations have no solutions because all the x-terms sum to zero and a contradictory (false) statement such as $0 = 5$ or $12 = 7$ is obtained. For instance, the equation

$$x = x + 1$$

has no solution. Watch for this type of equation in the exercises.

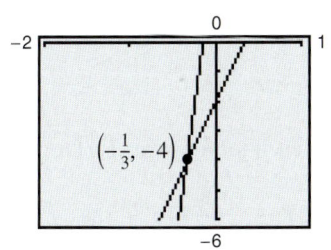
Example 2   Solving a Linear Equation

Solve

$$6(x - 1) + 4 = 3(7x + 1).$$

Solution

$$6(x - 1) + 4 = 3(7x + 1) \qquad \text{Write original equation.}$$

$$6x - 6 + 4 = 21x + 3 \qquad \text{Distributive Property}$$

$$6x - 2 = 21x + 3 \qquad \text{Simplify.}$$

$$-15x - 2 = 3 \qquad \text{Subtract } 21x \text{ from each side.}$$

$$-15x = 5 \qquad \text{Add 2 to each side.}$$

$$x = -\frac{1}{3} \qquad \text{Divide each side by } -15.$$

Check

Check this solution by substituting $-\frac{1}{3}$ for x in the original equation.

$$6(x - 1) + 4 = 3(7x + 1) \qquad \text{Write original equation.}$$

$$6\left(-\frac{1}{3} - 1\right) + 4 \overset{?}{=} 3\left[7\left(-\frac{1}{3}\right) + 1\right] \qquad \text{Substitute } -\frac{1}{3} \text{ for } x.$$

$$6\left(-\frac{4}{3}\right) + 4 \overset{?}{=} 3\left[-\frac{7}{3} + 1\right] \qquad \text{Simplify.}$$

$$6\left(-\frac{4}{3}\right) + 4 \overset{?}{=} 3\left(-\frac{4}{3}\right) \qquad \text{Simplify.}$$

$$-\frac{24}{3} + 4 \overset{?}{=} -\frac{12}{3} \qquad \text{Multiply.}$$

$$-8 + 4 \overset{?}{=} -4 \qquad \text{Simplify.}$$

$$-4 = -4 \qquad \text{Solution checks. } \checkmark$$

So, the solution is $x = -\frac{1}{3}$. Note that if you subtracted $6x$ from each side of the equation and then subtracted 3 from each side of the equation, you would still obtain the solution $x = -\frac{1}{3}$.

✓CHECKPOINT   Now try Exercise 29.

Equations That Lead to Linear Equations

To solve an equation involving fractional expressions, find the least common denominator (LCD) of all terms and multiply every term by the LCD. This process will clear the original equation of fractions and produce a simpler equation to work with.

STUDY TIP

An equation with a *single fraction* on each side can be cleared of denominators by **cross multiplying,** which is equivalent to multiplying by the LCD and then dividing out. To do this, multiply the left numerator by the right denominator and the right numerator by the left denominator as follows.

$$\frac{a}{b} = \frac{c}{d} \qquad \text{LCD is } bd.$$

$$\frac{a}{b} \cdot bd = \frac{c}{d} \cdot bd \qquad \text{Multiply by LCD.}$$

$$ad = cb \qquad \text{Divide out common factors.}$$

Example 3   An Equation Involving Fractional Expressions

Solve $\dfrac{x}{3} + \dfrac{3x}{4} = 2$.

Solution

$$\frac{x}{3} + \frac{3x}{4} = 2 \qquad\qquad \text{Write original equation.}$$

$$(12)\frac{x}{3} + (12)\frac{3x}{4} = (12)2 \qquad\qquad \text{Multiply each term by the LCD of 12.}$$

$$4x + 9x = 24 \qquad\qquad \text{Divide out and multiply.}$$

$$13x = 24 \qquad\qquad \text{Combine like terms.}$$

$$x = \frac{24}{13} \qquad\qquad \text{Divide each side by 13.}$$

The solution is $x = \frac{24}{13}$. Check this in the original equation.

✓**CHECKPOINT**   Now try Exercise 33.

When multiplying or dividing an equation by a *variable* quantity, it is possible to introduce an extraneous solution. An **extraneous solution** is one that does not satisfy the original equation. Therefore, it is essential that you check your solutions.

Example 4   An Equation with an Extraneous Solution

Solve $\dfrac{1}{x - 2} = \dfrac{3}{x + 2} - \dfrac{6x}{x^2 - 4}$.

Solution

STUDY TIP

Recall that the least common denominator of two or more fractions consists of the product of all prime factors in the denominators, with each factor given the highest power of its occurrence in any denominator. For instance, in Example 4, by factoring each denominator you can determine that the LCD is $(x + 2)(x - 2)$.

The LCD is $x^2 - 4$, or $(x + 2)(x - 2)$. Multiply each term by this LCD.

$$\frac{1}{x - 2}(x + 2)(x - 2) = \frac{3}{x + 2}(x + 2)(x - 2) - \frac{6x}{x^2 - 4}(x + 2)(x - 2)$$

$$x + 2 = 3(x - 2) - 6x, \qquad x \neq \pm 2$$

$$x + 2 = 3x - 6 - 6x$$

$$x + 2 = -3x - 6$$

$$4x = -8 \quad \Longrightarrow \quad x = -2 \qquad \text{Extraneous solution}$$

In the original equation, $x = -2$ yields a denominator of zero. So, $x = -2$ is an extraneous solution, and the original equation has *no solution*.

✓**CHECKPOINT**   Now try Exercise 53.

Finding Intercepts Algebraically

In Section 1.1, you learned to find x- and y-intercepts using a graphical approach. Because all the points on the x-axis have a y-coordinate equal to zero, and all the points on the y-axis have an x-coordinate equal to zero, you can use an algebraic approach to find x- and y-intercepts, as follows.

> ### Finding Intercepts Algebraically
>
> **1.** To find x-intercepts, set y equal to zero and solve the equation for x.
>
> **2.** To find y-intercepts, set x equal to zero and solve the equation for y.

Here is an example.

$$y = 4x + 1 \implies 0 = 4x + 1 \implies -1 = 4x \implies -\tfrac{1}{4} = x$$

$$y = 4x + 1 \implies y = 4(0) + 1 \implies y = 1$$

So, the x-intercept of $y = 4x + 1$ is $\left(-\tfrac{1}{4}, 0\right)$ and the y-intercept is $(0, 1)$.

Application

Female Participants in High School Athletics

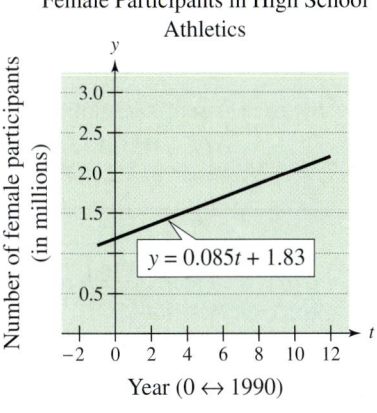

Number of female participants (in millions)

Year ($0 \leftrightarrow 1990$)

$y = 0.085t + 1.83$

FIGURE 1.14

Example 5   Female Participants in Athletic Programs

The number y (in millions) of female participants in high school athletic programs (in millions) in the United States from 1989 to 2002 can be approximated by the linear model

$$y = 0.085t + 1.83, \qquad -1 \le t \le 12$$

where $t = 0$ represents 1990. (a) Find the y-intercept of the graph of the linear model shown in Figure 1.14 algebraically. (b) Assuming that this linear pattern continues, find the year in which there will be 3.36 million female participants. (Source: National Federation of State High School Associations)

Solution

a. To find the y-intercept, let $t = 0$ and solve for y as follows.

$$y = 0.085t + 1.83 \qquad \text{Write original equation.}$$
$$= 0.085(0) + 1.83 \qquad \text{Substitute 0 for } t.$$
$$= 1.83 \qquad \text{Simplify.}$$

So, the y-intercept is $(0, 1.83)$.

b. Let $y = 3.36$ and solve the equation $3.36 = 0.085t + 1.83$ for t.

$$3.36 = 0.085t + 1.83 \qquad \text{Write original equation.}$$
$$1.53 = 0.085t \qquad \text{Subtract 1.83 from each side.}$$
$$18 = t \qquad \text{Divide each side by 0.085.}$$

Because $t = 0$ represents 1990, $t = 18$ must represent 2008. So, from this model, there will be 3.36 million female participants in 2008.

✓**CHECKPOINT**   Now try Exercise 97.

1.2   Exercises

VOCABULARY CHECK: Fill in the blanks.

1. An _____ is a statement that equates two algebraic expressions.

2. To find all values that satisfy an equation is to _____ the equation.

3. There are two types of equations, _____ and _____ equations.

4. A linear equation in one variable is an equation that can be written in the standard from _____.

5. When solving an equation, it is possible to introduce an _____ solution, which is a value that does not satisfy the original equation.

PREREQUISITE SKILLS REVIEW: Practice and review algebra skills needed for this section at **www.Eduspace.com.**

In Exercises 1–10, determine whether each value of *x* is a solution of the equation.

Equation	*Values*	
1. $5x - 3 = 3x + 5$	(a) $x = 0$	(b) $x = -5$
	(c) $x = 4$	(d) $x = 10$
2. $7 - 3x = 5x - 17$	(a) $x = -3$	(b) $x = 0$
	(c) $x = 8$	(d) $x = 3$
3. $3x^2 + 2x - 5$	(a) $x = -3$	(b) $x = 1$
$= 2x^2 - 2$	(c) $x = 4$	(d) $x = -5$
4. $5x^3 + 2x - 3$	(a) $x = 2$	(b) $x = -2$
$= 4x^3 + 2x - 11$	(c) $x = 0$	(d) $x = 10$
5. $\dfrac{5}{2x} - \dfrac{4}{x} = 3$	(a) $x = -\frac{1}{2}$	(b) $x = 4$
	(c) $x = 0$	(d) $x = \frac{1}{4}$
6. $3 + \dfrac{1}{x + 2} = 4$	(a) $x = -1$	(b) $x = -2$
	(c) $x = 0$	(d) $x = 5$
7. $\sqrt{3x - 2} = 4$	(a) $x = 3$	(b) $x = 2$
	(c) $x = 9$	(d) $x = -6$
8. $\sqrt[3]{x - 8} = 3$	(a) $x = 2$	(b) $x = -5$
	(c) $x = 35$	(d) $x = 8$
9. $6x^2 - 11x - 35 = 0$	(a) $x = -\frac{5}{3}$	(b) $x = -\frac{2}{7}$
	(c) $x = \frac{7}{2}$	(d) $x = \frac{5}{3}$
10. $10x^2 + 21x - 10 = 0$	(a) $x = \frac{2}{5}$	(b) $x = -\frac{5}{2}$
	(c) $x = -\frac{1}{3}$	(d) $x = -2$

In Exercises 11–20, determine whether the equation is an identity or a conditional equation.

11. $2(x - 1) = 2x - 2$ 12. $3(x + 2) = 5x + 4$

13. $-6(x - 3) + 5 = -2x + 10$

14. $3(x + 2) - 5 = 3x + 1$

15. $4(x + 1) - 2x = 2(x + 2)$

16. $-7(x - 3) + 4x = 3(7 - x)$

17. $x^2 - 8x + 5 = (x - 4)^2 - 11$

18. $x^2 + 2(3x - 2) = x^2 + 6x - 4$

19. $3 + \dfrac{1}{x + 1} = \dfrac{4x}{x + 1}$ 20. $\dfrac{5}{x} + \dfrac{3}{x} = 24$

In Exercises 21 and 22, justify each step of the solution.

21. $4x + 32 = 83$

$4x + 32 - 32 = 83 - 32$

$4x = 51$

$\dfrac{4x}{4} = \dfrac{51}{4}$

$x = \dfrac{51}{4}$

22. $3(x - 4) + 10 = 7$

$3x - 12 + 10 = 7$

$3x - 2 = 7$

$3x - 2 + 2 = 7 + 2$

$3x = 9$

$\dfrac{3x}{3} = \dfrac{9}{3}$

$x = 3$

In Exercises 23–38, solve the equation and check your solution.

23. $x + 11 = 15$ 24. $7 - x = 19$

25. $7 - 2x = 25$ 26. $7x + 2 = 23$

27. $8x - 5 = 3x + 20$ 28. $7x + 3 = 3x - 17$

29. $2(x + 5) - 7 = 3(x - 2)$

30. $3(x + 3) = 5(1 - x) - 1$

31. $x - 3(2x + 3) = 8 - 5x$

32. $9x - 10 = 5x + 2(2x - 5)$

33. $\dfrac{5x}{4} + \dfrac{1}{2} = x - \dfrac{1}{2}$

34. $\dfrac{x}{5} - \dfrac{x}{2} = 3 + \dfrac{3x}{10}$

35. $\frac{2}{3}(z + 5) - \frac{1}{4}(z + 24) = 0$

36. $\dfrac{3x}{2} + \dfrac{1}{4}(x - 2) = 10$

37. $0.25x + 0.75(10 - x) = 3$

38. $0.60x + 0.40(100 - x) = 50$

In Exercises 39–42, solve the equation using two different methods. Then explain which method is easier.

39. $3(x - 1) = 4$

40. $4(x + 3) = 15$

41. $\frac{1}{3}(x + 2) = 5$

42. $\frac{3}{4}(z - 4) = 6$

In Exercises 43–64, solve the equation and check your solution. (If not possible, explain why.)

43. $x + 8 = 2(x - 2) - x$

44. $8(x + 2) - 3(2x + 1) = 2(x + 5)$

45. $\dfrac{100 - 4x}{3} = \dfrac{5x + 6}{4} + 6$

46. $\dfrac{17 + y}{y} + \dfrac{32 + y}{y} = 100$

47. $\dfrac{5x - 4}{5x + 4} = \dfrac{2}{3}$

48. $\dfrac{10x + 3}{5x + 6} = \dfrac{1}{2}$

49. $10 - \dfrac{13}{x} = 4 + \dfrac{5}{x}$

50. $\dfrac{15}{x} - 4 = \dfrac{6}{x} + 3$

51. $3 = 2 + \dfrac{2}{z + 2}$

52. $\dfrac{1}{x} + \dfrac{2}{x - 5} = 0$

53. $\dfrac{x}{x + 4} + \dfrac{4}{x + 4} + 2 = 0$

54. $\dfrac{7}{2x + 1} - \dfrac{8x}{2x - 1} = -4$

55. $\dfrac{2}{(x - 4)(x - 2)} = \dfrac{1}{x - 4} + \dfrac{2}{x - 2}$

56. $\dfrac{4}{x - 1} + \dfrac{6}{3x + 1} = \dfrac{15}{3x + 1}$

57. $\dfrac{1}{x - 3} + \dfrac{1}{x + 3} - \dfrac{10}{x^2 - 9}$

58. $\dfrac{1}{x - 2} + \dfrac{3}{x + 3} = \dfrac{4}{x^2 + x - 6}$

59. $\dfrac{3}{x^2 - 3x} + \dfrac{4}{x} = \dfrac{1}{x - 3}$

60. $\dfrac{6}{x} - \dfrac{2}{x + 3} = \dfrac{3(x + 5)}{x^2 + 3x}$

61. $(x + 2)^2 + 5 = (x + 3)^2$

62. $(x + 1)^2 + 2(x - 2) = (x + 1)(x - 2)$

63. $(x + 2)^2 - x^2 = 4(x + 1)$

64. $(2x + 1)^2 = 4(x^2 + x + 1)$

Graphical Analysis   In Exercises 65–70, use a graphing utility to graph the equation and approximate any *x*-intercepts. Set $y = 0$ and solve the resulting equation. Compare the results with the graph's *x*-intercepts.

65. $y = 2(x - 1) - 4$

66. $y = \frac{4}{3}x + 2$

67. $y = 20 - (3x - 10)$

68. $y = 10 + 2(x - 2)$

69. $y = -38 + 5(9 - x)$

70. $y = 6x - 6\left(\frac{16}{11} + x\right)$

In Exercises 71–80, find the *x*- and *y*-intercepts of the graph of the equation algebraically.

71. $y = 12 - 5x$

72. $y = 16 - 3x$

73. $y = -3(2x + 1)$

74. $y = 5 - (6 - x)$

75. $2x + 3y = 10$

76. $4x - 5y = 12$

77. $\dfrac{2x}{5} + 8 - 3y = 0$

78. $\dfrac{8x}{3} + 5 - 2y = 0$

79. $4y - 0.75x + 1.2 = 0$

80. $3y + 2.5x - 3.4 = 0$

In Exercises 81–88, solve for *x*.

81. $4(x + 1) - ax = x + 5$

82. $4 - 2(x - 2b) = ax + 3$

83. $6x + ax = 2x + 5$

84. $5 + ax = 12 - bx$

85. $19x + \frac{1}{2}ax = x + 9$

86. $-5(3x - 6b) + 12 = 8 + 3ax$

87. $-2ax + 6(x + 3) = -4x + 1$

88. $\frac{4}{5}x - ax = 2\left(\frac{2}{5}x - 1\right) + 10$

In Exercises 89–92, solve the equation for *x*. (Round your solution to three decimal places.)

89. $0.275x + 0.725(500 - x) = 300$

90. $2.763 - 4.5(2.1x - 5.1432) = 6.32x + 5$

91. $\dfrac{2}{7.398} - \dfrac{4.405}{x} = \dfrac{1}{x}$

92. $\dfrac{3}{6.350} - \dfrac{6}{x} = 18$

93. *Geometry*   The surface area S of the circular cylinder shown in the figure is

$$S = 2\pi(25) + 2\pi(5h).$$

Find the height h of the cylinder if the surface area is 471 square feet. Use 3.14 for π.

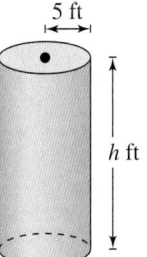

94. Geometry The surface area S of the rectangular solid in the figure is $S = 2(24) + 2(4x) + 2(6x)$. Find the length x of the box if the surface area is 248 square centimeters.

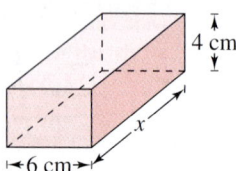

Model It

95. Anthropology The relationship between the length of an adult's femur (thigh bone) and the height of the adult can be approximated by the linear equations

$y = 0.432x - 10.44$ Female

$y = 0.449x - 12.15$ Male

where y is the length of the femur in inches and x is the height of the adult in inches (see figure).

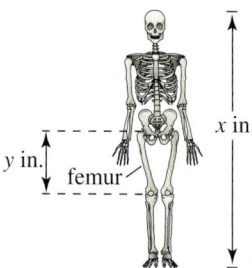

(a) An anthropologist discovers a femur belonging to an adult human female. The bone is 16 inches long. Estimate the height of the female.

(b) From the foot bones of an adult human male, an anthropologist estimates that the person's height was 69 inches. A few feet away from the site where the foot bones were discovered, the anthropologist discovers a male adult femur that is 19 inches long. Is it likely that both the foot bones and the thigh bone came from the same person?

(c) Complete the table to determine if there is a height of an adult for which an anthropologist would not be able to determine whether the femur belonged to a male or a female.

Model It (continued)

Height, x	Female femur length, y	Male femur length, y
60		
70		
80		
90		
100		
110		

(d) Solve part (c) algebraically by setting the two equations equal to each other and solving for x. Compare your solutions. Do you believe an anthropologist would ever have the problem of not being able to determine whether a femur belonged to a male or a female? Why or why not?

96. Tax Credits Use the following information about a possible tax credit for a family consisting of two adults and two children (see figure).

Earned income: E

Subsidy (a grant of money):
$S = 10,000 - \frac{1}{2}E, \quad 0 \le E \le 20,000$

Total income: $T = E + S$

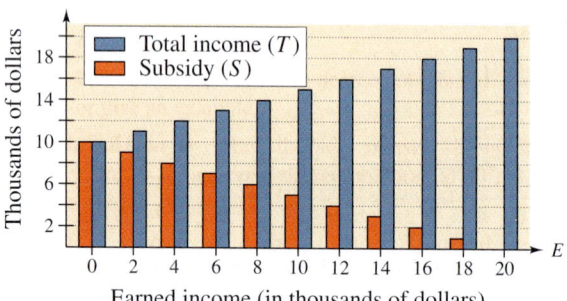

Earned income (in thousands of dollars)

(a) Write the total income T in terms of E.

(b) Find the earned income E if the subsidy is $6600.

(c) Find the earned income E if the total income is $13,800.

(d) Find the subsidy S if the total income is $12,500.

97. *Fuel Consumption*   The annual consumption y of gasoline (in billions of gallons) by vans, pickup trucks, and SUVs in the United States from 1989 to 2002 can be approximated by the model

$$y = 1.64t + 36.8, \quad -1 \le t \le 12$$

where t represents the year, with $t = 0$ corresponding to 1990.

(a) Sketch a graph of the model. Graphically estimate the y-intercept of the graph.

(b) Find the y-intercept of the graph algebraically.

(c) Assuming this linear pattern continues, find the year in which the annual consumption of gasoline will be 65 billion gallons. Does your answer seem reasonable? Explain.

98. *Labor Statistics*   The number of married women y (in millions) in the civilian work force in the United States from 1990 to 2002 (see figure) can be approximated by the model

$$y = 0.39t + 31.0, \quad 0 \le t \le 12$$

where t represents the year, with $t = 0$ corresponding to 1990. (Source: U.S. Bureau of Labor Statistics)

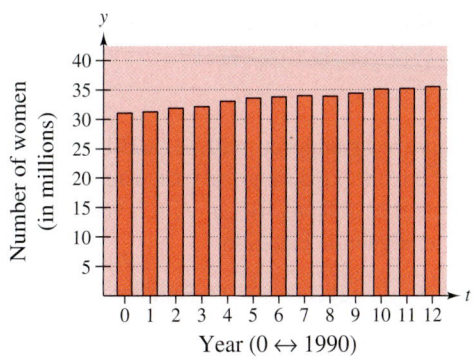

(a) According to this model, during which year did the number reach 33 million?

(b) Explain how you can solve part (a) graphically and algebraically.

99. *Operating Cost*   A delivery company has a fleet of vans. The annual operating cost C per van is

$$C = 0.32m + 2500$$

where m is the number of miles traveled by a van in a year. What number of miles will yield an annual operating cost of $10,000?

100. *Flood Control*   A river has risen 8 feet above its flood stage. The water begins to recede at a rate of 3 inches per hour. Write a mathematical model that shows the number of feet above flood stage after t hours. If the water continually recedes at this rate, when will the river be 1 foot above its flood stage?

Synthesis

True or False?   **In Exercises 101 and 102, determine whether the statement is true or false. Justify your answer.**

101. The equation $x(3 - x) = 10$ is a linear equation.

102. The equation $x^2 + 9x - 5 = 4 - x^3$ has no real solution.

103. *Think About It*   What is meant by *equivalent equations*? Give an example of two equivalent equations.

104. *Writing*   Describe the steps used to transform an equation into an equivalent equation.

105. *Exploration*

(a) Complete the table.

x	-1	0	1	2	3	4
$3.2x - 5.8$						

(b) Use the table in part (a) to determine the interval in which the solution to the equation $3.2x - 5.8 = 0$ is located. Explain your reasoning.

(c) Complete the table.

x	1.5	1.6	1.7	1.8	1.9	2.0
$3.2x - 5.8$						

(d) Use the table in part (c) to determine the interval in which the solution to the equation $3.2x - 5.8 = 0$ is located. Explain how this process can be used to approximate the solution to any desired degree of accuracy.

106. *Exploration*   Use the procedure in Exercise 105 to approximate the solution to the equation

$$0.3(x - 1.5) - 2 = 0$$

accurate to two decimal places.

Skills Review

In Exercises 107 and 108, simplify the expression.

107. $\dfrac{x^2 + 5x - 36}{2x^2 + 17x - 9}$

108. $\dfrac{x^2 - 49}{x^3 + x^2 + 3x - 21}$

In Exercises 109–112, sketch the graph of the equation.

109. $y = 3x - 5$

110. $y = -\frac{1}{2}x - \frac{9}{2}$

111. $y = -x^2 - 5x$

112. $y = \sqrt{5 - x}$

1.3   Modeling with Linear Equations

What you should learn

- Use a verbal model in a problem-solving plan.
- Write and use mathematical models to solve real-life problems.
- Solve mixture problems.
- Use common formulas to solve real-life problems.

Why you should learn it

You can use linear equations to determine the percents of income and of expenses of the federal government that come from various sources. See Exercise 42 on page 106.

© Bob Rowan; Progressive Image/Corbis

Introduction to Problem Solving

In this section, you will learn how algebra can be used to solve problems that occur in real-life situations. The process of translating phrases or sentences into algebraic expressions or equations is called **mathematical modeling.** A good approach to mathematical modeling is to use two stages. Begin by using the verbal description of the problem to form a *verbal model*. Then, after assigning labels to the quantities in the verbal model, form a *mathematical model* or *algebraic equation*.

$$\boxed{\text{Verbal Description}} \Rightarrow \boxed{\text{Verbal Model}} \Rightarrow \boxed{\text{Algebraic Equation}}$$

When you are trying to construct a verbal model, it is helpful to look for a *hidden equality*—a statement that two algebraic expressions are equal.

Example 1   Using a Verbal Model

You have accepted a job for which your annual salary will be $32,300. This salary includes a year-end bonus of $500. You will be paid twice a month. What will your gross pay (pay before taxes) be for each paycheck?

Solution

Because there are 12 months in a year and you will be paid twice a month, it follows that you will receive 24 paychecks during the year. You can construct an algebraic equation for this problem as follows. Begin with a verbal model, then assign labels, and finally form an algebraic equation.

Verbal Model:	Income for year = 24 paychecks + Bonus	
Labels:	Income for year = 32,300	(dollars)
	Amount of each paycheck = x	(dollars)
	Bonus = 500	(dollars)

Equation: $32{,}300 = 24x + 500$

The algebraic equation for this problem is a *linear equation* in the variable x, which you can solve as follows.

$$32{,}300 = 24x + 500 \qquad \text{Write original equation.}$$

$$32{,}300 - 500 = 24x + 500 - 500 \qquad \text{Subtract 500 from each side.}$$

$$31{,}800 = 24x \qquad \text{Simplify.}$$

$$\frac{31{,}800}{24} = \frac{24x}{24} \qquad \text{Divide each side by 24.}$$

$$1325 = x \qquad \text{Simplify.}$$

So, your gross pay for each paycheck will be $1325.

✔CHECKPOINT   Now try Exercise 29.

A fundamental step in writing a mathematical model to represent a real-life problem is translating key words and phrases into algebraic expressions and equations. The following list gives several examples.

Translating Key Words and Phrases

Key Words and Phrases	Verbal Description	Algebraic Expression or Equation
Equality:		
Equals, equal to, is, are, was, will be, represents	• The sale price S is $10 less than the list price L.	$S = L - 10$
Addition:		
Sum, plus, greater than, increased by, more than, exceeds, total of	• The sum of 5 and x • Seven more than y	$5 + x$ or $x + 5$ $7 + y$ or $y + 7$
Subtraction:		
Difference, minus, less than, decreased by, subtracted from, reduced by, the remainder	• The diffcrence of 4 and b • Three less than z	$4 - b$ $z - 3$
Multiplication:		
Product, multiplied by, twice, times, percent of	• Two times x • Three percent of t	$2x$ $0.03t$
Division:		
Quotient, divided by, ratio, per	• The ratio of x to 8	$\dfrac{x}{8}$

Using Mathematical Models

Example 2   Finding the Percent of a Raise

You have accepted a job that pays $8 an hour. You are told that after a two-month probationary period, your hourly wage will be increased to $9 an hour. What percent raise will you receive after the two-month period?

Solution

Verbal Model: Raise $=$ Percent $\cdot$ Old wage

Labels:
  Old wage $= 8$ (dollars per hour)
  New wage $= 9$ (dollars per hour)
  Raise $= 9 - 8 = 1$ (dollars per hour)
  Percent $= r$ (percent in decimal form)

Equation: $1 = r \cdot 8$

  $\dfrac{1}{8} = r$ Divide each side by 8.

  $0.125 = r$ Rewrite fraction as a decimal.

You will receive a raise of 0.125 or 12.5%.

✔ CHECKPOINT   Now try Exercise 41.

Example 3 **Finding the Percent of Monthly Expenses**

Your family has an annual income of \$57,000 and the following monthly expenses: mortgage (\$1100), car payment (\$375), food (\$295), utilities (\$240), and credit cards (\$220). The total value of the monthly expenses represents what percent of your family's annual income?

Solution

The total amount of your family's monthly expenses is \$2230. The total monthly expenses for 1 year are \$26,760.

Verbal Model: Monthly expenses = Percent · Income

Labels:
Income = 57,000   (dollars)
Monthly expenses = 26,760   (dollars)
Percent = r (in decimal form)

Equation:
$$26{,}760 = r \cdot 57{,}000$$

$$\frac{26{,}760}{57{,}000} = r \qquad \text{Divide each side by 57,000.}$$

$$0.469 \approx r \qquad \text{Use a calculator.}$$

Your family's monthly expenses are approximately 0.469 or 46.9% of your family's annual income.

✔**CHECKPOINT** Now try Exercise 47.

Example 4 **Finding the Dimensions of a Room**

A rectangular kitchen is twice as long as it is wide, and its perimeter is 84 feet. Find the dimensions of the kitchen.

Solution

For this problem, it helps to sketch a diagram, as shown in Figure 1.15.

Verbal Model: $2 \cdot$ Length $+ 2 \cdot$ Width $=$ Perimeter

Labels:
Perimeter = 84   (feet)
Width = w (feet)
Length = $l = 2w$ (feet)

Equation:
$$2(2w) + 2w = 84$$

$$6w = 84 \qquad \text{Group like terms.}$$

$$w = 14 \qquad \text{Divide each side by 6.}$$

Because the length is twice the width, you have

$$l = 2w \qquad \text{Length is twice width.}$$

$$= 2(14) = 28. \qquad \text{Substitute and simplify.}$$

So, the dimensions of the room are 14 feet by 28 feet.

✔**CHECKPOINT** Now try Exercise 49.

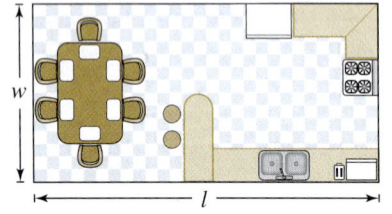

FIGURE 1.15

Example 5 A Distance Problem

A plane is flying nonstop from Atlanta to Portland, a distance of about 2700 miles, as shown in Figure 1.16. After 1.5 hours in the air, the plane flies over Kansas City (a distance of 820 miles from Atlanta). Estimate the time it will take the plane to fly from Atlanta to Portland.

Solution

Verbal Model:

$$\boxed{\text{Distance}} = \boxed{\text{Rate}} \cdot \boxed{\text{Time}}$$

Labels:

Distance = 2700   (miles)

Time = t (hours)

Rate = $\dfrac{\text{distance to Kansas City}}{\text{time to Kansas City}} = \dfrac{820}{1.5}$ (miles per hour)

Equation:

$$2700 = \frac{820}{1.5}t$$

$$4050 = 820t$$

$$\frac{4050}{820} = t$$

$$4.94 \approx t$$

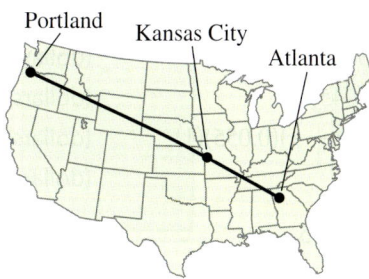

FIGURE 1.16

The trip will take about 4.94 hours, or about 4 hours and 56 minutes.

✔CHECKPOINT   Now try Exercise 53.

Example 6 An Application Involving Similar Triangles

To determine the height of the Aon Center Building (in Chicago), you measure the shadow cast by the building and find it to be 142 feet long, as shown in Figure 1.17. Then you measure the shadow cast by a four-foot post and find it to be 6 inches long. Estimate the building's height.

Solution

To solve this problem, you use a result from geometry that states that the ratios of corresponding sides of similar triangles are equal.

Verbal Model:

$$\frac{\text{Height of building}}{\text{Length of building's shadow}} = \frac{\text{Height of post}}{\text{Length of post's shadow}}$$

Labels:

Height of building = x (feet)

Length of building's shadow = 142   (feet)

Height of post = 4 feet = 48 inches   (inches)

Length of post's shadow = 6   (inches)

Equation:

$$\frac{x}{142} = \frac{48}{6}$$

$$x = 1136$$

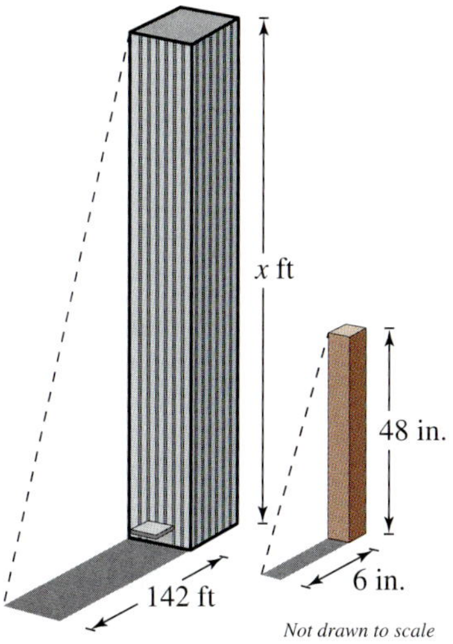

So, the Aon Center Building is about 1136 feet high.

✔CHECKPOINT   Now try Exercise 65.

x ft

48 in.

142 ft   6 in.

Not drawn to scale

FIGURE 1.17

When working with applied problems, you often need to rewrite a literal equation in terms of another variable. You can use the methods for solving linear equations to solve some literal equations for a specified variable. For instance, the formula for the perimeter of a rectangle, $P = 2l + 2w$, can be rewritten or solved for w as $w = \frac{1}{2}(P - 2l)$.

Example 9 **Using a Formula**

A cylindrical can has a volume of 200 cubic centimeters (cm^3) and a radius of 4 centimeters (cm), as shown in Figure 1.18. Find the height of the can.

Solution

The formula for the *volume of a cylinder* is $V = \pi r^2 h$. To find the height of the can, solve for h.

$$h = \frac{V}{\pi r^2}$$

Then, using $V = 200$ and $r = 4$, find the height.

$$h = \frac{200}{\pi (4)^2} \qquad \text{Substitute 200 for } V \text{ and 4 for } r.$$

$$= \frac{200}{16\pi} \qquad \text{Simplify denominator.}$$

$$\approx 3.98 \qquad \text{Use a calculator.}$$

You can use unit analysis to check that your answer is reasonable.

$$\frac{200 \text{ cm}^3}{16\pi \text{ cm}^2} \approx 3.98 \text{ cm}$$

✓**CHECKPOINT**   Now try Exercise 91.

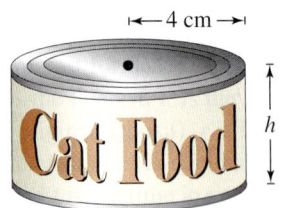

←4 cm→

h

FIGURE 1.18

𝒲RITING ABOUT 𝑀ATHEMATICS

Translating Algebraic Formulas Most people use algebraic formulas every day—sometimes without realizing it because they use a verbal form or think of an often-repeated calculation in steps. Translate each of the following verbal descriptions into an algebraic formula, and demonstrate the use of each formula.

a. **Designing Billboards** "The letters on a sign or billboard are designed to be readable at a certain distance. Take half the letter height in inches and multiply by 100 to find the readable distance in feet."—Thos. Hodgson, Hodgson Signs (Source: *Rules of Thumb* by Tom Parker)

b. **Percent of Calories from Fat** "To calculate percent of calories from fat, multiply grams of total fat per serving by 9, then divide by the number of calories per serving." (Source: *Good Housekeeping*)

c. **Building Stairs** "A set of steps will be comfortable to use if two times the height of one riser plus the width of one tread is equal to 26 inches." —Alice Lukens Bachelder, gardener (Source: *Rules of Thumb* by Tom Parker)

1.3 | Exercises

VOCABULARY CHECK:

In Exercises 1 and 2, fill in the blanks.

1. The process of translating phrases or sentences into algebraic expressions or equations is called _____ _____.

2. A good approach to mathematical modeling is a two-stage approach, using a verbal description to form a(n) _____ _____, and then after assigning labels to the quantities, form a(n) _____ _____.

In Exercises 3–8, write the formula for the given quantity.

3. Area of a circle: _____

4. Perimeter of a rectangle: _____

5. Volume of a cube: _____

6. Volume of a circular cylinder: _____

7. Balance if P dollars is invested at $r\%$ compounded monthly for t years: _____

8. Simple interest if P dollars is invested at $r\%$ for t years: _____

PREREQUISITE SKILLS REVIEW: Practice and review algebra skills needed for this section at **www.Eduspace.com.**

In Exercises 1–10, write a verbal description of the algebraic expression without using the variable.

1. $x + 4$

2. $t - 10$

3. $\dfrac{u}{5}$

4. $\dfrac{2}{3}x$

5. $\dfrac{y - 4}{5}$

6. $\dfrac{z + 10}{7}$

7. $-3(b + 2)$

8. $\dfrac{-5(x - 1)}{8}$

9. $12x(x - 5)$

10. $\dfrac{(q + 4)(3 - q)}{2q}$

In Exercises 11–22, write an algebraic expression for the verbal description.

11. The sum of two consecutive natural numbers

12. The product of two consecutive natural numbers

13. The product of two consecutive odd integers, the first of which is $2n - 1$

14. The sum of the squares of two consecutive even integers, the first of which is $2n$

15. The distance traveled in t hours by a car traveling at 50 miles per hour

16. The travel time for a plane traveling at a rate of r kilometers per hour for 200 kilometers

17. The amount of acid in x liters of a 20% acid solution

18. The sale price of an item that is discounted 20% of its list price L

19. The perimeter of a rectangle with a width x and a length that is twice the width

20. The area of a triangle with base 20 inches and height h inches

21. The total cost of producing x units for which the fixed costs are $1200 and the cost per unit is $25

22. The total revenue obtained by selling x units at $3.59 per unit

In Exercises 23–26, translate the statement into an algebraic expression or equation.

23. Thirty percent of the list price L

24. The amount of water in q quarts of a liquid that is 35% water

25. The percent of 500 that is represented by the number N

26. The percent change in sales from one month to the next if the monthly sales are S_1 and S_2, respectively

In Exercises 27 and 28, write an expression for the area of the region in the figure.

27.

28.

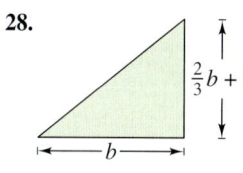

Number Problems In Exercises 29–34, write a mathematical model for the problem and solve.

29. The sum of two consecutive natural numbers is 525. Find the numbers.

30. The sum of three consecutive natural numbers is 804. Find the numbers.

31. One positive number is 5 times another number. The difference between the two numbers is 148. Find the numbers.

32. One positive number is $\frac{1}{5}$ of another number. The difference between the two numbers is 76. Find the numbers.

33. Find two consecutive integers whose product is 5 less than the square of the smaller number.

34. Find two consecutive natural numbers such that the difference of their reciprocals is $\frac{1}{4}$ the reciprocal of the smaller number.

In Exercises 35–40, solve the percent equation.

35. What is 30% of 45?   **36.** What is 175% of 360?

37. 432 is what percent of 1600?

38. 459 is what percent of 340?

39. 12 is $\frac{1}{2}$% of what number?

40. 70 is 40% of what number?

41. *Finance* A family has annual loan payments equaling 58.6% of their annual income. During the year, their loan payments total $13,077.75. What is their annual income?

Model It

42. *Government* The tables show the sources of income (in billions of dollars) and expenses (in billions of dollars) for the federal government in 2002. (Source: U.S. Office of Management and Budget)

Source of income	Income
Corporation taxes	148.0
Income tax	858.3
Social Security	700.8
Other	146.1

Source of expenses	Expense
Interest on debt	171.0
Health and human services	1317.9
Defense department	348.6
Other	173.5

Model It (continued)

(a) Find the percent of the total income for each category. Then use these percents to label the circle graph. To print an enlarged copy of the graph, go to the website *www.mathgraphs.com*.

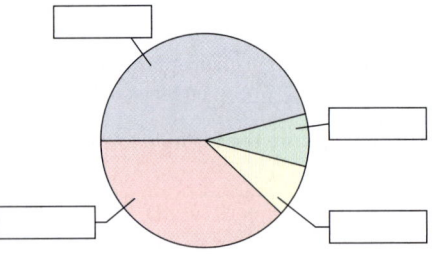

(b) Find the percent of the total expenses for each category. Then use these percents to label the circle graph. To print an enlarged copy of the graph, go to the website *www.mathgraphs.com*.

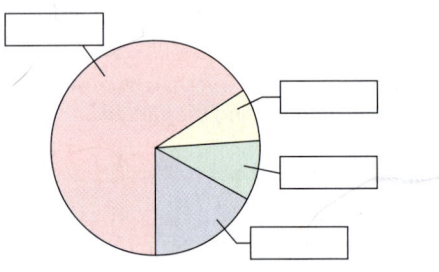

(c) Compare the total income and total expenses. How much of a surplus or deficit is there?

In Exercises 43–46, the prices of various items are given for 1990 and 2002. Find the percent change for each item. (Source: 2003 Statistical Abstract of the U.S.; IDC; Consumer Electronics Association)

	Item	1990	2002
43.	Gallon of regular unleaded gasoline	$1.16	$1.36
44.	Half-gallon of ice cream	$2.54	$3.76
45.	Pound of tomatoes	$0.86	$1.66
46.	Personal computer	$1050	$855

47. *Discount* The price of a swimming pool has been discounted 16.5%. The sale price is $1210.75. Find the original list price of the pool.

48. *Finance* A salesperson's weekly paycheck is 15% less than her coworker's paycheck. The two paychecks total $645. Find the amount of each paycheck.

49. *Dimensions of a Room*   A room is 1.5 times as long as it is wide, and its perimeter is 25 meters.

(a) Draw a diagram that represents the problem. Identify the length as *l* and the width as *w*.

(b) Write *l* in terms of *w* and write an equation for the perimeter in terms of *w*.

(c) Find the dimensions of the room.

50. *Dimensions of a Picture Frame*   A picture frame has a total perimeter of 2 meters. The height of the frame is 0.62 times its width.

(a) Draw a diagram that represents the problem. Identify the width as *w* and the height as *h*.

(b) Write *h* in terms of *w* and write an equation for the perimeter in terms of *w*.

(c) Find the dimensions of the picture frame.

51. *Course Grade*   To get an A in a course, you must have an average of at least 90 on four tests of 100 points each. The scores on your first three tests were 87, 92, and 84. What must you score on the fourth test to get an A for the course?

52. *Course Grade*   You are taking a course that has four tests. The first three tests are 100 points each and the fourth test is 200 points. To get an A in the course, you must have an average of at least 90% on the four tests. Your scores on the first three tests were 87, 92, and 84. What must you score on the fourth test to get an A for the course?

53. *Travel Time*   You are driving on a Canadian freeway to a town that is 300 kilometers from your home. After 30 minutes you pass a freeway exit that you know is 50 kilometers from your home. Assuming that you continue at the same constant speed, how long will it take for the entire trip?

54. *Travel Time*   Two cars start at an interstate interchange and travel in the same direction at average speeds of 40 miles per hour and 55 miles per hour. How much time must elapse before the two cars are 5 miles apart?

55. *Travel Time*   On the first part of a 317-mile trip, a salesperson averaged 58 miles per hour. He averaged only 52 miles per hour on the last part of the trip because of an increased volume of traffic. The total time of the trip was 5 hours and 45 minutes. Find the amount of time at each of the two speeds.

56. *Travel Time*   Students are traveling in two cars to a football game 135 miles away. The first car leaves on time and travels at an average speed of 45 miles per hour. The second car starts $\frac{1}{2}$ hour later and travels at an average speed of 55 miles per hour. How long will it take the second car to catch up to the first car? Will the second car catch up to the first car before the first car arrives at the game?

57. *Travel Time*   Two families meet at a park for a picnic. At the end of the day, one family travels east at an average speed of 42 miles per hour and the other travels west at an average speed of 50 miles per hour. Both families have approximately 160 miles to travel.

(a) Find the time it takes each family to get home.

(b) Find the time that will have elapsed when they are 100 miles apart.

(c) Find the distance the eastbound family has to travel after the westbound family has arrived home.

58. *Average Speed*   A truck driver traveled at an average speed of 55 miles per hour on a 200-mile trip to pick up a load of freight. On the return trip (with the truck fully loaded), the average speed was 40 miles per hour. What was the average speed for the round trip?

59. *Wind Speed*   An executive flew in the corporate jet to a meeting in a city 1500 kilometers away. After traveling the same amount of time on the return flight, the pilot mentioned that they still had 300 kilometers to go. The air speed of the plane was 600 kilometers per hour. How fast was the wind blowing? (Assume that the wind direction was parallel to the flight path and constant all day.)

60. *Physics*   Light travels at the speed of approximately 3.0×10^8 meters per second. Find the time in minutes required for light to travel from the sun to Earth (an approximate distance of 1.5×10^{11} meters).

61. *Radio Waves*   Radio waves travel at the same speed as light, approximately 3.0×10^8 meters per second. Find the time required for a radio wave to travel from Mission Control in Houston to NASA astronauts on the surface of the moon 3.84×10^8 meters away.

62. *Height of a Tree*   To obtain the height of a tree (see figure), you measure the tree's shadow and find that it is 8 meters long. You also measure the shadow of a two-meter lamppost and find that it is 75 centimeters long. How tall is the tree?

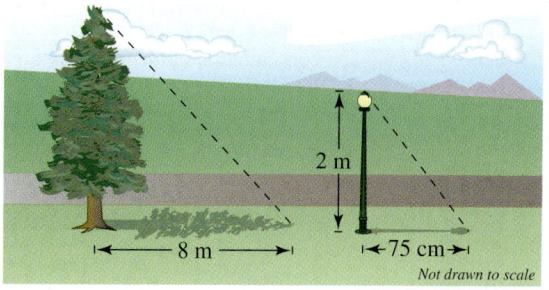

2 m

8 m   75 cm

Not drawn to scale

63. *Height of a Building*   To obtain the height of the Chrysler Building in New York, you measure the building's shadow and find that it is 87 feet long. You also measure the shadow of a four-foot stake and find that it is 4 inches long. How tall is the Chrysler Building?

64. Flagpole Height A person who is 6 feet tall walks away from a flagpole toward the tip of the shadow of the flagpole. When the person is 30 feet from the flagpole, the tips of the person's shadow and the shadow cast by the flagpole coincide at a point 5 feet in front of the person.

(a) Draw a diagram that gives a visual representation of the problem. Let h represent the height of the flagpole.

(b) Find the height of the flagpole.

65. Shadow Length A person who is 6 feet tall walks away from a 50-foot silo toward the tip of the silo's shadow. At a distance of 32 feet from the silo, the person's shadow begins to emerge beyond the silo's shadow. How much farther must the person walk to be completely out of the silo's shadow?

66. Investment You plan to invest $12,000 in two funds paying $4\frac{1}{2}\%$ and 5% simple interest. (There is more risk in the 5% fund.) Your goal is to obtain a total annual interest income of $580 from the investments. What is the smallest amount you can invest in the 5% fund and still meet your objective?

67. Investment You plan to invest $25,000 in two funds paying 3% and $4\frac{1}{2}\%$ simple interest. (There is more risk in the $4\frac{1}{2}\%$ fund.) Your goal is to obtain a total annual interest income of $1000 from the investments. What is the smallest amount you can invest in the $4\frac{1}{2}\%$ fund and still meet your objective?

68. Inventory A nursery has $20,000 of inventory in dogwood trees and red maple trees. The profit on a dogwood tree is 25% and the profit on a red maple tree is 17%. The profit for the entire stock is 20%. How much was invested in each type of tree?

69. Inventory An automobile dealer has $600,000 of inventory in minivans and SUVs. The profit on a minivan is 24% and the profit on an SUV is 28%. The profit for the entire stock is 25%. How much was invested in each type of vehicle?

70. Mixture Problem Using the values in the table, determine the amounts of solutions 1 and 2, needed to obtain the desired amount and concentration of the final mixture.

	Concentration			Amount of final solution
	Solution 1	Solution 2	Final solution	
(a)	10%	30%	25%	100 gal
(b)	25%	50%	30%	5 L
(c)	15%	45%	30%	10 qt
(d)	70%	90%	75%	25 gal

71. Mixture Problem A 100% concentrate is to be mixed with a mixture having a concentration of 40% to obtain 55 gallons of a mixture with a concentration of 75%. How much of the 100% concentrate will be needed?

72. Mixture Problem A forester mixes gasoline and oil to make 2 gallons of mixture for his two-cycle chain-saw engine. This mixture is 32 parts gasoline and 1 part two-cycle oil. How much gasoline must be added to bring the mixture to 40 parts gasoline and 1 part oil?

73. Mixture Problem A grocer mixes peanuts that cost $2.49 per pound and walnuts that cost $3.89 per pound to make 100 pounds of a mixture that costs $3.19 per pound. How much of each kind of nut is put into the mixture?

74. Company Costs An outdoor furniture manufacturer has fixed costs of $10,000 per month and average variable costs of $8.50 per unit manufactured. The company has $85,000 available to cover the monthly costs. How many units can the company manufacture? (*Fixed costs* are those that occur regardless of the level of production. *Variable costs* depend on the level of production.)

75. Company Costs A plumbing supply company has fixed costs of $10,000 per month and average variable costs of $9.30 per unit manufactured. The company has $85,000 available to cover the monthly costs. How many units can the company manufacture? (*Fixed costs* are those that occur regardless of the level of production. *Variable costs* depend on the level of production.)

76. Water Depth A trough is 12 feet long, 3 feet deep, and 3 feet wide (see figure). Find the depth of the water when the trough contains 70 gallons (1 gallon $\approx$ 0.13368 cubic foot).

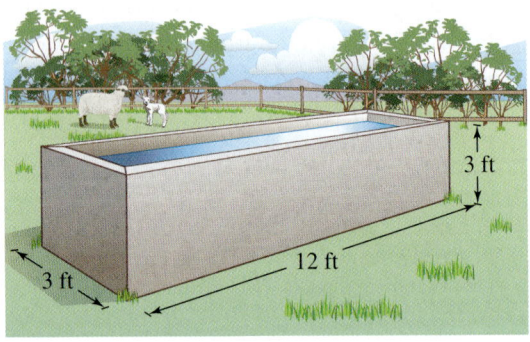

In Exercises 77–88, solve for the indicated variable.

77. Area of a Triangle
Solve for h: $A = \frac{1}{2}bh$

78. Area of a Trapezoid
Solve for b: $A = \frac{1}{2}(a + b)h$

79. Markup Solve for C: $S = C + RC$

80. Investment at Simple Interest
Solve for r: $A = P + Prt$

81. Volume of an Oblate Spheroid

Solve for b: $V = \frac{4}{3}\pi a^2 b$

82. Volume of a Spherical Segment

Solve for r: $V = \frac{1}{3}\pi h^2(3r - h)$

83. Free-Falling Body   Solve for a: $h = v_0 t + \frac{1}{2}at^2$

84. Lensmaker's Equation

Solve for R_1: $\dfrac{1}{f} = (n - 1)\left(\dfrac{1}{R_1} - \dfrac{1}{R_2}\right)$

85. Capacitance in Series Circuits

Solve for C_1: $C = \dfrac{1}{\dfrac{1}{C_1} + \dfrac{1}{C_2}}$

86. Arithmetic Progression

Solve for a: $S = \dfrac{n}{2}[2a + (n - 1)d]$

87. Arithmetic Progression

Solve for n: $L = a + (n - 1)d$

88. Geometric Progression   Solve for r: $S = \dfrac{rL - a}{r - 1}$

Physics   In Exercises 89 and 90, you have a uniform beam of length L with a fulcrum x feet from one end (see figure). Objects with weights W_1 and W_2 are placed at opposite ends of the beam. The beam will balance when $W_1 x = W_2(L - x)$. Find x such that the beam will balance.

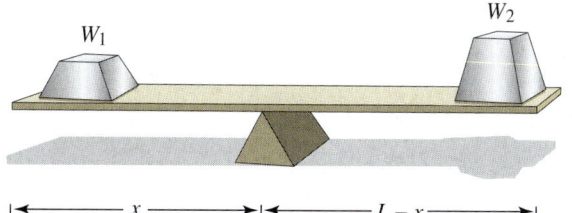

89. Two children weighing 50 pounds and 75 pounds are playing on a seesaw that is 10 feet long.

90. A person weighing 200 pounds is attempting to move a 550-pound rock with a bar that is 5 feet long.

91. Volume of a Billiard Ball   A billiard ball has a volume of 5.96 cubic inches. Find the radius of a billiard ball.

92. Length of a Tank   The diameter of a cylindrical propane gas tank is 4 feet. The total volume of the tank is 603.2 cubic feet. Find the length of the tank.

93. Temperature   The average daily temperature in San Diego, California is 64.4°F. What is San Diego's average daily temperature in degrees Celsius?   (Source: NOAA)

94. Temperature   The average daily temperature in Duluth, Minnesota is 39.1°F. What is Duluth's average daily temperature in degrees Celsius?   (Source: NOAA)

95. Temperature   The highest temperature ever recorded in Phoenix, Arizona was 50°C. What is this temperature in degrees Fahrenheit?   (Source: NOAA)

96. Temperature   The lowest temperature ever recorded in Louisville, Kentucky was -30°C. What is this temperature in degrees Fahrenheit?   (Source: NOAA)

Synthesis

True or False?   In Exercises 97 and 98, determine whether the statement is true or false. Justify your answer.

97. "8 less than z cubed divided by the difference of z squared and 9" can be written as

$$\frac{z^3 - 8}{(z - 9)^2}.$$

98. The volume of a cube with a side of length 9.5 inches is greater than the volume of a sphere with a radius of 5.9 inches.

99. Consider the linear equation $ax + b = 0$.

(a) What is the sign of the solution if $ab > 0$?

(b) What is the sign of the solution if $ab < 0$?

   In each case, explain your reasoning.

100. Write a linear equation that has the solution $x = -3$. (There are many correct answers.)

Skills Review

In Exercises 101–104, simplify the expression.

101. $(5x^4)(25x^2)^{-1}, \quad x \neq 0$

102. $\sqrt{150s^2t^3}$

103. $\dfrac{3}{x - 5} + \dfrac{2}{5 - x}$

104. $\dfrac{5}{x} + \dfrac{3x}{x^2 - 9} - \dfrac{10}{x + 3}$

In Exercises 105–108, rationalize the denominator.

105. $\dfrac{10}{7\sqrt{3}}$

106. $\dfrac{4}{\sqrt[3]{6}}$

107. $\dfrac{5}{\sqrt{6} + \sqrt{11}}$

108. $\dfrac{14}{3\sqrt{10} - 1}$

1.4 Quadratic Equations and Applications

What you should learn

- Solve quadratic equations by factoring.
- Solve quadratic equations by extracting square roots.
- Solve quadratic equations by completing the square.
- Use the Quadratic Formula to solve quadratic equations.
- Use quadratic equations to model and solve real-life problems.

Why you should learn it

Quadratic equations can be used to model and solve real-life problems. For instance, in Exercise 119 on page 123, you will use a quadratic equation to model average admission prices for movie theaters from 1997 to 2003.

© Indiapicture/Alamy

Factoring

A **quadratic equation** in x is an equation that can be written in the general form

$$ax^2 + bx + c = 0$$

where a, b, and c are real numbers with $a \neq 0$. A quadratic equation in x is also called a **second-degree polynomial equation** in x.

In this section, you will study four methods for solving quadratic equations: *factoring*, *extracting square roots*, *completing the square*, and the *Quadratic Formula*. The first method is based on the Zero-Factor Property from Section P.1.

If $ab = 0$, then $a = 0$ or $b = 0$.   Zero-Factor Property

To use this property, write the left side of the general form of a quadratic equation as the product of two linear factors. Then find the solutions of the quadratic equation by setting each linear factor equal to zero.

Example 1 Solving a Quadratic Equation by Factoring

a.

$2x^2 + 9x + 7 = 3$	Original equation
$2x^2 + 9x + 4 = 0$	Write in general form.
$(2x + 1)(x + 4) = 0$	Factor.
$2x + 1 = 0 \implies x = -\dfrac{1}{2}$	Set 1st factor equal to 0.
$x + 4 = 0 \implies x = -4$	Set 2nd factor equal to 0.

The solutions are $x = -\frac{1}{2}$ and $x = -4$. Check these in the original equation.

b.

$6x^2 - 3x = 0$	Original equation
$3x(2x - 1) = 0$	Factor.
$3x = 0 \implies x = 0$	Set 1st factor equal to 0.
$2x - 1 = 0 \implies x = \dfrac{1}{2}$	Set 2nd factor equal to 0.

The solutions are $x = 0$ and $x = \frac{1}{2}$. Check these in the original equation.

✓CHECKPOINT   Now try Exercise 9.

Be sure you see that the Zero-Factor Property works *only* for equations written in general form (in which the right side of the equation is zero). So, all terms must be collected on one side *before* factoring. For instance, in the equation $(x - 5)(x + 2) = 8$, it is *incorrect* to set each factor equal to 8. To solve this equation, you must multiply the binomials on the left side of the equation, and then subtract 8 from each side. After simplifying the left side of the equation, you can use the Zero-Factor Property to solve the equation. Try to solve this equation correctly.

Extracting Square Roots

There is a nice shortcut for solving quadratic equations of the form $u^2 = d$, where $d > 0$ and u is an algebraic expression. By factoring, you can see that this equation has two solutions.

$$u^2 = d \qquad \text{Write original equation.}$$

$$u^2 - d = 0 \qquad \text{Write in general form.}$$

$$\left(u + \sqrt{d}\right)\left(u - \sqrt{d}\right) = 0 \qquad \text{Factor.}$$

$$u + \sqrt{d} = 0 \quad \Longrightarrow \quad u = -\sqrt{d} \qquad \text{Set 1st factor equal to 0.}$$

$$u - \sqrt{d} = 0 \quad \Longrightarrow \quad u = \sqrt{d} \qquad \text{Set 2nd factor equal to 0.}$$

Because the two solutions differ only in sign, you can write the solutions together, using a "plus or minus sign," as

$$u = \pm\sqrt{d}.$$

This form of the solution is read as "u is equal to plus or minus the square root of d." Solving an equation of the form $u^2 = d$ without going through the steps of factoring is called **extracting square roots.**

> ## Extracting Square Roots
>
> The equation $u^2 = d$, where $d > 0$, has exactly two solutions:
>
> $$u = \sqrt{d} \qquad \text{and} \qquad u = -\sqrt{d}.$$
>
> These solutions can also be written as
>
> $$u = \pm\sqrt{d}.$$

Example 2   Extracting Square Roots

Solve each equation by extracting square roots.

a. $4x^2 = 12$ **b.** $(x - 3)^2 = 7$

Solution

a.
$$4x^2 = 12 \qquad \text{Write original equation.}$$
$$x^2 = 3 \qquad \text{Divide each side by 4.}$$
$$x = \pm\sqrt{3} \qquad \text{Extract square roots.}$$

When you take the square root of a variable expression, you must account for both positive and negative solutions. So, the solutions are $x = \sqrt{3}$ and $x = -\sqrt{3}$. Check these in the original equation.

b.
$$(x - 3)^2 = 7 \qquad \text{Write original equation.}$$
$$x - 3 = \pm\sqrt{7} \qquad \text{Extract square roots.}$$
$$x = 3 \pm \sqrt{7} \qquad \text{Add 3 to each side.}$$

The solutions are $x = 3 \pm \sqrt{7}$. Check these in the original equation.

✔CHECKPOINT   Now try Exercise 29.

Technology

You can use a graphing utility to check graphically the real solutions of a quadratic equation. Begin by writing the equation in general form. Then set y equal to the left side and graph the resulting equation. The x-intercepts of the equation represent the real solutions of the original equation. You can use the *zero* or *root* feature of a graphing utility to approximate the x-intercepts of the graph. For example, to check the solutions of $6x^2 - 3x = 0$, graph $y = 6x^2 - 3x$, and use the *zero* or *root* feature to approximate the x-intercepts to be $(0, 0)$ and $\left(\frac{1}{2}, 0\right)$, as shown below. These x-intercepts represent the solutions $x = 0$ and $x = \frac{1}{2}$, as found in Example 1(b).

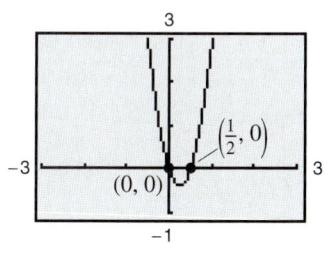

Completing the Square

The equation in Example 2(b) was given in the form $(x - 3)^2 = 7$ so that you could find the solution by extracting square roots. Suppose, however, that the equation had been given in the general form $x^2 - 6x + 2 = 0$. Because this equation is equivalent to the original, it has the same two solutions, $x = 3 \pm \sqrt{7}$. However, the left side of the equation is not factorable, and you cannot find its solutions unless you rewrite the equation by **completing the square.** Note that when you complete the square to solve a quadratic equation, you are just rewriting the equation so it can be solved by extracting square roots.

Completing the Square

To **complete the square** for the expression $x^2 + bx$, add $(b/2)^2$, which is the square of half the coefficient of x. Consequently,

$$x^2 + bx + \left(\frac{b}{2}\right)^2 = \left(x + \frac{b}{2}\right)^2.$$

Example 3   Completing the Square: Leading Coefficient Is 1

Solve $x^2 + 2x - 6 = 0$ by completing the square.

Solution

$x^2 + 2x - 6 = 0$	Write original equation.
$x^2 + 2x = 6$	Add 6 to each side.
$x^2 + 2x + 1^2 = 6 + 1^2$	Add 1^2 to each side.
	$(\text{half of } 2)^2$
$(x + 1)^2 = 7$	Simplify.
$x + 1 = \pm\sqrt{7}$	Take square root of each side.
$x = -1 \pm \sqrt{7}$	Subtract 1 from each side.

The solutions are $x = -1 \pm \sqrt{7}$. Check these in the original equation as follows.

Check

$x^2 + 2x - 6 = 0$	Write original equation.
$\left(-1 + \sqrt{7}\right)^2 + 2\left(-1 + \sqrt{7}\right) - 6 \overset{?}{=} 0$	Substitute $-1 + \sqrt{7}$ for x.
$8 - 2\sqrt{7} - 2 + 2\sqrt{7} - 6 \overset{?}{=} 0$	Multiply.
$8 - 2 - 6 = 0$	Solution checks ✓

Check the second solution in the original equation.

✓**CHECKPOINT**   Now try Exercise 37.

When solving quadratic equations by completing the square, you must add $(b/2)^2$ to *each side* in order to maintain equality. If the leading coefficient is *not* 1, you must divide each side of the equation by the leading coefficient *before* completing the square, as shown in Example 4.

Example 4   **Completing the Square: Leading Coefficient Is Not 1**

Solve $2x^2 + 8x + 3 = 0$ by completing the square.

Solution

$$2x^2 + 8x + 3 = 0 \qquad \text{Write original equation.}$$

$$2x^2 + 8x = -3 \qquad \text{Subtract 3 from each side.}$$

$$x^2 + 4x = -\frac{3}{2} \qquad \text{Divide each side by 2.}$$

$$x^2 + 4x + 2^2 = -\frac{3}{2} + 2^2 \qquad \text{Add } 2^2 \text{ to each side.}$$

$$\underbrace{\qquad}_{\text{(half of 4)}^2}$$

$$(x + 2)^2 = \frac{5}{2} \qquad \text{Simplify.}$$

$$x + 2 = \pm\sqrt{\frac{5}{2}} \qquad \text{Take square root of each side.}$$

$$x + 2 = \pm\frac{\sqrt{10}}{2} \qquad \text{Rationalize denominator.}$$

$$x = -2 \pm \frac{\sqrt{10}}{2} \qquad \text{Subtract 2 from each side.}$$

The solutions are $x = -2 \pm \dfrac{\sqrt{10}}{2}$. Check these in the original equation.

✓CHECKPOINT   Now try Exercise 39.

Example 5   **Completing the Square: Leading Coefficient Is Not 1**

$$3x^2 - 4x - 5 = 0 \qquad \text{Original equation}$$

$$3x^2 - 4x = 5 \qquad \text{Add 5 to each side.}$$

$$x^2 - \frac{4}{3}x = \frac{5}{3} \qquad \text{Divide each side by 3.}$$

$$x^2 - \frac{4}{3}x + \left(-\frac{2}{3}\right)^2 = \frac{5}{3} + \left(-\frac{2}{3}\right)^2 \qquad \text{Add } \left(-\frac{2}{3}\right)^2 \text{ to each side.}$$

$$x^2 - \frac{4}{3}x + \frac{4}{9} = \frac{19}{9} \qquad \text{Simplify.}$$

$$\left(x - \frac{2}{3}\right)^2 = \frac{19}{9} \qquad \text{Perfect square trinomial.}$$

$$x - \frac{2}{3} = \pm\frac{\sqrt{19}}{3} \qquad \text{Extract square roots.}$$

$$x = \frac{2}{3} \pm \frac{\sqrt{19}}{3} \qquad \text{Solutions}$$

✓CHECKPOINT   Now try Exercise 43.

The Quadratic Formula

Often in mathematics you are taught the long way of solving a problem first. Then, the longer method is used to develop shorter techniques. The long way stresses understanding and the short way stresses efficiency.

For instance, you can think of completing the square as a "long way" of solving a quadratic equation. When you use completing the square to solve quadratic equations, you must complete the square for *each* equation separately. In the following derivation, you complete the square *once* in a general setting to obtain the **Quadratic Formula**—a shortcut for solving quadratic equations.

$$ax^2 + bx + c = 0 \qquad \text{Write in general form, } a \neq 0.$$

$$ax^2 + bx = -c \qquad \text{Subtract } c \text{ from each side.}$$

$$x^2 + \frac{b}{a}x = -\frac{c}{a} \qquad \text{Divide each side by } a.$$

$$x^2 + \frac{b}{a}x + \left(\frac{b}{2a}\right)^2 = -\frac{c}{a} + \left(\frac{b}{2a}\right)^2 \qquad \text{Complete the square.}$$

$$\left(\text{half of } \frac{b}{a}\right)^2$$

$$\left(x + \frac{b}{2a}\right)^2 = \frac{b^2 - 4ac}{4a^2} \qquad \text{Simplify.}$$

$$x + \frac{b}{2a} = \pm\sqrt{\frac{b^2 - 4ac}{4a^2}} \qquad \text{Extract square roots.}$$

$$x = -\frac{b}{2a} \pm \frac{\sqrt{b^2 - 4ac}}{2|a|} \qquad \text{Solutions}$$

Note that because $\pm 2|a|$ represents the same numbers as $\pm 2a$, you can omit the absolute value sign. So, the formula simplifies to

$$x = \frac{-b \pm \sqrt{b^2 - 4ac}}{2a}.$$

STUDY TIP

You can solve every quadratic equation by completing the square or using the Quadratic Formula.

The Quadratic Formula

The solutions of a quadratic equation in the general form

$$ax^2 + bx + c = 0, \qquad a \neq 0$$

are given by the **Quadratic Formula**

$$x = \frac{-b \pm \sqrt{b^2 - 4ac}}{2a}.$$

The Quadratic Formula is one of the most important formulas in algebra. You should learn the verbal statement of the Quadratic Formula:

"Negative b, plus or minus the square root of b squared minus $4ac$, all divided by $2a$."

Exploration

From each graph, can you tell whether the discriminant is positive, zero, or negative? Explain your reasoning. Find each discriminant to verify your answers.

a. $x^2 - 2x = 0$

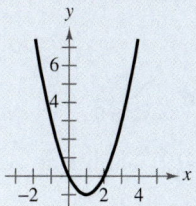

b. $x^2 - 2x + 1 = 0$

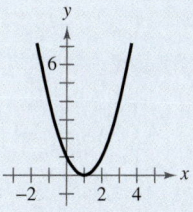

c. $x^2 - 2x + 2 = 0$

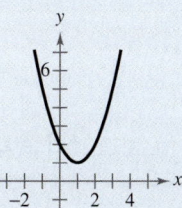

How many solutions would part (c) have if the linear term was $2x$? If the constant was -2?

In the Quadratic Formula, the quantity under the radical sign, $b^2 - 4ac$, is called the **discriminant** of the quadratic expression $ax^2 + bx + c$. It can be used to determine the nature of the solutions of a quadratic equation.

Solutions of a Quadratic Equation

The solutions of a quadratic equation $ax^2 + bx + c = 0$, $a \neq 0$, can be classified as follows. If the discriminant $b^2 - 4ac$ is

1. *positive*, then the quadratic equation has *two* distinct real solutions and its graph has *two* x-intercepts.

2. *zero*, then the quadratic equation has *one* repeated real solution and its graph has *one* x-intercept.

3. *negative*, then the quadratic equation has *no* real solutions and its graph has *no* x-intercepts.

If the discriminant of a quadratic equation is negative, as in case 3 above, then its square root is imaginary (not a real number) and the Quadratic Formula yields two complex solutions. You will study complex solutions in Section 1.5.

When using the Quadratic Formula, remember that *before* the formula can be applied, you must first write the quadratic equation in general form.

Example 6   The Quadratic Formula: Two Distinct Solutions

Use the Quadratic Formula to solve $x^2 + 3x = 9$.

Solution

The general form of the equation is $x^2 + 3x - 9 = 0$. The discriminant is $b^2 - 4ac = 9 + 36 = 45$, which is positive. So, the equation has two real solutions. You can solve the equation as follows.

$$x^2 + 3x - 9 = 0 \qquad \text{Write in general form.}$$

$$x = \frac{-b \pm \sqrt{b^2 - 4ac}}{2a} \qquad \text{Quadratic Formula}$$

$$x = \frac{-3 \pm \sqrt{(3)^2 - 4(1)(-9)}}{2(1)} \qquad \text{Substitute } a = 1, b = 3, \text{ and } c = -9.$$

$$x = \frac{-3 \pm \sqrt{45}}{2} \qquad \text{Simplify.}$$

$$x = \frac{-3 \pm 3\sqrt{5}}{2} \qquad \text{Simplify.}$$

The two solutions are:

$$x = \frac{-3 + 3\sqrt{5}}{2} \quad \text{and} \quad x = \frac{-3 - 3\sqrt{5}}{2}.$$

Check these in the original equation.

✓**CHECKPOINT**   Now try Exercise 75.

Applications

Quadratic equations often occur in problems dealing with area. Here is a simple example. "A square room has an area of 144 square feet. Find the dimensions of the room." To solve this problem, let x represent the length of each side of the room. Then, by solving the equation

$$x^2 = 144$$

you can conclude that each side of the room is 12 feet long. Note that although the equation $x^2 = 144$ has two solutions, $x = -12$ and $x = 12$, the negative solution does not make sense in the context of the problem, so you choose the positive solution.

Example 7   Finding the Dimensions of a Room

A bedroom is 3 feet longer than it is wide (see Figure 1.19) and has an area of 154 square feet. Find the dimensions of the room.

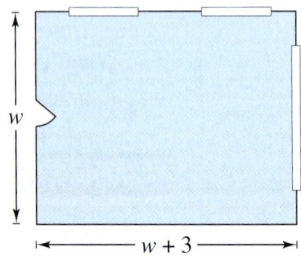

FIGURE **1.19**

Solution

Verbal Model:	Width of room	·	Length of room	=	Area of room

Labels: Width of room $= w$ (feet)
Length of room $= w + 3$ (feet)
Area of room $= 154$ (square feet)

Equation:

$$w(w + 3) = 154$$

$$w^2 + 3w - 154 = 0$$

$$(w - 11)(w + 14) = 0$$

$$w - 11 = 0 \quad \Longrightarrow \quad w = 11$$

$$w + 14 = 0 \quad \Longrightarrow \quad w = -14$$

Choosing the positive value, you find that the width is 11 feet and the length is $w + 3$, or 14 feet. You can check this solution by observing that the length is 3 feet longer than the width *and* that the product of the length and width is 154 square feet.

✓**CHECKPOINT**   Now try Exercise 109.

Another common application of quadratic equations involves an object that is falling (or projected into the air). The general equation that gives the height of such an object is called a **position equation,** and on Earth's surface it has the form

$$s = -16t^2 + v_0 t + s_0.$$

In this equation, s represents the height of the object (in feet), v_0 represents the initial velocity of the object (in feet per second), s_0 represents the initial height of the object (in feet), and t represents the time (in seconds).

Example 8   Falling Time

A construction worker on the 24th floor of a building project (see Figure 1.20) accidentally drops a wrench and yells "Look out below!" Could a person at ground level hear this warning in time to get out of the way? (*Note:* The speed of sound is about 1100 feet per second.)

Solution

Assume that each floor of the building is 10 feet high, so that the wrench is dropped from a height of 235 feet (the construction worker's hand is 5 feet below the ceiling of the 24th floor). Because sound travels at about 1100 feet per second, it follows that a person at ground level hears the warning within 1 second of the time the wrench is dropped. To set up a mathematical model for the height of the wrench, use the position equation

$$s = -16t^2 + v_0 t + s_0.$$

Because the object is dropped rather than thrown, the initial velocity is $v_0 = 0$ feet per second. Moreover, because the initial height is $s_0 = 235$ feet, you have the following model.

$$s = -16t^2 + (0)t + 235 = -16t^2 + 235$$

After the wrench has fallen for 1 second, its height is $-16(1)^2 + 235 = 219$ feet. After the wrench has fallen for 2 seconds, its height is $-16(2)^2 + 235 = 171$ feet. To find the number of seconds it takes the wrench to hit the ground, let the height s be zero and solve the equation for t.

$s = -16t^2 + 235$	Write position equation.
$0 = -16t^2 + 235$	Substitute 0 for height.
$16t^2 = 235$	Add $16t^2$ to each side.
$t^2 = \dfrac{235}{16}$	Divide each side by 16.
$t = \dfrac{\sqrt{235}}{4}$	Extract positive square root.
$t \approx 3.83$	Use a calculator.

The wrench will take about 3.83 seconds to hit the ground. If the person hears the warning 1 second after the wrench is dropped, the person still has almost 3 seconds to get out of the way.

✓**CHECKPOINT**   Now try Exercise 115.

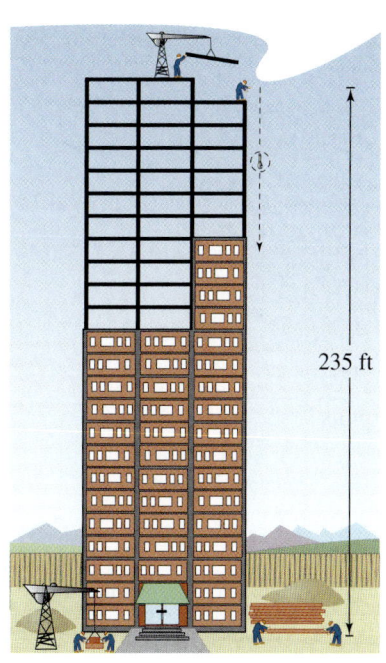

FIGURE **1.20**

235 ft

STUDY TIP

The position equation used in Example 8 ignores air resistance. This implies that it is appropriate to use the position equation only to model falling objects that have little air resistance and that fall over short distances.

A third type of application of a quadratic equation is one in which a quantity is changing over time t according to a quadratic model.

Example 9 Quadratic Modeling: Internet Users

From 2000 to 2007, the estimated number of Internet users I (in millions) in the United States can be modeled by the quadratic equation

$$I = -1.163t^2 + 17.19t + 125.9, \quad 0 \le t \le 7$$

where t represents the year, with $t = 0$ corresponding to 2000. According to this model, in which year will the number of Internet users reach or surpass 180 million? (Source: *eMarketer*)

Algebraic Solution

To find the year in which the number of Internet users will reach 180 million, you need to solve the equation

$$-1.163t^2 + 17.19t + 125.9 = 180.$$

To begin, write the equation in general form.

$$-1.163t^2 + 17.19t - 54.1 = 0$$

Then apply the Quadratic Formula.

$$t = \frac{-b \pm \sqrt{b^2 - 4ac}}{2a}$$

$$t = \frac{-17.19 \pm \sqrt{17.19^2 - 4(-1.163)(-54.1)}}{2(-1.163)}$$

$$\approx \frac{-17.19 \pm \sqrt{43.82}}{-2.326}$$

$$\approx 4.5 \text{ or } 10.2$$

Choose the smaller value $t \approx 4.5$. Because $t = 0$ corresponds to 2000, it follows that $t \approx 4.5$ must correspond to 2004. So, the number of Internet users should have reached 180 million during the year 2004.

✓**CHECKPOINT** Now try Exercise 119.

Numerical Solution

You can estimate the year in which the number of Internet users will reach or surpass 180 million by constructing a table of values. The table below shows the number of Internet users for each year from 2000 to 2007.

Year	t	I
2000	0	125.9
2001	1	141.9
2002	2	155.6
2003	3	167.0
2004	4	176.1
2005	5	182.8
2006	6	187.2
2007	7	189.2

From the table, you can see that sometime during 2004 the number of Internet users reached 180 million.

Technology

You can also use a graphical approach to solve Example 9. Use a graphing utility to graph

$$y_1 = -1.163t^2 + 17.19t + 125.9 \quad \text{and} \quad y_2 = 180$$

in the same viewing window. Then use the *intersect* feature to find the point(s) of intersection of the two graphs. You should obtain $t \approx 4.5$, which verifies the answer obtained algebraically.

A fourth type of application that often involves a quadratic equation is one dealing with the hypotenuse of a right triangle. In these types of applications, the **Pythagorean Theorem** is often used. The Pythagorean Theorem states that

$$a^2 + b^2 = c^2 \qquad \text{Pythagorean Theorem}$$

where a and b are the legs of a right triangle and c is the hypotenuse.

Example 10 An Application Involving the Pythagorean Theorem

An L-shaped sidewalk from the athletic center to the library on a college campus is shown in Figure 1.21. The sidewalk was constructed so that the length of one sidewalk forming the L was twice as long as the other. The length of the diagonal sidewalk that cuts across the grounds between the two buildings is 32 feet. How many feet does a person save by walking on the diagonal sidewalk?

Solution

Using the Pythagorean Theorem, you have the following.

$$
\begin{aligned}
x^2 + (2x)^2 &= 32^2 && \text{Pythagorean Theorem} \\
5x^2 &= 1024 && \text{Combine like terms.} \\
x^2 &= 204.8 && \text{Divide each side by 5.} \\
x &= \pm\sqrt{204.8} && \text{Take the square root of each side.} \\
x &= \sqrt{204.8} && \text{Extract positive square root.}
\end{aligned}
$$

The total distance covered by walking on the L-shaped sidewalk is

$$
\begin{aligned}
x + 2x &= 3x \\
&= 3\sqrt{204.8} \\
&\approx 42.9 \text{ feet.}
\end{aligned}
$$

Walking on the diagonal sidewalk saves a person about $42.9 - 32 = 10.9$ feet.

✓**CHECKPOINT**   Now try Exercise 123.

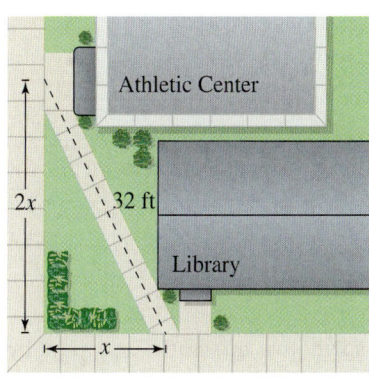

FIGURE **1.21**

---

*W*RITING ABOUT *M*ATHEMATICS

Comparing Solution Methods In this section, you studied four algebraic methods for solving quadratic equations. Solve each of the quadratic equations below in several different ways. Write a short paragraph explaining which method(s) you prefer. Does your preferred method depend on the equation?

a. $x^2 - 4x - 5 = 0$

b. $x^2 - 4x = 0$

c. $x^2 - 4x - 3 = 0$

d. $x^2 - 4x - 6 = 0$

1.4   Exercises

VOCABULARY CHECK: Fill in the blanks.

1. An equation of the form $ax^2 + bx + c = 0$, $a \neq 0$ is a _____ _____, or second-degree polynomial equation in x.

2. The four methods that can be used to solve a quadratic equation are _____, _____, _____, and the _____.

3. The part of the Quadratic Formula, $b^2 - 4ac$, known as the _____, determines the type of solutions of a quadratic equation.

4. The general equation that gives the height of an object (in feet) in terms of the time t (in seconds) is called the _____ equation, and has the form $s =$ _____, where v_0 represents the _____ and s_0 represents the _____.

5. An important theorem that is sometimes used in applications that require solving quadratic equations is the _____ _____.

PREREQUISITE SKILLS REVIEW: Practice and review algebra skills needed for this section at **www.Eduspace.com**.

In Exercises 1–6, write the quadratic equation in general form.

1. $2x^2 = 3 - 8x$
2. $x^2 = 16x$
3. $(x - 3)^2 = 3$
4. $13 - 3(x + 7)^2 = 0$
5. $\frac{1}{5}(3x^2 - 10) = 18x$
6. $x(x + 2) = 5x^2 + 1$

In Exercises 7–20, solve the quadratic equation by factoring.

7. $6x^2 + 3x = 0$
8. $9x^2 - 1 = 0$
9. $x^2 - 2x - 8 = 0$
10. $x^2 - 10x + 9 = 0$
11. $x^2 + 10x + 25 = 0$
12. $4x^2 + 12x + 9 = 0$
13. $3 + 5x - 2x^2 = 0$
14. $2x^2 = 19x + 33$
15. $x^2 + 4x = 12$
16. $-x^2 + 8x = 12$
17. $\frac{3}{4}x^2 + 8x + 20 = 0$
18. $\frac{1}{8}x^2 - x - 16 = 0$
19. $x^2 + 2ax + a^2 = 0$, a is a real number
20. $(x + a)^2 - b^2 = 0$, a and b are real numbers

In Exercises 21–34, solve the equation by extracting square roots.

21. $x^2 = 49$
22. $x^2 = 169$
23. $x^2 = 11$
24. $x^2 = 32$
25. $3x^2 = 81$
26. $9x^2 = 36$
27. $(x - 12)^2 = 16$
28. $(x + 13)^2 = 25$
29. $(x + 2)^2 = 14$
30. $(x - 5)^2 = 30$
31. $(2x - 1)^2 = 18$
32. $(4x + 7)^2 = 44$

33. $(x - 7)^2 = (x + 3)^2$
34. $(x + 5)^2 = (x + 4)^2$

In Exercises 35–44, solve the quadratic equation by completing the square.

35. $x^2 + 4x - 32 = 0$
36. $x^2 - 2x - 3 = 0$
37. $x^2 + 12x + 25 = 0$
38. $x^2 + 8x + 14 = 0$
39. $9x^2 - 18x = -3$
40. $9x^2 - 12x = 14$
41. $8 + 4x - x^2 = 0$
42. $-x^2 + x - 1 = 0$
43. $2x^2 + 5x - 8 = 0$
44. $4x^2 - 4x - 99 = 0$

In Exercises 45–50, rewrite the quadratic portion of the algebraic expression as the sum or difference of two squares by completing the square.

45. $\dfrac{1}{x^2 + 2x + 5}$
46. $\dfrac{1}{x^2 - 12x + 19}$
47. $\dfrac{4}{x^2 + 4x - 3}$
48. $\dfrac{5}{x^2 + 25x + 11}$
49. $\dfrac{1}{\sqrt{6x - x^2}}$
50. $\dfrac{1}{\sqrt{16 - 6x - x^2}}$

Graphical Analysis   In Exercises 51–58, (a) use a graphing utility to graph the equation, (b) use the graph to approximate any *x*-intercepts of the graph, (c) set $y = 0$ and solve the resulting equation, and (d) compare the result of part (c) with the *x*-intercepts of the graph.

51. $y = (x + 3)^2 - 4$

52. $y = (x - 4)^2 - 1$

53. $y = 1 - (x - 2)^2$

54. $y = 9 - (x - 8)^2$

55. $y = -4x^2 + 4x + 3$

56. $y = 4x^2 - 1$

57. $y = x^2 + 3x - 4$

58. $y = x^2 - 5x - 24$

In Exercises 59–66, use the discriminant to determine the number of real solutions of the quadratic equation.

59. $2x^2 - 5x + 5 = 0$

60. $-5x^2 - 4x + 1 = 0$

61. $2x^2 - x - 1 = 0$

62. $x^2 - 4x + 4 = 0$

63. $\frac{1}{3}x^2 - 5x + 25 = 0$

64. $\frac{4}{7}x^2 - 8x + 28 = 0$

65. $0.2x^2 + 1.2x - 8 = 0$

66. $9 + 2.4x - 8.3x^2 = 0$

In Exercises 67–90, use the Quadratic Formula to solve the equation.

67. $2x^2 + x - 1 = 0$

68. $2x^2 - x - 1 = 0$

69. $16x^2 + 8x - 3 = 0$

70. $25x^2 - 20x + 3 = 0$

71. $2 + 2x - x^2 = 0$

72. $x^2 - 10x + 22 = 0$

73. $x^2 + 14x + 44 = 0$

74. $6x = 4 - x^2$

75. $x^2 + 8x - 4 = 0$

76. $4x^2 - 4x - 4 = 0$

77. $12x - 9x^2 = -3$

78. $16x^2 + 22 = 40x$

79. $9x^2 + 24x + 16 = 0$

80. $36x^2 + 24x - 7 = 0$

81. $4x^2 + 4x = 7$

82. $16x^2 - 40x + 5 = 0$

83. $28x - 49x^2 = 4$

84. $3x + x^2 - 1 = 0$

85. $8t = 5 + 2t^2$

86. $25h^2 + 80h + 61 = 0$

87. $(y - 5)^2 = 2y$

88. $(z + 6)^2 = -2z$

89. $\frac{1}{2}x^2 + \frac{3}{8}x = 2$

90. $\left(\frac{5}{7}x - 14\right)^2 = 8x$

In Exercises 91–98, use the Quadratic Formula to solve the equation. (Round your answer to three decimal places.)

91. $5.1x^2 - 1.7x - 3.2 = 0$

92. $2x^2 - 2.50x - 0.42 = 0$

93. $-0.067x^2 - 0.852x + 1.277 = 0$

94. $-0.005x^2 + 0.101x - 0.193 = 0$

95. $422x^2 - 506x - 347 = 0$

96. $1100x^2 + 326x - 715 = 0$

97. $12.67x^2 + 31.55x + 8.09 = 0$

98. $-3.22x^2 - 0.08x + 28.651 = 0$

In Exercises 99–108, solve the equation using any convenient method.

99. $x^2 - 2x - 1 = 0$

100. $11x^2 + 33x = 0$

101. $(x + 3)^2 = 81$

102. $x^2 - 14x + 49 = 0$

103. $x^2 - x - \frac{11}{4} = 0$

104. $x^2 + 3x - \frac{3}{4} = 0$

105. $(x + 1)^2 = x^2$

106. $a^2x^2 - b^2 = 0$, *a* and *b* are real numbers

107. $3x + 4 = 2x^2 - 7$

108. $4x^2 + 2x + 4 = 2x + 8$

109. *Floor Space*   The floor of a one-story building is 14 feet longer than it is wide. The building has 1632 square feet of floor space.

(a) Draw a diagram that gives a visual representation of the floor space. Represent the width as *w* and show the length in terms of *w*.

(b) Write a quadratic equation in terms of *w*.

(c) Find the length and width of the floor of the building.

110. *Dimensions of a Corral*   A rancher has 100 meters of fencing to enclose two adjacent rectangular corrals (see figure). The rancher wants the enclosed area to be 350 square meters. What dimensions should the rancher use to obtain this area?

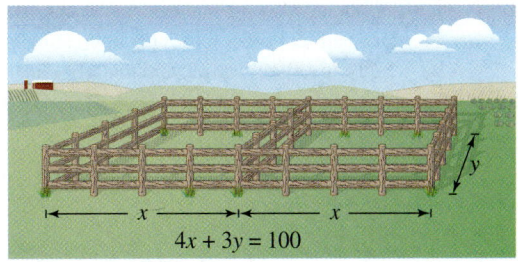

111. *Packaging*   An open box with a square base (see figure) is to be constructed from 84 square inches of material. The height of the box is 2 inches. What are the dimensions of the box? (*Hint:* The surface area is $S = x^2 + 4xh$.)

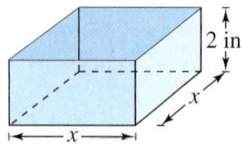

2 in.

112. *Packaging*   An open gift box is to be made from a square piece of material by cutting two-centimeter squares from the corners and turning up the sides (see figure). The volume of the finished box is to be 200 cubic centimeters. Find the size of the original piece of material.

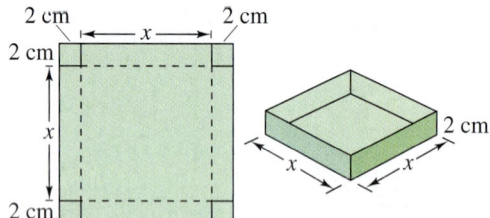

113. *Mowing the Lawn*   Two landscapers must mow a rectangular lawn that measures 100 feet by 200 feet. Each wants to mow no more than half of the lawn. The first starts by mowing around the outside of the lawn. The mower has a 24-inch cut. How wide a strip must the first landscaper mow on each of the four sides in order to mow no more than half of the lawn? Approximate the required number of trips around the lawn the first landscaper must take.

114. *Seating*   A rectangular classroom seats 72 students. If the seats were rearranged with three more seats in each row, the classroom would have two fewer rows. Find the original number of seats in each row.

In Exercises 115–118, use the position equation given in Example 8 as the model for the problem.

115. *Military*   A C-141 Starlifter flying at 32,000 feet over level terrain drops a 500-pound supply package.

(a) How long will it take until the supply package strikes the ground?

(b) The plane is flying at 500 miles per hour. How far will the supply package travel horizontally during its descent?

116. *Eiffel Tower*   You drop a coin from the top of the Eiffel Tower in Paris. The building has a height of 984 feet.

(a) Use the position equation to write a mathematical model for the height of the coin.

(b) Find the height of the coin after 4 seconds.

(c) How long will it take before the coin strikes the ground?

117. *Sports*   Some Major League Baseball pitchers can throw a fastball at speeds of up to and over 100 miles per hour. Assume a Major League Baseball pitcher throws a baseball straight up into the air at 100 miles per hour from a height of 6 feet 3 inches.

(a) Use the position equation to write a mathematical model for the height of the baseball.

(b) Find the height of the baseball after 3 seconds, 4 seconds, and 5 seconds. What must have occurred sometime in the interval $3 \le t \le 5$? Explain.

(c) How many seconds is the baseball in the air?

Model It

118. *CN Tower*   At 1815 feet tall, the CN Tower in Toronto, Ontario is the world's tallest self-supporting structure. An object is dropped from the top of the tower.

(a) Use the position equation to write a mathematical model for the height of the object.

(b) Complete the table.

Time, t	0	2	4	6	8	10	12
Height, s							

(c) From the table in part (b), determine the time interval during which the object reaches the ground. Numerically approximate the time it takes the object to reach the ground.

(d) Find the time it takes the object to reach the ground algebraically. How close was your numerical approximation?

 (e) Use a graphing utility with the appropriate viewing window to verify your answer(s) to parts (c) and (d).

119. *Data Analysis: Movie Tickets*   The average admission prices P for movie theaters from 1997 to 2003 can be approximated by the model

$P = -0.0081t^2 + 0.417t + 1.99, \quad 7 \le t \le 13$

where t represents the year, with $t = 7$ corresponding to 1997. (Source: Motion Picture Association of America, Inc.)

(a) Use the model to complete the table to determine when the average admission price reached or surpassed $5.

t	7	8	9	10	11	12	13
P							

(b) Verify your result from part (a) algebraically.

(c) Use the model to predict the average admission price for movie theaters in 2008. Is this prediction reasonable? How does this value compare with the admission price where you live?

120. *Data Analysis: Median Income*   The median incomes I (in dollars) of U.S. households from 1995 to 2002 can be approximated by the model

$I = -133.18t^2 + 3560.5t + 19{,}170, \quad 5 \le t \le 12$

where t represents the year, with $t = 5$ corresponding to 1995. (Source: U.S. Census Bureau)

(a) Use a graphing utility to graph the model. Then use the graph to determine in which year the median income reached or surpassed $40,000.

(b) Verify your result from part (a) algebraically.

(c) Use the model to predict the median incomes of U.S. households in 2008 and 2013. Can this model be used to predict the median income of U.S. households after 2002? Why or why not?

121. *Geometry*   The hypotenuse of an isosceles right triangle is 5 centimeters long. How long are its sides?

122. *Geometry*   An equilateral triangle has a height of 10 inches. How long is one of its sides? (*Hint:* Use the height of the triangle to partition the triangle into two congruent right triangles.)

123. *Flying Speed*   Two planes leave simultaneously from Chicago's O'Hare Airport, one flying due north and the other due east (see figure). The northbound plane is flying 50 miles per hour faster than the eastbound plane. After 3 hours, the planes are 2440 miles apart. Find the speed of each plane.

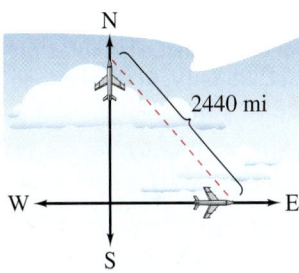

FIGURE FOR 123

124. *Boating*   A winch is used to tow a boat to a dock. The rope is attached to the boat at a point 15 feet below the level of the winch (see figure).

(a) Use the Pythagorean Theorem to write an equation giving the relationship between l and x.

(b) Find the distance from the boat to the dock when there is 75 feet of rope out.

125. *Revenue*   The demand equation for a product is $p = 20 - 0.0002x$, where p is the price per unit and x is the number of units sold. The total revenue for selling x units is

Revenue $= xp = x(20 - 0.0002x)$.

How many units must be sold to produce a revenue of $500,000?

126. *Revenue*   The demand equation for a product is $p = 60 - 0.0004x$, where p is the price per unit and x is the number of units sold. The total revenue for selling x units is

Revenue $= xp = x(60 - 0.0004x)$.

How many units must be sold to produce a revenue of $220,000?

Cost **In Exercises 127–130, use the cost equation to find the number of units *x* that a manufacturer can produce for the given cost *C*. Round your answer to the nearest positive integer.**

127. $C = 0.125x^2 + 20x + 500$ $C = \$14{,}000$

128. $C = 0.5x^2 + 15x + 5000$ $C = \$11{,}500$

129. $C = 800 + 0.04x + 0.002x^2$ $C = \$1680$

130. $C = 800 - 10x + \dfrac{x^2}{4}$ $C = \$896$

131. *Money in Circulation* The bar graph shows the total amounts of money *M* (in billions of dollars) in circulation in the United States from 1995 to 2003. These data can be approximated by the model

$$M - 1.835t^2 + 3.58t + 333.0, \quad 5 \leq t \leq 13$$

where *t* represents the year, with *t* = 5 corresponding to 1995. (Source: Financial Management Service, U.S. Department of the Treasury)

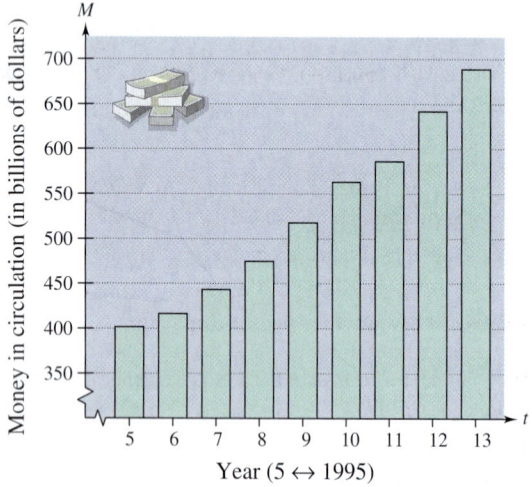

(a) Use the model to complete the table to determine when the total amount of money in circulation reached or surpassed $600 billion.

t	5	6	7	8	9	10	11	12	13
M									

(b) Verify your result from part (a) algebraically and graphically.

(c) Use the model to predict the total amount of money in circulation in 2008. Is this prediction reasonable? Explain.

132. *Biology* The metabolic rate of an ectothermic organism increases with increasing temperature within a certain range. The graph shows experimental data for the oxygen consumption *C* (in microliters per gram per hour) of a beetle at certain temperatures. The data can be approximated by the model

$$C = 0.45x^2 - 1.65x + 50.75, \quad 10 \leq x \leq 25$$

where *x* is the air temperature in degrees Celsius.

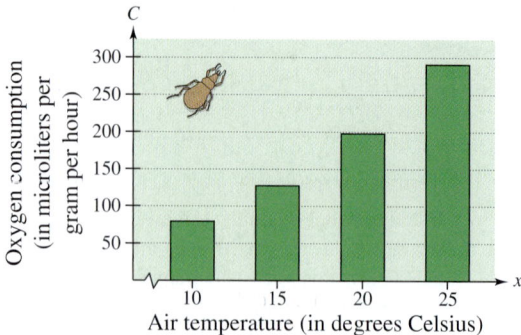

Air temperature (in degrees Celsius)

(a) The oxygen consumption is 150 microliters per gram per hour. What is the air temperature?

(b) The temperature is increased from 10°C to 20°C. The oxygen consumption is increased by approximately what factor?

133. *Flying Distance* A commercial jet flies to three cities whose locations form the vertices of a right triangle (see figure). The total flight distance (from Oklahoma City to Austin to New Orleans and back to Oklahoma City) is approximately 1348 miles. It is 560 miles between Oklahoma City and New Orleans. Approximate the other two distances.

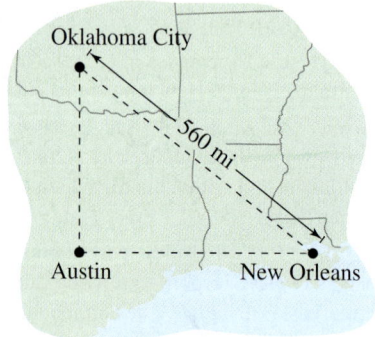

134. *Geometry*   An above ground swimming pool with the dimensions shown in the figure is to be constructed such that the volume of water in the pool is 1024 cubic feet. The height of the pool is to be 4 feet.

Not drawn to scale

(a) What are the possible dimensions of the base?

(b) One cubic foot of water weighs approximately 62.4 pounds. Find the total weight of the water in the pool.

(c) A water pump is filling the pool at a rate of 5 gallons per minute. Find the time that will be required for the pump to fill the pool. (*Hint:* One gallon of water is approximately 0.13368 cubic foot.)

Synthesis

True or False?   **In Exercises 135 and 136, determine whether the statement is true or false. Justify your answer.**

135. The quadratic equation $-3x^2 - x = 10$ has two real solutions.

136. If $(2x - 3)(x + 5) = 8$, then either $2x - 3 = 8$ or $x + 5 = 8$.

137. To solve the equation $2x^2 + 3x = 15x$, a student divides each side by x and solves the equation $2x + 3 = 15$. The resulting solution ($x = 6$) satisfies the original equation. Is there an error? Explain.

138. The graphs show the solutions of equations plotted on the real number line. In each case, determine whether the solution(s) is (are) for a linear equation, a quadratic equation, both, or neither. Explain.

(a) ●———●———●→ x
    a   b   c

(b) ———————●————→ x
            a

(c) ●———————●———→ x
    a   b

(d) ●———●———●———●→ x
    a   b   c   d

139. Solve $3(x + 4)^2 + (x + 4) - 2 = 0$ in two ways.

(a) Let $u = x + 4$, and solve the resulting equation for u. Then solve the u-solution for x.

(b) Expand and collect like terms in the equation, and solve the resulting equation for x.

(c) Which method is easier? Explain.

140. Solve each equation, given that a and b are not zero.

(a) $ax^2 + bx = 0$

(b) $ax^2 - ax = 0$

Think About It   **In Exercises 141–146, write a quadratic equation that has the given solutions. (There are many correct answers.)**

141. -3 and 6

142. -4 and -11

143. 8 and 14

144. $\frac{1}{6}$ and $-\frac{2}{5}$

145. $1 + \sqrt{2}$ and $1 - \sqrt{2}$

146. $-3 + \sqrt{5}$ and $-3 - \sqrt{5}$

Skills Review

In Exercises 147–150, identify the rule of algebra illustrated by the statement.

147. $(10x)y = 10(xy)$

148. $-4(x - 3) = -4x + 12$

149. $7x^4 + (-7x^4) = 0$

150. $(x + 4) + x^3 = x + (4 + x^3)$

In Exercises 151–154, find the product.

151. $(x + 3)(x - 6)$

152. $(x - 8)(x - 1)$

153. $(x + 4)(x^2 - x + 2)$

154. $(x + 9)(x^2 - 6x + 4)$

In Exercises 155–158, completely factor the expression.

155. $x^5 - 27x^2$

156. $x^3 - 5x^2 - 14x$

157. $x^3 + 5x^2 - 2x - 10$

158. $2x^3 + x^2 - 8x - 4$

159. Make a Decision   To work an extended application analyzing the population of the United States, visit this text's website at *college.hmco.com*. *(Data Source: U.S. Census Bureau)*

The *Make a Decision* exercise indicates a multi-part exercise using large data sets. Go to *college.hmco.com* to access these problems.

1.5 Complex Numbers

What you should learn

- Use the imaginary unit i to write complex numbers.
- Add, subtract, and multiply complex numbers.
- Use complex conjugates to write the quotient of two complex numbers in standard form.
- Find complex solutions of quadratic equations.

Why you should learn it

You can use complex numbers to model and solve real-life problems in electronics. For instance, in Exercise 83 on page 132, you will learn how to use complex numbers to find the impedance of an electrical circuit.

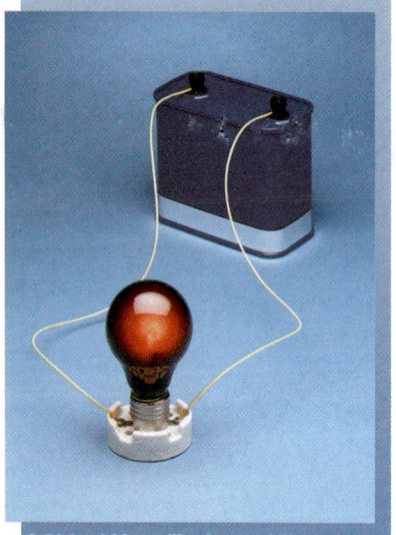

The Imaginary Unit i

In Section 1.4, you learned that some quadratic equations have no real solutions. For instance, the quadratic equation $x^2 + 1 = 0$ has no real solution because there is no real number x that can be squared to produce -1. To overcome this deficiency, mathematicians created an expanded system of numbers using the **imaginary unit i,** defined as

$$i = \sqrt{-1} \qquad \text{Imaginary unit}$$

where $i^2 = -1$. By adding real numbers to real multiples of this imaginary unit, the set of **complex numbers** is obtained. Each complex number can be written in the **standard form $a + bi$.** For instance, the standard form of the complex number $-5 + \sqrt{-9}$ is $-5 + 3i$ because

$$-5 + \sqrt{-9} = -5 + \sqrt{3^2(-1)} = -5 + 3\sqrt{-1} = -5 + 3i.$$

In the standard form $a + bi$, the real number a is called the **real part** of the **complex number $a + bi$,** and the number bi (where b is a real number) is called the **imaginary part** of the complex number.

Definition of a Complex Number

If a and b are real numbers, the number $a + bi$ is a **complex number,** and it is said to be written in **standard form.** If $b = 0$, the number $a + bi = a$ is a real number. If $b \neq 0$, the number $a + bi$ is called an **imaginary number.** A number of the form bi, where $b \neq 0$, is called a **pure imaginary number.**

The set of real numbers is a subset of the set of complex numbers, as shown in Figure 1.22. This is true because every real number a can be written as a complex number using $b = 0$. That is, for every real number a, you can write $a = a + 0i$.

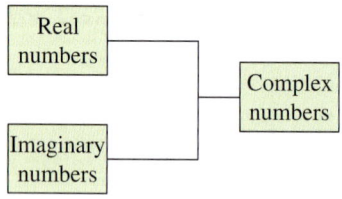

FIGURE 1.22

Equality of Complex Numbers

Two complex numbers $a + bi$ and $c + di$, written in standard form, are equal to each other

$$a + bi = c + di \qquad \text{Equality of two complex numbers}$$

if and only if $a = c$ and $b = d$.

Operations with Complex Numbers

To add (or subtract) two complex numbers, you add (or subtract) the real and imaginary parts of the numbers separately.

Addition and Subtraction of Complex Numbers

If $a + bi$ and $c + di$ are two complex numbers written in standard form, their sum and difference are defined as follows.

Sum: $(a + bi) + (c + di) = (a + c) + (b + d)i$

Difference: $(a + bi) - (c + di) = (a - c) + (b - d)i$

The **additive identity** in the complex number system is zero (the same as in the real number system). Furthermore, the **additive inverse** of the complex number $a + bi$ is

$$-(a + bi) = -a - bi. \qquad \text{Additive inverse}$$

So, you have

$$(a + bi) + (-a - bi) = 0 + 0i = 0.$$

Example 1   Adding and Subtracting Complex Numbers

a. $(4 + 7i) + (1 - 6i) = 4 + 7i + 1 - 6i$ Remove parentheses.

$$= (4 + 1) + (7i - 6i) \qquad \text{Group like terms.}$$

$$= 5 + i \qquad \text{Write in standard form.}$$

b. $(1 + 2i) - (4 + 2i) = 1 + 2i - 4 - 2i$ Remove parentheses.

$$= (1 - 4) + (2i - 2i) \qquad \text{Group like terms.}$$

$$= -3 + 0 \qquad \text{Simplify.}$$

$$= -3 \qquad \text{Write in standard form.}$$

c. $3i - (-2 + 3i) - (2 + 5i) = 3i + 2 - 3i - 2 - 5i$

$$= (2 - 2) + (3i - 3i - 5i)$$

$$= 0 - 5i$$

$$= -5i$$

d. $(3 + 2i) + (4 - i) - (7 + i) = 3 + 2i + 4 - i - 7 - i$

$$= (3 + 4 - 7) + (2i - i - i)$$

$$= 0 + 0i$$

$$= 0$$

✔**CHECKPOINT**   Now try Exercise 17.

Note in Examples 1(b) and 1(d) that the sum of two complex numbers can be a real number.

Many of the properties of real numbers are valid for complex numbers as well. Here are some examples.

Associative Properties of Addition and Multiplication

Commutative Properties of Addition and Multiplication

Distributive Property of Multiplication Over Addition

Notice below how these properties are used when two complex numbers are multiplied.

$$(a + bi)(c + di) = a(c + di) + bi(c + di) \qquad \text{Distributive Property}$$
$$= ac + (ad)i + (bc)i + (bd)i^2 \qquad \text{Distributive Property}$$
$$= ac + (ad)i + (bc)i + (bd)(-1) \qquad i^2 = -1$$
$$= ac - bd + (ad)i + (bc)i \qquad \text{Commutative Property}$$
$$= (ac - bd) + (ad + bc)i \qquad \text{Associative Property}$$

Rather than trying to memorize this multiplication rule, you should simply remember how the Distributive Property is used to multiply two complex numbers.

Example 2   Multiplying Complex Numbers

a. $4(-2 + 3i) = 4(-2) + 4(3i)$ Distributive Property

$= -8 + 12i$ Simplify.

b. $(2 - i)(4 + 3i) = 2(4 + 3i) - i(4 + 3i)$ Distributive Property

$= 8 + 6i - 4i - 3i^2$ Distributive Property

$= 8 + 6i - 4i - 3(-1)$ $i^2 = -1$

$= (8 + 3) + (6i - 4i)$ Group like terms.

$= 11 + 2i$ Write in standard form.

c. $(3 + 2i)(3 - 2i) = 3(3 - 2i) + 2i(3 - 2i)$ Distributive Property

$= 9 - 6i + 6i - 4i^2$ Distributive Property

$= 9 - 6i + 6i - 4(-1)$ $i^2 = -1$

$= 9 + 4$ Simplify.

$= 13$ Write in standard form.

d. $(3 + 2i)^2 = (3 + 2i)(3 + 2i)$ Square of a binomial

$= 3(3 + 2i) + 2i(3 + 2i)$ Distributive Property

$= 9 + 6i + 6i + 4i^2$ Distributive Property

$= 9 + 6i + 6i + 4(-1)$ $i^2 = -1$

$= 9 + 12i - 4$ Simplify.

$= 5 + 12i$ Write in standard form.

✓**CHECKPOINT**   Now try Exercise 27.

Complex Conjugates

Notice in Example 2(c) that the product of two complex numbers can be a real number. This occurs with pairs of complex numbers of the form $a + bi$ and $a - bi$, called **complex conjugates**.

$$(a + bi)(a - bi) = a^2 - abi + abi - b^2i^2$$
$$= a^2 - b^2(-1)$$
$$= a^2 + b^2$$

Example 3   Multiplying Conjugates

Multiply each complex number by its complex conjugate.

a. $1 + i$ **b.** $4 - 3i$

Solution

a. The complex conjugate of $1 + i$ is $1 - i$.

$$(1 + i)(1 - i) = 1^2 - i^2 = 1 - (-1) = 2$$

b. The complex conjugate of $4 - 3i$ is $4 + 3i$.

$$(4 - 3i)(4 + 3i) = 4^2 - (3i)^2 = 16 - 9i^2 = 16 - 9(-1) = 25$$

✔**CHECKPOINT**   Now try Exercise 37.

To write the quotient of $a + bi$ and $c + di$ in standard form, where c and d are not both zero, multiply the numerator and denominator by the complex conjugate of the *denominator* to obtain

$$\frac{a + bi}{c + di} = \frac{a + bi}{c + di}\left(\frac{c - di}{c - di}\right)$$

$$= \frac{(ac + bd) + (bc - ad)i}{c^2 + d^2}.$$ Standard form

STUDY TIP

Note that when you multiply the numerator and denominator of a quotient of complex numbers by

$$\frac{c - di}{c - di}$$

you are actually multiplying the quotient by a form of 1. You are not changing the original expression, you are only creating an expression that is equivalent to the original expression.

Example 4   Writing a Quotient of Complex Numbers in Standard Form

$$\frac{2 + 3i}{4 - 2i} = \frac{2 + 3i}{4 - 2i}\left(\frac{4 + 2i}{4 + 2i}\right)$$ Multiply numerator and denominator by complex conjugate of denominator.

$$= \frac{8 + 4i + 12i + 6i^2}{16 - 4i^2}$$ Expand.

$$= \frac{8 - 6 + 16i}{16 + 4}$$ $i^2 = -1$

$$= \frac{2 + 16i}{20}$$ Simplify.

$$= \frac{1}{10} + \frac{4}{5}i$$ Write in standard form.

✔**CHECKPOINT**   Now try Exercise 49.

Complex Solutions of Quadratic Equations

When using the Quadratic Formula to solve a quadratic equation, you often obtain a result such as $\sqrt{-3}$, which you know is not a real number. By factoring out $i = \sqrt{-1}$, you can write this number in standard form.

$$\sqrt{-3} = \sqrt{3(-1)} = \sqrt{3}\sqrt{-1} = \sqrt{3}i$$

The number $\sqrt{3}i$ is called the *principal square root* of -3.

STUDY TIP

The definition of principal square root uses the rule

$$\sqrt{ab} = \sqrt{a}\sqrt{b}$$

for $a > 0$ and $b < 0$. This rule is not valid if *both* a and b are negative. For example,

$$\sqrt{-5}\sqrt{-5} = \sqrt{5(-1)}\sqrt{5(-1)}$$
$$= \sqrt{5}i\sqrt{5}i$$
$$= \sqrt{25}i^2$$
$$= 5i^2 = -5$$

whereas

$$\sqrt{(-5)(-5)} = \sqrt{25} = 5.$$

To avoid problems with square roots of negative numbers, be sure to convert complex numbers to standard form *before* multiplying.

Principal Square Root of a Negative Number

If a is a positive number, the **principal square root** of the negative number $-a$ is defined as

$$\sqrt{-a} = \sqrt{a}i.$$

Example 5   **Writing Complex Numbers in Standard Form**

a. $\sqrt{-3}\sqrt{-12} = \sqrt{3}i\sqrt{12}i = \sqrt{36}i^2 = 6(-1) = -6$

b. $\sqrt{-48} - \sqrt{-27} = \sqrt{48}i - \sqrt{27}i = 4\sqrt{3}i - 3\sqrt{3}i = \sqrt{3}i$

c. $\left(-1 + \sqrt{-3}\right)^2 = \left(-1 + \sqrt{3}i\right)^2$

$$= (-1)^2 - 2\sqrt{3}i + \left(\sqrt{3}\right)^2(i^2)$$
$$= 1 - 2\sqrt{3}i + 3(-1)$$
$$= -2 - 2\sqrt{3}i$$

✓**CHECKPOINT**   Now try Exercise 59.

Example 6   **Complex Solutions of a Quadratic Equation**

Solve (a) $x^2 + 4 = 0$ and (b) $3x^2 - 2x + 5 = 0$.

Solution

a. $x^2 + 4 = 0$ Write original equation.

$\quad\quad x^2 = -4$ Subtract 4 from each side.

$\quad\quad x = \pm 2i$ Extract square roots.

b. $3x^2 - 2x + 5 = 0$ Write original equation.

$$x = \frac{-(-2) \pm \sqrt{(-2)^2 - 4(3)(5)}}{2(3)} \quad\quad \text{Quadratic Formula}$$

$$= \frac{2 \pm \sqrt{-56}}{6} \quad\quad \text{Simplify.}$$

$$= \frac{2 \pm 2\sqrt{14}i}{6} \quad\quad \text{Write } \sqrt{-56} \text{ in standard form.}$$

$$= \frac{1}{3} \pm \frac{\sqrt{14}}{3}i \quad\quad \text{Write in standard form.}$$

✓**CHECKPOINT**   Now try Exercise 65.

1.5 | Exercises

VOCABULARY CHECK:

1. Match the type of complex number with its definition.

(a) Real Number   (i) $a + bi$, $a \neq 0$, $b \neq 0$

(b) Imaginary number   (ii) $a + bi$, $a = 0$, $b \neq 0$

(c) Pure imaginary number (iii) $a + bi$, $b = 0$

In Exercises 2–4, fill in the blanks.

2. The imaginary unit i is defined as $i =$ _____, where $i^2 =$ _____.

3. If a is a positive number, the _____ _____ root of the negative number $-a$ is defined as $\sqrt{-a} = \sqrt{a}\,i$.

4. The numbers $a + bi$ and $a - bi$ are called _____ _____, and their product is a real number $a^2 + b^2$.

PREREQUISITE SKILLS REVIEW: Practice and review algebra skills needed for this section at **www.Eduspace.com.**

In Exercises 1–4, find real numbers a and b such that the equation is true.

1. $a + bi = -10 + 6i$ **2.** $a + bi = 13 + 4i$

3. $(a - 1) + (b + 3)i = 5 + 8i$

4. $(a + 6) + 2bi = 6 - 5i$

In Exercises 5–16, write the complex number in standard form.

5. $4 + \sqrt{-9}$ **6.** $3 + \sqrt{-16}$

7. $2 - \sqrt{-27}$ **8.** $1 + \sqrt{-8}$

9. $\sqrt{-75}$ **10.** $\sqrt{-4}$

11. 8 **12.** 45

13. $-6i + i^2$ **14.** $-4i^2 + 2i$

15. $\sqrt{-0.09}$ **16.** $\sqrt{-0.0004}$

In Exercises 17–26, perform the addition or subtraction and write the result in standard form.

17. $(5 + i) + (6 - 2i)$ **18.** $(13 - 2i) + (-5 + 6i)$

19. $(8 - i) - (4 - i)$ **20.** $(3 + 2i) - (6 + 13i)$

21. $\left(-2 + \sqrt{-8}\right) + \left(5 - \sqrt{-50}\right)$

22. $\left(8 + \sqrt{-18}\right) - \left(4 + 3\sqrt{2}\,i\right)$

23. $13i - (14 - 7i)$ **24.** $22 + (-5 + 8i) + 10i$

25. $-\left(\frac{3}{2} + \frac{5}{2}i\right) + \left(\frac{5}{3} + \frac{11}{3}i\right)$

26. $(1.6 + 3.2i) + (-5.8 + 4.3i)$

In Exercises 27–36, perform the operation and write the result in standard form.

27. $(1 + i)(3 - 2i)$ **28.** $(6 - 2i)(2 - 3i)$

29. $6i(5 - 2i)$ **30.** $-8i(9 + 4i)$

31. $\left(\sqrt{14} + \sqrt{10}i\right)\left(\sqrt{14} - \sqrt{10}i\right)$

32. $\left(\sqrt{3} + \sqrt{15}i\right)\left(\sqrt{3} - \sqrt{15}i\right)$

33. $(4 + 5i)^2$ **34.** $(2 - 3i)^2$

35. $(2 + 3i)^2 + (2 - 3i)^2$ **36.** $(1 - 2i)^2 - (1 + 2i)^2$

In Exercises 37–44, write the complex conjugate of the complex number. Then multiply the number by its complex conjugate.

37. $6 + 3i$ **38.** $7 - 12i$

39. $-1 - \sqrt{5}i$ **40.** $-3 + \sqrt{2}i$

41. $\sqrt{-20}$ **42.** $\sqrt{-15}$

43. $\sqrt{8}$ **44.** $1 + \sqrt{8}$

In Exercises 45–54, write the quotient in standard form.

45. $\dfrac{5}{i}$ **46.** $-\dfrac{14}{2i}$

47. $\dfrac{2}{4 - 5i}$ **48.** $\dfrac{5}{1 - i}$

49. $\dfrac{3 + i}{3 - i}$ **50.** $\dfrac{6 - 7i}{1 - 2i}$

51. $\dfrac{6 - 5i}{i}$ **52.** $\dfrac{8 + 16i}{2i}$

53. $\dfrac{3i}{(4 - 5i)^2}$ **54.** $\dfrac{5i}{(2 + 3i)^2}$

In Exercises 55–58, perform the operation and write the result in standard form.

55. $\dfrac{2}{1 + i} - \dfrac{3}{1 - i}$ **56.** $\dfrac{2i}{2 + i} + \dfrac{5}{2 - i}$

57. $\dfrac{i}{3 - 2i} + \dfrac{2i}{3 + 8i}$ **58.** $\dfrac{1 + i}{i} - \dfrac{3}{4 - i}$

In Exercises 59–64, write the complex number in standard form.

59. $\sqrt{-6} \cdot \sqrt{-2}$

60. $\sqrt{-5} \cdot \sqrt{-10}$

61. $\left(\sqrt{-10}\right)^2$

62. $\left(\sqrt{-75}\right)^2$

63. $\left(3 + \sqrt{-5}\right)\left(7 - \sqrt{-10}\right)$

64. $\left(2 - \sqrt{-6}\right)^2$

In Exercises 65–74, use the Quadratic Formula to solve the quadratic equation.

65. $x^2 - 2x + 2 = 0$

66. $x^2 + 6x + 10 = 0$

67. $4x^2 + 16x + 17 = 0$

68. $9x^2 - 6x + 37 = 0$

69. $4x^2 + 16x + 15 = 0$

70. $16t^2 - 4t + 3 = 0$

71. $\frac{3}{2}x^2 - 6x + 9 = 0$

72. $\frac{7}{8}x^2 - \frac{3}{4}x + \frac{5}{16} = 0$

73. $1.4x^2 - 2x - 10 = 0$

74. $4.5x^2 - 3x + 12 = 0$

In Exercises 75–82, simplify the complex number and write it in standard form.

75. $-6i^3 + i^2$

76. $4i^2 - 2i^3$

77. $-5i^5$

78. $(-i)^3$

79. $\left(\sqrt{-75}\right)^3$

80. $\left(\sqrt{-2}\right)^6$

81. $\dfrac{1}{i^3}$

82. $\dfrac{1}{(2i)^3}$

Model It

83. *Impedance*   The opposition to current in an electrical circuit is called its impedance. The impedance z in a parallel circuit with two pathways satisfies the equation

$$\frac{1}{z} = \frac{1}{z_1} + \frac{1}{z_2}$$

where z_1 is the impedance (in ohms) of pathway 1 and z_2 is the impedance of pathway 2.

(a) The impedance of each pathway in a parallel circuit is found by adding the impedances of all components in the pathway. Use the table to find z_1 and z_2.

(b) Find the impedance z.

	Resistor	Inductor	Capacitor
Symbol	$a\Omega$	$b\Omega$	$c\Omega$
Impedance	a	bi	$-ci$

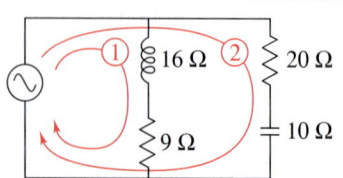

84. Cube each complex number.

    (a) 2   (b) $-1 + \sqrt{3}i$ (c) $-1 - \sqrt{3}i$

85. Raise each complex number to the fourth power.

    (a) 2   (b) -2 (c) $2i$ (d) $-2i$

86. Write each of the powers of i as i, $-i$, 1, or -1.

    (a) i^{40} (b) i^{25} (c) i^{50} (d) i^{67}

Synthesis

True or False?   In Exercises 87–89, determine whether the statement is true or false. Justify your answer.

87. There is no complex number that is equal to its complex conjugate.

88. $i\sqrt{6}$ is a solution of $x^4 - x^2 + 14 = 56$.

89. $i^{44} + i^{150} - i^{74} - i^{109} + i^{61} = -1$

90. *Error Analysis*   Describe the error.

$$\sqrt{-6}\sqrt{-6} = \sqrt{(-6)(-6)} = \sqrt{36} = 6$$

91. *Proof*   Prove that the complex conjugate of the product of two complex numbers $a_1 + b_1i$ and $a_2 + b_2i$ is the product of their complex conjugates.

92. *Proof*   Prove that the complex conjugate of the sum of two complex numbers $a_1 + b_1i$ and $a_2 + b_2i$ is the sum of their complex conjugates.

Skills Review

In Exercises 93–96, perform the operation and write the result in standard form.

93. $(4 + 3x) + (8 - 6x - x^2)$

94. $(x^3 - 3x^2) - (6 - 2x - 4x^2)$

95. $\left(3x - \frac{1}{2}\right)(x + 4)$

96. $(2x - 5)^2$

In Exercises 97–100, solve the equation and check your solution.

97. $-x - 12 = 19$

98. $8 - 3x = -34$

99. $4(5x - 6) - 3(6x + 1) = 0$

100. $5[x - (3x + 11)] = 20x - 15$

101. *Volume of an Oblate Spheroid*

Solve for a: $V = \frac{4}{3}\pi a^2 b$

102. *Newton's Law of Universal Gravitation*

Solve for r: $F = \alpha\dfrac{m_1 m_2}{r^2}$

103. *Mixture Problem*   A five-liter container contains a mixture with a concentration of 50%. How much of this mixture must be withdrawn and replaced by 100% concentrate to bring the mixture up to 60% concentration?

1.6   Other Types of Equations

What you should learn

- Solve polynomial equations of degree three or greater.
- Solve equations involving radicals.
- Solve equations involving fractions or absolute values.
- Use polynomial equations and equations involving radicals to model and solve real-life problems.

Why you should learn it

Polynomial equations, radical equations, and absolute value equations can be used to model and solve real-life problems. For instance, in Exercise 96 on page 142, a radical equation can be used to model the total monthly cost of airplane flights between Chicago and Denver.

© Austin Brown/Getty Images

Polynomial Equations

In this section you will extend the techniques for solving equations to nonlinear and nonquadratic equations. At this point in the text, you have only four basic methods for solving nonlinear equations—*factoring*, *extracting square roots*, *completing the square*, and the *Quadratic Formula*. So the main goal of this section is to learn to *rewrite* nonlinear equations in a form to which you can apply one of these methods.

Example 1 shows how to use factoring to solve a **polynomial equation,** which is an equation that can be written in the general form

$$a_n x^n + a_{n-1} x^{n-1} + \cdots + a_2 x^2 + a_1 x + a_0 = 0.$$

Example 1   Solving a Polynomial Equation by Factoring

Solve $3x^4 = 48x^2$.

Solution

First write the polynomial equation in general form with zero on one side, factor the other side, and then set each factor equal to zero and solve.

$3x^4 = 48x^2$		Write original equation.
$3x^4 - 48x^2 = 0$		Write in general form.
$3x^2(x^2 - 16) = 0$		Factor out common factor.
$3x^2(x + 4)(x - 4) = 0$		Write in factored form.
$3x^2 = 0$	$x = 0$	Set 1st factor equal to 0.
$x + 4 = 0$	$x = -4$	Set 2nd factor equal to 0.
$x - 4 = 0$	$x = 4$	Set 3rd factor equal to 0.

You can check these solutions by substituting in the original equation, as follows.

Check

$3(0)^4 = 48(0)^2$	0 checks. ✔
$3(-4)^4 = 48(-4)^2$	-4 checks. ✔
$3(4)^4 = 48(4)^2$	4 checks. ✔

So, you can conclude that the solutions are $x = 0$, $x = -4$, and $x = 4$.

✔**CHECKPOINT**   Now try Exercise 1.

A common mistake that is made in solving an equation like that in Example 1 is to divide each side of the equation by the variable factor x^2. This loses the solution $x = 0$. When solving an equation, always write the equation in general form, then factor the equation and set each factor equal to zero. Do not divide each side of an equation by a variable factor in an attempt to simplify the equation.

For a review of factoring special polynomial forms, see Section P.4.

Technology

You can use a graphing utility to check graphically the solutions of the equation in Example 2. To do this, graph the equation

$$y = x^3 - 3x^2 + 3x - 9.$$

Then use the *zero* or *root* feature to approximate any *x*-intercepts. As shown below, the *x*-intercept of the graph occurs at $(3, 0)$, confirming the *real* solution of $x = 3$ found in Example 2.

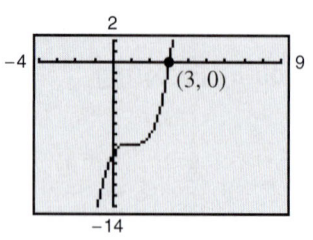

Try using a graphing utility to check the solutions found in Example 3.

Example 2   **Solving a Polynomial Equation by Factoring**

Solve $x^3 - 3x^2 + 3x - 9 = 0$.

Solution

$x^3 - 3x^2 + 3x - 9 = 0$	Write original equation.
$x^2(x - 3) + 3(x - 3) = 0$	Factor by grouping.
$(x - 3)(x^2 + 3) = 0$	Distributive Property
$x - 3 = 0 \quad\Longrightarrow\quad x = 3$	Set 1st factor equal to 0.
$x^2 + 3 = 0 \quad\Longrightarrow\quad x = \pm\sqrt{3}i$	Set 2nd factor equal to 0.

The solutions are $x = 3$, $x = \sqrt{3}i$, and $x = -\sqrt{3}i$.

✓CHECKPOINT   Now try Exercise 9.

Occasionally, mathematical models involve equations that are of **quadratic type.** In general, an equation is of quadratic type if it can be written in the form

$$au^2 + bu + c = 0$$

where $a \neq 0$ and u is an algebraic expression.

Example 3   **Solving an Equation of Quadratic Type**

Solve $x^4 - 3x^2 + 2 = 0$.

Solution

This equation is of quadratic type with $u = x^2$.

$$(x^2)^2 - 3(x^2) + 2 = 0$$

To solve this equation, you can factor the left side of the equation as the product of two second-degree polynomials.

$x^4 - 3x^2 + 2 = 0$	Write original equation.
$\overbrace{(x^2)^2}^{u^2} - \overbrace{3(x^2)}^{3u} + 2 = 0$	Quadratic form
$(x^2 - 1)(x^2 - 2) = 0$	Partially factor.
$(x + 1)(x - 1)(x^2 - 2) = 0$	Factor completely.
$x + 1 = 0 \quad\Longrightarrow\quad x = -1$	Set 1st factor equal to 0.
$x - 1 = 0 \quad\Longrightarrow\quad x = 1$	Set 2nd factor equal to 0.
$x^2 - 2 = 0 \quad\Longrightarrow\quad x = \pm\sqrt{2}$	Set 3rd factor equal to 0.

The solutions are $x = -1$, $x = 1$, $x = \sqrt{2}$, and $x = -\sqrt{2}$. Check these in the original equation.

✓CHECKPOINT   Now try Exercise 13.

Equations Involving Radicals

Operations such as squaring each side of an equation, raising each side of an equation to a rational power, and multiplying each side of an equation by a variable quantity all can introduce extraneous solutions. So, when you use any of these operations, checking your solutions is crucial.

Example 4   Solving Equations Involving Radicals

a.

$\sqrt{2x + 7} - x = 2$	Original equation
$\sqrt{2x + 7} = x + 2$	Isolate radical.
$2x + 7 = x^2 + 4x + 4$	Square each side.
$0 = x^2 + 2x - 3$	Write in general form.
$0 = (x + 3)(x - 1)$	Factor.
$x + 3 = 0 \implies x = -3$	Set 1st factor equal to 0.
$x - 1 = 0 \implies x = 1$	Set 2nd factor equal to 0.

By checking these values, you can determine that the only solution is $x = 1$.

b.

$\sqrt{2x - 5} - \sqrt{x - 3} = 1$	Original equation
$\sqrt{2x - 5} = \sqrt{x - 3} + 1$	Isolate $\sqrt{2x - 5}$.
$2x - 5 = x - 3 + 2\sqrt{x - 3} + 1$	Square each side.
$2x - 5 = x - 2 + 2\sqrt{x - 3}$	Combine like terms.
$x - 3 = 2\sqrt{x - 3}$	Isolate $2\sqrt{x - 3}$.
$x^2 - 6x + 9 = 4(x - 3)$	Square each side.
$x^2 - 10x + 21 = 0$	Write in general form.
$(x - 3)(x - 7) = 0$	Factor.
$x - 3 = 0 \implies x = 3$	Set 1st factor equal to 0.
$x - 7 = 0 \implies x = 7$	Set 2nd factor equal to 0.

The solutions are $x = 3$ and $x = 7$. Check these in the original equation.

✔**CHECKPOINT**   Now try Exercise 31.

STUDY TIP

When an equation contains two radicals, it may not be possible to isolate both. In such cases, you may have to raise each side of the equation to a power at *two* different stages in the solution, as shown in Example 4(b).

Example 5   Solving an Equation Involving a Rational Exponent

$(x - 4)^{2/3} = 25$	Original equation
$\sqrt[3]{(x - 4)^2} = 25$	Rewrite in radical form.
$(x - 4)^2 = 15{,}625$	Cube each side.
$x - 4 = \pm 125$	Take square root of each side.
$x = 129, \quad x = -121$	Add 4 to each side.

✔**CHECKPOINT**   Now try Exercise 45.

Equations with Fractions or Absolute Values

To solve an equation involving fractions, multiply each side of the equation by the least common denominator (LCD) of all terms in the equation. This procedure will "clear the equation of fractions." For instance, in the equation

$$\frac{2}{x^2 + 1} + \frac{1}{x} = \frac{2}{x}$$

you can multiply each side of the equation by $x(x^2 + 1)$. Try doing this and solve the resulting equation. You should obtain one solution: $x = 1$.

Example 6 Solving an Equation Involving Fractions

Solve $\dfrac{2}{x} = \dfrac{3}{x - 2} - 1$.

Solution

For this equation, the least common denominator of the three terms is $x(x - 2)$, so you begin by multiplying each term of the equation by this expression.

$$\frac{2}{x} = \frac{3}{x - 2} - 1 \qquad \text{Write original equation.}$$

$$x(x - 2)\frac{2}{x} = x(x - 2)\frac{3}{x - 2} - x(x - 2)(1) \qquad \text{Multiply each term by the LCD.}$$

$$2(x - 2) = 3x - x(x - 2) \qquad \text{Simplify.}$$

$$2x - 4 = -x^2 + 5x \qquad \text{Simplify.}$$

$$x^2 - 3x - 4 = 0 \qquad \text{Write in general form.}$$

$$(x - 4)(x + 1) = 0 \qquad \text{Factor.}$$

$$x - 4 = 0 \quad \Longrightarrow \quad x = 4 \qquad \text{Set 1st factor equal to 0.}$$

$$x + 1 = 0 \quad \Longrightarrow \quad x = -1 \qquad \text{Set 2nd factor equal to 0.}$$

Check $x = 4$

$$\frac{2}{x} = \frac{3}{x - 2} - 1$$

$$\frac{2}{4} \stackrel{?}{=} \frac{3}{4 - 2} - 1$$

$$\frac{1}{2} \stackrel{?}{=} \frac{3}{2} - 1$$

$$\frac{1}{2} = \frac{1}{2} \; \checkmark$$

Check $x = -1$

$$\frac{2}{x} = \frac{3}{x - 2} - 1$$

$$\frac{2}{-1} \stackrel{?}{=} \frac{3}{-1 - 2} - 1$$

$$-2 \stackrel{?}{=} -1 - 1$$

$$-2 = -2 \; \checkmark$$

So, the solutions are $x = 4$ and $x = -1$.

✓CHECKPOINT Now try Exercise 59.

To solve an equation involving an absolute value, remember that the expression inside the absolute value signs can be positive or negative. This results in *two* separate equations, each of which must be solved. For instance, the equation

$$|x - 2| = 3$$

results in the two equations $x - 2 = 3$ and $-(x - 2) = 3$, which implies that the equation has two solutions: $x = 5$ and $x = -1$.

Example 7   Solving an Equation Involving Absolute Value

Solve $|x^2 - 3x| = -4x + 6$.

Solution

Because the variable expression inside the absolute value signs can be positive or negative, you must solve the following two equations.

First Equation

$x^2 - 3x = -4x + 6$	Use positive expression.
$x^2 + x - 6 = 0$	Write in general form.
$(x + 3)(x - 2) = 0$	Factor.
$x + 3 = 0 \implies x = -3$	Set 1st factor equal to 0.
$x - 2 = 0 \implies x = 2$	Set 2nd factor equal to 0.

Second Equation

$-(x^2 - 3x) = -4x + 6$	Use negative expression.
$x^2 - 7x + 6 = 0$	Write in general form.
$(x - 1)(x - 6) = 0$	Factor.
$x - 1 = 0 \implies x = 1$	Set 1st factor equal to 0.
$x - 6 = 0 \implies x = 6$	Set 2nd factor equal to 0.

Check

$\left|(-3)^2 - 3(-3)\right| \overset{?}{=} -4(-3) + 6$	Substitute -3 for x.
$18 = 18$	-3 checks. ✔
$\left|(2)^2 - 3(2)\right| \overset{?}{=} -4(2) + 6$	Substitute 2 for x.
$2 \neq -2$	2 does not check.
$\left|(1)^2 - 3(1)\right| \overset{?}{=} -4(1) + 6$	Substitute 1 for x.
$2 = 2$	1 checks. ✔
$\left|(6)^2 - 3(6)\right| \overset{?}{=} -4(6) + 6$	Substitute 6 for x.
$18 \neq -18$	6 does not check.

The solutions are $x = -3$ and $x = 1$.

✓CHECKPOINT   Now try Exercise 67.

Applications

It would be impossible to categorize the many different types of applications that involve nonlinear and nonquadratic models. However, from the few examples and exercises that are given, you will gain some appreciation for the variety of applications that can occur.

Example 8   **Reduced Rates**

A ski club chartered a bus for a ski trip at a cost of $480. In an attempt to lower the bus fare per skier, the club invited nonmembers to go along. After five nonmembers joined the trip, the fare per skier decreased by $4.80. How many club members are going on the trip?

Solution

Begin the solution by creating a verbal model and assigning labels.

Verbal
Model: Cost per skier · Number of skiers = Cost of trip

Labels: Cost of trip $= 480$ (dollars)
Number of ski club members $= x$ (people)
Number of skiers $= x + 5$ (people)

Original cost per member $= \dfrac{480}{x}$ (dollars per person)

Cost per skier $= \dfrac{480}{x} - 4.80$ (dollars per person)

Equation:
$$\left(\frac{480}{x} - 4.80\right)(x + 5) = 480$$

$$\left(\frac{480 - 4.8x}{x}\right)(x + 5) = 480 \qquad \text{Write } \left(\frac{480}{x} - 4.80\right) \text{ as a fraction.}$$

$$(480 - 4.8x)(x + 5) = 480x \qquad \text{Multiply each side by } x.$$

$$480x + 2400 - 4.8x^2 - 24x = 480x \qquad \text{Multiply.}$$

$$-4.8x^2 - 24x + 2400 = 0 \qquad \text{Subtract } 480x \text{ from each side.}$$

$$x^2 + 5x - 500 = 0 \qquad \text{Divide each side by } -4.8.$$

$$(x + 25)(x - 20) = 0 \qquad \text{Factor.}$$

$$x + 25 = 0 \quad \Longrightarrow \quad x = -25$$

$$x - 20 = 0 \quad \Longrightarrow \quad x = 20$$

Choosing the positive value of x, you can conclude that 20 ski club members are going on the trip. Check this in the original statement of the problem, as follows.

$$\left(\frac{480}{20} - 4.80\right)(20 + 5) \stackrel{?}{=} 480 \qquad \text{Substitute 20 for } x.$$

$$(24 - 4.80)25 \stackrel{?}{=} 480 \qquad \text{Simplify.}$$

$$480 = 480 \qquad \text{20 checks. } \checkmark$$

✔CHECKPOINT   Now try Exercise 87.

Interest in a savings account is calculated by one of three basic methods: simple interest, interest compounded n times per year, and interest compounded continuously. The next example uses the formula for interest that is compounded n times per year.

$$A = P\left(1 + \frac{r}{n}\right)^{nt}$$

In this formula, A is the balance in the account, P is the principal (or original deposit), r is the annual interest rate (in decimal form), n is the number of compoundings per year, and t is the time in years. In Chapter 5, you will study a derivation of the formula above for interest compounded continuously.

Example 9   **Compound Interest**

When you were born, your grandparents deposited $5000 in a long-term investment in which the interest was compounded quarterly. Today, on your 25th birthday, the value of the investment is $25,062.59. What is the annual interest rate for this investment?

Solution

Formula: $A = P\left(1 + \frac{r}{n}\right)^{nt}$

Labels: Balance $= A = 25{,}062.59$ (dollars)
             Principal $= P = 5000$ (dollars)
             Time $= t = 25$ (years)
             Compoundings per year $= n = 4$ (compoundings per year)
             Annual interest rate $= r$ (percent in decimal form)

Equation: $25{,}062.59 = 5000\left(1 + \dfrac{r}{4}\right)^{4(25)}$

$\dfrac{25{,}062.59}{5000} = \left(1 + \dfrac{r}{4}\right)^{100}$ Divide each side by 5000.

$5.0125 \approx \left(1 + \dfrac{r}{4}\right)^{100}$ Use a calculator.

$(5.0125)^{1/100} = 1 + \dfrac{r}{4}$ Raise each side to reciprocal power.

$1.01625 \approx 1 + \dfrac{r}{4}$ Use a calculator.

$0.01625 = \dfrac{r}{4}$ Subtract 1 from each side.

$0.065 = r$ Multiply each side by 4.

The annual interest rate is about 0.065, or 6.5%. Check this in the original statement of the problem.

✓**CHECKPOINT**   Now try Exercise 91.

1.6 | Exercises

VOCABULARY CHECK: Fill in the blanks.

1. The equation $a_n x^n + a_{n-1} x^{n-1} + \cdots + a_2 x^2 + a_1 x + a_0 = 0$ is a _____ equation in x written in general form.

2. Squaring each side of an equation, multiplying each side of an equation by a variable quantity, and raising each side of an equation to a rational power are all operations that can introduce _____ solutions to a given equation.

3. The equation $2x^4 + x^2 + 1 = 0$ is of _____ _____.

PREREQUISITE SKILLS REVIEW: Practice and review algebra skills needed for this section at **www.Eduspace.com.**

In Exercises 1–24, find all solutions of the equation. Check your solutions in the original equation.

1. $4x^4 - 18x^2 = 0$
2. $20x^3 - 125x = 0$
3. $x^4 - 81 = 0$
4. $x^6 - 64 = 0$
5. $x^3 + 216 = 0$
6. $27x^3 - 512 = 0$
7. $5x^3 + 30x^2 + 45x = 0$
8. $9x^4 - 24x^3 + 16x^2 = 0$
9. $x^3 - 3x^2 - x + 3 = 0$
10. $x^3 + 2x^2 + 3x + 6 = 0$
11. $x^4 - x^3 + x - 1 = 0$
12. $x^4 + 2x^3 - 8x - 16 = 0$
13. $x^4 - 4x^2 + 3 = 0$
14. $x^4 + 5x^2 - 36 = 0$
15. $4x^4 - 65x^2 + 16 = 0$
16. $36t^4 + 29t^2 - 7 = 0$
17. $x^6 + 7x^3 - 8 = 0$
18. $x^6 + 3x^3 + 2 = 0$
19. $\dfrac{1}{x^2} + \dfrac{8}{x} + 15 = 0$
20. $6\left(\dfrac{x}{x+1}\right)^2 + 5\left(\dfrac{x}{x+1}\right) - 6 = 0$
21. $2x + 9\sqrt{x} = 5$
22. $6x - 7\sqrt{x} - 3 = 0$
23. $3x^{1/3} + 2x^{2/3} = 5$
24. $9t^{2/3} + 24t^{1/3} + 16 = 0$

Graphical Analysis In Exercises 25–28, (a) use a graphing utility to graph the equation, (b) use the graph to approximate any x-intercepts of the graph, (c) set $y = 0$ and solve the resulting equation, and (d) compare the result of part (c) with the x-intercepts of the graph.

25. $y = x^3 - 2x^2 - 3x$
26. $y = 2x^4 - 15x^3 + 18x^2$
27. $y = x^4 - 10x^2 + 9$
28. $y = x^4 - 29x^2 + 100$

In Exercises 29–52, find all solutions of the equation. Check your solutions in the original equation.

29. $\sqrt{2x} - 10 = 0$
30. $4\sqrt{x} - 3 = 0$
31. $\sqrt{x - 10} - 4 = 0$
32. $\sqrt{5 - x} - 3 = 0$
33. $\sqrt[3]{2x + 5} + 3 = 0$
34. $\sqrt[3]{3x + 1} - 5 = 0$
35. $-\sqrt{26 - 11x} + 4 = x$
36. $x + \sqrt{31 - 9x} = 5$
37. $\sqrt{x + 1} = \sqrt{3x + 1}$
38. $\sqrt{x + 5} = \sqrt{x - 5}$
39. $\sqrt{x} - \sqrt{x - 5} = 1$
40. $\sqrt{x} + \sqrt{x - 20} = 10$
41. $\sqrt{x + 5} + \sqrt{x - 5} = 10$
42. $2\sqrt{x + 1} - \sqrt{2x + 3} = 1$
43. $\sqrt{x + 2} - \sqrt{2x - 3} = -1$
44. $4\sqrt{x - 3} - \sqrt{6x - 17} = 3$
45. $(x - 5)^{3/2} = 8$
46. $(x + 3)^{3/2} = 8$
47. $(x + 3)^{2/3} = 8$
48. $(x + 2)^{2/3} = 9$
49. $(x^2 - 5)^{3/2} = 27$
50. $(x^2 - x - 22)^{3/2} = 27$
51. $3x(x - 1)^{1/2} + 2(x - 1)^{3/2} = 0$
52. $4x^2(x - 1)^{1/3} + 6x(x - 1)^{4/3} = 0$

Graphical Analysis In Exercises 53–56, (a) use a graphing utility to graph the equation, (b) use the graph to approximate any x-intercepts of the graph, (c) set $y = 0$ and solve the resulting equation, and (d) compare the result of part (c) with the x-intercepts of the graph.

53. $y = \sqrt{11x - 30} - x$
54. $y = 2x - \sqrt{15 - 4x}$
55. $y = \sqrt{7x + 36} - \sqrt{5x + 16} - 2$
56. $y = 3\sqrt{x} - \dfrac{4}{\sqrt{x}} - 4$

In Exercises 57–70, find all solutions of the equation. Check your solutions in the original equation.

57. $x = \dfrac{3}{x} + \dfrac{1}{2}$

58. $\dfrac{4}{x} - \dfrac{5}{3} = \dfrac{x}{6}$

59. $\dfrac{1}{x} - \dfrac{1}{x+1} = 3$

60. $\dfrac{4}{x+1} - \dfrac{3}{x+2} = 1$

61. $\dfrac{20-x}{x} = x$

62. $4x + 1 = \dfrac{3}{x}$

63. $\dfrac{x}{x^2-4} + \dfrac{1}{x+2} = 3$

64. $\dfrac{x+1}{3} - \dfrac{x+1}{x+2} = 0$

65. $|2x - 1| = 5$

66. $|3x + 2| = 7$

67. $|x| = x^2 + x - 3$

68. $|x^2 + 6x| = 3x + 18$

69. $|x + 1| = x^2 - 5$

70. $|x - 10| = x^2 - 10x$

 Graphical Analysis   In Exercises 71–74, (a) use a graphing utility to graph the equation, (b) use the graph to approximate any x-intercepts of the graph, (c) set $y = 0$ and solve the resulting equation, and (d) compare the result of part (c) with the x-intercepts of the graph.

71. $y = \dfrac{1}{x} - \dfrac{4}{x-1} - 1$ **72.** $y = x + \dfrac{9}{x+1} - 5$

73. $y = |x + 1| - 2$ **74.** $y = |x - 2| - 3$

In Exercises 75–78, find the real solutions of the equation algebraically. (Round your answers to three decimal places.)

75. $3.2x^4 - 1.5x^2 - 2.1 = 0$

76. $7.08x^6 + 4.15x^3 - 9.6 = 0$

77. $1.8x - 6\sqrt{x} - 5.6 = 0$

78. $4x^{2/3} + 8x^{1/3} + 3.6 = 0$

Think About It   In Exercises 79–86, find an equation that has the given solutions. (There are many correct answers.)

79. $-2, 5$ **80.** $0, 3, 5$

81. $-\dfrac{7}{3}, \dfrac{6}{7}$ **82.** $-\dfrac{1}{8}, -\dfrac{4}{5}$

83. $\sqrt{3}, -\sqrt{3}, 4$

84. $2\sqrt{7}, -\sqrt{7}$

85. $-1, 1, i, -i$

86. $4i, -4i, 6, -6$

87. *Chartering a Bus*   A college charters a bus for $1700 to take a group to a museum. When six more students join the trip, the cost per student drops by $7.50. How many students were in the original group?

88. *Renting an Apartment*   Three students are planning to rent an apartment for a year and share equally in the cost. By adding a fourth person, each person could save $75 a month. How much is the monthly rent?

89. *Airspeed*   An airline runs a commuter flight between Portland, Oregon and Seattle, Washington, which are 145 miles apart. If the average speed of the plane could be increased by 40 miles per hour, the travel time would be decreased by 12 minutes. What airspeed is required to obtain this decrease in travel time?

90. *Average Speed*   A family drove 1080 miles to their vacation lodge. Because of increased traffic density, their average speed on the return trip was decreased by 6 miles per hour and the trip took $2\frac{1}{2}$ hours longer. Determine their average speed on the way to the lodge.

91. *Mutual Funds*   A deposit of $2500 in a mutual fund reaches a balance of $3052.49 after 5 years. What annual interest rate on a certificate of deposit compounded monthly would yield an equivalent return?

92. *Mutual Funds*   A sales representative for a mutual funds company describes a "guaranteed investment fund" that the company is offering to new investors. You are told that if you deposit $10,000 in the fund you will be guaranteed a return of at least $25,000 after 20 years. (Assume the interest is compounded quarterly.)

(a) What is the annual interest rate if the investment only meets the minimum guaranteed amount?

(b) After 20 years, you receive $32,000. What is the annual interest rate?

93. *Number of Doctors*   The number of medical doctors D (in thousands) in the United States from 1994 to 2002 can be modeled by

$$D = 463.97 + 111.6\sqrt{t}, \quad 4 \le t \le 12$$

where t represents the year, with $t = 4$ corresponding to 1994.   (Source: American Medical Association)

(a) In which year did the number of medical doctors reach 816,000?

(b) Use the model to predict when the number of medical doctors will reach 900,000. Is this prediction reasonable? Explain.

94. *Voting Population*   The total voting-age population P (in millions) in the United States from 1990 to 2002 can be modeled by

$$P = \frac{182.45 - 3.189t}{1.00 - 0.026t}, \quad 0 \le t \le 12$$

where t represents the year, with $t = 0$ corresponding to 1990. (Source: U.S. Census Bureau)

(a) In which year did the total voting-age population reach 200 million?

(b) Use the model to predict when the total voting-age population will reach 230 million. Is this prediction reasonable? Explain.

95. *Saturated Steam*   The temperature T (in degrees Fahrenheit) of saturated steam increases as pressure increases. This relationship is approximated by the model

$$T = 75.82 - 2.11x + 43.51\sqrt{x}, \quad 5 \le x \le 40$$

where x is the absolute pressure (in pounds per square inch).

(a) Use the model to complete the table.

Absolute pressure, x	5	10	15	20
Temperature, T				

Absolute pressure, x	25	30	35	40
Temperature, T				

(b) The temperature of steam at sea level is 212°F. Use the table in part (a) to approximate the absolute pressure at this temperature.

(c) Solve part (b) algebraically.

 (d) Use a graphing utility to verify your solutions from parts (b) and (c).

96. *Airline Passengers*   An airline offers daily flights between Chicago and Denver. The total monthly cost C (in millions of dollars) of these flights is

$$C = \sqrt{0.2x + 1}$$

where x is the number of passengers (in thousands). The total cost of the flights for June is 2.5 million dollars. How many passengers flew in June?

97. *Demand*   The demand equation for a video game is modeled by

$$p = 40 - \sqrt{0.01x + 1}$$

where x is the number of units demanded per day and p is the price per unit. Approximate the demand when the price is $37.55.

98. *Demand*   The demand equation for a high definition television set is modeled by

$$p = 800 - \sqrt{0.01x + 1}$$

where x is the number of units demanded per month and p is the price per unit. Approximate the demand when the price is $750.

99. *Baseball*   A baseball diamond has the shape of a square in which the distance from home plate to second base is approximately $127\frac{1}{2}$ feet. Approximate the distance between the bases.

100. *Meteorology*   A meteorologist is positioned 100 feet from the point where a weather balloon is launched. When the balloon is at height h, the distance d (in feet) between the meteorologist and the balloon is $d = \sqrt{100^2 + h^2}$.

 (a) Use a graphing utility to graph the equation. Use the *trace* feature to approximate the value of h when $d = 200$.

(b) Complete the table. Use the table to approximate the value of h when $d = 200$.

h	160	165	170	175	180	185
d						

(c) Find h algebraically when $d = 200$.

(d) Compare the results of each method. In each case, what information did you gain that wasn't apparent in another solution method?

101. *Geometry*   You construct a cone with a base radius of 8 inches. The surface area S of the cone can be represented by the equation

$$S = 8\pi\sqrt{64 + h^2}$$

where h is the height of the cone.

 (a) Use a graphing utility to graph the equation. Use the *trace* feature to approximate the value of h when $S = 350$ square inches.

(b) Complete the table. Use the table to approximate the value of h when $S = 350$.

h	8	9	10	11	12	13
S						

(c) Find h algebraically when $S = 350$.

(d) Compare the results of each method. In each case, what information did you gain that wasn't apparent in another solution method?

102. *Labor*   Working together, two people can complete a task in 8 hours. Working alone, one person takes 2 hours longer than the other to complete the task. How long would it take for each person to complete the task?

103. *Labor*   Working together, two people can complete a task in 12 hours. Working alone, one person takes 3 hours longer than the other to complete the task. How long would it take for each person to complete the task?

Model It

104. *Power Line*   A power station is on one side of a river that is $\frac{3}{4}$ mile wide, and a factory is 8 miles downstream on the other side of the river, as shown in the figure. It costs $24 per foot to run power lines over land and $30 per foot to run them under water.

(a) Write the total cost C to run power lines in terms of x (see figure).

(b) Find the total cost when $x = 3$.

(c) Find the length x when $C = \$1,098,662.40$.

 (d) Use a graphing utility to graph the equation from part (a).

 (e) Use your graph from part (d) to find the value of x that minimizes the cost.

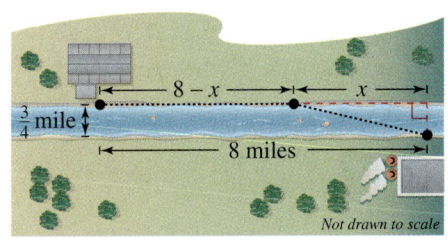

Not drawn to scale

In Exercises 105 and 106, solve for the indicated variable.

105. *A Person's Tangential Speed in a Rotor*

Solve for g: $v = \sqrt{\dfrac{gR}{\mu s}}$

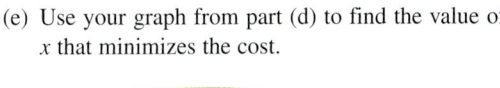

106. *Inductance*

Solve for Q: $i = \pm\sqrt{\dfrac{1}{LC}}\sqrt{Q^2 - q}$

Synthesis

True or False?   **In Exercises 107 and 108, determine whether the statement is true or false. Justify your answer.**

107. An equation can never have more than one extraneous solution.

108. When solving an absolute value equation, you will always have to check more than one solution.

In Exercises 109 and 110, find x such that the distance between the given points is 13. Explain your results.

109. $(1, 2), (x, -10)$

110. $(-8, 0), (x, 5)$ = distance formula

In Exercises 111 and 112, find y such that the distance between the given points is 17. Explain your results.

111. $(0, 0), (8, y)$

112. $(-8, 4), (7, y)$

In Exercises 113 and 114, consider an equation of the form $x + |x - a| = b$, where a and b are constants.

113. Find a and b when the solution of the equation is $x = 9$. (There are many correct answers.)

114. *Writing*   Write a short paragraph listing the steps required to solve this equation involving absolute values and explain why it is important to check your solutions.

In Exercises 115 and 116, consider an equation of the form $x + \sqrt{x - a} = b$, where a and b are constants.

115. Find a and b when the solution of the equation is $x = 20$. (There are many correct answers.)

116. *Writing*   Write a short paragraph listing the steps required to solve this equation involving radicals and explain why it is important to check your solutions.

Skills Review

In Exercises 117–120, perform the operation and simplify.

117. $\dfrac{8}{3x} + \dfrac{3}{2x}$

118. $\dfrac{2}{x^2 - 4} - \dfrac{1}{x^2 - 3x + 2}$

119. $\dfrac{2}{z + 2} - \left(3 - \dfrac{2}{z}\right)$

120. $25y^2 \div \dfrac{xy}{5}$

In Exercises 121 and 122, find all real solutions of the equation.

121. $x^2 - 22x + 121 = 0$

122. $x(x - 20) + 3(x - 20) = 0$

1.7   Linear Inequalities in One Variable

What you should learn

- Represent solutions of linear inequalities in one variable.
- Solve linear inequalities in one variable.
- Solve inequalities involving absolute values.
- Use inequalities to model and solve real-life problems.

Why you should learn it

Inequalities can be used to model and solve real-life problems. For instance, in Exercise 101 on page 152, you will use a linear inequality to analyze the average salary for elementary school teachers.

© Jose Luis Pelaez, Inc./Corbis

Introduction

Simple inequalities were reviewed in Section P.1. There, you used the inequality symbols $<$, $\leq$, $>$, and $\geq$ to compare two numbers and to denote subsets of real numbers. For instance, the simple inequality

$$x \geq 3$$

denotes all real numbers x that are greater than or equal to 3.

In this section, you will expand your work with inequalities to include more involved statements such as

$$5x - 7 < 3x + 9$$

and

$$-3 \leq 6x - 1 < 3.$$

As with an equation, you **solve an inequality** in the variable x by finding all values of x for which the inequality is true. Such values are **solutions** and are said to **satisfy** the inequality. The set of all real numbers that are solutions of an inequality is the **solution set** of the inequality. For instance, the solution set of

$$x + 1 < 4$$

is all real numbers that are less than 3.

The set of all points on the real number line that represent the solution set is the **graph of the inequality.** Graphs of many types of inequalities consist of intervals on the real number line. See Section P.1 to review the nine basic types of intervals on the real number line. Note that each type of interval can be classified as *bounded* or *unbounded*.

Example 1   Intervals and Inequalities

Write an inequality to represent each interval, and state whether the interval is bounded or unbounded.

a. $(-3, 5]$

b. $(-3, \infty)$

c. $[0, 2]$

d. $(-\infty, \infty)$

Solution

a. $(-3, 5]$ corresponds to $-3 < x \leq 5$.   Bounded

b. $(-3, \infty)$ corresponds to $-3 < x$.   Unbounded

c. $[0, 2]$ corresponds to $0 \leq x \leq 2$.   Bounded

d. $(-\infty, \infty)$ corresponds to $-\infty < x < \infty$.   Unbounded

✓CHECKPOINT   Now try Exercise 1.

Properties of Inequalities

The procedures for solving linear inequalities in one variable are much like those for solving linear equations. To isolate the variable, you can make use of the **Properties of Inequalities.** These properties are similar to the properties of equality, but there are two important exceptions. When each side of an inequality is multiplied or divided by a negative number, the direction of the inequality symbol must be reversed. Here is an example.

$$-2 < 5 \qquad \text{Original inequality}$$

$$(-3)(-2) > (-3)(5) \qquad \text{Multiply each side by } -3 \text{ and reverse inequality.}$$

$$6 > -15 \qquad \text{Simplify.}$$

Notice that if the inequality was not reversed you would obtain the false statement $6 < -15$.

Two inequalities that have the same solution set are **equivalent.** For instance, the inequalities

$$x + 2 < 5$$

and

$$x < 3$$

are equivalent. To obtain the second inequality from the first, you can subtract 2 from each side of the inequality. The following list describes the operations that can be used to create equivalent inequalities.

Properties of Inequalities

Let a, b, c, and d be real numbers.

1. Transitive Property

$$a < b \text{ and } b < c \quad \Longrightarrow \quad a < c$$

2. Addition of Inequalities

$$a < b \text{ and } c < d \quad \Longrightarrow \quad a + c < b + d$$

3. Addition of a Constant

$$a < b \quad \Longrightarrow \quad a + c < b + c$$

4. Multiplication by a Constant

$$\text{For } c > 0, a < b \quad \Longrightarrow \quad ac < bc$$

$$\text{For } c < 0, a < b \quad \Longrightarrow \quad ac > bc \qquad \text{Reverse the inequality.}$$

Each of the properties above is true if the symbol $<$ is replaced by $\leq$ and the symbol $>$ is replaced by $\geq$. For instance, another form of the multiplication property would be as follows.

$$\text{For } c > 0, a \leq b \quad \Longrightarrow \quad ac \leq bc$$

$$\text{For } c < 0, a \leq b \quad \Longrightarrow \quad ac \geq bc$$

Solving a Linear Inequality in One Variable

The simplest type of inequality is a **linear inequality** in one variable. For instance, $2x + 3 > 4$ is a linear inequality in x.

In the following examples, pay special attention to the steps in which the inequality symbol is reversed. Remember that when you multiply or divide by a negative number, you must reverse the inequality symbol.

Example 2 Solving Linear Inequalities

Solve each inequality.

a. $5x - 7 > 3x + 9$

b. $1 - \dfrac{3x}{2} \geq x - 4$

STUDY TIP

Checking the solution set of an inequality is not as simple as checking the solutions of an equation. You can, however, get an indication of the validity of a solution set by substituting a few convenient values of x.

Solution

a. $5x - 7 > 3x + 9$ Write original inequality.

$2x - 7 > 9$ Subtract $3x$ from each side.

$2x > 16$ Add 7 to each side.

$x > 8$ Divide each side by 2.

The solution set is all real numbers that are greater than 8, which is denoted by $(8, \infty)$. The graph of this solution set is shown in Figure 1.23. Note that a parenthesis at 8 on the real number line indicates that 8 *is not* part of the solution set.

Solution interval: $(8, \infty)$

FIGURE 1.23

b. $1 - \dfrac{3x}{2} \geq x - 4$ Write original inequality.

$2 - 3x \geq 2x - 8$ Multiply each side by 2.

$2 - 5x \geq -8$ Subtract $2x$ from each side.

$-5x \geq -10$ Subtract 2 from each side.

$x \leq 2$ Divide each side by -5 and reverse the inequality.

The solution set is all real numbers that are less than or equal to 2, which is denoted by $(-\infty, 2]$. The graph of this solution set is shown in Figure 1.24. Note that a bracket at 2 on the real number line indicates that 2 *is* part of the solution set.

Solution interval: $(-\infty, 2]$

FIGURE 1.24

CHECKPOINT Now try Exercise 25.

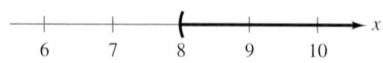

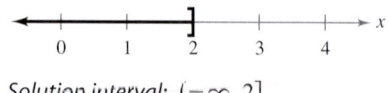

Sometimes it is possible to write two inequalities as a **double inequality.** For instance, you can write the two inequalities $-4 \le 5x - 2$ and $5x - 2 < 7$ more simply as

$$-4 \le 5x - 2 < 7.$$ Double inequality

This form allows you to solve the two inequalities together, as demonstrated in Example 3.

Example 3   Solving a Double Inequality

To solve a double inequality, you can isolate x as the middle term.

$$-3 \le 6x - 1 < 3$$ Original inequality

$$-3 + 1 \le 6x - 1 + 1 < 3 + 1$$ Add 1 to each part.

$$-2 \le 6x < 4$$ Simplify.

$$\frac{-2}{6} \le \frac{6x}{6} < \frac{4}{6}$$ Divide each part by 6.

$$-\frac{1}{3} \le x < \frac{2}{3}$$ Simplify.

The solution set is all real numbers that are greater than or equal to $-\frac{1}{3}$ and less than $\frac{2}{3}$, which is denoted by $\left[-\frac{1}{3}, \frac{2}{3}\right)$. The graph of this solution set is shown in Figure 1.25.

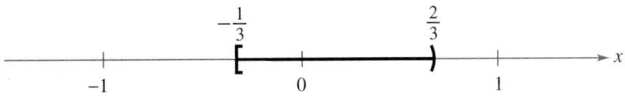

Solution interval: $\left[-\frac{1}{3}, \frac{2}{3}\right)$

FIGURE **1.**25

✓CHECKPOINT   Now try Exercise 37.

The double inequality in Example 3 could have been solved in two parts as follows.

$$-3 \le 6x - 1 \qquad \text{and} \qquad 6x - 1 < 3$$

$$-2 \le 6x \qquad\qquad\qquad 6x < 4$$

$$-\frac{1}{3} \le x \qquad\qquad\qquad x < \frac{2}{3}$$

The solution set consists of all real numbers that satisfy *both* inequalities. In other words, the solution set is the set of all values of x for which

$$-\frac{1}{3} \le x < \frac{2}{3}.$$

When combining two inequalities to form a double inequality, be sure that the inequalities satisfy the Transitive Property. For instance, it is *incorrect* to combine the inequalities $3 < x$ and $x \le -1$ as $3 < x \le -1$. This "inequality" is wrong because 3 is not less than -1.

Technology

A graphing utility can be used to identify the solution set of the graph of an inequality. For instance, to find the solution set of $|x - 5| < 2$ (see Example 4), rewrite the inequality as $|x - 5| - 2 < 0$, enter

$$Y1 = abs\,(X - 5) - 2,$$

and press the *graph* key. The graph should look like the one shown below.

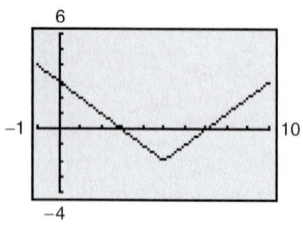

Notice that the graph is below the *x*-axis on the interval (3, 7).

Inequalities Involving Absolute Values

Solving an Absolute Value Inequality

Let x be a variable or an algebraic expression and let a be a real number such that $a \geq 0$.

1. The solutions of $|x| < a$ are all values of x that lie between $-a$ and a.

$$|x| < a \quad \text{if and only if} \quad -a < x < a. \qquad \text{Double inequality}$$

2. The solutions of $|x| > a$ are all values of x that are less than $-a$ or greater than a.

$$|x| > a \quad \text{if and only if} \quad x < -a \quad \text{or} \quad x > a. \qquad \text{Compound inequality}$$

These rules are also valid if < is replaced by ≤ and > is replaced by ≥.

Example 4   Solving an Absolute Value Inequality

Solve each inequality.

a. $|x - 5| < 2$ **b.** $|x + 3| \geq 7$

Solution

a.
$$|x - 5| < 2 \qquad \text{Write original inequality.}$$
$$-2 < x - 5 < 2 \qquad \text{Write equivalent inequalities.}$$
$$-2 + 5 < x - 5 + 5 < 2 + 5 \qquad \text{Add 5 to each part.}$$
$$3 < x < 7 \qquad \text{Simplify.}$$

The solution set is all real numbers that are greater than 3 and less than 7, which is denoted by (3, 7). The graph of this solution set is shown in Figure 1.26.

b.
$$|x + 3| \geq 7 \qquad \text{Write original inequality.}$$
$$x + 3 \leq -7 \quad \text{or} \quad x + 3 \geq 7 \qquad \text{Write equivalent inequalities.}$$
$$x + 3 - 3 \leq -7 - 3 \qquad x + 3 - 3 \geq 7 - 3 \qquad \text{Subtract 3 from each side.}$$
$$x \leq -10 \qquad x \geq 4 \qquad \text{Simplify.}$$

The solution set is all real numbers that are less than or equal to -10 *or* greater than or equal to 4. The interval notation for this solution set is $(-\infty, -10] \cup [4, \infty)$. The symbol $\cup$ is called a *union* symbol and is used to denote the combining of two sets. The graph of this solution set is shown in Figure 1.27.

STUDY TIP

Note that the graph of the inequality $|x - 5| < 2$ can be described as all real numbers *within* two units of 5, as shown in Figure 1.26.

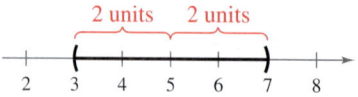

$|x - 5| < 2$: Solutions lie inside (3, 7)

FIGURE **1.26**

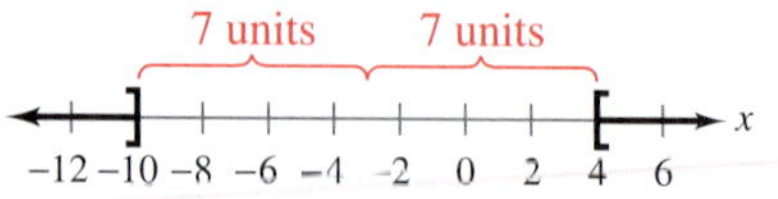

$|x + 3| \geq 7$: Solutions lie outside $(-10, 4)$

FIGURE **1.27**

✓**CHECKPOINT**   Now try Exercise 49.

Applications

The problem-solving plan described in Section 1.3 can be used to model and solve real-life problems that involve inequalities, as illustrated in Example 5.

Example 5   Comparative Shopping

You are choosing between two different cell phone plans. Plan A costs $49.99 per month for 500 minutes plus $0.40 for each additional minute. Plan B costs $45.99 per month for 500 minutes plus $0.45 for each additional minute. How many *additional* minutes must you use in one month for plan B to cost more than plan A?

Solution

Verbal Model: Monthly cost for plan B > Monthly cost for plan A

Labels: Minutes used (over 500) in one month $= m$ (minutes)
Monthly cost for plan A $= 0.40m + 49.99$ (dollars)
Monthly cost for plan B $= 0.45m + 45.99$ (dollars)

Inequality: $0.45m + 45.99 > 0.40m + 49.99$

$$0.05m > 4$$

$$m > 80 \text{ minutes}$$

Plan B costs more if you use more than 80 additional minutes in one month.

✔CHECKPOINT   Now try Exercise 91.

Example 6   Accuracy of a Measurement

You go to a candy store to buy chocolates that cost $9.89 per pound. The scale that is used in the store has a state seal of approval that indicates the scale is accurate to within half an ounce (or $\frac{1}{32}$ of a pound). According to the scale, your purchase weighs one-half pound and costs $4.95. How much might you have been undercharged or overcharged as a result of inaccuracy in the scale?

Solution

Let x represent the *true* weight of the candy. Because the scale is accurate to within half an ounce (or $\frac{1}{32}$ of a pound), the difference between the exact weight (x) and the scale weight $\left(\frac{1}{2}\right)$ is less than or equal to $\frac{1}{32}$ of a pound. That is, $\left| x - \frac{1}{2} \right| \leq \frac{1}{32}$. You can solve this inequality as follows.

$$-\tfrac{1}{32} \leq x - \tfrac{1}{2} \leq \tfrac{1}{32}$$

$$\tfrac{15}{32} \leq x \leq \tfrac{17}{32}$$

$$0.46875 \leq x \leq 0.53125$$

In other words, your "one-half pound" of candy could have weighed as little as 0.46875 pound (which would have cost $4.64) or as much as 0.53125 pound (which would have cost $5.25). So, you could have been overcharged by as much as $0.31 or undercharged by as much as $0.30.

✔CHECKPOINT   Now try Exercise 105.

1.7 Exercises

VOCABULARY CHECK: Fill in the blanks.

1. The set of all real numbers that are solutions to an inequality is the _____ _____ of the inequality.

2. The set of all points on the real number line that represent the solution set of an inequality is the _____ of the inequality.

3. To solve a linear inequality in one variable, you can use the properties of inequalities, which are identical to those used to solve equations, with the exception of multiplying or dividing each side by a _____ number.

4. Two inequalities that have the same solution set are _____ _____.

5. It is sometimes possible to write two inequalities as one inequality, called a _____ inequality.

6. The symbol $\cup$ is called a _____ symbol and is used to denote the combining of two sets.

PREREQUISITE SKILLS REVIEW: Practice and review algebra skills needed for this section at **www.Eduspace.com.**

In Exercises 1–6, (a) write an inequality that represents the interval and (b) state whether the interval is bounded or unbounded.

1. $[-1, 5]$
2. $(2, 10]$
3. $(11, \infty)$
4. $[-5, \infty)$
5. $(-\infty, -2)$
6. $(-\infty, 7]$

In Exercises 7–12, match the inequality with its graph. [The graphs are labeled (a), (b), (c), (d), (e), and (f).]

(a)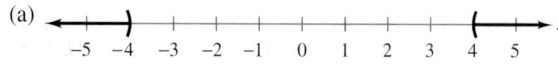

(b)

(c)

(d)

(e)

(f)

7. $x < 3$
8. $x \geq 5$
9. $-3 < x \leq 4$
10. $0 \leq x \leq \frac{9}{2}$
11. $|x| < 3$
12. $|x| > 4$

In Exercises 13–18, determine whether each value of x is a solution of the inequality.

Inequality	Values
13. $5x - 12 > 0$	(a) $x = 3$ (b) $x = -3$
	(c) $x = \frac{5}{2}$ (d) $x = \frac{3}{2}$
14. $2x + 1 < -3$	(a) $x = 0$ (b) $x = -\frac{1}{4}$
	(c) $x = -4$ (d) $x = -\frac{3}{2}$

Inequality	Values
15. $0 < \dfrac{x-2}{4} < 2$	(a) $x = 4$ (b) $x = 10$
	(c) $x = 0$ (d) $x = \frac{7}{2}$
16. $-1 < \dfrac{3-x}{2} \leq 1$	(a) $x = 0$ (b) $x = -5$
	(c) $x = 1$ (d) $x = 5$
17. $|x - 10| \geq 3$	(a) $x = 13$ (b) $x = -1$
	(c) $x = 14$ (d) $x = 9$
18. $|2x - 3| < 15$	(a) $x = -6$ (b) $x = 0$
	(c) $x = 12$ (d) $x = 7$

In Exercises 19–44, solve the inequality and sketch the solution on the real number line. (Some inequalities have no solutions.)

19. $4x < 12$
20. $10x < -40$
21. $-2x > -3$
22. $-6x > 15$
23. $x - 5 \geq 7$
24. $x + 7 \leq 12$
25. $2x + 7 < 3 + 4x$
26. $3x + 1 \geq 2 + x$
27. $2x - 1 \geq 1 - 5x$
28. $6x - 4 \leq 2 + 8x$
29. $4 - 2x < 3(3 - x)$
30. $4(x + 1) < 2x + 3$
31. $\frac{3}{4}x - 6 \leq x - 7$
32. $3 + \frac{2}{7}x > x - 2$
33. $\frac{1}{2}(8x + 1) \geq 3x + \frac{5}{2}$
34. $9x - 1 < \frac{3}{4}(16x - 2)$
35. $3.6x + 11 \geq -3.4$
36. $15.6 - 1.3x < -5.2$
37. $1 < 2x + 3 < 9$
38. $-8 \leq -(3x + 5) < 13$
39. $-4 < \dfrac{2x - 3}{3} < 4$
40. $0 \leq \dfrac{x + 3}{2} < 5$
41. $\frac{3}{4} > x + 1 > \frac{1}{4}$
42. $-1 < 2 - \dfrac{x}{3} < 1$
43. $3.2 \leq 0.4x - 1 \leq 4.4$
44. $4.5 > \dfrac{1.5x + 6}{2} > 10.5$

In Exercises 45–60, solve the inequality and sketch the solution on the real number line. (Some inequalities have no solution.)

45. $|x| < 6$

46. $|x| > 4$

47. $\left|\dfrac{x}{2}\right| > 1$

48. $\left|\dfrac{x}{5}\right| > 3$

49. $|x - 5| < -1$

50. $|x - 7| < -5$

51. $|x - 20| \le 6$

52. $|x - 8| \ge 0$

53. $|3 - 4x| \ge 9$

54. $|1 - 2x| < 5$

55. $\left|\dfrac{x - 3}{2}\right| \ge 4$

56. $\left|1 - \dfrac{2x}{3}\right| < 1$

57. $|9 - 2x| - 2 < -1$

58. $|x + 14| + 3 > 17$

59. $2|x + 10| \ge 9$

60. $3|4 - 5x| \le 9$

 Graphical Analysis   In Exercises 61–68, use a graphing utility to graph the inequality and identify the solution set.

61. $6x > 12$

62. $3x - 1 \le 5$

63. $5 - 2x \ge 1$

64. $3(x + 1) < x + 7$

65. $|x - 8| \le 14$

66. $|2x + 9| > 13$

67. $2|x + 7| \ge 13$

68. $\frac{1}{2}|x + 1| \le 3$

 Graphical Analysis   In Exercises 69–74, use a graphing utility to graph the equation. Use the graph to approximate the values of *x* that satisfy each inequality.

Equation	Inequalities	
69. $y = 2x - 3$	(a) $y \ge 1$	(b) $y \le 0$
70. $y = \frac{2}{3}x + 1$	(a) $y \le 5$	(b) $y \ge 0$
71. $y = -\frac{1}{2}x + 2$	(a) $0 \le y \le 3$	(b) $y \ge 0$
72. $y = -3x + 8$	(a) $-1 \le y \le 3$	(b) $y \le 0$
73. $y = |x - 3|$	(a) $y \le 2$	(b) $y \ge 4$
74. $y = \left|\frac{1}{2}x + 1\right|$	(a) $y \le 4$	(b) $y \ge 1$

In Exercises 75–80, find the interval(s) on the real number line for which the radicand is nonnegative.

75. $\sqrt{x - 5}$

76. $\sqrt{x - 10}$

77. $\sqrt{x + 3}$

78. $\sqrt{3 - x}$

79. $\sqrt[4]{7 - 2x}$

80. $\sqrt[4]{6x + 15}$

81. *Think About It*   The graph of $|x - 5| < 3$ can be described as all real numbers within three units of 5. Give a similar description of $|x - 10| < 8$.

82. *Think About It*   The graph of $|x - 2| > 5$ can be described as all real numbers more than five units from 2. Give a similar description of $|x - 8| > 4$.

In Exercises 83–90, use absolute value notation to define the interval (or pair of intervals) on the real number line.

83.

84.

85.

86.

87. All real numbers within 10 units of 12

88. All real numbers at least five units from 8

89. All real numbers more than four units from -3

90. All real numbers no more than seven units from -6

91. *Checking Account*   You can choose between two types of checking accounts at your local bank. Type A charges a monthly service fee of $6 plus $0.25 for each check written. Type B charges a monthly service fee of $4.50 plus $0.50 for each check written. How many checks must you write in a month in order for the monthly charges for type A to be less than that for type B?

92. *Copying Costs*   Your department sends its copying to the photocopy center of your company. The center bills your department $0.10 per page. You have investigated the possibility of buying a departmental copier for $3000. With your own copier, the cost per page would be $0.03. The expected life of the copier is 4 years. How many copies must you make in the four-year period to justify buying the copier?

93. *Investment*   In order for an investment of $1000 to grow to more than $1062.50 in 2 years, what must the annual interest rate be? $[A = P(1 + rt)]$

94. *Investment*   In order for an investment of $750 to grow to more than $825 in 2 years, what must the annual interest rate be? $[A = P(1 + rt)]$

95. *Cost, Revenue, and Profit*   The revenue for selling *x* units of a product is $R = 115.95x$. The cost of producing *x* units is

$$C = 95x + 750.$$

To obtain a profit, the revenue must be greater than the cost. For what values of *x* will this product return a profit?

96. *Cost, Revenue, and Profit*   The revenue for selling *x* units of a product is $R = 24.55x$. The cost of producing *x* units is

$$C = 15.4x + 150,000.$$

To obtain a profit, the revenue must be greater than the cost. For what values of *x* will this product return a profit?

97. **Daily Sales** A doughnut shop sells a dozen doughnuts for $2.95. Beyond the fixed costs (rent, utilities, and insurance) of $150 per day, it costs $1.45 for enough materials (flour, sugar, and so on) and labor to produce a dozen doughnuts. The daily profit from doughnut sales varies between $50 and $200. Between what levels (in dozens) do the daily sales vary?

98. **Weight Loss Program** A person enrolls in a diet and exercise program that guarantees a loss of at least $1\frac{1}{2}$ pounds per week. The person's weight at the beginning of the program is 164 pounds. Find the maximum number of weeks before the person attains a goal weight of 128 pounds.

99. **Data Analysis: IQ Scores and GPA** The admissions office of a college wants to determine whether there is a relationship between IQ scores x and grade-point averages y after the first year of school. An equation that models the data the admissions office obtained is

$$y = 0.067x - 5.638.$$

(a) Use a graphing utility to graph the model.

(b) Use the graph to estimate the values of x that predict a grade-point average of at least 3.0.

Model It

100. **Data Analysis: Weightlifting** You want to determine whether there is a relationship between an athlete's weight x (in pounds) and the athlete's maximum bench-press weight y (in pounds). The table shows a sample of data from 12 athletes.

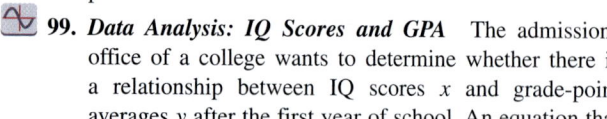

Athlete's weight, x	Bench-press weight, y
165	170
184	185
150	200
210	255
196	205
240	295
202	190
170	175
185	195
190	185
230	250
160	155

(a) Use a graphing utility to plot the data.

Model It (continued)

(b) A model for the data is $y = 1.3x - 36$. Use a graphing utility to graph the model in the same viewing window used in part (a).

(c) Use the graph to estimate the values of x that predict a maximum bench-press weight of at least 200 pounds.

(d) Verify your estimate from part (c) algebraically.

(e) Use the graph to write a statement about the accuracy of the model. If you think the graph indicates that an athlete's weight is not a particularly good indicator of the athlete's maximum bench-press weight, list other factors that might influence an individual's maximum bench-press weight.

101. **Teachers' Salaries** The average salary S (in thousands of dollars) for elementary school teachers in the United States from 1990 to 2002 is approximated by the model

$$S = 1.05t + 31.0, \qquad 0 \le t \le 12$$

where t represents the year, with $t = 0$ corresponding to 1990. (Source: National Education Association)

(a) According to this model, when was the average salary at least $32,000, but not more than $42,000?

(b) According to this model, when will the average salary exceed $48,000?

102. **Egg Production** The number of eggs E (in billions) produced in the United States from 1990 to 2002 can be modeled by

$$E = 1.64t + 67.2, \qquad 0 \le t \le 12$$

where t represents the year, with $t = 0$ corresponding to 1990. (Source: U.S. Department of Agriculture)

(a) According to this model, when was the annual egg production 70 billion, but no more than 80 billion?

(b) According to this model, when will the annual egg production exceed 95 billion?

103. **Geometry** The side of a square is measured as 10.4 inches with a possible error of $\frac{1}{16}$ inch. Using these measurements, determine the interval containing the possible areas of the square.

104. **Geometry** The side of a square is measured as 24.2 centimeters with a possible error of 0.25 centimeter. Using these measurements, determine the interval containing the possible areas of the square.

105. **Accuracy of Measurement** You stop at a self-service gas station to buy 15 gallons of 87-octane gasoline at $1.89 a gallon. The gas pump is accurate to within $\frac{1}{10}$ of a gallon. How much might you be undercharged or overcharged?

106. *Accuracy of Measurement* You buy six T-bone steaks that cost $14.99 per pound. The weight that is listed on the package is 5.72 pounds. The scale that weighed the package is accurate to within $\frac{1}{2}$ ounce. How much might you be undercharged or overcharged?

107. *Time Study* A time study was conducted to determine the length of time required to perform a particular task in a manufacturing process. The times required by approximately two-thirds of the workers in the study satisfied the inequality

$$\left| \frac{t - 15.6}{1.9} \right| < 1$$

where t is time in minutes. Determine the interval on the real number line in which these times lie.

108. *Height* The heights h of two-thirds of the members of a population satisfy the inequality

$$\left| \frac{h - 68.5}{2.7} \right| \le 1$$

where h is measured in inches. Determine the interval on the real number line in which these heights lie.

109. *Meteorology* An electronic device is to be operated in an environment with relative humidity h in the interval defined by $|h - 50| \le 30$. What are the minimum and maximum relative humidities for the operation of this device?

110. *Music* Michael Kasha of Florida State University used physics and mathematics to design a new classical guitar. He used the model for the frequency of the vibrations on a circular plate

$$v = \frac{2.6t}{d^2} \sqrt{\frac{E}{\rho}}$$

where v is the frequency (in vibrations per second), t is the plate thickness (in millimeters), d is the diameter of the plate, E is the elasticity of the plate material, and ρ is the density of the plate material. For fixed values of d, E, and ρ, the graph of the equation is a line (see figure).

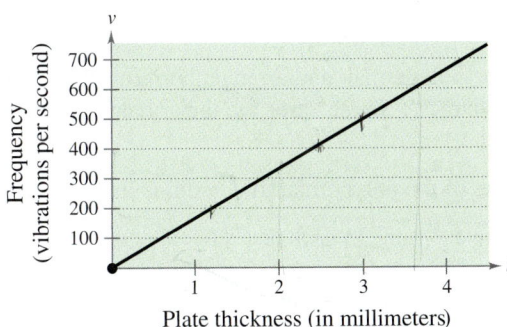

Plate thickness (in millimeters)

(a) Estimate the frequency when the plate thickness is 2 millimeters.

(b) Estimate the plate thickness when the frequency is 600 vibrations per second.

(c) Approximate the interval for the plate thickness when the frequency is between 200 and 400 vibrations per second.

(d) Approximate the interval for the frequency when the plate thickness is less than 3 millimeters.

Synthesis

True or False? **In Exercises 111 and 112, determine whether the statement is true or false. Justify your answer.**

111. If a, b, and c are real numbers, and $a \le b$, then $ac \le bc$.

112. If $-10 \le x \le 8$, then $-10 \ge -x$ and $-x \ge -8$.

113. Identify the graph of the inequality $|x - a| \ge 2$.

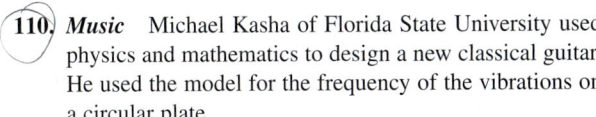

114. Find sets of values of a, b, and c such that $0 \le x \le 10$ is a solution of the inequality $|ax - b| \le c$.

Skills Review

In Exercises 115–118, find the distance between each pair of points. Then find the midpoint of the line segment joining the points.

115. $(-4, 2), (1, 12)$ **116.** $(1, -2), (10, 3)$

117. $(3, 6), (-5, -8)$ **118.** $(0, -3), (-6, 9)$

In Exercises 119–122, solve the equation.

119. $-6(2 - x) - 12 = 36$

120. $4(x + 7) - 9 = -6(-x - 1)$

121. $14x^2 + 5x - 1 = 0$

122. $x^3 + 5x^2 - 4x - 20 = 0$

123. Find the coordinates of the point located 3 units to the left of the y-axis and 10 units above the x-axis.

124. Determine the quadrant(s) in which the point (x, y) could be located if $y > 0$.

125. **Make a Decision** To work an extended application analyzing the number of heart disease deaths per 100,000 people in the United States, visit this text's website at *college.hmco.com*. *(Data Source: U.S. National Center for Health Statistics)*

1.8   Other Types of Inequalities

Polynomial Inequalities

To solve a polynomial inequality such as $x^2 - 2x - 3 < 0$, you can use the fact that a polynomial can change signs only at its zeros (the x-values that make the polynomial equal to zero). Between two consecutive zeros, a polynomial must be entirely positive or entirely negative. This means that when the real zeros of a polynomial are put in order, they divide the real number line into intervals in which the polynomial has no sign changes. These zeros are the **critical numbers** of the inequality, and the resulting intervals are the **test intervals** for the inequality. For instance, the polynomial above factors as

$$x^2 - 2x - 3 = (x + 1)(x - 3)$$

and has two zeros, $x = -1$ and $x = 3$. These zeros divide the real number line into three test intervals:

$$(-\infty, -1), \quad (-1, 3), \quad \text{and} \quad (3, \infty). \qquad \text{(See Figure 1.28.)}$$

So, to solve the inequality $x^2 - 2x - 3 < 0$, you need only test one value from each of these test intervals to determine whether the value satisfies the original inequality. If so, you can conclude that the interval is a solution of the inequality.

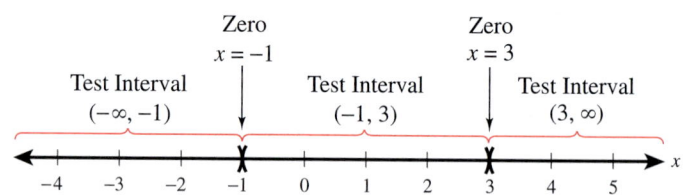

FIGURE **1.28**   *Three test intervals for $x^2 - 2x - 3$*

You can use the same basic approach to determine the test intervals for any polynomial.

Finding Test Intervals for a Polynomial

To determine the intervals on which the values of a polynomial are entirely negative or entirely positive, use the following steps.

1. Find all real zeros of the polynomial, and arrange the zeros in increasing order (from smallest to largest). These zeros are the critical numbers of the polynomial.

2. Use the critical numbers of the polynomial to determine its test intervals.

3. Choose one representative x-value in each test interval and evaluate the polynomial at that value. If the value of the polynomial is negative, the polynomial will have negative values for every x-value in the interval. If the value of the polynomial is positive, the polynomial will have positive values for every x-value in the interval.

Example 1   **Solving a Polynomial Inequality**

Solve

$$x^2 - x - 6 < 0.$$

Solution

By factoring the polynomial as

$$x^2 - x - 6 = (x + 2)(x - 3)$$

you can see that the critical numbers are $x = -2$ and $x = 3$. So, the polynomial's test intervals are

$$(-\infty, -2), \quad (-2, 3), \quad \text{and} \quad (3, \infty). \qquad \text{Test intervals}$$

In each test interval, choose a representative x-value and evaluate the polynomial.

Test Interval	x-Value	Polynomial Value	Conclusion
$(-\infty, -2)$	$x = -3$	$(-3)^2 - (-3) - 6 = 6$	Positive
$(-2, 3)$	$x = 0$	$(0)^2 - (0) - 6 = -6$	Negative
$(3, \infty)$	$x = 4$	$(4)^2 - (4) - 6 = 6$	Positive

From this you can conclude that the inequality is satisfied for all x-values in $(-2, 3)$. This implies that the solution of the inequality $x^2 - x - 6 < 0$ is the interval $(-2, 3)$, as shown in Figure 1.29. Note that the original inequality contains a less than symbol. This means that the solution set does not contain the endpoints of the test interval $(-2, 3)$.

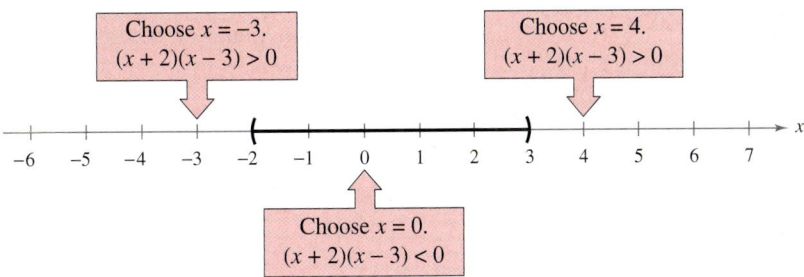

FIGURE **1.29**

✓**CHECKPOINT**   Now try Exercise 13.

As with linear inequalities, you can check the reasonableness of a solution by substituting x-values into the original inequality. For instance, to check the solution found in Example 1, try substituting several x-values from the interval $(-2, 3)$ into the inequality

$$x^2 - x - 6 < 0.$$

Regardless of which x-values you choose, the inequality should be satisfied.

You can also use a graph to check the result of Example 1. Sketch the graph of $y = x^2 - x - 6$, as shown in Figure 1.30. Notice that the graph is below the x-axis on the interval $(-2, 3)$.

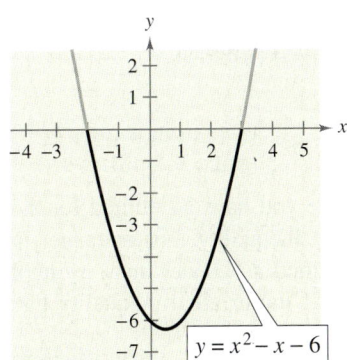

FIGURE **1.30**

In Example 1, the polynomial inequality was given in general form (with the polynomial on one side and zero on the other). Whenever this is not the case, you should begin the solution process by writing the inequality in general form.

Example 2 Solving a Polynomial Inequality

Solve $2x^3 - 3x^2 - 32x > -48$.

Solution

Begin by writing the inequality in general form.

$$2x^3 - 3x^2 - 32x > -48 \qquad \text{Write original inequality.}$$

$$2x^3 - 3x^2 - 32x + 48 > 0 \qquad \text{Write in general form.}$$

$$(x - 4)(x + 4)(2x - 3) > 0 \qquad \text{Factor.}$$

The critical numbers are $x = -4$, $x = \frac{3}{2}$, and $x = 4$, and the test intervals are $(-\infty, -4)$, $\left(-4, \frac{3}{2}\right)$, $\left(\frac{3}{2}, 4\right)$, and $(4, \infty)$.

Test Interval	x-Value	Polynomial Value	Conclusion
$(-\infty, -4)$	$x = -5$	$2(-5)^3 - 3(-5)^2 - 32(-5) + 48$	Negative
$\left(-4, \frac{3}{2}\right)$	$x = 0$	$2(0)^3 - 3(0)^2 - 32(0) + 48$	Positive
$\left(\frac{3}{2}, 4\right)$	$x = 2$	$2(2)^3 - 3(2)^2 - 32(2) + 48$	Negative
$(4, \infty)$	$x = 5$	$2(5)^3 - 3(5)^2 - 32(5) + 48$	Positive

From this you can conclude that the inequality is satisfied on the open intervals $\left(-4, \frac{3}{2}\right)$ and $(4, \infty)$. Therefore, the solution set consists of all real numbers in the intervals $\left(-4, \frac{3}{2}\right)$ and $(4, \infty)$, as shown in Figure 1.31.

STUDY TIP

You may find it easier to determine the sign of a polynomial from its *factored* form. For instance, in Example 2, if the test value $x = 2$ is substituted into the factored form

$$(x - 4)(x + 4)(2x - 3)$$

you can see that the sign pattern of the factors is

$$(-)(+)(+)$$

which yields a negative result. Try using the factored forms of the polynomials to determine the signs of the polynomials in the test intervals of the other examples in this section.

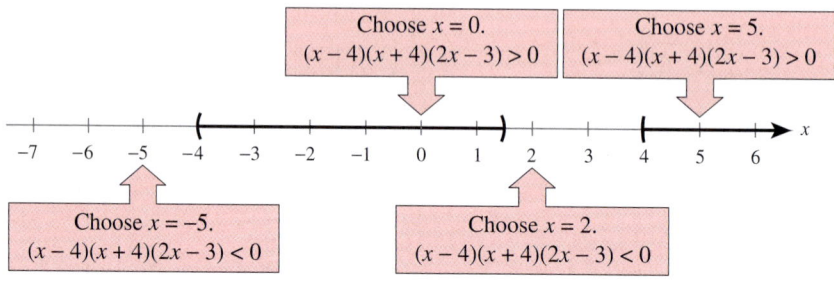

FIGURE 1.31

✔CHECKPOINT   Now try Exercise 21.

When solving a polynomial inequality, be sure you have accounted for the particular type of inequality symbol given in the inequality. For instance, in Example 2, note that the original inequality contained a "greater than" symbol and the solution consisted of two open intervals. If the original inequality had been

$$2x^3 - 3x^2 - 32x \geq -48$$

the solution would have consisted of the closed interval $\left[-4, \frac{3}{2}\right]$ and the interval $[4, \infty)$.

Each of the polynomial inequalities in Examples 1 and 2 has a solution set that consists of a single interval or the union of two intervals. When solving the exercises for this section, watch for unusual solution sets, as illustrated in Example 3.

Example 3   Unusual Solution Sets

a. The solution set of the following inequality consists of the entire set of real numbers, $(-\infty, \infty)$. In other words, the value of the quadratic $x^2 + 2x + 4$ is positive for every real value of x.

$$x^2 + 2x + 4 > 0$$

b. The solution set of the following inequality consists of the single real number $\{-1\}$, because the quadratic $x^2 + 2x + 1$ has only one critical number, $x = -1$, and it is the only value that satisfies the inequality.

$$x^2 + 2x + 1 \leq 0$$

c. The solution set of the following inequality is empty. In other words, the quadratic $x^2 + 3x + 5$ is not less than zero for any value of x.

$$x^2 + 3x + 5 < 0$$

d. The solution set of the following inequality consists of all real numbers except $x = 2$. In interval notation, this solution set can be written as $(-\infty, 2) \cup (2, \infty)$.

$$x^2 - 4x + 4 > 0$$

✓**CHECKPOINT**   Now try Exercise 25.

Exploration

You can use a graphing utility to verify the results in Example 3. For instance, the graph of $y = x^2 + 2x + 4$ is shown below. Notice that the y-values are greater than 0 for all values of x, as stated in Example 3(a). Use the graphing utility to graph the following:

$$y = x^2 + 2x + 1 \qquad y = x^2 + 3x + 5 \qquad y = x^2 - 4x + 4$$

Explain how you can use the graphs to verify the results of parts (b), (c), and (d) of Example 3.

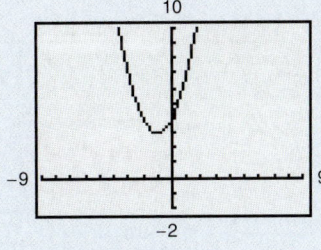

Rational Inequalities

The concepts of critical numbers and test intervals can be extended to rational inequalities. To do this, use the fact that the value of a rational expression can change sign only at its *zeros* (the x-values for which its numerator is zero) and its *undefined values* (the x-values for which its denominator is zero). These two types of numbers make up the *critical numbers* of a rational inequality. When solving a rational inequality, begin by writing the inequality in general form with the rational expression on the left and zero on the right.

Example 4 Solving a Rational Inequality

Solve $\dfrac{2x - 7}{x - 5} \le 3$.

Solution

$$\frac{2x - 7}{x - 5} \le 3 \qquad \text{Write original inequality.}$$

$$\frac{2x - 7}{x - 5} - 3 \le 0 \qquad \text{Write in general form.}$$

$$\frac{2x - 7 - 3x + 15}{x - 5} \le 0 \qquad \text{Find the LCD and add fractions.}$$

$$\frac{-x + 8}{x - 5} \le 0 \qquad \text{Simplify.}$$

Critical numbers: $x = 5, x = 8$ Zeros and undefined values of rational expression

Test intervals: $(-\infty, 5), (5, 8), (8, \infty)$

Test: Is $\dfrac{-x + 8}{x - 5} \le 0$?

After testing these intervals, as shown in Figure 1.32, you can see that the inequality is satisfied on the open intervals $(-\infty, 5)$ and $(8, \infty)$. Moreover, because $(-x + 8)/(x - 5) = 0$ when $x = 8$, you can conclude that the solution set consists of all real numbers in the intervals $(-\infty, 5) \cup [8, \infty)$. (Be sure to use a closed interval to indicate that x can equal 8.)

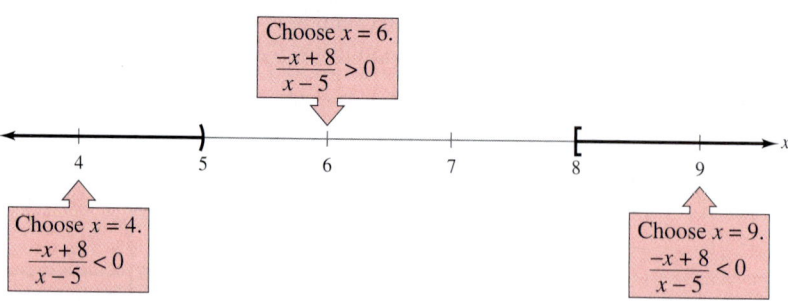

FIGURE **1.32**

✓CHECKPOINT   Now try Exercise 39.

Applications

One common application of inequalities comes from business and involves profit, revenue, and cost. The formula that relates these three quantities is

> Profit = Revenue − Cost
>
> $P = R - C.$

Example 5   Increasing the Profit for a Product

The marketing department of a calculator manufacturer has determined that the demand for a new model of calculator is

$$p = 100 - 0.00001x, \qquad 0 \le x \le 10{,}000{,}000 \qquad \text{\color{red}Demand equation}$$

where p is the price per calculator (in dollars) and x represents the number of calculators sold. (If this model is accurate, no one would be willing to pay \$100 for the calculator. At the other extreme, the company couldn't sell more than 10 million calculators.) The revenue for selling x calculators is

$$R = xp = x(100 - 0.00001x) \qquad \text{\color{red}Revenue equation}$$

as shown in Figure 1.33. The total cost of producing x calculators is \$10 per calculator plus a development cost of \$2,500,000. So, the total cost is

$$C = 10x + 2{,}500{,}000. \qquad \text{\color{red}Cost equation}$$

What price should the company charge per calculator to obtain a profit of at least \$190,000,000?

Solution

Verbal Model: Profit = Revenue − Cost

Equation: $P = R - C$

$$P = 100x - 0.00001x^2 - (10x + 2{,}500{,}000)$$

$$P = -0.00001x^2 + 90x - 2{,}500{,}000$$

To answer the question, solve the inequality

$$P \ge 190{,}000{,}000$$

$$-0.00001x^2 + 90x - 2{,}500{,}000 \ge 190{,}000{,}000.$$

When you write the inequality in general form, find the critical numbers and the test intervals, and then test a value in each test interval, you can find the solution to be

$$3{,}500{,}000 \le x \le 5{,}500{,}000$$

as shown in Figure 1.34. Substituting the x-values in the original price equation shows that prices of

$$\$45.00 \le p \le \$65.00$$

will yield a profit of at least \$190,000,000.

✓**CHECKPOINT**   Now try Exercise 71.

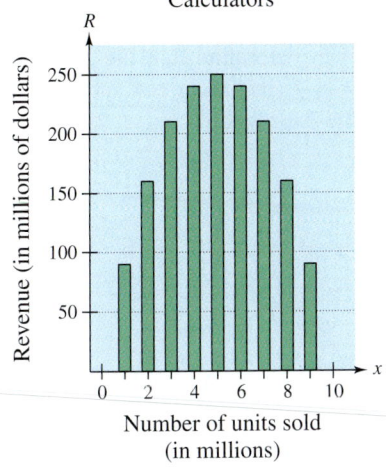

Calculators

FIGURE **1.33**

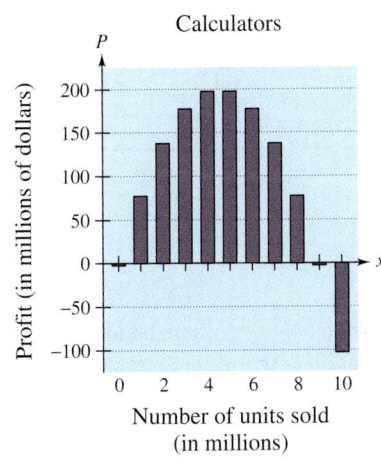

Calculators

FIGURE **1.34**

Another common application of inequalities is finding the domain of an expression that involves a square root, as shown in Example 6.

Example 6 Finding the Domain of an Expression

Find the domain of $\sqrt{64 - 4x^2}$.

Algebraic Solution

Remember that the domain of an expression is the set of all x-values for which the expression is defined. Because $\sqrt{64 - 4x^2}$ is defined (has real values) only if $64 - 4x^2$ is nonnegative, the domain is given by $64 - 4x^2 \geq 0$.

$$64 - 4x^2 \geq 0 \qquad \text{Write in general form.}$$

$$16 - x^2 \geq 0 \qquad \text{Divide each side by 4.}$$

$$(4 - x)(4 + x) \geq 0 \qquad \text{Write in factored form.}$$

So, the inequality has two critical numbers: $x = -4$ and $x = 4$. You can use these two numbers to test the inequality as follows.

Critical numbers: $\quad x = -4, x = 4$

Test intervals: $\quad (-\infty, -4), (-4, 4), (4, \infty)$

Test: $\qquad$ For what values of x is $\sqrt{64 - 4x^2} \geq 0$?

A test shows that the inequality is satisfied in the *closed interval* $[-4, 4]$. So, the domain of the expression $\sqrt{64 - 4x^2}$ is the interval $[-4, 4]$.

✓**CHECKPOINT** Now try Exercise 55.

Graphical Solution

Begin by sketching the graph of the equation $y = \sqrt{64 - 4x^2}$, as shown in Figure 1.35. From the graph, you can determine that the x-values extend from -4 to 4 (including -4 and 4). So, the domain of the expression $\sqrt{64 - 4x^2}$ is the interval $[-4, 4]$.

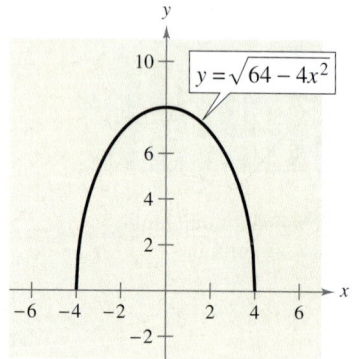

FIGURE **1.35**

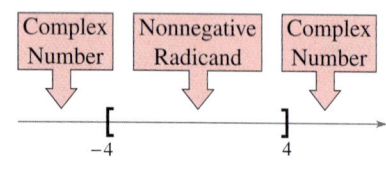

FIGURE **1.36**

To analyze a test interval, choose a representative x-value in the interval and evaluate the expression at that value. For instance, in Example 6, if you substitute any number from the interval $[-4, 4]$ into the expression $\sqrt{64 - 4x^2}$ you will obtain a nonnegative number under the radical symbol that simplifies to a real number. If you substitute any number from the intervals $(-\infty, -4)$ and $(4, \infty)$ you will obtain a complex number. It might be helpful to draw a visual representation of the intervals as shown in Figure 1.36.

𝒲RITING ABOUT 𝓜ATHEMATICS

Profit Analysis Consider the relationship

$$P = R - C$$

described on page 159. Write a paragraph discussing why it might be beneficial to solve $P < 0$ if you owned a business. Use the situation described in Example 5 to illustrate your reasoning.

1.8   Exercises

VOCABULARY CHECK: Fill in the blanks.

1. To solve a polynomial inequality, find the _____ numbers of the polynomial, and use these numbers to create _____ _____ for the inequality.

2. The critical numbers of a rational expression are its _____ and its _____ _____.

3. The formula that relates cost, revenue, and profit is _____.

PREREQUISITE SKILLS REVIEW: Practice and review algebra skills needed for this section at **www.Eduspace.com.**

In Exercises 1–4, determine whether each value of x is a solution of the inequality.

Inequality	Values
1. $x^2 - 3 < 0$	(a) $x = 3$ (b) $x = 0$
	(c) $x = \frac{3}{2}$ (d) $x = -5$
2. $x^2 - x - 12 \geq 0$	(a) $x = 5$ (b) $x = 0$
	(c) $x = -4$ (d) $x = -3$
3. $\dfrac{x + 2}{x - 4} \geq 3$	(a) $x = 5$ (b) $x = 4$
	(c) $x = -\frac{9}{2}$ (d) $x = \frac{9}{2}$
4. $\dfrac{3x^2}{x^2 + 4} < 1$	(a) $x = -2$ (b) $x = -1$
	(c) $x = 0$ (d) $x = 3$

In Exercises 5–8, find the critical numbers of the expression.

5. $2x^2 - x - 6$

6. $9x^3 - 25x^2$

7. $2 + \dfrac{3}{x - 5}$

8. $\dfrac{x}{x + 2} - \dfrac{2}{x - 1}$

In Exercises 9–26, solve the inequality and graph the solution on the real number line.

9. $x^2 \leq 9$

10. $x^2 < 36$

11. $(x + 2)^2 < 25$

12. $(x - 3)^2 \geq 1$

13. $x^2 + 4x + 4 \geq 9$

14. $x^2 - 6x + 9 < 16$

15. $x^2 + x < 6$

16. $x^2 + 2x > 3$

17. $x^2 + 2x - 3 < 0$

18. $x^2 - 4x - 1 > 0$

19. $x^2 + 8x - 5 \geq 0$

20. $-2x^2 + 6x + 15 \leq 0$

21. $x^3 - 3x^2 - x + 3 > 0$

22. $x^3 + 2x^2 - 4x - 8 \leq 0$

23. $x^3 - 2x^2 - 9x - 2 \geq -20$

24. $2x^3 + 13x^2 - 8x - 46 \geq 6$

25. $4x^2 - 4x + 1 \leq 0$

26. $x^2 + 3x + 8 > 0$

In Exercises 27–32, solve the inequality and write the solution set in interval notation.

27. $4x^3 - 6x^2 < 0$

28. $4x^3 - 12x^2 > 0$

29. $x^3 - 4x \geq 0$

30. $2x^3 - x^4 \leq 0$

31. $(x - 1)^2(x + 2)^3 \geq 0$

32. $x^4(x - 3) \leq 0$

Graphical Analysis   In Exercises 33–36, use a graphing utility to graph the equation. Use the graph to approximate the values of x that satisfy each inequality.

Equation	Inequalities
33. $y = -x^2 + 2x + 3$	(a) $y \leq 0$ (b) $y \geq 3$
34. $y = \frac{1}{2}x^2 - 2x + 1$	(a) $y \leq 0$ (b) $y \geq 7$
35. $y = \frac{1}{8}x^3 - \frac{1}{2}x$	(a) $y \geq 0$ (b) $y \leq 6$
36. $y = x^3 - x^2 - 16x + 16$	(a) $y \leq 0$ (b) $y \geq 36$

In Exercises 37–50, solve the inequality and graph the solution on the real number line.

37. $\dfrac{1}{x} - x > 0$

38. $\dfrac{1}{x} - 4 < 0$

39. $\dfrac{x + 6}{x + 1} - 2 < 0$

40. $\dfrac{x + 12}{x + 2} - 3 \geq 0$

41. $\dfrac{3x - 5}{x - 5} > 4$

42. $\dfrac{5 + 7x}{1 + 2x} < 4$

43. $\dfrac{4}{x + 5} > \dfrac{1}{2x + 3}$

44. $\dfrac{5}{x - 6} > \dfrac{3}{x + 2}$

45. $\dfrac{1}{x - 3} \leq \dfrac{9}{4x + 3}$

46. $\dfrac{1}{x} \geq \dfrac{1}{x + 3}$

47. $\dfrac{x^2 + 2x}{x^2 - 9} \leq 0$

48. $\dfrac{x^2 + x - 6}{x} \geq 0$

49. $\dfrac{5}{x - 1} - \dfrac{2x}{x + 1} < 1$

50. $\dfrac{3x}{x - 1} \leq \dfrac{x}{x + 4} + 3$

Graphical Analysis In Exercises 51–54, use a graphing utility to graph the equation. Use the graph to approximate the values of x that satisfy each inequality.

Equation	Inequalities
51. $y = \dfrac{3x}{x-2}$	(a) $y \le 0$ (b) $y \ge 6$
52. $y = \dfrac{2(x-2)}{x+1}$	(a) $y \le 0$ (b) $y \ge 8$
53. $y = \dfrac{2x^2}{x^2+4}$	(a) $y \ge 1$ (b) $y \le 2$
54. $y = \dfrac{5x}{x^2+4}$	(a) $y \ge 1$ (b) $y \le 0$

In Exercises 55–60, find the domain of x in the expression. Use a graphing utility to verify your result.

55. $\sqrt{4-x^2}$ **56.** $\sqrt{x^2-4}$

57. $\sqrt{x^2-7x+12}$ **58.** $\sqrt{144-9x^2}$

59. $\sqrt{\dfrac{x}{x^2-2x-35}}$ **60.** $\sqrt{\dfrac{x}{x^2-9}}$

In Exercises 61–66, solve the inequality. (Round your answers to two decimal places.)

61. $0.4x^2 + 5.26 < 10.2$

62. $-1.3x^2 + 3.78 > 2.12$

63. $-0.5x^2 + 12.5x + 1.6 > 0$

64. $1.2x^2 + 4.8x + 3.1 < 5.3$

65. $\dfrac{1}{2.3x-5.2} > 3.4$

66. $\dfrac{2}{3.1x-3.7} > 5.8$

67. *Height of a Projectile* A projectile is fired straight upward from ground level with an initial velocity of 160 feet per second.

(a) At what instant will it be back at ground level?

(b) When will the height exceed 384 feet?

68. *Height of a Projectile* A projectile is fired straight upward from ground level with an initial velocity of 128 feet per second.

(a) At what instant will it be back at ground level?

(b) When will the height be less than 128 feet?

69. *Geometry* A rectangular playing field with a perimeter of 100 meters is to have an area of at least 500 square meters. Within what bounds must the length of the rectangle lie?

70. *Geometry* A rectangular parking lot with a perimeter of 440 feet is to have an area of at least 8000 square feet. Within what bounds must the length of the rectangle lie?

71. *Cost, Revenue, and Profit* The revenue and cost equations for a product are

$$R = x(75 - 0.0005x) \quad \text{and} \quad C = 30x + 250,000$$

where R and C are measured in dollars and x represents the number of units sold. How many units must be sold to obtain a profit of at least $750,000? What is the price per unit?

72. *Cost, Revenue, and Profit* The revenue and cost equations for a product are

$$R = x(50 - 0.0002x) \quad \text{and} \quad C = 12x + 150,000$$

where R and C are measured in dollars and x represents the number of units sold. How many units must be sold to obtain a profit of at least $1,650,000? What is the price per unit?

Model It

73. *Cable Television* The percents C of households in the United States that owned a television and had cable from 1980 to 2003 can be modeled by

$$C = 0.0031t^3 - 0.216t^2 + 5.54t + 19.1, \quad 0 \le t \le 23$$

where t is the year, with $t = 0$ corresponding to 1980. (Source: Nielsen Media Research)

 (a) Use a graphing utility to graph the equation.

(b) Complete the table to determine the year in which the percent of households that own a television and have cable will exceed 75%.

t	24	26	28	30	32	34
C						

 (c) Use the *trace* feature of a graphing utility to verify your answer to part (b).

(d) Complete the table to determine the years during which the percent of households that own a television and have cable will be between 85% and 100%.

t	36	37	38	39	40	41	42	43
C								

 (e) Use the *trace* feature of a graphing utility to verify your answer to part (d).

(f) Explain why the model may give values greater than 100% even though such values are not reasonable.

$13.8 \leq L \leq 36.2$

74. Safe Load The maximum safe load uniformly distributed over a one-foot section of a two-inch-wide wooden beam is approximated by the model Load = $168.5d^2 - 472.1$, where d is the depth of the beam.

(a) Evaluate the model for $d = 4$, $d = 6$, $d = 8$, $d = 10$, and $d = 12$. Use the results to create a bar graph.

(b) Determine the minimum depth of the beam that will safely support a load of 2000 pounds.

75. Resistors When two resistors of resistances R_1 and R_2 are connected in parallel (see figure), the total resistance R satisfies the equation

$$\frac{1}{R} = \frac{1}{R_1} + \frac{1}{R_2}.$$

Find R_1 for a parallel circuit in which $R_2 = 2$ ohms and R must be at least 1 ohm.

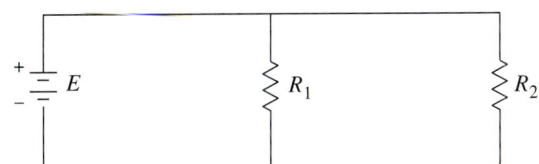

76. Education The numbers N (in thousands) of master's degrees earned by women in the United States from 1990 to 2002 are approximated by the model

$$N = -0.03t^2 + 9.6t + 172$$

where t represents the year, with $t = 0$ corresponding to 1990 (see figure). (Source: U.S. National Center for Education Statistics)

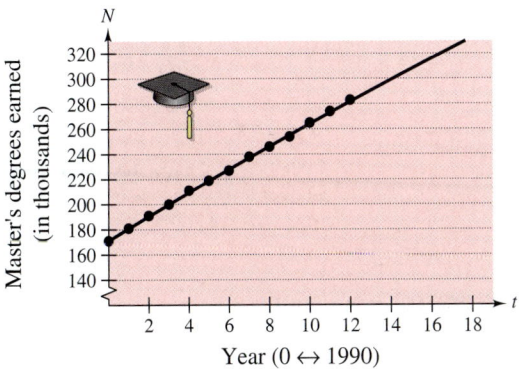

Year (0 ↔ 1990)

(a) According to the model, during what year did the number of master's degrees earned by women exceed 220,000?

(b) Use the graph to verify the result of part (a).

(c) According to the model, during what year will the number of master's degrees earned by women exceed 320,000?

(d) Use the graph to verify the result of part (c).

Synthesis

True or False? In Exercises 77 and 78, determine whether the statement is true or false. Justify your answer.

77. The zeros of the polynomial $x^3 - 2x^2 - 11x + 12 \geq 0$ divide the real number line into four test intervals.

78. The solution set of the inequality $\frac{3}{2}x^2 + 3x + 6 \geq 0$ is the entire set of real numbers.

Exploration In Exercises 79–82, find the interval for b such that the equation has at least one real solution.

79. $x^2 + bx + 4 = 0$

80. $x^2 + bx - 4 = 0$

81. $3x^2 + bx + 10 = 0$

82. $2x^2 + bx + 5 = 0$

83. (a) Write a conjecture about the intervals for b in Exercises 79–82. Explain your reasoning.

(b) What is the center of each interval for b in Exercises 79–82?

84. Consider the polynomial $(x - a)(x - b)$ and the real number line shown below.

(a) Identify the points on the line at which the polynomial is zero.

(b) In each of the three subintervals of the line, write the sign of each factor and the sign of the product.

(c) For what x-values does the polynomial change signs?

Skills Review

In Exercises 85–88, factor the expression completely.

85. $4x^2 + 20x + 25$

86. $(x + 3)^2 - 16$

87. $x^2(x + 3) - 4(x + 3)$

88. $2x^4 - 54x$

In Exercises 89 and 90, write an expression for the area of the region.

89.

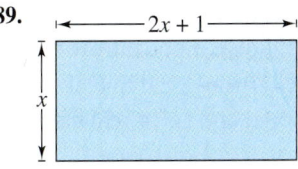

90.

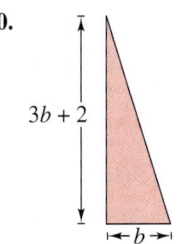

1 Chapter Summary

What did you learn?

Section 1.1
Review Exercises

☐ Sketch graphs of equations *(p. 78)* and find *x*- and *y*-intercepts of graphs of equations *(p. 81)*.   1–12

☐ Use symmetry to sketch graphs of equations *(p. 81)*.   13–20

☐ Find equations of, and sketch graphs of, circles *(p. 84)*.   21–28

☐ Use graphs of equations in solving real-life problems *(p. 85)*.   29, 30

Section 1.2

☐ Identify different types of equations and solve linear equations in one variable *(p. 89)*.   31–38

☐ Solve equations that lead to linear equations *(p. 92)*.   39–42

☐ Find *x*- and *y*-intercepts of graphs of equations algebraically *(p. 93)*.   43–50

☐ Use linear equations to model and solve real-life problems *(p. 93)*.   51, 52

Section 1.3

☐ Use a verbal model in a problem-solving plan *(p. 98)*.   53, 54

☐ Write and use mathematical models to solve real-life problems *(p. 99)*.   55–58

☐ Solve mixture problems *(p. 102)* and use common formulas to solve real-life problems *(p. 103)*.   59–64

Section 1.4

☐ Solve quadratic equations by factoring *(p. 110)*, by extracting square roots *(p. 111)*, by completing the square *(p. 112)*, and by using the Quadratic Formula *(p. 114)*.   65–74

☐ Use quadratic equations to model and solve real-life problems *(p. 116)*.   75, 76

Section 1.5

☐ Use the imaginary unit *i* to write complex numbers *(p. 126)*.   77–80

☐ Add, subtract, and multiply complex numbers *(p. 127)*.   81–86

☐ Use complex conjugates to write the quotient of two complex numbers in standard form *(p. 129)*.   87–90

☐ Find complex solutions of quadratic equations *(p. 130)*.   91–94

Section 1.6

☐ Solve polynomial equations of degree three or greater *(p. 133)* and solve equations involving radicals *(p. 135)*, fractions, and absolute values *(p. 136)*.   95–114

☐ Use different types of equations to model and solve real-life problems *(p. 138)*.   115, 116

Section 1.7

☐ Represent solutions of linear inequalities in one variable *(p. 144)*, solve linear inequalities in one variable *(p. 146)*, and solve inequalities involving absolute values *(p. 148)*.   117–130

☐ Use inequalities to model and solve real-life problems *(p. 149)*.   131, 132

Section 1.8

☐ Solve polynomial inequalities *(p. 154)* and rational inequalities *(p. 158)*.   133–144

☐ Use inequalities to model and solve real-life problems *(p. 159)*.   145, 146

1   Review Exercises

1.1 **In Exercises 1–4, complete a table of values. Use the solution points to sketch the graph of the equation.**

1. $y = 3x - 5$

2. $y = -\frac{1}{2}x + 2$

3. $y = x^2 - 3x$

4. $y = 2x^2 - x - 9$

In Exercises 5–10, sketch the graph *by hand*.

5. $y - 2x - 3 = 0$

6. $3x + 2y + 6 = 0$

7. $y = \sqrt{5 - x}$

8. $y = \sqrt{x + 2}$

9. $y + 2x^2 = 0$

10. $y = x^2 - 4x$

In Exercises 11 and 12, find the *x*- and *y*-intercepts of the graph of the equation.

11. $y = (x - 3)^2 - 4$

12. $y = |x + 1| - 3$

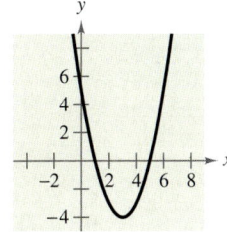

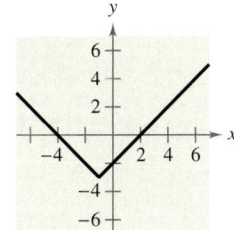

In Exercises 13–20, use the algebraic tests to check for symmetry with respect to both axes and the origin. Then sketch the graph of the equation.

13. $y = -4x + 1$

14. $y = 5x - 6$

15. $y = 5 - x^2$

16. $y = x^2 - 10$

17. $y = x^3 + 3$

18. $y = -6 - x^3$

19. $y = \sqrt{x + 5}$

20. $y = |x| + 9$

In Exercises 21–26, find the center and radius of the circle and sketch its graph.

21. $x^2 + y^2 = 9$

22. $x^2 + y^2 = 4$

23. $(x + 2)^2 + y^2 = 16$

24. $x^2 + (y - 8)^2 = 81$

25. $\left(x - \frac{1}{2}\right)^2 + (y + 1)^2 = 36$

26. $(x + 4)^2 + \left(y - \frac{3}{2}\right)^2 = 100$

27. Find the standard form of the equation of the circle for which the endpoints of a diameter are $(0, 0)$ and $(4, -6)$.

28. Find the standard form of the equation of the circle for which the endpoints of a diameter are $(-2, -3)$ and $(4, -10)$.

29. *Physics*   The force F (in pounds) required to stretch a spring x inches from its natural length (see figure) is

$$F = \frac{5}{4}x, \ 0 \le x \le 20.$$

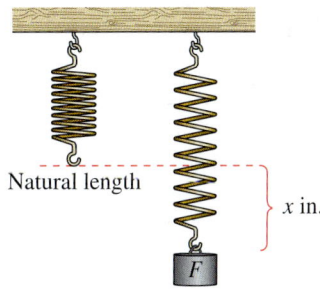

(a) Use the model to complete the table.

x	0	4	8	12	16	20
Force, F						

(b) Sketch a graph of the model.

(c) Use the graph to estimate the force necessary to stretch the spring 10 inches.

30. *Number of Stores*   The numbers N of Target stores for the years 1994 to 2003 can be approximated by the model

$$N = 3.69t^2 + 939, \quad 4 \le t \le 13$$

where t is the time (in years), with $t = 4$ corresponding to 1994.   (Source: Target Corp.)

(a) Sketch a graph of the model.

(b) Use the graph to estimate the year in which the number of stores was 1300.

1.2 **In Exercises 31–34, determine whether the equation is an identity or a conditional equation.**

31. $6 - (x - 2)^2 = 2 + 4x - x^2$

32. $3(x - 2) + 2x = 2(x + 3)$

33. $-x^3 + x(7 - x) + 3 = x(-x^2 - x) + 7(x + 1) - 4$

34. $3(x^2 - 4x + 8) = -10(x + 2) - 3x^2 + 6$

In Exercises 35–42, solve the equation (if possible) and check your solution.

35. $3x - 2(x + 5) = 10$

36. $4x + 2(7 - x) = 5$

37. $4(x + 3) - 3 = 2(4 - 3x) - 4$

38. $\frac{1}{2}(x - 3) - 2(x + 1) = 5$

39. $\frac{x}{5} - 3 = \frac{x}{3} + 1$

40. $\frac{4x - 3}{6} + \frac{x}{4} = x - 2$

41. $\frac{18}{x} = \frac{10}{x - 4}$

42. $\frac{5}{x - 2} = \frac{13}{2x - 3}$

In Exercises 43–50, find the x- and y-intercepts of the graph of the equation algebraically.

43. $y = 3x - 1$

44. $y = -5x + 6$

45. $y = 2(x - 4)$

46. $y = 4(7x + 1)$

47. $y = -\frac{1}{2}x + \frac{2}{3}$

48. $y = \frac{3}{4}x - \frac{1}{4}$

49. $3.8y - 0.5x + 1 = 0$

50. $1.5y + 2x - 1.2 = 0$

51. *Geometry*   The surface area S of the cylinder shown in the figure is approximated by

$$S = 2(3.14)(3)^2 + 2(3.14)(3)h.$$

The surface area is 244.92 square inches. Find the height h of the cylinder.

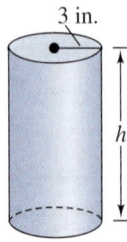

3 in.

h

52. *Temperature*   The Fahrenheit and Celsius temperature scales are related by the equation

$$C = \frac{5}{9}F - \frac{160}{9}.$$

Find the Fahrenheit temperature that corresponds to 100° Celsius.

1.3 **53.** *Profit*   In October, a greeting card company's total profit was 12% more than it was in September. The total profit for the two months was $689,000. Write a verbal model, assign labels, and write an algebraic equation to find the profit for each month.

54. *Discount*   The price of a digital camera has been discounted $85. The sale price is $340. Write a verbal model, assign labels, and write an algebraic equation to find the percent discount.

55. *Shadow Length*   A person who is 6 feet tall walks away from a streetlight toward the tip of the streetlight's shadow. When the person is 15 feet from the streetlight, the tip of the person's shadow and the shadow cast by the streetlight coincide at a point 5 feet in front of the person (see figure). How tall is the streetlight?

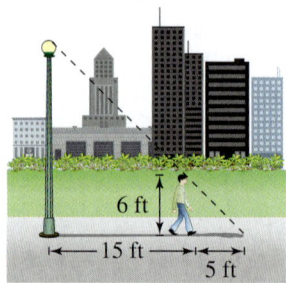

6 ft

15 ft

5 ft

56. *Finance*   A group agrees to share equally in the cost of a $48,000 piece of machinery. If it can find two more group members, each member's share will decrease by $4000. How many are presently in the group?

57. *Business Venture*   You are planning to start a small business that will require an investment of $90,000. You have found some people who are willing to share equally in the venture. If you can find three more people, each person's share will decrease by $2500. How many people have you found so far?

58. *Average Speed*   You commute 56 miles one way to work. The trip to work takes 10 minutes longer than the trip home. Your average speed on the trip home is 8 miles per hour faster. What is your average speed on the trip home?

59. *Mixture Problem*   A car radiator contains 10 liters of a 30% antifreeze solution. How many liters will have to be replaced with pure antifreeze if the resulting solution is to be 50% antifreeze?

60. *Investment*   You invested $6000 at $4\frac{1}{2}$% and $5\frac{1}{2}$% simple interest. During the first year, the two accounts earned $305. How much did you invest in each fund? (*Note:* The $5\frac{1}{2}$% account is more risky.)

In Exercises 61 and 62, solve for the indicated variable.

61. *Volume of a Cone*

Solve for h: $V = \frac{1}{3}\pi r^2 h$

62. *Kinetic Energy*

Solve for m: $E = \frac{1}{2}mv^2$

63. *Travel Time* Two cars start at a given time and travel in the same direction at average speeds of 40 miles per hour and 55 miles per hour. How much time will elapse before the two cars are 10 miles apart?

64. *Geometry* The volume of a circular cylinder is 81π cubic feet. The cylinder's radius is 3 feet. What is the height of the cylinder?

1.4 In Exercises 65–74, use any method to solve the quadratic equation.

65. $15 + x - 2x^2 = 0$

66. $2x^2 - x - 28 = 0$

67. $6 = 3x^2$

68. $16x^2 = 25$

69. $(x + 4)^2 = 18$

70. $(x - 8)^2 = 15$

71. $x^2 - 12x + 30 = 0$

72. $x^2 + 6x - 3 = 0$

73. $-2x^2 - 5x + 27 = 0$

74. $-20 - 3x + 3x^2 = 0$

75. *Simply Supported Beam* A simply supported 20-foot beam supports a uniformly distributed load of 1000 pounds per foot. The bending moment M (in foot-pounds) x feet from one end of the beam is given by $M = 500x(20 - x)$.

(a) Where is the bending moment zero?

 (b) Use a graphing utility to graph the equation.

 (c) Use the graph to determine the point on the beam where the bending moment is the greatest.

76. *Sports* You throw a softball straight up into the air at a velocity of 30 feet per second. You release the softball at a height of 5.8 feet and catch it when it falls back to a height of 6.2 feet.

(a) Use the position equation to write a mathematical model for the height of the softball.

(b) What is the height of the softball after 1 second?

(c) How many seconds is the softball in the air?

1.5 In Exercises 77–80, write the complex number in standard form.

77. $6 + \sqrt{-4}$

78. $3 - \sqrt{-25}$

79. $i^2 + 3i$

80. $-5i + i^2$

In Exercises 81–86, perform the operation and write the result in standard form.

81. $(7 + 5i) + (-4 + 2i)$

82. $\left(\frac{\sqrt{2}}{2} - \frac{\sqrt{2}}{2}i\right) - \left(\frac{\sqrt{2}}{2} + \frac{\sqrt{2}}{2}i\right)$

83. $5i(13 - 8i)$

84. $(1 + 6i)(5 - 2i)$

85. $(10 - 8i)(2 - 3i)$

86. $i(6 + i)(3 - 2i)$

In Exercises 87 and 88, write the quotient in standard form.

87. $\dfrac{6 + i}{4 - i}$

88. $\dfrac{3 + 2i}{5 + i}$

In Exercises 89 and 90, perform the operation and write the result in standard form.

89. $\dfrac{4}{2 - 3i} + \dfrac{2}{1 + i}$

90. $\dfrac{1}{2 + i} - \dfrac{5}{1 + 4i}$

In Exercises 91–94, find all solutions of the equation.

91. $3x^2 + 1 = 0$

92. $2 + 8x^2 = 0$

93. $x^2 - 2x + 10 = 0$

94. $6x^2 + 3x + 27 = 0$

1.6 In Exercises 95–114, find all solutions of the equation. Check your solutions in the original equation.

95. $5x^4 - 12x^3 = 0$

96. $4x^3 - 6x^2 = 0$

97. $x^4 - 5x^2 + 6 = 0$

98. $9x^4 + 27x^3 - 4x^2 - 12x = 0$

99. $\sqrt{x + 4} = 3$

100. $\sqrt{x - 2} - 8 = 0$

101. $\sqrt{2x + 3} + \sqrt{x - 2} = 2$

102. $5\sqrt{x} - \sqrt{x - 1} = 6$

103. $(x - 1)^{2/3} - 25 = 0$

104. $(x + 2)^{3/4} = 27$

105. $(x + 4)^{1/2} + 5x(x + 4)^{3/2} = 0$

106. $8x^2(x^2 - 4)^{1/3} + (x^2 - 4)^{4/3} = 0$

107. $\dfrac{5}{x} = 1 + \dfrac{3}{x + 2}$

108. $\dfrac{6}{x} + \dfrac{8}{x + 5} = 3$

109. $\dfrac{3}{x + 2} - \dfrac{1}{x} = \dfrac{1}{5x}$

110. $\dfrac{12}{x + 5} + \dfrac{5}{x} = \dfrac{20}{x}$

111. $|x - 5| = 10$

112. $|2x + 3| = 7$

113. $|x^2 - 3| = 2x$

114. $|x^2 - 6| = x$

115. *Demand* The demand equation for a hair dryer is

$$p = 42 - \sqrt{0.001x + 2}$$

where x is the number of units demanded per day and p is the price per unit. Find the demand if the price is set at $29.95.

116. Data Analysis: Newspapers The total numbers N of daily evening newspapers in the United States from 1970 to 2000 can be approximated by the model

$$N = 1481 - 4.6t^{3/2}, \quad 0 \le t \le 30$$

where t represents the year, with $t = 0$ corresponding to 1970. The actual numbers of newspapers for selected years are shown in the table.   (Source: Editor & Publisher Co.)

Year	Newspapers, N
1970	1429
1975	1436
1980	1388
1985	1220
1990	1084
1995	891
2000	727

(a) Use a graphing utility to plot the data and graph the model in the same viewing window. How well does the model fit the data?

(b) Use the graph in part (a) to estimate the year in which there were 800 daily evening newspapers.

(c) Use the model to verify algebraically the estimate from part (b).

1.7 **In Exercises 117–120, write an inequality that represents the interval and state whether the interval is bounded or unbounded.**

117. $(-7, 2]$ **118.** $(4, \infty)$

119. $(-\infty, -10]$ **120.** $[-2, 2]$

In Exercises 121–130, solve the inequality.

121. $9x - 8 \le 7x + 16$ **122.** $\frac{15}{2}x + 4 > 3x - 5$

123. $4(5 - 2x) \le \frac{1}{2}(8 - x)$

124. $\frac{1}{2}(3 - x) > \frac{1}{3}(2 - 3x)$

125. $-19 < 3x - 17 \le 34$ **126.** $-3 \le \frac{2x - 5}{3} < 5$

127. $|x| \le 4$ **128.** $|x - 2| < 1$

129. $|x - 3| > 4$ **130.** $\left| x - \frac{3}{2} \right| \ge \frac{3}{2}$

131. Geometry The side of a square is measured as 19.3 centimeters with a possible error of 0.5 centimeter. Using these measurements, determine the interval containing the area of the square.

132. Cost, Revenue, and Profit The revenue for selling x units of a product is $R = 125.33x$. The cost of producing x units is $C = 92x + 1200$. To obtain a profit, the revenue must be greater than the cost. Determine the smallest value of x for which this product returns a profit.

1.8 **In Exercises 133–144, solve the inequality.**

133. $x^2 - 6x - 27 < 0$ **134.** $x^2 - 2x \ge 3$

135. $6x^2 + 5x < 4$ **136.** $2x^2 + x \ge 15$

137. $x^3 - 16x \ge 0$ **138.** $12x^3 - 20x^2 < 0$

139. $\dfrac{x + 8}{x + 5} - 2 < 0$ **140.** $\dfrac{3x + 8}{x - 3} \le 4$

141. $\dfrac{2}{x + 1} \le \dfrac{3}{x - 1}$ **142.** $\dfrac{x - 5}{3 - x} < 0$

143. $\dfrac{x^2 + 7x + 12}{x} \ge 0$ **144.** $\dfrac{1}{x - 2} > \dfrac{1}{x}$

145. Investment P dollars invested at interest rate r compounded annually increases to an amount

$$A = P(1 + r)^2$$

in 2 years. An investment of \$5000 is to increase to an amount greater than \$5500 in 2 years. The interest rate must be greater than what percent?

146. Population of a Species A biologist introduces 200 ladybugs into a crop field. The population P of the ladybugs is approximated by the model

$$P = \frac{1000(1 + 3t)}{5 + t}$$

where t is the time in days. Find the time required for the population to increase to at least 2000 ladybugs.

Synthesis

True or False? **In Exercises 147 and 148, determine whether the statement is true or false. Justify your answer.**

147. $\sqrt{-18}\sqrt{-2} = \sqrt{(-18)(-2)}$

148. The equation $325x^2 - 717x + 398 = 0$ has no solution.

149. Writing Explain why it is essential to check your solutions to radical, absolute value, and rational equations.

150. Error Analysis What is wrong with the following solution?

$$|11x + 4| \ge 26$$

$11x + 4 \le 26$ or $11x + 4 \ge 26$
$11x \le 22$ $11x \ge 22$
$x \le 2$ $x \ge 2$

1 | Chapter Test

Take this test as you would take a test in class. When you are finished, check your work against the answers given in the back of the book.

In Exercises 1–6, check for symmetry with respect to both axes and the origin. Then sketch the graph of the equation. Identify any x- and y-intercepts.

1. $y = 4 - \frac{3}{4}x$

2. $y = 4 - \frac{3}{4}|x|$

3. $y = 4 - (x - 2)^2$

4. $y = x - x^3$

5. $y = \sqrt{3 - x}$

6. $(x - 3)^2 + y^2 = 9$

In Exercises 7–12, solve the equation (if possible).

7. $\frac{2}{3}(x - 1) + \frac{1}{4}x = 10$

8. $(x - 3)(x + 2) = 14$

9. $\dfrac{x - 2}{x + 2} + \dfrac{4}{x + 2} + 4 = 0$

10. $x^4 + x^2 - 6 = 0$

11. $2\sqrt{x} - \sqrt{2x + 1} = 1$

12. $|3x - 1| = 7$

In Exercises 13–16, solve the inequality. Sketch the solution set on the real number line.

13. $-3 \le 2(x + 4) < 14$

14. $\dfrac{2}{x} > \dfrac{5}{x + 6}$

15. $2x^2 + 5x > 12$

16. $|x - 15| \ge 5$

17. Perform each operation and write the result in standard form.

   (a) $10i - \left(3 + \sqrt{-25}\right)$ (b) $\left(2 + \sqrt{3}\,i\right)\left(2 - \sqrt{3}\,i\right)$

18. Write the quotient in standard form: $\dfrac{5}{2 + i}$.

19. The sales y (in billions of dollars) for Dell, Inc. from 1994 to 2003 can be approximated by the model

   $$y = 4.45t - 16.6, \quad 4 \le t \le 13$$

   where t is the time (in years), with $t = 4$ corresponding to 1994.   (Source: Dell, Inc.)

   (a) Sketch a graph of the model.

   (b) Assuming that the pattern continues, use the graph in part (a) to estimate the sales in 2008.

   (c) Use the model to verify algebraically the estimate from part (b).

20. A basketball has a volume of about 455.9 cubic inches. Find the radius of the basketball (accurate to three decimal places).

21. On the first part of a 350-kilometer trip, a salesperson travels 2 hours and 15 minutes at an average speed of 100 kilometers per hour. The salesperson needs to arrive at the destination in another hour and 20 minutes. Find the average speed required for the remainder of the trip.

22. The area of the ellipse in the figure at the left is $A = \pi ab$. If a and b satisfy the constraint $a + b = 100$, find a and b such that the area of the ellipse equals the area of the circle.

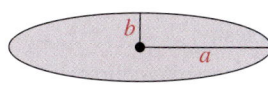

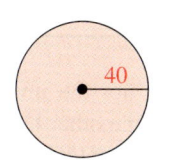

FIGURE FOR 22

7. A Pythagorean Triple is a group of three integers, such as 3, 4, and 5, that could be the lengths of the sides of a right triangle.

   (a) Find two other Pythagorean Triples.

   (b) Notice that $3 \cdot 4 \cdot 5 = 60$. Is the product of the three numbers in each Pythagorean Triple evenly divisible by 3? by 4? by 5?

   (c) Write a conjecture involving Pythagorean Triples and divisibility by 60.

8. Determine the solutions x_1 and x_2 of each quadratic equation. Use the values of x_1 and x_2 to fill in the boxes.

Equation	x_1, x_2	$x_1 + x_2$	$x_1 \cdot x_2$
(a) $x^2 - x - 6 = 0$			
(b) $2x^2 + 5x - 3 = 0$			
(c) $4x^2 - 9 = 0$			
(d) $x^2 - 10x + 34 = 0$			

9. Consider a general quadratic equation

   $$ax^2 + bx + c = 0$$

   whose solutions are x_1 and x_2. Use the results of Exercise 8 to determine a relationship among the coefficients a, b, and c and the sum $x_1 + x_2$ and the product $x_1 \cdot x_2$ of the solutions.

10. (a) The principal cube root of 125, $\sqrt[3]{125}$, is 5. Evaluate the expression x^3 for each value of x.

    (i) $x = \dfrac{-5 + 5\sqrt{3}i}{2}$

    (ii) $x = \dfrac{-5 - 5\sqrt{3}i}{2}$

    (b) The principal cube root of 27, $\sqrt[3]{27}$, is 3. Evaluate the expression x^3 for each value of x.

    (i) $x = \dfrac{-3 + 3\sqrt{3}i}{2}$

    (ii) $x = \dfrac{-3 - 3\sqrt{3}i}{2}$

    (c) Use the results of parts (a) and (b) to list possible cube roots of (i), 1, (ii) 8, and (iii) 64. Verify your results algebraically.

11. The multiplicative inverse of z is a complex number z_m such that $z \cdot z_m = 1$. Find the multiplicative inverse of each complex number.

    (a) $z = 1 + i$ (b) $z = 3 - i$ (c) $z = -2 + 8i$

12. Prove that the product of a complex number $a + bi$ and its complex conjugate is a real number.

13. A **fractal** is a geometric figure that consists of a pattern that is repeated infinitely on a smaller and smaller scale. The most famous fractal is called the **Mandelbrot Set,** named after the Polish-born mathematician Benoit Mandelbrot. To draw the Mandelbrot Set, consider the following sequence of numbers.

    $$c, c^2 + c, (c^2 + c)^2 + c, [(c^2 + c)^2 + c]^2 + c, \dots$$

    The behavior of this sequence depends on the value of the complex number c. If the sequence is bounded (the absolute value of each number in the sequence, $|a + bi| = \sqrt{a^2 + b^2}$, is less than some fixed number N), the complex number c is in the Mandelbrot Set, and if the sequence is unbounded (the absolute value of the terms of the sequence become infinitely large), the complex number c is not in the Mandelbrot Set. Determine whether the complex number c is in the Mandelbrot Set.

    (a) $c = i$ (b) $c = 1 + i$ (c) $c = -2$

    The figure below shows a black and yellow photo of the Mandelbrot Set.

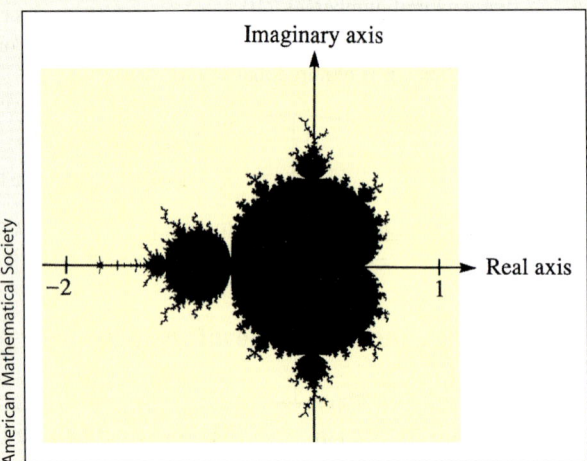

14. Use the equation $4\sqrt{x} = 2x + k$ to find three different values of k such that the equation has two solutions, one solution, and no solution. Describe the process you used to find the values.

15. Use the graph of $y = x^4 - x^3 - 6x^2 + 4x + 8$ to solve the inequality $x^4 - x^3 - 6x^2 + 4x + 8 > 0$.

16. When you buy a 16-ounce bag of chips, you expect to get precisely 16 ounces. The actual weight w (in ounces) of a "16-ounce" bag of chips is given by

    $$|w - 16| \le \frac{1}{2}.$$

    You buy four 16-ounce bags. What is the greatest amount you can expect to get? What is the smallest amount? Explain.

Functions and Their Graphs

2

Functions play a primary role in modeling real-life situations. The estimated growth in the number of digital music scales in the United States can be modeled by a cubic function.

© AP/Wide World Photos

SELECTED APPLICATIONS

Functions have many real-life applications. The applications listed below represent a small sample of the applications in this chapter.

2.1 Linear Equations in Two Variables

What you should learn

- Use slope to graph linear equations in two variables.
- Find slopes of lines.
- Write linear equations in two variables.
- Use slope to identify parallel and perpendicular lines.
- Use slope and linear equations in two variables to model and solve real-life problems.

Why you should learn it

Linear equations in two variables can be used to model and solve real-life problems. For instance, in Exercise 109 on page 186, you will use a linear equation to model student enrollment at the Pennsylvania State University.

The *HM mathSpace®* CD-ROM and *Eduspace®* contain additional resources related to the concepts discussed in this chapter.

Using Slope

The simplest mathematical model for relating two variables is the **linear equation in two variables** $y = mx + b$. The equation is called *linear* because its graph is a line. (In mathematics, the term *line* means *straight line*.) By letting $x = 0$, you can see that the line crosses the y-axis at $y = b$, as shown in Figure 2.1. In other words, the y-intercept is $(0, b)$. The steepness or slope of the line is m.

$$y = mx + b$$

Slope ⎦   ⎦ y-Intercept

The **slope** of a nonvertical line is the number of units the line rises (or falls) vertically for each unit of horizontal change from left to right, as shown in Figure 2.1 and Figure 2.2.

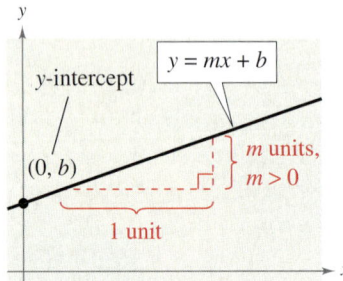

Positive slope, line rises.
FIGURE 2.1

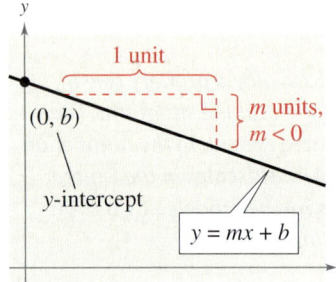

Negative slope, line falls.
FIGURE 2.2

A linear equation that is written in the form $y = mx + b$ is said to be written in **slope-intercept form.**

The Slope-Intercept Form of the Equation of a Line

The graph of the equation

$$y = mx + b$$

is a line whose slope is m and whose y-intercept is $(0, b)$.

Exploration

Use a graphing utility to compare the slopes of the lines $y = mx$, where $m = 0.5, 1, 2,$ and 4. Which line rises most quickly? Now, let $m = -0.5,$ $-1, -2,$ and -4. Which line falls most quickly? Use a square setting to obtain a true geometric perspective. What can you conclude about the slope and the "rate" at which the line rises or falls?

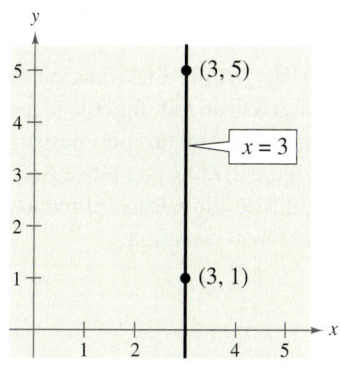

FIGURE **2.3**   *Slope is undefined.*

Once you have determined the slope and the *y*-intercept of a line, it is a relatively simple matter to sketch its graph. In the next example, note that none of the lines is vertical. A vertical line has an equation of the form

$$x = a.$$ Vertical line

The equation of a vertical line cannot be written in the form $y = mx + b$ because the slope of a vertical line is undefined, as indicated in Figure 2.3.

Example 1   Graphing a Linear Equation

Sketch the graph of each linear equation.

a. $y = 2x + 1$

b. $y = 2$

c. $x + y = 2$

Solution

a. Because $b = 1$, the *y*-intercept is $(0, 1)$. Moreover, because the slope is $m = 2$, the line *rises* two units for each unit the line moves to the right, as shown in Figure 2.4.

b. By writing this equation in the form $y = (0)x + 2$, you can see that the *y*-intercept is $(0, 2)$ and the slope is zero. A zero slope implies that the line is horizontal—that is, it doesn't rise *or* fall, as shown in Figure 2.5.

c. By writing this equation in slope-intercept form

$$x + y = 2$$ Write original equation.

$$y = -x + 2$$ Subtract *x* from each side.

$$y = (-1)x + 2$$ Write in slope-intercept form.

you can see that the *y*-intercept is $(0, 2)$. Moreover, because the slope is $m = -1$, the line *falls* one unit for each unit the line moves to the right, as shown in Figure 2.6.

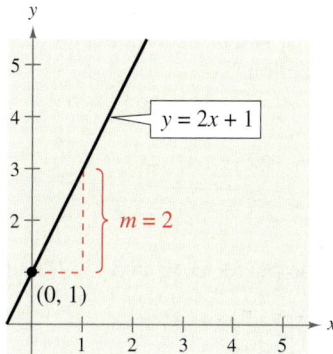

When m is positive, the line rises.
FIGURE **2.4**

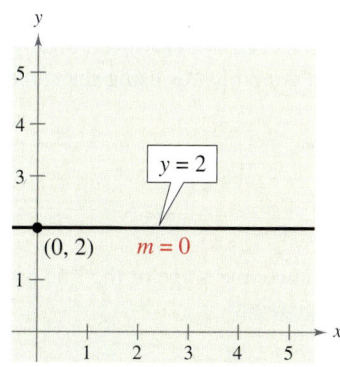

When m is 0, the line is horizontal.
FIGURE **2.5**

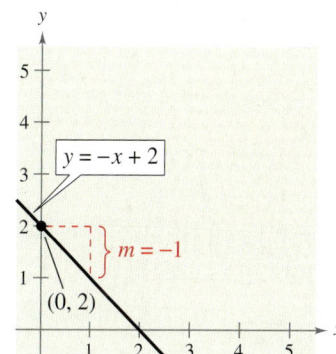

When m is negative, the line falls.
FIGURE **2.6**

✔**CHECKPOINT**   Now try Exercise 9.

Finding the Slope of a Line

Given an equation of a line, you can find its slope by writing the equation in slope-intercept form. If you are not given an equation, you can still find the slope of a line. For instance, suppose you want to find the slope of the line passing through the points (x_1, y_1) and (x_2, y_2), as shown in Figure 2.7. As you move from left to right along this line, a change of $(y_2 - y_1)$ units in the vertical direction corresponds to a change of $(x_2 - x_1)$ units in the horizontal direction.

$$y_2 - y_1 = \text{the change in } y = \text{rise}$$

and

$$x_2 - x_1 = \text{the change in } x = \text{run}$$

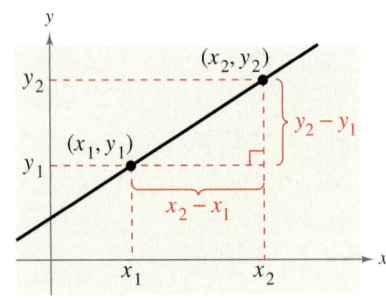

FIGURE **2.7**

The ratio of $(y_2 - y_1)$ to $(x_2 - x_1)$ represents the slope of the line that passes through the points (x_1, y_1) and (x_2, y_2).

$$\text{Slope} = \frac{\text{change in } y}{\text{change in } x}$$

$$= \frac{\text{rise}}{\text{run}}$$

$$= \frac{y_2 - y_1}{x_2 - x_1}$$

The Slope of a Line Passing Through Two Points

The **slope** m of the nonvertical line through (x_1, y_1) and (x_2, y_2) is

$$m = \frac{y_2 - y_1}{x_2 - x_1}$$

where $x_1 \neq x_2$.

When this formula is used for slope, the *order of subtraction* is important. Given two points on a line, you are free to label either one of them as (x_1, y_1) and the other as (x_2, y_2). However, once you have done this, you must form the numerator and denominator using the same order of subtraction.

$$m = \frac{y_2 - y_1}{x_2 - x_1} \qquad m = \frac{y_1 - y_2}{x_1 - x_2} \qquad m = \frac{y_2 - y_1}{x_1 - x_2}$$

Correct   Correct   Incorrect

For instance, the slope of the line passing through the points $(3, 4)$ and $(5, 7)$ can be calculated as

$$m = \frac{7 - 4}{5 - 3} = \frac{3}{2}$$

or, reversing the subtraction order in both the numerator and denominator, as

$$m = \frac{4 - 7}{3 - 5} = \frac{-3}{-2} = \frac{3}{2}.$$

Example 2 **Finding the Slope of a Line Through Two Points**

Find the slope of the line passing through each pair of points.

a. $(-2,\ 0)$ and $(3, 1)$ **b.** $(-1,\ 2)$ and $(2, 2)$

c. $(0, 4)$ and $(1,\ -1)$ **d.** $(3, 4)$ and $(3, 1)$

Solution

a. Letting $(x_1, y_1) = (-2,\ 0)$ and $(x_2, y_2) = (3, 1)$, you obtain a slope of

$$m = \frac{y_2 - y_1}{x_2 - x_1} = \frac{1 - 0}{3 - (-2)} = \frac{1}{5}. \qquad \text{See Figure 2.8.}$$

b. The slope of the line passing through $(-1, 2)$ and $(2, 2)$ is

$$m = \frac{2 - 2}{2 - (-1)} = \frac{0}{3} = 0. \qquad \text{See Figure 2.9.}$$

c. The slope of the line passing through $(0, 4)$ and $(1, -1)$ is

$$m = \frac{-1 - 4}{1 - 0} = \frac{-5}{1} = -5. \qquad \text{See Figure 2.10.}$$

d. The slope of the line passing through $(3, 4)$ and $(3, 1)$ is

$$m = \frac{1 - 4}{3 - 3} = \frac{-3}{0}. \qquad \text{See Figure 2.11.}$$

Because division by 0 is undefined, the slope is undefined and the line is vertical.

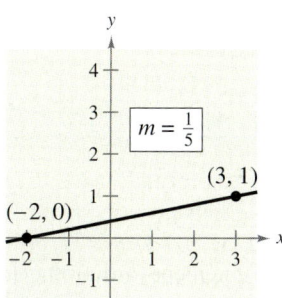

FIGURE 2.8

FIGURE 2.9

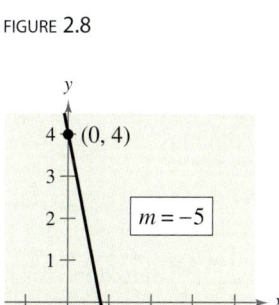

FIGURE 2.10

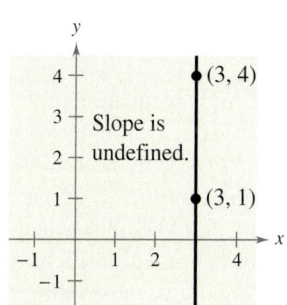

FIGURE 2.11

✔**CHECKPOINT**   Now try Exercise 21.

Writing Linear Equations in Two Variables

If (x_1, y_1) is a point on a line of slope m and (x, y) is *any other* point on the line, then

$$\frac{y - y_1}{x - x_1} = m.$$

This equation, involving the variables x and y, can be rewritten in the form

$$y - y_1 = m(x - x_1)$$

which is the **point-slope form** of the equation of a line.

Point-Slope Form of the Equation of a Line

The equation of the line with slope m passing through the point (x_1, y_1) is

$$y - y_1 = m(x - x_1).$$

The point-slope form is most useful for *finding* the equation of a line. You should remember this form.

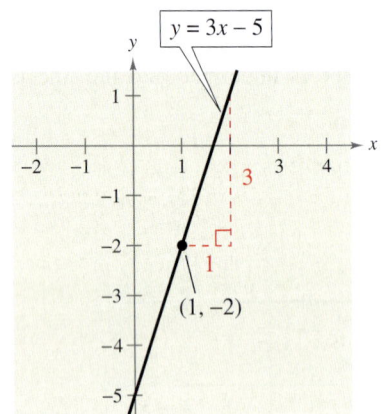

FIGURE 2.12

Example 3   Using the Point-Slope Form

Find the slope-intercept form of the equation of the line that has a slope of 3 and passes through the point $(1, -2)$.

Solution

Use the point-slope form with $m = 3$ and $(x_1, y_1) = (1, -2)$.

$$
\begin{aligned}
y - y_1 &= m(x - x_1) && \text{Point-slope form}\\
y - (-2) &= 3(x - 1) && \text{Substitute for } m, x_1, \text{ and } y_1.\\
y + 2 &= 3x - 3 && \text{Simplify.}\\
y &= 3x - 5 && \text{Write in slope-intercept form.}
\end{aligned}
$$

The slope-intercept form of the equation of the line is $y = 3x - 5$. The graph of this line is shown in Figure 2.12.

✓**CHECKPOINT**   Now try Exercise 39.

The point-slope form can be used to find an equation of the line passing through two points (x_1, y_1) and (x_2, y_2). To do this, first find the slope of the line

$$m = \frac{y_2 - y_1}{x_2 - x_1}, \qquad x_1 \neq x_2$$

and then use the point-slope form to obtain the equation

$$y - y_1 = \frac{y_2 - y_1}{x_2 - x_1}(x - x_1). \qquad \text{Two-point form}$$

This is sometimes called the **two-point form** of the equation of a line.

STUDY TIP

When you find an equation of the line that passes through two given points, you only need to substitute the coordinates of one of the points into the point-slope form. It does not matter which point you choose because both points will yield the same result.

Parallel and Perpendicular Lines

Slope can be used to decide whether two nonvertical lines in a plane are parallel, perpendicular, or neither.

> ### Parallel and Perpendicular Lines
>
> 1. Two distinct nonvertical lines are **parallel** if and only if their slopes are equal. That is, $m_1 = m_2$.
>
> 2. Two nonvertical lines are **perpendicular** if and only if their slopes are negative reciprocals of each other. That is, $m_1 = -1/m_2$.

Example 4   **Finding Parallel and Perpendicular Lines**

Find the slope-intercept forms of the equations of the lines that pass through the point $(2, -1)$ and are (a) parallel to and (b) perpendicular to the line $2x - 3y = 5$.

Solution

By writing the equation of the given line in slope-intercept form

$$2x - 3y = 5 \qquad \text{Write original equation.}$$

$$-3y = -2x + 5 \qquad \text{Subtract } 2x \text{ from each side.}$$

$$y = \tfrac{2}{3}x - \tfrac{5}{3} \qquad \text{Write in slope-intercept form.}$$

you can see that it has a slope of $m = \tfrac{2}{3}$, as shown in Figure 2.13.

a. Any line parallel to the given line must also have a slope of $\tfrac{2}{3}$. So, the line through $(2, -1)$ that is parallel to the given line has the following equation.

$$y - (-1) = \tfrac{2}{3}(x - 2) \qquad \text{Write in point-slope form.}$$

$$3(y + 1) = 2(x - 2) \qquad \text{Multiply each side by 3.}$$

$$3y + 3 = 2x - 4 \qquad \text{Distributive Property}$$

$$y = \tfrac{2}{3}x - \tfrac{7}{3} \qquad \text{Write in slope-intercept form.}$$

b. Any line perpendicular to the given line must have a slope of $-\tfrac{3}{2}$ $\left(\text{because } -\tfrac{3}{2}\right.$ is the negative reciprocal of $\tfrac{2}{3}\left.\right)$. So, the line through $(2, -1)$ that is perpendicular to the given line has the following equation.

$$y - (-1) = -\tfrac{3}{2}(x - 2) \qquad \text{Write in point-slope form.}$$

$$2(y + 1) = -3(x - 2) \qquad \text{Multiply each side by 2.}$$

$$2y + 2 = -3x + 6 \qquad \text{Distributive Property}$$

$$y = -\tfrac{3}{2}x + 2 \qquad \text{Write in slope-intercept form.}$$

✓**CHECKPOINT**   Now try Exercise 69.

Notice in Example 4 how the slope-intercept form is used to obtain information about the graph of a line, whereas the point-slope form is used to write the equation of a line.

Exploration

Find d_1 and d_2 in terms of m_1 and m_2, respectively (see figure). Then use the Pythagorean Theorem to find a relationship between m_1 and m_2.

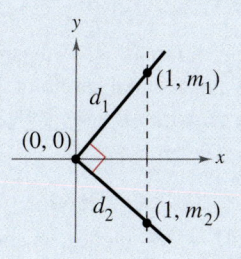

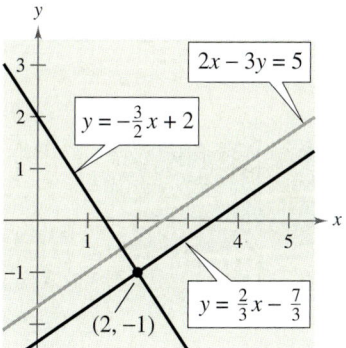

FIGURE 2.13

Technology

On a graphing utility, lines will not appear to have the correct slope unless you use a viewing window that has a square setting. For instance, try graphing the lines in Example 4 using the standard setting $-10 \le x \le 10$ and $-10 \le y \le 10$. Then reset the viewing window with the square setting $-9 \le x \le 9$ and $-6 \le y \le 6$. On which setting do the lines $y = \tfrac{2}{3}x - \tfrac{5}{3}$ and $y = -\tfrac{3}{2}x + 2$ appear to be perpendicular?

Applications

In real-life problems, the slope of a line can be interpreted as either a *ratio* or a *rate*. If the *x*-axis and *y*-axis have the same unit of measure, then the slope has no units and is a **ratio.** If the *x*-axis and *y*-axis have different units of measure, then the slope is a **rate** or **rate of change.**

Example 5   Using Slope as a Ratio

The maximum recommended slope of a wheelchair ramp is $\frac{1}{12}$. A business is installing a wheelchair ramp that rises 22 inches over a horizontal length of 24 feet. Is the ramp steeper than recommended?   (Source: *Americans with Disabilities Act Handbook*)

Solution

The horizontal length of the ramp is 24 feet or 12(24) = 288 inches, as shown in Figure 2.14. So, the slope of the ramp is

$$\text{Slope} = \frac{\text{vertical change}}{\text{horizontal change}} = \frac{22 \text{ in.}}{288 \text{ in.}} \approx 0.076.$$

Because $\frac{1}{12} \approx 0.083$, the slope of the ramp is not steeper than recommended.

FIGURE 2.14

✓CHECKPOINT   Now try Exercise 97.

Example 6   Using Slope as a Rate of Change

A kitchen appliance manufacturing company determines that the total cost in dollars of producing *x* units of a blender is

$$C = 25x + 3500. \qquad \text{Cost equation}$$

Describe the practical significance of the *y*-intercept and slope of this line.

Solution

The *y*-intercept (0, 3500) tells you that the cost of producing zero units is $3500. This is the *fixed cost* of production—it includes costs that must be paid regardless of the number of units produced. The slope of $m = 25$ tells you that the cost of producing each unit is $25, as shown in Figure 2.15. Economists call the cost per unit the *marginal cost*. If the production increases by one unit, then the "margin," or extra amount of cost, is $25. So, the cost increases at a rate of $25 per unit.

✓CHECKPOINT   Now try Exercise 101.

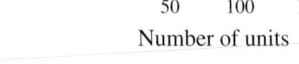

FIGURE 2.15   *Production cost*

Most business expenses can be deducted in the same year they occur. One exception is the cost of property that has a useful life of more than 1 year. Such costs must be *depreciated* (decreased in value) over the useful life of the property. If the *same amount* is depreciated each year, the procedure is called *linear* or *straight-line depreciation*. The *book value* is the difference between the original value and the total amount of depreciation accumulated to date.

Example 7   Straight-Line Depreciation

A college purchased exercise equipment worth $12,000 for the new campus fitness center. The equipment has a useful life of 8 years. The salvage value at the end of 8 years is $2000. Write a linear equation that describes the book value of the equipment each year.

Solution

Let V represent the value of the equipment at the end of year t. You can represent the initial value of the equipment by the data point $(0, 12{,}000)$ and the salvage value of the equipment by the data point $(8, 2000)$. The slope of the line is

$$m = \frac{2000 - 12{,}000}{8 - 0} = -\$1250$$

which represents the annual depreciation in *dollars per year*. Using the point-slope form, you can write the equation of the line as follows.

$$V - 12{,}000 = -1250(t - 0) \qquad \text{Write in point-slope form.}$$

$$V = -1250t + 12{,}000 \qquad \text{Write in slope-intercept form.}$$

The table shows the book value at the end of each year, and the graph of the equation is shown in Figure 2.16.

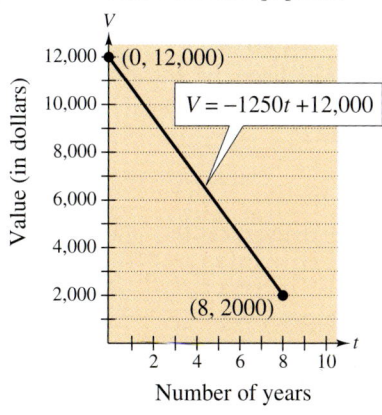

Useful Life of Equipment

FIGURE 2.16   *Straight-line depreciation*

Year, t	Value, V
0	12,000
1	10,750
2	9,500
3	8,250
4	7,000
5	5,750
6	4,500
7	3,250
8	2,000

✓CHECKPOINT   Now try Exercise 107.

In many real-life applications, the two data points that determine the line are often given in a disguised form. Note how the data points are described in Example 7.

Example 8   Predicting Sales per Share

The sales per share for Starbucks Corporation were $6.97 in 2001 and $8.47 in 2002. Using only this information, write a linear equation that gives the sales per share in terms of the year. Then predict the sales per share for 2003.   (Source: Starbucks Corporation)

Solution

Let $t = 1$ represent 2001. Then the two given values are represented by the data points $(1, 6.97)$ and $(2, 8.47)$. The slope of the line through these points is

$$m = \frac{8.47 - 6.97}{2 - 1}$$

$$= 1.5.$$

Using the point-slope form, you can find the equation that relates the sales per share y and the year t to be

$$y - 6.97 = 1.5(t - 1)$$ Write in point-slope form.

$$y = 1.5t + 5.47.$$ Write in slope-intercept form.

According to this equation, the sales per share in 2003 was $y = 1.5(3) + 5.47 = \$9.97$, as shown in Figure 2.17. (In this case, the prediction is quite good—the actual sales per share in 2003 was $10.35.)

✔**CHECKPOINT**   Now try Exercise 109.

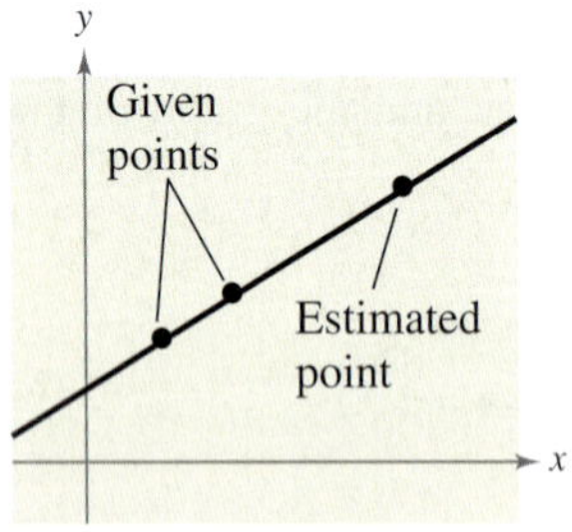

The prediction method illustrated in Example 8 is called **linear extrapolation.** Note in Figure 2.18 that an extrapolated point does not lie between the given points. When the estimated point lies between two given points, as shown in Figure 2.19, the procedure is called **linear interpolation.**

Because the slope of a vertical line is not defined, its equation cannot be written in slope-intercept form. However, every line has an equation that can be written in the **general form**

$$Ax + By + C = 0$$ General form

where A and B are not both zero. For instance, the vertical line given by $x = a$ can be represented by the general form $x - a = 0$.

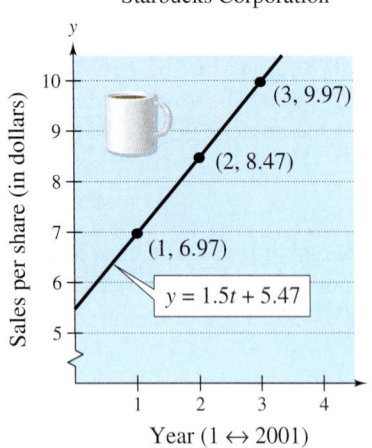

Starbucks Corporation

(3, 9.97)

(2, 8.47)

(1, 6.97)

$y = 1.5t + 5.47$

Year (1 ↔ 2001)

Sales per share (in dollars)

FIGURE 2.17

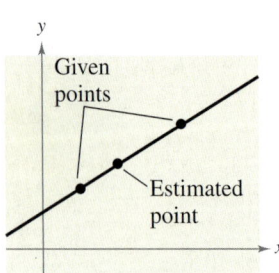

Given points

Estimated point

Linear extrapolation

FIGURE 2.18

Given points

Estimated point

Linear interpolation

FIGURE 2.19

Summary of Equations of Lines

1. General form: $Ax + By + C = 0$

2. Vertical line: $x = a$

3. Horizontal line: $y = b$

4. Slope-intercept form: $y = mx + b$

5. Point-slope form: $y - y_1 = m(x - x_1)$

6. Two-point form: $y - y_1 = \dfrac{y_2 - y_1}{x_2 - x_1}(x - x_1)$

2.1 | **Exercises**   The *HM mathSpace®* CD-ROM and *Eduspace®* for this text contain step-by-step solutions to all odd-numbered exercises. They also provide Tutorial Exercises for additional help.

VOCABULARY CHECK:

In Exercises 1–6, fill in the blanks.

1. The simplest mathematical model for relating two variables is the _____ equation in two variables $y = mx + b$.

2. For a line, the ratio of the change in y to the change in x is called the _____ of the line.

3. Two lines are _____ if and only if their slopes are equal.

4. Two lines are _____ if and only if their slopes are negative reciprocals of each other.

5. When the x-axis and y-axis have different units of measure, the slope can be interpreted as a _____.

6. The prediction method _____ _____ is the method used to estimate a point on a line that does not lie between the given points.

7. Match each equation of a line with its form.

   (a) $Ax + By + C = 0$ (i) Vertical line

   (b) $x = a$ (ii) Slope-intercept form

   (c) $y = b$ (iii) General form

   (d) $y = mx + b$ (iv) Point-slope form

   (e) $y - y_1 = m(x - x_1)$ (v) Horizontal line

PREREQUISITE SKILLS REVIEW: Practice and review algebra skills needed for this section at **www.Eduspace.com.**

In Exercises 1 and 2, identify the line that has each slope.

1. (a) $m = \frac{2}{3}$ 2. (a) $m = 0$

   (b) m is undefined.   (b) $m = -\frac{3}{4}$

   (c) $m = -2$ (c) $m = 1$

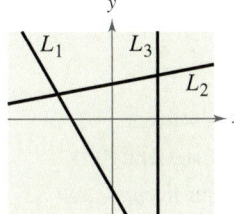   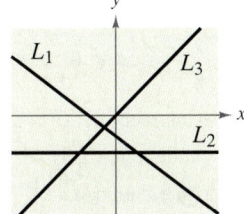

In Exercises 3 and 4, sketch the lines through the point with the indicated slopes on the same set of coordinate axes.

	Point			*Slopes*		

3. $(2, 3)$ (a) 0   (b) 1   (c) 2   (d) -3

4. $(-4, 1)$ (a) 3   (b) -3 (c) $\frac{1}{2}$ (d) Undefined

In Exercises 5–8, estimate the slope of the line.

5.   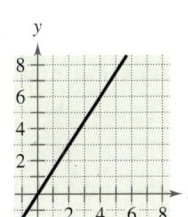   6.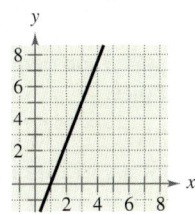

7.   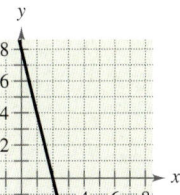   8.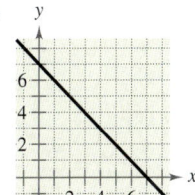

In Exercises 9–20, find the slope and y-intercept (if possible) of the equation of the line. Sketch the line.

9. $y = 5x + 3$ 10. $y = x - 10$

11. $y = -\frac{1}{2}x + 4$ 12. $y = -\frac{3}{2}x + 6$

13. $5x - 2 = 0$ 14. $3y + 5 = 0$

15. $7x + 6y = 30$ 16. $2x + 3y = 9$

17. $y - 3 = 0$ 18. $y + 4 = 0$

19. $x + 5 = 0$ 20. $x - 2 = 0$

In Exercises 21–28, plot the points and find the slope of the line passing through the pair of points.

21. $(-3, -2), (1, 6)$ 22. $(2, 4), (4, -4)$

23. $(-6, -1), (-6, 4)$ 24. $(0, -10), (-4, 0)$

25. $\left(\frac{11}{2}, -\frac{4}{3}\right), \left(-\frac{3}{2}, -\frac{1}{3}\right)$ 26. $\left(\frac{7}{8}, \frac{3}{4}\right), \left(\frac{5}{4}, -\frac{1}{4}\right)$

27. $(4.8, 3.1), (-5.2, 1.6)$

28. $(-1.75, -8.3), (2.25, -2.6)$

In Exercises 29–38, use the point on the line and the slope of the line to find three additional points through which the line passes. (There are many correct answers.)

Point	Slope
29. $(2, 1)$	$m = 0$
30. $(-4, 1)$	m is undefined.
31. $(5, -6)$	$m = 1$
32. $(10, -6)$	$m = -1$
33. $(-8, 1)$	m is undefined.
34. $(-3, -1)$	$m = 0$
35. $(-5, 4)$	$m = 2$
36. $(0, -9)$	$m = -2$
37. $(7, -2)$	$m = \frac{1}{2}$
38. $(-1, -6)$	$m = -\frac{1}{2}$

In Exercises 39–50, find the slope-intercept form of the equation of the line that passes through the given point and has the indicated slope. Sketch the line.

Point	Slope
39. $(0, -2)$	$m = 3$
40. $(0, 10)$	$m = -1$
41. $(-3, 6)$	$m = -2$
42. $(0, 0)$	$m = 4$
43. $(4, 0)$	$m = -\frac{1}{3}$
44. $(-2, -5)$	$m = \frac{3}{4}$
45. $(6, -1)$	m is undefined.
46. $(-10, 4)$	m is undefined.
47. $\left(4, \frac{5}{2}\right)$	$m = 0$
48. $\left(-\frac{1}{2}, \frac{3}{2}\right)$	$m = 0$
49. $(-5.1, 1.8)$	$m = 5$
50. $(2.3, -8.5)$	$m = -\frac{5}{2}$

In Exercises 51–64, find the slope-intercept form of the equation of the line passing through the points. Sketch the line.

51. $(5, -1), (-5, 5)$ **52.** $(4, 3), (-4, -4)$

53. $(-8, 1), (-8, 7)$ **54.** $(-1, 4), (6, 4)$

55. $\left(2, \frac{1}{2}\right), \left(\frac{1}{2}, \frac{5}{4}\right)$ **56.** $\left(1, 1\right), \left(6, -\frac{2}{3}\right)$

57. $\left(-\frac{1}{10}, -\frac{3}{5}\right), \left(\frac{9}{10}, -\frac{9}{5}\right)$ **58.** $\left(\frac{3}{4}, \frac{3}{2}\right), \left(-\frac{4}{3}, \frac{7}{4}\right)$

59. $(1, 0.6), (-2, -0.6)$

60. $(-8, 0.6), (2, -2.4)$

61. $(2, -1), \left(\frac{1}{3}, -1\right)$

62. $\left(\frac{1}{5}, -2\right), (-6, -2)$

63. $\left(\frac{7}{3}, -8\right), \left(\frac{7}{3}, 1\right)$

64. $(1.5, -2), (1.5, 0.2)$

In Exercises 65–68, determine whether the lines L_1 and L_2 passing through the pairs of points are parallel, perpendicular, or neither.

65. L_1: $(0, -1), (5, 9)$ **66.** L_1: $(-2, -1), (1, 5)$
    L_2: $(0, 3), (4, 1)$ L_2: $(1, 3), (5, -5)$

67. L_1: $(3, 6), (-6, 0)$ **68.** L_1: $(4, 8), (-4, 2)$
    L_2: $(0, -1), \left(5, \frac{7}{3}\right)$ L_2: $(3, -5), \left(-1, \frac{1}{3}\right)$

In Exercises 69–78, write the slope-intercept forms of the equations of the lines through the given point (a) parallel to the given line and (b) perpendicular to the given line.

Point	Line
69. $(2, 1)$	$4x - 2y = 3$
70. $(-3, 2)$	$x + y = 7$
71. $\left(-\frac{2}{3}, \frac{7}{8}\right)$	$3x + 4y = 7$
72. $\left(\frac{7}{8}, \frac{3}{4}\right)$	$5x + 3y = 0$
73. $(-1, 0)$	$y = -3$
74. $(4, -2)$	$y = 1$
75. $(2, 5)$	$x = 4$
76. $(-5, 1)$	$x = -2$
77. $(2.5, 6.8)$	$x - y = 4$
78. $(-3.9, -1.4)$	$6x + 2y = 9$

In Exercises 79–84, use the *intercept form* to find the equation of the line with the given intercepts. The intercept form of the equation of a line with intercepts $(a, 0)$ and $(0, b)$ is

$$\frac{x}{a} + \frac{y}{b} = 1, \ a \neq 0, \ b \neq 0.$$

79. x-intercept: $(2, 0)$ **80.** x-intercept: $(-3, 0)$
    y-intercept: $(0, 3)$ y-intercept: $(0, 4)$

81. x-intercept: $\left(-\frac{1}{6}, 0\right)$ **82.** x-intercept: $\left(\frac{2}{3}, 0\right)$
    y-intercept: $\left(0, -\frac{2}{3}\right)$ y-intercept: $(0, -2)$

83. Point on line: $(1, 2)$ **84.** Point on line: $(-3, 4)$
    x-intercept: $(c, 0)$ x-intercept: $(d, 0)$
    y-intercept: $(0, c), \ c \neq 0$ y-intercept: $(0, d), \ d \neq 0$

Graphical Interpretation In Exercises 85–88, identify any relationships that exist among the lines, and then use a graphing utility to graph the three equations in the same viewing window. Adjust the viewing window so that the slope appears visually correct—that is, so that parallel lines appear parallel and perpendicular lines appear to intersect at right angles.

85. (a) $y = 2x$ (b) $y = -2x$ (c) $y = \frac{1}{2}x$

86. (a) $y = \frac{2}{3}x$ (b) $y = -\frac{3}{2}x$ (c) $y = \frac{2}{3}x + 2$

87. (a) $y = -\frac{1}{2}x$ (b) $y = -\frac{1}{2}x + 3$ (c) $y = 2x - 4$

88. (a) $y = x - 8$ (b) $y = x + 1$ (c) $y = -x + 3$

In Exercises 89–92, find a relationship between x and y such that (x, y) is equidistant (the same distance) from the two points.

89. $(4, -1), (-2, 3)$

90. $(6, 5), (1, -8)$

91. $\left(3, \frac{5}{2}\right), (-7, 1)$

92. $\left(-\frac{1}{2}, -4\right), \left(\frac{7}{2}, \frac{5}{4}\right)$

93. Sales The following are the slopes of lines representing annual sales y in terms of time x in years. Use the slopes to interpret any change in annual sales for a one-year increase in time.

(a) The line has a slope of $m = 135$.

(b) The line has a slope of $m = 0$.

(c) The line has a slope of $m = -40$.

94. Revenue The following are the slopes of lines representing daily revenues y in terms of time x in days. Use the slopes to interpret any change in daily revenues for a one-day increase in time.

(a) The line has a slope of $m = 400$.

(b) The line has a slope of $m = 100$.

(c) The line has a slope of $m = 0$.

95. Average Salary The graph shows the average salaries for senior high school principals from 1990 through 2002. (Source: Educational Research Service)

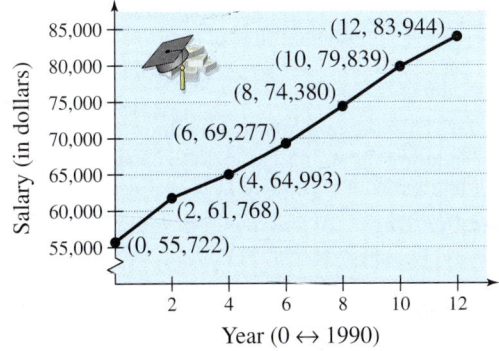

Year (0 ↔ 1990)

(a) Use the slopes to determine the time periods in which the average salary increased the greatest and the least.

(b) Find the slope of the line segment connecting the years 1990 and 2002.

(c) Interpret the meaning of the slope in part (b) in the context of the problem.

96. Net Profit The graph shows the net profits (in millions) for Applebee's International, Inc. for the years 1994 through 2003. (Source: Applebee's International, Inc.)

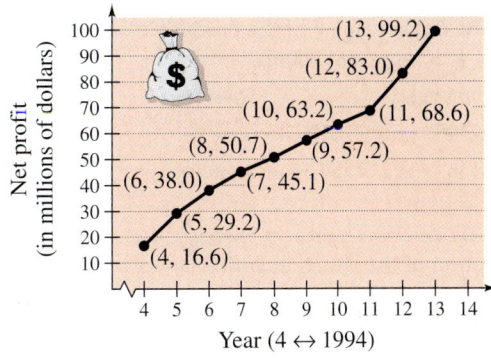

Year (4 ↔ 1994)

(a) Use the slopes to determine the years in which the net profit showed the greatest increase and the least increase.

(b) Find the slope of the line segment connecting the years 1994 and 2003.

(c) Interpret the meaning of the slope in part (b) in the context of the problem.

97. Road Grade You are driving on a road that has a 6% uphill grade (see figure). This means that the slope of the road is $\frac{6}{100}$. Approximate the amount of vertical change in your position if you drive 200 feet.

98. Road Grade From the top of a mountain road, a surveyor takes several horizontal measurements x and several vertical measurements y, as shown in the table (x and y are measured in feet).

x	300	600	900	1200	1500	1800	2100
y	-25	-50	-75	-100	-125	-150	-175

(a) Sketch a scatter plot of the data.

(b) Use a straightedge to sketch the line that you think best fits the data.

(c) Find an equation for the line you sketched in part (b).

(d) Interpret the meaning of the slope of the line in part (c) in the context of the problem.

(e) The surveyor needs to put up a road sign that indicates the steepness of the road. For instance, a surveyor would put up a sign that states "8% grade" on a road with a downhill grade that has a slope of $-\frac{8}{100}$. What should the sign state for the road in this problem?

Rate of Change   In Exercises 99 and 100, you are given the dollar value of a product in 2005 and the rate at which the value of the product is expected to change during the next 5 years. Use this information to write a linear equation that gives the dollar value V of the product in terms of the year t. (Let $t = 5$ represent 2005.)

	2005 Value	Rate
99.	$2540	$125 decrease per year
100.	$156	$4.50 increase per year

Graphical Interpretation   In Exercises 101–104, match the description of the situation with its graph. Also determine the slope and *y*-intercept of each graph and interpret the slope and *y*-intercept in the context of the situation. [The graphs are labeled (a), (b), (c), and (d).]

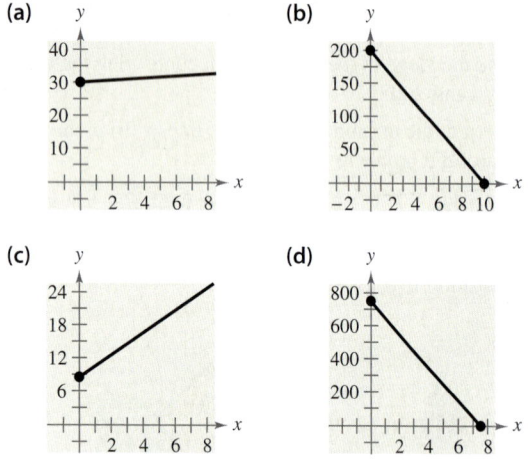

(a)   (b)   (c)   (d)

101. A person is paying $20 per week to a friend to repay a $200 loan.

102. An employee is paid $8.50 per hour plus $2 for each unit produced per hour.

103. A sales representative receives $30 per day for food plus $0.32 for each mile traveled.

104. A computer that was purchased for $750 depreciates $100 per year.

105. *Cash Flow per Share*   The cash flow per share for the Timberland Co. was $0.18 in 1995 and $4.04 in 2003. Write a linear equation that gives the cash flow per share in terms of the year. Let $t = 5$ represent 1995. Then predict the cash flows for the years 2008 and 2010. (Source: The Timberland Co.)

106. *Number of Stores*   In 1999 there were 4076 J.C. Penney stores and in 2003 there were 1078 stores. Write a linear equation that gives the number of stores in terms of the year. Let $t = 9$ represent 1999. Then predict the numbers of stores for the years 2008 and 2010. Are your answers reasonable? Explain. (Source: J.C. Penney Co.)

107. *Depreciation*   A sub shop purchases a used pizza oven for $875. After 5 years, the oven will have to be replaced. Write a linear equation giving the value V of the equipment during the 5 years it will be in use.

108. *Depreciation*   A school district purchases a high-volume printer, copier, and scanner for $25,000. After 10 years, the equipment will have to be replaced. Its value at that time is expected to be $2000. Write a linear equation giving the value V of the equipment during the 10 years it will be in use.

109. *College Enrollment*   The Pennsylvania State University had enrollments of 40,571 students in 2000 and 41,289 students in 2004 at its main campus in University Park, Pennsylvania. (Source: Penn State Fact Book)

   (a) Assuming the enrollment growth is linear, find a linear model that gives the enrollment in terms of the year t, where $t = 0$ corresponds to 2000.

   (b) Use your model from part (a) to predict the enrollments in 2008 and 2010.

   (c) What is the slope of your model? Explain its meaning in the context of the situation.

110. *College Enrollment*   The University of Florida had enrollments of 36,531 students in 1990 and 48,673 students in 2003. (Source: University of Florida)

   (a) What was the average annual change in enrollment from 1990 to 2003?

   (b) Use the average annual change in enrollment to estimate the enrollments in 1994, 1998, and 2002.

   (c) Write the equation of a line that represents the given data. What is its slope? Interpret the slope in the context of the problem.

   (d) Using the results of parts (a)–(c), write a short paragraph discussing the concepts of *slope* and *average rate of change*.

111. *Sales*   A discount outlet is offering a 15% discount on all items. Write a linear equation giving the sale price S for an item with a list price L.

112. *Hourly Wage*   A microchip manufacturer pays its assembly line workers $11.50 per hour. In addition, workers receive a piecework rate of $0.75 per unit produced. Write a linear equation for the hourly wage W in terms of the number of units x produced per hour.

113. *Cost, Revenue, and Profit*   A roofing contractor purchases a shingle delivery truck with a shingle elevator for $36,500. The vehicle requires an average expenditure of $5.25 per hour for fuel and maintenance, and the operator is paid $11.50 per hour.

   (a) Write a linear equation giving the total cost C of operating this equipment for t hours. (Include the purchase cost of the equipment.)

(b) Assuming that customers are charged $27 per hour of machine use, write an equation for the revenue R derived from t hours of use.

(c) Use the formula for profit $(P = R - C)$ to write an equation for the profit derived from t hours of use.

(d) Use the result of part (c) to find the break-even point—that is, the number of hours this equipment must be used to yield a profit of 0 dollars.

114. Rental Demand A real estate office handles an apartment complex with 50 units. When the rent per unit is $580 per month, all 50 units are occupied. However, when the rent is $625 per month, the average number of occupied units drops to 47. Assume that the relationship between the monthly rent p and the demand x is linear.

(a) Write the equation of the line giving the demand x in terms of the rent p.

(b) Use this equation to predict the number of units occupied when the rent is $655.

(c) Predict the number of units occupied when the rent is $595.

115. Geometry The length and width of a rectangular garden are 15 meters and 10 meters, respectively. A walkway of width x surrounds the garden.

(a) Draw a diagram that gives a visual representation of the problem.

(b) Write the equation for the perimeter y of the walkway in terms of x.

(c) Use a graphing utility to graph the equation for the perimeter.

(d) Determine the slope of the graph in part (c). For each additional one-meter increase in the width of the walkway, determine the increase in its perimeter.

116. Monthly Salary A pharmaceutical salesperson receives a monthly salary of $2500 plus a commission of 7% of sales. Write a linear equation for the salesperson's monthly wage W in terms of monthly sales S.

117. Business Costs A sales representative of a company using a personal car receives $120 per day for lodging and meals plus $0.38 per mile driven. Write a linear equation giving the daily cost C to the company in terms of x, the number of miles driven.

118. Sports The median salaries (in thousands of dollars) for players on the Los Angeles Dodgers from 1996 to 2003 are shown in the scatter plot. Find the equation of the line that you think best fits these data. (Let y represent the median salary and let t represent the year, with $t = 6$ corresponding to 1996.) (Source: *USA TODAY*)

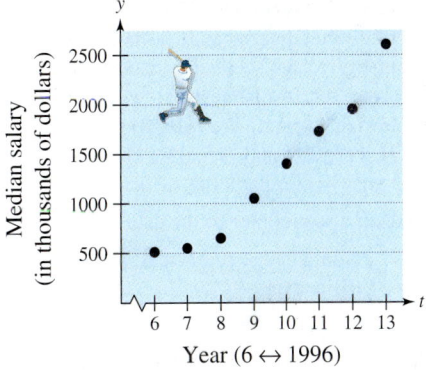

FIGURE FOR **118**

Model It

119. Data Analysis: Cell Phone Suscribers The numbers of cellular phone suscribers y (in millions) in the United States from 1990 through 2002, where x is the year, are shown as data points (x, y). (Source: Cellular Telecommunications & Internet Association)

(1990, 5.3)
(1991, 7.6)
(1992, 11.0)
(1993, 16.0)
(1994, 24.1)
(1995, 33.8)
(1996, 44.0)
(1997, 55.3)
(1998, 69.2)
(1999, 86.0)
(2000, 109.5)
(2001, 128.4)
(2002, 140.8)

(a) Sketch a scatter plot of the data. Let $x = 0$ correspond to 1990.

(b) Use a straightedge to sketch the line that you think best fits the data.

(c) Find the equation of the line from part (b). Explain the procedure you used.

(d) Write a short paragraph explaining the meanings of the slope and y-intercept of the line in terms of the data.

(e) Compare the values obtained using your model with the actual values.

(f) Use your model to estimate the number of cellular phone suscribers in 2008.

120. *Data Analysis: Average Scores* An instructor gives regular 20-point quizzes and 100-point exams in an algebra course. Average scores for six students, given as data points (x, y) where x is the average quiz score and y is the average test score, are $(18, 87)$, $(10, 55)$, $(19, 96)$, $(16, 79)$, $(13, 76)$, and $(15, 82)$. [*Note:* There are many correct answers for parts (b)–(d).]

  (a) Sketch a scatter plot of the data.

  (b) Use a straightedge to sketch the line that you think best fits the data.

  (c) Find an equation for the line you sketched in part (b).

  (d) Use the equation in part (c) to estimate the average test score for a person with an average quiz score of 17.

  (e) The instructor adds 4 points to the average test score of each student in the class. Describe the changes in the positions of the plotted points and the change in the equation of the line.

Synthesis

True or False? **In Exercises 121 and 122, determine whether the statement is true or false. Justify your answer.**

121. A line with a slope of $-\frac{5}{7}$ is steeper than a line with a slope of $-\frac{6}{7}$.

122. The line through $(-8, 2)$ and $(-1, 4)$ and the line through $(0, -4)$ and $(-7, 7)$ are parallel.

123. Explain how you could show that the points $A\,(2, 3)$, $B\,(2, 9)$, and $C\,(4, 3)$ are the vertices of a right triangle.

124. Explain why the slope of a vertical line is said to be undefined.

125. With the information shown in the graphs, is it possible to determine the slope of each line? Is it possible that the lines could have the same slope? Explain.

(a) 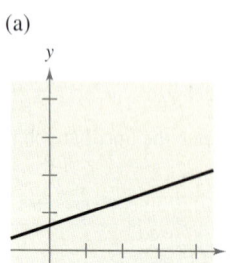 (b)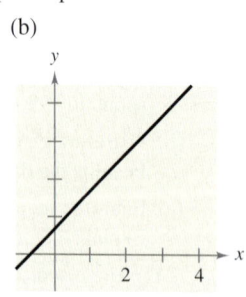

126. The slopes of two lines are -4 and $\frac{5}{2}$. Which is steeper? Explain.

127. The value V of a molding machine t years after it is purchased is

$$V = -4000t + 58{,}500, \quad 0 \le t \le 5.$$

Explain what the V-intercept and slope measure.

128. *Think About It* Is it possible for two lines with positive slopes to be perpendicular? Explain.

Skills Review

In Exercises 129–132, match the equation with its graph. [The graphs are labeled (a), (b), (c), and (d).]

(a)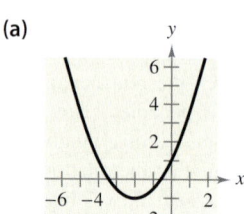

(b)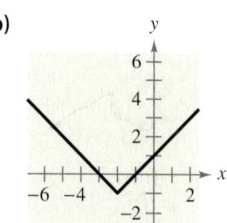

(c)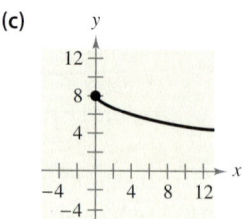

(d)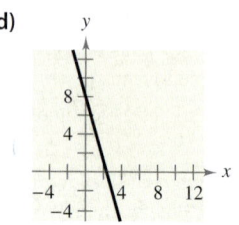

129. $y = 8 - 3x$

130. $y = 8 - \sqrt{x}$

131. $y = \frac{1}{2}x^2 + 2x + 1$

132. $y = |x + 2| - 1$

In Exercises 133–138, find all the solutions of the equation. Check your solution(s) in the original equation.

133. $-7(3 - x) = 14(x - 1)$

134. $\dfrac{8}{2x - 7} = \dfrac{4}{9 - 4x}$

135. $2x^2 - 21x + 49 = 0$

136. $x^2 - 8x + 3 = 0$

137. $\sqrt{x - 9} + 15 = 0$

138. $3x - 16\sqrt{x} + 5 = 0$

139. **Make a Decision** To work an extended application analyzing the numbers of bachelor's degrees earned by women in the United States from 1985 to 2002, visit this text's website at *college.hmco.com.* (*Data Source: U.S. National Center for Education Statistics*)

2.2 | Functions

What you should learn

- Determine whether relations between two variables are functions.
- Use function notation and evaluate functions.
- Find the domains of functions.
- Use functions to model and solve real-life problems.
- Evaluate difference quotients.

Why you should learn it

Functions can be used to model and solve real-life problems. For instance, in Exercise 100 on page 201, you will use a function to model the force of water against the face of a dam.

© Lester Lefkowitz/Corbis

Introduction to Functions

Many everyday phenomena involve two quantities that are related to each other by some rule of correspondence. The mathematical term for such a rule of correspondence is a **relation.** In mathematics, relations are often represented by mathematical equations and formulas. For instance, the simple interest I earned on $1000 for 1 year is related to the annual interest rate r by the formula $I = 1000r$.

The formula $I = 1000r$ represents a special kind of relation that matches each item from one set with *exactly one* item from a different set. Such a relation is called a **function.**

> ### Definition of Function
>
> A **function** f from a set A to a set B is a relation that assigns to each element x in the set A exactly one element y in the set B. The set A is the **domain** (or set of inputs) of the function f, and the set B contains the **range** (or set of outputs). *you can't have the same x and different ys*

To help understand this definition, look at the function that relates the time of day to the temperature in Figure 2.20.

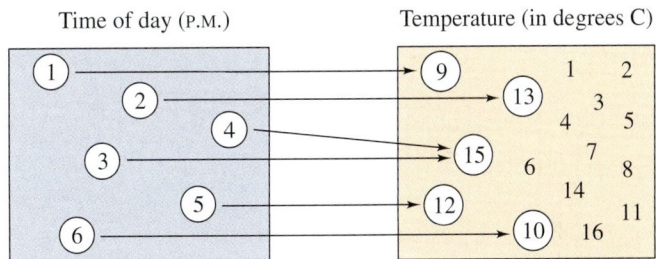

Set A is the domain.
Inputs: 1, 2, 3, 4, 5, 6

Set B contains the range.
Outputs: 9, 10, 12, 13, 15

FIGURE 2.20

This function can be represented by the following ordered pairs, in which the first coordinate (x-value) is the input and the second coordinate (y-value) is the output.

$$\{(1, 9°), (2, 13°), (3, 15°), (4, 15°), (5, 12°), (6, 10°)\}$$

> ### Characteristics of a Function from Set A to Set B
>
> 1. Each element in A must be matched with an element in B.
>
> 2. Some elements in B may not be matched with any element in A.
>
> 3. Two or more elements in A may be matched with the same element in B.
>
> 4. An element in A (the domain) cannot be matched with two different elements in B.

Functions are commonly represented in four ways.

> ## Four Ways to Represent a Function
>
> 1. *Verbally* by a sentence that describes how the input variable is related to the output variable
>
> 2. *Numerically* by a table or a list of ordered pairs that matches input values with output values
>
> 3. *Graphically* by points on a graph in a coordinate plane in which the input values are represented by the horizontal axis and the output values are represented by the vertical axis
>
> 4. *Algebraically* by an equation in two variables

To determine whether or not a relation is a function, you must decide whether each input value is matched with exactly one output value. If any input value is matched with two or more output values, the relation is not a function.

Example 1   Testing for Functions

Determine whether the relation represents y as a function of x.

a. The input value x is the number of representatives from a state, and the output value y is the number of senators.

b.

Input, x	Output, y
2	11
2	10
3	8
4	5
5	1

c.

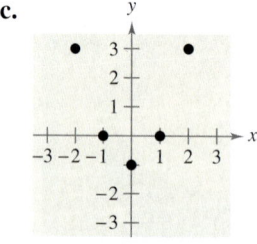

FIGURE 2.21

Solution

a. This verbal description *does* describe y as a function of x. Regardless of the value of x, the value of y is always 2. Such functions are called *constant functions*.

b. This table *does not* describe y as a function of x. The input value 2 is matched with two different y-values.

c. The graph in Figure 2.21 *does* describe y as a function of x. Each input value is matched with exactly one output value.

✓**CHECKPOINT**   Now try Exercise 5.

Representing functions by sets of ordered pairs is common in *discrete mathematics*. In algebra, however, it is more common to represent functions by equations or formulas involving two variables. For instance, the equation

$$y = x^2 \qquad \text{\color{red}{y is a function of x.}}$$

represents the variable y as a function of the variable x. In this equation, x is the **independent variable** and y is the **dependent variable.** The domain of the function is the set of all values taken on by the independent variable x, and the range of the function is the set of all values taken on by the dependent variable y.

Example 2   **Testing for Functions Represented Algebraically**

Which of the equations represent(s) y as a function of x?

a. $x^2 + y = 1$ **b.** $-x + y^2 = 1$

Solution

To determine whether y is a function of x, try to solve for y in terms of x.

a. Solving for y yields

$$x^2 + y = 1$$ Write original equation.

$$y = 1 - x^2.$$ Solve for y.

To each value of x there corresponds exactly one value of y. So, y is a function of x.

b. Solving for y yields

$$-x + y^2 = 1$$ Write original equation.

$$y^2 = 1 + x$$ Add x to each side.

$$y = \pm\sqrt{1 + x}.$$ Solve for y.

The $\pm$ indicates that to a given value of x there correspond two values of y. So, y is not a function of x.

✓**CHECKPOINT**   Now try Exercise 15.

Function Notation

When an equation is used to represent a function, it is convenient to name the function so that it can be referenced easily. For example, you know that the equation $y = 1 - x^2$ describes y as a function of x. Suppose you give this function the name "f." Then you can use the following **function notation.**

Input	Output	Equation
x	$f(x)$	$f(x) = 1 - x^2$

The symbol $f(x)$ is read as *the value of f at x* or simply f *of x*. The symbol $f(x)$ corresponds to the y-value for a given x. So, you can write $y = f(x)$. Keep in mind that f is the *name* of the function, whereas $f(x)$ is the *value* of the function at x. For instance, the function given by

$$f(x) = 3 - 2x$$

has *function values* denoted by $f(-1), f(0), f(2)$, and so on. To find these values, substitute the specified input values into the given equation.

For $x = -1$, $f(-1) = 3 - 2(-1) = 3 + 2 = 5$.

For $x = 0$, $f(0) = 3 - 2(0) = 3 - 0 = 3$.

For $x = 2$, $f(2) = 3 - 2(2) = 3 - 4 = -1$.

© Bettmann/Corbis

Historical Note
Leonhard Euler (1707–1783), a Swiss mathematician, is considered to have been the most prolific and productive mathematician in history. One of his greatest influences on mathematics was his use of symbols, or notation. The function notation $y = f(x)$ was introduced by Euler.

Although f is often used as a convenient function name and x is often used as the independent variable, you can use other letters. For instance,

$$f(x) = x^2 - 4x + 7, \quad f(t) = t^2 - 4t + 7, \quad \text{and} \quad g(s) = s^2 - 4s + 7$$

all define the same function. In fact, the role of the independent variable is that of a "placeholder." Consequently, the function could be described by

$$f(\ \ \ ) = (\ \ \ )^2 - 4(\ \ \ ) + 7.$$

> **STUDY TIP**
>
> In Example 3, note that $g(x + 2)$ is not equal to $g(x) + g(2)$. In general, $g(u + v) \neq g(u) + g(v)$.

Example 3   **Evaluating a Function**

Let $g(x) = -x^2 + 4x + 1$. Find each function value.

a. $g(2)$ **b.** $g(t)$ **c.** $g(x + 2)$

Solution

a. Replacing x with 2 in $g(x) = -x^2 + 4x + 1$ yields the following.

$$g(2) = -(2)^2 + 4(2) + 1 = -4 + 8 + 1 = 5$$

b. Replacing x with t yields the following.

$$g(t) = -(t)^2 + 4(t) + 1 = -t^2 + 4t + 1$$

c. Replacing x with $x + 2$ yields the following.

$$
\begin{aligned}
g(x + 2) &= -(x + 2)^2 + 4(x + 2) + 1 \\
&= -(x^2 + 4x + 4) + 4x + 8 + 1 \\
&= -x^2 - 4x - 4 + 4x + 8 + 1 \\
&= -x^2 + 5
\end{aligned}
$$

✓**CHECKPOINT**   Now try Exercise 29.

A function defined by two or more equations over a specified domain is called a **piecewise-defined function.**

Example 4   **A Piecewise-Defined Function**

Evaluate the function when $x = -1, 0$, and 1.

$$f(x) = \begin{cases} x^2 + 1, & x < 0 \\ x - 1, & x \geq 0 \end{cases}$$

Solution

Because $x = -1$ is less than 0, use $f(x) = x^2 + 1$ to obtain

$$f(-1) = (-1)^2 + 1 = 2.$$

For $x = 0$, use $f(x) = x - 1$ to obtain

$$f(0) = (0) - 1 = -1.$$

For $x = 1$, use $f(x) = x - 1$ to obtain

$$f(1) = (1) - 1 = 0.$$

✓**CHECKPOINT**   Now try Exercise 35.

Technology

Use a graphing utility to graph the functions given by $y = \sqrt{4 - x^2}$ and $y = \sqrt{x^2 - 4}$. What is the domain of each function? Do the domains of these two functions overlap? If so, for what values do the domains overlap?

The Domain of a Function

The domain of a function can be described explicitly or it can be *implied* by the expression used to define the function. The **implied domain** is the set of all real numbers for which the expression is defined. For instance, the function given by

$$f(x) = \frac{1}{x^2 - 4}$$ Domain excludes x-values that result in division by zero.

has an implied domain that consists of all real x other than $x = \pm 2$. These two values are excluded from the domain because division by zero is undefined. Another common type of implied domain is that used to avoid even roots of negative numbers. For example, the function given by

$$f(x) = \sqrt{x}$$ Domain excludes x-values that result in even roots of negative numbers.

is defined only for $x \geq 0$. So, its implied domain is the interval $[0, \infty)$. In general, the domain of a function *excludes* values that would cause division by zero *or* that would result in the even root of a negative number.

Example 5   Finding the Domain of a Function

Find the domain of each function.

a. f: $\{(-3, 0), (-1, 4), (0, 2), (2, 2), (4, -1)\}$ **b.** $g(x) = \dfrac{1}{x + 5}$

c. Volume of a sphere: $V = \frac{4}{3}\pi r^3$ **d.** $h(x) = \sqrt{4 - x^2}$

Solution

a. The domain of f consists of all first coordinates in the set of ordered pairs.

Domain $= \{-3, -1, 0, 2, 4\}$

b. Excluding x-values that yield zero in the denominator, the domain of g is the set of all real numbers x except $x = -5$.

c. Because this function represents the volume of a sphere, the values of the radius r must be positive. So, the domain is the set of all real numbers r such that $r > 0$.

d. This function is defined only for x-values for which

$$4 - x^2 \geq 0.$$

Using the methods described in Section 1.8, you can conclude that $-2 \leq x \leq 2$. So, the domain is the interval $[-2, 2]$.

✓CHECKPOINT   Now try Exercise 59.

In Example 5(c), note that the domain of a function may be implied by the physical context. For instance, from the equation

$$V = \frac{4}{3}\pi r^3$$

you would have no reason to restrict r to positive values, but the physical context implies that a sphere cannot have a negative or zero radius.

$$\boxed{\frac{h}{r} = 4}$$

FIGURE **2.22**

Applications

Example 6 The Dimensions of a Container

You work in the marketing department of a soft-drink company and are experimenting with a new can for iced tea that is slightly narrower and taller than a standard can. For your experimental can, the ratio of the height to the radius is 4, as shown in Figure 2.22.

a. Write the volume of the can as a function of the radius r.

b. Write the volume of the can as a function of the height h.

Solution

a. $V(r) = \pi r^2 h = \pi r^2(4r) = 4\pi r^3$ Write V as a function of r.

b. $V(h) = \pi \left(\dfrac{h}{4}\right)^2 h = \dfrac{\pi h^3}{16}$ Write V as a function of h.

✓**CHECKPOINT**   Now try Exercise 87.

Example 7 The Path of a Baseball

A baseball is hit at a point 3 feet above ground at a velocity of 100 feet per second and an angle of 45°. The path of the baseball is given by the function

$$f(x) = -0.0032x^2 + x + 3$$

where y and x are measured in feet, as shown in Figure 2.23. Will the baseball clear a 10-foot fence located 300 feet from home plate?

Baseball Path

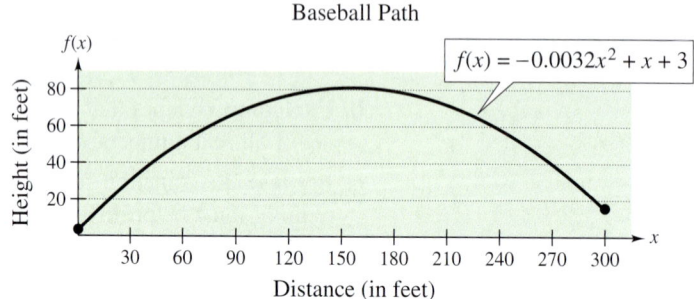

FIGURE **2.23**

Solution

When $x = 300$, the height of the baseball is

$$f(300) = -0.0032(300)^2 + 300 + 3$$
$$= 15 \text{ feet.}$$

So, the baseball will clear the fence.

✓**CHECKPOINT**   Now try Exercise 93.

In the equation in Example 7, the height of the baseball is a function of the distance from home plate.

Number of Alternative-Fueled
Vehicles in the U.S.

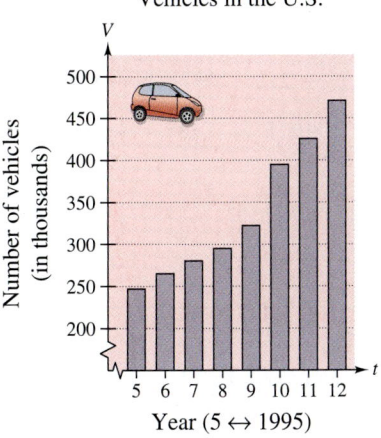

FIGURE **2.24**

Example 8   **Alternative-Fueled Vehicles**

The number V (in thousands) of alternative-fueled vehicles in the United States increased in a linear pattern from 1995 to 1999, as shown in Figure 2.24. Then, in 2000, the number of vehicles took a jump and, until 2002, increased in a different linear pattern. These two patterns can be approximated by the function

$$V(t) = \begin{cases} 18.08t + 155.3 & 5 \le t \le 9 \\ 38.20t + 10.2, & 10 \le t \le 12 \end{cases}$$

where t represents the year, with $t = 5$ corresponding to 1995. Use this function to approximate the number of alternative-fueled vehicles for each year from 1995 to 2002. (Source: Science Applications International Corporation; Energy Information Administration)

Solution

From 1995 to 1999, use $V(t) = 18.08t + 155.3$.

245.7	263.8	281.9	299.9	318.0
1995	1996	1997	1998	1999

From 2000 to 2002, use $V(t) = 38.20t + 10.2$.

392.2	430.4	468.6
2000	2001	2002

✓CHECKPOINT   Now try Exercise 95.

Difference Quotients

One of the basic definitions in calculus employs the ratio

$$\frac{f(x + h) - f(x)}{h}, \quad h \ne 0.$$

This ratio is called a **difference quotient,** as illustrated in Example 9.

Example 9   **Evaluating a Difference Quotient**

For $f(x) = x^2 - 4x + 7$, find $\dfrac{f(x + h) - f(x)}{h}$.

Solution

$$\frac{f(x + h) - f(x)}{h} = \frac{[(x + h)^2 - 4(x + h) + 7] - (x^2 - 4x + 7)}{h}$$

$$= \frac{x^2 + 2xh + h^2 - 4x - 4h + 7 - x^2 + 4x - 7}{h}$$

$$= \frac{2xh + h^2 - 4h}{h} = \frac{h(2x + h - 4)}{h} = 2x + h - 4, \ h \ne 0$$

✓CHECKPOINT   Now try Exercise 79.

The symbol ∫ indicates an example or exercise that highlights algebraic techniques specifically used in calculus.

You may find it easier to calculate the difference quotient in Example 9 by first finding $f(x + h)$, and then substituting the resulting expression into the difference quotient, as follows.

$$f(x + h) = (x + h)^2 - 4(x + h) + 7 = x^2 + 2xh + h^2 - 4x - 4h + 7$$

$$\frac{f(x + h) - f(x)}{h} = \frac{(x^2 + 2xh + h^2 - 4x - 4h + 7) - (x^2 - 4x + 7)}{h}$$

$$= \frac{2xh + h^2 - 4h}{h} = \frac{h(2x + h - 4)}{h} = 2x + h - 4, \quad h \neq 0$$

Summary of Function Terminology

Function: A **function** is a relationship between two variables such that to each value of the independent variable there corresponds exactly one value of the dependent variable.

Function Notation: $y = f(x)$

f is the *name* of the function.

y is the **dependent variable.**

x is the **independent variable.**

$f(x)$ is the *value of the function at x.*

Domain: The **domain** of a function is the set of all values (inputs) of the independent variable for which the function is defined. If x is in the domain of f, f is said to be *defined* at x. If x is not in the domain of f, f is said to be *undefined* at x.

Range: The **range** of a function is the set of all values (outputs) assumed by the dependent variable (that is, the set of all function values).

Implied Domain: If f is defined by an algebraic expression and the domain is not specified, the **implied domain** consists of all real numbers for which the expression is defined.

***W**RITING ABOUT **M**ATHEMATICS*

Everyday Functions In groups of two or three, identify common real-life functions. Consider everyday activities, events, and expenses, such as long distance telephone calls and car insurance. Here are two examples.

a. The statement, "Your happiness is a function of the grade you receive in this course" *is not* a correct mathematical use of the word "function." The word "happiness" is ambiguous.

b. The statement, "Your federal income tax is a function of your adjusted gross income" *is* a correct mathematical use of the word "function." Once you have determined your adjusted gross income, your income tax can be determined.

Describe your functions in words. Avoid using ambiguous words. Can you find an example of a piecewise-defined function?

2.2   Exercises

VOCABULARY CHECK: Fill in the blanks.

1. A relation that assigns to each element x from a set of inputs, or _____, exactly one element y in a set of outputs, or _____, is called a _____.

2. Functions are commonly represented in four different ways, _____, _____, _____, and _____.

3. For an equation that represents y as a function of x, the set of all values taken on by the _____ variable x is the domain, and the set of all values taken on by the _____ variable y is the range.

4. The function given by

$$f(x) = \begin{cases} 2x - 1, & x < 0 \\ x^2 + 4, & x \geq 0 \end{cases}$$

is an example of a _____ function.

5. If the domain of the function f is not given, then the set of values of the independent variable for which the expression is defined is called the _____ _____.

6. In calculus, one of the basic definitions is that of a _____ _____, given by $\dfrac{f(x + h) - f(x)}{h}$, $h \neq 0$.

PREREQUISITE SKILLS REVIEW: Practice and review algebra skills needed for this section at **www.Eduspace.com.**

In Exercises 1–4, is the relationship a function?

1.
Domain   *Range*

$-2 \longrightarrow 5$
-1 6
0 7
1 8
2

2.
Domain   *Range*

$-2 \longrightarrow 3$
-1 4
0 5
1
2

3.
Domain   *Range*

National League $\begin{array}{c} \text{Cubs} \\ \text{Pirates} \\ \text{Dodgers} \end{array}$

American League $\begin{array}{c} \text{Orioles} \\ \text{Yankees} \\ \text{Twins} \end{array}$

4.
Domain   *Range*
(Year)   (Number of North Atlantic tropical storms and hurricanes)

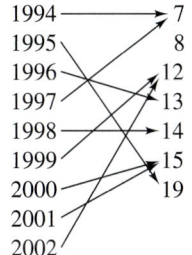

1994   7
1995   8
1996   12
1997   13
1998   14
1999   15
2000   19
2001
2002

In Exercises 5–8, does the table describe a function? Explain your reasoning.

5.

Input value	-2	-1	0	1	2
Output value	-8	-1	0	1	8

6.

Input value	0	1	2	1	0
Output value	-4	-2	0	2	4

7.

Input value	10	7	4	7	10
Output value	3	6	9	12	15

8.

Input value	0	3	9	12	15
Output value	3	3	3	3	3

In Exercises 9 and 10, which sets of ordered pairs represent functions from *A* to *B*? Explain.

9. $A = \{0, 1, 2, 3\}$ and $B = \{-2, -1, 0, 1, 2\}$

(a) $\{(0, 1), (1, -2), (2, 0), (3, 2)\}$

(b) $\{(0, -1), (2, 2), (1, -2), (3, 0), (1, 1)\}$

(c) $\{(0, 0), (1, 0), (2, 0), (3, 0)\}$

(d) $\{(0, 2), (3, 0), (1, 1)\}$

10. $A = \{a, b, c\}$ and $B = \{0, 1, 2, 3\}$

(a) $\{(a, 1), (c, 2), (c, 3), (b, 3)\}$

(b) $\{(a, 1), (b, 2), (c, 3)\}$

(c) $\{(1, a), (0, a), (2, c), (3, b)\}$

(d) $\{(c, 0), (b, 0), (a, 3)\}$

Circulation of Newspapers   **In Exercises 11 and 12, use the graph, which shows the circulation (in millions) of daily newspapers in the United States.**   (Source: Editor & Publisher Company)

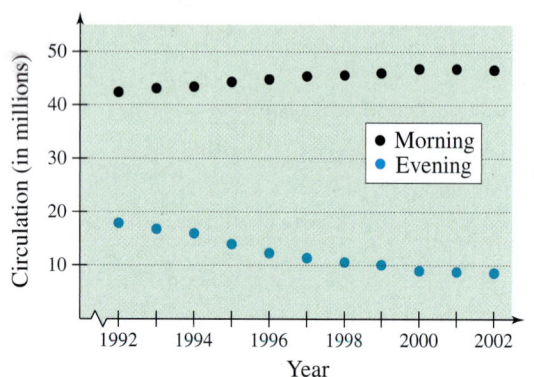

Year

11. Is the circulation of morning newspapers a function of the year? Is the circulation of evening newspapers a function of the year? Explain.

12. Let $f(x)$ represent the circulation of evening newspapers in year x. Find $f(1998)$.

In Exercises 13–24, determine whether the equation represents y as a function of x.

13. $x^2 + y^2 = 4$

14. $x = y^2$

15. $x^2 + y = 4$

16. $x + y^2 = 4$

17. $2x + 3y = 4$

18. $(x - 2)^2 + y^2 = 4$

19. $y^2 = x^2 - 1$

20. $y = \sqrt{x + 5}$

21. $y = |4 - x|$

22. $|y| = 4 - x$

23. $x = 14$

24. $y = -75$

In Exercises 25–38, evaluate the function at each specified value of the independent variable and simplify.

25. $f(x) = 2x - 3$

(a) $f(1)$ (b) $f(-3)$ (c) $f(x - 1)$

26. $g(y) = 7 - 3y$

(a) $g(0)$ (b) $g\left(\frac{7}{3}\right)$ (c) $g(s + 2)$

27. $V(r) = \frac{4}{3}\pi r^3$

(a) $V(3)$ (b) $V\left(\frac{3}{2}\right)$ (c) $V(2r)$

28. $h(t) = t^2 - 2t$

(a) $h(2)$ (b) $h(1.5)$ (c) $h(x + 2)$

29. $f(y) = 3 - \sqrt{y}$

(a) $f(4)$ (b) $f(0.25)$ (c) $f(4x^2)$

30. $f(x) = \sqrt{x + 8} + 2$

(a) $f(-8)$ (b) $f(1)$ (c) $f(x - 8)$

31. $q(x) = \dfrac{1}{x^2 - 9}$

(a) $q(0)$ (b) $q(3)$ (c) $q(y + 3)$

32. $q(t) = \dfrac{2t^2 + 3}{t^2}$

(a) $q(2)$ (b) $q(0)$ (c) $q(-x)$

33. $f(x) = \dfrac{|x|}{x}$

(a) $f(2)$ (b) $f(-2)$ (c) $f(x - 1)$

34. $f(x) = |x| + 4$

(a) $f(2)$ (b) $f(-2)$ (c) $f(x^2)$

35. $f(x) = \begin{cases} 2x + 1, & x < 0 \\ 2x + 2, & x \geq 0 \end{cases}$

(a) $f(-1)$ (b) $f(0)$ (c) $f(2)$

36. $f(x) = \begin{cases} x^2 + 2, & x \leq 1 \\ 2x^2 + 2, & x > 1 \end{cases}$

(a) $f(-2)$ (b) $f(1)$ (c) $f(2)$

37. $f(x) = \begin{cases} 3x - 1, & x < -1 \\ 4, & -1 \leq x \leq 1 \\ x^2, & x > 1 \end{cases}$

(a) $f(-2)$ (b) $f\left(-\frac{1}{2}\right)$ (c) $f(3)$

38. $f(x) = \begin{cases} 4 - 5x, & x \leq -2 \\ 0, & -2 < x < 2 \\ x^2 + 1, & x > 2 \end{cases}$

(a) $f(-3)$ (b) $f(4)$ (c) $f(-1)$

In Exercises 39–44, complete the table.

39. $f(x) = x^2 - 3$

x	-2	-1	0	1	2
$f(x)$					

40. $g(x) = \sqrt{x - 3}$

x	3	4	5	6	7
$g(x)$					

41. $h(t) = \frac{1}{2}|t + 3|$

t	-5	-4	-3	-2	-1
$h(t)$					

42. $f(s) = \dfrac{|s - 2|}{s - 2}$

s	0	1	$\frac{3}{2}$	$\frac{5}{2}$	4
$f(s)$					

43. $f(x) = \begin{cases} -\frac{1}{2}x + 4, & x \le 0 \\ (x - 2)^2, & x > 0 \end{cases}$

x	-2	-1	0	1	2
$f(x)$					

44. $f(x) = \begin{cases} 9 - x^2, & x < 3 \\ x - 3, & x \ge 3 \end{cases}$

x	1	2	3	4	5
$f(x)$					

In Exercises 45–52, find all real values of x such that $f(x) = 0$.

45. $f(x) = 15 - 3x$

46. $f(x) = 5x + 1$

47. $f(x) = \dfrac{3x - 4}{5}$

48. $f(x) = \dfrac{12 - x^2}{5}$

49. $f(x) = x^2 - 9$

50. $f(x) = x^2 - 8x + 15$

51. $f(x) = x^3 - x$

52. $f(x) = x^3 - x^2 - 4x + 4$

In Exercises 53–56, find the value(s) of x for which $f(x) = g(x)$.

53. $f(x) = x^2 + 2x + 1, \quad g(x) = 3x + 3$

54. $f(x) = x^4 - 2x^2, \quad g(x) = 2x^2$

55. $f(x) = \sqrt{3x} + 1, \quad g(x) = x + 1$

56. $f(x) = \sqrt{x} - 4, \quad g(x) = 2 - x$

In Exercises 57–70, find the domain of the function.

57. $f(x) = 5x^2 + 2x - 1$

58. $g(x) = 1 - 2x^2$

59. $h(t) = \dfrac{4}{t}$

60. $s(y) = \dfrac{3y}{y + 5}$

61. $g(y) = \sqrt{y - 10}$

62. $f(t) = \sqrt[3]{t + 4}$

63. $f(x) = \sqrt[4]{1 - x^2}$

64. $f(x) = \sqrt[4]{x^2 + 3x}$

65. $g(x) = \dfrac{1}{x} - \dfrac{3}{x + 2}$

66. $h(x) = \dfrac{10}{x^2 - 2x}$

67. $f(s) = \dfrac{\sqrt{s - 1}}{s - 4}$

68. $f(x) = \dfrac{\sqrt{x + 6}}{6 + x}$

69. $f(x) = \dfrac{x - 4}{\sqrt{x}}$

70. $f(x) = \dfrac{x - 5}{\sqrt{x^2 - 9}}$

In Exercises 71–74, assume that the domain of f is the set $A = \{-2, -1, 0, 1, 2\}$. Determine the set of ordered pairs that represents the function f.

71. $f(x) = x^2$

72. $f(x) = x^2 - 3$

73. $f(x) = |x| + 2$

74. $f(x) = |x + 1|$

Exploration   In Exercises 75–78, match the data with one of the following functions

$$f(x) = cx, \quad g(x) = cx^2, \quad h(x) = c\sqrt{|x|}, \quad \text{and} \quad r(x) = \dfrac{c}{x}$$

and determine the value of the constant c that will make the function fit the data in the table.

75.

x	-4	-1	0	1	4
y	-32	-2	0	-2	-32

76.

x	-4	-1	0	1	4
y	-1	$-\frac{1}{4}$	0	$\frac{1}{4}$	1

77.

x	-4	-1	0	1	4
y	-8	-32	Undef.	32	8

78.

x	-4	-1	0	1	4
y	6	3	0	3	6

∫ In Exercises 79–86, find the difference quotient and simplify your answer.

79. $f(x) = x^2 - x + 1, \quad \dfrac{f(2 + h) - f(2)}{h}, h \ne 0$

80. $f(x) = 5x - x^2, \quad \dfrac{f(5 + h) - f(5)}{h}, h \ne 0$

81. $f(x) = x^3 + 3x, \quad \dfrac{f(x + h) - f(x)}{h}, h \ne 0$

82. $f(x) = 4x^2 - 2x, \quad \dfrac{f(x + h) - f(x)}{h}, h \ne 0$

83. $g(x) = \dfrac{1}{x^2}, \quad \dfrac{g(x) - g(3)}{x - 3}, x \ne 3$

84. $f(t) = \dfrac{1}{t - 2}, \quad \dfrac{f(t) - f(1)}{t - 1}, t \ne 1$

85. $f(x) = \sqrt{5x}, \quad \dfrac{f(x) - f(5)}{x - 5}, x \ne 5$

86. $f(x) = x^{2/3} + 1, \quad \dfrac{f(x) - f(8)}{x - 8}, x \ne 8$

87. *Geometry*   Write the area A of a square as a function of its perimeter P.

88. *Geometry*   Write the area A of a circle as a function of its circumference C.

The symbol ∫ indicates an example or exercise that highlights algebraic techniques specifically used in calculus.

89. *Maximum Volume* An open box of maximum volume is to be made from a square piece of material 24 centimeters on a side by cutting equal squares from the corners and turning up the sides (see figure).

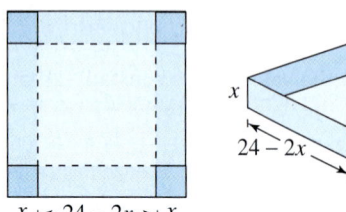

(a) The table shows the volume V (in cubic centimeters) of the box for various heights x (in centimeters). Use the table to estimate the maximum volume.

Height, x	1	2	3	4	5	6
Volume, V	484	800	972	1024	980	864

(b) Plot the points (x, V) from the table in part (a). Does the relation defined by the ordered pairs represent V as a function of x?

(c) If V is a function of x, write the function and determine its domain.

90. *Maximum Profit* The cost per unit in the production of a portable CD player is $60. The manufacturer charges $90 per unit for orders of 100 or less. To encourage large orders, the manufacturer reduces the charge by $0.15 per CD player for each unit ordered in excess of 100 (for example, there would be a charge of $87 per CD player for an order size of 120).

(a) The table shows the profit P (in dollars) for various numbers of units ordered, x. Use the table to estimate the maximum profit.

Units, x	110	120	130	140
Profit, P	3135	3240	3315	3360

Units, x	150	160	170
Profit, P	3375	3360	3315

(b) Plot the points (x, P) from the table in part (a). Does the relation defined by the ordered pairs represent P as a function of x?

(c) If P is a function of x, write the function and determine its domain.

91. *Geometry* A right triangle is formed in the first quadrant by the x- and y-axes and a line through the point $(2, 1)$ (see figure). Write the area A of the triangle as a function of x, and determine the domain of the function.

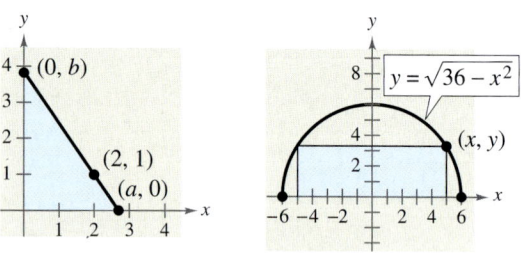

FIGURE FOR 91   FIGURE FOR 92

92. *Geometry* A rectangle is bounded by the x-axis and the semicircle $y = \sqrt{36 - x^2}$ (see figure). Write the area A of the rectangle as a function of x, and determine the domain of the function.

93. *Path of a Ball* The height y (in feet) of a baseball thrown by a child is

$$y = -\frac{1}{10}x^2 + 3x + 6$$

where x is the horizontal distance (in feet) from where the ball was thrown. Will the ball fly over the head of another child 30 feet away trying to catch the ball? (Assume that the child who is trying to catch the ball holds a baseball glove at a height of 5 feet.)

94. *Prescription Drugs* The amounts d (in billions of dollars) spent on prescription drugs in the United States from 1991 to 2002 (see figure) can be approximated by the model

$$d(t) = \begin{cases} 5.0t + 37, & 1 \le t \le 7 \\ 18.7t - 64, & 8 \le t \le 12 \end{cases}$$

where t represents the year, with $t = 1$ corresponding to 1991. Use this model to find the amount spent on prescription drugs in each year from 1991 to 2002. (Source: U.S. Centers for Medicare & Medicaid Services)

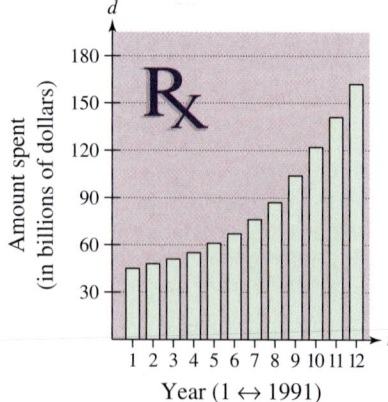

Year (1 ↔ 1991)

95. *Average Price* The average prices p (in thousands of dollars) of a new mobile home in the United States from 1990 to 2002 (see figure) can be approximated by the model

$$p(t) = \begin{cases} 0.182t^2 + 0.57t + 27.3, & 0 \le t \le 7 \\ 2.50t + 21.3, & 8 \le t \le 12 \end{cases}$$

where t represents the year, with $t = 0$ corresponding to 1990. Use this model to find the average price of a mobile home in each year from 1990 to 2002. (*Source: U.S. Census Bureau*)

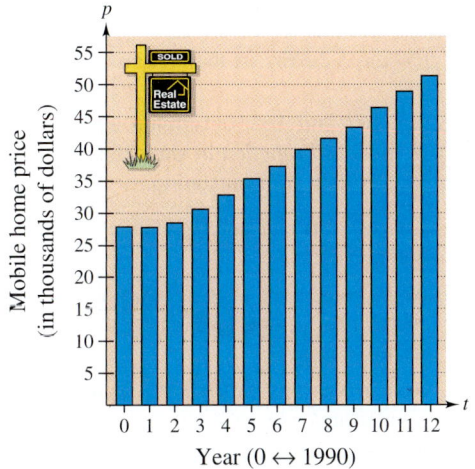

Year $(0 \leftrightarrow 1990)$

96. *Postal Regulations* A rectangular package to be sent by the U.S. Postal Service can have a maximum combined length and girth (perimeter of a cross section) of 108 inches (see figure).

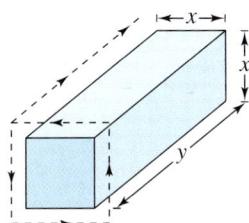

(a) Write the volume V of the package as a function of x. What is the domain of the function?

 (b) Use a graphing utility to graph your function. Be sure to use an appropriate window setting.

(c) What dimensions will maximize the volume of the package? Explain your answer.

97. *Cost, Revenue, and Profit* A company produces a product for which the variable cost is $12.30 per unit and the fixed costs are $98,000. The product sells for $17.98. Let x be the number of units produced and sold.

(a) The total cost for a business is the sum of the variable cost and the fixed costs. Write the total cost C as a function of the number of units produced.

(b) Write the revenue R as a function of the number of units sold.

(c) Write the profit P as a function of the number of units sold. (*Note:* $P = R - C$)

98. *Average Cost* The inventor of a new game believes that the variable cost for producing the game is $0.95 per unit and the fixed costs are $6000. The inventor sells each game for $1.69. Let x be the number of games sold.

(a) The total cost for a business is the sum of the variable cost and the fixed costs. Write the total cost C as a function of the number of games sold.

(b) Write the average cost per unit $\overline{C} = C/x$ as a function of x.

99. *Transportation* For groups of 80 or more people, a charter bus company determines the rate per person according to the formula

$$\text{Rate} = 8 - 0.05(n - 80), \quad n \ge 80$$

where the rate is given in dollars and n is the number of people.

(a) Write the revenue R for the bus company as a function of n.

(b) Use the function in part (a) to complete the table. What can you conclude?

n	90	100	110	120	130	140	150
$R(n)$							

100. *Physics* The force F (in tons) of water against the face of a dam is estimated by the function $F(y) = 149.76\sqrt{10}\,y^{5/2}$, where y is the depth of the water (in feet).

(a) Complete the table. What can you conclude from the table?

y	5	10	20	30	40
$F(y)$					

(b) Use the table to approximate the depth at which the force against the dam is 1,000,000 tons.

(c) Find the depth at which the force against the dam is 1,000,000 tons algebraically.

101. *Height of a Balloon* A balloon carrying a transmitter ascends vertically from a point 3000 feet from the receiving station.

(a) Draw a diagram that gives a visual representation of the problem. Let h represent the height of the balloon and let d represent the distance between the balloon and the receiving station.

(b) Write the height of the balloon as a function of d. What is the domain of the function?

Model It

102. *Wildlife*   The graph shows the numbers of threatened and endangered fish species in the world from 1996 through 2003. Let $f(t)$ represent the number of threatened and endangered fish species in the year t. (Source: U.S. Fish and Wildlife Service)

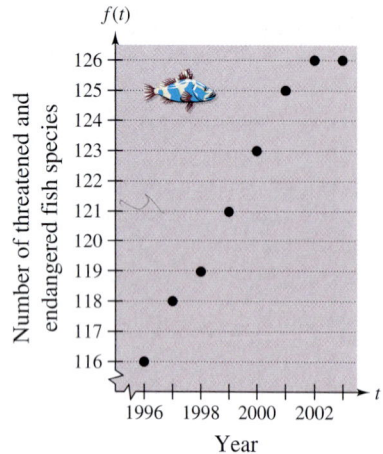

(a) Find $\dfrac{f(2003) - f(1996)}{2003 - 1996}$ and interpret the result in the context of the problem.

(b) Find a linear model for the data algebraically. Let N represent the number of threatened and endangered fish species and let $x = 6$ correspond to 1996.

(c) Use the model found in part (b) to complete the table.

x	6	7	8	9	10	11	12	13
N								

(d) Compare your results from part (c) with the actual data.

(e) Use a graphing utility to find a linear model for the data. Let $x = 6$ correspond to 1996. How does the model you found in part (b) compare with the model given by the graphing utility?

Synthesis

True or False? In Exercises 103 and 104, determine whether the statement is true or false. Justify your answer.

103. The domain of the function given by $f(x) = x^4 - 1$ is $(-\infty, \infty)$, and the range of $f(x)$ is $(0, \infty)$.

104. The set of ordered pairs $\{(-8, -2), (-6, 0), (-4, 0), (-2, 2), (0, 4), (2, -2)\}$ represents a function.

105. *Writing*   In your own words, explain the meanings of *domain* and *range*.

106. *Think About It*   Consider $f(x) = \sqrt{x - 2}$ and $g(x) = \sqrt[3]{x - 2}$. Why are the domains of f and g different?

In Exercises 107 and 108, determine whether the statements use the word *function* in ways that are mathematically correct. Explain your reasoning.

107. (a) The sales tax on a purchased item is a function of the selling price.

(b) Your score on the next algebra exam is a function of the number of hours you study the night before the exam.

108. (a) The amount in your savings account is a function of your salary.

(b) The speed at which a free-falling baseball strikes the ground is a function of the height from which it was dropped.

Skills Review

In Exercises 109–112, solve the equation.

109. $\dfrac{t}{3} + \dfrac{t}{5} = 1$ $\dfrac{5t + 3t}{15} = \dfrac{15}{15}$

110. $\dfrac{3}{t} + \dfrac{5}{t} = 1$

111. $\dfrac{3}{x(x + 1)} - \dfrac{4}{x} = \dfrac{1}{x + 1}$

112. $\dfrac{12}{x} - 3 = \dfrac{4}{x} + 9$

In Exercises 113–116, find the equation of the line passing through the pair of points.

113. $(-2, -5), (4, -1)$ **114.** $(10, 0), (1, 9)$

115. $(-6, 5), (3, -5)$ **116.** $\left(-\frac{1}{2}, 3\right), \left(\frac{11}{2}, -\frac{1}{3}\right)$

| 2.3 | **Analyzing Graphs of Functions** |

What you should learn

- Use the Vertical Line Test for functions.
- Find the zeros of functions.
- Determine intervals on which functions are increasing or decreasing and determine relative maximum and relative minimum values of functions.
- Determine the average rate of change of a function.
- Identify even and odd functions.

Why you should learn it

Graphs of functions can help you visualize relationships between variables in real life. For instance, in Exercise 86 on page 213, you will use the graph of a function to represent visually the temperature for a city over a 24–hour period.

The Graph of a Function

In Section 2.2, you studied functions from an algebraic point of view. In this section, you will study functions from a graphical perspective.

The **graph of a function** f is the collection of ordered pairs $(x, f(x))$ such that x is in the domain of f. As you study this section, remember that

$x =$ the directed distance from the y-axis

$y = f(x) =$ the directed distance from the x-axis

as shown in Figure 2.25.

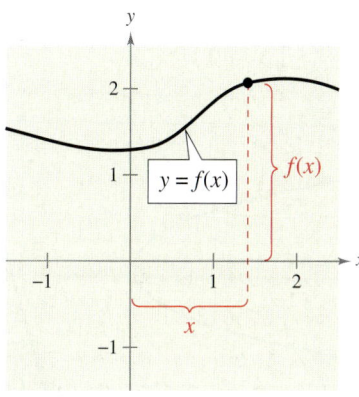

FIGURE 2.25

Example 1   **Finding the Domain and Range of a Function**

Use the graph of the function f, shown in Figure 2.26, to find (a) the domain of f, (b) the function values $f(-1)$ and $f(2)$, and (c) the range of f.

Solution

a. The closed dot at $(-1, 1)$ indicates that $x = -1$ is in the domain of f, whereas the open dot at $(5, 2)$ indicates that $x = 5$ is not in the domain. So, the domain of f is all x in the interval $[-1, 5)$.

b. Because $(-1, 1)$ is a point on the graph of f, it follows that $f(-1) = 1$. Similarly, because $(2, -3)$ is a point on the graph of f, it follows that $f(2) = -3$.

c. Because the graph does not extend below $f(2) = -3$ or above $f(0) = 3$, the range of f is the interval $[-3, 3]$.

✔CHECKPOINT   Now try Exercise 1.

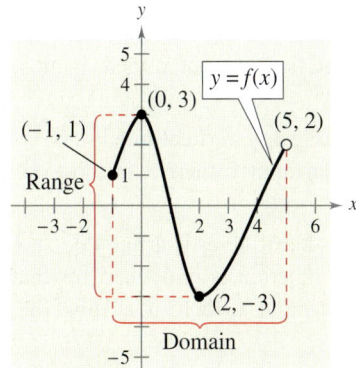

FIGURE 2.26

The use of dots (open or closed) at the extreme left and right points of a graph indicates that the graph does not extend beyond these points. If no such dots are shown, assume that the graph extends beyond these points.

By the definition of a function, at most one y-value corresponds to a given x-value. This means that the graph of a function cannot have two or more different points with the same x-coordinate, and no two points on the graph of a function can be vertically above or below each other. It follows, then, that a vertical line can intersect the graph of a function at most once. This observation provides a convenient visual test called the **Vertical Line Test** for functions.

> ### Vertical Line Test for Functions
>
> A set of points in a coordinate plane is the graph of y as a function of x if and only if no *vertical* line intersects the graph at more than one point.

Example 2 Vertical Line Test for Functions

Use the Vertical Line Test to decide whether the graphs in Figure 2.27 represent y as a function of x.

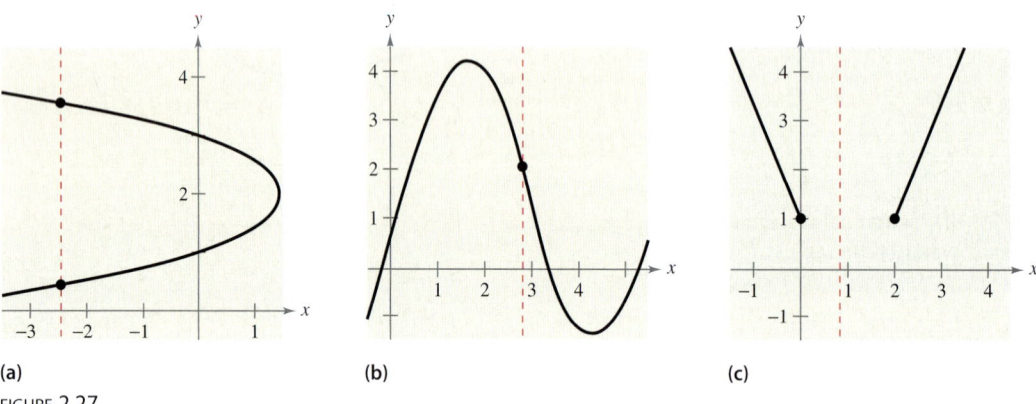

(a)   (b)   (c)

FIGURE 2.27

Solution

a. This *is not* a graph of y as a function of x, because you can find a vertical line that intersects the graph twice. That is, for a particular input x, there is more than one output y.

b. This *is* a graph of y as a function of x, because every vertical line intersects the graph at most once. That is, for a particular input x, there is at most one output y.

c. This *is* a graph of y as a function of x. (Note that if a vertical line does not intersect the graph, it simply means that the function is undefined for that particular value of x.) That is, for a particular input x, there is at most one output y.

✓**CHECKPOINT**   Now try Exercise 9.

Zeros of a Function

If the graph of a function of x has an x-intercept at $(a, 0)$, then a is a **zero** of the function.

> ### Zeros of a Function
>
> The **zeros of a function** f of x are the x-values for which $f(x) = 0$.

Example 3   Finding the Zeros of a Function

Find the zeros of each function.

a. $f(x) = 3x^2 + x - 10$ **b.** $g(x) = \sqrt{10 - x^2}$ **c.** $h(t) = \dfrac{2t - 3}{t + 5}$

Solution

To find the zeros of a function, set the function equal to zero and solve for the independent variable.

a. $3x^2 + x - 10 = 0$ Set $f(x)$ equal to 0.

$(3x - 5)(x + 2) = 0$ Factor.

$3x - 5 = 0$ $\Longrightarrow$ $x = \frac{5}{3}$ Set 1st factor equal to 0.

$x + 2 = 0$ $\Longrightarrow$ $x = -2$ Set 2nd factor equal to 0.

The zeros of f are $x = \frac{5}{3}$ and $x = -2$. In Figure 2.28, note that the graph of f has $\left(\frac{5}{3}, 0\right)$ and $(-2, 0)$ as its x-intercepts.

b. $\sqrt{10 - x^2} = 0$ Set $g(x)$ equal to 0.

$10 - x^2 = 0$ Square each side.

$10 = x^2$ Add x^2 to each side.

$\pm\sqrt{10} = x$ Extract square roots.

The zeros of g are $x = -\sqrt{10}$ and $x = \sqrt{10}$. In Figure 2.29, note that the graph of g has $\left(-\sqrt{10}, 0\right)$ and $\left(\sqrt{10}, 0\right)$ as its x-intercepts.

c. $\dfrac{2t - 3}{t + 5} = 0$ Set $h(t)$ equal to 0.

$2t - 3 = 0$ Multiply each side by $t + 5$.

$2t = 3$ Add 3 to each side.

$t = \dfrac{3}{2}$ Divide each side by 2.

The zero of h is $t = \frac{3}{2}$. In Figure 2.30, note that the graph of h has $\left(\frac{3}{2}, 0\right)$ as its t-intercept.

✔**CHECKPOINT**   Now try Exercise 15.

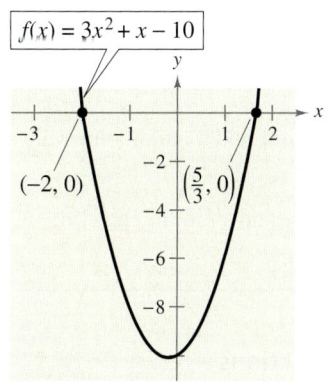

$f(x) = 3x^2 + x - 10$

Zeros of f: $x = -2, x = \frac{5}{3}$

FIGURE **2.28**

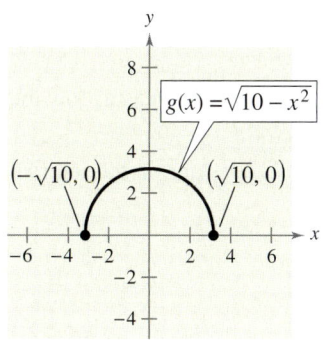

$g(x) = \sqrt{10 - x^2}$

Zeros of g: $x = \pm\sqrt{10}$

FIGURE **2.29**

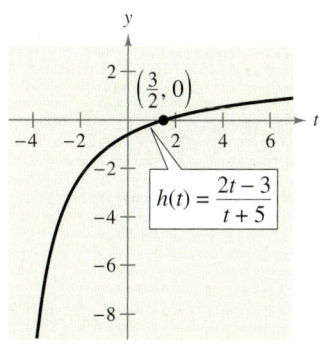

$h(t) = \dfrac{2t - 3}{t + 5}$

Zero of h: $t = \frac{3}{2}$

FIGURE **2.30**

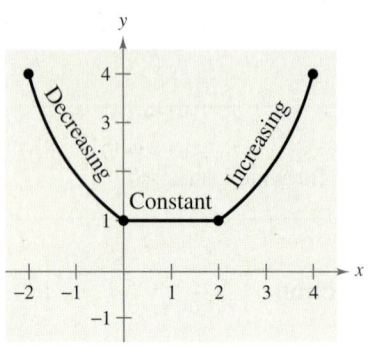

FIGURE **2.31**

Increasing and Decreasing Functions

The more you know about the graph of a function, the more you know about the function itself. Consider the graph shown in Figure 2.31. As you move from *left to right*, this graph falls from $x = -2$ to $x = 0$, is constant from $x = 0$ to $x = 2$, and rises from $x = 2$ to $x = 4$.

> ## Increasing, Decreasing, and Constant Functions
>
> A function f is **increasing** on an interval if, for any x_1 and x_2 in the interval, $x_1 < x_2$ implies $f(x_1) < f(x_2)$.
>
> A function f is **decreasing** on an interval if, for any x_1 and x_2 in the interval, $x_1 < x_2$ implies $f(x_1) > f(x_2)$.
>
> A function f is **constant** on an interval if, for any x_1 and x_2 in the interval, $f(x_1) = f(x_2)$.

Example 4   Increasing and Decreasing Functions

Use the graphs in Figure 2.32 to describe the increasing or decreasing behavior of each function.

Solution

a. This function is increasing over the entire real line.

b. This function is increasing on the interval $(-\infty, -1)$, decreasing on the interval $(-1, 1)$, and increasing on the interval $(1, \infty)$.

c. This function is increasing on the interval $(-\infty, 0)$, constant on the interval $(0, 2)$, and decreasing on the interval $(2, \infty)$.

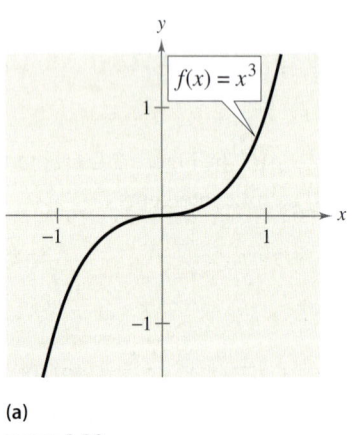

(a)

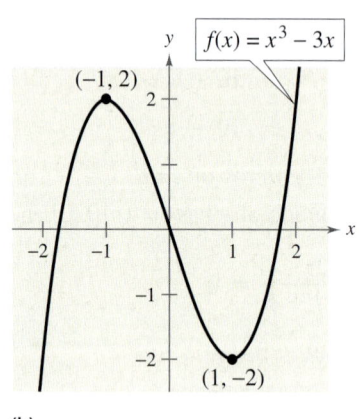

(b)

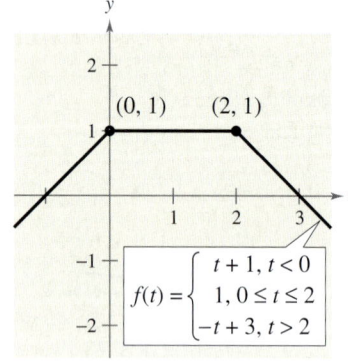

(c)

FIGURE **2.32**

✓CHECKPOINT   Now try Exercise 33.

To help you decide whether a function is increasing, decreasing, or constant on an interval, you can evaluate the function for several values of x. However, calculus is needed to determine, for certain, all intervals on which a function is increasing, decreasing, or constant.

STUDY TIP

A relative minimum or relative maximum is also referred to as a *local* minimum or *local* maximum.

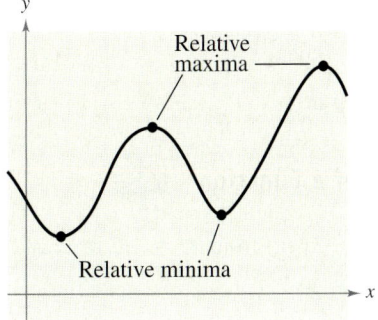

FIGURE 2.33

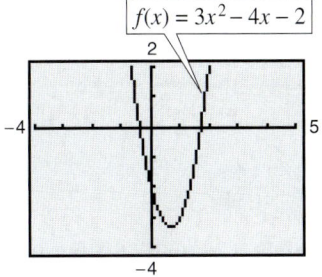

FIGURE 2.34

The points at which a function changes its increasing, decreasing, or constant behavior are helpful in determining the **relative minimum** or **relative maximum** values of the function.

Definitions of Relative Minimum and Relative Maximum

A function value $f(a)$ is called a **relative minimum** of f if there exists an interval (x_1, x_2) that contains a such that

$$x_1 < x < x_2 \quad \text{implies} \quad f(a) \le f(x).$$

A function value $f(a)$ is called a **relative maximum** of f if there exists an interval (x_1, x_2) that contains a such that

$$x_1 < x < x_2 \quad \text{implies} \quad f(a) \ge f(x).$$

Figure 2.33 shows several different examples of relative minima and relative maxima. In Section 3.1, you will study a technique for finding the *exact point* at which a second-degree polynomial function has a relative minimum or relative maximum. For the time being, however, you can use a graphing utility to find reasonable approximations of these points.

Example 5   Approximating a Relative Minimum

Use a graphing utility to approximate the relative minimum of the function given by $f(x) = 3x^2 - 4x - 2$.

Solution

The graph of f is shown in Figure 2.34. By using the *zoom* and *trace* features or the *minimum* feature of a graphing utility, you can estimate that the function has a relative minimum at the point

$$(0.67, -3.33). \qquad \text{Relative minimum}$$

Later, in Section 3.1, you will be able to determine that the exact point at which the relative minimum occurs is $\left(\frac{2}{3}, -\frac{10}{3}\right)$.

✔CHECKPOINT   Now try Exercise 49.

You can also use the *table* feature of a graphing utility to approximate numerically the relative minimum of the function in Example 5. Using a table that begins at 0.6 and increments the value of x by 0.01, you can approximate that the minimum of $f(x) = 3x^2 - 4x - 2$ occurs at the point $(0.67, -3.33)$.

Technology

If you use a graphing utility to estimate the *x*- and *y*-values of a relative minimum or relative maximum, the *zoom* feature will often produce graphs that are nearly flat. To overcome this problem, you can manually change the vertical setting of the viewing window. The graph will stretch vertically if the values of Ymin and Ymax are closer together.

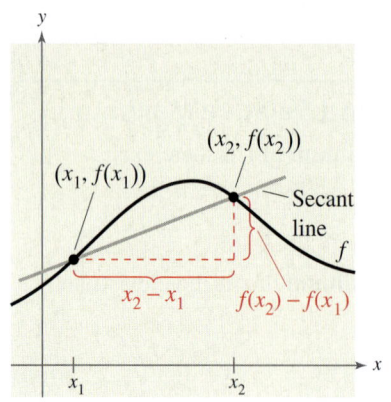

FIGURE 2.35

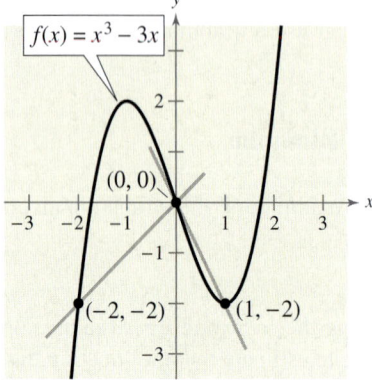

FIGURE 2.36

Average Rate of Change

In Section 2.1, you learned that the slope of a line can be interpreted as a *rate of change*. For a nonlinear graph whose slope changes at each point, the **average rate of change** between any two points $(x_1, f(x_1))$ and $(x_2, f(x_2))$ is the slope of the line through the two points (see Figure 2.35). The line through the two points is called the **secant line,** and the slope of this line is denoted as m_{sec}.

$$\text{Average rate of change of } f \text{ from } x_1 \text{ to } x_2 = \frac{f(x_2) - f(x_1)}{x_2 - x_1}$$

$$= \frac{\text{change in } y}{\text{change in } x}$$

$$= m_{sec}$$

Example 6   Average Rate of Change of a Function

Find the average rates of change of $f(x) = x^3 - 3x$ (a) from $x_1 = -2$ to $x_2 = 0$ and (b) from $x_1 = 0$ to $x_2 = 1$ (see Figure 2.36).

Solution

a. The average rate of change of f from $x_1 = -2$ to $x_2 = 0$ is

$$\frac{f(x_2) - f(x_1)}{x_2 - x_1} = \frac{f(0) - f(-2)}{0 - (-2)} = \frac{0 - (-2)}{2} = 1. \qquad \text{Secant line has positive slope.}$$

b. The average rate of change of f from $x_1 = 0$ to $x_2 = 1$ is

$$\frac{f(x_2) - f(x_1)}{x_2 - x_1} = \frac{f(1) - f(0)}{1 - 0} = \frac{-2 - 0}{1} = -2. \qquad \text{Secant line has negative slope.}$$

✓**CHECKPOINT**   Now try Exercise 63.

Example 7   Finding Average Speed

The distance s (in feet) a moving car is from a stoplight is given by the function $s(t) = 20t^{3/2}$, where t is the time (in seconds). Find the average speed of the car (a) from $t_1 = 0$ to $t_2 = 4$ seconds and (b) from $t_1 = 4$ to $t_2 = 9$ seconds.

Solution

a. The average speed of the car from $t_1 = 0$ to $t_2 = 4$ seconds is

$$\frac{s(t_2) - s(t_1)}{t_2 - t_1} = \frac{s(4) - s(0)}{4 - (0)} = \frac{160 - 0}{4} = 40 \text{ feet per second.}$$

b. The average speed of the car from $t_1 = 4$ to $t_2 = 9$ seconds is

$$\frac{s(t_2) - s(t_1)}{t_2 - t_1} = \frac{s(9) - s(4)}{9 - 4} = \frac{540 - 160}{5} = 76 \text{ feet per second.}$$

✓**CHECKPOINT**   Now try Exercise 89.

Exploration

Use the information in Example 7 to find the average speed of the car from $t_1 = 0$ to $t_2 = 9$ seconds. Explain why the result is less than the value obtained in part (b).

Even and Odd Functions

In Section 1.1, you studied different types of symmetry of a graph. In the terminology of functions, a function is said to be **even** if its graph is symmetric with respect to the y-axis and to be **odd** if its graph is symmetric with respect to the origin. The symmetry tests in Section 1.1 yield the following tests for even and odd functions.

> **Tests for Even and Odd Functions**
>
> A function $y = f(x)$ is **even** if, for each x in the domain of f,
>
> $$f(-x) = f(x).$$
>
> A function $y = f(x)$ is **odd** if, for each x in the domain of f,
>
> $$f(-x) = -f(x).$$

Exploration

Graph each of the functions with a graphing utility. Determine whether the function is *even*, *odd*, or *neither*.

$$f(x) = x^2 - x^4$$

$$g(x) = 2x^3 + 1$$

$$h(x) = x^5 - 2x^3 + x$$

$$j(x) = 2 - x^6 - x^8$$

$$k(x) = x^5 - 2x^4 + x - 2$$

$$p(x) = x^9 + 3x^5 - x^3 + x$$

What do you notice about the equations of functions that are odd? What do you notice about the equations of functions that are even? Can you describe a way to identify a function as odd or even by inspecting the equation? Can you describe a way to identify a function as neither odd nor even by inspecting the equation?

Example 8   Even and Odd Functions

a. The function $g(x) = x^3 - x$ is odd because $g(-x) = -g(x)$, as follows.

$$g(-x) = (-x)^3 - (-x) \qquad \text{Substitute } -x \text{ for } x.$$

$$= -x^3 + x \qquad \text{Simplify.}$$

$$= -(x^3 - x) \qquad \text{Distributive Property}$$

$$= -g(x) \qquad \text{Test for odd function}$$

b. The function $h(x) = x^2 + 1$ is even because $h(-x) = h(x)$, as follows.

$$h(-x) = (-x)^2 + 1 \qquad \text{Substitute } -x \text{ for } x.$$

$$= x^2 + 1 \qquad \text{Simplify.}$$

$$= h(x) \qquad \text{Test for even function}$$

The graphs and symmetry of these two functions are shown in Figure 2.37.

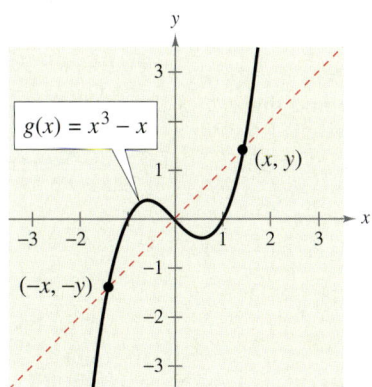

(a) **Symmetric to origin: Odd Function**

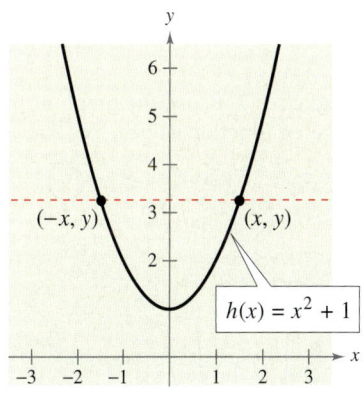

(b) **Symmetric to y-axis: Even Function**

FIGURE **2.37**

✔*CHECKPOINT*   Now try Exercise 71.

2.3   Exercises

VOCABULARY CHECK: Fill in the blanks.

1. The graph of a function f is the collection of _____ _____ or $(x, f(x))$ such that x is in the domain of f.

2. The _____ _____ _____ is used to determine whether the graph of an equation is a function of y in terms of x.

3. The _____ of a function f are the values of x for which $f(x) = 0$.

4. A function f is _____ on an interval if, for any x_1 and x_2 in the interval, $x_1 < x_2$ implies $f(x_1) > f(x_2)$.

5. A function value $f(a)$ is a relative _____ of f if there exists an interval (x_1, x_2) containing a such that $x_1 < x < x_2$ implies $f(a) \geq f(x)$.

6. The _____ _____ _____ _____ between any two points $(x_1, f(x_1))$ and $(x_2, f(x_2))$ is the slope of the line through the two points, and this line is called the _____ line.

7. A function f is _____ if for the each x in the domain of f, $f(-x) = -f(x)$.

8. A function f is _____ if its graph is symmetric with respect to the y-axis.

PREREQUISITE SKILLS REVIEW: Practice and review algebra skills needed for this section at **www.Eduspace.com**.

In Exercises 1–4, use the graph of the function to find the domain and range of f.

1.

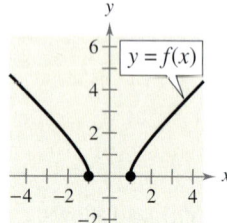

2.

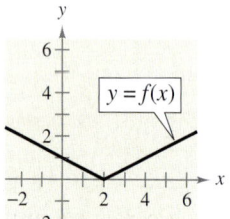

3.

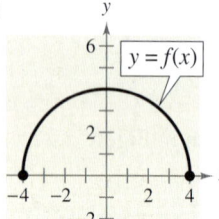

4.

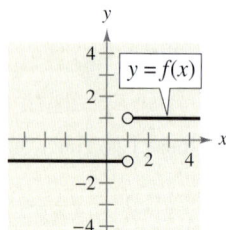

In Exercises 5–8, use the graph of the function to find the indicated function values.

5. (a) $f(-2)$ (b) $f(-1)$
    (c) $f\left(\frac{1}{2}\right)$ (d) $f(1)$

6. (a) $f(-1)$ (b) $f(2)$
    (c) $f(0)$ (d) $f(1)$

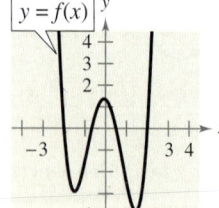

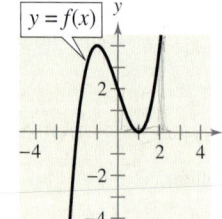

7. (a) $f(-2)$ (b) $f(1)$
    (c) $f(0)$ (d) $f(2)$

8. (a) $f(2)$ (b) $f(1)$
    (c) $f(3)$ (d) $f(-1)$

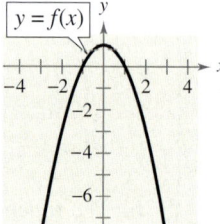

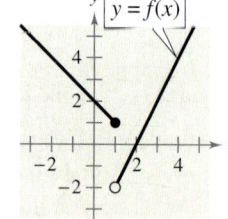

In Exercises 9–14, use the Vertical Line Test to determine whether y is a function of x. To print an enlarged copy of the graph, go to the website www.mathgraphs.com.

9. $y = \frac{1}{2}x^2$

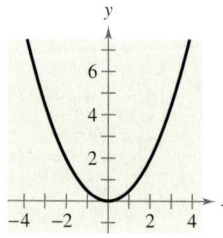

10. $y = \frac{1}{4}x^3$

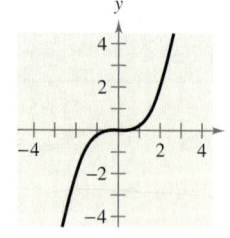

11. $x - y^2 = 1$

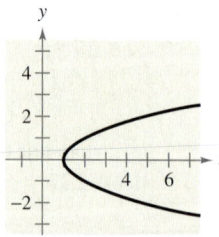

12. $x^2 + y^2 = 25$

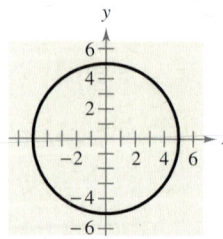

13. $x^2 = 2xy - 1$

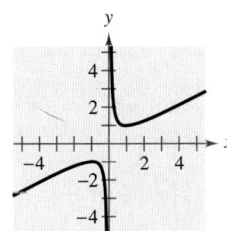

14. $x = |y + 2|$

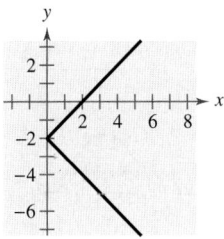

In Exercises 15–24, find the zeros of the function algebraically.

15. $f(x) = 2x^2 - 7x - 30$

16. $f(x) = 3x^2 + 22x - 16$

17. $f(x) = \dfrac{x}{9x^2 - 4}$

18. $f(x) = \dfrac{x^2 - 9x + 14}{4x}$

19. $f(x) = \frac{1}{2}x^3 - x$

20. $f(x) = x^3 - 4x^2 - 9x + 36$

21. $f(x) = 4x^3 - 24x^2 - x + 6$

22. $f(x) = 9x^4 - 25x^2$

23. $f(x) = \sqrt{2x} - 1$

24. $f(x) = \sqrt{3x + 2}$

In Exercises 25–30, (a) use a graphing utility to graph the function and find the zeros of the function and (b) verify your results from part (a) algebraically.

25. $f(x) = 3 + \dfrac{5}{x}$

26. $f(x) = x(x - 7)$

27. $f(x) = \sqrt{2x + 11}$

28. $f(x) = \sqrt{3x - 14} - 8$

29. $f(x) = \dfrac{3x - 1}{x - 6}$

30. $f(x) = \dfrac{2x^2 - 9}{3 - x}$

In Exercises 31–38, determine the intervals over which the function is increasing, decreasing, or constant.

31. $f(x) = \frac{3}{2}x$

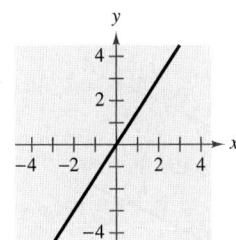

32. $f(x) = x^2 - 4x$

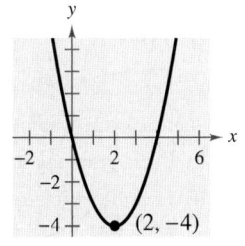

33. $f(x) = x^3 - 3x^2 + 2$

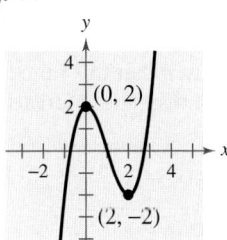

34. $f(x) = \sqrt{x^2 - 1}$

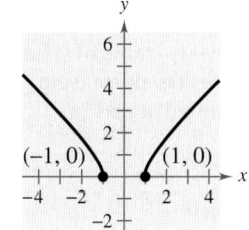

35. $f(x) = \begin{cases} x + 3, & x \le 0 \\ 3, & 0 < x \le 2 \\ 2x + 1, & x > 2 \end{cases}$

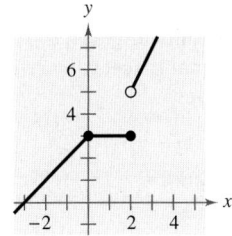

36. $f(x) = \begin{cases} 2x + 1, & x \le -1 \\ x^2 - 2, & x > -1 \end{cases}$

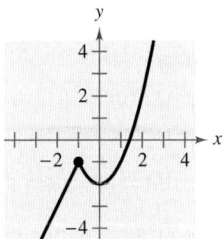

37. $f(x) = |x + 1| + |x - 1|$

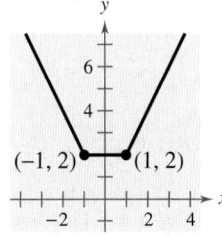

38. $f(x) = \dfrac{x^2 + x + 1}{x + 1}$

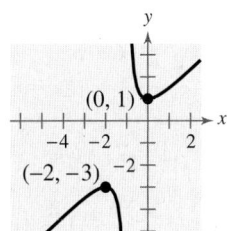

In Exercises 39–48, (a) use a graphing utility to graph the function and visually determine the intervals over which the function is increasing, decreasing, or constant, and (b) make a table of values to verify whether the function is increasing, decreasing, or constant over the intervals you identified in part (a).

39. $f(x) = 3$

40. $g(x) = x$

41. $g(s) = \dfrac{s^2}{4}$

42. $h(x) = x^2 - 4$

43. $f(t) = -t^4$

44. $f(x) = 3x^4 - 6x^2$

45. $f(x) = \sqrt{1 - x}$

46. $f(x) = x\sqrt{x + 3}$

47. $f(x) = x^{3/2}$

48. $f(x) = x^{2/3}$

In Exercises 49–54, use a graphing utility to graph the function and approximate (to two decimal places) any relative minimum or relative maximum values.

49. $f(x) = (x - 4)(x + 2)$

50. $f(x) = 3x^2 - 2x - 5$

51. $f(x) = -x^2 + 3x - 2$

52. $f(x) = -2x^2 + 9x$

53. $f(x) = x(x - 2)(x + 3)$

54. $f(x) = x^3 - 3x^2 - x + 1$

In Exercises 55–62, graph the function and determine the interval(s) for which $f(x) \geq 0$.

55. $f(x) = 4 - x$

56. $f(x) = 4x + 2$

57. $f(x) = x^2 + x$

58. $f(x) = x^2 - 4x$

59. $f(x) = \sqrt{x - 1}$

60. $f(x) = \sqrt{x + 2}$

61. $f(x) = -(1 + |x|)$

62. $f(x) = \frac{1}{2}(2 + |x|)$

In Exercises 63–70, find the average rate of change of the function from x_1 to x_2.

	Function	x-Values
63.	$f(x) = -2x + 15$	$x_1 = 0, x_2 = 3$
64.	$f(x) = 3x + 8$	$x_1 = 0, x_2 = 3$
65.	$f(x) = x^2 + 12x - 4$	$x_1 = 1, x_2 = 5$
66.	$f(x) = x^2 - 2x + 8$	$x_1 = 1, x_2 = 5$
67.	$f(x) = x^3 - 3x^2 - x$	$x_1 = 1, x_2 = 3$
68.	$f(x) = -x^3 + 6x^2 + x$	$x_1 = 1, x_2 = 6$
69.	$f(x) = -\sqrt{x - 2} + 5$	$x_1 = 3, x_2 = 11$
70.	$f(x) = -\sqrt{x + 1} + 3$	$x_1 = 3, x_2 = 8$

In Exercises 71–76, determine whether the function is even, odd, or neither. Then describe the symmetry.

71. $f(x) = x^6 - 2x^2 + 3$

72. $h(x) = x^3 - 5$

73. $g(x) = x^3 - 5x$

74. $f(x) = x\sqrt{1 - x^2}$

75. $f(t) = t^2 + 2t - 3$

76. $g(s) = 4s^{2/3}$

In Exercises 77–80, write the height h of the rectangle as a function of x.

77.

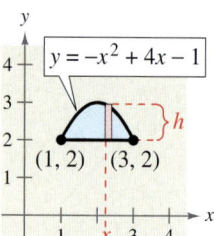

$y = -x^2 + 4x - 1$

78.

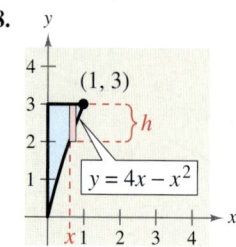

$y = 4x - x^2$

79.

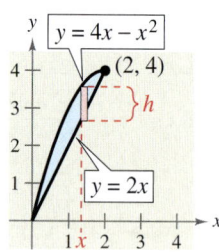

$y = 4x - x^2$

$y = 2x$

80.

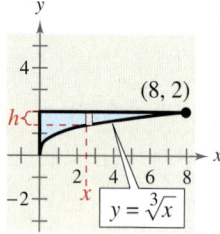

$y = \sqrt[3]{x}$

In Exercises 81–84, write the length L of the rectangle as a function of y.

81.

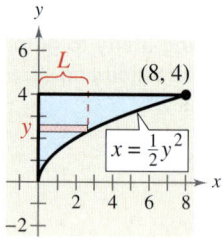

$x = \frac{1}{2}y^2$

82.

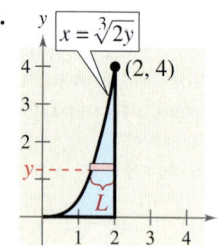

$x = \sqrt[3]{2y}$

83.

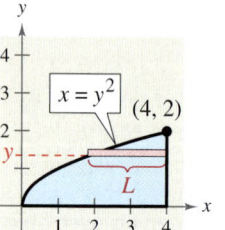

$x = y^2$

84.

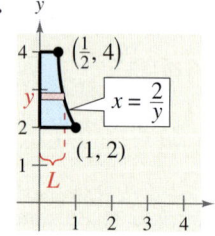

$x = \frac{2}{y}$

85. *Electronics* The number of lumens (time rate of flow of light) L from a fluorescent lamp can be approximated by the model

$$L = -0.294x^2 + 97.744x - 664.875, \quad 20 \leq x \leq 90$$

where x is the wattage of the lamp.

(a) Use a graphing utility to graph the function.

(b) Use the graph from part (a) to estimate the wattage necessary to obtain 2000 lumens.

Model It

86. Data Analysis: Temperature   The table shows the temperature y (in degrees Fahrenheit) of a certain city over a 24-hour period. Let x represent the time of day, where $x = 0$ corresponds to 6 A.M.

Time, x	Temperature, y
0	34
2	50
4	60
6	64
8	63
10	59
12	53
14	46
16	40
18	36
20	34
22	37
24	45

A model that represents these data is given by

$$y = 0.026x^3 - 1.03x^2 + 10.2x + 34, \quad 0 \le x \le 24.$$

(a) Use a graphing utility to create a scatter plot of the data. Then graph the model in the same viewing window.

(b) How well does the model fit the data?

(c) Use the graph to approximate the times when the temperature was increasing and decreasing.

(d) Use the graph to approximate the maximum and minimum temperatures during this 24-hour period.

(e) Could this model be used to predict the temperature for the city during the next 24-hour period? Why or why not?

87. Coordinate Axis Scale   Each function models the specified data for the years 1995 through 2005, with $t = 5$ corresponding to 1995. Estimate a reasonable scale for the vertical axis (e.g., hundreds, thousands, millions, etc.) of the graph and justify your answer. (There are many correct answers.)

(a) $f(t)$ represents the average salary of college professors.

(b) $f(t)$ represents the U.S. population.

(c) $f(t)$ represents the percent of the civilian work force that is unemployed.

88. Geometry   Corners of equal size are cut from a square with sides of length 8 meters (see figure).

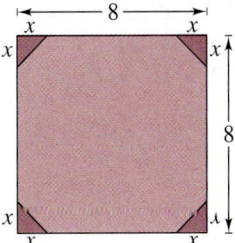

(a) Write the area A of the resulting figure as a function of x. Determine the domain of the function.

(b) Use a graphing utility to graph the area function over its domain. Use the graph to find the range of the function.

(c) Identify the figure that would result if x were chosen to be the maximum value in the domain of the function. What would be the length of each side of the figure?

89. Digital Music Sales   The estimated revenues r (in billions of dollars) from sales of digital music from 2002 to 2007 can be approximated by the model

$$r = 15.639t^3 - 104.75t^2 + 303.5t - 301, \quad 2 \le t \le 7$$

where t represents the year, with $t = 2$ corresponding to 2002.   (Source: Fortune)

(a) Use a graphing utility to graph the model.

(b) Find the average rate of change of the model from 2002 to 2007. Interpret your answer in the context of the problem.

90. Foreign College Students   The numbers of foreign students F (in thousands) enrolled in colleges in the United States from 1992 to 2002 can be approximated by the model.

$$F = 0.004t^4 + 0.46t^2 + 431.6, \quad 2 \le t \le 12$$

where t represents the year, with $t = 2$ corresponding to 1992.   (Source: Institute of International Education)

(a) Use a graphing utility to graph the model.

(b) Find the average rate of change of the model from 1992 to 2002. Interpret your answer in the context of the problem.

(c) Find the five-year time periods when the rate of change was the greatest and the least.

Physics In Exercises 91–96, (a) use the position equation $s = -16t^2 + v_0 t + s_0$ to write a function that represents the situation, (b) use a graphing utility to graph the function, (c) find the average rate of change of the function from t_1 to t_2, (d) interpret your answer to part (c) in the context of the problem, (e) find the equation of the secant line through t_1 and t_2, and (f) graph the secant line in the same viewing window as your position function.

91. An object is thrown upward from a height of 6 feet at a velocity of 64 feet per second.

$t_1 = 0, t_2 = 3$

92. An object is thrown upward from a height of 6.5 feet at a velocity of 72 feet per second.

$t_1 = 0, t_2 = 4$

93. An object is thrown upward from ground level at a velocity of 120 feet per second.

$t_1 = 3, t_2 = 5$

94. An object is thrown upward from ground level at a velocity of 96 feet per second.

$t_1 = 2, t_2 = 5$

95. An object is dropped from a height of 120 feet.

$t_1 = 0, t_2 = 2$

96. An object is dropped from a height of 80 feet.

$t_1 = 1, t_2 = 2$

Synthesis

True or False? In Exercises 97 and 98, determine whether the statement is true or false. Justify your answer.

97. A function with a square root cannot have a domain that is the set of real numbers.

98. It is possible for an odd function to have the interval $[0, \infty)$ as its domain.

99. If f is an even function, determine whether g is even, odd, or neither. Explain.

(a) $g(x) = -f(x)$

(b) $g(x) = f(-x)$

(c) $g(x) = f(x) - 2$

(d) $g(x) = f(x - 2)$

100. *Think About It* Does the graph in Exercise 11 represent x as a function of y? Explain.

Think About It In Exercises 101–104, find the coordinates of a second point on the graph of a function f if the given point is on the graph and the function is (a) even and (b) odd.

101. $\left(-\frac{3}{2}, 4\right)$

102. $\left(-\frac{5}{3}, -7\right)$

103. $(4, 9)$

104. $(5, -1)$

 105. *Writing* Use a graphing utility to graph each function. Write a paragraph describing any similarities and differences you observe among the graphs.

(a) $y = x$ (b) $y = x^2$

(c) $y = x^3$ (d) $y = x^4$

(e) $y = x^5$ (f) $y = x^6$

 106. *Conjecture* Use the results of Exercise 105 to make a conjecture about the graphs of $y = x^7$ and $y = x^8$. Use a graphing utility to graph the functions and compare the results with your conjecture.

Skills Review

In Exercises 107–110, solve the equation.

107. $x^2 - 10x = 0$

108. $100 - (x - 5)^2 = 0$

109. $x^3 - x = 0$

110. $16x^2 - 40x + 25 = 0$

In Exercises 111–114, evaluate the function at each specified value of the independent variable and simplify.

111. $f(x) = 5x - 8$

(a) $f(9)$ (b) $f(-4)$ (c) $f(x - 7)$

112. $f(x) = x^2 - 10x$

(a) $f(4)$ (b) $f(-8)$ (c) $f(x - 4)$

113. $f(x) = \sqrt{x - 12} - 9$

(a) $f(12)$ (b) $f(40)$ (c) $f\left(-\sqrt{36}\right)$

114. $f(x) = x^4 - x - 5$

(a) $f(-1)$ (b) $f\left(\frac{1}{2}\right)$ (c) $f\left(2\sqrt{3}\right)$

In Exercises 115 and 116, find the difference quotient and simplify your answer.

115. $f(x) = x^2 - 2x + 9, \quad \dfrac{f(3 + h) - f(3)}{h}, \ h \neq 0$

116. $f(x) = 5 + 6x - x^2, \quad \dfrac{f(6 + h) - f(6)}{h}, \ h \neq 0$

<table>
<tr><td>**2.4**</td><td>**A Library of Parent Functions**</td></tr>
</table>

What you should learn

- Identify and graph linear and squaring functions.
- Identify and graph cubic, square root, and reciprocal functions.
- Identify and graph step and other piecewise-defined functions.
- Recognize graphs of parent functions.

Why you should learn it

Step functions can be used to model real-life situations. For instance, in Exercise 63 on page 221, you will use a step function to model the cost of sending an overnight package from Los Angeles to Miami.

© Getty Images

Linear and Squaring Functions

One of the goals of this text is to enable you to recognize the basic shapes of the graphs of different types of functions. For instance, you know that the graph of the **linear function** $f(x) = ax + b$ is a line with slope $m = a$ and y-intercept at $(0, b)$. The graph of the linear function has the following characteristics.

- The domain of the function is the set of all real numbers.
- The range of the function is the set of all real numbers.
- The graph has an x-intercept of $(-b/m, 0)$ and a y-intercept of $(0, b)$.
- The graph is increasing if $m > 0$, decreasing if $m < 0$, and constant if $m = 0$.

Example 1   Writing a Linear Function

Write the linear function f for which $f(1) = 3$ and $f(4) = 0$.

Solution

To find the equation of the line that passes through $(x_1, y_1) = (1, 3)$ and $(x_2, y_2) = (4, 0)$, first find the slope of the line.

$$m = \frac{y_2 - y_1}{x_2 - x_1} = \frac{0 - 3}{4 - 1} = \frac{-3}{3} = -1$$

Next, use the point-slope form of the equation of a line.

$$y - y_1 = m(x - x_1) \qquad \text{Point-slope form}$$
$$y - 3 = -1(x - 1) \qquad \text{Substitute for } x_1, y_1, \text{ and } m.$$
$$y = -x + 4 \qquad \text{Simplify.}$$
$$f(x) = -x + 4 \qquad \text{Function notation}$$

The graph of this function is shown in Figure 2.38.

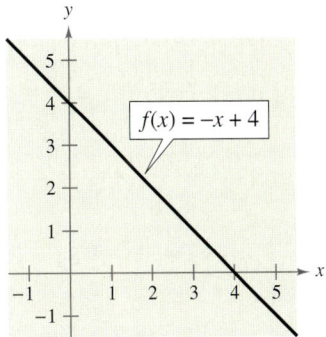

$f(x) = -x + 4$

FIGURE **2.38**

✓ **CHECKPOINT**   Now try Exercise 1.

There are two special types of linear functions, the **constant function** and the **identity function.** A constant function has the form

$$f(x) = c$$

and has the domain of all real numbers with a range consisting of a single real number c. The graph of a constant function is a horizontal line, as shown in Figure 2.39. The identity function has the form

$$f(x) = x.$$

Its domain and range are the set of all real numbers. The identity function has a slope of $m = 1$ and a y-intercept $(0, 0)$. The graph of the identity function is a line for which each x-coordinate equals the corresponding y-coordinate. The graph is always increasing, as shown in Figure 2.40.

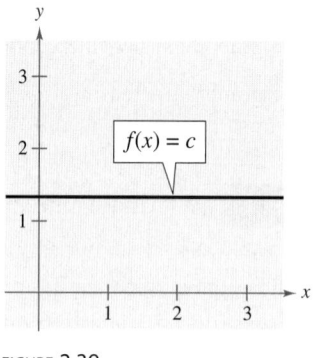

FIGURE 2.39   FIGURE 2.40

The graph of the **squaring function**

$$f(x) = x^2$$

is a U-shaped curve with the following characteristics.

- The domain of the function is the set of all real numbers.
- The range of the function is the set of all nonnegative real numbers.
- The function is even.
- The graph has an intercept at $(0, 0)$.
- The graph is decreasing on the interval $(-\infty, 0)$ and increasing on the interval $(0, \infty)$.
- The graph is symmetric with respect to the y-axis.
- The graph has a relative minimum at $(0, 0)$.

The graph of the squaring function is shown in Figure 2.41.

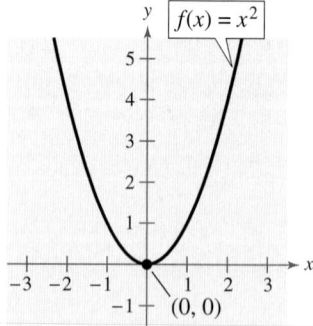

FIGURE 2.41

Cubic, Square Root, and Reciprocal Functions

The basic characteristics of the graphs of the **cubic, square root,** and **reciprocal functions** are summarized below.

1. The graph of the cubic function $f(x) = x^3$ has the following characteristics.

- The domain of the function is the set of all real numbers.
- The range of the function is the set of all real numbers.
- The function is odd.
- The graph has an intercept at $(0, 0)$.
- The graph is increasing on the interval $(-\infty, \infty)$.
- The graph is symmetric with respect to the origin.

The graph of the cubic function is shown in Figure 2.42.

2. The graph of the *square root* function $f(x) = \sqrt{x}$ has the following characteristics.

- The domain of the function is the set of all nonnegative real numbers.
- The range of the function is the set of all nonnegative real numbers.
- The graph has an intercept at $(0, 0)$.
- The graph is increasing on the interval $(0, \infty)$.

The graph of the square root function is shown in Figure 2.43.

3. The graph of the reciprocal function $f(x) = \dfrac{1}{x}$ has the following characteristics.

- The domain of the function is $(-\infty, 0) \cup (0, \infty)$.
- The range of the function is $(-\infty, 0) \cup (0, \infty)$.
- The function is odd.
- The graph does not have any intercepts.
- The graph is decreasing on the intervals $(-\infty, 0)$ and $(0, \infty)$.
- The graph is symmetric with respect to the origin.

The graph of the reciprocal function is shown in Figure 2.44.

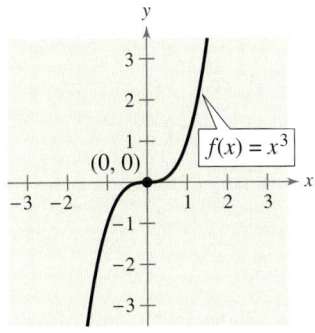

Cubic function

FIGURE 2.42

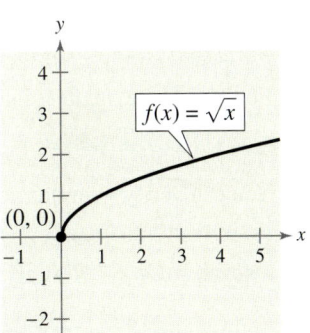

Square root function

FIGURE 2.43

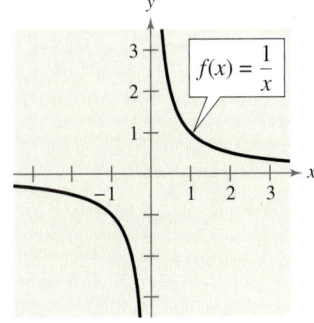

Reciprocal function

FIGURE 2.44

Step and Piecewise-Defined Functions

Functions whose graphs resemble sets of stairsteps are known as **step functions.** The most famous of the step functions is the **greatest integer function,** which is denoted by $[\![x]\!]$ and defined as

$$f(x) = [\![x]\!] = \text{the greatest integer less than or equal to } x.$$

Some values of the greatest integer function are as follows.

$$[\![-1]\!] = (\text{greatest integer} \le -1) = -1$$

$$\left[\!\left[-\tfrac{1}{2}\right]\!\right] = \left(\text{greatest integer} \le -\tfrac{1}{2}\right) = -1$$

$$\left[\!\left[\tfrac{1}{10}\right]\!\right] = \left(\text{greatest integer} \le \tfrac{1}{10}\right) = 0$$

$$[\![1.5]\!] = (\text{greatest integer} \le 1.5) = 1$$

The graph of the greatest integer function

$$f(x) = [\![x]\!]$$

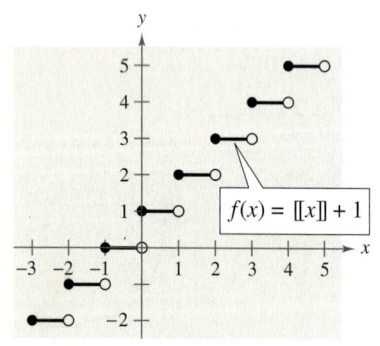

FIGURE 2.45

has the following characteristics, as shown in Figure 2.45.

- The domain of the function is the set of all real numbers.
- The range of the function is the set of all integers.
- The graph has a y-intercept at $(0, 0)$ and x-intercepts in the interval $[0, 1)$.
- The graph is constant between each pair of consecutive integers.
- The graph jumps vertically one unit at each integer value.

Example 2   Evaluating a Step Function

Evaluate the function when $x = -1, 2,$ and $\tfrac{3}{2}$.

$$f(x) = [\![x]\!] + 1$$

Solution

For $x = -1$, the greatest integer ≤ -1 is -1, so

$$f(-1) = [\![-1]\!] + 1 = -1 + 1 = 0.$$

For $x = 2$, the greatest integer ≤ 2 is 2, so

$$f(2) = [\![2]\!] + 1 = 2 + 1 = 3.$$

For $x = \tfrac{3}{2}$, the greatest integer $\le \tfrac{3}{2}$ is 1, so

$$f\left(\tfrac{3}{2}\right) = \left[\!\left[\tfrac{3}{2}\right]\!\right] + 1 = 1 + 1 = 2.$$

FIGURE 2.46

You can verify your answers by examining the graph of $f(x) = [\![x]\!] + 1$ shown in Figure 2.46.

✓CHECKPOINT   Now try Exercise 29.

Recall from Section 2.2 that a piecewise-defined function is defined by two or more equations over a specified domain. To graph a piecewise-defined function, graph each equation separately over the specified domain, as shown in Example 3.

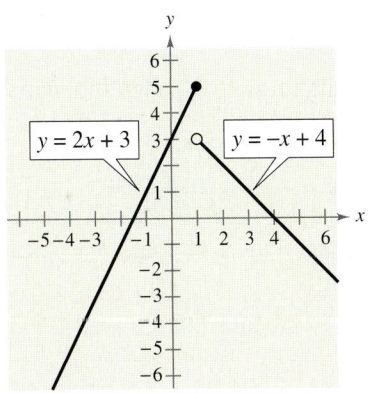

FIGURE 2.47

Example 3   **Graphing a Piecewise-Defined Function**

Sketch the graph of

$$f(x) = \begin{cases} 2x + 3, & x \le 1 \\ -x + 4, & x > 1 \end{cases}.$$

Solution

This piecewise-defined function is composed of two linear functions. At $x = 1$ and to the left of $x = 1$ the graph is the line $y = 2x + 3$, and to the right of $x = 1$ the graph is the line $y = -x + 4$, as shown in Figure 2.47. Notice that the point $(1, 5)$ is a solid dot and the point $(1, 3)$ is an open dot. This is because $f(1) = 2(1) + 3 = 5$.

✓**CHECKPOINT**   Now try Exercise 43.

Parent Functions

The eight graphs shown in Figure 2.48 represent the most commonly used functions in algebra. Familiarity with the basic characteristics of these simple graphs will help you analyze the shapes of more complicated graphs—in particular, graphs obtained from these graphs by the rigid and nonrigid transformations studied in the next section.

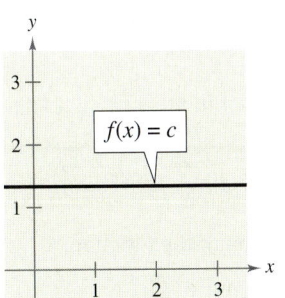

(a) Constant Function

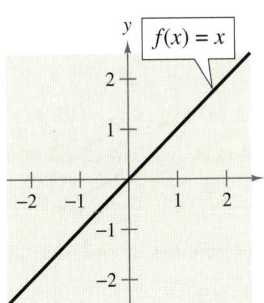

(b) Identity Function

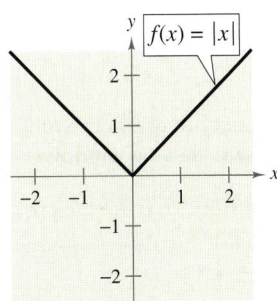

(c) Absolute Value Function

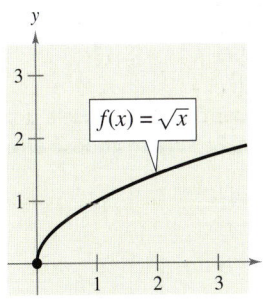

(d) Square Root Function

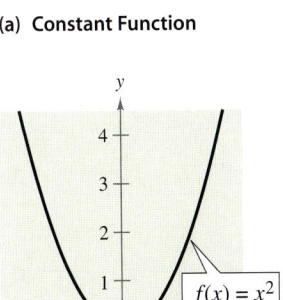

(e) Quadratic Function

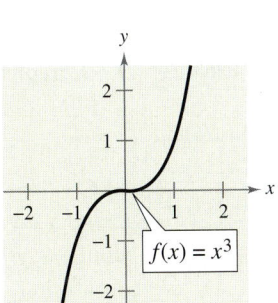

(f) Cubic Function

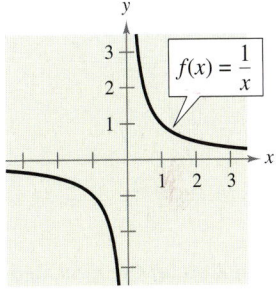

(g) Reciprocal Function

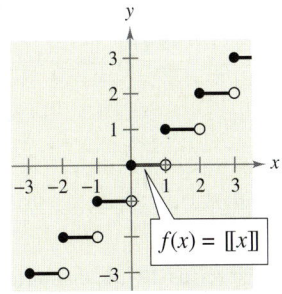

(h) Greatest Integer Function

FIGURE 2.48

2.4 Exercises

VOCABULARY CHECK: Match each function with its name.

1. $f(x) = [\![x]\!]$

2. $f(x) = x$

3. $f(x) = \dfrac{1}{x}$

4. $f(x) = x^2$

5. $f(x) = \sqrt{x}$

6. $f(x) = c$

7. $f(x) = |x|$

8. $f(x) = x^3$

9. $f(x) = ax + b$

(a) squaring function

(b) square root function

(c) cubic function

(d) linear function

(e) constant function

(f) absolute value function

(e) greatest integer function

(h) reciprocal function

(i) identity function

PREREQUISITE SKILLS REVIEW: Practice and review algebra skills needed for this section at **www.Eduspace.com.**

In Exercises 1–8, (a) write the linear function f such that it has the indicated function values and (b) sketch the graph of the function.

1. $f(1) = 4, f(0) = 6$

2. $f(-3) = -8, f(1) = 2$

3. $f(5) = -4, f(-2) = 17$

4. $f(3) = 9, f(-1) = -11$

5. $f(-5) = -1, f(5) = -1$

6. $f(-10) = 12, f(16) = -1$

7. $f\!\left(\tfrac{1}{2}\right) = -6, f(4) = -3$

8. $f\!\left(\tfrac{2}{3}\right) = -\tfrac{15}{2}, f(-4) = -11$

In Exercises 9–28, use a graphing utility to graph the function. Be sure to choose an appropriate viewing window.

9. $f(x) = -x - \tfrac{3}{4}$

10. $f(x) = 3x - \tfrac{5}{2}$

11. $f(x) = -\tfrac{1}{6}x - \tfrac{5}{2}$

12. $f(x) = \tfrac{5}{6} - \tfrac{2}{3}x$

13. $f(x) = x^2 - 2x$

14. $f(x) = -x^2 + 8x$

15. $h(x) = -x^2 + 4x + 12$

16. $g(x) = x^2 - 6x - 16$

17. $f(x) = x^3 - 1$

18. $f(x) = 8 - x^3$

19. $f(x) = (x - 1)^3 + 2$

20. $g(x) = 2(x + 3)^3 + 1$

21. $f(x) = 4\sqrt{x}$

22. $f(x) = 4 - 2\sqrt{x}$

23. $g(x) = 2 - \sqrt{x + 4}$

24. $h(x) = \sqrt{x + 2} + 3$

25. $f(x) = -\dfrac{1}{x}$

26. $f(x) = 4 + \dfrac{1}{x}$

27. $h(x) = \dfrac{1}{x + 2}$

28. $k(x) = \dfrac{1}{x - 3}$

In Exercises 29–36, evaluate the function for the indicated values.

29. $f(x) = [\![x]\!]$

(a) $f(2.1)$ (b) $f(2.9)$ (c) $f(-3.1)$ (d) $f\!\left(\tfrac{7}{2}\right)$

30. $g(x) = 2[\![x]\!]$

(a) $g(-3)$ (b) $g(0.25)$ (c) $g(9.5)$ (d) $g\!\left(\tfrac{11}{3}\right)$

31. $h(x) = [\![x + 3]\!]$

(a) $h(-2)$ (b) $h\!\left(\tfrac{1}{2}\right)$ (c) $h(4.2)$ (d) $h(-21.6)$

32. $f(x) = 4[\![x]\!] + 7$

(a) $f(0)$ (b) $f(-1.5)$ (c) $f(6)$ (d) $f\!\left(\tfrac{5}{3}\right)$

33. $h(x) = [\![3x - 1]\!]$

(a) $h(2.5)$ (b) $h(-3.2)$ (c) $h\!\left(\tfrac{7}{3}\right)$ (d) $h\!\left(-\tfrac{21}{3}\right)$

34. $k(x) = \left[\!\left[\tfrac{1}{2}x + 6\right]\!\right]$

(a) $k(5)$ (b) $k(-6.1)$ (c) $k(0.1)$ (d) $k(15)$

35. $g(x) = 3[\![x - 2]\!] + 5$

(a) $g(-2.7)$ (b) $g(-1)$ (c) $g(0.8)$ (d) $g(14.5)$

36. $g(x) = -7[\![x + 4]\!] + 6$

(a) $g\!\left(\tfrac{1}{8}\right)$ (b) $g(9)$ (c) $g(-4)$ (d) $g\!\left(\tfrac{3}{2}\right)$

In Exercises 37–42, sketch the graph of the function.

37. $g(x) = -[\![x]\!]$

38. $g(x) = 4[\![x]\!]$

39. $g(x) = [\![x]\!] - 2$

40. $g(x) = [\![x]\!] - 1$

41. $g(x) = [\![x + 1]\!]$

42. $g(x) = [\![x - 3]\!]$

In Exercises 43–50, graph the function.

43. $f(x) = \begin{cases} 2x + 3, & x < 0 \\ 3 - x, & x \geq 0 \end{cases}$

44. $g(x) = \begin{cases} x + 6, & x \leq -4 \\ \tfrac{1}{2}x - 4, & x > -4 \end{cases}$

45. $f(x) = \begin{cases} \sqrt{4 + x}, & x < 0 \\ \sqrt{4 - x}, & x \geq 0 \end{cases}$

46. $f(x) = \begin{cases} 1 - (x - 1)^2, & x \leq 2 \\ \sqrt{x - 2}, & x > 2 \end{cases}$

47. $f(x) = \begin{cases} x^2 + 5, & x \leq 1 \\ -x^2 + 4x + 3, & x > 1 \end{cases}$

48. $h(x) = \begin{cases} 3 - x^2, & x < 0 \\ x^2 + 2, & x \geq 0 \end{cases}$

49. $h(x) = \begin{cases} 4 - x^2, & x < -2 \\ 3 + x, & -2 \leq x < 0 \\ x^2 + 1, & x \geq 0 \end{cases}$

50. $k(x) = \begin{cases} 2x + 1, & x \leq -1 \\ 2x^2 - 1, & -1 < x \leq 1 \\ 1 - x^2, & x > 1 \end{cases}$

In Exercises 51 and 52, (a) use a graphing utility to graph the function, (b) state the domain and range of the function, and (c) describe the pattern of the graph.

51. $s(x) = 2\left(\frac{1}{4}x - \left[\!\left[\frac{1}{4}x\right]\!\right]\right)$ **52.** $g(x) = 2\left(\frac{1}{4}x - \left[\!\left[\frac{1}{4}x\right]\!\right]\right)^2$

In Exercises 53–60, (a) identify the parent function and the transformed parent function shown in the graph, (b) write an equation for the function shown in the graph, and (c) use a graphing utility to verify your answers in parts (a) and (b).

53.
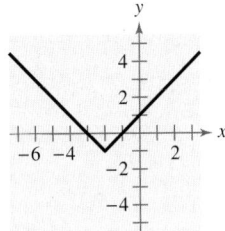

54.
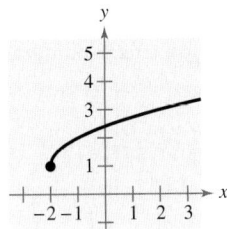

55.
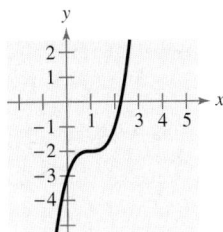

56.
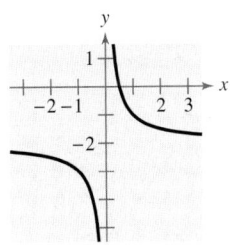

57.
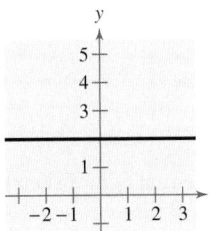

58.
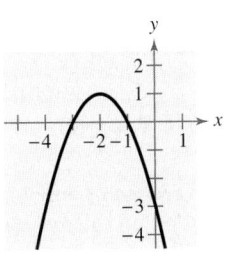

59.
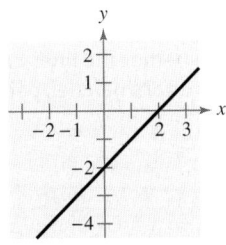

60.
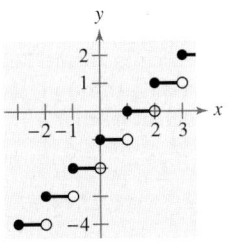

61. *Communications* The cost of a telephone call between Denver and Boise is \$0.60 for the first minute and \$0.42 for each additional minute or portion of a minute. A model for the total cost C (in dollars) of the phone call is $C = 0.60 - 0.42[\![1 - t]\!]$, $t > 0$ where t is the length of the phone call in minutes.

(a) Sketch the graph of the model.

(b) Determine the cost of a call lasting 12 minutes and 30 seconds.

62. *Communications* The cost of using a telephone calling card is \$1.05 for the first minute and \$0.38 for each additional minute or portion of a minute.

(a) A customer needs a model for the cost C of using a calling card for a call lasting t minutes. Which of the following is the appropriate model? Explain.
$$C_1(t) = 1.05 + 0.38[\![t - 1]\!]$$
$$C_2(t) = 1.05 - 0.38[\![-(t - 1)]\!]$$

(b) Graph the appropriate model. Determine the cost of a call lasting 18 minutes and 45 seconds.

63. *Delivery Charges* The cost of sending an overnight package from Los Angeles to Miami is \$10.75 for a package weighing up to but not including 1 pound and \$3.95 for each additional pound or portion of a pound. A model for the total cost C (in dollars) of sending the package is $C = 10.75 + 3.95[\![x]\!]$, $x > 0$ where x is the weight in pounds.

(a) Sketch a graph of the model.

(b) Determine the cost of sending a package that weighs 10.33 pounds.

64. *Delivery Charges* The cost of sending an overnight package from New York to Atlanta is \$9.80 for a package weighing up to but not including 1 pound and \$2.50 for each additional pound or portion of a pound.

(a) Use the greatest integer function to create a model for the cost C of overnight delivery of a package weighing x pounds, $x > 0$.

(b) Sketch the graph of the function.

65. *Wages* A mechanic is paid \$12.00 per hour for regular time and time-and-a-half for overtime. The weekly wage function is given by

$$W(h) = \begin{cases} 12h, & 0 < h \leq 40 \\ 18(h - 40) + 480, & h > 40 \end{cases}$$

where h is the number of hours worked in a week.

(a) Evaluate $W(30)$, $W(40)$, $W(45)$, and $W(50)$.

(b) The company increased the regular work week to 45 hours. What is the new weekly wage function?

66. *Snowstorm* During a nine-hour snowstorm, it snows at a rate of 1 inch per hour for the first 2 hours, at a rate of 2 inches per hour for the next 6 hours, and at a rate of 0.5 inch per hour for the final hour. Write and graph a piecewise-defined function that gives the depth of the snow during the snowstorm. How many inches of snow accumulated from the storm?

Model It

67. *Revenue* The table shows the monthly revenue y (in thousands of dollars) of a landscaping business for each month of the year 2005, with $x = 1$ representing January.

Month, x	Revenue, y
1	5.2
2	5.6
3	6.6
4	8.3
5	11.5
6	15.8
7	12.8
8	10.1
9	8.6
10	6.9
11	4.5
12	2.7

A mathematical model that represents these data is

$$f(x) = \begin{cases} -1.97x + 26.3 \\ 0.505x^2 - 1.47x + 6.3 \end{cases}.$$

(a) What is the domain of each part of the piecewise-defined function? How can you tell? Explain your reasoning.

(b) Sketch a graph of the model.

(c) Find $f(5)$ and $f(11)$, and interpret your results in the context of the problem.

(d) How do the values obtained from the model in part (b) compare with the actual data values?

68. *Fluid Flow* The intake pipe of a 100-gallon tank has a flow rate of 10 gallons per minute, and two drainpipes have flow rates of 5 gallons per minute each. The figure shows the volume V of fluid in the tank as a function of time t. Determine the combination of the input pipe and drain pipes in which the fluid is flowing in specific subintervals of the 1 hour of time shown on the graph. (There are many correct answers.)

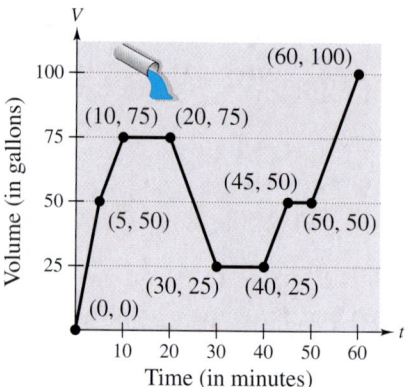

FIGURE FOR 68

Synthesis

True or False? In Exercises 69 and 70, determine whether the statement is true or false. Justify your answer.

69. A piecewise-defined function will always have at least one x-intercept or at least one y-intercept.

70. $f(x) = \begin{cases} 2, & 1 \le x < 2 \\ 4, & 2 \le x < 3 \\ 6, & 3 \le x < 4 \end{cases}$

can be rewritten as $f(x) = 2[\![x]\!]$, $1 \le x < 4$.

Exploration In Exercises 71 and 72, write equations for the piecewise-defined function shown in the graph.

71.

72.

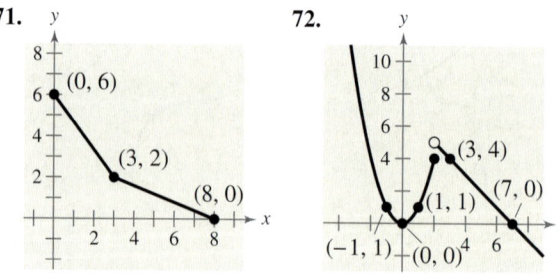

Skills Review

In Exercises 73 and 74, solve the inequality and sketch the solution on the real number line.

73. $3x + 4 \le 12 - 5x$ **74.** $2x + 1 > 6x - 9$

In Exercises 75 and 76, determine whether the lines L_1 and L_2 passing through the pairs of points are parallel, perpendicular, or neither.

75. L_1: $(-2, -2)$, $(2, 10)$ **76.** L_1: $(-1, -7)$, $(4, 3)$
  L_2: $(-1, 3)$, $(3, 9)$ L_2: $(1, 5)$, $(-2, -7)$

What you should learn

- Use vertical and horizontal shifts to sketch graphs of functions.
- Use reflections to sketch graphs of functions.
- Use nonrigid transformations to sketch graphs of functions.

Why you should learn it

Knowing the graphs of common functions and knowing how to shift, reflect, and stretch graphs of functions can help you sketch a wide variety of simple functions by hand. This skill is useful in sketching graphs of functions that model real-life data, such as in Exercise 68 on page 232, where you are asked to sketch the graph of a function that models the amounts of mortgage debt outstanding from 1990 through 2002.

© Ken Fisher/Getty Images

Shifting Graphs

Many functions have graphs that are simple transformations of the parent graphs summarized in Section 2.4. For example, you can obtain the graph of

$$h(x) = x^2 + 2$$

by shifting the graph of $f(x) = x^2$ *upward* two units, as shown in Figure 2.49. In function notation, h and f are related as follows.

$$h(x) = x^2 + 2 = f(x) + 2 \qquad \text{Upward shift of two units}$$

Similarly, you can obtain the graph of

$$g(x) = (x - 2)^2$$

by shifting the graph of $f(x) = x^2$ to the *right* two units, as shown in Figure 2.50. In this case, the functions g and f have the following relationship.

$$g(x) = (x - 2)^2 = f(x - 2) \qquad \text{Right shift of two units}$$

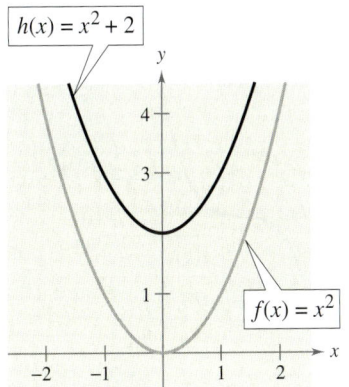

FIGURE 2.49

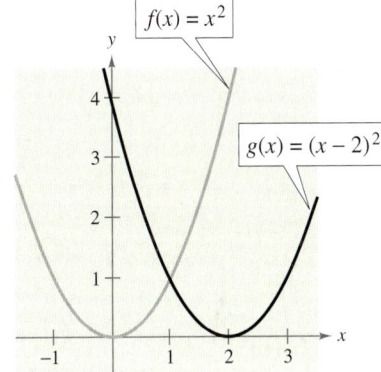

FIGURE 2.50

The following list summarizes this discussion about horizontal and vertical shifts.

Vertical and Horizontal Shifts

Let c be a positive real number. **Vertical and horizontal shifts** in the graph of $y = f(x)$ are represented as follows.

1. Vertical shift c units *upward*: $\qquad h(x) = f(x) + c$

2. Vertical shift c units *downward*: $\qquad h(x) = f(x) - c$

3. Horizontal shift c units to the *right*: $\qquad h(x) = f(x - c)$

4. Horizontal shift c units to the *left*: $\qquad h(x) = f(x + c)$

Some graphs can be obtained from combinations of vertical and horizontal shifts, as demonstrated in Example 1(b). Vertical and horizontal shifts generate a *family of functions*, each with the same shape but at different locations in the plane.

Example 1   Shifts in the Graphs of a Function

Use the graph of $f(x) = x^3$ to sketch the graph of each function.

a. $g(x) = x^3 - 1$

b. $h(x) = (x + 2)^3 + 1$

Solution

a. Relative to the graph of $f(x) = x^3$, the graph of $g(x) = x^3 - 1$ is a downward shift of one unit, as shown in Figure 2.51.

b. Relative to the graph of $f(x) = x^3$, the graph of $h(x) = (x + 2)^3 + 1$ involves a left shift of two units and an upward shift of one unit, as shown in Figure 2.52.

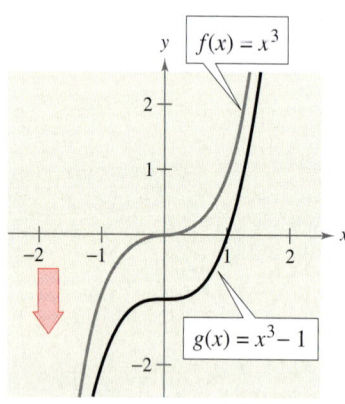

FIGURE 2.51

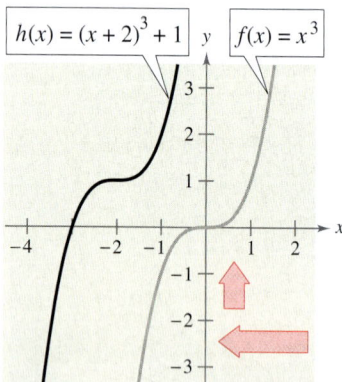

FIGURE 2.52

✓**CHECKPOINT**   Now try Exercise 1.

In Figure 2.52, notice that the same result is obtained if the vertical shift precedes the horizontal shift *or* if the horizontal shift precedes the vertical shift.

Exploration

Graphing utilities are ideal tools for exploring translations of functions. Graph f, g, and h in same viewing window. Before looking at the graphs, try to predict how the graphs of g and h relate to the graph of f.

a. $f(x) = x^2$, $g(x) = (x - 4)^2$, $h(x) = (x - 4)^2 + 3$

b. $f(x) = x^2$, $g(x) = (x + 1)^2$, $h(x) = (x + 1)^2 - 2$

c. $f(x) = x^2$, $g(x) = (x + 4)^2$, $h(x) = (x + 4)^2 + 2$

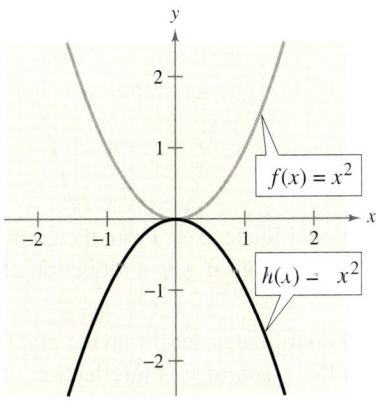

FIGURE **2.53**

Reflecting Graphs

The second common type of transformation is a **reflection.** For instance, if you consider the x-axis to be a mirror, the graph of

$$h(x) = -x^2$$

is the mirror image (or reflection) of the graph of

$$f(x) = x^2,$$

as shown in Figure 2.53.

Reflections in the Coordinate Axes

Reflections in the coordinate axes of the graph of $y = f(x)$ are represented as follows.

1. Reflection in the x-axis: $h(x) = -f(x)$

2. Reflection in the y-axis: $h(x) = f(-x)$

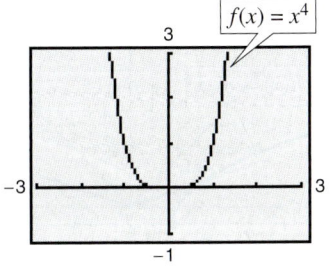

FIGURE **2.54**

Example 2   Finding Equations from Graphs

The graph of the function given by

$$f(x) = x^4$$

is shown in Figure 2.54. Each of the graphs in Figure 2.55 is a transformation of the graph of f. Find an equation for each of these functions.

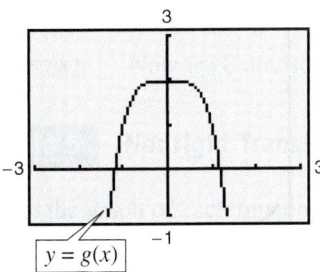

(a)

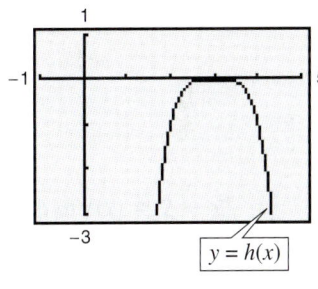

(b)

FIGURE **2.55**

Solution

a. The graph of g is a reflection in the x-axis *followed by* an upward shift of two units of the graph of $f(x) = x^4$. So, the equation for g is

$$g(x) = -x^4 + 2.$$

b. The graph of h is a horizontal shift of three units to the right *followed by* a reflection in the x-axis of the graph of $f(x) = x^4$. So, the equation for h is

$$h(x) = -(x - 3)^4.$$

✓CHECKPOINT   Now try Exercise 9.

Exploration

Reverse the order of transformations in Example 2(a). Do you obtain the same graph? Do the same for Example 2(b). Do you obtain the same graph? Explain.

2.5 Exercises

VOCABULARY CHECK:

In Exercises 1–5, fill in the blanks.

1. Horizontal shifts, vertical shifts, and reflections are called _____ transformations.

2. A reflection in the x-axis of $y = f(x)$ is represented by $h(x) =$ _____, while a reflection in the y-axis of $y = f(x)$ is represented by $h(x) =$ _____.

3. Transformations that cause a distortion in the shape of the graph of $y = f(x)$ are called _____ transformations.

4. A nonrigid transformation of $y = f(x)$ represented by $h(x) = f(cx)$ is a _____ _____ if $c > 1$ and a _____ _____ if $0 < c < 1$.

5. A nonrigid transformation of $y = f(x)$ represented by $g(x) = cf(x)$ is a _____ _____ if $c > 1$ and a _____ _____ if $0 < c < 1$.

6. Match the rigid transformation of $y = f(x)$ with the correct representation of the graph of h, where $c > 0$.
   (a) $h(x) = f(x) + c$ (i) A horizontal shift of f, c units to the right
   (b) $h(x) = f(x) - c$ (ii) A vertical shift of f, c units downward
   (c) $h(x) = f(x + c)$ (iii) A horizontal shift of f, c units to the left
   (d) $h(x) = f(x - c)$ (iv) A vertical shift of f, c units upward

PREREQUISITE SKILLS REVIEW: Practice and review algebra skills needed for this section at **www.Eduspace.com.**

1. For each function, sketch (on the same set of coordinate axes) a graph of each function for $c = -1, 1$, and 3.
   (a) $f(x) = |x| + c$
   (b) $f(x) = |x - c|$
   (c) $f(x) = |x + 4| + c$

2. For each function, sketch (on the same set of coordinate axes) a graph of each function for $c = -3, -1, 1$, and 3.
   (a) $f(x) = \sqrt{x} + c$
   (b) $f(x) = \sqrt{x - c}$
   (c) $f(x) = \sqrt{x - 3} + c$

3. For each function, sketch (on the same set of coordinate axes) a graph of each function for $c = -2, 0$, and 2.
   (a) $f(x) = [\![x]\!] + c$
   (b) $f(x) = [\![x + c]\!]$
   (c) $f(x) = [\![x - 1]\!] + c$

4. For each function, sketch (on the same set of coordinate axes) a graph of each function for $c = -3, -1, 1$, and 3.
   (a) $f(x) = \begin{cases} x^2 + c, & x < 0 \\ -x^2 + c, & x \geq 0 \end{cases}$
   (b) $f(x) = \begin{cases} (x + c)^2, & x < 0 \\ -(x + c)^2, & x \geq 0 \end{cases}$

In Exercises 5–8, use the graph of f to sketch each graph. To print an enlarged copy of the graph go to the website www.mathgraphs.com.

5. (a) $y = f(x) + 2$
   (b) $y = f(x - 2)$
   (c) $y = 2f(x)$
   (d) $y = -f(x)$
   (e) $y = f(x + 3)$
   (f) $y = f(-x)$
   (g) $y = f\left(\frac{1}{2}x\right)$

6. (a) $y = f(-x)$
   (b) $y = f(x) + 4$
   (c) $y = 2f(x)$
   (d) $y = -f(x - 4)$
   (e) $y = f(x) - 3$
   (f) $y = -f(x) - 1$
   (g) $y = f(2x)$

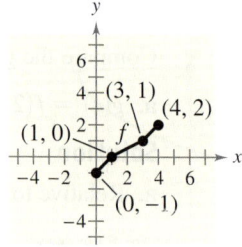

FIGURE FOR 5

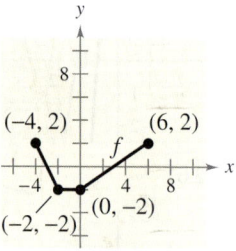

FIGURE FOR 6

7. (a) $y = f(x) - 1$
   (b) $y = f(x - 1)$
   (c) $y = f(-x)$
   (d) $y = f(x + 1)$
   (e) $y = -f(x - 2)$
   (f) $y = \frac{1}{2}f(x)$
   (g) $y = f(2x)$

8. (a) $y = f(x - 5)$
   (b) $y = -f(x) + 3$
   (c) $y = \frac{1}{3}f(x)$
   (d) $y = -f(x + 1)$
   (e) $y = f(-x)$
   (f) $y = f(x) - 10$
   (g) $y = f\left(\frac{1}{3}x\right)$

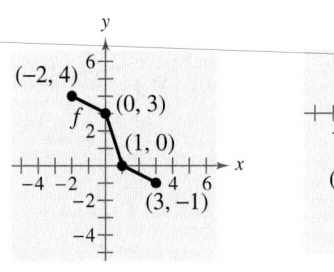

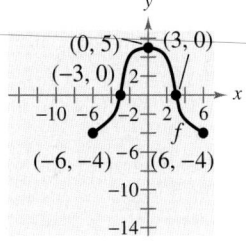

FIGURE FOR 7   FIGURE FOR 8

9. Use the graph of $f(x) = x^2$ to write an equation for each function whose graph is shown.

(a)

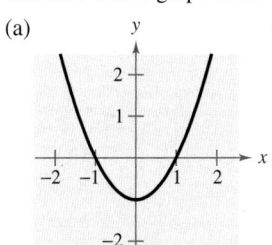

(b)

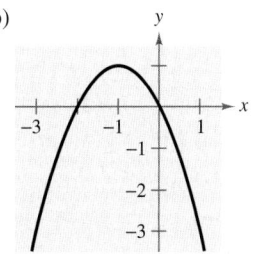

(c)

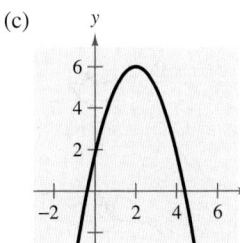

(d)

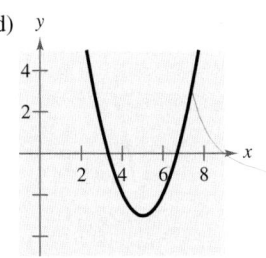

10. Use the graph of $f(x) = x^3$ to write an equation for each function whose graph is shown.

(a)

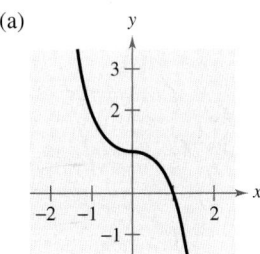

(b)

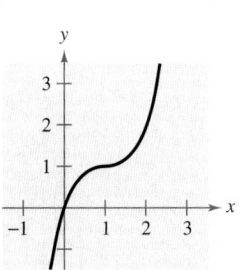

(c)

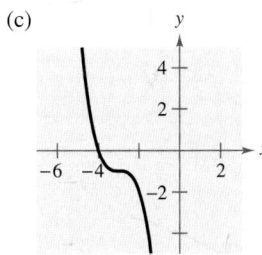

(d)

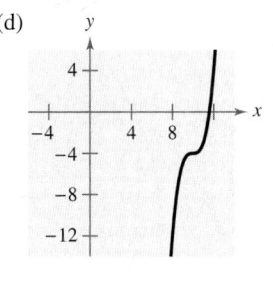

11. Use the graph of $f(x) = |x|$ to write an equation for each function whose graph is shown.

(a)

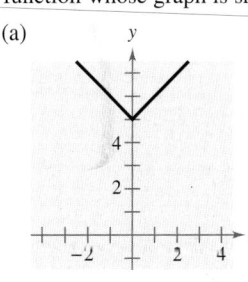

(b)

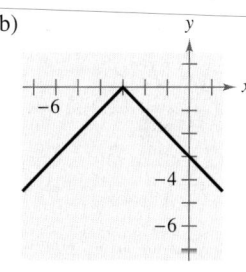

(c)

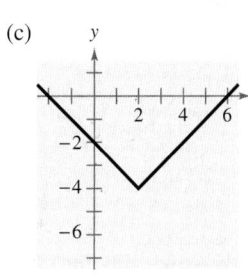

(d)

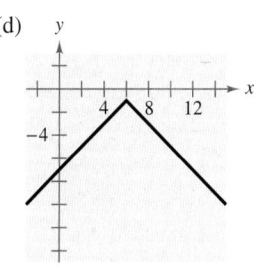

12. Use the graph of $f(x) = \sqrt{x}$ to write an equation for each function whose graph is shown.

(a)

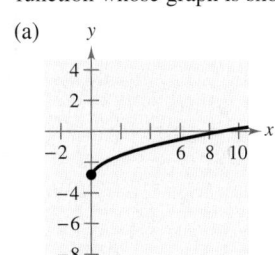

(b)

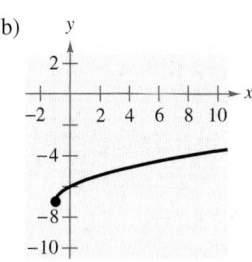

(c)

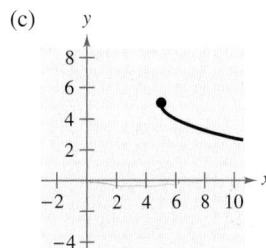

(d)

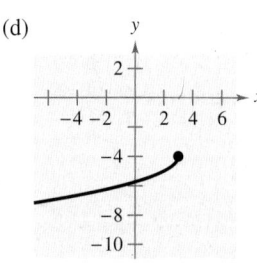

In Exercises 13–18, identify the parent function and the transformation shown in the graph. Write an equation for the function shown in the graph.

13.

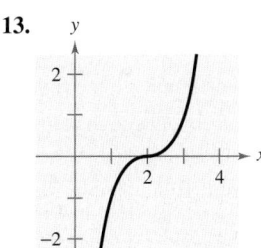

14.

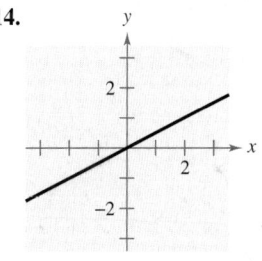

15.

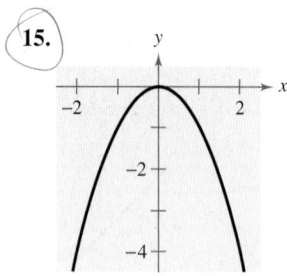

16.

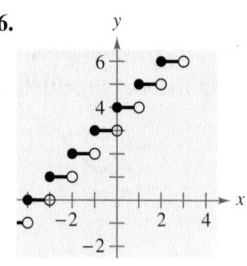

17.

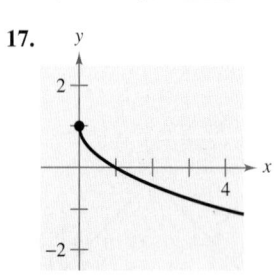

18.

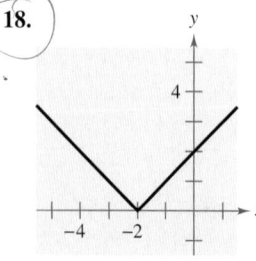

In Exercises 19–42, g is related to one of the parent functions described in this chapter. (a) Identify the parent function f. (b) Describe the sequence of tranformations from f to g. (c) Sketch the graph of g. (d) Use function notation to write g in terms of f.

19. $g(x) = 12 - x^2$

20. $g(x) = (x - 8)^2$

21. $g(x) = x^3 + 7$

22. $g(x) = -x^3 - 1$

23. $g(x) = \frac{2}{3}x^2 + 4$

24. $g(x) = 2(x - 7)^2$

25. $g(x) = 2 - (x + 5)^2$

26. $g(x) = -(x + 10)^2 + 5$

27. $g(x) = \sqrt{3x}$

28. $g(x) = \sqrt{\frac{1}{4}x}$

29. $g(x) = (x - 1)^3 + 2$

30. $g(x) = (x + 3)^3 - 10$

31. $g(x) = -|x| - 2$

32. $g(x) = 6 - |x + 5|$

33. $g(x) = -|x + 4| + 8$

34. $g(x) = |-x + 3| + 9$

35. $g(x) = 3 - [\![x]\!]$

36. $g(x) = 2[\![x + 5]\!]$

37. $g(x) = \sqrt{x - 9}$

38. $g(x) = \sqrt{x + 4} + 8$

39. $g(x) = \sqrt{7 - x} - 2$

40. $g(x) = -\sqrt{x + 1} - 6$

41. $g(x) = \sqrt{\frac{1}{2}x} - 4$

42. $g(x) = \sqrt{3x} + 1$

In Exercises 43–50, write an equation for the function that is described by the given characteristics.

43. The shape of $f(x) = x^2$, but moved two units to the right and eight units downward

44. The shape of $f(x) = x^2$, but moved three units to the left, seven units upward, and reflected in the x-axis

45. The shape of $f(x) = x^3$, but moved 13 units to the right

46. The shape of $f(x) = x^3$, but moved six units to the left, six units downward, and reflected in the y-axis

47. The shape of $f(x) = |x|$, but moved 10 units upward and reflected in the x-axis

48. The shape of $f(x) = |x|$, but moved one unit to the left and seven units downward

49. The shape of $f(x) = \sqrt{x}$, but moved six units to the left and reflected in both the x-axis and the y-axis

50. The shape of $f(x) = \sqrt{x}$, but moved nine units downward and reflected in both the x-axis and the y-axis

51. Use the graph of $f(x) = x^2$ to write an equation for each function whose graph is shown.

(a)

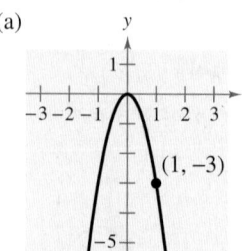

(b)

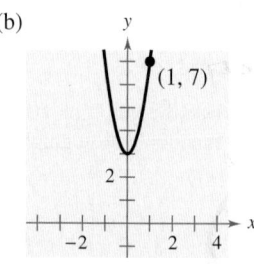

52. Use the graph of $f(x) = x^3$ to write an equation for each function whose graph is shown.

(a)

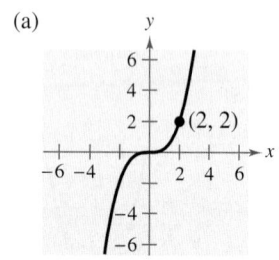

(b)

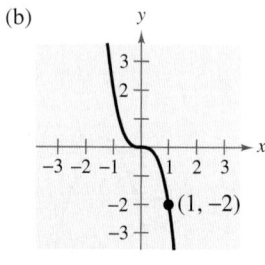

53. Use the graph of $f(x) = |x|$ to write an equation for each function whose graph is shown.

(a)

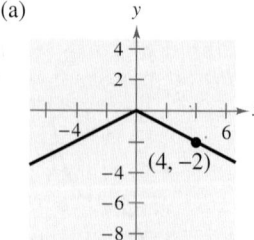

(b)

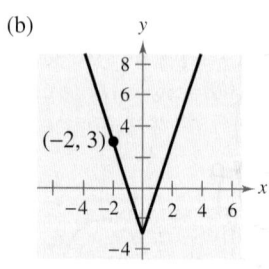

54. Use the graph of $f(x) = \sqrt{x}$ to write an equation for each function whose graph is shown.

(a)

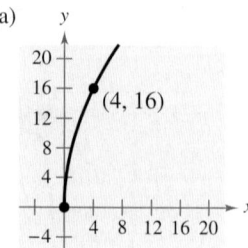

(b)

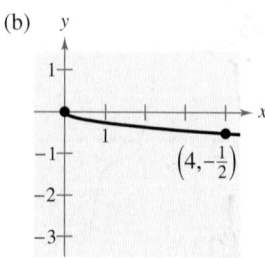

In Exercises 55–60, identify the parent function and the transformation shown in the graph. Write an equation for the function shown in the graph. Then use a graphing utility to verify your answer.

55.

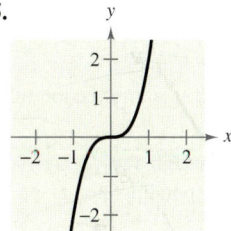

56.

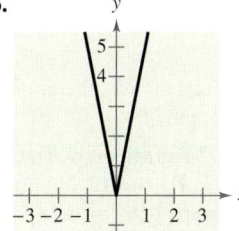

57.

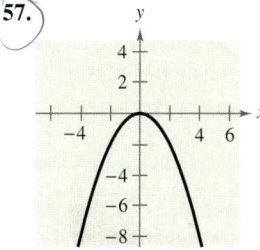

58.

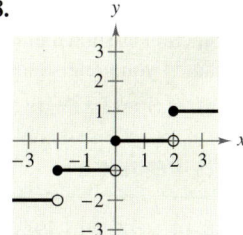

59.

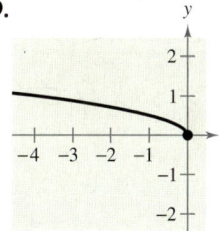

60.

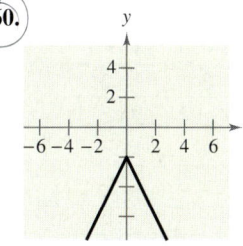

Graphical Analysis   In Exercises 61–64, use the viewing window shown to write a possible equation for the transformation of the parent function.

61.

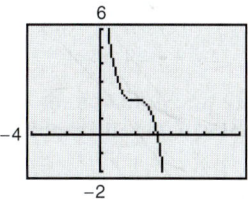

62.

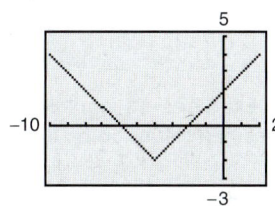

63.

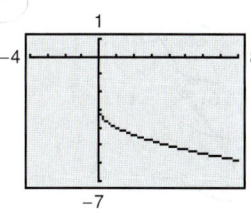

64.

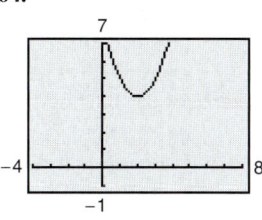

Graphical Reasoning   In Exercises 65 and 66, use the graph of f to sketch the graph of g. To print an enlarged copy of the graph, go to the website *www.mathgraphs.com*.

65.

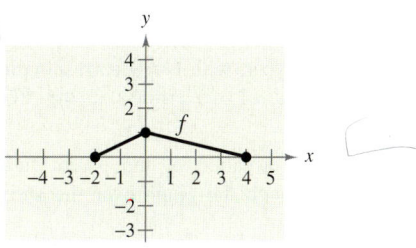

(a) $g(x) = f(x) + 2$ (b) $g(x) = f(x) - 1$

(c) $g(x) = f(-x)$ (d) $g(x) = -2f(x)$

(e) $g(x) = f(4x)$ (f) $g(x) = f\left(\frac{1}{2}x\right)$

66.

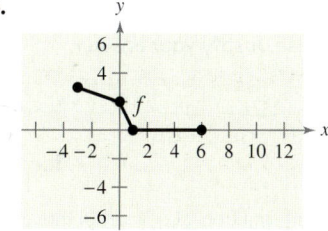

(a) $g(x) = f(x) - 5$ (b) $g(x) = f(x) + \frac{1}{2}$

(c) $g(x) = f(-x)$ (d) $g(x) = -4f(x)$

(e) $g(x) = f(2x) + 1$ (f) $g(x) = f\left(\frac{1}{4}x\right) - 2$

Model It

67. *Fuel Use*   The amounts of fuel F (in billions of gallons) used by trucks from 1980 through 2002 can be approximated by the function

$$F = f(t) = 20.6 + 0.035t^2, \quad 0 \le t \le 22$$

where t represents the year, with $t = 0$ corresponding to 1980.   (Source: U.S. Federal Highway Administration)

(a) Describe the transformation of the parent function $f(x) = x^2$. Then sketch the graph over the specified domain.

(b) Find the average rate of change of the function from 1980 to 2002. Interpret your answer in the context of the problem.

(c) Rewrite the function so that $t = 0$ represents 1990. Explain how you got your answer.

(d) Use the model from part (c) to predict the amount of fuel used by trucks in 2010. Does your answer seem reasonable? Explain.

68. *Finance* The amounts M (in trillions of dollars) of mortgage debt outstanding in the United States from 1990 through 2002 can be approximated by the function

$$M = f(t) = 0.0054(t + 20.396)^2, \quad 0 \le t \le 12$$

where t represents the year, with $t = 0$ corresponding to 1990. (Source: Board of Governors of the Federal Reserve System)

(a) Describe the transformation of the parent function $f(x) = x^2$. Then sketch the graph over the specified domain.

(b) Rewrite the function so that $t = 0$ represents 2000. Explain how you got your answer.

Synthesis

True or False? **In Exercises 69 and 70, determine whether the statement is true or false. Justify your answer.**

69. The graphs of

$$f(x) = |x| + 6 \quad \text{and} \quad f(x) = |-x| + 6$$

are identical.

70. If the graph of the parent function $f(x) = x^2$ is moved six units to the right, three units upward, and reflected in the x-axis, then the point $(-2, 19)$ will lie on the graph of the transformation.

71. *Describing Profits* Management originally predicted that the profits from the sales of a new product would be approximated by the graph of the function f shown. The actual profits are shown by the function g along with a verbal description. Use the concepts of transformations of graphs to write g in terms of f.

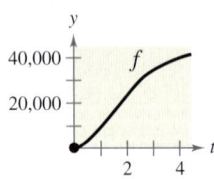

(a) The profits were only three-fourths as large as expected.

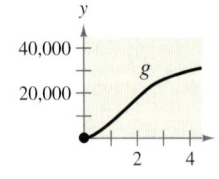

(b) The profits were consistently $10,000 greater than predicted.

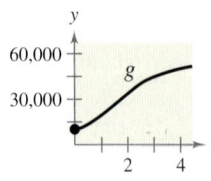

(c) There was a two-year delay in the introduction of the product. After sales began, profits grew as expected.

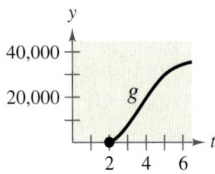

72. Explain why the graph of $y = -f(x)$ is a reflection of the graph of $y = f(x)$ about the x-axis.

73. The graph of $y = f(x)$ passes through the points $(0, 1)$, $(1, 2)$, and $(2, 3)$. Find the corresponding points on the graph of $y = f(x + 2) - 1$.

74. *Think About It* You can use either of two methods to graph a function: plotting points or translating a parent function as shown in this section. Which method of graphing do you prefer to use for each function? Explain.

(a) $f(x) = 3x^2 - 4x + 1$ (b) $f(x) = 2(x - 1)^2 - 6$

Skills Review

In Exercises 75–82, perform the operation and simplify.

75. $\dfrac{4}{x} + \dfrac{4}{1 - x}$

76. $\dfrac{2}{x + 5} - \dfrac{2}{x - 5}$

77. $\dfrac{3}{x - 1} - \dfrac{2}{x(x - 1)}$

78. $\dfrac{x}{x - 5} + \dfrac{1}{2}$

79. $(x - 4)\left(\dfrac{1}{\sqrt{x^2 - 4}}\right)$

80. $\left(\dfrac{x}{x^2 - 4}\right)\left(\dfrac{x^2 - x - 2}{x^2}\right)$

81. $(x^2 - 9) \div \left(\dfrac{x + 3}{5}\right)$

82. $\left(\dfrac{x}{x^2 - 3x - 28}\right) \div \left(\dfrac{x^2 + 3x}{x^2 + 5x + 4}\right)$

In Exercises 83 and 84, evaluate the function at the specified values of the independent variable and simplify.

83. $f(x) = x^2 - 6x + 11$

  (a) $f(-3)$ (b) $f\left(-\frac{1}{2}\right)$ (c) $f(x - 3)$

84. $f(x) = \sqrt{x + 10} - 3$

  (a) $f(-10)$ (b) $f(26)$ (c) $f(x - 10)$

In Exercises 85–88, find the domain of the function.

85. $f(x) = \dfrac{2}{11 - x}$

86. $f(x) = \dfrac{\sqrt{x - 3}}{x - 8}$

87. $f(x) = \sqrt{81 - x^2}$

88. $f(x) = \sqrt[3]{4 - x^2}$

2.6   Combinations of Functions: Composite Functions

Arithmetic Combinations of Functions

Just as two real numbers can be combined by the operations of addition, subtraction, multiplication, and division to form other real numbers, two *functions* can be combined to create new functions. For example, the functions given by $f(x) = 2x - 3$ and $g(x) = x^2 - 1$ can be combined to form the sum, difference, product, and quotient of f and g.

$$f(x) + g(x) = (2x - 3) + (x^2 - 1)$$
$$= x^2 + 2x - 4 \qquad \text{Sum}$$
$$f(x) - g(x) = (2x - 3) - (x^2 - 1)$$
$$= -x^2 + 2x - 2 \qquad \text{Difference}$$
$$f(x)g(x) = (2x - 3)(x^2 - 1)$$
$$= 2x^3 - 3x^2 - 2x + 3 \qquad \text{Product}$$
$$\frac{f(x)}{g(x)} = \frac{2x - 3}{x^2 - 1}, \quad x \neq \pm 1 \qquad \text{Quotient}$$

The domain of an **arithmetic combination** of functions f and g consists of all real numbers that are common to the domains of f and g. In the case of the quotient $f(x)/g(x)$, there is the further restriction that $g(x) \neq 0$.

Sum, Difference, Product, and Quotient of Functions

Let f and g be two functions with overlapping domains. Then, for all x common to both domains, the *sum, difference, product,* and *quotient* of f and g are defined as follows.

1. *Sum:* $\quad (f + g)(x) = f(x) + g(x)$

2. *Difference:* $\quad (f - g)(x) = f(x) - g(x)$

3. *Product:* $\quad (fg)(x) = f(x) \cdot g(x)$

4. *Quotient:* $\quad \left(\dfrac{f}{g}\right)(x) = \dfrac{f(x)}{g(x)}, \qquad g(x) \neq 0$

Example 1   Finding the Sum of Two Functions

Given $f(x) = 2x + 1$ and $g(x) = x^2 + 2x - 1$, find $(f + g)(x)$.

Solution

$$(f + g)(x) = f(x) + g(x) = (2x + 1) + (x^2 + 2x - 1) = x^2 + 4x$$

✔**CHECKPOINT**   Now try Exercise 5(a).

| Example 2 | **Finding the Difference of Two Functions** |

Given $f(x) = 2x + 1$ and $g(x) = x^2 + 2x - 1$, find $(f - g)(x)$. Then evaluate the difference when $x = 2$.

Solution

The difference of f and g is

$$(f - g)(x) = f(x) - g(x)$$
$$= (2x + 1) - (x^2 + 2x - 1)$$
$$= -x^2 + 2.$$

When $x = 2$, the value of this difference is

$$(f - g)(2) = -(2)^2 + 2$$
$$= -2.$$

✔**CHECKPOINT** Now try Exercise 5(b).

In Examples 1 and 2, both f and g have domains that consist of all real numbers. So, the domains of $(f + g)$ and $(f - g)$ are also the set of all real numbers. Remember that any restrictions on the domains of f and g must be considered when forming the sum, difference, product, or quotient of f and g.

| Example 3 | **Finding the Domains of Quotients of Functions** |

Find $\left(\dfrac{f}{g}\right)(x)$ and $\left(\dfrac{g}{f}\right)(x)$ for the functions given by

$$f(x) = \sqrt{x} \qquad \text{and} \qquad g(x) = \sqrt{4 - x^2}.$$

Then find the domains of f/g and g/f.

Solution

The quotient of f and g is

$$\left(\frac{f}{g}\right)(x) = \frac{f(x)}{g(x)} = \frac{\sqrt{x}}{\sqrt{4 - x^2}}$$

and the quotient of g and f is

$$\left(\frac{g}{f}\right)(x) = \frac{g(x)}{f(x)} = \frac{\sqrt{4 - x^2}}{\sqrt{x}}.$$

The domain of f is $[0, \infty)$ and the domain of g is $[-2, 2]$. The intersection of these domains is $[0, 2]$. So, the domains of $\left(\dfrac{f}{g}\right)$ and $\left(\dfrac{g}{f}\right)$ are as follows.

$$\text{Domain of } \left(\frac{f}{g}\right): [0, 2) \qquad \text{Domain of } \left(\frac{g}{f}\right): (0, 2]$$

Note that the domain of (f/g) includes $x = 0$, but not $x = 2$, because $x = 2$ yields a zero in the denominator, whereas the domain of (g/f) includes $x = 2$, but not $x = 0$, because $x = 0$ yields a zero in the denominator.

✔**CHECKPOINT** Now try Exercise 5(d).

Composition of Functions

Another way of combining two functions is to form the **composition** of one with the other. For instance, if $f(x) = x^2$ and $g(x) = x + 1$, the composition of f with g is

$$f(g(x)) = f(x + 1)$$
$$= (x + 1)^2.$$

This composition is denoted as $(f \circ g)$ and reads as "f composed with g."

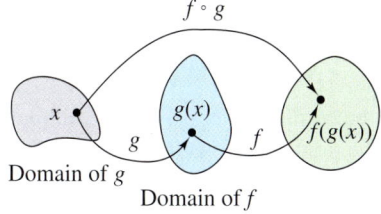

Domain of g

Domain of f

FIGURE 2.63

> ### Definition of Composition of Two Functions
>
> The **composition** of the function f with the function g is
>
> $$(f \circ g)(x) = f(g(x)).$$
>
> The domain of $(f \circ g)$ is the set of all x in the domain of g such that $g(x)$ is in the domain of f. (See Figure 2.63.)

Example 4   Composition of Functions

Given $f(x) = x + 2$ and $g(x) = 4 - x^2$, find the following.

a. $(f \circ g)(x)$ **b.** $(g \circ f)(x)$ **c.** $(g \circ f)(-2)$

Solution

a. The composition of f with g is as follows.

$$(f \circ g)(x) = f(g(x))$$ Definition of $f \circ g$
$$= f(4 - x^2)$$ Definition of $g(x)$
$$= (4 - x^2) + 2$$ Definition of $f(x)$
$$= -x^2 + 6$$ Simplify.

b. The composition of g with f is as follows.

$$(g \circ f)(x) = g(f(x))$$ Definition of $g \circ f$
$$= g(x + 2)$$ Definition of $f(x)$
$$= 4 - (x + 2)^2$$ Definition of $g(x)$
$$= 4 - (x^2 + 4x + 4)$$ Expand.
$$= -x^2 - 4x$$ Simplify.

Note that, in this case, $(f \circ g)(x) \neq (g \circ f)(x)$.

c. Using the result of part (b), you can write the following.

$$(g \circ f)(-2) = -(-2)^2 - 4(-2)$$ Substitute.
$$= -4 + 8$$ Simplify.
$$= 4$$ Simplify.

✓**CHECKPOINT**   Now try Exercise 31.

STUDY TIP

The following tables of values help illustrate the composition $(f \circ g)(x)$ given in Example 4.

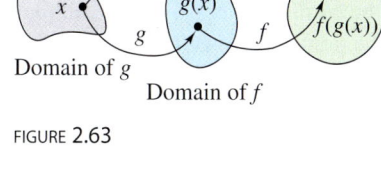

x	0	1	2	3
$g(x)$	4	3	0	-5

$g(x)$	4	3	0	-5
$f(g(x))$	6	5	2	-3

x	0	1	2	3
$f(g(x))$	6	5	2	-3

Note that the first two tables can be combined (or "composed") to produce the values given in the third table.

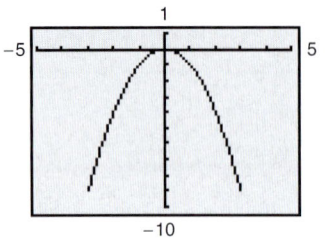
Example 5   Finding the Domain of a Composite Function

Given $f(x) = x^2 - 9$ and $g(x) = \sqrt{9 - x^2}$, find the composition $(f \circ g)(x)$. Then find the domain of $(f \circ g)$.

Solution

$$(f \circ g)(x) = f(g(x))$$

$$= f\left(\sqrt{9 - x^2}\right)$$

$$= \left(\sqrt{9 - x^2}\right)^2 - 9$$

$$= 9 - x^2 - 9$$

$$= -x^2$$

From this, it might appear that the domain of the composition is the set of all real numbers. This, however is not true. Because the domain of f is the set of all real numbers and the domain of g is $-3 \le x \le 3$, the domain of $(f \circ g)$ is $-3 \le x \le 3$.

✓**CHECKPOINT**   Now try Exercise 35.

In Examples 4 and 5, you formed the composition of two given functions. In calculus, it is also important to be able to identify two functions that make up a given composite function. For instance, the function h given by

$$h(x) = (3x - 5)^3$$

is the composition of f with g, where $f(x) = x^3$ and $g(x) = 3x - 5$. That is,

$$h(x) = (3x - 5)^3 = [g(x)]^3 = f(g(x)).$$

Basically, to "decompose" a composite function, look for an "inner" function and an "outer" function. In the function h above, $g(x) = 3x - 5$ is the inner function and $f(x) = x^3$ is the outer function.

Example 6   Decomposing a Composite Function

Write the function given by $h(x) = \dfrac{1}{(x - 2)^2}$ as a composition of two functions.

Solution

One way to write h as a composition of two functions is to take the inner function to be $g(x) = x - 2$ and the outer function to be

$$f(x) = \frac{1}{x^2} = x^{-2}.$$

Then you can write

$$h(x) = \frac{1}{(x - 2)^2} = (x - 2)^{-2} = f(x - 2) = f(g(x)).$$

✓**CHECKPOINT**   Now try Exercise 47.

Application

Example 7 **Bacteria Count**

The number N of bacteria in a refrigerated food is given by

$$N(T) = 20T^2 - 80T + 500, \qquad 2 \le T \le 14$$

where T is the temperature of the food in degrees Celsius. When the food is removed from refrigeration, the temperature of the food is given by

$$T(t) = 4t + 2, \qquad 0 \le t \le 3$$

where t is the time in hours. (a) Find the composition $N(T(t))$ and interpret its meaning in context. (b) Find the time when the bacterial count reaches 2000.

Solution

a. $N(T(t)) = 20(4t + 2)^2 - 80(4t + 2) + 500$

$$= 20(16t^2 + 16t + 4) - 320t - 160 + 500$$

$$= 320t^2 + 320t + 80 - 320t - 160 + 500$$

$$= 320t^2 + 420$$

The composite function $N(T(t))$ represents the number of bacteria in the food as a function of the amount of time the food has been out of refrigeration.

b. The bacterial count will reach 2000 when $320t^2 + 420 = 2000$. Solve this equation to find that the count will reach 2000 when $t \approx 2.2$ hours. When you solve this equation, note that the negative value is rejected because it is not in the domain of the composite function.

✓*CHECKPOINT*   Now try Exercise 65.

---

*W*RITING ABOUT *M*ATHEMATICS

Analyzing Arithmetic Combinations of Functions

a. Use the graphs of f and $(f + g)$ in Figure 2.64 to make a table showing the values of $g(x)$ when $x = 1, 2, 3, 4, 5,$ and 6. Explain your reasoning.

b. Use the graphs of f and $(f - h)$ in Figure 2.64 to make a table showing the values of $h(x)$ when $x = 1, 2, 3, 4, 5,$ and 6. Explain your reasoning.

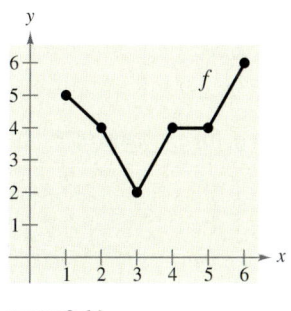
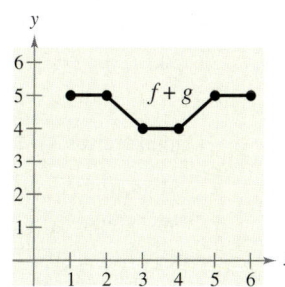
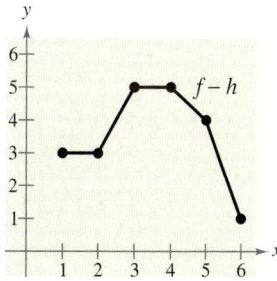

FIGURE **2.64**

2.6 | Exercises

VOCABULARY CHECK: Fill in the blanks.

1. Two functions f and g can be combined by the arithmetic operations of _____, _____, _____, and _____ to create new functions.

2. The _____ of the function f with g is $(f \circ g)(x) = f(g(x))$.

3. The domain of $(f \circ g)$ is all x in the domain of g such that _____ is in the domain of f.

4. To decompose a composite function, look for an _____ function and an _____ function.

PREREQUISITE SKILLS REVIEW: Practice and review algebra skills needed for this section at **www.Eduspace.com.**

In Exercises 1–4, use the graphs of f and g to graph $h(x) = (f + g)(x)$. To print an enlarged copy of the graph, go to the website **www.mathgraphs.com.**

1.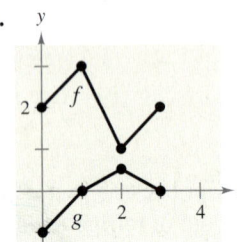

2.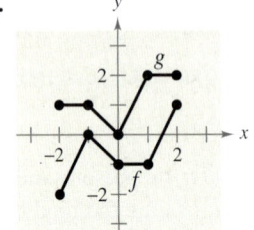

3.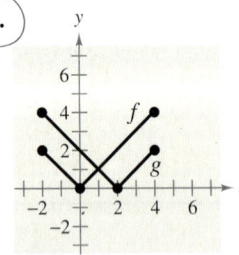

4.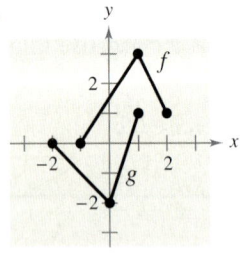

In Exercises 5–12, find (a) $(f + g)(x)$, (b) $(f - g)(x)$, (c) $(fg)(x)$, and (d) $(f/g)(x)$. What is the domain of f/g?

5. $f(x) = x + 2,$ $\qquad g(x) = x - 2$
6. $f(x) = 2x - 5,$ $\qquad g(x) = 2 - x$
7. $f(x) = x^2,$ $\qquad g(x) = 4x - 5$
8. $f(x) = 2x - 5,$ $\qquad g(x) = 4$
9. $f(x) = x^2 + 6,$ $\qquad g(x) = \sqrt{1 - x}$
10. $f(x) = \sqrt{x^2 - 4},$ $\qquad g(x) = \dfrac{x^2}{x^2 + 1}$
11. $f(x) = \dfrac{1}{x},$ $\qquad g(x) = \dfrac{1}{x^2}$
12. $f(x) = \dfrac{x}{x + 1},$ $\qquad g(x) = x^3$

In Exercises 13–24, evaluate the indicated function for $f(x) = x^2 + 1$ and $g(x) = x - 4$.

13. $(f + g)(2)$ $\qquad$ 14. $(f - g)(-1)$
15. $(f - g)(0)$ $\qquad$ 16. $(f + g)(1)$
17. $(f - g)(3t)$ $\qquad$ 18. $(f + g)(t - 2)$
19. $(fg)(6)$ $\qquad$ 20. $(fg)(-6)$
21. $\left(\dfrac{f}{g}\right)(5)$ $\qquad$ 22. $\left(\dfrac{f}{g}\right)(0)$
23. $\left(\dfrac{f}{g}\right)(-1) - g(3)$ $\qquad$ 24. $(fg)(5) + f(4)$

In Exercises 25–28, graph the functions f, g, and $f + g$ on the same set of coordinate axes.

25. $f(x) = \frac{1}{2}x,$ $\qquad g(x) = x - 1$
26. $f(x) = \frac{1}{3}x,$ $\qquad g(x) = -x + 4$
27. $f(x) = x^2,$ $\qquad g(x) = -2x$
28. $f(x) = 4 - x^2,$ $\qquad g(x) = x$

 Graphical Reasoning In Exercises 29 and 30, use a graphing utility to graph f, g, and $f + g$ in the same viewing window. Which function contributes most to the magnitude of the sum when $0 \le x \le 2$? Which function contributes most to the magnitude of the sum when $x > 6$?

29. $f(x) = 3x,$ $\qquad g(x) = -\dfrac{x^3}{10}$
30. $f(x) = \dfrac{x}{2},$ $\qquad g(x) = \sqrt{x}$

In Exercises 31–34, find (a) $f \circ g$, (b) $g \circ f$, and (c) $f \circ f$.

31. $f(x) = x^2,$ $\qquad g(x) = x - 1$
32. $f(x) = 3x + 5,$ $\qquad g(x) = 5 - x$
33. $f(x) = \sqrt[3]{x - 1},$ $\qquad g(x) = x^3 + 1$
34. $f(x) = x^3,$ $\qquad g(x) = \dfrac{1}{x}$

In Exercises 35–42, find (a) $f \circ g$ and (b) $g \circ f$. Find the domain of each function and each composite function.

35. $f(x) = \sqrt{x + 4}$, $g(x) = x^2$
36. $f(x) = \sqrt[3]{x - 5}$, $g(x) = x^3 + 1$
37. $f(x) = x^2 + 1$, $g(x) = \sqrt{x}$
38. $f(x) = x^{2/3}$, $g(x) = x^6$
39. $f(x) = |x|$, $g(x) = x + 6$
40. $f(x) = |x - 4|$, $g(x) = 3 - x$
41. $f(x) = \dfrac{1}{x}$, $g(x) = x + 3$
42. $f(x) = \dfrac{3}{x^2 - 1}$, $g(x) = x + 1$

In Exercises 43–46, use the graphs of f and g to evaluate the functions.

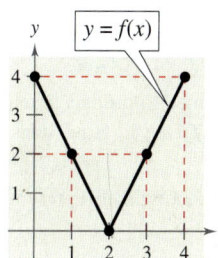
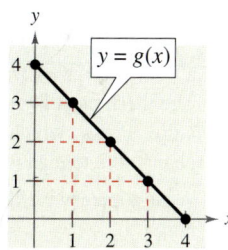

43. (a) $(f + g)(3)$ (b) $(f/g)(2)$
44. (a) $(f - g)(1)$ (b) $(fg)(4)$
45. (a) $(f \circ g)(2)$ (b) $(g \circ f)(2)$
46. (a) $(f \circ g)(1)$ (b) $(g \circ f)(3)$

In Exercises 47–54, find two functions f and g such that $(f \circ g)(x) = h(x)$. (There are many correct answers.)

47. $h(x) = (2x + 1)^2$ 48. $h(x) = (1 - x)^3$
49. $h(x) = \sqrt[3]{x^2 - 4}$ 50. $h(x) = \sqrt{9 - x}$
51. $h(x) = \dfrac{1}{x + 2}$ 52. $h(x) = \dfrac{4}{(5x + 2)^2}$
53. $h(x) = \dfrac{-x^2 + 3}{4 - x^2}$ 54. $h(x) = \dfrac{27x^3 + 6x}{10 - 27x^3}$

55. **Stopping Distance**   The research and development department of an automobile manufacturer has determined that when a driver is required to stop quickly to avoid an accident, the distance (in feet) the car travels during the driver's reaction time is given by $R(x) = \frac{3}{4}x$, where x is the speed of the car in miles per hour. The distance (in feet) traveled while the driver is braking is given by $B(x) = \frac{1}{15}x^2$. Find the function that represents the total stopping distance T. Graph the functions R, B, and T on the same set of coordinate axes for $0 \le x \le 60$.

56. **Sales**   From 2000 to 2005, the sales R_1 (in thousands of dollars) for one of two restaurants owned by the same parent company can be modeled by

$$R_1 = 480 - 8t - 0.8t^2, \qquad t = 0, 1, 2, 3, 4, 5$$

where $t = 0$ represents 2000. During the same six-year period, the sales R_2 (in thousands of dollars) for the second restaurant can be modeled by

$$R_2 = 254 + 0.78t, \qquad t = 0, 1, 2, 3, 4, 5.$$

(a) Write a function R_3 that represents the total sales of the two restaurants owned by the same parent company.

 (b) Use a graphing utility to graph R_1, R_2, and R_3 in the same viewing window.

57. **Vital Statistics**   Let $b(t)$ be the number of births in the United States in year t, and let $d(t)$ represent the number of deaths in the United States in year t, where $t = 0$ corresponds to 2000.

(a) If $p(t)$ is the population of the United States in year t, find the function $c(t)$ that represents the percent change in the population of the United States.

(b) Interpret the value of $c(5)$.

58. **Pets**   Let $d(t)$ be the number of dogs in the United States in year t, and let $c(t)$ be the number of cats in the United States in year t, where $t = 0$ corresponds to 2000.

(a) Find the function $p(t)$ that represents the total number of dogs and cats in the United States.

(b) Interpret the value of $p(5)$.

(c) Let $n(t)$ represent the population of the United States in year t, where $t = 0$ corresponds to 2000. Find and interpret

$$h(t) = \frac{p(t)}{n(t)}.$$

59. **Military Personnel**   The total numbers of Army personnel (in thousands) A and Navy personnel (in thousands) N from 1990 to 2002 can be approximated by the models

$$A(t) = 3.36t^2 - 59.8t + 735$$

and

$$N(t) = 1.95t^2 - 42.2t + 603$$

where t represents the year, with $t = 0$ corresponding to 1990. (Source: Department of Defense)

(a) Find and interpret $(A + N)(t)$. Evaluate this function for $t = 4, 8$, and 12.

(b) Find and interpret $(A - N)(t)$. Evaluate this function for $t = 4, 8$, and 12.

60. *Sales* The sales of exercise equipment E (in millions of dollars) in the United States from 1997 to 2003 can be approximated by the function

$$E(t) = 25.95t^2 - 231.2t + 3356$$

and the U.S. population P (in millions) from 1997 to 2003 can be approximated by the function

$$P(t) = 3.02t + 252.0$$

where t represents the year, with $t = 7$ corresponding to 1997. (Source: National Sporting Goods Association, U.S. Census Bureau)

(a) Find and interpret $h(t) = \dfrac{E(t)}{P(t)}$.

(b) Evaluate the function in part (a) for $t = 7$, 10, and 12.

Model It

61. *Health Care Costs* The table shows the total amounts (in billions of dollars) spent on health services and supplies in the United States (including Puerto Rico) for the years 1995 through 2001. The variables y_1, y_2, and y_3 represent out-of-pocket payments, insurance premiums, and other types of payments, respectively. (Source: Centers for Medicare and Medicaid Services)

Year	y_1	y_2	y_3
1995	146.2	329.1	44.8
1996	152.0	344.1	48.1
1997	162.2	359.9	52.1
1998	175.2	382.0	55.6
1999	184.4	412.1	57.8
2000	194.7	449.0	57.4
2001	205.5	496.1	57.8

(a) Use the *regression* feature of a graphing utility to find a linear model for y_1 and quadratic models for y_2 and y_3. Let $t = 5$ represent 1995.

(b) Find $y_1 + y_2 + y_3$. What does this sum represent?

(c) Use a graphing utility to graph y_1, y_2, y_3, and $y_1 + y_2 + y_3$ in the same viewing window.

(d) Use the model from part (b) to estimate the total amounts spent on health services and supplies in the years 2008 and 2010.

62. *Graphical Reasoning* An electronically controlled thermostat in a home is programmed to lower the temperature automatically during the night. The temperature in the house T (in degrees Fahrenheit) is given in terms of t, the time in hours on a 24-hour clock (see figure).

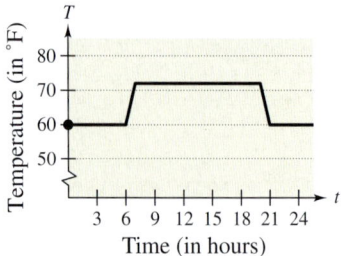

(a) Explain why T is a function of t.

(b) Approximate $T(4)$ and $T(15)$.

(c) The thermostat is reprogrammed to produce a temperature H for which $H(t) = T(t - 1)$. How does this change the temperature?

(d) The thermostat is reprogrammed to produce a temperature H for which $H(t) = T(t) - 1$. How does this change the temperature?

(e) Write a piecewise-defined function that represents the graph.

63. *Geometry* A square concrete foundation is prepared as a base for a cylindrical tank (see figure).

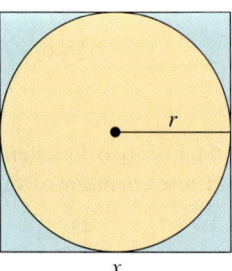

(a) Write the radius r of the tank as a function of the length x of the sides of the square.

(b) Write the area A of the circular base of the tank as a function of the radius r.

(c) Find and interpret $(A \circ r)(x)$.

64. *Physics*   A pebble is dropped into a calm pond, causing ripples in the form of concentric circles (see figure). The radius r (in feet) of the outer ripple is $r(t) = 0.6t$, where t is the time in seconds after the pebble strikes the water. The area A of the circle is given by the function $A(r) = \pi r^2$. Find and interpret $(A \circ r)(t)$.

65. *Bacteria Count*   The number N of bacteria in a refrigerated food is given by

$$N(T) = 10T^2 - 20T + 600, \quad 1 \le T \le 20$$

where T is the temperature of the food in degrees Celsius. When the food is removed from refrigeration, the temperature of the food is given by

$$T(t) = 3t + 2, \quad 0 \le t \le 6$$

where t is the time in hours.

(a) Find the composition $N(T(t))$ and interpret its meaning in context.

(b) Find the time when the bacterial count reaches 1500.

66. *Cost*   The weekly cost C of producing x units in a manufacturing process is given by

$$C(x) = 60x + 750.$$

The number of units x produced in t hours is given by

$$x(t) = 50t.$$

(a) Find and interpret $(C \circ x)(t)$.

(b) Find the time that must elapse in order for the cost to increase to \$15,000.

67. *Salary*   You are a sales representative for a clothing manufacturer. You are paid an annual salary, plus a bonus of 3% of your sales over \$500,000. Consider the two functions given by

$$f(x) = x - 500,000 \quad \text{and} \quad g(x) = 0.03x.$$

If x is greater than \$500,000, which of the following represents your bonus? Explain your reasoning.

(a) $f(g(x))$ (b) $g(f(x))$

68. *Consumer Awareness*   The suggested retail price of a new hybrid car is p dollars. The dealership advertises a factory rebate of \$2000 and a 10% discount.

(a) Write a function R in terms of p giving the cost of the hybrid car after receiving the rebate from the factory.

(b) Write a function S in terms of p giving the cost of the hybrid car after receiving the dealership discount.

(c) Form the composite functions $(R \circ S)(p)$ and $(S \circ R)(p)$ and interpret each.

(d) Find $(R \circ S)(20{,}500)$ and $(S \circ R)(20{,}500)$. Which yields the lower cost for the hybrid car? Explain.

Synthesis

True or False?   **In Exercises 69 and 70, determine whether the statement is true or false. Justify your answer.**

69. If $f(x) = x + 1$ and $g(x) = 6x$, then

$$(f \circ g)(x) = (g \circ f)(x).$$

70. If you are given two functions $f(x)$ and $g(x)$, you can calculate $(f \circ g)(x)$ if and only if the range of g is a subset of the domain of f.

71. *Proof*   Prove that the product of two odd functions is an even function, and that the product of two even functions is an even function.

72. *Conjecture*   Use examples to hypothesize whether the product of an odd function and an even function is even or odd. Then prove your hypothesis.

Skills Review

Average Rate of Change   **In Exercises 73–76, find the difference quotient**

$$\frac{f(x+h) - f(x)}{h}$$

and simplify your answer.

73. $f(x) = 3x - 4$ **74.** $f(x) = 1 - x^2$

75. $f(x) = \dfrac{4}{x}$ **76.** $f(x) = \sqrt{2x + 1}$

In Exercises 77–80, find an equation of the line that passes through the given point and has the indicated slope. Sketch the line.

77. $(2, -4), m = 3$ **78.** $(-6, 3), m = -1$

79. $(8, -1), m = -\frac{3}{2}$ **80.** $(7, 0), m = \frac{5}{7}$

2.7   Inverse Functions

What you should learn

- Find inverse functions informally and verify that two functions are inverse functions of each other.
- Use graphs of functions to determine whether functions have inverse functions.
- Use the Horizontal Line Test to determine if functions are one-to-one.
- Find inverse functions algebraically.

Why you should learn it

Inverse functions can be used to model and solve real-life problems. For instance, in Exercise 80 on page 250, an inverse function can be used to determine the year in which there was a given dollar amount of sales of digital cameras in the United States.

© Tim Boyle/Getty Images

Inverse Functions

Recall from Section 2.2, that a function can be represented by a set of ordered pairs. For instance, the function $f(x) = x + 4$ from the set $A = \{1, 2, 3, 4\}$ to the set $B = \{5, 6, 7, 8\}$ can be written as follows.

$$f(x) = x + 4: \ \{(1, 5), (2, 6), (3, 7), (4, 8)\}$$

In this case, by interchanging the first and second coordinates of each of these ordered pairs, you can form the **inverse function** of f, which is denoted by f^{-1}. It is a function from the set B to the set A, and can be written as follows.

$$f^{-1}(x) = x - 4: \ \{(5, 1), (6, 2), (7, 3), (8, 4)\}$$

Note that the domain of f is equal to the range of f^{-1}, and vice versa, as shown in Figure 2.65. Also note that the functions f and f^{-1} have the effect of "undoing" each other. In other words, when you form the composition of f with f^{-1} or the composition of f^{-1} with f, you obtain the identity function.

$$f(f^{-1}(x)) = f(x - 4) = (x - 4) + 4 = x$$

$$f^{-1}(f(x)) = f^{-1}(x + 4) = (x + 4) - 4 = x$$

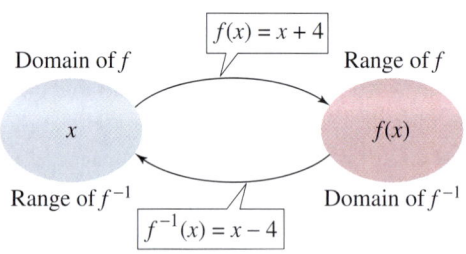

FIGURE 2.65

Example 1   Finding Inverse Functions Informally

Find the inverse function of $f(x) = 4x$. Then verify that both $f(f^{-1}(x))$ and $f^{-1}(f(x))$ are equal to the identity function.

Solution

The function f *multiplies* each input by 4. To "undo" this function, you need to *divide* each input by 4. So, the inverse function of $f(x) = 4x$ is

$$f^{-1}(x) = \frac{x}{4}.$$

You can verify that both $f(f^{-1}(x)) = x$ and $f^{-1}(f(x)) = x$ as follows.

$$f(f^{-1}(x)) = f\left(\frac{x}{4}\right) = 4\left(\frac{x}{4}\right) = x \qquad f^{-1}(f(x)) = f^{-1}(4x) = \frac{4x}{4} = x$$

✓**CHECKPOINT**   Now try Exercise 1.

Consider the functions given by

$$f(x) = x + 2$$

and

$$f^{-1}(x) = x - 2.$$

Evaluate $f(f^{-1}(x))$ and $f^{-1}(f(x))$ for the indicated values of x. What can you conclude about the functions?

x	-10	0	7	45
$f(f^{-1}(x))$				
$f^{-1}(f(x))$				

Definition of Inverse Function

Let f and g be two functions such that

$$f(g(x)) = x \qquad \text{for every } x \text{ in the domain of } g$$

and

$$g(f(x)) = x \qquad \text{for every } x \text{ in the domain of } f.$$

Under these conditions, the function g is the **inverse function** of the function f. The function g is denoted by f^{-1} (read "f-inverse"). So,

$$f(f^{-1}(x)) = x \qquad \text{and} \qquad f^{-1}(f(x)) = x.$$

The domain of f must be equal to the range of f^{-1}, and the range of f must be equal to the domain of f^{-1}.

Don't be confused by the use of -1 to denote the inverse function f^{-1}. In this text, whenever f^{-1} is written, it *always* refers to the inverse function of the function f and *not* to the reciprocal of $f(x)$.

If the function g is the inverse function of the function f, it must also be true that the function f is the inverse function of the function g. For this reason, you can say that the functions f and g are *inverse functions of each other*.

Example 2   Verifying Inverse Functions

Which of the functions is the inverse function of $f(x) = \dfrac{5}{x - 2}$?

$$g(x) = \frac{x - 2}{5} \qquad h(x) = \frac{5}{x} + 2$$

Solution

By forming the composition of f with g, you have

$$f(g(x)) = f\left(\frac{x - 2}{5}\right)$$

$$= \frac{5}{\left(\dfrac{x - 2}{5}\right) - 2} \qquad \text{Substitute } \frac{x - 2}{5} \text{ for } x.$$

$$= \frac{25}{x - 12} \neq x.$$

Because this composition is not equal to the identity function x, it follows that g *is not* the inverse function of f. By forming the composition of f with h, you have

$$f(h(x)) = f\left(\frac{5}{x} + 2\right) = \frac{5}{\left(\dfrac{5}{x} + 2\right) - 2} = \frac{5}{\left(\dfrac{5}{x}\right)} = x.$$

So, it appears that h *is* the inverse function of f. You can confirm this by showing that the composition of h with f is also equal to the identity function.

✓**CHECKPOINT**   Now try Exercise 5.

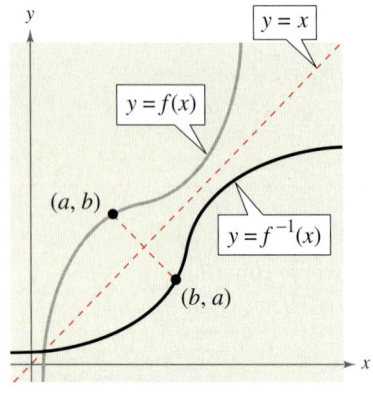

FIGURE **2.66**

The Graph of an Inverse Function

The graphs of a function f and its inverse function f^{-1} are related to each other in the following way. If the point (a, b) lies on the graph of f, then the point (b, a) must lie on the graph of f^{-1}, and vice versa. This means that the graph of f^{-1} is a *reflection* of the graph of f in the line $y = x$, as shown in Figure 2.66.

Example 3   Finding Inverse Functions Graphically

Sketch the graphs of the inverse functions $f(x) = 2x - 3$ and $f^{-1}(x) = \frac{1}{2}(x + 3)$ on the same rectangular coordinate system and show that the graphs are reflections of each other in the line $y = x$.

Solution

The graphs of f and f^{-1} are shown in Figure 2.67. It appears that the graphs are reflections of each other in the line $y = x$. You can further verify this reflective property by testing a few points on each graph. Note in the following list that if the point (a, b) is on the graph of f, the point (b, a) is on the graph of f^{-1}.

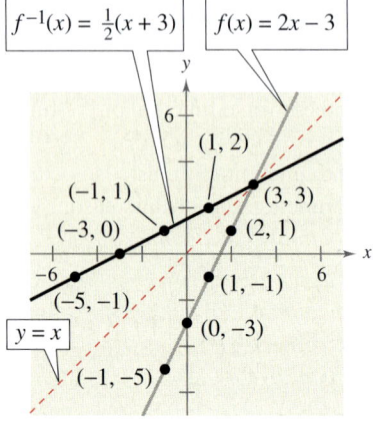

FIGURE **2.67**

Graph of $f(x) = 2x - 3$	Graph of $f^{-1}(x) = \frac{1}{2}(x + 3)$
$(-1, -5)$	$(-5, -1)$
$(0, -3)$	$(-3, 0)$
$(1, -1)$	$(-1, 1)$
$(2, 1)$	$(1, 2)$
$(3, 3)$	$(3, 3)$

✓**CHECKPOINT**   Now try Exercise 15.

Example 4   Finding Inverse Functions Graphically

Sketch the graphs of the inverse functions $f(x) = x^2$ ($x \geq 0$) and $f^{-1}(x) = \sqrt{x}$ on the same rectangular coordinate system and show that the graphs are reflections of each other in the line $y = x$.

Solution

The graphs of f and f^{-1} are shown in Figure 2.68. It appears that the graphs are reflections of each other in the line $y = x$. You can further verify this reflective property by testing a few points on each graph. Note in the following list that if the point (a, b) is on the graph of f, the point (b, a) is on the graph of f^{-1}.

Graph of $f(x) = x^2$, $x \geq 0$	Graph of $f^{-1}(x) = \sqrt{x}$
$(0, 0)$	$(0, 0)$
$(1, 1)$	$(1, 1)$
$(2, 4)$	$(4, 2)$
$(3, 9)$	$(9, 3)$

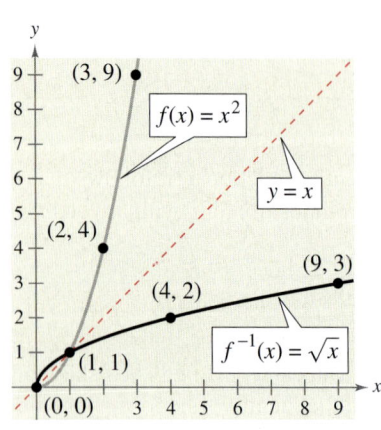

FIGURE **2.68**

Try showing that $f(f^{-1}(x)) = x$ and $f^{-1}(f(x)) = x$.

✓**CHECKPOINT**   Now try Exercise 17.

One-to-One Functions

The reflective property of the graphs of inverse functions gives you a nice *geometric* test for determining whether a function has an inverse function. This test is called the **Horizontal Line Test** for inverse functions.

> ### Horizontal Line Test for Inverse Functions
>
> A function f has an inverse function if and only if no *horizontal* line intersects the graph of f at more than one point.

If no horizontal line intersects the graph of f at more than one point, then no y-value is matched with more than one x-value. This is the essential characteristic of what are called **one-to-one functions.**

> ### One-to-One Functions
>
> A function f is **one-to-one** if each value of the dependent variable corresponds to exactly one value of the independent variable. A function f has an inverse function if and only if f is one-to-one.

Consider the function given by $f(x) = x^2$. The table on the left is a table of values for $f(x) = x^2$. The table of values on the right is made up by interchanging the columns of the first table. The table on the right does not represent a function because the input $x = 4$ is matched with two different outputs: $y = -2$ and $y = 2$. So, $f(x) = x^2$ is not one-to-one and does not have an inverse function.

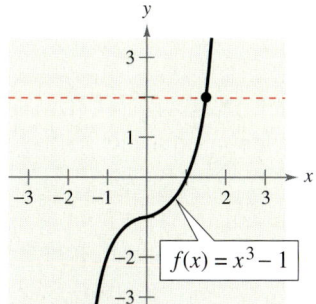

FIGURE 2.69

x	$f(x) = x^2$
-2	4
-1	1
0	0
1	1
2	4
3	9

x	y
4	-2
1	-1
0	0
1	1
4	2
9	3

FIGURE 2.70

Example 5   **Applying the Horizontal Line Test**

a. The graph of the function given by $f(x) = x^3 - 1$ is shown in Figure 2.69. Because no horizontal line intersects the graph of f at more than one point, you can conclude that f *is* a one-to-one function and *does* have an inverse function.

b. The graph of the function given by $f(x) = x^2 - 1$ is shown in Figure 2.70. Because it is possible to find a horizontal line that intersects the graph of f at more than one point, you can conclude that f *is not* a one-to-one function and *does not* have an inverse function.

✓CHECKPOINT   Now try Exercise 29.

Finding Inverse Functions Algebraically

For simple functions (such as the one in Example 1), you can find inverse functions by inspection. For more complicated functions, however, it is best to use the following guidelines. The key step in these guidelines is Step 3—interchanging the roles of x and y. This step corresponds to the fact that inverse functions have ordered pairs with the coordinates reversed.

Finding an Inverse Function

1. Use the Horizontal Line Test to decide whether f has an inverse function.
2. In the equation for $f(x)$, replace $f(x)$ by y.
3. Interchange the roles of x and y, and solve for y.
4. Replace y by $f^{-1}(x)$ in the new equation.
5. Verify that f and f^{-1} are inverse functions of each other by showing that the domain of f is equal to the range of f^{-1}, the range of f is equal to the domain of f^{-1}, and $f(f^{-1}(x)) = x$ and $f^{-1}(f(x)) = x$.

STUDY TIP

Note what happens when you try to find the inverse function of a function that is not one-to-one.

$f(x) = x^2 + 1$ Original function

$y = x^2 + 1$ Replace $f(x)$ by y.

$x = y^2 + 1$ Interchange x and y.

$x - 1 = y^2$ Isolate y-term.

$y = \pm\sqrt{x - 1}$ Solve for y.

You obtain two y-values for each x.

Example 6 Finding an Inverse Function Algebraically

Find the inverse function of

$$f(x) = \frac{5 - 3x}{2}.$$

Solution

The graph of f is a line, as shown in Figure 2.71. This graph passes the Horizontal Line Test. So, you know that f is one-to-one and has an inverse function.

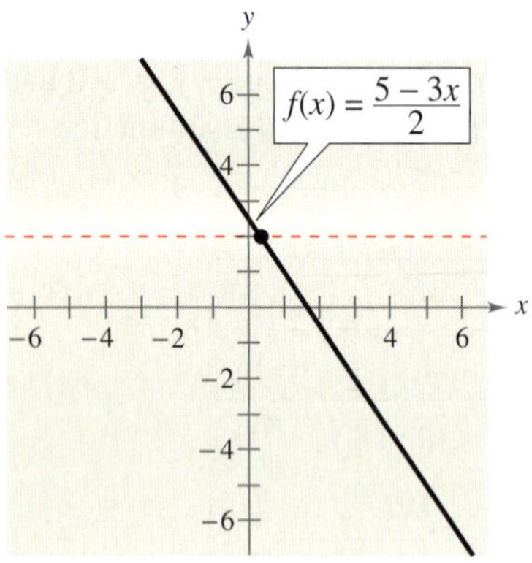

FIGURE 2.71

$$f(x) = \frac{5 - 3x}{2} \quad \text{Write original function.}$$

$$y = \frac{5 - 3x}{2} \quad \text{Replace } f(x) \text{ by } y.$$

$$x = \frac{5 - 3y}{2} \quad \text{Interchange } x \text{ and } y.$$

$$2x = 5 - 3y \quad \text{Multiply each side by 2.}$$

$$3y = 5 - 2x \quad \text{Isolate the } y\text{-term.}$$

$$y = \frac{5 - 2x}{3} \quad \text{Solve for } y.$$

$$f^{-1}(x) = \frac{5 - 2x}{3} \quad \text{Replace } y \text{ by } f^{-1}(x).$$

Note that both f and f^{-1} have domains and ranges that consist of the entire set of real numbers. Check that $f(f^{-1}(x)) = x$ and $f^{-1}(f(x)) = x$.

✓CHECKPOINT Now try Exercise 55.

Exploration

Restrict the domain of $f(x) = x^2 + 1$ to $x \geq 0$. Use a graphing utility to graph the function. Does the restricted function have an inverse function? Explain.

Example 7   **Finding an Inverse Function**

Find the inverse function of

$$f(x) = \sqrt[3]{x + 1}.$$

Solution

FIGURE **2.72**

The graph of f is a curve, as shown in Figure 2.72. Because this graph passes the Horizontal Line Test, you know that f is one-to-one and has an inverse function.

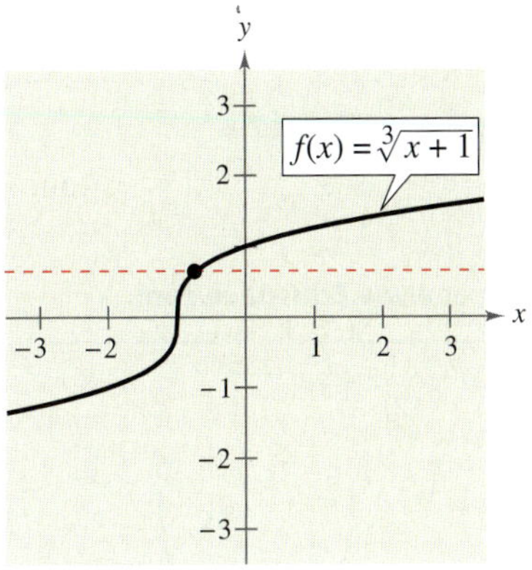

$f(x) = \sqrt[3]{x + 1}$	Write original function.
$y = \sqrt[3]{x + 1}$	Replace $f(x)$ by y.
$x = \sqrt[3]{y + 1}$	Interchange x and y.
$x^3 = y + 1$	Cube each side.
$x^3 - 1 = y$	Solve for y.
$x^3 - 1 = f^{-1}(x)$	Replace y by $f^{-1}(x)$.

Both f and f^{-1} have domains and ranges that consist of the entire set of real numbers. You can verify this result numerically as shown in the tables below.

x	$f(x)$
-28	-3
-9	-2
-2	-1
-1	0
0	1
7	2
26	3

x	$f^{-1}(x)$
-3	-28
-2	-9
-1	-2
0	-1
1	0
2	7
3	26

✓CHECKPOINT   Now try Exercise 61.

WRITING ABOUT MATHEMATICS

The Existence of an Inverse Function Write a short paragraph describing why the following functions do or do not have inverse functions.

a. Let x represent the retail price of an item (in dollars), and let $f(x)$ represent the sales tax on the item. Assume that the sales tax is 6% of the retail price *and* that the sales tax is rounded to the nearest cent. Does this function have an inverse function? (*Hint:* Can you undo this function?

For instance, if you know that the sales tax is $0.12, can you determine exactly what the retail price is?)

b. Let x represent the temperature in degrees Celsius, and let $f(x)$ represent the temperature in degrees Fahrenheit. Does this function have an inverse function? (*Hint:* The formula for converting from degrees Celsius to degrees Fahrenheit is $F = \frac{9}{5}C + 32$.)

2.7 | Exercises

VOCABULARY CHECK: Fill in the blanks.

1. If the composite functions $f(g(x)) = x$ and $g(f(x)) = x$ then the function g is the _____ function of f.
2. The domain of f is the _____ of f^{-1}, and the _____ of f^{-1} is the range of f.
3. The graphs of f and f^{-1} are reflections of each other in the line _____.
4. A function f is _____ if each value of the dependent variable corresponds to exactly one value of the independent variable.
5. A graphical test for the existence of an inverse function of f is called the _____ Line Test.

PREREQUISITE SKILLS REVIEW: Practice and review algebra skills needed for this section at **www.Eduspace.com**.

In Exercises 1–8, find the inverse function of f informally.
Verify that $f(f^{-1}(x)) = x$ and $f^{-1}(f(x)) = x$.

1. $f(x) = 6x$
2. $f(x) = \frac{1}{3}x$
3. $f(x) = x + 9$
4. $f(x) = x - 4$
5. $f(x) = 3x + 1$
6. $f(x) = \dfrac{x - 1}{5}$
7. $f(x) = \sqrt[3]{x}$
8. $f(x) = x^5$

11.

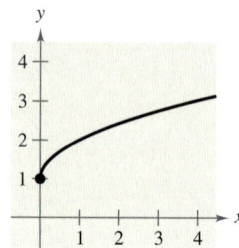

12.

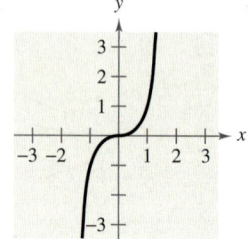

In Exercises 9–12, match the graph of the function with the graph of its inverse function. [The graphs of the inverse functions are labeled (a), (b), (c), and (d).]

(a)

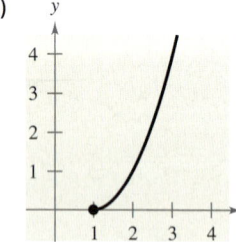

(b)

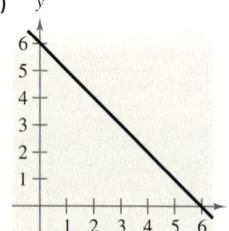

(c)

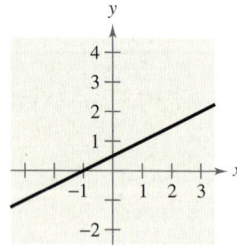

(d)

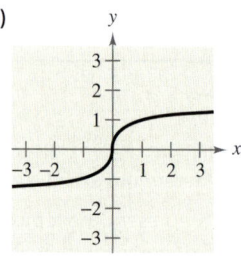

9.

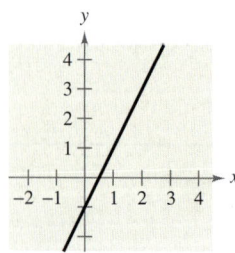

10.

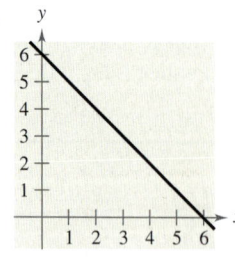

In Exercises 13–24, show that f and g are inverse functions
(a) algebraically and (b) graphically.

13. $f(x) = 2x$, $\qquad g(x) = \dfrac{x}{2}$

14. $f(x) = x - 5$, $\qquad g(x) = x + 5$

15. $f(x) = 7x + 1$, $\qquad g(x) = \dfrac{x - 1}{7}$

16. $f(x) = 3 - 4x$, $\qquad g(x) = \dfrac{3 - x}{4}$

17. $f(x) = \dfrac{x^3}{8}$, $\qquad g(x) = \sqrt[3]{8x}$

18. $f(x) = \dfrac{1}{x}$, $\qquad g(x) = \dfrac{1}{x}$

19. $f(x) = \sqrt{x - 4}$, $\qquad g(x) = x^2 + 4$, $\quad x \geq 0$

20. $f(x) = 1 - x^3$, $\qquad g(x) = \sqrt[3]{1 - x}$

21. $f(x) = 9 - x^2$, $\quad x \geq 0$, $\qquad g(x) = \sqrt{9 - x}$, $\quad x \leq 9$

22. $f(x) = \dfrac{1}{1 + x}$, $\quad x \geq 0$, $\qquad g(x) = \dfrac{1 - x}{x}$, $\quad 0 < x \leq 1$

23. $f(x) = \dfrac{x - 1}{x + 5}$, $\qquad g(x) = -\dfrac{5x + 1}{x - 1}$

24. $f(x) = \dfrac{x + 3}{x - 2}$, $\qquad g(x) = \dfrac{2x + 3}{x - 1}$

In Exercises 25 and 26, does the function have an inverse function?

25.

x	-1	0	1	2	3	4
$f(x)$	-2	1	2	1	-2	-6

26.

x	-3	-2	-1	0	2	3
$f(x)$	10	6	4	1	-3	-10

In Exercises 27 and 28, use the table of values for $y = f(x)$ to complete a table for $y = f^{-1}(x)$.

27.

x	-2	-1	0	1	2	3
$f(x)$	-2	0	2	4	6	8

28.

x	-3	-2	-1	0	1	2
$f(x)$	-10	-7	-4	-1	2	5

In Exercises 29–32, does the function have an inverse function?

29.
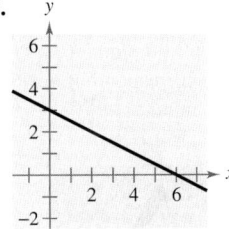

30.
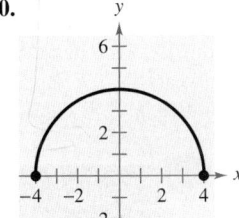

31.
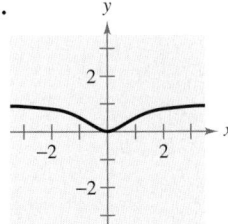

32.
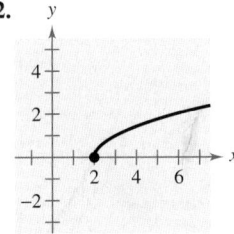

In Exercises 33–38, use a graphing utility to graph the function, and use the Horizontal Line Test to determine whether the function is one-to-one and so has an inverse function.

33. $g(x) = \dfrac{4 - x}{6}$

34. $f(x) = 10$

35. $h(x) = |x + 4| - |x - 4|$

36. $g(x) = (x + 5)^3$

37. $f(x) = -2x\sqrt{16 - x^2}$

38. $f(x) = \frac{1}{8}(x + 2)^2 - 1$

In Exercises 39–54, (a) find the inverse function of f, (b) graph both f and f^{-1} on the same set of coordinate axes, (c) describe the relationship between the graphs of f and f^{-1}, and (d) state the domain and range of f and f^{-1}.

39. $f(x) = 2x - 3$

40. $f(x) = 3x + 1$

41. $f(x) = x^5 - 2$

42. $f(x) = x^3 + 1$

43. $f(x) = \sqrt{x}$

44. $f(x) = x^2, \quad x \geq 0$

45. $f(x) = \sqrt{4 - x^2}, \quad 0 \leq x \leq 2$

46. $f(x) = x^2 - 2, \quad x \leq 0$

47. $f(x) = \dfrac{4}{x}$

48. $f(x) = -\dfrac{2}{x}$

49. $f(x) = \dfrac{x + 1}{x - 2}$

50. $f(x) = \dfrac{x - 3}{x + 2}$

51. $f(x) = \sqrt[3]{x - 1}$

52. $f(x) = x^{3/5}$

53. $f(x) = \dfrac{6x + 4}{4x + 5}$

54. $f(x) = \dfrac{8x - 4}{2x + 6}$

In Exercises 55–68, determine whether the function has an inverse function. If it does, find the inverse function.

55. $f(x) = x^4$

56. $f(x) = \dfrac{1}{x^2}$

57. $g(x) = \dfrac{x}{8}$

58. $f(x) = 3x + 5$

59. $p(x) = -4$

60. $f(x) = \dfrac{3x + 4}{5}$

61. $f(x) = (x + 3)^2, \quad x \geq -3$

62. $q(x) = (x - 5)^2$

63. $f(x) = \begin{cases} x + 3, & x < 0 \\ 6 - x, & x \geq 0 \end{cases}$

64. $f(x) = \begin{cases} -x, & x \leq 0 \\ x^2 - 3x, & x > 0 \end{cases}$

65. $h(x) = -\dfrac{4}{x^2}$

66. $f(x) = |x - 2|, \quad x \leq 2$

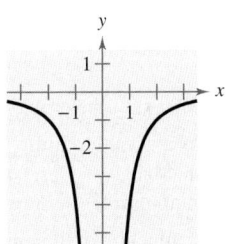

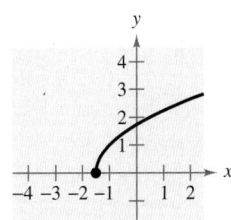

67. $f(x) = \sqrt{2x + 3}$

68. $f(x) = \sqrt{x - 2}$

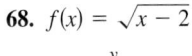

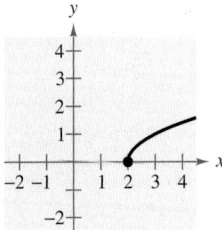

In Exercises 69–74, use the functions given by $f(x) = \frac{1}{8}x - 3$ and $g(x) = x^3$ to find the indicated value or function.

69. $(f^{-1} \circ g^{-1})(1)$

70. $(g^{-1} \circ f^{-1})(-3)$

71. $(f^{-1} \circ f^{-1})(6)$

72. $(g^{-1} \circ g^{-1})(-4)$

73. $(f \circ g)^{-1}$

74. $g^{-1} \circ f^{-1}$

In Exercises 75–78, use the functions given by $f(x) = x + 4$ and $g(x) = 2x - 5$ to find the specified function.

75. $g^{-1} \circ f^{-1}$

76. $f^{-1} \circ g^{-1}$

77. $(f \circ g)^{-1}$

78. $(g \circ f)^{-1}$

Model It

79. *U.S. Households* The numbers of households f (in thousands) in the United States from 1995 to 2003 are shown in the table. The time (in years) is given by t, with $t = 5$ corresponding to 1995. (Source: U.S. Census Bureau)

Year, t	Households, $f(t)$
5	98,990
6	99,627
7	101,018
8	102,528
9	103,874
10	104,705
11	108,209
12	109,297
13	111,278

(a) Find $f^{-1}(108,209)$.

(b) What does f^{-1} mean in the context of the problem?

 (c) Use the *regression* feature of a graphing utility to find a linear model for the data, $y = mx + b$. (Round m and b to two decimal places.)

 (d) Algebraically find the inverse function of the linear model in part (c).

 (e) Use the inverse function of the linear model you found in part (d) to approximate $f^{-1}(117, 022)$.

 (f) Use the inverse function of the linear model you found in part (d) to approximate $f^{-1}(108,209)$. How does this value compare with the original data shown in the table?

80. *Digital Camera Sales* The factory sales f (in millions of dollars) of digital cameras in the United States from 1998 through 2003 are shown in the table. The time (in years) is given by t, with $t = 8$ corresponding to 1998. (Source: Consumer Electronincs Association)

Year, t	Sales, $f(t)$
8	519
9	1209
10	1825
11	1972
12	2794
13	3421

(a) Does f^{-1} exist?

(b) If f^{-1} exists, what does it represent in the context of the problem?

(c) If f^{-1} exists, find $f^{-1}(1825)$.

(d) If the table was extended to 2004 and if the factory sales of digital cameras for that year was $2794 million, would f^{-1} exist? Explain.

81. *Miles Traveled* The total numbers f (in billions) of miles traveled by motor vehicles in the United States from 1995 through 2002 are shown in the table. The time (in years) is given by t, with $t = 5$ corresponding to 1995. (Source: U.S. Federal Highway Administration)

Year, t	Miles traveled, $f(t)$
5	2423
6	2486
7	2562
8	2632
9	2691
10	2747
11	2797
12	2856

(a) Does f^{-1} exist?

(b) If f^{-1} exists, what does it mean in the context of the problem?

(c) If f^{-1} exists, find $f^{-1}(2632)$.

(d) If the table was extended to 2003 and if the total number of miles traveled by motor vehicles for that year was 2747 billion, would f^{-1} exist? Explain.

82. *Hourly Wage*   Your wage is $8.00 per hour plus $0.75 for each unit produced per hour. So, your hourly wage y in terms of the number of units produced is

$$y = 8 + 0.75x.$$

(a) Find the inverse function.

(b) What does each variable represent in the inverse function?

(c) Determine the number of units produced when your hourly wage is $22.25.

83. *Diesel Mechanics*   The function given by

$$y = 0.03x^2 + 245.50, \qquad 0 < x < 100$$

approximates the exhaust temperature y in degrees Fahrenheit, where x is the percent load for a diesel engine.

(a) Find the inverse function. What does each variable represent in the inverse function?

 (b) Use a graphing utility to graph the inverse function.

(c) The exhaust temperature of the engine must not exceed 500 degrees Fahrenheit. What is the percent load interval?

84. *Cost*   You need a total of 50 pounds of two types of ground beef costing $1.25 and $1.60 per pound, respectively. A model for the total cost y of the two types of beef is

$$y = 1.25x + 1.60(50 - x)$$

where x is the number of pounds of the less expensive ground beef.

(a) Find the inverse function of the cost function. What does each variable represent in the inverse function?

(b) Use the context of the problem to determine the domain of the inverse function.

(c) Determine the number of pounds of the less expensive ground beef purchased when the total cost is $73.

Synthesis

True or False?   **In Exercises 85 and 86, determine whether the statement is true or false. Justify your answer.**

85. If f is an even function, f^{-1} exists.

86. If the inverse function of f exists and the graph of f has a y-intercept, the y-intercept of f is an x-intercept of f^{-1}.

87. *Proof*   Prove that if f and g are one-to-one functions, then $(f \circ g)^{-1}(x) = (g^{-1} \circ f^{-1})(x)$.

88. *Proof*   Prove that if f is a one-to-one odd function, then f^{-1} is an odd function.

In Exercises 89–92, use the graph of the function f to create a table of values for the given points. Then create a second table that can be used to find f^{-1}, and sketch the graph of f^{-1} if possible.

89.

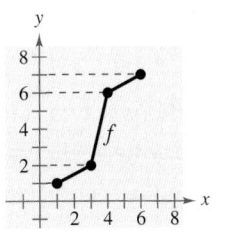

90.

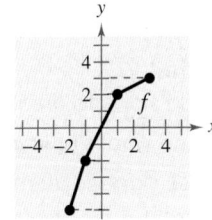

91.

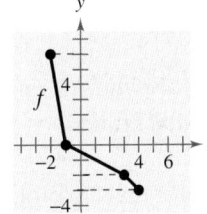

92.

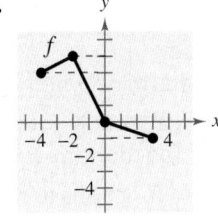

93. *Think About It*   The function given by

$$f(x) = k(2 - x - x^3)$$

has an inverse function, and $f^{-1}(3) = -2$. Find k.

94. *Think About It*   The function given by

$$f(x) = k(x^3 + 3x - 4)$$

has an inverse function, and $f^{-1}(-5) = 2$. Find k.

Skills Review

In Exercises 95–102, solve the equation using any convenient method.

95. $x^2 = 64$

96. $(x - 5)^2 = 8$

97. $4x^2 - 12x + 9 = 0$

98. $9x^2 + 12x + 3 = 0$

99. $x^2 - 6x + 4 = 0$

100. $2x^2 - 4x - 6 = 0$

101. $50 + 5x = 3x^2$

102. $2x^2 + 4x - 9 = 2(x - 1)^2$

103. Find two consecutive positive even integers whose product is 288.

104. *Geometry*   A triangular sign has a height that is twice its base. The area of the sign is 10 square feet. Find the base and height of the sign.

2 Chapter Summary

What did you learn?

2 Review Exercises

2.1 **In Exercises 1 and 2, identify the line that has each slope.**

1. (a) $m = \frac{3}{2}$
   (b) $m = 0$
   (c) $m = -3$
   (d) $m = -\frac{1}{5}$

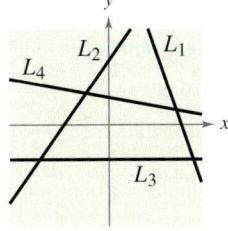

2. (a) m is undefined.
   (b) $m = -1$
   (c) $m = \frac{5}{2}$
   (d) $m = \frac{1}{2}$

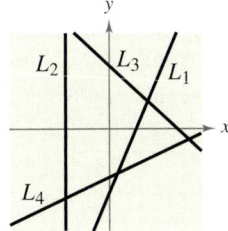

In Exercises 3–10, find the slope and y-intercept (if possible) of the equation of the line. Sketch the line.

3. $y = -2x - 7$
4. $y = 4x - 3$
5. $y = 6$
6. $x = -3$
7. $y = 3x + 13$
8. $y = -10x + 9$
9. $y = -\frac{5}{2}x - 1$
10. $y = \frac{5}{6}x + 5$

In Exercises 11 and 12, use the concept of slope to find t such that the three points lie on the same line.

11. $(-2, 5), (0, t), (1, 1)$
12. $(-6, 1), (1, t), (10, 5)$

In Exercises 13 and 14, use the point on the line and the slope of the line to find three additional points through which the line passes. (There are many correct answers.)

Point	Slope
13. $(2, -1)$	$m = \frac{1}{4}$
14. $(-3, 5)$	$m = -\frac{3}{2}$

In Exercises 15–18, plot the points and find the slope of the line passing through the pair of points.

15. $(3, -4), (-7, 1)$
16. $(-1, 8), (6, 5)$
17. $(-4.5, 6), (2.1, 3)$
18. $(-3, 2), (8, 2)$

In Exercises 19–22, find the slope-intercept form of the equation of the line that passes through the given point and has the indicated slope. Sketch the line.

Point	Slope
19. $(0, -5)$	$m = \frac{3}{2}$
20. $(-2, 6)$	$m = 0$
21. $(10, -3)$	$m = -\frac{1}{2}$
22. $(-8, 5)$	m is undefined.

In Exercises 23–26, find the slope-intercept form of the equation of the line passing through the points.

23. $(0, 0), (0, 10)$
24. $(2, 5), (-2, -1)$
25. $(-1, 4), (2, 0)$
26. $(11, -2), (6, -1)$

In Exercises 27 and 28, write the slope-intercept forms of the equations of the lines through the given point (a) parallel to the given line and (b) perpendicular to the given line.

Point	Line
27. $(3, -2)$	$5x - 4y = 8$
28. $(-8, 3)$	$2x + 3y = 5$

Rate of Change **In Exercises 29 and 30, you are given the dollar value of a product in 2006 and the rate at which the value of the product is expected to change during the next 5 years. Use this information to write a linear equation that gives the dollar value V of the product in terms of the year t. (Let $t = 6$ represent 2006.)**

2006 Value	Rate
29. $12,500	$850 increase per year
30. $72.95	$5.15 increase per year

31. *Sales* During the second and third quarters of the year, a salvage yard had sales of $160,000 and $185,000, respectively. The growth of sales follows a linear pattern. Estimate sales during the fourth quarter.

32. *Inflation* The dollar value of a product in 2005 is $85, and the product is expected to increase in value at a rate of $3.75 per year.

(a) Write a linear equation that gives the dollar value V of the product in terms of the year t. (Let $t = 5$ represent 2005.)

 (b) Use a graphing utility to graph the equation found in part (a).

(c) Move the cursor along the graph of the sales model to estimate the dollar value of the product in 2010.

2.2 In Exercises 33 and 34, determine which of the sets of ordered pairs represents a function from *A* to *B*. Explain your reasoning.

33. $A = \{10, 20, 30, 40\}$ and $B = \{0, 2, 4, 6\}$

   (a) $\{(20, 4), (40, 0), (20, 6), (30, 2)\}$

   (b) $\{(10, 4), (20, 4), (30, 4), (40, 4)\}$

   (c) $\{(40, 0), (30, 2), (20, 4), (10, 6)\}$

   (d) $\{(20, 2), (10, 0), (40, 4)\}$

34. $A = \{u, v, w\}$ and $B = \{-2, -1, 0, 1, 2\}$

   (a) $\{(v, -1), (u, 2), (w, 0), (u, -2)\}$

   (b) $\{(u, -2), (v, 2), (w, 1)\}$

   (c) $\{(u, 2), (v, 2), (w, 1), (w, 1)\}$

   (d) $\{(w, -2), (v, 0), (w, 2)\}$

In Exercises 35–38, determine whether the equation represents *y* as a function of *x*.

35. $16x - y^4 = 0$ **36.** $2x - y - 3 = 0$

37. $y = \sqrt{1 - x}$ **38.** $|y| = x + 2$

In Exercises 39–42, evaluate the function at each specified value of the independent variable and simplify.

39. $f(x) = x^2 + 1$

   (a) $f(2)$ (b) $f(-4)$ (c) $f(t^2)$ (d) $f(t + 1)$

40. $g(x) = x^{4/3}$

   (a) $g(8)$ (b) $g(t + 1)$ (c) $g(-27)$ (d) $g(-x)$

41. $h(x) = \begin{cases} 2x + 1, & x \le -1 \\ x^2 + 2, & x > -1 \end{cases}$

   (a) $h(-2)$ (b) $h(-1)$ (c) $h(0)$ (d) $h(2)$

42. $f(x) = \dfrac{4}{x^2 + 1}$

   (a) $f(1)$ (b) $f(-5)$ (c) $f(-t)$ (d) $f(0)$

In Exercises 43–48, find the domain of the function. Verify your result with a graph.

43. $f(x) = \sqrt{25 - x^2}$ **44.** $f(x) = 3x + 4$

45. $g(s) = \dfrac{5}{3s - 9}$ **46.** $f(x) = \sqrt{x^2 + 8x}$

47. $h(x) = \dfrac{x}{x^2 - x - 6}$ **48.** $h(t) = |t + 1|$

49. *Physics* The velocity of a ball projected upward from ground level is given by $v(t) = -32t + 48$, where *t* is the time in seconds and *v* is the velocity in feet per second.

   (a) Find the velocity when $t = 1$.

   (b) Find the time when the ball reaches its maximum height. [*Hint:* Find the time when $v(t) = 0$.]

   (c) Find the velocity when $t = 2$.

50. *Total Cost* A hand tool manufacturer produces a product for which the variable cost is $5.35 per unit and the fixed costs are $16,000. The company sells the product for $8.20 and can sell all that it produces.

   (a) Find the total cost as a function of *x*, the number of units produced.

   (b) Find the profit as a function of *x*.

51. *Geometry* A wire 24 inches long is to be cut into four pieces to form a rectangle with one side of length *x*.

   (a) Write the area *A* of the rectangle as a function of *x*.

   (b) Determine the domain of the function.

52. *Mixture Problem* From a full 50-liter container of a 40% concentration of acid, *x* liters is removed and replaced with 100% acid.

   (a) Write the amount of acid in the final mixture as a function of *x*.

   (b) Determine the domain and range of the function.

   (c) Determine *x* if the final mixture is 50% acid.

f In Exercises 53 and 54, find the difference quotient and simplify your answer.

53. $f(x) = 2x^2 + 3x - 1$, $\dfrac{f(x + h) - f(x)}{h}$, $h \ne 0$

54. $f(x) = x^3 - 5x^2 + x$, $\dfrac{f(x + h) - f(x)}{h}$, $h \ne 0$

2.3 In Exercises 55–58, use the Vertical Line Test to determine whether *y* is a function of *x*. To print an enlarged copy of the graph, go to the website *www.mathgraphs.com*.

55. $y = (x - 3)^2$ **56.** $y = -\frac{3}{5}x^3 - 2x + 1$

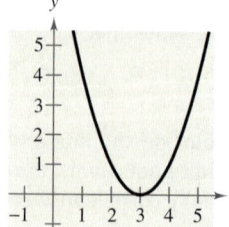
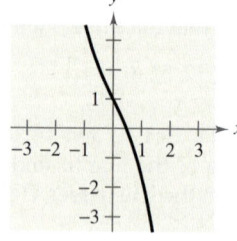

57. $x - 4 = y^2$ **58.** $x = -|4 - y|$

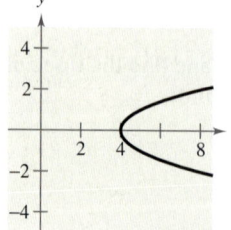
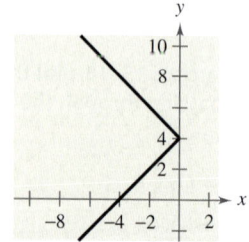

In Exercises 59–62, find the zeros of the function algebraically.

59. $f(x) = 3x^2 - 16x + 21$

60. $f(x) = 5x^2 + 4x - 1$

61. $f(x) = \dfrac{8x + 3}{11 - x}$

62. $f(x) = x^3 - x^2 - 25x + 25$

In Exercises 63 and 64, determine the intervals over which the function is increasing, decreasing, or constant.

63. $f(x) = |x| + |x + 1|$ **64.** $f(x) = (x^2 - 4)^2$

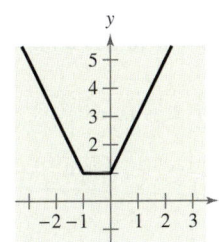   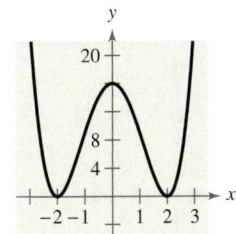

 In Exercises 65–68, use a graphing utility to graph the function and approximate (to two decimal places) any relative minimum or relative maximum values.

65. $f(x) = -x^2 + 2x + 1$ **66.** $f(x) = x^4 - 4x^2 - 2$

67. $f(x) = x^3 - 6x^4$

68. $f(x) = x^3 - 4x^2 + x - 1$

 In Exercises 69–72, find the average rate of change of the function from x_1 to x_2.

Function	x-Values
69. $f(x) = -x^2 + 8x - 4$	$x_1 = 0, x_2 = 4$
70. $f(x) = x^3 + 12x - 2$	$x_1 = 0, x_2 = 4$
71. $f(x) = 2 - \sqrt{x + 1}$	$x_1 = 3, x_2 = 7$
72. $f(x) = 1 - \sqrt{x + 3}$	$x_1 = 1, x_2 = 6$

In Exercises 73–76, determine whether the function is even, odd, or neither.

73. $f(x) = x^5 + 4x - 7$ **74.** $f(x) = x^4 - 20x^2$

75. $f(x) = 2x\sqrt{x^2 + 3}$ **76.** $f(x) = \sqrt[5]{6x^2}$

2.4 In Exercises 77–80, write the linear function f such that it has the indicated function values. Then sketch the graph of the function.

77. $f(2) = -6, \; f(-1) = 3$

78. $f(0) = -5, \; f(4) = -8$

79. $f\left(-\frac{4}{5}\right) = 2, \; f\left(\frac{11}{5}\right) = 7$

80. $f(3.3) = 5.6, \; f(-4.7) = -1.4$

In Exercises 81–90, graph the function.

81. $f(x) = 3 - x^2$ **82.** $h(x) = x^3 - 2$

83. $f(x) = -\sqrt{x}$

84. $f(x) = \sqrt{x + 1}$

85. $g(x) = \dfrac{3}{x}$

86. $g(x) = \dfrac{1}{x + 5}$

87. $f(x) = [\![x]\!] - 2$

88. $g(x) = [\![x + 4]\!]$

89. $f(x) = \begin{cases} 5x - 3, & x \ge -1 \\ -4x + 5, & x < -1 \end{cases}$

90. $f(x) = \begin{cases} x^2 - 2, & x < -2 \\ 5, & -2 \le x \le 0 \\ 8x - 5, & x > 0 \end{cases}$

In Exercises 91 and 92, the figure shows the graph of a transformed parent function. Identify the parent function.

91.   **92.**

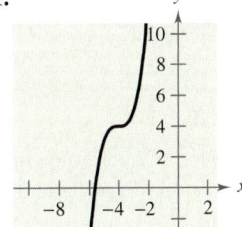   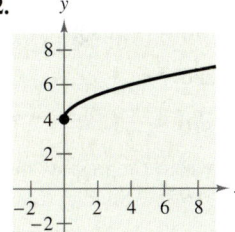

2.5 In Exercises 93–106, h is related to one of the parent functions described in this chapter. (a) Identify the parent function f. (b) Describe the sequence of transformations from f to h. (c) Sketch the graph of h. (d) Use function notation to write h in terms of f.

93. $h(x) = x^2 - 9$

94. $h(x) = (x - 2)^3 + 2$

95. $h(x) = \sqrt{x - 7}$

96. $h(x) = |x + 3| - 5$

97. $h(x) = -(x + 3)^2 + 1$

98. $h(x) = -(x - 5)^3 - 5$

99. $h(x) = -[\![x]\!] + 6$

100. $h(x) = -\sqrt{x + 1} + 9$

101. $h(x) = -|-x + 4| + 6$

102. $h(x) = -(x + 1)^2 - 3$

103. $h(x) = 5[\![x - 9]\!]$

104. $h(x) = -\frac{1}{3}x^3$

105. $h(x) = -2\sqrt{x - 4}$

106. $h(x) = \frac{1}{2}|x| - 1$

2.6 In Exercises 107 and 108, find (a) $(f + g)(x)$, (b) $(f - g)(x)$, (c) $(fg)(x)$, and (d) $(f/g)(x)$. What is the domain of f/g?

107. $f(x) = x^2 + 3$, $g(x) = 2x - 1$

108. $f(x) = x^2 - 4$, $g(x) = \sqrt{3 - x}$

In Exercises 109 and 110, find (a) $f \circ g$ and (b) $g \circ f$. Find the domain of each function and each composite function.

109. $f(x) = \frac{1}{3}x - 3$, $g(x) = 3x + 1$

110. $f(x) = x^3 - 4$, $g(x) = \sqrt[3]{x + 7}$

 In Exercises 111 and 112, find two functions f and g such that $(f \circ g)(x) = h(x)$. (There are many correct answers.)

111. $h(x) = (6x - 5)^3$

112. $h(x) = \sqrt[3]{x + 2}$

113. *Electronics Sales* The factory sales (in millions of dollars) for VCRs $v(t)$ and DVD players $d(t)$ from 1997 to 2003 can be approximated by the functions

$$v(t) = -31.86t^2 + 233.6t + 2594$$

and

$$d(t) = -4.18t^2 + 571.0t - 3706$$

where t represents the year, with $t = 7$ corresponding to 1997. (Source: Consumer Electronics Association)

(a) Find and interpret $(v + d)(t)$.

 (b) Use a graphing utility to graph $v(t)$, $d(t)$, and the function from part (a) in the same viewing window.

(c) Find $(v + d)(10)$. Use the graph in part (b) to verify your result.

114. *Bacteria Count* The number N of bacteria in a refrigerated food is given by

$$N(T) = 25T^2 - 50T + 300, \quad 2 \le T \le 20$$

where T is the temperature of the food in degrees Celsius. When the food is removed from refrigeration, the temperature of the food is given by

$$T(t) = 2t + 1, \quad 0 \le t \le 9$$

where t is the time in hours (a) Find the composition $N(T(t))$ and interpret its meaning in context and (b) find the time when the bacterial count reaches 750.

2.7 In Exercises 115 and 116, find the inverse function of f informally. Verify that $f(f^{-1}(x)) = x$ and $f^{-1}(f(x)) = x$.

115. $f(x) = x - 7$

116. $f(x) = x + 5$

In Exercises 117 and 118, determine whether the function has an inverse function.

117.

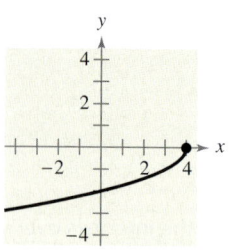

118.

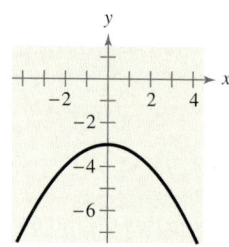

In Exercises 119–122, use a graphing utility to graph the function, and use the Horizontal Line Test to determine whether the function is one-to-one and so has an inverse function.

119. $f(x) = 4 - \frac{1}{3}x$

120. $f(x) = (x - 1)^2$

121. $h(t) = \dfrac{2}{t - 3}$

122. $g(x) = \sqrt{x + 6}$

In Exercises 123–126, (a) find the inverse function of f, (b) graph both f and f^{-1} on the same set of coordinate axes, (c) describe the relationship between the graphs of f and f^{-1}, and (d) state the domain and range of f and f^{-1}.

123. $f(x) = \frac{1}{2}x - 3$

124. $f(x) = 5x - 7$

125. $f(x) = \sqrt{x + 1}$

126. $f(x) = x^3 + 2$

In Exercises 127 and 128, restrict the domain of the function f to an interval over which the function is increasing and determine f^{-1} over that interval.

127. $f(x) = 2(x - 4)^2$

128. $f(x) = |x - 2|$

Synthesis

True or False? In Exercises 129 and 130, determine whether the statement is true or false. Justify your answer.

129. Relative to the graph of $f(x) = \sqrt{x}$, the function given by $h(x) = -\sqrt{x + 9} - 13$ is shifted 9 units to the left and 13 units downward, then reflected in the x-axis.

130. If f and g are two inverse functions, then the domain of g is equal to the range of f.

131. *Writing* Explain how to tell whether a relation between two variables is a function.

132. *Writing* Explain the difference between the Vertical Line Test and the Horizontal Line Test.

133. *Writing* Describe the basic characteristics of the cubic function. Describe the basic characteristics of $f(x) = x^3 + 1$.

2 Chapter Test

Take this test as you would take a test in class. When you are finished, check your work against the answers given in the back of the book.

In Exercises 1 and 2, find an equation of the line passing through the points. Then sketch the line.

1. $(2, -3), (-4, 9)$

2. $(3, 0.8), (7, -6)$

3. Find an equation of the line that passes through the point $(3, 8)$ and is (a) parallel to and (b) perpendicular to the line $-4x + 7y = -5$.

In Exercises 4 and 5, evaluate the function at each specified value of the independent variable and simplify.

4. $f(x) = |x + 2| - 15$

 (a) $f(-8)$ (b) $f(14)$ (c) $f(x - 6)$

5. $f(x) = \dfrac{\sqrt{x + 9}}{x^2 - 81}$

 (a) $f(7)$ (b) $f(-5)$ (c) $f(x - 9)$

In Exercises 6 and 7, determine the domain of the function.

6. $f(x) = \sqrt{100 - x^2}$

7. $f(x) = |-x + 6| + 2$

In Exercises 8–10, (a) use a graphing utility to graph the function, (b) approximate the intervals over which the function is increasing, decreasing, or constant, and (c) determine whether the function is even, odd, or neither.

8. $f(x) = 2x^6 + 5x^4 - x^2$

9. $f(x) = 4x\sqrt{3 - x}$

10. $f(x) = |x + 5|$

11. Use a graphing utility to approximate any relative minimum or maximum values of $f(x) = -x^3 + 2x - 1$.

12. Find the average rate of change of $f(x) = -2x^2 + 5x - 3$ from $x_1 = 1$ to $x_2 = 3$.

13. Sketch the graph of $f(x) = \begin{cases} 3x + 7, & x \le -3 \\ 4x^2 - 1, & x > -3 \end{cases}$.

In Exercises 14–16, (a) identify the parent function in the transformation, (b) describe the sequence of transformations from f to h, and (c) sketch the graph of h.

14. $h(x) = -[\![x]\!]$

15. $h(x) = -\sqrt{x + 5} + 8$

16. $h(x) = \frac{1}{4}|x + 1| - 3$

In Exercises 17 and 18, find (a) $(f + g)(x)$, (b) $(f - g)(x)$, (c) $(fg)(x)$, (d) $(f/g)(x)$, (e) $(f \circ g)(x)$, and (f) $(g \circ f)(x)$.

17. $f(x) = 3x^2 - 7, \quad g(x) = -x^2 - 4x + 5$

18. $f(x) = \dfrac{1}{x}, \quad g(x) = 2\sqrt{x}$

In Exercises 19–21, determine whether the function has an inverse function, and if so, find the inverse function.

19. $f(x) = x^3 + 8$

20. $f(x) = |x^2 - 3| + 6$

21. $f(x) = \dfrac{3x\sqrt{x}}{8}$

22. It costs a company \$58 to produce 6 units of a product and \$78 to produce 10 units. How much does it cost to produce 25 units, assuming that the cost function is linear?

2 Cumulative Test for Chapters P–2

Take this test to review the material from earlier chapters. When you are finished, check your work against the answers given in the back of the book.

In Exercises 1 and 2, simplify the expression.

1. $\dfrac{8x^2y^{-3}}{30x^{-1}y^2}$

2. $\sqrt{24x^4y^3}$

In Exercises 3–5, perform the operation and simplify the result.

3. $4x - [2x + 3(2 - x)]$ **4.** $(x - 2)(x^2 + x - 3)$ **5.** $\dfrac{2}{s + 3} - \dfrac{1}{s + 1}$

In Exercises 6–8, factor the expression completely.

6. $25 - (x - 2)^2$ **7.** $x - 5x^2 - 6x^3$ **8.** $54 - 16x^3$

In Exercises 9 and 10, write an expression for the area of the region.

9.

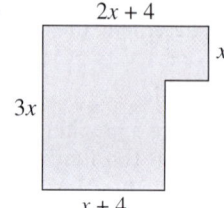

10.

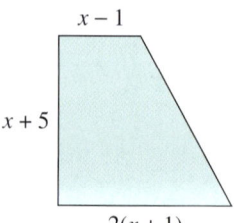

In Exercises 11–13, graph the equation without using a graphing utility.

11. $x - 3y + 12 = 0$ **12.** $y = x^2 - 9$ **13.** $y = \sqrt{4 - x}$

In Exercises 14–16, solve the equation and check your solution.

14. $2x - 4 = 7x + 5$ **15.** $-(x + 3) = 14(x - 6)$ **16.** $\dfrac{1}{x - 2} = \dfrac{10}{4x + 3}$

In Exercises 17–22, solve the equation using any convenient method and check your solutions. State the method you used.

17. $x^2 - 4x + 3 = 0$ **18.** $-2x^2 + 8x + 12 = 0$

19. $\frac{3}{4}x^2 = 12$ **20.** $3x^2 + 5x - 6 = 0$

21. $3x^2 + 9x + 1 = 0$ **22.** $\frac{1}{2}x^2 - 7 = 25$

In Exercises 23–28, solve the equation (if possible).

23. $x^4 + 12x^3 + 4x^2 + 48x = 0$ **24.** $8x^3 - 48x^2 + 72x = 0$

25. $x^{2/3} + 13 = 17$ **26.** $\sqrt{x + 10} = x - 2$

27. $|4(x - 2)| = 28$ **28.** $|x - 12| = -2$

In Exercises 29 and 30, determine whether each value of x is a solution of the inequality.

29. $4x + 2 > 7$

(a) $x = -1$ (b) $x = \frac{1}{2}$

(c) $x = \frac{3}{2}$ (d) $x = 2$

30. $|5x - 1| < 4$

(a) $x = -1$ (b) $x = -\frac{1}{2}$

(c) $x = 1$ (d) $x = 2$

In Exercises 31–34, solve the inequality and sketch the solution on the real number line.

31. $|x + 1| \le 6$

33. $5x^2 + 12x + 7 \ge 0$

32. $|7 + 8x| > 5$

34. $-x^2 + x + 4 < 0$

35. Find an equation of the line passing through $\left(-\frac{1}{2}, 1\right)$ and $(3, 8)$.

36. Explain why the graph at the left does not represent y as a function of x.

37. Evaluate (if possible) the function given by $f(x) = \dfrac{x}{x - 2}$ for each value.

(a) $f(6)$ (b) $f(2)$ (c) $f(s + 2)$

In Exercises 38–40, determine whether the function is even, odd, or neither.

38. $f(x) = 10 - \sqrt{3 - x}$ **39.** $f(x) = x^5 - x^3 + 2$ **40.** $f(x) = 2x^4 - 4$

41. Compare the graph of each function with the graph of $y = \sqrt[3]{x}$. (*Note:* It is not necessary to sketch the graphs.)

(a) $r(x) = \frac{1}{2}\sqrt[3]{x}$ (b) $h(x) = \sqrt[3]{x} + 2$ (c) $g(x) = \sqrt[3]{x + 2}$

In Exercises 42 and 43, find (a) $(f + g)(x)$, (b) $(f - g)(x)$, (c) $(fg)(x)$, and (d) $(f/g)(x)$. What is the domain of f/g?

42. $f(x) = x - 3, \quad g(x) = 4x + 1$ **43.** $f(x) = \sqrt{x - 1}, \quad g(x) = x^2 + 1$

In Exercises 44 and 45, find (a) $f \circ g$ and (b) $g \circ f$. Find the domain of each composite function.

44. $f(x) = 2x^2, \quad g(x) = \sqrt{x + 6}$ **45.** $f(x) = x - 2, \quad g(x) = |x|$

46. Determine whether $h(x) = 5x - 2$ has an inverse function. If so, find the inverse function.

47. A group of n people decide to buy a \$36,000 minibus. Each person will pay an equal share of the cost. If three additional people join the group, the cost per person will decrease by \$1000. Find n.

48. For groups of 80 or more, a charter bus company determines the rate per person according to the formula

Rate $= \$8.00 - \$0.05(n - 80), \quad n \ge 80$.

(a) Write the revenue R as a function of n.

(b) Use a graphing utility to graph the revenue function. Move the cursor along the function to estimate the number of passengers that will maximize the revenue.

49. The height of an object thrown vertically upward from a height of 6 feet at a velocity of 64 feet per second can be modeled by $s(t) = -16t^2 + 64t + 6$, where s is the height (in feet) and t is the time (in seconds). Find the average rate of change of the function from $t_1 = 0$ to $t_2 = 2$. Interpret your answer in the context of the problem.

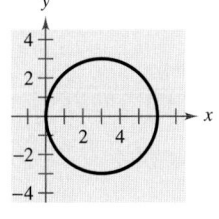

FIGURE FOR **36**

Proofs in Mathematics

Biconditional Statements

Recall from the Proofs in Mathematics in Chapter 1 that a conditional statement is a statement of the form "if p, then q." A statement of the form "p if and only if q" is called a **biconditional statement.** A biconditional statement, denoted by

$$p \leftrightarrow q \qquad \text{Biconditional statement}$$

is the conjunction of the conditional statement $p \rightarrow q$ and its converse $q \rightarrow p$.

A biconditional statement can be either true or false. To be true, *both* the conditional statement and its converse must be true.

Example 1 Analyzing a Biconditional Statement

Consider the statement $x = 3$ if and only if $x^2 = 9$.

a. Is the statement a biconditional statement? **b.** Is the statement true?

Solution

a. The statement is a biconditional statement because it is of the form "p if and only if q."

b. The statement can be rewritten as the following conditional statement and its converse.

> *Conditional statement:* If $x = 3$, then $x^2 = 9$.
> *Converse:* If $x^2 = 9$, then $x = 3$.

The first of these statements is true, but the second is false because x could also equal -3. So, the biconditional statement is false.

Knowing how to use biconditional statements is an important tool for reasoning in mathematics.

Example 2 Analyzing a Biconditional Statement

Determine whether the biconditional statement is true or false. If it is false, provide a counterexample.

A number is divisible by 5 if and only if it ends in 0.

Solution

The biconditional statement can be rewritten as the following conditional statement and its converse.

> *Conditional statement:* If a number is divisible by 5, then it ends in 0.
> *Converse:* If a number ends in 0, then it is divisible by 5.

The conditional statement is false. A counterexample is the number 15, which is divisible by 5 but does not end in 0.

This collection of thought-provoking and challenging exercises further explores and expands upon concepts learned in this chapter.

1. As a salesperson, you receive a monthly salary of $2000, plus a commission of 7% of sales. You are offered a new job at $2300 per month, plus a commission of 5% of sales.

   (a) Write a linear equation for your current monthly wage W_1 in terms of your monthly sales S.

   (b) Write a linear equation for the monthly wage W_2 of your new job offer in terms of the monthly sales S.

    (c) Use a graphing utility to graph both equations in the same viewing window. Find the point of intersection. What does it signify?

   (d) You think you can sell $20,000 per month. Should you change jobs? Explain.

2. For the numbers 2 through 9 on a telephone keypad (see figure), create two relations: one mapping numbers onto letters, and the other mapping letters onto numbers. Are both relations functions? Explain.

3. What can be said about the sum and difference of each of the following?

   (a) Two even functions   (b) Two odd functions

   (c) An odd function and an even function

4. The two functions given by

   $$f(x) = x \quad \text{and} \quad g(x) = -x$$

   are their own inverse functions. Graph each function and explain why this is true. Graph other linear functions that are their own inverse functions. Find a general formula for a family of linear functions that are their own inverse functions.

5. Prove that a function of the following form is even.

   $$y = a_{2n}x^{2n} + a_{2n-2}x^{2n-2} + \cdots + a_2x^2 + a_0$$

6. A miniature golf professional is trying to make a hole-in-one on the miniature golf green shown. A coordinate plane is placed over the golf green. The golf ball is at the point $(2.5, 2)$ and the hole is at the point $(9.5, 2)$. The professional wants to bank the ball off the side wall of the green at the point (x, y). Find the coordinates of the point (x, y). Then write an equation for the path of the ball.

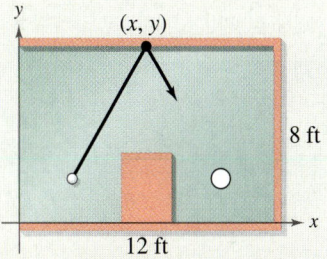

FIGURE FOR 6

7. At 2:00 P.M. on April 11, 1912, the *Titanic* left Cobh, Ireland, on her voyage to New York City. At 11:40 P.M. on April 14, the *Titanic* struck an iceberg and sank, having covered only about 2100 miles of the approximately 3400-mile trip.

   (a) What was the total length of the *Titantic's* voyage in hours?

   (b) What was the *Titanic's* average speed in miles per hour?

   (c) Write a function relating the *Titanic's* distance from New York City and the number of hours traveled. Find the domain and range of the function.

   (d) Graph the function from part (c).

8. Consider the function given by $f(x) = -x^2 + 4x - 3$. Find the average rate of change of the function from x_1 to x_2.

   (a) $x_1 = 1, x_2 = 2$ (b) $x_1 = 1, x_2 = 1.5$

   (c) $x_1 = 1, x_2 = 1.25$

   (d) $x_1 = 1, x_2 = 1.125$

   (e) $x_1 = 1, x_2 = 1.0625$

   (f) Does the average rate of change seem to be approaching one value? If so, what value?

   (g) Find the equation of the secant line through the points $(x_1, f(x_1))$ and $(x_2, f(x_2))$ for parts (a)–(e).

   (h) Find the equation of the line though the point $(1, f(1))$ using your answer from part (f) as the slope of the line.

9. Consider the functions given by $f(x) = 4x$ and $g(x) = x + 6$.

   (a) Find $(f \circ g)(x)$.

   (b) Find $(f \circ g)^{-1}(x)$.

   (c) Find $f^{-1}(x)$ and $g^{-1}(x)$.

   (d) Find $(g^{-1} \circ f^{-1})(x)$ and compare the result with that of part (b).

   (e) Repeat parts (a) through (d) for $f(x) = x^3 + 1$ and $g(x) = 2x$.

   (f) Write two one-to-one functions f and g, and repeat parts (a) through (d) for these functions.

   (g) Make a conjecture about $(f \circ g)^{-1}(x)$ and $(g^{-1} \circ f^{-1})(x)$.

261

10. You are in a boat 2 miles from the nearest point on the coast. You are to travel to a point Q, 3 miles down the coast and 1 mile inland (see figure). You can row at 2 miles per hour and walk at 4 miles per hour.

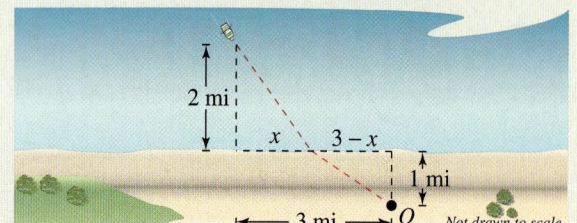

(a) Write the total time T of the trip as a function of x.

(b) Determine the domain of the function.

 (c) Use a graphing utility to graph the function. Be sure to choose an appropriate viewing window.

 (d) Use the *zoom* and *trace* features to find the value of x that minimizes T.

 (e) Write a brief paragraph interpreting these values.

11. The **Heaviside function** $H(x)$ is widely used in engineering applications. (See figure.) To print an enlarged copy of the graph, go to the website *www.mathgraphs.com*.

$$H(x) = \begin{cases} 1, & x \geq 0 \\ 0, & x < 0 \end{cases}$$

Sketch the graph of each function by hand.

(a) $H(x) - 2$ (b) $H(x - 2)$ (c) $-H(x)$

(d) $H(-x)$ (e) $\frac{1}{2}H(x)$ (f) $-H(x - 2) + 2$

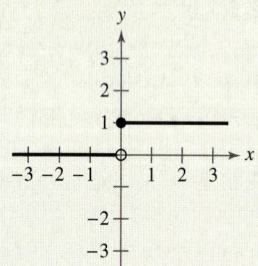

12. Let $f(x) = \dfrac{1}{1 - x}$.

(a) What are the domain and range of f?

(b) Find $f(f(x))$. What is the domain of this function?

(c) Find $f(f(f(x)))$. Is the graph a line? Why or why not?

13. Show that the Associative Property holds for compositions of functions—that is,

$$(f \circ (g \circ h))(x) = ((f \circ g) \circ h)(x).$$

14. Consider the graph of the function f shown in the figure. Use this graph to sketch the graph of each function. To print an enlarged copy of the graph, go to the website *www.mathgraphs.com*.

(a) $f(x + 1)$ (b) $f(x) + 1$ (c) $2f(x)$ (d) $f(-x)$

(e) $-f(x)$ (f) $|f(x)|$ (g) $f(|x|)$

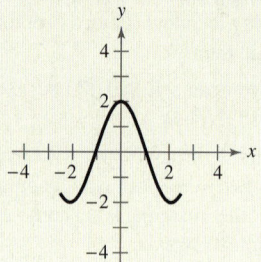

15. Use the graphs of f and f^{-1} to complete each table of function values.

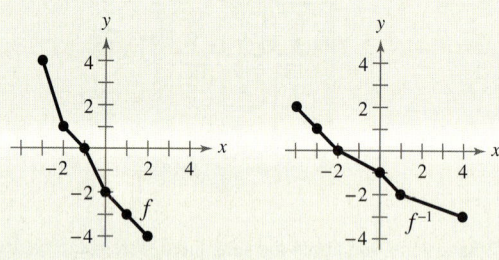

(a)

x	-4	-2	0	4
$(f(f^{-1}(x))$				

(b)

x	-3	-2	0	1
$(f + f^{-1})(x)$				

(c)

x	-3	-2	0	1
$(f \cdot f^{-1})(x)$				

(d)

x	-4	-3	0	4
$|f^{-1}(x)|$				

Polynomial Functions

3

Quadratic functions are often used to model real-life phenomena, such as the path of a diver.

© Martin Rose/Bongarts/Getty Images

SELECTED APPLICATIONS

Polynomial functions have many applications in real life, as can be seen by the examples below, which represent a small sample of the applications in this chapter.

3.1 Quadratic Functions and Models

What you should learn

- Analyze graphs of quadratic functions.
- Write quadratic functions in standard form and use the results to sketch graphs of functions.
- Use quadratic functions to model and solve real-life problems.

Why you should learn it

Quadratic functions can be used to model data to analyze consumer behavior. For instance, in Exercise 83 on page 273, you will use a quadratic function to model the revenue earned from manufacturing handheld video games.

© John Henley/Corbis

The *HM mathSpace*® CD-ROM and *Eduspace*® for this text contain additional resources related to the concepts discussed in this chapter.

The Graph of a Quadratic Function

In this and the next section, you will study the graphs of polynomial functions. In Section 2.4, you were introduced to the following basic functions.

$f(x) = ax + b$ Linear function

$f(x) = c$ Constant function

$f(x) = x^2$ Squaring function

These functions are examples of **polynomial functions.**

Definition of Polynomial Function

Let n be a nonnegative integer and let $a_n, a_{n-1}, \ldots, a_2, a_1, a_0$ be real numbers with $a_n \neq 0$. The function given by

$$f(x) = a_n x^n + a_{n-1} x^{n-1} + \cdots + a_2 x^2 + a_1 x + a_0$$

is called a **polynomial function of x with degree n.**

Polynomial functions are classified by degree. For instance, a constant function has degree 0 and a linear function has degree 1. In this section, you will study second-degree polynomial functions, which are called **quadratic functions.**

For instance, each of the following functions is a quadratic function.

$f(x) = x^2 + 6x + 2$

$g(x) = 2(x + 1)^2 - 3$

$h(x) = 9 + \frac{1}{4}x^2$

$k(x) = -3x^2 + 4$

$m(x) = (x - 2)(x + 1)$

Note that the squaring function is a simple quadratic function that has degree 2.

Definition of Quadratic Function

Let a, b, and c be real numbers with $a \neq 0$. The function given by

$$f(x) = ax^2 + bx + c \qquad \text{Quadratic function}$$

is called a **quadratic function.**

The graph of a quadratic function is a special type of "U"-shaped curve called a **parabola.** Parabolas occur in many real-life applications—especially those involving reflective properties of satellite dishes and flashlight reflectors. You will study these properties in Section 4.3.

All parabolas are symmetric with respect to a line called the **axis of symmetry,** or simply the **axis** of the parabola. The point where the axis intersects the parabola is the **vertex** of the parabola, as shown in Figure 3.1. If the leading coefficient is positive, the graph of

$$f(x) = ax^2 + bx + c$$

is a parabola that opens upward. If the leading coefficient is negative, the graph of

$$f(x) = ax^2 + bx + c$$

is a parabola that opens downward.

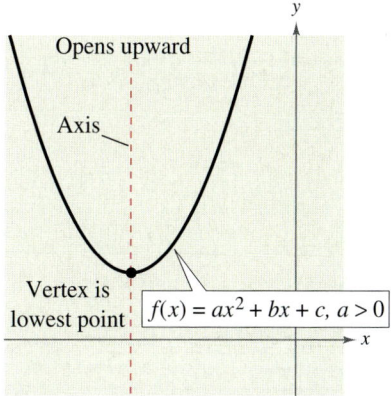

Leading coefficient is positive.   Leading coefficient is negative.

FIGURE 3.1

The simplest type of quadratic function is

$$f(x) = ax^2.$$

Its graph is a parabola whose vertex is $(0, 0)$. If $a > 0$, the vertex is the point with the *minimum* y-value on the graph, and if $a < 0$, the vertex is the point with the *maximum* y-value on the graph, as shown in Figure 3.2.

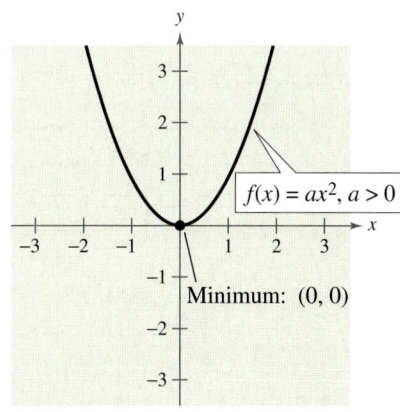
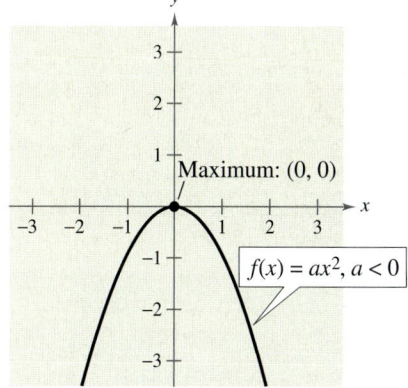

Leading coefficient is positive.   Leading coefficient is negative.

FIGURE 3.2

When sketching the graph of $f(x) = ax^2$, it is helpful to use the graph of $y = x^2$ as a reference, as discussed in Section 2.5.

Example 1 **Sketching Graphs of Quadratic Functions**

a. Compare the graphs of $y = x^2$ and $f(x) = \frac{1}{3}x^2$.

b. Compare the graphs of $y = x^2$ and $g(x) = 2x^2$.

Solution

a. Compared with $y = x^2$, each output of $f(x) = \frac{1}{3}x^2$ "shrinks" by a factor of $\frac{1}{3}$, creating the broader parabola shown in Figure 3.3.

b. Compared with $y = x^2$, each output of $g(x) = 2x^2$ "stretches" by a factor of 2, creating the narrower parabola shown in Figure 3.4.

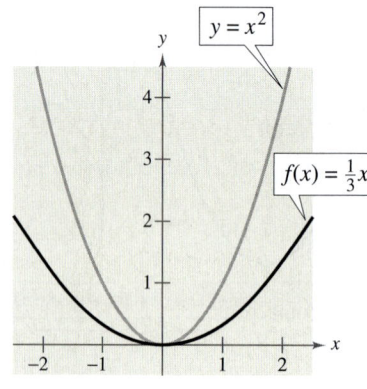

FIGURE **3.3**   FIGURE **3.4**

✓**CHECKPOINT**   Now try Exercise 9.

In Example 1, note that the coefficient a determines how widely the parabola given by $f(x) = ax^2$ opens. If $|a|$ is small, the parabola opens more widely than if $|a|$ is large.

Recall from Section 2.5 that the graphs of $y = f(x \pm c)$, $y = f(x) \pm c$, $y = f(-x)$, and $y = -f(x)$ are rigid transformations of the graph of $y = f(x)$. For instance, in Figure 3.5, notice how the graph of $y = x^2$ can be transformed to produce the graphs of $f(x) = -x^2 + 1$ and $g(x) = (x + 2)^2 - 3$.

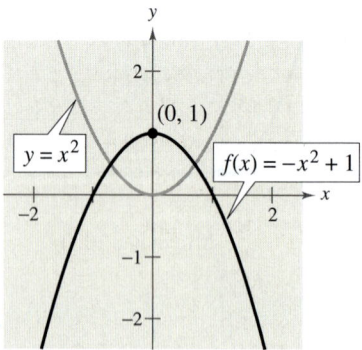

Reflection in x-axis followed by an upward shift of one unit

FIGURE **3.5**

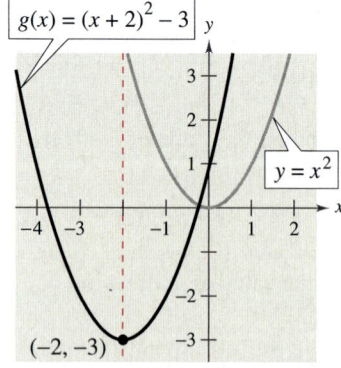

Left shift of two units followed by a downward shift of three units

The Standard Form of a Quadratic Function

The **standard form** of a quadratic function is $f(x) = a(x - h)^2 + k$. This form is especially convenient for sketching a parabola because it identifies the vertex of the parabola as (h, k).

> ## Standard Form of a Quadratic Function
>
> The quadratic function given by
>
> $$f(x) = a(x - h)^2 + k, \qquad a \neq 0$$
>
> is in **standard form.** The graph of f is a parabola whose axis is the vertical line $x = h$ and whose vertex is the point (h, k). If $a > 0$, the parabola opens upward, and if $a < 0$, the parabola opens downward.

To graph a parabola, it is helpful to begin by writing the quadratic function in standard form using the process of completing the square, as illustrated in Example 2. In this example, notice that when completing the square, you *add and subtract* the square of half the coefficient of x within the parentheses instead of adding the value to each side of the equation as was done in Section 1.4.

Example 2   Graphing a Parabola in Standard Form

Sketch the graph of $f(x) = 2x^2 + 8x + 7$ and identify the vertex and the axis of the parabola.

Solution

Begin by writing the quadratic function in standard form. Notice that the first step in completing the square is to factor out any coefficient of x^2 that is not 1.

$$f(x) = 2x^2 + 8x + 7 \qquad \text{Write original function.}$$
$$= 2(x^2 + 4x) + 7 \qquad \text{Factor 2 out of } x\text{-terms.}$$
$$= 2(x^2 + 4x + 4 - 4) + 7 \qquad \text{Add and subtract 4 within parentheses.}$$

$$(4/2)^2$$

After adding and subtracting 4 within the parentheses, you must now regroup the terms to form a perfect square trinomial. The -4 can be removed from inside the parentheses; however, because of the 2 outside of the parentheses, you must multiply -4 by 2, as shown below.

$$f(x) = 2(x^2 + 4x + 4) - 2(4) + 7 \qquad \text{Regroup terms.}$$
$$= 2(x^2 + 4x + 4) - 8 + 7 \qquad \text{Simplify.}$$
$$= 2(x + 2)^2 - 1 \qquad \text{Write in standard form.}$$

From this form, you can see that the graph of f is a parabola that opens upward and has its vertex at $(-2, -1)$. This corresponds to a left shift of two units and a downward shift of one unit relative to the graph of $y = 2x^2$, as shown in Figure 3.6. In the figure, you can see that the axis of the parabola is the vertical line through the vertex, $x = -2$.

✓CHECKPOINT   Now try Exercise 13.

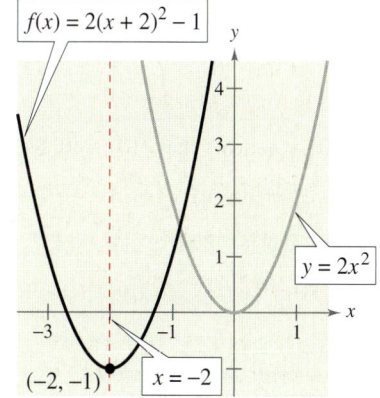

$f(x) = 2(x + 2)^2 - 1$

$y = 2x^2$

$(-2, -1)$ $x = -2$

FIGURE 3.6

To find the x-intercepts of the graph of $f(x) = ax^2 + bx + c$, you must solve the equation $ax^2 + bx + c = 0$. If $ax^2 + bx + c$ does not factor, you can use the Quadratic Formula to find the x-intercepts. Remember, however, that a parabola may not have x-intercepts.

Example 3 Finding the Vertex and x-Intercepts of a Parabola

Sketch the graph of $f(x) = -x^2 + 6x - 8$ and identify the vertex and x-intercepts.

Solution

$$f(x) = -x^2 + 6x - 8 \qquad \text{Write original function.}$$
$$= -(x^2 - 6x) - 8 \qquad \text{Factor } -1 \text{ out of } x\text{-terms.}$$
$$= -(x^2 - 6x + 9 - 9) - 8 \qquad \begin{array}{l}\text{Add and subtract 9 within} \\ \text{parentheses.}\end{array}$$

$$(-6/2)^2$$

$$= -(x^2 - 6x + 9) - (-9) - 8 \qquad \text{Regroup terms.}$$
$$= -(x - 3)^2 + 1 \qquad \text{Write in standard form.}$$

From this form, you can see that f is a parabola that opens downward with vertex $(3, 1)$. The x-intercepts of the graph are determined as follows.

$$-(x^2 - 6x + 8) = 0 \qquad \text{Factor out } -1.$$
$$-(x - 2)(x - 4) = 0 \qquad \text{Factor.}$$
$$x - 2 = 0 \quad \Longrightarrow \quad x = 2 \qquad \text{Set 1st factor equal to 0.}$$
$$x - 4 = 0 \quad \Longrightarrow \quad x = 4 \qquad \text{Set 2nd factor equal to 0.}$$

So, the x-intercepts are $(2, 0)$ and $(4, 0)$, as shown in Figure 3.7.

✓CHECKPOINT Now try Exercise 23.

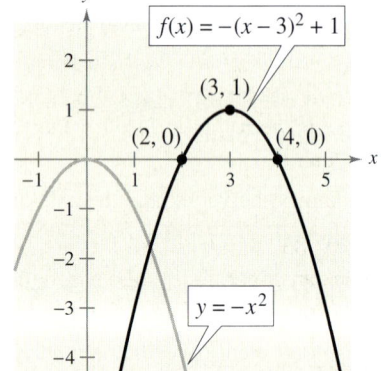

$f(x) = -(x - 3)^2 + 1$

$(3, 1)$

$(2, 0)$ $(4, 0)$

$y = -x^2$

FIGURE 3.7

Example 4 Writing the Equation of a Parabola

Write the standard form of the equation of the parabola whose vertex is $(1, 2)$ and that passes through the point $(0, 0)$, as shown in Figure 3.8.

Solution

Because the vertex of the parabola is at $(h, k) = (1, 2)$, the equation has the form

$$f(x) = a(x - 1)^2 + 2. \qquad \text{Substitute for } h \text{ and } k \text{ in standard form.}$$

Because the parabola passes through the point $(0, 0)$, it follows that $f(0) = 0$. So,

$$0 = a(0 - 1)^2 + 2 \quad \Longrightarrow \quad a = -2 \qquad \text{Substitute 0 for } x; \text{ solve for } a.$$

which implies that the equation in standard form is $f(x) = -2(x - 1)^2 + 2$.

✓CHECKPOINT Now try Exercise 43.

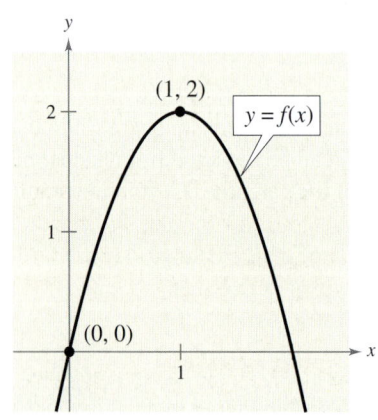

$(1, 2)$

$y = f(x)$

$(0, 0)$

FIGURE 3.8

Applications

Many applications involve finding the maximum or minimum value of a quadratic function. You can find the maximum or minimum value of a quadratic function by locating the vertex of the graph of the function.

Vertex of a Parabola

The vertex of the graph of $f(x) = ax^2 + bx + c$ is $\left(-\dfrac{b}{2a},\ f\left(-\dfrac{b}{2a}\right)\right)$.

1. If $a > 0$, has a *minimum* at $x = -\dfrac{b}{2a}$.

2. If $a < 0$, has a *maximum* at $x = -\dfrac{b}{2a}$.

Example 5   **The Maximum Height of a Baseball**

A baseball is hit at a point 3 feet above the ground at a velocity of 100 feet per second and at an angle of 45° with respect to the ground. The path of the baseball is given by the function $f(x) = -0.0032x^2 + x + 3$, where $f(x)$ is the height of the baseball (in feet) and x is the horizontal distance from home plate (in feet). What is the maximum height reached by the baseball?

Solution

From the given function, you can see that $a = -0.0032$ and $b = 1$. Because the function has a maximum when $x = -b/(2a)$, you can conclude that the baseball reaches its maximum height when it is x feet from home plate, where x is

$$x = -\frac{b}{2a} x = -\frac{b}{2a} = -\frac{1}{2(-0.0032)} = 156.25 \text{ feet.}$$

At this distance, the maximum height is $f(156.25) = -0.0032(156.25)^2 + 156.25 + 3 = 81.125$ feet. The path of the baseball is shown in Figure 3.9.

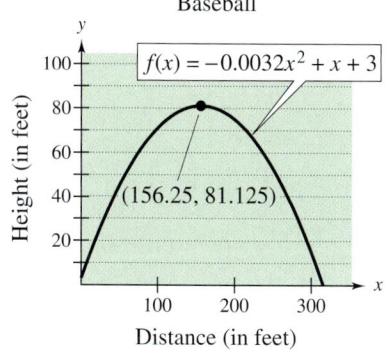

Baseball

$f(x) = -0.0032x^2 + x + 3$

(156.25, 81.125)

Height (in feet)

Distance (in feet)

FIGURE 3.9

✔CHECKPOINT   Now try Exercise 77.

Example 6   **Minimizing Cost**

A small local soft-drink manufacturer has daily production costs of $C = 70{,}000 - 120x + 0.075x^2$, where C is the total cost (in dollars) and x is the number of units produced. How many units should be produced each day to yield a minimum cost?

Solution

Use the fact that the function has a minimum when $x = -b/(2a)$. From the given function you can see that $a = 0.075$ and $b = -120$. So, producing

$$x = -\frac{b}{2a} = -\frac{-120}{2(0.075)} = 800 \text{ units}$$

each day will yield a minimum cost.

✔CHECKPOINT   Now try Exercise 83.

3.1   Exercises

The *HM mathSpace*® CD-ROM and *Eduspace*® for this text contain step-by-step solutions to all odd-numbered exercises. They also provide Tutorial Exercises for additional help.

VOCABULARY CHECK: Fill in the blanks.

1. A polynomial function of degree n and leading coefficient a_n is a function of the form
   $f(x) = a_n x^n + a_{n-1} x^{n-1} + \cdots + a_1 x + a_0$ $(a_n \neq 0)$ where n is a _____ _____ and
   $a_n, a_{n-1}, \ldots, a_1, a_0$ are _____ numbers.

2. A _____ function is a second-degree polynomial function, and its graph is called a _____.

3. The graph of a quadratic function is symmetric about its _____.

4. If the graph of a quadratic function opens upward, then its leading coefficient is _____ and the vertex of the graph is a _____.

5. If the graph of a quadratic function opens downward, then its leading coefficient is _____ and the vertex of the graph is a _____.

PREREQUISITE SKILLS REVIEW: Practice and review algebra skills needed for this section at **www.Eduspace.com**.

In Exercises 1–8, match the quadratic function with its graph. [The graphs are labeled (a), (b), (c), (d), (e), (f), (g), and (h).]

1. $f(x) = (x - 2)^2$
2. $f(x) = (x + 4)^2$
3. $f(x) = x^2 - 2$
4. $f(x) = 3 - x^2$
5. $f(x) = 4 - (x - 2)^2$
6. $f(x) = (x + 1)^2 - 2$
7. $f(x) = -(x - 3)^2 - 2$
8. $f(x) = -(x - 4)^2$

(a)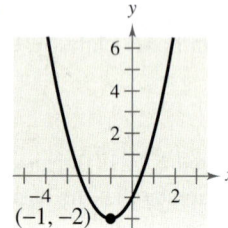

(b)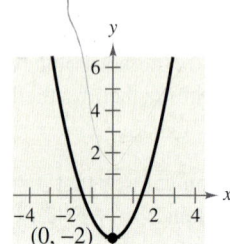

(c)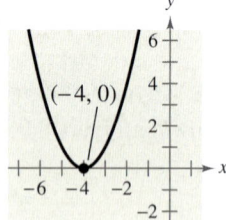

(d)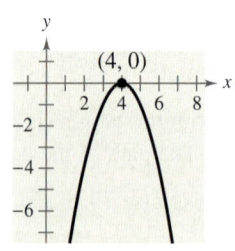

(e)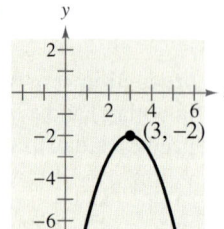

(f)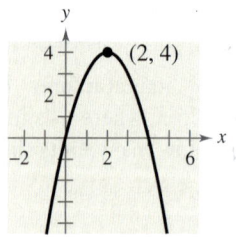

(g)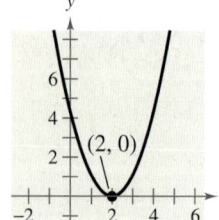

(h)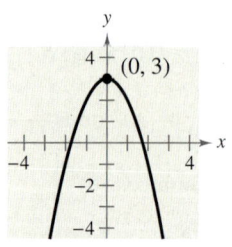

In Exercises 9–12, graph each function. Compare the graph of each function with the graph of $y = x^2$.

9. (a) $f(x) = \frac{1}{2}x^2$
    (b) $g(x) = -\frac{1}{8}x^2$
    (c) $h(x) = \frac{3}{2}x^2$
    (d) $k(x) = -3x^2$

10. (a) $f(x) = x^2 + 1$
    (b) $g(x) = x^2 - 1$
    (c) $h(x) = x^2 + 3$
    (d) $k(x) = x^2 - 3$

11. (a) $f(x) = (x - 1)^2$
    (b) $g(x) = (3x)^2 + 1$
    (c) $h(x) = \left(\frac{1}{3}x\right)^2 - 3$
    (d) $k(x) = (x + 3)^2$

12. (a) $f(x) = -\frac{1}{2}(x - 2)^2 + 1$
    (b) $g(x) = \left[\frac{1}{2}(x - 1)\right]^2 - 3$
    (c) $h(x) = -\frac{1}{2}(x + 2)^2 - 1$
    (d) $k(x) = [2(x + 1)]^2 + 4$

In Exercises 13–28, sketch the graph of the quadratic function without using a graphing utility. Identify the vertex, axis of symmetry, and x-intercept(s).

13. $f(x) = x^2 - 5$
14. $h(x) = 25 - x^2$
15. $f(x) = \frac{1}{2}x^2 - 4$
16. $f(x) = 16 - \frac{1}{4}x^2$
17. $f(x) = (x + 5)^2 - 6$
18. $f(x) = (x - 6)^2 + 3$
19. $h(x) = x^2 - 8x + 16$
20. $g(x) = x^2 + 2x + 1$
21. $f(x) = x^2 - x + \frac{5}{4}$
22. $f(x) = x^2 + 3x + \frac{1}{4}$
23. $f(x) = -x^2 + 2x + 5$
24. $f(x) = -x^2 - 4x + 1$
25. $h(x) = 4x^2 - 4x + 21$
26. $f(x) = 2x^2 - x + 1$
27. $f(x) = \frac{1}{4}x^2 - 2x - 12$
28. $f(x) = -\frac{1}{3}x^2 + 3x - 6$

In Exercises 29–36, use a graphing utility to graph the quadratic function. Identify the vertex, axis of symmetry, and *x*-intercepts. Then check your results algebraically by writing the quadratic function in standard form.

29. $f(x) = -(x^2 + 2x - 3)$ **30.** $f(x) = -(x^2 + x - 30)$

31. $g(x) = x^2 + 8x + 11$ **32.** $f(x) = x^2 + 10x + 14$

33. $f(x) = 2x^2 - 16x + 31$ **34.** $f(x) = -4x^2 + 24x - 41$

35. $g(x) = \frac{1}{2}(x^2 + 4x - 2)$ **36.** $f(x) = \frac{3}{5}(x^2 + 6x - 5)$

In Exercises 37–42, find the standard form of the quadratic function.

37.
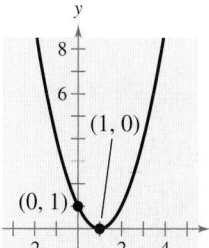

38.
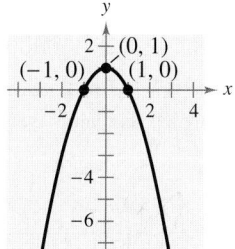

39.
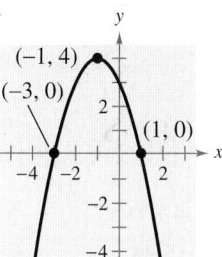

40.
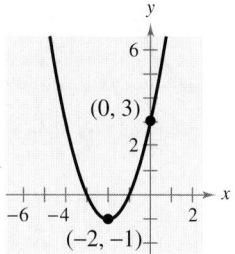

41.
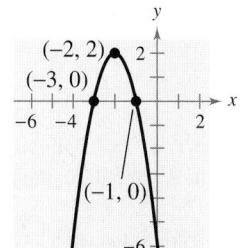

42.
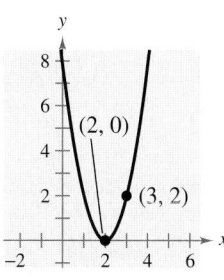

In Exercises 43–52, write the standard form of the equation of the parabola that has the indicated vertex and whose graph passes through the given point.

43. Vertex: $(-2, 5)$; point: $(0, 9)$

44. Vertex: $(4, -1)$; point: $(2, 3)$

45. Vertex: $(3, 4)$; point: $(1, 2)$

46. Vertex: $(2, 3)$; point: $(0, 2)$

47. Vertex: $(5, 12)$; point: $(7, 15)$

48. Vertex: $(-2, -2)$; point: $(-1, 0)$

49. Vertex: $\left(-\frac{1}{4}, \frac{3}{2}\right)$; point: $(-2, 0)$

50. Vertex: $\left(\frac{5}{2}, -\frac{3}{4}\right)$; point: $(-2, 4)$

51. Vertex: $\left(-\frac{5}{2}, 0\right)$; point: $\left(-\frac{7}{2}, -\frac{16}{3}\right)$

52. Vertex: $(6, 6)$; point: $\left(\frac{61}{10}, \frac{3}{2}\right)$

Graphical Reasoning In Exercises 53–56, determine the *x*-intercept(s) of the graph visually. Then find the *x*-intercepts algebraically to confirm your results.

53. $y = x^2 - 16$

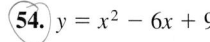
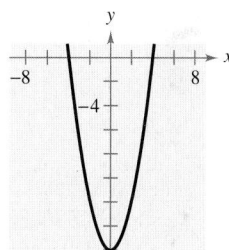

54. $y = x^2 - 6x + 9$

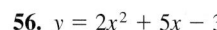
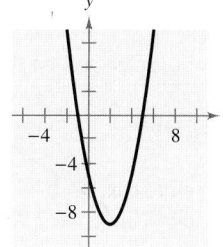

55. $y = x^2 - 4x - 5$

56. $y = 2x^2 + 5x - 3$

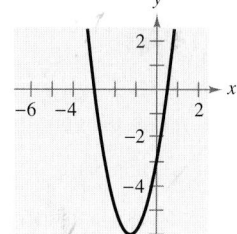

 In Exercises 57–64, use a graphing utility to graph the quadratic function. Find the *x*-intercepts of the graph and compare them with the solutions of the corresponding quadratic equation when $f(x) = 0$.

57. $f(x) = x^2 - 4x$

58. $f(x) = -2x^2 + 10x$

59. $f(x) = x^2 - 9x + 18$

60. $f(x) = x^2 - 8x - 20$

61. $f(x) = 2x^2 - 7x - 30$

62. $f(x) = 4x^2 + 25x - 21$

63. $f(x) = -\frac{1}{2}(x^2 - 6x - 7)$

64. $f(x) = \frac{7}{10}(x^2 + 12x - 45)$

In Exercises 65–70, find two quadratic functions, one that opens upward and one that opens downward, whose graphs have the given *x*-intercepts. (There are many correct answers.)

65. $(-1, 0), (3, 0)$ **66.** $(-5, 0), (5, 0)$ put in factored form & You

67. $(0, 0), (10, 0)$ **68.** $(4, 0), (8, 0)$

69. $(-3, 0), \left(-\frac{1}{2}, 0\right)$ **70.** $\left(-\frac{5}{2}, 0\right), (2, 0)$

In Exercises 71–74, find two positive real numbers whose product is a maximum.

71. The sum is 110.

72. The sum is S.

73. The sum of the first and twice the second is 24.

74. The sum of the first and three times the second is 42.

75. *Numerical, Graphical, and Analytical Analysis* A rancher has 200 feet of fencing to enclose two adjacent rectangular corrals (see figure).

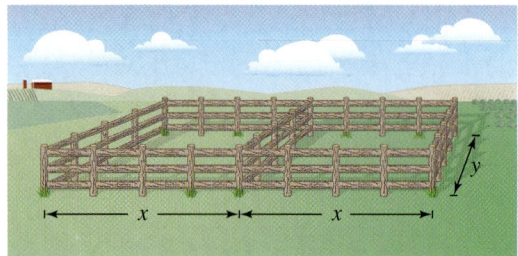

(a) Write the area A of the corral as a function of x.

(b) Create a table showing possible values of x and the corresponding areas of the corral. Use the table to estimate the dimensions that will produce the maximum enclosed area.

 (c) Use a graphing utility to graph the area function. Use the graph to approximate the dimensions that will produce the maximum enclosed area.

(d) Write the area function in standard form to find analytically the dimensions that will produce the maximum area.

(e) Compare your results from parts (b), (c), and (d).

76. *Geometry* An indoor physical fitness room consists of a rectangular region with a semicircle on each end (see figure). The perimeter of the room is to be a 200-meter single-lane running track.

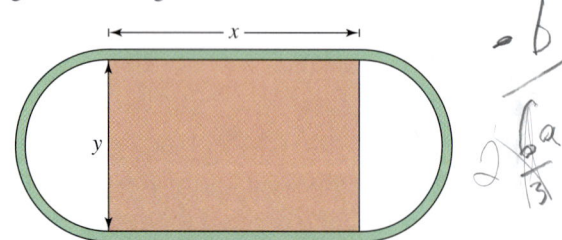

(a) Determine the radius of the semicircular ends of the room. Determine the distance, in terms of y, around the inside edge of the two semicircular parts of the track.

(b) Use the result of part (a) to write an equation, in terms of x and y, for the distance traveled in one lap around the track. Solve for y.

(c) Use the result of part (b) to write the area A of the rectangular region as a function of x. What dimensions will produce a maximum area of the rectangle?

77. *Path of a Diver* The path of a diver is given by

$$y = -\frac{4}{9}x^2 + \frac{24}{9}x + 12$$

where y is the height (in feet) and x is the horizontal distance from the end of the diving board (in feet). What is the maximum height of the diver?

78. *Height of a Ball* The height y (in feet) of a punted football is given by

$$y = -\frac{16}{2025}x^2 + \frac{9}{5}x + 1.5$$

where x is the horizontal distance (in feet) from the point at which the ball is punted (see figure).

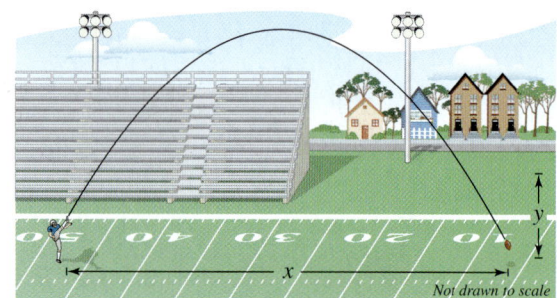

(a) How high is the ball when it is punted?

(b) What is the maximum height of the punt?

(c) How long is the punt?

79. *Minimum Cost* A manufacturer of lighting fixtures has daily production costs of

$$C = 800 - 10x + 0.25x^2$$

where C is the total cost (in dollars) and x is the number of units produced. How many fixtures should be produced each day to yield a minimum cost?

80. *Minimum Cost* A textile manufacturer has daily production costs of

$$C = 100,000 - 110x + 0.045x^2$$

where C is the total cost (in dollars) and x is the number of units produced. How many units should be produced each day to yield a minimum cost?

81. *Maximum Profit* The profit P (in dollars) for a company that produces antivirus and system utilities software is

$$P = -0.0002x^2 + 140x - 250,000$$

where x is the number of units sold. What sales level will yield a maximum profit?

82. Maximum Profit   The profit P (in hundreds of dollars) that a company makes depends on the amount x (in hundreds of dollars) the company spends on advertising according to the model

$$P = 230 + 20x - 0.5x^2.$$

What expenditure for advertising will yield a maximum profit?

83. Maximum Revenue   The total revenue R earned (in thousands of dollars) from manufacturing handheld video games is given by

$$R(p) = -25p^2 + 1200p$$

where p is the price per unit (in dollars).

(a) Find the revenue earned for each price per unit given below.

   $20

   $25

   $30

(b) Find the unit price that will yield a maximum revenue. What is the maximum revenue? Explain your results.

84. Maximum Revenue   The total revenue R earned per day (in dollars) from a pet-sitting service is given by

$$R(p) = -12p^2 + 150p$$

where p is the price charged per pet (in dollars).

(a) Find the revenue earned for each price per pet given below.

   $4

   $6

   $8

(b) Find the price that will yield a maximum revenue. What is the maximum revenue? Explain your results.

85. Graphical Analysis   From 1960 to 2003, the per capita consumption C of cigarettes by Americans (age 18 and older) can be modeled by

$$C = 4299 - 1.8t - 1.36t^2, \quad 0 \le t \le 43$$

where t is the year, with $t = 0$ corresponding to 1960. (Source: *Tobacco Outlook Report*)

 (a) Use a graphing utility to graph the model.

(b) Use the graph of the model to approximate the maximum average annual consumption. Beginning in 1966, all cigarette packages were required by law to carry a health warning. Do you think the warning had any effect? Explain.

(c) In 2000, the U.S. population (age 18 and over) was 209,128,094. Of those, about 48,308,590 were smokers. What was the average annual cigarette consumption *per smoker* in 2000? What was the average daily cigarette consumption *per smoker*?

Model It

86. Data Analysis   The numbers y (in thousands) of hairdressers and cosmetologists in the United States for the years 1994 through 2002 are shown in the table. (Source: U.S. Bureau of Labor Statistics)

Year	Number of hairdressers and cosmetologists, y
1994	753
1995	750
1996	737
1997	748
1998	763
1999	784
2000	820
2001	854
2002	908

(a) Use a graphing utility to create a scatter plot of the data. Let x represent the year, with $x = 4$ corresponding to 1994.

(b) Use the *regression* feature of a graphing utility to find a quadratic model for the data.

(c) Use a graphing utility to graph the model in the same viewing window as the scatter plot. How well does the model fit the data?

(d) Use the *trace* feature of the graphing utility to approximate the year in which the number of hairdressers and cosmetologists was the least.

(e) Verify your answer to part (d) algebraically.

(f) Use the model to predict the number of hairdressers and cosmetologists in 2008.

 87. Wind Drag   The number of horsepower y required to overcome wind drag on an automobile is approximated by

$$y = 0.002s^2 + 0.005s - 0.029, \quad 0 \le s \le 100$$

where s is the speed of the car (in miles per hour).

(a) Use a graphing utility to graph the function.

(b) Graphically estimate the maximum speed of the car if the power required to overcome wind drag is not to exceed 10 horsepower. Verify your estimate algebraically.

88. *Maximum Fuel Economy*   A study was done to compare the speed x (in miles per hour) with the mileage y (in miles per gallon) of an automobile. The results are shown in the table.   (Source: Federal Highway Administration)

Speed, x	Mileage, y
15	22.3
20	25.5
25	27.5
30	29.0
35	28.8
40	30.0
45	29.9
50	30.2
55	30.4
60	28.8
65	27.4
70	25.3
75	23.3

(a) Use a graphing utility to create a scatter plot of the data.

(b) Use the *regression* feature of a graphing utility to find a quadratic model for the data.

(c) Use a graphing utility to graph the model in the same viewing window as the scatter plot.

(d) Estimate the speed for which the miles per gallon is greatest.

Synthesis

True or False?   In Exercises 89 and 90, determine whether the statement is true or false. Justify your answer.

89. The function given by $f(x) = -12x^2 - 1$ has no x-intercepts.

90. The graphs of

$$f(x) = -4x^2 - 10x + 7$$

and

$$g(x) = 12x^2 + 30x + 1$$

have the same axis of symmetry.

91. Write the quadratic function

$$f(x) = ax^2 + bx + c$$

in standard form to verify that the vertex occurs at

$$\left(-\frac{b}{2a}, f\left(-\frac{b}{2a}\right)\right).$$

92. *Profit*   The profit P (in millions of dollars) for a recreational vehicle retailer is modeled by a quadratic function of the form

$$P = at^2 + bt + c$$

where t represents the year. If you were president of the company, which of the models below would you prefer? Explain your reasoning.

(a) a is positive and $-b/(2a) \le t$.

(b) a is positive and $t \le -b/(2a)$.

(c) a is negative and $-b/(2a) \le t$.

(d) a is negative and $t \le -b/(2a)$.

93. Is it possible for a quadratic equation to have only one x-intercept? Explain.

94. Assume that the function given by

$$f(x) = ax^2 + bx + c, \quad a \ne 0$$

has two real zeros. Show that the x-coordinate of the vertex of the graph is the average of the zeros of f. (*Hint:* Use the Quadratic Formula.)

Skills Review

In Exercises 95–98, find the equation of the line in slope-intercept form that has the given characteristics.

95. Passes through the points $(-4, 3)$ and $(2, 1)$

96. Passes through the point $\left(\frac{7}{2}, 2\right)$ and has a slope of $\frac{3}{2}$

97. Passes through the point $(0, 3)$ and is perpendicular to the line $4x + 5y = 10$

98. Passes through the point $(-8, 4)$ and is parallel to the line $y = -3x + 2$

In Exercises 99–104, let $f(x) = 14x - 3$ and let $g(x) = 8x^2$. Find the indicated value.

99. $(f + g)(-3)$

100. $(g - f)(2)$

101. $(fg)\left(-\frac{4}{7}\right)$

102. $\left(\dfrac{f}{g}\right)(-1.5)$

103. $(f \circ g)(-1)$

104. $(g \circ f)(0)$

105. **Make a Decision**   To work an extended application analyzing the height of a basketball after it has been dropped, visit this text's website at *college.hmco.com*.

What you should learn

- Use transformations to sketch graphs of polynomial functions.
- Use the Leading Coefficient Test to determine the end behavior of graphs of polynomial functions.
- Find and use zeros of polynomial functions as sketching aids.
- Use the Intermediate Value Theorem to help locate zeros of polynomial functions.

Why you should learn it

You can use polynomial functions to analyze business situations such as how revenue is related to advertising expenses, as discussed in Exercise 98 on page 287.

Graphs of Polynomial Functions

In this section, you will study basic features of the graphs of polynomial functions. The first feature is that the graph of a polynomial function is **continuous.** Essentially, this means that the graph of a polynomial function has no breaks, holes, or gaps, as shown in Figure 3.10(a). The graph shown in Figure 3.10(b) is an example of a piecewise-defined function that is not continuous.

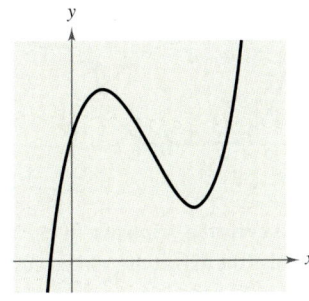

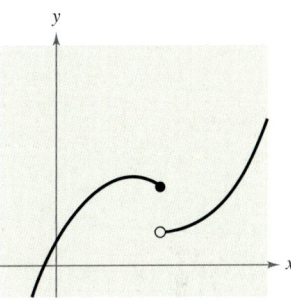

(a) **Polynomial functions have continuous graphs.**

(b) **Functions with graphs that are not continuous are not polynomial functions.**

FIGURE 3.10

The second feature is that the graph of a polynomial function has only smooth, rounded turns, as shown in Figure 3.11. A polynomial function cannot have a sharp turn. For instance, the function given by $f(x) = |x|$, which has a sharp turn at the point $(0, 0)$, as shown in Figure 3.12, is not a polynomial function.

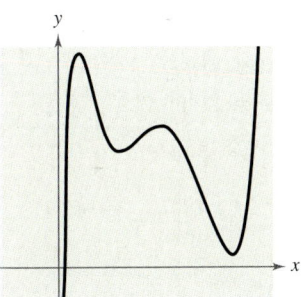

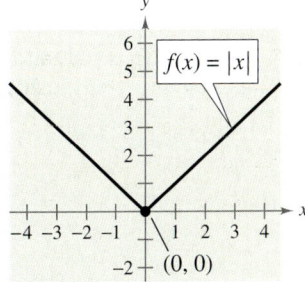

Polynomial functions have graphs with smooth rounded turns.
FIGURE 3.11

Graphs of polynomial functions cannot have sharp turns.
FIGURE 3.12

The graphs of polynomial functions of degree greater than 2 are more difficult to analyze than the graphs of polynomials of degree 0, 1, or 2. However, using the features presented in this section, coupled with your knowledge of point plotting, intercepts, and symmetry, you should be able to make reasonably accurate sketches *by hand.*

The polynomial functions that have the simplest graphs are monomials of the form $f(x) = x^n$, where n is an integer greater than zero. From Figure 3.13, you can see that when n is *even*, the graph is similar to the graph of $f(x) = x^2$, and when n is *odd*, the graph is similar to the graph of $f(x) = x^3$. Moreover, the greater the value of n, the flatter the graph near the origin. Polynomial functions of the form $f(x) = x^n$ are often referred to as **power functions.**

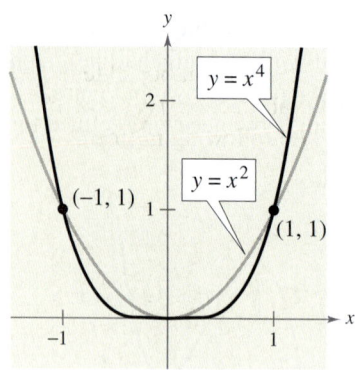

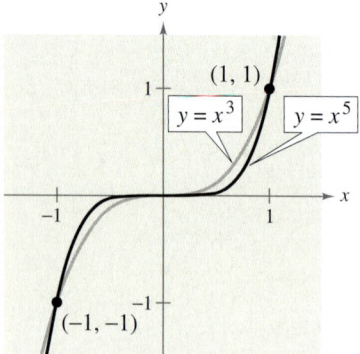

(a) If n is even, the graph of $y = x^n$ touches the axis at the x-intercept.

(b) If n is odd, the graph of $y = x^n$ crosses the axis at the x-intercept.

FIGURE 3.13

Example 1 Sketching Transformations of Monomial Functions

Sketch the graph of each function.

a. $f(x) = -x^5$ **b.** $h(x) = (x + 1)^4$

Solution

a. Because the degree of $f(x) = -x^5$ is odd, its graph is similar to the graph of $y = x^3$. In Figure 3.14, note that the negative coefficient has the effect of reflecting the graph in the x-axis.

b. The graph of $h(x) = (x + 1)^4$, as shown in Figure 3.15, is a left shift by one unit of the graph of $y = x^4$.

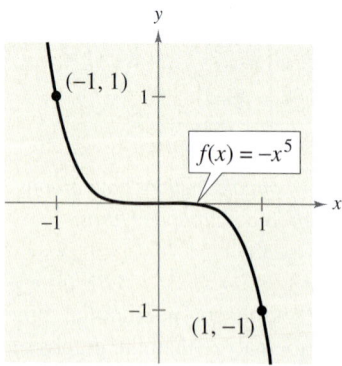

FIGURE 3.14

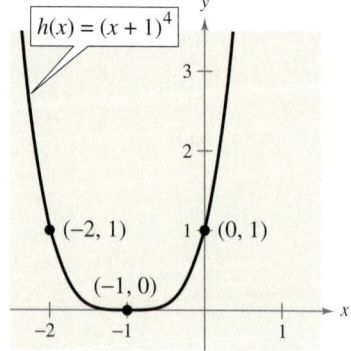

FIGURE 3.15

✔CHECKPOINT Now try Exercise 9.

The Leading Coefficient Test

In Example 1, note that both graphs eventually rise or fall without bound as x moves to the right. Whether the graph of a polynomial function eventually rises or falls can be determined by the function's degree (even or odd) and by its leading coefficient, as indicated in the **Leading Coefficient Test.**

Leading Coefficient Test

As x moves without bound to the left or to the right, the graph of the polynomial function $f(x) = a_n x^n + \cdots + a_1 x + a_0$ eventually rises or falls in the following manner.

1. When n is *odd:*

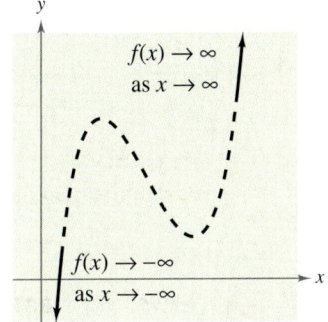 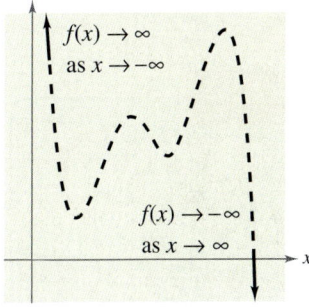

If the leading coefficient is positive ($a_n > 0$), the graph falls to the left and rises to the right.

If the leading coefficient is negative ($a_n < 0$), the graph rises to the left and falls to the right.

2. When n is *even:*

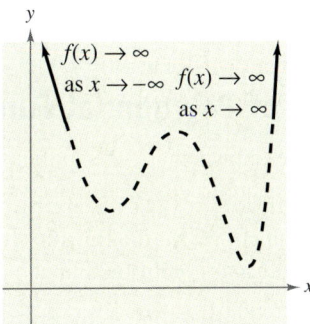 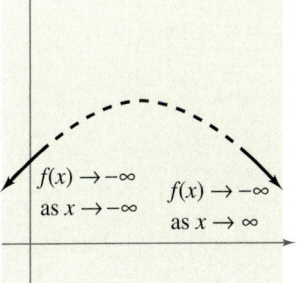

If the leading coefficient is positive ($a_n > 0$), the graph rises to the left and right.

If the leading coefficient is negative ($a_n < 0$), the graph falls to the left and right.

The dashed portions of the graphs indicate that the test determines *only* the right-hand and left-hand behavior of the graph.

Example 2 **Applying the Leading Coefficient Test**

Describe the right-hand and left-hand behavior of the graph of each function.

a. $f(x) = -x^3 + 4x$ **b.** $f(x) = x^4 - 5x^2 + 4$ **c.** $f(x) = x^5 - x$

Solution

a. Because the degree is odd and the leading coefficient is negative, the graph rises to the left and falls to the right, as shown in Figure 3.16.

b. Because the degree is even and the leading coefficient is positive, the graph rises to the left and right, as shown in Figure 3.17.

c. Because the degree is odd and the leading coefficient is positive, the graph falls to the left and rises to the right, as shown in Figure 3.18.

<table>
<tr><td>

Exploration

For each of the graphs in Example 2, count the number of zeros of the polynomial function and the number of relative minima and relative maxima. Compare these numbers with the degree of the polynomial. What do you observe?

</td></tr>
</table>

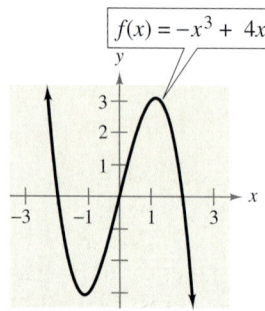

FIGURE **3.16**

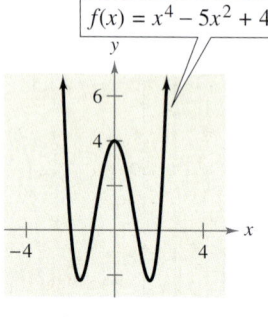

FIGURE **3.17**

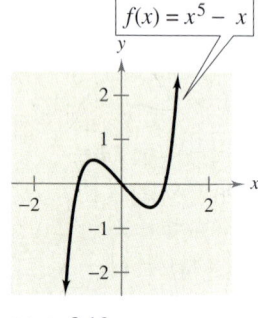

FIGURE **3.18**

✓*CHECKPOINT* Now try Exercise 15.

In Example 2, note that the Leading Coefficient Test tells you only whether the graph *eventually* rises or falls to the right or left. Other characteristics of the graph, such as intercepts and minimum and maximum points, must be determined by other tests.

Zeros of Polynomial Functions

It can be shown that for a polynomial function f of degree n, the following statements are true.

1. The function f has, at most, n real zeros. (You will study this result in detail in the discussion of the Fundamental Theorem of Algebra in Section 3.4.)

2. The graph of f has, at most, $n - 1$ turning points. (Turning points, also called relative minima or relative maxima, are points at which the graph changes from increasing to decreasing or vice versa.)

STUDY TIP

Remember that the *zeros* of a function of x are the x-values for which the function is zero.

Finding the zeros of polynomial functions is one of the most important problems in algebra. There is a strong interplay between graphical and algebraic approaches to this problem. Sometimes you can use information about the graph of a function to help find its zeros, and in other cases you can use information about the zeros of a function to help sketch its graph. Finding zeros of polynomial functions is closely related to factoring and finding x-intercepts.

Real Zeros of Polynomial Functions

If f is a polynomial function and a is a real number, the following statements are equivalent.

1. $x = a$ is a *zero* of the function f.

2. $x = a$ is a *solution* of the polynomial equation $f(x) = 0$.

3. $(x - a)$ is a *factor* of the polynomial $f(x)$.

4. $(a, 0)$ is an *x-intercept* of the graph of f.

Example 3   **Finding the Zeros of a Polynomial Function**

Find all real zeros of

$$f(x) = -2x^4 + 2x^2.$$

Then determine the number of turning points of the graph of the function.

Algebraic Solution

To find the real zeros of the function, set $f(x)$ equal to zero and solve for x.

$$-2x^4 + 2x^2 = 0 \qquad \text{Set } f(x) \text{ equal to 0.}$$

$$-2x^2(x^2 - 1) = 0 \qquad \text{Remove common monomial factor.}$$

$$-2x^2(x - 1)(x + 1) = 0 \qquad \text{Factor completely.}$$

So, the real zeros are $x = 0$, $x = 1$, and $x = -1$. Because the function is a fourth-degree polynomial, the graph of f can have at most $4 - 1 = 3$ turning points.

Graphical Solution

Use a graphing utility to graph $y = -2x^4 + 2x^2$. In Figure 3.19, the graph appears to have zeros at $(0, 0)$, $(1, 0)$, and $(-1, 0)$. Use the *zero* or *root* feature, or the *zoom* and *trace* features, of the graphing utility to verify these zeros. So, the real zeros are $x = 0$, $x = 1$, and $x = -1$. From the figure, you can see that the graph has three turning points. This is consistent with the fact that a fourth-degree polynomial can have at most three turning points.

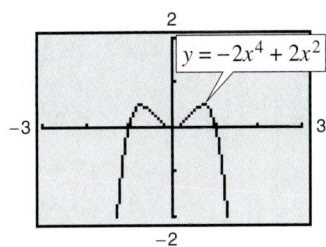

FIGURE **3.19**

✓CHECKPOINT   Now try Exercise 27.

In Example 3, note that because k is even, the factor $-2x^2$ yields the *repeated* zero $x = 0$. The graph touches the x-axis at $x = 0$, as shown in Figure 3.19.

Repeated Zeros

A factor $(x - a)^k$, $k > 1$, yields a **repeated zero** $x = a$ of **multiplicity** k.

1. If k is odd, the graph *crosses* the x-axis at $x = a$.

2. If k is even, the graph *touches* the x-axis (but does not cross the x-axis) at $x = a$.

Technology

Example 4 uses an *algebraic approach* to describe the graph of the function. A graphing utility is a complement to this approach. Remember that an important aspect of using a graphing utility is to find a viewing window that shows all significant features of the graph. For instance, the viewing window in part (a) illustrates all of the significant features of the function in Example 4.

a.

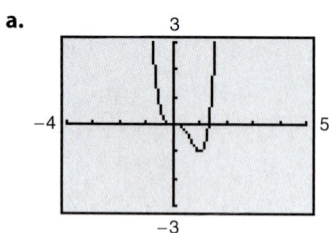

b.

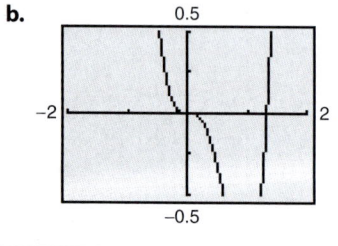

A polynomial function is written in **standard form** if its terms are written in descending order of exponents from left to right. Before applying the Leading Coefficient Test to a polynomial function, it is a good idea to check that the polynomial function is written in standard form.

Example 4   **Sketching the Graph of a Polynomial Function**

Sketch the graph of $f(x) = 3x^4 - 4x^3$.

Solution

1. *Apply the Leading Coefficient Test.* Because the leading coefficient is positive and the degree is even, you know that the graph eventually rises to the left and to the right (see Figure 3.20).

2. *Find the Zeros of the Polynomial.* By factoring

$$f(x) = 3x^4 - 4x^3 = x^3(3x - 4) \qquad \text{Remove common factor.}$$

you can see that the zeros of f are $x = 0$ and $x = \frac{4}{3}$ (both of odd multiplicity). So, the x-intercepts occur at $(0, 0)$ and $\left(\frac{4}{3}, 0\right)$. Add these points to your graph, as shown in Figure 3.20.

3. *Plot a Few Additional Points.* To sketch the graph by hand, find a few additional points, as shown in the table. Then plot the points (see Figure 3.21).

x	-1	0.5	1	1.5
$f(x)$	7	-0.3125	-1	1.6875

4. *Draw the Graph.* Draw a continuous curve through the points, as shown in Figure 3.21. Because both zeros are of odd multiplicity, you know that the graph should cross the x-axis at $x = 0$ and $x = \frac{4}{3}$. If you are unsure of the shape of that portion of the graph, plot some additional points.

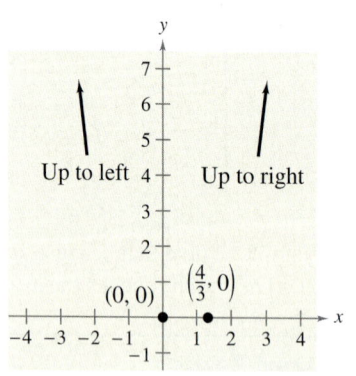

FIGURE **3.20**

FIGURE **3.21**

✓**CHECKPOINT**   Now try Exercise 67.

| Example 5 | **Sketching the Graph of a Polynomial Function** |

Sketch the graph of $f(x) = -2x^3 + 6x^2 - \frac{9}{2}x$.

Solution

1. *Apply the Leading Coefficient Test.*   Because the leading coefficient is negative and the degree is odd, you know that the graph eventually rises to the left and falls to the right (see Figure 3.22).

2. *Find the Zeros of the Polynomial.*   By factoring

$$f(x) = -2x^3 + 6x^2 - \frac{9}{2}x$$

$$= -\frac{1}{2}x(4x^2 - 12x + 9)$$

$$= -\frac{1}{2}x(2x - 3)^2$$

you can see that the zeros of f are $x = 0$ (odd multiplicity) and $x = \frac{3}{2}$ (even multiplicity). So, the x-intercepts occur at $(0, 0)$ and $\left(\frac{3}{2}, 0\right)$. Add these points to your graph, as shown in Figure 3.22.

3. *Plot a Few Additional Points.*   To sketch the graph by hand, find a few additional points, as shown in the table. Then plot the points (see Figure 3.23).

x	-0.5	0.5	1	2
$f(x)$	4	-1	-0.5	-1

4. *Draw the Graph.*   Draw a continuous curve through the points, as shown in Figure 3.23. As indicated by the multiplicities of the zeros, the graph crosses the x-axis at $(0, 0)$ but does not cross the x-axis at $\left(\frac{3}{2}, 0\right)$.

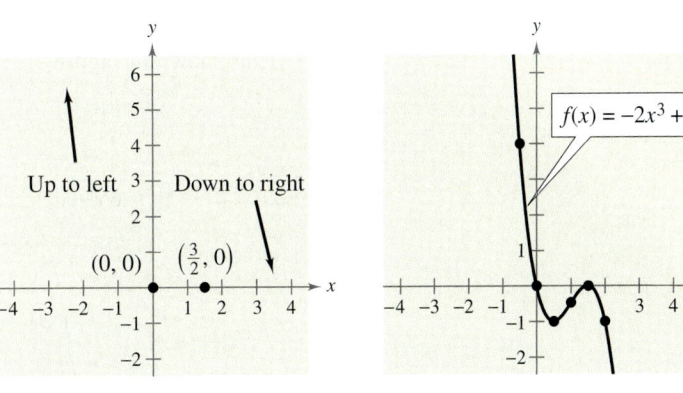

FIGURE 3.22   FIGURE 3.23

✓CHECKPOINT   Now try Exercise 69.

The Intermediate Value Theorem

The next theorem, called the **Intermediate Value Theorem,** illustrates the existence of real zeros of polynomial functions. This theorem implies that if $(a, f(a))$ and $(b, f(b))$ are two points on the graph of a polynomial function such that $f(a) \neq f(b)$, then for any number d between $f(a)$ and $f(b)$ there must be a number c between a and b such that $f(c) = d$. (See Figure 3.24.)

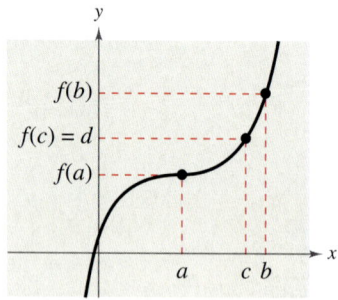

FIGURE 3.24

Intermediate Value Theorem

Let a and b be real numbers such that $a < b$. If f is a polynomial function such that $f(a) \neq f(b)$, then, in the interval $[a, b]$, f takes on every value between $f(a)$ and $f(b)$.

The Intermediate Value Theorem helps you locate the real zeros of a polynomial function in the following way. If you can find a value $x = a$ at which a polynomial function is positive, and another value $x = b$ at which it is negative, you can conclude that the function has at least one real zero between these two values. For example, the function given by $f(x) = x^3 + x^2 + 1$ is negative when $x = -2$ and positive when $x = -1$. Therefore, it follows from the Intermediate Value Theorem that f must have a real zero somewhere between -2 and -1, as shown in Figure 3.25.

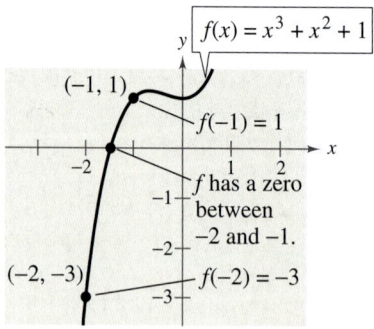

FIGURE 3.25

By continuing this line of reasoning, you can approximate any real zeros of a polynomial function to any desired accuracy. This concept is further demonstrated in Example 6.

| Example 6 | **Approximating a Zero of a Polynomial Function** |

Use the Intermediate Value Theorem to approximate the real zero of

$$f(x) = x^3 - x^2 + 1.$$

Solution

Begin by computing a few function values, as follows.

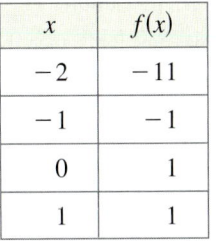

x	$f(x)$
-2	-11
-1	-1
0	1
1	1

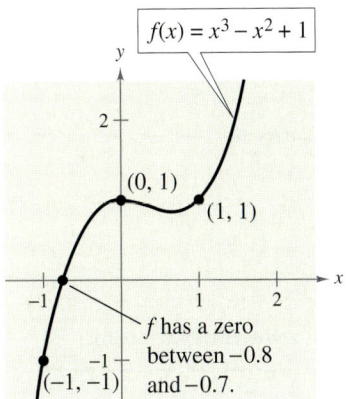

FIGURE 3.26

Because $f(-1)$ is negative and $f(0)$ is positive, you can apply the Intermediate Value Theorem to conclude that the function has a zero between -1 and 0. To pinpoint this zero more closely, divide the interval $[-1, 0]$ into tenths and evaluate the function at each point. When you do this, you will find that

$$f(-0.8) = -0.152 \qquad \text{and} \qquad f(-0.7) = 0.167.$$

So, f must have a zero between -0.8 and -0.7, as shown in Figure 3.26. For a more accurate approximation, compute function values between $f(-0.8)$ and $f(-0.7)$ and apply the Intermediate Value Theorem again. By continuing this process, you can approximate this zero to any desired accuracy.

✓**CHECKPOINT**   Now try Exercise 85.

Technology

You can use the *table* feature of a graphing utility to approximate the zeros of a polynomial function. For instance, for the function given by

$$f(x) = -2x^3 - 3x^2 + 3$$

create a table that shows the function values for $-20 \leq x \leq 20$, as shown in the first table at the right. Scroll through the table looking for consecutive function values that differ in sign. From the table, you can see that $f(0)$ and $f(1)$ differ in sign. So, you can conclude from the Intermediate Value Theorem that the function has a zero between 0 and 1. You can adjust your table to show function values for $0 \leq x \leq 1$ using increments of 0.1, as shown in the second table at the right. By scrolling through the table you can see that $f(0.8)$ and $f(0.9)$ differ in sign. So, the function has a zero between 0.8 and 0.9. If you repeat this process several times, you should obtain $x \approx 0.806$ as the zero of the function. Use the *zero* or *root* feature of a graphing utility to confirm this result.

X	Y₁	
-2	7	
-1	2	
0	3	
1	-2	
2	-25	
3	-78	
4	-173	
X=1		

X	Y₁	
.4	2.392	
.5	2	
.6	1.488	
.7	.844	
.8	.056	
.9	-.888	
1	-2	
X=.9		

3.2 | Exercises

VOCABULARY CHECK: Fill in the blanks.

1. The graphs of all polynomial functions are _____, which means that the graphs have no breaks, holes, or gaps.

2. The _____ _____ _____ is used to determine the left-hand and right-hand behavior of the graph of a polynomial function.

3. A polynomial function of degree n has at most _____ real zeros and at most _____ turning points.

4. If $x = a$ is a zero of a polynomial function f, then the following three statements are true.

   (a) $x = a$ is a _____ of the polynomial equation $f(x) = 0$.

   (b) _____ is a factor of the polynomial $f(x)$.

   (c) $(a, 0)$ is an _____ of the graph f.

5. If a real zero of a polynomial function is of even multiplicity, then the graph of f _____ the x-axis at $x = a$, and if it is of odd multiplicity then the graph of f _____ the x-axis at $x = a$.

6. A polynomial function is written in _____ form if its terms are written in descending order of exponents from left to right.

7. The _____ _____ Theorem states that if f is a polynomial function such that $f(a) \neq f(b)$, then in the interval $[a, b]$, f takes on every value between $f(a)$ and $f(b)$.

PREREQUISITE SKILLS REVIEW: Practice and review algebra skills needed for this section at **www.Eduspace.com.**

In Exercises 1–8, match the polynomial function with its graph. [The graphs are labeled (a), (b), (c), (d), (e), (f), (g), and (h).]

(a)
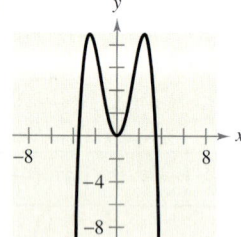

(b)
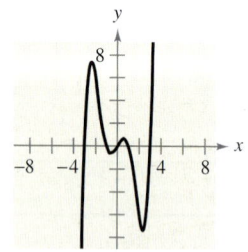

(g)
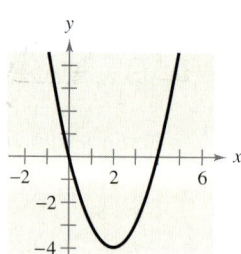

(h)
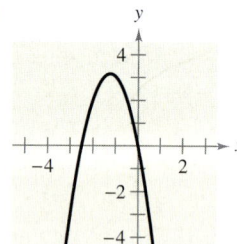

(c)
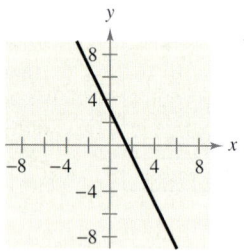

(d)
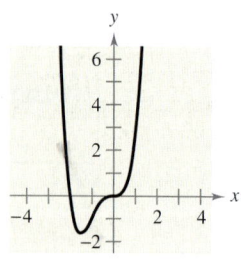

1. $f(x) = -2x + 3$
2. $f(x) = x^2 - 4x$
3. $f(x) = -2x^2 - 5x$
4. $f(x) = 2x^3 - 3x + 1$
5. $f(x) = -\frac{1}{4}x^4 + 3x^2$
6. $f(x) = -\frac{1}{3}x^3 + x^2 - \frac{4}{3}$
7. $f(x) = x^4 + 2x^3$
8. $f(x) = \frac{1}{5}x^5 - 2x^3 + \frac{9}{5}x$

In Exercises 9–12, sketch the graph of $y = x^n$ and each transformation.

9. $y = x^3$

   (a) $f(x) = (x - 2)^3$
   (b) $f(x) = x^3 - 2$
   (c) $f(x) = -\frac{1}{2}x^3$
   (d) $f(x) = (x - 2)^3 - 2$

10. $y = x^5$

   (a) $f(x) = (x + 1)^5$
   (b) $f(x) = x^5 + 1$
   (c) $f(x) = 1 - \frac{1}{2}x^5$
   (d) $f(x) = -\frac{1}{2}(x + 1)^5$

(e)
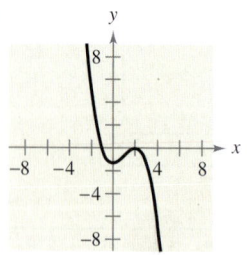

(f)
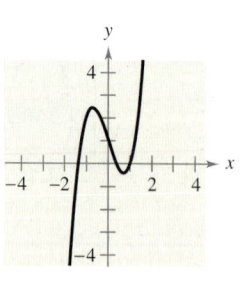

11. $y = x^4$

   (a) $f(x) = (x + 3)^4$
   (b) $f(x) = x^4 - 3$
   (c) $f(x) = 4 - x^4$
   (d) $f(x) = \frac{1}{2}(x - 1)^4$
   (e) $f(x) = (2x)^4 + 1$
   (f) $f(x) = \left(\frac{1}{2}x\right)^4 - 2$

12. $y = x^6$

(a) $f(x) = -\frac{1}{8}x^6$ (b) $f(x) = (x + 2)^6 - 4$

(c) $f(x) = x^6 - 4$ (d) $f(x) = -\frac{1}{4}x^6 + 1$

(e) $f(x) = \left(\frac{1}{4}x\right)^6 - 2$ (f) $f(x) = (2x)^6 - 1$

In Exercises 13–22, describe the right-hand and left-hand behavior of the graph of the polynomial function.

13. $f(x) = \frac{1}{3}x^3 + 5x$

14. $f(x) = 2x^2 - 3x + 1$

15. $g(x) = 5 - \frac{7}{2}x - 3x^2$

16. $h(x) = 1 - x^6$

17. $f(x) = -2.1x^5 + 4x^3 - 2$

18. $f(x) = 2x^5 - 5x + 7.5$

19. $f(x) = 6 - 2x + 4x^2 - 5x^3$

20. $f(x) = \dfrac{3x^4 - 2x + 5}{4}$

21. $h(t) = -\frac{2}{3}(t^2 - 5t + 3)$

22. $f(s) = -\frac{7}{8}(s^3 + 5s^2 - 7s + 1)$

 Graphical Analysis   In Exercises 23–26, use a graphing utility to graph the functions f and g in the same viewing window. Zoom out sufficiently far to show that the right-hand and left-hand behaviors of f and g appear identical.

23. $f(x) = 3x^3 - 9x + 1$, $g(x) = 3x^3$

24. $f(x) = -\frac{1}{3}(x^3 - 3x + 2)$, $g(x) = -\frac{1}{3}x^3$

25. $f(x) = -(x^4 - 4x^3 + 16x)$, $g(x) = -x^4$

26. $f(x) = 3x^4 - 6x^2$, $g(x) = 3x^4$

In Exercises 27–42, (a) find all the real zeros of the polynomial function, (b) determine the multiplicity of each zero and the number of turning points of the graph of the function, and (c) use a graphing utility to graph the function and verify your answers.

27. $f(x) = x^2 - 25$

28. $f(x) = 49 - x^2$

29. $h(t) = t^2 - 6t + 9$

30. $f(x) = x^2 + 10x + 25$

31. $f(x) = \frac{1}{3}x^2 + \frac{1}{3}x - \frac{2}{3}$

32. $f(x) = \frac{1}{2}x^2 + \frac{5}{2}x - \frac{3}{2}$

33. $f(x) = 3x^3 - 12x^2 + 3x$

34. $g(x) = 5x(x^2 - 2x - 1)$

35. $f(t) = t^3 - 4t^2 + 4t$

36. $f(x) = x^4 - x^3 - 20x^2$

37. $g(t) = t^5 - 6t^3 + 9t$

38. $f(x) = x^5 + x^3 - 6x$

39. $f(x) = 5x^4 + 15x^2 + 10$

40. $f(x) = 2x^4 - 2x^2 - 40$

41. $g(x) = x^3 + 3x^2 - 4x - 12$

42. $f(x) = x^3 - 4x^2 - 25x + 100$

 Graphical Analysis   In Exercises 43–46, (a) use a graphing utility to graph the function, (b) use the graph to approximate any x-intercepts of the graph, (c) set $y = 0$ and solve the resulting equation, and (d) compare the results of part (c) with any x-intercepts of the graph.

43. $y = 4x^3 - 20x^2 + 25x$

44. $y = 4x^3 + 4x^2 - 8x + 8$

45. $y = x^5 - 5x^3 + 4x$

46. $y = \frac{1}{4}x^3(x^2 - 9)$

In Exercises 47–56, find a polynomial function that has the given zeros. (There are many correct answers.)

47. $0, 10$ **48.** $0, -3$

49. $2, -6$ **50.** $-4, 5$

51. $0, -2, -3$ **52.** $0, 2, 5$

53. $4, -3, 3, 0$ **54.** $-2, -1, 0, 1, 2$

55. $1 + \sqrt{3}, 1 - \sqrt{3}$ **56.** $2, 4 + \sqrt{5}, 4 - \sqrt{5}$

In Exercises 57–66, find a polynomial of degree n that has the given zero(s). (There are many correct answers.)

Zero(s)	Degree
57. $x = -2$	$n = 2$
58. $x = -8, -4$	$n = 2$
59. $x = -3, 0, 1$	$n = 3$
60. $x = -2, 4, 7$	$n = 3$
61. $x = 0, \sqrt{3}, -\sqrt{3}$	$n = 3$
62. $x = 9$	$n = 3$
63. $x = -5, 1, 2$	$n = 4$
64. $x = -4, -1, 3, 6$	$n = 4$
65. $x = 0, -4$	$n = 5$
66. $x = -3, 1, 5, 6$	$n = 5$

In Exercises 67–80, sketch the graph of the function by (a) applying the Leading Coefficient Test, (b) finding the zeros of the polynomial, (c) plotting sufficient solution points, and (d) drawing a continuous curve through the points.

67. $f(x) = x^3 - 9x$ **68.** $g(x) = x^4 - 4x^2$

69. $f(t) = \frac{1}{4}(t^2 - 2t + 15)$

70. $g(x) = -x^2 + 10x - 16$

71. $f(x) = x^3 - 3x^2$ **72.** $f(x) = 1 - x^3$

73. $f(x) = 3x^3 - 15x^2 + 18x$

74. $f(x) = -4x^3 + 4x^2 + 15x$

75. $f(x) = -5x^2 - x^3$ **76.** $f(x) = -48x^2 + 3x^4$

77. $f(x) = x^2(x - 4)$ **78.** $h(x) = \frac{1}{3}x^3(x - 4)^2$

79. $g(t) = -\frac{1}{4}(t - 2)^2(t + 2)^2$

80. $g(x) = \frac{1}{10}(x + 1)^2(x - 3)^3$

In Exercises 81–84, use a graphing utility to graph the function. Use the *zero* or *root* feature to approximate the real zeros of the function. Then determine the multiplicity of each zero.

81. $f(x) = x^3 - 4x$

82. $f(x) = \frac{1}{4}x^4 - 2x^2$

83. $g(x) = \frac{1}{5}(x + 1)^2(x - 3)(2x - 9)$

84. $h(x) = \frac{1}{5}(x + 2)^2(3x - 5)^2$

 In Exercises 85–88, use the Intermediate Value Theorem and the *table* feature of a graphing utility to find intervals one unit in length in which the polynomial function is guaranteed to have a zero. Adjust the table to approximate the zeros of the function. Use the *zero* or *root* feature of a graphing utility to verify your results.

85. $f(x) = x^3 - 3x^2 + 3$

86. $f(x) = 0.11x^3 - 2.07x^2 + 9.81x - 6.88$

87. $g(x) = 3x^4 + 4x^3 - 3$

88. $h(x) = x^4 - 10x^2 + 3$

89. *Numerical and Graphical Analysis* An open box is to be made from a square piece of material, 36 inches on a side, by cutting equal squares with sides of length x from the corners and turning up the sides (see figure).

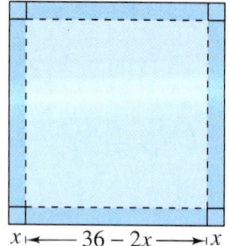 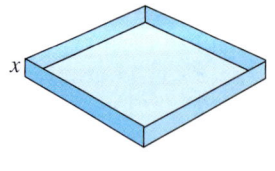

 $x \longleftarrow 36 - 2x \longrightarrow x$

(a) Verify that the volume of the box is given by the function

$$V(x) = x(36 - 2x)^2.$$

(b) Determine the domain of the function.

 (c) Use a graphing utility to create a table that shows the box height x and the corresponding volumes V. Use the table to estimate the dimensions that will produce a maximum volume.

(d) Use a graphing utility to graph V and use the graph to estimate the value of x for which $V(x)$ is maximum. Compare your result with that of part (c).

90. *Maximum Volume* An open box with locking tabs is to be made from a square piece of material 24 inches on a side. This is to be done by cutting equal squares from the corners and folding along the dashed lines shown in the figure.

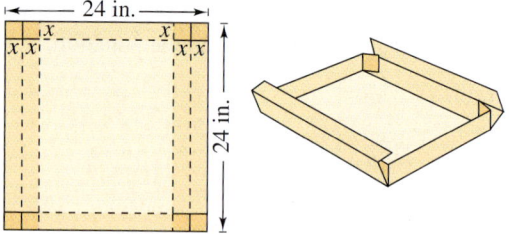

(a) Verify that the volume of the box is given by the function

$$V(x) = 8x(6 - x)(12 - x).$$

(b) Determine the domain of the function V.

(c) Sketch a graph of the function and estimate the value of x for which $V(x)$ is maximum.

91. *Construction* A roofing contractor is fabricating gutters from 12-inch aluminum sheeting. The contractor plans to use an aluminum siding folding press to create the gutter by creasing equal lengths for the sidewalls (see figure).

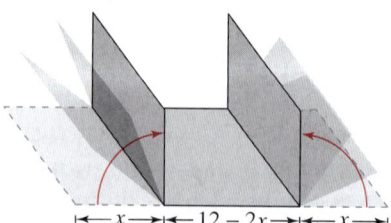

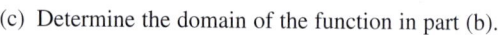 $x \longrightarrow \longleftarrow 12 - 2x \longrightarrow \longleftarrow x \longrightarrow$

(a) Let x represent the height of the sidewall of the gutter. Write a function A that represents the cross-sectional area of the gutter.

(b) The length of the aluminum sheeting is 16 feet. Write a function V that represents the volume of one run of gutter in terms of x.

(c) Determine the domain of the function in part (b).

 (d) Use a graphing utility to create a table that shows the sidewall height x and the corresponding volumes V. Use the table to estimate the dimensions that will produce a maximum volume.

 (e) Use a graphing utility to graph V. Use the graph to estimate the value of x for which $V(x)$ is a maximum. Compare your result with that of part (d).

(f) Would the value of x change if the aluminum sheeting were of different lengths? Explain.

92. *Construction*   An industrial propane tank is formed by adjoining two hemispheres to the ends of a right circular cylinder. The length of the cylindrical portion of the tank is four times the radius of the hemispherical components (see figure).

   (a) Write a function that represents the total volume V of the tank in terms of r.

   (b) Find the domain of the function.

   (c) Use a graphing utility to graph the function.

   (d) The total volume of the tank is to be 120 cubic feet. Use the graph from part (c) to estimate the radius and length of the cylindrical portion of the tank.

Data Analysis: Home Prices   In Exercise 93–96, use the table, which shows the median prices (in thousands of dollars) of new privately owned U.S. homes in the Midwest y_1 and in the South y_2 for the years 1997 through 2003. The data can be approximated by the following models.

$$y_1 = 0.139t^3 - 4.42t^2 + 51.1t - 39$$

$$y_2 = 0.056t^3 - 1.73t^2 + 23.8t + 29$$

In the models, t represents the year, with $t = 7$ corresponding to 1997. (Source: U.S. Census Bureau; U.S. Department of Housing and Urban Development)

Year, t	y_1	y_2
7	150	130
8	158	136
9	164	146
10	170	148
11	173	155
12	178	163
13	184	168

93. Use a graphing utility to plot the data and graph the model for y_1 in the same viewing window. How closely does the model represent the data?

94. Use a graphing utility to plot the data and graph the model for y_2 in the same viewing window. How closely does the model represent the data?

95. Use the models to predict the median prices of a new privately owned home in both regions in 2008. Do your answers seem reasonable? Explain.

96. Use the graphs of the models in Exercises 93 and 94 to write a short paragraph about the relationship between the median prices of homes in the two regions.

Model It

97. *Tree Growth*   The growth of a red oak tree is approximated by the function

$$G = -0.003t^3 + 0.137t^2 + 0.458t - 0.839$$

where G is the height of the tree (in feet) and t ($2 \leq t \leq 34$) is its age (in years).

   (a) Use a graphing utility to graph the function. (*Hint:* Use a viewing window in which $-10 \leq x \leq 45$ and $-5 \leq y \leq 60$.)

   (b) Estimate the age of the tree when it is growing most rapidly. This point is called the *point of diminishing returns* because the increase in size will be less with each additional year.

   (c) Using calculus, the point of diminishing returns can also be found by finding the vertex of the parabola given by

$$y = -0.009t^2 + 0.274t + 0.458.$$

   Find the vertex of this parabola.

   (d) Compare your results from parts (b) and (c).

---

98. *Revenue*   The total revenue R (in millions of dollars) for a company is related to its advertising expense by the function

$$R = \frac{1}{100{,}000}(-x^3 + 600x^2), \qquad 0 \leq x \leq 400$$

where x is the amount spent on advertising (in tens of thousands of dollars). Use the graph of this function, shown in the figure, to estimate the point on the graph at which the function is increasing most rapidly. This point is called the *point of diminishing returns* because any expense above this amount will yield less return per dollar invested in advertising.

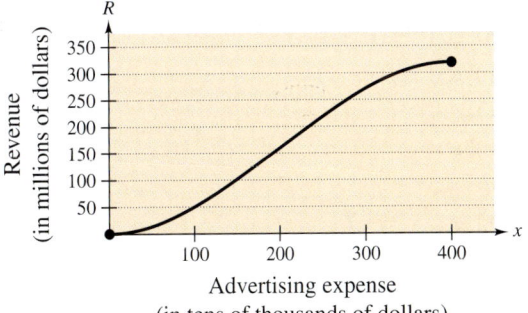

Synthesis

True or False? In Exercises 99–101, determine whether the statement is true or false. Justify your answer.

99. A fifth-degree polynomial can have five turning points in its graph.

100. It is possible for a sixth-degree polynomial to have only one solution.

101. The graph of the function given by

$$f(x) = 2 + x - x^2 + x^3 - x^4 + x^5 + x^6 - x^7$$

rises to the left and falls to the right.

102. *Graphical Analysis* For each graph, describe a polynomial function that could represent the graph. (Indicate the degree of the function and the sign of its leading coefficient.)

(a)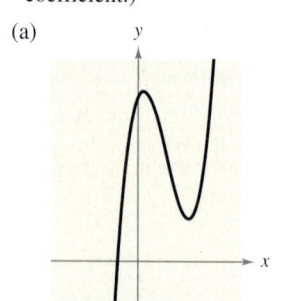

(b)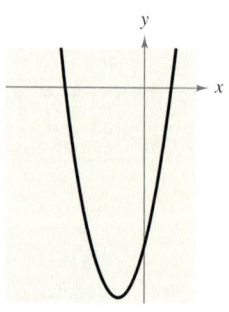

(c)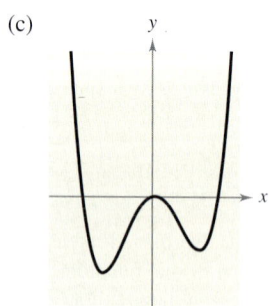

(d)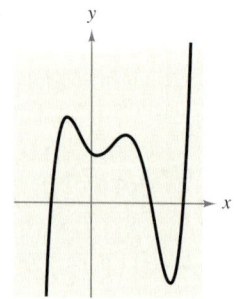

103. *Graphical Reasoning* Sketch a graph of the function given by $f(x) = x^4$. Explain how the graph of each function g differs (if it does) from the graph of each function f. Determine whether g is odd, even, or neither.

(a) $g(x) = f(x) + 2$

(b) $g(x) = f(x + 2)$

(c) $g(x) = f(-x)$

(d) $g(x) = -f(x)$

(e) $g(x) = f\left(\frac{1}{2}x\right)$

(f) $g(x) = \frac{1}{2}f(x)$

(g) $g(x) = f\left(x^{3/4}\right)$

(h) $g(x) = (f \circ f)(x)$

104. *Exploration* Explore the transformations of the form $g(x) = a(x - h)^5 + k$.

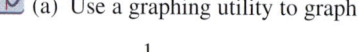 (a) Use a graphing utility to graph the functions given by

$$y_1 = -\frac{1}{3}(x - 2)^5 + 1$$

and

$$y_2 = \frac{3}{5}(x + 2)^5 - 3.$$

Determine whether the graphs are increasing or decreasing. Explain.

(b) Will the graph of g always be increasing or decreasing? If so, is this behavior determined by a, h, or k? Explain.

(c) Use a graphing utility to graph the function given by

$$H(x) = x^5 - 3x^3 + 2x + 1.$$

Use the graph and the result of part (b) to determine whether H can be written in the form $H(x) = a(x - h)^5 + k$. Explain.

Skills Review

In Exercises 105–108, factor the expression completely.

105. $5x^2 + 7x - 24$

106. $6x^3 - 61x^2 + 10x$

107. $4x^4 - 7x^3 - 15x^2$

108. $y^3 + 216$

In Exercises 109–112, solve the equation by factoring.

109. $2x^2 - x - 28 = 0$

110. $3x^2 - 22x - 16 = 0$

111. $12x^2 + 11x - 5 = 0$

112. $x^2 + 24x + 144 = 0$

In Exercises 113–116, solve the equation by completing the square.

113. $x^2 - 2x - 21 = 0$

114. $x^2 - 8x + 2 = 0$

115. $2x^2 + 5x - 20 = 0$

116. $3x^2 + 4x - 9 = 0$

In Exercises 117–122, describe the transformation from a common function that occurs in $f(x)$. Then sketch its graph.

117. $f(x) = (x + 4)^2$

118. $f(x) = 3 - x^2$

119. $f(x) = \sqrt{x + 1} - 5$

120. $f(x) = 7 - \sqrt{x - 6}$

121. $f(x) = 2[\![x]\!] + 9$

122. $f(x) = 10 - \frac{1}{3}[\![x + 3]\!]$

<table>
<tr><td>**3.3**</td><td>**Polynomial and Synthetic Division**</td></tr>
</table>

What you should learn

- Use long division to divide polynomials by other polynomials.
- Use synthetic division to divide polynomials by binomials of the form $(x - k)$.
- Use the Remainder Theorem and the Factor Theorem.

Why you should learn it

Synthetic division can help you evaluate polynomial functions. For instance, in Exercise 75 on page 296, you will use synthetic division to determine the number of U.S. military personnel in 2008.

© Kevin Fleming/Corbis

Long Division of Polynomials

In this section, you will study two procedures for *dividing* polynomials. These procedures are especially valuable in factoring and finding the zeros of polynomial functions. To begin, suppose you are given the graph of

$$f(x) = 6x^3 - 19x^2 + 16x - 4.$$

Notice that a zero of f occurs at $x = 2$, as shown in Figure 3.27. Because $x = 2$ is a zero of f, you know that $(x - 2)$ is a factor of $f(x)$. This means that there exists a second-degree polynomial $q(x)$ such that

$$f(x) = (x - 2) \cdot q(x).$$

To find $q(x)$, you can use **long division,** as illustrated in Example 1.

Example 1   Long Division of Polynomials

Divide $6x^3 - 19x^2 + 16x - 4$ by $x - 2$, and use the result to factor the polynomial completely.

Solution

Think $\dfrac{6x^3}{x} = 6x^2$.

Think $\dfrac{-7x^2}{x} = -7x$.

Think $\dfrac{2x}{x} = 2$.

$$
\begin{array}{r}
6x^2 - 7x + 2 \\
x - 2 \overline{\big)\ 6x^3 - 19x^2 + 16x - 4} \\
\underline{6x^3 - 12x^2} \\
-7x^2 + 16x \\
\underline{-7x^2 + 14x} \\
2x - 4 \\
\underline{2x - 4} \\
0
\end{array}
$$

Multiply: $6x^2(x - 2)$.
Subtract.
Multiply: $-7x(x - 2)$.
Subtract.
Multiply: $2(x - 2)$.
Subtract.

From this division, you can conclude that

$$6x^3 - 19x^2 + 16x - 4 = (x - 2)(6x^2 - 7x + 2)$$

and by factoring the quadratic $6x^2 - 7x + 2$, you have

$$6x^3 - 19x^2 + 16x - 4 = (x - 2)(2x - 1)(3x - 2).$$

Note that this factorization agrees with the graph shown in Figure 3.27 in that the three x-intercepts occur at $x = 2$, $x = \frac{1}{2}$, and $x = \frac{2}{3}$.

✓ *CHECKPOINT*   Now try Exercise 5.

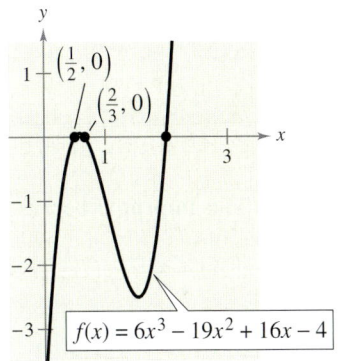

$f(x) = 6x^3 - 19x^2 + 16x - 4$

FIGURE 3.27

In Example 1, $x - 2$ is a factor of the polynomial $6x^3 - 19x^2 + 16x - 4$, and the long division process produces a remainder of zero. Often, long division will produce a nonzero remainder. For instance, if you divide $x^2 + 3x + 5$ by $x + 1$, you obtain the following.

$$
\begin{array}{r}
x + 2 \quad \longleftarrow \text{Quotient} \\
x + 1 \overline{)\, x^2 + 3x + 5} \quad \longleftarrow \text{Dividend} \\
\underline{x^2 + x} \\
2x + 5 \\
\underline{2x + 2} \\
3 \quad \longleftarrow \text{Remainder}
\end{array}
$$

Divisor $\longrightarrow$

In fractional form, you can write this result as follows.

$$
\underbrace{\frac{\overbrace{x^2 + 3x + 5}^{\text{Dividend}}}{\underbrace{x + 1}_{\text{Divisor}}}}_{} = \overbrace{x + 2}^{\text{Quotient}} + \frac{\overset{\text{Remainder}}{3}}{\underbrace{x + 1}_{\text{Divisor}}}
$$

This implies that

$$
x^2 + 3x + 5 = (x + 1)(x + 2) + 3 \qquad \text{Multiply each side by } (x + 1).
$$

which illustrates the following theorem, called the **Division Algorithm.**

The Division Algorithm

If $f(x)$ and $d(x)$ are polynomials such that $d(x) \neq 0$, and the degree of $d(x)$ is less than or equal to the degree of $f(x)$, there exist unique polynomials $q(x)$ and $r(x)$ such that

$$
f(x) = d(x)q(x) + r(x)
$$

Dividend $\;$ Divisor $\;$ Quotient $\;$ Remainder

where $r(x) = 0$ *or* the degree of $r(x)$ is less than the degree of $d(x)$. If the remainder $r(x)$ is zero, $d(x)$ *divides evenly* into $f(x)$.

The Division Algorithm can also be written as

$$
\frac{f(x)}{d(x)} = q(x) + \frac{r(x)}{d(x)}.
$$

In the Division Algorithm, the rational expression $f(x)/d(x)$ is **improper** because the degree of $f(x)$ is greater than or equal to the degree of $d(x)$. On the other hand, the rational expression $r(x)/d(x)$ is **proper** because the degree of $r(x)$ is less than the degree of $d(x)$.

Before you apply the Division Algorithm, follow these steps.

1. Write the dividend and divisor in descending powers of the variable.

2. Insert placeholders with zero coefficients for missing powers of the variable.

Example 2   Long Division of Polynomials

Divide $x^3 - 1$ by $x - 1$.

Solution

Because there is no x^2-term or x-term in the dividend, you need to line up the subtraction by using zero coefficients (or leaving spaces) for the missing terms.

$$
\begin{array}{r}
x^2 + x + 1 \\
x - 1 \overline{\smash{)}\, x^3 + 0x^2 + 0x - 1} \\
\underline{x^3 - x^2} \\
x^2 + 0x \\
\underline{x^2 - x} \\
x - 1 \\
\underline{x - 1} \\
0
\end{array}
$$

So, $x - 1$ divides evenly into $x^3 - 1$, and you can write

$$
\frac{x^3 - 1}{x - 1} = x^2 + x + 1, \quad x \neq 1.
$$

✔CHECKPOINT   Now try Exercise 13.

You can check the result of Example 2 by multiplying.

$$(x - 1)(x^2 + x + 1) = x^3 + x^2 + x - x^2 - x - 1 = x^3 - 1$$

Example 3   Long Division of Polynomials

Divide $2x^4 + 4x^3 - 5x^2 + 3x - 2$ by $x^2 + 2x - 3$.

Solution

$$
\begin{array}{r}
2x^2 + 1 \\
x^2 + 2x - 3 \overline{\smash{)}\, 2x^4 + 4x^3 - 5x^2 + 3x - 2} \\
\underline{2x^4 + 4x^3 - 6x^2} \\
x^2 + 3x - 2 \\
\underline{x^2 + 2x - 3} \\
x + 1
\end{array}
$$

Note that the first subtraction eliminated two terms from the dividend. When this happens, the quotient skips a term. You can write the result as

$$
\frac{2x^4 + 4x^3 - 5x^2 + 3x - 2}{x^2 + 2x - 3} = 2x^2 + 1 + \frac{x + 1}{x^2 + 2x - 3}.
$$

✔CHECKPOINT   Now try Exercise 15.

Synthetic Division

There is a nice shortcut for long division of polynomials when dividing by divisors of the form $x - k$. This shortcut is called **synthetic division.** The pattern for synthetic division of a cubic polynomial is summarized as follows. (The pattern for higher-degree polynomials is similar.)

Synthetic Division (for a Cubic Polynomial)

To divide $ax^3 + bx^2 + cx + d$ by $x - k$, use the following pattern.

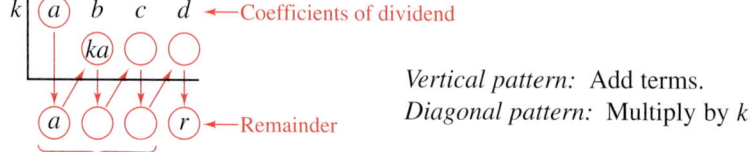

Vertical pattern: Add terms.
Diagonal pattern: Multiply by k.

Synthetic division works only for divisors of the form $x - k$. [Remember that $x + k = x - (-k)$.] You cannot use synthetic division to divide a polynomial by a quadratic such as $x^2 - 3$.

Example 4 Using Synthetic Division

Use synthetic division to divide $x^4 - 10x^2 - 2x + 4$ by $x + 3$.

Solution

You should set up the array as follows. Note that a zero is included for the missing x^3-term in the dividend.

$$-3 \quad | \quad 1 \quad 0 \quad -10 \quad -2 \quad 4$$

Then, use the synthetic division pattern by adding terms in columns and multiplying the results by -3.

Divisor: $x + 3$ Dividend: $x^4 - 10x^2 - 2x + 4$

$$
\begin{array}{c|ccccc}
-3 & 1 & 0 & -10 & -2 & 4 \\
   &   & -3 & 9 & 3 & -3 \\
\hline
   & 1 & -3 & -1 & 1 & 1 \quad \longleftarrow \text{Remainder: } 1
\end{array}
$$

Quotient: $x^3 - 3x^2 - x + 1$

So, you have

$$\frac{x^4 - 10x^2 - 2x + 4}{x + 3} = x^3 - 3x^2 - x + 1 + \frac{1}{x + 3}.$$

✔CHECKPOINT   Now try Exercise 19.

The Remainder and Factor Theorems

The remainder obtained in the synthetic division process has an important interpretation, as described in the **Remainder Theorem.**

> ### The Remainder Theorem
>
> If a polynomial $f(x)$ is divided by $x - k$, the remainder is
>
> $$r = f(k).$$

For a proof of the Remainder Theorem, see Proofs in Mathematics on page 331.

The Remainder Theorem tells you that synthetic division can be used to evaluate a polynomial function. That is, to evaluate a polynomial function $f(x)$ when $x = k$, divide $f(x)$ by $x - k$. The remainder will be $f(k)$, as illustrated in Example 5.

Example 5   Using the Remainder Theorem

Use the Remainder Theorem to evaluate the following function at $x = -2$.

$$f(x) = 3x^3 + 8x^2 + 5x - 7$$

Solution

Using synthetic division, you obtain the following.

$$
\begin{array}{r|rrrr}
-2 & 3 & 8 & 5 & -7 \\
   &   & -6 & -4 & -2 \\
\hline
   & 3 & 2 & 1 & -9
\end{array}
$$

Because the remainder is $r = -9$, you can conclude that

$$f(-2) = -9. \qquad \textcolor{red}{r = f(k)}$$

This means that $(-2, -9)$ is a point on the graph of f. You can check this by substituting $x = -2$ in the original function.

Check

$$f(\textcolor{red}{-2}) = 3(\textcolor{red}{-2})^3 + 8(\textcolor{red}{-2})^2 + 5(\textcolor{red}{-2}) - 7$$

$$= 3(-8) + 8(4) - 10 - 7 = -9$$

✔**CHECKPOINT**   Now try Exercise 45.

Another important theorem is the **Factor Theorem,** stated below. This theorem states that you can test to see whether a polynomial has $(x - k)$ as a factor by evaluating the polynomial at $x = k$. If the result is 0, $(x - k)$ is a factor.

> ### The Factor Theorem
>
> A polynomial $f(x)$ has a factor $(x - k)$ if and only if $f(k) = 0$.

For a proof of the Factor Theorem, see Proofs in Mathematics on page 331.

Example 6 **Factoring a Polynomial: Repeated Division**

Show that $(x - 2)$ and $(x + 3)$ are factors of

$$f(x) = 2x^4 + 7x^3 - 4x^2 - 27x - 18.$$

Then find the remaining factors of $f(x)$.

Solution

Using synthetic division with the factor $(x - 2)$, you obtain the following.

$$
\begin{array}{r|rrrrr}
2 & 2 & 7 & -4 & -27 & -18 \\
  &   & 4 & 22 & 36 & 18 \\
\hline
  & 2 & 11 & 18 & 9 & 0
\end{array}
$$

0 remainder, so $f(2) = 0$ and $(x - 2)$ is a factor.

Take the result of this division and perform synthetic division again using the factor $(x + 3)$.

$$
\begin{array}{r|rrrr}
-3 & 2 & 11 & 18 & 9 \\
   &   & -6 & -15 & -9 \\
\hline
   & 2 & 5 & 3 & 0
\end{array}
$$

0 remainder, so $f(-3) = 0$ and $(x + 3)$ is a factor.

Because the resulting quadratic expression factors as

$$2x^2 + 5x + 3 = (2x + 3)(x + 1)$$

the complete factorization of $f(x)$ is

$$f(x) = (x - 2)(x + 3)(2x + 3)(x + 1).$$

Note that this factorization implies that f has four real zeros:

$$x = 2, \ x = -3, \ x = -\tfrac{3}{2}, \text{ and } x = -1.$$

This is confirmed by the graph of f, which is shown in Figure 3.28.

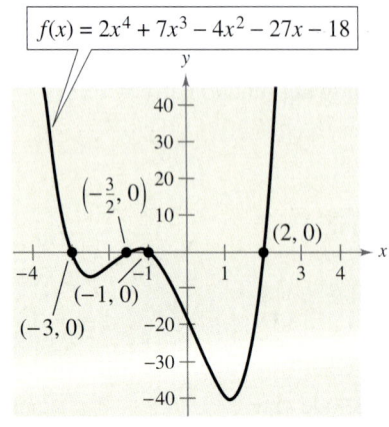

$f(x) = 2x^4 + 7x^3 - 4x^2 - 27x - 18$

FIGURE 3.28

✓CHECKPOINT Now try Exercise 57.

Uses of the Remainder in Synthetic Division

The remainder r, obtained in the synthetic division of $f(x)$ by $x - k$, provides the following information.

1. The remainder r gives the value of f at $x = k$. That is, $r = f(k)$.

2. If $r = 0$, $(x - k)$ is a factor of $f(x)$.

3. If $r = 0$, $(k, 0)$ is an x-intercept of the graph of f.

Throughout this text, the importance of developing several problem-solving strategies is emphasized. In the exercises for this section, try using more than one strategy to solve several of the exercises. For instance, if you find that $x - k$ divides evenly into $f(x)$ (with no remainder), try sketching the graph of f. You should find that $(k, 0)$ is an x-intercept of the graph.

3.3 | Exercises

VOCABULARY CHECK:

1. Two forms of the Division Algorithm are shown below. Identify and label each term or function.

$$f(x) = d(x)q(x) + r(x) \qquad \frac{f(x)}{d(x)} = q(x) + \frac{r(x)}{d(x)}$$

In Exercises 2–5, fill in the blanks.

2. The rational expression $p(x)/q(x)$ is called _____ if the degree of the numerator is greater than or equal to that of the denominator, and is called _____ if the degree of the numerator is less than that of the denominator.

3. An alternative method to long division of polynomials is called _____ _____, in which the divisor must be of the form $x - k$.

4. The _____ Theorem states that a polynomial $f(x)$ has a factor $(x - k)$ if and only if $f(k) = 0$.

5. The _____ Theorem states that if a polynomial $f(x)$ is divided by $x - k$, the remainder is $r = f(k)$.

PREREQUISITE SKILLS REVIEW: Practice and review algebra skills needed for this section at **www.Eduspace.com.**

Analytical Analysis   **In Exercises 1 and 2, use long division to verify that $y_1 = y_2$.**

1. $y_1 = \dfrac{x^2}{x + 2}, \quad y_2 = x - 2 + \dfrac{4}{x + 2}$

2. $y_1 = \dfrac{x^4 - 3x^2 - 1}{x^2 + 5}, \quad y_2 = x^2 - 8 + \dfrac{39}{x^2 + 5}$

 Graphical Analysis   **In Exercises 3 and 4, (a) use a graphing utility to graph the two equations in the same viewing window, (b) use the graphs to verify that the expressions are equivalent, and (c) use long division to verify the results algebraically.**

3. $y_1 = \dfrac{x^5 - 3x^3}{x^2 + 1}, \quad y_2 = x^3 - 4x + \dfrac{4x}{x^2 + 1}$

4. $y_1 = \dfrac{x^3 - 2x^2 + 5}{x^2 + x + 1}, \quad y_2 = x - 3 + \dfrac{2(x + 4)}{x^2 + x + 1}$

In Exercises 5–18, use long division to divide.

5. $(2x^2 + 10x + 12) \div (x + 3)$

6. $(5x^2 - 17x - 12) \div (x - 4)$

7. $(4x^3 - 7x^2 - 11x + 5) \div (4x + 5)$

8. $(6x^3 - 16x^2 + 17x - 6) \div (3x - 2)$

9. $(x^4 + 5x^3 + 6x^2 - x - 2) \div (x + 2)$

10. $(x^3 + 4x^2 - 3x - 12) \div (x - 3)$

11. $(7x + 3) \div (x + 2)$ 12. $(8x - 5) \div (2x + 1)$

13. $(6x^3 + 10x^2 + x + 8) \div (2x^2 + 1)$

14. $(x^3 - 9) \div (x^2 + 1)$

15. $(x^4 + 3x^2 + 1) \div (x^2 - 2x + 3)$

16. $(x^5 + 7) \div (x^3 - 1)$

17. $\dfrac{x^4}{(x - 1)^3}$ 18. $\dfrac{2x^3 - 4x^2 - 15x + 5}{(x - 1)^2}$

In Exercises 19–36, use synthetic division to divide.

19. $(3x^3 - 17x^2 + 15x - 25) \div (x - 5)$

20. $(5x^3 + 18x^2 + 7x - 6) \div (x + 3)$

21. $(4x^3 - 9x + 8x^2 - 18) \div (x + 2)$

22. $(9x^3 - 16x - 18x^2 + 32) \div (x - 2)$

23. $(-x^3 + 75x - 250) \div (x + 10)$

24. $(3x^3 - 16x^2 - 72) \div (x - 6)$

25. $(5x^3 - 6x^2 + 8) \div (x - 4)$

26. $(5x^3 + 6x + 8) \div (x + 2)$

27. $\dfrac{10x^4 - 50x^3 - 800}{x - 6}$ 28. $\dfrac{x^5 - 13x^4 - 120x + 80}{x + 3}$

29. $\dfrac{x^3 + 512}{x + 8}$ 30. $\dfrac{x^3 - 729}{x - 9}$

31. $\dfrac{-3x^4}{x - 2}$ 32. $\dfrac{-3x^4}{x + 2}$

33. $\dfrac{180x - x^4}{x - 6}$ 34. $\dfrac{5 - 3x + 2x^2 - x^3}{x + 1}$

35. $\dfrac{4x^3 + 16x^2 - 23x - 15}{x + \frac{1}{2}}$ 36. $\dfrac{3x^3 - 4x^2 + 5}{x - \frac{3}{2}}$

In Exercises 37–44, write the function in the form $f(x) = (x - k)q(x) + r$ for the given value of k, and demonstrate that $f(k) = r$.

Function	Value of k
37. $f(x) = x^3 - x^2 - 14x + 11$	$k = 4$
38. $f(x) = x^3 - 5x^2 - 11x + 8$	$k = -2$

Function	Value of k
39. $f(x) = 15x^4 + 10x^3 - 6x^2 + 14$	$k = -\frac{2}{3}$
40. $f(x) = 10x^3 - 22x^2 - 3x + 4$	$k = \frac{1}{5}$
41. $f(x) = x^3 + 3x^2 - 2x - 14$	$k = \sqrt{2}$
42. $f(x) = x^3 + 2x^2 - 5x - 4$	$k = -\sqrt{5}$
43. $f(x) = -4x^3 + 6x^2 + 12x + 4$	$k = 1 - \sqrt{3}$
44. $f(x) = -3x^3 + 8x^2 + 10x - 8$	$k = 2 + \sqrt{2}$

In Exercises 45–48, use synthetic division to find each function value. Verify your answers using another method.

45. $f(x) = 4x^3 - 13x + 10$

    (a) $f(1)$ (b) $f(-2)$ (c) $f\left(\frac{1}{2}\right)$ (d) $f(8)$

46. $g(x) = x^6 - 4x^4 + 3x^2 + 2$

    (a) $g(2)$ (b) $g(-4)$ (c) $g(3)$ (d) $g(-1)$

47. $h(x) = 3x^3 + 5x^2 - 10x + 1$

    (a) $h(3)$ (b) $h\left(\frac{1}{3}\right)$ (c) $h(-2)$ (d) $h(-5)$

48. $f(x) = 0.4x^4 - 1.6x^3 + 0.7x^2 - 2$

    (a) $f(1)$ (b) $f(-2)$ (c) $f(5)$ (d) $f(-10)$

In Exercises 49–56, use synthetic division to show that x is a solution of the third-degree polynomial equation, and use the result to factor the polynomial completely. List all real solutions of the equation.

Polynomial Equation	Value of x
49. $x^3 - 7x + 6 = 0$	$x = 2$
50. $x^3 - 28x - 48 = 0$	$x = -4$
51. $2x^3 - 15x^2 + 27x - 10 = 0$	$x = \frac{1}{2}$
52. $48x^3 - 80x^2 + 41x - 6 = 0$	$x = \frac{2}{3}$
53. $x^3 + 2x^2 - 3x - 6 = 0$	$x = \sqrt{3}$
54. $x^3 + 2x^2 - 2x - 4 = 0$	$x = \sqrt{2}$
55. $x^3 - 3x^2 + 2 = 0$	$x = 1 + \sqrt{3}$
56. $x^3 - x^2 - 13x - 3 = 0$	$x = 2 - \sqrt{5}$

In Exercises 57–64, (a) verify the given factors of the function f, (b) find the remaining factors of f, (c) use your results to write the complete factorization of f, (d) list all real zeros of f, and (e) confirm your results by using a graphing utility to graph the function.

Function	Factors
57. $f(x) = 2x^3 + x^2 - 5x + 2$	$(x + 2), (x - 1)$
58. $f(x) = 3x^3 + 2x^2 - 19x + 6$	$(x + 3), (x - 2)$
59. $f(x) = x^4 - 4x^3 - 15x^2$ $\qquad + 58x - 40$	$(x - 5), (x + 4)$
60. $f(x) = 8x^4 - 14x^3 - 71x^2$ $\qquad - 10x + 24$	$(x + 2), (x - 4)$

Function	Factors
61. $f(x) = 6x^3 + 41x^2 - 9x - 14$	$(2x + 1), (3x - 2)$
62. $f(x) = 10x^3 - 11x^2 - 72x + 45$	$(2x + 5), (5x - 3)$
63. $f(x) = 2x^3 - x^2 - 10x + 5$	$(2x - 1), (x + \sqrt{5})$
64. $f(x) = x^3 + 3x^2 - 48x - 144$	$(x + 4\sqrt{3}), (x + 3)$

 Graphical Analysis **In Exercises 65–68, (a) use the *zero* or *root* feature of a graphing utility to approximate the zeros of the function accurate to three decimal places, (b) determine one of the exact zeros, and (c) use synthetic division to verify your result from part (b), and then factor the polynomial completely.**

65. $f(x) = x^3 - 2x^2 - 5x + 10$

66. $g(x) = x^3 - 4x^2 - 2x + 8$

67. $h(t) = t^3 - 2t^2 - 7t + 2$

68. $f(s) = s^3 - 12s^2 + 40s - 24$

In Exercises 69–72, simplify the rational expression by using long division or synthetic division.

69. $\dfrac{4x^3 - 8x^2 + x + 3}{2x - 3}$ **70.** $\dfrac{x^3 + x^2 - 64x - 64}{x + 8}$

71. $\dfrac{x^4 + 6x^3 + 11x^2 + 6x}{x^2 + 3x + 2}$ **72.** $\dfrac{x^4 + 9x^3 - 5x^2 - 36x + 4}{x^2 - 4}$

Model It

73. *Data Analysis: Military Personnel* The numbers M (in thousands) of United States military personnel on active duty for the years 1993 through 2003 are shown in the table, where t represents the year, with $t = 3$ corresponding to 1993. (Source: U.S. Department of Defense)

Year, t	Military personnel, M
3	1705
4	1611
5	1518
6	1472
7	1439
8	1407
9	1386
10	1384
11	1385
12	1412
13	1434

Model It (continued)

(a) Use a graphing utility to create a scatter plot of the data.

(b) Use the *regression* feature of the graphing utility to find a cubic model for the data. Graph the model in the same viewing window as the scatter plot.

(c) Use the model to create a table of estimated values of M. Compare the model with the original data.

(d) Use synthetic division to evaluate the model for the year 2008. Even though the model is relatively accurate for estimating the given data, would you use this model to predict the number of military personnel in the future? Explain.

74. Data Analysis: Cable Television   The average monthly basic rates R (in dollars) for cable television in the United States for the years 1992 through 2002 are shown in the table, where t represents the year, with $t = 2$ corresponding to 1992.   (Source: Kagan Research LLC)

Year, t	Basic rate, R
2	19.08
3	19.39
4	21.62
5	23.07
6	24.41
7	26.48
8	27.81
9	28.92
10	30.37
11	32.87
12	34.71

(a) Use a graphing utility to create a scatter plot of the data.

(b) Use the *regression* feature of the graphing utility to find a cubic model for the data. Then graph the model in the same viewing window as the scatter plot. Compare the model with the data.

(c) Use synthetic division to evaluate the model for the year 2008.

Synthesis

True or False?   In Exercises 75–77, determine whether the statement is true or false. Justify your answer.

75. If $(7x + 4)$ is a factor of some polynomial function f, then $\frac{4}{7}$ is a zero of f.

76. $(2x - 1)$ is a factor of the polynomial
$$6x^6 + x^5 - 92x^4 + 45x^3 + 184x^2 + 4x - 48.$$

77. The rational expression
$$\frac{x^3 + 2x^2 - 13x + 10}{x^2 - 4x - 12}$$
is improper.

78. Exploration   Use the form $f(x) = (x - k)q(x) + r$ to create a cubic function that (a) passes through the point $(2, 5)$ and rises to the right, and (b) passes through the point $(-3, 1)$ and falls to the right. (There are many correct answers.)

Think About It   In Exercises 79 and 80, perform the division by assuming that *n* is a positive integer.

79. $\dfrac{x^{3n} + 9x^{2n} + 27x^n + 27}{x^n + 3}$ **80.** $\dfrac{x^{3n} - 3x^{2n} + 5x^n - 6}{x^n - 2}$

81. Writing   Briefly explain what it means for a divisor to divide evenly into a dividend.

82. Writing   Briefly explain how to check polynomial division, and justify your reasoning. Give an example.

Exploration   In Exercises 83 and 84, find the constant *c* such that the denominator will divide evenly into the numerator.

83. $\dfrac{x^3 + 4x^2 - 3x + c}{x - 5}$ **84.** $\dfrac{x^5 - 2x^2 + x + c}{x + 2}$

Think About It   In Exercises 85 and 86, answer the questions about the division $f(x) \div (x - k)$, where $f(x) = (x + 3)^2(x - 3)(x + 1)^3$.

85. What is the remainder when $k = -3$? Explain.

86. If it is necessary to find $f(2)$, is it easier to evaluate the function directly or to use synthetic division? Explain.

Skills Review

In Exercises 87–92, use any method to solve the quadratic equation.

87. $9x^2 - 25 = 0$ **88.** $16x^2 - 21 = 0$

89. $5x^2 - 3x - 14 = 0$ **90.** $8x^2 - 22x + 15 = 0$

91. $2x^2 + 6x + 3 = 0$ **92.** $x^2 + 3x - 3 = 0$

In Exercises 93–96, find a polynomial function that has the given zeros. (There are many correct answers.)

93. $0, 3, 4$ **94.** $-6, 1$

95. $-3, 1 + \sqrt{2}, 1 - \sqrt{2}$ **96.** $1, -2, 2 + \sqrt{3}, 2 - \sqrt{3}$

3.4 Zeros of Polynomial Functions

What you should learn

- Use the Fundamental Theorem of Algebra to determine the number of zeros of polynomial functions.
- Find rational zeros of polynomial functions.
- Find conjugate pairs of complex zeros.
- Find zeros of polynomials by factoring.
- Use Descartes's Rule of Signs and the Upper and Lower Bound Rules to find zeros of polynomials.

Why you should learn it

Finding zeros of polynomial functions is an important part of solving real-life problems. For instance, in Exercise 109 on page 311, the zeros of a polynomial function can help you analyze the attendance at women's college basketball games.

The Fundamental Theorem of Algebra

You know that an nth-degree polynomial can have at most n real zeros. In the complex number system, this statement can be improved. That is, in the complex number system, every nth-degree polynomial function has *precisely* n zeros. This important result is derived from the **Fundamental Theorem of Algebra,** first proved by the German mathematician Carl Friedrich Gauss (1777–1855).

> **The Fundamental Theorem of Algebra**
>
> If $f(x)$ is a polynomial of degree n, where $n > 0$, then f has at least one zero in the complex number system.

Using the Fundamental Theorem of Algebra and the equivalence of zeros and factors, you obtain the **Linear Factorization Theorem.**

> **Linear Factorization Theorem**
>
> If $f(x)$ is a polynomial of degree n, where $n > 0$, then f has precisely n linear factors
>
> $$f(x) = a_n(x - c_1)(x - c_2) \cdots (x - c_n)$$
>
> where $c_1, c_2, \ldots, c_n$ are complex numbers.

For a proof of the Linear Factorization Theorem, see Proofs in Mathematics on page 332.

Note that the Fundamental Theorem of Algebra and the Linear Factorization Theorem tell you only that the zeros or factors of a polynomial exist, not how to find them. Such theorems are called *existence theorems*.

Example 1   Zeros of Polynomial Functions

a. The first-degree polynomial $f(x) = x - 2$ has exactly *one* zero: $x = 2$.

b. Counting multiplicity, the second-degree polynomial function

$$f(x) = x^2 - 6x + 9 = (x - 3)(x - 3)$$

has exactly *two* zeros: $x = 3$ and $x = 3$. (This is called a *repeated zero*.)

c. The third-degree polynomial function

$$f(x) = x^3 + 4x = x(x^2 + 4) = x(x - 2i)(x + 2i)$$

has exactly *three* zeros: $x = 0$, $x = 2i$, and $x = -2i$.

d. The fourth-degree polynomial function

$$f(x) = x^4 - 1 = (x - 1)(x + 1)(x - i)(x + i)$$

has exactly *four* zeros: $x = 1$, $x = -1$, $x = i$, and $x = -i$.

✔**CHECKPOINT**   Now try Exercise 1.

STUDY TIP

Recall that in order to find the zeros of a function $f(x)$, set $f(x)$ equal to 0 and solve the resulting equation for x. For instance, the function in Example 1(a) has a zero at $x = 2$ because

$$x - 2 = 0$$
$$x = 2.$$

The Rational Zero Test

The **Rational Zero Test** relates the possible rational zeros of a polynomial (having integer coefficients) to the leading coefficient and to the constant term of the polynomial.

> ## The Rational Zero Test
>
> If the polynomial $f(x) = a_n x^n + a_{n-1} x^{n-1} + \cdots + a_2 x^2 + a_1 x + a_0$ has *integer* coefficients, every rational zero of f has the form
>
> $$\text{Rational zero} = \frac{p}{q}$$
>
> where p and q have no common factors other than 1, and
>
> $p = $ a factor of the constant term a_0
>
> $q = $ a factor of the leading coefficient a_n.

To use the Rational Zero Test, you should first list all rational numbers whose numerators are factors of the constant term and whose denominators are factors of the leading coefficient.

$$\text{Possible rational zeros} = \frac{\text{factors of constant term}}{\text{factors of leading coefficient}}$$

Having formed this list of *possible rational zeros*, use a trial-and-error method to determine which, if any, are actual zeros of the polynomial. Note that when the leading coefficient is 1, the possible rational zeros are simply the factors of the constant term.

Example 2   Rational Zero Test with Leading Coefficient of 1

Find the rational zeros of

$$f(x) = x^3 + x + 1.$$

Solution

Because the leading coefficient is 1, the possible rational zeros are ± 1, the factors of the constant term. By testing these possible zeros, you can see that neither works.

$$f(1) = (1)^3 + 1 + 1$$
$$= 3$$
$$f(-1) = (-1)^3 + (-1) + 1$$
$$= -1$$

So, you can conclude that the given polynomial has *no* rational zeros. Note from the graph of f in Figure 3.29 that f does have one real zero between -1 and 0. However, by the Rational Zero Test, you know that this real zero is *not* a rational number.

✔**CHECKPOINT**   Now try Exercise 7.

Historical Note
Although they were not contemporaries, Jean Le Rond d'Alembert (1717–1783) worked independently of Carl Gauss in trying to prove the Fundamental Theorem of Algebra. His efforts were such that, in France, the Fundamental Theorem of Algebra is frequently known as the Theorem of d'Alembert.

Fogg Art Museum

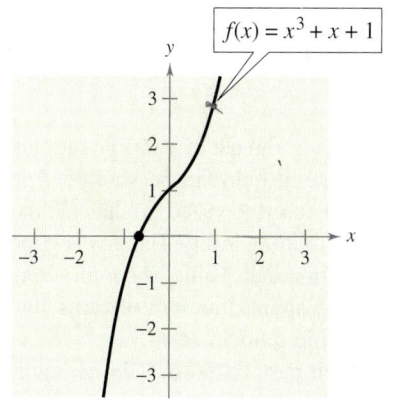

$f(x) = x^3 + x + 1$

FIGURE **3.29**

Example 3   Rational Zero Test with Leading Coefficient of 1

Find the rational zeros of $f(x) = x^4 - x^3 + x^2 - 3x - 6$.

Solution

Because the leading coefficient is 1, the possible rational zeros are the factors of the constant term.

Possible rational zeros: $\pm 1, \pm 2, \pm 3, \pm 6$

By applying synthetic division successively, you can determine that $x = -1$ and $x = 2$ are the only two rational zeros.

$$
\begin{array}{r|rrrr}
-1 & 1 & -1 & 1 & -3 & -6 \\
   &   & -1 & 2 & -3 & 6 \\
\hline
   & 1 & -2 & 3 & -6 & 0
\end{array}
$$ ⟶ 0 remainder, so $x = -1$ is a zero.

$$
\begin{array}{r|rrrr}
2 & 1 & -2 & 3 & -6 \\
  &   & 2 & 0 & 6 \\
\hline
  & 1 & 0 & 3 & 0
\end{array}
$$ ⟶ 0 remainder, so $x = 2$ is a zero.

So, $f(x)$ factors as

$$f(x) = (x + 1)(x - 2)(x^2 + 3).$$

Because the factor $(x^2 + 3)$ produces no real zeros, you can conclude that $x = -1$ and $x = 2$ are the only *real* zeros of f, which is verified in Figure 3.30.

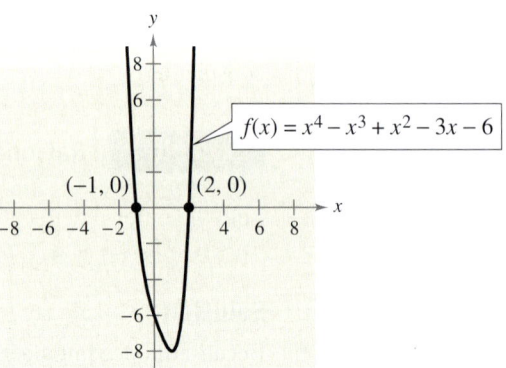

FIGURE 3.30

✔CHECKPOINT   Now try Exercise 11.

If the leading coefficient of a polynomial is not 1, the list of possible rational zeros can increase dramatically. In such cases, the search can be shortened in several ways: (1) a programmable calculator can be used to speed up the calculations; (2) a graph, drawn either by hand or with a graphing utility, can give a good estimate of the locations of the zeros; (3) the Intermediate Value Theorem along with a table generated by a graphing utility can give approximations of zeros; and (4) synthetic division can be used to test the possible rational zeros.

Finding the first zero is often the most difficult part. After that, the search is simplified by working with the lower-degree polynomial obtained in synthetic division, as shown in Example 3.

| Example 4 | Using the Rational Zero Test |

Find the rational zeros of $f(x) = 2x^3 + 3x^2 - 8x + 3$.

Solution

The leading coefficient is 2 and the constant term is 3.

$$Possible\ rational\ zeros: \frac{Factors\ of\ 3}{Factors\ of\ 2} = \frac{\pm 1, \pm 3}{\pm 1, \pm 2} = \pm 1, \pm 3, \pm \frac{1}{2}, \pm \frac{3}{2}$$

By synthetic division, you can determine that $x = 1$ is a rational zero.

$$\begin{array}{r|rrrr} 1 & 2 & 3 & -8 & 3 \\ & & 2 & 5 & -3 \\ \hline & 2 & 5 & -3 & 0 \end{array}$$

So, $f(x)$ factors as

$$f(x) = (x - 1)(2x^2 + 5x - 3)$$
$$= (x - 1)(2x - 1)(x + 3)$$

and you can conclude that the rational zeros of f are $x = 1$, $x = \frac{1}{2}$, and $x = -3$.

✓**CHECKPOINT**   Now try Exercise 17.

Recall from Section 3.2 that if $x = a$ is a zero of the polynomial function f, then $x = a$ is a solution of the polynomial equation $f(x) = 0$.

| Example 5 | Solving a Polynomial Equation |

Find all the real solutions of $-10x^3 + 15x^2 + 16x - 12 = 0$.

Solution

The leading coefficient is -10 and the constant term is -12.

$$Possible\ rational\ solutions: \frac{Factors\ of\ -12}{Factors\ of\ -10} = \frac{\pm 1, \pm 2, \pm 3, \pm 4, \pm 6, \pm 12}{\pm 1, \pm 2, \pm 5, \pm 10}$$

With so many possibilities (32, in fact), it is worth your time to stop and sketch a graph. From Figure 3.31, it looks like three reasonable solutions would be $x = -\frac{6}{5}$, $x = \frac{1}{2}$, and $x = 2$. Testing these by synthetic division shows that $x = 2$ is the only rational solution. So, you have

$$(x - 2)(-10x^2 - 5x + 6) = 0.$$

Using the Quadratic Formula for the second factor, you find that the two additional solutions are irrational numbers.

$$x = \frac{-5 - \sqrt{265}}{20} \approx -1.0639$$

and

$$x = \frac{-5 + \sqrt{265}}{20} \approx 0.5639$$

✓**CHECKPOINT**   Now try Exercise 23.

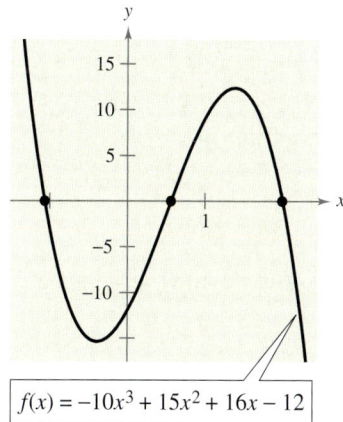

$f(x) = -10x^3 + 15x^2 + 16x - 12$

FIGURE 3.31

Conjugate Pairs

In Example 1(c) and (d), note that the pairs of complex zeros are **conjugates.** That is, they are of the form $a + bi$ and $a - bi$.

> ### Complex Zeros Occur in Conjugate Pairs
>
> Let $f(x)$ be a polynomial function that has *real coefficients*. If $a + bi$, where $b \neq 0$, is a zero of the function, the conjugate $a - bi$ is also a zero of the function.

Be sure you see that this result is true only if the polynomial function has *real coefficients*. For instance, the result applies to the function given by $f(x) = x^2 + 1$ but not to the function given by $g(x) = x - i$.

Example 6 Finding a Polynomial with Given Zeros

Find a fourth-degree polynomial function with real coefficients that has -1, -1, and $3i$ as zeros.

Solution

Because $3i$ is a zero *and* the polynomial is stated to have real coefficients, you know that the conjugate $-3i$ must also be a zero. So, from the Linear Factorization Theorem, $f(x)$ can be written as

$$f(x) = a(x + 1)(x + 1)(x - 3i)(x + 3i).$$

For simplicity, let $a = 1$ to obtain

$$f(x) = (x^2 + 2x + 1)(x^2 + 9)$$

$$= x^4 + 2x^3 + 10x^2 + 18x + 9.$$

✔**CHECKPOINT**   Now try Exercise 37.

Factoring a Polynomial

The Linear Factorization Theorem shows that you can write any nth-degree polynomial as the product of n linear factors.

$$f(x) = a_n(x - c_1)(x - c_2)(x - c_3) \cdots (x - c_n)$$

However, this result includes the possibility that some of the values of c_i are complex. The following theorem says that even if you do not want to get involved with "complex factors," you can still write $f(x)$ as the product of linear and/or quadratic factors. For a proof of this theorem, see Proofs in Mathematics on page 332.

> ### Factors of a Polynomial
>
> Every polynomial of degree $n > 0$ with real coefficients can be written as the product of linear and quadratic factors with real coefficients, where the quadratic factors have no real zeros.

A quadratic factor with no real zeros is said to be *prime* or **irreducible over the reals.** Be sure you see that this is not the same as being *irreducible over the rationals.* For example, the quadratic $x^2 + 1 = (x - i)(x + i)$ is irreducible over the reals (and therefore over the rationals). On the other hand, the quadratic $x^2 - 2 = \left(x - \sqrt{2}\right)\left(x + \sqrt{2}\right)$ is irreducible over the rationals but *reducible* over the reals.

Example 7   Finding the Zeros of a Polynomial Function

Find all the zeros of $f(x) = x^4 - 3x^3 + 6x^2 + 2x - 60$ given that $1 + 3i$ is a zero of f.

Algebraic Solution

Because complex zeros occur in conjugate pairs, you know that $1 - 3i$ is also a zero of f. This means that both

$$[x - (1 + 3i)] \quad \text{and} \quad [x - (1 - 3i)]$$

are factors of f. Multiplying these two factors produces

$$[x - (1 + 3i)][x - (1 - 3i)] = [(x - 1) - 3i][(x - 1) + 3i]$$

$$= (x - 1)^2 - 9i^2$$

$$= x^2 - 2x + 10.$$

Using long division, you can divide $x^2 - 2x + 10$ into f to obtain the following.

$$
\begin{array}{r}
x^2 - x - 6 \\
x^2 - 2x + 10 \overline{\smash{\big)}\, x^4 - 3x^3 + 6x^2 + 2x - 60} \\
\underline{x^4 - 2x^3 + 10x^2} \\
-x^3 - 4x^2 + 2x \\
\underline{-x^3 + 2x^2 - 10x} \\
-6x^2 + 12x - 60 \\
\underline{-6x^2 + 12x - 60} \\
0
\end{array}
$$

So, you have

$$f(x) = (x^2 - 2x + 10)(x^2 - x - 6)$$

$$= (x^2 - 2x + 10)(x - 3)(x + 2)$$

and you can conclude that the zeros of f are $x = 1 + 3i$, $x = 1 - 3i$, $x = 3$, and $x = -2$.

✔**CHECKPOINT**   Now try Exercise 47.

Graphical Solution

Because complex zeros always occur in conjugate pairs, you know that $1 - 3i$ is also a zero of f. Because the polynomial is a fourth-degree polynomial, you know that there are at most two other zeros of the function. Use a graphing utility to graph

$$y = x^4 - 3x^3 + 6x^2 + 2x - 60$$

as shown in Figure 3.32.

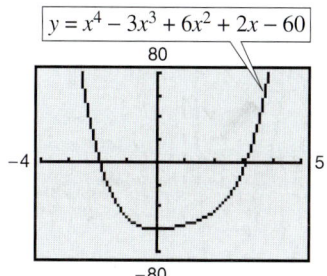

FIGURE 3.32

You can see that -2 and 3 appear to be zeros of the graph of the function. Use the *zero* or *root* feature or the *zoom* and *trace* features of the graphing utility to confirm that $x = -2$ and $x = 3$ are zeros of the graph. So, you can conclude that the zeros of f are $x = 1 + 3i$, $x = 1 - 3i$, $x = 3$, and $x = -2$.

In Example 7, if you were not told that $1 + 3i$ is a zero of f, you could still find all zeros of the function by using synthetic division to find the real zeros -2 and 3. Then you could factor the polynomial as $(x + 2)(x - 3)(x^2 - 2x + 10)$. Finally, by using the Quadratic Formula, you could determine that the zeros are $x = -2$, $x = 3$, $x = 1 + 3i$, and $x = 1 - 3i$.

STUDY TIP

In Example 8, the fifth-degree polynomial function has three real zeros. In such cases, you can use the *zoom* and *trace* features or the *zero* or *root* feature of a graphing utility to approximate the real zeros. You can then use these real zeros to determine the complex zeros algebraically.

Example 8 shows how to find all the zeros of a polynomial function, including complex zeros.

Example 8   **Finding the Zeros of a Polynomial Function**

Write $f(x) = x^5 + x^3 + 2x^2 - 12x + 8$ as the product of linear factors, and list all of its zeros.

Solution

The possible rational zeros are $\pm 1, \pm 2, \pm 4,$ and ± 8. Synthetic division produces the following.

$$
\begin{array}{r|rrrrrr}
1 & 1 & 0 & 1 & 2 & -12 & 8 \\
  &   & 1 & 1 & 2 & 4 & -8 \\
\hline
  & 1 & 1 & 2 & 4 & -8 & 0 \\
\end{array}
$$
$\longrightarrow$ 1 is a zero.

$$
\begin{array}{r|rrrrr}
-2 & 1 & 1 & 2 & 4 & -8 \\
   &   & -2 & 2 & -8 & 8 \\
\hline
   & 1 & -1 & 4 & -4 & 0 \\
\end{array}
$$
$\longrightarrow$ -2 is a zero.

So, you have

$$f(x) = x^5 + x^3 + 2x^2 - 12x + 8$$
$$= (x - 1)(x + 2)(x^3 - x^2 + 4x - 4).$$

You can factor $x^3 - x^2 + 4x - 4$ as $(x - 1)(x^2 + 4)$, and by factoring $x^2 + 4$ as

$$x^2 - (-4) = \left(x - \sqrt{-4}\right)\left(x + \sqrt{-4}\right)$$
$$= (x - 2i)(x + 2i)$$

you obtain

$$f(x) = (x - 1)(x - 1)(x + 2)(x - 2i)(x + 2i)$$

which gives the following five zeros of f.

$$x = 1, \ x = 1, \ x = -2, \ x = 2i, \quad \text{and} \quad x = -2i$$

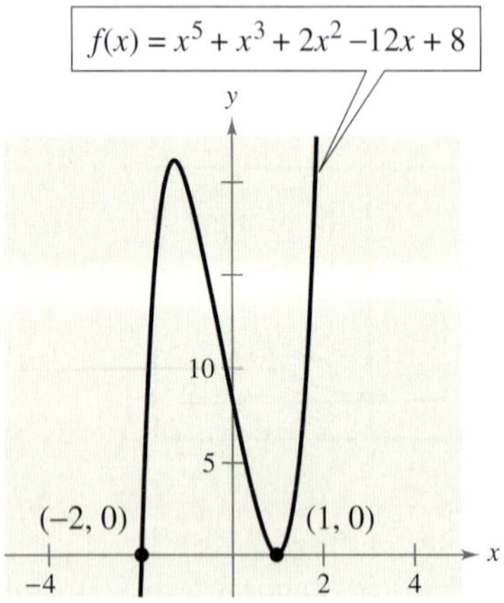

From the graph of f shown in Figure 3.33, you can see that the *real* zeros are the only ones that appear as x-intercepts. Note that $x = 1$ is a repeated zero.

✓**CHECKPOINT**   Now try Exercise 63.

$f(x) = x^5 + x^3 + 2x^2 - 12x + 8$

FIGURE 3.33

Technology

You can use the *table* feature of a graphing utility to help you determine which of the possible rational zeros are zeros of the polynomial in Example 8. The table should be set to *ask* mode. Then enter each of the possible rational zeros in the table. When you do this, you will see that there are two rational zeros, -2 and 1, as shown at the right.

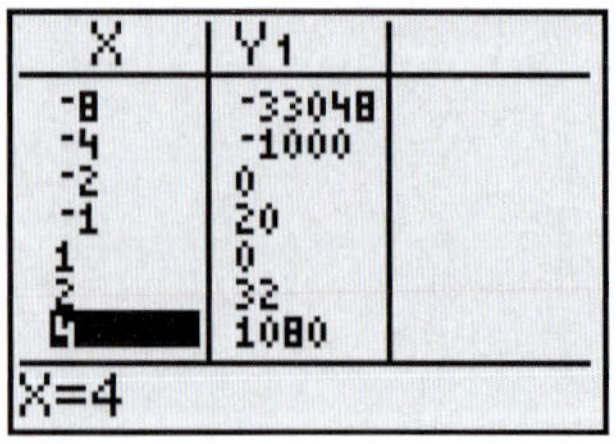

X	Y1
-8	-33048
-4	-1000
-2	0
-1	20
1	0
2	32
4	1080

X=4

Other Tests for Zeros of Polynomials

You know that an nth-degree polynomial function can have *at most n* real zeros. Of course, many nth-degree polynomials do not have that many real zeros. For instance, $f(x) = x^2 + 1$ has no real zeros, and $f(x) = x^3 + 1$ has only one real zero. The following theorem, called **Descartes's Rule of Signs,** sheds more light on the number of real zeros of a polynomial.

Descartes's Rule of Signs

Let $f(x) = a_n x^n + a_{n-1} x^{n-1} + \cdots + a_2 x^2 + a_1 x + a_0$ be a polynomial with real coefficients and $a_0 \neq 0$.

1. The number of *positive real zeros* of f is either equal to the number of variations in sign of $f(x)$ or less than that number by an even integer.

2. The number of *negative real zeros* of f is either equal to the number of variations in sign of $f(-x)$ or less than that number by an even integer.

A **variation in sign** means that two consecutive coefficients have opposite signs.

When using Descartes's Rule of Signs, a zero of multiplicity k should be counted as k zeros. For instance, the polynomial $x^3 - 3x + 2$ has two variations in sign, and so has either two positive or no positive real zeros. Because

$$x^3 - 3x + 2 = (x - 1)(x - 1)(x + 2)$$

you can see that the two positive real zeros are $x = 1$ of multiplicity 2.

Example 9   **Using Descartes's Rule of Signs**

Describe the possible real zeros of

$$f(x) = 3x^3 - 5x^2 + 6x - 4.$$

Solution

The original polynomial has *three* variations in sign.

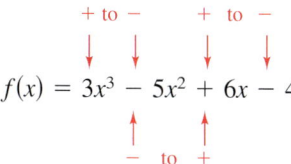

The polynomial

$$f(-x) = 3(-x)^3 - 5(-x)^2 + 6(-x) - 4$$

$$= -3x^3 - 5x^2 - 6x - 4$$

has no variations in sign. So, from Descartes's Rule of Signs, the polynomial $f(x) = 3x^3 - 5x^2 + 6x - 4$ has either three positive real zeros or one positive real zero, and has no negative real zeros. From the graph in Figure 3.34, you can see that the function has only one real zero (it is a positive number, near $x = 1$).

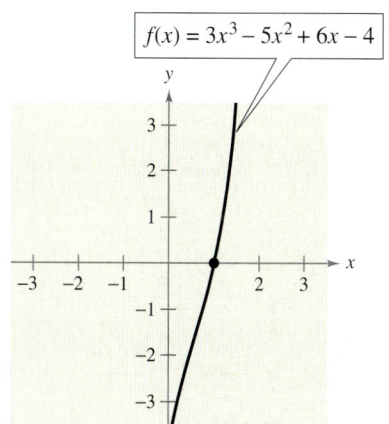

FIGURE **3.34**

✓**CHECKPOINT**   Now try Exercise 79.

Another test for zeros of a polynomial function is related to the sign pattern in the last row of the synthetic division array. This test can give you an upper or lower bound of the real zeros of f. A real number b is an **upper bound** for the real zeros of f if no zeros are greater than b. Similarly, b is a **lower bound** if no real zeros of f are less than b.

Upper and Lower Bound Rules

Let $f(x)$ be a polynomial with real coefficients and a positive leading coefficient. Suppose $f(x)$ is divided by $x - c$, using synthetic division.

1. If $c > 0$ and each number in the last row is either positive or zero, c is an **upper bound** for the real zeros of f.

2. If $c < 0$ and the numbers in the last row are alternately positive and negative (zero entries count as positive or negative), c is a **lower bound** for the real zeros of f.

Example 10   Finding the Zeros of a Polynomial Function

Find the real zeros of $f(x) = 6x^3 - 4x^2 + 3x - 2$.

Solution

The possible real zeros are as follows.

$$\frac{\text{Factors of 2}}{\text{Factors of 6}} = \frac{\pm 1, \pm 2}{\pm 1, \pm 2, \pm 3, \pm 6} = \pm 1, \pm \frac{1}{2}, \pm \frac{1}{3}, \pm \frac{1}{6}, \pm \frac{2}{3}, \pm 2$$

The original polynomial $f(x)$ has three variations in sign. The polynomial

$$f(-x) = 6(-x)^3 - 4(-x)^2 + 3(-x) - 2$$
$$= -6x^3 - 4x^2 - 3x - 2$$

has no variations in sign. As a result of these two findings, you can apply Descartes's Rule of Signs to conclude that there are three positive real zeros or one positive real zero, and no negative zeros. Trying $x = 1$ produces the following.

$$
\begin{array}{r|rrrr}
1 & 6 & -4 & 3 & -2 \\
  &   & 6 & 2 & 5 \\
\hline
  & 6 & 2 & 5 & 3 \\
\end{array}
$$

So, $x = 1$ is not a zero, but because the last row has all positive entries, you know that $x = 1$ is an upper bound for the real zeros. So, you can restrict the search to zeros between 0 and 1. By trial and error, you can determine that $x = \frac{2}{3}$ is a zero. So,

$$f(x) = \left(x - \frac{2}{3}\right)(6x^2 + 3).$$

Because $6x^2 + 3$ has no real zeros, it follows that $x = \frac{2}{3}$ is the only real zero.

✔**CHECKPOINT**   Now try Exercise 87.

Before concluding this section, here are two additional hints that can help you find the real zeros of a polynomial.

1. If the terms of $f(x)$ have a common monomial factor, it should be factored out before applying the tests in this section. For instance, by writing

$$f(x) = x^4 - 5x^3 + 3x^2 + x$$
$$= x(x^3 - 5x^2 + 3x + 1)$$

you can see that $x = 0$ is a zero of f and that the remaining zeros can be obtained by analyzing the cubic factor.

2. If you are able to find all but two zeros of $f(x)$, you can always use the Quadratic Formula on the remaining quadratic factor. For instance, if you succeeded in writing

$$f(x) = x^4 - 5x^3 + 3x^2 + x$$
$$= x(x - 1)(x^2 - 4x - 1)$$

you can apply the Quadratic Formula to $x^2 - 4x - 1$ to conclude that the two remaining zeros are $x = 2 + \sqrt{5}$ and $x = 2 - \sqrt{5}$.

Example 11   **Using a Polynomial Model**

You are designing candle-making kits. Each kit contains 25 cubic inches of candle wax and a mold for making a pyramid-shaped candle. You want the height of the candle to be 2 inches less than the length of each side of the candle's square base. What should the dimensions of your candle mold be?

Solution

The volume of a pyramid is $V = \frac{1}{3}Bh$, where B is the area of the base and h is the height. The area of the base is x^2 and the height is $(x - 2)$. So, the volume of the pyramid is $V = \frac{1}{3}x^2(x - 2)$. Substituting 25 for the volume yields the following.

$$25 = \frac{1}{3}x^2(x - 2) \qquad \text{Substitute 25 for } V.$$

$$75 = x^3 - 2x^2 \qquad \text{Multiply each side by 3.}$$

$$0 = x^3 - 2x^2 - 75 \qquad \text{Write in general form.}$$

The possible rational solutions are $x = \pm 1, \pm 3, \pm 5, \pm 15, \pm 25, \pm 75$. Use synthetic division to test some of the possible solutions. Note that in this case, it makes sense to test only positive x-values. Using synthetic division, you can determine that $x = 5$ is a solution.

$$
\begin{array}{r|rrrr}
5 & 1 & -2 & 0 & -75 \\
  &   & 5 & 15 & 75 \\
\hline
  & 1 & 3 & 15 & 0 \\
\end{array}
$$

The other two solutions, which satisfy $x^2 + 3x + 15 = 0$, are imaginary and can be discarded. You can conclude that the base of the candle mold should be 5 inches by 5 inches and the height of the mold should be $5 - 2 = 3$ inches.

✔**CHECKPOINT**   Now try Exercise 107.

3.4   Exercises

VOCABULARY CHECK: Fill in the blanks.

1. The _____ _____ of _____ states that if $f(x)$ is a polynomial of degree n $(n > 0)$, then f has at least one zero in the complex number system.

2. The _____ _____ _____ states that if $f(x)$ is a polynomial of degree n $(n > 0)$, then f has precisely n linear factors $f(x) = a_n(x - c_1)(x - c_2) \cdots (x - c_n)$ where $c_1, c_2, \ldots, c_n$ are complex numbers.

3. The test that gives a list of the possible rational zeros of a polynomial function is called the _____ _____ Test.

4. If $a + bi$ is a complex zero of a polynomial with real coefficients, then so is its _____, $a - bi$.

5. A quadratic factor that cannot be factored further as a product of linear factors containing real numbers is said to be _____ over the _____.

6. The theorem that can be used to determine the possible numbers of positive real zeros and negative real zeros of a function is called _____ _____ of _____.

7. A real number b is a(n) _____ bound for the real zeros of f if no real zeros are less than b, and is a(n) _____ bound if no real zeros are greater than b.

PREREQUISITE SKILLS REVIEW: Practice and review algebra skills needed for this section at **www.Eduspace.com.**

In Exercises 1–6, find all the zeros of the function.

1. $f(x) = x(x - 6)^2$

2. $f(x) = x^2(x + 3)(x^2 - 1)$

3. $g(x) = (x - 2)(x + 4)^3$

4. $f(x) = (x + 5)(x - 8)^2$

5. $f(x) = (x + 6)(x + i)(x - i)$

6. $h(t) = (t - 3)(t - 2)(t - 3i)(t + 3i)$

In Exercises 7–10, use the Rational Zero Test to list all possible rational zeros of f. Verify that the zeros of f shown on the graph are contained in the list.

7. $f(x) = x^3 + 3x^2 - x - 3$

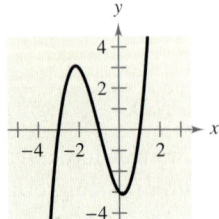

8. $f(x) = x^3 - 4x^2 - 4x + 16$

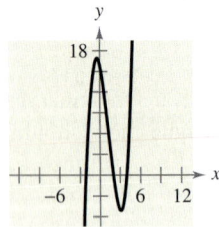

9. $f(x) = 2x^4 - 17x^3 + 35x^2 + 9x - 45$

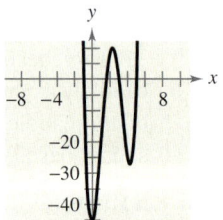

10. $f(x) = 4x^5 - 8x^4 - 5x^3 + 10x^2 + x - 2$

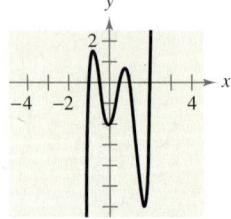

In Exercises 11–20, find all the rational zeros of the function.

11. $f(x) = x^3 - 6x^2 + 11x - 6$

12. $f(x) = x^3 - 7x - 6$

13. $g(x) = x^3 - 4x^2 - x + 4$

14. $h(x) = x^3 - 9x^2 + 20x - 12$

15. $h(t) = t^3 + 12t^2 + 21t + 10$

16. $p(x) = x^3 - 9x^2 + 27x - 27$

17. $C(x) = 2x^3 + 3x^2 - 1$

18. $f(x) = 3x^3 - 19x^2 + 33x - 9$

19. $f(x) = 9x^4 - 9x^3 - 58x^2 + 4x + 24$

20. $f(x) = 2x^4 - 15x^3 + 23x^2 + 15x - 25$

In Exercises 21–24, find all real solutions of the polynomial equation.

21. $z^4 - z^3 - 2z - 4 = 0$

22. $x^4 - 13x^2 - 12x = 0$

23. $2y^4 + 7y^3 - 26y^2 + 23y - 6 = 0$

24. $x^5 - x^4 - 3x^3 + 5x^2 - 2x = 0$

In Exercises 25–28, (a) list the possible rational zeros of f, (b) sketch the graph of f so that some of the possible zeros in part (a) can be disregarded, and then (c) determine all real zeros of f.

25. $f(x) = x^3 + x^2 - 4x - 4$

26. $f(x) = -3x^3 + 20x^2 - 36x + 16$

27. $f(x) = -4x^3 + 15x^2 - 8x - 3$

28. $f(x) = 4x^3 - 12x^2 - x + 15$

 In Exercises 29–32, (a) list the possible rational zeros of f, (b) use a graphing utility to graph f so that some of the possible zeros in part (a) can be disregarded, and then (c) determine all real zeros of f.

29. $f(x) = -2x^4 + 13x^3 - 21x^2 + 2x + 8$

30. $f(x) = 4x^4 - 17x^2 + 4$

31. $f(x) = 32x^3 - 52x^2 + 17x + 3$

32. $f(x) = 4x^3 + 7x^2 - 11x - 18$

 Graphical Analysis   In Exercises 33–36, (a) use the *zero* or *root* feature of a graphing utility to approximate the zeros of the function accurate to three decimal places, (b) determine one of the exact zeros (use synthetic division to verify your result), and (c) factor the polynomial completely.

33. $f(x) = x^4 - 3x^2 + 2$ **34.** $P(t) = t^4 - 7t^2 + 12$

35. $h(x) = x^5 - 7x^4 + 10x^3 + 14x^2 - 24x$

36. $g(x) = 6x^4 - 11x^3 - 51x^2 + 99x - 27$

In Exercises 37–42, find a polynomial function with real coefficients that has the given zeros. (There are many correct answers.)

37. $1, 5i, -5i$ **38.** $4, 3i, -3i$

39. $6, -5 + 2i, -5 - 2i$ **40.** $2, 4 + i, 4 - i$

41. $\frac{2}{3}, -1, 3 + \sqrt{2}i$ **42.** $-5, -5, 1 + \sqrt{3}i$

In Exercises 43–46, write the polynomial (a) as the product of factors that are irreducible over the *rationals*, (b) as the product of linear and quadratic factors that are irreducible over the *reals*, and (c) in completely factored form.

43. $f(x) = x^4 + 6x^2 - 27$

44. $f(x) = x^4 - 2x^3 - 3x^2 + 12x - 18$
    (*Hint:* One factor is $x^2 - 6$.)

45. $f(x) = x^4 - 4x^3 + 5x^2 - 2x - 6$
    (*Hint:* One factor is $x^2 - 2x - 2$.)

46. $f(x) = x^4 - 3x^3 - x^2 - 12x - 20$
    (*Hint:* One factor is $x^2 + 4$.)

In Exercises 47–54, use the given zero to find all the zeros of the function.

Function	Zero
47. $f(x) = 2x^3 + 3x^2 + 50x + 75$	$5i$
48. $f(x) = x^3 + x^2 + 9x + 9$	$3i$
49. $f(x) = 2x^4 - x^3 + 7x^2 - 4x - 4$	$2i$
50. $g(x) = x^3 - 7x^2 - x + 87$	$5 + 2i$
51. $g(x) = 4x^3 + 23x^2 + 34x - 10$	$-3 + i$
52. $h(x) = 3x^3 - 4x^2 + 8x + 8$	$1 - \sqrt{3}i$
53. $f(x) = x^4 + 3x^3 - 5x^2 - 21x + 22$	$-3 + \sqrt{2}i$
54. $f(x) = x^3 + 4x^2 + 14x + 20$	$-1 - 3i$

In Exercises 55–72, find all the zeros of the function and write the polynomial as a product of linear factors.

55. $f(x) = x^2 + 25$ **56.** $f(x) = x^2 - x + 56$

57. $h(x) = x^2 - 4x + 1$ **58.** $g(x) = x^2 + 10x + 23$

59. $f(x) = x^4 - 81$

60. $f(y) = y^4 - 625$

61. $f(z) = z^2 - 2z + 2$

62. $h(x) = x^3 - 3x^2 + 4x - 2$

63. $g(x) = x^3 - 6x^2 + 13x - 10$

64. $f(x) = x^3 - 2x^2 - 11x + 52$

65. $h(x) = x^3 - x + 6$

66. $h(x) = x^3 + 9x^2 + 27x + 35$

67. $f(x) = 5x^3 - 9x^2 + 28x + 6$

68. $g(x) = 3x^3 - 4x^2 + 8x + 8$

69. $g(x) = x^4 - 4x^3 + 8x^2 - 16x + 16$

70. $h(x) = x^4 + 6x^3 + 10x^2 + 6x + 9$

71. $f(x) = x^4 + 10x^2 + 9$ **72.** $f(x) = x^4 + 29x^2 + 100$

 In Exercises 73–78, find all the zeros of the function. When there is an extended list of possible rational zeros, use a graphing utility to graph the function in order to discard any rational zeros that are obviously not zeros of the function.

73. $f(x) = x^3 + 24x^2 + 214x + 740$

74. $f(s) = 2s^3 - 5s^2 + 12s - 5$

75. $f(x) = 16x^3 - 20x^2 - 4x + 15$

76. $f(x) = 9x^3 - 15x^2 + 11x - 5$

77. $f(x) = 2x^4 + 5x^3 + 4x^2 + 5x + 2$

78. $g(x) = x^5 - 8x^4 + 28x^3 - 56x^2 + 64x - 32$

In Exercises 79–86, use Descartes's Rule of Signs to determine the possible numbers of positive and negative zeros of the function.

79. $g(x) = 5x^5 + 10x$ **80.** $h(x) = 4x^2 - 8x + 3$

81. $h(x) = 3x^4 + 2x^2 + 1$ **82.** $h(x) = 2x^4 - 3x + 2$

83. $g(x) = 2x^3 - 3x^2 - 3$

84. $f(x) = 4x^3 - 3x^2 + 2x - 1$

85. $f(x) = -5x^3 + x^2 - x + 5$

86. $f(x) = 3x^3 + 2x^2 + x + 3$

In Exercises 87–90, use synthetic division to verify the upper and lower bounds of the real zeros of f.

87. $f(x) = x^4 - 4x^3 + 15$

    (a) Upper: $x = 4$ (b) Lower: $x = -1$

88. $f(x) = 2x^3 - 3x^2 - 12x + 8$

    (a) Upper: $x = 4$ (b) Lower: $x = -3$

89. $f(x) = x^4 - 4x^3 + 16x - 16$

    (a) Upper: $x = 5$ (b) Lower: $x = -3$

90. $f(x) = 2x^4 - 8x + 3$

    (a) Upper: $x = 3$ (b) Lower: $x = -4$

In Exercises 91–94, find all the real zeros of the function.

91. $f(x) = 4x^3 - 3x - 1$

92. $f(z) = 12z^3 - 4z^2 - 27z + 9$

93. $f(y) = 4y^3 + 3y^2 + 8y + 6$

94. $g(x) = 3x^3 - 2x^2 + 15x - 10$

In Exercises 95–98, find all the rational zeros of the polynomial function.

95. $P(x) = x^4 - \frac{25}{4}x^2 + 9 = \frac{1}{4}(4x^4 - 25x^2 + 36)$

96. $f(x) = x^3 - \frac{3}{2}x^2 - \frac{23}{2}x + 6 = \frac{1}{2}(2x^3 - 3x^2 - 23x + 12)$

97. $f(x) = x^3 - \frac{1}{4}x^2 - x + \frac{1}{4} = \frac{1}{4}(4x^3 - x^2 - 4x + 1)$

98. $f(z) = z^3 + \frac{11}{6}z^2 - \frac{1}{2}z - \frac{1}{3} = \frac{1}{6}(6z^3 + 11z^2 - 3z - 2)$

In Exercises 99–102, match the cubic function with the numbers of rational and irrational zeros.

(a) Rational zeros: 0; irrational zeros: 1

(b) Rational zeros: 3; irrational zeros: 0

(c) Rational zeros: 1; irrational zeros: 2

(d) Rational zeros: 1; irrational zeros: 0

99. $f(x) = x^3 - 1$ **100.** $f(x) = x^3 - 2$

101. $f(x) = x^3 - x$ **102.** $f(x) = x^3 - 2x$

103. Geometry An open box is to be made from a rectangular piece of material, 15 centimeters by 9 centimeters, by cutting equal squares from the corners and turning up the sides.

(a) Let x represent the length of the sides of the squares removed. Draw a diagram showing the squares removed from the original piece of material and the resulting dimensions of the open box.

(b) Use the diagram to write the volume V of the box as a function of x. Determine the domain of the function.

(c) Sketch the graph of the function and approximate the dimensions of the box that will yield a maximum volume.

(d) Find values of x such that $V = 56$. Which of these values is a physical impossibility in the construction of the box? Explain.

104. Geometry A rectangular package to be sent by a delivery service (see figure) can have a maximum combined length and girth (perimeter of a cross section) of 120 inches.

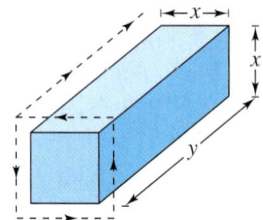

(a) Show that the volume of the package is

$$V(x) = 4x^2(30 - x).$$

 (b) Use a graphing utility to graph the function and approximate the dimensions of the package that will yield a maximum volume.

(c) Find values of x such that $V = 13,500$. Which of these values is a physical impossibility in the construction of the package? Explain.

105. Advertising Cost A company that produces MP3 players estimates that the profit P (in dollars) for selling a particular model is given by

$$P = -76x^3 + 4830x^2 - 320,000, \quad 0 \le x \le 60$$

where x is the advertising expense (in tens of thousands of dollars). Using this model, find the smaller of two advertising amounts that will yield a profit of $2,500,000.

106. Advertising Cost A company that manufactures bicycles estimates that the profit P (in dollars) for selling a particular model is given by

$$P = -45x^3 + 2500x^2 - 275,000, \quad 0 \le x \le 50$$

where x is the advertising expense (in tens of thousands of dollars). Using this model, find the smaller of two advertising amounts that will yield a profit of $800,000.

107. Geometry A bulk food storage bin with dimensions 2 feet by 3 feet by 4 feet needs to be increased in size to hold five times as much food as the current bin. (Assume each dimension is increased by the same amount.)

(a) Write a function that represents the volume V of the new bin.

(b) Find the dimensions of the new bin.

108. Geometry A rancher wants to enlarge an existing rectangular corral such that the total area of the new corral is 1.5 times that of the original corral. The current corral's dimensions are 250 feet by 160 feet. The rancher wants to increase each dimension by the same amount.

(a) Write a function that represents the area A of the new corral.

(b) Find the dimensions of the new corral.

(c) A rancher wants to add a length to the sides of the corral that are 160 feet, and twice the length to the sides that are 250 feet, such that the total area of the new corral is 1.5 times that of the original corral. Repeat parts (a) and (b). Explain your results.

109. Cost The ordering and transportation cost C (in thousands of dollars) for the components used in manufacturing a product is given by

$$C = 100\left(\frac{200}{x^2} + \frac{x}{x + 30}\right), \quad x \geq 1$$

where x is the order size (in hundreds). In calculus, it can be shown that the cost is a minimum when

$$3x^3 - 40x^2 - 2400x - 36{,}000 = 0.$$

Use a calculator to approximate the optimal order size to the nearest hundred units.

110. Height of a Baseball A baseball is thrown upward from a height of 6 feet with an initial velocity of 48 feet per second, and its height h (in feet) is

$$h(t) = -16t^2 + 48t + 6, \quad 0 \leq t \leq 3$$

where t is the time (in seconds). You are told the ball reaches a height of 64 feet. Is this possible?

111. Profit The demand equation for a certain product is $p = 140 - 0.0001x$, where p is the unit price (in dollars) of the product and x is the number of units produced and sold. The cost equation for the product is $C = 80x + 150{,}000$, where C is the total cost (in dollars) and x is the number of units produced. The total profit obtained by producing and selling x units is

$$P = R - C = xp - C.$$

You are working in the marketing department of the company that produces this product, and you are asked to determine a price p that will yield a profit of 9 million dollars. Is this possible? Explain.

Model It

112. Athletics The attendance A (in millions) at NCAA women's college basketball games for the years 1997 through 2003 is shown in the table, where t represents the year, with $t = 7$ corresponding to 1997. (Source: National Collegiate Athletic Association)

Year, t	Attendance, A
7	6.7
8	7.4
9	8.0
10	8.7
11	8.8
12	9.5
13	10.2

(a) Use the *regression* feature of a graphing utility to find a cubic model for the data.

(b) Use the graphing utility to create a scatter plot of the data. Then graph the model and the scatter plot in the same viewing window. How do they compare?

(c) According to the model found in part (a), in what year did attendance reach 8.5 million?

(d) According to the model found in part (a), in what year did attendance reach 9 million?

(e) According to the right-hand behavior of the model, will the attendance continue to increase? Explain.

Synthesis

True or False? **In Exercises 113 and 114, decide whether the statement is true or false. Justify your answer.**

113. It is possible for a third-degree polynomial function with integer coefficients to have no real zeros.

114. If $x = -i$ is a zero of the function given by

$$f(x) = x^3 + ix^2 + ix - 1$$

then $x = i$ must also be a zero of f.

Think About It **In Exercises 115–120, determine (if possible) the zeros of the function g if the function f has zeros at $x = r_1, x = r_2,$ and $x = r_3$.**

115. $g(x) = -f(x)$ **116.** $g(x) = 3f(x)$

117. $g(x) = f(x - 5)$ **118.** $g(x) = f(2x)$

119. $g(x) = 3 + f(x)$ **120.** $g(x) = f(-x)$

121. ***Exploration***   Use a graphing utility to graph the function given by $f(x) = x^4 - 4x^2 + k$ for different values of k. Find values of k such that the zeros of f satisfy the specified characteristics. (Some parts do not have unique answers.)

   (a) Four real zeros

   (b) Two real zeros, each of multiplicity 2

   (c) Two real zeros and two complex zeros

   (d) Four complex zeros

122. ***Think About It***   Will the answers to Exercise 121 change for the function g?

   (a) $g(x) = f(x - 2)$ (b) $g(x) = f(2x)$

123. ***Think About It***   A third-degree polynomial function f has real zeros $-2, \frac{1}{2}$, and 3, and its leading coefficient is negative. Write an equation for f. Sketch the graph of f. How many different polynomial functions are possible for f?

124. ***Think About It***   Sketch the graph of a fifth-degree polynomial function whose leading coefficient is positive and that has one zero at $x = 3$ of multiplicity 2.

125. ***Writing***   Compile a list of all the various techniques for factoring a polynomial that have been covered so far in the text. Give an example illustrating each technique, and write a paragraph discussing when the use of each technique is appropriate.

126. Use the information in the table to answer each question.

Interval	Value of $f(x)$
$(-\infty, -2)$	Positive
$(-2, 1)$	Negative
$(1, 4)$	Negative
$(4, \infty)$	Positive

   (a) What are the three real zeros of the polynomial function f?

   (b) What can be said about the behavior of the graph of f at $x = 1$?

   (c) What is the least possible degree of f? Explain. Can the degree of f ever be odd? Explain.

   (d) Is the leading coefficient of f positive or negative? Explain.

   (e) Write an equation for f. (There are many correct answers.)

   (f) Sketch a graph of the equation you wrote in part (e).

127. (a) Find a quadratic function f (with integer coefficients) that has $\pm \sqrt{b} i$ as zeros. Assume that b is a positive integer.

   (b) Find a quadratic function f (with integer coefficients) that has $a \pm bi$ as zeros. Assume that b is a positive integer.

128. ***Graphical Reasoning***   The graph of one of the following functions is shown below. Identify the function shown in the graph. Explain why each of the others is not the correct function. Use a graphing utility to verify your result.

   (a) $f(x) = x^2(x + 2)(x - 3.5)$

   (b) $g(x) = (x + 2)(x - 3.5)$

   (c) $h(x) = (x + 2)(x - 3.5)(x^2 + 1)$

   (d) $k(x) = (x + 1)(x + 2)(x - 3.5)$

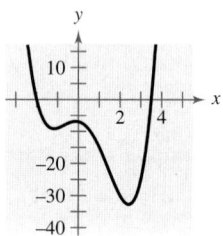

Skills Review

In Exercises 129–132, perform the operation and simplify.

129. $(-3 + 6i) - (8 - 3i)$

130. $(12 - 5i) + 16i$

131. $(6 - 2i)(1 + 7i)$

132. $(9 - 5i)(9 + 5i)$

In Exercises 133–138, use the graph of f to sketch the graph of g. To print an enlarged copy of the graph, go to the website *www.mathgraphs.com*.

133. $g(x) = f(x - 2)$

134. $g(x) = f(x) - 2$

135. $g(x) = 2f(x)$

136. $g(x) = f(-x)$

137. $g(x) = f(2x)$

138. $g(x) = f\left(\frac{1}{2}x\right)$

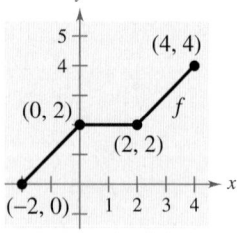

3.5   Mathematical Modeling and Variation

What you should learn

- Use mathematical models to approximate sets of data points.
- Use the *regression* feature of a graphing utility to find the equation of a least squares regression line.
- Write mathematical models for direct variation.
- Write mathematical models for direct variation as an *n*th power.
- Write mathematical models for inverse variation.
- Write mathematical models for joint variation.

Why you should learn it

You can use functions as models to represent a wide variety of real-life data sets. For instance, in Exercise 71 on page 323, a variation model can be used to model the water temperature of the ocean at various depths.

Introduction

You have already studied some techniques for fitting models to data. For instance, in Section 2.1, you learned how to find the equation of a line that passes through two points. In this section, you will study other techniques for fitting models to data: *least squares regression* and *direct and inverse variation*. The resulting models are either polynomial functions or rational functions. (Rational functions will be studied in Chapter 4.)

Example 1   A Mathematical Model

The numbers of insured commercial banks y (in thousands) in the United States for the years 1996 to 2001 are shown in the table.   (Source: Federal Deposit Insurance Corporation)

Year	Insured commercial banks, y
1996	9.53
1997	9.14
1998	8.77
1999	8.58
2000	8.32
2001	8.08

A linear model that approximates the data is $y = -0.283t + 11.14$ for $6 \leq t \leq 11$, where t is the year, with $t = 6$ corresponding to 1996. Plot the actual data *and* the model on the same graph. How closely does the model represent the data?

Solution

The actual data are plotted in Figure 3.35, along with the graph of the linear model. From the graph, it appears that the model is a "good fit" for the actual data. You can see how well the model fits by comparing the actual values of y with the values of y given by the model. The values given by the model are labeled $y*$ in the table below.

t	6	7	8	9	10	11
y	9.53	9.14	8.77	8.58	8.32	8.08
$y*$	9.44	9.16	8.88	8.59	8.31	8.03

✓**CHECKPOINT**   Now try Exercise 1.

Note in Example 1 that you could have chosen any two points to find a line that fits the data. However, the given linear model was found using the *regression* feature of a graphing utility and is the line that *best* fits the data. This concept of a "best-fitting" line is discussed on the next page.

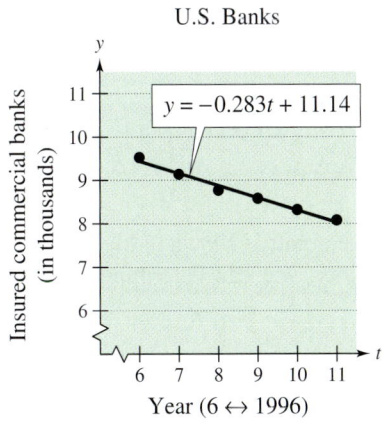

U.S. Banks

$y = -0.283t + 11.14$

Year (6 ↔ 1996)

FIGURE 3.35

Least Squares Regression and Graphing Utilities

So far in this text, you have worked with many different types of mathematical models that approximate real-life data. In some instances the model was given (as in Example 1), whereas in other instances you were asked to find the model using simple algebraic techniques or a graphing utility.

To find a model that approximates the data most accurately, statisticians use a measure called the **sum of square differences,** which is the sum of the squares of the differences between actual data values and model values. The "best-fitting" linear model, called the **least squares regression line,** is the one with the least sum of square differences. Recall that you can approximate this line visually by plotting the data points and drawing the line that appears to fit best—or you can enter the data points into a calculator or computer and use the *linear regression* feature of the calculator or computer. When you use the *regression* feauture of a graphing calculator or computer program, you will notice that the program may also output an "*r*-value." This *r*-value is the **correlation coefficient** of the data and gives a measure of how well the model fits the data. The closer the value of $|r|$ is to 1, the better the fit.

| Example 2 | Finding a Least Squares Regression Line |

The amounts p (in millions of dollars) of total annual prize money awarded at the Indianapolis 500 race from 1995 to 2004 are shown in the table. Construct a scatter plot that represents the data and find the least squares regression line for the data. (Source: indy500.com)

Year	Prize money, p
1995	8.06
1996	8.11
1997	8.61
1998	8.72
1999	9.05
2000	9.48
2001	9.61
2002	10.03
2003	10.15
2004	10.25

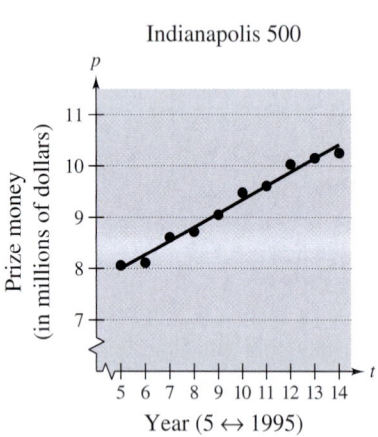

Indianapolis 500

FIGURE 3.36

t	p	$p*$
5	8.06	8.00
6	8.11	8.27
7	8.61	8.54
8	8.72	8.80
9	9.05	9.07
10	9.48	9.34
11	9.61	9.61
12	10.03	9.88
13	10.15	10.14
14	10.25	10.41

Solution

Let $t = 5$ represent 1995. The scatter plot for the points is shown in Figure 3.36. Using the *regression* feature of a graphing utility, you can determine that the equation of the least squares regression line is

$$p = 0.268t + 6.66.$$

To check this model, compare the actual p-values with the p-values given by the model, which are labeled $p*$ in the table at the left. The correlation coefficient for this model is $r \approx 0.991$, which implies that the model is a good fit.

✓**CHECKPOINT** Now try Exercise 7.

Direct Variation

There are two basic types of linear models. The more general model has a y-intercept that is nonzero.

$$y = mx + b, \quad b \neq 0$$

The simpler model

$$y = kx$$

has a y-intercept that is zero. In the simpler model, y is said to **vary directly** as x, or to be **directly proportional** to x.

Direct Variation

The following statements are equivalent.

1. y **varies directly** as x.

2. y is **directly proportional** to x.

3. $y = kx$ for some nonzero constant k.

k is the **constant of variation** or the **constant of proportionality.**

Example 3   **Direct Variation**

In Pennsylvania, the state income tax is directly proportional to *gross income*. You are working in Pennsylvania and your state income tax deduction is $46.05 for a gross monthly income of $1500. Find a mathematical model that gives the Pennsylvania state income tax in terms of gross income.

Solution

Verbal Model: State income tax $= k \cdot$ Gross income

Labels: State income tax $= y$ (dollars)
Gross income $= x$ (dollars)
Income tax rate $= k$ (percent in decimal form)

Equation: $y = kx$

To solve for k, substitute the given information into the equation $y = kx$, and then solve for k.

$$y = kx \qquad \text{\textcolor{red}{Write direct variation model.}}$$

$$46.05 = k(1500) \qquad \text{\textcolor{red}{Substitute } y = 46.05 \text{ and } x = 1500.}$$

$$0.0307 = k \qquad \text{\textcolor{red}{Simplify.}}$$

So, the equation (or model) for state income tax in Pennsylvania is

$$y = 0.0307x.$$

In other words, Pennsylvania has a state income tax rate of 3.07% of gross income. The graph of this equation is shown in Figure 3.37.

✔**CHECKPOINT**   Now try Exercise 33.

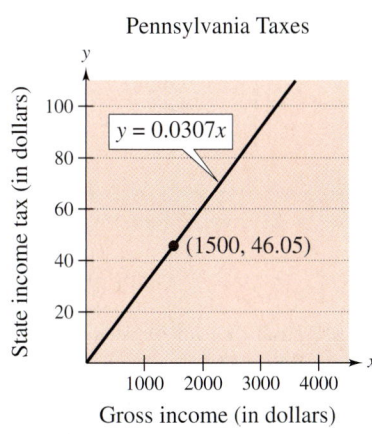

Pennsylvania Taxes

$y = 0.0307x$

(1500, 46.05)

State income tax (in dollars)

Gross income (in dollars)

FIGURE **3.37**

Direct Variation as an *n*th Power

Another type of direct variation relates one variable to a *power* of another variable. For example, in the formula for the area of a circle

$$A = \pi r^2$$

the area A is directly proportional to the square of the radius r. Note that for this formula, π is the constant of proportionality.

Direct Variation as an *n*th Power

The following statements are equivalent.

1. y **varies directly as the *n*th power** of x.

2. y is **directly proportional to the *n*th power** of x.

3. $y = kx^n$ for some constant k.

Example 4   **Direct Variation as *n*th Power**

The distance a ball rolls down an inclined plane is directly proportional to the square of the time it rolls. During the first second, the ball rolls 8 feet. (See Figure 3.38.)

a. Write an equation relating the distance traveled to the time.

b. How far will the ball roll during the first 3 seconds?

Solution

a. Letting d be the distance (in feet) the ball rolls and letting t be the time (in seconds), you have

$$d = kt^2.$$

Now, because $d = 8$ when $t = 1$, you can see that $k = 8$, as follows.

$$d = kt^2$$
$$8 = k(1)^2$$
$$8 = k$$

So, the equation relating distance to time is

$$d = 8t^2.$$

b. When $t = 3$, the distance traveled is $d = 8(3)^2 = 8(9) = 72$ feet.

✔**CHECKPOINT**   Now try Exercise 63.

In Examples 3 and 4, the direct variations are such that an *increase* in one variable corresponds to an *increase* in the other variable. This is also true in the model $d = \frac{1}{5}F, F > 0$, where an increase in F results in an increase in d. You should not, however, assume that this always occurs with direct variation. For example, in the model $y = -3x$, an increase in x results in a *decrease* in y, and yet y is said to vary directly as x.

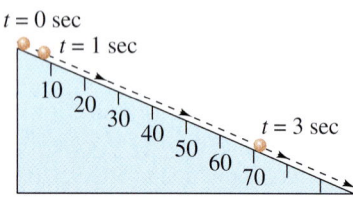

$t = 0$ sec
$t = 1$ sec
10
20
30
40
50
60
70
$t = 3$ sec

FIGURE 3.38

Inverse Variation

> ### Inverse Variation
>
> The following statements are equivalent.
>
> **1.** y **varies inversely** as x.   **2.** y is **inversely proportional** to x.
>
> **3.** $y = \dfrac{k}{x}$ for some constant k.

If x and y are related by an equation of the form $y = k/x^n$, then y varies inversely as the nth power of x (or y is inversely proportional to the nth power of x).

Some applications of variation involve problems with *both* direct and inverse variation in the same model. These types of models are said to have **combined variation.**

Example 5   Direct and Inverse Variation

A gas law states that the volume of an enclosed gas varies directly as the temperature *and* inversely as the pressure, as shown in Figure 3.39. The pressure of a gas is 0.75 kilogram per square centimeter when the temperature is 294 K and the volume is 8000 cubic centimeters. (a) Write an equation relating pressure, temperature, and volume. (b) Find the pressure when the temperature is 300 K and the volume is 7000 cubic centimeters.

P_1

P_2

V_1

V_2

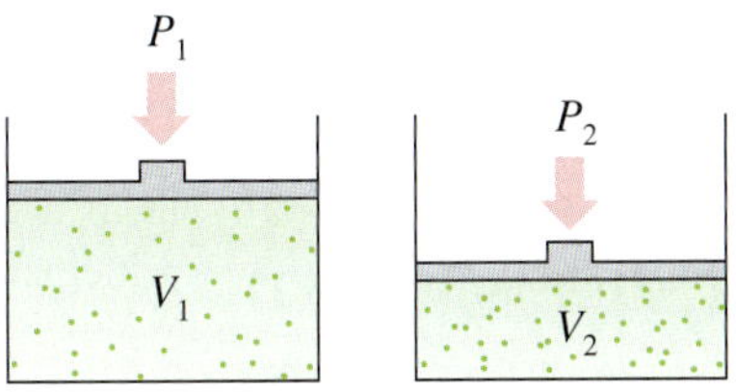

$P_2 > P_1$ then $V_2 < V_1$

FIGURE 3.39   *If the temperature is held constant and pressure increases, volume decreases.*

Solution

a. Let V be volume (in cubic centimeters), let P be pressure (in kilograms per square centimeter), and let T be temperature (in Kelvin). Because V varies directly as T and inversely as P, you have

$$V = \frac{kT}{P}.$$

Now, because $P = 0.75$ when $T = 294$ and $V = 8000$, you have

$$8000 = \frac{k(294)}{0.75}$$

$$k = \frac{6000}{294} = \frac{1000}{49}.$$

So, the equation relating pressure, temperature, and volume is

$$V = \frac{1000}{49}\left(\frac{T}{P}\right).$$

b. When $T = 300$ and $V = 7000$, the pressure is

$$P = \frac{1000}{49}\left(\frac{300}{7000}\right) = \frac{300}{343} \approx 0.87 \text{ kilogram per square centimeter.}$$

✔**CHECKPOINT**   Now try Exercise 65.

Joint Variation

In Example 5, note that when a direct variation and an inverse variation occur in the same statement, they are coupled with the word "and." To describe two different *direct* variations in the same statement, the word **jointly** is used.

Joint Variation

The following statements are equivalent.

1. z **varies jointly** as x and y.

2. z is **jointly proportional** to x and y.

3. $z = kxy$ for some constant k.

If x, y, and z are related by an equation of the form

$$z = kx^n y^m$$

then z varies jointly as the nth power of x and the mth power of y.

Example 6 **Joint Variation**

The *simple* interest for a certain savings account is jointly proportional to the time and the principal. After one quarter (3 months), the interest on a principal of $5000 is $43.75.

a. Write an equation relating the interest, principal, and time.

b. Find the interest after three quarters.

Solution

a. Let I = interest (in dollars), P = principal (in dollars), and t = time (in years). Because I is jointly proportional to P and t, you have

$$I = kPt.$$

For $I = 43.75$, $P = 5000$, and $t = \frac{1}{4}$, you have

$$43.75 = k(5000)\left(\frac{1}{4}\right)$$

which implies that $k = 4(43.75)/5000 = 0.035$. So, the equation relating interest, principal, and time is

$$I = 0.035Pt$$

which is the familiar equation for simple interest where the constant of proportionality, 0.035, represents an annual interest rate of 3.5%.

b. When $P = \$5000$ and $t = \frac{3}{4}$, the interest is

$$I = (0.035)(5000)\left(\frac{3}{4}\right)$$

$$= \$131.25.$$

✓**CHECKPOINT** Now try Exercise 67.

3.5   Exercises

VOCABULARY CHECK: Fill in the blanks.

1. Two techniques for fitting models to data are called direct _____ and least squares _____.

2. Statisticians use a measure called _____ of_____ _____ to find a model that approximates a set of data most accurately.

3. An *r*-value of a set of data, also called a _____ _____, gives a measure of how well a model fits a set of data.

4. Direct variation models can be described as *y* varies directly as *x*, or *y* is _____ _____ to *x*.

5. In direct variation models of the form $y = kx$, *k* is called the _____ of _____.

6. The direct variation model $y = kx^n$ can be described as *y* varies directly as the *n*th power of *x*, or *y* is _____ _____ to the *n*th power of *x*.

7. The mathematical model $y = \dfrac{k}{x}$ is an example of _____ variation.

8. Mathematical models that involve both direct and inverse variation are said to have _____ variation.

9. The joint variation model $z = kxy$ can be described as *z* varies jointly as *x* and *y*, or *z* is _____ _____ to *x* and *y*.

PREREQUISITE SKILLS REVIEW: Practice and review algebra skills needed for this section at **www.Eduspace.com.**

---

1. **Employment** The total numbers of employees (in thousands) in the United States from 1992 to 2002 are given by the following ordered pairs.

(1992, 128,105)   (1998, 137,673)
(1993, 129,200)   (1999, 139,368)
(1994, 131,056)   (2000, 142,583)
(1995, 132,304)   (2001, 143,734)
(1996, 133,943)   (2002, 144,683)
(1997, 136,297)

A linear model that approximates the data is $y = 1767.0t + 123{,}916$, where *y* represents the number of employees (in thousands) and $t = 2$ represents 1992. Plot the actual data and the model on the same set of coordinate axes. How closely does the model represent the data? (Source: U.S. Bureau of Labor Statistics)

2. **Sports** The winning times (in minutes) in the women's 400-meter freestyle swimming event in the Olympics from 1948 to 2004 are given by the following ordered pairs.

(1948, 5.30)   (1980, 4.15)
(1952, 5.20)   (1984, 4.12)
(1956, 4.91)   (1988, 4.06)
(1960, 4.84)   (1992, 4.12)
(1964, 4.72)   (1996, 4.12)
(1968, 4.53)   (2000, 4.10)
(1972, 4.32)   (2004, 4.09)
(1976, 4.16)

A linear model that approximates the data is $y = -0.022t + 5.03$, where *y* represents the winning time (in minutes) and $t = 0$ represents 1950. Plot the actual data and the model on the same set of coordinate axes. How closely does the model represent the data? Does it appear that another type of model may be a better fit? Explain. (Source: *The World Almanac and Book of Facts*)

In Exercises 3–6, sketch the line that you think best approximates the data in the scatter plot. Then find an equation of the line. To print an enlarged copy of the graph, go to the website *www.mathgraphs.com.*

3.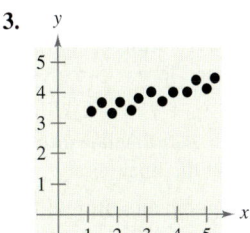

4.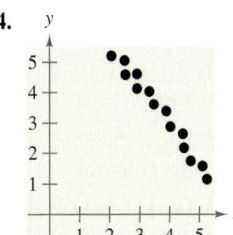

5.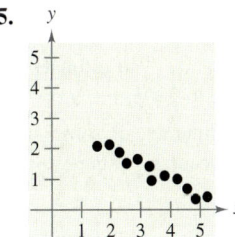

6.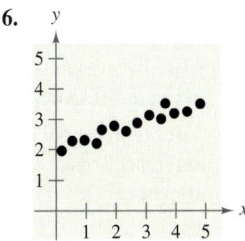

7. Sports The lengths (in feet) of the winning men's discus throws in the Olympics from 1912 to 2004 are listed below. (Source: *The World Almanac and Book of Facts*)

1912	148.3	1952	180.5	1980	218.7
1920	146.6	1956	184.9	1984	218.5
1924	151.3	1960	194.2	1988	225.8
1928	155.3	1964	200.1	1992	213.7
1932	162.3	1968	212.5	1996	227.7
1936	165.6	1972	211.3	2000	227.3
1948	173.2	1976	221.5	2004	229.3

(a) Sketch a scatter plot of the data. Let y represent the length of the winning discus throw (in feet) and let $t = 12$ represent 1912.

(b) Use a straightedge to sketch the best-fitting line through the points and find an equation of the line.

(c) Use the *regression* feature of a graphing utility to find the least squares regression line that fits the data.

(d) Compare the linear model you found in part (b) with the linear model given by the graphing utility in part (c).

(e) Use the models from parts (b) and (c) to estimate the winning men's discus throw in the year 2008.

(f) Use your school's library, the Internet, or some other reference source to analyze the accuracy of the estimate in part (e).

8. Revenue The total revenues (in millions of dollars) for Outback Steakhouse from 1995 to 2003 are listed below. (Source: Outback Steakhouse, Inc.)

1995	664.0	2000	1906.0
1996	937.4	2001	2127.0
1997	1151.6	2002	2362.1
1998	1358.9	2003	2744.4
1999	1646.0		

(a) Sketch a scatter plot of the data. Let y represent the total revenue (in millions of dollars) and let $t = 5$ represent 1995.

(b) Use a straightedge to sketch the best-fitting line through the points and find an equation of the line.

(c) Use the *regression* feature of a graphing utility to find the least squares regression line that fits the data.

(d) Compare the linear model you found in part (b) with the linear model given by the graphing utility in part (c).

(e) Use the models from parts (b) and (c) to estimate the revenues of Outback Steakhouse in 2005.

(f) Use your school's library, the Internet, or some other reference source to analyze the accuracy of the estimate in part (e).

9. Data Analysis: Broadway Shows The table shows the annual gross ticket sales S (in millions of dollars) for Broadway shows in New York City from 1995 through 2004. (Source: The League of American Theatres and Producers, Inc.)

Year	Sales, S
1995	406
1996	436
1997	499
1998	558
1999	588
2000	603
2001	666
2002	643
2003	721
2004	771

(a) Use a graphing utility to create a scatter plot of the data. Let $t = 5$ represent 1995.

(b) Use the *regression* feature of a graphing utility to find the equation of the least squares regression line that fits the data.

(c) Use the graphing utility to graph the scatter plot you found in part (a) and the model you found in part (b) in the same viewing window. How closely does the model represent the data?

(d) Use the model to estimate the annual gross ticket sales in 2005 and 2007.

(e) Interpret the meaning of the slope of the linear model in the context of the problem.

10. Data Analysis: Television Households The table shows the numbers x (in millions) of households with cable television and the numbers y (in millions) of households with color television sets in the United States from 1995 through 2002. (Source: Nielson Media Research; Television Bureau of Advertising, Inc.)

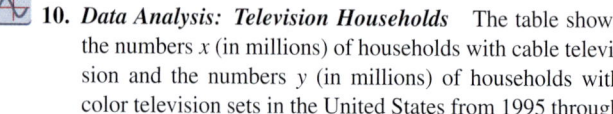

Households with cable, x	Households with color TV, y
63	94
65	95
66	97
67	98
75	99
77	101
80	102
86	105

(a) Use the *regression* feature of a graphing utility to find the equation of the least squares regression line that fits the data.

(b) Use the graphing utility to create a scatter plot of the data. Then graph the model you found in part (a) and the scatter plot in the same viewing window. How closely does the model represent the data?

(c) Use the model to estimate the number of households with color television sets if the number of households with cable television is 90 million.

(d) Interpret the meaning of the slope of the linear model in the context of the problem.

Think About It   **In Exercises 11 and 12, use the graph to determine whether y varies directly as some power of x or inversely as some power of x. Explain.**

11.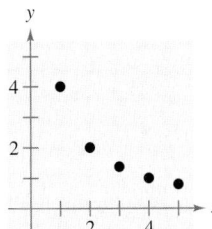

12.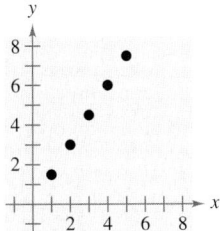

In Exercises 13–16, use the given value of k to complete the table for the direct variation model $y = kx^2$. Plot the points on a rectangular coordinate system.

x	2	4	6	8	10
$y = kx^2$					

13. $k = 1$

15. $k = \frac{1}{2}$

14. $k = 2$

16. $k = \frac{1}{4}$

In Exercises 17–20, use the given value of k to complete the table for the inverse variation model

$$y = \frac{k}{x^2}.$$

Plot the points on a rectangular coordinate system.

x	2	4	6	8	10
$y = \dfrac{k}{x^2}$					

17. $k = 2$

19. $k = 10$

18. $k = 5$

20. $k = 20$

In Exercises 21–24, determine whether the variation model is of the form $y = kx$ or $y = k/x$, and find k.

21.

x	5	10	15	20	25
y	1	$\frac{1}{2}$	$\frac{1}{3}$	$\frac{1}{4}$	$\frac{1}{5}$

22.

x	5	10	15	20	25
y	2	4	6	8	10

23.

x	5	10	15	20	25
y	-3.5	-7	-10.5	-14	-17.5

24.

x	5	10	15	20	25
y	24	12	8	6	$\frac{24}{5}$

Direct Variation   **In Exercises 25–28, assume that y is directly proportional to x. Use the given x-value and y-value to find a linear model that relates y and x.**

25. $x = 5,\ y = 12$

26. $x = 2,\ y = 14$

27. $x = 10,\ y = 2050$

28. $x = 6,\ y = 580$

29. *Simple Interest*   The simple interest on an investment is directly proportional to the amount of the investment. By investing $2500 in a certain bond issue, you obtained an interest payment of $87.50 after 1 year. Find a mathematical model that gives the interest I for this bond issue after 1 year in terms of the amount invested P.

30. *Simple Interest*   The simple interest on an investment is directly proportional to the amount of the investment. By investing $5000 in a municipal bond, you obtained an interest payment of $187.50 after 1 year. Find a mathematical model that gives the interest I for this municipal bond after 1 year in terms of the amount invested P.

31. *Measurement*   On a yardstick with scales in inches and centimeters, you notice that 13 inches is approximately the same length as 33 centimeters. Use this information to find a mathematical model that relates centimeters to inches. Then use the model to find the numbers of centimeters in 10 inches and 20 inches.

32. *Measurement*   When buying gasoline, you notice that 14 gallons of gasoline is approximately the same amount of gasoline as 53 liters. Use this information to find a linear model that relates gallons to liters. Then use the model to find the numbers of liters in 5 gallons and 25 gallons.

33. *Taxes* Property tax is based on the assessed value of a property. A house that has an assessed value of $150,000 has a property tax of $5520. Find a mathematical model that gives the amount of property tax y in terms of the assessed value x of the property. Use the model to find the property tax on a house that has an assessed value of $200,000.

34. *Taxes* State sales tax is based on retail price. An item that sells for $145.99 has a sales tax of $10.22. Find a mathematical model that gives the amount of sales tax y in terms of the retail price x. Use the model to find the sales tax on a $540.50 purchase.

Hooke's Law **In Exercises 35–38, use Hooke's Law for springs, which states that the distance a spring is stretched (or compressed) varies directly as the force on the spring.**

35. A force of 265 newtons stretches a spring 0.15 meter (see figure).

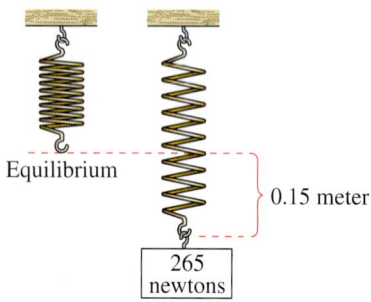

Equilibrium

0.15 meter

265 newtons

(a) How far will a force of 90 newtons stretch the spring?

(b) What force is required to stretch the spring 0.1 meter?

36. A force of 220 newtons stretches a spring 0.12 meter. What force is required to stretch the spring 0.16 meter?

37. The coiled spring of a toy supports the weight of a child. The spring is compressed a distance of 1.9 inches by the weight of a 25-pound child. The toy will not work properly if its spring is compressed more than 3 inches. What is the weight of the heaviest child who should be allowed to use the toy?

38. An overhead garage door has two springs, one on each side of the door (see figure). A force of 15 pounds is required to stretch each spring 1 foot. Because of a pulley system, the springs stretch only one-half the distance the door travels. The door moves a total of 8 feet, and the springs are at their natural length when the door is open. Find the combined lifting force applied to the door by the springs when the door is closed.

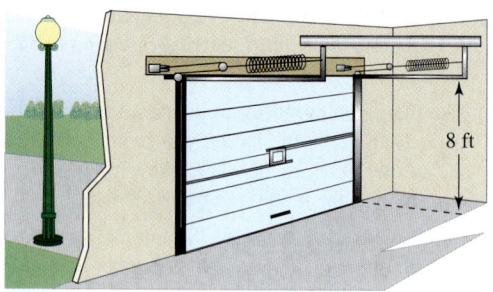

8 ft

FIGURE FOR **38**

In Exercises 39–48, find a mathematical model for the verbal statement.

39. A varies directly as the square of r.

40. V varies directly as the cube of e.

41. y varies inversely as the square of x.

42. h varies inversely as the square root of s.

43. F varies directly as g and inversely as r^2.

44. z is jointly proportional to the square of x and the cube of y.

45. *Boyle's Law:* For a constant temperature, the pressure P of a gas is inversely proportional to the volume V of the gas.

46. *Newton's Law of Cooling:* The rate of change R of the temperature of an object is proportional to the difference between the temperature T of the object and the temperature T_e of the environment in which the object is placed.

47. *Newton's Law of Universal Gravitation:* The gravitational attraction F between two objects of masses m_1 and m_2 is proportional to the product of the masses and inversely proportional to the square of the distance r between the objects.

48. *Logistic Growth:* The rate of growth R of a population is jointly proportional to the size S of the population and the difference between S and the maximum population size L that the environment can support.

In Exercises 49–54, write a sentence using the variation terminology of this section to describe the formula.

49. *Area of a triangle:* $A = \frac{1}{2}bh$

50. *Surface area of a sphere:* $S = 4\pi r^2$

51. *Volume of a sphere:* $V = \frac{4}{3}\pi r^3$

52. *Volume of a right circular cylinder:* $V = \pi r^2 h$

53. *Average speed:* $r = \dfrac{d}{t}$

54. *Free vibrations:* $\omega = \sqrt{\dfrac{kg}{W}}$

In Exercises 55–62, find a mathematical model representing the statement. (In each case, determine the constant of proportionality.)

55. A varies directly as r^2. ($A = 9\pi$ when $r = 3$.)

56. y varies inversely as x. ($y = 3$ when $x = 25$.)

57. y is inversely proportional to x. ($y = 7$ when $x = 4$.)

58. z varies jointly as x and y. ($z = 64$ when $x = 4$ and $y = 8$.)

59. F is jointly proportional to r and the third power of s. ($F = 4158$ when $r = 11$ and $s = 3$.)

60. P varies directly as x and inversely as the square of y. $\left(P = \frac{28}{3}\text{ when }x = 42\text{ and }y = 9.\right)$

61. z varies directly as the square of x and inversely as y. ($z = 6$ when $x = 6$ and $y = 4$.)

62. v varies jointly as p and q and inversely as the square of s. ($v = 1.5$ when $p = 4.1$, $q = 6.3$, and $s = 1.2$.)

Ecology   **In Exercises 63 and 64, use the fact that the diameter of the largest particle that can be moved by a stream varies approximately directly as the square of the velocity of the stream.**

63. A stream with a velocity of $\frac{1}{4}$ mile per hour can move coarse sand particles about 0.02 inch in diameter. Approximate the velocity required to carry particles 0.12 inch in diameter.

64. A stream of velocity v can move particles of diameter d or less. By what factor does d increase when the velocity is doubled?

Resistance   **In Exercises 65 and 66, use the fact that the resistance of a wire carrying an electrical current is directly proportional to its length and inversely proportional to its cross-sectional area.**

65. If #28 copper wire (which has a diameter of 0.0126 inch) has a resistance of 66.17 ohms per thousand feet, what length of #28 copper wire will produce a resistance of 33.5 ohms?

66. A 14-foot piece of copper wire produces a resistance of 0.05 ohm. Use the constant of proportionality from Exercise 65 to find the diameter of the wire.

67. *Work*   The work W (in joules) done when lifting an object varies jointly with the mass m (in kilograms) of the object and the height h (in meters) that the object is lifted. The work done when a 120-kilogram object is lifted 1.8 meters is 2116.8 joules. How much work is done when lifting a 100-kilogram object 1.5 meters?

68. *Spending*   The prices of three sizes of pizza at a pizza shop are as follows.

9-inch: $8.78, 12-inch: $11.78, 15-inch: $14.18

You would expect that the price of a certain size of pizza would be directly proportional to its surface area. Is that the case for this pizza shop? If not, which size of pizza is the best buy?

69. *Fluid Flow*   The velocity v of a fluid flowing in a conduit is inversely proportional to the cross-sectional area of the conduit. (Assume that the volume of the flow per unit of time is held constant.) Determine the change in the velocity of water flowing from a hose when a person places a finger over the end of the hose to decrease its cross-sectional area by 25%.

70. *Beam Load*   The maximum load that can be safely supported by a horizontal beam varies jointly as the width of the beam and the square of its depth, and inversely as the length of the beam. Determine the changes in the maximum safe load under the following conditions.

    (a) The width and length of the beam are doubled.

    (b) The width and depth of the beam are doubled.

    (c) All three of the dimensions are doubled.

    (d) The depth of the beam is halved.

Model It

71. *Data Analysis: Ocean Temperatures*   An oceanographer took readings of the water temperatures C (in degrees Celsius) at several depths d (in meters). The data collected are shown in the table.

Depth, d	Temperature, C
1000	4.2°
2000	1.9°
3000	1.4°
4000	1.2°
5000	0.9°

(a) Sketch a scatter plot of the data.

(b) Does it appear that the data can be modeled by the inverse variation model $C = k/d$? If so, find k for each pair of coordinates.

(c) Determine the mean value of k from part (b) to find the inverse variation model $C = k/d$.

 (d) Use a graphing utility to plot the data points and the inverse model in part (c).

(e) Use the model to approximate the depth at which the water temperature is 3°C.

72. *Data Analysis: Physics Experiment* An experiment in a physics lab requires a student to measure the compressed lengths y (in centimeters) of a spring when various forces of F pounds are applied. The data are shown in the table.

Force, F	Length, y
0	0
2	1.15
4	2.3
6	3.45
8	4.6
10	5.75
12	6.9

(a) Sketch a scatter plot of the data.

(b) Does it appear that the data can be modeled by Hooke's Law? If so, estimate k. (See Exercises 35–38.)

(c) Use the model in part (b) to approximate the force required to compress the spring 9 centimeters.

73. *Data Analysis: Light Intensity* A light probe is located x centimeters from a light source, and the intensity y (in microwatts per square centimeter) of the light is measured. The results are shown as ordered pairs (x, y).

(30, 0.1881)   (34, 0.1543)   (38, 0.1172)

(42, 0.0998)   (46, 0.0775)   (50, 0.0645)

A model for the data is $y = 262.76/x^{2.12}$.

 (a) Use a graphing utility to plot the data points and the model in the same viewing window.

(b) Use the model to approximate the light intensity 25 centimeters from the light source.

74. *Illumination* The illumination from a light source varies inversely as the square of the distance from the light source. When the distance from a light source is doubled, how does the illumination change? Discuss this model in terms of the data given in Exercise 73. Give a possible explanation of the difference.

Synthesis

True or False? **In Exercises 75–77, decide whether the statement is true or false. Justify your answer.**

75. If y varies directly as x, then if x increases, y will increase as well.

76. In the equation for kinetic energy, $E = \frac{1}{2}mv^2$, the amount of kinetic energy E is directly proportional to the mass m of an object and the square of its velocity v.

77. If the correlation coefficient for a least squares regression line is close to -1, the regression line cannot be used to describe the data.

78. Discuss how well the data shown in each scatter plot can be approximated by a linear model.

(a)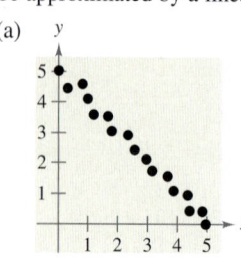

(b)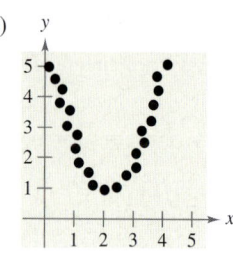

(c)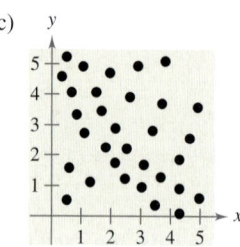

(d)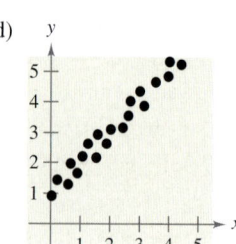

79. *Writing* A linear mathematical model for predicting prize winnings at a race is based on data for 3 years. Write a paragraph discussing the potential accuracy or inaccuracy of such a model.

80. *Research Project* Use your school's library, the Internet, or some other reference source to find data that you think describe a linear relationship. Create a scatter plot of the data and find the least squares regression line that represents the data points. Interpret the slope and y-intercept in the context of the data. Write a summary of your findings.

Skills Review

In Exercises 81–84, solve the inequality and graph the solution on the real number line.

81. $3x + 2 > 17$

82. $-7x + 10 \le -1 + x$

83. $|2x - 1| < 9$ **84.** $|4 - 3x| + 7 \ge 12$

In Exercises 85 and 86, evaluate the function at each value of the independent variable and simplify.

85. $f(x) = \dfrac{x^2 + 5}{x - 3}$

(a) $f(0)$ (b) $f(-3)$ (c) $f(4)$

86. $f(x) = \begin{cases} -x^2 + 10, & x \ge -2 \\ 6x^2 - 1, & x < -2 \end{cases}$

(a) $f(-2)$ (b) $f(1)$ (c) $f(-8)$

87. **Make a Decision** To work an extended application analyzing registered voters in the United States, visit this text's website at *college.hmco.com*. *(Data Source: U.S. Census Bureau)*

3   Chapter Summary

What did you learn?

3.1 In Exercises 1 and 2, graph each function. Compare the graph of each function with the graph of $y = x^2$.

1. (a) $f(x) = 2x^2$
   (b) $g(x) = -2x^2$
   (c) $h(x) = x^2 + 2$
   (d) $k(x) = (x + 2)^2$

2. (a) $f(x) = x^2 - 4$
   (b) $g(x) = 4 - x^2$
   (c) $h(x) = (x - 3)^2$
   (d) $k(x) = \frac{1}{2}x^2 - 1$

In Exercises 3–14, write the quadratic function in standard form and sketch its graph. Identify the vertex, axis of symmetry, and x-intercept(s).

3. $g(x) = x^2 - 2x$

4. $f(x) = 6x - x^2$

5. $f(x) = x^2 + 8x + 10$

6. $h(x) = 3 + 4x - x^2$

7. $f(t) = -2t^2 + 4t + 1$

8. $f(x) = x^2 - 8x + 12$

9. $h(x) = 4x^2 + 4x + 13$

10. $f(x) = x^2 - 6x + 1$

11. $h(x) = x^2 + 5x - 4$

12. $f(x) = 4x^2 + 4x + 5$

13. $f(x) = \frac{1}{3}(x^2 + 5x - 4)$

14. $f(x) = \frac{1}{2}(6x^2 - 24x + 22)$

In Exercises 15–18, write the standard form of the equation of the parabola that has the indicated vertex and whose graph passes through the given point.

15.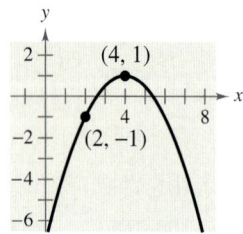

16.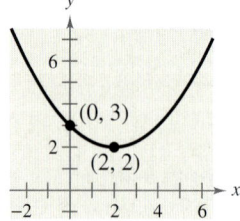

17. Vertex: $(1, -4)$; point: $(2, -3)$

18. Vertex: $(2, 3)$; point: $(-1, 6)$

19. *Numerical, Graphical, and Analytical Analysis* A rectangle is inscribed in the region bounded by the x-axis, the y-axis, and the graph of $x + 2y - 8 = 0$, as shown in the figure.

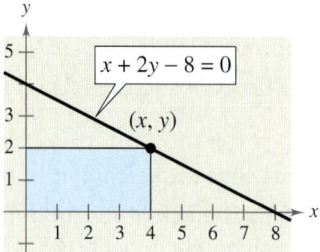

(a) Write the area A of the rectangle as a function of x.

(b) Determine the domain of the function in the context of the problem.

(c) Create a table showing possible values of x and the corresponding area of the rectangle.

 (d) Use a graphing utility to graph the area function. Use the graph to approximate the dimensions that will produce the maximum area.

(e) Write the area function in standard form to find analytically the dimensions that will produce the maximum area.

20. *Geometry* The perimeter of a rectangle is 200 meters.

(a) Draw a diagram that gives a visual representation of the problem. Label the length and width as x and y, respectively.

(b) Write y as a function of x. Use the result to write the area as a function of x.

(c) Of all possible rectangles with perimeters of 200 meters, find the dimensions of the one with the maximum area.

21. *Maximum Revenue* The total revenue R earned (in dollars) from producing a gift box of candles is given by

$$R(p) = -10p^2 + 800p$$

where p is the price per unit (in dollars).

(a) Find the revenues when the prices per box are $20, $25, and $30.

(b) Find the unit price that will yield a maximum revenue. What is the maximum revenue? Explain your results.

22. *Maximum Profit* A real estate office handles an apartment building that has 50 units. When the rent is $540 per month, all units are occupied. However, for each $30 increase in rent, one unit becomes vacant. Each occupied unit requires an average of $18 per month for service and repairs. What rent should be charged to obtain the maximum profit?

23. *Minimum Cost* A soft-drink manufacturer has daily production costs of

$$C = 70,000 - 120x + 0.055x^2$$

where C is the total cost (in dollars) and x is the number of units produced. How many units should be produced each day to yield a minimum cost?

24. *Sociology* The average age of the groom at a first marriage for a given age of the bride can be approximated by the model

$$y = -0.107x^2 + 5.68x - 48.5, \quad 20 \le x \le 25$$

where y is the age of the groom and x is the age of the bride. Sketch a graph of the model. For what age of the bride is the average age of the groom 26? (Source: U.S. Census Bureau)

3.2 In Exercises 25–30, sketch the graphs of $y = x^n$ and the transformation.

25. $y = x^3$, $f(x) = -(x - 4)^3$
26. $y = x^3$, $f(x) = -4x^3$
27. $y = x^4$, $f(x) = 2 - x^4$
28. $y = x^4$, $f(x) = 2(x - 2)^4$
29. $y = x^5$, $f(x) = (x - 3)^5$
30. $y = x^5$, $f(x) = \frac{1}{2}x^5 + 3$

In Exercises 31–34, describe the right-hand and left-hand behavior of the graph of the polynomial function.

31. $f(x) = -x^2 + 6x + 9$
32. $f(x) = \frac{1}{2}x^3 + 2x$
33. $g(x) = \frac{3}{4}(x^4 + 3x^2 + 2)$
34. $h(x) = -x^5 - 7x^2 + 10x$

In Exercises 35–40, find all the real zeros of the polynomial function. Determine the multiplicity of each zero and the number of turning points of the graph of the function. Use a graphing utility to verify your answers.

35. $f(x) = 2x^2 + 11x - 21$
36. $f(x) = x(x + 3)^2$
37. $f(t) = t^3 - 3t$
38. $f(x) = x^3 - 8x^2$
39. $f(x) = -12x^3 + 20x^2$
40. $g(x) = x^4 - x^3 - 2x^2$

In Exercises 41–44, sketch the graph of the function by (a) applying the Leading Coefficient Test, (b) finding the zeros of the polynomial, (c) plotting sufficient solution points, and (d) drawing a continuous curve through the points.

41. $f(x) = -x^3 + x^2 - 2$
42. $g(x) = 2x^3 + 4x^2$
43. $f(x) = x(x^3 + x^2 - 5x + 3)$
44. $h(x) = 3x^2 - x^4$

In Exercises 45–48, use the Intermediate Value Theorem and the *table* feature of a graphing utility to find intervals one unit in length in which the polynomial function is guaranteed to have a zero. Adjust the table to approximate the zeros of the function. Use the *zero* or *root* feature of the graphing utility to verify your results.

45. $f(x) = 3x^3 - x^2 + 3$
46. $f(x) = 0.25x^3 - 3.65x + 6.12$
47. $f(x) = x^4 - 5x - 1$
48. $f(x) = 7x^4 + 3x^3 - 8x^2 + 2$

3.3 In Exercises 49–54, use long division to divide.

49. $\dfrac{24x^2 - x - 8}{3x - 2}$

50. $\dfrac{4x + 7}{3x - 2}$

51. $\dfrac{5x^3 - 13x^2 - x + 2}{x^2 - 3x + 1}$

52. $\dfrac{3x^4}{x^2 - 1}$

53. $\dfrac{x^4 - 3x^3 + 4x^2 - 6x + 3}{x^2 + 2}$

54. $\dfrac{6x^4 + 10x^3 + 13x^2 - 5x + 2}{2x^2 - 1}$

In Exercises 55–58, use synthetic division to divide.

55. $\dfrac{6x^4 - 4x^3 - 27x^2 + 18x}{x - 2}$

56. $\dfrac{0.1x^3 + 0.3x^2 - 0.5}{x - 5}$

57. $\dfrac{2x^3 - 19x^2 + 38x + 24}{x - 4}$

58. $\dfrac{3x^3 + 20x^2 + 29x - 12}{x + 3}$

In Exercises 59 and 60, use synthetic division to determine whether the given values of x are zeros of the function.

59. $f(x) = 20x^4 + 9x^3 - 14x^2 - 3x$

  (a) $x = -1$ (b) $x = \frac{3}{4}$ (c) $x = 0$ (d) $x = 1$

60. $f(x) = 3x^3 - 8x^2 - 20x + 16$

  (a) $x = 4$ (b) $x = -4$ (c) $x = \frac{2}{3}$ (d) $x = -1$

In Exercises 61 and 62, use synthetic division to find each function value.

61. $f(x) = x^4 + 10x^3 - 24x^2 + 20x + 44$

  (a) $f(-3)$ (b) $f(-1)$

62. $g(t) = 2t^5 - 5t^4 - 8t + 20$

  (a) $g(-4)$ (b) $g(\sqrt{2})$

In Exercises 63–66, (a) verify the given factor(s) of the function f, (b) find the remaining factors of f, (c) use your results to write the complete factorization of f, (d) list all real zeros of f, and (e) confirm your results by using a graphing utility to graph the function.

Function	Factor(s)
63. $f(x) = x^3 + 4x^2 - 25x - 28$	$(x - 4)$
64. $f(x) = 2x^3 + 11x^2 - 21x - 90$	$(x + 6)$
65. $f(x) = x^4 - 4x^3 - 7x^2 + 22x + 24$	$(x + 2)(x - 3)$
66. $f(x) = x^4 - 11x^3 + 41x^2 - 61x + 30$	$(x - 2)(x - 5)$

3.4 In Exercises 67–72, find all the zeros of the function.

67. $f(x) = 3x(x - 2)^2$

68. $f(x) = (x - 4)(x + 9)^2$

69. $f(x) = x^2 - 9x + 8$

70. $f(x) = x^3 + 6x$

71. $f(x) = (x + 4)(x - 6)(x - 2i)(x + 2i)$

72. $f(x) = (x - 8)(x - 5)^2(x - 3 + i)(x - 3 - i)$

In Exercises 73 and 74, use the Rational Zero Test to list all possible rational zeros of f.

73. $f(x) = -4x^3 + 8x^2 - 3x + 15$

74. $f(x) = 3x^4 + 4x^3 - 5x^2 - 8$

In Exercises 75–80, find all the rational zeros of the function.

75. $f(x) = x^3 - 2x^2 - 21x - 18$

76. $f(x) = 3x^3 - 20x^2 + 7x + 30$

77. $f(x) = x^3 - 10x^2 + 17x - 8$

78. $f(x) = x^3 + 9x^2 + 24x + 20$

79. $f(x) = x^4 + x^3 - 11x^2 + x - 12$

80. $f(x) = 25x^4 + 25x^3 - 154x^2 - 4x + 24$

In Exercises 81 and 82, find a polynomial function with real coefficients that has the given zeros. (There are many correct answers.)

81. $\frac{2}{3}, 4, \sqrt{3}i$ **82.** $2, -3, 1 - 2i$

In Exercises 83–86, use the given zero to find all the zeros of the function.

Function	Zero
83. $f(x) = x^3 - 4x^2 + x - 4$	i
84. $h(x) = -x^3 + 2x^2 - 16x + 32$	$-4i$
85. $g(x) = 2x^4 - 3x^3 - 13x^2 + 37x - 15$	$2 + i$
86. $f(x) = 4x^4 - 11x^3 + 14x^2 - 6x$	$1 - i$

In Exercises 87–90, find all the zeros of the function and write the polynomial as a product of linear factors.

87. $f(x) = x^3 + 4x^2 - 5x$

88. $g(x) = x^3 - 7x^2 + 36$

89. $g(x) = x^4 + 4x^3 - 3x^2 + 40x + 208$

90. $f(x) = x^4 + 8x^3 + 8x^2 - 72x - 153$

 In Exercises 91–94, use a graphing utility to (a) graph the function, (b) determine the number of real zeros of the function, and (c) approximate the real zeros of the function to the nearest hundredth.

91. $f(x) = x^4 + 2x + 1$

92. $g(x) = x^3 - 3x^2 + 3x + 2$

93. $h(x) = x^3 - 6x^2 + 12x - 10$

94. $f(x) = x^5 + 2x^3 - 3x - 20$

In Exercises 95 and 96, use Descartes's Rule of Signs to determine the possible numbers of positive and negative zeros of the function.

95. $g(x) = 5x^3 + 3x^2 - 6x + 9$

96. $h(x) = -2x^5 + 4x^3 - 2x^2 + 5$

In Exercises 97 and 98, use synthetic division to verify the upper and lower bounds of the real zeros of f.

97. $f(x) = 4x^3 - 3x^2 + 4x - 3$

  (a) Upper: $x = 1$

  (b) Lower: $x = -\frac{1}{4}$

98. $f(x) = 2x^3 - 5x^2 - 14x + 8$

  (a) Upper: $x = 8$

  (b) Lower: $x = -4$

3.5 **99. *Median Income*** The median incomes I (in thousands of dollars) for married-couple families in the United States from 1995 through 2002 are shown in the table. A linear model that approximates these data is

$$I = 2.09t + 37.2$$

where t represents the year, with $t = 5$ corresponding to 1995. (Source: U.S. Census Bureau)

Year	Median income, I
1995	47.1
1996	49.7
1997	51.6
1998	54.2
1999	56.5
2000	59.1
2001	60.3
2002	61.1

(a) Plot the actual data and the model on the same set of coordinate axes.

(b) How closely does the model represent the data?

 100. *Data Analysis: Electronic Games* The table shows the factory sales S (in millions of dollars) of electronic gaming software in the United States from 1995 through 2003. (Source: Consumer Electronics Association)

Year	Sales, S
1995	3000
1996	3500
1997	3900
1998	4480
1999	5100
2000	5850
2001	6725
2002	7375
2003	7744

(a) Use a graphing utility to create a scatter plot of the data. Let t represent the year, with $t = 5$ corresponding to 1995.

(b) Use the *regression* feature of the graphing utility to find the equation of the least squares regression line that fits the data. Then graph the model and the scatter plot you found in part (a) in the same viewing window. How closely does the model represent the data?

(c) Use the model to estimate the factory sales of electronic gaming software in the year 2008.

(d) Interpret the meaning of the slope of the linear model in the context of the problem.

101. *Measurement* You notice a billboard indicating that it is 2.5 miles or 4 kilometers to the next restaurant of a national fast-food chain. Use this information to find a mathematical model that relates miles to kilometers. Then use the model to find the numbers of kilometers in 2 miles and 10 miles.

102. *Energy* The power P produced by a wind turbine is proportional to the cube of the wind speed S. A wind speed of 27 miles per hour produces a power output of 750 kilowatts. Find the output for a wind speed of 40 miles per hour.

103. *Frictional Force* The frictional force F between the tires and the road required to keep a car on a curved section of a highway is directly proportional to the square of the speed s of the car. If the speed of the car is doubled, the force will change by what factor?

104. *Demand* A company has found that the daily demand x for its boxes of chocolates is inversely proportional to the price p. When the price is $5, the demand is 800 boxes. Approximate the demand when the price is increased to $6.

105. *Travel Time* The travel time between two cities is inversely proportional to the average speed. A train travels between the cities in 3 hours at an average speed of 65 miles per hour. How long would it take to travel between the cities at an average speed of 80 miles per hour?

106. *Cost* The cost of constructing a wooden box with a square base varies jointly as the height of the box and the square of the width of the box. A box of height 16 inches and of width 6 inches costs $28.80. How much would a box of height 14 inches and of width 8 inches cost?

Synthesis

True or False? **In Exercises 107 and 108, determine whether the statement is true or false. Justify your answer.**

107. A fourth-degree polynomial with real coefficients can have -5, $-8i$, $4i$, and 5 as its zeros.

108. If y is directly proportional to x, then x is directly proportional to y.

109. *Writing* Explain how to determine the maximum or minimum value of a quadratic function.

110. *Writing* Explain the connections between factors of a polynomial, zeros of a polynomial function, and solutions of a polynomial equation.

3 | Chapter Test

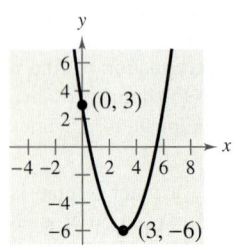

FIGURE FOR **3**

Take this test as you would take a test in class. When you are finished, check your work against the answers given in the back of the book.

1. Describe how the graph of g differs from the graph of $f(x) = x^2$.

    (a) $g(x) = 2 - x^2$ (b) $g(x) = \left(x - \frac{3}{2}\right)^2$

2. Identify the vertex and intercepts of the graph of $y = x^2 + 4x + 3$.

3. Find an equation of the parabola shown in the figure at the left.

4. The path of a ball is given by $y = -\frac{1}{20}x^2 + 3x + 5$, where y is the height (in feet) of the ball and x is the horizontal distance (in feet) from where the ball was thrown.

    (a) Find the maximum height of the ball.

    (b) Which number determines the height at which the ball was thrown? Does changing this value change the coordinates of the maximum height of the ball? Explain.

5. Determine the right-hand and left-hand behavior of the graph of the function $h(t) = -\frac{3}{4}t^5 + 2t^2$. Then sketch its graph.

6. Divide using long division.   **7.** Divide using synthetic division.

$$\frac{3x^3 + 4x - 1}{x^2 + 1}$$ $$\frac{2x^4 - 5x^2 - 3}{x - 2}$$

8. Use synthetic division to show that $x = \sqrt{3}$ is a zero of the function given by

$$f(x) = 4x^3 - x^2 - 12x + 3.$$

Use the result to factor the polynomial function completely and list all the real zeros of the function.

In Exercises 9 and 10, find all the rational zeros of the function.

9. $g(t) = 2t^4 - 3t^3 + 16t - 24$ **10.** $h(x) = 3x^5 + 2x^4 - 3x - 2$

In Exercises 11 and 12, find a polynomial function with real coefficients that has the given zeros. (There are many correct answers.)

11. $0, 3, 3 + i, 3 - i$ **12.** $1 + \sqrt{3}\,i, 1 - \sqrt{3}\,i, 2, 2$

In Exercises 13 and 14, find all the zeros of the function.

13. $f(x) = x^3 + 2x^2 + 5x + 10$ **14.** $f(x) = x^4 - 9x^2 - 22x - 24$

In Exercises 15–17, find a mathematical model that represents the statement. (In each case, determine the constant of proportionality.)

15. v varies directly as the square root of s. ($v = 24$ when $s = 16$.)

16. A varies jointly as x and y. ($A = 500$ when $x = 15$ and $y = 8$.)

17. b varies inversely as a. ($b = 32$ when $a = 1.5$.)

18. The table at the left shows the median salaries S (in thousands of dollars) for hockey players on the Philadelphia Flyers from 2001 through 2004, where $t = 1$ represents 2001. Use the *regression* feature of a graphing utility to find the equation of the least squares regression line that fits the data. How well does the model represent the data? (Source: USA Today)

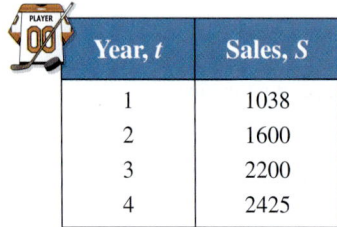

Year, t	Sales, S
1	1038
2	1600
3	2200
4	2425

TABLE FOR **18**

Proofs in Mathematics

These two pages contain proofs of four important theorems about polynomial functions. The first two theorems are from Section 3.3, and the second two theorems are from Section 3.4.

> ## The Remainder Theorem   *(p. 293)*
>
> If a polynomial $f(x)$ is divided by $x - k$, the remainder is
>
> $$r = f(k).$$

Proof

From the Division Algorithm, you have

$$f(x) = (x - k)q(x) + r(x)$$

and because either $r(x) = 0$ or the degree of $r(x)$ is less than the degree of $x - k$, you know that $r(x)$ must be a constant. That is, $r(x) = r$. Now, by evaluating $f(x)$ at $x = k$, you have

$$f(k) = (k - k)q(k) + r$$

$$= (0)q(k) + r = r.$$

To be successful in algebra, it is important that you understand the connection among *factors* of a polynomial, *zeros* of a polynomial function, and *solutions* or *roots* of a polynomial equation. The Factor Theorem is the basis for this connection.

> ## The Factor Theorem   *(p. 293)*
>
> A polynomial $f(x)$ has a factor $(x - k)$ if and only if $f(k) = 0$.

Proof

Using the Division Algorithm with the factor $(x - k)$, you have

$$f(x) = (x - k)q(x) + r(x).$$

By the Remainder Theorem, $r(x) = r = f(k)$, and you have

$$f(x) = (x - k)q(x) + f(k)$$

where $q(x)$ is a polynomial of lesser degree than $f(x)$. If $f(k) = 0$, then

$$f(x) = (x - k)q(x)$$

and you see that $(x - k)$ is a factor of $f(x)$. Conversely, if $(x - k)$ is a factor of $f(x)$, division of $f(x)$ by $(x - k)$ yields a remainder of 0. So, by the Remainder Theorem, you have $f(k) = 0$.

The Linear Factorization Theorem is closely related to the Fundamental Theorem of Algebra. The Fundamental Theorem of Algebra has a long and interesting history. In the early work with polynomial equations, The Fundamental Theorem of Algebra was thought to have been not true, because imaginary solutions were not considered. In fact, in the very early work by mathematicians such as Abu al-Khwarizmi (c. 800 A.D.), negative solutions were also not considered.

Once imaginary numbers were accepted, several mathematicians attempted to give a general proof of the Fundamental Theorem of Algebra. These included Gottfried von Leibniz (1702), Jean d'Alembert (1746), Leonhard Euler (1749), Joseph-Louis Lagrange (1772), and Pierre Simon Laplace (1795). The mathematician usually credited with the first correct proof of the Fundamental Theorem of Algebra is Carl Friedrich Gauss, who published the proof in his doctoral thesis in 1799.

Linear Factorization Theorem   *(p. 298)*

If $f(x)$ is a polynomial of degree n, where $n > 0$, then f has precisely n linear factors

$$f(x) = a_n(x - c_1)(x - c_2) \cdots (x - c_n)$$

where $c_1, c_2, \ldots, c_n$ are complex numbers.

Proof

Using the Fundamental Theorem of Algebra, you know that f must have at least one zero, c_1. Consequently, $(x - c_1)$ is a factor of $f(x)$, and you have

$$f(x) = (x - c_1)f_1(x).$$

If the degree of $f_1(x)$ is greater than zero, you again apply the Fundamental Theorem to conclude that f_1 must have a zero c_2, which implies that

$$f(x) = (x - c_1)(x - c_2)f_2(x).$$

It is clear that the degree of $f_1(x)$ is $n - 1$, that the degree of $f_2(x)$ is $n - 2$, and that you can repeatedly apply the Fundamental Theorem n times until you obtain

$$f(x) = a_n(x - c_1)(x - c_2) \cdots (x - c_n)$$

where a_n is the leading coefficient of the polynomial $f(x)$.

Factors of a Polynomial   *(p. 302)*

Every polynomial of degree $n > 0$ with real coefficients can be written as the product of linear and quadratic factors with real coefficients, where the quadratic factors have no real zeros.

Proof

To begin, you use the Linear Factorization Theorem to conclude that $f(x)$ can be *completely* factored in the form

$$f(x) = d(x - c_1)(x - c_2)(x - c_3) \cdots (x - c_n).$$

If each c_i is real, there is nothing more to prove. If any c_i is complex ($c_i = a + bi$, $b \neq 0$), then, because the coefficients of $f(x)$ are real, you know that the conjugate $c_j = a - bi$ is also a zero. By multiplying the corresponding factors, you obtain

$$(x - c_i)(x - c_j) = [x - (a + bi)][x - (a - bi)]$$
$$= x^2 - 2ax + (a^2 + b^2)$$

where each coefficient is real.

This collection of thought-provoking and challenging exercises further explores and expands upon concepts learned in this chapter.

1. (a) Find the zeros of each quadratic function $g(x)$.

   (i) $g(x) = x^2 - 4x - 12$

   (ii) $g(x) = x^2 + 5x$

   (iii) $g(x) = x^2 + 3x - 10$

   (iv) $g(x) = x^2 - 4x + 4$

   (v) $g(x) = x^2 - 2x - 6$

   (vi) $g(x) = x^2 + 3x + 4$

    (b) For each function in part (a), use a graphing utility to graph $f(x) = (x - 2) \cdot g(x)$. Verify that $(2, 0)$ is an x-intercept of the graph of $f(x)$. Describe any similarities or differences in the behavior of the six functions at this x-intercept.

    (c) For each function in part (b), use the graph of $f(x)$ to approximate the other x-intercepts of the graph.

    (d) Describe the connections that you find among the results of parts (a), (b), and (c).

2. Quonset huts were developed during World War II. They were temporary housing structures that could be assembled quickly and easily. A Quonset hut is shaped like a half cylinder. A manufacturer has 600 square feet of material with which to build a Quonset hut.

   (a) The formula for the surface area of half a cylinder is $S = \pi r^2 + \pi r l$, where r is the radius and l is the length of the hut. Solve this equation for l when $S = 600$.

   (b) The formula for the volume of the hut is $V = \frac{1}{2}\pi r^2 l$. Write the volume V of the Quonset hut as a polynomial function of r.

   (c) Use the function you wrote in part (b) to find the maximum volume of a Quonset hut with a surface area of 600 square feet. What are the dimensions of the hut?

3. Show that if $f(x) = ax^3 + bx^2 + cx + d$ then $f(k) = r$, where $r = ak^3 + bk^2 + ck + d$ using long division. In other words, verify the Remainder Theorem for a third-degree polynomial function.

4. In 2000 B.C., the Babylonians solved polynomial equations by referring to tables of values. One such table gave the values of $y^3 + y^2$. To be able to use this table, the Babylonians sometimes had to manipulate the equation as shown below.

   $ax^3 + bx^2 = c$ Original equation

   $\dfrac{a^3 x^3}{b^3} + \dfrac{a^2 x^2}{b^2} = \dfrac{a^2 c}{b^3}$ Multiply each side by $\dfrac{a^2}{b^3}$.

   $\left(\dfrac{ax}{b}\right)^3 + \left(\dfrac{ax}{b}\right)^2 = \dfrac{a^2 c}{b^3}$ Rewrite.

Then they would find $(a^2 c)/b^3$ in the $y^3 + y^2$ column of the table. Because they knew that the corresponding y-value was equal to $(ax)/b$, they could conclude that $x = (by)/a$.

   (a) Calculate $y^3 + y^2$ for $y = 1, 2, 3, \ldots, 10$. Record the values in a table.

   Use the table from part (a) and the method above to solve each equation.

   (b) $x^3 + x^2 = 252$

   (c) $x^3 + 2x^2 = 288$

   (d) $3x^3 + x^2 = 90$

   (e) $2x^3 + 5x^2 = 2500$

   (f) $7x^3 + 6x^2 = 1728$

   (g) $10x^3 + 3x^2 = 297$

   Using the methods from this chapter, verify your solution to each equation.

5. At a glassware factory, molten cobalt glass is poured into molds to make paperweights. Each mold is a rectangular prism whose height is 3 inches greater than the length of each side of the square base. A machine pours 20 cubic inches of liquid glass into each mold. What are the dimensions of the mold?

6. (a) Complete the table.

Function	Zeros	Sum of zeros	Product of zeros
$f_1(x) = x^2 - 5x + 6$			
$f_2(x) = x^3 - 7x + 6$			
$f_3(x) = x^4 + 2x^3 + x^2 + 8x - 12$			
$f_4(x) = x^5 - 3x^4 - 9x^3 + 25x^2 - 6x$			

   (b) Use the table to make a conjecture relating the sum of the zeros of a polynomial function to the coefficients of the polynomial function.

   (c) Use the table to make a conjecture relating the product of the zeros of a polynomial function to the coefficients of the polynomial function.

333

7. Determine whether the statement is true or false. If false, provide one or more reasons why the statement is false and correct the statement. Let $f(x) = ax^3 + bx^2 + cx + d$, $a \neq 0$, and let $f(2) = -1$. Then

$$\frac{f(x)}{x+1} = q(x) + \frac{2}{x+1}$$

where $q(x)$ is a second-degree polynomial.

8. The parabola shown in the figure has an equation of the form $y = ax^2 + bx + c$. Find the equation for this parabola by the following methods. (a) Find the equation analytically. (b) Use the *regression* feature of a graphing utility to find the equation.

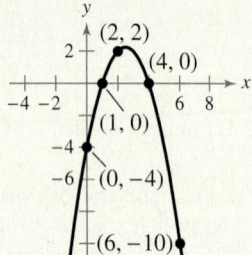

9. One of the fundamental themes of calculus is to find the slope of the tangent line to a curve at a point. To see how this can be done, consider the point $(2, 4)$ on the graph of the quadratic function $f(x) = x^2$.

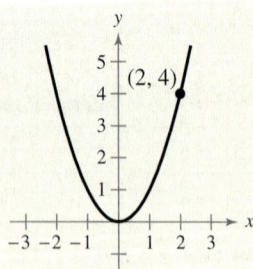

(a) Find the slope of the line joining $(2, 4)$ and $(3, 9)$. Is the slope of the tangent line at $(2, 4)$ greater than or less than the slope of the line through $(2, 4)$ and $(3, 9)$?

(b) Find the slope of the line joining $(2, 4)$ and $(1, 1)$. Is the slope of the tangent line at $(2, 4)$ greater than or less than the slope of the line through $(2, 4)$ and $(1, 1)$?

(c) Find the slope of the line joining $(2, 4)$ and $(2.1, 4.41)$. Is the slope of the tangent line at $(2, 4)$ greater than or less than the slope of the line through $(2, 4)$ and $(2.1, 4.41)$?

(d) Find the slope of the line joining $(2, 4)$ and $(2 + h, f(2 + h))$ in terms of the nonzero number h.

(e) Evaluate the slope formula from part (d) for $h = -1$, 1, and 0.1. Compare these values with those in parts (a)–(c).

(f) What can you conclude the slope of the tangent line at $(2, 4)$ to be? Explain your answer.

10. A rancher plans to fence a rectangular pasture adjacent to a river (see figure). The rancher has 100 meters of fencing, and no fencing is needed along the river.

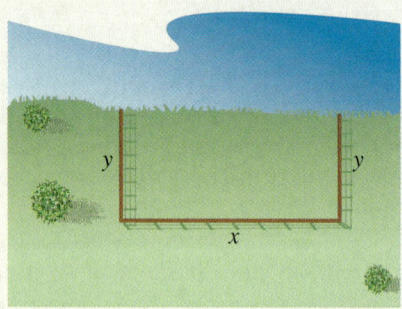

(a) Write the are A as a function of x, the length of the side of the pasture parallel to the river. What is the domain of $A(x)$?

(b) Graph the function $A(x)$ and estimate the dimensions that yield the maximum area of the pasture.

(c) Find the exact dimensions that yield the maximum area of the pasture by writing the quadratic function in standard form.

11. A wire 100 centimeters in length is cut into two pieces. One piece is bent to form a square and the other to form a circle. Let x equal the length of the wire used to form the square.

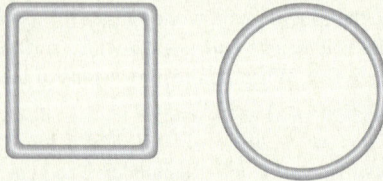

(a) Write the function that represents the combined area of the two figures.

(b) Determine the domain of the function.

(c) Find the value(s) of x that yield a maximum area and a minimum area.

(d) Explain your reasoning.

12. Find a formula for the polynomial division: $\dfrac{x^n - 1}{x - 1}$.

334

Rational Functions and Conics

4

The nine planets move about the sun in elliptical orbits. You can use the techniques presented in this chapter to determine the distances between the planets and the center of the sun.

Kauko Helavuo/Getty Images

SELECTED APPLICATIONS

Rational functions and conics have many real-life applications. The applications listed below represent a small sample of the applications in this chapter.

4.1 Rational Functions and Asymptotes

Introduction

A **rational function** can be written in the form

$$f(x) = \frac{N(x)}{D(x)}$$

where $N(x)$ and $D(x)$ are polynomials and $D(x)$ is not the zero polynomial.

In general, the *domain* of a rational function of x includes all real numbers except x-values that make the denominator zero. Much of the discussion of rational functions will focus on their graphical behavior near the x-values excluded from the domain.

Example 1 **Finding the Domain of a Rational Function**

Find the domain of $f(x) = \dfrac{1}{x}$ and discuss the behavior of f near any excluded x-values.

Solution

Because the denominator is zero when $x = 0$, the domain of f is all real numbers except $x = 0$. To determine the behavior of f near this excluded value, evaluate $f(x)$ to the left and right of $x = 0$, as indicated in the following tables.

x	-1	-0.5	-0.1	-0.01	-0.001	$\longrightarrow 0$
$f(x)$	-1	-2	-10	-100	-1000	$\longrightarrow -\infty$

x	$0 \longleftarrow$	0.001	0.01	0.1	0.5	1
$f(x)$	$\infty \longleftarrow$	1000	100	10	2	1

Note that as x approaches 0 *from the left*, $f(x)$ decreases without bound. In contrast, as x approaches 0 *from the right*, $f(x)$ increases without bound. The graph of f is shown in Figure 4.1.

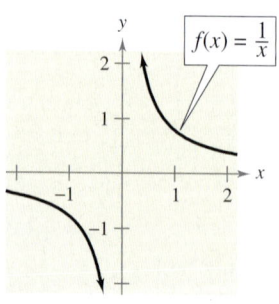

$f(x) = \dfrac{1}{x}$

FIGURE 4.1

✔CHECKPOINT   Now try Exercise 1.

Horizontal and Vertical Asymptotes

In Example 1, the behavior of f near $x = 0$ is denoted as follows.

$$f(x) \longrightarrow -\infty \text{ as } x \longrightarrow 0^- \qquad f(x) \longrightarrow \infty \text{ as } x \longrightarrow 0^+$$

$f(x)$ decreases without bound as x approaches 0 from the left. $f(x)$ increases without bound as x approaches 0 from the right.

The line $x = 0$ is a **vertical asymptote** of the graph of f, as shown in Figure 4.2. From this figure, you can see that the graph of f also has a **horizontal asymptote**— the line $y = 0$. This means that the values of $f(x) = 1/x$ approach zero as x increases or decreases without bound.

$$f(x) \longrightarrow 0 \text{ as } x \longrightarrow -\infty \qquad f(x) \longrightarrow 0 \text{ as } x \longrightarrow \infty$$

$f(x)$ approaches 0 as x decreases without bound. $f(x)$ approaches 0 as x increases without bound.

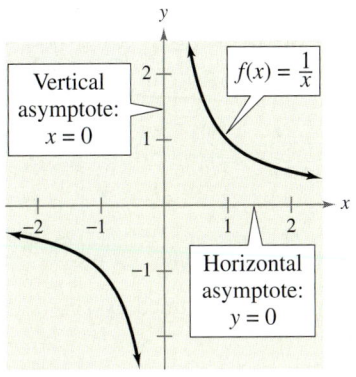

Vertical asymptote: $x = 0$

$f(x) = \frac{1}{x}$

Horizontal asymptote: $y = 0$

FIGURE 4.2

Definitions of Vertical and Horizontal Asymptotes

1. The line $x = a$ is a **vertical asymptote** of the graph of f if

$$f(x) \longrightarrow \infty \quad \text{or} \quad f(x) \longrightarrow -\infty$$

as $x \longrightarrow a$, either from the right or from the left.

2. The line $y = b$ is a **horizontal asymptote** of the graph of f if

$$f(x) \longrightarrow b$$

as $x \longrightarrow \infty$ or $x \longrightarrow -\infty$.

Eventually (as $x \longrightarrow \infty$ or $x \longrightarrow -\infty$), the distance between the horizontal asymptote and the points on the graph must approach zero. Figure 4.3 shows the horizontal and vertical asymptotes of the graphs of three rational functions.

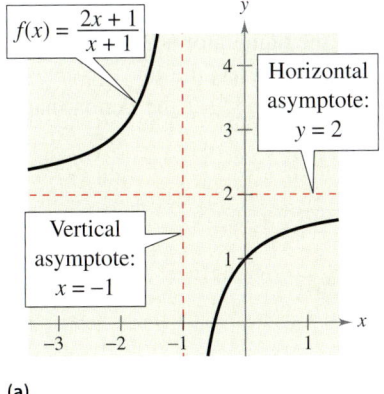

$f(x) = \dfrac{2x + 1}{x + 1}$

Horizontal asymptote: $y = 2$

Vertical asymptote: $x = -1$

(a)

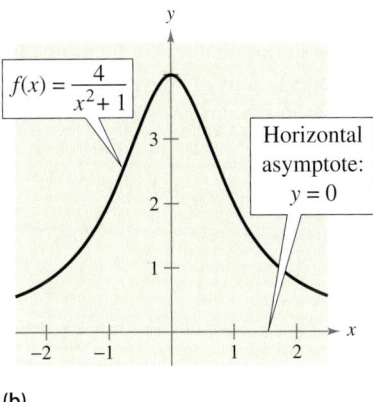

$f(x) = \dfrac{4}{x^2 + 1}$

Horizontal asymptote: $y = 0$

(b)

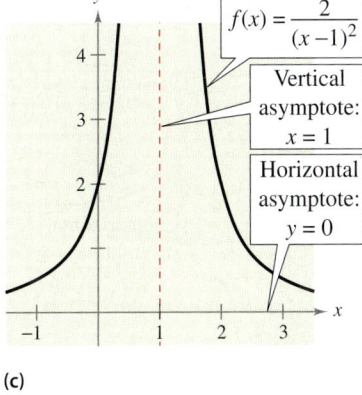

$f(x) = \dfrac{2}{(x-1)^2}$

Vertical asymptote: $x = 1$

Horizontal asymptote: $y = 0$

(c)

FIGURE 4.3

The graphs of $f(x) = 1/x$ in Figure 4.2 and $f(x) = (2x + 1)/(x + 1)$ in Figure 4.3(a) are **hyperbolas.** You will study hyperbolas in Sections 4.3 and 4.4.

Asymptotes of a Rational Function

Let f be the rational function given by

$$f(x) = \frac{N(x)}{D(x)} = \frac{a_n x^n + a_{n-1} x^{n-1} + \cdots + a_1 x + a_0}{b_m x^m + b_{m-1} x^{m-1} + \cdots + b_1 x + b_0}$$

where $N(x)$ and $D(x)$ have no common factors.

1. The graph of f has *vertical* asymptotes at the zeros of $D(x)$.

2. The graph of f has one or no *horizontal* asymptote determined by comparing the degrees of $N(x)$ and $D(x)$.

   a. If $n < m$, the graph of f has the line $y = 0$ (the x-axis) as a horizontal asymptote.

   b. If $n = m$, the graph of f has the line $y = a_n/b_m$ (ratio of the leading coefficients) as a horizontal asymptote.

   c. If $n > m$, the graph of f has no horizontal asymptote.

Example 2   Finding Horizontal and Vertical Asymptotes

Find all horizontal and vertical asymptotes of the graph of each rational function.

a. $f(x) = \dfrac{2x}{3x^2 + 1}$ **b.** $f(x) = \dfrac{2x^2}{x^2 - 1}$

Solution

a. For this rational function, the degree of the numerator is *less than* the degree of the denominator, so the graph has the line $y = 0$ as a horizontal asymptote. To find any vertical asymptotes, set the denominator equal to zero and solve the resulting equation for x. Because the equation $3x^2 + 1 = 0$ has no real solutions, you can conclude that the graph has no vertical asymptote. The graph of the function is shown in Figure 4.4.

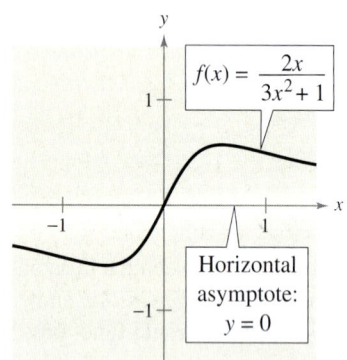

$$f(x) = \frac{2x}{3x^2 + 1}$$

Horizontal asymptote: $y = 0$

FIGURE **4.4**

b. For this rational function, the degree of the numerator is *equal to* the degree of the denominator. The leading coefficient of the numerator is 2 and the leading coefficient of the denominator is 1, so the graph has the line $y = 2$ as a horizontal asymptote. To find any vertical asymptotes, set the denominator equal to zero and solve the resulting equation for x.

$$x^2 - 1 = 0 \qquad \text{Set denominator equal to zero.}$$

$$(x + 1)(x - 1) = 0 \qquad \text{Factor.}$$

$$x + 1 = 0 \quad \Longrightarrow \quad x = -1 \qquad \text{Set 1st factor equal to 0.}$$

$$x - 1 = 0 \quad \Longrightarrow \quad x = 1 \qquad \text{Set 2nd factor equal to 0.}$$

This equation has two real solutions $x = -1$ and $x = 1$, so the graph has the lines $x = -1$ and $x = 1$ as vertical asymptotes. The graph of the function is shown in Figure 4.5.

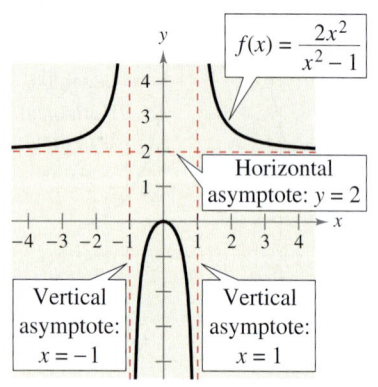

$$f(x) = \frac{2x^2}{x^2 - 1}$$

Horizontal asymptote: $y = 2$

Vertical asymptote: $x = -1$

Vertical asymptote: $x = 1$

FIGURE **4.5**

✓CHECKPOINT   Now try Exercise 9.

| Example 3 | **Finding Horizontal and Vertical Asymptotes** |

Find all horizontal and vertical asymptotes of the graph of $f(x) = \dfrac{x^2 + x - 2}{x^2 - x - 6}$.

Solution

For this rational function, the degree of the numerator is *equal to* the degree of the denominator. The leading coefficient of both the numerator and denominator is 1, so the graph has the line $y = 1$ as a horizontal asymptote. To find any vertical asymptotes, first factor the numerator and denominator as follows.

$$f(x) = \frac{x^2 + x - 2}{x^2 - x - 6} = \frac{(x - 1)(x + 2)}{(x + 2)(x - 3)} = \frac{x - 1}{x - 3}, \quad x \neq 2$$

By setting the denominator $x - 3$ (of the simplified function) equal to zero, you can determine that the graph has the line $x = 3$ as a vertical asymptote.

✓**CHECKPOINT**   Now try Exercise 25.

Applications

There are many examples of asymptotic behavior in real life. For instance, Example 4 shows how a vertical asymptote can be used to analyze the cost of removing pollutants from smokestack emissions.

| Example 4 | **Cost-Benefit Model** | |

A utility company burns coal to generate electricity. The cost C (in dollars) of removing $p\%$ of the smokestack pollutants is given by $C = 80{,}000p/(100 - p)$ for $0 \leq p < 100$. Sketch the graph of this function. You are a member of a state legislature considering a law that would require utility companies to remove 90% of the pollutants from their smokestack emissions. The current law requires 85% removal. How much additional cost would the utility company incur as a result of the new law?

Solution

The graph of this function is shown in Figure 4.6. Note that the graph has a vertical asymptote at $p = 100$. Because the current law requires 85% removal, the current cost to the utility company is

$$C = \frac{80{,}000(85)}{100 - 85} \approx \$453{,}333. \qquad \text{Evaluate } C \text{ when } p = 85.$$

If the new law increases the percent removal to 90%, the cost will be

$$C = \frac{80{,}000(90)}{100 - 90} = \$720{,}000. \qquad \text{Evaluate } C \text{ when } p = 90.$$

So, the new law would require the utility company to spend an additional

$$720{,}000 - 453{,}333 = \$266{,}667. \qquad \begin{array}{l}\text{Subtract 85\% removal cost}\\ \text{from 90\% removal cost.}\end{array}$$

✓**CHECKPOINT**   Now try Exercise 45.

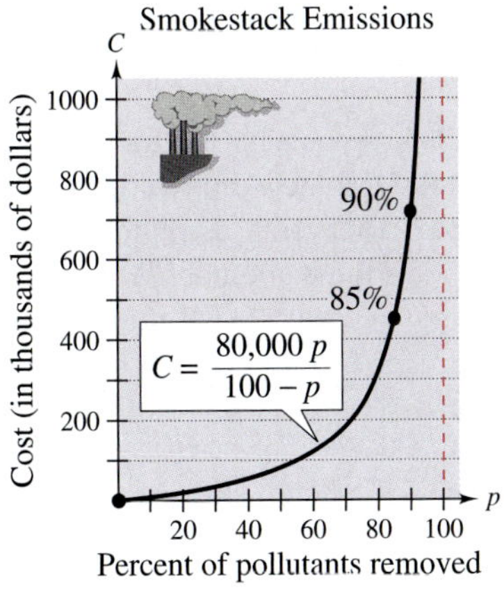

Smokestack Emissions

Cost (in thousands of dollars)

$C = \dfrac{80{,}000\,p}{100 - p}$

90%

85%

Percent of pollutants removed

FIGURE 4.6

Ultraviolet Radiation

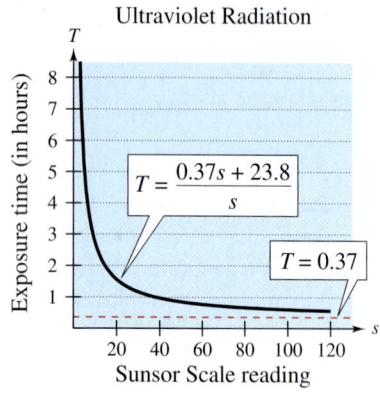

$$T = \frac{0.37s + 23.8}{s}$$

$T = 0.37$

FIGURE 4.7

Example 5 **Ultraviolet Radiation**

For a person with sensitive skin, the amount of time T (in hours) the person can be exposed to the sun with minimal burning can be modeled by

$$T = \frac{0.37s + 23.8}{s}, \qquad 0 < s \le 120$$

where s is the Sunsor Scale reading. The Sunsor Scale is based on the level of intensity of UVB rays. (Source: Sunsor, Inc.)

a. Find the amounts of time a person with sensitive skin can be exposed to the sun with minimal burning when $s = 10$, $s = 25$, and $s = 100$.

b. If the model were valid for all $s > 0$, what would be the horizontal asymptote of this function, and what would it represent?

Solution

a. When $s = 10$, $T = \dfrac{0.37(10) + 23.8}{10}$

$$= 2.75 \text{ hours.}$$

When $s = 25$, $T = \dfrac{0.37(25) + 23.8}{25}$

$$\approx 1.32 \text{ hours.}$$

When $s = 100$, $T = \dfrac{0.37(100) + 23.8}{100}$

$$\approx 0.61 \text{ hour.}$$

b. As shown in Figure 4.7, the horizontal asymptote is the line $T = 0.37$. This line represents the shortest possible exposure time with minimal burning.

✓CHECKPOINT Now try Exercise 41.

---

*W*RITING ABOUT *M*ATHEMATICS

Asymptotes of Graphs of Rational Functions Do you think it is possible for the graph of a rational function to cross its horizontal asymptote? If so, how can you determine when the graph of a rational function will cross its horizontal asymptote? Use the graphs of the following functions to investigate these questions. Write a summary of your conclusions. Explain your reasoning.

a. $f(x) = \dfrac{x}{x^2 + 1}$

b. $g(x) = \dfrac{x}{x^2 - 3}$

c. $h(x) = \dfrac{x^2}{2x^3 - x}$

VOCABULARY CHECK: Fill in the blanks.

1. Functions of the form $f(x) = N(x)/D(x)$, where $N(x)$ and $D(x)$ are polynomials and $D(x)$ is not the zero polynomial, are called _____ _____.

2. If $f(x) \to \pm\infty$ as $x \to a$ from the left or the right, then $x = a$ is a _____ _____ of the graph of f.

3. If $f(x) \to b$ as $x \to \pm\infty$, then $y = b$ is a _____ _____ of the graph of f.

PREREQUISITE SKILLS REVIEW: Practice and review algebra skills needed for this section at **www.Eduspace.com.**

In Exercises 1–4, (a) complete each table for the function, (b) determine the vertical and horizontal asymptotes of the graph of the function, and (c) find the domain of the function.

x	$f(x)$
0.5	
0.9	
0.99	
0.999	

x	$f(x)$
1.5	
1.1	
1.01	
1.001	

x	$f(x)$
5	
10	
100	
1000	

1. $f(x) = \dfrac{1}{x-1}$

2. $f(x) = \dfrac{5x}{x-1}$

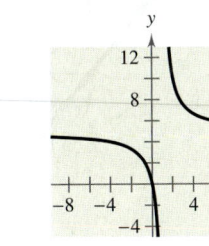

3. $f(x) = \dfrac{3x^2}{x^2-1}$

4. $f(x) = \dfrac{4x}{x^2-1}$

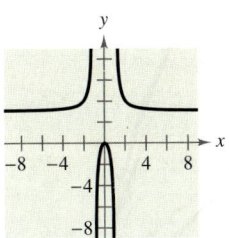
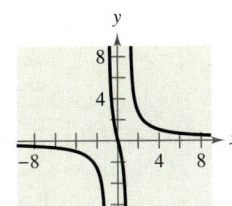

In Exercises 5–12, find the domain of the function and identify any horizontal and vertical asymptotes.

5. $f(x) = \dfrac{1}{x^2}$

6. $f(x) = \dfrac{4}{(x-2)^3}$

7. $f(x) = \dfrac{2+x}{2-x}$

8. $f(x) = \dfrac{1-5x}{1+2x}$

9. $f(x) = \dfrac{x^3}{x^2-1}$

10. $f(x) = \dfrac{2x^2}{x+1}$

11. $f(x) = \dfrac{3x^2+1}{x^2+x+9}$

12. $f(x) = \dfrac{3x^2+x-5}{x^2+1}$

In Exercises 13–16, match the rational function with its graph. [The graphs are labeled (a), (b), (c), and (d).]

(a)
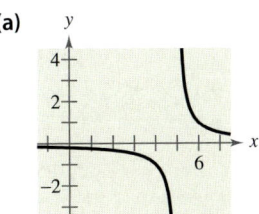

(b)
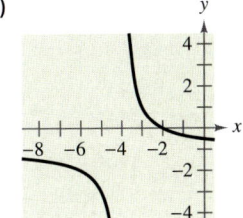

(c)
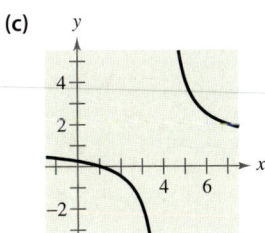

(d)
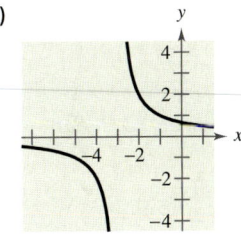

13. $f(x) = \dfrac{2}{x+3}$

14. $f(x) = \dfrac{1}{x-5}$

15. $f(x) = \dfrac{x-1}{x-4}$

16. $f(x) = -\dfrac{x+2}{x+4}$

In Exercises 17–24, find the zeros (if any) of the rational function.

17. $g(x) = \dfrac{x^2-1}{x+1}$

18. $f(x) = \dfrac{x^2-2}{x-3}$

19. $h(x) = 2 + \dfrac{5}{x^2+2}$

20. $f(x) = 1 + \dfrac{3}{x^2-4}$

21. $f(x) = 1 - \dfrac{3}{x-3}$

22. $g(x) = 4 - \dfrac{2}{x+5}$

23. $g(x) = \dfrac{x^3-8}{x^2+1}$

24. $f(x) = \dfrac{x^3-1}{x^2+6}$

In Exercises 25–32, find the domain of the function and identify any horizontal and vertical asymptotes.

25. $f(x) = \dfrac{x - 4}{x^2 - 16}$

26. $f(x) = \dfrac{x + 3}{x^2 - 9}$

27. $f(x) = \dfrac{x^2 - 1}{x^2 - 2x - 3}$

28. $f(x) = \dfrac{x^2 - 4}{x^2 - 3x + 2}$

29. $f(x) = \dfrac{x^2 - 3x - 4}{2x^2 + x - 1}$

30. $f(x) = \dfrac{x^2 + x - 2}{2x^2 + 5x + 2}$

31. $f(x) = \dfrac{6x^2 + 5x - 6}{3x^2 - 8x + 4}$

32. $f(x) = \dfrac{6x^2 - 11x + 3}{6x^2 - 7x - 3}$

Analytical and Numerical Analysis In Exercises 33–36, (a) determine the domains of f and g, (b) simplify f and find any vertical asymptotes of f, (c) complete the table, and (d) explain how the two functions differ.

33. $f(x) = \dfrac{x^2 - 4}{x + 2}, \quad g(x) = x - 2$

x	-4	-3	-2.5	-2	-1.5	-1	0
$f(x)$							
$g(x)$							

34. $f(x) = \dfrac{x^2(x + 3)}{x^2 + 3x}, \quad g(x) = x$

x	-3	-2	-1	0	1	2	3
$f(x)$							
$g(x)$							

35. $f(x) = \dfrac{2x - 1}{2x^2 - x}, \quad g(x) = \dfrac{1}{x}$

x	-1	-0.5	0	0.5	2	3	4
$f(x)$							
$g(x)$							

36. $f(x) = \dfrac{2x - 8}{x^2 - 9x + 20}, \quad g(x) = \dfrac{2}{x - 5}$

x	0	1	2	3	4	5	6
$f(x)$							
$g(x)$							

Exploration In Exercises 37–40, (a) determine the value that the function f approaches as the magnitude of x increases. Is $f(x)$ greater than or less than this functional value when (b) x is positive and large in magnitude and (c) x is negative and large in magnitude?

37. $f(x) = 4 - \dfrac{1}{x}$

38. $f(x) = 2 + \dfrac{1}{x - 3}$

39. $f(x) = \dfrac{2x - 1}{x - 3}$

40. $f(x) = \dfrac{2x - 1}{x^2 + 1}$

Data Analysis: Physics Experiment In Exercises 41 and 42, consider a physics laboratory experiment designed to determine an unknown mass. A flexible metal meter stick is clamped to a table with 50 centimeters overhanging the edge (see figure). Known masses M ranging from 200 grams to 2000 grams are attached to the end of the meter stick. For each mass, the meter stick is displaced vertically and then allowed to oscillate. The average time t of one oscillation (in seconds) for each mass is recorded in the table.

Mass, M	Time, t
200	0.450
400	0.597
600	0.721
800	0.831
1000	0.906
1200	1.003
1400	1.008
1600	1.168
1800	1.218
2000	1.338

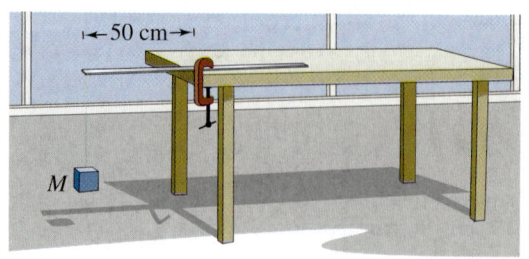

41. A model for the data that can be used to predict the time of one oscillation is

$$t = \frac{38M + 16{,}965}{10(M + 5000)}.$$

(a) Use this model to create a table showing the predicted time for each of the masses shown in the table.

(b) Compare the predicted times with the experimental times. What can you conclude?

42. Use the model in Exercise 41 to approximate the mass of an object for which $t = 1.056$ seconds.

43. *Pollution*   The cost C (in millions of dollars) of removing $p\%$ of the industrial and municipal pollutants discharged into a river is given by

$$C = \frac{255p}{100 - p}, \quad 0 \le p < 100.$$

(a) Use a graphing utility to graph the cost function.

(b) Find the costs of removing 10%, 40%, and 75% of the pollutants.

(c) According to this model, would it be possible to remove 100% of the pollutants? Explain.

44. *Recycling*   In a pilot project, a rural township is given recycling bins for separating and storing recyclable products. The cost C (in dollars) for supplying bins to $p\%$ of the population is given by

$$C = \frac{25{,}000p}{100 - p}, \quad 0 \le p < 100.$$

(a) Use a graphing utility to graph the cost function.

(b) Find the costs of supplying bins to 15%, 50%, and 90% of the population.

(c) According to this model, would it be possible to supply bins to 100% of the residents? Explain.

45. *Population Growth*   The game commission introduces 100 deer into newly acquired state game lands. The population N of the herd is modeled by

$$N = \frac{20(5 + 3t)}{1 + 0.04t}, \quad t \ge 0$$

where t is the time in years (see figure).

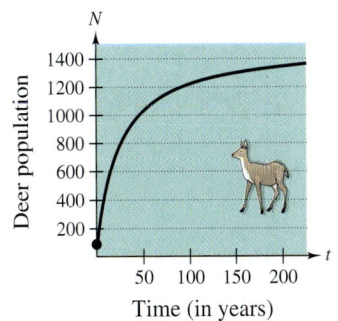

Time (in years)

(a) Find the populations when $t = 5$, $t = 10$, and $t = 25$.

(b) What is the limiting size of the herd as time increases?

46. *Food Consumption*   A biology class performs an experiment comparing the quantity of food consumed by a certain kind of moth with the quantity supplied. The model for the experimental data is given by

$$y = \frac{1.568x - 0.001}{6.360x + 1}, \quad x > 0$$

where x is the quantity (in milligrams) of food supplied and y is the quantity (in milligrams) of food consumed (see figure). At what level of consumption will the moth become satiated?

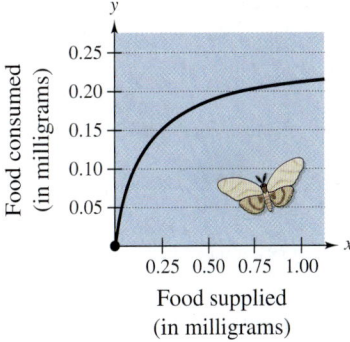

Food supplied
(in milligrams)

47. *Human Memory Model*   Psychologists have developed mathematical models to predict memory performance as a function of the number of trials n of a certain task. Consider the learning curve

$$P = \frac{0.5 + 0.9(n - 1)}{1 + 0.9(n - 1)}, \quad n > 0$$

where P is the fraction of correct responses after n trials.

(a) Complete the table for this model. What does it suggest?

n	1	2	3	4	5	6	7	8	9	10
P										

(b) According to this model, what is the limiting percent of correct responses as n increases?

48. *Human Memory Model*   How would the limiting percent of correct responses change if the human memory model in Exercise 47 were changed to

$$P = \frac{0.5 + 0.6(n - 1)}{1 + 0.8(n - 1)}, \quad n > 0?$$

Model It

49. Data Analysis: DVD Sales The values D (in millions of dollars) of DVDs shipped in the United States in the years 1998 through 2003 are shown in the table. (Source: Recording Industry Association of America)

Year	Value, D
1998	12.2
1999	66.3
2000	80.3
2001	190.7
2002	236.3
2003	369.6

A model for these data is given by

$$D = \frac{1.391t^2 - 73.37}{-0.003t^2 + 1.00}, \quad 8 \le t \le 13$$

where t represents the year, with $t = 8$ corresponding to 1998.

 (a) Use a graphing utility to plot the data and graph the model in the same viewing window. How well does the model fit the data?

(b) Use the model to estimate the value of DVDs shipped in 2006. Is this estimate reasonable?

(c) Would this model be useful for predicting the value of DVDs shipped after 2003? Explain.

 (d) Use the *regression* feature of a graphing utility to find linear and quadratic models for the data.

 (e) Use your models from part (d) to estimate the value of DVDs shipped in 2006. How do these values compare with your answer to part (b)?

(f) Which model do you think is the best fit for the data and the best for predicting future values of DVDs shipped?

50. Sales The sales S (in millions of dollars) for the Yankee Candle Company in the years 1998 through 2003 are shown in the table. (Source: The Yankee Candle Company)

1998 184.5	1999 256.6	2000 338.8
2001 379.8	2002 444.8	2003 508.6

A model for these data is given by

$$S = \frac{5.816t^2 - 130.68}{0.004t^2 + 1.00}, \quad 8 \le t \le 13$$

where t represents the year, with $t = 8$ corresponding to 1998.

 (a) Use a graphing utility to plot the data and graph the model in the same viewing window. How well does the model fit the data?

(b) Use the model to estimate the sales for the Yankee Candle Company in 2008.

(c) Would this model be useful for estimating sales after 2008? Explain.

Synthesis

True or False? In Exercises 51 and 52, determine whether the statement is true or false. Justify your answer.

51. A polynomial function can have infinitely many vertical asymptotes.

52. $f(x) = x^3 - 2x^2 - 5x + 6$ is a rational function.

Think About It In Exercises 53–56, write a rational function f that has the specified characteristics. (There are many correct answers.)

53. Vertical asymptote: None

Horizontal asymptote: $y = 2$

54. Vertical asymptotes: $x = 0, x = \frac{5}{2}$

Horizontal asymptote: $y = -3$

55. Vertical asymptote: $x = -2, x = 1$

Horizontal asymptote: None

56. Vertical asymptote: $x = 3$

Horizontal asymptote: x-axis

57. Think About It Give an example of a rational function whose domain is the set of all real numbers. Give an example of a rational function whose domain is the set of all real numbers except $x = 20$.

58. Writing Describe what is meant by an asymptote of a graph.

Skills Review

In Exercises 59 and 60, find the inverse function of f. Then graph both f and f^{-1} in the same coordinate plane.

59. $f(x) = 8x - 7$

60. $f(x) = \frac{1}{9}x$

In Exercises 61–64, divide using long division.

61. $(x^2 + 5x + 6) \div (x - 4)$

62. $(x^2 - 10x + 15) \div (x - 3)$

63. $(2x^2 + x - 11) \div (x + 5)$

64. $(4x^2 + 3x - 10) \div (x + 6)$

<table>
<tr><td>**4.2**</td><td>**Graphs of Rational Functions**</td></tr>
</table>

What you should learn

- Analyze and sketch graphs of rational functions.
- Sketch graphs of rational functions that have slant asymptotes.
- Use graphs of rational functions to model and real-life problems.

Why you should learn it

You can use rational functions to model average speed over a distance. For example, see Exercise 83 on page 353.

Mike Powell/Getty Images

Analyzing Graphs of Rational Functions

To sketch the graph of a rational function, use the following guidelines.

Guidelines for Analyzing Graphs of Rational Functions

Let $f(x) = N(x)/D(x)$, where $N(x)$ and $D(x)$ are polynomials.

1. Simplify f, if possible.

2. Find and plot the y-intercept (if any) by evaluating $f(0)$.

3. Find the zeros of the numerator (if any) by solving the equation $N(x) = 0$. Then plot the corresponding x-intercepts.

4. Find the zeros of the denominator (if any) by solving the equation $D(x) = 0$. Then sketch the corresponding vertical asymptotes.

5. Find and sketch the horizontal asymptote (if any) by using the rule for finding the horizontal asymptote of a rational function.

6. Plot at least one point *between* and one point *beyond* each x-intercept and vertical asymptote.

7. Use smooth curves to complete the graph between and beyond the vertical asymptotes.

You may also want to test for symmetry when graphing rational functions, especially for simple rational functions. Recall from Section 2.4 that the graph of $f(x) - 1/x$ is symmetric with respect to the origin.

Technology

Some graphing utilities have difficulty graphing rational functions that have vertical asymptotes. Often, the utility will connect parts of the graph that are not supposed to be connected. For instance, the screen below on the left shows the graph of $f(x) = 1/(x - 2)$. Notice that the graph should consist of two unconnected portions—one to the left of $x = 2$ and the other to the right of $x = 2$. To eliminate this problem, you can try changing the mode of the graphing utility to *dot mode*. The problem with this is that the graph is then represented as a collection of dots (as shown in the screen on the right) rather than as a smooth curve.

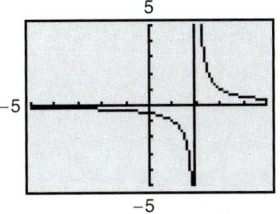

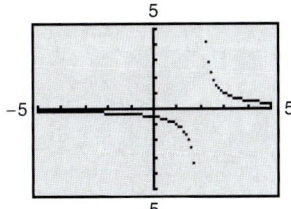

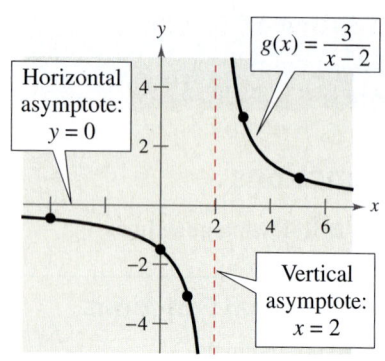

Horizontal asymptote: $y = 0$

$g(x) = \dfrac{3}{x-2}$

Vertical asymptote: $x = 2$

FIGURE **4.8**

Example 1   **Sketching the Graph of a Rational Function**

Sketch the graph of $g(x) = \dfrac{3}{x-2}$ and state its domain.

Solution

y-intercept:	$\left(0, -\frac{3}{2}\right)$, because $g(0) = -\frac{3}{2}$
x-intercept:	None, because $3 \neq 0$
Vertical asymptote:	$x = 2$, zero of denominator
Horizontal asymptote:	$y = 0$, because degree of $N(x) <$ degree of $D(x)$
Additional points:	

x	-4	1	2	3	5
$g(x)$	-0.5	-3	Undefined	3	1

By plotting the intercepts, asymptotes, and a few additional points, you can obtain the graph shown in Figure 4.8. The domain of g is all real numbers except $x = 2$.

✓**CHECKPOINT**   Now try Exercise 13.

The graph of g in Example 1 is a vertical stretch and a right shift of the graph of $f(x) = 1/x$, because

$$g(x) = \frac{3}{x-2}$$

$$= 3\left(\frac{1}{x-2}\right)$$

$$= 3f(x-2).$$

STUDY TIP

Note in the examples in this section that the vertical asymptotes are included in the table of additional points. This is done to emphasize numerically the behavior of the graph of the function.

Example 2   **Sketching the Graph of a Rational Function**

Sketch the graph of $f(x) = \dfrac{2x-1}{x}$ and state its domain.

Solution

y-intercept:	None, because $x = 0$ is not in the domain
x-intercept:	$\left(\frac{1}{2}, 0\right)$, because $f\left(\frac{1}{2}\right) = 0$
Vertical asymptote:	$x = 0$, zero of denominator
Horizontal asymptote:	$y = 2$, because degree of $N(x) =$ degree of $D(x)$
Additional points:	

x	-4	-1	0	$\frac{1}{4}$	4
$f(x)$	2.25	3	Undefined	-2	1.75

Horizontal asymptote: $y = 2$

Vertical asymptote: $x = 0$

$f(x) = \dfrac{2x-1}{x}$

FIGURE **4.9**

By plotting the intercepts, asymptotes, and a few additional points, you can obtain the graph shown in Figure 4.9. The domain of f is all real numbers except $x = 0$.

✓**CHECKPOINT**   Now try Exercise 17.

| Example 3 | Sketching the Graph of a Rational Function |

Sketch the graph of $f(x) = \dfrac{x}{x^2 - x - 2}$.

Solution

Factor the denominator to determine more easily the zeros of the denominator.

$$f(x) = \frac{x}{x^2 - x - 2} = \frac{x}{(x + 1)(x - 2)}$$

y-intercept: $(0, 0)$, because $f(0) = 0$

x-intercept: $(0, 0)$, because $f(0) = 0$

Vertical asymptotes: $x = -1, x = 2$, zeros of denominator

Horizontal asymptote: $y = 0$, because degree of $N(x)$ < degree of $D(x)$

Additional points:

x	-3	-1	-0.5	1	2	3
$f(x)$	-0.3	Undefined	0.4	-0.5	Undefined	0.75

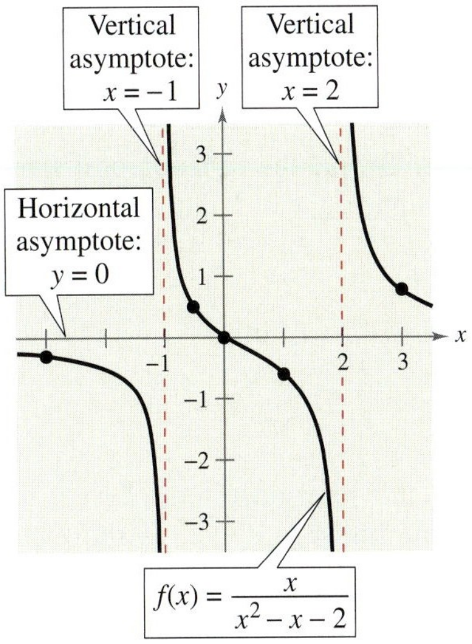

The graph is shown in Figure 4.10.

✓**CHECKPOINT**   Now try Exercise 29.

Vertical asymptote: $x = -1$

Vertical asymptote: $x = 2$

Horizontal asymptote: $y = 0$

$f(x) = \dfrac{x}{x^2 - x - 2}$

FIGURE **4.10**

| Example 4 | Sketching the Graph of a Rational Function |

Sketch the graph of $f(x) = \dfrac{x^2 - 9}{x^2 - 2x - 3}$.

Solution

By factoring the numerator and denominator, you have

$$f(x) = \frac{x^2 - 9}{x^2 - 2x - 3} = \frac{(x - 3)(x + 3)}{(x - 3)(x + 1)} = \frac{x + 3}{x + 1}, \quad x \neq 3.$$

y-intercept: $(0, 3)$, because $f(0) = 3$

x-intercept: $(-3, 0)$, because $f(-3) = 0$

Vertical asymptote: $x = -1$, zero of (simplified) denominator

Horizontal asymptote: $y = 1$, because degree of $N(x)$ = degree of $D(x)$

Additional points:

x	-5	-2	-1	-0.5	1	3	4
$f(x)$	0.5	-1	Undefined	5	2	Undefined	1.4

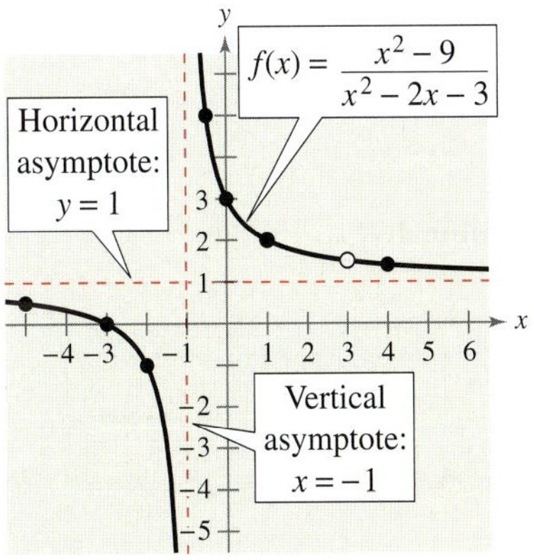

The graph is shown in Figure 4.11. Notice there is a hole in the graph at $x = 3$ because the function is not defined when $x = 3$.

✓**CHECKPOINT**   Now try Exercise 37.

$f(x) = \dfrac{x^2 - 9}{x^2 - 2x - 3}$

Horizontal asymptote: $y = 1$

Vertical asymptote: $x = -1$

FIGURE **4.11**   *HOLE AT $x = 3$*

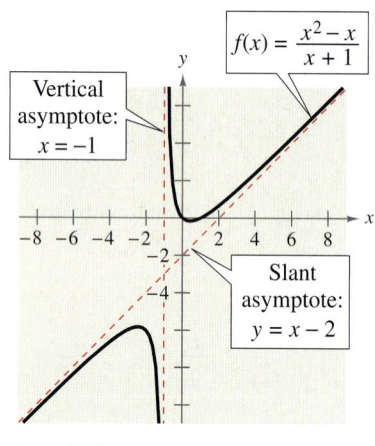

Vertical asymptote: $x = -1$

$f(x) = \dfrac{x^2 - x}{x + 1}$

Slant asymptote: $y = x - 2$

FIGURE **4.12**

Slant Asymptotes

Consider a rational function whose denominator is of degree 1 or greater. If the degree of the numerator is exactly *one more* than the degree of the denominator, the graph of the function has a **slant** (or **oblique**) **asymptote.** For example, the graph of

$$f(x) = \frac{x^2 - x}{x + 1}$$

has a slant asymptote, as shown in Figure 4.12. To find the equation of a slant asymptote, use long division. For instance, by dividing $x + 1$ into $x^2 - x$, you obtain

$$f(x) = \frac{x^2 - x}{x + 1} = \underbrace{x - 2}_{\substack{\text{Slant asymptote} \\ (y = x - 2)}} + \frac{2}{x + 1}.$$

As x increases or decreases without bound, the remainder term $2/(x + 1)$ approaches 0, so the graph of f approaches the line $y = x - 2$, as shown in Figure 4.12.

Example 5   A Rational Function with a Slant Asymptote

Sketch the graph of $f(x) = \dfrac{x^2 - x - 2}{x - 1}$.

Solution

First write $f(x)$ in two different ways. Factoring the numerator

$$f(x) = \frac{x^2 - x - 2}{x - 1} = \frac{(x - 2)(x + 1)}{x - 1}$$

allows you to recognize the x-intercepts. Long division

$$f(x) = \frac{x^2 - x - 2}{x - 1} = x - \frac{2}{x - 1}$$

allows you to recognize that the line $y = x$ is a slant asymptote of the graph.

y-intercept:	$(0, 2)$, because $f(0) = 2$
x-intercepts:	$(-1, 0)$ and $(2, 0)$
Vertical asymptote:	$x = 1$, zero of denominator
Slant asymptote:	$y = x$
Additional points:	

x	-2	0.5	1	1.5	3
$f(x)$	-1.33	4.5	Undefined	-2.5	2

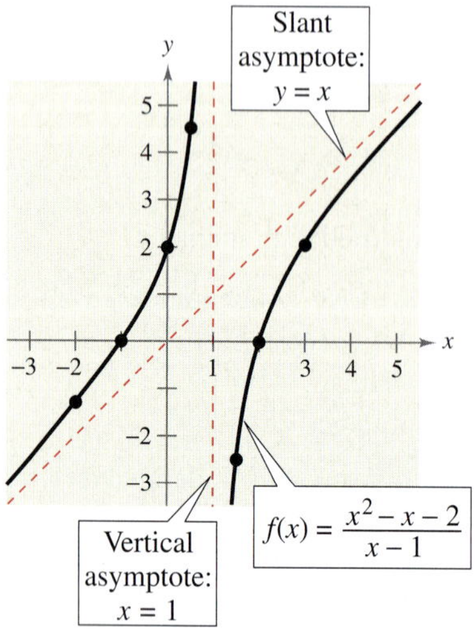

Slant asymptote: $y = x$

$f(x) = \dfrac{x^2 - x - 2}{x - 1}$

Vertical asymptote: $x = 1$

FIGURE **4.13**

The graph is shown in Figure 4.13.

✔**CHECKPOINT**   Now try Exercise 59.

Application

Example 6   **Finding a Minimum Area**

A rectangular page is designed to contain 48 square inches of print. The margins at the top and bottom of the page are each 1 inch deep. The margins on each side are $1\frac{1}{2}$ inches wide. What should the dimensions of the page be so that the least amount of paper is used?

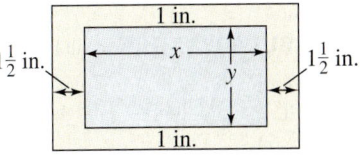

FIGURE **4.14**

Graphical Solution

Let A be the area to be minimized. From Figure 4.14, you can write

$$A = (x + 3)(y + 2).$$

The printed area inside the margins is modeled by $48 = xy$ or $y = 48/x$. To find the minimum area, rewrite the equation for A in terms of just one variable by substituting $48/x$ for y.

$$A = (x + 3)\left(\frac{48}{x} + 2\right)$$

$$= \frac{(x + 3)(48 + 2x)}{x}, \quad x > 0$$

The graph of this rational function is shown in Figure 4.15. Because x represents the width of the printed area, you need consider only the portion of the graph for which x is positive. Using a graphing utility, you can approximate the minimum value of A to occur when $x \approx 8.5$ inches. The corresponding value of y is $48/8.5 \approx 5.6$ inches. So, the dimensions should be

$$x + 3 \approx 11.5 \text{ inches} \quad \text{by} \quad y + 2 \approx 7.6 \text{ inches}.$$

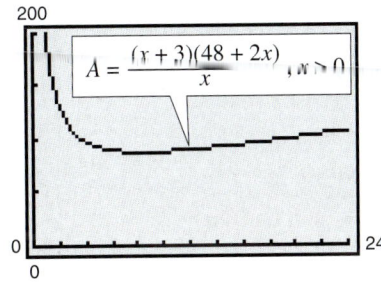

FIGURE **4.15**

✓CHECKPOINT   Now try Exercise 77.

Numerical Solution

Let A be the area to be minimized. From Figure 4.14, you can write

$$A = (x + 3)(y + 2).$$

The printed area inside the margins is modeled by $48 = xy$ or $y = 48/x$. To find the minimum area, rewrite the equation for A in terms of just one variable by substituting $48/x$ for y.

$$A = (x + 3)\left(\frac{48}{x} + 2\right) = \frac{(x + 3)(48 + 2x)}{x}, \quad x > 0$$

Use the *table* feature of a graphing utility to create a table of values for the function

$$y_1 = \frac{(x + 3)(48 + 2x)}{x}$$

beginning at $x = 1$. From the table, you can see that the minimum value of y_1 occurs when x is somewhere between 8 and 9, as shown in Figure 4.16. To approximate the minimum value of y_1 to one decimal place, change the table so that it starts at $x = 8$ and increases by 0.1. The minimum value of y_1 occurs when $x \approx 8.5$, as shown in Figure 4.17. The corresponding value of y is $48/8.5 \approx 5.6$ inches. So, the dimensions should be $x + 3 \approx 11.5$ inches by $y + 2 \approx 7.6$ inches.

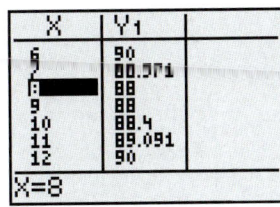

FIGURE **4.16**

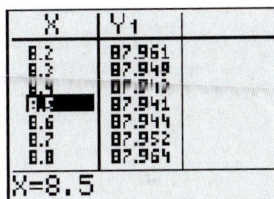

FIGURE **4.17**

If you go on to take a course in calculus, you will learn an analytic technique for finding the exact value of x that produces a minimum area. In this case, that value is $x = 6\sqrt{2} \approx 8.485$.

4.2 | Exercises

VOCABULARY CHECK: Fill in the blanks.

1. For the rational function given by $f(x) = N(x)/D(x)$, if the degree of $N(x)$ is exactly one more than the degree of $D(x)$, then the graph of f has a _____ (or oblique) _____.

2. The graph of $f(x) = 1/x$ has a _____ asymptote at $x = 0$.

PREREQUISITE SKILLS REVIEW: Practice and review algebra skills needed for this section at **www.Eduspace.com.**

In Exercises 1–4, use the graph of $f(x) = 2/x$ to sketch the graph of g.

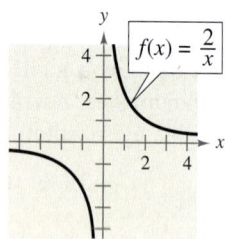

1. $g(x) = \dfrac{2}{x} + 3$

2. $g(x) = \dfrac{2}{x - 3}$

3. $g(x) = -\dfrac{2}{x}$

4. $g(x) = \dfrac{1}{x + 2}$

In Exercises 5–8, use the graph of $f(x) = 2/x^2$ to sketch the graph of g.

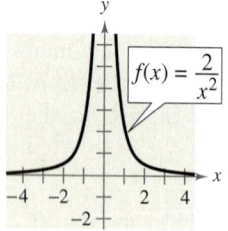

5. $g(x) = \dfrac{2}{x^2} - 1$

6. $g(x) = -\dfrac{2}{x^2}$

7. $g(x) = \dfrac{2}{(x - 1)^2}$

8. $g(x) = \dfrac{1}{2x^2}$

In Exercises 9–12, use the graph of $f(x) = 4/x^3$ to sketch the graph of g.

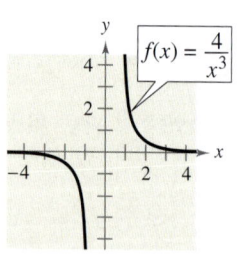

9. $g(x) = \dfrac{4}{(x + 3)^3}$

10. $g(x) = \dfrac{4}{x^3} + 3$

11. $g(x) = -\dfrac{4}{x^3}$

12. $g(x) = \dfrac{1}{x^3}$

In Exercises 13–42, (a) state the domain of the function, (b) identify all intercepts, (c) find any vertical and horizontal asymptotes, and (d) plot additional solution points as needed to sketch the graph of the rational function.

13. $f(x) = \dfrac{1}{x + 2}$

14. $f(x) = \dfrac{1}{x - 3}$

15. $h(x) = \dfrac{-1}{x + 2}$

16. $g(x) = \dfrac{1}{3 - x}$

17. $C(x) = \dfrac{5 + 2x}{1 + x}$

18. $P(x) = \dfrac{1 - 3x}{1 - x}$

19. $g(x) = \dfrac{1}{x + 2} + 2$

20. $f(x) = 2 - \dfrac{3}{x^2}$

21. $f(x) = \dfrac{x^2}{x^2 + 9}$

22. $f(t) = \dfrac{1 - 2t}{t}$

23. $h(x) = \dfrac{x^2}{x^2 - 9}$

24. $g(x) = \dfrac{x}{x^2 - 9}$

25. $g(s) = \dfrac{s}{s^2 + 1}$

26. $f(x) = -\dfrac{1}{(x - 2)^2}$

27. $g(x) = \dfrac{4(x + 1)}{x(x - 4)}$

28. $h(x) = \dfrac{2}{x^2(x - 2)}$

29. $f(x) = \dfrac{3x}{x^2 - x - 2}$

30. $f(x) = \dfrac{2x}{x^2 + x - 2}$

31. $h(x) = \dfrac{x^2 - 5x + 4}{x^2 - 4}$

32. $g(x) = \dfrac{x^2 - 2x - 8}{x^2 - 9}$

33. $f(x) = \dfrac{6x}{x^2 - 5x - 14}$

34. $f(x) = \dfrac{3(x^2 + 1)}{x^2 + 2x - 15}$

35. $f(x) = \dfrac{2x^2 - 5x - 3}{x^3 - 2x^2 - x + 2}$

36. $f(x) = \dfrac{x^2 - x - 2}{x^3 - 2x^2 - 5x + 6}$

37. $f(x) = \dfrac{x^2 + 3x}{x^2 + x - 6}$

38. $f(x) = \dfrac{5(x + 4)}{x^2 + x - 12}$

39. $f(x) = \dfrac{2x^2 - 5x + 2}{2x^2 - x - 6}$

40. $f(x) = \dfrac{3x^2 - 8x + 4}{2x^2 - 3x - 2}$

41. $f(t) = \dfrac{t^2 - 1}{t + 1}$

42. $f(x) = \dfrac{x^2 - 16}{x - 4}$

Analytical, Numerical, and Graphical Analysis   In Exercises 43–46, do the following.

(a) Determine the domains of f and g.

(b) Simplify f and find any vertical asymptotes of the graph of f.

(c) Compare the functions by completing the table.

 (d) Use a graphing utility to graph f and g in the same viewing window.

 (e) Explain why the graphing utility may not show the difference in the domains of f and g.

43. $f(x) = \dfrac{x^2 - 1}{x + 1}, \quad g(x) = x - 1$

x	-3	-2	-1.5	-1	-0.5	0	1
$f(x)$							
$g(x)$							

44. $f(x) = \dfrac{x^2(x - 2)}{x^2 - 2x}, \quad g(x) = x$

x	-1	0	1	1.5	2	2.5	3
$f(x)$							
$g(x)$							

45. $f(x) = \dfrac{x - 2}{x^2 - 2x}, \quad g(x) = \dfrac{1}{x}$

x	-0.5	0	0.5	1	1.5	2	3
$f(x)$							
$g(x)$							

46. $f(x) = \dfrac{2x - 6}{x^2 - 7x + 12}, \quad g(x) = \dfrac{2}{x - 4}$

x	0	1	2	3	4	5	6
$f(x)$							
$g(x)$							

In Exercises 47–62, (a) state the domain of the function, (b) identify all intercepts, (c) identify any vertical and slant asymptotes, and (d) plot additional solution points as needed to sketch the graph of the rational function.

47. $h(x) = \dfrac{x^2 - 4}{x}$

48. $g(x) = \dfrac{x^2 + 5}{x}$

49. $f(x) = \dfrac{2x^2 + 1}{x}$

50. $f(x) = \dfrac{1 - x^2}{x}$

51. $g(x) = \dfrac{x^2 + 1}{x}$

52. $h(x) = \dfrac{x^2}{x - 1}$

53. $f(t) = -\dfrac{t^2 + 1}{t + 5}$

54. $f(x) = \dfrac{x^2}{3x + 1}$

55. $f(x) = \dfrac{x^3}{x^2 - 1}$

56. $g(x) = \dfrac{x^3}{2x^2 - 8}$

57. $f(x) = \dfrac{x^3 - 1}{x^2 - x}$

58. $f(x) = \dfrac{x^4 + x}{x^3}$

59. $f(x) = \dfrac{x^2 - x + 1}{x - 1}$

60. $f(x) = \dfrac{2x^2 - 5x + 5}{x - 2}$

61. $f(x) = \dfrac{2x^3 - x^2 - 2x + 1}{x^2 + 3x + 2}$

62. $f(x) = \dfrac{2x^3 + x^2 - 8x - 4}{x^2 - 3x + 2}$

 In Exercises 63–66, use a graphing utility to graph the rational function. Give the domain of the function and identify any asymptotes. Then zoom out sufficiently far so that the graph appears as a line. Identify the line.

63. $f(x) = \dfrac{x^2 + 5x + 8}{x + 3}$

64. $f(x) = \dfrac{2x^2 + x}{x + 1}$

65. $g(x) = \dfrac{1 + 3x^2 - x^3}{x^2}$

66. $h(x) = \dfrac{12 - 2x - x^2}{2(4 + x)}$

Graphical Reasoning In Exercises 67–70, (a) use the graph to determine any *x*-intercepts of the graph of the rational function and (b) set $y = 0$ and solve the resulting equation to confirm your result in part (a).

67. $y = \dfrac{x + 1}{x - 3}$

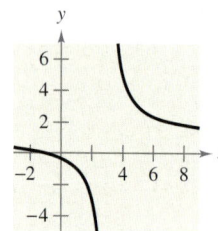

68. $y = \dfrac{2x}{x - 3}$

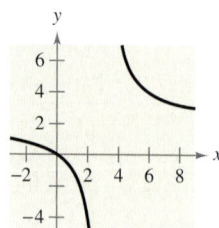

69. $y = \dfrac{1}{x} - x$

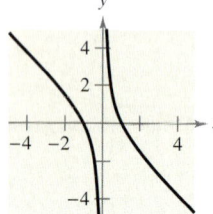

70. $y = x - 3 + \dfrac{2}{x}$

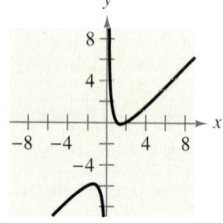

 Graphical Reasoning In Exercises 71–74, (a) use a graphing utility to graph the rational function and determine any *x*-intercepts of the graph and (b) set $y = 0$ and solve the resulting equation to confirm your result in part (a).

71. $y = \dfrac{1}{x + 5} + \dfrac{4}{x}$

72. $y = 20\left(\dfrac{2}{x + 1} - \dfrac{3}{x}\right)$

73. $y = x - \dfrac{6}{x - 1}$

74. $y = x - \dfrac{9}{x}$

75. **Concentration of a Mixture** A 1000-liter tank contains 50 liters of a 25% brine solution. You add *x* liters of a 75% brine solution to the tank.

(a) Show that the concentration *C*, the proportion of brine to total solution, in the final mixture is

$$C = \dfrac{3x + 50}{4(x + 50)}.$$

(b) Determine the domain of the function based on the physical constraints of the problem.

(c) Sketch a graph of the concentration function.

(d) As the tank is filled, what happens to the rate at which the concentration of brine is increasing? What percent does the concentration of brine appear to approach?

76. **Geometry** A rectangular region of length *x* and width *y* has an area of 500 square meters.

(a) Write the width *y* as a function of *x*.

(b) Determine the domain of the function based on the physical constraints of the problem.

(c) Sketch a graph of the function and determine the width of the rectangle when $x = 30$ meters.

77. **Page Design** A page that is *x* inches wide and *y* inches high contains 30 square inches of print. The top and bottom margins are 1 inch deep and the margins on each side are 2 inches wide (see figure).

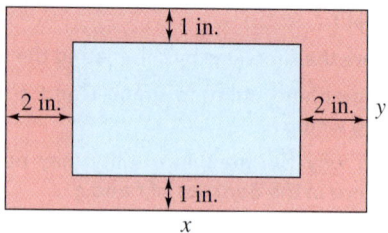

(a) Show that the total area *A* on the page is

$$A = \dfrac{2x(x + 11)}{x - 4}.$$

(b) Determine the domain of the function based on the physical constraints of the problem.

 (c) Use a graphing utility to graph the area function and approximate the page size for which the least amount of paper will be used. Verify your answer numerically using the *table* feature of the graphing utility.

78. **Page Design** A rectangular page is designed to contain 64 square inches of print. The margins at the top and bottom of the page are each 1 inch deep. The margins on each side are $1\frac{1}{2}$ inches wide. What should the dimensions of the page be so that the least amount of paper is used?

 In Exercises 79 and 80, use a graphing utility to graph the function and locate any relative maximum or minimum points on the graph.

79. $f(x) = \dfrac{3(x + 1)}{x^2 + x + 1}$

80. $C(x) = x + \dfrac{32}{x}$

 81. **Minimum Cost** The ordering and transportation cost *C* (in thousands of dollars) for the components used in manufacturing a product is given by

$$C = 100\left(\dfrac{200}{x^2} + \dfrac{x}{x + 30}\right), \quad x \geq 1$$

where *x* is the order size (in hundreds). Use a graphing utility to graph the cost function. From the graph, estimate the order size that minimizes cost.

82. *Minimum Cost* The cost C of producing x units of a product is given by

$$C = 0.2x^2 + 10x + 5$$

and the average cost per unit is given by

$$\overline{C} = \frac{C}{x} = \frac{0.2x^2 + 10x + 5}{x}, \quad x > 0.$$

Sketch the graph of the average cost function and estimate the number of units that should be produced to minimize the average cost per unit.

Model It

83. *Average Speed* A driver averaged 50 miles per hour on the round trip between Akron, Ohio, and Columbus, Ohio, 100 miles away. The average speeds for going and returning were x and y miles per hour, respectively.

(a) Show that $y = \dfrac{25x}{x - 25}$.

(b) Determine the vertical and horizontal asymptotes of the graph of the function.

 (c) Use a graphing utility to graph the function.

(d) Complete the table.

x	30	35	40	45	50	55	60
y							

(e) Are the results in the table what you expected? Explain.

(f) Is it possible to average 20 miles per hour in one direction and still average 50 miles per hour on the round trip? Explain.

---

84. *Medicine* The concentration C of a chemical in the bloodstream t hours after injection into muscle tissue is given by

$$C = \frac{3t^2 + t}{t^3 + 50}, \quad t > 0.$$

(a) Determine the horizontal asymptote of the graph of the function and interpret its meaning in the context of the problem.

 (b) Use a graphing utility to graph the function and approximate the time when the bloodstream concentration is greatest.

 (c) Use a graphing utility to determine when the concentration is less than 0.345.

Synthesis

True or False? **In Exercises 85–88, determine whether the statement is true or false. Justify your answer.**

85. If the graph of a rational function f has a vertical asymptote at $x = 5$, it is possible to sketch the graph without lifting your pencil from the paper.

86. The graph of a rational function can never cross one of its asymptotes.

87. The graph of $f(x) = \dfrac{2x^3}{x + 1}$ has a slant asymptote.

88. Every rational function has a horizontal asymptote.

 Think About It **In Exercises 89 and 90, use a graphing utility to graph the function. Explain why there is no vertical asymptote when a superficial examination of the function may indicate that there should be one.**

89. $h(x) = \dfrac{6 - 2x}{3 - x}$ 90. $g(x) = \dfrac{x^2 + x - 2}{x - 1}$

91. ***Think About It*** Write a rational function satisfying the following criteria.

Vertical asymptote: $x = 2$

Slant asymptote: $y = x + 1$

Zero of the function: $x = -2$

92. ***Think About It*** Write a rational function satisfying the following criteria.

Vertical asymptote: $x = -1$

Slant asymptote: $y = x + 2$

Zero of the function: $x = 3$

Skills Review

In Exercises 93–96, completely factor the expression.

93. $x^2 - 15x + 56$ 94. $3x^2 + 23x - 36$

95. $x^3 - 5x^2 + 4x - 20$ 96. $x^3 + 6x^2 - 2x - 12$

In Exercises 97–100, solve the inequality and graph the solution on the real number line.

97. $10 - 3x \le 0$ 98. $5 - 2x > 5(x + 1)$

99. $|4(x - 2)| < 20$ 100. $\frac{1}{2}|2x + 3| \ge 5$

---

101. **Make a Decision** To work an extended application analyzing the total manpower of the Department of Defense, visit this text's website at *college.hmco.com*. *(Data Source: U.S. Department of Defense)*

4.3 Conics

What you should learn

- Recognize the four basic conics: circles, ellipses, parabolas, and hyperbolas.
- Recognize, graph, and write equations of parabolas (vertex at origin).
- Recognize, graph, and write equations of ellipses (center at origin).
- Recognize, graph, and write equations of hyperbolas (center at origin).

Why you should learn it

Conics have been used for hundreds of years to model and solve engineering problems. For instance, in Exercise 33 on page 364, a parabola can be used to model the cables of the Golden Gate Bridge.

Cosmo Condina/Getty Images

Introduction

Conic sections were discovered during the classical Greek period, 600 to 300 B.C. This early Greek study was largely concerned with the geometric properties of conics. It was not until the early 17th century that the broad applicability of conics became apparent and played a prominent role in the early development of calculus.

A **conic section** (or simply **conic**) is the intersection of a plane and a double-napped cone. Notice in Figure 4.18 that in the formation of the four basic conics, the intersecting plane does not pass through the vertex of the cone. When the plane does pass through the vertex, the resulting figure is a **degenerate conic,** as shown in Figure 4.19.

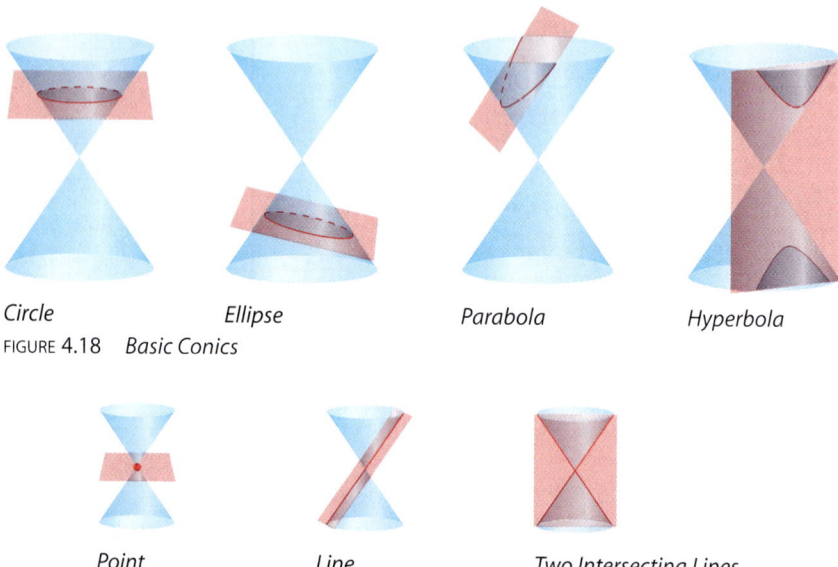

Circle   Ellipse   Parabola   Hyperbola

FIGURE 4.18   *Basic Conics*

Point   Line   Two Intersecting Lines

FIGURE 4.19   *Degenerate Conics*

There are several ways to approach the study of conics. You could begin by defining conics in terms of the intersections of planes and cones, as the Greeks did, or you could define them algebraically, in terms of the general second-degree equation

$$Ax^2 + Bxy + Cy^2 + Dx + Ey + F = 0.$$

However, you will study a third approach, in which each of the conics is defined as a *locus* (collection) of points satisfying a certain geometric property. For example, in Section 1.1 you saw how the definition of a circle as *the collection of all points (x, y) that are equidistant from a fixed point (h, k)* led easily to the standard form of the equation of a circle

$$(x - h)^2 + (y - k)^2 = r^2. \qquad \text{Equation of a circle}$$

Recall from Section 1.1 that the center of a circle is at (h, k) and that the radius of the circle is r.

Parabolas

In Section 3.1, you learned that the graph of the quadratic function

$$f(x) = ax^2 + bx + c$$

is a parabola that opens upward or downward. The following definition of a parabola is more general in the sense that it is independent of the orientation of the parabola.

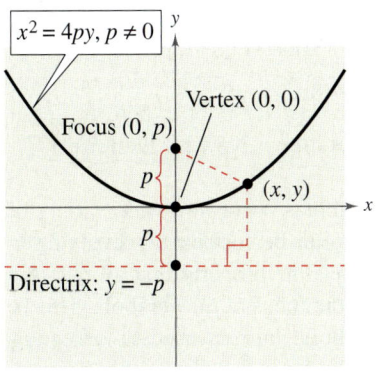

FIGURE 4.20   *Parabola*

> ## Definition of a Parabola
>
> A **parabola** is the set of all points (x, y) in a plane that are equidistant from a fixed line, the **directrix,** and a fixed point, the **focus,** not on the line. (See Figure 4.20.) The **vertex** is the midpoint between the focus and the directrix. The **axis** of the parabola is the line passing through the focus and the vertex.

> ## Standard Equation of a Parabola (Vertex at Origin)
>
> The **standard form of the equation of a parabola** with vertex at $(0, 0)$ and directrix $y = -p$ is
>
> $$x^2 = 4py, \qquad p \neq 0. \qquad \text{Vertical axis}$$
>
> For directrix $x = -p$, the equation is
>
> $$y^2 = 4px, \qquad p \neq 0. \qquad \text{Horizontal axis}$$
>
> The focus is on the axis p units (directed distance) from the vertex.

For a proof of the standard form of the equation of a parabola, see Proofs in Mathematics on page 380.

Notice that a parabola can have a vertical or a horizontal axis and that a parabola is symmetric with respect to its axis. Examples of each are shown in Figure 4.21.

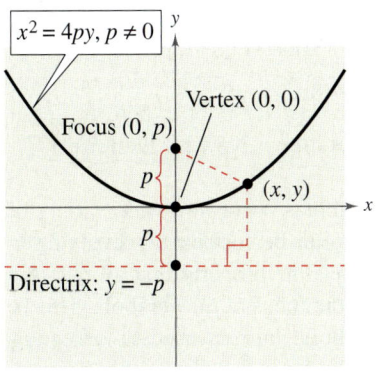
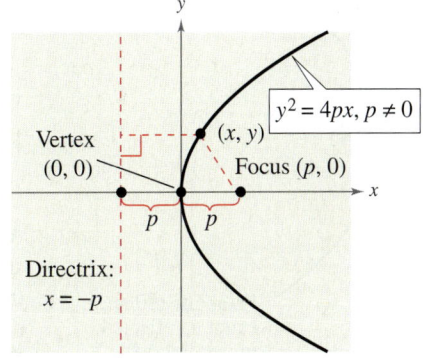

(a) Parabola with vertical axis   **(b)** Parabola with horizontal axis

FIGURE 4.21

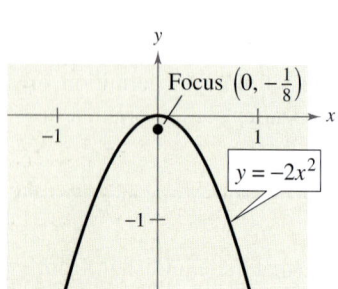

FIGURE 4.22

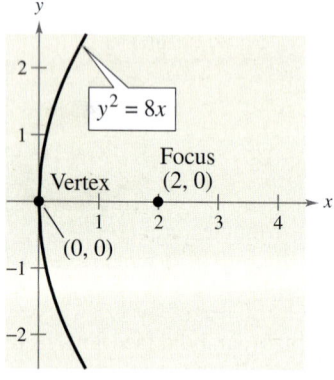

FIGURE 4.23

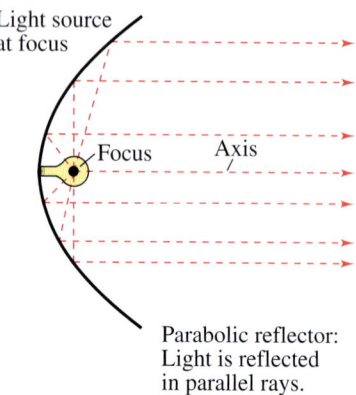

Parabolic reflector:
Light is reflected
in parallel rays.

FIGURE 4.24

Example 1 Finding the Focus of a Parabola

Find the focus of the parabola whose equation is $y = -2x^2$.

Solution

Because the squared term in the equation involves x, you know that the axis is vertical, and the equation is of the form

$$x^2 = 4py.$$

You can write the original equation in this form as follows.

$$x^2 = -\frac{1}{2}y$$

$$x^2 = 4\left(-\frac{1}{8}\right)y \qquad \text{Write in standard form.}$$

So, $p = -\frac{1}{8}$. Because p is negative, the parabola opens downward (see Figure 4.22), and the focus of the parabola is

$$(0, p) = \left(0, -\frac{1}{8}\right). \qquad \text{Focus}$$

✓**CHECKPOINT** Now try Exercise 11.

Example 2 A Parabola with a Horizontal Axis

Write the standard form of the equation of the parabola with vertex at the origin and focus at $(2, 0)$.

Solution

The axis of the parabola is horizontal, passing through $(0, 0)$ and $(2, 0)$, as shown in Figure 4.23. So, the standard form is

$$y^2 = 4px.$$

Because the focus is $p = 2$ units from the vertex, the equation is

$$y^2 = 4(2)x$$

$$y^2 = 8x.$$

✓**CHECKPOINT** Now try Exercise 17.

Parabolas occur in a wide variety of applications. For instance, a parabolic reflector can be formed by revolving a parabola about its axis. The resulting surface has the property that all incoming rays parallel to the axis are reflected through the focus of the parabola. This is the principle behind the construction of the parabolic mirrors used in reflecting telescopes. Conversely, the light rays emanating from the focus of a parabolic reflector used in a flashlight are all parallel to one another, as shown in Figure 4.24.

Ellipses

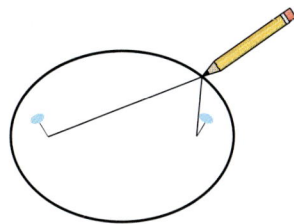

$d_1 + d_2$ is constant.

FIGURE **4.25**

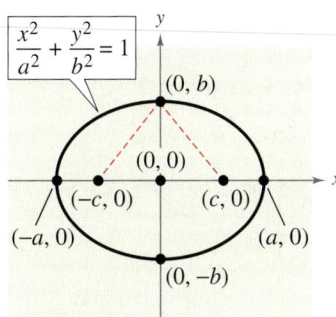 <!-- FIGURE 4.26 thumbtack/pencil illustration -->

FIGURE **4.26**

Definition of an Ellipse

An **ellipse** is the set of all points (x, y) in a plane the sum of whose distances from two distinct fixed points (**foci**) is constant. See Figure 4.25.

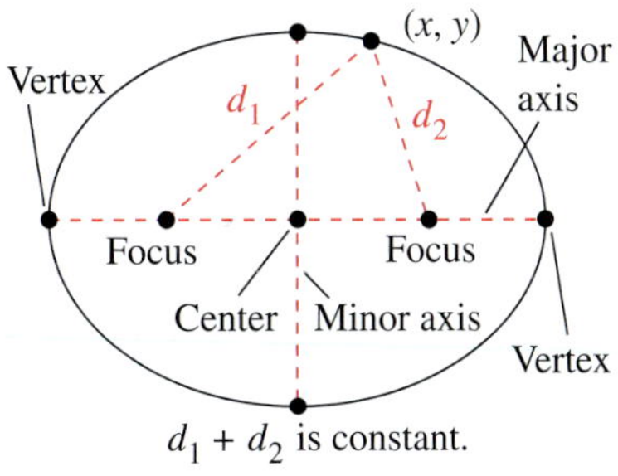

The line through the foci intersects the ellipse at two points (**vertices**). The chord joining the vertices is the **major axis,** and its midpoint is the **center** of the ellipse. The chord perpendicular to the major axis at the center is the **minor axis.** (See Figure 4.25).

You can visualize the definition of an ellipse by imagining two thumbtacks placed at the foci, as shown in Figure 4.26. If the ends of a fixed length of string are fastened to the thumbtacks and the string is *drawn taut* with a pencil, the path traced by the pencil will be an ellipse.

The standard form of the equation of an ellipse takes one of two forms, depending on whether the major axis is horizontal or vertical.

Standard Equation of an Ellipse (Center at Origin)

The **standard form of the equation of an ellipse** centered at the origin with major and minor axes of lengths $2a$ and $2b$ (where $0 < b < a$) is

$$\frac{x^2}{a^2} + \frac{y^2}{b^2} = 1 \qquad \text{or} \qquad \frac{x^2}{b^2} + \frac{y^2}{a^2} = 1.$$

The vertices and foci lie on the major axis, a and c units, respectively, from the center, as shown in Figure 4.27. Moreover, a, b, and c are related by the equation $c^2 = a^2 - b^2$.

Exploration

An ellipse can be drawn using two thumbtacks placed at the foci of the ellipse, a string of fixed length (greater than the distance between the tacks), and a pencil, as shown in Figure 4.26. Try doing this. Vary the length of the string and the distance between the thumbtacks. Explain how to obtain ellipses that are almost circular. Explain how to obtain ellipses that are long and narrow.

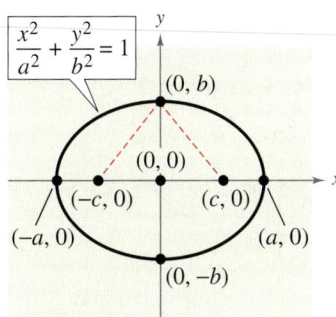 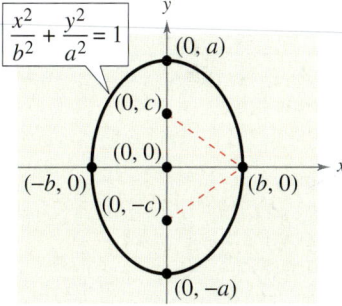

(a) Major axis is horizontal; minor axis is vertical.

(b) Major axis is vertical; minor axis is horizontal.

FIGURE **4.27**

In Figure 4.27(a), note that because the sum of the distances from a point on the ellipse to the two foci is constant, it follows that

(Sum of distances from $(0, b)$ to foci) = (sum of distances from $(a, 0)$ to foci)

$$2\sqrt{b^2 + c^2} = (a + c) + (a - c)$$
$$\sqrt{b^2 + c^2} = a$$
$$c^2 = a^2 - b^2.$$

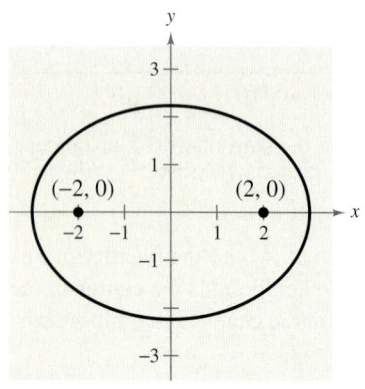

FIGURE 4.28

Example 3 **Finding the Standard Equation of an Ellipse**

Find the standard form of the equation of the ellipse shown in Figure 4.28.

Solution

From Figure 4.28, the foci occur at $(-2, 0)$ and $(2, 0)$. So, the center of the ellipse is $(0, 0)$, the major axis is horizontal, and the ellipse has an equation of the form

$$\frac{x^2}{a^2} + \frac{y^2}{b^2} = 1. \qquad \text{Standard form}$$

Also from Figure 4.28, the length of the major axis is $2a = 6$. This implies that $a = 3$. Moreover, the distance from the center to either focus is $c = 2$. Finally,

$$b^2 = a^2 - c^2 = 3^2 - 2^2 = 9 - 4 = 5.$$

Substituting $a^2 = 3^2$ and $b^2 = \left(\sqrt{5}\right)^2$ yields the following equation in standard form.

$$\frac{x^2}{3^2} + \frac{y^2}{\left(\sqrt{5}\right)^2} = 1$$

This equation simplifies to

$$\frac{x^2}{9} + \frac{y^2}{5} = 1.$$

✔**CHECKPOINT** Now try Exercise 49.

Technology

Conics can be graphed using a graphing calculator by first solving for y. You may have to graph the conic using two separate equations. For example, you can graph the ellipse from Example 4 by graphing both

$$y_1 = \sqrt{36 - 4x^2}$$

and

$$y_2 = -\sqrt{36 - 4x^2}$$

in the same viewing window.

Example 4 **Sketching an Ellipse**

Sketch the ellipse given by $4x^2 + y^2 = 36$, and identify the vertices.

Solution

$$4x^2 + y^2 = 36 \qquad \text{Write original equation.}$$

$$\frac{4x^2}{36} + \frac{y^2}{36} = \frac{36}{36} \qquad \text{Divide each side by 36.}$$

$$\frac{x^2}{3^2} + \frac{y^2}{6^2} = 1 \qquad \text{Write in standard form.}$$

$$\frac{x^2}{9} + \frac{y^2}{36} = 1 \qquad \text{Simplify.}$$

Because the denominator of the y^2-term is larger than the denominator of the x^2-term, you can conclude that the major axis is vertical. Moreover, because $a = 6$, the vertices are $(0, -6)$ and $(0, 6)$. Finally, because $b = 3$, the endpoints of the minor axis (or co-vertices) are $(-3, 0)$ and $(3, 0)$, as shown in Figure 4.29. Note that you can sketch the ellipse by locating the endpoints of the two axes. Because 3^2 is the denominator of the x^2-term, move three units to the *right and left* of the center to locate the endpoints of the horizontal axis. Similarly, because 6^2 is the denominator of the y^2-term, move six units *up and down* from the center to locate the endpoints of the vertical axis.

✔**CHECKPOINT** Now try Exercise 41.

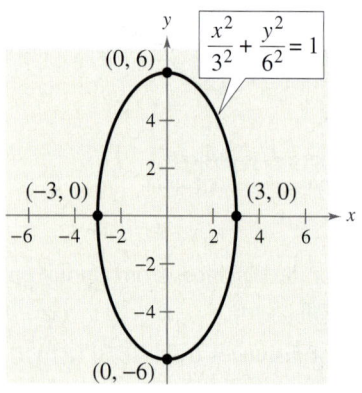

FIGURE 4.29

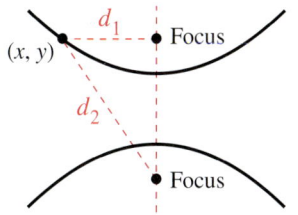

$d_2 - d_1$ is a positive constant.

(a)

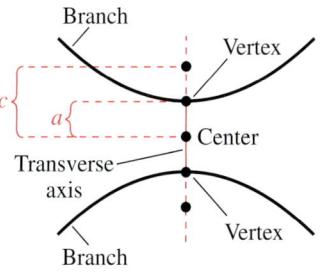

(b)

FIGURE 4.30

Hyperbolas

The definition of a **hyperbola** is similar to that of an ellipse. The difference is that for an ellipse the *sum* of the distances between the foci and a point on the ellipse is constant, whereas for a hyperbola the *difference* of the distances between the foci and a point on the hyperbola is constant.

> ## Definition of a Hyperbola
>
> A **hyperbola** is the set of all points (x, y) in a plane the difference of whose distances from two distinct fixed points **(foci)** is a positive constant. See Figure 4.30(a).

The graph of a hyperbola has two disconnected parts **(branches).** The line through the two foci intersects the hyperbola at two points **(vertices).** The line segment connecting the vertices is the **transverse axis,** and the midpoint of the transverse axis is the **center** of the hyperbola. See Figure 4.30(b).

> ## Standard Equation of a Hyperbola (Center at Origin)
>
> The **standard form of the equation of a hyperbola** with center at the origin (where $a \neq 0$ and $b \neq 0$) is
>
> $$\frac{x^2}{a^2} - \frac{y^2}{b^2} = 1 \qquad \text{Transverse axis is horizontal.}$$
>
> or
>
> $$\frac{y^2}{a^2} - \frac{x^2}{b^2} = 1. \qquad \text{Transverse axis is vertical.}$$
>
> The vertices and foci are, respectively, a and c units from the center. Moreover, a, b, and c are related by the equation $b^2 = c^2 - a^2$. See Figure 4.31.

STUDY TIP

When finding the foci of ellipses and hyperbolas, notice that the relationships between a, b, and c differ slightly.

Finding the foci of an ellipse:

$$c^2 = a^2 - b^2$$

Finding the foci of a hyperbola:

$$c^2 = a^2 + b^2$$

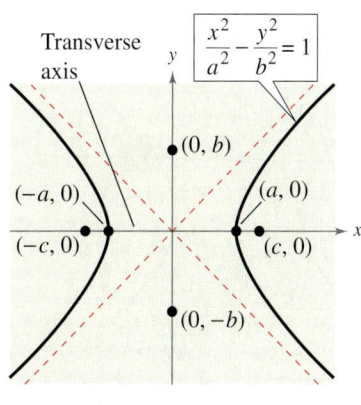

(a)

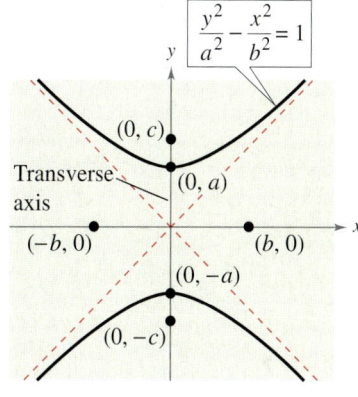

(b)

FIGURE 4.31

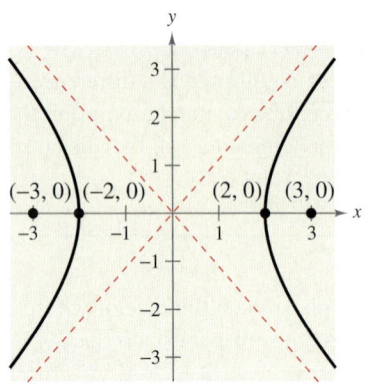

FIGURE **4.32**

Example 5   Finding the Standard Equation of a Hyperbola

Find the standard form of the equation of the hyperbola with foci at $(-3, 0)$ and $(3, 0)$ and vertices at $(-2, 0)$ and $(2, 0)$, as shown in Figure 4.32.

Solution

From the graph, you can determine that $c = 3$, because the foci are three units from the center. Moreover, $a = 2$ because the vertices are two units from the center. So, it follows that

$$b^2 = c^2 - a^2$$
$$= 3^2 - 2^2$$
$$= 9 - 4$$
$$= 5.$$

Because the transverse axis is horizontal, the standard form of the equation is

$$\frac{x^2}{a^2} - \frac{y^2}{b^2} = 1.$$

Finally, substitute $a^2 = 2^2$ and $b^2 = \left(\sqrt{5}\right)^2$ to obtain

$$\frac{x^2}{2^2} - \frac{y^2}{\left(\sqrt{5}\right)^2} = 1 \qquad \text{Write in standard form.}$$

$$\frac{x^2}{4} - \frac{y^2}{5} = 1. \qquad \text{Simplify.}$$

✓**CHECKPOINT**   Now try Exercise 69.

An important aid in sketching the graph of a hyperbola is the determination of its *asymptotes*, as shown in Figure 4.33. Each hyperbola has two asymptotes that intersect at the center of the hyperbola. Furthermore, the asymptotes pass through the corners of a rectangle of dimensions $2a$ by $2b$. The line segment of length $2b$ joining $(0, b)$ and $(0, -b)$ [or $(-b, 0)$ and $(b, 0)$] is the **conjugate axis** of the hyperbola.

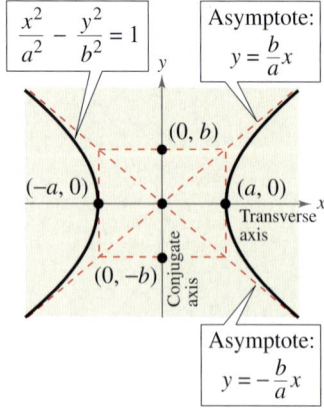

(a) **Transverse axis is horizontal;**
     conjugate axis is vertical.

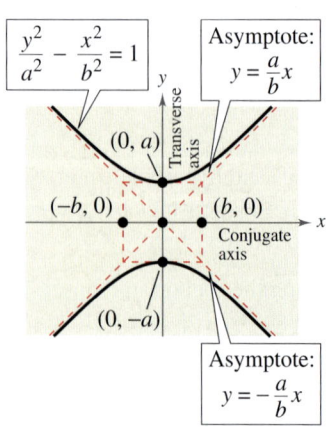

(b) **Transverse axis is vertical;**
     conjugate axis is horizontal.

FIGURE **4.33**

Asymptotes of a Hyperbola (Center at Origin)

The **asymptotes of a hyperbola** with center at $(0, 0)$ are

$$y = \frac{b}{a}x \quad \text{and} \quad y = -\frac{b}{a}x$$ Transverse axis is horizontal.

or

$$y = \frac{a}{b}x \quad \text{and} \quad y = -\frac{a}{b}x.$$ Transverse axis is vertical.

Exploration

Use a graphing utility to graph the hyperbola in Example 6. Does your graph look like the one shown in Figure 4.35? If not, what must you do to obtain both the upper and lower portions of the hyperbola? Explain your reasoning.

Example 6   Sketching a Hyperbola

Sketch the hyperbola whose equation is

$$4x^2 - y^2 = 16.$$

Solution

$$4x^2 - y^2 = 16$$ Write original equation.

$$\frac{4x^2}{16} - \frac{y^2}{16} = \frac{16}{16}$$ Divide each side by 16.

$$\frac{x^2}{2^2} - \frac{y^2}{4^2} = 1$$ Write in standard form.

$$\frac{x^2}{4} - \frac{y^2}{16} = 1$$ Simplify.

Because the x^2-term is positive, you can conclude that the transverse axis is horizontal and the vertices occur at $(-2, 0)$ and $(2, 0)$. Moreover, the endpoints of the conjugate axis occur at $(0, -4)$ and $(0, 4)$, and you can sketch the rectangle shown in Figure 4.34. Finally, by drawing the asymptotes through the corners of this rectangle, you can complete the sketch shown in Figure 4.35. Note that the asymptotes are $y = 2x$ and $y = -2x$.

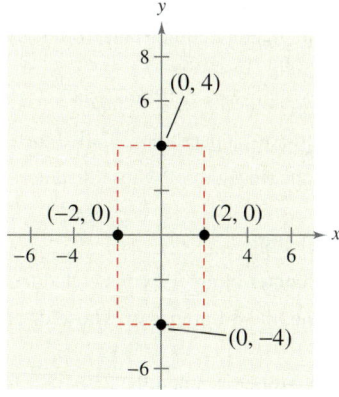

FIGURE 4.34   FIGURE 4.35

✓**CHECKPOINT**   Now try Exercise 67.

Example 7 **Finding the Standard Equation of a Hyperbola**

Find the standard form of the equation of the hyperbola that has vertices at $(0, -3)$ and $(0, 3)$ and asymptotes $y = -2x$ and $y = 2x$, as shown in Figure 4.36.

Solution

Because the transverse axis is vertical, the asymptotes are of the forms

$$y = \frac{a}{b}x \quad \text{and} \quad y = -\frac{a}{b}x.$$

Using the fact that $y = 2x$ and $y = -2x$, you can determine that

$$\frac{a}{b} = 2.$$

Because $a = 3$, you can determine that $b = \frac{3}{2}$. Finally, you can conclude that the hyperbola has the following equation.

$$\frac{y^2}{3^2} - \frac{x^2}{\left(\frac{3}{2}\right)^2} = 1 \qquad \text{Write in standard form.}$$

$$\frac{y^2}{9} - \frac{x^2}{\dfrac{9}{4}} = 1 \qquad \text{Simplify.}$$

✓**CHECKPOINT** Now try Exercise 71.

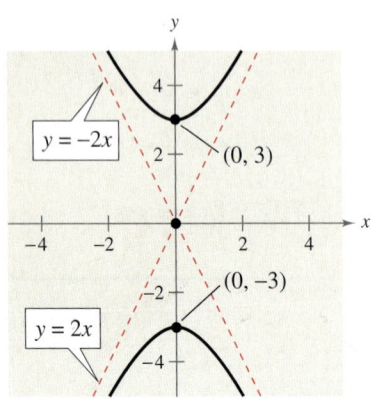

FIGURE 4.36

4.3 Exercises

VOCABULARY CHECK: Fill in the blanks.

1. A _____ is the intersection of a plane and a double-napped cone.

2. The equation $(x - h)^2 + (y - k)^2 = r^2$ is the standard form of the equation of a _____ with center _____ and radius _____.

3. A _____ is the set of all points (x, y) in a plane that are equidistant from a fixed line, called the _____, and a fixed point, called the _____, not on the line.

4. The _____ of a parabola is the midpoint between the focus and the directrix.

5. The line that passes through the focus and the vertex of a parabola is called the _____ of the parabola.

6. An _____ is the set of all points (x, y) in a plane, the sum of whose distances from two distinct fixed points, called_____, is constant.

7. The chord joining the vertices of an ellipse is called the _____ _____, and its midpoint is the _____ of the ellipse.

8. The chord perpendicular to the major axis at the center of an ellipse is called the _____ _____ of the ellipse.

9. A _____ is the set of all points (x, y) in a plane, the difference of whose distances from two distinct fixed points, called _____, is a positive constant.

10. The line segment connecting the vertices of a hyperbola is called the _____ _____, and the midpoint of the line segment is the _____ of the hyperbola.

PREREQUISITE SKILLS REVIEW: Practice and review algebra skills needed for this section at **www.Eduspace.com.**

In Exercises 1–10, match the equation with its graph. If the graph of an equation is not shown, write "not shown." [The graphs are labeled (a), (b), (c), (d), (e), (f), (g), and (h).]

(a)
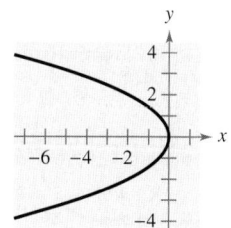

(b)
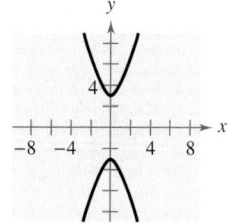

(c)
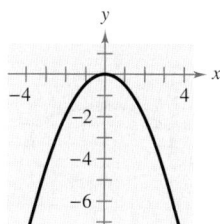

(d)
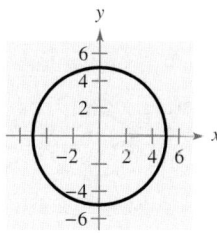

(e)
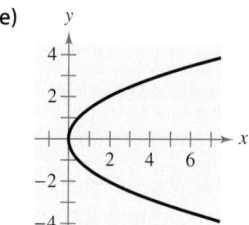

(f)
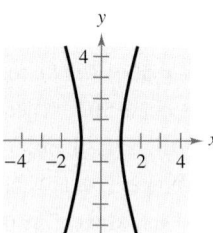

(g)
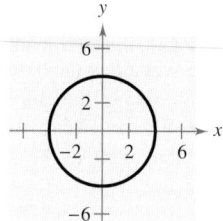

(h)
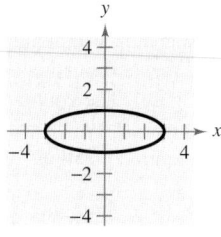

1. $x^2 = 2y$

2. $x^2 = -2y$

3. $y^2 = 2x$

4. $y^2 = -2x$

5. $9x^2 + y^2 = 9$

6. $x^2 + 9y^2 = 9$

7. $9x^2 - y^2 = 9$

8. $y^2 - 9x^2 = 9$

9. $x^2 + y^2 = 25$

10. $x^2 + y^2 = 16$

In Exercises 11–16, find the vertex and focus of the parabola and sketch its graph.

11. $y = \frac{1}{2}x^2$

12. $y = 2x^2$

13. $y^2 = -6x$

14. $y^2 = 3x$

15. $x^2 + 8y = 0$

16. $x + y^2 = 0$

In Exercises 17–26, find the standard form of the equation of the parabola with the given characteristic(s) and vertex at the origin.

17. Focus: $(-2, 0)$

18. Focus: $(0, -2)$

19. Focus: $\left(0, -\frac{3}{2}\right)$

20. Focus: $\left(\frac{5}{2}, 0\right)$

21. Directrix: $y = -1$

22. Directrix: $y = 2$

23. Directrix: $x = 3$

24. Directrix: $x = -2$

25. Passes through the point $(4, 6)$; horizontal axis

26. Passes through the point $(-2, -2)$; vertical axis

In Exercises 27–30, find the standard form of the equation of the parabola and determine the coordinates of the focus.

27.
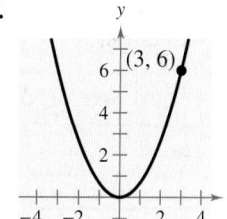

28.
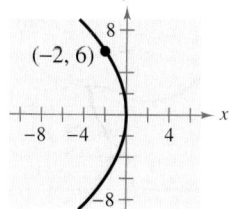

29.
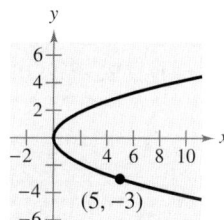

30.
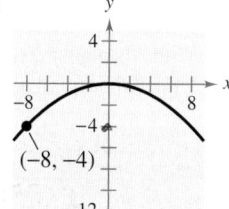

31. *Flashlight* The light bulb in a flashlight is at the focus of the parabolic reflector, 1.5 centimeters from the vertex of the reflector (see figure). Write an equation for a cross section of the flashlight's reflector with its focus on the positive x-axis and its vertex at the origin.

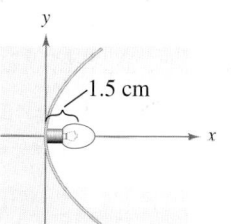

FIGURE FOR 31   FIGURE FOR 32

32. *Satellite Antenna* Write an equation for a cross section of the parabolic television dish antenna shown in the figure.

Model It

33. Suspension Bridge Each cable of the Golden Gate Bridge is suspended (in the shape of a parabola) between two towers that are 1280 meters apart. The top of each tower is 152 meters above the roadway. The cables touch the roadway at the midpoint between the towers.

(a) Draw a sketch of the bridge. Locate the origin of a rectangular coordinate system at the center of the roadway. Label the coordinates of the known points.

(b) Write an equation that models the cables.

(c) Complete the table by finding the height y of the suspension cables over the roadway at a distance of x meters from the center of the bridge.

Distance, x	0	200	400	500	600
Height, y					

34. Beam Deflection A simply supported beam (see figure) is 64 feet long and has a load at the center. The deflection of the beam at its center is 1 inch. The shape of the deflected beam is parabolic.

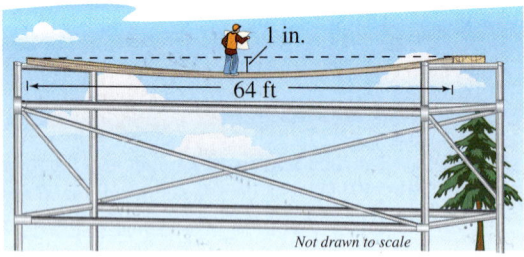

Not drawn to scale

(a) Find an equation of the parabola. (Assume that the origin is at the center of the beam.)

(b) How far from the center of the beam is the deflection $\frac{1}{2}$ inch?

In Exercises 35–42, find the center and vertices of the ellipse and sketch its graph.

35. $\dfrac{x^2}{25} + \dfrac{y^2}{16} = 1$

36. $\dfrac{x^2}{144} + \dfrac{y^2}{169} = 1$

37. $\dfrac{x^2}{25/9} + \dfrac{y^2}{16/9} = 1$

38. $\dfrac{x^2}{4} + \dfrac{y^2}{1/4} = 1$

39. $\dfrac{x^2}{9} + \dfrac{y^2}{5} = 1$

40. $\dfrac{x^2}{28} + \dfrac{y^2}{64} = 1$

41. $4x^2 + y^2 = 1$

42. $4x^2 + 9y^2 = 36$

In Exercises 43–52, find the standard form of the equation of the ellipse with the given characteristics and center at the origin.

43.

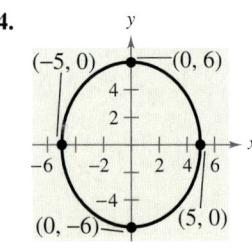

44.

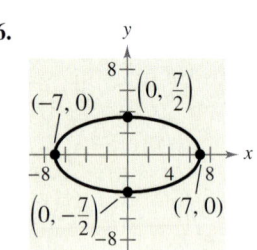

45.

(0, 3/2), (−2, 0), (2, 0), (0, −3/2)

46.

(−7, 0), (0, 7/2), (0, −7/2), (7, 0)

47. Vertices: $(\pm 5, 0)$; foci: $(\pm 2, 0)$

48. Vertices: $(0, \pm 8)$; foci: $(0, \pm 4)$

49. Foci: $(\pm 5, 0)$; major axis of length 12

50. Foci: $(\pm 2, 0)$; major axis of length 8

51. Vertices: $(0, \pm 5)$; passes through the point $(4, 2)$

52. Major axis vertical; passes through the points $(0, 4)$ and $(2, 0)$

53. Architecture A fireplace arch is to be constructed in the shape of a semiellipse. The opening is to have a height of 2 feet at the center and a width of 6 feet along the base (see figure). The contractor draws the outline of the ellipse on the wall by the method shown in Figure 4.26. Give the required positions of the tacks and the length of the string.

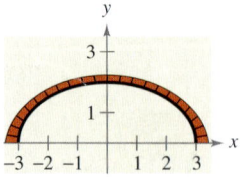

54. Architecture A semielliptical arch over a tunnel for a one-way road through a mountain has a major axis of 50 feet and a height at the center of 10 feet.

(a) Draw a rectangular coordinate system on a sketch of the tunnel with the center of the road entering the tunnel at the origin. Identify the coordinates of the known points.

(b) Find an equation of the semielliptical arch over the tunnel.

(c) You are driving a moving truck that has a width of 8 feet and a height of 9 feet. Will the moving truck clear the opening of the arch?

55. *Architecture*   Repeat Exercise 54 for a semielliptical arch with a major axis of 40 feet and a height at the center of 15 feet. The dimensions of the truck are 10 feet wide by 14 feet high.

56. *Geometry*   A line segment through a focus of an ellipse with endpoints on the ellipse and perpendicular to the major axis is called a **latus rectum** of the ellipse. Therefore, an ellipse has two latera recta. Knowing the length of the latera recta is helpful in sketching an ellipse because it yields other points on the curve (see figure). Show that the length of each latus rectum is $2b^2/a$.

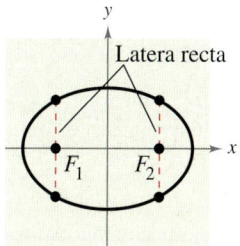

In Exercises 57–60, sketch the graph of the ellipse, using the latera recta (see Exercise 56).

57. $\dfrac{x^2}{4} + \dfrac{y^2}{1} = 1$ **58.** $\dfrac{x^2}{9} + \dfrac{y^2}{16} = 1$

59. $9x^2 + 4y^2 = 36$ **60.** $5x^2 + 3y^2 = 15$

In Exercises 61–68, find the center and vertices of the hyperbola and sketch its graph, using asymptotes as sketching aids.

61. $x^2 - y^2 = 1$ **62.** $\dfrac{x^2}{9} - \dfrac{y^2}{16} = 1$

63. $\dfrac{y^2}{1} - \dfrac{x^2}{4} = 1$ **64.** $\dfrac{y^2}{9} - \dfrac{x^2}{1} = 1$

65. $\dfrac{y^2}{25} - \dfrac{x^2}{144} = 1$ **66.** $\dfrac{x^2}{36} - \dfrac{y^2}{4} = 1$

67. $4y^2 - x^2 = 1$ **68.** $4y^2 - 9x^2 = 36$

In Exercises 69–76, find the standard form of the equation of the hyperbola with the given characteristics and center at the origin.

69. Vertices: $(0, \pm 2)$; foci: $(0, \pm 4)$

70. Vertices: $(\pm 3, 0)$; foci: $(\pm 5, 0)$

71. Vertices: $(\pm 1, 0)$; asymptotes: $y = \pm 3x$

72. Vertices: $(0, \pm 3)$; asymptotes: $y = \pm 3x$

73. Foci: $(0, \pm 8)$; asymptotes: $y = \pm 4x$

74. Foci: $(\pm 10, 0)$; asymptotes: $y = \pm \frac{3}{4}x$

75. Vertices: $(0, \pm 3)$; passes through the point: $(-2, 5)$

76. Vertices: $(\pm 2, 0)$; passes through the point: $\left(3, \sqrt{3}\right)$

77. *Art*   A sculpture has a hyperbolic cross section (see figure).

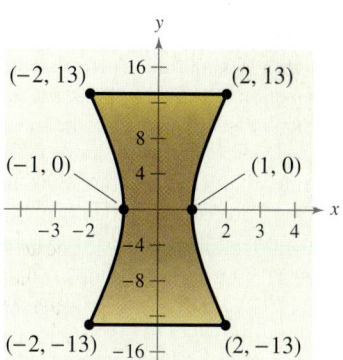

(a) Write an equation that models the curved sides of the sculpture.

(b) Each unit on the coordinate plane represents 1 foot. Find the width of the sculpture at a height of 5 feet.

78. *Optics*   A hyperbolic mirror (used in some telescopes) has the property that a light ray directed at the focus will be reflected to the other focus. The focus of a hyperbolic mirror (see figure) has coordinates $(24, 0)$. Find the vertex of the mirror if its mount at the top edge of the mirror has coordinates $(24, 24)$.

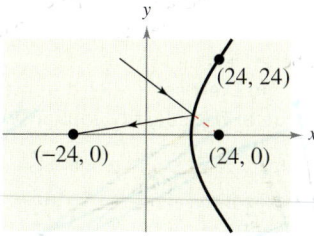

79. *Aeronautics*   When an airplane travels faster than the speed of sound, the sound waves form a cone behind the airplane. If the airplane is flying parallel to the ground, the sound waves intersect the ground in a hyperbola with the airplane directly above its center (see figure). A sonic boom is heard along the hyperbola. You hear a sonic boom that is audible along a hyperbola with the equation

$$\frac{x^2}{100} - \frac{y^2}{4} = 1$$

where x and y are measured in miles. What is the shortest horizontal distance you could be to the airplane?

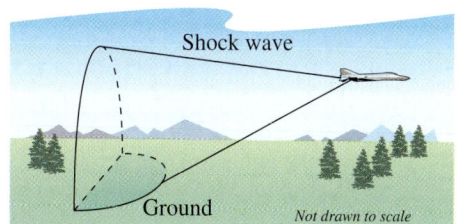

80. *Navigation* Long distance radio navigation for aircraft and ships uses synchronized pulses transmitted by widely separated transmitting stations. These pulses travel at the speed of light (186,000 miles per second). The difference in the times of arrival of these pulses at an aircraft or ship is constant on a hyperbola having the transmitting stations as foci.

Assume that two stations 300 miles apart are positioned on a rectangular coordinate system at points with coordinates $(-150, 0)$ and $(150, 0)$ and that a ship is traveling on a path with coordinates $(x, 75)$, as shown in the figure. Find the x-coordinate of the position of the ship if the time difference between the pulses from the transmitting stations is 1000 microseconds (0.001 second).

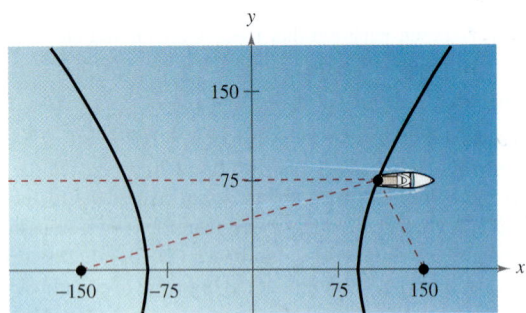

Synthesis

True or False? In Exercises 81–83, determine whether the statement is true or false. Justify your answer.

81. The equation $x^2 - y^2 = 144$ represents a circle.

82. The major axis of the ellipse $y^2 + 16x^2 = 64$ is vertical.

83. It is possible for a parabola to intersect its directrix.

84. *Exploration* Consider the ellipse

$$\frac{x^2}{a^2} + \frac{y^2}{b^2} = 1, \quad a + b = 20.$$

(a) The area of the ellipse is given by $A = \pi ab$. Write the area of the ellipse as a function of a.

(b) Find the equation of an ellipse with an area of 264 square centimeters.

(c) Complete the table using your equation from part (a), and make a conjecture about the shape of the ellipse with maximum area.

a	8	9	10	11	12	13
A						

 (d) Use a graphing utility to graph the area function and use the graph to support your conjecture in part (c).

In Exercises 85–90, identify the conic. Explain your reasoning.

85. $4x^2 + 4y^2 - 16 = 0$

86. $4y^2 - 5x^2 + 20 = 0$

87. $3y^2 - 6x = 0$

88. $2x^2 + 4y^2 - 12 = 0$

89. $4x^2 + y^2 - 16 = 0$

90. $2x^2 - 12y = 0$

91. *Think About It* How can you tell if an ellipse is a circle from the equation?

92. *Think About It* Is the graph of $x^2 + 4y^4 = 4$ an ellipse? Explain.

93. *Think About It* The graph of $x^2 - y^2 = 0$ is a degenerate conic. Sketch this graph and identify the degenerate conic.

94. *Think About It* Which part of the graph of the ellipse $4x^2 + 9y^2 = 36$ is represented by each equation? (Do not graph.)

(a) $x = -\frac{3}{2}\sqrt{4 - y^2}$ (b) $y = \frac{2}{3}\sqrt{9 - x^2}$

95. *Writing* At the beginning of this section, you learned that each type of conic section can be formed by the intersection of a plane and a double-napped cone. Write a short paragraph describing examples of physical situations in which hyperbolas are formed.

96. *Writing* Write a paragraph discussing the changes in the shape and orientation of the graph of the ellipse

$$\frac{x^2}{a^2} + \frac{y^2}{4^2} = 1$$

as a increases from 1 to 8.

97. Use the definition of an ellipse to derive the standard form of the equation of an ellipse.

98. Use the definition of a hyperbola to derive the standard form of the equation of a hyperbola.

Skills Review

In Exercises 99–102, sketch the graph of the quadratic function without using a graphing utility. Identify the vertex.

99. $f(x) = x^2 - 8$

100. $f(x) = 25 - x^2$

101. $f(x) = x^2 + 8x + 12$

102. $f(x) = x^2 - 4x - 21$

103. Find a polynomial with integer coefficients that has the zeros 3, $2 + i$, and $2 - i$.

104. Find all the zeros of $f(x) = 2x^3 - 3x^2 + 50x - 75$ if one of the zeros is $x = \frac{3}{2}$.

105. List the possible rational zeros of the function given by $g(x) = 6x^4 + 7x^3 - 29x^2 - 28x + 20$.

 106. Use a graphing utility to graph the function given by $h(x) = 2x^4 + x^3 - 19x^2 - 9x + 9$. Use the graph and the Rational Zero Test to find the zeros of h.

4.4 Translations of Conics

What you should learn

- Recognize equations of conics that have been shifted vertically or horizontally in the plane.
- Write and graph equations of conics that have been shifted vertically or horizontally in the plane.

Why you should learn it

In some real-life applications, it is not convenient to use conics whose centers or vertices are at the origin. For instance, in Exercise 31 on page 373, a parabola can be used to model the maximum revenue of Papa John's International.

Vertical and Horizontal Shifts of Conics

In Section 4.3 you looked at conic sections whose graphs were in *standard position*. In this section you will study the equations of conic sections that have been shifted vertically or horizontally in the plane.

Standard Forms of Equations of Conics

Circle: Center $= (h, k)$; radius $= r$

$$(x - h)^2 + (y - k)^2 = r^2$$

Ellipse: Center $= (h, k)$

Major axis length $= 2a$; minor axis length $= 2b$

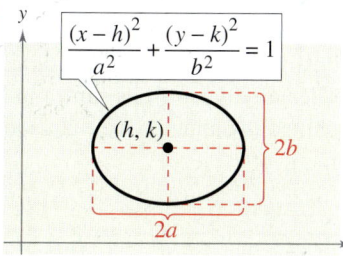

$$\frac{(x-h)^2}{a^2} + \frac{(y-k)^2}{b^2} = 1$$

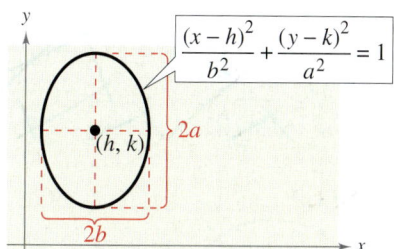

$$\frac{(x-h)^2}{b^2} + \frac{(y-k)^2}{a^2} = 1$$

Hyperbola: Center $= (h, k)$

Transverse axis length $= 2a$; conjugate axis length $= 2b$

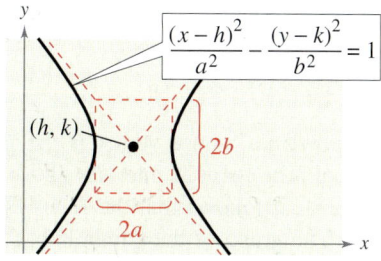

$$\frac{(x-h)^2}{a^2} - \frac{(y-k)^2}{b^2} = 1$$

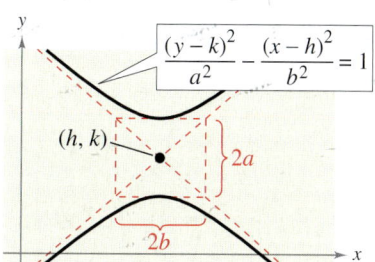

$$\frac{(y-k)^2}{a^2} - \frac{(x-h)^2}{b^2} = 1$$

Parabola: Vertex $= (h, k)$

Directed distance from vertex to focus $= p$

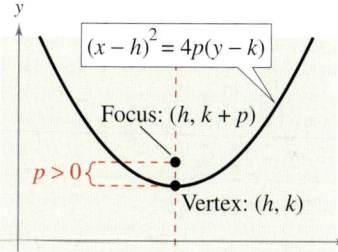

$(x - h)^2 = 4p(y - k)$

Focus: $(h, k + p)$

$p > 0$

Vertex: (h, k)

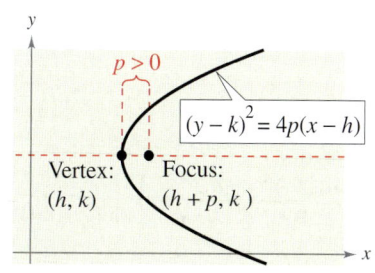

$(y - k)^2 = 4p(x - h)$

$p > 0$

Vertex: (h, k) Focus: $(h + p, k)$

STUDY TIP

Consider the equation of the ellipse

$$\frac{(x-h)^2}{a^2} + \frac{(y-k)^2}{b^2} = 1.$$

If you let $a = b$, then the equation can be rewritten as

$$(x - h)^2 + (y - k)^2 = a^2$$

which is the standard form of the equation of a circle with radius $r = a$ (see Section 1.1). Geometrically, when $a = b$ for an ellipse, the major and minor axes are of equal length, and so the graph is a circle [see Example 1(a)].

Example 1   Equations of Conic Sections

Identify each conic. Then describe the translation of the graph of the conic.

a. $(x - 1)^2 + (y + 2)^2 = 3^2$ **b.** $\dfrac{(x - 2)^2}{3^2} + \dfrac{(y - 1)^2}{2^2} = 1$

c. $\dfrac{(x - 3)^2}{1^2} - \dfrac{(y - 2)^2}{3^2} = 1$ **d.** $(x - 2)^2 = 4(-1)(y - 3)$

Solution

a. The graph of $(x - 1)^2 + (y + 2)^2 = 3^2$ is a *circle* whose center is the point $(1, -2)$ and whose radius is 3, as shown in Figure 4.37. The graph of the circle has been shifted one unit to the right and two units downward from standard position.

b. The graph of

$$\frac{(x - 2)^2}{3^2} + \frac{(y - 1)^2}{2^2} = 1$$

is an *ellipse* whose center is the point $(2, 1)$. The major axis of the ellipse is horizontal and of length $2(3) = 6$, and the minor axis of the ellipse is vertical and of length $2(2) = 4$, as shown in Figure 4.38. The graph of the ellipse has been shifted two units to the right and one unit upward from standard position.

c. The graph of

$$\frac{(x - 3)^2}{1^2} - \frac{(y - 2)^2}{3^2} = 1$$

is a *hyperbola* whose center is the point $(3, 2)$. The transverse axis is horizontal and of length $2(1) = 2$, and the conjugate axis is vertical and of length $2(3) = 6$, as shown in Figure 4.39. The graph of the hyperbola has been shifted three units to the right and two units upward from standard position.

d. The graph of

$$(x - 2)^2 = 4(-1)(y - 3)$$

is a *parabola* whose vertex is the point $(2, 3)$. The axis of the parabola is vertical. The focus is one unit above or below the vertex. Moreover, because $p = -1$, it follows that the focus lies *below* the vertex, as shown in Figure 4.40. The graph of the parabola has been reflected in the x-axis, shifted two units to the left, and shifted three units upward from standard position.

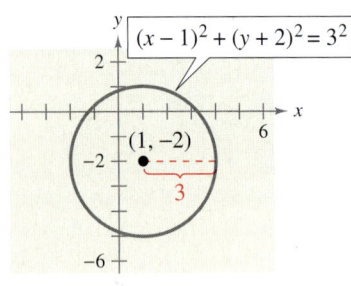

FIGURE **4.37**   *Circle*

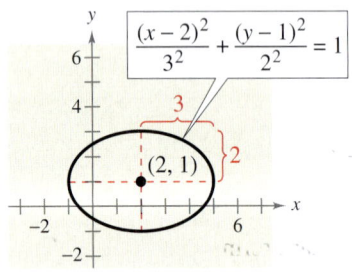

FIGURE **4.38**   *Ellipse*

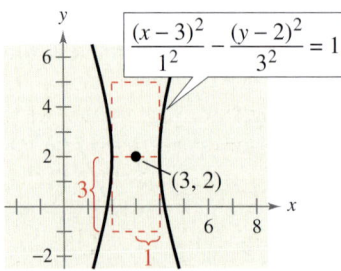

FIGURE **4.39**   *Hyperbola*

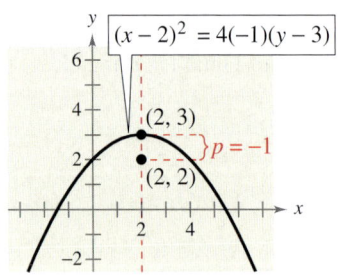

FIGURE **4.40**   *Parabola*

✔**CHECKPOINT**   Now try Exercise 1.

Equations of Conics in Standard Form

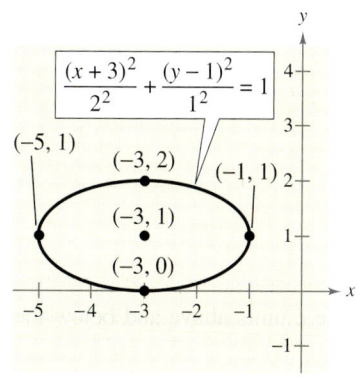

FIGURE **4.41**

Example 2   Finding the Standard Form of a Parabola

Find the vertex and focus of the parabola $x^2 - 2x + 4y - 3 = 0$.

Solution

Complete the square to write the equation in standard form.

$x^2 - 2x + 4y - 3 = 0$	Write original equation.
$x^2 - 2x = -4y + 3$	Group terms.
$x^2 - 2x + 1 = -4y + 3 + 1$	Add 1 to each side.
$(x - 1)^2 = -4y + 4$	Write in completed square form.
$(x - 1)^2 = 4(-1)(y - 1)$	Write in standard form, $(x - h)^2 = 4p(y - k)$.

From this standard form, it follows that $h = 1$, $k = 1$, and $p = -1$. Because the axis is vertical and p is negative, the parabola opens downward. The vertex is $(h, k) = (1, 1)$ and the focus is $(h, k + p) = (1, 0)$. (See Figure 4.41.)

✔CHECKPOINT   Now try Exercise 23.

STUDY TIP

Note in Example 2 that p is the directed distance from the vertex to the focus. Because the axis of the parabola is vertical and $p = -1$, the focus is one unit below the vertex, and the parabola opens downward.

Example 3   Sketching an Ellipse

Sketch the ellipse $x^2 + 4y^2 + 6x - 8y + 9 = 0$.

Solution

Complete the square to write the equation in standard form.

$x^2 + 4y^2 + 6x - 8y + 9 = 0$	Write original equation.
$(x^2 + 6x + \quad) + (4y^2 - 8y + \quad) = -9$	Group terms.
$(x^2 + 6x + \quad) + 4(y^2 - 2y + \quad) = -9$	Factor 4 out of y-terms.
$(x^2 + 6x + 9) + 4(y^2 - 2y + 1) = -9 + 9 + 4(1)$	Add 9 and $4(1) = 4$ to each side.
$(x + 3)^2 + 4(y - 1)^2 = 4$	Write in completed square form.
$\dfrac{(x + 3)^2}{4} + \dfrac{4(y - 1)^2}{4} = 1$	Divide each side by 4.
$\dfrac{(x + 3)^2}{2^2} + \dfrac{(y - 1)^2}{1^2} = 1$	$\dfrac{(x - h)^2}{a^2} + \dfrac{(y - k)^2}{b^2} = 1$

From this standard form, it follows that the center is $(h, k) = (-3, 1)$. Because the denominator of the x-term is $a^2 = 2^2$, the endpoints of the major axis lie two units to the right and left of the center. Similarly, because the denominator of the y-term is $b^2 = 1^2$, the endpoints of the minor axis lie one unit up and down from the center. The ellipse is shown in Figure 4.42.

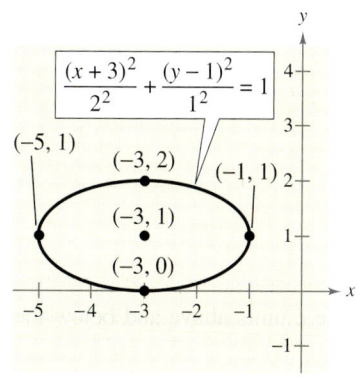

FIGURE **4.42**

✔CHECKPOINT   Now try Exercise 39.

4.4 | Exercises

VOCABULARY CHECK: Match the description of the conic with its standard equation. The equations are labeled (a), (b), (c), (d), (e), (f), and g.

(a) $\dfrac{(x-h)^2}{a^2} + \dfrac{(y-k)^2}{b^2} = 1$ 　　 **(b)** $\dfrac{(x-h)^2}{a^2} - \dfrac{(y-k)^2}{b^2} = 1$ 　　 **(c)** $\dfrac{(y-k)^2}{a^2} - \dfrac{(x-h)^2}{b^2} = 1$

(d) $\dfrac{(x-h)^2}{b^2} + \dfrac{(y-k)^2}{a^2} = 1$ 　　 **(e)** $(x-h)^2 = 4p(y-k)$ 　　 **(f)** $(y-k)^2 = 4p(x-h)$

(g) $(x-h)^2 + (y-k)^2 = r^2$

1. Circle 　　　　 **2.** Ellipse with vertical major axis 　　　 **3.** Parabola with vertical axis

4. Hyperbola with horizontal transverse axis 　　 **5.** Ellipse with horizontal major axis 　　 **6.** Parabola with horizontal axis

7. Hyperbola with vertical transverse axis

PREREQUISITE SKILLS REVIEW: Practice and review algebra skills needed for this section at **www.Eduspace.com.**

In Exercises 1–6, describe the translation of the graph of the conic.

1. $(x+2)^2 + (y-1)^2 = 4$ 　　 **2.** $(y-1)^2 = 4(2)(x+2)$

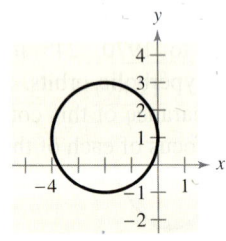 　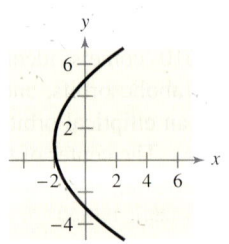

3. $\dfrac{(y+3)^2}{4} - (x-1)^2 = 1$ 　　 **4.** $\dfrac{(x-2)^2}{9} + \dfrac{(y+1)^2}{4} = 1$

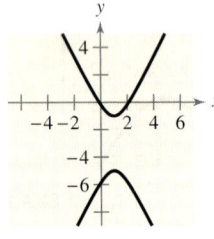 　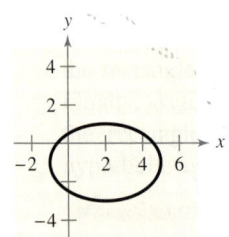

5. $\dfrac{(x+4)^2}{9} + \dfrac{(y+2)^2}{16} = 1$ 　　 **6.** $\dfrac{(x+2)^2}{4} - \dfrac{(y-3)^2}{9} = 1$

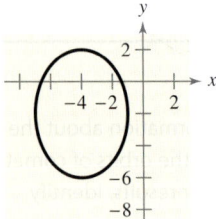 　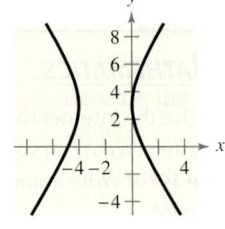

In Exercises 7–12, identify the center and radius of the circle.

7. $x^2 + y^2 = 49$ 　　　　　 **8.** $x^2 + y^2 = 1$

9. $(x+3)^2 + (y-8)^2 = 16$

10. $(x+9)^2 + (y+1)^2 = 36$

11. $(x-1)^2 + y^2 = 10$ 　　 **12.** $x^2 + (y+12)^2 = 24$

In Exercises 13–16, write the equation of the circle in standard form, and then identify its center and radius.

13. $x^2 + y^2 - 2x + 6y + 9 = 0$

14. $x^2 + y^2 - 10x - 6y + 25 = 0$

15. $4x^2 + 4y^2 + 12x - 24y + 41 = 0$

16. $9x^2 + 9y^2 + 54x - 36y + 17 = 0$

In Exercises 17–24, find the vertex, focus, and directrix of the parabola, and sketch its graph.

17. $(x-1)^2 + 8(y+2) = 0$ 　　 **18.** $(x+3) + (y-2)^2 = 0$

19. $\left(y+\tfrac{1}{2}\right)^2 = 2(x-5)$ 　　 **20.** $\left(x+\tfrac{1}{2}\right)^2 = 4(y-3)$

21. $y = \tfrac{1}{4}(x^2 - 2x + 5)$ 　　 **22.** $4x - y^2 - 2y - 33 = 0$

23. $y^2 + 6y + 8x + 25 = 0$ 　　 **24.** $y^2 - 4y - 4x = 0$

In Exercises 25–30, find the standard form of the equation of the parabola with the given characteristics.

25. Vertex: $(3, 2)$; focus: $(1, 2)$

26. Vertex: $(-1, 2)$; focus: $(-1, 0)$

27. Vertex: $(0, 4)$; directrix: $y = 2$

28. Vertex: $(-2, 1)$; directrix: $x = 1$

29. Focus: $(2, 2)$; directrix: $x = -2$

30. Focus: $(0, 0)$; directrix: $y = 4$

Model It

31. Revenue The revenues R (in millions of dollars) for Papa John's International for the years 1997 through 2003 are shown in the table. (Source: Papa John's International)

Year	Revenue, R
1997	508.8
1998	669.8
1999	805.3
2000	944.7
2001	971.2
2002	946.2
2003	917.4

(a) Use a graphing utility to find an equation of the parabola $y = at^2 + bt + c$ that models the data. Write the equation in standard form. Let t represent the year, with $t = 7$ corresponding to 1997.

(b) Find the coordinates of the vertex and interpret its meaning in the context of the problem.

(c) Use a graphing utility to graph the function.

(d) Use the *trace* feature of the graphing utility to approximate *graphically* the year in which revenue was maximum.

(e) Use the *table* feature of the graphing utility to approximate *numerically* the year in which revenue was maximum.

(f) Compare the results of parts (b), (d), and (e). What did you learn by using all three approaches?

32. Satellite Orbit A satellite in a 100-mile-high circular orbit around Earth has a velocity of approximately 17,500 miles per hour (see figure). If this velocity is multiplied by $\sqrt{2}$, the satellite will have the minimum velocity necessary to escape Earth's gravity and it will follow a parabolic path with the center of Earth as the focus.

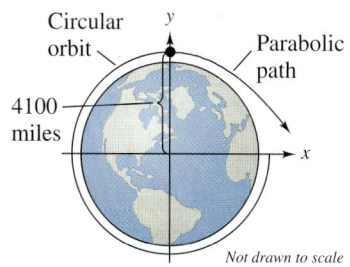

Not drawn to scale

(a) Find the escape velocity of the satellite.

(b) Find an equation of its path (assume that the radius of Earth is 4000 miles).

33. Projectile Motion A cargo plane is flying at an altitude of 30,000 feet and a speed of 540 miles per hour (792 feet per second). How many feet will a supply crate dropped from the plane travel horizontally before it hits the ground if the path of the crate is modeled by

$$x^2 = -39,204(y - 30,000)?$$

34. Path of a Projectile The path of a softball is modeled by $-12.5(y - 7.125) = (x - 6.25)^2$. The coordinates x and y are measured in feet, with $x = 0$ corresponding to the position from which the ball was thrown.

(a) Use a graphing utility to graph the trajectory of the softball.

(b) Use the *trace* feature of the graphing utility to approximate the highest point and the range of the trajectory.

In Exercises 35–42, find the center, foci, and vertices of the ellipse, and sketch its graph.

35. $\dfrac{(x-1)^2}{9} + \dfrac{(y-5)^2}{25} = 1$ **36.** $\dfrac{(x-6)^2}{4} + \dfrac{(y+7)^2}{16} = 1$

37. $(x+2)^2 + \dfrac{(y+4)^2}{1/4} = 1$ **38.** $\dfrac{(x-3)^2}{25/9} + (y-8)^2 = 1$

39. $9x^2 + 4y^2 + 36x - 24y + 36 = 0$

40. $9x^2 + 4y^2 - 36x + 8y + 31 = 0$

41. $16x^2 + 25y^2 - 32x + 50y + 16 = 0$

42. $9x^2 + 25y^2 - 36x - 50y + 61 = 0$

In Exercises 43–52, find the standard form of the equation of the ellipse with the given characteristics.

43. Vertices: $(4, 4), (4, -4)$; minor axis of length 6

44. Vertices: $(-1, 2), (5, 2)$; minor axis of length 4

45. Vertices: $(0, 2), (4, 2)$; minor axis of length 2

46. Foci: $(0, 0), (4, 0)$; major axis of length 8

47. Foci: $(0, 0), (0, 8)$; major axis of length 16

48. Center: $(2, -1)$; vertex: $\left(2, \frac{1}{2}\right)$; Minor axis of length 2

49. Center: $(0, 4)$; $a = 2c$; vertices: $(-4, 4), (4, 4)$

50. Center: $(3, 2)$; $a = 3c$; foci: $(1, 2), (5, 2)$

51. Vertices: $(0, 2), (4, 2)$; endpoints of the minor axis: $(2, 3), (2, 1)$

52. Vertices: $(5, 0), (5, 12)$; endpoints of the minor axis: $(0, 6), (10, 6)$

In Exercises 53 and 54, e is called the *eccentricity* of an ellipse, and is defined by $e = c/a$. It measures the flatness of the ellipse.

53. Find the standard form of the equation of the ellipse with vertices $(\pm 5, 0)$ and eccentricity $e = \frac{3}{5}$.

54. Find the standard form of the equation of the ellipse with vertices $(0, \pm 8)$ and eccentricity $e = \frac{1}{2}$.

55. *Planetary Motion* The planet Pluto moves in an elliptical orbit with the sun at one of the foci, as shown in the figure. The length of half of the major axis, a, is 3.67×10^9 miles, and the eccentricity is 0.249. Find the smallest distance (*perihelion*) and the greatest distance (*aphelion*) of Pluto from the center of the sun.

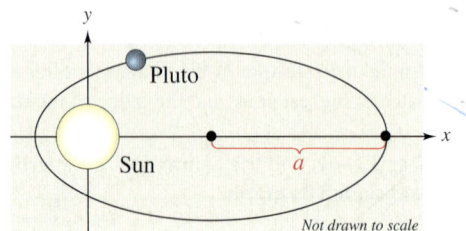

Not drawn to scale

56. *Australian Football* In Australia, football by Australian Rules (or rugby) is played on elliptical fields. The field can be a maximum of 170 yards wide and a maximum of 200 yards long. Let the center of a field of maximum size be represented by the point $(0, 85)$. Write the standard form of the equation of the ellipse that represents this field. (Source: Australian Football League)

In Exercises 57–66, find the center, foci, and vertices of the hyperbola, and sketch its graph, using the asymptotes as an aid.

57. $\dfrac{(x-1)^2}{4} - \dfrac{(y+2)^2}{1} = 1$ **58.** $\dfrac{(x-1)^2}{144} - \dfrac{(y-4)^2}{25} = 1$

59. $(y+6)^2 - (x-2)^2 = 1$ **60.** $\dfrac{(y-1)^2}{1/4} - \dfrac{(x+3)^2}{1/9} = 1$

61. $9x^2 - y^2 - 36x - 6y + 18 = 0$

62. $x^2 - 9y^2 + 36y - 72 = 0$

63. $x^2 - 9y^2 + 2x - 54y - 80 = 0$

64. $16y^2 - x^2 + 2x + 64y + 63 = 0$

65. $9y^2 - 4x^2 + 8x + 18y + 41 = 0$

66. $11y^2 - 3x^2 + 12x + 44y + 48 = 0$

In Exercises 67–76, find the standard form of the equation of the hyperbola with the given characteristics.

67. Vertices: $(0, 2), (0, 0)$; foci: $(0, 3), (0, -1)$

68. Vertices: $(1, 2), (5, 2)$; foci: $(0, 2), (6, 2)$

69. Vertices: $(2, 0), (6, 0)$; foci: $(0, 0), (8, 0)$

70. Vertices: $(2, 3), (2, -3)$; foci: $(2, 5), (2, -5)$

71. Vertices: $(4, 1), (4, 9)$; foci: $(4, 0), (4, 10)$

72. Vertices: $(-2, 1), (2, 1)$; foci: $(-3, 1), (3, 1)$

73. Vertices: $(2, 3), (2, -3)$; passes through the point $(0, 5)$

74. Vertices: $(-2, 1), (2, 1)$; passes through the point $(4, 3)$

75. Vertices: $(0, 2), (6, 2)$; asymptotes: $y = \frac{2}{3}x, y = 4 - \frac{2}{3}x$

76. Vertices: $(3, 0), (3, 4)$; asymptotes: $y = \frac{2}{3}x, y = 4 - \frac{2}{3}x$

In Exercises 77–84, identify the conic by writing its equation in standard form. Then sketch its graph.

77. $x^2 + y^2 - 6x + 4y + 9 = 0$

78. $x^2 + 4y^2 - 6x + 16y + 21 = 0$

79. $4x^2 - y^2 - 4x - 3 = 0$

80. $y^2 - 4y - 4x = 0$

81. $4x^2 + 3y^2 + 8x - 24y + 51 = 0$

82. $4y^2 - 2x^2 - 4y - 8x - 15 = 0$

83. $25x^2 - 10x - 200y - 119 = 0$

84. $4x^2 + 4y^2 - 16y + 15 = 0$

Synthesis

True or False? In Exercises 85 and 86, determine whether the statement is true or false. Justify your answer.

85. The conic represented by the equation $3x^2 + 2y^2 - 18x - 16y + 58 = 0$ is an ellipse.

86. The graphs of $x^2 + 10y - 10x + 5 = 0$ and $x^2 + 16y^2 + 10x - 32y - 23 = 0$ do not intersect.

Exploration In Exercises 87 and 88, consider the ellipse

$$\frac{x^2}{a^2} + \frac{y^2}{b^2} = 1.$$

87. Show that the equation of the ellipse can be written as

$$\frac{(x-h)^2}{a^2} + \frac{(y-k)^2}{a^2(1-e^2)} = 1$$

where e is the eccentricity.

88. Use a graphing utility to graph the ellipse

$$\frac{(x-2)^2}{4} + \frac{(y-3)^2}{4(1-e^2)} = 1$$

for $e = 0.95, 0.75, 0.5, 0.25,$ and 0. Make a conjecture about the change in the shape of the ellipse as e approaches 0.

Skills Review

In Exercises 89–92, find the inverse function of f.

89. $f(x) = 10 - 7x$ **90.** $f(x) = 2 - x^3$

91. $f(x) = \sqrt{x+8}$ **92.** $f(x) = \sqrt[3]{x^3 + 4}$

4 Chapter Summary

What did you learn?

4 Review Exercises

4.1 In Exercises 1–4, find the domain of the rational function.

1. $f(x) = \dfrac{5x}{x + 12}$

2. $f(x) = \dfrac{3x^2}{1 + 3x}$

3. $f(x) = \dfrac{8}{x^2 - 10x + 24}$

4. $f(x) = \dfrac{x^2 + x - 2}{x^2 + 4}$

In Exercises 5–10, identify any horizontal or vertical asymptotes.

5. $f(x) = \dfrac{4}{x + 3}$

6. $f(x) = \dfrac{2x^2 + 5x - 3}{x^2 + 2}$

7. $g(x) = \dfrac{x^2}{x^2 - 4}$

8. $g(x) = \dfrac{1}{(x - 3)^2}$

9. $h(x) = \dfrac{2x - 10}{x^2 - 2x - 15}$

10. $h(x) = \dfrac{x^3 - 4x^2}{x^2 + 3x + 2}$

11. *Average Cost* A business has a production cost of $C = 0.5x + 500$ for producing x units of a product. The average cost per unit, $\overline{C}$, is given by

$$\overline{C} = \frac{C}{x} = \frac{0.5x + 500}{x}, \quad x > 0.$$

Determine the average cost per unit as x increases without bound. (Find the horizontal asymptote.)

12. *Seizure of Illegal Drugs* The cost C (in millions of dollars) for the federal government to seize $p\%$ of an illegal drug as it enters the country is given by

$$C = \frac{528p}{100 - p}, \quad 0 \le p < 100.$$

 (a) Use a graphing utility to graph the cost function.

(b) Find the costs of seizing 25%, 50%, and 75% of the drug.

(c) According to this model, would it be possible to seize 100% of the drug?

4.2 In Exercises 13–24, (a) state the domain of the function, (b) identify all intercepts, (c) find any vertical and horizontal asymptotes, and (d) plot additional solution points as needed to sketch the graph of the rational function.

13. $f(x) = \dfrac{-5}{x^2}$

14. $f(x) = \dfrac{4}{x}$

15. $g(x) = \dfrac{2 + x}{1 - x}$

16. $h(x) = \dfrac{x - 3}{x - 2}$

17. $p(x) = \dfrac{x^2}{x^2 + 1}$

18. $f(x) = \dfrac{2x}{x^2 + 4}$

19. $f(x) = \dfrac{x}{x^2 + 1}$

20. $h(x) = \dfrac{4}{(x - 1)^2}$

21. $f(x) = \dfrac{-6x^2}{x^2 + 1}$

22. $y = \dfrac{2x^2}{x^2 - 4}$

23. $f(x) = \dfrac{6x^2 - 11x + 3}{3x^2 - x}$

24. $f(x) = \dfrac{6x^2 - 7x + 2}{4x^2 - 1}$

In Exercises 25–30, (a) state the domain of the function, (b) identify all intercepts, (c) identify any vertical and slant asymptotes, and (d) plot additional solution points as needed to sketch the graph of the rational function.

25. $f(x) = \dfrac{2x^3}{x^2 + 1}$

26. $f(x) = \dfrac{x^2 + 1}{x + 1}$

27. $f(x) = \dfrac{x^2 + 3x - 10}{x + 2}$

28. $f(x) = \dfrac{x^3}{x^2 - 4}$

29. $f(x) = \dfrac{3x^3 - 2x^2 - 3x + 2}{3x^2 - x - 4}$

30. $f(x) = \dfrac{3x^3 - 4x^2 - 12x + 16}{3x^2 + 5x - 2}$

31. *Average Cost* The cost of producing x units of a product is C, and the average cost per unit $\overline{C}$ is given by

$$\overline{C} = \frac{C}{x} = \frac{100,000 + 0.9x}{x}, \quad x > 0.$$

(a) Graph the average cost function.

(b) Find the average costs of producing $x = 1000, 10,000,$ and $100,000$ units.

(c) By increasing the level of production, what is the smallest average cost per unit you can obtain? Explain your reasoning.

32. *Page Design* A page that is x inches wide and y inches high contains 30 square inches of print. The top and bottom margins are 2 inches deep and the margins on each side are 2 inches wide.

(a) Draw a diagram that gives a visual representation of the problem.

(b) Show that the total area A on the page is

$$A = \frac{2x(2x + 7)}{x - 4}.$$

(c) Determine the domain of the function based on the physical constraints of the problem.

 (d) Use a graphing utility to graph the area function and approximate the page size for which the least amount of paper will be used. Verify your answer numerically using the *table* feature of the graphing utility.

33. *Photosynthesis* The amount y of CO_2 uptake (in milligrams per square decimeter per hour) at optimal temperatures and with the natural supply of CO_2 is approximated by the model

$$y = \frac{18.47x - 2.96}{0.23x + 1}, \quad x > 0$$

where x is the light intensity (in watts per square meter). Use a graphing utility to graph the function and determine the limiting amount of CO_2 uptake.

34. *Medicine* The concentration C of a medication in the bloodstream t hours after injection into muscle tissue is given by

$$C(t) = \frac{2t + 1}{t^2 + 4}, \quad t > 0.$$

(a) Determine the horizontal asymptote of the graph of the function and interpret its meaning in the context of the problem.

 (b) Use a graphing utility to graph the function and approximate the time when the bloodstream concentration is greatest.

4.3 In Exercises 35–42, identify the conic.

35. $y^2 = -12x$

36. $16x^2 + y^2 = 16$

37. $\dfrac{x^2}{9} - \dfrac{y^2}{1} = 1$

38. $\dfrac{x^2}{1} + \dfrac{y^2}{9} = 1$

39. $x^2 + 20y = 0$

40. $x^2 + y^2 = 100$

41. $\dfrac{y^2}{49} - \dfrac{x^2}{144} = 1$

42. $\dfrac{x^2}{49} + \dfrac{y^2}{144} = 1$

In Exercises 43–46, find the standard form of the equation of the parabola with the given focus and vertex at the origin.

43. Passes through the point $(1, 2)$; horizontal axis
44. Passes through the point $(4, -2)$; vertical axis

45. Focus: $(-6, 0)$

46. Focus: $(0, 3)$

47. *Satellite Antenna* A cross section of a large parabolic antenna (see figure) is modeled by

$$y = \frac{x^2}{200}, \quad -100 \le x \le 100.$$

The receiving and transmitting equipment is positioned at the focus. Find the coordinates of the focus.

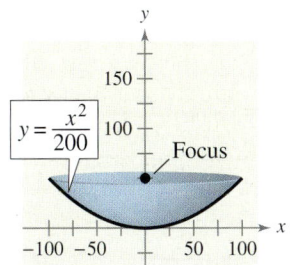

48. *Suspension Bridge* Each cable of a suspension bridge is suspended (in the shape of a parabola) between two towers (see figure).

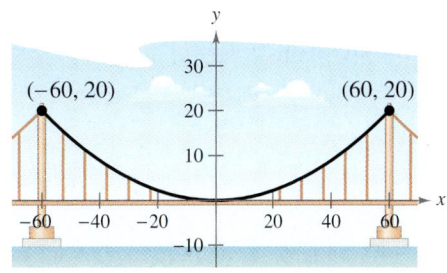

(a) Find the coordinates of the focus.

(b) Write an equation that models the cables.

In Exercises 49–54, find the standard form of the equation of the ellipse with the given characteristics and center at the origin.

49. Vertices: $(\pm 5, 0)$; minor axis of length 6

50. Vertices: $(0, \pm 10)$; minor axis of length 8

51. Vertices: $(0, \pm 6)$; passes through the point $(2, 2)$

52. Vertices: $(\pm 7, 0)$; foci: $(\pm 6, 0)$

53. Foci: $(\pm 6, 0)$; minor axis of length 10

54. Foci: $(\pm 3, 0)$; major axis of length 12

55. *Architecture* A semielliptical archway is to be formed over the entrance to an estate (see figure). The arch is to be set on pillars that are 10 feet apart and is to have a height (atop the pillars) of 4 feet. Where should the foci be placed in order to sketch the arch?

56. *Wading Pool* You are building a wading pool that is in the shape of an ellipse. Your plans give an equation for the elliptical shape of the pool measured in feet as

$$\frac{x^2}{324} + \frac{y^2}{196} = 1.$$

Find the longest distance across the pool, the shortest distance, and the distance between the foci.

In Exercises 57–60, find the standard form of the equation of the hyperbola with the given characteristics and center at the origin.

57. Vertices: $(0, \pm 1)$; foci: $(0, \pm 3)$

58. Vertices: $(\pm 4, 0)$; foci: $(\pm 6, 0)$

59. Vertices: $(\pm 1, 0)$; asymptotes: $y = \pm 2x$

60. Vertices: $(0, \pm 2)$; asymptotes: $y = \pm \dfrac{2}{\sqrt{5}} x$

4.4 **In Exercises 61–68, identify the conic by writing its equation in standard form. Then sketch its graph and describe the translation.**

61. $x^2 - 6x + 2y + 9 = 0$

62. $y^2 - 12y - 8x + 20 = 0$

63. $x^2 + y^2 - 2x - 4y + 5 = 0$

64. $16x^2 + 16y^2 - 16x + 24y - 3 = 0$

65. $x^2 + 9y^2 + 10x - 18y + 25 = 0$

66. $4x^2 + y^2 - 16x + 15 = 0$

67. $9x^2 - y^2 - 72x + 8y + 119 = 0$

68. $x^2 - 9y^2 + 10x + 18y + 7 = 0$

In Exercises 69–72, find the standard form of the equation of the parabola with the given characteristics.

69. Vertex: $(-6, 4)$; passes through the point: $(0, 0)$

70. Vertex: $(0, 5)$; passes through the point: $(6, 0)$

71. Vertex: $(4, 2)$; focus: $(4, 0)$

72. Vertex: $(2, 0)$; focus: $(0, 0)$

In Exercises 73–76, find the standard form of the equation of the ellipse with the given characteristics.

73. Vertices: $(0, 3)$, $(10, 3)$; passes through the point: $(5, 0)$

74. Center: $(0, 4)$; vertices: $(0, 0)$, $(0, 8)$

75. Vertices: $(-3, 0)$, $(7, 0)$; foci: $(0, 0)$, $(4, 0)$

76. Vertices: $(2, 0)$, $(2, 4)$; foci: $(2, 1)$, $(2, 3)$

In Exercises 77–82, find the standard form of the equation of the hyperbola with the given characteristics.

77. Vertices: $(\pm 6, 7)$; asymptotes: $y = -\frac{1}{2}x + 7$, $y = \frac{1}{2}x + 7$

78. Vertices: $(0, 0)$, $(0, -4)$; passes through the point: $\left(2, 2\left(\sqrt{5} - 1\right)\right)$

79. Vertices: $(-10, 3)$, $(6, 3)$; foci: $(-12, 3)$, $(8, 3)$

80. Vertices: $(2, 2)$, $(-2, 2)$; foci: $(4, 2)$, $(-4, 2)$

81. Foci: $(0, 0)$, $(8, 0)$; asymptotes: $y = \pm 2(x - 4)$

82. Foci: $(3, \pm 2)$; asymptotes: $y = \pm 2(x - 3)$

83. *Architecture*   A parabolic archway is 12 meters high at the vertex. At a height of 10 meters, the width of the archway is 8 meters (see figure). How wide is the archway at ground level?

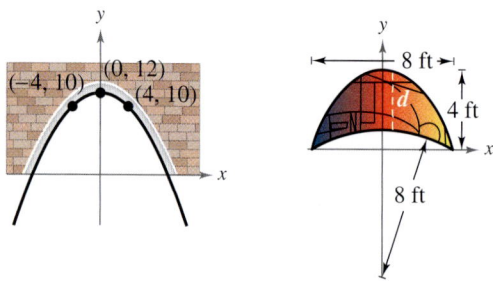

FIGURE FOR **83**   FIGURE FOR **84**

84. *Architecture*   A church window (see figure) is bounded above by a parabola and below by the arc of a circle.

(a) Find equations for the parabola and the circle.

(b) Complete the table by filling in the vertical distance d between the circle and the parabola for each given value of x.

x	0	1	2	3	4
d					

85. *Running Path*   Let $(0, 0)$ represent a water fountain located in a city park. Each day you run through the park along a path given by

$$x^2 + y^2 - 200x - 52{,}500 = 0$$

where x and y are measured in meters.

(a) What type of conic is your path? Explain your reasoning.

(b) Write the equation of the path in standard form. Sketch a graph of the equation.

(c) After you run, you walk to the water fountain. If you stop running at $(-100, 150)$, how far must you walk for a drink of water?

Synthesis

True or False?   **In Exercises 86 and 87, determine whether the statement is true or false. Justify your answer.**

86. The domain of a rational function can never be the set of all real numbers.

87. The graph of the equation

$$Ax^2 + Bxy + Cy^2 + Dx + Ey + F = 0$$

can be a single point.

4 Chapter Test

Take this test as you would take a test in class. When you are finished, check your work against the answers given in the back of the book.

In Exercises 1–3, find the domain of the function and identify any asymptotes.

1. $y = \dfrac{2}{4 - x}$

2. $f(x) = \dfrac{3 - x^2}{3 + x^2}$

3. $g(x) = \dfrac{x^2 + 2x - 3}{x - 2}$

In Exercises 4–9, identify any intercepts and asymptotes of the graph the function. Then sketch a graph of the function.

4. $h(x) = \dfrac{4}{x^2} - 1$

5. $g(x) = \dfrac{x^2 + 2}{x - 1}$

6. $f(x) = \dfrac{x + 1}{x^2 + x - 12}$

7. $f(x) = \dfrac{2x^2 - 5x - 12}{x^2 - 16}$

8. $f(x) = \dfrac{x^2 - 9}{2x - 6}$

9. $g(x) = \dfrac{2x^3 - 7x^2 + 4x + 4}{x^2 - x - 2}$

10. A rectangular page is designed to contain 36 square inches of print. The margins at the top and bottom of the page are 2 inches deep. The margins on each side are 1 inch wide. What should the dimensions of the page be so that the least amount of paper is used?

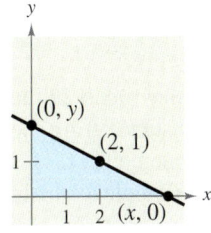

FIGURE FOR 11

11. A triangle is formed by the coordinate axes and a line through the point $(2, 1)$, as shown in the figure.

(a) Verify that $y = 1 + \dfrac{2}{x - 2}$.

(b) Write the area A of the triangle as a function of x. Determine the domain of the function in the context of the problem.

(c) Graph the area function. Estimate the minimum area of the triangle from the graph.

In Exercises 12–17, graph the conic and identify the center, vertices, and foci, if applicable.

12. $y^2 - 8x = 0$

13. $x^2 + y^2 - 10x + 4y + 4 = 0$

14. $x^2 - 10x - 2y + 19 = 0$

15. $x^2 - \dfrac{y^2}{4} = 1$

16. $x^2 - 4y^2 - 4x = 0$

17. $x^2 + 3y^2 - 2x + 36y + 100 = 0$

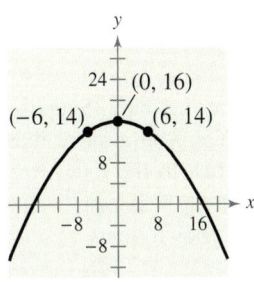

FIGURE FOR 20

18. Find an equation of the ellipse with vertices $(0, 2)$ and $(8, 2)$ and minor axis of length 4.

19. Find an equation of the hyperbola with vertices $(0, \pm 3)$ and asymptotes $y = \pm \frac{3}{2}x$.

20. A parabolic archway is 16 meters high at the vertex. At a height of 14 meters, the width of the archway is 12 meters, as shown in the figure. How wide is the archway at ground level?

21. The moon orbits Earth in an elliptical path with the center of Earth at one focus, as shown in the figure. The major and minor axes of the orbit have lengths of 768,800 kilometers and 767,640 kilometers, respectively. Find the smallest distance (perigee) and the greatest distance (apogee) from the center of the moon to the center of Earth.

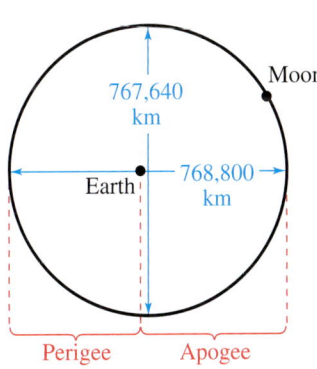

FIGURE FOR 21

Proofs in Mathematics

You can use the definition of a parabola to derive the standard form of the equation of a parabola whose directrix is parallel to the x-axis or to the y-axis.

> ### Standard Equation of a Parabola (Vertex at Origin) *(p. 355)*
>
> The standard form of the equation of a parabola with vertex at $(0, 0)$ and directrix $y = -p$ is
>
> $$x^2 = 4py, \qquad p \neq 0. \qquad \text{Vertical axis}$$
>
> For directrix $x = -p$, the equation is
>
> $$y^2 = 4px, \qquad p \neq 0. \qquad \text{Horizontal axis}$$
>
> The focus is on the axis p units (directed distance) from the vertex.

Proof

For the first case, suppose the directrix $(y = -p)$ is parallel to the x-axis. In the figure, you assume that $p > 0$, and because p is the directed distance from the vertex to the focus, the focus must lie above the vertex. Because the point (x, y) is equidistant from $(0, p)$ and $y = -p$, you can apply the Distance Formula to obtain

$$\sqrt{(x - 0)^2 + (y - p)^2} = y + p \qquad \text{Distance Formula}$$

$$x^2 + (y - p)^2 = (y + p)^2 \qquad \text{Square each side.}$$

$$x^2 + y^2 - 2py + p^2 = y^2 + 2py + p^2 \qquad \text{Expand.}$$

$$x^2 = 4py. \qquad \text{Simplify.}$$

A proof of the second case is similar to the proof of the first case. Suppose the directrix $(x = -p)$ is parallel to the y-axis. In the figure, you assume that $p > 0$, and because p is the directed distance from the vertex to the focus, the focus must lie to the right of the vertex. Because the point (x, y) is equidistant from $(p, 0)$ and $x = -p$, you can apply the Distance Formula as follows.

$$\sqrt{(x - p)^2 + (y - 0)^2} = x + p \qquad \text{Distance Formula}$$

$$(x - p)^2 + y^2 = (x + p)^2 \qquad \text{Square each side.}$$

$$x^2 - 2px + p^2 + y^2 = x^2 + 2px + p^2 \qquad \text{Expand.}$$

$$y^2 = 4px \qquad \text{Simplify.}$$

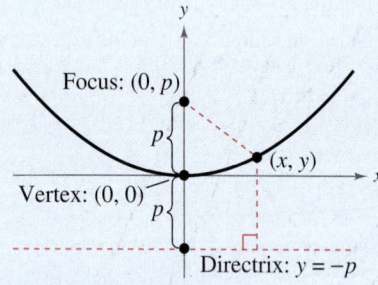

Parabola with vertical axis

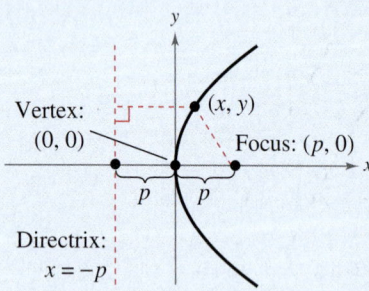

Parabola with horizontal axis

P.S. Problem Solving

This collection of thought-provoking and challenging exercises further explores and expands upon concepts learned in this chapter.

1. Match the graph of the rational function given by

$$f(x) = \frac{ax + b}{cx + d}$$

with the given conditions.

(a)

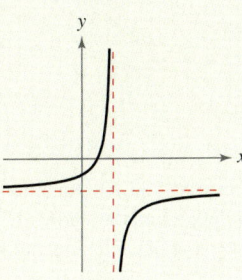

(b)

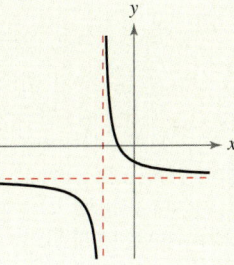

(c)

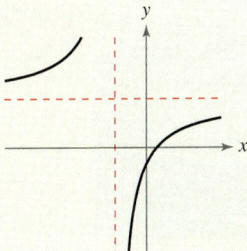

(d)

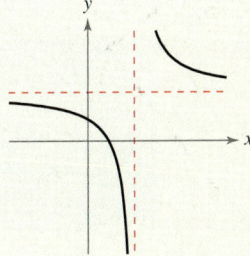

(i) $a > 0$ (ii) $a > 0$ (iii) $a < 0$ (iv) $a > 0$
   $b < 0$ $b > 0$ $b > 0$ $b < 0$
   $c > 0$ $c < 0$ $c > 0$ $c > 0$
   $d < 0$ $d < 0$ $d < 0$ $d > 0$

2. Consider the function given by

$$f(x) = \frac{ax}{(x - b)^2}.$$

(a) Determine the effect on the graph of f if $b \neq 0$ and a is varied. Consider cases in which a is positive and a is negative.

(b) Determine the effect on the graph of f if $a \neq 0$ and b is varied.

 3. The endpoints of the interval over which distinct vision is possible is called the *near point* and *far point* of the eye (see figure). With increasing age, these points normally change. The table shows the approximate near points y (in inches) for various ages x (in years).

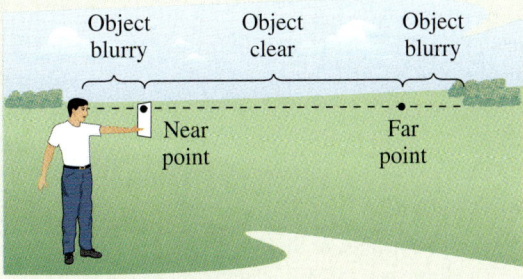

Age, x	Near point, y
16	3.0
32	4.7
44	9.8
50	19.7
60	39.4

(a) Use the *regression* feature of a graphing utility to find a quadratic model for the data. Use a graphing utility to plot the data and graph the model in the same viewing window.

(b) Find a rational model for the data. Take the reciprocals of the near points to generate the points $(x, 1/y)$. Use the *regression* feature of a graphing utility to find a linear model for the data. The resulting line has the form

$$\frac{1}{y} = ax + b.$$

Solve for y. Use a graphing utility to plot the data and graph the model in the same viewing window.

(c) Use the *table* feature of a graphing utility to create a table showing the predicted near point based on each model for each of the ages in the original table. How well do the models fit the original data?

(d) Use both models to estimate the near point for a person who is 25 years old. Which model is a better fit?

(e) Do you think either model can be used to predict the near point for a person who is 70 years old? Explain.

4. Statuary Hall is an elliptical room in the United States Capitol in Washington D.C. The room is also called the Whispering Gallery because a person standing at one focus of the room can hear even a whisper spoken by a person standing at the other focus. This occurs because any sound that is emitted from one focus of an ellipse will reflect off the side of the ellipse to the other focus. Statuary Hall is 46 feet wide and 97 feet long.

(a) Find an equation that models the shape of the room.

(b) How far apart are the two foci?

(c) What is the area of the floor of the room? (The area of an ellipse is $A = \pi ab$.)

5. Use the figure to show that $|d_2 - d_1| = 2a$.

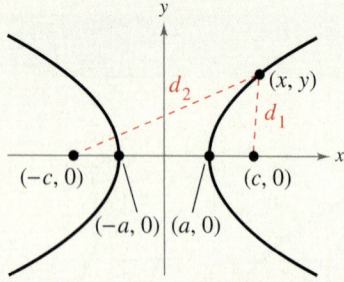

6. Find an equation of a hyperbola such that for any point on the hyperbola, the difference between its distances from the points $(2, 2)$ and $(10, 2)$ is 6.

7. The filament of a lightbulb is a thin wire that glows when electricity passes through it. The filament of a car headlight is at the focus of a parabolic reflector, which sends light out in a straight beam. Given that the filament is 1.5 inches from the vertex, find an equation for the cross section of the reflector. A reflector is 7 inches wide. How deep is it?

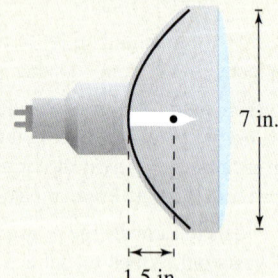

7 in.

1.5 in.

8. Consider the parabola $x^2 = 4py$.

(a) Use a graphing utility to graph the parabola for $p = 1$, $p = 2$, $p = 3$, and $p = 4$. Describe the effect on the graph when p increases.

(b) Locate the focus for each parabola in part (a).

(c) For each parabola in part (a), find the length of the chord passing through the focus and parallel to the directrix. How can the length of this chord be determined directly from the standard form of the equation of the parabola?

(d) Explain how the result of part (c) can be used as a sketching aid when graphing parabolas.

9. Let (x_1, y_1) be the coordinates of a point on the parabola $x^2 = 4py$. The equation of the line that just touches the parabola at the point (x_1, y_1), called a *tangent line*, is given by

$$y - y_1 = \frac{x_1}{2p}(x - x_1).$$

(a) What is the slope of the tangent line?

(b) For each parabola in Exercise 8, find the equations of the tangent lines at the endpoints of the chord. Use a graphing utility to graph the parabola and tangent lines.

10. A tour boat travels between two islands that are 12 miles apart (see figure). For each trip between the islands, there is enough fuel for a 20-mile trip.

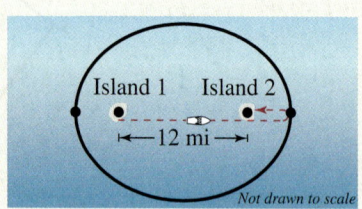

Island 1   Island 2

←—12 mi—→

Not drawn to scale

(a) Explain why the region in which the boat can travel is bounded by an ellipse.

(b) Let $(0, 0)$ represent the center of the ellipse. Find the coordinates of the center of each island.

(c) The boat travels from one island, straight past the other island to one vertex of the ellipse, and back to the second island. How many miles does the boat travel? Use your answer to find the coordinates of the vertex.

(d) Use the results of parts (b) and (c) to write an equation of the ellipse that bounds the region in which the boat can travel.

11. Prove that the graph of the equation

$$Ax^2 + Cy^2 + Dx + Ey + F = 0$$

is one of the following (except in degenerate cases).

Conic	Condition
(a) Circle	$A = C$
(b) Parabola	$A = 0$ or $C = 0$ (but not both)
(c) Ellipse	$AC > 0$
(d) Hyperbola	$AC < 0$

Exponential and Logarithmic Functions

5

Carbon dating is a method used to determine the ages of archeological artifacts up to 50,000 years old. For example, archeologists are using carbon dating to determine the ages of the great pyramids of Egypt.

© Sylvain Grandadam/Getty Images

SELECTED APPLICATIONS

Exponential and logarithmic functions have many real-life applications. The applications listed below represent a small sample of the applications in this chapter.

5.1 Exponential Functions and Their Graphs

What you should learn

- Recognize and evaluate exponential functions with base a.
- Graph exponential functions and use the One-to-One Property.
- Recognize, evaluate, and graph exponential functions with base e.
- Use exponential functions to model and solve real-life problems.

Why you should learn it

Exponential functions can be used to model and solve real-life problems. For instance, in Exercise 70 on page 394, an exponential function is used to model the atmospheric pressure at different altitudes.

© Comstock Images/Alamy

Exponential Functions

So far, this text has dealt mainly with **algebraic functions,** which include polynomial functions and rational functions. In this chapter, you will study two types of nonalgebraic functions—*exponential functions* and *logarithmic functions.* These functions are examples of **transcendental functions.**

> ### Definition of Exponential Function
>
> The **exponential function** f **with base** a is denoted by
>
> $$f(x) = a^x$$
>
> where $a > 0$, $a \neq 1$, and x is any real number.

The base $a = 1$ is excluded because it yields $f(x) = 1^x = 1$. This is a constant function, not an exponential function.

You have evaluated a^x for integer and rational values of x. For example, you know that $4^3 = 64$ and $4^{1/2} = 2$. However, to evaluate 4^x for any real number x, you need to interpret forms with *irrational* exponents. For the purposes of this text, it is sufficient to think of

$$a^{\sqrt{2}} \quad (\text{where } \sqrt{2} \approx 1.41421356)$$

as the number that has the successively closer approximations

$$a^{1.4}, a^{1.41}, a^{1.414}, a^{1.4142}, a^{1.41421}, \ldots$$

Example 1 Evaluating Exponential Functions

Use a calculator to evaluate each function at the indicated value of x.

Function	Value
a. $f(x) = 2^x$	$x = -3.1$
b. $f(x) = 2^{-x}$	$x = \pi$
c. $f(x) = 0.6^x$	$x = \frac{3}{2}$

Solution

Function Value	Graphing Calculator Keystrokes	Display
a. $f(-3.1) = 2^{-3.1}$	2 [∧] [(−)] 3.1 [ENTER]	0.1166291
b. $f(\pi) = 2^{-\pi}$	2 [∧] [(−)] π [ENTER]	0.1133147
c. $f\left(\frac{3}{2}\right) = (0.6)^{3/2}$	.6 [∧] [(] 3 [÷] 2 [)] [ENTER]	0.4647580

✓**CHECKPOINT**   Now try Exercise 1.

When evaluating exponential functions with a calculator, remember to enclose fractional exponents in parentheses. Because the calculator follows the order of operations, parentheses are crucial in order to obtain the correct result.

The *HM mathSpace®* CD-ROM and *Eduspace®* for this text contain additional resources related to the concepts discussed in this chapter.

Graphs of Exponential Functions

The graphs of all exponential functions have similar characteristics, as shown in Examples 2, 3, and 5.

Exploration

Note that an exponential function $f(x) = a^x$ is a constant raised to a variable power, whereas a power function $g(x) = x^n$ is a variable raised to a constant power. Use a graphing utility to graph each pair of functions in the same viewing window. Describe any similarities and differences in the graphs.

a. $y_1 = 2^x$, $y_2 = x^2$

b. $y_1 = 3^x$, $y_2 = x^3$

Example 2   Graphs of $y = a^x$

In the same coordinate plane, sketch the graph of each function.

a. $f(x) = 2^x$ **b.** $g(x) = 4^x$

Solution

The table below lists some values for each function, and Figure 5.1 shows the graphs of the two functions. Note that both graphs are increasing. Moreover, the graph of $g(x) = 4^x$ is increasing more rapidly than the graph of $f(x) = 2^x$.

x	-3	-2	-1	0	1	2
2^x	$\frac{1}{8}$	$\frac{1}{4}$	$\frac{1}{2}$	1	2	4
4^x	$\frac{1}{64}$	$\frac{1}{16}$	$\frac{1}{4}$	1	4	16

✓**CHECKPOINT**   Now try Exercise 11.

The table in Example 2 was evaluated by hand. You could, of course, use a graphing utility to construct tables with even more values.

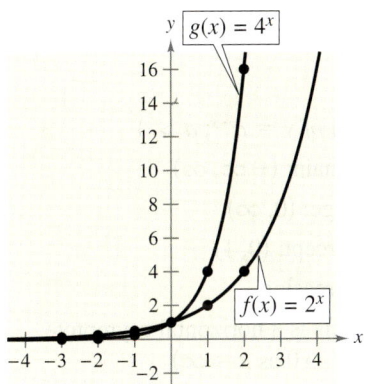

FIGURE 5.1

Example 3   Graphs of $y = a^{-x}$

In the same coordinate plane, sketch the graph of each function.

a. $F(x) = 2^{-x}$ **b.** $G(x) = 4^{-x}$

Solution

The table below lists some values for each function, and Figure 5.2 shows the graphs of the two functions. Note that both graphs are decreasing. Moreover, the graph of $G(x) = 4^{-x}$ is decreasing more rapidly than the graph of $F(x) = 2^{-x}$.

x	-2	-1	0	1	2	3
2^{-x}	4	2	1	$\frac{1}{2}$	$\frac{1}{4}$	$\frac{1}{8}$
4^{-x}	16	4	1	$\frac{1}{4}$	$\frac{1}{16}$	$\frac{1}{64}$

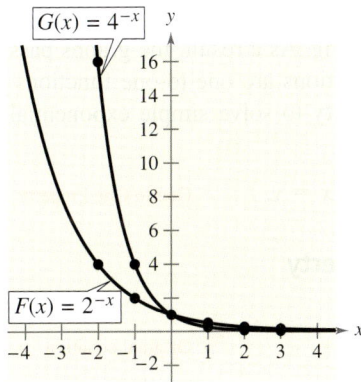

FIGURE 5.2

✓**CHECKPOINT**   Now try Exercise 13.

In Example 3, note that by using one of the properties of exponents, the functions $F(x) = 2^{-x}$ and $G(x) = 4^{-x}$ can be rewritten with positive exponents.

$$F(x) = 2^{-x} = \frac{1}{2^x} = \left(\frac{1}{2}\right)^x \quad \text{and} \quad G(x) = 4^{-x} = \frac{1}{4^x} = \left(\frac{1}{4}\right)^x$$

5.1   Exercises

VOCABULARY CHECK: Fill in the blanks.

1. Polynomials and rational functions are examples of _____ functions.

2. Exponential and logarithmic functions are examples of nonalgebraic functions, also called _____ functions.

3. The exponential function given by $f(x) = e^x$ is called the _____ _____ function, and the base e is called the _____ base.

4. To find the amount A in an account after t years with principal P and an annual interest rate r compounded n times per year, you can use the formula _____.

5. To find the amount A in an account after t years with principal P and an annual interest rate r compounded continuously, you can use the formula _____.

PREREQUISITE SKILLS REVIEW: Practice and review algebra skills needed for this section at **www.Eduspace.com.**

In Exercises 1–6, evaluate the function at the indicated value of *x*. Round your result to three decimal places.

Function	Value
1. $f(x) = 3.4^x$	$x = 5.6$
2. $f(x) = 2.3^x$	$x = \frac{3}{2}$
3. $f(x) = 5^x$	$x = -\pi$
4. $f(x) = \left(\frac{2}{3}\right)^{5x}$	$x = \frac{3}{10}$
5. $g(x) = 5000(2^x)$	$x = -1.5$
6. $f(x) = 200(1.2)^{12x}$	$x = 24$

In Exercises 7–10, match the exponential function with its graph. [The graphs are labeled (a), (b), (c), and (d).]

(a)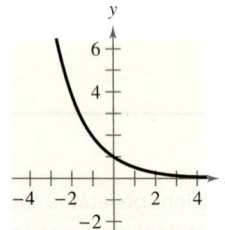

(b)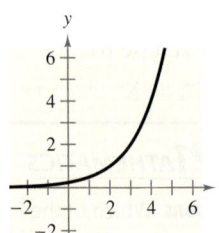

(c)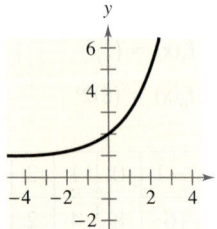

(d)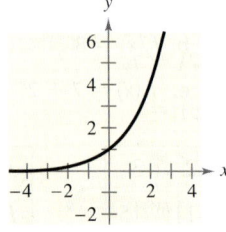

7. $f(x) = 2^x$ **8.** $f(x) = 2^x + 1$

9. $f(x) = 2^{-x}$ **10.** $f(x) = 2^{x-2}$

 In Exercises 11–16, use a graphing utility to construct a table of values for the function. Then sketch the graph of the function.

11. $f(x) = \left(\frac{1}{2}\right)^x$ **12.** $f(x) = \left(\frac{1}{2}\right)^{-x}$

13. $f(x) = 6^{-x}$ **14.** $f(x) = 6^x$

15. $f(x) = 2^{x-1}$ **16.** $f(x) = 4^{x-3} + 3$

In Exercises 17–22, use the graph of *f* to describe the transformation that yields the graph of *g*.

17. $f(x) = 3^x$, $g(x) = 3^{x-4}$

18. $f(x) = 4^x$, $g(x) = 4^x + 1$

19. $f(x) = -2^x$, $g(x) = 5 - 2^x$

20. $f(x) = 10^x$, $g(x) = 10^{-x+3}$

21. $f(x) = \left(\frac{7}{2}\right)^x$, $g(x) = -\left(\frac{7}{2}\right)^{-x+6}$

22. $f(x) = 0.3^x$, $g(x) = -0.3^x + 5$

 In Exercises 23–26, use a graphing utility to graph the exponential function.

23. $y = 2^{-x^2}$ **24.** $y = 3^{-|x|}$

25. $y = 3^{x-2} + 1$ **26.** $y = 4^{x+1} - 2$

In Exercises 27–32, evaluate the function at the indicated value of *x*. Round your result to three decimal places.

Function	Value
27. $h(x) = e^{-x}$	$x = \frac{3}{4}$
28. $f(x) = e^x$	$x = 3.2$
29. $f(x) = 2e^{-5x}$	$x = 10$
30. $f(x) = 1.5e^{x/2}$	$x = 240$
31. $f(x) = 5000e^{0.06x}$	$x = 6$
32. $f(x) = 250e^{0.05x}$	$x = 20$

In Exercises 33–38, use a graphing utility to construct a table of values for the function. Then sketch the graph of the function.

33. $f(x) = e^x$

34. $f(x) = e^{-x}$

35. $f(x) = 3e^{x+4}$

36. $f(x) = 2e^{-0.5x}$

37. $f(x) = 2e^{x-2} + 4$

38. $f(x) = 2 + e^{x-5}$

 In Exercises 39–44, use a graphing utility to graph the exponential function.

39. $y = 1.08^{-5x}$

40. $y = 1.08^{5x}$

41. $s(t) = 2e^{0.12t}$

42. $s(t) = 3e^{-0.2t}$

43. $g(x) = 1 + e^{-x}$

44. $h(x) = e^{x-2}$

In Exercise 45–52, use the One-to-One Property to solve the equation for x.

45. $3^{x+1} = 27$

46. $2^{x-3} = 16$

47. $2^{x-2} = \dfrac{1}{32}$

48. $\left(\dfrac{1}{5}\right)^{x+1} = 125$

49. $e^{3x+2} = e^3$

50. $e^{2x-1} = e^4$

51. $e^{x^2-3} = e^{2x}$

52. $e^{x^2+6} = e^{5x}$

Compound Interest   In Exercises 53–56, complete the table to determine the balance A for P dollars invested at rate r for t years and compounded n times per year.

n	1	2	4	12	365	Continuous
A						

53. $P = \$2500$, $r = 2.5\%$, $t = 10$ years

54. $P = \$1000$, $r = 4\%$, $t = 10$ years

55. $P = \$2500$, $r = 3\%$, $t = 20$ years

56. $P = \$1000$, $r = 6\%$, $t = 40$ years

Compound Interest   In Exercises 57–60, complete the table to determine the balance A for $\$12{,}000$ invested at rate r for t years, compounded continuously.

t	10	20	30	40	50
A					

57. $r = 4\%$

58. $r = 6\%$

59. $r = 6.5\%$

60. $r = 3.5\%$

61. *Trust Fund*   On the day of a child's birth, a deposit of $\$25{,}000$ is made in a trust fund that pays 8.75% interest, compounded continuously. Determine the balance in this account on the child's 25th birthday.

62. *Trust Fund*   A deposit of $\$5000$ is made in a trust fund that pays 7.5% interest, compounded continuously. It is specified that the balance will be given to the college from which the donor graduated after the money has earned interest for 50 years. How much will the college receive?

63. *Inflation*   If the annual rate of inflation averages 4% over the next 10 years, the approximate costs C of goods or services during any year in that decade will be modeled by $C(t) = P(1.04)^t$, where t is the time in years and P is the present cost. The price of an oil change for your car is presently $\$23.95$. Estimate the price 10 years from now.

64. *Demand*   The demand equation for a product is given by

$$p = 5000\left(1 - \dfrac{4}{4 + e^{-0.002x}}\right)$$

where p is the price and x is the number of units.

 (a) Use a graphing utility to graph the demand function for $x > 0$ and $p > 0$.

(b) Find the price p for a demand of $x = 500$ units.

 (c) Use the graph in part (a) to approximate the greatest price that will still yield a demand of at least 600 units.

65. *Computer Virus*   The number V of computers infected by a computer virus increases according to the model $V(t) = 100e^{4.6052t}$, where t is the time in hours. Find (a) $V(1)$, (b) $V(1.5)$, and (c) $V(2)$.

66. *Population*   The population P (in millions) of Russia from 1996 to 2004 can be approximated by the model $P = 152.26e^{-0.0039t}$, where t represents the year, with $t = 6$ corresponding to 1996. (Source: Census Bureau, International Data Base)

(a) According to the model, is the population of Russia increasing or decreasing? Explain.

(b) Find the population of Russia in 1998 and 2000.

(c) Use the model to predict the population of Russia in 2010.

67. *Radioactive Decay*   Let Q represent a mass of radioactive radium (^{226}Ra) (in grams), whose half-life is 1599 years. The quantity of radium present after t years is $Q = 25\left(\frac{1}{2}\right)^{t/1599}$.

(a) Determine the initial quantity (when $t = 0$).

(b) Determine the quantity present after 1000 years.

 (c) Use a graphing utility to graph the function over the interval $t = 0$ to $t = 5000$.

68. *Radioactive Decay*   Let Q represent a mass of carbon 14 (^{14}C) (in grams), whose half-life is 5715 years. The quantity of carbon 14 present after t years is $Q = 10\left(\frac{1}{2}\right)^{t/5715}$.

(a) Determine the initial quantity (when $t = 0$).

(b) Determine the quantity present after 2000 years.

(c) Sketch the graph of this function over the interval $t = 0$ to $t = 10{,}000$.

Model It

69. *Data Analysis: Biology*   To estimate the amount of defoliation caused by the gypsy moth during a given year, a forester counts the number x of egg masses on $\frac{1}{40}$ of an acre (circle of radius 18.6 feet) in the fall. The percent of defoliation y the next spring is shown in the table.   (Source: USDA, Forest Service)

Egg masses, x	Percent of defoliation, y
0	12
25	44
50	81
75	96
100	99

A model for the data is given by

$$y = \frac{100}{1 + 7e^{-0.069x}}.$$

 (a) Use a graphing utility to create a scatter plot of the data and graph the model in the same viewing window.

(b) Create a table that compares the model with the sample data.

(c) Estimate the percent of defoliation if 36 egg masses are counted on $\frac{1}{40}$ acre.

 (d) You observe that $\frac{2}{3}$ of a forest is defoliated the following spring. Use the graph in part (a) to estimate the number of egg masses per $\frac{1}{40}$ acre.

70. *Data Analysis: Meteorology*   A meteorologist measures the atmospheric pressure P (in pascals) at altitude h (in kilometers). The data are shown in the table.

Altitude, h	Pressure, P
0	101,293
5	54,735
10	23,294
15	12,157
20	5,069

A model for the data is given by $P = 107{,}428e^{-0.150h}$.

(a) Sketch a scatter plot of the data and graph the model on the same set of axes.

(b) Estimate the atmospheric pressure at a height of 8 kilometers.

Synthesis

True or False?   In Exercises 71 and 72, determine whether the statement is true or false. Justify your answer.

71. The line $y = -2$ is an asymptote for the graph of $f(x) = 10^x - 2$.

72. $e = \dfrac{271{,}801}{99{,}990}$.

Think About It   In Exercises 73–76, use properties of exponents to determine which functions (if any) are the same.

73. $f(x) = 3^{x-2}$
$g(x) = 3^x - 9$
$h(x) = \frac{1}{9}(3^x)$

74. $f(x) = 4^x + 12$
$g(x) = 2^{2x+6}$
$h(x) = 64(4^x)$

75. $f(x) = 16(4^{-x})$
$g(x) = \left(\frac{1}{4}\right)^{x-2}$
$h(x) = 16(2^{-2x})$

76. $f(x) = e^{-x} + 3$
$g(x) = e^{3-x}$
$h(x) = -e^{x-3}$

77. Graph the functions given by $y = 3^x$ and $y = 4^x$ and use the graphs to solve each inequality.

(a) $4^x < 3^x$ (b) $4^x > 3^x$

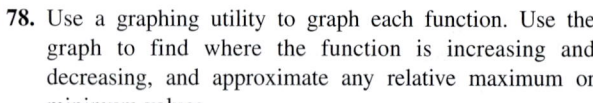 **78.** Use a graphing utility to graph each function. Use the graph to find where the function is increasing and decreasing, and approximate any relative maximum or minimum values.

(a) $f(x) = x^2 e^{-x}$ (b) $g(x) = x2^{3-x}$

 79. *Graphical Analysis*   Use a graphing utility to graph

$$f(x) = \left(1 + \frac{0.5}{x}\right)^x \quad \text{and} \quad g(x) = e^{0.5}$$

in the same viewing window. What is the relationship between f and g as x increases and decreases without bound?

80. *Think About It*   Which functions are exponential?

(a) $3x$ (b) $3x^2$ (c) 3^x (d) 2^{-x}

Skills Review

In Exercises 81 and 82, solve for y.

81. $x^2 + y^2 = 25$

82. $x - |y| = 2$

In Exercises 83 and 84, sketch the graph of the function.

83. $f(x) = \dfrac{2}{9 + x}$

84. $f(x) = \sqrt{7 - x}$

85. **Make a Decision**   To work an extended application analyzing the population per square mile of the United States, visit this text's website at *college.hmco.com*.   (Data Source: U.S. Census Bureau)

What you should learn

- Recognize and evaluate logarithmic functions with base a.
- Graph logarithmic functions.
- Recognize, evaluate, and graph natural logarithmic functions.
- Use logarithmic functions to model and solve real-life problems.

Why you should learn it

Logarithmic functions are often used to model scientific observations. For instance, in Exercise 89 on page 404, a logarithmic function is used to model human memory.

© Ariel Skelley/Corbis

Logarithmic Functions

In Section 2.7, you studied the concept of an inverse function. There, you learned that if a function is one-to-one—that is, if the function has the property that no horizontal line intersects the graph of the function more than once—the function must have an inverse function. By looking back at the graphs of the exponential functions introduced in Section 5.1, you will see that every function of the form $f(x) = a^x$ passes the Horizontal Line Test and therefore must have an inverse function. This inverse function is called the **logarithmic function with base a.**

> ### Definition of Logarithmic Function with Base a
>
> For $x > 0$, $a > 0$, and $a \neq 1$,
>
> $\qquad y = \log_a x$ if and only if $x = a^y$.
>
> The function given by
>
> $\qquad f(x) = \log_a x$ Read as "log base a of x."
>
> is called the **logarithmic function with base a.**

The equations

$\qquad y = \log_a x \qquad$ and $\qquad x = a^y$

are equivalent. The first equation is in logarithmic form and the second is in exponential form. For example, the logarithmic equation $2 = \log_3 9$ can be rewritten in exponential form as $9 = 3^2$. The exponential equation $5^3 = 125$ can be rewritten in logarithmic form as $\log_5 125 = 3$.

When evaluating logarithms, remember that *a logarithm is an exponent.* This means that $\log_a x$ is the exponent to which a must be raised to obtain x. For instance, $\log_2 8 = 3$ because 2 must be raised to the third power to get 8.

Example 1   **Evaluating Logarithms**

Use the definition of logarithmic function to evaluate each logarithm at the indicated value of x.

a. $f(x) = \log_2 x, \quad x = 32$ **b.** $f(x) = \log_3 x, \quad x = 1$

c. $f(x) = \log_4 x, \quad x = 2$ **d.** $f(x) = \log_{10} x, \quad x = \frac{1}{100}$

Solution

a. $f(32) = \log_2 32 = 5$ because $2^5 = 32$.

b. $f(1) = \log_3 1 = 0$ because $3^0 = 1$.

c. $f(2) = \log_4 2 = \frac{1}{2}$ because $4^{1/2} = \sqrt{4} = 2$.

d. $f\left(\frac{1}{100}\right) = \log_{10} \frac{1}{100} = -2$ because $10^{-2} = \frac{1}{10^2} = \frac{1}{100}$.

✓**CHECKPOINT**   Now try Exercise 17.

STUDY TIP

Remember that a logarithm is an exponent. So, to evaluate the logarithmic expression $\log_a x$, you need to ask the question, "To what power must a be raised to obtain x?"

Complete the table for $f(x) = 10^x$.

x	-2	-1	0	1	2
$f(x)$					

Complete the table for $f(x) = \log x$.

x	$\frac{1}{100}$	$\frac{1}{10}$	1	10	100
$f(x)$					

Compare the two tables. What is the relationship between $f(x) = 10^x$ and $f(x) = \log x$?

The logarithmic function with base 10 is called the **common logarithmic function.** It is denoted by $\log_{10}$ or simply by log. On most calculators, this function is denoted by $\boxed{\text{LOG}}$. Example 2 shows how to use a calculator to evaluate common logarithmic functions. You will learn how to use a calculator to calculate logarithms to any base in the next section.

Example 2 Evaluating Common Logarithms on a Calculator

Use a calculator to evaluate the function given by $f(x) = \log x$ at each value of x.

a. $x = 10$ **b.** $x = \frac{1}{3}$ **c.** $x = 2.5$ **d.** $x = -2$

Solution

Function Value	Graphing Calculator Keystrokes	Display
a. $f(10) = \log 10$	$\boxed{\text{LOG}}$ 10 $\boxed{\text{ENTER}}$	1
b. $f(\frac{1}{3}) = \log\frac{1}{3}$	$\boxed{\text{LOG}}$ $\boxed{(}$ 1 $\boxed{\div}$ 3 $\boxed{)}$ $\boxed{\text{ENTER}}$	-0.4771213
c. $f(2.5) = \log 2.5$	$\boxed{\text{LOG}}$ 2.5 $\boxed{\text{ENTER}}$	0.3979400
d. $f(-2) = \log(-2)$	$\boxed{\text{LOG}}$ $\boxed{(-)}$ 2 $\boxed{\text{ENTER}}$	ERROR

Note that the calculator displays an error message (or a complex number) when you try to evaluate $\log(-2)$. The reason for this is that there is no real number power to which 10 can be raised to obtain -2.

✓**CHECKPOINT** Now try Exercise 23.

The following properties follow directly from the definition of the logarithmic function with base a.

Properties of Logarithms

1. $\log_a 1 = 0$ because $a^0 = 1$.

2. $\log_a a = 1$ because $a^1 = a$.

3. $\log_a a^x = x$ and $a^{\log_a x} = x$ Inverse Properties

4. If $\log_a x = \log_a y$, then $x = y$.   One-to-One Property

---

Example 3 Using Properties of Logarithms

a. Simplify: $\log_4 1$ **b.** Simplify: $\log_{\sqrt{7}} \sqrt{7}$ **c.** Simplify: $6^{\log_6 20}$

Solution

a. Using Property 1, it follows that $\log_4 1 = 0$.

b. Using Property 2, you can conclude that $\log_{\sqrt{7}} \sqrt{7} = 1$.

c. Using the Inverse Property (Property 3), it follows that $6^{\log_6 20} = 20$.

✓**CHECKPOINT** Now try Exercise 27.

You can use the One-to-One Property (Property 4) to solve simple logarithmic equations, as shown in Example 4.

Example 4   **Using the One-to-One Property**

a. $\log_3 x = \log_3 12$ Original equation

     $x = 12$ One-to-One Property

b. $\log(2x + 1) = \log x \implies 2x + 1 = x \implies x = -1$

c. $\log_4(x^2 - 6) = \log_4 10 \implies x^2 - 6 = 10 \implies x^2 = 16 \implies x = \pm 4$

✔CHECKPOINT   Now try Exercise 79.

Graphs of Logarithmic Functions

To sketch the graph of $y = \log_a x$, you can use the fact that the graphs of inverse functions are reflections of each other in the line $y = x$.

Example 5   **Graphs of Exponential and Logarithmic Functions**

In the same coordinate plane, sketch the graph of each function.

a. $f(x) = 2^x$ **b.** $g(x) = \log_2 x$

Solution

a. For $f(x) = 2^x$, construct a table of values. By plotting these points and connecting them with a smooth curve, you obtain the graph shown in Figure 5.13.

x	-2	-1	0	1	2	3
$f(x) = 2^x$	$\frac{1}{4}$	$\frac{1}{2}$	1	2	4	8

b. Because $g(x) = \log_2 x$ is the inverse function of $f(x) = 2^x$, the graph of g is obtained by plotting the points $(f(x), x)$ and connecting them with a smooth curve. The graph of g is a reflection of the graph of f in the line $y = x$, as shown in Figure 5.13.

✔CHECKPOINT   Now try Exercise 31.

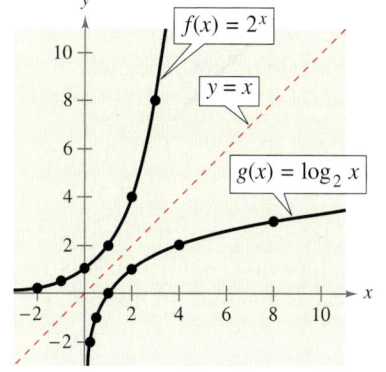

FIGURE 5.13

Example 6   **Sketching the Graph of a Logarithmic Function**

Sketch the graph of the common logarithmic function $f(x) = \log x$. Identify the vertical asymptote.

Solution

Begin by constructing a table of values. Note that some of the values can be obtained without a calculator by using the Inverse Property of Logarithms. Others require a calculator. Next, plot the points and connect them with a smooth curve, as shown in Figure 5.14. The vertical asymptote is $x = 0$ (y-axis).

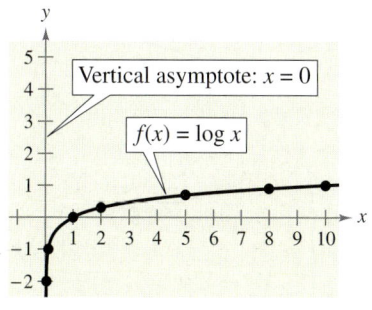

FIGURE 5.14

	Without calculator				With calculator		
x	$\frac{1}{100}$	$\frac{1}{10}$	1	10	2	5	8
$f(x) = \log x$	-2	-1	0	1	0.301	0.699	0.903

✔CHECKPOINT   Now try Exercise 37.

The nature of the graph in Figure 5.14 is typical of functions of the form $f(x) = \log_a x, a > 1$. They have one x-intercept and one vertical asymptote. Notice how slowly the graph rises for $x > 1$. The basic characteristics of logarithmic graphs are summarized in Figure 5.15.

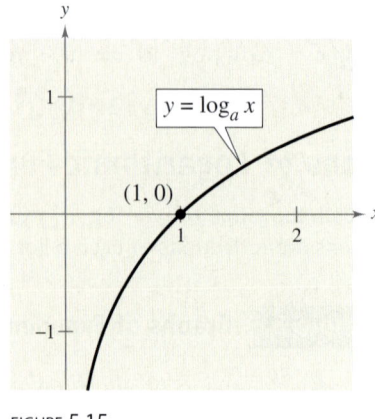

FIGURE 5.15

Graph of $y = \log_a x, a > 1$

- Domain: $(0, \infty)$
- Range: $(-\infty, \infty)$
- x-intercept: $(1, 0)$
- Increasing
- One-to-one, therefore has an inverse function
- y-axis is a vertical asymptote $(\log_a x \to -\infty$ as $x \to 0^+)$.
- Continuous
- Reflection of graph of $y = a^x$ about the line $y = x$

The basic characteristics of the graph of $f(x) = a^x$ are shown below to illustrate the inverse relation between $f(x) = a^x$ and $g(x) = \log_a x$.

- Domain: $(-\infty, \infty)$ • Range: $(0, \infty)$
- y-intercept: $(0,1)$ • x-axis is a horizontal asymptote $(a^x \to 0$ as $x \to -\infty)$.

In the next example, the graph of $y = \log_a x$ is used to sketch the graphs of functions of the form $f(x) = b \pm \log_a(x + c)$. Notice how a horizontal shift of the graph results in a horizontal shift of the vertical asymptote.

Example 7 **Shifting Graphs of Logarithmic Functions**

The graph of each of the functions is similar to the graph of $f(x) = \log x$.

a. Because $g(x) = \log(x - 1) = f(x - 1)$, the graph of g can be obtained by shifting the graph of f one unit to the right, as shown in Figure 5.16.

b. Because $h(x) = 2 + \log x = 2 + f(x)$, the graph of h can be obtained by shifting the graph of f two units upward, as shown in Figure 5.17.

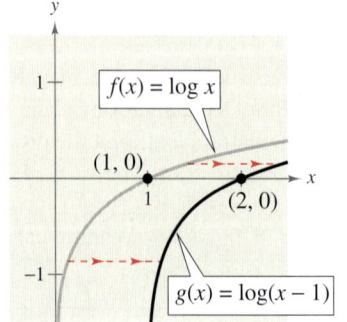

FIGURE 5.16

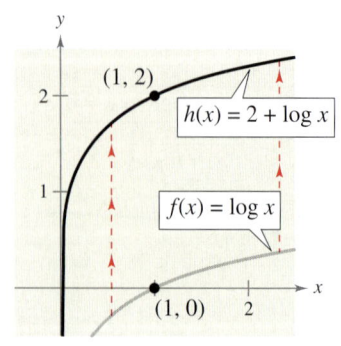

FIGURE 5.17

✓**CHECKPOINT** Now try Exercise 39.

The Natural Logarithmic Function

By looking back at the graph of the natural exponential function introduced in Section 5.1 on page 388, you will see that $f(x) = e^x$ is one-to-one and so has an inverse function. This inverse function is called the **natural logarithmic function** and is denoted by the special symbol ln x, read as "the natural log of x" or "el en of x." Note that the natural logarithm is written without a base. The base is understood to be e.

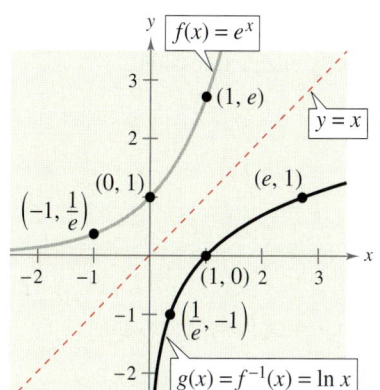

$f(x) = e^x$

$y = x$

$(1, e)$

$(0, 1)$ $(e, 1)$

$\left(-1, \dfrac{1}{e}\right)$

$(1, 0)$

$\left(\dfrac{1}{e}, -1\right)$

$g(x) = f^{-1}(x) = \ln x$

Reflection of graph of $f(x) = e^x$ about the line $y = x$

FIGURE 5.18

> ### The Natural Logarithmic Function
>
> The function defined by
>
> $$f(x) = \log_e x = \ln x, \quad x > 0$$
>
> is called the **natural logarithmic function.**

The definition above implies that the natural logarithmic function and the natural exponential function are inverse functions of each other. So, every logarithmic equation can be written in an equivalent exponential form and every exponential equation can be written in logarithmic form. That is, $y = \ln x$ and $x = e^y$ are equivalent equations.

Because the functions given by $f(x) = e^x$ and $g(x) = \ln x$ are inverse functions of each other, their graphs are reflections of each other in the line $y = x$. This reflective property is illustrated in Figure 5.18.

On most calculators, the natural logarithm is denoted by $\boxed{\text{LN}}$, as illustrated in Example 8.

Example 8   Evaluating the Natural Logarithmic Function

Use a calculator to evaluate the function given by $f(x) = \ln x$ for each value of x.

a. $x = 2$ **b.** $x = 0.3$ **c.** $x = -1$ **d.** $x = 1 + \sqrt{2}$

Solution

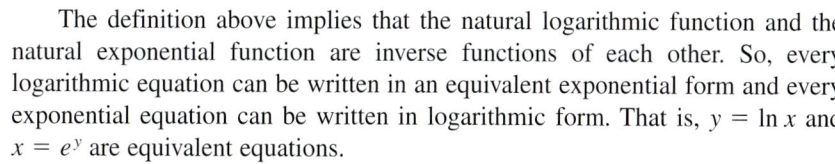

Function Value	Graphing Calculator Keystrokes	Display
a. $f(2) = \ln 2$	$\boxed{\text{LN}}$ 2 $\boxed{\text{ENTER}}$	0.6931472
b. $f(0.3) = \ln 0.3$	$\boxed{\text{LN}}$ .3 $\boxed{\text{ENTER}}$	−1.2039728
c. $f(-1) = \ln(-1)$	$\boxed{\text{LN}}$ $\boxed{(-)}$ 1 $\boxed{\text{ENTER}}$	ERROR
d. $f(1 + \sqrt{2}) = \ln(1 + \sqrt{2})$	$\boxed{\text{LN}}$ $\boxed{(}$ 1 $\boxed{+}$ $\boxed{\sqrt{\ }}$ 2 $\boxed{)}$ $\boxed{\text{ENTER}}$	0.8813736

✓**CHECKPOINT**   Now try Exercise 61.

In Example 8, be sure you see that $\ln(-1)$ gives an error message on most calculators. (Some calculators may display a complex number.) This occurs because the domain of $\ln x$ is the set of positive real numbers (see Figure 5.18). So, $\ln(-1)$ is undefined.

The four properties of logarithms listed on page 396 are also valid for natural logarithms.

STUDY TIP

Notice that as with every other logarithmic function, the domain of the natural logarithmic function is the set of *positive real numbers*—be sure you see that ln x is not defined for zero or for negative numbers.

Properties of Natural Logarithms

1. $\ln 1 = 0$ because $e^0 = 1$.
2. $\ln e = 1$ because $e^1 = e$.
3. $\ln e^x = x$ and $e^{\ln x} = x$ Inverse Properties
4. If $\ln x = \ln y$, then $x = y$.   One-to-One Property

Example 9   Using Properties of Natural Logarithms

Use the properties of natural logarithms to simplify each expression.

a. $\ln \dfrac{1}{e}$ **b.** $e^{\ln 5}$ **c.** $\dfrac{\ln 1}{3}$ **d.** $2 \ln e$

Solution

a. $\ln \dfrac{1}{e} = \ln e^{-1} = -1$ Inverse Property   **b.** $e^{\ln 5} = 5$ Inverse Property

c. $\dfrac{\ln 1}{3} = \dfrac{0}{3} = 0$ Property 1   **d.** $2 \ln e = 2(1) = 2$ Property 2

✓**CHECKPOINT**   Now try Exercise 65.

Example 10   Finding the Domains of Logarithmic Functions

Find the domain of each function.

a. $f(x) = \ln(x - 2)$ **b.** $g(x) = \ln(2 - x)$ **c.** $h(x) = \ln x^2$

Solution

a. Because $\ln(x - 2)$ is defined only if $x - 2 > 0$, it follows that the domain of f is $(2, \infty)$. The graph of f is shown in Figure 5.19.

b. Because $\ln(2 - x)$ is defined only if $2 - x > 0$, it follows that the domain of g is $(-\infty, 2)$. The graph of g is shown in Figure 5.20.

c. Because $\ln x^2$ is defined only if $x^2 > 0$, it follows that the domain of h is all real numbers except $x = 0$. The graph of h is shown in Figure 5.21.

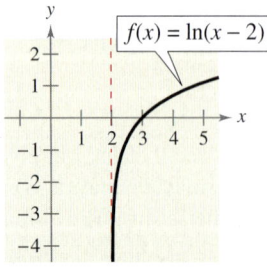

FIGURE 5.19

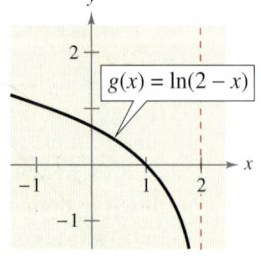

FIGURE 5.20

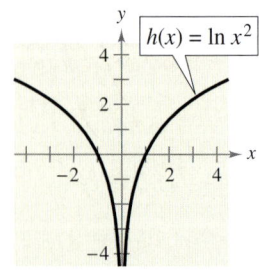

FIGURE 5.21

✓**CHECKPOINT**   Now try Exercise 69.

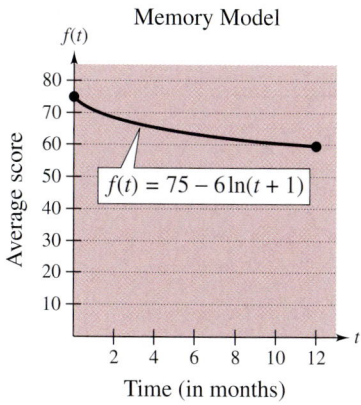

Memory Model

FIGURE 5.22

Application

Example 11   Human Memory Model

Students participating in a psychology experiment attended several lectures on a subject and were given an exam. Every month for a year after the exam, the students were retested to see how much of the material they remembered. The average scores for the group are given by the *human memory model*

$$f(t) = 75 - 6 \ln(t + 1), \quad 0 \le t \le 12$$

where t is the time in months. The graph of f is shown in Figure 5.22.

a. What was the average score on the original ($t = 0$) exam?

b. What was the average score at the end of $t = 2$ months?

c. What was the average score at the end of $t = 6$ months?

Solution

a. The original average score was

$$f(0) = 75 - 6 \ln(0 + 1) \qquad \text{Substitute 0 for } t.$$

$$= 75 - 6 \ln 1 \qquad \text{Simplify.}$$

$$= 75 - 6(0) \qquad \text{Property of natural logarithms}$$

$$= 75. \qquad \text{Solution}$$

b. After 2 months, the average score was

$$f(2) = 75 - 6 \ln(2 + 1) \qquad \text{Substitute 2 for } t.$$

$$= 75 - 6 \ln 3 \qquad \text{Simplify.}$$

$$\approx 75 - 6(1.0986) \qquad \text{Use a calculator.}$$

$$\approx 68.4. \qquad \text{Solution}$$

c. After 6 months, the average score was

$$f(6) = 75 - 6 \ln(6 + 1) \qquad \text{Substitute 6 for } t.$$

$$= 75 - 6 \ln 7 \qquad \text{Simplify.}$$

$$\approx 75 - 6(1.9459) \qquad \text{Use a calculator.}$$

$$\approx 63.3. \qquad \text{Solution}$$

✔CHECKPOINT   Now try Exercise 89.

---

*W*RITING ABOUT *M*ATHEMATICS

Analyzing a Human Memory Model Use a graphing utility to determine the time in months when the average score in Example 11 was 60. Explain your method of solving the problem. Describe another way that you can use a graphing utility to determine the answer.

5.2 | Exercises

VOCABULARY CHECK: Fill in the blanks.

1. The inverse function of the exponential function given by $f(x) = a^x$ is called the _____ function with base a.

2. The common logarithmic function has base _____ .

3. The logarithmic function given by $f(x) = \ln x$ is called the _____ logarithmic function and has base _____.

4. The Inverse Property of logarithms and exponentials states that $\log_a a^x = x$ and _____.

5. The One-to-One Property of natural logarithms states that if $\ln x = \ln y$, then _____.

PREREQUISITE SKILLS REVIEW: Practice and review algebra skills needed for this section at **www.Eduspace.com**.

In Exercises 1–8, write the logarithmic equation in exponential form. For example, the exponential form of $\log_5 25 = 2$ is $5^2 = 25$.

1. $\log_4 64 = 3$
2. $\log_3 81 = 4$
3. $\log_7 \frac{1}{49} = -2$
4. $\log \frac{1}{1000} = -3$
5. $\log_{32} 4 = \frac{2}{5}$
6. $\log_{16} 8 = \frac{3}{4}$
7. $\log_{36} 6 = \frac{1}{2}$
8. $\log_8 4 = \frac{2}{3}$

In Exercises 9–16, write the exponential equation in logarithmic form. For example, the logarithmic form of $2^3 = 8$ is $\log_2 8 = 3$.

9. $5^3 = 125$
10. $8^2 = 64$
11. $81^{1/4} = 3$
12. $9^{3/2} = 27$
13. $6^{-2} = \frac{1}{36}$
14. $4^{-3} = \frac{1}{64}$
15. $7^0 = 1$
16. $10^{-3} = 0.001$

In Exercises 17–22, evaluate the function at the indicated value of x without using a calculator.

Function	Value
17. $f(x) = \log_2 x$	$x = 16$
18. $f(x) = \log_{16} x$	$x = 4$
19. $f(x) = \log_7 x$	$x = 1$
20. $f(x) = \log x$	$x = 10$
21. $g(x) = \log_a x$	$x = a^2$
22. $g(x) = \log_b x$	$x = b^{-3}$

 In Exercises 23–26, use a calculator to evaluate $f(x) = \log x$ at the indicated value of x. Round your result to three decimal places.

23. $x = \frac{4}{5}$
24. $x = \frac{1}{500}$
25. $x = 12.5$
26. $x = 75.25$

In Exercises 27–30, use the properties of logarithms to simplify the expression.

27. $\log_3 3^4$
28. $\log_{1.5} 1$
29. $\log_\pi \pi$
30. $9^{\log_9 15}$

In Exercises 31–38, find the domain, x-intercept, and vertical asymptote of the logarithmic function and sketch its graph.

31. $f(x) = \log_4 x$
32. $g(x) = \log_6 x$
33. $y = -\log_3 x + 2$
34. $h(x) = \log_4(x - 3)$
35. $f(x) = -\log_6(x + 2)$
36. $y = \log_5(x - 1) + 4$
37. $y = \log\left(\frac{x}{5}\right)$
38. $y = \log(-x)$

In Exercises 39–44, use the graph of $g(x) = \log_3 x$ to match the given function with its graph. Then describe the relationship between the graphs of f and g. [The graphs are labeled (a), (b), (c), (d), (e), and (f).]

(a)
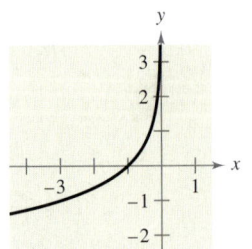

(b)
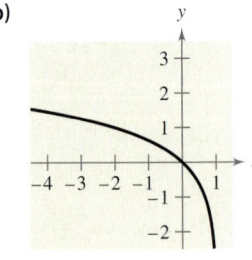

(c)
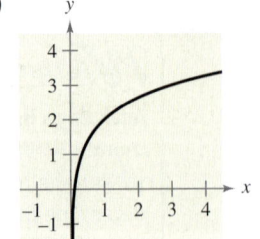

(d)
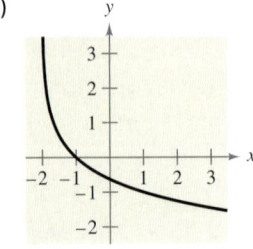

(e)

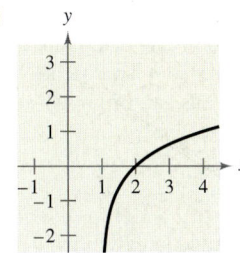

(f)

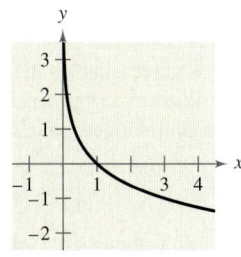

39. $f(x) = \log_3 x + 2$

40. $f(x) = -\log_3 x$

41. $f(x) = -\log_3(x + 2)$

42. $f(x) = \log_3(x - 1)$

43. $f(x) = \log_3(1 - x)$

44. $f(x) = -\log_3(-x)$

In Exercises 45–52, write the logarithmic equation in exponential form.

45. $\ln \frac{1}{2} = -0.693 \ldots$

46. $\ln \frac{2}{5} = -0.916 \ldots$

47. $\ln 4 = 1.386 \ldots$

48. $\ln 10 = 2.302 \ldots$

49. $\ln 250 = 5.521 \ldots$

50. $\ln 679 = 6.520 \ldots$

51. $\ln 1 = 0$

52. $\ln e = 1$

In Exercises 53–60, write the exponential equation in logarithmic form.

53. $e^3 = 20.0855 \ldots$

54. $e^2 = 7.3890 \ldots$

55. $e^{1/2} = 1.6487 \ldots$

56. $e^{1/3} = 1.3956 \ldots$

57. $e^{-0.5} = 0.6065 \ldots$

58. $e^{-4.1} = 0.0165 \ldots$

59. $e^x = 4$

60. $e^{2x} = 3$

 In Exercises 61–64, use a calculator to evaluate the function at the indicated value of x. Round your result to three decimal places.

Function	Value
61. $f(x) = \ln x$	$x = 18.42$
62. $f(x) = 3 \ln x$	$x = 0.32$
63. $g(x) = 2 \ln x$	$x = 0.75$
64. $g(x) = -\ln x$	$x = \frac{1}{2}$

In Exercises 65–68, evaluate $g(x) = \ln x$ at the indicated value of x without using a calculator.

65. $x = e^3$

66. $x = e^{-2}$

67. $x = e^{-2/3}$

68. $x = e^{-5/2}$

In Exercises 69–72, find the domain, x-intercept, and vertical asymptote of the logarithmic function and sketch its graph.

69. $f(x) = \ln(x - 1)$

70. $h(x) = \ln(x + 1)$

71. $g(x) = \ln(-x)$

72. $f(x) = \ln(3 - x)$

 In Exercises 73–78, use a graphing utility to graph the function. Be sure to use an appropriate viewing window.

73. $f(x) = \log(x + 1)$

74. $f(x) = \log(x - 1)$

75. $f(x) = \ln(x - 1)$

76. $f(x) = \ln(x + 2)$

77. $f(x) = \ln x + 2$

78. $f(x) = 3 \ln x - 1$

In Exercises 79–86, use the One-to-One Property to solve the equation for x.

79. $\log_2(x + 1) = \log_2 4$

80. $\log_2(x - 3) = \log_2 9$

81. $\log(2x + 1) = \log 15$

82. $\log(5x + 3) = \log 12$

83. $\ln(x + 2) = \ln 6$

84. $\ln(x - 4) = \ln 2$

85. $\ln(x^2 - 2) = \ln 23$

86. $\ln(x^2 - x) = \ln 6$

Model It

87. *Monthly Payment*   The model

$$t = 12.542 \ln\left(\frac{x}{x - 1000}\right), \quad x > 1000$$

approximates the length of a home mortgage of $150,000 at 8% in terms of the monthly payment. In the model, t is the length of the mortgage in years and x is the monthly payment in dollars (see figure).

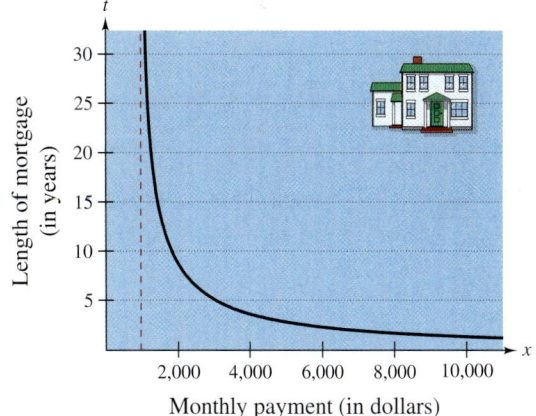

Monthly payment (in dollars)

(a) Use the model to approximate the lengths of a $150,000 mortgage at 8% when the monthly payment is $1100.65 and when the monthly payment is $1254.68.

(b) Approximate the total amounts paid over the term of the mortgage with a monthly payment of $1100.65 and with a monthly payment of $1254.68.

(c) Approximate the total interest charges for a monthly payment of $1100.65 and for a monthly payment of $1254.68.

(d) What is the vertical asymptote for the model? Interpret its meaning in the context of the problem.

88. *Compound Interest* A principal P, invested at $9\frac{1}{2}\%$ and compounded continuously, increases to an amount K times the original principal after t years, where t is given by $t = (\ln K)/0.095$.

(a) Complete the table and interpret your results.

K	1	2	4	6	8	10	12
t							

(b) Sketch a graph of the function.

89. *Human Memory Model* Students in a mathematics class were given an exam and then retested monthly with an equivalent exam. The average scores for the class are given by the human memory model $f(t) = 80 - 17 \log(t + 1)$, $0 \le t \le 12$ where t is the time in months.

 (a) Use a graphing utility to graph the model over the specified domain.

(b) What was the average score on the original exam $(t = 0)$?

(c) What was the average score after 4 months?

(d) What was the average score after 10 months?

90. *Sound Intensity* The relationship between the number of decibels β and the intensity of a sound I in watts per square meter is

$$\beta = 10 \log\left(\frac{I}{10^{-12}}\right).$$

(a) Determine the number of decibels of a sound with an intensity of 1 watt per square meter.

(b) Determine the number of decibels of a sound with an intensity of 10^{-2} watt per square meter.

(c) The intensity of the sound in part (a) is 100 times as great as that in part (b). Is the number of decibels 100 times as great? Explain.

Synthesis

True or False? In Exercises 91 and 92, determine whether the statement is true or false. Justify your answer.

91. You can determine the graph of $f(x) = \log_6 x$ by graphing $g(x) = 6^x$ and reflecting it about the x-axis.

92. The graph of $f(x) = \log_3 x$ contains the point $(27, 3)$.

In Exercises 93–96, sketch the graph of f and g and describe the relationship between the graphs of f and g. What is the relationship between the functions f and g?

93. $f(x) = 3^x$, $g(x) = \log_3 x$

94. $f(x) = 5^x$, $g(x) = \log_5 x$

95. $f(x) = e^x$, $g(x) = \ln x$

96. $f(x) = 10^x$, $g(x) = \log x$

 97. *Graphical Analysis* Use a graphing utility to graph f and g in the same viewing window and determine which is increasing at the greater rate as x approaches $+\infty$. What can you conclude about the rate of growth of the natural logarithmic function?

(a) $f(x) = \ln x$, $g(x) = \sqrt{x}$

(b) $f(x) = \ln x$, $g(x) = \sqrt[4]{x}$

98. (a) Complete the table for the function given by

$$f(x) = \frac{\ln x}{x}.$$

x	1	5	10	10^2	10^4	10^6
f(x)						

(b) Use the table in part (a) to determine what value $f(x)$ approaches as x increases without bound.

(c) Use a graphing utility to confirm the result of part (b).

99. *Think About It* The table of values was obtained by evaluating a function. Determine which of the statements may be true and which must be false.

x	1	2	8
y	0	1	3

(a) y is an exponential function of x.

(b) y is a logarithmic function of x.

(c) x is an exponential function of y.

(d) y is a linear function of x.

100. *Writing* Explain why $\log_a x$ is defined only for $0 < a < 1$ and $a > 1$.

 In Exercises 101 and 102, (a) use a graphing utility to graph the function, (b) use the graph to determine the intervals in which the function is increasing and decreasing, and (c) approximate any relative maximum or minimum values of the function.

101. $f(x) = |\ln x|$ **102.** $h(x) = \ln(x^2 + 1)$

Skills Review

In Exercises 103–108, evaluate the function for $f(x) = 3x + 2$ and $g(x) = x^3 - 1$.

103. $(f + g)(2)$ **104.** $(f - g)(-1)$

105. $(fg)(6)$ **106.** $\left(\dfrac{f}{g}\right)(0)$

107. $(f \circ g)(7)$ **108.** $(g \circ f)(-3)$

5.3   Properties of Logarithms

What you should learn

- Use the change-of-base formula to rewrite and evaluate logarithmic expressions.
- Use properties of logarithms to evaluate or rewrite logarithmic expressions.
- Use properties of logarithms to expand or condense logarithmic expressions.
- Use logarithmic functions to model and solve real-life problems.

Why you should learn it

Logarithmic functions can be used to model and solve real-life problems. For instance, in Exercises 81–83 on page 410, a logarithmic function is used to model the relationship between the number of decibels and the intensity of a sound.

AP Photo/Stephen Chernin

Change of Base

Most calculators have only two types of log keys, one for common logarithms (base 10) and one for natural logarithms (base e). Although common logs and natural logs are the most frequently used, you may occasionally need to evaluate logarithms to other bases. To do this, you can use the following **change-of-base formula.**

Change-of-Base Formula

Let a, b, and x be positive real numbers such that $a \neq 1$ and $b \neq 1$. Then $\log_a x$ can be converted to a different base as follows.

Base b	Base 10	Base e
$\log_a x = \dfrac{\log_b x}{\log_b a}$	$\log_a x = \dfrac{\log x}{\log a}$	$\log_a x = \dfrac{\ln x}{\ln a}$

One way to look at the change-of-base formula is that logarithms to base a are simply *constant multiples* of logarithms to base b. The constant multiplier is $1/(\log_b a)$.

Example 1   Changing Bases Using Common Logarithms

a. $\log_4 25 = \dfrac{\log 25}{\log 4}$ $\log_a x = \dfrac{\log x}{\log a}$

$\approx \dfrac{1.39794}{0.60206}$ Use a calculator.

≈ 2.3219 Simplify.

b. $\log_2 12 = \dfrac{\log 12}{\log 2} \approx \dfrac{1.07918}{0.30103} \approx 3.5850$

✔CHECKPOINT   Now try Exercise 1(a).

Example 2   Changing Bases Using Natural Logarithms

a. $\log_4 25 = \dfrac{\ln 25}{\ln 4}$ $\log_a x = \dfrac{\ln x}{\ln a}$

$\approx \dfrac{3.21888}{1.38629}$ Use a calculator.

≈ 2.3219 Simplify.

b. $\log_2 12 = \dfrac{\ln 12}{\ln 2} \approx \dfrac{2.48491}{0.69315} \approx 3.5850$

✔CHECKPOINT   Now try Exercise 1(b).

Properties of Logarithms

You know from the preceding section that the logarithmic function with base a is the *inverse function* of the exponential function with base a. So, it makes sense that the properties of exponents should have corresponding properties involving logarithms. For instance, the exponential property $a^0 = 1$ has the corresponding logarithmic property $\log_a 1 = 0$.

Properties of Logarithms

Let a be a positive number such that $a \neq 1$, and let n be a real number. If u and v are positive real numbers, the following properties are true.

	Logarithm with Base a	Natural Logarithm
1. Product Property:	$\log_a(uv) = \log_a u + \log_a v$	$\ln(uv) = \ln u + \ln v$
2. Quotient Property:	$\log_a \dfrac{u}{v} = \log_a u - \log_a v$	$\ln \dfrac{u}{v} = \ln u - \ln v$
3. Power Property:	$\log_a u^n = n \log_a u$	$\ln u^n = n \ln u$

For proofs of the properties listed above, see Proofs in Mathematics on page 444.

Example 3   **Using Properties of Logarithms**

Write each logarithm in terms of $\ln 2$ and $\ln 3$.

a. $\ln 6$ **b.** $\ln \dfrac{2}{27}$

Solution

a. $\ln 6 = \ln(2 \cdot 3)$ Rewrite 6 as $2 \cdot 3$.

$\qquad = \ln 2 + \ln 3$ Product Property

b. $\ln \dfrac{2}{27} = \ln 2 - \ln 27$ Quotient Property

$\qquad = \ln 2 - \ln 3^3$ Rewrite 27 as 3^3.

$\qquad = \ln 2 - 3 \ln 3$ Power Property

 **Now try Exercise 17.**

Example 4   **Using Properties of Logarithms**

Find the exact value of each expression without using a calculator.

a. $\log_5 \sqrt[3]{5}$ **b.** $\ln e^6 - \ln e^2$

Solution

a. $\log_5 \sqrt[3]{5} = \log_5 5^{1/3} = \tfrac{1}{3} \log_5 5 = \tfrac{1}{3}(1) = \tfrac{1}{3}$

b. $\ln e^6 - \ln e^2 = \ln \dfrac{e^6}{e^2} = \ln e^4 = 4 \ln e = 4(1) = 4$

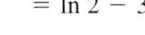 **Now try Exercise 23.**

The Granger Collection

Historical Note
John Napier, a Scottish mathematician, developed logarithms as a way to simplify some of the tedious calculations of his day. Beginning in 1594, Napier worked about 20 years on the invention of logarithms. Napier was only partially successful in his quest to simplify tedious calculations. Nonetheless, the development of logarithms was a step forward and received immediate recognition.

Rewriting Logarithmic Expressions

The properties of logarithms are useful for rewriting logarithmic expressions in forms that simplify the operations of algebra. This is true because these properties convert complicated products, quotients, and exponential forms into simpler sums, differences, and products, respectively.

Example 5   **Expanding Logarithmic Expressions**

Expand each logarithmic expression.

a. $\log_4 5x^3y$ **b.** $\ln \dfrac{\sqrt{3x-5}}{7}$

Solution

a. $\log_4 5x^3y = \log_4 5 + \log_4 x^3 + \log_4 y$ Product Property

$\qquad\qquad\quad = \log_4 5 + 3\log_4 x + \log_4 y$ Power Property

b. $\ln \dfrac{\sqrt{3x-5}}{7} = \ln \dfrac{(3x-5)^{1/2}}{7}$ Rewrite using rational exponent.

$\qquad\qquad\quad = \ln(3x-5)^{1/2} - \ln 7$ Quotient Property

$\qquad\qquad\quad = \dfrac{1}{2}\ln(3x-5) - \ln 7$ Power Property

✓**CHECKPOINT**   Now try Exercise 47.

In Example 5, the properties of logarithms were used to *expand* logarithmic expressions. In Example 6, this procedure is reversed and the properties of logarithms are used to *condense* logarithmic expressions.

Example 6   **Condensing Logarithmic Expressions**

Condense each logarithmic expression.

a. $\frac{1}{2}\log x + 3\log(x+1)$ **b.** $2\ln(x+2) - \ln x$

c. $\frac{1}{3}[\log_2 x + \log_2(x+1)]$

Solution

a. $\frac{1}{2}\log x + 3\log(x+1) = \log x^{1/2} + \log(x+1)^3$ Power Property

$\qquad\qquad\qquad\qquad = \log\left[\sqrt{x}(x+1)^3\right]$ Product Property

b. $2\ln(x+2) - \ln x = \ln(x+2)^2 - \ln x$ Power Property

$\qquad\qquad\qquad = \ln \dfrac{(x+2)^2}{x}$ Quotient Property

c. $\frac{1}{3}[\log_2 x + \log_2(x+1)] = \frac{1}{3}\{\log_2[x(x+1)]\}$ Product Property

$\qquad\qquad\qquad\qquad = \log_2[x(x+1)]^{1/3}$ Power Property

$\qquad\qquad\qquad\qquad = \log_2 \sqrt[3]{x(x+1)}$ Rewrite with a radical.

✓**CHECKPOINT**   Now try Exercise 69.

Exploration

Use a graphing utility to graph the functions given by

$$y_1 = \ln x - \ln(x-3)$$

and

$$y_2 = \ln \dfrac{x}{x-3}$$

in the same viewing window. Does the graphing utility show the functions with the same domain? If so, should it? Explain your reasoning.

Application

One method of determining how the x- and y-values for a set of nonlinear data are related is to take the natural logarithm of each of the x- and y-values. If the points are graphed and fall on a line, then you can determine that the x- and y-values are related by the equation

$$\ln y = m \ln x$$

where m is the slope of the line.

Example 7 **Finding a Mathematical Model**

The table shows the mean distance x and the period (the time it takes a planet to orbit the sun) y for each of the six planets that are closest to the sun. In the table, the mean distance is given in terms of astronomical units (where Earth's mean distance is defined as 1.0), and the period is given in years. Find an equation that relates y and x.

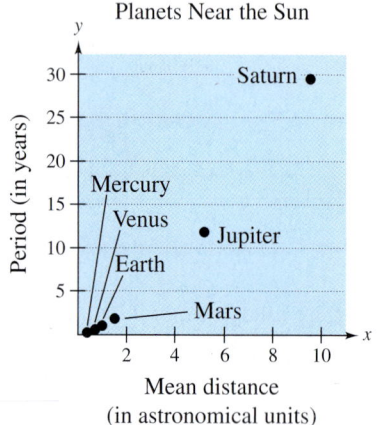

Planets Near the Sun

FIGURE 5.23

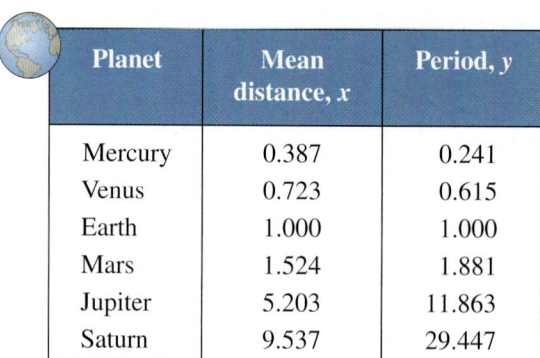

Planet	Mean distance, x	Period, y
Mercury	0.387	0.241
Venus	0.723	0.615
Earth	1.000	1.000
Mars	1.524	1.881
Jupiter	5.203	11.863
Saturn	9.537	29.447

Solution

The points in the table above are plotted in Figure 5.23. From this figure it is not clear how to find an equation that relates y and x. To solve this problem, take the natural logarithm of each of the x- and y-values in the table. This produces the following results.

Planet	Mercury	Venus	Earth	Mars	Jupiter	Saturn
$\ln x$	-0.949	-0.324	0.000	0.421	1.649	2.255
$\ln y$	-1.423	-0.486	0.000	0.632	2.473	3.383

Now, by plotting the points in the second table, you can see that all six of the points appear to lie in a line (see Figure 5.24). Choose any two points to determine the slope of the line. Using the two points $(0.421, 0.632)$ and $(0, 0)$, you can determine that the slope of the line is

$$m = \frac{0.632 - 0}{0.421 - 0} \approx 1.5 = \frac{3}{2}.$$

By the point-slope form, the equation of the line is $Y = \frac{3}{2}X$, where $Y = \ln y$ and $X = \ln x$. You can therefore conclude that $\ln y = \frac{3}{2} \ln x$.

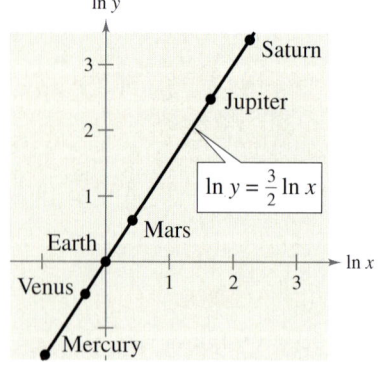

FIGURE 5.24

✔**CHECKPOINT** Now try Exercise 85.

5.3 | Exercises

VOCABULARY CHECK:

In Exercises 1 and 2, fill in the blanks.

1. To evaluate a logarithm to any base, you can use the _____ formula.

2. The change-of-base formula for base e is given by $\log_a x =$ _____.

In Exercises 3–5, match the property of logarithms with its name.

3. $\log_a(uv) = \log_a u + \log_a v$ (a) Power Property

4. $\ln u^n = n \ln u$ (b) Quotient Property

5. $\log_a \dfrac{u}{v} = \log_a u - \log_a v$ (c) Product Property

PREREQUISITE SKILLS REVIEW: Practice and review algebra skills needed for this section at **www.Eduspace.com.**

In Exercises 1–8, rewrite the logarithm as a ratio of (a) common logarithms and (b) natural logarithms.

1. $\log_5 x$ **2.** $\log_3 x$

3. $\log_{1/5} x$ **4.** $\log_{1/3} x$

5. $\log_x \frac{3}{10}$ **6.** $\log_x \frac{3}{4}$

7. $\log_{2.6} x$ **8.** $\log_{7.1} x$

In Exercises 9–16, evaluate the logarithm using the change-of-base formula. Round your result to three decimal places.

9. $\log_3 7$ **10.** $\log_7 4$

11. $\log_{1/2} 4$ **12.** $\log_{1/4} 5$

13. $\log_9 0.4$ **14.** $\log_{20} 0.125$

15. $\log_{15} 1250$ **16.** $\log_3 0.015$

In Exercises 17–22, use the properties of logarithms to rewrite and simplify the logarithmic expression.

17. $\log_4 8$ **18.** $\log_2(4^2 \cdot 3^4)$

19. $\log_5 \frac{1}{250}$ **20.** $\log \frac{9}{300}$

21. $\ln(5e^6)$ **22.** $\ln \dfrac{6}{e^2}$

In Exercises 23–38, find the exact value of the logarithmic expression without using a calculator. (If this is not possible, state the reason.)

23. $\log_3 9$ **24.** $\log_5 \frac{1}{125}$

25. $\log_2 \sqrt[4]{8}$ **26.** $\log_6 \sqrt[3]{6}$

27. $\log_4 16^{1.2}$ **28.** $\log_3 81^{-0.2}$

29. $\log_3(-9)$ **30.** $\log_2(-16)$

31. $\ln e^{4.5}$

32. $3 \ln e^4$

33. $\ln \dfrac{1}{\sqrt{e}}$

34. $\ln \sqrt[4]{e^3}$

35. $\ln e^2 + \ln e^5$

36. $2 \ln e^6 - \ln e^5$

37. $\log_5 75 - \log_5 3$

38. $\log_4 2 + \log_4 32$

In Exercises 39–60, use the properties of logarithms to expand the expression as a sum, difference, and/or constant multiple of logarithms. (Assume all variables are positive.)

39. $\log_4 5x$ **40.** $\log_3 10z$

41. $\log_8 x^4$ **42.** $\log_{10} \dfrac{y}{2}$

43. $\log_5 \dfrac{5}{x}$ **44.** $\log_6 \dfrac{1}{z^3}$

45. $\ln \sqrt{z}$ **46.** $\ln \sqrt[3]{t}$

47. $\ln xyz^2$ **48.** $\log 4x^2 y$

49. $\ln z(z - 1)^2, \; z > 1$ **50.** $\ln\left(\dfrac{x^2 - 1}{x^3}\right), \; x > 1$

51. $\log_2 \dfrac{\sqrt{a - 1}}{9}, \; a > 1$ **52.** $\ln \dfrac{6}{\sqrt{x^2 + 1}}$

53. $\ln \sqrt[3]{\dfrac{x}{y}}$ **54.** $\ln \sqrt{\dfrac{x^2}{y^3}}$

55. $\ln \dfrac{x^4 \sqrt{y}}{z^5}$ **56.** $\log_2 \dfrac{\sqrt{x}\, y^4}{z^4}$

57. $\log_5 \dfrac{x^2}{y^2 z^3}$ **58.** $\log_{10} \dfrac{xy^4}{z^5}$

59. $\ln \sqrt[4]{x^3(x^2 + 3)}$ **60.** $\ln \sqrt{x^2(x + 2)}$

In Exercises 61–78, condense the expression to the logarithm of a single quantity.

61. $\ln x + \ln 3$

62. $\ln y + \ln t$

63. $\log_4 z - \log_4 y$

64. $\log_5 8 - \log_5 t$

65. $2 \log_2(x + 4)$

66. $\frac{2}{3} \log_7(z - 2)$

67. $\frac{1}{4} \log_3 5x$

68. $-4 \log_6 2x$

69. $\ln x - 3 \ln(x + 1)$

70. $2 \ln 8 + 5 \ln (z - 4)$

71. $\log x - 2 \log y + 3 \log z$

72. $3 \log_3 x + 4 \log_3 y - 4 \log_3 z$

73. $\ln x - 4[\ln(x + 2) + \ln(x - 2)]$

74. $4[\ln z + \ln(z + 5)] - 2 \ln(z - 5)$

75. $\frac{1}{3}[2 \ln(x + 3) + \ln x - \ln(x^2 - 1)]$

76. $2[3 \ln x - \ln(x + 1) - \ln(x - 1)]$

77. $\frac{1}{3}[\log_8 y + 2 \log_8(y + 4)] - \log_8(y - 1)$

78. $\frac{1}{2}[\log_4(x + 1) + 2 \log_4(x - 1)] + 6 \log_4 x$

In Exercises 79 and 80, compare the logarithmic quantities. If two are equal, explain why.

79. $\dfrac{\log_2 32}{\log_2 4}$, $\quad \log_2 \dfrac{32}{4}$, $\quad \log_2 32 - \log_2 4$

80. $\log_7 \sqrt{70}$, $\quad \log_7 35$, $\quad \frac{1}{2} + \log_7 \sqrt{10}$

Sound Intensity In Exercises 81–83, use the following information. The relationship between the number of decibels β and the intensity of a sound I in watts per square meter is given by

$$\beta = 10 \log\left(\frac{I}{10^{-12}}\right).$$

81. Use the properties of logarithms to write the formula in simpler form, and determine the number of decibels of a sound with an intensity of 10^{-6} watt per square meter.

82. Find the difference in loudness between an average office with an intensity of 1.26×10^{-7} watt per square meter and a broadcast studio with an intensity of 3.16×10^{-5} watt per square meter.

83. You and your roommate are playing your stereos at the same time and at the same intensity. How much louder is the music when both stereos are playing compared with just one stereo playing?

Model It

84. *Human Memory Model* Students participating in a psychology experiment attended several lectures and were given an exam. Every month for a year after the exam, the students were retested to see how much of the material they remembered. The average scores for the group can be modeled by the human memory model

$$f(t) = 90 - 15 \log(t + 1), \quad 0 \le t \le 12$$

where t is the time in months.

(a) Use the properties of logarithms to write the function in another form.

(b) What was the average score on the original exam $(t = 0)$?

(c) What was the average score after 4 months?

(d) What was the average score after 12 months?

 (e) Use a graphing utility to graph the function over the specified domain.

(f) Use the graph in part (e) to determine when the average score will decrease to 75.

(g) Verify your answer to part (f) numerically.

85. *Galloping Speeds of Animals* Four-legged animals run with two different types of motion: trotting and galloping. An animal that is trotting has at least one foot on the ground at all times, whereas an animal that is galloping has all four feet off the ground at some point in its stride. The number of strides per minute at which an animal breaks from a trot to a gallop depends on the weight of the animal. Use the table to find a logarithmic equation that relates an animal's weight x (in pounds) and its lowest galloping speed y (in strides per minute).

Weight, x	Galloping Speed, y
25	191.5
35	182.7
50	173.8
75	164.2
500	125.9
1000	114.2

86. *Comparing Models*   A cup of water at an initial temperature of 78° C is placed in a room at a constant temperature of 21° C. The temperature of the water is measured every 5 minutes during a half-hour period. The results are recorded as ordered pairs of the form (t, T), where t is the time (in minutes) and T is the temperature (in degrees Celsius).

(0, 78.0°), (5, 66.0°), (10, 57.5°), (15, 51.2°), (20, 46.3°), (25, 42.4°), (30, 39.6°)

(a) The graph of the model for the data should be asymptotic with the graph of the temperature of the room. Subtract the room temperature from each of the temperatures in the ordered pairs. Use a graphing utility to plot the data points (t, T) and $(t, T - 21)$.

(b) An exponential model for the data $(t, T - 21)$ is given by

$$T - 21 = 54.4(0.964)^t.$$

Solve for T and graph the model. Compare the result with the plot of the original data.

(c) Take the natural logarithms of the revised temperatures. Use a graphing utility to plot the points $(t, \ln(T - 21))$ and observe that the points appear to be linear. Use the *regression* feature of the graphing utility to fit a line to these data. This resulting line has the form

$$\ln(T - 21) = at + b.$$

Use the properties of the logarithms to solve for T. Verify that the result is equivalent to the model in part (b).

(d) Fit a rational model to the data. Take the reciprocals of the y-coordinates of the revised data points to generate the points

$$\left(t, \frac{1}{T - 21}\right).$$

Use a graphing utility to graph these points and observe that they appear to be linear. Use the *regression* feature of a graphing utility to fit a line to these data. The resulting line has the form

$$\frac{1}{T - 21} = at + b.$$

Solve for T, and use a graphing utility to graph the rational function and the original data points.

(e) Write a short paragraph explaining why the transformations of the data were necessary to obtain each model. Why did taking the logarithms of the temperatures lead to a linear scatter plot? Why did taking the reciprocals of the temperature lead to a linear scatter plot?

Synthesis

True or False?   In Exercises 87–92, determine whether the statement is true or false given that $f(x) = \ln x$. Justify your answer.

87. $f(0) = 0$

88. $f(ax) = f(a) + f(x), \quad a > 0, x > 0$

89. $f(x - 2) = f(x) - f(2), \quad x > 2$

90. $\sqrt{f(x)} = \frac{1}{2}f(x)$ *false not sqrt of whole thing*

91. If $f(u) = 2f(v)$, then $v = u^2$.

92. If $f(x) < 0$, then $0 < x < 1$.

93. *Proof*   Prove that $\log_b \dfrac{u}{v} = \log_b u - \log_b v$.

94. *Proof*   Prove that $\log_b u^n = n \log_b u$.

 In Exercises 95–100, use the change-of-base formula to rewrite the logarithm as a ratio of logarithms. Then use a graphing utility to graph both functions in the same viewing window to verify that the functions are equivalent.

95. $f(x) = \log_2 x$

96. $f(x) = \log_4 x$

97. $f(x) = \log_{1/2} x$

98. $f(x) = \log_{1/4} x$

99. $f(x) = \log_{11.8} x$

100. $f(x) = \log_{12.4} x$

101. *Think About It*   Consider the functions below.

$$f(x) = \ln \frac{x}{2}, \quad g(x) = \frac{\ln x}{\ln 2}, \quad h(x) = \ln x - \ln 2$$

Which two functions should have identical graphs? Verify your answer by sketching the graphs of all three functions on the same set of coordinate axes.

102. *Exploration*   For how many integers between 1 and 20 can the natural logarithms be approximated given that $\ln 2 \approx 0.6931$, $\ln 3 \approx 1.0986$,   and $\ln 5 \approx 1.6094$? Approximate these logarithms (do not use a calculator).

Skills Review

In Exercises 103–106, simplify the expression.

103. $\dfrac{24xy^{-2}}{16x^{-3}y}$

104. $\left(\dfrac{2x^2}{3y}\right)^{-3}$

105. $(18x^3y^4)^{-3}(18x^3y^4)^3$

106. $xy(x^{-1} + y^{-1})^{-1}$

In Exercises 107–110, solve the equation.

107. $3x^2 + 2x - 1 = 0$

108. $4x^2 - 5x + 1 = 0$

109. $\dfrac{2}{3x + 1} = \dfrac{x}{4}$

110. $\dfrac{5}{x - 1} = \dfrac{2x}{3}$

5.4 Exponential and Logarithmic Equations

What you should learn

- Solve simple exponential and logarithmic equations.
- Solve more complicated exponential equations.
- Solve more complicated logarithmic equations.
- Use exponential and logarithmic equations to model and solve real-life problems.

Why you should learn it

Exponential and logarithmic equations are used to model and solve life science applications. For instance, in Exercise 112, on page 421, a logarithmic function is used to model the number of trees per acre given the average diameter of the trees.

© James Marshall/Corbis

Introduction

So far in this chapter, you have studied the definitions, graphs, and properties of exponential and logarithmic functions. In this section, you will study procedures for *solving equations* involving these exponential and logarithmic functions.

There are two basic strategies for solving exponential or logarithmic equations. The first is based on the One-to-One Properties and was used to solve simple exponential and logarithmic equations in Sections 5.1 and 5.2. The second is based on the Inverse Properties. For $a > 0$ and $a \neq 1$, the following properties are true for all x and y for which $\log_a x$ and $\log_a y$ are defined.

One-to-One Properties

$a^x = a^y$ if and only if $x = y$.

$\log_a x = \log_a y$ if and only if $x = y$.

Inverse Properties

$a^{\log_a x} = x$

$\log_a a^x = x$

Example 1 Solving Simple Equations

Original Equation	*Rewritten Equation*	*Solution*	*Property*
a. $2^x = 32$	$2^x = 2^5$	$x = 5$	One-to-One
b. $\ln x - \ln 3 = 0$	$\ln x = \ln 3$	$x = 3$	One-to-One
c. $\left(\frac{1}{3}\right)^x = 9$	$3^{-x} = 3^2$	$x = -2$	One-to-One
d. $e^x = 7$	$\ln e^x = \ln 7$	$x = \ln 7$	Inverse
e. $\ln x = -3$	$e^{\ln x} = e^{-3}$	$x = e^{-3}$	Inverse
f. $\log x = -1$	$10^{\log x} = 10^{-1}$	$x = 10^{-1} = \frac{1}{10}$	Inverse

✓**CHECKPOINT** Now try Exercise 13.

The strategies used in Example 1 are summarized as follows.

Strategies for Solving Exponential and Logarithmic Equations

1. Rewrite the original equation in a form that allows the use of the One-to-One Properties of exponential or logarithmic functions.

2. Rewrite an *exponential* equation in logarithmic form and apply the Inverse Property of logarithmic functions.

3. Rewrite a *logarithmic* equation in exponential form and apply the Inverse Property of exponential functions.

Solving Exponential Equations

Example 2   **Solving Exponential Equations**

Solve each equation and approximate the result to three decimal places if necessary.

a. $e^{-x^2} = e^{-3x-4}$ **b.** $3(2^x) = 42$

Solution

a.

$$e^{-x^2} = e^{-3x-4}$$ Write original equation.

$$-x^2 = -3x - 4$$ One-to-One Property

$$x^2 - 3x - 4 = 0$$ Write in general form.

$$(x + 1)(x - 4) = 0$$ Factor.

$$(x + 1) = 0 \Longrightarrow x = -1$$ Set 1st factor equal to 0.

$$(x - 4) = 0 \Longrightarrow x = 4$$ Set 2nd factor equal to 0.

The solutions are $x = -1$ and $x = 4$. Check these in the original equation.

b.

$$3(2^x) = 42$$ Write original equation.

$$2^x = 14$$ Divide each side by 3.

$$\log_2 2^x = \log_2 14$$ Take log (base 2) of each side.

$$x = \log_2 14$$ Inverse Property

$$x = \frac{\ln 14}{\ln 2} \approx 3.807$$ Change-of-base formula

The solution is $x = \log_2 14 \approx 3.807$. Check this in the original equation.

✓**CHECKPOINT**   Now try Exercise 25.

In Example 2(b), the exact solution is $x = \log_2 14$ and the approximate solution is $x \approx 3.807$. An exact answer is preferred when the solution is an intermediate step in a larger problem. For a final answer, an approximate solution is easier to comprehend.

Example 3   **Solving an Exponential Equation**

Solve $e^x + 5 = 60$ and approximate the result to three decimal places.

Solution

$$e^x + 5 = 60$$ Write original equation.

$$e^x = 55$$ Subtract 5 from each side.

$$\ln e^x = \ln 55$$ Take natural log of each side.

$$x = \ln 55 \approx 4.007$$ Inverse Property

The solution is $x = \ln 55 \approx 4.007$. Check this in the original equation.

✓**CHECKPOINT**   Now try Exercise 51.

STUDY TIP

Remember that the natural logarithmic function has a base of e.

Example 4 Solving an Exponential Equation

Solve $2(3^{2t-5}) - 4 = 11$ and approximate the result to three decimal places.

Solution

$2(3^{2t-5}) - 4 = 11$	Write original equation.
$2(3^{2t-5}) = 15$	Add 4 to each side.
$3^{2t-5} = \dfrac{15}{2}$	Divide each side by 2.
$\log_3 3^{2t-5} = \log_3 \dfrac{15}{2}$	Take log (base 3) of each side.
$2t - 5 = \log_3 \dfrac{15}{2}$	Inverse Property
$2t = 5 + \log_3 7.5$	Add 5 to each side.
$t = \dfrac{5}{2} + \dfrac{1}{2}\log_3 7.5$	Divide each side by 2.
$t \approx 3.417$	Use a calculator.

The solution is $t = \frac{5}{2} + \frac{1}{2}\log_3 7.5 \approx 3.417$. Check this in the original equation.

✓CHECKPOINT Now try Exercise 53.

> **STUDY TIP**
>
> Remember that to evaluate a logarithm such as $\log_3 7.5$, you need to use the change-of-base formula.
>
> $\log_3 7.5 = \dfrac{\ln 7.5}{\ln 3} \approx 1.834$

When an equation involves two or more exponential expressions, you can still use a procedure similar to that demonstrated in Examples 2, 3, and 4. However, the algebra is a bit more complicated.

Example 5 Solving an Exponential Equation of Quadratic Type

Solve $e^{2x} - 3e^x + 2 = 0$.

Algebraic Solution

$e^{2x} - 3e^x + 2 = 0$	Write original equation.
$(e^x)^2 - 3e^x + 2 = 0$	Write in quadratic form.
$(e^x - 2)(e^x - 1) = 0$	Factor.
$e^x - 2 = 0$	Set 1st factor equal to 0.
$x = \ln 2$	Solution
$e^x - 1 = 0$	Set 2nd factor equal to 0.
$x = 0$	Solution

The solutions are $x = \ln 2 \approx 0.693$ and $x = 0$. Check these in the original equation.

✓CHECKPOINT Now try Exercise 67.

Graphical Solution

Use a graphing utility to graph $y = e^{2x} - 3e^x + 2$. Use the *zero* or *root* feature or the *zoom* and *trace* features of the graphing utility to approximate the values of x for which $y = 0$. In Figure 5.25, you can see that the zeros occur at $x = 0$ and at $x \approx 0.693$. So, the solutions are $x = 0$ and $x \approx 0.693$.

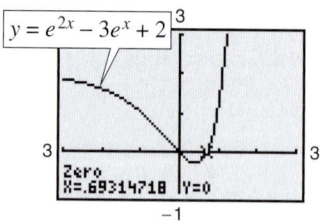

FIGURE 5.25

Solving Logarithmic Equations

To solve a logarithmic equation, you can write it in exponential form.

$\ln x = 3$	Logarithmic form
$e^{\ln x} = e^3$	Exponentiate each side.
$x = e^3$	Exponential form

This procedure is called *exponentiating* each side of an equation.

Example 6   Solving Logarithmic Equations

a.

$\ln x = 2$	Original equation
$e^{\ln x} = e^2$	Exponentiate each side.
$x = e^2$	Inverse Property

b.

$\log_3(5x - 1) = \log_3(x + 7)$	Original equation
$5x - 1 = x + 7$	One-to-One Property
$4x = 8$	Add $-x$ and 1 to each side.
$x = 2$	Divide each side by 4.

c.

$\log_6(3x + 14) - \log_6 5 = \log_6 2x$	Original equation
$\log_6\left(\dfrac{3x + 14}{5}\right) = \log_6 2x$	Quotient Property of Logarithms
$\dfrac{3x + 14}{5} = 2x$	One-to-One Property
$3x + 14 = 10x$	Cross multiply.
$-7x = -14$	Isolate x.
$x = 2$	Divide each side by -7.

✔**CHECKPOINT**   Now try Exercise 77.

Example 7   Solving a Logarithmic Equation

Solve $5 + 2 \ln x = 4$ and approximate the result to three decimal places.

Solution

$5 + 2 \ln x = 4$	Write original equation.
$2 \ln x = -1$	Subtract 5 from each side.
$\ln x = -\dfrac{1}{2}$	Divide each side by 2.
$e^{\ln x} = e^{-1/2}$	Exponentiate each side.
$x = e^{-1/2}$	Inverse Property
$x \approx 0.607$	Use a calculator.

✔**CHECKPOINT**   Now try Exercise 85.

Example 8   **Solving a Logarithmic Equation**

Solve $2 \log_5 3x = 4$.

Solution

$2 \log_5 3x = 4$	Write original equation.
$\log_5 3x = 2$	Divide each side by 2.
$5^{\log_5 3x} = 5^2$	Exponentiate each side (base 5).
$3x = 25$	Inverse Property
$x = \dfrac{25}{3}$	Divide each side by 3.

The solution is $x = \frac{25}{3}$. Check this in the original equation.

✓**CHECKPOINT**   Now try Exercise 87.

STUDY TIP

Notice in Example 9 that the logarithmic part of the equation is condensed into a single logarithm before exponentiating each side of the equation.

Because the domain of a logarithmic function generally does not include all real numbers, you should be sure to check for extraneous solutions of logarithmic equations.

Example 9   **Checking for Extraneous Solutions**

Solve $\log 5x + \log(x - 1) = 2$.

Algebraic Solution

$\log 5x + \log(x - 1) = 2$	Write original equation.
$\log[5x(x - 1)] = 2$	Product Property of Logarithms
$10^{\log(5x^2 - 5x)} = 10^2$	Exponentiate each side (base 10).
$5x^2 - 5x = 100$	Inverse Property
$x^2 - x - 20 = 0$	Write in general form.
$(x - 5)(x + 4) = 0$	Factor.
$x - 5 = 0$	Set 1st factor equal to 0.
$x = 5$	Solution
$x + 4 = 0$	Set 2nd factor equal to 0.
$x = -4$	Solution

The solutions appear to be $x = 5$ and $x = -4$. However, when you check these in the original equation, you can see that $x = 5$ is the only solution.

✓**CHECKPOINT**   Now try Exercise 99.

Graphical Solution

Use a graphing utility to graph $y_1 = \log 5x + \log(x - 1)$ and $y_2 = 2$ in the same viewing window. From the graph shown in Figure 5.26, it appears that the graphs intersect at one point. Use the *intersect* feature or the *zoom* and *trace* features to determine that the graphs intersect at approximately $(5, 2)$. So, the solution is $x = 5$. Verify that 5 is an exact solution algebraically.

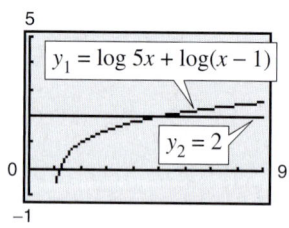

FIGURE 5.26

In Example 9, the domain of $\log 5x$ is $x > 0$ and the domain of $\log(x - 1)$ is $x > 1$, so the domain of the original equation is $x > 1$. Because the domain is all real numbers greater than 1, the solution $x = -4$ is extraneous. The graph in Figure 5.26 verifies this concept.

Applications

Example 10   **Doubling an Investment**

You have deposited $500 in an account that pays 6.75% interest, compounded continuously. How long will it take your money to double?

Solution

Using the formula for continuous compounding, you can find that the balance in the account is

$$A = Pe^{rt}$$

$$A = 500e^{0.0675t}.$$

To find the time required for the balance to double, let $A = 1000$ and solve the resulting equation for t.

$500e^{0.0675t} = 1000$	Let $A = 1000$.
$e^{0.0675t} = 2$	Divide each side by 500.
$\ln e^{0.0675t} = \ln 2$	Take natural log of each side.
$0.0675t = \ln 2$	Inverse Property
$t = \dfrac{\ln 2}{0.0675}$	Divide each side by 0.0675.
$t \approx 10.27$	Use a calculator.

The balance in the account will double after approximately 10.27 years. This result is demonstrated graphically in Figure 5.27.

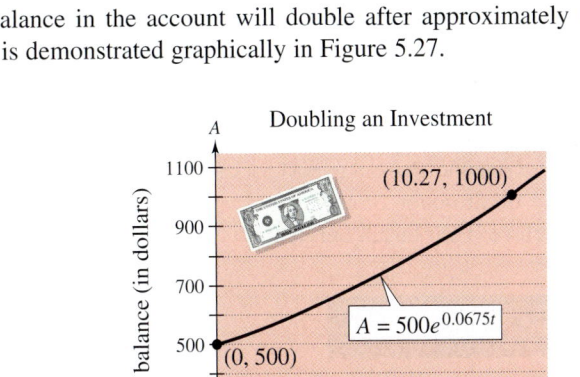

FIGURE 5.27

✓**CHECKPOINT**   Now try Exercise 107.

In Example 10, an approximate answer of 10.27 years is given. Within the context of the problem, the exact solution, $(\ln 2)/0.0675$ years, does not make sense as an answer.

Exploration

The *effective yield* of a savings plan is the percent increase in the balance after 1 year. Find the effective yield for each savings plan when $1000 is deposited in a savings account.

a. 7% annual interest rate, compounded annually

b. 7% annual interest rate, compounded continuously

c. 7% annual interest rate, compounded quarterly

d. 7.25% annual interest rate, compounded quarterly

Which savings plan has the greatest effective yield? Which savings plan will have the highest balance after 5 years?

Endangered Animal Species

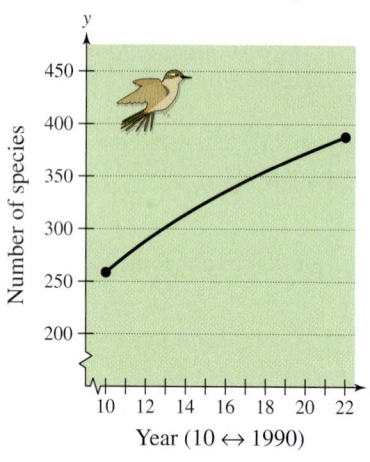

Number of species

Year (10 ↔ 1990)

FIGURE **5.28**

Example 11 **Endangered Animals**

The number y of endangered animal species in the United States from 1990 to 2002 can be modeled by

$$y = -119 + 164 \ln t, \quad 10 \leq t \leq 22$$

where t represents the year, with $t = 10$ corresponding to 1990 (see Figure 5.28). During which year did the number of endangered animal species reach 357? (Source: U.S. Fish and Wildlife Service)

Solution

$-119 + 164 \ln t = y$	Write original equation.
$-119 + 164 \ln t = 357$	Substitute 357 for y.
$164 \ln t = 476$	Add 119 to each side.
$\ln t = \dfrac{476}{164}$	Divide each side by 164.
$e^{\ln t} \approx e^{476/164}$	Exponentiate each side.
$t \approx e^{476/164}$	Inverse Property
$t \approx 18$	Use a calculator.

The solution is $t \approx 18$. Because $t = 10$ represents 1990, it follows that the number of endangered animals reached 357 in 1998.

✓**CHECKPOINT** Now try Exercise 113.

𝒲RITING ABOUT 𝓜ATHEMATICS

Comparing Mathematical Models The table shows the U.S. Postal Service rates y for sending an express mail package for selected years from 1985 through 2002, where $x = 5$ represents 1985. (Source: U.S. Postal Service)

Year, x	Rate, y
5	10.75
8	12.00
11	13.95
15	15.00
19	15.75
21	16.00
22	17.85

a. Create a scatter plot of the data. Find a linear model for the data, and add its graph to your scatter plot. According to this model, when will the rate for sending an express mail package reach $19.00?

b. Create a new table showing values for ln x and ln y and create a scatter plot of these transformed data. Use the method illustrated in Example 7 in Section 5.3 to find a model for the transformed data, and add its graph to your scatter plot. According to this model, when will the rate for sending an express mail package reach $19.00?

c. Solve the model in part (b) for y, and add its graph to your scatter plot in part (a). Which model better fits the original data? Which model will better predict future rates? Explain.

5.4 | Exercises

VOCABULARY CHECK: Fill in the blanks.

1. To _____ an equation in x means to find all values of x for which the equation is true.

2. To solve exponential and logarithmic equations, you can use the following One-to-One and Inverse Properties.

    (a) $a^x = a^y$ if and only if _____.

    (b) $\log_a x = \log_a y$ if and only if _____.

    (c) $a^{\log_a x} =$ _____

    (d) $\log_a a^x =$ _____

3. An _____ solution does not satisfy the original equation.

PREREQUISITE SKILLS REVIEW: Practice and review algebra skills needed for this section at **www.Eduspace.com.**

In Exercises 1–8, determine whether each x-value is a solution (or an approximate solution) of the equation.

1. $4^{2x-7} = 64$

    (a) $x = 5$

    (b) $x = 2$

2. $2^{3x+1} = 32$

    (a) $x = -1$

    (b) $x = 2$

3. $3e^{x+2} = 75$

    (a) $x = -2 + e^{25}$

    (b) $x = -2 + \ln 25$

    (c) $x \approx 1.219$

4. $2e^{5x+2} = 12$

    (a) $x = \frac{1}{5}(-2 + \ln 6)$

    (b) $x = \dfrac{\ln 6}{5 \ln 2}$

    (c) $x \approx -0.0416$

5. $\log_4(3x) = 3$

    (a) $x \approx 21.333$

    (b) $x = -4$

    (c) $x = \frac{64}{3}$

6. $\log_2(x + 3) = 10$

    (a) $x = 1021$

    (b) $x = 17$

    (c) $x = 10^2 - 3$

7. $\ln(2x + 3) = 5.8$

    (a) $x = \frac{1}{2}(-3 + \ln 5.8)$

    (b) $x = \frac{1}{2}(-3 + e^{5.8})$

    (c) $x \approx 163.650$

8. $\ln(x - 1) = 3.8$

    (a) $x = 1 + e^{3.8}$

    (b) $x \approx 45.701$

    (c) $x = 1 + \ln 3.8$

In Exercises 9–20, solve for x.

9. $4^x = 16$

10. $3^x = 243$

11. $\left(\frac{1}{2}\right)^x = 32$

12. $\left(\frac{1}{4}\right)^x = 64$

13. $\ln x - \ln 2 = 0$

14. $\ln x - \ln 5 = 0$

15. $e^x = 2$

16. $e^x = 4$

17. $\ln x = -1$

18. $\ln x = -7$

19. $\log_4 x = 3$

20. $\log_5 x = -3$

In Exercises 21–24, approximate the point of intersection of the graphs of f and g. Then solve the equation $f(x) = g(x)$ algebraically to verify your approximation.

21. $f(x) = 2^x$

    $g(x) = 8$

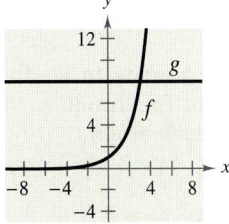

22. $f(x) = 27^x$

    $g(x) = 9$

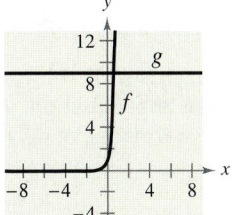

23. $f(x) = \log_3 x$

    $g(x) = 2$

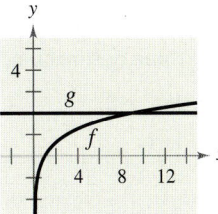

24. $f(x) = \ln(x - 4)$

    $g(x) = 0$

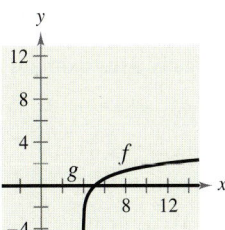

In Exercises 25–66, solve the exponential equation algebraically. Approximate the result to three decimal places.

25. $e^x = e^{x^2-2}$

26. $e^{2x} = e^{x^2-8}$

27. $e^{x^2-3} = e^{x-2}$

28. $e^{-x^2} = e^{x^2-2x}$

29. $4(3^x) = 20$

30. $2(5^x) = 32$

31. $2e^x = 10$

32. $4e^x = 91$

33. $e^x - 9 = 19$

34. $6^x + 10 = 47$

35. $3^{2x} = 80$

36. $6^{5x} = 3000$

37. $5^{-t/2} = 0.20$

38. $4^{-3t} = 0.10$

39. $3^{x-1} = 27$

40. $2^{x-3} = 32$

41. $2^{3-x} = 565$

42. $8^{-2-x} = 431$

43. $8(10^{3x}) = 12$

44. $5(10^{x-6}) = 7$

45. $3(5^{x-1}) = 21$

46. $8(3^{6-x}) = 40$

47. $e^{3x} = 12$

48. $e^{2x} = 50$

49. $500e^{-x} = 300$

50. $1000e^{-4x} = 75$

51. $7 - 2e^x = 5$

52. $-14 + 3e^x = 11$

53. $6(2^{3x-1}) - 7 = 9$

54. $8(4^{6-2x}) + 13 = 41$

55. $e^{2x} - 4e^x - 5 = 0$

56. $e^{2x} - 5e^x + 6 = 0$

57. $e^{2x} - 3e^x - 4 = 0$

58. $e^{2x} + 9e^x + 36 = 0$

59. $\dfrac{500}{100 - e^{x/2}} = 20$

60. $\dfrac{400}{1 + e^{-x}} = 350$

61. $\dfrac{3000}{2 + e^{2x}} = 2$

62. $\dfrac{119}{e^{6x} - 14} = 7$

63. $\left(1 + \dfrac{0.065}{365}\right)^{365t} = 4$

64. $\left(4 - \dfrac{2.471}{40}\right)^{9t} = 21$

65. $\left(1 + \dfrac{0.10}{12}\right)^{12t} = 2$

66. $\left(16 - \dfrac{0.878}{26}\right)^{3t} = 30$

 In Exercises 67–74, use a graphing utility to graph and solve the equation. Approximate the result to three decimal places. Verify your result algebraically.

67. $6e^{1-x} = 25$

68. $-4e^{-x-1} + 15 = 0$

69. $3e^{3x/2} = 962$

70. $8e^{-2x/3} = 11$

71. $e^{0.09t} = 3$

72. $-e^{1.8x} + 7 = 0$

73. $e^{0.125t} - 8 = 0$

74. $e^{2.724x} = 29$

In Exercises 75–102, solve the logarithmic equation algebraically. Approximate the result to three decimal places.

75. $\ln x = -3$

76. $\ln x = 2$

77. $\ln 2x = 2.4$

78. $\ln 4x = 1$

79. $\log x = 6$

80. $\log 3z = 2$

81. $3 \ln 5x = 10$

82. $2 \ln x = 7$

83. $\ln \sqrt{x + 2} = 1$

84. $\ln \sqrt{x - 8} = 5$

85. $7 + 3 \ln x = 5$

86. $2 - 6 \ln x = 10$

87. $6 \log_3(0.5x) = 11$

88. $5 \log_{10}(x - 2) = 11$

89. $\ln x - \ln(x + 1) = 2$

90. $\ln x + \ln(x + 1) = 1$

91. $\ln x + \ln(x - 2) = 1$

92. $\ln x + \ln(x + 3) = 1$

93. $\ln(x + 5) = \ln(x - 1) - \ln(x + 1)$

94. $\ln(x + 1) - \ln(x - 2) = \ln x$

95. $\log_2(2x - 3) = \log_2(x + 4)$

96. $\log(x - 6) = \log(2x + 1)$

97. $\log(x + 4) - \log x = \log(x + 2)$

98. $\log_2 x + \log_2(x + 2) = \log_2(x + 6)$

99. $\log_4 x - \log_4(x - 1) = \frac{1}{2}$

100. $\log_3 x + \log_3(x - 8) = 2$

101. $\log 8x - \log(1 + \sqrt{x}) = 2$

102. $\log 4x - \log(12 + \sqrt{x}) = 2$

 In Exercises 103–106, use a graphing utility to graph and solve the equation. Approximate the result to three decimal places. Verify your result algebraically.

103. $7 = 2^x$

104. $500 = 1500e^{-x/2}$

105. $3 - \ln x = 0$

106. $10 - 4 \ln(x - 2) = 0$

Compound Interest **In Exercises 107 and 108, $2500 is invested in an account at interest rate r, compounded continuously. Find the time required for the amount to (a) double and (b) triple.**

107. $r = 0.085$

108. $r = 0.12$

109. *Demand* The demand equation for a microwave oven is given by

$$p = 500 - 0.5(e^{0.004x}).$$

Find the demand x for a price of (a) $p = \$350$ and (b) $p = \$300$.

110. *Demand* The demand equation for a hand-held electronic organizer is

$$p = 5000\left(1 - \frac{4}{4 + e^{-0.002x}}\right).$$

Find the demand x for a price of (a) $p = \$600$ and (b) $p = \$400$.

111. *Forest Yield* The yield V (in millions of cubic feet per acre) for a forest at age t years is given by

$$V = 6.7e^{-48.1/t}.$$

 (a) Use a graphing utility to graph the function.

(b) Determine the horizontal asymptote of the function. Interpret its meaning in the context of the problem.

(c) Find the time necessary to obtain a yield of 1.3 million cubic feet.

112. *Trees per Acre*   The number N of trees of a given species per acre is approximated by the model $N = 68(10^{-0.04x})$, $5 \le x \le 40$ where x is the average diameter of the trees (in inches) 3 feet above the ground. Use the model to approximate the average diameter of the trees in a test plot when $N = 21$.

113. *Medicine*   The number y of hospitals in the United States from 1995 to 2002 can be modeled by

$$y = 7312 - 630.0 \ln t, \quad 5 \le t \le 12$$

where t represents the year, with $t = 5$ corresponding to 1995. During which year did the number of hospitals reach 5800?   (Source: Health Forum)

114. *Sports*   The number y of daily fee golf facilities in the United States from 1995 to 2003 can be modeled by $y = 4381 + 1883.6 \ln t$, $5 \le t \le 13$ where t represents the year, with $t = 5$ corresponding to 1995. During which year did the number of daily fee golf facilities reach 9000?   (Source: National Golf Foundation)

115. *Average Heights*   The percent m of American males between the ages of 18 and 24 who are no more than x inches tall is modeled by

$$m(x) = \frac{100}{1 + e^{-0.6114(x - 69.71)}}$$

and the percent f of American females between the ages of 18 and 24 who are no more than x inches tall is modeled by

$$f(x) = \frac{100}{1 + e^{-0.66607(x - 64.51)}}.$$

(Source: U.S. National Center for Health Statistics)

(a) Use the graph to determine any horizontal asymptotes of the graphs of the functions. Interpret the meaning in the context of the problem.

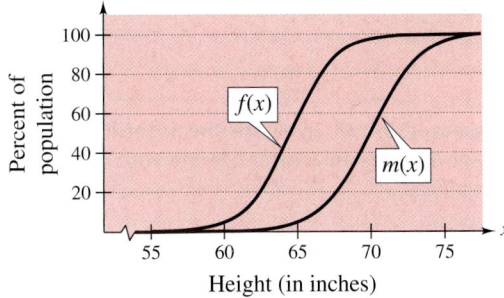

(b) What is the average height of each sex?

116. *Learning Curve*   In a group project in learning theory, a mathematical model for the proportion P of correct responses after n trials was found to be

$$P = \frac{0.83}{1 + e^{-0.2n}}.$$

 (a) Use a graphing utility to graph the function.

 (b) Use the graph to determine any horizontal asymptotes of the graph of the function. Interpret the meaning of the upper asymptote in the context of this problem.

(c) After how many trials will 60% of the responses be correct?

Model It

117. *Automobiles*   Automobiles are designed with crumple zones that help protect their occupants in crashes. The crumple zones allow the occupants to move short distances when the automobiles come to abrupt stops. The greater the distance moved, the fewer g's the crash victims experience. (One g is equal to the acceleration due to gravity. For very short periods of time, humans have withstood as much as 40 g's.) In crash tests with vehicles moving at 90 kilometers per hour, analysts measured the numbers of g's experienced during deceleration by crash dummies that were permitted to move x meters during impact. The data are shown in the table.

x	g's
0.2	158
0.4	80
0.6	53
0.8	40
1.0	32

A model for the data is given by

$$y = -3.00 + 11.88 \ln x + \frac{36.94}{x}$$

where y is the number of g's.

(a) Complete the table using the model.

x	0.2	0.4	0.6	0.8	1.0
y					

 (b) Use a graphing utility to graph the data points and the model in the same viewing window. How do they compare?

(c) Use the model to estimate the distance traveled during impact if the passenger deceleration must not exceed 30 g's.

(d) Do you think it is practical to lower the number of g's experienced during impact to fewer than 23? Explain your reasoning.

118. *Data Analysis*   An object at a temperature of 160°C was removed from a furnace and placed in a room at 20°C. The temperature T of the object was measured each hour h and recorded in the table. A model for the data is given by $T = 20[1 + 7(2^{-h})]$. The graph of this model is shown in the figure.

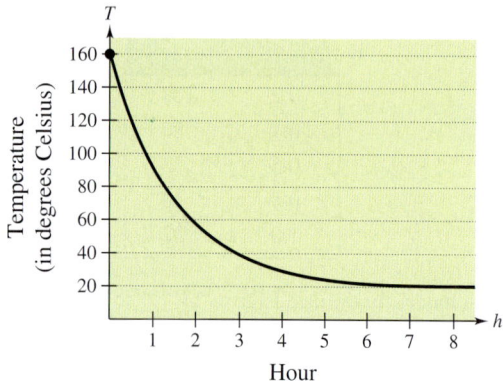

Hour, h	Temperature, T
0	160°
1	90°
2	56°
3	38°
4	29°
5	24°

(a) Use the graph to identify the horizontal asymptote of the model and interpret the asymptote in the context of the problem.

(b) Use the model to approximate the time when the temperature of the object was 100°C.

Synthesis

True or False?   In Exercises 119–122, rewrite each verbal statement as an equation. Then decide whether the statement is true or false. Justify your answer.

119. The logarithm of the product of two numbers is equal to the sum of the logarithms of the numbers.

120. The logarithm of the sum of two numbers is equal to the product of the logarithms of the numbers.

121. The logarithm of the difference of two numbers is equal to the difference of the logarithms of the numbers.

122. The logarithm of the quotient of two numbers is equal to the difference of the logarithms of the numbers.

123. *Think About It*   Is it possible for a logarithmic equation to have more than one extraneous solution? Explain.

124. *Finance*   You are investing P dollars at an annual interest rate of r, compounded continuously, for t years. Which of the following would result in the highest value of the investment? Explain your reasoning.

(a) Double the amount you invest.

(b) Double your interest rate.

(c) Double the number of years.

125. *Think About It*   Are the times required for the investments in Exercises 107 and 108 to quadruple twice as long as the times for them to double? Give a reason for your answer and verify your answer algebraically.

126. *Writing*   Write two or three sentences stating the general guidelines that you follow when solving (a) exponential equations and (b) logarithmic equations.

Skills Review

In Exercises 127–130, simplify the expression.

127. $\sqrt{48x^2y^5}$

128. $\sqrt{32} - 2\sqrt{25}$

129. $\sqrt[3]{25} \cdot \sqrt[3]{15}$

130. $\dfrac{3}{\sqrt{10} - 2}$

In Exercises 131–134, sketch a graph of the function.

131. $f(x) = |x| + 9$

132. $f(x) = |x + 2| - 8$

133. $g(x) = \begin{cases} 2x, & x < 0 \\ -x^2 + 4, & x \ge 0 \end{cases}$

134. $g(x) = \begin{cases} x - 3, & x \le -1 \\ x^2 + 1, & x > -1 \end{cases}$

In Exercises 135–138, evaluate the logarithm using the change-of-base formula. Approximate your result to three decimal places.

135. $\log_6 9$

136. $\log_3 4$

137. $\log_{3/4} 5$

138. $\log_8 22$

5.5   Exponential and Logarithmic Models

Alan Becker/Getty Images

Introduction

The five most common types of mathematical models involving exponential functions and logarithmic functions are as follows.

1. **Exponential growth model:** $y = ae^{bx}, \quad b > 0$
2. **Exponential decay model:** $y = ae^{-bx}, \quad b > 0$
3. **Gaussian model:** $y = ae^{-(x-b)^2/c}$
4. **Logistic growth model:** $y = \dfrac{a}{1 + be^{-rx}}$
5. **Logarithmic models:** $y = a + b \ln x, \quad y = a + b \log x$

The basic shapes of the graphs of these functions are shown in Figure 5.29.

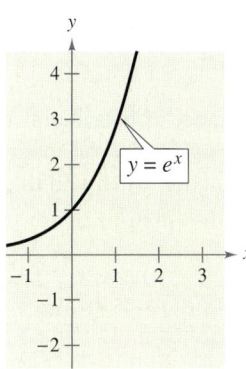

EXPONENTIAL GROWTH MODEL

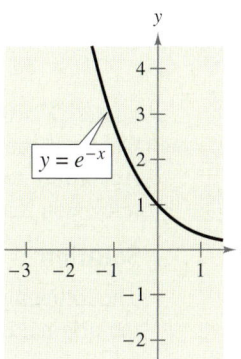

EXPONENTIAL DECAY MODEL

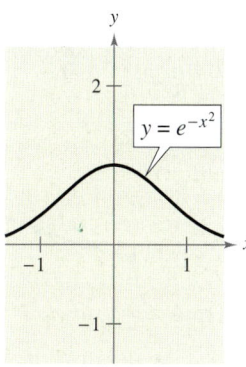

GAUSSIAN MODEL

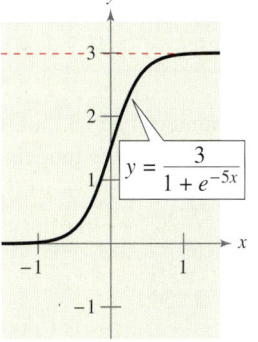

LOGISTIC GROWTH MODEL

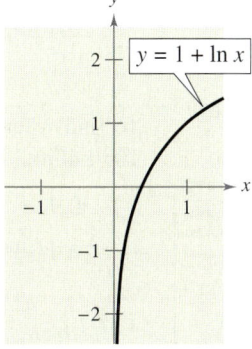

NATURAL LOGARITHMIC MODEL

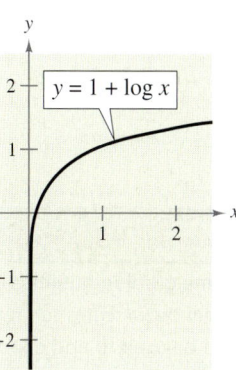

COMMON LOGARITHMIC MODEL

FIGURE 5.29

You can often gain quite a bit of insight into a situation modeled by an exponential or logarithmic function by identifying and interpreting the function's asymptotes. Use the graphs in Figure 5.29 to identify the asymptotes of the graph of each function.

Exponential Growth and Decay

Example 1 Digital Television

Estimates of the numbers (in millions) of U.S. households with digital television from 2003 through 2007 are shown in the table. The scatter plot of the data is shown in Figure 5.30. (Source: eMarketer)

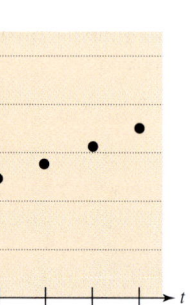

Digital Television

FIGURE 5.30

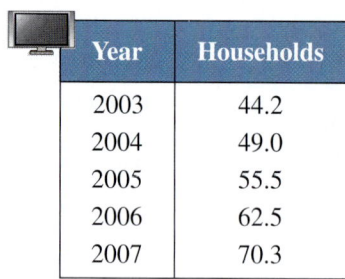

Year	Households
2003	44.2
2004	49.0
2005	55.5
2006	62.5
2007	70.3

An exponential growth model that approximates these data is given by

$$D = 30.92e^{0.1171t}, \quad 3 \le t \le 7$$

where D is the number of households (in millions) and $t = 3$ represents 2003. Compare the values given by the model with the estimates shown in the table. According to this model, when will the number of U.S. households with digital television reach 100 million?

Solution

The following table compares the two sets of figures. The graph of the model and the original data are shown in Figure 5.31.

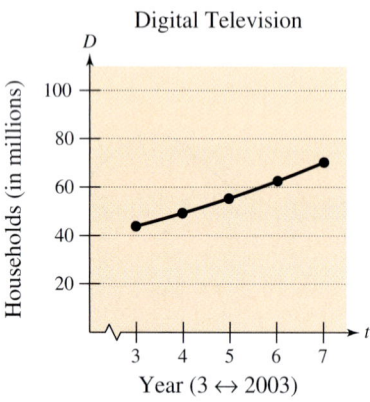

Digital Television

FIGURE 5.31

Year	2003	2004	2005	2006	2007
Households	44.2	49.0	55.5	62.5	70.3
Model	43.9	49.4	55.5	62.4	70.2

To find when the number of U.S. households with digital television will reach 100 million, let $D = 100$ in the model and solve for t.

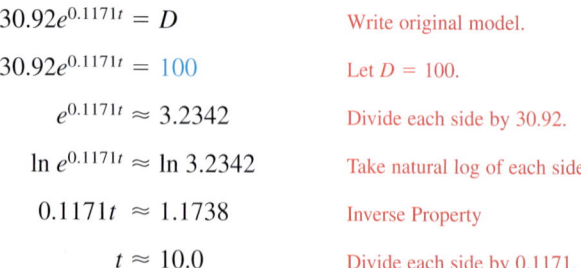

$30.92e^{0.1171t} = D$	Write original model.
$30.92e^{0.1171t} = 100$	Let $D = 100$.
$e^{0.1171t} \approx 3.2342$	Divide each side by 30.92.
$\ln e^{0.1171t} \approx \ln 3.2342$	Take natural log of each side.
$0.1171t \approx 1.1738$	Inverse Property
$t \approx 10.0$	Divide each side by 0.1171.

According to the model, the number of U.S. households with digital television will reach 100 million in 2010.

✓CHECKPOINT   Now try Exercise 35.

Technology

Some graphing utilities have an *exponential regression* feature that can be used to find exponential models that represent data. If you have such a graphing utility, try using it to find an exponential model for the data given in Example 1. How does your model compare with the model given in Example 1?

In Example 1, you were given the exponential growth model. But suppose this model were not given; how could you find such a model? One technique for doing this is demonstrated in Example 2.

Example 2   Modeling Population Growth

In a research experiment, a population of fruit flies is increasing according to the law of exponential growth. After 2 days there are 100 flies, and after 4 days there are 300 flies. How many flies will there be after 5 days?

Solution

Let y be the number of flies at time t. From the given information, you know that $y = 100$ when $t = 2$ and $y = 300$ when $t = 4$. Substituting this information into the model $y = ae^{bt}$ produces

$$100 = ae^{2b} \quad \text{and} \quad 300 = ae^{4b}.$$

To solve for b, solve for a in the first equation.

$$100 = ae^{2b} \quad \implies \quad a = \frac{100}{e^{2b}} \qquad \text{Solve for } a \text{ in the first equation.}$$

Then substitute the result into the second equation.

$$300 = ae^{4b} \qquad \text{Write second equation.}$$

$$300 = \left(\frac{100}{e^{2b}}\right)e^{4b} \qquad \text{Substitute } 100/e^{2b} \text{ for } a.$$

$$\frac{300}{100} = e^{2b} \qquad \text{Divide each side by 100.}$$

$$\ln 3 = 2b \qquad \text{Take natural log of each side.}$$

$$\frac{1}{2}\ln 3 = b \qquad \text{Solve for } b.$$

Using $b = \frac{1}{2}\ln 3$ and the equation you found for a, you can determine that

$$a = \frac{100}{e^{2[(1/2)\ln 3]}} \qquad \text{Substitute } \tfrac{1}{2}\ln 3 \text{ for } b.$$

$$= \frac{100}{e^{\ln 3}} \qquad \text{Simplify.}$$

$$= \frac{100}{3} \qquad \text{Inverse Property}$$

$$\approx 33.33. \qquad \text{Simplify.}$$

So, with $a \approx 33.33$ and $b = \frac{1}{2}\ln 3 \approx 0.5493$, the exponential growth model is

$$y = 33.33e^{0.5493t}$$

as shown in Figure 5.32. This implies that, after 5 days, the population will be

$$y = 33.33e^{0.5493(5)} \approx 520 \text{ flies.}$$

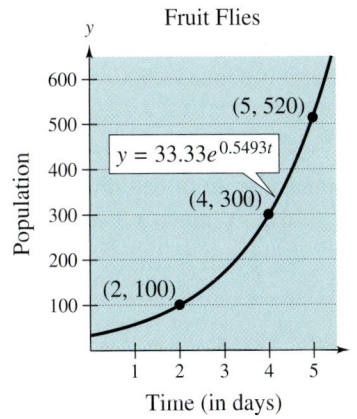

Fruit Flies

$y = 33.33e^{0.5493t}$

(5, 520)
(4, 300)
(2, 100)

Population

Time (in days)

FIGURE 5.32

✔CHECKPOINT   Now try Exercise 37.

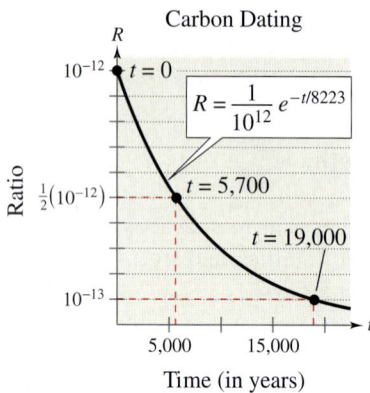

Carbon Dating

$$R = \frac{1}{10^{12}} e^{-t/8223}$$

FIGURE 5.33

In living organic material, the ratio of the number of radioactive carbon isotopes (carbon 14) to the number of nonradioactive carbon isotopes (carbon 12) is about 1 to 10^{12}. When organic material dies, its carbon 12 content remains fixed, whereas its radioactive carbon 14 begins to decay with a half-life of about 5700 years. To estimate the age of dead organic material, scientists use the following formula, which denotes the ratio of carbon 14 to carbon 12 present at any time t (in years).

$$R = \frac{1}{10^{12}} e^{-t/8223} \qquad \text{Carbon dating model}$$

The graph of R is shown in Figure 5.33. Note that R decreases as t increases.

Example 3   Carbon Dating

Estimate the age of a newly discovered fossil in which the ratio of carbon 14 to carbon 12 is

$$R = \frac{1}{10^{13}}.$$

Solution

In the carbon dating model, substitute the given value of R to obtain the following.

$$\frac{1}{10^{12}} e^{-t/8223} = R \qquad \text{Write original model.}$$

$$\frac{e^{-t/8223}}{10^{12}} = \frac{1}{10^{13}} \qquad \text{Let } R = \frac{1}{10^{13}}.$$

$$e^{-t/8223} = \frac{1}{10} \qquad \text{Multiply each side by } 10^{12}.$$

$$\ln e^{-t/8223} = \ln \frac{1}{10} \qquad \text{Take natural log of each side.}$$

$$-\frac{t}{8223} \approx -2.3026 \qquad \text{Inverse Property}$$

$$t \approx 18{,}934 \qquad \text{Multiply each side by } -8223.$$

So, to the nearest thousand years, the age of the fossil is about 19,000 years.

✓CHECKPOINT   Now try Exercise 41.

The value of b in the exponential decay model $y = ae^{-bt}$ determines the *decay* of radioactive isotopes. For instance, to find how much of an initial 10 grams of ^{226}Ra isotope with a half-life of 1599 years is left after 500 years, substitute this information into the model $y = ae^{-bt}$.

$$\frac{1}{2}(10) = 10e^{-b(1599)} \implies \ln\frac{1}{2} = -1599b \implies b = -\frac{\ln\frac{1}{2}}{1599}$$

Using the value of b found above and $a = 10$, the amount left is

$$y = 10e^{-[-\ln(1/2)/1599](500)} \approx 8.05 \text{ grams.}$$

STUDY TIP

The carbon dating model in Example 3 assumed that the carbon 14 to carbon 12 ratio was one part in 10,000,000,000,000. Suppose an error in measurement occurred and the actual ratio was one part in 8,000,000,000,000. The fossil age corresponding to the actual ratio would then be approximately 17,000 years. Try checking this result.

Gaussian Models

As mentioned at the beginning of this section, Gaussian models are of the form

$$y = ae^{-(x-b)^2/c}.$$

This type of model is commonly used in probability and statistics to represent populations that are **normally distributed.** The graph of a Gaussian model is called a **bell-shaped curve.** Try graphing the normal distribution with a graphing utility. Can you see why it is called a bell-shaped curve?

For *standard* normal distributions, the model takes the form

$$y = \frac{1}{\sqrt{2\pi}}e^{-x^2/2}.$$

The **average value** for a population can be found from the bell-shaped curve by observing where the maximum y-value of the function occurs. The x-value corresponding to the maximum y-value of the function represents the average value of the independent variable—in this case, x.

Example 4   **SAT Scores**

In 2004, the Scholastic Aptitude Test (SAT) math scores for college-bound seniors roughly followed the normal distribution given by

$$y = 0.0035e^{-(x-518)^2/25,992}, \quad 200 \le x \le 800$$

where x is the SAT score for mathematics. Sketch the graph of this function. From the graph, estimate the average SAT score.   (Source: College Board)

Solution

The graph of the function is shown in Figure 5.34. On this bell-shaped curve, the maximum value of the curve represents the average score. From the graph, you can estimate that the average mathematics score for college-bound seniors in 2004 was 518.

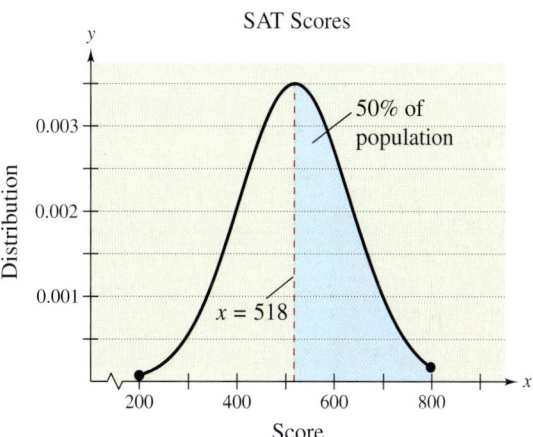

FIGURE **5.34**

✓CHECKPOINT   Now try Exercise 47.

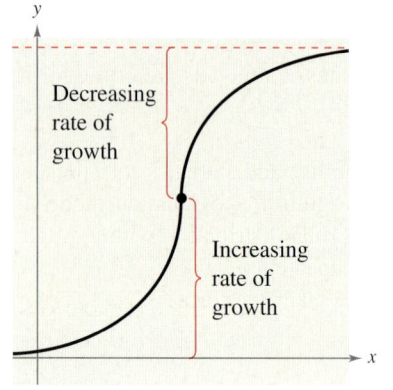

FIGURE 5.35

Logistic Growth Models

Some populations initially have rapid growth, followed by a declining rate of growth, as indicated by the graph in Figure 5.35. One model for describing this type of growth pattern is the **logistic curve** given by the function

$$y = \frac{a}{1 + be^{-rx}}$$

where y is the population size and x is the time. An example is a bacteria culture that is initially allowed to grow under ideal conditions, and then under less favorable conditions that inhibit growth. A logistic growth curve is also called a **sigmoidal curve.**

Example 5   Spread of a Virus

On a college campus of 5000 students, one student returns from vacation with a contagious and long-lasting flu virus. The spread of the virus is modeled by

$$y = \frac{5000}{1 + 4999e^{-0.8t}}, \quad t \geq 0$$

where y is the total number of students infected after t days. The college will cancel classes when 40% or more of the students are infected.

a. How many students are infected after 5 days?

b. After how many days will the college cancel classes?

Solution

a. After 5 days, the number of students infected is

$$y = \frac{5000}{1 + 4999e^{-0.8(5)}} = \frac{5000}{1 + 4999e^{-4}} \approx 54.$$

b. Classes are canceled when the number infected is $(0.40)(5000) = 2000$.

$$2000 = \frac{5000}{1 + 4999e^{-0.8t}}$$

$$1 + 4999e^{-0.8t} = 2.5$$

$$e^{-0.8t} = \frac{1.5}{4999}$$

$$\ln e^{-0.8t} = \ln \frac{1.5}{4999}$$

$$-0.8t = \ln \frac{1.5}{4999}$$

$$t = -\frac{1}{0.8} \ln \frac{1.5}{4999}$$

$$t \approx 10.1$$

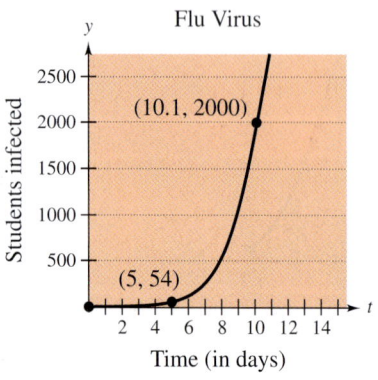

FIGURE 5.36

So, after about 10 days, at least 40% of the students will be infected, and the college will cancel classes. The graph of the function is shown in Figure 5.36.

✓**CHECKPOINT**   Now try Exercise 49.

Logarithmic Models

On December 26, 2004, an earthquake of magnitude 9.0 struck northern Sumatra and many other Asian countries. This earthquake caused a deadly tsunami and was the fourth largest earthquake in the world since 1900.

Example 6   **Magnitudes of Earthquakes**

On the Richter scale, the magnitude R of an earthquake of intensity I is given by

$$R = \log \frac{I}{I_0}$$

where $I_0 = 1$ is the minimum intensity used for comparison. Find the intensities per unit of area for each earthquake. (Intensity is a measure of the wave energy of an earthquake.)

a. Northern Sumatra in 2004: $R = 9.0$

b. Southeastern Alaska in 2004: $R = 6.8$

Solution

a. Because $I_0 = 1$ and $R = 9.0$, you have

$$9.0 = \log \frac{I}{1} \qquad \text{Substitute 1 for } I_0 \text{ and 9.0 for } R.$$

$$10^{9.0} = 10^{\log I} \qquad \text{Exponentiate each side.}$$

$$I = 10^{9.0} \approx 100{,}000{,}000. \qquad \text{Inverse Property}$$

b. For $R = 6.8$, you have

$$6.8 = \log \frac{I}{1} \qquad \text{Substitute 1 for } I_0 \text{ and 6.8 for } R.$$

$$10^{6.8} = 10^{\log I} \qquad \text{Exponentiate each side.}$$

$$I = 10^{6.8} \approx 6{,}310{,}000. \qquad \text{Inverse Property}$$

Note that an increase of 2.2 units on the Richter scale (from 6.8 to 9.0) represents an increase in intensity by a factor of

$$\frac{1{,}000{,}000{,}000}{6{,}310{,}000} \approx 158.$$

In other words, the intensity of the earthquake in Sumatra was about 158 times greater than that of the earthquake in Alaska.

✓CHECKPOINT   Now try Exercise 51.

t	Year	Population, P
1	1910	92.23
2	1920	106.02
3	1930	123.20
4	1940	132.16
5	1950	151.33
6	1960	179.32
7	1970	203.30
8	1980	226.54
9	1990	248.72
10	2000	281.42

𝑊RITING ABOUT 𝑀ATHEMATICS

Comparing Population Models The populations P (in millions) of the United States for the census years from 1910 to 2000 are shown in the table at the left. Least squares regression analysis gives the best quadratic model for these data as $P = 1.0328t^2 + 9.607t + 81.82$, and the best exponential model for these data as $P = 82.677e^{0.124t}$. Which model better fits the data? Describe how you reached your conclusion. (Source: U.S. Census Bureau)

5.5 | Exercises

VOCABULARY CHECK: Fill in the blanks.

1. An exponential growth model has the form _____ and an exponential decay model has the form _____.
2. A logarithmic model has the form _____ or _____.
3. Gaussian models are commonly used in probability and statistics to represent populations that are _____ _____.
4. The graph of a Gaussian model is _____ shaped, where the _____ _____ is the maximum y-value of the graph.
5. A logistic curve is also called a _____ curve.

PREREQUISITE SKILLS REVIEW: Practice and review algebra skills needed for this section at **www.Eduspace.com.**

In Exercises 1–6, match the function with its graph. [The graphs are labeled (a), (b), (c), (d), (e), and (f).]

(a)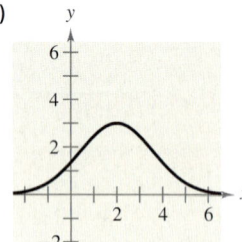

(b)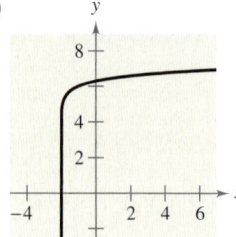

(c)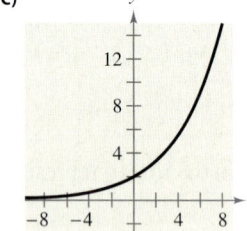

(d)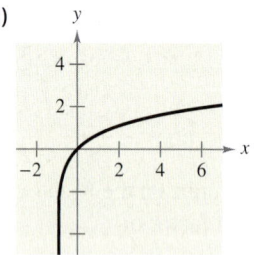

(e)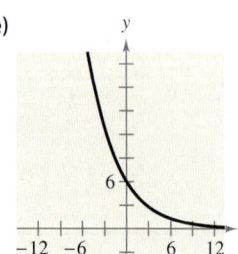

(f)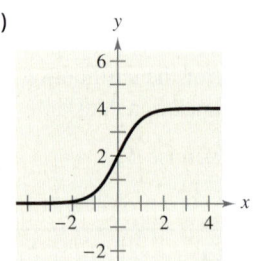

1. $y = 2e^{x/4}$

2. $y = 6e^{-x/4}$

3. $y = 6 + \log(x + 2)$

4. $y = 3e^{-(x-2)^2/5}$

5. $y = \ln(x + 1)$

6. $y = \dfrac{4}{1 + e^{-2x}}$

Compound Interest In Exercises 7–14, complete the table for a savings account in which interest is compounded continuously.

	Initial Investment	Annual % Rate	Time to Double	Amount After 10 Years
7.	$1000	3.5%		
8.	$750	$10\frac{1}{2}\%$		
9.	$750		$7\frac{3}{4}$ yr	
10.	$10,000		12 yr	
11.	$500			$1505.00
12.	$600			$19,205.00
13.		4.5%		$10,000.00
14.		2%		$2000.00

Compound Interest In Exercises 15 and 16, determine the principal P that must be invested at rate r, compounded monthly, so that $500,000 will be available for retirement in t years.

15. $r = 7\frac{1}{2}\%, t = 20$ 16. $r = 12\%, t = 40$

Compound Interest In Exercises 17 and 18, determine the time necessary for $1000 to double if it is invested at interest rate r compounded (a) annually, (b) monthly, (c) daily, and (d) continuously.

17. $r = 11\%$ 18. $r = 10\frac{1}{2}\%$

19. *Compound Interest* Complete the table for the time t necessary for P dollars to triple if interest is compounded continuously at rate r.

r	2%	4%	6%	8%	10%	12%
t						

 20. *Modeling Data* Draw a scatter plot of the data in Exercise 19. Use the *regression* feature of a graphing utility to find a model for the data.

21. *Compound Interest*   Complete the table for the time t necessary for P dollars to triple if interest is compounded annually at rate r.

r	2%	4%	6%	8%	10%	12%
t						

 22. *Modeling Data*   Draw a scatter plot of the data in Exercise 21. Use the *regression* feature of a graphing utility to find a model for the data.

23. *Comparing Models*   If $1 is invested in an account over a 10-year period, the amount in the account, where t represents the time in years, is given by $A = 1 + 0.075[\![t]\!]$ or $A = e^{0.07t}$ depending on whether the account pays simple interest at $7\frac{1}{2}$% or continuous compound interest at 7%. Graph each function on the same set of axes. Which grows at a higher rate? (Remember that $[\![t]\!]$ is the greatest integer function discussed in Section 2.4.)

 24. *Comparing Models*   If $1 is invested in an account over a 10-year period, the amount in the account, where t represents the time in years, is given by

$$A = 1 + 0.06[\![t]\!] \quad \text{or} \quad A = \left(1 + \frac{0.055}{365}\right)^{[\![365t]\!]}$$

depending on whether the account pays simple interest at 6% or compound interest at $5\frac{1}{2}$% compounded daily. Use a graphing utility to graph each function in the same viewing window. Which grows at a higher rate?

Radioactive Decay   In Exercises 25–30, complete the table for the radioactive isotope.

Isotope	Half-life (years)	Initial Quantity	Amount After 1000 Years
25. ^{226}Ra	1599	10 g	
26. ^{226}Ra	1599		1.5 g
27. ^{14}C	5715		2 g
28. ^{14}C	5715	3 g	
29. ^{239}Pu	24,100		2.1 g
30. ^{239}Pu	24,100		0.4 g

In Exercises 31–34, find the exponential model $y = ae^{bx}$ that fits the points shown in the graph or table.

31.

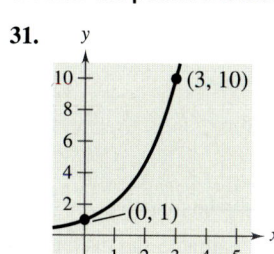

32.

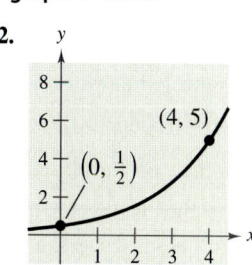

33.

x	0	4
y	5	1

34.

x	0	3
y	1	$\frac{1}{4}$

35. *Population*   The population P (in thousands) of Pittsburgh, Pennsylvania from 2000 through 2003 can be modeled by $P = 2430e^{-0.0029t}$, where t represents the year, with $t = 0$ corresponding to 2000.   (Source: U.S. Census Bureau)

(a) According to the model, was the population of Pittsburgh increasing or decreasing from 2000 to 2003? Explain your reasoning.

(b) What were the populations of Pittsburgh in 2000 and 2003?

(c) According to the model, when will the population be approximately 2.3 million?

Model It

36. *Population*   The table shows the populations (in millions) of five countries in 2000 and the projected populations (in millions) for the year 2010.   (Source: U.S. Census Bureau)

Country	2000	2010
Bulgaria	7.8	7.1
Canada	31.3	34.3
China	1268.9	1347.6
United Kingdom	59.5	61.2
United States	282.3	309.2

(a) Find the exponential growth or decay model $y = ae^{bt}$ or $y = ae^{-bt}$ for the population of each country by letting $t = 0$ correspond to 2000. Use the model to predict the population of each country in 2030.

(b) You can see that the populations of the United States and the United Kingdom are growing at different rates. What constant in the equation $y = ae^{bt}$ is determined by these different growth rates? Discuss the relationship between the different growth rates and the magnitude of the constant.

(c) You can see that the population of China is increasing while the population of Bulgaria is decreasing. What constant in the equation $y = ae^{bt}$ reflects this difference? Explain.

37. Website Growth The number y of hits a new search-engine website receives each month can be modeled by

$$y = 4080e^{kt}$$

where t represents the number of months the website has been operating. In the website's third month, there were 10,000 hits. Find the value of k, and use this result to predict the number of hits the website will receive after 24 months.

38. Value of a Painting The value V (in millions of dollars) of a famous painting can be modeled by

$$V = 10e^{kt}$$

where t represents the year, with $t = 0$ corresponding to 1990. In 2004, the same painting was sold for $65 million. Find the value of k, and use this result to predict the value of the painting in 2010.

39. Bacteria Growth The number N of bacteria in a culture is modeled by

$$N = 100e^{kt}$$

where t is the time in hours. If $N = 300$ when $t = 5$, estimate the time required for the population to double in size.

40. Bacteria Growth The number N of bacteria in a culture is modeled by

$$N = 250e^{kt}$$

where t is the time in hours. If $N = 280$ when $t = 10$, estimate the time required for the population to double in size.

41. Carbon Dating

(a) The ratio of carbon 14 to carbon 12 in a piece of wood discovered in a cave is $R = 1/8^{14}$. Estimate the age of the piece of wood.

(b) The ratio of carbon 14 to carbon 12 in a piece of paper buried in a tomb is $R = 1/13^{11}$. Estimate the age of the piece of paper.

42. Radioactive Decay Carbon 14 dating assumes that the carbon dioxide on Earth today has the same radioactive content as it did centuries ago. If this is true, the amount of ^{14}C absorbed by a tree that grew several centuries ago should be the same as the amount of ^{14}C absorbed by a tree growing today. A piece of ancient charcoal contains only 15% as much radioactive carbon as a piece of modern charcoal. How long ago was the tree burned to make the ancient charcoal if the half-life of ^{14}C is 5715 years?

43. Depreciation A 2005 Jeep Wrangler that costs $30,788 new has a book value of $18,000 after 2 years.

(a) Find the linear model $V = mt + b$.

(b) Find the exponential model $V = ae^{kt}$.

(c) Use a graphing utility to graph the two models in the same viewing window. Which model depreciates faster in the first 2 years?

(d) Find the book values of the vehicle after 1 year and after 3 years using each model.

(e) Explain the advantages and disadvantages of using each model to a buyer and a seller.

44. Depreciation A Dell Inspiron 8600 laptop computer that costs $1150 new has a book value of $550 after 2 years.

(a) Find the linear model $V = mt + b$.

(b) Find the exponential model $V = ae^{kt}$.

(c) Use a graphing utility to graph the two models in the same viewing window. Which model depreciates faster in the first 2 years?

(d) Find the book values of the computer after 1 year and after 3 years using each model.

(e) Explain the advantages and disadvantages to a buyer and a seller of using each model.

45. Sales The sales S (in thousands of units) of a new CD burner after it has been on the market for t years are modeled by

$$S(t) = 100(1 - e^{kt}).$$

Fifteen thousand units of the new product were sold the first year.

(a) Complete the model by solving for k.

(b) Sketch the graph of the model.

(c) Use the model to estimate the number of units sold after 5 years.

46. Learning Curve The management at a plastics factory has found that the maximum number of units a worker can produce in a day is 30. The learning curve for the number N of units produced per day after a new employee has worked t days is modeled by

$$N = 30(1 - e^{kt}).$$

After 20 days on the job, a new employee produces 19 units.

(a) Find the learning curve for this employee (first, find the value of k).

(b) How many days should pass before this employee is producing 25 units per day?

47. IQ Scores The IQ scores from a sample of a class of returning adult students at a small northeastern college roughly follow the normal distribution

$$y = 0.0266e^{-(x-100)^2/450}, \quad 70 \le x \le 115$$

where x is the IQ score.

(a) Use a graphing utility to graph the function.

(b) From the graph in part (a), estimate the average IQ score of an adult student.

48. Education   The time (in hours per week) a student utilizes a math-tutoring center roughly follows the normal distribution

$$y = 0.7979e^{-(x-5.4)^2/0.5}, \quad 4 \le x \le 7$$

where x is the number of hours.

(a) Use a graphing utility to graph the function.

(b) From the graph in part (a), estimate the average number of hours per week a student uses the tutor center.

49. Population Growth   A conservation organization releases 100 animals of an endangered species into a game preserve. The organization believes that the preserve has a carrying capacity of 1000 animals and that the growth of the pack will be modeled by the logistic curve

$$p(t) = \frac{1000}{1 + 9e^{-0.1656t}}$$

where t is measured in years (see figure).

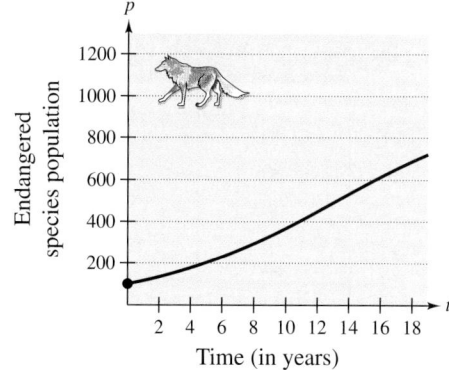

Time (in years)

(a) Estimate the population after 5 years.

(b) After how many years will the population be 500?

 (c) Use a graphing utility to graph the function. Use the graph to determine the horizontal asymptotes, and interpret the meaning of the larger p-value in the context of the problem.

50. Sales   After discontinuing all advertising for a tool kit in 2000, the manufacturer noted that sales began to drop according to the model

$$S = \frac{500,000}{1 + 0.6e^{kt}}$$

where S represents the number of units sold and $t = 0$ represents 2000. In 2004, the company sold 300,000 units.

(a) Complete the model by solving for k.

(b) Estimate sales in 2008.

Geology   **In Exercises 51 and 52, use the Richter scale**

$$R = \log \frac{I}{I_0}$$

for measuring the magnitudes of earthquakes.

51. Find the intensity I of an earthquake measuring R on the Richter scale (let $I_0 = 1$).

(a) Central Alaska in 2002, $R = 7.9$

(b) Hokkaido, Japan in 2003, $R = 8.3$

(c) Illinois in 2004, $R = 4.2$

52. Find the magnitude R of each earthquake of intensity I (let $I_0 = 1$).

(a) $I = 80,500,000$ (b) $I = 48,275,000$

(c) $I = 251,200$

Intensity of Sound   **In Exercises 53–56, use the following information for determining sound intensity. The level of sound β, in decibels, with an intensity of I, is given by**

$$\beta = 10 \log \frac{I}{I_0}$$

where I_0 is an intensity of 10^{-12} watt per square meter, corresponding roughly to the faintest sound that can be heard by the human ear. In Exercises 53 and 54, find the level of sound β.

53. (a) $I = 10^{-10}$ watt per m² (quiet room)

(b) $I = 10^{-5}$ watt per m² (busy street corner)

(c) $I = 10^{-8}$ watt per m² (quiet radio)

(d) $I = 10^0$ watt per m² (threshold of pain)

54. (a) $I = 10^{-11}$ watt per m² (rustle of leaves)

(b) $I = 10^2$ watt per m² (jet at 30 meters)

(c) $I = 10^{-4}$ watt per m² (door slamming)

(d) $I = 10^{-2}$ watt per m² (siren at 30 meters)

55. Due to the installation of noise suppression materials, the noise level in an auditorium was reduced from 93 to 80 decibels. Find the percent decrease in the intensity level of the noise as a result of the installation of these materials.

56. Due to the installation of a muffler, the noise level of an engine was reduced from 88 to 72 decibels. Find the percent decrease in the intensity level of the noise as a result of the installation of the muffler.

pH Levels   **In Exercises 57–62, use the acidity model given by pH = $-\log[\text{H}^+]$, where acidity (pH) is a measure of the hydrogen ion concentration $[\text{H}^+]$ (measured in moles of hydrogen per liter) of a solution.**

57. Find the pH if $[\text{H}^+] = 2.3 \times 10^{-5}$.

58. Find the pH if $[\text{H}^+] = 11.3 \times 10^{-6}$.

59. Compute $[H^+]$ for a solution in which pH $= 5.8$.

60. Compute $[H^+]$ for a solution in which pH $= 3.2$.

61. Apple juice has a pH of 2.9 and drinking water has a pH of 8.0. The hydrogen ion concentration of the apple juice is how many times the concentration of drinking water?

62. The pH of a solution is decreased by one unit. The hydrogen ion concentration is increased by what factor?

63. *Forensics* At 8:30 A.M., a coroner was called to the home of a person who had died during the night. In order to estimate the time of death, the coroner took the person's temperature twice. At 9:00 A.M. the temperature was 85.7°F, and at 11:00 a.m. the temperature was 82.8°F. From these two temperatures, the coroner was able to determine that the time elapsed since death and the body temperature were related by the formula

$$t = -10 \ln \frac{T - 70}{98.6 - 70}$$

where t is the time in hours elapsed since the person died and T is the temperature (in degrees Fahrenheit) of the person's body. Assume that the person had a normal body temperature of 98.6°F at death, and that the room temperature was a constant 70°F. (This formula is derived from a general cooling principle called *Newton's Law of Cooling*.) Use the formula to estimate the time of death of the person.

64. *Home Mortgage* A \$120,000 home mortgage for 35 years at $7\frac{1}{2}\%$ has a monthly payment of \$809.39. Part of the monthly payment is paid toward the interest charge on the unpaid balance, and the remainder of the payment is used to reduce the principal. The amount that is paid toward the interest is

$$u = M - \left(M - \frac{Pr}{12}\right)\left(1 + \frac{r}{12}\right)^{12t}$$

and the amount that is paid toward the reduction of the principal is

$$v = \left(M - \frac{Pr}{12}\right)\left(1 + \frac{r}{12}\right)^{12t}.$$

In these formulas, P is the size of the mortgage, r is the interest rate, M is the monthly payment, and t is the time in years.

(a) Use a graphing utility to graph each function in the same viewing window. (The viewing window should show all 35 years of mortgage payments.)

(b) In the early years of the mortgage, is the larger part of the monthly payment paid toward the interest or the principal? Approximate the time when the monthly payment is evenly divided between interest and principal reduction.

(c) Repeat parts (a) and (b) for a repayment period of 20 years ($M = \$966.71$). What can you conclude?

65. *Home Mortgage* The total interest u paid on a home mortgage of P dollars at interest rate r for t years is

$$u = P\left[\frac{rt}{1 - \left(\dfrac{1}{1 + r/12}\right)^{12t}} - 1\right].$$

Consider a \$120,000 home mortgage at $7\frac{1}{2}\%$.

 (a) Use a graphing utility to graph the total interest function.

(b) Approximate the length of the mortgage for which the total interest paid is the same as the size of the mortgage. Is it possible that some people are paying twice as much in interest charges as the size of the mortgage?

66. *Data Analysis* The table shows the time t (in seconds) required to attain a speed of s miles per hour from a standing start for a car.

Speed, s	Time, t
30	3.4
40	5.0
50	7.0
60	9.3
70	12.0
80	15.8
90	20.0

Two models for these data are as follows.

$$t_1 = 40.757 + 0.556s - 15.817 \ln s$$

$$t_2 = 1.2259 + 0.0023s^2$$

(a) Use the *regression* feature of a graphing utility to find a linear model t_3 and an exponential model t_4 for the data.

(b) Use a graphing utility to graph the data and each model in the same viewing window.

(c) Create a table comparing the data with estimates obtained from each model.

(d) Use the results of part (c) to find the sum of the absolute values of the differences between the data and the estimated values given by each model. Based on the four sums, which model do you think better fits the data? Explain.

Synthesis

True or False? **In Exercises 67–70, determine whether the statement is true or false. Justify your answer.**

67. The domain of a logistic growth function cannot be the set of real numbers.

68. A logistic growth function will always have an x-intercept.

69. The graph of

$$f(x) = \frac{4}{1 + 6e^{-2x}} + 5$$

is the graph of

$$g(x) = \frac{4}{1 + 6e^{-2x}}$$

shifted to the right five units.

70. The graph of a Gaussian model will never have an x-intercept.

71. Identify each model as linear, logarithmic, exponential, logistic, or none of the above. Explain your reasoning.

(a) y

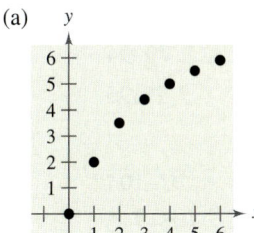

(b) y

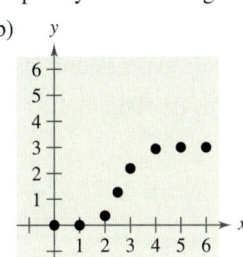

(c) y

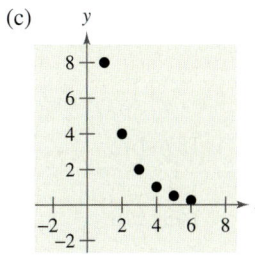

(d) y

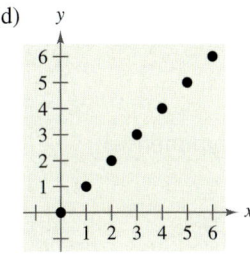

(e) y

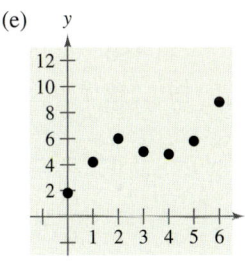

(f) y

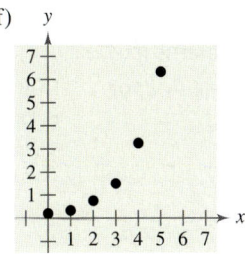

72. ***Writing*** Use your school's library, the Internet, or some other reference source to write a paper describing John Napier's work with logarithms.

Skills Review

In Exercises 73–78, (a) plot the points, (b) find the distance between the points, (c) find the midpoint of the line segment joining the points, and (d) find the slope of the line passing through the points.

73. $(-1, 2), (0, 5)$

74. $(4, -3), (-6, 1)$

75. $(3, 3), (14, -2)$

76. $(7, 0), (10, 4)$

77. $\left(\frac{1}{2}, -\frac{1}{4}\right), \left(\frac{3}{4}, 0\right)$

78. $\left(\frac{7}{3}, \frac{1}{6}\right), \left(-\frac{2}{3}, -\frac{1}{3}\right)$

In Exercises 79–88, sketch the graph of the equation.

79. $y = 10 - 3x$

80. $y = -4x - 1$

81. $y = -2x^2 - 3$

82. $y = 2x^2 - 7x - 30$

83. $3x^2 - 4y = 0$

84. $-x^2 - 8y = 0$

85. $y = \dfrac{4}{1 - 3x}$

86. $y = \dfrac{x^2}{-x - 2}$

87. $x^2 + (y - 8)^2 = 25$

88. $(x - 4)^2 + (y + 7) = 4$

In Exercises 89–92, graph the exponential function.

89. $f(x) = 2^{x-1} + 5$

90. $f(x) = -2^{-x-1} - 1$

91. $f(x) = 3^x - 4$

92. $f(x) = -3^x + 4$

93. Make a Decision To work an extended application analyzing the net sales for Kohl's Corporation from 1992 to 2004, visit this text's website at *college.hmco.com*. *(Data Source: Kohl's Illinois, Inc.)*

5	**Chapter Summary**

What did you learn?

5   Review Exercises

5.1 In Exercises 1–6, evaluate the function at the indicated value of *x*. Round your result to three decimal places.

Function	Value
1. $f(x) = 6.1^x$	$x = 2.4$
2. $f(x) = 30^x$	$x = \sqrt{3}$
3. $f(x) = 2^{-0.5x}$	$x = \pi$
4. $f(x) = 1278^{x/5}$	$x = 1$
5. $f(x) = 7(0.2^x)$	$x = -\sqrt{11}$
6. $f(x) = -14(5^x)$	$x = -0.8$

In Exercises 7–10, match the function with its graph. [The graphs are labeled (a), (b), (c), and (d).]

(a)
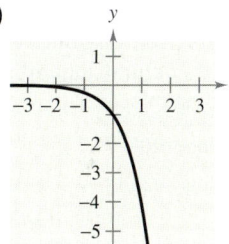

(b)
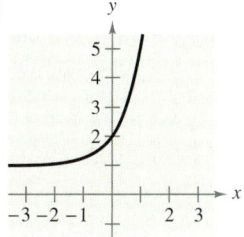

(c)
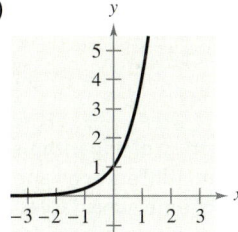

(d)
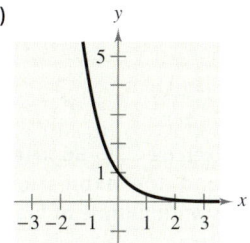

7. $f(x) = 4^x$

8. $f(x) = 4^{-x}$

9. $f(x) = -4^x$

10. $f(x) = 4^x + 1$

In Exercises 11–14, use the graph of *f* to describe the transformation that yields the graph of *g*.

11. $f(x) = 5^x$, $g(x) = 5^{x-1}$

12. $f(x) = 4^x$, $g(x) = 4^x - 3$

13. $f(x) = \left(\frac{1}{2}\right)^x$, $g(x) = -\left(\frac{1}{2}\right)^{x+2}$

14. $f(x) = \left(\frac{2}{3}\right)^x$, $g(x) = 8 - \left(\frac{2}{3}\right)^x$

 In Exercises 15–22, use a graphing utility to construct a table of values for the function. Then sketch the graph of the function.

15. $f(x) = 4^{-x} + 4$

16. $f(x) = -4^x - 3$

17. $f(x) = -2.65^{x+1}$

18. $f(x) = 2.65^{x-1}$

19. $f(x) = 5^{x-2} + 4$

20. $f(x) = 2^{x-6} - 5$

21. $f(x) = \left(\frac{1}{2}\right)^{-x} + 3$

22. $f(x) = \left(\frac{1}{8}\right)^{x+2} - 5$

In Exercises 23–26, use the One-to-One Property to solve the equation for *x*.

23. $3^{x+2} = \dfrac{1}{9}$

24. $\left(\dfrac{1}{3}\right)^{x-2} = 81$

25. $e^{5x-7} = e^{15}$

26. $e^{8-2x} = e^{-3}$

In Exercises 27–30, evaluate the function given by $f(x) = e^x$ at the indicated value of *x*. Round your result to three decimal places.

27. $x = 8$

28. $x = \frac{5}{8}$

29. $x = -1.7$

30. $x = 0.278$

 In Exercises 31–34, use a graphing utility to construct a table of values for the function. Then sketch the graph of the function.

31. $h(x) = e^{-x/2}$

32. $h(x) = 2 - e^{-x/2}$

33. $f(x) = e^{x+2}$

34. $s(t) = 4e^{-2/t}, \quad t > 0$

Compound Interest   In Exercises 35 and 36, complete the table to determine the balance *A* for *P* dollars invested at rate *r* for *t* years and compounded *n* times per year.

n	1	2	4	12	365	Continuous
A						

35. $P = \$3500, \ r = 6.5\%, \ t = 10$ years

36. $P = \$2000, \ r = 5\%, \ t = 30$ years

37. *Waiting Times*   The average time between incoming calls at a switchboard is 3 minutes. The probability *F* of waiting less than *t* minutes until the next incoming call is approximated by the model $F(t) = 1 - e^{-t/3}$. A call has just come in. Find the probability that the next call will be within

(a) $\frac{1}{2}$ minute.   (b) 2 minutes.   (c) 5 minutes.

38. *Depreciation*   After *t* years, the value *V* of a car that originally cost \$14,000 is given by $V(t) = 14,000\left(\frac{3}{4}\right)^t$.

 (a) Use a graphing utility to graph the function.

(b) Find the value of the car 2 years after it was purchased.

(c) According to the model, when does the car depreciate most rapidly? Is this realistic? Explain.

39. *Trust Fund* On the day a person is born, a deposit of $50,000 is made in a trust fund that pays 8.75% interest, compounded continuously.

(a) Find the balance on the person's 35th birthday.

(b) How much longer would the person have to wait for the balance in the trust fund to double?

40. *Radioactive Decay* Let Q represent a mass of plutonium 241 (^{241}Pu) (in grams), whose half-life is 14.4 years. The quantity of plutonium 241 present after t years is given by $Q = 100\left(\frac{1}{2}\right)^{t/14.4}$.

(a) Determine the initial quantity (when $t = 0$).

(b) Determine the quantity present after 10 years.

(c) Sketch the graph of this function over the interval $t = 0$ to $t = 100$.

5.2 In Exercises 41–44, write the exponential equation in logarithmic form.

41. $4^3 = 64$

42. $25^{3/2} = 125$

43. $e^{0.8} = 2.2255\ldots$

44. $e^0 = 1$

In Exercises 45–48, evaluate the function at the indicated value of x without using a calculator.

Function	Value
45. $f(x) = \log x$	$x = 1000$
46. $g(x) = \log_9 x$	$x = 3$
47. $g(x) = \log_2 x$	$x = \frac{1}{8}$
48. $f(x) = \log_4 x$	$x = \frac{1}{4}$

In Exercises 49–52, use the One-to-One Property to solve the equation for x.

49. $\log_4 (x + 7) = \log_4 14$

50. $\log_8 (3x - 10) = \log_8 5$

51. $\ln(x + 9) = \ln 4$

52. $\ln(2x - 1) = \ln 11$

In Exercises 53–58, find the domain, x-intercept, and vertical asymptote of the logarithmic function and sketch its graph.

53. $g(x) = \log_7 x$

54. $g(x) = \log_5 x$

55. $f(x) = \log\left(\frac{x}{3}\right)$

56. $f(x) = 6 + \log x$

57. $f(x) = 4 - \log(x + 5)$

58. $f(x) = \log(x - 3) + 1$

 In Exercises 59–64, use a calculator to evaluate the function given by $f(x) = \ln x$ at the indicated value of x. Round your result to three decimal places if necessary.

59. $x = 22.6$

60. $x = 0.98$

61. $x = e^{-12}$

62. $x = e^7$

63. $x = \sqrt{7} + 5$

64. $x = \dfrac{\sqrt{3}}{8}$

In Exercises 65–68, find the domain, x-intercept, and vertical asymptote of the logarithmic function and sketch its graph.

65. $f(x) = \ln x + 3$

66. $f(x) = \ln(x - 3)$

67. $h(x) = \ln(x^2)$

68. $f(x) = \frac{1}{4} \ln x$

69. *Antler Spread* The antler spread a (in inches) and shoulder height h (in inches) of an adult male American elk are related by the model $h = 116 \log(a + 40) - 176$. Approximate the shoulder height of a male American elk with an antler spread of 55 inches.

70. *Snow Removal* The number of miles s of roads cleared of snow is approximated by the model

$$s = 25 - \frac{13 \ln(h/12)}{\ln 3}, \quad 2 \le h \le 15$$

where h is the depth of the snow in inches. Use this model to find s when $h = 10$ inches.

5.3 In Exercises 71–74, evaluate the logarithm using the change-of-base formula. Do each exercise twice, once with common logarithms and once with natural logarithms. Round your the results to three decimal places.

71. $\log_4 9$

72. $\log_{12} 200$

73. $\log_{1/2} 5$

74. $\log_3 0.28$

In Exercises 75–78, use the properties of logarithms to rewrite and simplify the logarithmic expression.

75. $\log 18$

76. $\log_2 \left(\frac{1}{12}\right)$

77. $\ln 20$

78. $\ln(3e^{-4})$

In Exercises 79–86, use the properties of logarithms to expand the expression as a sum, difference, and/or constant multiple of logarithms. (Assume all variables are positive.)

79. $\log_5 5x^2$

80. $\log 7x^4$

81. $\log_3 \dfrac{6}{\sqrt[3]{x}}$

82. $\log_7 \dfrac{\sqrt{x}}{4}$

83. $\ln x^2 y^2 z$

84. $\ln 3xy^2$

85. $\ln\left(\dfrac{x + 3}{xy}\right)$

86. $\ln\left(\dfrac{y - 1}{4}\right)^2, \quad y > 1$

In Exercises 87–94, condense the expression to the logarithm of a single quantity.

87. $\log_2 5 + \log_2 x$

88. $\log_6 y - 2 \log_6 z$

89. $\ln x - \frac{1}{4} \ln y$

90. $3 \ln x + 2 \ln(x + 1)$

91. $\frac{1}{3} \log_8(x + 4) + 7 \log_8 y$

92. $-2 \log x - 5 \log(x + 6)$

93. $\frac{1}{2} \ln(2x - 1) - 2 \ln(x + 1)$

94. $5 \ln(x - 2) - \ln(x + 2) - 3 \ln x$

95. *Climb Rate* The time t (in minutes) for a small plane to climb to an altitude of h feet is modeled by

$$t = 50 \log \frac{18{,}000}{18{,}000 - h}$$

where 18,000 feet is the plane's absolute ceiling.

(a) Determine the domain of the function in the context of the problem.

 (b) Use a graphing utility to graph the function and identify any asymptotes.

(c) As the plane approaches its absolute ceiling, what can be said about the time required to increase its altitude?

(d) Find the time for the plane to climb to an altitude of 4000 feet.

96. *Human Memory Model* Students in a learning theory study were given an exam and then retested monthly for 6 months with an equivalent exam. The data obtained in the study are given as the ordered pairs (t, s), where t is the time in months after the initial exam and s is the average score for the class. Use these data to find a logarithmic equation that relates t and s.

(1, 84.2), (2, 78.4), (3, 72.1),

(4, 68.5), (5, 67.1), (6, 65.3)

5.4 In Exercises 97–104, solve for x.

97. $8^x = 512$ **98.** $6^x = \frac{1}{216}$

99. $e^x = 3$ **100.** $e^x = 6$

101. $\log_4 x = 2$ **102.** $\log_6 x = -1$

103. $\ln x = 4$ **104.** $\ln x = -3$

In Exercises 105–114, solve the exponential equation algebraically. Approximate your result to three decimal places.

105. $e^x = 12$ **106.** $e^{3x} = 25$

107. $e^{4x} = e^{x^2 + 3}$ **108.** $14e^{3x+2} = 560$

109. $2^x + 13 = 35$ **110.** $6^x - 28 = -8$

111. $-4(5^x) = -68$ **112.** $2(12^x) = 190$

113. $e^{2x} - 7e^x + 10 = 0$ **114.** $e^{2x} - 6e^x + 8 = 0$

 In Exercises 115–118, use a graphing utility to graph and solve the equation. Approximate the result to three decimal places.

115. $2^{0.6x} - 3x = 0$ **116.** $4^{-0.2x} + x = 0$

117. $25e^{-0.3x} = 12$ **118.** $4e^{1.2x} = 9$

In Exercises 119–130, solve the logarithmic equation algebraically. Approximate the result to three decimal places.

119. $\ln 3x = 8.2$ **120.** $\ln 5x = 7.2$

121. $2 \ln 4x = 15$ **122.** $4 \ln 3x = 15$

123. $\ln x - \ln 3 = 2$ **124.** $\ln \sqrt{x + 8} = 3$

125. $\ln \sqrt{x + 1} = 2$ **126.** $\ln x - \ln 5 = 4$

127. $\log_8 (x - 1) = \log_8 (x - 2) - \log_8 (x + 2)$

128. $\log_6 (x + 2) - \log_6 x = \log_6 (x + 5)$

129. $\log(1 - x) = -1$

130. $\log(-x - 4) = 2$

 In Exercises 131–134, use a graphing utility to graph and solve the equation. Approximate the result to three decimal places.

131. $2 \ln(x + 3) + 3x = 8$ **132.** $6 \log(x^2 + 1) - x = 0$

133. $4 \ln(x + 5) - x = 10$ **134.** $x - 2 \log(x + 4) = 0$

135. *Compound Interest* You deposit $7550 in an account that pays 7.25% interest, compounded continuously. How long will it take for the money to triple?

136. *Meteorology* The speed of the wind S (in miles per hour) near the center of a tornado and the distance d (in miles) the tornado travels are related by the model $S = 93 \log d + 65$. On March 18, 1925, a large tornado struck portions of Missouri, Illinois, and Indiana with a wind speed at the center of about 283 miles per hour. Approximate the distance traveled by this tornado.

5.5 In Exercises 137–142, match the function with its graph. [The graphs are labeled (a), (b), (c), (d), (e), and (f).]

(a)
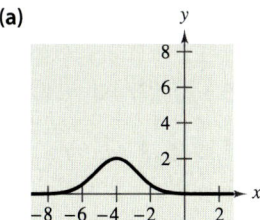

(b)
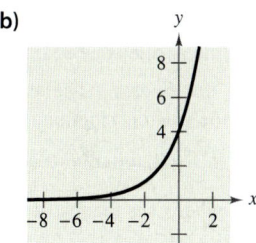

(c)
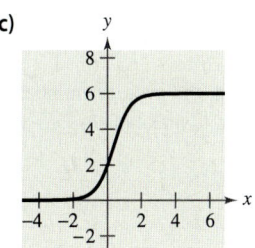

(d)
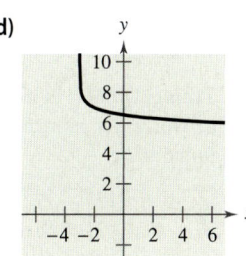

(e)
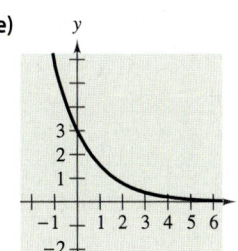

(f)
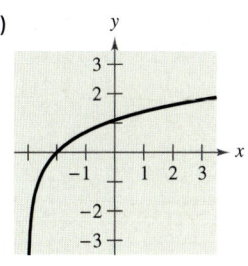

137. $y = 3e^{-2x/3}$

138. $y = 4e^{2x/3}$

139. $y = \ln(x + 3)$

140. $y = 7 - \log(x + 3)$

141. $y = 2e^{-(x+4)^2/3}$

142. $y = \dfrac{6}{1 + 2e^{-2x}}$

In Exercises 143 and 144, find the exponential model $y = ae^{bx}$ that passes through the points.

143. $(0, 2), (4, 3)$

144. $\left(0, \frac{1}{2}\right), (5, 5)$

145. *Population* The population P of South Carolina (in thousands) from 1990 through 2003 can be modeled by $P = 3499e^{0.0135t}$, where t represents the year, with $t = 0$ corresponding to 1990. According to this model, when will the population reach 4.5 million? (Source: U.S. Census Bureau)

146. *Radioactive Decay* The half-life of radioactive uranium II (^{234}U) is about 250,000 years. What percent of a present amount of radioactive uranium II will remain after 5000 years?

147. *Compound Interest* A deposit of $10,000 is made in a savings account for which the interest is compounded continuously. The balance will double in 5 years.

(a) What is the annual interest rate for this account?

(b) Find the balance after 1 year.

148. *Wildlife Population* A species of bat is in danger of becoming extinct. Five years ago, the total population of the species was 2000. Two years ago, the total population of the species was 1400. What was the total population of the species one year ago?

 149. *Test Scores* The test scores for a biology test follow a normal distribution modeled by

$$y = 0.0499e^{-(x-71)^2/128}, \quad 40 \le x \le 100$$

where x is the test score.

(a) Use a graphing utility to graph the equation.

(b) From the graph in part (a), estimate the average test score.

150. *Typing Speed* In a typing class, the average number N of words per minute typed after t weeks of lessons was found to be

$$N = \frac{157}{1 + 5.4e^{-0.12t}}.$$

Find the time necessary to type (a) 50 words per minute and (b) 75 words per minute.

151. *Sound Intensity* The relationship between the number of decibels β and the intensity of a sound I in watts per square centimeter is

$$\beta = 10 \log\left(\frac{I}{10^{-16}}\right).$$

Determine the intensity of a sound in watts per square centimeter if the decibel level is 125.

152. *Geology* On the Richter scale, the magnitude R of an earthquake of intensity I is given by

$$R = \log \frac{I}{I_0}$$

where $I_0 = 1$ is the minimum intensity used for comparison. Find the intensity per unit of area for each value of R.

(a) $R = 8.4$ (b) $R = 6.85$ (c) $R = 9.1$

Synthesis

True or False? **In Exercises 153 and 154, determine whether the equation is true or false. Justify your answer.**

153. $\log_b b^{2x} = 2x$

154. $\ln(x + y) = \ln x + \ln y$

155. The graphs of $y = e^{kt}$ are shown where $k = a, b, c,$ and d. Which of the four values are negative? Which are positive? Explain your reasoning.

(a)

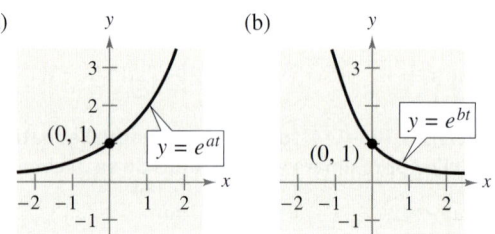

(b)

(c)

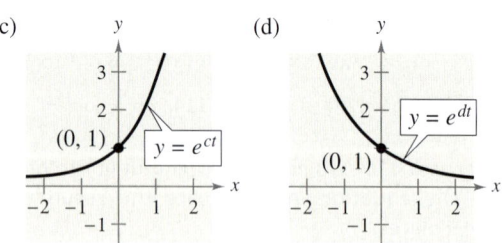

(d)

5   Chapter Test

Take this test as you would take a test in class. When you are finished, check your work against the answers given in the back of the book.

In Exercises 1–4, evaluate the expression. Approximate your result to three decimal places.

1. $12.4^{2.79}$ **2.** $4^{3\pi/2}$ **3.** $e^{-7/10}$ **4.** $e^{3.1}$

In Exercises 5–7, construct a table of values. Then sketch the graph of the function.

5. $f(x) = 10^{-x}$ **6.** $f(x) = -6^{x-2}$ **7.** $f(x) = 1 - e^{2x}$

8. Evaluate (a) $\log_7 7^{-0.89}$ and (b) $4.6 \ln e^2$.

In Exercises 9–11, construct a table of values. Then sketch the graph of the function. Identify any asymptotes.

9. $f(x) = -\log x - 6$ **10.** $f(x) = \ln(x - 4)$ **11.** $f(x) = 1 + \ln(x + 6)$

In Exercises 12–14, evaluate the logarithm using the change-of-base formula. Round your result to three decimal places.

12. $\log_7 44$ **13.** $\log_{2/5} 0.9$ **14.** $\log_{24} 68$

In Exercises 15–17, use the properties of logarithms to expand the expression as a sum, difference, and/or constant multiple of logarithms.

15. $\log_2 3a^4$ **16.** $\ln \dfrac{5\sqrt{x}}{6}$ **17.** $\log \dfrac{7x^2}{yz^3}$

In Exercises 18–20, condense the expression to the logarithm of a single quantity.

18. $\log_3 13 + \log_3 y$ **19.** $4 \ln x - 4 \ln y$

20. $2 \ln x + \ln(x - 5) - 3 \ln y$

In Exercises 21–26, solve the equation algebraically. Approximate your result to three decimal places.

21. $5^x = \dfrac{1}{25}$ **22.** $3e^{-5x} = 132$

23. $\dfrac{1025}{8 + e^{4x}} = 5$ **24.** $\ln x = \dfrac{1}{2}$

25. $18 + 4 \ln x = 7$ **26.** $\log x - \log(8 - 5x) = 2$

27. Find an exponential growth model for the graph shown in the figure.

28. The half-life of radioactive actinium (^{227}Ac) is 21.77 years. What percent of a present amount of radioactive actinium will remain after 19 years?

29. A model that can be used for predicting the height H (in centimeters) of a child based on his or her age is $H = 70.228 + 5.104x + 9.222 \ln x$, $\frac{1}{4} \le x \le 6$, where x is the age of the child in years. (Source: Snapshots of Applications in Mathematics)

(a) Construct a table of values. Then sketch the graph of the model.

(b) Use the graph from part (a) to estimate the height of a four-year-old child. Then calculate the actual height using the model.

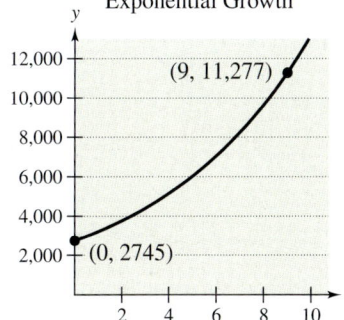

Exponential Growth

(9, 11,277)

(0, 2745)

FIGURE FOR 27

5 Cumulative Test for Chapters 3–5

Take this test to review the material from earlier chapters. When you are finished, check your work against the answers given in the back of the book.

1. Find the quadratic function whose graph has a vertex at $(-8, 5)$ and passes through the point $(-4, -7)$.

In Exercises 2–4, sketch the graph of the function without the aid of a graphing utility.

2. $h(x) = -(x^2 + 4x)$
3. $f(t) = \frac{1}{4}t(t - 2)^2$
4. $g(s) = s^2 + 4s + 10$

In Exercises 5 and 6, find all the zeros of the function.

5. $f(x) = x^3 + 2x^2 + 4x + 8$
6. $f(x) = x^4 + 4x^3 - 21x^2$

7. Divide: $\dfrac{6x^3 - 4x^2}{2x^2 + 1}$.

8. Use synthetic division to divide $2x^4 + 3x^3 - 6x + 5$ by $x + 2$.

9. Use a graphing utility to approximate (to the nearest hundredth) the real zero of the function given by $g(x) = x^3 + 3x^2 - 6$.

10. Find a polynomial with real coefficients that has -5, -2, and $2 + \sqrt{3}i$ as its zeros.

In Exercises 11 and 12, find the domain of the function, and identify all asymptotes. Sketch the graph of the function.

11. $f(x) = \dfrac{2x}{x - 3}$
12. $f(x) = \dfrac{4x^2}{x - 5}$

In Exercises 13–15, sketch the graph of the rational function by hand. Be sure to identify all intercepts and asymptotes.

13. $f(x) = \dfrac{2x}{x^2 - 9}$
14. $f(x) = \dfrac{x^2 - 4x + 3}{x^2 - 2x - 3}$
15. $f(x) = \dfrac{x^3 + 3x^2 - 4x - 12}{x^2 - x - 2}$

In Exercises 16 and 17, sketch a graph of the conic.

16. $\dfrac{(x + 3)^2}{16} - \dfrac{(y + 4)^2}{25} = 1$
17. $\dfrac{(x - 2)^2}{4} + \dfrac{(y + 1)^2}{9} = 1$

FIGURE FOR 18

18. Find an equation of the parabola shown in the figure.

19. Find an equation of the hyperbola with foci $(0, 0)$ and $(0, 4)$ and asymptotes $y = \pm\frac{1}{2}x + 2$.

In Exercises 20 and 21, use the graph of f to describe the transformation that yields the graph of g. Use a graphing utility to graph both equations in the same viewing window.

20. $f(x) = \left(\frac{2}{5}\right)^x$, $g(x) = -\left(\frac{2}{5}\right)^{-x+3}$
21. $f(x) = 2.2^x$, $g(x) = -2.2^x + 4$

In Exercises 22–25, use a calculator to evaluate each expression. Round your result to three decimal places.

22. $\log 98$
23. $\log\left(\frac{6}{7}\right)$
24. $\ln\sqrt{31}$
25. $\ln\left(\sqrt{40} - 5\right)$

In Exercises 26–28, evaluate the logarithm using the change-of-base formula. Round your answer to three decimal places.

26. $\log_7 1.8$ **27.** $\log_3 0.149$ **28.** $\log_{1/2} 17$

29. Use the properties of logarithms to expand $\ln\left(\dfrac{x^2 - 16}{x^4}\right)$, where $x > 4$.

30. Write $2\ln x - \frac{1}{2}\ln(x + 5)$ as a logarithm of a single quantity.

In Exercises 31–36, solve the equation algebraically. Approximate the result to three decimal places.

31. $6e^{2x} = 72$ **32.** $4^{x-5} + 21 = 30$

33. $e^{2x} - 11e^x + 24 = 0$ **34.** $\log_2 x + \log_2 5 = 6$

35. $\ln 4x - \ln 2 = 8$ **36.** $\ln\sqrt{x + 2} = 3$

37. Use a graphing utility to graph

$$f(x) = \frac{1000}{1 + 4e^{-0.2x}}$$

and determine the horizontal asymptotes.

38. Let x be the amount (in hundreds of dollars) that an online stock-trading company spends on advertising, and let P be the profit (in thousands of dollars), where $P = 230 + 20x - \frac{1}{2}x^2$. What amount of advertising will yield a maximum profit?

39. The sales S (in billions of dollars) of lottery tickets in the United States from 1997 through 2003 are shown in the table. (Source: TLF Publications, Inc.)

(a) Use a graphing utility to create a scatter plot of the data. Let t represent the year, with $t = 7$ corresponding to 1997.

(b) Use the *regression* feature of the graphing utility to find a quadratic model for the data.

(c) Use the graphing utility to graph the model in the same viewing window used for the scatter plot. How well does the model fit the data?

(d) Use the model to predict the sales of lottery tickets in 2008. Does your answer seem reasonable? Explain.

Year	Sales, S
1997	35.5
1998	35.6
1999	36.0
2000	37.2
2001	38.4
2002	42.0
2003	43.5

TABLE FOR 39

40. On the day a grandchild is born, a grandparent deposits $2500 in a fund earning 7.5%, compounded continuously. Determine the balance in the account at the time of the grandchild's 25th birthday.

41. The number N of bacteria in a culture is given by the model $N = 175e^{kt}$, where t is the time in hours. If $N = 420$ when $t = 8$, estimate the time required for the population to double in size.

42. The population P of Texas (in thousands) from 1990 through 2003 can be modeled by $P = 16{,}989e^{0.0207t}$, where t represents the year, with $t = 0$ corresponding to 1990. According to this model, when will the population reach 24 million?   (Source: U.S. Census Bureau)

43. The population p of a species of bird t years after it is introduced into a new habitat is given by

$$p = \frac{1200}{1 + 3e^{-t/5}}.$$

(a) Determine the population size that was introduced into the habitat.

(b) Determine the population after 5 years.

(c) After how many years will the population be 800?

Proofs in Mathematics

Each of the following three properties of logarithms can be proved by using properties of exponential functions.

Properties of Logarithms *(p. 406)*

Let a be a positive number such that $a \neq 1$, and let n be a real number. If u and v are positive real numbers, the following properties are true.

	Logarithm with Base a	*Natural Logarithm*
1. Product Property:	$\log_a(uv) = \log_a u + \log_a v$	$\ln(uv) = \ln u + \ln v$
2. Quotient Property:	$\log_a \dfrac{u}{v} = \log_a u - \log_a v$	$\ln \dfrac{u}{v} = \ln u - \ln v$
3. Power Property:	$\log_a u^n = n \log_a u$	$\ln u^n = n \ln u$

Proof

Let

$$x = \log_a u \quad \text{and} \quad y = \log_a v.$$

The corresponding exponential forms of these two equations are

$$a^x = u \quad \text{and} \quad a^y = v.$$

To prove the Product Property, multiply u and v to obtain

$$uv = a^x a^y = a^{x+y}.$$

The corresponding logarithmic form of $uv = a^{x+y}$ is $\log_a(uv) = x + y$. So,

$$\log_a(uv) = \log_a u + \log_a v.$$

To prove the Quotient Property, divide u by v to obtain

$$\frac{u}{v} = \frac{a^x}{a^y} = a^{x-y}.$$

The corresponding logarithmic form of $u/v = a^{x-y}$ is $\log_a(u/v) = x - y$. So,

$$\log_a \frac{u}{v} = \log_a u - \log_a v.$$

To prove the Power Property, substitute a^x for u in the expression $\log_a u^n$, as follows.

$$
\begin{aligned}
\log_a u^n &= \log_a (a^x)^n && \text{Substitute } a^x \text{ for } u. \\
&= \log_a a^{nx} && \text{Property of exponents} \\
&= nx && \text{Inverse Property of Logarithms} \\
&= n \log_a u && \text{Substitute } \log_a u \text{ for } x.
\end{aligned}
$$

So, $\log_a u^n = n \log_a u$.

This collection of thought-provoking and challenging exercises further explores and expands upon concepts learned in this chapter.

1. Graph the exponential function given by $y = a^x$ for $a = 0.5$, 1.2, and 2.0. Which of these curves intersects the line $y = x$? Determine all positive numbers a for which the curve $y = a^x$ intersects the line $y = x$.

 2. Use a graphing utility to graph $y_1 = e^x$ and each of the functions $y_2 = x^2$, $y_3 = x^3$, $y_4 = \sqrt{x}$, and $y_5 = |x|$. Which function increases at the greatest rate as x approaches $+\infty$?

 3. Use the result of Exercise 2 to make a conjecture about the rate of growth of $y_1 = e^x$ and $y = x^n$, where n is a natural number and x approaches $+\infty$.

 4. Use the results of Exercises 2 and 3 to describe what is implied when it is stated that a quantity is growing exponentially.

5. Given the exponential function

$$f(x) = a^x$$

show that

(a) $f(u + v) = f(u) \cdot f(v)$.

(b) $f(2x) = [f(x)]^2$.

6. Given that

$$f(x) = \frac{e^x + e^{-x}}{2} \text{ and } g(x) = \frac{e^x - e^{-x}}{2}$$

show that

$$[f(x)]^2 - [g(x)]^2 = 1.$$

 7. Use a graphing utility to compare the graph of the function given by $y = e^x$ with the graph of each given function. [$n!$ (read "n factorial") is defined as $n! = 1 \cdot 2 \cdot 3 \cdots (n - 1) \cdot n$.]

(a) $y_1 = 1 + \dfrac{x}{1!}$

(b) $y_2 = 1 + \dfrac{x}{1!} + \dfrac{x^2}{2!}$

(c) $y_3 = 1 + \dfrac{x}{1!} + \dfrac{x^2}{2!} + \dfrac{x^3}{3!}$

8. Identify the pattern of successive polynomials given in Exercise 7. Extend the pattern one more term and compare the graph of the resulting polynomial function with the graph of $y = e^x$. What do you think this pattern implies?

9. Graph the function given by

$$f(x) = e^x - e^{-x}.$$

From the graph, the function appears to be one-to-one. Assuming that the function has an inverse function, find $f^{-1}(x)$.

10. Find a pattern for $f^{-1}(x)$ if

$$f(x) = \frac{a^x + 1}{a^x - 1}$$

where $a > 0$, $a \neq 1$.

11. By observation, identify the equation that corresponds to the graph. Explain your reasoning.

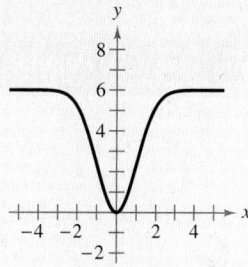

(a) $y = 6e^{-x^2/2}$

(b) $y = \dfrac{6}{1 + e^{-x/2}}$

(c) $y = 6(1 - e^{-x^2/2})$

12. You have two options for investing \$500. The first earns 7% compounded annually and the second earns 7% simple interest. The figure shows the growth of each investment over a 30-year period.

(a) Identify which graph represents each type of investment. Explain your reasoning.

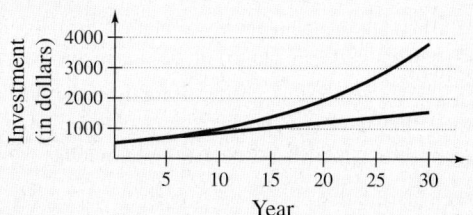

(b) Verify your answer in part (a) by finding the equations that model the investment growth and graphing the models.

(c) Which option would you choose? Explain your reasoning.

13. Two different samples of radioactive isotopes are decaying. The isotopes have initial amounts of c_1 and c_2, as well as half-lives of k_1 and k_2, respectively. Find the time required for the samples to decay to equal amounts.

14. A lab culture initially contains 500 bacteria. Two hours later, the number of bacteria has decreased to 200. Find the exponential decay model of the form

$$B = B_0 a^{kt}$$

that can be used to approximate the number of bacteria after t hours.

15. The table shows the colonial population estimates of the American colonies from 1700 to 1780. (Source: U.S. Census Bureau)

Year	Population
1700	250,900
1710	331,700
1720	466,200
1730	629,400
1740	905,600
1750	1,170,800
1760	1,593,600
1770	2,148,100
1780	2,780,400

In each of the following, let y represent the population in the year t, with $t = 0$ corresponding to 1700.

(a) Use the *regression* feature of a graphing utility to find an exponential model for the data.

(b) Use the *regression* feature of the graphing utility to find a quadratic model for the data.

(c) Use the graphing utility to plot the data and the models from parts (a) and (b) in the same viewing window.

(d) Which model is a better fit for the data? Would you use this model to predict the population of the United States in 2010? Explain your reasoning.

16. Show that $\dfrac{\log_a x}{\log_{a/b} x} = 1 + \log_a \dfrac{1}{b}$.

17. Solve $(\ln x)^2 = \ln x^2$.

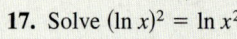**18.** Use a graphing utility to compare the graph of the function given by $y = \ln x$ with the graph of each given function.

(a) $y_1 = x - 1$

(b) $y_2 = (x - 1) - \frac{1}{2}(x - 1)^2$

(c) $y_3 = (x - 1) - \frac{1}{2}(x - 1)^2 + \frac{1}{3}(x - 1)^3$

19. Identify the pattern of successive polynomials given in Exercise 18. Extend the pattern one more term and compare the graph of the resulting polynomial function with the graph of $y = \ln x$. What do you think the pattern implies?

20. Using

$$y = ab^x \qquad \text{and} \qquad y = ax^b$$

take the natural logarithm of each side of each equation. What are the slope and y-intercept of the line relating x and $\ln y$ for $y = ab^x$? What are the slope and y-intercept of the line relating $\ln x$ and $\ln y$ for $y = ax^b$?

In Exercises 21 and 22, use the model

$$y = 80.4 - 11 \ln x, \quad 100 \le x \le 1500$$

which approximates the minimum required ventilation rate in terms of the air space per child in a public school classroom. In the model, x is the air space per child in cubic feet and y is the ventilation rate per child in cubic feet per minute.

21. Use a graphing utility to graph the model and approximate the required ventilation rate if there is 300 cubic feet of air space per child.

22. A classroom is designed for 30 students. The air conditioning system in the room has the capacity of moving 450 cubic feet of air per minute.

(a) Determine the ventilation rate per child, assuming that the room is filled to capacity.

(b) Estimate the air space required per child.

(c) Determine the minimum number of square feet of floor space required for the room if the ceiling height is 30 feet.

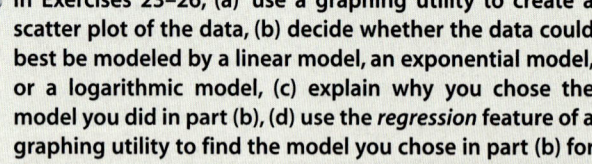**In Exercises 23–26, (a) use a graphing utility to create a scatter plot of the data, (b) decide whether the data could best be modeled by a linear model, an exponential model, or a logarithmic model, (c) explain why you chose the model you did in part (b), (d) use the *regression* feature of a graphing utility to find the model you chose in part (b) for the data and graph the model with the scatter plot, and (e) determine how well the model you chose fits the data.**

23. $(1, 2.0), (1.5, 3.5), (2, 4.0), (4, 5.8), (6, 7.0), (8, 7.8)$

24. $(1, 4.4), (1.5, 4.7), (2, 5.5), (4, 9.9), (6, 18.1), (8, 33.0)$

25. $(1, 7.5), (1.5, 7.0), (2, 6.8), (4, 5.0), (6, 3.5), (8, 2.0)$

26. $(1, 5.0), (1.5, 6.0), (2, 6.4), (4, 7.8), (6, 8.6), (8, 9.0)$

Systems of Equations and Inequalities

6

Systems of equations can be used to determine the combinations of scoring plays for different sports, such as football.

© Robert Galbraith/Reuters/Corbis

SELECTED APPLICATIONS

Systems of equations and inequalities have many real-life applications. The applications listed below represent a small sample of the applications in this chapter.

- Break-Even Analysis,
  Exercises 61–64, page 456

- Data Analysis: Renewable Energy,
  Exercise 71, page 457

- Acid Mixture,
  Exercise 51, page 468

- Sports,
  Exercise 51, page 481

- Electrical Network,
  Exercise 65, page 482

- Thermodynamics,
  Exercise 57, page 492

- Data Analysis: Prescription Drugs,
  Exercise 77, page 502

- Investment Portfolio,
  Exercises 47 and 48, page 513

- Supply and Demand,
  Exercises 75 and 76, page 517

447

6.1 Linear and Nonlinear Systems of Equations

What you should learn

• Use the method of substitution to solve systems of linear equations in two variables.
• Use the method of substitution to solve systems of nonlinear equations in two variables.
• Use a graphical approach to solve systems of equations in two variables.
• Use systems of equations to model and solve real-life problems.

Why you should learn it

Graphs of systems of equations help you solve real-life problems. For instance, in Exercise 71 on page 457, you can use the graph of a system of equations to approximate when the consumption of wind energy exceeded the consumption of solar energy.

© ML Sinibaldi/Corbis

The *HM mathSpace®* CD-ROM and *Eduspace®* for this text contain additional resources related to the concepts discussed in this chapter.

The Method of Substitution

Up to this point in the text, most problems have involved either a function of one variable or a single equation in two variables. However, many problems in science, business, and engineering involve two or more equations in two or more variables. To solve such problems, you need to find solutions of a **system of equations.** Here is an example of a system of two equations in two unknowns.

$$\begin{cases} 2x + y = 5 & \text{Equation 1} \\ 3x - 2y = 4 & \text{Equation 2} \end{cases}$$

A **solution** of this system is an ordered pair that satisfies each equation in the system. Finding the set of all solutions is called **solving the system of equations.** For instance, the ordered pair (2, 1) is a solution of this system. To check this, you can substitute 2 for x and 1 for y in *each* equation.

Check (2, 1) in Equation 1 and Equation 2:

$$2x + y = 5 \qquad \text{Write Equation 1.}$$
$$2(2) + 1 \stackrel{?}{=} 5 \qquad \text{Substitute 2 for } x \text{ and 1 for } y.$$
$$4 + 1 = 5 \qquad \text{Solution checks in Equation 1. } \checkmark$$

$$3x - 2y = 4 \qquad \text{Write Equation 2.}$$
$$3(2) - 2(1) \stackrel{?}{=} 4 \qquad \text{Substitute 2 for } x \text{ and 1 for } y.$$
$$6 - 2 = 4 \qquad \text{Solution checks in Equation 2. } \checkmark$$

In this chapter, you will study four ways to solve systems of equations, beginning with the **method of substitution.**

Method	Section	Type of System
1. Substitution	6.1	Linear or nonlinear, two variables
2. Graphical method	6.1	Linear or nonlinear, two variables
3. Elimination	6.2	Linear, two variables
4. Gaussian elimination	6.3	Linear, three or more variables

Method of Substitution

1. *Solve* one of the equations for one variable in terms of the other.

2. *Substitute* the expression found in Step 1 into the other equation to obtain an equation in one variable.

3. *Solve* the equation obtained in Step 2.

4. *Back-substitute* the value obtained in Step 3 into the expression obtained in Step 1 to find the value of the other variable.

5. *Check* that the solution satisfies *each* of the original equations.

Exploration

Use a graphing utility to graph $y_1 = 4 - x$ and $y_2 = x - 2$ in the same viewing window. Use the *zoom* and *trace* features to find the coordinates of the point of intersection. What is the relationship between the point of intersection and the solution found in Example 1?

Example 1   **Solving a System of Equations by Substitution**

Solve the system of equations.

$$\begin{cases} x + y = 4 & \text{Equation 1} \\ x - y = 2 & \text{Equation 2} \end{cases}$$

Solution

Begin by solving for y in Equation 1.

$$y = 4 - x \qquad \text{Solve for } y \text{ in Equation 1.}$$

Next, substitute this expression for y into Equation 2 and solve the resulting single-variable equation for x.

$x - y = 2$	Write Equation 2.
$x - (4 - x) = 2$	Substitute $4 - x$ for y.
$x - 4 + x = 2$	Distributive Property
$2x = 6$	Combine like terms.
$x = 3$	Divide each side by 2.

Finally, you can solve for y by *back-substituting* $x = 3$ into the equation $y = 4 - x$, to obtain

$y = 4 - x$	Write revised Equation 1.
$y = 4 - 3$	Substitute 3 for x.
$y = 1.$	Solve for y.

The solution is the ordered pair $(3, 1)$. You can check this solution as follows.

Check

Substitute $(3, 1)$ into Equation 1:

$x + y = 4$	Write Equation 1.
$3 + 1 \overset{?}{=} 4$	Substitute for x and y.
$4 = 4$	Solution checks in Equation 1. ✓

Substitute $(3, 1)$ into Equation 2:

$x - y = 2$	Write Equation 2.
$3 - 1 \overset{?}{=} 2$	Substitute for x and y.
$2 = 2$	Solution checks in Equation 2. ✓

Because $(3, 1)$ satisfies both equations in the system, it is a solution of the system of equations.

✓**CHECKPOINT**   Now try Exercise 5.

The term *back-substitution* implies that you work *backwards*. First you solve for one of the variables, and then you substitute that value *back* into one of the equations in the system to find the value of the other variable.

STUDY TIP

Because many steps are required to solve a system of equations, it is very easy to make errors in arithmetic. So, you should always check your solution by substituting it into *each* equation in the original system.

Example 2   Solving a System by Substitution

A total of $12,000 is invested in two funds paying 5% and 3% simple interest. (Recall that the formula for simple interest is $I = Prt$, where P is the principal, r is the annual interest rate, and t is the time.) The yearly interest is $500. How much is invested at each rate?

Solution

Verbal Model:

5% fund	+	3% fund	=	Total investment
5% interest	+	3% interest	=	Total interest

Labels:

Amount in 5% fund $= x$	(dollars)
Interest for 5% fund $= 0.05x$	(dollars)
Amount in 3% fund $= y$	(dollars)
Interest for 3% fund $= 0.03y$	(dollars)
Total investment $= 12{,}000$	(dollars)
Total interest $= 500$	(dollars)

System:

$$\begin{cases} x + \; y = 12{,}000 & \text{Equation 1} \\ 0.05x + 0.03y = 500 & \text{Equation 2} \end{cases}$$

To begin, it is convenient to multiply each side of Equation 2 by 100. This eliminates the need to work with decimals.

$$100(0.05x + 0.03y) = 100(500) \qquad \text{Multiply each side by 100.}$$

$$5x + 3y = 50{,}000 \qquad \text{Revised Equation 2}$$

To solve this system, you can solve for x in Equation 1.

$$x = 12{,}000 - y \qquad \text{Revised Equation 1}$$

Then, substitute this expression for x into revised Equation 2 and solve the resulting equation for y.

$$5x + 3y = 50{,}000 \qquad \text{Write revised Equation 2.}$$

$$5(12{,}000 - y) + 3y = 50{,}000 \qquad \text{Substitute } 12{,}000 - y \text{ for } x.$$

$$60{,}000 - 5y + 3y = 50{,}000 \qquad \text{Distributive Property}$$

$$-2y = -10{,}000 \qquad \text{Combine like terms.}$$

$$y = 5000 \qquad \text{Divide each side by } -2.$$

Next, back-substitute the value $y = 5000$ to solve for x.

$$x = 12{,}000 - y \qquad \text{Write revised Equation 1.}$$

$$x = 12{,}000 - 5000 \qquad \text{Substitute 5000 for } y.$$

$$x = 7000 \qquad \text{Simplify.}$$

The solution is $(7000, 5000)$. So, $7000 is invested at 5% and $5000 is invested at 3%. Check this in the original system.

✔**CHECKPOINT**   Now try Exercise 19.

STUDY TIP

When using the method of substitution, it does not matter which variable you choose to solve for first. Whether you solve for y first or x first, you will obtain the same solution. When making your choice, you should choose the variable and equation that are easier to work with. For instance, in Example 2, solving for x in Equation 1 is easier than solving for x in Equation 2.

Technology

One way to check the answers you obtain in this section is to use a graphing utility. For instance, enter the two equations in Example 2

$$y_1 = 12{,}000 - x$$

$$y_2 = \frac{500 - 0.05x}{0.03}$$

and find an appropriate viewing window that shows where the two lines intersect. Then use the *intersect* feature or the *zoom* and *trace* features to find the point of intersection. Does this point agree with the solution obtained at the right?

Nonlinear Systems of Equations

The equations in Examples 1 and 2 are linear. The method of substitution can also be used to solve systems in which one or both of the equations are nonlinear.

Example 3   Substitution: Two-Solution Case

Solve the system of equations.

$$\begin{cases} x^2 + 4x - y = 7 & \text{Equation 1} \\ 2x - y = -1 & \text{Equation 2} \end{cases}$$

Solution

Begin by solving for y in Equation 2 to obtain $y = 2x + 1$. Next, substitute this expression for y into Equation 1 and solve for x.

$x^2 + 4x - (2x + 1) = 7$	Substitute $2x + 1$ for y into Equation 1.
$x^2 + 2x - 1 = 7$	Simplify.
$x^2 + 2x - 8 = 0$	Write in general form.
$(x + 4)(x - 2) = 0$	Factor.
$x = -4, 2$	Solve for x.

Back-substituting these values of x to solve for the corresponding values of y produces the solutions $(-4, -7)$ and $(2, 5)$. Check these in the original system.

✓**CHECKPOINT**   Now try Exercise 25.

When using the method of substitution, you may encounter an equation that has no solution, as shown in Example 4.

Example 4   Substitution: No-Real-Solution Case

Solve the system of equations.

$$\begin{cases} -x + y = 4 & \text{Equation 1} \\ x^2 + y = 3 & \text{Equation 2} \end{cases}$$

Solution

Begin by solving for y in Equation 1 to obtain $y = x + 4$. Next, substitute this expression for y into Equation 2 and solve for x.

$x^2 + (x + 4) = 3$	Substitute $x + 4$ for y into Equation 2.
$x^2 + x + 1 = 0$	Simplify.
$x = \dfrac{-1 \pm \sqrt{-3}}{2}$	Use the Quadratic Formula.

Because the discriminant is negative, the equation $x^2 + x + 1 = 0$ has no (real) solution. So, the original system has no (real) solution.

✓**CHECKPOINT**   Now try Exercise 27.

Exploration

Use a graphing utility to graph the two equations in Example 3

$$y_1 = x^2 + 4x - 7$$

$$y_2 = 2x + 1$$

in the same viewing window. How many solutions do you think this system has? Repeat this experiment for the equations in Example 4. How many solutions does this system have? Explain your reasoning.

Graphical Approach to Finding Solutions

From Examples 2, 3, and 4, you can see that a system of two equations in two unknowns can have exactly one solution, more than one solution, or no solution. By using a **graphical method,** you can gain insight about the number of solutions and the location(s) of the solution(s) of a system of equations by graphing each of the equations in the same coordinate plane. The solutions of the system correspond to the **points of intersection** of the graphs. For instance, the two equations in Figure 6.1 graph as two lines with a *single point* of intersection; the two equations in Figure 6.2 graph as a parabola and a line with *two points* of intersection; and the two equations in Figure 6.3 graph as a line and a parabola that have *no points* of intersection.

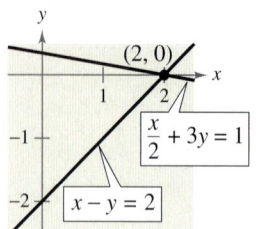

One intersection point
FIGURE 6.1

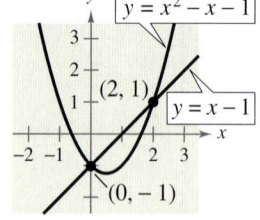

Two intersection points
FIGURE 6.2

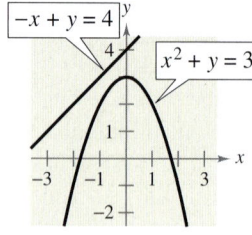

No intersection points
FIGURE 6.3

Example 5   Solving a System of Equations Graphically

Solve the system of equations.

$$\begin{cases} y = \ln x & \text{Equation 1} \\ x + y = 1 & \text{Equation 2} \end{cases}$$

Solution

Sketch the graphs of the two equations. From the graphs of these equations, it is clear that there is only one point of intersection and that $(1, 0)$ is the solution point (see Figure 6.4). You can confirm this by substituting 1 for x and 0 for y in *both* equations.

Check (1, 0) in Equation 1:

$y = \ln x$ Write Equation 1.

$0 = \ln 1$ Equation 1 checks. ✓

Check (1, 0) in Equation 2:

$x + y = 1$ Write Equation 2.

$1 + 0 = 1$ Equation 2 checks. ✓

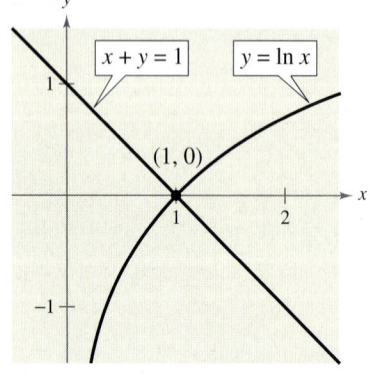

FIGURE 6.4

✓CHECKPOINT   Now try Exercise 33.

Example 5 shows the value of a graphical approach to solving systems of equations in two variables. Notice what would happen if you tried only the substitution method in Example 5. You would obtain the equation $x + \ln x = 1$. It would be difficult to solve this equation for x using standard algebraic techniques.

Applications

The total cost C of producing x units of a product typically has two components—the initial cost and the cost per unit. When enough units have been sold so that the total revenue R equals the total cost C, the sales are said to have reached the **break-even point.** You will find that the break-even point corresponds to the point of intersection of the cost and revenue curves.

Example 6	Break-Even Analysis	

A shoe company invests \$300,000 in equipment to produce a new line of athletic footwear. Each pair of shoes costs \$5 to produce and is sold for \$60. How many pairs of shoes must be sold before the business breaks even?

Solution

The total cost of producing x units is

$$\text{Total cost} = \text{Cost per unit} \cdot \text{Number of units} + \text{Initial cost}$$

$$C = 5x + 300{,}000. \qquad \textcolor{orange}{\text{Equation 1}}$$

The revenue obtained by selling x units is

$$\text{Total revenue} = \text{Price per unit} \cdot \text{Number of units}$$

$$R = 60x. \qquad \textcolor{orange}{\text{Equation 2}}$$

Because the break-even point occurs when $R = C$, you have $C = 60x$, and the system of equations to solve is

$$\begin{cases} C = 5x + 300{,}000 \\ C = 60x \end{cases}.$$

Now you can solve by substitution.

$60x = 5x + 300{,}000$	Substitute $60x$ for C in Equation 1.
$55x = 300{,}000$	Subtract $5x$ from each side.
$x \approx 5455$	Divide each side by 55.

So, the company must sell about 5455 pairs of shoes to break even. Note in Figure 6.5 that revenue less than the break-even point corresponds to an overall loss, whereas revenue greater than the break-even point corresponds to a profit.

✓**CHECKPOINT**   Now try Exercise 63.

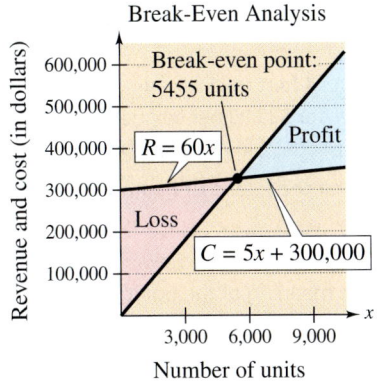

Break-Even Analysis

Break-even point: 5455 units

$R = 60x$

Profit

Loss

$C = 5x + 300{,}000$

Revenue and cost (in dollars)

Number of units

FIGURE **6.5**

Another way to view the solution in Example 6 is to consider the profit function

$$P = R - C.$$

The break-even point occurs when the profit is 0, which is the same as saying that $R = C$.

Example 7 Movie Ticket Sales

The weekly ticket sales for a new comedy movie decreased each week. At the same time, the weekly ticket sales for a new drama movie increased each week. Models that approximate the weekly ticket sales S (in millions of dollars) for each movie are

$$\begin{cases} S = 60 - 8x & \text{Comedy} \\ S = 10 + 4.5x & \text{Drama} \end{cases}$$

where x represents the number of weeks each movie was in theaters, with $x = 0$ corresponding to the ticket sales during the opening weekend. After how many weeks will the ticket sales for the two movies be equal?

Algebraic Solution

Because the second equation has already been solved for S in terms of x, substitute this value into the first equation and solve for x, as follows.

$10 + 4.5x = 60 - 8x$ Substitute for S in Equation 1.

$4.5x + 8x = 60 - 10$ Add $8x$ and -10 to each side.

$12.5x = 50$ Combine like terms.

$x = 4$ Divide each side by 12.5.

So, the weekly ticket sales for the two movies will be equal after 4 weeks.

 **CHECKPOINT**   Now try Exercise 65.

Numerical Solution

You can create a table of values for each model to determine when the ticket sales for the two movies will be equal.

Number of weeks, x	0	1	2	3	4	5	6
Sales, S (comedy)	60	52	44	36	28	20	12
Sales, S (drama)	10	14.5	19	23.5	28	32.5	37

So, from the table above, you can see that the weekly ticket sales for the two movies will be equal after 4 weeks.

WRITING ABOUT MATHEMATICS

Interpreting Points of Intersection You plan to rent a 14-foot truck for a two-day local move. At truck rental agency A, you can rent a truck for $29.95 per day plus $0.49 per mile. At agency B, you can rent a truck for $50 per day plus $0.25 per mile.

a. Write a total cost equation in terms of x and y for the total cost of renting the truck from each agency.

b. Use a graphing utility to graph the two equations in the same viewing window and find the point of intersection. Interpret the meaning of the point of intersection in the context of the problem.

c. Which agency should you choose if you plan to travel a total of 100 miles during the two-day move? Why?

d. How does the situation change if you plan to drive 200 miles during the two-day move?

6.1 | Exercises

The *HM mathSpace*® CD-ROM and *Eduspace*® for this text contain step-by-step solutions to all odd-numbered exercises. They also provide Tutorial Exercises for additional help.

VOCABULARY CHECK: Fill in the blanks.

1. A set of two or more equations in two or more variables is called a _____ of _____.

2. A _____ of a system of equations is an ordered pair that satisfies each equation in the system.

3. Finding the set of all solutions to a system of equations is called _____ the system of equations.

4. The first step in solving a system of equations by the method of _____ is to solve one of the equations for one variable in terms of the other variable.

5. Graphically, the solution of a system of two equations is the _____ of _____ of the graphs of the two equations.

6. In business applications, the point at which the revenue equals costs is called the _____ point.

PREREQUISITE SKILLS REVIEW: Practice and review algebra skills needed for this section at **www.Eduspace.com.**

In Exercises 1–4, determine whether each ordered pair is a solution of the system of equations.

1. $\begin{cases} 4x - y = 1 \\ 6x + y = -6 \end{cases}$ (a) $(0, -3)$ (b) $(-1, -4)$
   (c) $\left(-\frac{3}{2}, -2\right)$ (d) $\left(-\frac{1}{2}, -3\right)$

2. $\begin{cases} 4x^2 + y = 3 \\ -x - y = 11 \end{cases}$ (a) $(2, -13)$ (b) $(2, -9)$
   (c) $\left(-\frac{3}{2}, -\frac{31}{3}\right)$ (d) $\left(-\frac{7}{4}, -\frac{37}{4}\right)$

3. $\begin{cases} y = -2e^x \\ 3x - y = 2 \end{cases}$ (a) $(-2, 0)$ (b) $(0, -2)$
   (c) $(0, -3)$ (d) $(-1, 2)$

4. $\begin{cases} -\log x + 3 = y \\ \frac{1}{9}x + y = \frac{28}{9} \end{cases}$ (a) $\left(9, \frac{37}{9}\right)$ (b) $(10, 2)$
   (c) $(1, 3)$ (d) $(2, 4)$

In Exercises 5–14, solve the system by the method of substitution. Check your solution graphically.

5. $\begin{cases} 2x + y = 6 \\ -x + y = 0 \end{cases}$

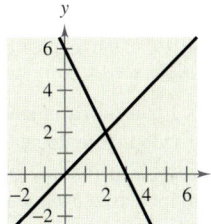

6. $\begin{cases} x - y = -4 \\ x + 2y = 5 \end{cases}$

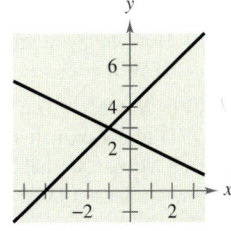

7. $\begin{cases} x - y = -4 \\ x^2 - y = -2 \end{cases}$

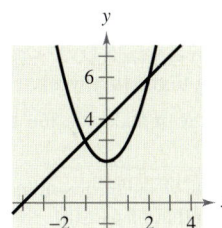

8. $\begin{cases} 3x + y = 2 \\ x^3 - 2 + y = 0 \end{cases}$

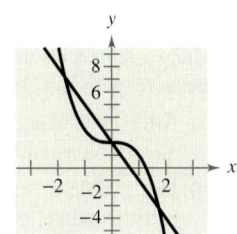

9. $\begin{cases} -2x + y = -5 \\ x^2 + y^2 = 25 \end{cases}$

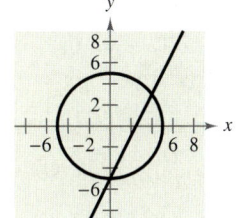

10. $\begin{cases} x + y = 0 \\ x^3 - 5x - y = 0 \end{cases}$

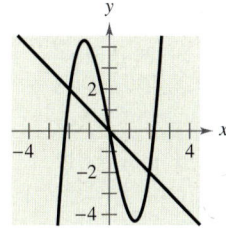

11. $\begin{cases} x^2 + y = 0 \\ x^2 - 4x - y = 0 \end{cases}$

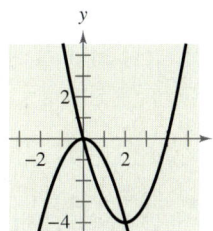

12. $\begin{cases} y = -2x^2 + 2 \\ y = 2(x^4 - 2x^2 + 1) \end{cases}$

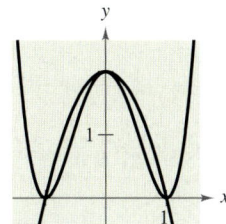

13. $\begin{cases} y = x^3 - 3x^2 + 1 \\ y = x^2 - 3x + 1 \end{cases}$

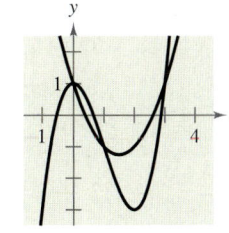

14. $\begin{cases} y = x^3 - 3x^2 + 4 \\ y = -2x + 4 \end{cases}$

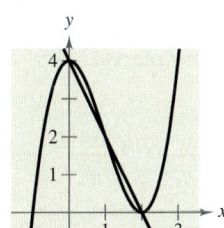

In Exercises 15–28, solve the system by the method of substitution.

15. $\begin{cases} x - y = 0 \\ 5x - 3y = 10 \end{cases}$

16. $\begin{cases} x + 2y = 1 \\ 5x - 4y = -23 \end{cases}$

17. $\begin{cases} 2x - y + 2 = 0 \\ 4x + y - 5 = 0 \end{cases}$

18. $\begin{cases} 6x - 3y - 4 = 0 \\ x + 2y - 4 = 0 \end{cases}$

19. $\begin{cases} 1.5x + 0.8y = 2.3 \\ 0.3x - 0.2y = 0.1 \end{cases}$

20. $\begin{cases} 0.5x + 3.2y = 9.0 \\ 0.2x - 1.6y = -3.6 \end{cases}$

21. $\begin{cases} \frac{1}{5}x + \frac{1}{2}y = 8 \\ x + y = 20 \end{cases}$

22. $\begin{cases} \frac{1}{2}x + \frac{3}{4}y = 10 \\ \frac{3}{4}x - y = 4 \end{cases}$

23. $\begin{cases} 6x + 5y = -3 \\ -x - \frac{5}{6}y = -7 \end{cases}$

24. $\begin{cases} -\frac{2}{3}x + y = 2 \\ 2x - 3y = 6 \end{cases}$

25. $\begin{cases} x^2 - y = 0 \\ 2x + y = 0 \end{cases}$

26. $\begin{cases} x - 2y = 0 \\ 3x - y^2 = 0 \end{cases}$

27. $\begin{cases} x - y = -1 \\ x^2 - y = -4 \end{cases}$

28. $\begin{cases} y = -x \\ y = x^3 + 3x^2 + 2x \end{cases}$

In Exercises 29–42, solve the system graphically.

29. $\begin{cases} -x + 2y = 2 \\ 3x + y = 15 \end{cases}$

30. $\begin{cases} x + y = 0 \\ 3x - 2y = 10 \end{cases}$

31. $\begin{cases} x - 3y = -2 \\ 5x + 3y = 17 \end{cases}$

32. $\begin{cases} -x + 2y = 1 \\ x - y = 2 \end{cases}$

33. $\begin{cases} x + y = 4 \\ x^2 + y^2 - 4x = 0 \end{cases}$

34. $\begin{cases} -x + y = 3 \\ x^2 - 6x - 27 + y^2 = 0 \end{cases}$

35. $\begin{cases} x - y + 3 = 0 \\ x^2 - 4x + 7 = y \end{cases}$

36. $\begin{cases} y^2 - 4x + 11 = 0 \\ -\frac{1}{2}x + y = -\frac{1}{2} \end{cases}$

37. $\begin{cases} 7x + 8y = 24 \\ x - 8y = 8 \end{cases}$

38. $\begin{cases} x - y = 0 \\ 5x - 2y = 6 \end{cases}$

39. $\begin{cases} 3x - 2y = 0 \\ x^2 - y^2 = 4 \end{cases}$

40. $\begin{cases} 2x - y + 3 = 0 \\ x^2 + y^2 - 4x = 0 \end{cases}$

41. $\begin{cases} x^2 + y^2 = 25 \\ 3x^2 - 16y = 0 \end{cases}$

42. $\begin{cases} x^2 + y^2 = 25 \\ (x - 8)^2 + y^2 = 41 \end{cases}$

In Exercises 43–48, use a graphing utility to solve the system of equations. Find the solution accurate to two decimal places.

43. $\begin{cases} y = e^x \\ x - y + 1 = 0 \end{cases}$

44. $\begin{cases} y = -4e^{-x} \\ y + 3x + 8 = 0 \end{cases}$

45. $\begin{cases} x + 2y = 8 \\ y = \log_2 x \end{cases}$

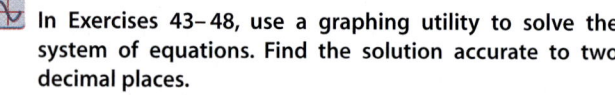

46. $\begin{cases} y = -2 + \ln(x - 1) \\ 3y + 2x = 9 \end{cases}$

47. $\begin{cases} x^2 + y^2 = 169 \\ x^2 - 8y = 104 \end{cases}$

48. $\begin{cases} x^2 + y^2 = 4 \\ 2x^2 - y = 2 \end{cases}$

In Exercises 49–60, solve the system graphically or algebraically. Explain your choice of method.

49. $\begin{cases} y = 2x \\ y = x^2 + 1 \end{cases}$

50. $\begin{cases} x + y = 4 \\ x^2 + y = 2 \end{cases}$

51. $\begin{cases} 3x - 7y + 6 = 0 \\ x^2 - y^2 = 4 \end{cases}$

52. $\begin{cases} x^2 + y^2 = 25 \\ 2x + y = 10 \end{cases}$

53. $\begin{cases} x - 2y = 4 \\ x^2 - y = 0 \end{cases}$

54. $\begin{cases} y = (x + 1)^3 \\ y = \sqrt{x - 1} \end{cases}$

55. $\begin{cases} y - e^{-x} = 1 \\ y - \ln x = 3 \end{cases}$

56. $\begin{cases} x^2 + y = 4 \\ e^x - y = 0 \end{cases}$

57. $\begin{cases} y = x^4 - 2x^2 + 1 \\ y = 1 - x^2 \end{cases}$

58. $\begin{cases} y = x^3 - 2x^2 + x - 1 \\ y = -x^2 + 3x - 1 \end{cases}$

59. $\begin{cases} xy - 1 = 0 \\ 2x - 4y + 7 = 0 \end{cases}$

60. $\begin{cases} x - 2y = 1 \\ y = \sqrt{x - 1} \end{cases}$

Break-Even Analysis **In Exercises 61 and 62, find the sales necessary to break even ($R = C$) for the cost C of producing x units and the revenue R obtained by selling x units. (Round to the nearest whole unit.)**

61. $C = 8650x + 250{,}000, \quad R = 9950x$

62. $C = 5.5\sqrt{x} + 10{,}000, \quad R = 3.29x$

63. ***Break-Even Analysis*** A small software company invests $16,000 to produce a software package that will sell for $55.95. Each unit can be produced for $35.45.

(a) How many units must be sold to break even?

(b) How many units must be sold to make a profit of $60,000?

64. ***Break-Even Analysis*** A small fast-food restaurant invests $5000 to produce a new food item that will sell for $3.49. Each item can be produced for $2.16.

(a) How many items must be sold to break even?

(b) How many items must be sold to make a profit of $8500?

65. ***DVD Rentals*** The weekly rentals for a newly released DVD of an animated film at a local video store decreased each week. At the same time, the weekly rentals for a newly released DVD of a horror film increased each week. Models that approximate the weekly rentals R for each DVD are

$$\begin{cases} R = 360 - 24x \quad \text{Animated film} \\ R = 24 + 18x \quad \text{Horror film} \end{cases}$$

where x represents the number of weeks each DVD was in the store, with $x = 1$ corresponding to the first week.

(a) After how many weeks will the rentals for the two movies be equal?

(b) Use a table to solve the system of equations numerically. Compare your result with that of part (a).

66. *CD Sales* The total weekly sales for a newly released rock CD increased each week. At the same time, the total weekly sales for a newly released rap CD decreased each week. Models that approximate the total weekly sales S (in thousands of units) for each CD are

$$\begin{cases} S = 25x + 100 & \text{Rock CD} \\ S = -50x + 475 & \text{Rap CD} \end{cases}$$

where x represents the number of weeks each CD was in stores, with $x = 0$ corresponding to the CD sales on the day each CD was first released in stores.

(a) After how many weeks will the sales for the two CDs be equal?

(b) Use a table to solve the system of equations numerically. Compare your result with that of part (a).

67. *Choice of Two Jobs* You are offered two jobs selling dental supplies. One company offers a straight commission of 6% of sales. The other company offers a salary of $350 per week plus 3% of sales. How much would you have to sell in a week in order to make the straight commission offer better?

 68. *Supply and Demand* The supply and demand curves for a business dealing with wheat are

Supply: $p = 1.45 + 0.00014x^2$

Demand: $p = (2.388 - 0.007x)^2$

where p is the price in dollars per bushel and x is the quantity in bushels per day. Use a graphing utility to graph the supply and demand equations and find the market equilibrium. (The market equilibrium is the point of intersection of the graphs for $x > 0$.)

69. *Investment Portfolio* A total of $25,000 is invested in two funds paying 6% and 8.5% simple interest. (The 6% investment has a lower risk.) The investor wants a yearly interest income of $2000 from the two investments.

(a) Write a system of equations in which one equation represents the total amount invested and the other equation represents the $2000 required in interest. Let x and y represent the amounts invested at 6% and 8.5%, respectively.

 (b) Use a graphing utility to graph the two equations in the same viewing window. As the amount invested at 6% increases, how does the amount invested at 8.5% change? How does the amount of interest income change? Explain.

(c) What amount should be invested at 6% to meet the requirement of $2000 per year in interest?

70. *Log Volume* You are offered two different rules for estimating the number of board feet in a 16-foot log. (A board foot is a unit of measure for lumber equal to a board 1 foot square and 1 inch thick.) The first rule is the *Doyle Log Rule* and is modeled by

$$V_1 = (D - 4)^2, \quad 5 \le D \le 40$$

and the other is the *Scribner Log Rule* and is modeled by

$$V_2 = 0.79D^2 - 2D - 4, \quad 5 \le D \le 40$$

where D is the diameter (in inches) of the log and V is its volume (in board feet).

 (a) Use a graphing utility to graph the two log rules in the same viewing window.

(b) For what diameter do the two scales agree?

(c) You are selling large logs by the board foot. Which scale would you use? Explain your reasoning.

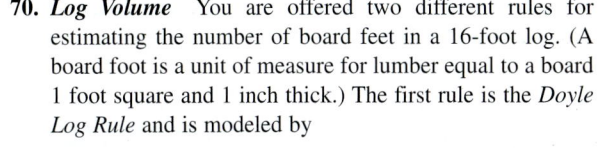

71. *Data Analysis: Renewable Energy* The table shows the consumption C (in trillions of Btus) of solar energy and wind energy in the United States from 1998 to 2003. (Source: Energy Information Administration)

Year	Solar, C	Wind, C
1998	70	31
1999	69	46
2000	66	57
2001	65	68
2002	64	105
2003	63	108

 (a) Use the *regression* feature of a graphing utility to find a quadratic model for the solar energy consumption data and a linear model for the wind energy consumption data. Let t represent the year, with $t = 8$ corresponding to 1998.

 (b) Use a graphing utility to graph the data and the two models in the same viewing window.

 (c) Use the graph from part (b) to approximate the point of intersection of the graphs of the models. Interpret your answer in the context of the problem.

(d) Approximate the point of intersection of the graphs of the models algebraically.

 (e) Compare your results from parts (c) and (d).

(f) Use your school's library, the Internet, or some other reference source to research the advantages and disadvantages of using renewable energy.

72. *Data Analysis: Population* The table shows the populations P (in thousands) of Alabama and Colorado from 1999 to 2003. (Source: U.S. Census Bureau)

Year	Alabama, P	Colorado, P
1999	4430	4226
2000	4447	4302
2001	4466	4429
2002	4479	4501
2003	4501	4551

(a) Use the *regression* feature of a graphing utility to find linear models for each set of data. Graph the models in the same viewing window. Let t represent the year, with $t = 9$ corresponding to 1999.

(b) Use your graph from part (a) to approximate when the population of Colorado exceeded the population of Alabama.

(c) Verify your answer from part (b) algebraically.

Geometry **In Exercises 73–76, find the dimensions of the rectangle meeting the specified conditions.**

73. The perimeter is 30 meters and the length is 3 meters greater than the width.

74. The perimeter is 280 centimeters and the width is 20 centimeters less than the length.

75. The perimeter is 42 inches and the width is three-fourths the length.

76. The perimeter is 210 feet and the length is $1\frac{1}{2}$ times the width.

77. *Geometry* What are the dimensions of a rectangular tract of land if its perimeter is 40 kilometers and its area is 96 square kilometers?

78. *Geometry* What are the dimensions of an isosceles right triangle with a two-inch hypotenuse and an area of 1 square inch?

Synthesis

True or False? **In Exercises 79 and 80, determine whether the statement is true or false. Justify your answer.**

79. In order to solve a system of equations by substitution, you must always solve for y in one of the two equations and then back-substitute.

80. If a system consists of a parabola and a circle, then the system can have at most two solutions.

81. *Writing* List and explain the steps used to solve a system of equations by the method of substitution.

82. *Think About It* When solving a system of equations by substitution, how do you recognize that the system has no solution?

83. *Exploration* Find an equation of a line whose graph intersects the graph of the parabola $y = x^2$ at (a) two points, (b) one point, and (c) no points. (There is more than one correct answer.)

84. *Conjecture* Consider the system of equations

$$\begin{cases} y = b^x \\ y = x^b \end{cases}.$$

(a) Use a graphing utility to graph the system for $b = 1, 2, 3,$ and 4.

(b) For a fixed even value of $b > 1$, make a conjecture about the number of points of intersection of the graphs in part (a).

Skills Review

In Exercises 85–90, find the general form of the equation of the line passing through the two points.

85. $(-2, 7), (5, 5)$

86. $(3.5, 4), (10, 6)$

87. $(6, 3), (10, 3)$

88. $(4, -2), (4, 5)$

89. $\left(\frac{3}{5}, 0\right), (4, 6)$

90. $\left(-\frac{7}{3}, 8\right), \left(\frac{5}{2}, \frac{1}{2}\right)$

In Exercises 91–94, find the domain of the function and identify any horizontal or vertical asymptotes.

91. $f(x) = \dfrac{5}{x - 6}$

92. $f(x) = \dfrac{2x - 7}{3x + 2}$

93. $f(x) = \dfrac{x^2 + 2}{x^2 - 16}$

94. $f(x) = 3 - \dfrac{2}{x^2}$

6.2 Two-Variable Linear Systems

What you should learn

- Use the method of elimination to solve systems of linear equations in two variables.
- Interpret graphically the numbers of solutions of systems of linear equations in two variables.
- Use systems of linear equations in two variables to model and solve real-life problems.

Why you should learn it

You can use systems of equations in two variables to model and solve real-life problems. For instance, in Exercise 63 on page 469, you will solve a system of equations to find a linear model that represents the relationship between wheat yield and amount of fertilizer applied.

© Bill Stormont/Corbis

The Method of Elimination

In Section 6.1, you studied two methods for solving a system of equations: substitution and graphing. Now you will study the **method of elimination.** The key step in this method is to obtain, for one of the variables, coefficients that differ only in sign so that *adding* the equations eliminates the variable.

$$3x + 5y = 7 \qquad \text{Equation 1}$$
$$\underline{-3x - 2y = -1} \qquad \text{Equation 2}$$
$$3y = 6 \qquad \text{Add equations.}$$

Note that by adding the two equations, you eliminate the x-terms and obtain a single equation in y. Solving this equation for y produces $y = 2$, which you can then back-substitute into one of the original equations to solve for x.

Example 1 **Solving a System of Equations by Elimination**

Solve the system of linear equations.

$$\begin{cases} 3x + 2y = 4 & \text{Equation 1} \\ 5x - 2y = 8 & \text{Equation 2} \end{cases}$$

Solution

Because the coefficients of y differ only in sign, you can eliminate the y-terms by adding the two equations.

$$3x + 2y = 4 \qquad \text{Write Equation 1.}$$
$$\underline{5x - 2y = 8} \qquad \text{Write Equation 2.}$$
$$8x = 12 \qquad \text{Add equations.}$$

So, $x = \frac{3}{2}$. By back-substituting this value into Equation 1, you can solve for y.

$$3x + 2y = 4 \qquad \text{Write Equation 1.}$$
$$3\left(\tfrac{3}{2}\right) + 2y = 4 \qquad \text{Substitute } \tfrac{3}{2} \text{ for } x.$$
$$\tfrac{9}{2} + 2y = 4 \qquad \text{Simplify.}$$
$$y = -\tfrac{1}{4} \qquad \text{Solve for } y.$$

The solution is $\left(\frac{3}{2}, -\frac{1}{4}\right)$. Check this in the original system, as follows.

Check

$$3\left(\tfrac{3}{2}\right) + 2\left(-\tfrac{1}{4}\right) \overset{?}{=} 4 \qquad \text{Substitute into Equation 1.}$$
$$\tfrac{9}{2} - \tfrac{1}{2} = 4 \qquad \text{Equation 1 checks. } \checkmark$$
$$5\left(\tfrac{3}{2}\right) - 2\left(-\tfrac{1}{4}\right) \overset{?}{=} 8 \qquad \text{Substitute into Equation 2.}$$
$$\tfrac{15}{2} + \tfrac{1}{2} = 8 \qquad \text{Equation 2 checks. } \checkmark$$

✓**CHECKPOINT**   Now try Exercise 11.

Exploration

Use the method of substitution to solve the system in Example 1. Which method is easier?

> ### Method of Elimination
>
> To use the **method of elimination** to solve a system of two linear equations in x and y, perform the following steps.
>
> 1. *Obtain coefficients* for x (or y) that differ only in sign by multiplying all terms of one or both equations by suitably chosen constants.
>
> 2. *Add* the equations to eliminate one variable, and solve the resulting equation.
>
> 3. *Back-substitute* the value obtained in Step 2 into either of the original equations and solve for the other variable.
>
> 4. *Check* your solution in both of the original equations.

Example 2 Solving a System of Equations by Elimination

Solve the system of linear equations.

$$\begin{cases} 2x - 3y = -7 & \text{Equation 1} \\ 3x + \ y = -5 & \text{Equation 2} \end{cases}$$

Solution

For this system, you can obtain coefficients that differ only in sign by multiplying Equation 2 by 3.

$2x - 3y = -7$	$2x - 3y = \ -7$	Write Equation 1.
$3x + \ y = -5$	$9x + 3y = -15$	Multiply Equation 2 by 3.
	$11x \qquad = -22$	Add equations.

So, you can see that $x = -2$. By back-substituting this value of x into Equation 1, you can solve for y.

$2x - 3y = -7$	Write Equation 1.
$2(-2) - 3y = -7$	Substitute -2 for x.
$-3y = -3$	Combine like terms.
$y = 1$	Solve for y.

The solution is $(-2, 1)$. Check this in the original system, as follows.

Check

$2x - 3y = -7$	Write original Equation 1.
$2(-2) - 3(1) \overset{?}{=} -7$	Substitute into Equation 1.
$-4 - 3 = -7$	Equation 1 checks. ✔
$3x + y = -5$	Write original Equation 2.
$3(-2) + 1 \overset{?}{=} -5$	Substitute into Equation 2.
$-6 + 1 = -5$	Equation 2 checks. ✔

✔**CHECKPOINT** Now try Exercise 13.

Exploration

Rewrite each system of equations in slope-intercept form and sketch the graph of each system. What is the relationship between the slopes of the two lines and the number of points of intersection?

a. $\begin{cases} 5x - y = -1 \\ -x + y = -5 \end{cases}$

b. $\begin{cases} 4x - 3y = \ \ 1 \\ -8x + 6y = -2 \end{cases}$

c. $\begin{cases} x + 2y = \ \ 3 \\ x + 2y = -8 \end{cases}$

In Example 2, the two systems of linear equations (the original system and the system obtained by multiplying by constants)

$$\begin{cases} 2x - 3y = -7 \\ 3x + y = -5 \end{cases} \quad \text{and} \quad \begin{cases} 2x - 3y = -7 \\ 9x + 3y = -15 \end{cases}$$

are called **equivalent systems** because they have precisely the same solution set. The operations that can be performed on a system of linear equations to produce an equivalent system are (1) interchanging any two equations, (2) multiplying an equation by a nonzero constant, and (3) adding a multiple of one equation to any other equation in the system.

Example 3   Solving the System of Equations by Elimination

Solve the system of linear equations.

$$\begin{cases} 5x + 3y = 9 & \text{Equation 1} \\ 2x - 4y = 14 & \text{Equation 2} \end{cases}$$

Algebraic Solution

You can obtain coefficients that differ only in sign by multiplying Equation 1 by 4 and multiplying Equation 2 by 3.

$$5x + 3y = 9 \implies 20x + 12y = 36 \quad \text{Multiply Equation 1 by 4.}$$
$$2x - 4y = 14 \implies \underline{6x - 12y = 42} \quad \text{Multiply Equation 2 by 3.}$$
$$26x \qquad = 78 \quad \text{Add equations.}$$

From this equation, you can see that $x = 3$. By back-substituting this value of x into Equation 2, you can solve for y.

$$2x - 4y = 14 \qquad \text{Write Equation 2.}$$
$$2(3) - 4y = 14 \qquad \text{Substitute 3 for } x.$$
$$-4y = 8 \qquad \text{Combine like terms.}$$
$$y = -2 \qquad \text{Solve for } y.$$

The solution is $(3, -2)$. Check this in the original system.

Graphical Solution

Solve each equation for y. Then use a graphing utility to graph $y_1 = -\frac{5}{3}x + 3$ and $y_2 = \frac{1}{2}x - \frac{7}{2}$ in the same viewing window. Use the *intersect* feature or the *zoom* and *trace* features to approximate the point of intersection of the graphs. From the graph in Figure 6.6, you can see that the point of intersection is $(3, -2)$. You can determine that this is the exact solution by checking $(3, -2)$ in both equations.

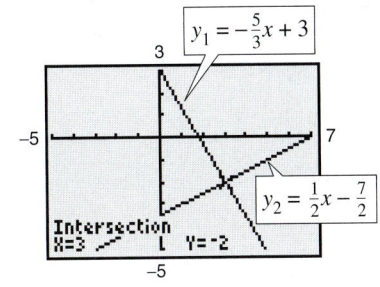

FIGURE 6.6

✓CHECKPOINT   Now try Exercise 15.

You can check the solution from Example 3 as follows.

$$5(3) + 3(-2) \stackrel{?}{=} 9 \qquad \text{Substitute 3 for } x \text{ and } -2 \text{ for } y \text{ in Equation 1.}$$
$$15 - 6 = 9 \qquad \text{Equation 1 checks. } ✓$$
$$2(3) - 4(-2) \stackrel{?}{=} 14 \qquad \text{Substitute 3 for } x \text{ and } -2 \text{ for } y \text{ in Equation 2.}$$
$$6 + 8 = 14 \qquad \text{Equation 2 checks. } ✓$$

Keep in mind that the terminology and methods discussed in this section apply only to systems of *linear* equations.

Graphical Interpretation of Solutions

It is possible for a *general* system of equations to have exactly one solution, two or more solutions, or no solution. If a system of *linear* equations has two different solutions, it must have an *infinite* number of solutions.

Graphical Interpretations of Solutions

For a system of two linear equations in two variables, the number of solutions is one of the following.

Number of Solutions	Graphical Interpretation	Slopes of Lines
1. Exactly one solution	The two lines intersect at one point.	The slopes of the two lines are not equal.
2. Infinitely many solutions	The two lines coincide (are identical).	The slopes of the two lines are equal.
3. No solution	The two lines are parallel.	The slopes of the two lines are equal.

A system of linear equations is **consistent** if it has at least one solution. A consistent system with exactly one solution is *independent*, whereas a consistent system with infinitely many solutions is *dependent*. A system is **inconsistent** if it has no solution.

Example 4 Recognizing Graphs of Linear Systems

Match each system of linear equations with its graph in Figure 6.7. Describe the number of solutions and state whether the system is consistent or inconsistent.

a. $\begin{cases} 2x - 3y = 3 \\ -4x + 6y = 6 \end{cases}$ **b.** $\begin{cases} 2x - 3y = 3 \\ x + 2y = 5 \end{cases}$ **c.** $\begin{cases} 2x - 3y = 3 \\ -4x + 6y = -6 \end{cases}$

i.

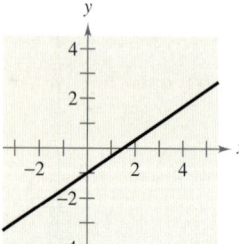

ii.

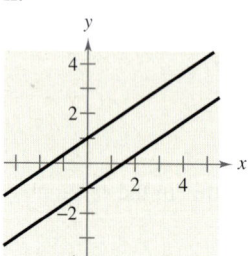

iii.

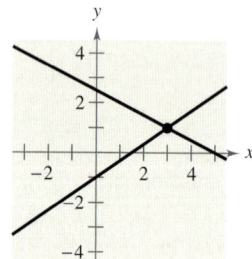

FIGURE 6.7

STUDY TIP

A comparison of the slopes of two lines gives useful information about the number of solutions of the corresponding system of equations. To solve a system of equations graphically, it helps to begin by writing the equations in slope-intercept form. Try doing this for the systems in Example 4.

Solution

a. The graph of system (a) is a pair of parallel lines (ii). The lines have no point of intersection, so the system has no solution. The system is inconsistent.

b. The graph of system (b) is a pair of intersecting lines (iii). The lines have one point of intersection, so the system has exactly one solution. The system is consistent.

c. The graph of system (c) is a pair of lines that coincide (i). The lines have infinitely many points of intersection, so the system has infinitely many solutions. The system is consistent.

✓CHECKPOINT Now try Exercises 31–34.

In Examples 5 and 6, note how you can use the method of elimination to determine that a system of linear equations has no solution or infinitely many solutions.

Example 5   No-Solution Case: Method of Elimination

Solve the system of linear equations.

$$\begin{cases} x - 2y = 3 & \text{Equation 1} \\ -2x + 4y = 1 & \text{Equation 2} \end{cases}$$

Solution

To obtain coefficients that differ only in sign, multiply Equation 1 by 2.

$x - 2y = 3$	$2x - 4y = 6$	Multiply Equation 1 by 2.
$-2x + 4y = 1$	$-2x + 4y = 1$	Write Equation 2.
	$0 = 7$	False statement

Because there are no values of x and y for which $0 = 7$, you can conclude that the system is inconsistent and has no solution. The lines corresponding to the two equations in this system are shown in Figure 6.8. Note that the two lines are parallel and therefore have no point of intersection.

✓**CHECKPOINT**   Now try Exercise 19.

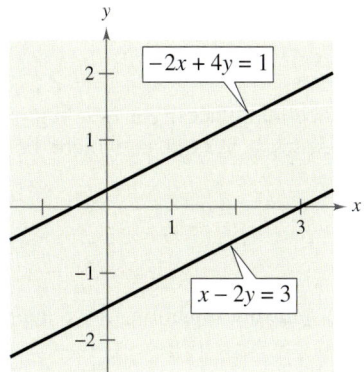

FIGURE **6.8**

In Example 5, note that the occurrence of a false statement, such as $0 = 7$, indicates that the system has no solution. In the next example, note that the occurrence of a statement that is true for all values of the variables, such as $0 = 0$, indicates that the system has infinitely many solutions.

Example 6   Many-Solution Case: Method of Elimination

Solve the system of linear equations.

$$\begin{cases} 2x - y = 1 & \text{Equation 1} \\ 4x - 2y = 2 & \text{Equation 2} \end{cases}$$

Solution

To obtain coefficients that differ only in sign, multiply Equation 2 by $-\frac{1}{2}$.

$2x - y = 1$	$2x - y = 1$	Write Equation 1.
$4x - 2y = 2$	$-2x + y = -1$	Multiply Equation 2 by $-\frac{1}{2}$.
	$0 = 0$	Add equations.

Because the two equations turn out to be equivalent (have the same solution set), you can conclude that the system has infinitely many solutions. The solution set consists of all points (x, y) lying on the line $2x - y = 1$, as shown in Figure 6.9. Letting $x = a$, where a is any real number, you can see that the solutions to the system are $(a, 2a - 1)$.

✓**CHECKPOINT**   Now try Exercise 23.

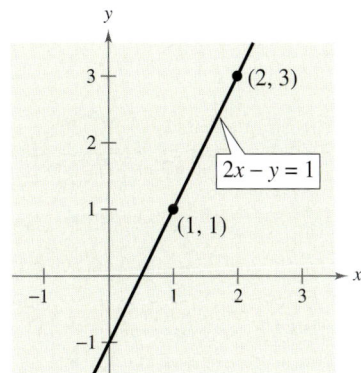

FIGURE **6.9**

Example 7 illustrates a strategy for solving a system of linear equations that has decimal coefficients.

Example 7 **A Linear System Having Decimal Coefficients**

Solve the system of linear equations.

$$\begin{cases} 0.02x - 0.05y = -0.38 & \text{Equation 1} \\ 0.03x + 0.04y = 1.04 & \text{Equation 2} \end{cases}$$

Solution

Because the coefficients in this system have two decimal places, you can begin by multiplying each equation by 100. This produces a system in which the coefficients are all integers.

$$\begin{cases} 2x - 5y = -38 & \text{Revised Equation 1} \\ 3x + 4y = 104 & \text{Revised Equation 2} \end{cases}$$

Now, to obtain coefficients that differ only in sign, multiply Equation 1 by 3 and multiply Equation 2 by -2.

$$
\begin{array}{ll}
2x - 5y = -38 & \quad 6x - 15y = -114 \quad \text{Multiply Equation 1 by 3.} \\
3x + 4y = 104 & \quad \underline{-6x - 8y = -208} \quad \text{Multiply Equation 2 by } -2. \\
& \qquad\quad\; - 23y = -322 \quad \text{Add equations.}
\end{array}
$$

So, you can conclude that

$$y = \frac{-322}{-23}$$
$$= 14.$$

Back-substituting this value into revised Equation 2 produces the following.

$$
\begin{array}{ll}
3x + 4y = 104 & \text{Write revised Equation 2.} \\
3x + 4(14) = 104 & \text{Substitute 14 for } y. \\
3x = 48 & \text{Combine like terms.} \\
x = 16 & \text{Solve for } x.
\end{array}
$$

The solution is $(16, 14)$. Check this in the original system, as follows.

Check

$$
\begin{array}{ll}
0.02x - 0.05y = -0.38 & \text{Write original Equation 1.} \\
0.02(16) - 0.05(14) \stackrel{?}{=} -0.38 & \text{Substitute into Equation 1.} \\
0.32 - 0.70 = -0.38 & \text{Equation 1 checks. } \checkmark \\
0.03x + 0.04y = 1.04 & \text{Write original Equation 2.} \\
0.03(16) + 0.04(14) \stackrel{?}{=} 1.04 & \text{Substitute into Equation 2.} \\
0.48 + 0.56 = 1.04 & \text{Equation 2 checks. } \checkmark
\end{array}
$$

✓CHECKPOINT Now try Exercise 25.

Applications

At this point, you may be asking the question "How can I tell which application problems can be solved using a system of linear equations?" The answer comes from the following considerations.

1. Does the problem involve more than one unknown quantity?

2. Are there two (or more) equations or conditions to be satisfied?

If one or both of these situations occur, the appropriate mathematical model for the problem may be a system of linear equations.

Example 8   An Application of a Linear System

An airplane flying into a headwind travels the 2000-mile flying distance between Chicopee, Massachusetts and Salt Lake City, Utah in 4 hours and 24 minutes. On the return flight, the same distance is traveled in 4 hours. Find the airspeed of the plane and the speed of the wind, assuming that both remain constant.

Solution

The two unknown quantities are the speeds of the wind and the plane. If r_1 is the speed of the plane and r_2 is the speed of the wind, then

$$r_1 - r_2 = \text{speed of the plane against the wind}$$

$$r_1 + r_2 = \text{speed of the plane with the wind}$$

as shown in Figure 6.10. Using the formula distance $= (\text{rate})(\text{time})$ for these two speeds, you obtain the following equations.

$$2000 = (r_1 - r_2)\left(4 + \frac{24}{60}\right)$$

$$2000 = (r_1 + r_2)(4)$$

These two equations simplify as follows.

$$\begin{cases} 5000 = 11r_1 - 11r_2 & \text{Equation 1} \\ 500\ \ =\ \ r_1 + \ \ r_2 & \text{Equation 2} \end{cases}$$

To solve this system by elimination, multiply Equation 2 by 11.

$$5000 = 11r_1 - 11r_2 \quad\Longrightarrow\quad 5000 = 11r_1 - 11r_2 \qquad \text{Write Equation 1.}$$

$$\underline{500 = \ \ r_1 + \ \ r_2} \quad\Longrightarrow\quad \underline{5500 = 11r_1 + 11r_2} \qquad \text{Multiply Equation 2 by 11.}$$

$$10{,}500 = 22r_1 \qquad \text{Add equations.}$$

So,

$$r_1 = \frac{10{,}500}{22} = \frac{5250}{11} \approx 477.27 \text{ miles per hour} \qquad \text{Speed of plane}$$

$$r_2 = 500 - \frac{5250}{11} = \frac{250}{11} \approx 22.73 \text{ miles per hour.} \qquad \text{Speed of wind}$$

Check this solution in the original statement of the problem.

✔**CHECKPOINT**   Now try Exercise 43.

Original flight

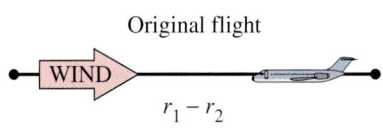

$r_1 - r_2$

Return flight

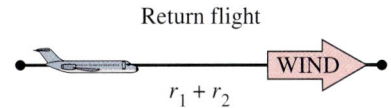

$r_1 + r_2$

FIGURE **6.10**

In a free market, the demands for many products are related to the prices of the products. As the prices decrease, the demands by consumers increase and the amounts that producers are able or willing to supply decrease.

Example 9   Finding the Equilibrium Point

The demand and supply functions for a new type of personal digital assistant are

$$\begin{cases} p = 150 - 0.00001x & \text{Demand equation} \\ p = 60 + 0.00002x & \text{Supply equation} \end{cases}$$

where p is the price in dollars and x represents the number of units. Find the equilibrium point for this market. The **equilibrium point** is the price p and number of units x that satisfy both the demand and supply equations.

Solution

Because p is written in terms of x, begin by substituting the value of p given in the supply equation into the demand equation.

$$p = 150 - 0.00001x \qquad \text{Write demand equation.}$$
$$60 + 0.00002x = 150 - 0.00001x \qquad \text{Substitute } 60 + 0.00002x \text{ for } p.$$
$$0.00003x = 90 \qquad \text{Combine like terms.}$$
$$x = 3,000,000 \qquad \text{Solve for } x.$$

So, the equilibrium point occurs when the demand and supply are each 3 million units. (See Figure 6.11.) The price that corresponds to this x-value is obtained by back-substituting $x = 3,000,000$ into either of the original equations. For instance, back-substituting into the demand equation produces

$$p = 150 - 0.00001(3,000,000)$$
$$= 150 - 30$$
$$= \$120.$$

The solution is $(3,000,000, 120)$. You can check this as follows.

Check

Substitute $(3,000,000, 120)$ into the demand equation.

$$p = 150 - 0.00001x \qquad \text{Write demand equation.}$$
$$120 \overset{?}{=} 150 - 0.00001(3,000,000) \qquad \text{Substitute 120 for } p \text{ and 3,000,000 for } x.$$
$$120 = 120 \qquad \text{Solution checks in demand equation.} ✓$$

Substitute $(3,000,000, 120)$ into the supply equation.

$$p = 60 + 0.00002x \qquad \text{Write supply equation.}$$
$$120 \overset{?}{=} 60 + 0.00002(3,000,000) \qquad \text{Substitute 120 for } p \text{ and 3,000,000 for } x.$$
$$120 = 120 \qquad \text{Solution checks in supply equation.} ✓$$

✔CHECKPOINT   Now try Exercise 45.

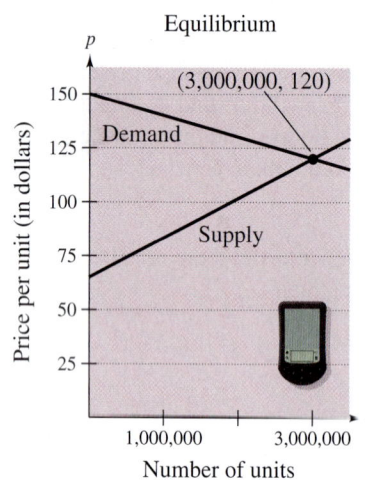

Equilibrium
(3,000,000, 120)
Demand
Supply
Price per unit (in dollars)
150
125
100
75
50
25
1,000,000   3,000,000
Number of units

FIGURE **6.11**

6.2 | Exercises

VOCABULARY CHECK: Fill in the blanks.

1. The first step in solving a system of equations by the method of _____ is to obtain coefficients for x (or y) that differ only in sign.

2. Two systems of equations that have the same solution set are called _____ systems.

3. A system of linear equations that has at least one solution is called _____, whereas a system of linear equations that has no solution is called _____.

4. In business applications, the _____ _____ is defined as the price p and the number of units x that satisfy both the demand and supply equations.

PREREQUISITE SKILLS REVIEW: Practice and review algebra skills needed for this section at **www.Eduspace.com.**

In Exercises 1–10, solve the system by the method of elimination. Label each line with its equation. To print an enlarged copy of the graph, go to the website **www.mathgraphs.com.**

1. $\begin{cases} 2x + y = 5 \\ x - y = 1 \end{cases}$

2. $\begin{cases} x + 3y = 1 \\ -x + 2y = 4 \end{cases}$

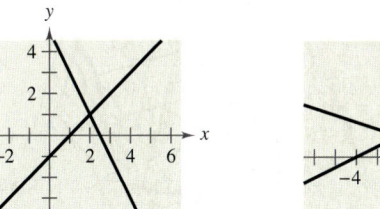

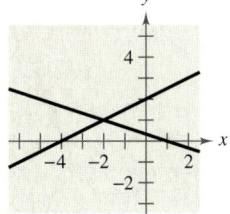

3. $\begin{cases} x + y = 0 \\ 3x + 2y = 1 \end{cases}$

4. $\begin{cases} 2x - y = 3 \\ 4x + 3y = 21 \end{cases}$

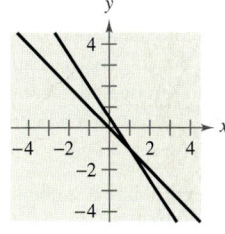

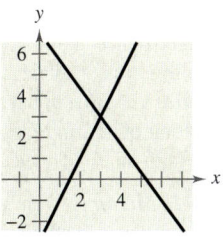

5. $\begin{cases} x - y = 2 \\ -2x + 2y = 5 \end{cases}$

6. $\begin{cases} 3x + 2y = 3 \\ 6x + 4y = 14 \end{cases}$

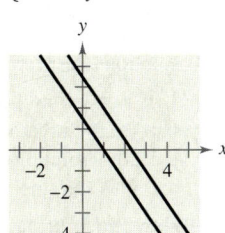

7. $\begin{cases} 3x - 2y = 5 \\ -6x + 4y = -10 \end{cases}$

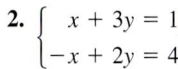

8. $\begin{cases} 9x - 3y = -15 \\ -3x + y = 5 \end{cases}$

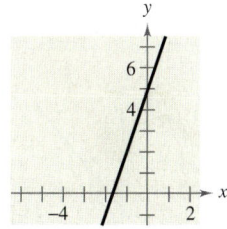

9. $\begin{cases} 9x + 3y = 1 \\ 3x - 6y = 5 \end{cases}$

10. $\begin{cases} 5x + 3y = -18 \\ 2x - 6y = 1 \end{cases}$

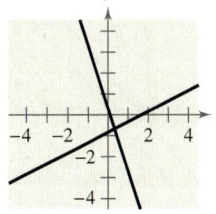

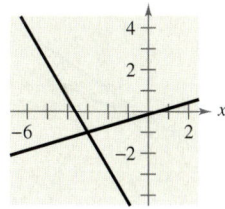

In Exercises 11–30, solve the system by the method of elimination and check any solutions algebraically.

11. $\begin{cases} x + 2y = 4 \\ x - 2y = 1 \end{cases}$

12. $\begin{cases} 3x - 5y = 2 \\ 2x + 5y = 13 \end{cases}$

13. $\begin{cases} 2x + 3y = 18 \\ 5x - y = 11 \end{cases}$

14. $\begin{cases} x + 7y = 12 \\ 3x - 5y = 10 \end{cases}$

15. $\begin{cases} 3x + 2y = 10 \\ 2x + 5y = 3 \end{cases}$

16. $\begin{cases} 2r + 4s = 5 \\ 16r + 50s = 55 \end{cases}$

17. $\begin{cases} 5u + 6v = 24 \\ 3u + 5v = 18 \end{cases}$

18. $\begin{cases} 3x + 11y = 4 \\ -2x - 5y = 9 \end{cases}$

19. $\begin{cases} \frac{9}{5}x + \frac{6}{5}y = 4 \\ 9x + 6y = 3 \end{cases}$

20. $\begin{cases} \frac{3}{4}x + y = \frac{1}{8} \\ \frac{9}{4}x + 3y = \frac{3}{8} \end{cases}$

21. $\begin{cases} \dfrac{x}{4} + \dfrac{y}{6} = 1 \\ x - y = 3 \end{cases}$

22. $\begin{cases} \dfrac{2}{3}x + \dfrac{1}{6}y = \dfrac{2}{3} \\ 4x + \ y = 4 \end{cases}$

23. $\begin{cases} -5x + \ 6y = -3 \\ 20x - 24y = \ 12 \end{cases}$

24. $\begin{cases} 7x + \ 8y = \ \ 6 \\ -14x - 16y = -12 \end{cases}$

25. $\begin{cases} 0.05x - 0.03y = 0.21 \\ 0.07x + 0.02y = 0.16 \end{cases}$

26. $\begin{cases} 0.2x - 0.5y = -27.8 \\ 0.3x + 0.4y = \ \ 68.7 \end{cases}$

27. $\begin{cases} 4b + \ 3m = \ 3 \\ 3b + 11m = 13 \end{cases}$

28. $\begin{cases} 2x + 5y = \ 8 \\ 5x + 8y = 10 \end{cases}$

29. $\begin{cases} \dfrac{x + 3}{4} + \dfrac{y - 1}{3} = \ 1 \\ 2x - y = 12 \end{cases}$

30. $\begin{cases} \dfrac{x - 1}{2} + \dfrac{y + 2}{3} = 4 \\ x - 2y = 5 \end{cases}$

In Exercises 31–34, match the system of linear equations with its graph. Describe the number of solutions and state whether the system is consistent or inconsistent. [The graphs are labeled (a), (b), (c) and (d).]

(a)

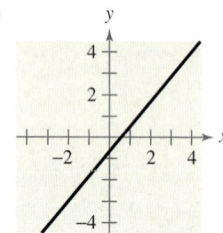

(b)

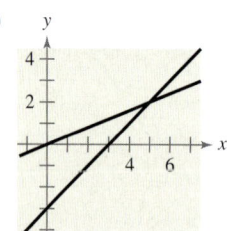

(c)

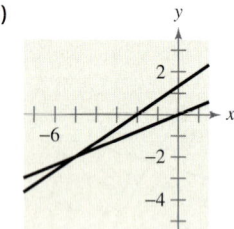

(d)

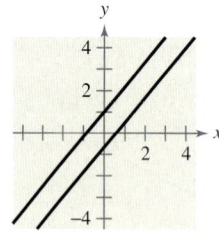

31. $\begin{cases} 2x - 5y = 0 \\ x - \ y = 3 \end{cases}$

32. $\begin{cases} -7x + \ 6y = -4 \\ 14x - 12y = \ \ 8 \end{cases}$

33. $\begin{cases} 2x - 5y = \ \ 0 \\ 2x - 3y = -4 \end{cases}$

34. $\begin{cases} 7x - 6y = -6 \\ -7x + 6y = -4 \end{cases}$

In Exercises 35–42, use any method to solve the system.

35. $\begin{cases} 3x - 5y = 7 \\ 2x + \ y = 9 \end{cases}$

36. $\begin{cases} -x + 3y = 17 \\ 4x + 3y = \ 7 \end{cases}$

37. $\begin{cases} y = 2x - \ 5 \\ y = 5x - 11 \end{cases}$

38. $\begin{cases} 7x + 3y = 16 \\ y = x + 2 \end{cases}$

39. $\begin{cases} x - 5y = 21 \\ 6x + 5y = 21 \end{cases}$

40. $\begin{cases} y = -3x - 8 \\ y = 15 - 2x \end{cases}$

41. $\begin{cases} -2x + 8y = 19 \\ y = x - 3 \end{cases}$

42. $\begin{cases} 4x - 3y = \ \ 6 \\ -5x + 7y = -1 \end{cases}$

43. *Airplane Speed* An airplane flying into a headwind travels the 1800-mile flying distance between Pittsburgh, Pennsylvania and Phoenix, Arizona in 3 hours and 36 minutes. On the return flight, the distance is traveled in 3 hours. Find the airspeed of the plane and the speed of the wind, assuming that both remain constant.

44. *Airplane Speed* Two planes start from Los Angeles International Airport and fly in opposite directions. The second plane starts $\frac{1}{2}$ hour after the first plane, but its speed is 80 kilometers per hour faster. Find the airspeed of each plane if 2 hours after the first plane departs the planes are 3200 kilometers apart.

Supply and Demand In Exercises 45–48, find the equilibrium point of the demand and supply equations. The equilibrium point is the price *p* and number of units *x* that satisfy both the demand and supply equations.

Demand	Supply
45. $p = 50 - 0.5x$	$p = 0.125x$
46. $p = 100 - 0.05x$	$p = 25 + 0.1x$
47. $p - 140 - 0.00002x$	$p = 80 + 0.00001x$
48. $p = 400 - 0.0002x$	$p = 225 + 0.0005x$

49. *Nutrition* Two cheeseburgers and one small order of French fries from a fast-food restaurant contain a total of 850 calories. Three cheeseburgers and two small orders of French fries contain a total of 1390 calories. Find the caloric content of each item.

50. *Nutrition* One eight-ounce glass of apple juice and one eight-ounce glass of orange juice contain a total of 185 milligrams of vitamin C. Two eight-ounce glasses of apple juice and three eight-ounce glasses of orange juice contain a total of 452 milligrams of vitamin C. How much vitamin C is in an eight-ounce glass of each type of juice?

51. *Acid Mixture* Ten liters of a 30% acid solution is obtained by mixing a 20% solution with a 50% solution.

(a) Write a system of equations in which one equation represents the amount of final mixture required and the other represents the percent of acid in the final mixture. Let *x* and *y* represent the amounts of the 20% and 50% solutions, respectively.

 (b) Use a graphing utility to graph the two equations in part (a) in the same viewing window. As the amount of the 20% solution increases, how does the amount of the 50% solution change?

(c) How much of each solution is required to obtain the specified concentration of the final mixture?

52. *Fuel Mixture*   Five hundred gallons of 89 octane gasoline is obtained by mixing 87 octane gasoline with 92 octane gasoline.

(a) Write a system of equations in which one equation represents the amount of final mixture required and the other represents the amounts of 87 and 92 octane gasolines in the final mixture. Let x and y represent the numbers of gallons of 87 octane and 92 octane gasolines, respectively.

(b) Use a graphing utility to graph the two equations in part (a) in the same viewing window. As the amount of 87 octane gasoline increases, how does the amount of 92 octane gasoline change?

(c) How much of each type of gasoline is required to obtain the 500 gallons of 89 octane gasoline?

53. *Investment Portfolio*   A total of $12,000 is invested in two corporate bonds that pay 7.5% and 9% simple interest. The investor wants an annual interest income of $990 from the investments. What amount should be invested in the 7.5% bond?

54. *Investment Portfolio*   A total of $32,000 is invested in two municipal bonds that pay 5.75% and 6.25% simple interest. The investor wants an annual interest income of $1900 from the investments. What amount should be invested in the 5.75% bond?

55. *Ticket Sales*   At a local high school city championship basketball game, 1435 tickets were sold. A student admission ticket cost $1.50 and an adult admission ticket cost $5.00. The sum of all the total ticket receipts for the basketball game were $3552.50. How many of each type of ticket were sold?

56. *Consumer Awareness*   A department store held a sale to sell all of the 214 winter jackets that remained after the season ended. Until noon, each jacket in the store was priced at $31.95. At noon, the price of the jackets was further reduced to $18.95. After the last jacket was sold, total receipts for the clearance sale were $5108.30. How many jackets were sold before noon and how many were sold after noon?

Fitting a Line to Data   **In Exercises 57–62, find the least squares regression line $y = ax + b$ for the points**

$$(x_1, y_1), (x_2, y_2), \ldots, (x_n, y_n)$$

by solving the system for a and b.

$$nb + \left(\sum_{i=1}^{n} x_i \right) a = \left(\sum_{i=1}^{n} y_i \right)$$

$$\left(\sum_{i=1}^{n} x_i \right) b + \left(\sum_{i=1}^{n} x_i^2 \right) a = \left(\sum_{i=1}^{n} x_i y_i \right)$$

Then use a graphing utility to confirm the result. (If you are unfamiliar with summation notation, look at the discussion in Section 8.1 or in Appendix A at the website for this text at *college.hmco.com*.)

57. $\begin{cases} 5b + 10a = 20.2 \\ 10b + 30a = 50.1 \end{cases}$ **58.** $\begin{cases} 5b + 10a = 11.7 \\ 10b + 30a = 25.6 \end{cases}$

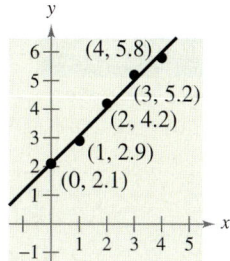
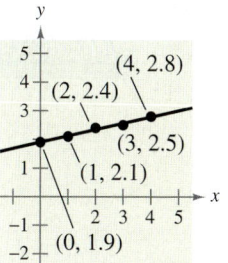

59. $\begin{cases} 7b + 21a = 35.1 \\ 21b + 91a = 114.2 \end{cases}$ **60.** $\begin{cases} 6b + 15a = 23.6 \\ 15b + 55a = 48.8 \end{cases}$

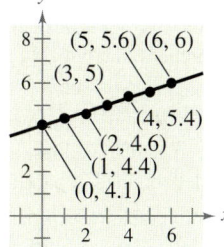
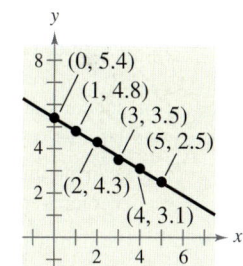

61. $(0, 4), (1, 3), (1, 1), (2, 0)$

62. $(1, 0), (2, 0), (3, 0), (3, 1), (4, 1), (4, 2), (5, 2), (6, 2)$

63. *Data Analysis*   A farmer used four test plots to determine the relationship between wheat yield y (in bushels per acre) and the amount of fertilizer x (in hundreds of pounds per acre). The results are shown in the table.

Fertilizer, x	Yield, y
1.0	32
1.5	41
2.0	48
2.5	53

(a) Use the technique demonstrated in Exercises 57–62 to set up a system of equations for the data and to find the least squares regression line $y = ax + b$.

(b) Use the linear model to predict the yield for a fertilizer application of 160 pounds per acre.

Model It

64. *Data Analysis* The table shows the average room rates y for a hotel room in the United States for the years 1995 through 2001. (Source: American Hotel & Motel Association)

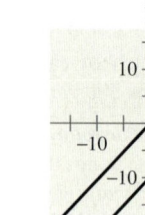

Year	Average room rate, y
1995	$66.65
1996	$70.93
1997	$75.31
1998	$78.62
1999	$81.33
2000	$85.89
2001	$88.27

(a) Use the technique demonstrated in Exercises 57–62 to set up a system of equations for the data and to find the least squares regression line $y = at + b$. Let t represent the year, with $t = 5$ corresponding to 1995.

 (b) Use the *regression* feature of a graphing utility to find a linear model for the data. How does this model compare with the model obtained in part (a)?

(c) Use the linear model to create a table of estimated values of y. Compare the estimated values with the actual data.

(d) Use the linear model to predict the average room rate in 2002. The actual average room rate in 2002 was $83.54. How does this value compare with your prediction?

(e) Use the linear model to predict when the average room rate will be $100.00. Using your result from part (d), do you think this prediction is accurate?

Synthesis

True or False? In Exercises 65 and 66, determine whether the statement is true or false. Justify your answer.

65. If two lines do not have exactly one point of intersection, then they must be parallel.

66. Solving a system of equations graphically will always give an exact solution.

67. *Writing* Briefly explain whether or not it is possible for a consistent system of linear equations to have exactly two solutions.

68. *Think About It* Give examples of a system of linear equations that has (a) no solution and (b) an infinite number of solutions.

Think About It In Exercises 69 and 70, the graphs of the two equations appear to be parallel. Yet, when the system is solved algebraically, you find that the system does have a solution. Find the solution and explain why it does not appear on the portion of the graph that is shown.

69. $\begin{cases} 100y - x = 200 \\ 99y - x = -198 \end{cases}$

70. $\begin{cases} 21x - 20y = 0 \\ 13x - 12y = 120 \end{cases}$

In Exercises 71 and 72, find the value of k such that the system of linear equations is inconsistent.

71. $\begin{cases} 4x - 8y = -3 \\ 2x + ky = 16 \end{cases}$

72. $\begin{cases} 15x + 3y = 6 \\ -10x + ky = 9 \end{cases}$

Skills Review

In Exercises 73–80, solve the inequality and graph the solution on the real number line.

73. $-11 - 6x \geq 33$

74. $2(x - 3) > -5x + 1$

75. $8x - 15 \leq -4(2x - 1)$

76. $-6 \leq 3x - 10 < 6$

77. $|x - 8| < 10$

78. $|x + 10| \geq -3$

79. $2x^2 + 3x - 35 < 0$

80. $3x^2 + 12x > 0$

In Exercises 81–84, write the expression as the logarithm of a single quantity.

81. $\ln x + \ln 6$

82. $\ln x - 5 \ln(x + 3)$

83. $\log_9 12 - \log_9 x$

84. $\frac{1}{4} \log_6 3x$

In Exercises 85 and 86, solve the system by the method of substitution.

85. $\begin{cases} 2x - y = 4 \\ -4x + 2y = -12 \end{cases}$

86. $\begin{cases} 30x - 40y - 33 = 0 \\ 10x + 20y - 21 = 0 \end{cases}$

87. Make a Decision To work an extended application analyzing the average undergraduate tuition, room, and board charges at private colleges in the United States from 1985 to 2003, visit this text's website at *college.hmco.com*. (Data Source: U.S. Dept. of Education)

6.3 Multivariable Linear Systems

What you should learn

- Use back-substitution to solve linear systems in row-echelon form.
- Use Gaussian elimination to solve systems of linear equations.
- Solve nonsquare systems of linear equations.
- Use systems of linear equations in three or more variables to model and solve real-life problems.

Why you should learn it

Systems of linear equations in three or more variables can be used to model and solve real-life problems. For instance, in Exercise 71 on page 483, a system of linear equations can be used to analyze the reproduction rates of deer in a wildlife preserve.

Jeanne Drake/Tony Stone Images

Row-Echelon Form and Back-Substitution

The method of elimination can be applied to a system of linear equations in more than two variables. In fact, this method easily adapts to computer use for solving linear systems with dozens of variables.

When elimination is used to solve a system of linear equations, the goal is to rewrite the system in a form to which back-substitution can be applied. To see how this works, consider the following two systems of linear equations.

System of Three Linear Equations in Three Variables: (See Example 3.)

$$\begin{cases} x - 2y + 3z = 9 \\ -x + 3y = -4 \\ 2x - 5y + 5z = 17 \end{cases}$$

Equivalent System in Row-Echelon Form: (See Example 1.)

$$\begin{cases} x - 2y + 3z = 9 \\ y + 3z = 5 \\ z = 2 \end{cases}$$

The second system is said to be in **row-echelon form,** which means that it has a "stair-step" pattern with leading coefficients of 1. After comparing the two systems, it should be clear that it is easier to solve the system in row-echelon form, using back-substitution.

Example 1   Using Back-Substitution in Row-Echelon Form

Solve the system of linear equations.

$$\begin{cases} x - 2y + 3z = 9 & \text{Equation 1} \\ y + 3z = 5 & \text{Equation 2} \\ z = 2 & \text{Equation 3} \end{cases}$$

Solution

From Equation 3, you know the value of z. To solve for y, substitute $z = 2$ into Equation 2 to obtain

$$y + 3(2) = 5 \qquad \text{Substitute 2 for } z.$$
$$y = -1. \qquad \text{Solve for } y.$$

Finally, substitute $y = -1$ and $z = 2$ into Equation 1 to obtain

$$x - 2(-1) + 3(2) = 9 \qquad \text{Substitute } -1 \text{ for } y \text{ and 2 for } z.$$
$$x = 1. \qquad \text{Solve for } x.$$

The solution is $x = 1$, $y = -1$, and $z = 2$, which can be written as the **ordered triple** $(1, -1, 2)$. Check this in the original system of equations.

✓**CHECKPOINT**   Now try Exercise 5.

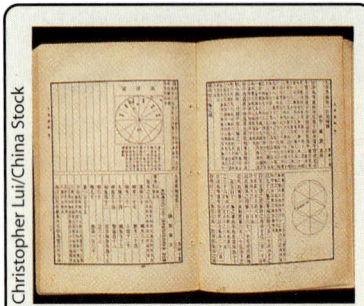

Christopher Lui/China Stock

Historical Note

One of the most influential Chinese mathematics books was the *Chui-chang suan-shu* or *Nine Chapters on the Mathematical Art* (written in approximately 250 B.C.). Chapter Eight of the *Nine Chapters* contained solutions of systems of linear equations using positive and negative numbers. One such system was as follows.

$$\begin{cases} 3x + 2y + z = 39 \\ 2x + 3y + z = 34 \\ x + 2y + 3z = 26 \end{cases}$$

This system was solved using column operations on a matrix. Matrices (plural for matrix) will be discussed in the next chapter.

Gaussian Elimination

Two systems of equations are *equivalent* if they have the same solution set. To solve a system that is not in row-echelon form, first convert it to an *equivalent* system that is in row-echelon form by using the following operations.

Operations That Produce Equivalent Systems

Each of the following **row operations** on a system of linear equations produces an *equivalent* system of linear equations.

1. Interchange two equations.

2. Multiply one of the equations by a nonzero constant.

3. Add a multiple of one of the equations to another equation to replace the latter equation.

To see how this is done, take another look at the method of elimination, as applied to a system of two linear equations.

Example 2 Using Gaussian Elimination to Solve a System

Solve the system of linear equations.

$$\begin{cases} 3x - 2y = -1 & \text{Equation 1} \\ x - y = 0 & \text{Equation 2} \end{cases}$$

Solution

There are two strategies that seem reasonable: eliminate the variable x or eliminate the variable y. The following steps show how to use the first strategy.

$$\begin{cases} x - y = 0 \\ 3x - 2y = -1 \end{cases}$$ Interchange the two equations in the system.

$$\begin{cases} -3x + 3y = 0 \\ 3x - 2y = -1 \end{cases}$$ Multiply the first equation by -3.

$$\begin{aligned} -3x + 3y &= 0 \\ \underline{3x - 2y} &= \underline{-1} \\ y &= -1 \end{aligned}$$ Add the multiple of the first equation to the second equation to obtain a new second equation.

$$\begin{cases} x - y = 0 \\ y = -1 \end{cases}$$ New system in row-echelon form

Now, using back-substitution, you can determine that the solution is $y = -1$ and $x = -1$, which can be written as the ordered pair $(-1, -1)$. Check this solution in the original system of equations.

STUDY TIP

As demonstrated in the first step in the solution of Example 2, interchanging rows is an easy way of obtaining a leading coefficient of 1.

CHECK Point Now try Exercise 13.

As shown in Example 2, rewriting a system of linear equations in row-echelon form usually involves a chain of equivalent systems, each of which is obtained by using one of the three basic row operations listed on the previous page. This process is called **Gaussian elimination,** after the German mathematician Carl Friedrich Gauss (1777–1855).

Example 3   Using Gaussian Elimination to Solve a System

Solve the system of linear equations.

$$\begin{cases} x - 2y + 3z = 9 & \text{Equation 1} \\ -x + 3y \quad\;\; = -4 & \text{Equation 2} \\ 2x - 5y + 5z = 17 & \text{Equation 3} \end{cases}$$

STUDY TIP

Arithmetic errors are often made when performing elementary row operations. You should note the operation performed in each step so that you can go back and check your work.

Solution

Because the leading coefficient of the first equation is 1, you can begin by saving the x at the upper left and eliminating the other x-terms from the first column.

$$\begin{array}{ll} x - 2y + 3z = 9 & \text{Write Equation 1.} \\ \underline{-x + 3y \quad\;\; = -4} & \text{Write Equation 2.} \\ y + 3z = 5 & \text{Add Equation 1 to Equation 2.} \end{array}$$

$$\begin{cases} x - 2y + 3z = 9 \\ \quad\quad y + 3z = 5 \\ 2x - 5y + 5z = 17 \end{cases}$$

Adding the first equation to the second equation produces a new second equation.

$$\begin{array}{ll} -2x + 4y - 6z = -18 & \text{Multiply Equation 1 by } -2. \\ \underline{2x - 5y + 5z = \quad 17} & \text{Write Equation 3.} \\ -y - z = -1 & \text{Add revised Equation 1 to Equation 3.} \end{array}$$

$$\begin{cases} x - 2y + 3z = 9 \\ \quad\quad y + 3z = 5 \\ \quad\; -y - z = -1 \end{cases}$$

Adding -2 times the first equation to the third equation produces a new third equation.

Now that all but the first x have been eliminated from the first column, go to work on the second column. (You need to eliminate y from the third equation.)

$$\begin{cases} x - 2y + 3z = 9 \\ \quad\quad y + 3z = 5 \\ \quad\quad\quad 2z = 4 \end{cases}$$

Adding the second equation to the third equation produces a new third equation.

Finally, you need a coefficient of 1 for z in the third equation.

$$\begin{cases} x - 2y + 3z = 9 \\ \quad\quad y + 3z = 5 \\ \quad\quad\quad z = 2 \end{cases}$$

Multiplying the third equation by $\frac{1}{2}$ produces a new third equation.

This is the same system that was solved in Example 1, and, as in that example, you can conclude that the solution is

$$x = 1, \qquad y = -1, \qquad \text{and} \qquad z = 2.$$

✔**CHECKPOINT**   Now try Exercise 15.

The next example involves an inconsistent system—one that has no solution. The key to recognizing an inconsistent system is that at some stage in the elimination process you obtain a false statement such as $0 = -2$.

Example 4 An Inconsistent System

Solve the system of linear equations.

$$\begin{cases} x - 3y + z = 1 & \text{Equation 1} \\ 2x - y - 2z = 2 & \text{Equation 2} \\ x + 2y - 3z = -1 & \text{Equation 3} \end{cases}$$

Solution

$$\begin{cases} x - 3y + z = 1 \\ 5y - 4z = 0 \\ x + 2y - 3z = -1 \end{cases}$$

Adding -2 times the first equation to the second equation produces a new second equation.

$$\begin{cases} x - 3y + z = 1 \\ 5y - 4z = 0 \\ 5y - 4z = -2 \end{cases}$$

Adding -1 times the first equation to the third equation produces a new third equation.

$$\begin{cases} x - 3y + z = 1 \\ 5y - 4z = 0 \\ 0 = -2 \end{cases}$$

Adding -1 times the second equation to the third equation produces a new third equation.

Because $0 = -2$ is a false statement, you can conclude that this system is inconsistent and so has no solution. Moreover, because this system is equivalent to the original system, you can conclude that the original system also has no solution.

✔**CHECKPOINT**   Now try Exercise 19.

As with a system of linear equations in two variables, the solution(s) of a system of linear equations in more than two variables must fall into one of three categories.

> ### The Number of Solutions of a Linear System
>
> For a system of linear equations, exactly one of the following is true.
>
> **1.** There is exactly one solution.
>
> **2.** There are infinitely many solutions.
>
> **3.** There is no solution.

In Section 6.2, you learned that a system of two linear equations in two variables can be represented graphically as a pair of lines that are intersecting, coincident, or parallel. A system of three linear equations in three variables has a similar graphical representation—it can be represented as three planes in space that intersect in one point (exactly one solution) [see Figure 6.12], intersect in a line or a plane (infinitely many solutions) [see Figures 6.13 and 6.14], or have no points common to all three planes (no solution) [see Figures 6.15 and 6.16].

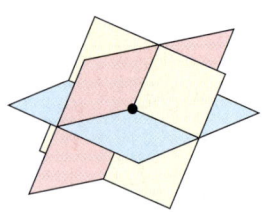

FIGURE 6.12   *Solution: one point*

FIGURE 6.13   *Solution: one line*

FIGURE 6.14   *Solution: one plane*

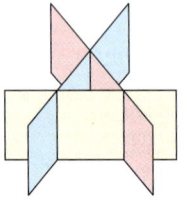

FIGURE 6.15   *Solution: none*

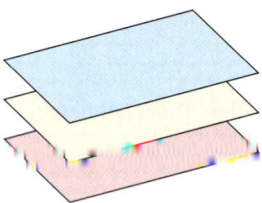

FIGURE 6.16   *Solution: none*

Example 5   **A System with Infinitely Many Solutions**

Solve the system of linear equations.

$$\begin{cases} x + y - 3z = -1 & \text{Equation 1} \\ \quad\quad y - z = 0 & \text{Equation 2} \\ -x + 2y = 1 & \text{Equation 3} \end{cases}$$

Solution

$$\begin{cases} x + y - 3z = -1 \\ \quad\quad y - z = 0 \\ \quad\quad 3y - 3z = 0 \end{cases}$$

> Adding the first equation to the third equation produces a new third equation.

$$\begin{cases} x + y - 3z = -1 \\ \quad\quad y - z = 0 \\ \quad\quad\quad 0 = 0 \end{cases}$$

> Adding -3 times the second equation to the third equation produces a new third equation.

This result means that Equation 3 depends on Equations 1 and 2 in the sense that it gives no additional information about the variables. Because $0 = 0$ is a true statement, you can conclude that this system will have infinitely many solutions. However, it is incorrect to say simply that the solution is "infinite." You must also specify the correct form of the solution. So, the original system is equivalent to the system

$$\begin{cases} x + y - 3z = -1 \\ \quad\quad y - z = 0 \end{cases}.$$

In the last equation, solve for y in terms of z to obtain $y = z$. Back-substituting for y in the first equation produces $x = 2z - 1$. Finally, letting $z = a$, where a is a real number, the solutions to the given system are all of the form $x = 2a - 1$, $y = a$, and $z = a$. So, every ordered triple of the form

$$(2a - 1, a, a), \quad a \text{ is a real number}$$

is a solution of the system.

✔**CHECKPOINT**   Now try Exercise 23.

In Example 5, there are other ways to write the same infinite set of solutions. For instance, letting $x = b$, the solutions could have been written as

$$\left(b, \tfrac{1}{2}(b + 1), \tfrac{1}{2}(b + 1)\right), \quad b \text{ is a real number.}$$

To convince yourself that this description produces the same set of solutions, consider the following.

Substitution	*Solution*	
$a = 0$	$(2(0) - 1, 0, 0) = (-1, 0, 0)$	Same solution
$b = -1$	$\left(-1, \tfrac{1}{2}(-1 + 1), \tfrac{1}{2}(-1 + 1)\right) = (-1, 0, 0)$	
$a = 1$	$(2(1) - 1, 1, 1) = (1, 1, 1)$	Same solution
$b = 1$	$\left(1, \tfrac{1}{2}(1 + 1), \tfrac{1}{2}(1 + 1)\right) = (1, 1, 1)$	
$a = 2$	$(2(2) - 1, 2, 2) = (3, 2, 2)$	Same solution
$b = 3$	$\left(3, \tfrac{1}{2}(3 + 1), \tfrac{1}{2}(3 + 1)\right) = (3, 2, 2)$	

STUDY TIP

In Example 5, x and y are solved in terms of the third variable z. To write the correct form of the solution to the system that does not use any of the three variables of the system, let a represent any real number and let $z = a$. Then solve for x and y. The solution can then be written in terms of a, which is not one of the variables of the system.

STUDY TIP

When comparing descriptions of an infinite solution set, keep in mind that there is more than one way to describe the set.

Nonsquare Systems

So far, each system of linear equations you have looked at has been *square*, which means that the number of equations is equal to the number of variables. In a **nonsquare** system, the number of equations differs from the number of variables. A system of linear equations cannot have a unique solution unless there are at least as many equations as there are variables in the system.

Example 6 **A System with Fewer Equations than Variables**

Solve the system of linear equations.

$$\begin{cases} x - 2y + z = 2 & \text{Equation 1} \\ 2x - y - z = 1 & \text{Equation 2} \end{cases}$$

Solution

Begin by rewriting the system in row-echelon form.

$$\begin{cases} x - 2y + z = 2 \\ 3y - 3z = -3 \end{cases}$$

> Adding -2 times the first equation to the second equation produces a new second equation.

$$\begin{cases} x - 2y + z = 2 \\ y - z = -1 \end{cases}$$

> Multiplying the second equation by $\frac{1}{3}$ produces a new second equation.

Solve for y in terms of z, to obtain

$$y = z - 1.$$

By back-substituting into Equation 1, you can solve for x, as follows.

$$x - 2y + z = 2 \qquad \text{Write Equation 1.}$$
$$x - 2(z - 1) + z = 2 \qquad \text{Substitute for } y \text{ in Equation 1.}$$
$$x - 2z + 2 + z = 2 \qquad \text{Distributive Property}$$
$$x = z \qquad \text{Solve for } x.$$

Finally, by letting $z = a$, where a is a real number, you have the solution

$$x = a, \qquad y = a - 1, \qquad \text{and} \qquad z = a.$$

So, every ordered triple of the form

$$(a, a - 1, a), \qquad a \text{ is a real number}$$

is a solution of the system. Because there were originally three variables and only two equations, the system cannot have a unique solution.

✓CHECKPOINT Now try Exercise 27.

In Example 6, try choosing some values of a to obtain different solutions of the system, such as $(1, 0, 1)$, $(2, 1, 2)$, and $(3, 2, 3)$. Then check each of the solutions in the original system to verify that they are solutions of the original system.

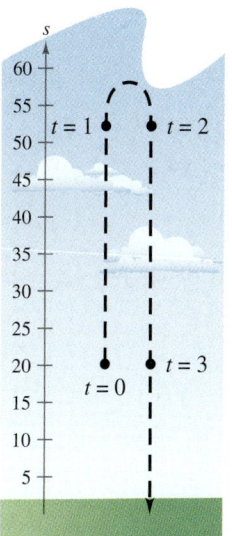

FIGURE **6.17**

Applications

Example 7   **Vertical Motion**

The height at time t of an object that is moving in a (vertical) line with constant acceleration a is given by the **position equation**

$$s = \tfrac{1}{2}at^2 + v_0 t + s_0.$$

The height s is measured in feet, the acceleration a is measured in feet per second squared, t is measured in seconds, v_0 is the initial velocity (at $t = 0$), and s_0 is the initial height. Find the values of a, v_0, and s_0 if $s = 52$ at $t = 1$, $s = 52$ at $t = 2$, and $s = 20$ at $t = 3$, and interpret the result. (See Figure 6.17.)

Solution

By substituting the three values of t and s into the position equation, you can obtain three linear equations in a, v_0, and s_0.

When $t = 1$: $\tfrac{1}{2}a(1)^2 + v_0(1) + s_0 = 52$ ⟹ $a + 2v_0 + 2s_0 = 104$

When $t = 2$: $\tfrac{1}{2}a(2)^2 + v_0(2) + s_0 = 52$ ⟹ $2a + 2v_0 + s_0 = 52$

When $t = 3$: $\tfrac{1}{2}a(3)^2 + v_0(3) + s_0 = 20$ ⟹ $9a + 6v_0 + 2s_0 = 40$

This produces the following system of linear equations.

$$\begin{cases} a + 2v_0 + 2s_0 = 104 \\ 2a + 2v_0 + s_0 = 52 \\ 9a + 6v_0 + 2s_0 = 40 \end{cases}$$

Now solve the system using Gaussian elimination.

$$\begin{cases} a + 2v_0 + 2s_0 = 104 \\ - 2v_0 - 3s_0 = -156 \\ 9a + 6v_0 + 2s_0 = 40 \end{cases}$$

> Adding -2 times the first equation to the second equation produces a new second equation.

$$\begin{cases} a + 2v_0 + 2s_0 = 104 \\ - 2v_0 - 3s_0 = -156 \\ - 12v_0 - 16s_0 = -896 \end{cases}$$

> Adding -9 times the first equation to the third equation produces a new third equation.

$$\begin{cases} a + 2v_0 + 2s_0 = 104 \\ - 2v_0 - 3s_0 = -156 \\ 2s_0 = 40 \end{cases}$$

> Adding -6 times the second equation to the third equation produces a new third equation.

$$\begin{cases} a + 2v_0 + 2s_0 = 104 \\ v_0 + \tfrac{3}{2}s_0 = 78 \\ s_0 = 20 \end{cases}$$

> Multiplying the second equation by $-\tfrac{1}{2}$ produces a new second equation and multiplying the third equation by $\tfrac{1}{2}$ produces a new third equation.

So, the solution of this system is $a = -32$, $v_0 = 48$, and $s_0 = 20$. This solution results in a position equation of $s = -16t^2 + 48t + 20$ and implies that the object was thrown upward at a velocity of 48 feet per second from a height of 20 feet.

✓**CHECKPOINT**   Now try Exercise 39.

Example 8   **Data Analysis: Curve-Fitting**

Find a quadratic equation

$$y = ax^2 + bx + c$$

whose graph passes through the points $(-1, 3)$, $(1, 1)$, and $(2, 6)$.

Solution

Because the graph of $y = ax^2 + bx + c$ passes through the points $(-1, 3)$, $(1, 1)$, and $(2, 6)$, you can write the following.

When $x = -1$, $y = 3$: $a(-1)^2 + b(-1) + c = 3$

When $x = 1$, $y = 1$: $a(1)^2 + b(1) + c = 1$

When $x = 2$, $y = 6$: $a(2)^2 + b(2) + c = 6$

This produces the following system of linear equations.

$$\begin{cases} a - b + c = 3 & \text{Equation 1} \\ a + b + c = 1 & \text{Equation 2} \\ 4a + 2b + c = 6 & \text{Equation 3} \end{cases}$$

The solution of this system is $a = 2$, $b = -1$, and $c = 0$. So, the equation of the parabola is $y = 2x^2 - x$, as shown in Figure 6.18.

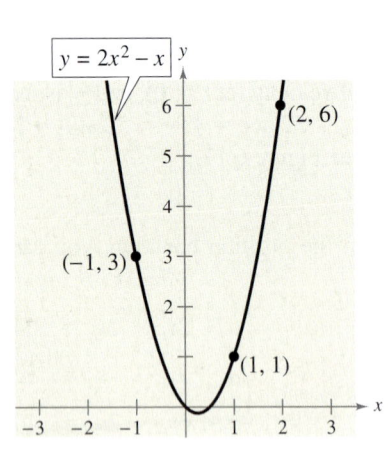

FIGURE 6.18

✓CHECKPOINT   Now try Exercise 43.

Example 9   **Investment Analysis**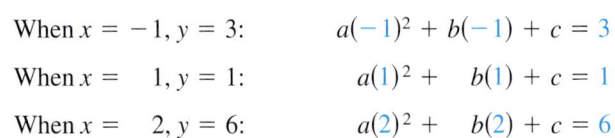

An inheritance of $12,000 was invested among three funds: a money-market fund that paid 5% annually, municipal bonds that paid 6% annually, and mutual funds that paid 12% annually. The amount invested in mutual funds was $4000 more than the amount invested in municipal bonds. The total interest earned during the first year was $1120. How much was invested in each type of fund?

Solution

Let x, y, and z represent the amounts invested in the money-market fund, municipal bonds, and mutual funds, respectively. From the given information, you can write the following equations.

$$\begin{cases} x + y + z = 12{,}000 & \text{Equation 1} \\ z = y + 4000 & \text{Equation 2} \\ 0.05x + 0.06y + 0.12z = 1120 & \text{Equation 3} \end{cases}$$

Rewriting this system in standard form without decimals produces the following.

$$\begin{cases} x + y + z = 12{,}000 & \text{Equation 1} \\ -y + z = 4{,}000 & \text{Equation 2} \\ 5x + 6y + 12z = 112{,}000 & \text{Equation 3} \end{cases}$$

Using Gaussian elimination to solve this system yields $x = 2000$, $y = 3000$, and $z = 7000$. So, $2000 was invested in the money-market fund, $3000 was invested in municipal bonds, and $7000 was invested in mutual funds.

✓CHECKPOINT   Now try Exercise 53.

6.3   Exercises

VOCABULARY CHECK: Fill in the blanks.

1. A system of equations that is in _____ form has a "stair-step" pattern with leading coefficients of 1.

2. A solution to a system of three linear equations in three unknowns can be written as an _____ _____, which has the form (x, y, z).

3. The process used to write a system of linear equations in row-echelon form is called _____ elimination.

4. Interchanging two equations of a system of linear equations is a _____ _____ that produces an equivalent system.

5. A system of equations is called _____ if the number of equations differs from the number of variables in the system.

6. The equation $s = \frac{1}{2}at^2 + v_0t + s_0$ is called the _____ equation, and it models the height s of an object at time t that is moving in a vertical line with a constant acceleration a.

PREREQUISITE SKILLS REVIEW: Practice and review algebra skills needed for this section at **www.Eduspace.com.**

In Exercises 1–4, determine whether each ordered triple is a solution of the system of equations.

1. $\begin{cases} 3x - y + z = 1 \\ 2x \quad\;\; - 3z = -14 \\ \quad\;\; 5y + 2z = 8 \end{cases}$

   (a) $(2, 0, -3)$ (b) $(-2, 0, 8)$
   (c) $(0, -1, 3)$ (d) $(-1, 0, 4)$

2. $\begin{cases} 3x + 4y - z = 17 \\ 5x - y + 2z = -2 \\ 2x - 3y + 7z = -21 \end{cases}$

   (a) $(3, -1, 2)$ (b) $(1, 3, -2)$
   (c) $(4, 1, -3)$ (d) $(1, -2, 2)$

3. $\begin{cases} 4x + y - z = 0 \\ -8x - 6y + z = -\frac{7}{4} \\ 3x - y \quad\;\; = -\frac{9}{4} \end{cases}$

   (a) $\left(\frac{1}{2}, -\frac{3}{4}, -\frac{7}{4}\right)$ (b) $\left(-\frac{3}{2}, \frac{5}{4}, -\frac{5}{4}\right)$
   (c) $\left(-\frac{1}{2}, \frac{3}{4}, -\frac{5}{4}\right)$ (d) $\left(-\frac{1}{2}, \frac{1}{6}, -\frac{3}{4}\right)$

4. $\begin{cases} -4x - y - 8z = -6 \\ \quad\;\; y + z = 0 \\ 4x - 7y \quad\;\; = 6 \end{cases}$

   (a) $(-2, -2, 2)$ (b) $\left(-\frac{33}{2}, -10, 10\right)$
   (c) $\left(\frac{1}{8}, -\frac{1}{2}, \frac{1}{2}\right)$ (d) $\left(-\frac{11}{2}, -4, 4\right)$

In Exercises 5–10, use back-substitution to solve the system of linear equations.

5. $\begin{cases} 2x - y + 5z = 24 \\ \quad\;\; y + 2z = 6 \\ \quad\;\; z = 4 \end{cases}$

6. $\begin{cases} 4x - 3y - 2z = 21 \\ \quad\;\; 6y - 5z = -8 \\ \quad\;\; z = -2 \end{cases}$

7. $\begin{cases} 2x + y - 3z = 10 \\ \quad\;\; y + z = 12 \\ \quad\;\; z = 2 \end{cases}$

8. $\begin{cases} x - y + 2z = 22 \\ \quad\;\; 3y - 8z = -9 \\ \quad\;\; z = -3 \end{cases}$

9. $\begin{cases} 4x - 2y + z = 8 \\ \quad\;\; -y + z = 4 \\ \quad\;\; z = 2 \end{cases}$

10. $\begin{cases} 5x \quad\;\; - 8z = 22 \\ \quad\;\; 3y - 5z = 10 \\ \quad\;\; z = -4 \end{cases}$

In Exercises 11 and 12, perform the row operation and write the equivalent system.

11. Add Equation 1 to Equation 2.

   $\begin{cases} x - 2y + 3z = 5 & \text{Equation 1} \\ -x + 3y - 5z = 4 & \text{Equation 2} \\ 2x \quad\;\; - 3z = 0 & \text{Equation 3} \end{cases}$

   What did this operation accomplish?

12. Add -2 times Equation 1 to Equation 3.

   $\begin{cases} x - 2y + 3z = 5 & \text{Equation 1} \\ -x + 3y - 5z = 4 & \text{Equation 2} \\ 2x \quad\;\; - 3z = 0 & \text{Equation 3} \end{cases}$

   What did this operation accomplish?

In Exercises 13–38, solve the system of linear equations and check any solution algebraically.

13. $\begin{cases} x + y + z = 6 \\ 2x - y + z = 3 \\ 3x \quad\;\; - z = 0 \end{cases}$

14. $\begin{cases} x + y + z = 3 \\ x - 2y + 4z = 5 \\ \quad\;\; 3y + 4z = 5 \end{cases}$

15. $\begin{cases} 2x \quad\quad + 2z = 2 \\ 5x + 3y \quad\;\; = 4 \\ \quad\;\; 3y - 4z = 4 \end{cases}$

16. $\begin{cases} 2x + 4y + z = 1 \\ x - 2y - 3z = 2 \\ x + y - z = -1 \end{cases}$

17. $\begin{cases} \quad\;\; 6y + 4z = -12 \\ 3x + 3y \quad\;\; = 9 \\ 2x \quad\quad - 3z = 10 \end{cases}$

18. $\begin{cases} 2x + 4y - z = 7 \\ 2x - 4y + 2z = -6 \\ x + 4y + z = 0 \end{cases}$

19. $\begin{cases} 2x + y - z = 7 \\ x - 2y + 2z = -9 \\ 3x - y + z = 5 \end{cases}$

20. $\begin{cases} 5x - 3y + 2z = 3 \\ 2x + 4y - z = 7 \\ x - 11y + 4z = 3 \end{cases}$

21. $\begin{cases} 3x - 5y + 5z = 1 \\ 5x - 2y + 3z = 0 \\ 7x - y + 3z = 0 \end{cases}$

22. $\begin{cases} 2x + y + 3z = 1 \\ 2x + 6y + 8z = 3 \\ 6x + 8y + 18z = 5 \end{cases}$

23. $\begin{cases} x + 2y - 7z = -4 \\ 2x + y + z = 13 \\ 3x + 9y - 36z = -33 \end{cases}$

24. $\begin{cases} 2x + y - 3z = 4 \\ 4x \quad\;\; + 2z = 10 \\ -2x + 3y - 13z = -8 \end{cases}$

25. $\begin{cases} 3x - 3y + 6z = 6 \\ x + 2y - z = 5 \\ 5x - 8y + 13z = 7 \end{cases}$

26. $\begin{cases} x \quad\quad + 2z = 5 \\ 3x - y - z = 1 \\ 6x - y + 5z = 16 \end{cases}$

27. $\begin{cases} x - 2y + 5z = 2 \\ 4x \quad\quad - z = 0 \end{cases}$

28. $\begin{cases} x - 3y + 2z = 18 \\ 5x - 13y + 12z = 80 \end{cases}$

29. $\begin{cases} 2x - 3y + z = -2 \\ -4x + 9y \quad\;\; = 7 \end{cases}$

30. $\begin{cases} 2x + 3y + 3z = 7 \\ 4x + 18y + 15z = 44 \end{cases}$

31. $\begin{cases} x \quad\quad\quad + 3w = 4 \\ 2y - z - w = 0 \\ 3y \quad\;\; - 2w = 1 \\ 2x - y + 4z \quad\;\; = 5 \end{cases}$

32. $\begin{cases} x + y + z + w = 6 \\ 2x + 3y \quad\;\; - w = 0 \\ -3x + 4y + z + 2w = 4 \\ x + 2y - z + w = 0 \end{cases}$

33. $\begin{cases} x \quad\quad + 4z = 1 \\ x + y + 10z = 10 \\ 2x - y + 2z = -5 \end{cases}$

34. $\begin{cases} 2x - 2y - 6z = -4 \\ -3x + 2y + 6z = 1 \\ x - y - 5z = -3 \end{cases}$

35. $\begin{cases} 2x + 3y \quad\;\; = 0 \\ 4x + 3y - z = 0 \\ 8x + 3y + 3z = 0 \end{cases}$

36. $\begin{cases} 4x + 3y + 17z = 0 \\ 5x + 4y + 22z = 0 \\ 4x + 2y + 19z = 0 \end{cases}$

37. $\begin{cases} 12x + 5y + z = 0 \\ 23x + 4y - z = 0 \end{cases}$

38. $\begin{cases} 2x - y - z = 0 \\ -2x + 6y + 4z = 2 \end{cases}$

Vertical Motion **In Exercises 39–42, an object moving vertically is at the given heights at the specified times. Find the position equation** $s = \frac{1}{2}at^2 + v_0t + s_0$ **for the object.**

39. At $t = 1$ second, $s = 128$ feet
    At $t = 2$ seconds, $s = 80$ feet
    At $t = 3$ seconds, $s = 0$ feet

40. At $t = 1$ second, $s = 48$ feet
    At $t = 2$ seconds, $s = 64$ feet
    At $t = 3$ seconds, $s = 48$ feet

41. At $t = 1$ second, $s = 452$ feet
    At $t = 2$ seconds, $s = 372$ feet
    At $t = 3$ seconds, $s = 260$ feet

42. At $t = 1$ second, $s = 132$ feet
    At $t = 2$ seconds, $s = 100$ feet
    At $t = 3$ seconds, $s = 36$ feet

In Exercises 43–46, find the equation of the parabola

$$y = ax^2 + bx + c$$

that passes through the points. To verify your result, use a graphing utility to plot the points and graph the parabola.

43. $(0, 0), (2, -2), (4, 0)$
44. $(0, 3), (1, 4), (2, 3)$
45. $(2, 0), (3, -1), (4, 0)$
46. $(1, 3), (2, 2), (3, -3)$

In Exercises 47–50, find the equation of the circle

$$x^2 + y^2 + Dx + Ey + F = 0$$

that passes through the points. To verify your result, use a graphing utility to plot the points and graph the circle.

47. $(0, 0), (2, 2), (4, 0)$
48. $(0, 0), (0, 6), (3, 3)$
49. $(-3, -1), (2, 4), (-6, 8)$
50. $(0, 0), (0, -2), (3, 0)$

51. *Sports*   In Super Bowl I, on January 15, 1967, the Green Bay Packers defeated the Kansas City Chiefs by a score of 35 to 10. The total points scored came from 13 different scoring plays, which were a combination of touchdowns, extra-point kicks, and field goals, worth 6, 1, and 3 points respectively. The same number of touchdowns and extra point kicks were scored. There were six times as many touchdowns as field goals. How many touchdowns, extra-point kicks, and field goals were scored during the game?   (Source: SuperBowl.com)

52. *Sports*   In the 2004 Women's NCAA Final Four Championship game, the University of Connecticut Huskies defeated the University of Tennessee Lady Volunteers by a score of 70 to 61. The Huskies won by scoring a combination of two-point baskets, three-point baskets, and one-point free throws. The number of two-point baskets was two more than the number of free throws. The number of free throws was one more than two times the number of three-point baskets. What combination of scoring accounted for the Huskies' 70 points?   (Source: National Collegiate Athletic Association)

53. *Finance*   A small corporation borrowed $775,000 to expand its clothing line. Some of the money was borrowed at 8%, some at 9%, and some at 10%. How much was borrowed at each rate if the annual interest owed was $67,500 and the amount borrowed at 8% was four times the amount borrowed at 10%?

54. *Finance*   A small corporation borrowed $800,000 to expand its line of toys. Some of the money was borrowed at 8%, some at 9%, and some at 10%. How much was borrowed at each rate if the annual interest owed was $67,000 and the amount borrowed at 8% was five times the amount borrowed at 10%?

Investment Portfolio   In Exercises 55 and 56, consider an investor with a portfolio totaling $500,000 that is invested in certificates of deposit, municipal bonds, blue-chip stocks, and growth or speculative stocks. How much is invested in each type of investment?

55. The certificates of deposit pay 10% annually, and the municipal bonds pay 8% annually. Over a five-year period, the investor expects the blue-chip stocks to return 12% annually and the growth stocks to return 13% annually. The investor wants a combined annual return of 10% and also wants to have only one-fourth of the portfolio invested in stocks.

56. The certificates of deposit pay 9% annually, and the municipal bonds pay 5% annually. Over a five-year period, the investor expects the blue-chip stocks to return 12% annually and the growth stocks to return 14% annually. The investor wants a combined annual return of 10% and also wants to have only one-fourth of the portfolio invested in stocks.

57. *Agriculture*   A mixture of 5 pounds of fertilizer A, 13 pounds of fertilizer B, and 4 pounds of fertilizer C provides the optimal nutrients for a plant. Commercial brand X contains equal parts of fertilizer B and fertilizer C. Commercial brand Y contains one part of fertilizer A and two parts of fertilizer B. Commercial brand Z contains two parts of fertilizer A, five parts of fertilizer B, and two parts of fertilizer C. How much of each fertilizer brand is needed to obtain the desired mixture?

58. *Agriculture*   A mixture of 12 liters of chemical A, 16 liters of chemical B, and 26 liters of chemical C is required to kill a destructive crop insect. Commercial spray X contains 1, 2, and 2 parts, respectively, of these chemicals. Commercial spray Y contains only chemical C. Commercial spray Z contains only chemicals A and B in equal amounts. How much of each type of commercial spray is needed to get the desired mixture?

59. *Coffee Mixture*   A coffee manufacturer sells a 10-pound package of coffee that consists of three flavors of coffee. Vanilla-flavored coffee costs $2 per pound, hazelnut-flavored coffee costs $2.50 per pound, and mocha-flavored coffee costs $3 per pound. The package contains the same amount of hazelnut coffee as mocha coffee. The cost of the 10-pound package is $26. How many pounds of each type of coffee are in the package?

60. *Floral Arrangements*   A florist is creating 10 centerpieces for a wedding. The florist can use roses that cost $2.50 each, lilies that cost $4 each, and irises that cost $2 each to make the bouquets. The customer has a budget of $300 and wants each bouquet to contain 12 flowers, with twice as many roses used as the other two types of flowers combined. How many of each type of flower should be in each centerpiece?

61. *Advertising* A health insurance company advertises on television, radio, and in the local newspaper. The marketing department has an advertising budget of $42,000 per month. A television ad costs $1000, a radio ad costs $200, and a newspaper ad costs $500. The department wants to run 60 ads per month, and have as many television ads as radio and newspaper ads combined. How many of each type of ad can the department run each month?

62. *Radio* You work as a disc jockey at your college radio station. You are supposed to play 32 songs within two hours. You are to choose the songs from the latest rock, dance, and pop albums. You want to play twice as many rock songs as pop songs and four more pop songs than dance songs. How many of each type of song will you play?

63. *Acid Mixture* A chemist needs 10 liters of a 25% acid solution. The solution is to be mixed from three solutions whose concentrations are 10%, 20%, and 50%. How many liters of each solution will satisfy each condition?

(a) Use 2 liters of the 50% solution.

(b) Use as little as possible of the 50% solution.

(c) Use as much as possible of the 50% solution.

64. *Acid Mixture* A chemist needs 12 gallons of a 20% acid solution. The solution is to be mixed from three solutions whose concentrations are 10%, 15%, and 25%. How many gallons of each solution will satisfy each condition?

(a) Use 4 gallons of the 25% solution.

(b) Use as little as possible of the 25% solution.

(c) Use as much as possible of the 25% solution.

65. *Electrical Network* Applying Kirchhoff's Laws to the electrical network in the figure, the currents I_1, I_2, and I_3 are the solution of the system

$$\begin{cases} I_1 - I_2 + I_3 = 0 \\ 3I_1 + 2I_2 \quad\;\; = 7 \\ \quad\;\; 2I_2 + 4I_3 = 8 \end{cases}$$

find the currents.

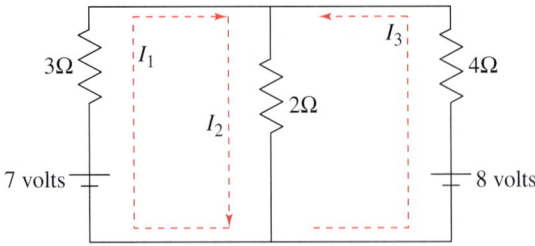

66. *Pulley System* A system of pulleys is loaded with 128-pound and 32-pound weights (see figure). The tensions t_1 and t_2 in the ropes and the acceleration a of the 32-pound weight are found by solving the system of equations

$$\begin{cases} t_1 - 2t_2 \qquad\;\; = 0 \\ t_1 \qquad\;\; - 2a = 128 \\ \qquad t_2 + a = 32 \end{cases}$$

where t_1 and t_2 are measured in pounds and a is measured in feet per second squared.

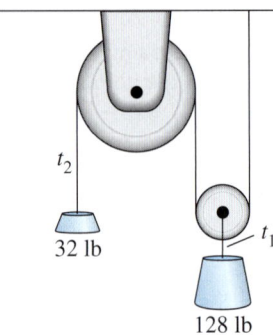

(a) Solve this system.

(b) The 32-pound weight in the pulley system is replaced by a 64-pound weight. The new pulley system will be modeled by the following system of equations.

$$\begin{cases} t_1 - 2t_2 \qquad\;\; = 0 \\ t_1 \qquad\;\; - 2a = 128 \\ \qquad t_2 + a = 64 \end{cases}$$

Solve this system and use your answer for the acceleration to describe what (if anything) is happening in the pulley system.

Fitting a Parabola In Exercises 67–70, find the least squares regression parabola $y = ax^2 + bx + c$ for the points $(x_1, y_1), (x_2, y_2), \ldots, (x_n, y_n)$ by solving the following system of linear equations for a, b, and c. Then use the *regression* feature of a graphing utility to confirm the result. (If you are unfamiliar with summation notation, look at the discussion in Section 8.1 or in Appendix A at the website for this text at *college.hmco.com*.)

$$nc + \left(\sum_{i=1}^{n} x_i\right)b + \left(\sum_{i=1}^{n} x_i^2\right)a = \sum_{i=1}^{n} y_i$$

$$\left(\sum_{i=1}^{n} x_i\right)c + \left(\sum_{i=1}^{n} x_i^2\right)b + \left(\sum_{i=1}^{n} x_i^3\right)a = \sum_{i=1}^{n} x_i y_i$$

$$\left(\sum_{i=1}^{n} x_i^2\right)c + \left(\sum_{i=1}^{n} x_i^3\right)b + \left(\sum_{i=1}^{n} x_i^4\right)a = \sum_{i=1}^{n} x_i^2 y_i$$

67.

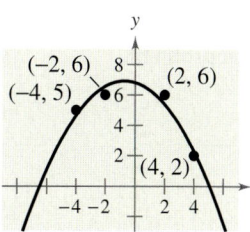

68.

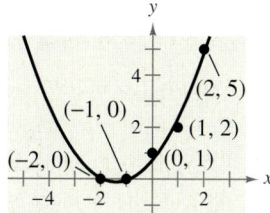

69.

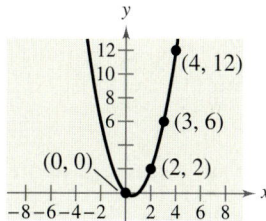

70.

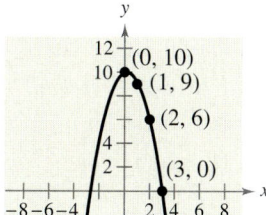

Model It

71. *Data Analysis: Wildlife* A wildlife management team studied the reproduction rates of deer in three tracts of a wildlife preserve. Each tract contained 5 acres. In each tract, the number of females x, and the percent of females y that had offspring the following year, were recorded. The results are shown in the table.

Number, x	Percent, y
100	75
120	68
140	55

(a) Use the technique demonstrated in Exercises 67–70 to set up a system of equations for the data and to find a least squares regression parabola that models the data.

 (b) Use a graphing utility to graph the parabola and the data in the same viewing window.

(c) Use the model to create a table of estimated values of y. Compare the estimated values with the actual data.

(d) Use the model to estimate the percent of females that had offspring when there were 170 females.

(e) Use the model to estimate the number of females when 40% of the females had offspring.

---

72. *Data Analysis: Stopping Distance* In testing a new automobile braking system, the speed x (in miles per hour) and the stopping distance y (in feet) were recorded in the table.

Speed, x	Stopping distance, y
30	55
40	105
50	188

(a) Use the technique demonstrated in Exercises 67–70 to set up a system of equations for the data and to find a least squares regression parabola that models the data.

(b) Graph the parabola and the data on the same set of axes.

(c) Use the model to estimate the stopping distance when the speed is 70 miles per hour.

73. *Sports* In Super Bowl XXXVIII, on February 1, 2004, the New England Patriots beat the Carolina Panthers by a score of 32 to 29. The total points scored came from 16 different scoring plays, which were a combination of touchdowns, extra-point kicks, two-point conversions, and field goals, worth 6, 1, 2, and 3 points, respectively. There were four times as many touchdowns as field goals and two times as many field goals as two-point conversions. How many touchdowns, extra-point kicks, two-point conversions, and field goals were scored during the game? (*Source: SuperBowl.com*)

74. *Sports* In the 2005 Orange Bowl, the University of Southern California won the National Championship by defeating the University of Oklahoma by a score of 55 to 19. The total points scored came from 22 different scoring plays, which were a combination of touchdowns, extra-point kicks, field goals and safeties, worth 6, 1, 3, and 2 points respectively. The same number of touchdowns and extra-point kicks were scored, and there were three times as many field goals as safeties. How many touchdowns, extra-point kicks, field goals, and safeties were scored? (*Source: ESPN.com*)

$\textit{f}$ *Advanced Applications* In Exercises 75–78, find values of x, y, and λ that satisfy the system. These systems arise in certain optimization problems in calculus, and λ is called a Lagrange multiplier.

75. $\begin{cases} y + \lambda = 0 \\ x + \lambda = 0 \\ x + y - 10 = 0 \end{cases}$

76. $\begin{cases} 2x + \lambda = 0 \\ 2y + \lambda = 0 \\ x + y - 4 = 0 \end{cases}$

77. $\begin{cases} 2x - 2x\lambda = 0 \\ -2y + \lambda = 0 \\ y - x^2 = 0 \end{cases}$

78. $\begin{cases} 2 + 2y + 2\lambda = 0 \\ 2x + 1 + \lambda = 0 \\ 2x + y - 100 = 0 \end{cases}$

Synthesis

True or False? In Exercises 79 and 80, determine whether the statement is true or false. Justify your answer.

79. The system

$\begin{cases} x + 3y - 6z = -16 \\ 2y - z = -1 \\ z = 3 \end{cases}$

is in row-echelon form.

80. If a system of three linear equations is inconsistent, then its graph has no points common to all three equations.

81. *Think About It* Are the following two systems of equations equivalent? Give reasons for your answer.

$\begin{cases} x + 3y - z = 6 \\ 2x - y + 2z = 1 \\ 3x + 2y - z = 2 \end{cases}$ $\begin{cases} x + 3y - z = 6 \\ -7y + 4z = 1 \\ -7y - 4z = -16 \end{cases}$

82. *Writing* When using Gaussian elimination to solve a system of linear equations, explain how you can recognize that the system has no solution. Give an example that illustrates your answer.

In Exercises 83–86, find two systems of linear equations that have the ordered triple as a solution. (There are many correct answers.)

83. $(4, -1, 2)$

84. $(-5, -2, 1)$

85. $\left(3, -\frac{1}{2}, \frac{7}{4}\right)$

86. $\left(-\frac{3}{2}, 4, -7\right)$

Skills Review

In Exercises 87–90, solve the percent problem.

87. What is $7\frac{1}{2}\%$ of 85?

88. 225 is what percent of 150?

89. 0.5% of what number is 400?

90. 48% of what number is 132?

In Exercises 91–96, perform the operation and write the result in standard form.

91. $(7 - i) + (4 + 2i)$

92. $(-6 + 3i) - (1 + 6i)$

93. $(4 - i)(5 + 2i)$

94. $(1 + 2i)(3 - 4i)$

95. $\dfrac{i}{1 + i} + \dfrac{6}{1 - i}$

96. $\dfrac{i}{4 + i} - \dfrac{2i}{8 - 3i}$

In Exercises 97–100, (a) determine the real zeros of f and (b) sketch the graph of f.

97. $f(x) = x^3 + x^2 - 12x$

98. $f(x) = -8x^4 + 32x^2$

99. $f(x) = 2x^3 + 5x^2 - 21x - 36$

100. $f(x) = 6x^3 - 29x^2 - 6x + 5$

In Exercises 101–104, use a graphing utility to construct a table of values for the equation. Then sketch the graph of the equation by hand.

101. $y = 4^{x-4} - 5$

102. $y = \left(\frac{5}{2}\right)^{-x+1} - 4$

103. $y = 1.9^{-0.8x} + 3$

104. $y = 3.5^{-x+2} + 6$

In Exercises 105 and 106, solve the system by elimination.

105. $\begin{cases} 2x + y = 120 \\ x + 2y = 120 \end{cases}$

106. $\begin{cases} 6x - 5y = 3 \\ 10x - 12y = 5 \end{cases}$

107. **Make a Decision** To work an extended application analyzing the earnings per share for Wal-Mart Stores, Inc. from 1988 to 2003, visit this text's website at *college.hmco.com*. *(Data Source: Wal-Mart Stores, Inc.)*

6.4   Partial Fractions

What you should learn

- Recognize partial fraction decompositions of rational expressions.
- Find partial fraction decompositions of rational expressions.

Why you should learn it

Partial fractions can help you analyze the behavior of a rational function. For instance, in Exercise 57 on page 492, you can analyze the exhaust temperatures of a diesel engine using partial fractions.

© Michael Rosenfeld/Getty Images

STUDY TIP

In Section P.5, you learned how to combine expressions such as

$$\frac{1}{x-2} + \frac{-1}{x+3} = \frac{5}{(x-2)(x+3)}.$$

The method of partial fractions shows you how to reverse this process.

$$\frac{5}{(x-2)(x+3)} = \frac{?}{x-2} + \frac{?}{x+3}$$

Introduction

In this section, you will learn to write a rational expression as the sum of two or more simpler rational expressions. For example, the rational expression

$$\frac{x+7}{x^2-x-6}$$

can be written as the sum of two fractions with first-degree denominators. That is,

Partial fraction decomposition
of $\dfrac{x+7}{x^2-x-6}$

$$\underbrace{\frac{x+7}{x^2-x-6}}_{} = \underbrace{\frac{2}{x-3}}_{\substack{\text{Partial}\\\text{fraction}}} + \underbrace{\frac{-1}{x+2}}_{\substack{\text{Partial}\\\text{fraction}}}.$$

Each fraction on the right side of the equation is a **partial fraction,** and together they make up the **partial fraction decomposition** of the left side.

Decomposition of $N(x)/D(x)$ into Partial Fractions

1. *Divide if improper:* If $N(x)/D(x)$ is an improper fraction [degree of $N(x) \geq$ degree of $D(x)$], divide the denominator into the numerator to obtain

$$\frac{N(x)}{D(x)} = (\text{polynomial}) + \frac{N_1(x)}{D(x)}$$

and apply Steps 2, 3, and 4 below to the proper rational expression $N_1(x)/D(x)$. Note that $N_1(x)$ is the remainder from the division of $N(x)$ by $D(x)$.

2. *Factor the denominator:* Completely factor the denominator into factors of the form

$$(px+q)^m \quad \text{and} \quad (ax^2+bx+c)^n$$

where (ax^2+bx+c) is irreducible.

3. *Linear factors:* For *each* factor of the form $(px+q)^m$, the partial fraction decomposition must include the following sum of m fractions.

$$\frac{A_1}{(px+q)} + \frac{A_2}{(px+q)^2} + \cdots + \frac{A_m}{(px+q)^m}$$

4. *Quadratic factors:* For *each* factor of the form $(ax^2+bx+c)^n$, the partial fraction decomposition must include the following sum of n fractions.

$$\frac{B_1x+C_1}{ax^2+bx+c} + \frac{B_2x+C_2}{(ax^2+bx+c)^2} + \cdots + \frac{B_nx+C_n}{(ax^2+bx+c)^n}$$

The inequality in Example 1 is a nonlinear inequality in two variables. Most of the following examples involve **linear inequalities** such as $ax + by < c$ (a and b are not both zero). The graph of a linear inequality is a half-plane lying on one side of the line $ax + by = c$.

Example 2 Sketching the Graph of a Linear Inequality

Sketch the graph of each linear inequality.

a. $x > -2$ **b.** $y \le 3$

Solution

a. The graph of the corresponding equation $x = -2$ is a vertical line. The points that satisfy the inequality $x > -2$ are those lying to the right of this line, as shown in Figure 6.20.

b. The graph of the corresponding equation $y = 3$ is a horizontal line. The points that satisfy the inequality $y \le 3$ are those lying below (or on) this line, as shown in Figure 6.21.

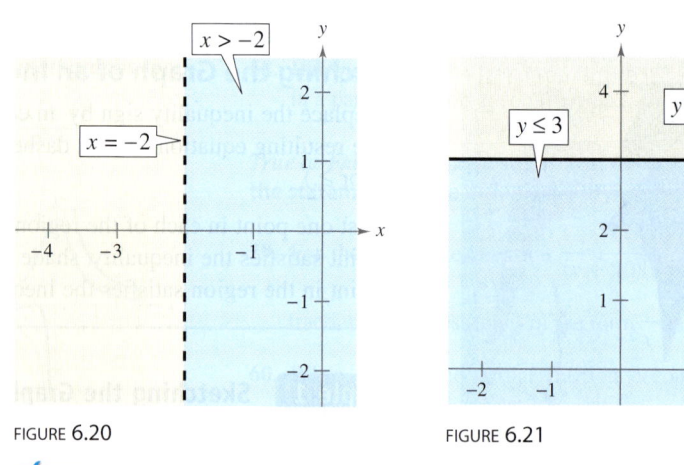

FIGURE 6.20   FIGURE 6.21

✓CHECKPOINT   Now try Exercise 3.

Example 3 Sketching the Graph of a Linear Inequality

Sketch the graph of $x - y < 2$.

Solution

The graph of the corresponding equation $x - y = 2$ is a line, as shown in Figure 6.22. Because the origin $(0, 0)$ satisfies the inequality, the graph consists of the half-plane lying above the line. (Try checking a point below the line. Regardless of which point you choose, you will see that it does not satisfy the inequality.)

✓CHECKPOINT   Now try Exercise 9.

To graph a linear inequality, it can help to write the inequality in slope-intercept form. For instance, by writing $x - y < 2$ in the form

$$y > x - 2$$

you can see that the solution points lie *above* the line $x - y = 2$ (or $y = x - 2$), as shown in Figure 6.22.

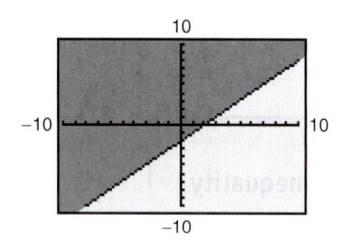
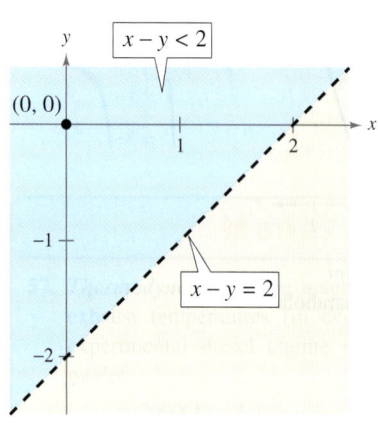

FIGURE 6.22

Systems of Inequalities

Many practical problems in business, science, and engineering involve systems of linear inequalities. A **solution** of a system of inequalities in x and y is a point (x, y) that satisfies each inequality in the system.

To sketch the graph of a system of inequalities in two variables, first sketch the graph of each individual inequality (on the same coordinate system) and then find the region that is *common* to every graph in the system. This region represents the **solution set** of the system. For systems of *linear inequalities*, it is helpful to find the vertices of the solution region.

Example 4   Solving a System of Inequalities

Sketch the graph (and label the vertices) of the solution set of the system.

$$\begin{cases} x - y < 2 & \text{Inequality 1} \\ \quad x > -2 & \text{Inequality 2} \\ \quad\quad y \le 3 & \text{Inequality 3} \end{cases}$$

Solution

The graphs of these inequalities are shown in Figures 6.22, 6.20, and 6.21, respectively, on page 494. The triangular region common to all three graphs can be found by superimposing the graphs on the same coordinate system, as shown in Figure 6.23. To find the vertices of the region, solve the three systems of corresponding equations obtained by taking *pairs* of equations representing the boundaries of the individual regions.

> ## STUDY TIP
>
> Using different colored pencils to shade the solution of each inequality in a system will make identifying the solution of the system of inequalities easier.

Vertex A: $(-2, -4)$	*Vertex B:* $(5, 3)$	*Vertex C:* $(-2, 3)$
$\begin{cases} x - y = 2 \\ \quad x = -2 \end{cases}$	$\begin{cases} x - y = 2 \\ \quad y = 3 \end{cases}$	$\begin{cases} x = -2 \\ y = 3 \end{cases}$

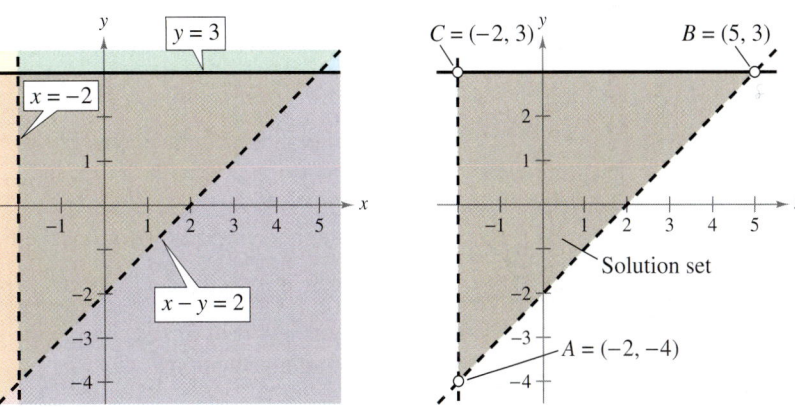

FIGURE **6.23**

Note in Figure 6.23 that the vertices of the region are represented by open dots. This means that the vertices *are not* solutions of the system of inequalities.

✓**CHECKPOINT**   Now try Exercise 35.

For the triangular region shown in Figure 6.23, each point of intersection of a pair of boundary lines corresponds to a vertex. With more complicated regions, two border lines can sometimes intersect at a point that is not a vertex of the region, as shown in Figure 6.24. To keep track of which points of intersection are actually vertices of the region, you should sketch the region and refer to your sketch as you find each point of intersection.

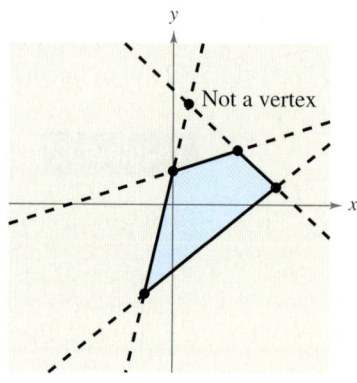

FIGURE 6.24

Example 5   Solving a System of Inequalities

Sketch the region containing all points that satisfy the system of inequalities.

$$\begin{cases} x^2 - y \le 1 & \text{Inequality 1} \\ -x + y \le 1 & \text{Inequality 2} \end{cases}$$

Solution

As shown in Figure 6.25, the points that satisfy the inequality

$$x^2 - y \le 1 \qquad \text{Inequality 1}$$

are the points lying above (or on) the parabola given by

$$y = x^2 - 1. \qquad \text{Parabola}$$

The points satisfying the inequality

$$-x + y \le 1 \qquad \text{Inequality 2}$$

are the points lying below (or on) the line given by

$$y = x + 1. \qquad \text{Line}$$

To find the points of intersection of the parabola and the line, solve the system of corresponding equations.

$$\begin{cases} x^2 - y = 1 \\ -x + y = 1 \end{cases}$$

Using the method of substitution, you can find the solutions to be $(-1, 0)$ and $(2, 3)$. So, the region containing all points that satisfy the system is indicated by the shaded region in Figure 6.25.

✔CHECKPOINT   Now try Exercise 37.

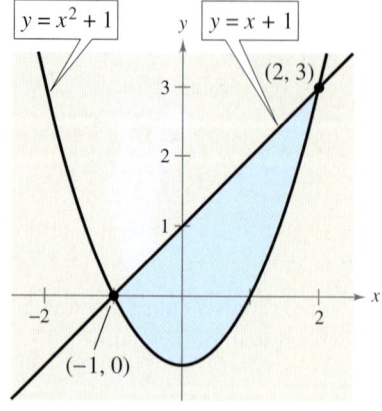

FIGURE 6.25

When solving a system of inequalities, you should be aware that the system might have no solution *or* it might be represented by an unbounded region in the plane. These two possibilities are shown in Examples 6 and 7.

Example 6   A System with No Solution

Sketch the solution set of the system of inequalities.

$$\begin{cases} x + y > 3 & \text{Inequality 1} \\ x + y < -1 & \text{Inequality 2} \end{cases}$$

Solution

From the way the system is written, it is clear that the system has no solution, because the quantity $(x + y)$ cannot be both less than -1 and greater than 3. Graphically, the inequality $x + y > 3$ is represented by the half-plane lying above the line $x + y = 3$, and the inequality $x + y < -1$ is represented by the half-plane lying below the line $x + y = -1$, as shown in Figure 6.26. These two half-planes have no points in common. So, the system of inequalities has no solution.

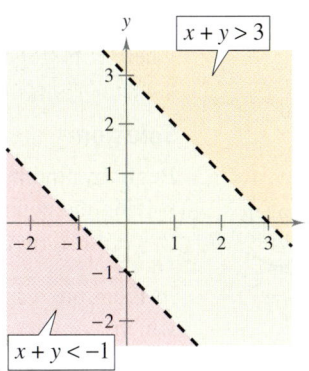

FIGURE **6.26**

✔CHECKPOINT   Now try Exercise 39.

Example 7   An Unbounded Solution Set

Sketch the solution set of the system of inequalities.

$$\begin{cases} x + y < 3 & \text{Inequality 1} \\ x + 2y > 3 & \text{Inequality 2} \end{cases}$$

Solution

The graph of the inequality $x + y < 3$ is the half-plane that lies below the line $x + y = 3$, as shown in Figure 6.27. The graph of the inequality $x + 2y > 3$ is the half-plane that lies above the line $x + 2y = 3$. The intersection of these two half-planes is an *infinite wedge* that has a vertex at $(3, 0)$. So, the solution set of the system of inequalities is unbounded.

✔CHECKPOINT   Now try Exercise 41.

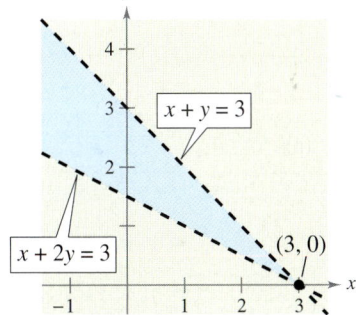

FIGURE **6.27**

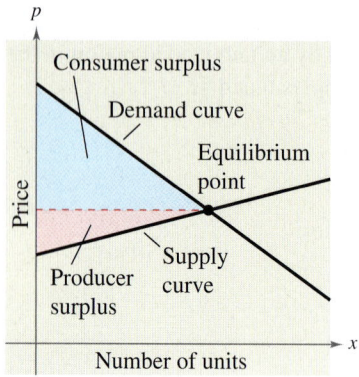

FIGURE **6.28**

Applications

Example 9 in Section 6.2 discussed the *equilibrium point* for a system of demand and supply functions. The next example discusses two related concepts that economists call **consumer surplus** and **producer surplus.** As shown in Figure 6.28, the consumer surplus is defined as the area of the region that lies *below* the demand curve, *above* the horizontal line passing through the equilibrium point, and to the right of the *p*-axis. Similarly, the producer surplus is defined as the area of the region that lies *above* the supply curve, *below* the horizontal line passing through the equilibrium point, and to the right of the *p*-axis. The consumer surplus is a measure of the amount that consumers would have been willing to pay *above what they actually paid*, whereas the producer surplus is a measure of the amount that producers would have been willing to receive *below what they actually received*.

Example 8   Consumer Surplus and Producer Surplus

The demand and supply functions for a new type of personal digital assistant are given by

$$\begin{cases} p = 150 - 0.00001x & \text{Demand equation} \\ p = 60 + 0.00002x & \text{Supply equation} \end{cases}$$

where p is the price (in dollars) and x represents the number of units. Find the consumer surplus and producer surplus for these two equations.

Solution

Begin by finding the equilibrium point (when supply and demand are equal) by solving the equation

$$60 + 0.00002x = 150 - 0.00001x.$$

In Example 9 in Section 6.2, you saw that the solution is $x = 3,000,000$ units, which corresponds to an equilibrium price of $p = \$120$. So, the consumer surplus and producer surplus are the areas of the following triangular regions.

Consumer Surplus	Producer Surplus
$\begin{cases} p \le 150 - 0.00001x \\ p \ge 120 \\ x \ge 0 \end{cases}$	$\begin{cases} p \ge 60 + 0.00002x \\ p \le 120 \\ x \ge 0 \end{cases}$

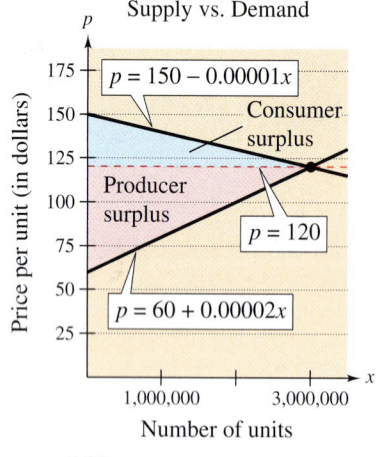

FIGURE **6.29**

In Figure 6.29, you can see that the consumer and producer surpluses are defined as the areas of the shaded triangles.

$$\begin{aligned} \text{Consumer surplus} &= \frac{1}{2}(\text{base})(\text{height}) \\ &= \frac{1}{2}(3,000,000)(30) = \$45,000,000 \end{aligned}$$

$$\begin{aligned} \text{Producer surplus} &= \frac{1}{2}(\text{base})(\text{height}) \\ &= \frac{1}{2}(3,000,000)(60) = \$90,000,000 \end{aligned}$$

✓**CHECKPOINT**   Now try Exercise 65.

Example 9   **Nutrition**

The liquid portion of a diet is to provide at least 300 calories, 36 units of vitamin A, and 90 units of vitamin C. A cup of dietary drink X provides 60 calories, 12 units of vitamin A, and 10 units of vitamin C. A cup of dietary drink Y provides 60 calories, 6 units of vitamin A, and 30 units of vitamin C. Set up a system of linear inequalities that describes how many cups of each drink should be consumed each day to meet or exceed the minimum daily requirements for calories and vitamins.

Solution

Begin by letting x and y represent the following.

x = number of cups of dietary drink X

y = number of cups of dietary drink Y

To meet or exceed the minimum daily requirements, the following inequalities must be satisfied.

$$\begin{cases} 60x + 60y \geq 300 & \text{\textcolor{red}{Calories}} \\ 12x + 6y \geq 36 & \text{\textcolor{red}{Vitamin A}} \\ 10x + 30y \geq 90 & \text{\textcolor{red}{Vitamin C}} \\ x \geq 0 \\ y \geq 0 \end{cases}$$

The last two inequalities are included because x and y cannot be negative. The graph of this system of inequalities is shown in Figure 6.30. (More is said about this application in Example 6 in Section 6.6.)

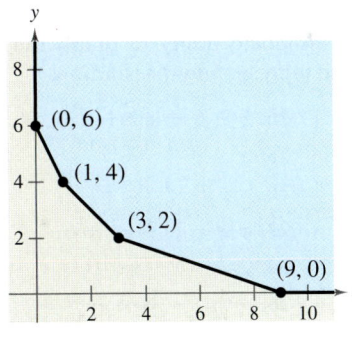

FIGURE 6.30

✓CHECKPOINT   Now try Exercise 69.

***W**RITING ABOUT **M**ATHEMATICS*

Creating a System of Inequalities Plot the points $(0, 0)$, $(4, 0)$, $(3, 2)$, and $(0, 2)$ in a coordinate plane. Draw the quadrilateral that has these four points as its vertices. Write a system of linear inequalities that has the quadrilateral as its solution. Explain how you found the system of inequalities.

6.5 | Exercises

VOCABULARY CHECK: Fill in the blanks.

1. An ordered pair (a, b) is a _____ of an inequality in x and y if the inequality is true when a and b are substituted for x and y, respectively.

2. The _____ of an inequality is the collection of all solutions of the inequality.

3. The graph of a _____ inequality is a half-plane lying on one side of the line $ax + by = c$.

4. A _____ of a system of inequalities in x and y is a point (x, y) that satisfies each inequality in the system.

5. The area of the region that lies below the demand curve, above the horizontal line passing through the equilibrium point, to the right of the p-axis is called the _____ _____.

PREREQUISITE SKILLS REVIEW: Practice and review algebra skills needed for this section at **www.Eduspace.com.**

In Exercises 1–14, sketch the graph of the inequality.

1. $y < 2 - x^2$
2. $y^2 - x < 0$
3. $x \geq 2$
4. $x \leq 4$
5. $y \geq -1$
6. $y \leq 3$
7. $y < 2 - x$
8. $y > 2x - 4$
9. $2y - x \geq 4$
10. $5x + 3y \geq -15$
11. $(x + 1)^2 + (y - 2)^2 < 9$
12. $(x - 1)^2 + (y - 4)^2 > 9$
13. $y \leq \dfrac{1}{1 + x^2}$
14. $y > \dfrac{-15}{x^2 + x + 4}$

 In Exercises 15–26, use a graphing utility to graph the inequality. Shade the region representing the solution.

15. $y < \ln x$
16. $y \geq 6 - \ln(x + 5)$
17. $y < 3^{-x-4}$
18. $y \leq 2^{2x-0.5} - 7$
19. $y \geq \frac{2}{3}x - 1$
20. $y \leq 6 - \frac{3}{2}x$
21. $y < -3.8x + 1.1$
22. $y \geq -20.74 + 2.66x$
23. $x^2 + 5y - 10 \leq 0$
24. $2x^2 - y - 3 > 0$
25. $\frac{5}{2}y - 3x^2 - 6 \geq 0$
26. $-\frac{1}{10}x^2 - \frac{3}{8}y < -\frac{1}{4}$

In Exercises 27–30, write an inequality for the shaded region shown in the figure.

27.
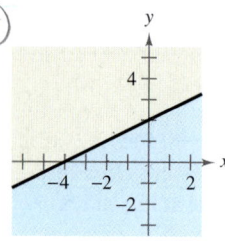

28.
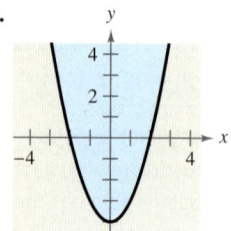

29.
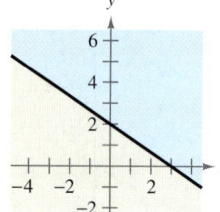

30.
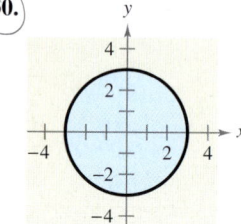

In Exercises 31–34, determine whether each ordered pair is a solution of the system of linear inequalities.

31. $\begin{cases} x \geq -4 \\ y > -3 \\ y \leq -8x - 3 \end{cases}$
(a) $(0, 0)$ (b) $(-1, -3)$
(c) $(-4, 0)$ (d) $(-3, 11)$

32. $\begin{cases} -2x + 5y \geq 3 \\ y < 4 \\ -4x + 2y < 7 \end{cases}$
(a) $(0, 2)$ (b) $(-6, 4)$
(c) $(-8, -2)$ (d) $(-3, 2)$

33. $\begin{cases} 3x + y > 1 \\ -y - \frac{1}{2}x^2 \leq -4 \\ -15x + 4y > 0 \end{cases}$
(a) $(0, 10)$ (b) $(0, -1)$
(c) $(2, 9)$ (d) $(-1, 6)$

34. $\begin{cases} x^2 + y^2 \geq 36 \\ -3x + y \leq 10 \\ \frac{2}{3}x - y \geq 5 \end{cases}$
(a) $(-1, 7)$ (b) $(-5, 1)$
(c) $(6, 0)$ (d) $(4, -8)$

In Exercises 35–48, sketch the graph and label the vertices of the solution set of the system of inequalities.

35. $\begin{cases} x + y \leq 1 \\ -x + y \leq 1 \\ y \geq 0 \end{cases}$

36. $\begin{cases} 3x + 2y < 6 \\ x > 0 \\ y > 0 \end{cases}$

37. $\begin{cases} x^2 + y \leq 5 \\ x \geq -1 \\ y \geq 0 \end{cases}$

38. $\begin{cases} 2x^2 + y \geq 2 \\ x \leq 2 \\ y \leq 1 \end{cases}$

39. $\begin{cases} 2x + y > 2 \\ 6x + 3y < 2 \end{cases}$

40. $\begin{cases} x - 7y > -36 \\ 5x + 2y > 5 \\ 6x - 5y > 6 \end{cases}$

41. $\begin{cases} -3x + 2y < 6 \\ x - 4y > -2 \\ 2x + y < 3 \end{cases}$

42. $\begin{cases} x - 2y < -6 \\ 5x - 3y > -9 \end{cases}$

43. $\begin{cases} x > y^2 \\ x < y + 2 \end{cases}$

44. $\begin{cases} x - y^2 > 0 \\ x - y > 2 \end{cases}$

45. $\begin{cases} x^2 + y^2 \le 9 \\ x^2 + y^2 \ge 1 \end{cases}$

46. $\begin{cases} x^2 + y^2 \le 25 \\ 4x - 3y \le 0 \end{cases}$

47. $\begin{cases} 3x + 4 \ge y^2 \\ x - y < 0 \end{cases}$ $y7x$

48. $\begin{cases} x < 2y - y^2 \\ 0 < x + y \end{cases}$

 In Exercises 49–54, use a graphing utility to graph the inequalities. Shade the region representing the solution set of the system.

49. $\begin{cases} y \le \sqrt{3x} + 1 \\ y \ge x^2 + 1 \end{cases}$

50. $\begin{cases} y < -x^2 + 2x + 3 \\ y > x^2 - 4x + 3 \end{cases}$

51. $\begin{cases} y < x^3 - 2x + 1 \\ y > -2x \\ x \le 1 \end{cases}$

52. $\begin{cases} y \ge x^4 - 2x^2 + 1 \\ y \le 1 - x^2 \end{cases}$

53. $\begin{cases} x^2 y \ge 1 \\ 0 < x \le 4 \\ y \le 4 \end{cases}$

54. $\begin{cases} y \le e^{-x^2/2} \\ y \ge 0 \\ -2 \le x \le 2 \end{cases}$

In Exercises 55–64, derive a set of inequalities to describe the region.

55.

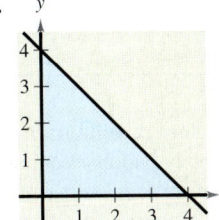

56.

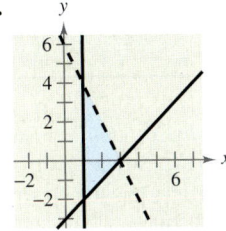

57.

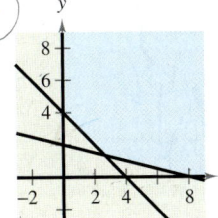

58.

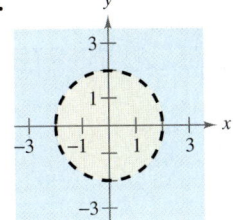

59.

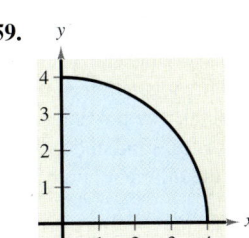

60.

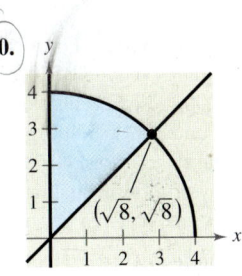

$(\sqrt{8}, \sqrt{8})$

61. Rectangle: vertices at $(2, 1), (5, 1), (5, 7), (2, 7)$

62. Parallelogram: vertices at $(0, 0), (4, 0), (1, 4), (5, 4)$

63. Triangle: vertices at $(0, 0), (5, 0), (2, 3)$

64. Triangle: vertices at $(-1, 0), (1, 0), (0, 1)$

Supply and Demand   In Exercises 65–68, (a) graph the systems representing the consumer surplus and producer surplus for the supply and demand equations and (b) find the consumer surplus and producer surplus.

	Demand	*Supply*
65.	$p = 50 - 0.5x$	$p = 0.125x$
66.	$p = 100 - 0.05x$	$p = 25 + 0.1x$
67.	$p = 140 - 0.00002x$	$p = 80 + 0.00001x$
68.	$p = 400 - 0.0002x$	$p = 225 + 0.0005x$

69. *Production*   A furniture company can sell all the tables and chairs it produces. Each table requires 1 hour in the assembly center and $1\frac{1}{3}$ hours in the finishing center. Each chair requires $1\frac{1}{2}$ hours in the assembly center and $1\frac{1}{2}$ hours in the finishing center. The company's assembly center is available 12 hours per day, and its finishing center is available 15 hours per day. Find and graph a system of inequalities describing all possible production levels.

70. *Inventory*   A store sells two models of computers. Because of the demand, the store stocks at least twice as many units of model A as of model B. The costs to the store for the two models are $800 and $1200, respectively. The management does not want more than $20,000 in computer inventory at any one time, and it wants at least four model A computers and two model B computers in inventory at all times. Find and graph a system of inequalities describing all possible inventory levels.

71. *Investment Analysis*   A person plans to invest up to $20,000 in two different interest-bearing accounts. Each account is to contain at least $5000. Moreover, the amount in one account should be at least twice the amount in the other account. Find and graph a system of inequalities to describe the various amounts that can be deposited in each account.

72. Ticket Sales For a concert event, there are $30 reserved seat tickets and $20 general admission tickets. There are 2000 reserved seats available, and fire regulations limit the number of paid ticket holders to 3000. The promoter must take in at least $75,000 in ticket sales. Find and graph a system of inequalities describing the different numbers of tickets that can be sold.

73. Shipping A warehouse supervisor is told to ship at least 50 packages of gravel that weigh 55 pounds each and at least 40 bags of stone that weigh 70 pounds each. The maximum weight capacity in the truck he is loading is 7500 pounds. Find and graph a system of inequalities describing the numbers of bags of stone and gravel that he can send.

74. Truck Scheduling A small company that manufactures two models of exercise machines has an order for 15 units of the standard model and 16 units of the deluxe model. The company has trucks of two different sizes that can haul the products, as shown in the table.

Truck	Standard	Deluxe
Large	6	3
Medium	4	6

Find and graph a system of inequalities describing the numbers of trucks of each size that are needed to deliver the order.

75. Nutrition A dietitian is asked to design a special dietary supplement using two different foods. Each ounce of food X contains 20 units of calcium, 15 units of iron, and 10 units of vitamin B. Each ounce of food Y contains 10 units of calcium, 10 units of iron, and 20 units of vitamin B. The minimum daily requirements of the diet are 300 units of calcium, 150 units of iron, and 200 units of vitamin B.

(a) Write a system of inequalities describing the different amounts of food X and food Y that can be used.

(b) Sketch a graph of the region corresponding to the system in part (a).

(c) Find two solutions of the system and interpret their meanings in the context of the problem.

76. Health A person's maximum heart rate is $220 - x$, where x is the person's age in years for $20 \leq x \leq 70$. When a person exercises, it is recommended that the person strive for a heart rate that is at least 50% of the maximum and at most 75% of the maximum. (Source: American Heart Association)

(a) Write a system of inequalities that describes the exercise target heart rate region.

(b) Sketch a graph of the region in part (a).

(c) Find two solutions to the system and interpret their meanings in the context of the problem.

Model It

77. Data Analysis: Prescription Drugs The table shows the retail sales y (in billions of dollars) of prescription drugs in the United States from 1999 to 2003. (Source: National Association of Chain Drug Stores)

Year	Retail sales, y
1999	125.8
2000	145.6
2001	164.1
2002	182.7
2003	203.1

(a) Use the *regression* feature of a graphing utility to find a linear model for the data. Let t represent the year, with $t = 9$ corresponding to 1999.

(b) The total retail sales of prescription drugs in the United States during this five-year period can be approximated by finding the area of the trapezoid bounded by the linear model you found in part (a) and the lines $y = 0$, $t = 8.5$, and $t = 13.5$. Use a graphing utility to graph this region.

(c) Use the formula for the area of a trapezoid to approximate the total retail sales of prescription drugs.

78. Physical Fitness Facility An indoor running track is to be constructed with a space for body-building equipment inside the track (see figure). The track must be at least 125 meters long, and the body-building space must have an area of at least 500 square meters.

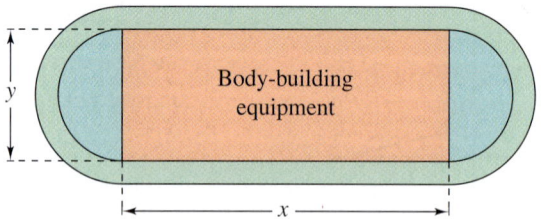

(a) Find a system of inequalities describing the requirements of the facility.

(b) Graph the system from part (a).

Synthesis

True or False?   **In Exercises 79 and 80, determine whether the statement is true or false. Justify your answer.**

79. The area of the figure defined by the system

$$\begin{cases} x \geq -3 \\ x \leq 6 \\ y \leq 5 \\ y \geq -6 \end{cases}$$

is 99 square units.

80. The graph below shows the solution of the system

$$\begin{cases} y \leq 6 \\ -4x - 9y > 6. \\ 3x + y^2 \geq 2 \end{cases}$$

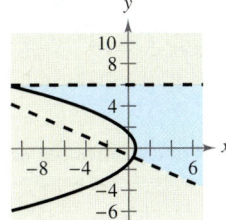

81. *Writing*   Explain the difference between the graphs of the inequality $x \leq 4$ on the real number line and on the rectangular coordinate system.

82. *Think About It*   After graphing the boundary of an inequality in x and y, how do you decide on which side of the boundary the solution set of the inequality lies?

83. *Graphical Reasoning*   Two concentric circles have radii x and y, where $y > x$. The area between the circles must be at least 10 square units.

(a) Find a system of inequalities describing the constraints on the circles.

 (b) Use a graphing utility to graph the system of inequalities in part (a). Graph the line $y = x$ in the same viewing window.

(c) Identify the graph of the line in relation to the boundary of the inequality. Explain its meaning in the context of the problem.

84. The graph of the solution of the inequality $x + 2y < 6$ is shown in the figure. Describe how the solution set would change for each of the following.

(a) $x + 2y \leq 6$ (b) $x + 2y > 6$

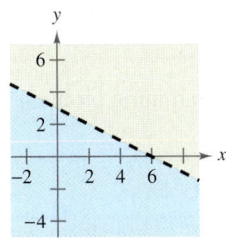

In Exercises 85–88, match the system of inequalities with the graph of its solution. [The graphs are labeled (a), (b), (c), and (d).]

(a)

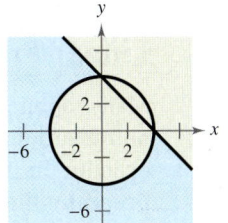

(b)

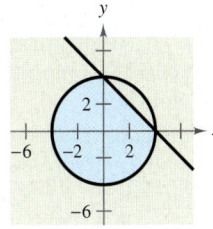

(c)

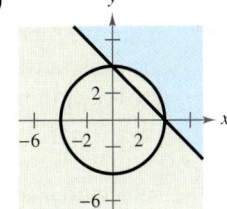

(d)

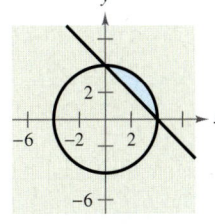

85. $\begin{cases} x^2 + y^2 \leq 16 \\ x + y \geq 4 \end{cases}$

86. $\begin{cases} x^2 + y^2 \leq 16 \\ x + y \leq 4 \end{cases}$

87. $\begin{cases} x^2 + y^2 \geq 16 \\ x + y \geq 4 \end{cases}$

88. $\begin{cases} x^2 + y^2 \geq 16 \\ x + y \leq 4 \end{cases}$

Skills Review

In Exercises 89–94, find the equation of the line passing through the two points.

89. $(-2, 6), (4, -4)$

90. $(-8, 0), (3, -1)$

91. $\left(\frac{3}{4}, -2\right), \left(-\frac{7}{2}, 5\right)$

92. $\left(-\frac{1}{2}, 0\right), \left(\frac{11}{2}, 12\right)$

93. $(3.4, -5.2), (-2.6, 0.8)$

94. $(-4.1, -3.8), (2.9, 8.2)$

 95. *Data Analysis: Cell Phone Bills*   The average monthly cell phone bills y (in dollars) in the United States from 1998 to 2003, where t is the year, are shown as data points (t, y).   (Source: Cellular Telecommunications & Internet Association)

(1998, 39.43),   (1999, 41.24),   (2000, 45.27)

(2001, 47.37),   (2002, 48.40),   (2003, 49.91)

(a) Use the *regression* feature of a graphing utility to find a linear model, a quadratic model, and an exponential model for the data. Let $t = 8$ correspond to 1998.

(b) Use a graphing utility to plot the data and the models in the same viewing window.

(c) Which model is the best fit for the data?

(d) Use the model from part (c) to predict the average monthly cell phone bill in 2008.

6.6   Linear Programming

Linear Programming: A Graphical Approach

Many applications in business and economics involve a process called **optimization,** in which you are asked to find the minimum or maximum of a quantity. In this section, you will study an optimization strategy called **linear programming.**

A two-dimensional linear programming problem consists of a linear **objective function** and a system of linear inequalities called **constraints.** The objective function gives the quantity that is to be maximized (or minimized), and the constraints determine the set of **feasible solutions.** For example, suppose you are asked to maximize the value of

$$z = ax + by \qquad \text{Objective function}$$

subject to a set of constraints that determines the shaded region in Figure 6.31.

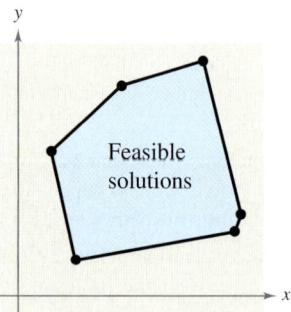

FIGURE 6.31

Because every point in the shaded region satisfies each constraint, it is not clear how you should find the point that yields a maximum value of z. Fortunately, it can be shown that if there is an optimal solution, it must occur at one of the vertices. This means that *you can find the maximum value of z by testing z at each of the vertices.*

> **Optimal Solution of a Linear Programming Problem**
>
> If a linear programming problem has a solution, it must occur at a vertex of the set of feasible solutions. If there is more than one solution, at least one of them must occur at such a vertex. In either case, the value of the objective function is unique.

Some guidelines for solving a linear programming problem in two variables are listed at the top of the next page.

> ### Solving a Linear Programming Problem
>
> 1. Sketch the region corresponding to the system of constraints. (The points inside or on the boundary of the region are *feasible solutions*.)
>
> 2. Find the vertices of the region.
>
> 3. Test the objective function at each of the vertices and select the values of the variables that optimize the objective function. For a bounded region, both a minimum and a maximum value will exist. (For an unbounded region, *if* an optimal solution exists, it will occur at a vertex.)

Example 1   Solving a Linear Programming Problem

Find the maximum value of

$$z = 3x + 2y$$ Objective function

subject to the following constraints.

$$\left.\begin{array}{r} x \geq 0 \\ y \geq 0 \\ x + 2y \leq 4 \\ x - y \leq 1 \end{array}\right\}$$ Constraints

Solution

The constraints form the region shown in Figure 6.32. At the four vertices of this region, the objective function has the following values.

At $(0, 0)$: $z = 3(0) + 2(0) = 0$
At $(1, 0)$: $z = 3(1) + 2(0) = 3$
At $(2, 1)$: $z = 3(2) + 2(1) = 8$ Maximum value of z
At $(0, 2)$: $z = 3(0) + 2(2) = 4$

So, the maximum value of z is 8, and this occurs when $x = 2$ and $y = 1$.

 **CHECKPOINT**   Now try Exercise 5.

In Example 1, try testing some of the *interior* points in the region. You will see that the corresponding values of z are less than 8. Here are some examples.

At $(1, 1)$: $z = 3(1) + 2(1) = 5$ At $\left(\frac{1}{2}, \frac{3}{2}\right)$: $z = 3\left(\frac{1}{2}\right) + 2\left(\frac{3}{2}\right) = \frac{9}{2}$

To see why the maximum value of the objective function in Example 1 must occur at a vertex, consider writing the objective function in slope-intercept form

$$y = -\frac{3}{2}x + \frac{z}{2}$$ Family of lines

where $z/2$ is the y-intercept of the objective function. This equation represents a family of lines, each of slope $-\frac{3}{2}$. Of these infinitely many lines, you want the one that has the largest z-value while still intersecting the region determined by the constraints. In other words, of all the lines whose slope is $-\frac{3}{2}$, you want the one that has the largest y-intercept *and* intersects the given region, as shown in Figure 6.33. From the graph you can see that such a line will pass through one (or more) of the vertices of the region.

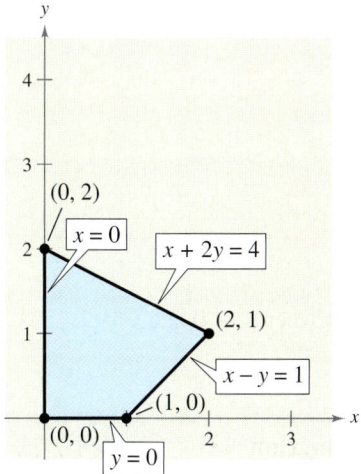

FIGURE 6.32

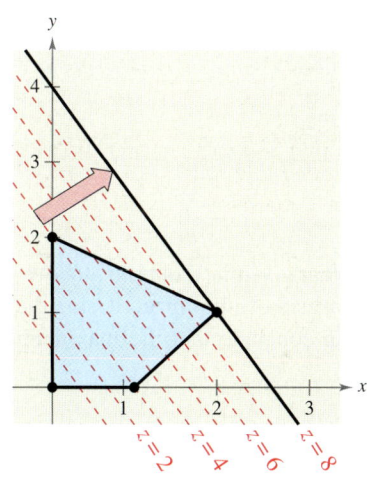

FIGURE 6.33

The next example shows that the same basic procedure can be used to solve a problem in which the objective function is to be *minimized*.

Example 2 Minimizing an Objective Function

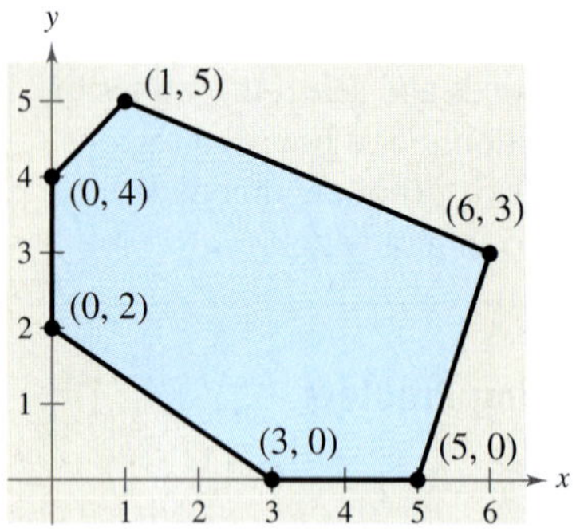

Find the minimum value of

$$z = 5x + 7y \qquad \text{Objective function}$$

where $x \geq 0$ and $y \geq 0$, subject to the following constraints.

$$\left.\begin{array}{r} 2x + 3y \geq 6 \\ 3x - y \leq 15 \\ -x + y \leq 4 \\ 2x + 5y \leq 27 \end{array}\right\} \qquad \text{Constraints}$$

Solution

The region bounded by the constraints is shown in Figure 6.34. By testing the objective function at each vertex, you obtain the following.

At $(0, 2)$: $z = 5(0) + 7(2) = 14$ Minimum value of z

At $(0, 4)$: $z = 5(0) + 7(4) = 28$

At $(1, 5)$: $z = 5(1) + 7(5) = 40$

At $(6, 3)$: $z = 5(6) + 7(3) = 51$

At $(5, 0)$: $z = 5(5) + 7(0) = 25$

At $(3, 0)$: $z = 5(3) + 7(0) = 15$

So, the minimum value of z is 14, and this occurs when $x = 0$ and $y = 2$.

 **CHECKPOINT** Now try Exercise 13.

Example 3 Maximizing an Objective Function

Find the maximum value of

$$z = 5x + 7y \qquad \text{Objective function}$$

where $x \geq 0$ and $y \geq 0$, subject to the following constraints.

$$\left.\begin{array}{r} 2x + 3y \geq 6 \\ 3x - y \leq 15 \\ -x + y \leq 4 \\ 2x + 5y \leq 27 \end{array}\right\} \qquad \text{Constraints}$$

Solution

This linear programming problem is identical to that given in Example 2 above, *except* that the objective function is maximized instead of minimized. Using the values of z at the vertices shown above, you can conclude that the maximum value of z is

$$z = 5(6) + 7(3) = 51$$

and occurs when $x = 6$ and $y = 3$.

 **CHECKPOINT** Now try Exercise 15.

FIGURE 6.34

(1, 5)
(0, 4)
(6, 3)
(0, 2)
(3, 0)
(5, 0)

Edward W. Souza/News Service/Stanford University

Historical Note
George Dantzig (1914–) was the first to propose the simplex method, or linear programming, in 1947. This technique defined the steps needed to find the optimal solution to a complex multivariable problem.

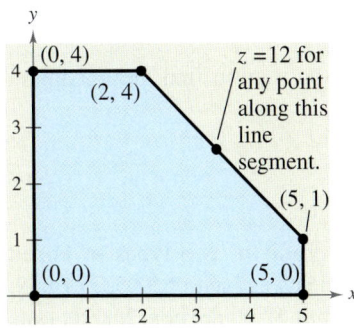

$z = 12$ for any point along this line segment.

FIGURE **6.35**

It is possible for the maximum (or minimum) value in a linear programming problem to occur at *two* different vertices. For instance, at the vertices of the region shown in Figure 6.35, the objective function

$$z = 2x + 2y \qquad \text{Objective function}$$

has the following values.

At $(0, 0)$: $z = 2(0) + 2(0) = 0$
At $(0, 4)$: $z = 2(0) + 2(4) = 8$
At $(2, 4)$: $z = 2(2) + 2(4) = 12$ Maximum value of z
At $(5, 1)$: $z = 2(5) + 2(1) = 12$ Maximum value of z
At $(5, 0)$: $z = 2(5) + 2(0) = 10$

In this case, you can conclude that the objective function has a maximum value not only at the vertices $(2, 4)$ and $(5, 1)$; it also has a maximum value (of 12) at *any point on the line segment connecting these two vertices*. Note that the objective function in slope-intercept form $y = -x + \frac{1}{2}z$ has the same slope as the line through the vertices $(2, 4)$ and $(5, 1)$.

Some linear programming problems have no optimal solutions. This can occur if the region determined by the constraints is *unbounded*. Example 4 illustrates such a problem.

Example 4   An Unbounded Region

Find the maximum value of

$$z = 4x + 2y \qquad \text{Objective function}$$

where $x \geq 0$ and $y \geq 0$, subject to the following constraints.

$$\left. \begin{array}{r} x + 2y \geq 4 \\ 3x + y \geq 7 \\ -x + 2y \leq 7 \end{array} \right\} \qquad \text{Constraints}$$

Solution

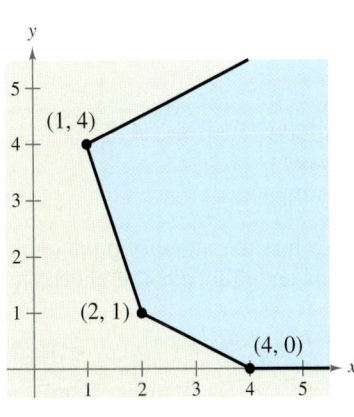

FIGURE **6.36**

The region determined by the constraints is shown in Figure 6.36. For this unbounded region, there is no maximum value of z. To see this, note that the point $(x, 0)$ lies in the region for all values of $x \geq 4$. Substituting this point into the objective function, you get

$$z = 4(x) + 2(0) = 4x.$$

By choosing x to be large, you can obtain values of z that are as large as you want. So, there is no maximum value of z. However, there *is* a minimum value of z.

At $(1, 4)$: $z = 4(1) + 2(4) = 12$
At $(2, 1)$: $z = 4(2) + 2(1) = 10$ Minimum value of z
At $(4, 0)$: $z = 4(4) + 2(0) = 16$

So, the minimum value of z is 10, and this occurs when $x = 2$ and $y = 1$.

✓CHECKPOINT   Now try Exercise 17.

Applications

Example 5 shows how linear programming can be used to find the maximum profit in a business application.

Example 5   Optimal Profit

A candy manufacturer wants to maximize the profit for two types of boxed chocolates. A box of chocolate covered creams yields a profit of $1.50 per box, and a box of chocolate covered nuts yields a profit of $2.00 per box. Market tests and available resources have indicated the following constraints.

1. The combined production level should not exceed 1200 boxes per month.
2. The demand for a box of chocolate covered nuts is no more than half the demand for a box of chocolate covered creams.
3. The production level for chocolate covered creams should be less than or equal to 600 boxes plus three times the production level for chocolate covered nuts.

Solution

Let x be the number of boxes of chocolate covered creams and let y be the number of boxes of chocolate covered nuts. So, the objective function (for the combined profit) is given by

$$P = 1.5x + 2y. \qquad \text{Objective function}$$

The three constraints translate into the following linear inequalities.

1. $x + y \le 1200$ ⟹ $x + y \le 1200$
2. $y \le \tfrac{1}{2}x$ ⟹ $-x + 2y \le 0$
3. $x \le 600 + 3y$ ⟹ $x - 3y \le 600$

Because neither x nor y can be negative, you also have the two additional constraints of $x \ge 0$ and $y \ge 0$. Figure 6.37 shows the region determined by the constraints. To find the maximum profit, test the values of P at the vertices of the region.

$$
\begin{aligned}
\text{At } (0, 0):\ & P = 1.5(0) + 2(0) = 0 \\
\text{At } (800, 400):\ & P = 1.5(800) + 2(400) = 2000 \qquad \text{Maximum profit} \\
\text{At } (1050, 150):\ & P = 1.5(1050) + 2(150) = 1875 \\
\text{At } (600, 0):\ & P = 1.5(600) + 2(0) = 900
\end{aligned}
$$

So, the maximum profit is $2000, and it occurs when the monthly production consists of 800 boxes of chocolate covered creams and 400 boxes of chocolate covered nuts.

✔**CHECKPOINT**   Now try Exercise 39.

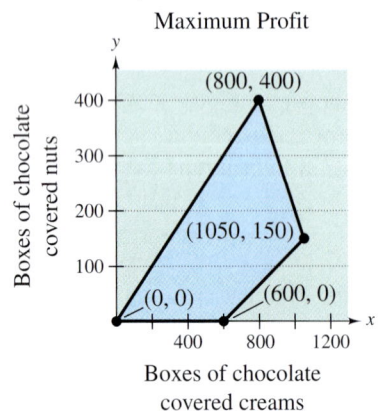

Maximum Profit

Boxes of chocolate covered nuts

Boxes of chocolate covered creams

FIGURE **6.37**

In Example 5, if the manufacturer improved the production of chocolate covered creams so that they yielded a profit of $2.50 per unit, the maximum profit could then be found using the objective function $P = 2.5x + 2y$. By testing the values of P at the vertices of the region, you would find that the maximum profit was $2925 and that it occurred when $x = 1050$ and $y = 150$.

Example 6   **Optimal Cost**

The liquid portion of a diet is to provide at least 300 calories, 36 units of vitamin A, and 90 units of vitamin C. A cup of dietary drink X costs $0.12 and provides 60 calories, 12 units of vitamin A, and 10 units of vitamin C. A cup of dietary drink Y costs $0.15 and provides 60 calories, 6 units of vitamin A, and 30 units of vitamin C. How many cups of each drink should be consumed each day to obtain an optimal cost and still meet the daily requirements?

Solution

As in Example 9 in Section 6.5, let x be the number of cups of dietary drink X and let y be the number of cups of dietary drink Y.

$$\left.\begin{array}{lrcl}
\text{For calories:} & 60x + 60y & \geq & 300 \\
\text{For vitamin A:} & 12x + 6y & \geq & 36 \\
\text{For vitamin C:} & 10x + 30y & \geq & 90 \\
& x & \geq & 0 \\
& y & \geq & 0
\end{array}\right\} \quad \text{Constraints}$$

The cost C is given by $C = 0.12x + 0.15y$.   Objective function

The graph of the region corresponding to the constraints is shown in Figure 6.38. Because you want to incur as little cost as possible, you want to determine the *minimum* cost. To determine the minimum cost, test C at each vertex of the region.

At $(0, 6)$: $C = 0.12(0) + 0.15(6) = 0.90$

At $(1, 4)$: $C = 0.12(1) + 0.15(4) = 0.72$

At $(3, 2)$: $C = 0.12(3) + 0.15(2) = 0.66$ Minimum value of C

At $(9, 0)$: $C = 0.12(9) + 0.15(0) = 1.08$

So, the minimum cost is $0.66 per day, and this occurs when 3 cups of drink X and 2 cups of drink Y are consumed each day.

✔**CHECKPOINT**   Now try Exercise 43.

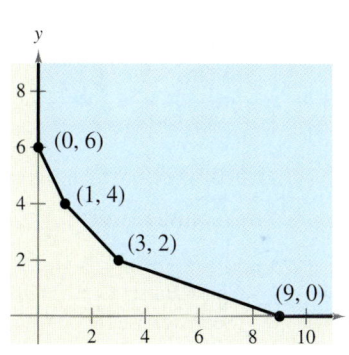

FIGURE **6.38**

𝒲RITING ABOUT 𝓜ATHEMATICS

Creating a Linear Programming Problem   Sketch the region determined by the following constraints.

$$\left.\begin{array}{rcl}
x + 2y & \leq & 8 \\
x + y & \leq & 5 \\
x & \geq & 0 \\
y & \geq & 0
\end{array}\right\} \quad \text{Constraints}$$

Find, if possible, an objective function of the form $z = ax + by$ that has a maximum at each indicated vertex of the region.

a. $(0, 4)$ **b.** $(2, 3)$ **c.** $(5, 0)$ **d.** $(0, 0)$

Explain how you found each objective function.

6.6 Exercises

VOCABULARY CHECK: Fill in the blanks.

1. In the process called _____, you are asked to find the maximum or minimum value of a quantity.

2. One type of optimization strategy is called _____ _____.

3. The _____ function of a linear programming problem gives the quantity that is to be maximized or minimized.

4. The _____ of a linear programming problem determine the set of _____ _____.

5. If a linear programming problem has a solution, it must occur at a _____ of the set of feasible solutions.

PREREQUISITE SKILLS REVIEW: Practice and review algebra skills needed for this section at **www.Eduspace.com.**

In Exercises 1–12, find the minimum and maximum values of the objective function and where they occur, subject to the indicated constraints. (For each exercise, the graph of the region determined by the constraints is provided.)

1. Objective function:

   $z = 4x + 3y$

   Constraints:

   $x \geq 0$

   $y \geq 0$

   $x + y \leq 5$

   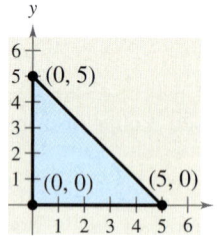

2. Objective function:

   $z = 2x + 8y$

   Constraints:

   $x \geq 0$

   $y \geq 0$

   $2x + y \leq 4$

   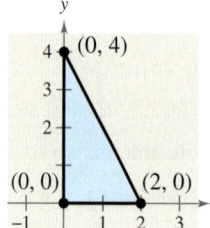

3. Objective function:

   $z = 3x + 8y$

   Constraints:
   (See Exercise 1.)

4. Objective function:

   $z = 7x + 3y$

   Constraints:
   (See Exercise 2.)

5. Objective function:

   $z = 3x + 2y$

   Constraints:

   $x \geq 0$

   $y \geq 0$

   $x + 3y \leq 15$

   $4x + y \leq 16$

6. Objective function:

   $z = 4x + 5y$

   Constraints:

   $x \geq 0$

   $2x + 3y \geq 6$

   $3x - y \leq 9$

   $x + 4y \leq 16$

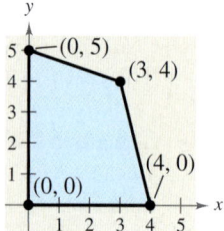

FIGURE FOR **5**

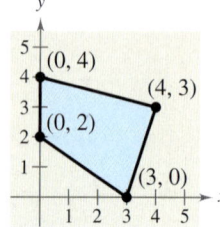

FIGURE FOR **6**

7. Objective function:

   $z = 5x + 0.5y$

   Constraints:
   (See Exercise 5.)

8. Objective function:

   $z = 2x + y$

   Constraints:
   (See Exercise 6.)

9. Objective function:

   $z = 10x + 7y$

   Constraints:

   $0 \leq x \leq 60$

   $0 \leq y \leq 45$

   $5x + 6y \leq 420$

10. Objective function:

    $z = 25x + 35y$

    Constraints:

    $x \geq 0$

    $y \geq 0$

    $8x + 9y \leq 7200$

    $8x + 9y \geq 3600$

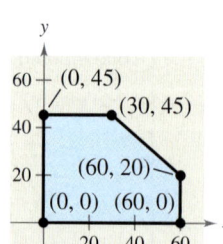

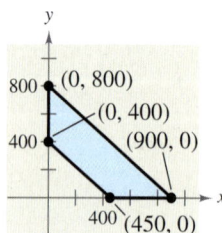

11. Objective function:

    $z = 25x + 30y$

    Constraints:
    (See Exercise 9.)

12. Objective function:

    $z = 15x + 20y$

    Constraints:
    (See Exercise 10.)

In Exercises 13–20, sketch the region determined by the constraints. Then find the minimum and maximum values of the objective function and where they occur, subject to the indicated constraints.

13. Objective function:

    $z = 6x + 10y$

    Constraints:

    $x \geq 0$

    $y \geq 0$

    $2x + 5y \leq 10$

14. Objective function:

    $z = 7x + 8y$

    Constraints:

    $x \geq 0$

    $y \geq 0$

    $x + \frac{1}{2}y \leq 4$

15. Objective function:

    $z = 9x + 24y$

    Constraints:

    (See Exercise 13.)

16. Objective function:

    $z = 7x + 2y$

    Constraints:

    (See Exercise 14.)

17. Objective function:

    $z = 4x + 5y$

    Constraints:

    $x \geq 0$

    $y \geq 0$

    $x + y \geq 8$

    $3x + 5y \geq 30$

18. Objective function:

    $z = 4x + 5y$

    Constraints:

    $x \geq 0$

    $y \geq 0$

    $2x + 2y \leq 10$

    $x + 2y \leq 6$

19. Objective function:

    $z = 2x + 7y$

    Constraints:

    (See Exercise 17.)

20. Objective function:

    $z = 2x - y$

    Constraints:

    (See Exercise 18.)

In Exercises 21–24, use a graphing utility to graph the region determined by the constraints. Then find the minimum and maximum values of the objective function and where they occur, subject to the constraints.

21. Objective function:

    $z = 4x + y$

    Constraints:

    $x \geq 0$

    $y \geq 0$

    $x + 2y \leq 40$

    $2x + 3y \geq 72$

22. Objective function:

    $z = x$

    Constraints:

    $x \geq 0$

    $y \geq 0$

    $2x + 3y \leq 60$

    $2x + y \leq 28$

    $4x + y \leq 48$

23. Objective function:

    $z = x + 4y$

    Constraints:

    (See Exercise 21.)

24. Objective function:

    $z = y$

    Constraints:

    (See Exercise 22.)

In Exercises 25–28, find the maximum value of the objective function and where it occurs, subject to the constraints $x \geq 0$, $y \geq 0$, $3x + y \leq 15$, and $4x + 3y \leq 30$.

25. $z = 2x + y$

26. $z = 5x + y$

27. $z = x + y$

28. $z = 3x + y$

In Exercises 29–32, find the maximum value of the objective function and where it occurs, subject to the constraints $x \geq 0$, $y \geq 0$, $x + 4y \leq 20$, $x + y \leq 18$, and $2x + 2y \leq 21$.

29. $z = x + 5y$

30. $z = 2x + 4y$

31. $z = 4x + 5y$

32. $z = 4x + y$

In Exercises 33–38, the linear programming problem has an unusual characteristic. Sketch a graph of the solution region for the problem and describe the unusual characteristic. Find the maximum value of the objective function and where it occurs.

33. Objective function:

    $z = 2.5x + y$

    Constraints:

    $x \geq 0$

    $y \geq 0$

    $3x + 5y \leq 15$

    $5x + 2y \leq 10$

34. Objective function:

    $z = x + y$

    Constraints:

    $x \geq 0$

    $y \geq 0$

    $-x + y \leq 1$

    $-x + 2y \leq 4$

35. Objective function:

    $z = -x + 2y$

    Constraints:

    $x \geq 0$

    $y \geq 0$

    $x \leq 10$

    $x + y \leq 7$

36. Objective function:

    $z = x + y$

    Constraints:

    $x \geq 0$

    $y \geq 0$

    $-x + y \leq 0$

    $-3x + y \geq 3$

37. Objective function:

    $z = 3x + 4y$

    Constraints:

    $x \geq 0$

    $y \geq 0$

    $x + y \leq 1$

    $2x + y \leq 4$

38. Objective function:

    $z = x + 2y$

    Constraints:

    $x \geq 0$

    $y \geq 0$

    $x + 2y \leq 4$

    $2x + y \leq 4$

39. *Optimal Profit* A manufacturer produces two models of bicycles. The times (in hours) required for assembling, painting, and packaging each model are shown in the table.

Process	Hours, model A	Hours, model B
Assembling	2	2.5
Painting	4	1
Packaging	1	0.75

The total times available for assembling, painting, and packaging are 4000 hours, 4800 hours, and 1500 hours, respectively. The profits per unit are $45 for model A and $50 for model B. What is the optimal production level for each model? What is the optimal profit?

40. *Optimal Profit* A manufacturer produces two models of bicycles. The times (in hours) required for assembling, painting, and packaging each model are shown in the table.

Process	Hours, model A	Hours, model B
Assembling	2.5	3
Painting	2	1
Packaging	0.75	1.25

The total times available for assembling, painting, and packaging are 4000 hours, 2500 hours, and 1500 hours, respectively. The profits per unit are $50 for model A and $52 for model B. What is the optimal production level for each model? What is the optimal profit?

41. *Optimal Profit* A merchant plans to sell two models of MP3 players at costs of $250 and $300. The $250 model yields a profit of $25 per unit and the $300 model yields a profit of $40 per unit. The merchant estimates that the total monthly demand will not exceed 250 units. The merchant does not want to invest more than $65,000 in inventory for these products. What is the optimal inventory level for each model? What is the optimal profit?

42. *Optimal Profit* A fruit grower has 150 acres of land available to raise two crops, A and B. It takes 1 day to trim an acre of crop A and 2 days to trim an acre of crop B, and there are 240 days per year available for trimming. It takes 0.3 day to pick an acre of crop A and 0.1 day to pick an acre of crop B, and there are 30 days available for picking. The profit is $140 per acre for crop A and $235 per acre for crop B. What is the optimal acreage for each fruit? What is the optimal profit?

43. *Optimal Cost* A farming cooperative mixes two brands of cattle feed. Brand X costs $25 per bag and contains two units of nutritional element A, two units of element B, and two units of element C. Brand Y costs $20 per bag and contains one unit of nutritional element A, nine units of element B, and three units of element C. The minimum requirements of nutrients A, B, and C are 12 units, 36 units, and 24 units, respectively. What is the optimal number of bags of each brand that should be mixed? What is the optimal cost?

Model It

44. *Optimal Cost* According to AAA (Automobile Association of America), on January 24, 2005, the national average price per gallon for regular unleaded (87-octane) gasoline was $1.84, and the price for premium unleaded (93-octane) gasoline was $2.03.

(a) Write an objective function that models the cost of the blend of mid-grade unleaded gasoline (89-octane).

(b) Determine the constraints for the objective function in part (a).

(c) Sketch a graph of the region determined by the constraints from part (b).

(d) Determine the blend of regular and premium unleaded gasoline that results in an optimal cost of mid-grade unleaded gasoline.

(e) What is the optimal cost?

(f) Is the cost lower than the national average of $1.96 per gallon for mid-grade unleaded gasoline?

45. *Optimal Revenue* An accounting firm has 900 hours of staff time and 155 hours of reviewing time available each week. The firm charges $2500 for an audit and $350 for a tax return. Each audit requires 75 hours of staff time and 10 hours of review time. Each tax return requires 12.5 hours of staff time and 2.5 hours of review time. What numbers of audits and tax returns will yield an optimal revenue? What is the optimal revenue?

46. *Optimal Revenue* The accounting firm in Exercise 45 lowers its charge for an audit to $2000. What numbers of audits and tax returns will yield an optimal revenue? What is the optimal revenue?

47. *Investment Portfolio* An investor has up to $250,000 to invest in two types of investments. Type A pays 8% annually and type B pays 10% annually. To have a well-balanced portfolio, the investor imposes the following conditions. At least one-fourth of the total portfolio is to be allocated to type A investments and at least one-fourth of the portfolio is to be allocated to type B investments. What is the optimal amount that should be invested in each type of investment? What is the optimal return?

48. *Investment Portfolio* An investor has up to $450,000 to invest in two types of investments. Type A pays 6% annually and type B pays 10% annually. To have a well-balanced portfolio, the investor imposes the following conditions. At least one-half of the total portfolio is to be allocated to type A investments and at least one-fourth of the portfolio is to be allocated to type B investments. What is the optimal amount that should be invested in each type of investment? What is the optimal return?

Synthesis

True or False? **In Exercises 49 and 50, determine whether the statement is true or false. Justify your answer.**

49. If an objective function has a maximum value at the vertices $(4, 7)$ and $(8, 3)$, you can conclude that it also has a maximum value at the points $(4.5, 6.5)$ and $(7.8, 3.2)$.

50. When solving a linear programming problem, if the objective function has a maximum value at more than one vertex, you can assume that there are an infinite number of points that will produce the maximum value.

In Exercises 51 and 52, determine values of t such that the objective function has maximum values at the indicated vertices.

51. Objective function:
$$z = 3x + ty$$
Constraints:
$$x \geq 0$$
$$y \geq 0$$
$$x + 3y \leq 15$$
$$4x + y \leq 16$$
(a) $(0, 5)$
(b) $(3, 4)$

52. Objective function:
$$z = 3x + ty$$
Constraints:
$$x \geq 0$$
$$y \geq 0$$
$$x + 2y \leq 4$$
$$x - y \leq 1$$
(a) $(2, 1)$
(b) $(0, 2)$

Think About It **In Exercises 53–56, find an objective function that has a maximum or minimum value at the indicated vertex of the constraint region shown below. (There are many correct answers.)**

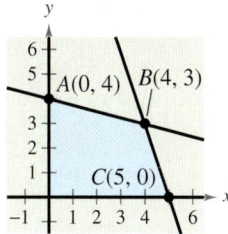

53. The maximum occurs at vertex A.

54. The maximum occurs at vertex B.

55. The maximum occurs at vertex C.

56. The minimum occurs at vertex C.

Skills Review

In Exercises 57–60, simplify the complex fraction.

57. $\dfrac{\left(\dfrac{9}{x}\right)}{\left(\dfrac{6}{x} + 2\right)}$

58. $\dfrac{\left(1 + \dfrac{2}{x}\right)}{\left(x - \dfrac{4}{x}\right)}$

59. $\dfrac{\left(\dfrac{4}{x^2 - 9} + \dfrac{2}{x - 2}\right)}{\left(\dfrac{1}{x + 3} + \dfrac{1}{x - 3}\right)}$

60. $\dfrac{\left(\dfrac{1}{x + 1} + \dfrac{1}{2}\right)}{\left(\dfrac{3}{2x^2 + 4x + 2}\right)}$

In Exercises 61–66, solve the equation algebraically. Round the result to three decimal places.

61. $e^{2x} + 2e^x - 15 = 0$

62. $e^{2x} - 10e^x + 24 = 0$

63. $8(62 - e^{x/4}) = 192$

64. $\dfrac{150}{e^{-x} - 4} = 75$

65. $7 \ln 3x = 12$

66. $\ln(x + 9)^2 = 2$

In Exercises 67 and 68, solve the system of linear equations and check any solution algebraically.

67. $\begin{cases} -x - 2y + 3z = -23 \\ 2x + 6y - z = 17 \\ 5y + z = 8 \end{cases}$

68. $\begin{cases} 7x - 3y + 5z = -28 \\ 4x + 4z = -16 \\ 7x + 2y - z = 0 \end{cases}$

6 Chapter Summary

What did you learn?

6   Review Exercises

6.1 **In Exercises 1–8, solve the system by the method of substitution.**

1. $\begin{cases} x + y = 2 \\ x - y = 0 \end{cases}$

2. $\begin{cases} 2x - 3y = 3 \\ x - y = 0 \end{cases}$

3. $\begin{cases} 0.5x + y = 0.75 \\ 1.25x - 4.5y = -2.5 \end{cases}$

4. $\begin{cases} -x + \frac{2}{5}y = \frac{3}{5} \\ -x + \frac{1}{5}y = -\frac{4}{5} \end{cases}$

5. $\begin{cases} x^2 - y^2 = 9 \\ x - y = 1 \end{cases}$

6. $\begin{cases} x^2 + y^2 = 169 \\ 3x + 2y = 39 \end{cases}$

7. $\begin{cases} y = 2x^2 \\ y = x^4 - 2x^2 \end{cases}$

8. $\begin{cases} x = y + 3 \\ x = y^2 + 1 \end{cases}$

In Exercises 9–12, solve the system graphically.

9. $\begin{cases} 2x - y = 10 \\ x + 5y = -6 \end{cases}$

10. $\begin{cases} 8x - 3y = -3 \\ 2x + 5y = 28 \end{cases}$

11. $\begin{cases} y = 2x^2 - 4x + 1 \\ y = x^2 - 4x + 3 \end{cases}$

12. $\begin{cases} y^2 - 2y + x = 0 \\ x + y = 0 \end{cases}$

 In Exercises 13 and 14, use a graphing utility to solve the system of equations. Find the solution accurate to two decimal places.

13. $\begin{cases} y = -2e^{-x} \\ 2e^x + y = 0 \end{cases}$

14. $\begin{cases} y = \ln(x - 1) - 3 \\ y = 4 - \frac{1}{2}x \end{cases}$

15. *Break-Even Analysis* You set up a scrapbook business and make an initial investment of \$50,000. The unit cost of a scrapbook kit is \$12 and the selling price is \$25. How many kits must you sell to break even?

16. *Choice of Two Jobs* You are offered two sales jobs at a pharmaceutical company. One company offers an annual salary of \$35,000 plus a year-end bonus of 1.5% of your total sales. The other company offers an annual salary of \$32,000 plus a year-end bonus of 2% of your total sales. What amount of sales will make the second offer better? Explain.

17. *Geometry* The perimeter of a rectangle is 480 meters and its length is 150% of its width. Find the dimensions of the rectangle.

18. *Geometry* The perimeter of a rectangle is 68 feet and its width is $\frac{8}{9}$ times its length. Find the dimensions of the rectangle.

6.2 **In Exercises 19–26, solve the system by the method of elimination.**

19. $\begin{cases} 2x - y = 2 \\ 6x + 8y = 39 \end{cases}$

20. $\begin{cases} 40x + 30y = 24 \\ 20x - 50y = -14 \end{cases}$

21. $\begin{cases} 0.2x + 0.3y = 0.14 \\ 0.4x + 0.5y = 0.20 \end{cases}$

22. $\begin{cases} 12x + 42y = -17 \\ 30x - 18y = 19 \end{cases}$

23. $\begin{cases} 3x - 2y = 0 \\ 3x + 2(y + 5) = 10 \end{cases}$

24. $\begin{cases} 7x + 12y = 63 \\ 2x + 3(y + 2) = 21 \end{cases}$

25. $\begin{cases} 1.25x - 2y = 3.5 \\ 5x - 8y = 14 \end{cases}$

26. $\begin{cases} 1.5x + 2.5y = 8.5 \\ 6x + 10y = 24 \end{cases}$

In Exercises 27–30, match the system of linear equations with its graph. Describe the number of solutions and state whether the system is consistent or inconsistent. [The graphs are labeled (a), (b), (c), and (d).]

(a)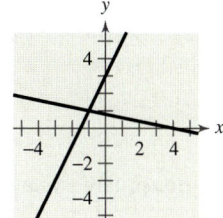

(b)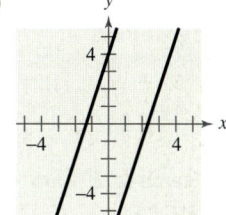

(c)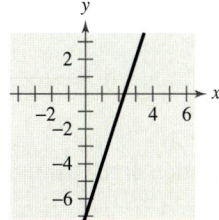

(d)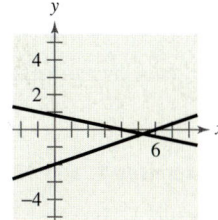

27. $\begin{cases} x + 5y = 4 \\ x - 3y = 6 \end{cases}$

28. $\begin{cases} -3x + y = -7 \\ 9x - 3y = 21 \end{cases}$

29. $\begin{cases} 3x - y = 7 \\ -6x + 2y = 8 \end{cases}$

30. $\begin{cases} 2x - y = -3 \\ x + 5y = 4 \end{cases}$

Supply and Demand **In Exercises 31 and 32, find the equilibrium point of demand and supply equations.**

	Demand	*Supply*
31.	$p = 37 - 0.0002x$	$p = 22 + 0.00001x$
32.	$p = 120 - 0.0001x$	$p = 45 + 0.0002x$

6.3 In Exercises 33 and 34, use back-substitution to solve the system of linear equations.

33. $\begin{cases} x - 4y + 3z = 3 \\ -y + z = -1 \\ z = -5 \end{cases}$

34. $\begin{cases} x - 7y + 8z = 85 \\ y - 9z = -35 \\ z = 3 \end{cases}$

In Exercises 35–38, use Gaussian elimination to solve the system of equations.

35. $\begin{cases} x + 2y + 6z = 4 \\ -3x + 2y - z = -4 \\ 4x + 2z = 16 \end{cases}$

36. $\begin{cases} x + 3y - z = 13 \\ 2x - 5z = 23 \\ 4x - y - 2z = 14 \end{cases}$

37. $\begin{cases} x - 2y + z = -6 \\ 2x - 3y = -7 \\ -x + 3y - 3z = 11 \end{cases}$

38. $\begin{cases} 2x + 6z = -9 \\ 3x - 2y + 11z = -16 \\ 3x - y + 7z = -11 \end{cases}$

In Exercises 39 and 40, solve the nonsquare system of equations.

39. $\begin{cases} 5x - 12y + 7z = 16 \\ 3x - 7y + 4z = 9 \end{cases}$

40. $\begin{cases} 2x + 5y - 19z = 34 \\ 3x + 8y - 31z = 54 \end{cases}$

In Exercises 41 and 42, find the equation of the parabola

$y = ax^2 + bx + c$

that passes through the points. To verify your result, use a graphing utility to plot the points and graph the parabola.

41.
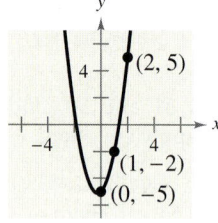

42.
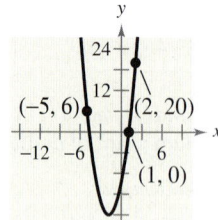

In Exercises 43 and 44, find the equation of the circle

$x^2 + y^2 + Dx + Ey + F = 0$

that passes through the points. To verify your result, use a graphing utility to plot the points and graph the circle.

43.
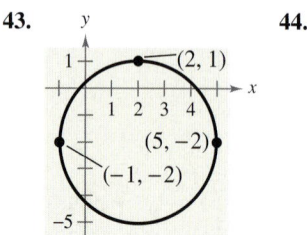

44.
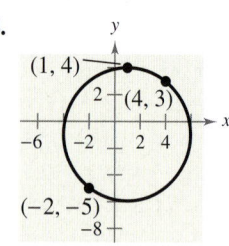

45. *Data Analysis: Online Shopping* The table shows the projected numbers y (in millions) of people shopping online in the United States from 2003 to 2005. (Source: eMarketer)

Year	Online shoppers, y
2003	101.7
2004	108.4
2005	121.1

(a) Use the technique demonstrated in Exercises 67–70 in Section 6.3 to set up a system of equations for the data and to find a least squares regression parabola that models the data. Let x represent the year, with $x = 3$ corresponding to 2003.

(b) Use a graphing utility to graph the parabola and the data in the same viewing window. How well does the model fit the data?

(c) Use the model to estimate the number of online shoppers in 2008. Does your answer seem reasonable?

46. *Agriculture* A mixture of 6 gallons of chemical A, 8 gallons of chemical B, and 13 gallons of chemical C is required to kill a destructive crop insect. Commercial spray X contains 1, 2, and 2 parts, respectively, of these chemicals. Commercial spray Y contains only chemical C. Commercial spray Z contains chemicals A, B, and C in equal amounts. How much of each type of commercial spray is needed to get the desired mixture?

47. *Investment Analysis* An inheritance of $40,000 was divided among three investments yielding $3500 in interest per year. The interest rates for the three investments were 7%, 9%, and 11%. Find the amount placed in each investment if the second and third were $3000 and $5000 less than the first, respectively.

48. *Vertical Motion*   An object moving vertically is at the given heights at the specified times. Find the position equation $s = \frac{1}{2}at^2 + v_0t + s_0$ for the object.

(a) At $t = 1$ second, $s = 134$ feet

At $t = 2$ seconds, $s = 86$ feet

At $t = 3$ seconds, $s = 6$ feet

(b) At $t = 1$ second, $s = 184$ feet

At $t = 2$ seconds, $s = 116$ feet

At $t = 3$ seconds, $s = 16$ feet

6.4 In Exercises 49–52, write the form of the partial fraction decomposition for the rational expression. Do not solve for the constants.

49. $\dfrac{3}{x^2 + 20x}$

50. $\dfrac{x - 8}{x^2 - 3x - 28}$

51. $\dfrac{3x - 4}{x^3 - 5x^2}$

52. $\dfrac{x - 2}{x(x^2 + 2)^2}$

In Exercises 53–60, write the partial fraction decomposition of the rational expression.

53. $\dfrac{4 - x}{x^2 + 6x + 8}$

54. $\dfrac{-x}{x^2 + 3x + 2}$

55. $\dfrac{x^2}{x^2 + 2x - 15}$

56. $\dfrac{9}{x^2 - 9}$

57. $\dfrac{x^2 + 2x}{x^3 - x^2 + x - 1}$

58. $\dfrac{4x}{3(x - 1)^2}$

59. $\dfrac{3x^2 + 4x}{(x^2 + 1)^2}$

60. $\dfrac{4x^2}{(x - 1)(x^2 + 1)}$

6.5 In Exercises 61–64, sketch the graph of the inequality.

61. $y \le 5 - \frac{1}{2}x$

62. $3y - x \ge 7$

63. $y - 4x^2 > -1$

64. $y \le \dfrac{3}{x^2 + 2}$

In Exercises 65–72, sketch the graph and label the vertices of the solution set of the system of inequalities.

65. $\begin{cases} x + 2y \le 160 \\ 3x + y \le 180 \\ x \ge 0 \\ y \ge 0 \end{cases}$

66. $\begin{cases} 2x + 3y \le 24 \\ 2x + y \le 16 \\ x \ge 0 \\ y \ge 0 \end{cases}$

67. $\begin{cases} 3x + 2y \ge 24 \\ x + 2y \ge 12 \\ 2 \le x \le 15 \\ y \le 15 \end{cases}$

68. $\begin{cases} 2x + y \ge 16 \\ x + 3y \ge 18 \\ 0 \le x \le 25 \\ 0 \le y \le 25 \end{cases}$

69. $\begin{cases} y < x + 1 \\ y > x^2 - 1 \end{cases}$

70. $\begin{cases} y \le 6 - 2x - x^2 \\ y \ge x + 6 \end{cases}$

71. $\begin{cases} 2x - 3y \ge 0 \\ 2x - y \le 8 \\ y \ge 0 \end{cases}$

72. $\begin{cases} x^2 + y^2 \le 9 \\ (x - 3)^2 + y^2 \le 9 \end{cases}$

73. *Inventory Costs*   A warehouse operator has 24,000 square feet of floor space in which to store two products. Each unit of product I requires 20 square feet of floor space and costs \$12 per day to store. Each unit of product II requires 30 square feet of floor space and costs \$8 per day to store. The total storage cost per day cannot exceed \$12,400. Find and graph a system of inequalities describing all possible inventory levels.

74. *Nutrition*   A dietitian is asked to design a special dietary supplement using two different foods. Each ounce of food X contains 12 units of calcium, 10 units of iron, and 20 units of vitamin B. Each ounce of food Y contains 15 units of calcium, 20 units of iron, and 12 units of vitamin B. The minimum daily requirements of the diet are 300 units of calcium, 280 units of iron, and 300 units of vitamin B.

(a) Write a system of inequalities describing the different amounts of food X and food Y that can be used.

(b) Sketch a graph of the region in part (a).

(c) Find two solutions to the system and interpret their meanings in the context of the problem.

Supply and Demand   In Exercises 75 and 76, (a) graph the systems representing the consumer surplus and producer surplus for the supply and demand equations and (b) find the consumer surplus and producer surplus.

Demand	*Supply*
75. $p = 160 - 0.0001x$	$p = 70 + 0.0002x$
76. $p = 130 - 0.0002x$	$p = 30 + 0.0003x$

6.6 In Exercises 77–82, sketch the region determined by the constraints. Then find the minimum and maximum values of the objective function and where they occur, subject to the indicated restraints.

77. Objective function:

$z = 3x + 4y$

Constraints:

$x \ge 0$

$y \ge 0$

$2x + 5y \le 50$

$4x + y \le 28$

78. Objective function:

$z = 10x + 7y$

Constraints:

$x \ge 0$

$y \ge 0$

$2x + y \ge 100$

$x + y \ge 75$

79. Objective function:

$z = 1.75x + 2.25y$

Constraints:

$$x \geq 0$$
$$y \geq 0$$
$$2x + y \geq 25$$
$$3x + 2y \geq 45$$

80. Objective function:

$z = 50x + 70y$

Constraints:

$$x \geq 0$$
$$y \geq 0$$
$$x + 2y \leq 1500$$
$$5x + 2y \leq 3500$$

81. Objective function:

$z = 5x + 11y$

Constraints:

$$x \geq 0$$
$$y \geq 0$$
$$x + 3y \leq 12$$
$$3x + 2y \leq 15$$

82. Objective function:

$z = -2x + y$

Constraints:

$$x \geq 0$$
$$y \geq 0$$
$$x + y \geq 7$$
$$5x + 2y \geq 20$$

83. *Optimal Revenue* A student is working part time as a hairdresser to pay college expenses. The student may work no more than 24 hours per week. Haircuts cost $25 and require an average of 20 minutes, and permanents cost $70 and require an average of 1 hour and 10 minutes. What combination of haircuts and/or permanents will yield an optimal revenue? What is the optimal revenue?

84. *Optimal Profit* A shoe manufacturer produces a walking shoe and a running shoe yielding profits of $18 and $24, respectively. Each shoe must go through three processes, for which the required times per unit are shown in the table.

	Process I	Process II	Process III
Hours for walking shoe	4	1	1
Hours for running shoe	2	2	1
Hours available per day	24	9	8

What is the optimal production level for each type of shoe? What is the optimal profit?

85. *Optimal Cost* A pet supply company mixes two brands of dry dog food. Brand X costs $15 per bag and contains eight units of nutritional element A, one unit of nutritional element B, and two units of nutritional element C. Brand Y costs $30 per bag and contains two units of nutritional element A, one unit of nutritional element B, and seven units of nutritional element C. Each bag of mixed dog food must contain at least 16 units, 5 units, and 20 units of nutritional elements A, B, and C, respectively. Find the numbers of bags of brands X and Y that should be mixed to produce a mixture meeting the minimum nutritional requirements and having an optimal cost. What is the optimal cost?

86. *Optimal Cost* Regular unleaded gasoline and premium unleaded gasoline have octane ratings of 87 and 93, respectively. For the week of January 3, 2005, regular unleaded gasoline in Houston, Texas averaged $1.63 per gallon. For the same week, premium unleaded gasoline averaged $1.83 per gallon. Determine the blend of regular and premium unleaded gasoline that results in an optimal cost of midgrade unleaded (89-octane) gasoline. What is the optimal cost? (Source: Energy Information Administration)

Synthesis

True or False? In Exercises 87 and 88, determine whether the statement is true or false. Justify your answer.

87. The system

$$\begin{cases} y \leq 5 \\ y \geq -2 \\ y \geq \frac{7}{2}x - 9 \\ y \geq -\frac{7}{2}x + 26 \end{cases}$$

represents the region covered by an isosceles trapezoid.

88. It is possible for an objective function of a linear programming problem to have exactly 10 maximum value points.

In Exercises 89–92, find a system of linear equations having the ordered pair as a solution. (There are many correct answers.)

89. $(-6, 8)$

90. $(5, -4)$

91. $\left(\frac{4}{3}, 3\right)$

92. $\left(-1, \frac{9}{4}\right)$

In Exercises 93–96, find a system of linear equations having the ordered triple as a solution. (There are many answers.)

93. $(4, -1, 3)$

94. $(-3, 5, 6)$

95. $\left(5, \frac{3}{2}, 2\right)$

96. $\left(\frac{3}{4}, -2, 8\right)$

97. *Writing* Explain what is meant by an inconsistent system of linear equations.

98. How can you tell graphically that a system of linear equations in two variables has no solution? Give an example.

99. *Writing* Write a brief paragraph describing any advantages of substitution over the graphical method of solving a system of equations.

6 | Chapter Test

Take this test as you would take a test in class. When you are finished, check your work against the answers given in the back of the book.

In Exercises 1–3, solve the system by the method of substitution.

1. $\begin{cases} x - y = -7 \\ 4x + 5y = 8 \end{cases}$

2. $\begin{cases} y = x - 1 \\ y = (x - 1)^3 \end{cases}$

3. $\begin{cases} 2x - y^2 = 0 \\ x - y = 4 \end{cases}$

In Exercises 4–6, solve the system graphically.

4. $\begin{cases} 2x - 3y = 0 \\ 2x + 3y = 12 \end{cases}$

5. $\begin{cases} y = 9 - x^2 \\ y = x + 3 \end{cases}$

6. $\begin{cases} y - \ln x = 12 \\ 7x - 2y + 11 = -6 \end{cases}$

In Exercises 7–10, solve the linear system by the method of elimination.

7. $\begin{cases} 2x + 3y = 17 \\ 5x - 4y = -15 \end{cases}$

8. $\begin{cases} 2.5x - y = 6 \\ 3x + 4y = 2 \end{cases}$

9. $\begin{cases} x - 2y + 3z = 11 \\ 2x \quad\quad - z = 3 \\ \quad\quad 3y + z = -8 \end{cases}$

10. $\begin{cases} 3x + 2y + z = 17 \\ -x + y + z = 4 \\ x - y - z = 3 \end{cases}$

In Exercises 11–14, write the partial fraction decomposition of the rational expression.

11. $\dfrac{2x + 5}{x^2 - x - 2}$

12. $\dfrac{3x^2 - 2x + 4}{x^2(2 - x)}$

13. $\dfrac{x^2 + 5}{x^3 - x}$

14. $\dfrac{x^2 - 4}{x^3 + 2x}$

In Exercises 15–17, sketch the graph and label the vertices of the solution of the system of inequalities.

15. $\begin{cases} 2x + y \le 4 \\ 2x - y \ge 0 \\ x \ge 0 \end{cases}$

16. $\begin{cases} y < -x^2 + x + 4 \\ y > 4x \end{cases}$

17. $\begin{cases} x^2 + y^2 \le 16 \\ x \ge 1 \\ y \ge -3 \end{cases}$

18. Find the maximum and minimum values of the objective function $z = 20x + 12y$ and where they occur, subject to the following constraints.

$$\left. \begin{array}{r} x \ge 0 \\ y \ge 0 \\ x + 4y \le 32 \\ 3x + 2y \le 36 \end{array} \right\} \quad \text{Constraints}$$

19. A total of \$50,000 is invested in two funds paying 8% and 8.5% simple interest. The yearly interest is \$4150. How much is invested at each rate?

20. Find the equation of the parabola $y = ax^2 + bx + c$ passing through the points $(0, 6), (-2, 2),$ and $\left(3, \frac{9}{2}\right)$.

21. A manufacturer produces two types of television stands. The amounts (in hours) of time for assembling, staining, and packaging the two models are shown in the table at the left. The total amounts of time available for assembling, staining, and packaging are 4000, 8950, and 2650 hours, respectively. The profits per unit are \$30 (model I) and \$40 (model II). What is the optimal inventory level for each model? What is the optimal profit?

	Model I	Model II
Assembling	0.5	0.75
Staining	2.0	1.5
Packaging	0.5	0.5

TABLE FOR 21

Proofs in Mathematics

An **indirect proof** can be useful in proving statements of the form "*p* implies *q*." Recall that the conditional statement $p \rightarrow q$ is false only when *p* is true and *q* is false. To prove a conditional statement indirectly, assume that *p* is true and *q* is false. If this assumption leads to an impossibility, then you have proved that the conditional statement is true. An indirect proof is also called a **proof by contradiction.**

You can use an indirect proof to prove the following conditional statement,

"If *a* is a positive integer and a^2 is divisible by 2, then *a* is divisible by 2,"

as follows. First, assume that *p*, "*a* is a positive integer and a^2 is divisible by 2," is true and *q*, "*a* is divisible by 2," is false. This means that *a* is not divisible by 2. If so, *a* is odd and can be written as $a = 2n + 1$, where *n* is an integer.

$a = 2n + 1$	Definition of an odd integer
$a^2 = 4n^2 + 4n + 1$	Square each side.
$a^2 = 2(2n^2 + 2n) + 1$	Distributive Property

So, by the definition of an odd integer, a^2 is odd. This contradicts the assumption, and you can conclude that *a* is divisible by 2.

Example Using an Indirect Proof

Use an indirect proof to prove that $\sqrt{2}$ is an irrational number.

Solution

Begin by assuming that $\sqrt{2}$ is *not* an irrational number. Then $\sqrt{2}$ can be written as the quotient of two integers *a* and *b* $(b \neq 0)$ that have no common factors.

$\sqrt{2} = \dfrac{a}{b}$	Assume that $\sqrt{2}$ is a rational number.
$2 = \dfrac{a^2}{b^2}$	Square each side.
$2b^2 = a^2$	Multiply each side by b^2.

This implies that 2 is a factor of a^2. So, 2 is also a factor of *a*, and *a* can be written as 2*c*, where *c* is an integer.

$2b^2 = (2c)^2$	Substitute 2*c* for *a*.
$2b^2 = 4c^2$	Simplify.
$b^2 = 2c^2$	Divide each side by 2.

This implies that 2 is a factor of b^2 and also a factor of *b*. So, 2 is a factor of both *a* and *b*. This contradicts the assumption that *a* and *b* have no common factors. So, you can conclude that $\sqrt{2}$ is an irrational number.

This collection of thought-provoking and challenging exercises further explores and expands upon concepts learned in this chapter.

1. A theorem from geometry states that if a triangle is inscribed in a circle such that one side of the triangle is a diameter of the circle, then the triangle is a right triangle. Show that this theorem is true for the circle

   $$x^2 + y^2 = 100$$

   and the triangle formed by the lines

   $$y = 0, \quad y = \tfrac{1}{2}x + 5, \quad \text{and} \quad y = -2x + 20.$$

2. Find k_1 and k_2 such that the system of equations has an infinite number of solutions.

   $$\begin{cases} 3x - 5y = 8 \\ 2x + k_1 y = k_2 \end{cases}$$

3. Consider the following system of linear equations in x and y.

   $$\begin{cases} ax + by = e \\ cx + dy = f \end{cases}$$

   Under what conditions will the system have exactly one solution?

4. Graph the lines determined by each system of linear equations. Then use Gaussian elimination to solve each system. At each step of the elimination process, graph the corresponding lines. What do you observe?

   (a) $\begin{cases} x - 4y = -3 \\ 5x - 6y = 13 \end{cases}$

   (b) $\begin{cases} 2x - 3y = 7 \\ -4x + 6y = -14 \end{cases}$

5. A system of two equations in two unknowns is solved and has a finite number of solutions. Determine the maximum number of solutions of the system satisfying each condition.

   (a) Both equations are linear.

   (b) One equation is linear and the other is quadratic.

   (c) Both equations are quadratic.

6. In the 2004 presidential election, approximately 118.304 million voters divided their votes among three presidential candidates. George W. Bush received 3,320,000 votes more than John Kerry. Ralph Nader received 0.3% of the votes. Write and solve a system of equations to find the total number of votes cast for each candidate. Let B represent the total votes cast for Bush, K the total votes cast for Kerry, and N the total votes cast for Nader. (Source: CNN.com)

7. The Vietnam Veterans Memorial (or "The Wall") in Washington, D.C. was designed by Maya Ying Lin when she was a student at Yale University. This monument has two vertical, triangular sections of black granite with a common side (see figure). The bottom of each section is level with the ground. The tops of the two sections can be approximately modeled by the equations

   $$-2x + 50y = 505 \quad \text{and} \quad 2x + 50y = 505$$

   when the x-axis is superimposed at the base of the wall. Each unit in the coordinate system represents 1 foot. How high is the memorial at the point where the two sections meet? How long is each section?

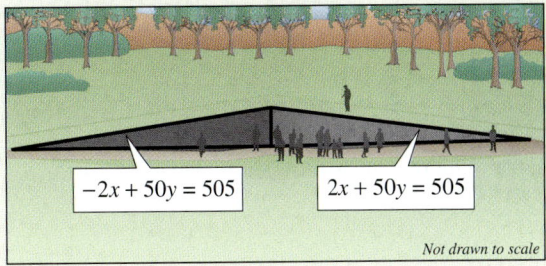

$-2x + 50y = 505$ $2x + 50y = 505$

Not drawn to scale

8. Weights of atoms and molecules are measured in atomic mass units (u). A molecule of C_2H_6 (ethane) is made up of two carbon atoms and six hydrogen atoms and weighs 30.07 u. A molecule of C_3H_8 (propane) is made up of three carbon atoms and eight hydrogen atoms and weighs 44.097 u. Find the weights of a carbon atom and a hydrogen atom.

9. To connect a DVD player to a television set, a cable with special connectors is required at both ends. You buy a six-foot cable for $15.50 and a three-foot cable for $10.25. Assuming that the cost of a cable is the sum of the cost of the two connectors and the cost of the cable itself, what is the cost of a four-foot cable? Explain your reasoning.

10. A hotel 35 miles from an airport runs a shuttle service to and from the airport. The 9:00 A.M. bus leaves for the airport traveling at 30 miles per hour. The 9:15 A.M. bus leaves for the airport traveling at 40 miles per hour. Write a system of linear equations that represents distance as a function of time for each bus. Graph and solve the system. How far from the airport will the 9:15 A.M. bus catch up to the 9:00 A.M. bus?

11. Solve each system of equations by letting $X = 1/x$, $Y = 1/y$, and $Z = 1/z$.

(a) $\begin{cases} \dfrac{12}{x} - \dfrac{12}{y} = 7 \\ \dfrac{3}{x} + \dfrac{4}{y} = 0 \end{cases}$

(b) $\begin{cases} \dfrac{2}{x} + \dfrac{1}{y} - \dfrac{3}{z} = 4 \\ \dfrac{4}{x} \qquad + \dfrac{2}{z} = 10 \\ -\dfrac{2}{x} + \dfrac{3}{y} - \dfrac{13}{z} = -8 \end{cases}$

12. What values should be given to a, b, and c so that the linear system shown has $(-1, 2, -3)$ as its only solution?

$\begin{cases} x + 2y - 3z = a \qquad \text{Equation 1} \\ -x - y + z = b \qquad \text{Equation 2} \\ 2x + 3y - 2z = c \qquad \text{Equation 3} \end{cases}$

13. The following system has one solution: $x = 1$, $y = -1$, and $z = 2$.

$\begin{cases} 4x - 2y + 5z = 16 \\ x + y \qquad = 0 \\ -x - 3y + 2z = 6 \end{cases}$

Solve the system given by (a) Equation 1 and Equation 2, (b) Equation 1 and Equation 3, and (c) Equation 2 and Equation 3. (d) How many solutions does each of these systems have?

14. Solve the system of linear equations algebraically.

$\begin{cases} x_1 - x_2 + 2x_3 + 2x_4 + 6x_5 = 6 \\ 3x_1 - 2x_2 + 4x_3 + 4x_4 + 12x_5 = 14 \\ -x_2 - x_3 - x_4 - 3x_5 = -3 \\ 2x_1 - 2x_2 + 4x_3 + 5x_4 + 15x_5 = 10 \\ 2x_1 - 2x_2 + 4x_3 + 4x_4 + 13x_5 = 13 \end{cases}$

15. Each day, an average adult moose can process about 32 kilograms of terrestrial vegetation (twigs and leaves) and aquatic vegetation. From this food, it needs to obtain about 1.9 grams of sodium and 11,000 calories of energy. Aquatic vegetation has about 0.15 gram of sodium per kilogram and about 193 calories of energy per kilogram, whereas terrestrial vegetation has minimal sodium and about four times more energy than aquatic vegetation. Write and graph a system of inequalities that describes the amounts t and a of terrestrial and aquatic vegetation, respectively, for the daily diet of an average adult moose. (Source: Biology by Numbers)

16. For a healthy person who is 4 feet 10 inches tall, the recommended minimum weight is about 91 pounds and increases by about 3.7 pounds for each additional inch of height. The recommended maximum weight is about 119 pounds and increases by about 4.8 pounds for each additional inch of height. (Source: Dietary Guidelines Advisory Committee)

(a) Let x be the number of inches by which a person's height exceeds 4 feet 10 inches and let y be the person's weight in pounds. Write a system of inequalities that describes the possible values of x and y for a healthy person.

(b) Use a graphing utility to graph the system of inequalities from part (a).

(c) What is the recommended weight range for someone 6 feet tall?

17. The cholesterol in human blood is necessary, but too much cholesterol can lead to health problems. A blood cholesterol test gives three readings: LDL ("bad") cholesterol, HDL ("good") cholesterol, and total cholesterol (LDL + HDL). It is recommended that your LDL cholesterol level be less than 130 milligrams per deciliter, your HDL cholesterol level be at least 35 milligrams per deciliter, and your total cholesterol level be no more than 200 milligrams per deciliter. (Source: WebMD, Inc.)

(a) Write a system of linear inequalities for the recommended cholesterol levels. Let x represent HDL cholesterol and let y represent LDL cholesterol.

(b) Graph the system of inequalities from part (a). Label any vertices of the solution region.

(c) Are the following cholesterol levels within recommendations? Explain your reasoning.

LDL: 120 milligrams per deciliter

HDL: 90 milligrams per deciliter

Total: 210 milligrams per deciliter

(d) Give an example of cholesterol levels in which the LDL cholesterol level is too high but the HDL and total cholesterol levels are acceptable.

(e) Another recommendation is that the ratio of total cholesterol to HDL cholesterol be less than 4. Find a point in your solution region from part (b) that meets this recommendation, and explain why it meets the recommendation.

Matrices and Determinants

7

Matrices can be used to analyze financial information such as the profit a fruit farmer makes on two fruit crops.

Darren McCollester/Getty Images

SELECTED APPLICATIONS

Matrices have many real-life applications. The applications listed below represent a small sample of the applications in this chapter.

7.1 Matrices and Systems of Equations

What you should learn

• Write matrices and identify their orders.
• Perform elementary row operations on matrices.
• Use matrices and Gaussian elimination to solve systems of linear equations.
• Use matrices and Gauss-Jordan elimination to solve systems of linear equations.

Why you should learn it

You can use matrices to solve systems of linear equations in two or more variables. For instance, in Exercise 90 on page 537, you will use a matrix to find a model for the number of people who participated in snowboarding in the United States from 1997 to 2001.

Matrices

In this section, you will study a streamlined technique for solving systems of linear equations. This technique involves the use of a rectangular array of real numbers called a **matrix.** The plural of matrix is *matrices.*

Definition of Matrix

If m and n are positive integers, an $m \times n$ (read "m by n") matrix is a rectangular array

$$
\begin{array}{c}
\quad\quad \text{Column 1} \quad \text{Column 2} \quad \text{Column 3} \quad \cdots \quad \text{Column } n \\
\begin{array}{l}
\text{Row 1} \\
\text{Row 2} \\
\text{Row 3} \\
\vdots \\
\text{Row } m
\end{array}
\begin{bmatrix}
a_{11} & a_{12} & a_{13} & \cdots & a_{1n} \\
a_{21} & a_{22} & a_{23} & \cdots & a_{2n} \\
a_{31} & a_{32} & a_{33} & \cdots & a_{3n} \\
\vdots & \vdots & \vdots & & \vdots \\
a_{m1} & a_{m2} & a_{m3} & \cdots & a_{mn}
\end{bmatrix}
\end{array}
$$

in which each **entry,** a_{ij}, of the matrix is a number. An $m \times n$ matrix has m rows and n columns. Matrices are usually denoted by capital letters.

The entry in the ith row and jth column is denoted by the *double subscript* notation a_{ij}. For instance, a_{23} refers to the entry in the second row, third column. A matrix having m rows and n columns is said to be of **order** $m \times n$. If $m = n$, the matrix is **square** of order n. For a square matrix, the entries $a_{11}, a_{22}, a_{33}, \ldots$ are the **main diagonal** entries.

Example 1 Order of Matrices

Determine the order of each matrix.

a. $\begin{bmatrix} 2 \end{bmatrix}$ **b.** $\begin{bmatrix} 1 & -3 & 0 & \frac{1}{2} \end{bmatrix}$

c. $\begin{bmatrix} 0 & 0 \\ 0 & 0 \end{bmatrix}$ **d.** $\begin{bmatrix} 5 & 0 \\ 2 & -2 \\ -7 & 4 \end{bmatrix}$

Solution

a. This matrix has *one* row and *one* column. The order of the matrix is 1×1.

b. This matrix has *one* row and *four* columns. The order of the matrix is 1×4.

c. This matrix has *two* rows and *two* columns. The order of the matrix is 2×2.

d. This matrix has *three* rows and *two* columns. The order of the matrix is 3×2.

✓**CHECKPOINT**   Now try Exercise 1.

A matrix that has only one row is called a **row matrix,** and a matrix that has only one column is called a **column matrix.**

A matrix derived from a system of linear equations (each written in standard form with the constant term on the right) is the **augmented matrix** of the system. Moreover, the matrix derived from the coefficients of the system (but not including the constant terms) is the **coefficient matrix** of the system.

$$\textit{System:} \quad \begin{cases} x - 4y + 3z = 5 \\ -x + 3y - z = -3 \\ 2x \qquad - 4z = 6 \end{cases}$$

$$\textit{Augmented Matrix:} \quad \begin{bmatrix} 1 & -4 & 3 & \vdots & 5 \\ -1 & 3 & -1 & \vdots & -3 \\ 2 & 0 & -4 & \vdots & 6 \end{bmatrix}$$

$$\textit{Coefficient Matrix:} \quad \begin{bmatrix} 1 & -4 & 3 \\ -1 & 3 & -1 \\ 2 & 0 & -4 \end{bmatrix}$$

Note the use of 0 for the missing coefficient of the y-variable in the third equation, and also note the fourth column of constant terms in the augmented matrix.

When forming either the coefficient matrix or the augmented matrix of a system, you should begin by vertically aligning the variables in the equations and using zeros for the coefficients of the missing variables.

Example 2   Writing an Augmented Matrix

Write the augmented matrix for the system of linear equations.

$$\begin{cases} x + 3y - w = 9 \\ -y + 4z + 2w = -2 \\ x - 5z - 6w = 0 \\ 2x + 4y - 3z = 4 \end{cases}$$

What is the order of the augmented matrix?

Solution

Begin by rewriting the linear system and aligning the variables.

$$\begin{cases} x + 3y \qquad - w = 9 \\ \quad -y + 4z + 2w = -2 \\ x \qquad - 5z - 6w = 0 \\ 2x + 4y - 3z \qquad = 4 \end{cases}$$

Next, use the coefficients and constant terms as the matrix entries. Include zeros for the coefficients of the missing variables.

$$\begin{matrix} R_1 \\ R_2 \\ R_3 \\ R_4 \end{matrix} \begin{bmatrix} 1 & 3 & 0 & -1 & \vdots & 9 \\ 0 & -1 & 4 & 2 & \vdots & -2 \\ 1 & 0 & -5 & -6 & \vdots & 0 \\ 2 & 4 & -3 & 0 & \vdots & 4 \end{bmatrix}$$

The augmented matrix has four rows and five columns, so it is a 4×5 matrix. The notation R_n is used to designate each row in the matrix. For example, Row 1 is represented by R_1.

✓CHECKPOINT   Now try Exercise 9.

Elementary Row Operations

In Section 6.3, you studied three operations that can be used on a system of linear equations to produce an equivalent system.

1. Interchange two equations.

2. Multiply an equation by a nonzero constant.

3. Add a multiple of an equation to another equation.

In matrix terminology, these three operations correspond to **elementary row operations.** An elementary row operation on an augmented matrix of a given system of linear equations produces a new augmented matrix corresponding to a new (but equivalent) system of linear equations. Two matrices are **row-equivalent** if one can be obtained from the other by a sequence of elementary row operations.

> ### Elementary Row Operations
>
> 1. Interchange two rows.
>
> 2. Multiply a row by a nonzero constant.
>
> 3. Add a multiple of a row to another row.

Although elementary row operations are simple to perform, they involve a lot of arithmetic. Because it is easy to make a mistake, you should get in the habit of noting the elementary row operations performed in each step so that you can go back and check your work.

Example 3   Elementary Row Operations

a. Interchange the first and second rows of the original matrix.

Original Matrix

$$\begin{bmatrix} 0 & 1 & 3 & 4 \\ -1 & 2 & 0 & 3 \\ 2 & -3 & 4 & 1 \end{bmatrix}$$

New Row-Equivalent Matrix

$$\begin{matrix} R_2 \\ R_1 \end{matrix} \begin{bmatrix} -1 & 2 & 0 & 3 \\ 0 & 1 & 3 & 4 \\ 2 & -3 & 4 & 1 \end{bmatrix}$$

b. Multiply the first row of the original matrix by $\frac{1}{2}$.

Original Matrix

$$\begin{bmatrix} 2 & -4 & 6 & -2 \\ 1 & 3 & -3 & 0 \\ 5 & -2 & 1 & 2 \end{bmatrix}$$

New Row-Equivalent Matrix

$$\frac{1}{2}R_1 \rightarrow \begin{bmatrix} 1 & -2 & 3 & -1 \\ 1 & 3 & -3 & 0 \\ 5 & -2 & 1 & 2 \end{bmatrix}$$

c. Add -2 times the first row of the original matrix to the third row.

Original Matrix

$$\begin{bmatrix} 1 & 2 & -4 & 3 \\ 0 & 3 & -2 & -1 \\ 2 & 1 & 5 & -2 \end{bmatrix}$$

New Row-Equivalent Matrix

$$\begin{matrix} \\ \\ -2R_1 + R_3 \rightarrow \end{matrix} \begin{bmatrix} 1 & 2 & -4 & 3 \\ 0 & 3 & -2 & -1 \\ 0 & -3 & 13 & -8 \end{bmatrix}$$

Note that the elementary row operation is written beside the row that is *changed*.

✓**CHECKPOINT**   Now try Exercise 25.

In Example 3 in Section 6.3, you used Gaussian elimination with back-substitution to solve a system of linear equations. The next example demonstrates the matrix version of Gaussian elimination. The two methods are essentially the same. The basic difference is that with matrices you do not need to keep writing the variables.

Example 4   Comparing Linear Systems and Matrix Operations

Linear System

$$\begin{cases} x - 2y + 3z = 9 \\ -x + 3y \quad\;\; = -4 \\ 2x - 5y + 5z = 17 \end{cases}$$

Add the first equation to the second equation.

$$\begin{cases} x - 2y + 3z = 9 \\ \quad\;\; y + 3z = 5 \\ 2x - 5y + 5z = 17 \end{cases}$$

Add -2 times the first equation to the third equation.

$$\begin{cases} x - 2y + 3z = 9 \\ \quad\;\; y + 3z = 5 \\ \quad -y - z = -1 \end{cases}$$

Add the second equation to the third equation.

$$\begin{cases} x - 2y + 3z = 9 \\ \quad\;\; y + 3z = 5 \\ \quad\quad\;\; 2z = 4 \end{cases}$$

Multiply the third equation by $\frac{1}{2}$.

$$\begin{cases} x - 2y + 3z = 9 \\ \quad\;\; y + 3z = 5 \\ \quad\quad\;\; z = 2 \end{cases}$$

Associated Augmented Matrix

$$\begin{bmatrix} 1 & -2 & 3 & \vdots & 9 \\ -1 & 3 & 0 & \vdots & -4 \\ 2 & -5 & 5 & \vdots & 17 \end{bmatrix}$$

Add the first row to the second row $(R_1 + R_2)$.

$$R_1 + R_2 \rightarrow \begin{bmatrix} 1 & -2 & 3 & \vdots & 9 \\ 0 & 1 & 3 & \vdots & 5 \\ 2 & -5 & 5 & \vdots & 17 \end{bmatrix}$$

Add -2 times the first row to the third row $(-2R_1 + R_3)$.

$$-2R_1 + R_3 \rightarrow \begin{bmatrix} 1 & -2 & 3 & \vdots & 9 \\ 0 & 1 & 3 & \vdots & 5 \\ 0 & -1 & -1 & \vdots & -1 \end{bmatrix}$$

Add the second row to the third row $(R_2 + R_3)$.

$$R_2 + R_3 \rightarrow \begin{bmatrix} 1 & -2 & 3 & \vdots & 9 \\ 0 & 1 & 3 & \vdots & 5 \\ 0 & 0 & 2 & \vdots & 4 \end{bmatrix}$$

Multiply the third row by $\frac{1}{2}$ $\left(\frac{1}{2}R_3\right)$.

$$\tfrac{1}{2}R_3 \rightarrow \begin{bmatrix} 1 & -2 & 3 & \vdots & 9 \\ 0 & 1 & 3 & \vdots & 5 \\ 0 & 0 & 1 & \vdots & 2 \end{bmatrix}$$

At this point, you can use back-substitution to find x and y.

$$y + 3(2) = 5 \qquad \text{\color{red}{Substitute 2 for }} z.$$
$$y = -1 \qquad \text{\color{red}{Solve for }} y.$$
$$x - 2(-1) + 3(2) = 9 \qquad \text{\color{red}{Substitute }} -1 \text{ for } y \text{ and 2 for } z.$$
$$x = 1 \qquad \text{\color{red}{Solve for }} x.$$

The solution is $x = 1$, $y = -1$, and $z = 2$.

✓CHECKPOINT   Now try Exercise 27.

STUDY TIP

Remember that you should check a solution by substituting the values of x, y, and z into each equation of the original system. For example, you can check the solution to Example 4 as follows.

Equation 1:
$1 - 2(-1) + 3(2) = 9$ ✓

Equation 2:
$-1 + 3(-1) = -4$ ✓

Equation 3:
$2(1) - 5(-1) + 5(2) = 17$ ✓

The last matrix in Example 4 is said to be in **row-echelon form.** The term *echelon* refers to the stair-step pattern formed by the nonzero elements of the matrix. To be in this form, a matrix must have the following properties.

> **Row-Echelon Form and Reduced Row-Echelon Form**
>
> A matrix in **row-echelon form** has the following properties.
>
> **1.** Any rows consisting entirely of zeros occur at the bottom of the matrix.
>
> **2.** For each row that does not consist entirely of zeros, the first nonzero entry is 1 (called a **leading 1**).
>
> **3.** For two successive (nonzero) rows, the leading 1 in the higher row is farther to the left than the leading 1 in the lower row.
>
> A matrix in *row-echelon form* is in **reduced row-echelon form** if every column that has a leading 1 has zeros in every position above and below its leading 1.

Example 5   Row-Echelon Form

Determine whether each matrix is in row-echelon form. If it is, determine whether the matrix is in reduced row-echelon form.

a. $\begin{bmatrix} 1 & 2 & -1 & 4 \\ 0 & 1 & 0 & 3 \\ 0 & 0 & 1 & -2 \end{bmatrix}$ b. $\begin{bmatrix} 1 & 2 & -1 & 2 \\ 0 & 0 & 0 & 0 \\ 0 & 1 & 2 & -4 \end{bmatrix}$

c. $\begin{bmatrix} 1 & -5 & 2 & -1 & 3 \\ 0 & 0 & 1 & 3 & -2 \\ 0 & 0 & 0 & 1 & 4 \\ 0 & 0 & 0 & 0 & 1 \end{bmatrix}$ d. $\begin{bmatrix} 1 & 0 & 0 & -1 \\ 0 & 1 & 0 & 2 \\ 0 & 0 & 1 & 3 \\ 0 & 0 & 0 & 0 \end{bmatrix}$

e. $\begin{bmatrix} 1 & 2 & -3 & 4 \\ 0 & 2 & 1 & -1 \\ 0 & 0 & 1 & -3 \end{bmatrix}$ f. $\begin{bmatrix} 0 & 1 & 0 & 5 \\ 0 & 0 & 1 & 3 \\ 0 & 0 & 0 & 0 \end{bmatrix}$

Solution

The matrices in (a), (c), (d), and (f) are in row-echelon form. The matrices in (d) and (f) are in *reduced* row-echelon form because every column that has a leading 1 has zeros in every position above and below its leading 1. The matrix in (b) is not in row-echelon form because a row of all zeros does not occur at the bottom of the matrix. The matrix in (e) is not in row-echelon form because the first nonzero entry in Row 2 is not a leading 1.

✓**CHECKPOINT**   Now try Exercise 29.

Every matrix is row-equivalent to a matrix in row-echelon form. For instance, in Example 5, you can change the matrix in part (e) to row-echelon form by multiplying its second row by $\frac{1}{2}$.

Gaussian Elimination with Back-Substitution

Gaussian elimination with back-substitution works well for solving systems of linear equations by hand or with a computer. For this algorithm, the order in which the elementary row operations are performed is important. You should operate from left to right by columns, using elementary row operations to obtain zeros in all entries directly below the leading 1's.

Example 6   **Gaussian Elimination with Back-Substitution**

Solve the system
$$\begin{cases} \quad\quad\; y + \; z - 2w = \; -3 \\ x + 2y - \; z \quad\quad\;\; = \quad\; 2 \\ 2x + 4y + \; z - 3w = \; -2 \\ x - 4y - 7z - \; w = -19 \end{cases}.$$

Solution

$$\begin{bmatrix} 0 & 1 & 1 & -2 & \vdots & -3 \\ 1 & 2 & -1 & 0 & \vdots & 2 \\ 2 & 4 & 1 & -3 & \vdots & -2 \\ 1 & -4 & -7 & -1 & \vdots & -19 \end{bmatrix}$$

Write augmented matrix.

$$\begin{matrix} R_2 \\ R_1 \end{matrix} \begin{bmatrix} 1 & 2 & -1 & 0 & \vdots & 2 \\ 0 & 1 & 1 & -2 & \vdots & -3 \\ 2 & 4 & 1 & -3 & \vdots & -2 \\ 1 & -4 & -7 & -1 & \vdots & -19 \end{bmatrix}$$

Interchange R_1 and R_2 so first column has leading 1 in upper left corner.

$$\begin{matrix} \\ \\ -2R_1 + R_3 \rightarrow \\ -R_1 + R_4 \rightarrow \end{matrix} \begin{bmatrix} 1 & 2 & -1 & 0 & \vdots & 2 \\ 0 & 1 & 1 & -2 & \vdots & -3 \\ 0 & 0 & 3 & -3 & \vdots & -6 \\ 0 & -6 & -6 & -1 & \vdots & -21 \end{bmatrix}$$

Perform operations on R_3 and R_4 so first column has zeros below its leading 1.

$$\begin{matrix} \\ \\ \\ 6R_2 + R_4 \rightarrow \end{matrix} \begin{bmatrix} 1 & 2 & -1 & 0 & \vdots & 2 \\ 0 & 1 & 1 & -2 & \vdots & -3 \\ 0 & 0 & 3 & -3 & \vdots & -6 \\ 0 & 0 & 0 & -13 & \vdots & -39 \end{bmatrix}$$

Perform operations on R_4 so second column has zeros below its leading 1.

$$\begin{matrix} \\ \\ \frac{1}{3}R_3 \rightarrow \\ -\frac{1}{13}R_4 \rightarrow \end{matrix} \begin{bmatrix} 1 & 2 & -1 & 0 & \vdots & 2 \\ 0 & 1 & 1 & -2 & \vdots & -3 \\ 0 & 0 & 1 & -1 & \vdots & -2 \\ 0 & 0 & 0 & 1 & \vdots & 3 \end{bmatrix}$$

Perform operations on R_3 and R_4 so third and fourth columns have leading 1's.

The matrix is now in row-echelon form, and the corresponding system is

$$\begin{cases} x + 2y - z \quad\quad\;\; = \quad\; 2 \\ \quad\quad\; y + z - 2w = -3 \\ \quad\quad\quad\;\; z - \; w = -2 \\ \quad\quad\quad\quad\quad\;\; w = \quad\; 3 \end{cases}.$$

Using back-substitution, the solution is $x = -1$, $y = 2$, $z = 1$, and $w = 3$.

✓**CHECKPOINT**   Now try Exercise 51.

The procedure for using Gaussian elimination with back-substitution is summarized below.

> ## Gaussian Elimination with Back-Substitution
>
> **1.** Write the augmented matrix of the system of linear equations.
>
> **2.** Use elementary row operations to rewrite the augmented matrix in row-echelon form.
>
> **3.** Write the system of linear equations corresponding to the matrix in row-echelon form, and use back-substitution to find the solution.

When solving a system of linear equations, remember that it is possible for the system to have no solution. If, in the elimination process, you obtain a row with zeros except for the last entry, it is unnecessary to continue the elimination process. You can simply conclude that the system has no solution, or is *inconsistent*.

Example 7 A System with No Solution

Solve the system $\begin{cases} x - y + 2z = 4 \\ x \quad\quad + z = 6 \\ 2x - 3y + 5z = 4 \\ 3x + 2y - z = 1 \end{cases}$.

Solution

$$\begin{bmatrix} 1 & -1 & 2 & \vdots & 4 \\ 1 & 0 & 1 & \vdots & 6 \\ 2 & -3 & 5 & \vdots & 4 \\ 3 & 2 & -1 & \vdots & 1 \end{bmatrix}$$
Write augmented matrix.

$$\begin{matrix} \\ -R_1 + R_2 \rightarrow \\ -2R_1 + R_3 \rightarrow \\ -3R_1 + R_4 \rightarrow \end{matrix} \begin{bmatrix} 1 & -1 & 2 & \vdots & 4 \\ 0 & 1 & -1 & \vdots & 2 \\ 0 & -1 & 1 & \vdots & -4 \\ 0 & 5 & -7 & \vdots & -11 \end{bmatrix}$$
Perform row operations.

$$\begin{matrix} \\ \\ R_2 + R_3 \rightarrow \\ \end{matrix} \begin{bmatrix} 1 & -1 & 2 & \vdots & 4 \\ 0 & 1 & -1 & \vdots & 2 \\ 0 & 0 & 0 & \vdots & -2 \\ 0 & 5 & -7 & \vdots & -11 \end{bmatrix}$$
Perform row operations.

Note that the third row of this matrix consists of zeros except for the last entry. This means that the original system of linear equations is inconsistent. You can see why this is true by converting back to a system of linear equations.

$$\begin{cases} x - y + 2z = 4 \\ y - z = 2 \\ 0 = -2 \\ 5y - 7z = -11 \end{cases}$$

Because the third equation is not possible, the system has no solution.

✓**CHECKPOINT** Now try Exercise 57.

Gauss-Jordan Elimination

With Gaussian elimination, elementary row operations are applied to a matrix to obtain a (row-equivalent) row-echelon form of the matrix. A second method of elimination, called **Gauss-Jordan elimination,** after Carl Friedrich Gauss and Wilhelm Jordan (1842–1899), continues the reduction process until a *reduced* row-echelon form is obtained. This procedure is demonstrated in Example 8.

Example 8   Gauss-Jordan Elimination

Use Gauss-Jordan elimination to solve the system $\begin{cases} x - 2y + 3z = 9 \\ -x + 3y = -4. \\ 2x - 5y + 5z = 17 \end{cases}$

Solution

In Example 4, Gaussian elimination was used to obtain the row-echelon form of the linear system above.

$$\begin{bmatrix} 1 & -2 & 3 & \vdots & 9 \\ 0 & 1 & 3 & \vdots & 5 \\ 0 & 0 & 1 & \vdots & 2 \end{bmatrix}$$

Now, apply elementary row operations until you obtain zeros above each of the leading 1's, as follows.

$$\begin{array}{c} 2R_2 + R_1 \rightarrow \\ \\ \\ \end{array} \begin{bmatrix} 1 & 0 & 9 & \vdots & 19 \\ 0 & 1 & 3 & \vdots & 5 \\ 0 & 0 & 1 & \vdots & 2 \end{bmatrix}$$

Perform operations on R_1 so second column has a zero above its leading 1.

$$\begin{array}{c} -9R_3 + R_1 \rightarrow \\ -3R_3 + R_2 \rightarrow \\ \\ \end{array} \begin{bmatrix} 1 & 0 & 0 & \vdots & 1 \\ 0 & 1 & 0 & \vdots & -1 \\ 0 & 0 & 1 & \vdots & 2 \end{bmatrix}$$

Perform operations on R_1 and R_2 so third column has zeros above its leading 1.

The matrix is now in reduced row-echelon form. Converting back to a system of linear equations, you have

$$\begin{cases} x = 1 \\ y = -1. \\ z = 2 \end{cases}$$

Now you can simply read the solution, $x = 1$, $y = -1$, and $z = 2$, which can be written as the ordered triple $(1, -1, 2)$.

✓**CHECKPOINT**   Now try Exercise 59.

The elimination procedures described in this section sometimes result in fractional coefficients. For instance, in the elimination procedure for the system

$$\begin{cases} 2x - 5y + 5z = 17 \\ 3x - 2y + 3z = 11 \\ -3x + 3y = -6 \end{cases}$$

you may be inclined to multiply the first row by $\frac{1}{2}$ to produce a leading 1, which will result in working with fractional coefficients. You can sometimes avoid fractions by judiciously choosing the order in which you apply elementary row operations.

Technology

For a demonstration of a graphical approach to Gauss-Jordan elimination on a 2×3 matrix, see the Visualizing Row Operations Program available for several models of graphing calculators at our website *college.hmco.com*.

STUDY TIP

The advantage of using Gauss-Jordan elimination to solve a system of linear equations is that the solution of the system is easily found without using back-substitution, as illustrated in Example 8.

Recall from Chapter 6 that when there are fewer equations than variables in a system of equations, then the system has either no solution or infinitely many solutions.

Example 9 A System with an Infinite Number of Solutions

Solve the system.

$$\begin{cases} 2x + 4y - 2z = 0 \\ 3x + 5y \quad\quad = 1 \end{cases}$$

Solution

$$\begin{bmatrix} 2 & 4 & -2 & \vdots & 0 \\ 3 & 5 & 0 & \vdots & 1 \end{bmatrix}$$

$$\frac{1}{2}R_1 \rightarrow \begin{bmatrix} 1 & 2 & -1 & \vdots & 0 \\ 3 & 5 & 0 & \vdots & 1 \end{bmatrix}$$

$$-3R_1 + R_2 \rightarrow \begin{bmatrix} 1 & 2 & -1 & \vdots & 0 \\ 0 & -1 & 3 & \vdots & 1 \end{bmatrix}$$

$$-R_2 \rightarrow \begin{bmatrix} 1 & 2 & -1 & \vdots & 0 \\ 0 & 1 & -3 & \vdots & -1 \end{bmatrix}$$

$$-2R_2 + R_1 \rightarrow \begin{bmatrix} 1 & 0 & 5 & \vdots & 2 \\ 0 & 1 & -3 & \vdots & -1 \end{bmatrix}$$

The corresponding system of equations is

$$\begin{cases} x + 5z = \quad 2 \\ y - 3z = -1 \end{cases}.$$

Solving for x and y in terms of z, you have

$$x = -5z + 2 \quad \text{and} \quad y = 3z - 1.$$

To write a solution to the system that does not use any of the three variables of the system, let a represent any real number and let

$$z = a.$$

Now substitute a for z in the equations for x and y.

$$x = -5z + 2 = -5a + 2$$

$$y = 3z - 1 = 3a - 1$$

So, the solution set can be written as an ordered triple with the form

$$(-5a + 2, 3a - 1, a)$$

where a is any real number. Remember that a solution set of this form represents an infinite number of solutions. Try substituting values for a to obtain a few solutions. Then check each solution in the original equation.

✓CHECKPOINT   Now try Exercise 65.

STUDY TIP

In Example 9, x and y are solved for in terms of the third variable z. To write a solution to the system that does not use any of the three variables of the system, let a represent any real number and let $z = a$. Then solve for x and y. The solution can then be written in terms of a, which is not one of the variables of the system.

It is worth noting that the row-echelon form of a matrix is not unique. That is, two different sequences of elementary row operations may yield different row-echelon forms. This is demonstrated in Example 10.

Example 10 **Comparing Row-Echelon Forms**

Compare the following row-echelon form with the one found in Example 4. Is it the same? Does it yield the same solution?

$$\begin{cases} x - 2y + 3z = 9 \\ -x + 3y = -4 \\ 2x - 5y + 5z = 17 \end{cases}$$

$$\begin{bmatrix} 1 & -2 & 3 & \vdots & 9 \\ -1 & 3 & 0 & \vdots & -4 \\ 2 & -5 & 5 & \vdots & 17 \end{bmatrix}$$

$$\begin{matrix} R_2 \\ R_1 \end{matrix} \begin{bmatrix} -1 & 3 & 0 & \vdots & -4 \\ 1 & -2 & 3 & \vdots & 9 \\ 2 & -5 & 5 & \vdots & 17 \end{bmatrix}$$

$$-R_1 \rightarrow \begin{bmatrix} 1 & -3 & 0 & \vdots & 4 \\ 1 & -2 & 3 & \vdots & 9 \\ 2 & -5 & 5 & \vdots & 17 \end{bmatrix}$$

$$\begin{matrix} -R_1 + R_2 \rightarrow \\ -2R_1 + R_3 \rightarrow \end{matrix} \begin{bmatrix} 1 & -3 & 0 & \vdots & 4 \\ 0 & 1 & 3 & \vdots & 5 \\ 0 & 1 & 5 & \vdots & 9 \end{bmatrix}$$

$$-R_2 + R_3 \rightarrow \begin{bmatrix} 1 & -3 & 0 & \vdots & 4 \\ 0 & 1 & 3 & \vdots & 5 \\ 0 & 0 & 2 & \vdots & 4 \end{bmatrix}$$

$$\tfrac{1}{2}R_3 \rightarrow \begin{bmatrix} 1 & -3 & 0 & \vdots & 4 \\ 0 & 1 & 3 & \vdots & 5 \\ 0 & 0 & 1 & \vdots & 2 \end{bmatrix}$$

Solution

This row-echelon form is different from that obtained in Example 4. The corresponding system of linear equations for this row-echelon matrix is

$$\begin{cases} x - 3y = 4 \\ y + 3z = 5. \\ z = 2 \end{cases}$$

Using back-substitution on this system, you obtain the solution

$$x = 1, \, y = -1, \text{ and } z = 2$$

which is the same solution that was obtained in Example 4.

✓*CHECKPOINT*   Now try Exercise 77.

   You have seen that the row-echelon form of a given matrix *is not* unique; however, the *reduced* row-echelon form of a given matrix *is* unique. Try applying Gauss-Jordan elimination to the row-echelon matrix in Example 10 to see that you obtain the same reduced row-echelon form as in Example 8.

VOCABULARY CHECK: Fill in the blanks.

1. A rectangular array of real numbers than can be used to solve a system of linear equations is called a _____.

2. A matrix is _____ if the number of rows equals the number of columns.

3. For a square matrix, the entries $a_{11}, a_{22}, a_{33}, \ldots, a_{nn}$ are the _____ _____ entries.

4. A matrix with only one row is called a _____ matrix and a matrix with only one column is called a _____ matrix.

5. The matrix derived from a system of linear equations is called the _____ matrix of the system.

6. The matrix derived from the coefficients of a system of linear equations is called the _____ matrix of the system.

7. Two matrices are called _____ if one of the matrices can be obtained from the other by a sequence of elementary row operations.

8. A matrix in row-echelon form is in _____ _____ _____ if every column that has a leading 1 has zeros in every position above and below its leading 1.

9. The process of using row operations to write a matrix in reduced row-echelon form is called _____ _____.

PREREQUISITE SKILLS REVIEW: Practice and review algebra skills needed for this section at **www.Eduspace.com.**

In Exercises 1–6, determine the order of the matrix.

1. $\begin{bmatrix} 7 & 0 \end{bmatrix}$

2. $\begin{bmatrix} 5 & -3 & 8 & 7 \end{bmatrix}$

3. $\begin{bmatrix} 2 \\ 36 \\ 3 \end{bmatrix}$

4. $\begin{bmatrix} -3 & 7 & 15 & 0 \\ 0 & 0 & 3 & 3 \\ 1 & 1 & 6 & 7 \end{bmatrix}$

5. $\begin{bmatrix} 33 & 45 \\ -9 & 20 \end{bmatrix}$

6. $\begin{bmatrix} -7 & 6 & 4 \\ 0 & -5 & 1 \end{bmatrix}$

In Exercises 7–12, write the augmented matrix for the system of linear equations.

7. $\begin{cases} 4x - 3y = -5 \\ -x + 3y = 12 \end{cases}$

8. $\begin{cases} 7x + 4y = 22 \\ 5x - 9y = 15 \end{cases}$

9. $\begin{cases} x + 10y - 2z = 2 \\ 5x - 3y + 4z = 0 \\ 2x + y = 6 \end{cases}$

10. $\begin{cases} -x - 8y + 5z = 8 \\ -7x - 15z = -38 \\ 3x - y + 8z = 20 \end{cases}$

11. $\begin{cases} 7x - 5y + z = 13 \\ 19x - 8z = 10 \end{cases}$

12. $\begin{cases} 9x + 2y - 3z = 20 \\ -25y + 11z = -5 \end{cases}$

In Exercises 13–18, write the system of linear equations represented by the augmented matrix. (Use variables x, y, z, and w, if applicable.)

13. $\begin{bmatrix} 1 & 2 & \vdots & 7 \\ 2 & -3 & \vdots & 4 \end{bmatrix}$

14. $\begin{bmatrix} 7 & -5 & \vdots & 0 \\ 8 & 3 & \vdots & -2 \end{bmatrix}$

15. $\begin{bmatrix} 2 & 0 & 5 & \vdots & -12 \\ 0 & 1 & -2 & \vdots & 7 \\ 6 & 3 & 0 & \vdots & 2 \end{bmatrix}$

16. $\begin{bmatrix} 4 & -5 & -1 & \vdots & 18 \\ -11 & 0 & 6 & \vdots & 25 \\ 3 & 8 & 0 & \vdots & -29 \end{bmatrix}$

17. $\begin{bmatrix} 9 & 12 & 3 & 0 & \vdots & 0 \\ -2 & 18 & 5 & 2 & \vdots & 10 \\ 1 & 7 & -8 & 0 & \vdots & -4 \\ 3 & 0 & 2 & 0 & \vdots & -10 \end{bmatrix}$

18. $\begin{bmatrix} 6 & 2 & -1 & -5 & \vdots & -25 \\ -1 & 0 & 7 & 3 & \vdots & 7 \\ 4 & -1 & -10 & 6 & \vdots & 23 \\ 0 & 8 & 1 & -11 & \vdots & -21 \end{bmatrix}$

In Exercises 19–22, fill in the blank(s) using elementary row operations to form a row-equivalent matrix.

19. $\begin{bmatrix} 1 & 4 & 3 \\ 2 & 10 & 5 \end{bmatrix}$

$\begin{bmatrix} 1 & 4 & 3 \\ 0 & & -1 \end{bmatrix}$

20. $\begin{bmatrix} 3 & 6 & 8 \\ 4 & -3 & 6 \end{bmatrix}$

$\begin{bmatrix} 1 & & \frac{8}{3} \\ 4 & -3 & 6 \end{bmatrix}$

21. $\begin{bmatrix} 1 & 1 & 4 & -1 \\ 3 & 8 & 10 & 3 \\ -2 & 1 & 12 & 6 \end{bmatrix}$

$\begin{bmatrix} 1 & 1 & 4 & -1 \\ 0 & 5 & & \\ 0 & 3 & & \end{bmatrix}$

$\begin{bmatrix} 1 & 1 & 4 & -1 \\ 0 & 1 & -\frac{2}{5} & \frac{6}{5} \\ 0 & 3 & & \end{bmatrix}$

22. $\begin{bmatrix} 2 & 4 & 8 & 3 \\ 1 & -1 & -3 & 2 \\ 2 & 6 & 4 & 9 \end{bmatrix}$

$\begin{bmatrix} 1 & & & \\ 1 & -1 & -3 & 2 \\ 2 & 6 & 4 & 9 \end{bmatrix}$

$\begin{bmatrix} 1 & 2 & 4 & \frac{3}{2} \\ 0 & & -7 & \frac{1}{2} \\ 0 & 2 & & \end{bmatrix}$

In Exercises 23–26, identify the elementary row operation(s) being performed to obtain the new row-equivalent matrix.

23.
| Original Matrix | New Row-Equivalent Matrix |

$$\begin{bmatrix} -2 & 5 & 1 \\ 3 & -1 & -8 \end{bmatrix} \qquad \begin{bmatrix} 13 & 0 & -39 \\ 3 & -1 & -8 \end{bmatrix}$$

24.
Original Matrix $\qquad$ New Row-Equivalent Matrix

$$\begin{bmatrix} 3 & -1 & -4 \\ -4 & 3 & 7 \end{bmatrix} \qquad \begin{bmatrix} 3 & -1 & -4 \\ 5 & 0 & -5 \end{bmatrix}$$

25.
Original Matrix $\qquad$ New Row-Equivalent Matrix

$$\begin{bmatrix} 0 & -1 & -5 & 5 \\ -1 & 3 & -7 & 6 \\ 4 & -5 & 1 & 3 \end{bmatrix} \qquad \begin{bmatrix} -1 & 3 & -7 & 6 \\ 0 & -1 & -5 & 5 \\ 0 & 7 & -27 & 27 \end{bmatrix}$$

26.
Original Matrix $\qquad$ New Row-Equivalent Matrix

$$\begin{bmatrix} -1 & -2 & 3 & -2 \\ 2 & -5 & 1 & -7 \\ 5 & 4 & -7 & 6 \end{bmatrix} \qquad \begin{bmatrix} -1 & -2 & 3 & -2 \\ 0 & -9 & 7 & -11 \\ 0 & -6 & 8 & -4 \end{bmatrix}$$

27. Perform the sequence of row operations on the matrix. What did the operations accomplish?

$$\begin{bmatrix} 1 & 2 & 3 \\ 2 & -1 & -4 \\ 3 & 1 & -1 \end{bmatrix}$$

(a) Add -2 times R_1 to R_2.

(b) Add -3 times R_1 to R_3.

(c) Add -1 times R_2 to R_3.

(d) Multiply R_2 by $-\frac{1}{5}$.

(e) Add -2 times R_2 to R_1.

28. Perform the sequence of row operations on the matrix. What did the operations accomplish?

$$\begin{bmatrix} 7 & 1 \\ 0 & 2 \\ -3 & 4 \\ 4 & 1 \end{bmatrix}$$

(a) Add R_3 to R_4.

(b) Interchange R_1 and R_4.

(c) Add 3 times R_1 to R_3.

(d) Add -7 times R_1 to R_4.

(e) Multiply R_2 by $\frac{1}{2}$.

(f) Add the appropriate multiples of R_2 to R_1, R_3, and R_4.

In Exercises 29–32, determine whether the matrix is in row-echelon form. If it is, determine if it is also in reduced row-echelon form.

29. $\begin{bmatrix} 1 & 0 & 0 & 0 \\ 0 & 1 & 1 & 5 \\ 0 & 0 & 0 & 0 \end{bmatrix}$

30. $\begin{bmatrix} 1 & 3 & 0 & 0 \\ 0 & 0 & 1 & 8 \\ 0 & 0 & 0 & 0 \end{bmatrix}$

31. $\begin{bmatrix} 2 & 0 & 4 & 0 \\ 0 & -1 & 3 & 6 \\ 0 & 0 & 1 & 5 \end{bmatrix}$

32. $\begin{bmatrix} 1 & 0 & 2 & 1 \\ 0 & 1 & -3 & 10 \\ 0 & 0 & 1 & 0 \end{bmatrix}$

In Exercises 33–36, write the matrix in row-echelon form. (Remember that the row-echelon form of a matrix is not unique.)

33. $\begin{bmatrix} 1 & 1 & 0 & 5 \\ -2 & -1 & 2 & -10 \\ 3 & 6 & 7 & 14 \end{bmatrix}$

34. $\begin{bmatrix} 1 & 2 & -1 & 3 \\ 3 & 7 & -5 & 14 \\ -2 & -1 & -3 & 8 \end{bmatrix}$

35. $\begin{bmatrix} 1 & -1 & -1 & 1 \\ 5 & -4 & 1 & 8 \\ -6 & 8 & 18 & 0 \end{bmatrix}$

36. $\begin{bmatrix} 1 & -3 & 0 & -7 \\ -3 & 10 & 1 & 23 \\ 4 & -10 & 2 & -24 \end{bmatrix}$

 In Exercises 37–42, use the matrix capabilities of a graphing utility to write the matrix in *reduced* row-echelon form.

37. $\begin{bmatrix} 3 & 3 & 3 \\ -1 & 0 & -4 \\ 2 & 4 & -2 \end{bmatrix}$

38. $\begin{bmatrix} 1 & 3 & 2 \\ 5 & 15 & 9 \\ 2 & 6 & 10 \end{bmatrix}$

39. $\begin{bmatrix} 1 & 2 & 3 & -5 \\ 1 & 2 & 4 & -9 \\ -2 & -4 & -4 & 3 \\ 4 & 8 & 11 & -14 \end{bmatrix}$

40. $\begin{bmatrix} -2 & 3 & -1 & -2 \\ 4 & -2 & 5 & 8 \\ 1 & 5 & -2 & 0 \\ 3 & 8 & -10 & -30 \end{bmatrix}$

41. $\begin{bmatrix} -3 & 5 & 1 & 12 \\ 1 & -1 & 1 & 4 \end{bmatrix}$

42. $\begin{bmatrix} 5 & 1 & 2 & 4 \\ -1 & 5 & 10 & -32 \end{bmatrix}$

In Exercises 43–46, write the system of linear equations represented by the augmented matrix. Then use back-substitution to solve. (Use variables *x*, *y*, and *z*, if applicable.)

43. $\left[\begin{array}{cc:c} 1 & -2 & 4 \\ 0 & 1 & -3 \end{array}\right]$

44. $\left[\begin{array}{cc:c} 1 & 5 & 0 \\ 0 & 1 & -1 \end{array}\right]$

45. $\left[\begin{array}{ccc:c} 1 & -1 & 2 & 4 \\ 0 & 1 & -1 & 2 \\ 0 & 0 & 1 & -2 \end{array}\right]$

46. $\left[\begin{array}{ccc:c} 1 & 2 & -2 & -1 \\ 0 & 1 & 1 & 9 \\ 0 & 0 & 1 & -3 \end{array}\right]$

In Exercises 47–50, an augmented matrix that represents a system of linear equations (in variables x, y, and z, if applicable) has been reduced using Gauss-Jordan elimination. Write the solution represented by the augmented matrix.

47. $\begin{bmatrix} 1 & 0 & \vdots & 3 \\ 0 & 1 & \vdots & -4 \end{bmatrix}$

48. $\begin{bmatrix} 1 & 0 & \vdots & -6 \\ 0 & 1 & \vdots & 10 \end{bmatrix}$

49. $\begin{bmatrix} 1 & 0 & 0 & \vdots & -4 \\ 0 & 1 & 0 & \vdots & -10 \\ 0 & 0 & 1 & \vdots & 4 \end{bmatrix}$

50. $\begin{bmatrix} 1 & 0 & 0 & \vdots & 5 \\ 0 & 1 & 0 & \vdots & -3 \\ 0 & 0 & 1 & \vdots & 0 \end{bmatrix}$

In Exercises 51–70, use matrices to solve the system of equations (if possible). Use Gaussian elimination with back-substitution or Gauss-Jordan elimination.

51. $\begin{cases} x + 2y = 7 \\ 2x + y = 8 \end{cases}$

52. $\begin{cases} 2x + 6y = 16 \\ 2x + 3y = 7 \end{cases}$

53. $\begin{cases} 3x - 2y = -27 \\ x + 3y = 13 \end{cases}$

54. $\begin{cases} -x + y = 4 \\ 2x - 4y = -34 \end{cases}$

55. $\begin{cases} -2x + 6y = 22 \\ x + 2y = -9 \end{cases}$

56. $\begin{cases} 5x - 5y = -5 \\ -2x - 3y = 7 \end{cases}$

57. $\begin{cases} -x + 2y = 1.5 \\ 2x - 4y = 3 \end{cases}$

58. $\begin{cases} x - 3y = 5 \\ -2x + 6y = -10 \end{cases}$

59. $\begin{cases} x - 3z = -2 \\ 3x + y - 2z = 5 \\ 2x + 2y + z = 4 \end{cases}$

60. $\begin{cases} 2x - y + 3z = 24 \\ 2y - z = 14 \\ 7x - 5y = 6 \end{cases}$

61. $\begin{cases} -x + y - z = -14 \\ 2x - y + z = 21 \\ 3x + 2y + z = 19 \end{cases}$

62. $\begin{cases} 2x + 2y - z = 2 \\ x - 3y + z = -28 \\ -x + y = 14 \end{cases}$

63. $\begin{cases} x + 2y - 3z = -28 \\ 4y + 2z = 0 \\ -x + y - z = -5 \end{cases}$

64. $\begin{cases} 3x - 2y + z = 15 \\ -x + y + 2z = -10 \\ x - y - 4z = 14 \end{cases}$

65. $\begin{cases} x + y - 5z = 3 \\ x - 2z = 1 \\ 2x - y - z = 0 \end{cases}$

66. $\begin{cases} 2x + 3z = 3 \\ 4x - 3y + 7z = 5 \\ 8x - 9y + 15z = 9 \end{cases}$

67. $\begin{cases} x + 2y + z + 2w = 8 \\ 3x + 7y + 6z + 9w = 26 \end{cases}$

68. $\begin{cases} 4x + 12y - 7z - 20w = 22 \\ 3x + 9y - 5z - 28w = 30 \end{cases}$

69. $\begin{cases} -x + y = -22 \\ 3x + 4y = 4 \\ 4x - 8y = 32 \end{cases}$

70. $\begin{cases} x + 2y = 0 \\ x + y = 6 \\ 3x - 2y = 8 \end{cases}$

In Exercises 71–76, use the matrix capabilities of a graphing utility to reduce the augmented matrix corresponding to the system of equations, and solve the system.

71. $\begin{cases} 3x + 3y + 12z = 6 \\ x + y + 4z = 2 \\ 2x + 5y + 20z = 10 \\ -x + 2y + 8z = 4 \end{cases}$

72. $\begin{cases} 2x + 10y + 2z = 6 \\ x + 5y + 2z = 6 \\ x + 5y + z = 3 \\ -3x - 15y - 3z = -9 \end{cases}$

73. $\begin{cases} 2x + y - z + 2w = -6 \\ 3x + 4y + w = 1 \\ x + 5y + 2z + 6w = -3 \\ 5x + 2y - z - w = 3 \end{cases}$

74. $\begin{cases} x + 2y + 2z + 4w = 11 \\ 3x + 6y + 5z + 12w = 30 \\ x + 3y - 3z + 2w = -5 \\ 6x - y - z + w = -9 \end{cases}$

75. $\begin{cases} x + y + z + w = 0 \\ 2x + 3y + z - 2w = 0 \\ 3x + 5y + z = 0 \end{cases}$

76. $\begin{cases} x + 2y + z + 3w = 0 \\ x - y + w = 0 \\ y - z + 2w = 0 \end{cases}$

In Exercises 77–80, determine whether the two systems of linear equations yield the same solution. If so, find the solution using matrices.

77. (a) $\begin{cases} x - 2y + z = -6 \\ y - 5z = 16 \\ z = -3 \end{cases}$ (b) $\begin{cases} x + y - 2z = 6 \\ y + 3z = -8 \\ z = -3 \end{cases}$

78. (a) $\begin{cases} x - 3y + 4z = -11 \\ y - z = -4 \\ z = 2 \end{cases}$ (b) $\begin{cases} x + 4y = -11 \\ y + 3z = 4 \\ z = 2 \end{cases}$

79. (a) $\begin{cases} x - 4y + 5z = 27 \\ y - 7z = -54 \\ z = 8 \end{cases}$ (b) $\begin{cases} x - 6y + z = 15 \\ y + 5z = 42 \\ z = 8 \end{cases}$

80. (a) $\begin{cases} x + 3y - z = 19 \\ y + 6z = -18 \\ z = -4 \end{cases}$ (b) $\begin{cases} x - y + 3z = -15 \\ y - 2z = 14 \\ z = -4 \end{cases}$

81. Use the system

$\begin{cases} x + 3y + z = 3 \\ x + 5y + 5z = 1 \\ 2x + 6y + 3z = 8 \end{cases}$

to write two different matrices in row-echelon form that yield the same solution.

82. *Electrical Network*   The currents in an electrical network are given by the solution of the system

$$\begin{cases} I_1 - I_2 + I_3 = 0 \\ 3I_1 + 4I_2 \quad\quad = 18 \\ \quad\quad I_2 + 3I_3 = 6 \end{cases}$$

where $I_1, I_2,$ and I_3 are measured in amperes. Solve the system of equations using matrices.

83. *Partial Fractions*   Use a system of equations to write the partial fraction decomposition of the rational expression. Solve the system using matrices.

$$\frac{4x^2}{(x+1)^2(x-1)} = \frac{A}{x-1} + \frac{B}{x+1} + \frac{C}{(x+1)^2}$$

84. *Partial Fractions*   Use a system of equations to write the partial fraction decomposition of the rational expression. Solve the system using matrices.

$$\frac{8x^2}{(x-1)^2(x+1)} = \frac{A}{x+1} + \frac{B}{x-1} + \frac{C}{(x-1)^2}$$

85. *Finance*   A small shoe corporation borrowed $1,500,000 to expand its line of shoes. Some of the money was borrowed at 7%, some at 8%, and some at 10%. Use a system of equations to determine how much was borrowed at each rate if the annual interest was $130,500 and the amount borrowed at 10% was 4 times the amount borrowed at 7%. Solve the system using matrices.

86. *Finance*   A small software corporation borrowed $500,000 to expand its software line. Some of the money was borrowed at 9%, some at 10%, and some at 12%. Use a system of equations to determine how much was borrowed at each rate if the annual interest was $52,000 and the amount borrowed at 10% was $2\frac{1}{2}$ times the amount borrowed at 9%. Solve the system using matrices.

In Exercises 87 and 88, use a system of equations to find the specified equation that passes through the points. Solve the system using matrices. Use a graphing utility to verify your results.

87. Parabola:

$$y = ax^2 + bx + c$$

(3, 20), (2, 13), (1, 8)

88. Parabola:

$$y = ax^2 + bx + c$$

(1, 9), (2, 8), (3, 5)

89. *Mathematical Modeling*   A videotape of the path of a ball thrown by a baseball player was analyzed with a grid covering the TV screen. The tape was paused three times, and the position of the ball was measured each time. The coordinates obtained are shown in the table. (x and y are measured in feet.)

Horizontal distance, x	Height, y
0	5.0
15	9.6
30	12.4

(a) Use a system of equations to find the equation of the parabola $y = ax^2 + bx + c$ that passes through the three points. Solve the system using matrices.

(b) Use a graphing utility to graph the parabola.

(c) Graphically approximate the maximum height of the ball and the point at which the ball struck the ground.

(d) Analytically find the maximum height of the ball and the point at which the ball struck the ground.

(e) Compare your results from parts (c) and (d).

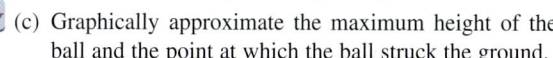

Model It

90. *Data Analysis: Snowboarders*   The table shows the numbers of people y (in millions) in the United States who participated in snowboarding for selected years from 1997 to 2001. (Source: National Sporting Goods Association)

Year	Number, y
1997	2.8
1999	3.3
2001	5.3

(a) Use a system of equations to find the equation of the parabola $y = at^2 + bt + c$ that passes through the points. Let t represent the year, with $t = 7$ corresponding to 1997. Solve the system using matrices.

(b) Use a graphing utility to graph the parabola.

(c) Use the equation in part (a) to estimate the number of people who participated in snowboarding in 2003. How does this value compare with the actual 2003 value of 6.3 million?

(d) Use the equation in part (a) to estimate y in the year 2008. Is the estimate reasonable? Explain.

Network Analysis In Exercises 91 and 92, answer the questions about the specified network. (In a network it is assumed that the total flow into each junction is equal to the total flow out of each junction.)

91. Water flowing through a network of pipes (in thousands of cubic meters per hour) is shown in the figure.

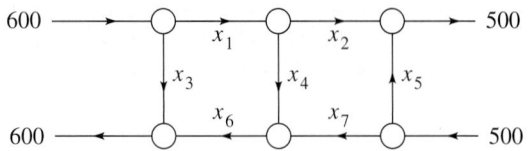

(a) Solve this system using matrices for the water flow represented by x_i, $i = 1, 2, \ldots, 7$.

(b) Find the network flow pattern when $x_6 = 0$ and $x_7 = 0$.

(c) Find the network flow pattern when $x_5 = 1000$ and $x_6 = 0$.

92. The flow of traffic (in vehicles per hour) through a network of streets is shown in the figure.

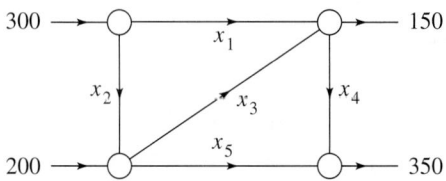

(a) Solve this system using matrices for the traffic flow represented by x_i, $i = 1, 2, \ldots, 5$.

(b) Find the traffic flow when $x_2 = 200$ and $x_3 = 50$.

(c) Find the traffic flow when $x_2 = 150$ and $x_3 = 0$.

Synthesis

True or False? In Exercises 93–95, determine whether the statement is true or false. Justify your answer.

93. $\begin{bmatrix} 5 & 0 & -2 & 7 \\ -1 & 3 & -6 & 0 \end{bmatrix}$ is a 4×2 matrix.

94. The matrix

$$\begin{bmatrix} 0 & 0 & 0 & 0 \\ 0 & 0 & 1 & -4 \\ 0 & 1 & 0 & 2 \\ 1 & 0 & 0 & 5 \end{bmatrix}$$

is in reduced row-echelon form.

95. The method of Gaussian elimination reduces a matrix until a reduced row-echelon form is obtained.

96. *Think About It* The augmented matrix represents a system of linear equations (in variables x, y, and z) that has been reduced using Gauss-Jordan elimination. Write a system of equations with nonzero coefficients that is represented by the reduced matrix. (There are many correct answers.)

$$\begin{bmatrix} 1 & 0 & 3 & \vdots & -2 \\ 0 & 1 & 4 & \vdots & 1 \\ 0 & 0 & 0 & \vdots & 0 \end{bmatrix}$$

97. *Think About It*

(a) Describe the row-echelon form of an augmented matrix that corresponds to a system of linear equations that is inconsistent.

(b) Describe the row-echelon form of an augmented matrix that corresponds to a system of linear equations that has an infinite number of solutions.

98. Describe the three elementary row operations that can be performed on an augmented matrix.

99. What is the relationship between the three elementary row operations performed on an augmented matrix and the operations that lead to equivalent systems of equations?

100. *Writing* In your own words, describe the difference between a matrix in row-echelon form and a matrix in reduced row-echelon form.

Skills Review

In Exercises 101–106, sketch the graph of the function. Do not use a graphing utility.

101. $f(x) = \dfrac{2x^2 - 4x}{3x - x^2}$

102. $f(x) = \dfrac{x^2 - 2x + 1}{x^2 - 1}$

103. $f(x) = 2^{x-1}$

104. $g(x) = 3^{-x+2}$

105. $h(x) = \ln(x - 1)$

106. $f(x) = 3 + \ln x$

What you should learn

- Decide whether two matrices are equal.
- Add and subtract matrices and multiply matrices by scalars.
- Multiply two matrices.
- Use matrix operations to model and solve real-life problems.

Why you should learn it

Matrix operations can be used to model and solve real-life problems. For instance, in Exercise 70 on page 553, matrix operations are used to analyze annual health care costs.

© Royalty-Free/Corbis

Equality of Matrices

In Section 7.1, you used matrices to solve systems of linear equations. There is a rich mathematical theory of matrices, and its applications are numerous. This section and the next two introduce some fundamentals of matrix theory. It is standard mathematical convention to represent matrices in any of the following three ways.

Representation of Matrices

1. A matrix can be denoted by an uppercase letter such as A, B, or C.
2. A matrix can be denoted by a representative element enclosed in brackets, such as $[a_{ij}]$, $[b_{ij}]$, or $[c_{ij}]$.
3. A matrix can be denoted by a rectangular array of numbers such as

$$A = [a_{ij}] = \begin{bmatrix} a_{11} & a_{12} & a_{13} & \cdots & a_{1n} \\ a_{21} & a_{22} & a_{23} & \cdots & a_{2n} \\ a_{31} & a_{32} & a_{33} & \cdots & a_{3n} \\ \vdots & \vdots & \vdots & & \vdots \\ a_{m1} & a_{m2} & a_{m3} & \cdots & a_{mn} \end{bmatrix}.$$

Two matrices $A = [a_{ij}]$ and $B = [b_{ij}]$ are **equal** if they have the same order $(m \times n)$ and $a_{ij} = b_{ij}$ for $1 \le i \le m$ and $1 \le j \le n$. In other words, two matrices are equal if their corresponding entries are equal.

Example 1   Equality of Matrices

Solve for a_{11}, a_{12}, a_{21}, and a_{22} in the following matrix equation.

$$\begin{bmatrix} a_{11} & a_{12} \\ a_{21} & a_{22} \end{bmatrix} = \begin{bmatrix} 2 & -1 \\ -3 & 0 \end{bmatrix}$$

Solution

Because two matrices are equal only if their corresponding entries are equal, you can conclude that

$$a_{11} = 2, \quad a_{12} = -1, \quad a_{21} = -3, \quad \text{and} \quad a_{22} = 0.$$

✓CHECKPOINT   Now try Exercise 1.

Be sure you see that for two matrices to be equal, they must have the same order *and* their corresponding entries must be equal. For instance,

$$\begin{bmatrix} 2 & -1 \\ \sqrt{4} & \frac{1}{2} \end{bmatrix} = \begin{bmatrix} 2 & -1 \\ 2 & 0.5 \end{bmatrix} \quad \text{but} \quad \begin{bmatrix} 2 & -1 \\ 3 & 4 \\ 0 & 0 \end{bmatrix} \ne \begin{bmatrix} 2 & -1 \\ 3 & 4 \end{bmatrix}.$$

Matrix Addition and Scalar Multiplication

In this section, three basic matrix operations will be covered. The first two are matrix addition and scalar multiplication. With matrix addition, you can add two matrices (of the same order) by adding their corresponding entries.

Definition of Matrix Addition

If $A = [a_{ij}]$ and $B = [b_{ij}]$ are matrices of order $m \times n$, their sum is the $m \times n$ matrix given by

$$A + B = [a_{ij} + b_{ij}].$$

The sum of two matrices of different orders is undefined.

Historical Note
Arthur Cayley (1821–1895), a British mathematician, invented matrices around 1858. Cayley was a Cambridge University graduate and a lawyer by profession. His groundbreaking work on matrices was begun as he studied the theory of transformations. Cayley also was instrumental in the development of determinants. Cayley and two American mathematicians, Benjamin Peirce (1809–1880) and his son Charles S. Peirce (1839–1914), are credited with developing "matrix algebra."

Example 2 **Addition of Matrices**

a. $\begin{bmatrix} -1 & 2 \\ 0 & 1 \end{bmatrix} + \begin{bmatrix} 1 & 3 \\ -1 & 2 \end{bmatrix} = \begin{bmatrix} -1+1 & 2+3 \\ 0+(-1) & 1+2 \end{bmatrix}$

$\qquad\qquad\qquad\qquad = \begin{bmatrix} 0 & 5 \\ -1 & 3 \end{bmatrix}$

b. $\begin{bmatrix} 0 & 1 & -2 \\ 1 & 2 & 3 \end{bmatrix} + \begin{bmatrix} 0 & 0 & 0 \\ 0 & 0 & 0 \end{bmatrix} = \begin{bmatrix} 0 & 1 & -2 \\ 1 & 2 & 3 \end{bmatrix}$

c. $\begin{bmatrix} 1 \\ -3 \\ -2 \end{bmatrix} + \begin{bmatrix} -1 \\ 3 \\ 2 \end{bmatrix} = \begin{bmatrix} 0 \\ 0 \\ 0 \end{bmatrix}$

d. The sum of

$$A = \begin{bmatrix} 2 & 1 & 0 \\ 4 & 0 & -1 \\ 3 & -2 & 2 \end{bmatrix} \qquad \text{and}$$

$$B = \begin{bmatrix} 0 & 1 \\ -1 & 3 \\ 2 & 4 \end{bmatrix}$$

is undefined because A is of order 3×3 and B is of order 3×2.

✓**CHECKPOINT**　　Now try Exercise 7(a).

In operations with matrices, numbers are usually referred to as **scalars.** In this text, scalars will always be real numbers. You can multiply a matrix A by a scalar c by multiplying each entry in A by c.

Definition of Scalar Multiplication

If $A = [a_{ij}]$ is an $m \times n$ matrix and c is a scalar, the **scalar multiple** of A by c is the $m \times n$ matrix given by

$$cA = [ca_{ij}].$$

The symbol $-A$ represents the negation of A, which is the scalar product $(-1)A$. Moreover, if A and B are of the same order, then $A - B$ represents the sum of A and $(-1)B$. That is,

$$A - B = A + (-1)B. \qquad \text{Subtraction of matrices}$$

The order of operations for matrix expressions is similar to that for real numbers. In particular, you perform scalar multiplication before matrix addition and subtraction, as shown in Example 3(c).

Example 3   Scalar Multiplication and Matrix Subtraction

For the following matrices, find (a) $3A$, (b) $-B$, and (c) $3A - B$.

$$A = \begin{bmatrix} 2 & 2 & 4 \\ -3 & 0 & -1 \\ 2 & 1 & 2 \end{bmatrix} \quad \text{and} \quad B = \begin{bmatrix} 2 & 0 & 0 \\ 1 & -4 & 3 \\ -1 & 3 & 2 \end{bmatrix}$$

Solution

a. $3A = 3\begin{bmatrix} 2 & 2 & 4 \\ -3 & 0 & -1 \\ 2 & 1 & 2 \end{bmatrix}$ Scalar multiplication

$$= \begin{bmatrix} 3(2) & 3(2) & 3(4) \\ 3(-3) & 3(0) & 3(-1) \\ 3(2) & 3(1) & 3(2) \end{bmatrix}$$ Multiply each entry by 3.

$$= \begin{bmatrix} 6 & 6 & 12 \\ -9 & 0 & -3 \\ 6 & 3 & 6 \end{bmatrix}$$ Simplify.

b. $-B = (-1)\begin{bmatrix} 2 & 0 & 0 \\ 1 & -4 & 3 \\ -1 & 3 & 2 \end{bmatrix}$ Definition of negation

$$= \begin{bmatrix} -2 & 0 & 0 \\ -1 & 4 & -3 \\ 1 & -3 & -2 \end{bmatrix}$$ Multiply each entry by -1.

c. $3A - B = \begin{bmatrix} 6 & 6 & 12 \\ -9 & 0 & -3 \\ 6 & 3 & 6 \end{bmatrix} - \begin{bmatrix} 2 & 0 & 0 \\ 1 & -4 & 3 \\ -1 & 3 & 2 \end{bmatrix}$ Matrix subtraction

$$= \begin{bmatrix} 4 & 6 & 12 \\ -10 & 4 & -6 \\ 7 & 0 & 4 \end{bmatrix}$$ Subtract corresponding entries.

✓CHECKPOINT   Now try Exercises 7(b), (c), and (d).

It is often convenient to rewrite the scalar multiple cA by factoring c out of every entry in the matrix. For instance, in the following example, the scalar $\frac{1}{2}$ has been factored out of the matrix.

$$\begin{bmatrix} \frac{1}{2} & -\frac{3}{2} \\ \frac{5}{2} & \frac{1}{2} \end{bmatrix} = \begin{bmatrix} \frac{1}{2}(1) & \frac{1}{2}(-3) \\ \frac{1}{2}(5) & \frac{1}{2}(1) \end{bmatrix} = \frac{1}{2}\begin{bmatrix} 1 & -3 \\ 5 & 1 \end{bmatrix}$$

Exploration

Consider matrices A, B, and C below. Perform the indicated operations and compare the results.

$$A = \begin{bmatrix} 3 & -1 \\ 4 & 7 \end{bmatrix}, \quad B = \begin{bmatrix} -2 & 0 \\ 8 & 1 \end{bmatrix},$$

$$C = \begin{bmatrix} 5 & 2 \\ 2 & -6 \end{bmatrix}$$

a. Find $A + B$ and $B + A$.

b. Find $A + B$, then add C to the resulting matrix. Find $B + C$, then add A to the resulting matrix.

c. Find $2A$ and $2B$, then add the two resulting matrices. Find $A + B$, then multiply the resulting matrix by 2.

The properties of matrix addition and scalar multiplication are similar to those of addition and multiplication of real numbers.

Properties of Matrix Addition and Scalar Multiplication

Let A, B, and C be $m \times n$ matrices and let c and d be scalars.

1. $A + B = B + A$ Commutative Property of Matrix Addition

2. $A + (B + C) = (A + B) + C$ Associative Property of Matrix Addition

3. $(cd)A = c(dA)$ Associative Property of Scalar Multiplication

4. $1A = A$ Scalar Identity Property

5. $c(A + B) = cA + cB$ Distributive Property

6. $(c + d)A = cA + dA$ Distributive Property

Note that the Associative Property of Matrix Addition allows you to write expressions such as $A + B + C$ without ambiguity because the same sum occurs no matter how the matrices are grouped. This same reasoning applies to sums of four or more matrices.

Example 4   Addition of More than Two Matrices

By adding corresponding entries, you obtain the following sum of four matrices.

$$\begin{bmatrix} 1 \\ 2 \\ -3 \end{bmatrix} + \begin{bmatrix} -1 \\ -1 \\ 2 \end{bmatrix} + \begin{bmatrix} 0 \\ 1 \\ 4 \end{bmatrix} + \begin{bmatrix} 2 \\ -3 \\ -2 \end{bmatrix} = \begin{bmatrix} 2 \\ -1 \\ 1 \end{bmatrix}$$

✓**CHECKPOINT**   Now try Exercise 13.

Example 5   Using the Distributive Property

Perform the indicated matrix operations.

$$3\left(\begin{bmatrix} -2 & 0 \\ 4 & 1 \end{bmatrix} + \begin{bmatrix} 4 & -2 \\ 3 & 7 \end{bmatrix} \right)$$

Solution

$$3\left(\begin{bmatrix} -2 & 0 \\ 4 & 1 \end{bmatrix} + \begin{bmatrix} 4 & -2 \\ 3 & 7 \end{bmatrix} \right) = 3\begin{bmatrix} -2 & 0 \\ 4 & 1 \end{bmatrix} + 3\begin{bmatrix} 4 & -2 \\ 3 & 7 \end{bmatrix}$$

$$= \begin{bmatrix} -6 & 0 \\ 12 & 3 \end{bmatrix} + \begin{bmatrix} 12 & -6 \\ 9 & 21 \end{bmatrix}$$

$$= \begin{bmatrix} 6 & -6 \\ 21 & 24 \end{bmatrix}$$

✓**CHECKPOINT**   Now try Exercise 15.

In Example 5, you could add the two matrices first and then multiply the matrix by 3, as follows. Notice that you obtain the same result.

$$3\left(\begin{bmatrix} -2 & 0 \\ 4 & 1 \end{bmatrix} + \begin{bmatrix} 4 & -2 \\ 3 & 7 \end{bmatrix} \right) = 3\begin{bmatrix} 2 & -2 \\ 7 & 8 \end{bmatrix} = \begin{bmatrix} 6 & -6 \\ 21 & 24 \end{bmatrix}$$

Technology

Most graphing utilities have the capability of performing matrix operations. Consult the user's guide for your graphing utility for specific keystrokes. Try using a graphing utility to find the sum of the matrices

$$A = \begin{bmatrix} 2 & -3 \\ -1 & 0 \end{bmatrix}$$

and

$$B = \begin{bmatrix} -1 & 4 \\ 2 & -5 \end{bmatrix}.$$

One important property of addition of real numbers is that the number 0 is the additive identity. That is, $c + 0 = c$ for any real number c. For matrices, a similar property holds. That is, if A is an $m \times n$ matrix and O is the $m \times n$ **zero matrix** consisting entirely of zeros, then $A + O = A$.

In other words, O is the **additive identity** for the set of all $m \times n$ matrices. For example, the following matrices are the additive identities for the set of all 2×3 and 2×2 matrices.

$$O = \begin{bmatrix} 0 & 0 & 0 \\ 0 & 0 & 0 \end{bmatrix} \quad \text{and} \quad O = \begin{bmatrix} 0 & 0 \\ 0 & 0 \end{bmatrix}$$

$\underbrace{\hspace{3cm}}$ $\underbrace{\hspace{3cm}}$
2×3 zero matrix 2×2 zero matrix

The algebra of real numbers and the algebra of matrices have many similarities. For example, compare the following solutions.

Real Numbers	$m \times n$ Matrices
(Solve for x.)	(Solve for X.)
$x + a = b$	$X + A = B$
$x + a + (-a) = b + (-a)$	$X + A + (-A) = B + (-A)$
$x + 0 = b - a$	$X + O = B - A$
$x = b - a$	$X = B - A$

The algebra of real numbers and the algebra of matrices also have important differences, which will be discussed later.

STUDY TIP

Remember that matrices are denoted by capital letters. So, when you solve for X, you are solving for a *matrix* that makes the matrix equation true.

Example 6   Solving a Matrix Equation

Solve for X in the equation $3X + A = B$, where

$$A = \begin{bmatrix} 1 & -2 \\ 0 & 3 \end{bmatrix} \quad \text{and} \quad B = \begin{bmatrix} -3 & 4 \\ 2 & 1 \end{bmatrix}.$$

Solution

Begin by solving the equation for X to obtain

$$3X = B - A$$

$$X = \frac{1}{3}(B - A).$$

Now, using the matrices A and B, you have

$$X = \frac{1}{3}\left(\begin{bmatrix} -3 & 4 \\ 2 & 1 \end{bmatrix} - \begin{bmatrix} 1 & -2 \\ 0 & 3 \end{bmatrix} \right) \qquad \text{Substitute the matrices.}$$

$$= \frac{1}{3}\begin{bmatrix} -4 & 6 \\ 2 & -2 \end{bmatrix} \qquad \text{Subtract matrix } A \text{ from matrix } B.$$

$$= \begin{bmatrix} -\frac{4}{3} & 2 \\ \frac{2}{3} & -\frac{2}{3} \end{bmatrix}. \qquad \text{Multiply the matrix by } \frac{1}{3}.$$

✓**CHECKPOINT**   Now try Exercise 25.

Matrix Multiplication

The third basic matrix operation is **matrix multiplication.** At first glance, the definition may seem unusual. You will see later, however, that this definition of the product of two matrices has many practical applications.

Definition of Matrix Multiplication

If $A = [a_{ij}]$ is an $m \times n$ matrix and $B = [b_{ij}]$ is an $n \times p$ matrix, the product AB is an $m \times p$ matrix

$$AB = [c_{ij}]$$

where $c_{ij} = a_{i1}b_{1j} + a_{i2}b_{2j} + a_{i3}b_{3j} + \cdots + a_{in}b_{nj}.$

The definition of matrix multiplication indicates a *row-by-column* multiplication, where the entry in the ith row and jth column of the product AB is obtained by multiplying the entries in the ith row of A by the corresponding entries in the jth column of B and then adding the results. The general pattern for matrix multiplication is as follows.

$$\begin{bmatrix} a_{11} & a_{12} & a_{13} & \cdots & a_{1n} \\ a_{21} & a_{22} & a_{23} & \cdots & a_{2n} \\ a_{31} & a_{32} & a_{33} & \cdots & a_{3n} \\ \vdots & \vdots & \vdots & & \vdots \\ a_{i1} & a_{i2} & a_{i3} & \cdots & a_{in} \\ \vdots & \vdots & \vdots & & \vdots \\ a_{m1} & a_{m2} & a_{m3} & \cdots & a_{mn} \end{bmatrix} \begin{bmatrix} b_{11} & b_{12} & \cdots & b_{1j} & \cdots & b_{1p} \\ b_{21} & b_{22} & \cdots & b_{2j} & \cdots & b_{2p} \\ b_{31} & b_{32} & \cdots & b_{3j} & \cdots & b_{3p} \\ \vdots & \vdots & & \vdots & & \vdots \\ b_{n1} & b_{n2} & \cdots & b_{nj} & \cdots & b_{np} \end{bmatrix} = \begin{bmatrix} c_{11} & c_{12} & \cdots & c_{1j} & \cdots & c_{1p} \\ c_{21} & c_{22} & \cdots & c_{2j} & \cdots & c_{2p} \\ \vdots & \vdots & & \vdots & & \vdots \\ c_{i1} & c_{i2} & \cdots & c_{ij} & \cdots & c_{ip} \\ \vdots & \vdots & & \vdots & & \vdots \\ c_{m1} & c_{m2} & \cdots & c_{mj} & \cdots & c_{mp} \end{bmatrix}$$

$$a_{i1}b_{1j} + a_{i2}b_{2j} + a_{i3}b_{3j} + \cdots + a_{in}b_{nj} = c_{ij}$$

Example 7 **Finding the Product of Two Matrices**

First, note that the product AB is defined because the number of columns of A is equal to the number of rows of B. Moreover, the product AB has order 3×2. To find the entries of the product, multiply each row of A by each column of B, as follows.

$$AB = \begin{bmatrix} -1 & 3 \\ 4 & -2 \\ 5 & 0 \end{bmatrix} \begin{bmatrix} -3 & 2 \\ -4 & 1 \end{bmatrix}$$

$$= \begin{bmatrix} (-1)(-3) + (3)(-4) & (-1)(2) + (3)(1) \\ (4)(-3) + (-2)(-4) & (4)(2) + (-2)(1) \\ (5)(-3) + (0)(-4) & (5)(2) + (0)(1) \end{bmatrix}$$

$$= \begin{bmatrix} -9 & 1 \\ -4 & 6 \\ -15 & 10 \end{bmatrix}$$

✓**CHECKPOINT** Now try Exercise 29.

Be sure you understand that for the product of two matrices to be defined, the number of *columns* of the first matrix must equal the number of *rows* of the second matrix. That is, the middle two indices must be the same. The outside two indices give the order of the product, as shown below.

$$\underset{m \times n}{A} \quad \times \quad \underset{n \times p}{B} \quad = \quad \underset{m \times p}{AB}$$

Equal

Order of AB

Example 8   Finding the Product of Two Matrices

Find the product AB where

$$A = \begin{bmatrix} 1 & 0 & 3 \\ 2 & -1 & -2 \end{bmatrix} \quad \text{and} \quad B = \begin{bmatrix} -2 & 4 \\ 1 & 0 \\ -1 & 1 \end{bmatrix}.$$

Solution

Note that the order of A is 2×3 and the order of B is 3×2. So, the product AB has order 2×2.

$$AB = \begin{bmatrix} 1 & 0 & 3 \\ 2 & -1 & -2 \end{bmatrix} \begin{bmatrix} -2 & 4 \\ 1 & 0 \\ -1 & 1 \end{bmatrix}$$

$$= \begin{bmatrix} 1(-2) + & 0(1) + & 3(-1) & 1(4) + & 0(0) + & 3(1) \\ 2(-2) + & (-1)(1) + & (-2)(-1) & 2(4) + & (-1)(0) + & (-2)(1) \end{bmatrix}$$

$$= \begin{bmatrix} -5 & 7 \\ -3 & 6 \end{bmatrix}$$

✓**CHECKPOINT**   Now try Exercise 31.

Example 9   Patterns in Matrix Multiplication

a. $\begin{bmatrix} 3 & 4 \\ -2 & 5 \end{bmatrix} \begin{bmatrix} 1 & 0 \\ 0 & 1 \end{bmatrix} = \begin{bmatrix} 3 & 4 \\ -2 & 5 \end{bmatrix}$

       2×2 2×2 2×2

b. $\begin{bmatrix} 6 & 2 & 0 \\ 3 & -1 & 2 \\ 1 & 4 & 6 \end{bmatrix} \begin{bmatrix} 1 \\ 2 \\ -3 \end{bmatrix} = \begin{bmatrix} 10 \\ -5 \\ -9 \end{bmatrix}$

       3×3 3×1 3×1

c. The product AB for the following matrices is not defined.

$$A = \begin{bmatrix} -2 & 1 \\ 1 & -3 \\ 1 & 4 \end{bmatrix} \quad \text{and} \quad B = \begin{bmatrix} -2 & 3 & 1 & 4 \\ 0 & 1 & -1 & 2 \\ 2 & -1 & 0 & 1 \end{bmatrix}$$

       3×2 3×4

✓**CHECKPOINT**   Now try Exercise 33.

Exploration

Use the following matrices to find AB, BA, $(AB)C$, and $A(BC)$. What do your results tell you about matrix multiplication, commutativity, and associativity?

$$A = \begin{bmatrix} 1 & 2 \\ 3 & 4 \end{bmatrix},$$

$$B = \begin{bmatrix} 0 & 1 \\ 2 & 3 \end{bmatrix},$$

$$C = \begin{bmatrix} 3 & 0 \\ 0 & 1 \end{bmatrix}$$

S1

The notat
the augm
when ma
matrix A.
represent
echelon f
matrix th
the syste

Example 12   **Softball Team Expenses**

Two softball teams submit equipment lists to their sponsors.

	Women's Team	Men's Team
Bats	12	15
Balls	45	38
Gloves	15	17

Each bat costs $80, each ball costs $6, and each glove costs $60. Use matrices to find the total cost of equipment for each team.

Solution

The equipment lists E and the costs per item C can be written in matrix form as

$$E = \begin{bmatrix} 12 & 15 \\ 45 & 38 \\ 15 & 17 \end{bmatrix}$$

and

$$C = \begin{bmatrix} 80 & 6 & 60 \end{bmatrix}.$$

The total cost of equipment for each team is given by the product

$$CE = \begin{bmatrix} 80 & 6 & 60 \end{bmatrix} \begin{bmatrix} 12 & 15 \\ 45 & 38 \\ 15 & 17 \end{bmatrix}$$

$$= \begin{bmatrix} 80(12) + 6(45) + 60(15) & 80(15) + 6(38) + 60(17) \end{bmatrix}$$

$$= \begin{bmatrix} 2130 & 2448 \end{bmatrix}.$$

So, the total cost of equipment for the women's team is $2130 and the total cost of equipment for the men's team is $2448. Notice that you cannot find the total cost using the product EC because EC is not defined. That is, the number of columns of E (2 columns) does not equal the number of rows of C (1 row).

✔**CHECKPOINT**   Now try Exercise 63.

---

*W*RITING ABOUT *M*ATHEMATICS

Problem Posing   Write a matrix multiplication application problem that uses the matrix

$$A = \begin{bmatrix} 20 & 42 & 33 \\ 17 & 30 & 50 \end{bmatrix}.$$

Exchange problems with another student in your class. Form the matrices that represent the problem, and solve the problem. Interpret your solution in the context of the problem. Check with the creator of the problem to see if you are correct. Discuss other ways to represent and/or approach the problem.

7.2   Exercises

VOCABULARY CHECK:

In Exercises 1–4, fill in the blanks.

1. Two matrices are _____ if all of their corresponding entries are equal.

2. When performing matrix operations, real numbers are often referred to as _____.

3. A matrix consisting entirely of zeros is called a _____ matrix and is denoted by _____.

4. The $n \times n$ matrix consisting of 1's on its main diagonal and 0's elsewhere is called the _____ matrix of order n.

In Exercises 5 and 6, match the matrix property with the correct form. A, B, and C are matrices of order $m \times n$, and c and d are scalars.

5. (a) $1A = A$ (i) Distributive Property

(b) $A + (B + C) = (A + B) + C$ (ii) Commutative Property of Matrix Addition

(c) $(c + d)A = cA + dA$ (iii) Scalar Identity Property

(d) $(cd)A = c(dA)$ (iv) Associative Property of Matrix Addition

(e) $A + B = B + A$ (v) Associative Property of Scalar Multiplication

6. (a) $A + O = A$ (i) Distributive Property

(b) $c(AB) = A(cB)$ (ii) Additive Identity of Matrix Addition

(c) $A(B + C) = AB + AC$ (iii) Associative Property of Multiplication

(d) $A(BC) = (AB)C$ (iv) Associative Property of Scalar Multiplication

PREREQUISITE SKILLS REVIEW: Practice and review algebra skills needed for this section at **www.Eduspace.com.**

In Exercises 1–4, find x and y.

1. $\begin{bmatrix} x & -2 \\ 7 & y \end{bmatrix} = \begin{bmatrix} -4 & -2 \\ 7 & 22 \end{bmatrix}$

2. $\begin{bmatrix} -5 & x \\ y & 8 \end{bmatrix} = \begin{bmatrix} -5 & 13 \\ 12 & 8 \end{bmatrix}$

3. $\begin{bmatrix} 16 & 4 & 5 & 4 \\ -3 & 13 & 15 & 6 \\ 0 & 2 & 4 & 0 \end{bmatrix} = \begin{bmatrix} 16 & 4 & 2x+1 & 4 \\ -3 & 13 & & 15 & 3x \\ 0 & 2 & 3y-5 & 0 \end{bmatrix}$

4. $\begin{bmatrix} x+2 & 8 & -3 \\ 1 & 2y & 2x \\ 7 & -2 & y+2 \end{bmatrix} = \begin{bmatrix} 2x+6 & 8 & -3 \\ 1 & 18 & -8 \\ 7 & -2 & 11 \end{bmatrix}$

In Exercises 5–12, if possible, find (a) $A + B$, (b) $A - B$, (c) $3A$, and (d) $3A - 2B$.

5. $A = \begin{bmatrix} 1 & -1 \\ 2 & -1 \end{bmatrix}$, $B = \begin{bmatrix} 2 & -1 \\ -1 & 8 \end{bmatrix}$

6. $A = \begin{bmatrix} 1 & 2 \\ 2 & 1 \end{bmatrix}$, $B = \begin{bmatrix} -3 & -2 \\ 4 & 2 \end{bmatrix}$

7. $A = \begin{bmatrix} 6 & -1 \\ 2 & 4 \\ -3 & 5 \end{bmatrix}$, $B = \begin{bmatrix} 1 & 4 \\ -1 & 5 \\ 1 & 10 \end{bmatrix}$

8. $A = \begin{bmatrix} 2 & 1 & 1 \\ -1 & -1 & 4 \end{bmatrix}$, $B = \begin{bmatrix} 2 & -3 & 4 \\ -3 & 1 & -2 \end{bmatrix}$

9. $A = \begin{bmatrix} 2 & 2 & -1 & 0 & 1 \\ 1 & 1 & -2 & 0 & -1 \end{bmatrix}$,
$B = \begin{bmatrix} 1 & 1 & -1 & 1 & 0 \\ -3 & 4 & 9 & -6 & -7 \end{bmatrix}$

10. $A = \begin{bmatrix} -1 & 4 & 0 \\ 3 & -2 & 2 \\ 5 & 4 & -1 \\ 0 & 8 & -6 \\ -4 & -1 & 0 \end{bmatrix}$, $B = \begin{bmatrix} -3 & 5 & 1 \\ 2 & -4 & -7 \\ 10 & -9 & -1 \\ 3 & 2 & -4 \\ 0 & 1 & -2 \end{bmatrix}$

11. $A = \begin{bmatrix} 6 & 0 & 3 \\ -1 & -4 & 0 \end{bmatrix}$, $B = \begin{bmatrix} 8 & -1 \\ 4 & -3 \end{bmatrix}$

12. $A = \begin{bmatrix} 3 \\ 2 \\ -1 \end{bmatrix}$, $B = \begin{bmatrix} -4 & 6 & 2 \end{bmatrix}$

In Exercises 13–18, evaluate the expression.

13. $\begin{bmatrix} -5 & 0 \\ 3 & -6 \end{bmatrix} + \begin{bmatrix} 7 & 1 \\ -2 & -1 \end{bmatrix} + \begin{bmatrix} -10 & -8 \\ 14 & 6 \end{bmatrix}$

14. $\begin{bmatrix} 6 & 8 \\ -1 & 0 \end{bmatrix} + \begin{bmatrix} 0 & 5 \\ -3 & -1 \end{bmatrix} + \begin{bmatrix} -11 & -7 \\ 2 & -1 \end{bmatrix}$

15. $4\left(\begin{bmatrix} -4 & 0 & 1 \\ 0 & 2 & 3 \end{bmatrix} - \begin{bmatrix} 2 & 1 & -2 \\ 3 & -6 & 0 \end{bmatrix}\right)$

16. $\frac{1}{2}([5 \quad -2 \quad 4 \quad 0] + [14 \quad 6 \quad -18 \quad 9])$

17. $-3\left(\begin{bmatrix} 0 & -3 \\ 7 & 2 \end{bmatrix} + \begin{bmatrix} -6 & 3 \\ 8 & 1 \end{bmatrix}\right) - 2\begin{bmatrix} 4 & -4 \\ 7 & -9 \end{bmatrix}$

18. $-1\begin{bmatrix} 4 & 11 \\ -2 & -1 \\ 9 & 3 \end{bmatrix} + \frac{1}{6}\left(\begin{bmatrix} -5 & -1 \\ 3 & 4 \\ 0 & 13 \end{bmatrix} + \begin{bmatrix} 7 & 5 \\ -9 & -1 \\ 6 & -1 \end{bmatrix}\right)$

 In Exercises 19–22, use the matrix capabilities of a graphing utility to evaluate the expression. Round your results to three decimal places, if necessary.

19. $\frac{3}{7}\begin{bmatrix} 2 & 5 \\ -1 & -4 \end{bmatrix} + 6\begin{bmatrix} -3 & 0 \\ 2 & 2 \end{bmatrix}$

20. $55\left(\begin{bmatrix} 14 & -11 \\ -22 & 19 \end{bmatrix} + \begin{bmatrix} -22 & 20 \\ 13 & 6 \end{bmatrix}\right)$

21. $-\begin{bmatrix} 3.211 & 6.829 \\ -1.004 & 4.914 \\ 0.055 & -3.889 \end{bmatrix} - \begin{bmatrix} -1.630 & -3.090 \\ 5.256 & 8.335 \\ -9.768 & 4.251 \end{bmatrix}$

22. $-12\left(\begin{bmatrix} 6 & 20 \\ 1 & -9 \\ -2 & 5 \end{bmatrix} + \begin{bmatrix} 14 & -15 \\ -8 & -6 \\ 7 & 0 \end{bmatrix} + \begin{bmatrix} -31 & -19 \\ 16 & 10 \\ 24 & -10 \end{bmatrix}\right)$

In Exercises 23–26, solve fo X in the equation given

$$A = \begin{bmatrix} -2 & -1 \\ 1 & 0 \\ 3 & -4 \end{bmatrix} \quad \text{and} \quad B = \begin{bmatrix} 0 & 3 \\ 2 & 0 \\ -4 & -1 \end{bmatrix}.$$

23. $X = 3A - 2B$

24. $2X = 2A - B$

25. $2X + 3A = B$

26. $2A + 4B = -2X$

In Exercises 27–34, if possible, find AB and state the order of the result.

27. $A = \begin{bmatrix} 2 & 1 \\ -3 & 4 \\ 1 & 6 \end{bmatrix}$, $B = \begin{bmatrix} 0 & -1 & 0 \\ 4 & 0 & 2 \\ 8 & -1 & 7 \end{bmatrix}$

28. $A = \begin{bmatrix} 1 & 0 & 3 & -2 \\ 6 & 13 & 8 & -17 \end{bmatrix}$, $B = \begin{bmatrix} 1 & 6 \\ 4 & 2 \end{bmatrix}$

29. $A = \begin{bmatrix} 0 & -1 & 0 \\ 4 & 0 & 2 \\ 8 & -1 & 7 \end{bmatrix}$, $B = \begin{bmatrix} 2 & 1 \\ -3 & 4 \\ 1 & 6 \end{bmatrix}$

30. $A = \begin{bmatrix} -1 & 3 \\ 4 & -5 \\ 0 & 2 \end{bmatrix}$, $B = \begin{bmatrix} 1 & 2 \\ 0 & 7 \end{bmatrix}$

31. $A = \begin{bmatrix} 1 & 0 & 0 \\ 0 & 4 & 0 \\ 0 & 0 & -2 \end{bmatrix}$, $B = \begin{bmatrix} 3 & 0 & 0 \\ 0 & -1 & 0 \\ 0 & 0 & 5 \end{bmatrix}$

32. $A = \begin{bmatrix} 5 & 0 & 0 \\ 0 & -8 & 0 \\ 0 & 0 & 7 \end{bmatrix}$, $B = \begin{bmatrix} \frac{1}{5} & 0 & 0 \\ 0 & -\frac{1}{8} & 0 \\ 0 & 0 & \frac{1}{2} \end{bmatrix}$

33. $A = \begin{bmatrix} 0 & 0 & 5 \\ 0 & 0 & -3 \\ 0 & 0 & 4 \end{bmatrix}$, $B = \begin{bmatrix} 6 & -11 & 4 \\ 8 & 16 & 4 \\ 0 & 0 & 0 \end{bmatrix}$

34. $A = \begin{bmatrix} 10 \\ 12 \end{bmatrix}$, $B = [6 \quad -2 \quad 1 \quad 6]$

 In Exercises 35–40, use the matrix capabilities of a graphing utility to find AB, if possible.

35. $A = \begin{bmatrix} 5 & 6 & -3 \\ -2 & 5 & 1 \\ 10 & -5 & 5 \end{bmatrix}$, $B = \begin{bmatrix} 1 & -1 & 2 \\ 8 & 1 & 4 \\ 4 & -2 & 9 \end{bmatrix}$

36. $A = \begin{bmatrix} 11 & -12 & 4 \\ 14 & 10 & 12 \\ 6 & -2 & 9 \end{bmatrix}$, $B = \begin{bmatrix} 12 & 10 \\ -5 & 12 \\ 15 & 16 \end{bmatrix}$

37. $A = \begin{bmatrix} -3 & 8 & -6 & 8 \\ -12 & 15 & 9 & 6 \\ 5 & -1 & 1 & 5 \end{bmatrix}$, $B = \begin{bmatrix} 3 & 1 & 6 \\ 24 & 15 & 14 \\ 16 & 10 & 21 \\ 8 & -4 & 10 \end{bmatrix}$

38. $A = \begin{bmatrix} -2 & 4 & 8 \\ 21 & 5 & 6 \\ 13 & 2 & 6 \end{bmatrix}$, $B = \begin{bmatrix} 2 & 0 \\ -7 & 15 \\ 32 & 14 \\ 0.5 & 1.6 \end{bmatrix}$

39. $A = \begin{bmatrix} 9 & 10 & -38 & 18 \\ 100 & -50 & 250 & 75 \end{bmatrix}$,

$B = \begin{bmatrix} 52 & -85 & 27 & 45 \\ 40 & -35 & 60 & 82 \end{bmatrix}$

40. $A = \begin{bmatrix} 15 & -18 \\ -4 & 12 \\ -8 & 22 \end{bmatrix}$, $B = \begin{bmatrix} -7 & 22 & 1 \\ 8 & 16 & 24 \end{bmatrix}$

In Exercises 41–46, if possible, find (a) AB, (b) BA, and (c) A^2. (*Note:* $A^2 = AA$.)

41. $A = \begin{bmatrix} 1 & 2 \\ 4 & 2 \end{bmatrix}$, $B = \begin{bmatrix} 2 & -1 \\ -1 & 8 \end{bmatrix}$

42. $A = \begin{bmatrix} 2 & -1 \\ 1 & 4 \end{bmatrix}$, $B = \begin{bmatrix} 0 & 0 \\ 3 & -3 \end{bmatrix}$

43. $A = \begin{bmatrix} 3 & -1 \\ 1 & 3 \end{bmatrix}$, $B = \begin{bmatrix} 1 & -3 \\ 3 & 1 \end{bmatrix}$

44. $A = \begin{bmatrix} 1 & -1 \\ 1 & 1 \end{bmatrix}$, $B = \begin{bmatrix} 1 & 3 \\ -3 & 1 \end{bmatrix}$

45. $A = \begin{bmatrix} 7 \\ 8 \\ -1 \end{bmatrix}$, $B = [1 \quad 1 \quad 2]$

46. $A = [3 \quad 2 \quad 1]$, $B = \begin{bmatrix} 2 \\ 3 \\ 0 \end{bmatrix}$

In Exercises 47–50, evaluate the expression. Use the matrix capabilities of a graphing utility to verify your answer.

47. $\begin{bmatrix} 3 & 1 \\ 0 & -2 \end{bmatrix}\begin{bmatrix} 1 & 0 \\ -2 & 2 \end{bmatrix}\begin{bmatrix} 1 & 0 \\ 2 & 4 \end{bmatrix}$

48. $-3\left(\begin{bmatrix} 6 & 5 & -1 \\ 1 & -2 & 0 \end{bmatrix}\begin{bmatrix} 0 & 3 \\ -1 & -3 \\ 4 & 1 \end{bmatrix}\right)$

49. $\begin{bmatrix} 0 & 2 & -2 \\ 4 & 1 & 2 \end{bmatrix}\left(\begin{bmatrix} 4 & 0 \\ 0 & -1 \\ -1 & 2 \end{bmatrix}+\begin{bmatrix} -2 & 3 \\ -3 & 5 \\ 0 & -3 \end{bmatrix}\right)$

50. $\begin{bmatrix} 3 \\ -1 \\ 5 \\ 7 \end{bmatrix}\left([5 \quad -6]+[7 \quad -1]+[-8 \quad 9]\right)$

In Exercises 51–58, (a) write the system of linear equations as a matrix equation, $AX = B$, and (b) use Gauss-Jordan elimination on the augmented matrix $[A \vdots B]$ to solve for the matrix X.

51. $\begin{cases} -x_1 + x_2 = 4 \\ -2x_1 + x_2 = 0 \end{cases}$
52. $\begin{cases} 2x_1 + 3x_2 = 5 \\ x_1 + 4x_2 = 10 \end{cases}$

53. $\begin{cases} -2x_1 - 3x_2 = -4 \\ 6x_1 + x_2 = -36 \end{cases}$
54. $\begin{cases} -4x_1 + 9x_2 = -13 \\ x_1 - 3x_2 = 12 \end{cases}$

55. $\begin{cases} x_1 - 2x_2 + 3x_3 = 9 \\ -x_1 + 3x_2 - x_3 = -6 \\ 2x_1 - 5x_2 + 5x_3 = 17 \end{cases}$

56. $\begin{cases} x_1 + x_2 - 3x_3 = 9 \\ -x_1 + 2x_2 = 6 \\ x_1 - x_2 + x_3 = -5 \end{cases}$

57. $\begin{cases} x_1 - 5x_2 + 2x_3 = -20 \\ -3x_1 + x_2 - x_3 = 8 \\ -2x_2 + 5x_3 = -16 \end{cases}$

58. $\begin{cases} x_1 - x_2 + 4x_3 = 17 \\ x_1 + 3x_2 = -11 \\ -6x_2 + 5x_3 = 40 \end{cases}$

59. *Manufacturing*   A corporation has three factories, each of which manufactures acoustic guitars and electric guitars. The number of units of guitars produced at factory j in one day is represented by a_{ij} in the matrix

$$A = \begin{bmatrix} 70 & 50 & 25 \\ 35 & 100 & 70 \end{bmatrix}.$$

Find the production levels if production is increased by 20%.

60. *Manufacturing*   A corporation has four factories, each of which manufactures sport utility vehicles and pickup trucks. The number of units of vehicle i produced at factory j in one day is represented by a_{ij} in the matrix

$$A = \begin{bmatrix} 100 & 90 & 70 & 30 \\ 40 & 20 & 60 & 60 \end{bmatrix}.$$

Find the production levels if production is increased by 10%.

61. *Agriculture*   A fruit grower raises two crops, apples and peaches. Each of these crops is sent to three different outlets for sale. These outlets are The Farmer's Market, The Fruit Stand, and The Fruit Farm. The numbers of bushels of apples sent to the three outlets are 125, 100, and 75, respectively. The numbers of bushels of peaches sent to the three outlets are 100, 175, and 125, respectively. The profit per bushel for apples is \$3.50 and the profit per bushel for peaches is \$6.00.

(a) Write a matrix A that represents the number of bushels of each crop i that are shipped to each outlet j. State what each entry a_{ij} of the matrix represents.

(b) Write a matrix B that represents the profit per bushel of each fruit. State what each entry b_{ij} of the matrix represents.

(c) Find the product BA and state what each entry of the matrix represents.

62. *Revenue*   A manufacturer of electronics produces three models of portable CD players, which are shipped to two warehouses. The number of units of model i that are shipped to warehouse j is represented by a_{ij} in the matrix

$$A = \begin{bmatrix} 5{,}000 & 4{,}000 \\ 6{,}000 & 10{,}000 \\ 8{,}000 & 5{,}000 \end{bmatrix}.$$

The prices per unit are represented by the matrix

$$B = [\$39.50 \quad \$44.50 \quad \$56.50].$$

Compute BA and interpret the result.

63. *Inventory*   A company sells five models of computers through three retail outlets. The inventories are represented by S.

$$\begin{matrix} & \text{Model} \\ & \overbrace{\begin{matrix} \text{A} & \text{B} & \text{C} & \text{D} & \text{E} \end{matrix}} \\ S = & \begin{bmatrix} 3 & 2 & 2 & 3 & 0 \\ 0 & 2 & 3 & 4 & 3 \\ 4 & 2 & 1 & 3 & 2 \end{bmatrix} \begin{matrix} 1 \\ 2 \\ 3 \end{matrix} \end{matrix} \text{Outlet}$$

The wholesale and retail prices are represented by T.

$$\begin{matrix} & \text{Price} \\ & \overbrace{\begin{matrix} \text{Wholesale} & \text{Retail} \end{matrix}} \\ T = & \begin{bmatrix} \$840 & \$1100 \\ \$1200 & \$1350 \\ \$1450 & \$1650 \\ \$2650 & \$3000 \\ \$3050 & \$3200 \end{bmatrix} \begin{matrix} \text{A} \\ \text{B} \\ \text{C} \\ \text{D} \\ \text{E} \end{matrix} \end{matrix} \text{Model}$$

Compute ST and interpret the result.

64. Voting Preferences The matrix

From

$$P = \begin{bmatrix} 0.6 & 0.1 & 0.1 \\ 0.2 & 0.7 & 0.1 \\ 0.2 & 0.2 & 0.8 \end{bmatrix} \begin{matrix} R \\ D \\ I \end{matrix} \Bigg\} \text{ To}$$

with column headings R D I

is called a *stochastic matrix*. Each entry $p_{ij} (i \neq j)$ represents the proportion of the voting population that changes from party i to party j, and p_{ii} represents the proportion that remains loyal to the party from one election to the next. Compute and interpret P^2.

 65. Voting Preferences Use a graphing utility to find P^3, P^4, P^5, P^6, P^7, and P^8 for the matrix given in Exercise 64. Can you detect a pattern as P is raised to higher powers?

66. Labor/Wage Requirements A company that manufactures boats has the following labor-hour and wage requirements.

Labor per boat

Department

$$S = \begin{bmatrix} 1.0 \text{ hr} & 0.5 \text{ hr} & 0.2 \text{ hr} \\ 1.6 \text{ hr} & 1.0 \text{ hr} & 0.2 \text{ hr} \\ 2.5 \text{ hr} & 2.0 \text{ hr} & 1.4 \text{ hr} \end{bmatrix} \begin{matrix} \text{Small} \\ \text{Medium} \\ \text{Large} \end{matrix} \Bigg\} \text{ Boat size}$$

with column headings Cutting Assembly Packaging

Wages per hour

Plant

$$T = \begin{bmatrix} \$12 & \$10 \\ \$9 & \$8 \\ \$8 & \$7 \end{bmatrix} \begin{matrix} \text{Cutting} \\ \text{Assembly} \\ \text{Packaging} \end{matrix} \Bigg\} \text{ Department}$$

with column headings A B

Compute ST and interpret the result.

67. Profit At a local dairy mart, the numbers of gallons of skim milk, 2% milk, and whole milk sold over the weekend are represented by A.

$$A = \begin{bmatrix} 40 & 64 & 52 \\ 60 & 82 & 76 \\ 76 & 96 & 84 \end{bmatrix} \begin{matrix} \text{Friday} \\ \text{Saturday} \\ \text{Sunday} \end{matrix}$$

with column headings Skim milk, 2% milk, Whole milk

The selling prices (in dollars per gallon) and the profits (in dollars per gallon) for the three types of milk sold by the dairy mart are represented by B.

$$B = \begin{bmatrix} 2.65 & 0.65 \\ 2.85 & 0.70 \\ 3.05 & 0.85 \end{bmatrix} \begin{matrix} \text{Skim milk} \\ \text{2% milk} \\ \text{Whole milk} \end{matrix}$$

with column headings Selling price, Profit

(a) Compute AB and interpret the result.

(b) Find the dairy mart's total profit from milk sales for the weekend.

68. Profit At a convenience store, the numbers of gallons of 87-octane, 89-octane, and 93-octane gasoline sold over the weekend are represented by A.

Octane

$$A = \begin{bmatrix} 580 & 840 & 320 \\ 560 & 420 & 160 \\ 860 & 1020 & 540 \end{bmatrix} \begin{matrix} \text{Friday} \\ \text{Saturday} \\ \text{Sunday} \end{matrix}$$

with column headings 87 89 93

The selling prices per gallon and the profits per gallon for the three grades of gasoline sold by the convenience store are represents by B.

$$B = \begin{bmatrix} 1.95 & 0.32 \\ 2.05 & 0.36 \\ 2.15 & 0.40 \end{bmatrix} \begin{matrix} 87 \\ 89 \\ 93 \end{matrix} \Bigg\} \text{ Octane}$$

with column headings Selling price, Profit

(a) Compute AB and interpret the result.

(b) Find the convenience store's profit from gasoline sales for the weekend.

69. Exercise The numbers of calories burned by individuals of different body weights performing different types of aerobic exercises for a 20-minute time period are shown in matrix A.

Calories burned

$$A = \begin{bmatrix} 109 & 136 \\ 127 & 159 \\ 64 & 79 \end{bmatrix} \begin{matrix} \text{Bicycling} \\ \text{Jogging} \\ \text{Walking} \end{matrix}$$

with column headings 120-lb person, 150-lb person

(a) A 120-pound person and a 150-pound person bicycled for 40 minutes, jogged for 10 minutes, and walked for 60 minutes. Organize the time spent exercising in a matrix B.

(b) Compute BA and interpret the result.

Model It

70. Health Care   The health care plans offered this year by a local manufacturing plant are as follows. For individuals, the comprehensive plan costs $694.32, the HMO standard plan costs $451.80, and the HMO Plus plan costs $489.48. For families, the comprehensive plan costs $1725.36, the HMO standard plan costs $1187.76 and the HMO Plus plan costs $1248.12. The plant expects the costs of the plans to change next year as follows. For individuals, the costs for the comprehensive, HMO standard, and HMO Plus plans will be $683.91, $463.10, and $499.27, respectively. For families, the costs for the comprehensive, HMO standard, and HMO Plus plans will be $1699.48, $1217.45, and $1273.08, respectively.

(a) Organize the information using two matrices A and B, where A represents the health care plan costs for this year and B represents the health care plan costs for next year. State what each entry of each matrix represents.

(b) Compute $A - B$ and interpret the result.

(c) The employees receive monthly paychecks from which the health care plan costs are deducted. Use the matrices from part (a) to write matrices that show how much will be deducted from each employees' paycheck this year and next year.

(d) Suppose the costs of each plan instead increase by 4% next year. Write a matrix that shows the new monthly payment.

Synthesis

True or False?   In Exercises 71 and 72, determine whether the statement is true or false. Justify your answer.

71. Two matrices can be added only if they have the same order.

72. $\begin{bmatrix} -6 & -2 \\ 2 & -6 \end{bmatrix}\begin{bmatrix} 4 & 0 \\ 0 & -1 \end{bmatrix} = \begin{bmatrix} 4 & 0 \\ 0 & -1 \end{bmatrix}\begin{bmatrix} -6 & -2 \\ 2 & -6 \end{bmatrix}$

Think About It   In Exercises 73–80, let matrices A, B, C, and D be of orders 2×3, 2×3, 3×2, and 2×2, respectively. Determine whether the matrices are of proper order to perform the operation(s). If so, give the order of the answer.

73. $A + 2C$

74. $B - 3C$

75. AB

76. BC

77. $BC - D$

78. $CB - D$

79. $D(A - 3B)$

80. $(BC - D)A$

81. *Think About It*   If a, b, and c are real numbers such that $c \neq 0$ and $ac = bc$, then $a = b$. However, if A, B, and C are nonzero matrices such that $AC = BC$, then A is *not* necessarily equal to B. Illustrate this using the following matrices.

$$A = \begin{bmatrix} 0 & 1 \\ 0 & 1 \end{bmatrix}, \quad B = \begin{bmatrix} 1 & 0 \\ 1 & 0 \end{bmatrix}, \quad C = \begin{bmatrix} 2 & 3 \\ 2 & 3 \end{bmatrix}$$

82. *Think About It*   If a and b are real numbers such that $ab = 0$, then $a = 0$ or $b = 0$. However, if A and B are matrices such that $AB = O$, it is *not* necessarily true that $A = O$ or $B = O$. Illustrate this using the following matrices.

$$A = \begin{bmatrix} 3 & 3 \\ 4 & 4 \end{bmatrix}, \quad B = \begin{bmatrix} 1 & -1 \\ -1 & 1 \end{bmatrix}$$

83. *Exploration*   Let A and B be unequal diagonal matrices of the same order. (A **diagonal matrix** is a square matrix in which each entry not on the main diagonal is zero.) Determine the products AB for several pairs of such matrices. Make a conjecture about a quick rule for such products.

84. *Exploration*   Let $i = \sqrt{-1}$ and let

$$A = \begin{bmatrix} i & 0 \\ 0 & i \end{bmatrix} \quad \text{and} \quad B = \begin{bmatrix} 0 & -i \\ i & 0 \end{bmatrix}.$$

(a) Find A^2, A^3, and A^4. Identify any similarities with i^2, i^3, and i^4.

(b) Find and identify B^2.

Skills Review

In Exercises 85–90, solve the equation.

85. $3x^2 + 20x - 32 = 0$

86. $8x^2 - 10x - 3 = 0$

87. $4x^3 + 10x^2 - 3x = 0$

88. $3x^3 + 22x^2 - 45x = 0$

89. $3x^3 - 12x^2 + 5x - 20 = 0$

90. $2x^3 - 5x^2 - 12x + 30 = 0$

In Exercises 91–94, solve the system of linear equations both graphically and algebraically.

91. $\begin{cases} -x + 4y = -9 \\ 5x - 8y = 39 \end{cases}$

92. $\begin{cases} 8x - 3y = -17 \\ -6x + 7y = 27 \end{cases}$

93. $\begin{cases} -x + 2y = -5 \\ -3x - y = -8 \end{cases}$

94. $\begin{cases} 6x - 13y = 11 \\ 9x + 5y = 41 \end{cases}$

What you should learn

- Verify that two matrices are inverses of each other.
- Use Gauss-Jordan elimination to find the inverses of matrices.
- Use a formula to find the inverses of 2×2 matrices.
- Use inverse matrices to solve systems of linear equations.

Why you should learn it

You can use inverse matrices to model and solve real-life problems. For instance, in Exercise 72 on page 562, an inverse matrix is used to find a linear model for the number of licensed drivers in the United States.

Jon Love/Getty Images

The Inverse of a Matrix

This section further develops the algebra of matrices. To begin, consider the real number equation $ax = b$. To solve this equation for x, multiply each side of the equation by a^{-1} (provided that $a \neq 0$).

$$ax = b$$

$$(a^{-1}a)x = a^{-1}b$$

$$(1)x = a^{-1}b$$

$$x = a^{-1}b$$

The number a^{-1} is called the *multiplicative inverse of a* because $a^{-1}a = 1$. The definition of the multiplicative **inverse of a matrix** is similar.

Definition of the Inverse of a Square Matrix

Let A be an $n \times n$ matrix and let I_n be the $n \times n$ identity matrix. If there exists a matrix A^{-1} such that

$$AA^{-1} = I_n = A^{-1}A$$

then A^{-1} is called the **inverse** of A. The symbol A^{-1} is read "A inverse."

Example 1 The Inverse of a Matrix

Show that B is the inverse of A, where

$$A = \begin{bmatrix} -1 & 2 \\ -1 & 1 \end{bmatrix} \quad \text{and} \quad B = \begin{bmatrix} 1 & -2 \\ 1 & -1 \end{bmatrix}.$$

Solution

To show that B is the inverse of A, show that $AB = I = BA$, as follows.

$$AB = \begin{bmatrix} -1 & 2 \\ -1 & 1 \end{bmatrix} \begin{bmatrix} 1 & -2 \\ 1 & -1 \end{bmatrix} = \begin{bmatrix} -1+2 & 2-2 \\ -1+1 & 2-1 \end{bmatrix} = \begin{bmatrix} 1 & 0 \\ 0 & 1 \end{bmatrix}$$

$$BA = \begin{bmatrix} 1 & -2 \\ 1 & -1 \end{bmatrix} \begin{bmatrix} -1 & 2 \\ -1 & 1 \end{bmatrix} = \begin{bmatrix} -1+2 & 2-2 \\ -1+1 & 2-1 \end{bmatrix} = \begin{bmatrix} 1 & 0 \\ 0 & 1 \end{bmatrix}$$

As you can see, $AB = I = BA$. This is an example of a square matrix that has an inverse. Note that not all square matrices have an inverse.

✓CHECKPOINT Now try Exercise 1.

Recall that it is not always true that $AB = BA$, even if both products are defined. However, if A and B are both square matrices and $AB = I_n$, it can be shown that $BA = I_n$. So, in Example 1, you need only to check that $AB = I_2$.

Finding Inverse Matrices

If a matrix A has an inverse, A is called **invertible** (or **nonsingular**); otherwise, A is called **singular.** A nonsquare matrix cannot have an inverse. To see this, note that if A is of order $m \times n$ and B is of order $n \times m$ (where $m \neq n$), the products AB and BA are of different orders and so cannot be equal to each other. Not all square matrices have inverses (see the matrix at the bottom of page 557). If, however, a matrix does have an inverse, that inverse is unique. Example 2 shows how to use a system of equations to find the inverse of a matrix.

Example 2   Finding the Inverse of a Matrix

Find the inverse of

$$A = \begin{bmatrix} 1 & 4 \\ -1 & -3 \end{bmatrix}.$$

Solution

To find the inverse of A, try to solve the matrix equation $AX = I$ for X.

$$\begin{array}{ccc} A & X & I \end{array}$$

$$\begin{bmatrix} 1 & 4 \\ -1 & -3 \end{bmatrix}\begin{bmatrix} x_{11} & x_{12} \\ x_{21} & x_{22} \end{bmatrix} = \begin{bmatrix} 1 & 0 \\ 0 & 1 \end{bmatrix}$$

$$\begin{bmatrix} x_{11} + 4x_{21} & x_{12} + 4x_{22} \\ -x_{11} - 3x_{21} & -x_{12} - 3x_{22} \end{bmatrix} = \begin{bmatrix} 1 & 0 \\ 0 & 1 \end{bmatrix}$$

Equating corresponding entries, you obtain two systems of linear equations.

$$\begin{cases} x_{11} + 4x_{21} = 1 \\ -x_{11} - 3x_{21} = 0 \end{cases}$$ Linear system with two variables, x_{11} and x_{21}.

$$\begin{cases} x_{12} + 4x_{22} = 0 \\ -x_{12} - 3x_{22} = 1 \end{cases}$$ Linear system with two variables, x_{12} and x_{22}.

Solve the first system using elementary row operations to determine that $x_{11} = -3$ and $x_{21} = 1$. From the second system you can determine that $x_{12} = -4$ and $x_{22} = 1$. Therefore, the inverse of A is

$$X = A^{-1}$$

$$= \begin{bmatrix} -3 & -4 \\ 1 & 1 \end{bmatrix}.$$

You can use matrix multiplication to check this result.

Check

$$AA^{-1} = \begin{bmatrix} 1 & 4 \\ -1 & -3 \end{bmatrix}\begin{bmatrix} -3 & -4 \\ 1 & 1 \end{bmatrix} = \begin{bmatrix} 1 & 0 \\ 0 & 1 \end{bmatrix} \checkmark$$

$$A^{-1}A = \begin{bmatrix} -3 & -4 \\ 1 & 1 \end{bmatrix}\begin{bmatrix} 1 & 4 \\ -1 & -3 \end{bmatrix} = \begin{bmatrix} 1 & 0 \\ 0 & 1 \end{bmatrix} \checkmark$$

✓CHECKPOINT   Now try Exercise 13.

In Example 2, note that the two systems of linear equations have the *same coefficient matrix A.* Rather than solve the two systems represented by

$$\begin{bmatrix} 1 & 4 & \vdots & 1 \\ -1 & -3 & \vdots & 0 \end{bmatrix}$$

and

$$\begin{bmatrix} 1 & 4 & \vdots & 0 \\ -1 & -3 & \vdots & 1 \end{bmatrix}$$

separately, you can solve them *simultaneously* by *adjoining* the identity matrix to the coefficient matrix to obtain

$$\begin{matrix} A & & I \\ \begin{bmatrix} 1 & 4 & \vdots & 1 & 0 \\ -1 & -3 & \vdots & 0 & 1 \end{bmatrix} \end{matrix}.$$

This "doubly augmented" matrix can be represented as $[A \ \vdots \ I]$. By applying Gauss-Jordan elimination to this matrix, you can solve *both* systems with a single elimination process.

$$\begin{bmatrix} 1 & 4 & \vdots & 1 & 0 \\ -1 & -3 & \vdots & 0 & 1 \end{bmatrix}$$

$$\begin{matrix} R_1 + R_2 \rightarrow \end{matrix} \begin{bmatrix} 1 & 4 & \vdots & 1 & 0 \\ 0 & 1 & \vdots & 1 & 1 \end{bmatrix}$$

$$\begin{matrix} -4R_2 + R_1 \rightarrow \end{matrix} \begin{bmatrix} 1 & 0 & \vdots & -3 & -4 \\ 0 & 1 & \vdots & 1 & 1 \end{bmatrix}$$

So, from the "doubly augmented" matrix $[A \ \vdots \ I]$, you obtain the matrix $[I \ \vdots \ A^{-1}]$.

$$\begin{matrix} A & & I \\ \begin{bmatrix} 1 & 4 & \vdots & 1 & 0 \\ -1 & -3 & \vdots & 0 & 1 \end{bmatrix} \end{matrix} \implies \begin{matrix} I & & A^{-1} \\ \begin{bmatrix} 1 & 0 & \vdots & -3 & -4 \\ 0 & 1 & \vdots & 1 & 1 \end{bmatrix} \end{matrix}$$

This procedure (or algorithm) works for any square matrix that has an inverse.

Technology

Most graphing utilities can find the inverse of a square matrix. To do so, you may have to use the inverse key $\boxed{x^{-1}}$. Consult the user's guide for your graphing utility for specific keystrokes.

Finding an Inverse Matrix

Let A be a square matrix of order n.

1. Write the $n \times 2n$ matrix that consists of the given matrix A on the left and the $n \times n$ identity matrix I on the right to obtain $[A \ \vdots \ I]$.

2. If possible, row reduce A to I using elementary row operations on the *entire* matrix $[A \ \vdots \ I]$. The result will be the matrix $[I \ \vdots \ A^{-1}]$. If this is not possible, A is not invertible.

3. Check your work by multiplying to see that $AA^{-1} = I = A^{-1}A$.

| Example 3 | **Finding the Inverse of a Matrix** |

Find the inverse of $A = \begin{bmatrix} 1 & -1 & 0 \\ 1 & 0 & -1 \\ 6 & -2 & -3 \end{bmatrix}$.

Solution

Begin by adjoining the identity matrix to A to form the matrix

$$[A \vdots I] = \begin{bmatrix} 1 & -1 & 0 & \vdots & 1 & 0 & 0 \\ 1 & 0 & -1 & \vdots & 0 & 1 & 0 \\ 6 & -2 & -3 & \vdots & 0 & 0 & 1 \end{bmatrix}.$$

Use elementary row operations to obtain the form $[I \vdots A^{-1}]$, as follows.

$$\begin{array}{c} \\ -R_1 + R_2 \rightarrow \\ -6R_1 + R_3 \rightarrow \end{array} \begin{bmatrix} 1 & -1 & 0 & \vdots & 1 & 0 & 0 \\ 0 & 1 & -1 & \vdots & -1 & 1 & 0 \\ 0 & 4 & -3 & \vdots & -6 & 0 & 1 \end{bmatrix}$$

$$\begin{array}{c} R_2 + R_1 \rightarrow \\ \\ -4R_2 + R_3 \rightarrow \end{array} \begin{bmatrix} 1 & 0 & -1 & \vdots & 0 & 1 & 0 \\ 0 & 1 & -1 & \vdots & -1 & 1 & 0 \\ 0 & 0 & 1 & \vdots & -2 & -4 & 1 \end{bmatrix}$$

$$\begin{array}{c} R_3 + R_1 \rightarrow \\ R_3 + R_2 \rightarrow \\ \\ \end{array} \begin{bmatrix} 1 & 0 & 0 & \vdots & -2 & -3 & 1 \\ 0 & 1 & 0 & \vdots & -3 & -3 & 1 \\ 0 & 0 & 1 & \vdots & -2 & -4 & 1 \end{bmatrix} = [I \vdots A^{-1}]$$

So, the matrix A is invertible and its inverse is

$$A^{-1} = \begin{bmatrix} -2 & -3 & 1 \\ -3 & -3 & 1 \\ -2 & -4 & 1 \end{bmatrix}.$$

Confirm this result by multiplying A and A^{-1} to obtain I, as follows.

Check

$$AA^{-1} = \begin{bmatrix} 1 & -1 & 0 \\ 1 & 0 & -1 \\ 6 & -2 & -3 \end{bmatrix} \begin{bmatrix} -2 & -3 & 1 \\ -3 & -3 & 1 \\ -2 & -4 & 1 \end{bmatrix} = \begin{bmatrix} 1 & 0 & 0 \\ 0 & 1 & 0 \\ 0 & 0 & 1 \end{bmatrix} = I$$

✔CHECKPOINT   Now try Exercise 21.

STUDY TIP

Be sure to check your solution because it is easy to make algebraic errors when using elementary row operations.

The process shown in Example 3 applies to any $n \times n$ matrix A. When using this algorithm, if the matrix A does not reduce to the identity matrix, then A does not have an inverse. For instance, the following matrix has no inverse.

$$A = \begin{bmatrix} 1 & 2 & 0 \\ 3 & -1 & 2 \\ -2 & 3 & -2 \end{bmatrix}$$

To confirm that matrix A above has no inverse, adjoin the identity matrix to A to form $[A: I]$ and perform elementary row operations on the matrix. After doing so, you will see that it is impossible to obtain the identity matrix I on the left. Therefore, A is not invertible.

The Inverse of a 2 × 2 Matrix

Using Gauss-Jordan elimination to find the inverse of a matrix works well (even as a computer technique) for matrices of order 3×3 or greater. For 2×2 matrices, however, many people prefer to use a formula for the inverse rather than Gauss-Jordan elimination. This simple formula, which works *only* for 2×2 matrices, is explained as follows. If A is a 2×2 matrix given by

$$A = \begin{bmatrix} a & b \\ c & d \end{bmatrix}$$

then A is invertible if and only if $ad - bc \neq 0$. Moreover, if $ad - bc \neq 0$, the inverse is given by

$$A^{-1} = \frac{1}{ad - bc} \begin{bmatrix} d & -b \\ -c & a \end{bmatrix}. \qquad \text{Formula for inverse of matrix } A$$

The denominator $ad - bc$ is called the **determinant** of the 2×2 matrix A. You will study determinants in the next section.

Example 4   Finding the Inverse of a 2 × 2 Matrix

If possible, find the inverse of each matrix.

a. $A = \begin{bmatrix} 3 & -1 \\ -2 & 2 \end{bmatrix}$

b. $B = \begin{bmatrix} 3 & -1 \\ -6 & 2 \end{bmatrix}$

Solution

a. For the matrix A, apply the formula for the inverse of a 2×2 matrix to obtain

$$ad - bc = (3)(2) - (-1)(-2)$$
$$= 4.$$

Because this quantity is not zero, the inverse is formed by interchanging the entries on the main diagonal, changing the signs of the other two entries, and multiplying by the scalar $\frac{1}{4}$, as follows.

$$A^{-1} = \frac{1}{4}\begin{bmatrix} 2 & 1 \\ 2 & 3 \end{bmatrix} \qquad \text{Substitute for } a, b, c, d, \text{ and the determinant.}$$

$$= \begin{bmatrix} \frac{1}{2} & \frac{1}{4} \\ \frac{1}{2} & \frac{3}{4} \end{bmatrix} \qquad \text{Multiply by the scalar } \frac{1}{4}.$$

b. For the matrix B, you have

$$ad - bc = (3)(2) - (-1)(-6)$$
$$= 0$$

which means that B is not invertible.

✓**CHECKPOINT**   Now try Exercise 39.

Systems of Linear Equations

You know that a system of linear equations can have exactly one solution, infinitely many solutions, or no solution. If the coefficient matrix A of a *square* system (a system that has the same number of equations as variables) is invertible, the system has a unique solution, which is defined as follows.

> ## A System of Equations with a Unique Solution
>
> If A is an invertible matrix, the system of linear equations represented by $AX = B$ has a unique solution given by
>
> $$X = A^{-1}B.$$

Technology

To solve a system of equations with a graphing utility, enter the matrices A and B in the matrix editor. Then, using the inverse key, solve for X.

A $\boxed{x^{-1}}$ B $\boxed{\text{ENTER}}$

The screen will display the solution, matrix X.

Example 5   **Solving a System Using an Inverse**

You are going to invest \$10,000 in AAA-rated bonds, AA-rated bonds, and B-rated bonds and want an annual return of \$730. The average yields are 6% on AAA bonds, 7.5% on AA bonds, and 9.5% on B bonds. You will invest twice as much in AAA bonds as in B bonds. Your investment can be represented as

$$\begin{cases} x + \quad y + \quad\quad z = 10,000 \\ 0.06x + 0.075y + 0.095z = \quad 730 \\ x \quad\quad\quad - \quad 2z = \quad\quad 0 \end{cases}$$

where x, y, and z represent the amounts invested in AAA, AA, and B bonds, respectively. Use an inverse matrix to solve the system.

Solution

Begin by writing the system in the matrix form $AX = B$.

$$\begin{bmatrix} 1 & 1 & 1 \\ 0.06 & 0.075 & 0.095 \\ 1 & 0 & -2 \end{bmatrix} \begin{bmatrix} x \\ y \\ z \end{bmatrix} = \begin{bmatrix} 10,000 \\ 730 \\ 0 \end{bmatrix}$$

Then, use Gauss-Jordan elimination to find A^{-1}.

$$A^{-1} = \begin{bmatrix} 15 & -200 & -2 \\ -21.5 & 300 & 3.5 \\ 7.5 & -100 & -1.5 \end{bmatrix}$$

Finally, multiply B by A^{-1} on the left to obtain the solution.

$$X = A^{-1}B$$

$$= \begin{bmatrix} 15 & -200 & -2 \\ -21.5 & 300 & 3.5 \\ 7.5 & -100 & -1.5 \end{bmatrix} \begin{bmatrix} 10,000 \\ 730 \\ 0 \end{bmatrix} = \begin{bmatrix} 4000 \\ 4000 \\ 2000 \end{bmatrix}$$

The solution to the system is $x = 4000$, $y = 4000$, and $z = 2000$. So, you will invest \$4000 in AAA bonds, \$4000 in AA bonds, and \$2000 in B bonds.

✓**CHECKPOINT**   Now try Exercise 67.

7.3 Exercises

VOCABULARY CHECK: Fill in the blanks.

1. In a _____ matrix, the number of rows equals the number of columns.

2. If there exists an $n \times n$ matrix A^{-1} such that $AA^{-1} = I_n = A^{-1}A$, then A^{-1} is called the _____ of A.

3. If a matrix A has an inverse, it is called invertible or _____; if it does not have an inverse, it is called _____.

4. If A is an invertible matrix, the system of linear equations represented by $AX = B$ has a unique solution given by $X =$ _____.

PREREQUISITE SKILLS REVIEW: Practice and review algebra skills needed for this section at **www.Eduspace.com**.

In Exercises 1–10, show that B is the inverse of A.

1. $A = \begin{bmatrix} 2 & 1 \\ 5 & 3 \end{bmatrix}$, $B = \begin{bmatrix} 3 & -1 \\ -5 & 2 \end{bmatrix}$

2. $A = \begin{bmatrix} 1 & -1 \\ -1 & 2 \end{bmatrix}$, $B = \begin{bmatrix} 2 & 1 \\ 1 & 1 \end{bmatrix}$

3. $A = \begin{bmatrix} 1 & 2 \\ 3 & 4 \end{bmatrix}$, $B = \begin{bmatrix} -2 & 1 \\ \frac{3}{2} & -\frac{1}{2} \end{bmatrix}$

4. $A = \begin{bmatrix} 1 & -1 \\ 2 & 3 \end{bmatrix}$, $B = \begin{bmatrix} \frac{3}{5} & \frac{1}{5} \\ -\frac{2}{5} & \frac{1}{5} \end{bmatrix}$

5. $A = \begin{bmatrix} 2 & -17 & 11 \\ -1 & 11 & -7 \\ 0 & 3 & -2 \end{bmatrix}$, $B = \begin{bmatrix} 1 & 1 & 2 \\ 2 & 4 & -3 \\ 3 & 6 & -5 \end{bmatrix}$

6. $A = \begin{bmatrix} -4 & 1 & 5 \\ -1 & 2 & 4 \\ 0 & -1 & -1 \end{bmatrix}$, $B = \begin{bmatrix} -\frac{1}{2} & 1 & \frac{3}{2} \\ \frac{1}{4} & -1 & -\frac{11}{4} \\ -\frac{1}{4} & 1 & \frac{7}{4} \end{bmatrix}$

7. $A = \begin{bmatrix} 2 & 0 & 1 & 1 \\ 3 & 0 & 0 & 1 \\ -1 & 1 & -2 & 1 \\ 4 & -1 & 1 & 0 \end{bmatrix}$, $B = \begin{bmatrix} -1 & 2 & -1 & -1 \\ -4 & 9 & -5 & -6 \\ 0 & 1 & -1 & -1 \\ 3 & -5 & 3 & 3 \end{bmatrix}$

8. $A = \begin{bmatrix} -2 & 0 & 1 & 0 \\ 1 & -1 & -3 & 0 \\ -2 & -1 & 0 & -2 \\ 0 & 1 & 3 & -1 \end{bmatrix}$, $B = \begin{bmatrix} -3 & -3 & 1 & -2 \\ 12 & 14 & -5 & 10 \\ -5 & -6 & 2 & -4 \\ -3 & -4 & 1 & -3 \end{bmatrix}$

9. $A = \begin{bmatrix} -2 & 2 & 3 \\ 1 & -1 & 0 \\ 0 & 1 & 4 \end{bmatrix}$, $B = \frac{1}{3}\begin{bmatrix} -4 & -5 & 3 \\ -4 & -8 & 3 \\ 1 & 2 & 0 \end{bmatrix}$

10. $A = \begin{bmatrix} -1 & 1 & 0 & -1 \\ 1 & -1 & 1 & 0 \\ -1 & 1 & 2 & 0 \\ 0 & -1 & 1 & 1 \end{bmatrix}$,
$B = \frac{1}{3}\begin{bmatrix} -3 & 1 & 1 & -3 \\ -3 & -1 & 2 & -3 \\ 0 & 1 & 1 & 0 \\ -3 & -2 & 1 & 0 \end{bmatrix}$

In Exercises 11–26, find the inverse of the matrix (if it exists).

11. $\begin{bmatrix} 2 & 0 \\ 0 & 3 \end{bmatrix}$

12. $\begin{bmatrix} 1 & 2 \\ 3 & 7 \end{bmatrix}$

13. $\begin{bmatrix} 1 & -2 \\ 2 & -3 \end{bmatrix}$

14. $\begin{bmatrix} -7 & 33 \\ 4 & -19 \end{bmatrix}$

15. $\begin{bmatrix} -1 & 1 \\ -2 & 1 \end{bmatrix}$

16. $\begin{bmatrix} 11 & 1 \\ -1 & 0 \end{bmatrix}$

17. $\begin{bmatrix} 2 & 4 \\ 4 & 8 \end{bmatrix}$

18. $\begin{bmatrix} 2 & 3 \\ 1 & 4 \end{bmatrix}$

19. $\begin{bmatrix} 2 & 7 & 1 \\ -3 & -9 & 2 \end{bmatrix}$

20. $\begin{bmatrix} -2 & 5 \\ 6 & -15 \\ 0 & 1 \end{bmatrix}$

21. $\begin{bmatrix} 1 & 1 & 1 \\ 3 & 5 & 4 \\ 3 & 6 & 5 \end{bmatrix}$

22. $\begin{bmatrix} 1 & 2 & 2 \\ 3 & 7 & 9 \\ -1 & -4 & -7 \end{bmatrix}$

23. $\begin{bmatrix} 1 & 0 & 0 \\ 3 & 4 & 0 \\ 2 & 5 & 5 \end{bmatrix}$

24. $\begin{bmatrix} 1 & 0 & 0 \\ 3 & 0 & 0 \\ 2 & 5 & 5 \end{bmatrix}$

25. $\begin{bmatrix} -8 & 0 & 0 & 0 \\ 0 & 1 & 0 & 0 \\ 0 & 0 & 4 & 0 \\ 0 & 0 & 0 & -5 \end{bmatrix}$

26. $\begin{bmatrix} 1 & 3 & -2 & 0 \\ 0 & 2 & 4 & 6 \\ 0 & 0 & -2 & 1 \\ 0 & 0 & 0 & 5 \end{bmatrix}$

In Exercises 27–38, use the matrix capabilities of a graphing utility to find the inverse of the matrix (if it exists).

27. $\begin{bmatrix} 1 & 2 & -1 \\ 3 & 7 & -10 \\ -5 & -7 & -15 \end{bmatrix}$

28. $\begin{bmatrix} 10 & 5 & -7 \\ -5 & 1 & 4 \\ 3 & 2 & -2 \end{bmatrix}$

29. $\begin{bmatrix} 1 & 1 & 2 \\ 3 & 1 & 0 \\ -2 & 0 & 3 \end{bmatrix}$

30. $\begin{bmatrix} 3 & 2 & 2 \\ 2 & 2 & 2 \\ -4 & 4 & 3 \end{bmatrix}$

31. $\begin{bmatrix} -\frac{1}{2} & \frac{3}{4} & \frac{1}{4} \\ 1 & 0 & -\frac{3}{2} \\ 0 & -1 & \frac{1}{2} \end{bmatrix}$

32. $\begin{bmatrix} -\frac{5}{6} & \frac{1}{3} & \frac{11}{6} \\ 0 & \frac{2}{3} & 2 \\ 1 & -\frac{1}{2} & -\frac{5}{2} \end{bmatrix}$

33. $\begin{bmatrix} 0.1 & 0.2 & 0.3 \\ -0.3 & 0.2 & 0.2 \\ 0.5 & 0.4 & 0.4 \end{bmatrix}$ **34.** $\begin{bmatrix} 0.6 & 0 & -0.3 \\ 0.7 & -1 & 0.2 \\ 1 & 0 & -0.9 \end{bmatrix}$

35. $\begin{bmatrix} 1 & 0 & 3 & 0 \\ 0 & 2 & 0 & 4 \\ 1 & 0 & 3 & 0 \\ 0 & 2 & 0 & 4 \end{bmatrix}$ **36.** $\begin{bmatrix} 4 & 8 & -7 & 14 \\ 2 & 5 & -4 & 6 \\ 0 & 2 & 1 & -7 \\ 3 & 6 & -5 & 10 \end{bmatrix}$

37. $\begin{bmatrix} -1 & 0 & 1 & 0 \\ 0 & 2 & 0 & -1 \\ 2 & 0 & -1 & 0 \\ 0 & -1 & 0 & 1 \end{bmatrix}$ **38.** $\begin{bmatrix} 1 & -2 & -1 & -2 \\ 3 & -5 & -2 & -3 \\ 2 & -5 & -2 & -5 \\ -1 & 4 & 4 & 11 \end{bmatrix}$

In Exercises 39–44, use the formula on page 558 to find the inverse of the 2×2 matrix (if it exists).

39. $\begin{bmatrix} 5 & -2 \\ 2 & 3 \end{bmatrix}$ **40.** $\begin{bmatrix} 7 & 12 \\ -8 & -5 \end{bmatrix}$

41. $\begin{bmatrix} -4 & -6 \\ 2 & 3 \end{bmatrix}$ **42.** $\begin{bmatrix} -12 & 3 \\ 5 & -2 \end{bmatrix}$

43. $\begin{bmatrix} \frac{7}{2} & -\frac{3}{4} \\ \frac{1}{5} & \frac{4}{5} \end{bmatrix}$ **44.** $\begin{bmatrix} -\frac{1}{4} & \frac{9}{4} \\ \frac{5}{3} & \frac{8}{9} \end{bmatrix}$

In Exercises 45–48, use the inverse matrix found in Exercise 13 to solve the system of linear equations.

45. $\begin{cases} x - 2y = 5 \\ 2x - 3y = 10 \end{cases}$ **46.** $\begin{cases} x - 2y = 0 \\ 2x - 3y = 3 \end{cases}$

47. $\begin{cases} x - 2y = 4 \\ 2x - 3y = 2 \end{cases}$ **48.** $\begin{cases} x - 2y = 1 \\ 2x - 3y = -2 \end{cases}$

In Exercises 49 and 50, use the inverse matrix found in Exercise 21 to solve the system of linear equations.

49. $\begin{cases} x + y + z = 0 \\ 3x + 5y + 4z = 5 \\ 3x + 6y + 5z = 2 \end{cases}$ **50.** $\begin{cases} x + y + z = -1 \\ 3x + 5y + 4z = 2 \\ 3x + 6y + 5z = 0 \end{cases}$

In Exercises 51 and 52, use the inverse matrix found in Exercise 38 to solve the system of linear equations.

51. $\begin{cases} x_1 - 2x_2 - x_3 - 2x_4 = 0 \\ 3x_1 - 5x_2 - 2x_3 - 3x_4 = 1 \\ 2x_1 - 5x_2 - 2x_3 - 5x_4 = -1 \\ -x_1 + 4x_2 + 4x_3 + 11x_4 = 2 \end{cases}$

52. $\begin{cases} x_1 - 2x_2 - x_3 - 2x_4 = 1 \\ 3x_1 - 5x_2 - 2x_3 - 3x_4 = -2 \\ 2x_1 - 5x_2 - 2x_3 - 5x_4 = 0 \\ -x_1 + 4x_2 + 4x_3 + 11x_4 = -3 \end{cases}$

In Exercises 53–60, use an inverse matrix to solve (if possible) the system of linear equations.

53. $\begin{cases} 3x + 4y = -2 \\ 5x + 3y = 4 \end{cases}$ **54.** $\begin{cases} 18x + 12y = 13 \\ 30x + 24y = 23 \end{cases}$

55. $\begin{cases} -0.4x + 0.8y = 1.6 \\ 2x - 4y = 5 \end{cases}$ **56.** $\begin{cases} 0.2x - 0.6y = 2.4 \\ -x + 1.4y = -8.8 \end{cases}$

57. $\begin{cases} -\frac{1}{4}x + \frac{3}{8}y = -2 \\ \frac{3}{2}x + \frac{3}{4}y = -12 \end{cases}$ **58.** $\begin{cases} \frac{5}{6}x - y = -20 \\ \frac{4}{3}x - \frac{7}{2}y = -51 \end{cases}$

59. $\begin{cases} 4x - y + z = -5 \\ 2x + 2y + 3z = 10 \\ 5x - 2y + 6z = 1 \end{cases}$ **60.** $\begin{cases} 4x - 2y + 3z = -2 \\ 2x + 2y + 5z = 16 \\ 8x - 5y - 2z = 4 \end{cases}$

In Exercises 61–66, use the matrix capabilities of a graphing utility to solve (if possible) the system of linear equations.

61. $\begin{cases} 5x - 3y + 2z = 2 \\ 2x + 2y - 3z = 3 \\ x - 7y + 8z = -4 \end{cases}$ **62.** $\begin{cases} 2x + 3y + 5z = 4 \\ 3x + 5y + 9z = 7 \\ 5x + 9y + 17z = 13 \end{cases}$

63. $\begin{cases} 3x - 2y + z = -29 \\ -4x + y - 3z = 37 \\ x - 5y + z = -24 \end{cases}$

64. $\begin{cases} -8x + 7y - 10z = -151 \\ 12x + 3y - 5z = 86 \\ 15x - 9y + 2z = 187 \end{cases}$

65. $\begin{cases} 7x - 3y + 2w = 41 \\ -2x + y - w = -13 \\ 4x + z - 2w = 12 \\ -x + y - w = -8 \end{cases}$

66. $\begin{cases} 2x + 5y + w = 11 \\ x + 4y + 2z - 2w = -7 \\ 2x - 2y + 5z + w = 3 \\ x - 3w = -1 \end{cases}$

Investment Portfolio   In Exercises 67–70, consider a person who invests in AAA-rated bonds, A-rated bonds, and B-rated bonds. The average yields are 6.5% on AAA bonds, 7% on A bonds, and 9% on B bonds. The person invests twice as much in B bonds as in A bonds. Let *x*, *y*, and *z* represent the amounts invested in AAA, A, and B bonds, respectively.

$\begin{cases} x + y + z = \text{(total investment)} \\ 0.065x + 0.07y + 0.09z = \text{(annual return)} \\ 2y - z = 0 \end{cases}$

Use the inverse of the coefficient matrix of this system to find the amount invested in each type of bond.

Total Investment	Annual Return
67. $10,000	$705
68. $10,000	$760
69. $12,000	$835
70. $500,000	$38,000

71. Circuit Analysis Consider the circuit shown in the figure. The currents I_1, I_2, and I_3, in amperes, are the solution of the system of linear equations

$$\begin{cases} 2I_1 \quad\quad + 4I_3 = E_1 \\ \quad\quad I_2 + 4I_3 = E_2 \\ I_1 + I_2 - \ I_3 = 0 \end{cases}$$

where E_1 and E_2 are voltages. Use the inverse of the coefficient matrix of this system to find the unknown currents for the voltages.

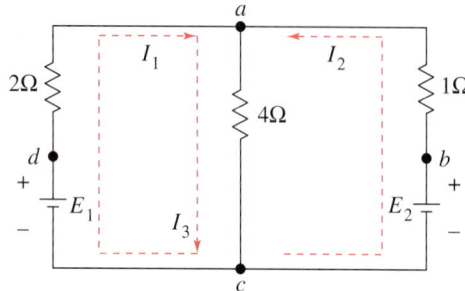

(a) $E_1 = 14$ volts, $E_2 = 28$ volts

(b) $E_1 = 24$ volts, $E_2 = 23$ volts

Model It

72. Data Analysis: Licensed Drivers The table shows the numbers y (in millions) of licensed drivers in the United States for selected years 1997 to 2001. (Source: U.S. Federal Highway Administration)

Year	Drivers, y
1997	182.7
1999	187.2
2001	191.3

(a) Use the technique demonstrated in Exercises 57–62 in Section 6.2 to create a system of linear equations for the data. Let t represent the year, with $t = 7$ corresponding to 1997.

(b) Use the matrix capabilities of a graphing utility to find an inverse matrix to solve the system from part (a) and find the least squares regression line $y = at + b$.

(c) Use the result of part (b) to estimate the number of licensed drivers in 2003.

(d) The actual number of licensed drivers in 2003 was 196.2 million. How does this value compare with your estimate from part (c)?

Model It (continued)

(e) Use the result of part (b) to estimate when the number of licensed drivers will reach 208 million.

Synthesis

True or False? In Exercises 73 and 74, determine whether the statement is true or false. Justify your answer.

73. Multiplication of an invertible matrix and its inverse is commutative.

74. If you multiply two square matrices and obtain the identity matrix, you can assume that the matrices are inverses of one another.

75. If A is a 2×2 matrix $A = \begin{bmatrix} a & b \\ c & d \end{bmatrix}$, then A is invertible if and only if $ad - bc \neq 0$. If $ad - bc \neq 0$, verify that the inverse is

$$A^{-1} = \frac{1}{ad - bc} \begin{bmatrix} d & -b \\ -c & a \end{bmatrix}.$$

76. Exploration Consider matrices of the form

$$A = \begin{bmatrix} a_{11} & 0 & 0 & 0 & \cdots & 0 \\ 0 & a_{22} & 0 & 0 & \cdots & 0 \\ 0 & 0 & a_{33} & 0 & \cdots & 0 \\ \vdots & \vdots & \vdots & \vdots & \cdots & \vdots \\ 0 & 0 & 0 & 0 & \cdots & a_{nn} \end{bmatrix}.$$

(a) Write a 2×2 matrix and a 3×3 matrix in the form of A. Find the inverse of each.

(b) Use the result of part (a) to make a conjecture about the inverses of matrices in the form of A.

Skills Review

In Exercises 77 and 78, solve the inequality and sketch the solution on the real number line.

77. $|x + 7| \geq 2$ **78.** $|2x - 1| < 3$

In Exercises 79–82, solve the equation. Approximate the result to three decimal places.

79. $3^{x/2} = 315$ **80.** $2000e^{-x/5} = 400$

81. $\log_2 x - 2 = 4.5$ **82.** $\ln x + \ln(x - 1) = 0$

83. Make a Decision To work an extended application analyzing the number of U.S. households with color televisions from 1985 to 2005, visit this text's website at *college.hmco.com*. (Data Source: Nielsen Media Research)

7.4 The Determinant of a Square Matrix

Why you should learn it

Determinants are often used in other branches of mathematics. For instance, Exercises 79–84 on page 570 show some types of determinants that are useful when changes in variables are made in calculus.

The Determinant of a 2 × 2 Matrix

Every *square* matrix can be associated with a real number called its **determinant.** Determinants have many uses, and several will be discussed in this and the next section. Historically, the use of determinants arose from special number patterns that occur when systems of linear equations are solved. For instance, the system

$$\begin{cases} a_1 x + b_1 y = c_1 \\ a_2 x + b_2 y = c_2 \end{cases}$$

has a solution

$$x = \frac{c_1 b_2 - c_2 b_1}{a_1 b_2 - a_2 b_1} \quad \text{and} \quad y = \frac{a_1 c_2 - a_2 c_1}{a_1 b_2 - a_2 b_1}$$

provided that $a_1 b_2 - a_2 b_1 \neq 0$. Note that the denominators of the two fractions are the same. This denominator is called the *determinant* of the coefficient matrix of the system.

Coefficient Matrix   *Determinant*

$$A = \begin{bmatrix} a_1 & b_1 \\ a_2 & b_2 \end{bmatrix} \qquad \det(A) = a_1 b_2 - a_2 b_1$$

The determinant of the matrix A can also be denoted by vertical bars on both sides of the matrix, as indicated in the following definition.

Definition of the Determinant of a 2 × 2 Matrix

The **determinant** of the matrix

$$A = \begin{bmatrix} a_1 & b_1 \\ a_2 & b_2 \end{bmatrix}$$

is given by

$$\det(A) = |A| = \begin{vmatrix} a_1 & b_1 \\ a_2 & b_2 \end{vmatrix} = a_1 b_2 - a_2 b_1.$$

In this text, $\det(A)$ and $|A|$ are used interchangeably to represent the determinant of A. Although vertical bars are also used to denote the absolute value of a real number, the context will show which use is intended.

A convenient method for remembering the formula for the determinant of a 2×2 matrix is shown in the following diagram.

$$\det(A) = \begin{vmatrix} a_1 & b_1 \\ a_2 & b_2 \end{vmatrix} = a_1 b_2 - a_2 b_1$$

Note that the determinant is the difference of the products of the two diagonals of the matrix.

Example 1   **The Determinant of a 2 × 2 Matrix**

Find the determinant of each matrix.

a. $A = \begin{bmatrix} 2 & -3 \\ 1 & 2 \end{bmatrix}$

b. $B = \begin{bmatrix} 2 & 1 \\ 4 & 2 \end{bmatrix}$

c. $C = \begin{bmatrix} 0 & \frac{3}{2} \\ 2 & 4 \end{bmatrix}$

Solution

a. $\det(A) = \begin{vmatrix} 2 & -3 \\ 1 & 2 \end{vmatrix}$

$= 2(2) - 1(-3)$

$= 4 + 3 = 7$

b. $\det(B) = \begin{vmatrix} 2 & 1 \\ 4 & 2 \end{vmatrix}$

$= 2(2) - 4(1)$

$= 4 - 4 = 0$

c. $\det(C) = \begin{vmatrix} 0 & \frac{3}{2} \\ 2 & 4 \end{vmatrix}$

$= 0(4) - 2\left(\frac{3}{2}\right)$

$= 0 - 3 = -3$

✓**CHECKPOINT**   Now try Exercise 5.

Notice in Example 1 that the determinant of a matrix can be positive, zero, or negative.

The determinant of a matrix of order 1×1 is defined simply as the entry of the matrix. For instance, if $A = [-2]$, then $\det(A) = -2$.

Exploration

Use a graphing utility with matrix capabilities to find the determinant of the following matrix.

$A = \begin{bmatrix} 1 & 2 \\ -1 & 0 \\ 3 & -2 \end{bmatrix}$

What message appears on the screen? Why does the graphing utility display this message?

Technology

Most graphing utilities can evaluate the determinant of a matrix. For instance, you can evaluate the determinant of

$A = \begin{bmatrix} 2 & -3 \\ 1 & 2 \end{bmatrix}$

by entering the matrix as $[A]$ and then choosing the *determinant* feature. The result should be 7, as in Example 1(a). Try evaluating the determinants of other matrices. Consult the user's guide for your graphing utility for specific keystrokes.

Minors and Cofactors

To define the determinant of a square matrix of order 3×3 or higher, it is convenient to introduce the concepts of **minors** and **cofactors.**

Sign Pattern for Cofactors

$$\begin{bmatrix} + & - & + \\ - & + & - \\ + & - & + \end{bmatrix}$$

3×3 matrix

$$\begin{bmatrix} + & - & + & - \\ - & + & - & + \\ + & - & + & - \\ - & + & - & + \end{bmatrix}$$

4×4 matrix

$$\begin{bmatrix} + & - & + & - & + & \cdots \\ - & + & - & + & - & \cdots \\ + & - & + & - & + & \cdots \\ - & + & - & + & - & \cdots \\ + & - & + & - & + & \cdots \\ \vdots & \vdots & \vdots & \vdots & \vdots & \end{bmatrix}$$

$n \times n$ matrix

> ### Minors and Cofactors of a Square Matrix
>
> If A is a square matrix, the **minor** M_{ij} of the entry a_{ij} is the determinant of the matrix obtained by deleting the ith row and jth column of A. The **cofactor** C_{ij} of the entry a_{ij} is
>
> $$C_{ij} = (-1)^{i+j} M_{ij}.$$

In the sign pattern for cofactors at the left, notice that *odd* positions (where $i + j$ is odd) have negative signs and *even* positions (where $i + j$ is even) have positive signs.

Example 2   Finding the Minors and Cofactors of a Matrix

Find all the minors and cofactors of

$$A = \begin{bmatrix} 0 & 2 & 1 \\ 3 & -1 & 2 \\ 4 & 0 & 1 \end{bmatrix}.$$

Solution

To find the minor M_{11}, delete the first row and first column of A and evaluate the determinant of the resulting matrix.

$$\begin{bmatrix} 0 & 2 & 1 \\ 3 & -1 & 2 \\ 4 & 0 & 1 \end{bmatrix}, \quad M_{11} = \begin{vmatrix} -1 & 2 \\ 0 & 1 \end{vmatrix} = -1(1) - 0(2) = -1$$

Similarly, to find M_{12}, delete the first row and second column.

$$\begin{bmatrix} 0 & 2 & 1 \\ 3 & -1 & 2 \\ 4 & 0 & 1 \end{bmatrix}, \quad M_{12} = \begin{vmatrix} 3 & 2 \\ 4 & 1 \end{vmatrix} = 3(1) - 4(2) = -5$$

Continuing this pattern, you obtain the minors.

$$M_{11} = -1 \qquad M_{12} = -5 \qquad M_{13} = 4$$
$$M_{21} = 2 \qquad M_{22} = -4 \qquad M_{23} = -8$$
$$M_{31} = 5 \qquad M_{32} = -3 \qquad M_{33} = -6$$

Now, to find the cofactors, combine these minors with the checkerboard pattern of signs for a 3×3 matrix shown at the upper left.

$$C_{11} = -1 \qquad C_{12} = 5 \qquad C_{13} = 4$$
$$C_{21} = -2 \qquad C_{22} = -4 \qquad C_{23} = 8$$
$$C_{31} = 5 \qquad C_{32} = 3 \qquad C_{33} = -6$$

✔**CHECKPOINT**   Now try Exercise 27.

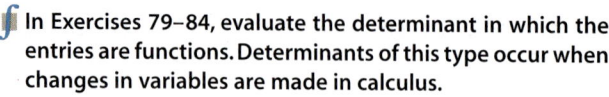 In Exercises 79–84, evaluate the determinant in which the entries are functions. Determinants of this type occur when changes in variables are made in calculus.

79. $\begin{vmatrix} 4u & -1 \\ -1 & 2v \end{vmatrix}$

80. $\begin{vmatrix} 3x^2 & -3y^2 \\ 1 & 1 \end{vmatrix}$

81. $\begin{vmatrix} e^{2x} & e^{3x} \\ 2e^{2x} & 3e^{3x} \end{vmatrix}$

82. $\begin{vmatrix} e^{-x} & xe^{-x} \\ -e^{-x} & (1-x)e^{-x} \end{vmatrix}$

83. $\begin{vmatrix} x & \ln x \\ 1 & 1/x \end{vmatrix}$

84. $\begin{vmatrix} x & x\ln x \\ 1 & 1+\ln x \end{vmatrix}$

Synthesis

True or False? In Exercises 85 and 86, determine whether the statement is true or false. Justify your answer.

85. If a square matrix has an entire row of zeros, the determinant will always be zero.

86. If two columns of a square matrix are the same, the determinant of the matrix will be zero.

87. *Exploration* Find square matrices A and B to demonstrate that $|A + B| \neq |A| + |B|$.

88. *Exploration* Consider square matrices in which the entries are consecutive integers. An example of such a matrix is

$$\begin{bmatrix} 4 & 5 & 6 \\ 7 & 8 & 9 \\ 10 & 11 & 12 \end{bmatrix}.$$

 (a) Use a graphing utility to evaluate the determinants of four matrices of this type. Make a conjecture based on the results.

(b) Verify your conjecture.

89. *Writing* Write a brief paragraph explaining the difference between a square matrix and its determinant.

90. *Think About It* If A is a matrix of order 3×3 such that $|A| = 5$, is it possible to find $|2A|$? Explain.

Properties of Determinants In Exercises 91–93, a property of determinants is given (A and B are square matrices). State how the property has been applied to the given determinants and use a graphing utility to verify the results.

91. If B is obtained from A by interchanging two rows of A or interchanging two columns of A, then $|B| = -|A|$.

(a) $\begin{vmatrix} 1 & 3 & 4 \\ -7 & 2 & -5 \\ 6 & 1 & 2 \end{vmatrix} = -\begin{vmatrix} 1 & 4 & 3 \\ -7 & -5 & 2 \\ 6 & 2 & 1 \end{vmatrix}$

(b) $\begin{vmatrix} 1 & 3 & 4 \\ -2 & 2 & 0 \\ 1 & 6 & 2 \end{vmatrix} = -\begin{vmatrix} 1 & 6 & 2 \\ -2 & 2 & 0 \\ 1 & 3 & 4 \end{vmatrix}$

92. If B is obtained from A by adding a multiple of a row of A to another row of A or by adding a multiple of a column of A to another column of A, then $|B| = |A|$.

(a) $\begin{vmatrix} 1 & -3 \\ 5 & 2 \end{vmatrix} = \begin{vmatrix} 1 & -3 \\ 0 & 17 \end{vmatrix}$

(b) $\begin{vmatrix} 5 & 4 & 2 \\ 2 & -3 & 4 \\ 7 & 6 & 3 \end{vmatrix} = \begin{vmatrix} 1 & 10 & -6 \\ 2 & -3 & 4 \\ 7 & 6 & 3 \end{vmatrix}$

93. If B is obtained from A by multiplying a row by a nonzero constant c or by multiplying a column by a nonzero constant c, then $|B| = c|A|$.

(a) $\begin{vmatrix} 5 & 10 \\ 2 & -3 \end{vmatrix} = 5\begin{vmatrix} 1 & 2 \\ 2 & -3 \end{vmatrix}$

(b) $\begin{vmatrix} 1 & 8 & -3 \\ 3 & -12 & 6 \\ 7 & 4 & 3 \end{vmatrix} = 12\begin{vmatrix} 1 & 2 & -1 \\ 3 & -3 & 2 \\ 7 & 1 & 3 \end{vmatrix}$

94. *Exploration* A **diagonal matrix** is a square matrix with all zero entries above and below its main diagonal. Evaluate the determinant of each diagonal matrix. Make a conjecture based on your results.

(a) $\begin{vmatrix} 7 & 0 \\ 0 & 4 \end{vmatrix}$

(b) $\begin{vmatrix} -1 & 0 & 0 \\ 0 & 5 & 0 \\ 0 & 0 & 2 \end{vmatrix}$

(c) $\begin{vmatrix} 2 & 0 & 0 & 0 \\ 0 & -2 & 0 & 0 \\ 0 & 0 & 1 & 0 \\ 0 & 0 & 0 & 3 \end{vmatrix}$

Skills Review

In Exercises 95–100, find the domain of the function.

95. $f(x) = x^3 - 2x$

96. $g(x) = \sqrt[3]{x}$

97. $h(x) = \sqrt{16 - x^2}$

98. $A(x) = \dfrac{3}{36 - x^2}$

99. $g(t) = \ln(t - 1)$

100. $f(s) = 625e^{-0.5s}$

In Exercises 101 and 102, sketch the graph of the solution of the system of inequalities.

101. $\begin{cases} x + y \leq 8 \\ x \geq -3 \\ 2x - y < 5 \end{cases}$

102. $\begin{cases} -x - y > 4 \\ y \leq 1 \\ 7x + 4y \leq -10 \end{cases}$

In Exercises 103–106, find the inverse of the matrix (if it exists).

103. $\begin{bmatrix} -4 & 1 \\ 8 & -1 \end{bmatrix}$

104. $\begin{bmatrix} -5 & -8 \\ 3 & 6 \end{bmatrix}$

105. $\begin{bmatrix} -7 & 2 & 9 \\ 2 & -4 & -6 \\ 3 & 5 & 2 \end{bmatrix}$

106. $\begin{bmatrix} -6 & 2 & 0 \\ 1 & 3 & -2 \\ -2 & 0 & 1 \end{bmatrix}$

7.5 Applications of Matrices and Determinants

What you should learn

- Use Cramer's Rule to solve systems of linear equations.
- Use determinants to find the areas of triangles.
- Use a determinant to test for collinear points and find an equation of a line passing through two points.
- Use matrices to encode and decode messages.

Why you should learn it

You can use Cramer's Rule to solve real-life problems. For instance, in Exercise 58 on page 582, Cramer's Rule is used to find a quadratic model for the number of U.S. Supreme Court cases waiting to be tried.

© Lester Lefkowitz/Corbis

Cramer's Rule

So far, you have studied three methods for solving a system of linear equations: substitution, elimination with equations, and elimination with matrices. In this section, you will study one more method, **Cramer's Rule,** named after Gabriel Cramer (1704–1752). This rule uses determinants to write the solution of a system of linear equations. To see how Cramer's Rule works, take another look at the solution described at the beginning of Section 7.4. There, it was pointed out that the system

$$\begin{cases} a_1x + b_1y = c_1 \\ a_2x + b_2y = c_2 \end{cases}$$

has a solution

$$x = \frac{c_1b_2 - c_2b_1}{a_1b_2 - a_2b_1} \quad \text{and} \quad y = \frac{a_1c_2 - a_2c_1}{a_1b_2 - a_2b_1}$$

provided that $a_1b_2 - a_2b_1 \neq 0$. Each numerator and denominator in this solution can be expressed as a determinant, as follows.

$$x = \frac{c_1b_2 - c_2b_1}{a_1b_2 - a_2b_1} = \frac{\begin{vmatrix} c_1 & b_1 \\ c_2 & b_2 \end{vmatrix}}{\begin{vmatrix} a_1 & b_1 \\ a_2 & b_2 \end{vmatrix}} \qquad y = \frac{a_1c_2 - a_2c_1}{a_1b_2 - a_2b_1} = \frac{\begin{vmatrix} a_1 & c_1 \\ a_2 & c_2 \end{vmatrix}}{\begin{vmatrix} a_1 & b_1 \\ a_2 & b_2 \end{vmatrix}}$$

Relative to the original system, the denominator for x and y is simply the determinant of the *coefficient* matrix of the system. This determinant is denoted by D. The numerators for x and y are denoted by D_x and D_y, respectively. They are formed by using the column of constants as replacements for the coefficients of x and y, as follows.

Coefficient Matrix	D	D_x	D_y
$\begin{bmatrix} a_1 & b_1 \\ a_2 & b_2 \end{bmatrix}$	$\begin{vmatrix} a_1 & b_1 \\ a_2 & b_2 \end{vmatrix}$	$\begin{vmatrix} c_1 & b_1 \\ c_2 & b_2 \end{vmatrix}$	$\begin{vmatrix} a_1 & c_1 \\ a_2 & c_2 \end{vmatrix}$

For example, given the system

$$\begin{cases} 2x - 5y = 3 \\ -4x + 3y = 8 \end{cases}$$

the coefficient matrix, D, D_x, and D_y are as follows.

Coefficient Matrix	D	D_x	D_y
$\begin{bmatrix} 2 & -5 \\ -4 & 3 \end{bmatrix}$	$\begin{vmatrix} 2 & -5 \\ -4 & 3 \end{vmatrix}$	$\begin{vmatrix} 3 & -5 \\ 8 & 3 \end{vmatrix}$	$\begin{vmatrix} 2 & 3 \\ -4 & 8 \end{vmatrix}$

Cramer's Rule generalizes easily to systems of n equations in n variables. The value of each variable is given as the quotient of two determinants. The denominator is the determinant of the coefficient matrix, and the numerator is the determinant of the matrix formed by replacing the column corresponding to the variable (being solved for) with the column representing the constants. For instance, the solution for x_3 in the following system is shown.

$$\begin{cases} a_{11}x_1 + a_{12}x_2 + a_{13}x_3 = b_1 \\ a_{21}x_1 + a_{22}x_2 + a_{23}x_3 = b_2 \\ a_{31}x_1 + a_{32}x_2 + a_{33}x_3 = b_3 \end{cases} \qquad x_3 = \frac{|A_3|}{|A|} = \frac{\begin{vmatrix} a_{11} & a_{12} & b_1 \\ a_{21} & a_{22} & b_2 \\ a_{31} & a_{32} & b_3 \end{vmatrix}}{\begin{vmatrix} a_{11} & a_{12} & a_{13} \\ a_{21} & a_{22} & a_{23} \\ a_{31} & a_{32} & a_{33} \end{vmatrix}}$$

Cramer's Rule

If a system of n linear equations in n variables has a coefficient matrix A with a nonzero determinant $|A|$, the solution of the system is

$$x_1 = \frac{|A_1|}{|A|}, \quad x_2 = \frac{|A_2|}{|A|}, \quad \cdots, \quad x_n = \frac{|A_n|}{|A|}$$

where the ith column of A_i is the column of constants in the system of equations. If the determinant of the coefficient matrix is zero, the system has either no solution or infinitely many solutions.

Example 1   Using Cramer's Rule for a 2 × 2 System

Use Cramer's Rule to solve the system of linear equations.

$$\begin{cases} 4x - 2y = 10 \\ 3x - 5y = 11 \end{cases}$$

Solution

To begin, find the determinant of the coefficient matrix.

$$D = \begin{vmatrix} 4 & -2 \\ 3 & -5 \end{vmatrix} = -20 - (-6) = -14$$

Because this determinant is not zero, you can apply Cramer's Rule.

$$x = \frac{D_x}{D} = \frac{\begin{vmatrix} 10 & -2 \\ 11 & -5 \end{vmatrix}}{-14} = \frac{-50 - (-22)}{-14} = \frac{-28}{-14} = 2$$

$$y = \frac{D_y}{D} = \frac{\begin{vmatrix} 4 & 10 \\ 3 & 11 \end{vmatrix}}{-14} = \frac{44 - 30}{-14} = \frac{14}{-14} = -1$$

So, the solution is $x = 2$ and $y = -1$. Check this in the original system.

✓CHECKPOINT   Now try Exercise 1.

Example 2   **Using Cramer's Rule for a 3 × 3 System**

Use Cramer's Rule to solve the system of linear equations.

$$\begin{cases} -x + 2y - 3z = 1 \\ 2x \quad\quad\; + \;\; z = 0 \\ 3x - 4y + 4z = 2 \end{cases}$$

Solution

To find the determinant of the coefficient matrix

$$\begin{bmatrix} -1 & 2 & -3 \\ 2 & 0 & 1 \\ 3 & -4 & 4 \end{bmatrix}$$

expand along the second row, as follows.

$$D = 2(-1)^3 \begin{vmatrix} 2 & -3 \\ -4 & 4 \end{vmatrix} + 0(-1)^4 \begin{vmatrix} -1 & -3 \\ 3 & 4 \end{vmatrix} + 1(-1)^5 \begin{vmatrix} -1 & 2 \\ 3 & -4 \end{vmatrix}$$

$$= -2(-4) + 0 - 1(-2)$$

$$= 10$$

Because this determinant is not zero, you can apply Cramer's Rule.

$$x = \frac{D_x}{D} = \frac{\begin{vmatrix} 1 & 2 & -3 \\ 0 & 0 & 1 \\ 2 & -4 & 4 \end{vmatrix}}{10} = \frac{8}{10} = \frac{4}{5}$$

$$y = \frac{D_y}{D} = \frac{\begin{vmatrix} -1 & 1 & -3 \\ 2 & 0 & 1 \\ 3 & 2 & 4 \end{vmatrix}}{10} = \frac{-15}{10} = -\frac{3}{2}$$

$$z = \frac{D_z}{D} = \frac{\begin{vmatrix} -1 & 2 & 1 \\ 2 & 0 & 0 \\ 3 & -4 & 2 \end{vmatrix}}{10} = \frac{-16}{10} = -\frac{8}{5}$$

The solution is $\left(\frac{4}{5}, -\frac{3}{2}, -\frac{8}{5}\right)$. Check this in the original system as follows.

Check

$$-\left(\tfrac{4}{5}\right) + 2\left(-\tfrac{3}{2}\right) - 3\left(-\tfrac{8}{5}\right) \stackrel{?}{=} 1 \qquad \text{\textcolor{red}{Substitute into Equation 1.}}$$

$$-\tfrac{4}{5} - 3 + \tfrac{24}{5} = 1 \qquad \text{\textcolor{red}{Equation 1 checks. ✓}}$$

$$2\left(\tfrac{4}{5}\right) + \left(-\tfrac{8}{5}\right) \stackrel{?}{=} 0 \qquad \text{\textcolor{red}{Substitute into Equation 2.}}$$

$$\tfrac{8}{5} - \tfrac{8}{5} = 0 \qquad \text{\textcolor{red}{Equation 2 checks. ✓}}$$

$$3\left(\tfrac{4}{5}\right) - 4\left(-\tfrac{3}{2}\right) + 4\left(-\tfrac{8}{5}\right) \stackrel{?}{=} 2 \qquad \text{\textcolor{red}{Substitute into Equation 3.}}$$

$$\tfrac{12}{5} + 6 - \tfrac{32}{5} = 2 \qquad \text{\textcolor{red}{Equation 3 checks. ✓}}$$

✓**CHECKPOINT**   Now try Exercise 7.

Remember that Cramer's Rule does not apply when the determinant of the coefficient matrix is zero. This would create division by zero, which is undefined.

Area of a Triangle

Another application of matrices and determinants is finding the area of a triangle whose vertices are given as points in a coordinate plane.

> ### Area of a Triangle
>
> The area of a triangle with vertices (x_1, y_1), (x_2, y_2), and (x_3, y_3) is
>
> $$\text{Area} = \pm \frac{1}{2} \begin{vmatrix} x_1 & y_1 & 1 \\ x_2 & y_2 & 1 \\ x_3 & y_3 & 1 \end{vmatrix}$$
>
> where the symbol $\pm$ indicates that the appropriate sign should be chosen to yield a positive area.

Example 3   Finding the Area of a Triangle

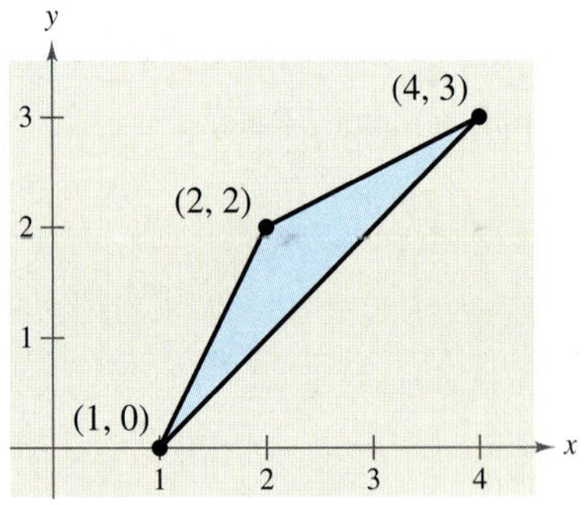

Find the area of a triangle whose vertices are $(1, 0)$, $(2, 2)$, and $(4, 3)$, as shown in Figure 7.1.

Solution

Let $(x_1, y_1) = (1, 0)$, $(x_2, y_2) = (2, 2)$, and $(x_3, y_3) = (4, 3)$. Then, to find the area of the triangle, evaluate the determinant.

$$\begin{vmatrix} x_1 & y_1 & 1 \\ x_2 & y_2 & 1 \\ x_3 & y_3 & 1 \end{vmatrix} = \begin{vmatrix} 1 & 0 & 1 \\ 2 & 2 & 1 \\ 4 & 3 & 1 \end{vmatrix}$$

$$= 1(-1)^2 \begin{vmatrix} 2 & 1 \\ 3 & 1 \end{vmatrix} + 0(-1)^3 \begin{vmatrix} 2 & 1 \\ 4 & 1 \end{vmatrix} + 1(-1)^4 \begin{vmatrix} 2 & 2 \\ 4 & 3 \end{vmatrix}$$

$$= 1(-1) + 0 + 1(-2) = -3.$$

Using this value, you can conclude that the area of the triangle is

$$\text{Area} = -\frac{1}{2} \begin{vmatrix} 1 & 0 & 1 \\ 2 & 2 & 1 \\ 4 & 3 & 1 \end{vmatrix} \qquad \text{Choose } (-) \text{ so that the area is positive.}$$

$$= -\frac{1}{2}(-3) = \frac{3}{2} \text{ square units.}$$

✔CHECKPOINT   Now try Exercise 19.

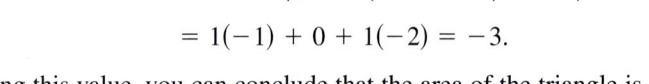

y

3 ⊸ (4, 3)

2 ⊸ (2, 2)

1 ⊸

(1, 0)
————————→ **x**
1   2   3   4

FIGURE **7.1**

> *Exploration*
>
> Use determinants to find the area of a triangle with vertices $(3, -1)$, $(7, -1)$, and $(7, 5)$. Confirm your answer by plotting the points in a coordinate plane and using the formula
>
> $$\text{Area} = \tfrac{1}{2}(\text{base})(\text{height}).$$

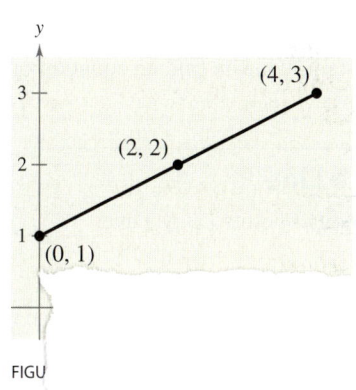

FIGU

Lines in a Plane

What if the three points in Example 3 had been on the same line? What would have happened had the area formula been applied to three such points? The answer is that the determinant would have been zero. Consider, for instance, the three collinear points $(0, 1)$, $(2, 2)$, and $(4, 3)$, as shown in Figure 7.2. The area of the "triangle" that has these three points as vertices is

$$\begin{vmatrix} 0 & 1 & 1 \\ & 1 & 1 \\ & 1 & 1 \end{vmatrix} = \frac{1}{2}\left[0(-1)^2 \begin{vmatrix} 2 & 1 \\ 3 & 1 \end{vmatrix} + 1(-1)^3 \begin{vmatrix} 2 & 1 \\ 4 & 1 \end{vmatrix} + 1(-1)^4 \begin{vmatrix} 2 & 2 \\ 4 & 3 \end{vmatrix} \right]$$

$$= \frac{1}{2}[0 - 1(-2) + 1(-2)]$$

$$= 0.$$

generalized as follows.

Collinear Points

s (x_1, y_1), (x_2, y_2), and (x_3, y_3) are **collinear** (lie on the same line) if

$$\begin{vmatrix} & y_1 & 1 \\ & y_2 & 1 \\ & y_3 & 1 \end{vmatrix} = 0.$$

4 ### Testing for Collinear Points

whether the points $(-2, -2)$, $(1, 1)$, and $(7, 5)$ are collinear. (See

Letting $(x_1, y_1) = (-2, -2)$, $(x_2, y_2) = (1, 1)$, and $(x_3, y_3) = (7, 5)$, you have

$$\begin{vmatrix} x_1 & y_1 & 1 \\ x_2 & y_2 & 1 \\ x_3 & y_3 & 1 \end{vmatrix} = \begin{vmatrix} -2 & -2 & 1 \\ 1 & 1 & 1 \\ 7 & 5 & 1 \end{vmatrix}$$

$$= -2(-1)^2 \begin{vmatrix} 1 & 1 \\ 5 & 1 \end{vmatrix} + (-2)(-1)^3 \begin{vmatrix} 1 & 1 \\ 7 & 1 \end{vmatrix} + 1(-1)^4 \begin{vmatrix} 1 & 1 \\ 7 & 5 \end{vmatrix}$$

$$= -2(-4) + 2(-6) + 1(-2)$$

$$= -6.$$

FIGURE 7.3

Because the value of this determinant is *not* zero, you can conclude that the three points do not lie on the same line. Moreover, the area of the triangle with vertices at these points is $\left(-\frac{1}{2}\right)(-6) = 3$ square units.

✓CHECKPOINT   Now try Exercise 31.

The test for collinear points can be adapted to another use. That is, if you are given two points on a rectangular coordinate system, you can find an equation of the line passing through the two points, as follows.

Two-Point Form of the Equation of a Line

An equation of the line passing through the distinct points (x_1, y_1) and (x_2, y_2) is given by

$$\begin{vmatrix} x & y & 1 \\ x_1 & y_1 & 1 \\ x_2 & y_2 & 1 \end{vmatrix} = 0.$$

Example 5   Finding an Equation of a Line

Find an equation of the line passing through the two points $(2, 4)$ and $(-1, 3)$, as shown in Figure 7.4.

Solution

Let $(x_1, y_1) = (2, 4)$ and $(x_2, y_2) = (-1, 3)$. Applying the determinant formula for the equation of a line produces

$$\begin{vmatrix} x & y & 1 \\ 2 & 4 & 1 \\ -1 & 3 & 1 \end{vmatrix} = 0.$$

To evaluate this determinant, you can expand by cofactors along the first row to obtain the following.

$$x(-1)^2 \begin{vmatrix} 4 & 1 \\ 3 & 1 \end{vmatrix} + y(-1)^3 \begin{vmatrix} 2 & 1 \\ -1 & 1 \end{vmatrix} + 1(-1)^4 \begin{vmatrix} 2 & 4 \\ -1 & 3 \end{vmatrix} = 0$$

$$x(1)(1) + y(-1)(3) + (1)(1)(10) = 0$$

$$x - 3y + 10 = 0$$

So, an equation of the line is

$$x - 3y + 10 = 0.$$

✓**CHECKPOINT**   Now try Exercise 39.

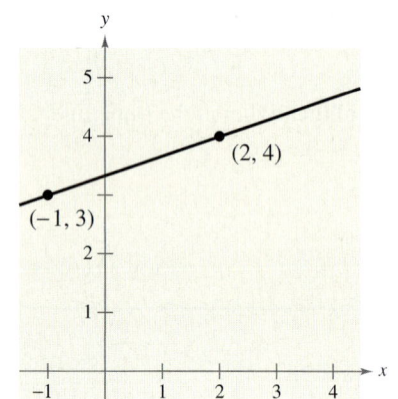

FIGURE **7.4**

Note that this method of finding the equation of a line works for all lines, including horizontal and vertical lines. For instance, the equation of the vertical line through $(2, 0)$ and $(2, 2)$ is

$$\begin{vmatrix} x & y & 1 \\ 2 & 0 & 1 \\ 2 & 2 & 1 \end{vmatrix} = 0$$

$$4 - 2x = 0$$

$$x = 2.$$

Cryptography

A **cryptogram** is a message written according to a secret code. (The Greek word *kryptos* means "hidden.") Matrix multiplication can be used to encode and decode messages. To begin, you need to assign a number to each letter in the alphabet (with 0 assigned to a blank space), as follows.

0 = _	9 = I	18 = R
1 = A	10 = J	19 = S
2 = B	11 = K	20 = T
3 = C	12 = L	21 = U
4 = D	13 = M	22 = V
5 = E	14 = N	23 = W
6 = F	15 = O	24 = X
7 = G	16 = P	25 = Y
8 = H	17 = Q	26 = Z

Then the message is converted to numbers and partitioned into **uncoded row matrices,** each having n entries, as demonstrated in Example 6.

Example 6   Forming Uncoded Row Matrices

Write the uncoded row matrices of order 1×3 for the message

MEET ME MONDAY.

Solution

Partitioning the message (including blank spaces, but ignoring punctuation) into groups of three produces the following uncoded row matrices.

$$[13 \quad 5 \quad 5] \quad [20 \quad 0 \quad 13] \quad [5 \quad 0 \quad 13] \quad [15 \quad 14 \quad 4] \quad [1 \quad 25 \quad 0]$$
$$\text{M} \quad \text{E} \quad \text{E} \quad \text{T} \qquad \text{M} \quad \text{E} \qquad \text{M} \quad \text{O} \quad \text{N} \quad \text{D} \quad \text{A} \quad \text{Y}$$

Note that a blank space is used to fill out the last uncoded row matrix.

✓CHECKPOINT   Now try Exercise 45.

To encode a message, use the techniques demonstrated in Section 7.3 to choose an $n \times n$ invertible matrix such as

$$A = \begin{bmatrix} 1 & -2 & 2 \\ -1 & 1 & 3 \\ 1 & -1 & -4 \end{bmatrix}$$

and multiply the uncoded row matrices by A (on the right) to obtain **coded row matrices.** Here is an example.

Uncoded Matrix   Encoding Matrix A   Coded Matrix

$$[13 \quad 5 \quad 5] \begin{bmatrix} 1 & -2 & 2 \\ -1 & 1 & 3 \\ 1 & -1 & -4 \end{bmatrix} = [13 \quad -26 \quad 21]$$

Example 7 **Encoding a Message**

Use the following invertible matrix to encode the message MEET ME MONDAY.

$$A = \begin{bmatrix} 1 & -2 & 2 \\ -1 & 1 & 3 \\ 1 & -1 & -4 \end{bmatrix}$$

Solution

The coded row matrices are obtained by multiplying each of the uncoded row matrices found in Example 6 by the matrix A, as follows.

Uncoded Matrix   *Encoding Matrix A*   *Coded Matrix*

$$\begin{bmatrix} 13 & 5 & 5 \end{bmatrix} \begin{bmatrix} 1 & -2 & 2 \\ -1 & 1 & 3 \\ 1 & -1 & -4 \end{bmatrix} = \begin{bmatrix} 13 & -26 & 21 \end{bmatrix}$$

$$\begin{bmatrix} 20 & 0 & 13 \end{bmatrix} \begin{bmatrix} 1 & -2 & 2 \\ -1 & 1 & 3 \\ 1 & -1 & -4 \end{bmatrix} = \begin{bmatrix} 33 & -53 & -12 \end{bmatrix}$$

$$\begin{bmatrix} 5 & 0 & 13 \end{bmatrix} \begin{bmatrix} 1 & -2 & 2 \\ -1 & 1 & 3 \\ 1 & -1 & -4 \end{bmatrix} = \begin{bmatrix} 18 & -23 & -42 \end{bmatrix}$$

$$\begin{bmatrix} 15 & 14 & 4 \end{bmatrix} \begin{bmatrix} 1 & -2 & 2 \\ -1 & 1 & 3 \\ 1 & -1 & -4 \end{bmatrix} = \begin{bmatrix} 5 & -20 & 56 \end{bmatrix}$$

$$\begin{bmatrix} 1 & 25 & 0 \end{bmatrix} \begin{bmatrix} 1 & -2 & 2 \\ -1 & 1 & 3 \\ 1 & -1 & -4 \end{bmatrix} = \begin{bmatrix} -24 & 23 & 77 \end{bmatrix}$$

So, the sequence of coded row matrices is

$$\begin{bmatrix} 13 & -26 & 21 \end{bmatrix}\begin{bmatrix} 33 & -53 & -12 \end{bmatrix}\begin{bmatrix} 18 & -23 & -42 \end{bmatrix}\begin{bmatrix} 5 & -20 & 56 \end{bmatrix}\begin{bmatrix} -24 & 23 & 77 \end{bmatrix}.$$

Finally, removing the matrix notation produces the following cryptogram.

$$13 \ -26 \ 21 \ 33 \ -53 \ -12 \ 18 \ -23 \ -42 \ 5 \ -20 \ 56 \ -24 \ 23 \ 77$$

✔**CHECKPOINT**   Now try Exercise 47.

For those who do not know the encoding matrix A, decoding the cryptogram found in Example 7 is difficult. But for an authorized receiver who knows the encoding matrix A, decoding is simple. The receiver just needs to multiply the coded row matrices by A^{-1} (on the right) to retrieve the uncoded row matrices. Here is an example.

$$\underbrace{\begin{bmatrix} 13 & -26 & 21 \end{bmatrix}}_{\text{Coded}} \underbrace{\begin{bmatrix} -1 & -10 & -8 \\ -1 & -6 & -5 \\ 0 & -1 & -1 \end{bmatrix}}_{A^{-1}} = \underbrace{\begin{bmatrix} 13 & 5 & 5 \end{bmatrix}}_{\text{Uncoded}}$$

Example 8   **Decoding a Message**

Use the inverse of the matrix

$$A = \begin{bmatrix} 1 & -2 & 2 \\ -1 & 1 & 3 \\ 1 & -1 & -4 \end{bmatrix}$$

to decode the cryptogram

$$13 \ -26 \ 21 \ 33 \ -53 \ -12 \ 18 \ -23 \ -42 \ 5 \ -20 \ 56 \ -24 \ 23 \ 77.$$

Solution

First find A^{-1} by using the techniques demonstrated in Section 7.3. A^{-1} is the decoding matrix. Then partition the message into groups of three to form the coded row matrices. Finally, multiply each coded row matrix by A^{-1} (on the right).

Coded Matrix   Decoding Matrix A^{-1} Decoded Matrix

$$\begin{bmatrix} 13 & -26 & 21 \end{bmatrix} \begin{bmatrix} -1 & -10 & -8 \\ -1 & -6 & -5 \\ 0 & -1 & -1 \end{bmatrix} = \begin{bmatrix} 13 & 5 & 5 \end{bmatrix}$$

$$\begin{bmatrix} 33 & -53 & -12 \end{bmatrix} \begin{bmatrix} -1 & -10 & -8 \\ -1 & -6 & -5 \\ 0 & -1 & -1 \end{bmatrix} = \begin{bmatrix} 20 & 0 & 13 \end{bmatrix}$$

$$\begin{bmatrix} 18 & -23 & -42 \end{bmatrix} \begin{bmatrix} -1 & -10 & -8 \\ -1 & -6 & -5 \\ 0 & -1 & -1 \end{bmatrix} = \begin{bmatrix} 5 & 0 & 13 \end{bmatrix}$$

$$\begin{bmatrix} 5 & -20 & 56 \end{bmatrix} \begin{bmatrix} -1 & -10 & -8 \\ -1 & -6 & -5 \\ 0 & -1 & -1 \end{bmatrix} = \begin{bmatrix} 15 & 14 & 4 \end{bmatrix}$$

$$\begin{bmatrix} -24 & 23 & 77 \end{bmatrix} \begin{bmatrix} -1 & -10 & -8 \\ -1 & -6 & -5 \\ 0 & -1 & -1 \end{bmatrix} = \begin{bmatrix} 1 & 25 & 0 \end{bmatrix}$$

So, the message is as follows.

$$\begin{bmatrix} 13 & 5 & 5 \end{bmatrix} \begin{bmatrix} 20 & 0 & 13 \end{bmatrix} \begin{bmatrix} 5 & 0 & 13 \end{bmatrix} \begin{bmatrix} 15 & 14 & 4 \end{bmatrix} \begin{bmatrix} 1 & 25 & 0 \end{bmatrix}$$

M E E T   M E   M O N D A Y

✓CHECKPOINT   Now try Exercise 53.

© Corbis

Historical Note

During World War II, Navajo soldiers created a code using their native language to send messages between battalions. Native words were assigned to represent characters in the English alphabet, and they created a number of expressions for important military terms, like *iron-fish* to mean *submarine*. Without the Navajo Code Talkers, the Second World War might have had a very different outcome.

𝒲RITING ABOUT 𝓜ATHEMATICS

Cryptography   Use your school's library, the Internet, or some other reference source to research information about another type of cryptography. Write a short paragraph describing how mathematics is used to code and decode messages.

7.5 | Exercises

VOCABULARY CHECK: Fill in the blanks.

1. The method of using determinants to solve a system of linear equations is called _____ _____.

2. Three points are _____ if the points lie on the same line.

3. The area A of a triangle with vertices (x_1, y_1), (x_2, y_2), and (x_3, y_3) is given by _____.

4. A message written according to a secret code is called a _____.

5. To encode a message, choose an invertible matrix A and multiply the _____ row matrices by A (on the right) to obtain _____ row matrices.

PREREQUISITE SKILLS REVIEW: Practice and review algebra skills needed for this section at **www.Eduspace.com**.

In Exercises 1–10, use Cramer's Rule to solve (if possible) the system of equations.

1. $\begin{cases} 3x + 4y = -2 \\ 5x + 3y = 4 \end{cases}$
2. $\begin{cases} -4x - 7y = 47 \\ -x + 6y = -27 \end{cases}$

3. $\begin{cases} 3x + 2y = -2 \\ 6x + 4y = 4 \end{cases}$
4. $\begin{cases} 6x - 5y = 17 \\ -13x + 3y = -76 \end{cases}$

5. $\begin{cases} -0.4x + 0.8y = 1.6 \\ 0.2x + 0.3y = 2.2 \end{cases}$
6. $\begin{cases} 2.4x - 1.3y = 14.63 \\ -4.6x + 0.5y = -11.51 \end{cases}$

7. $\begin{cases} 4x - y + z = -5 \\ 2x + 2y + 3z = 10 \\ 5x - 2y + 6z = 1 \end{cases}$
8. $\begin{cases} 4x - 2y + 3z = -2 \\ 2x + 2y + 5z = 16 \\ 8x - 5y - 2z = 4 \end{cases}$

9. $\begin{cases} x + 2y + 3z = -3 \\ -2x + y - z = 6 \\ 3x - 3y + 2z = -11 \end{cases}$
10. $\begin{cases} 5x - 4y + z = -14 \\ -x + 2y - 2z = 10 \\ 3x + y + z = 1 \end{cases}$

In Exercises 11–14, use a graphing utility and Cramer's Rule to solve (if possible) the system of equations.

11. $\begin{cases} 3x + 3y + 5z = 1 \\ 3x + 5y + 9z = 2 \\ 5x + 9y + 17z = 4 \end{cases}$
12. $\begin{cases} x + 2y - z = -7 \\ 2x - 2y - 2z = -8 \\ -x + 3y + 4z = 8 \end{cases}$

13. $\begin{cases} 2x + y + 2z = 6 \\ -x + 2y - 3z = 0 \\ 3x + 2y - z = 6 \end{cases}$
14. $\begin{cases} 2x + 3y + 5z = 4 \\ 3x + 5y + 9z = 7 \\ 5x + 9y + 17z = 13 \end{cases}$

In Exercises 15–24, use a determinant and the given vertices of a triangle to find the area of the triangle.

15.

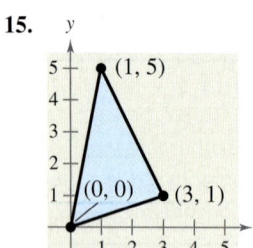

16.

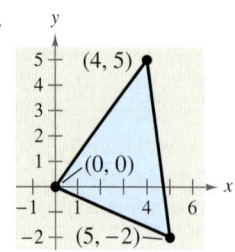

17.

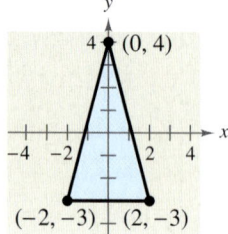

18.

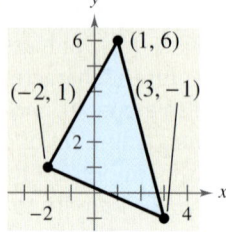

19.

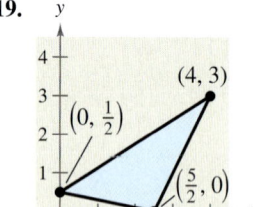

20.

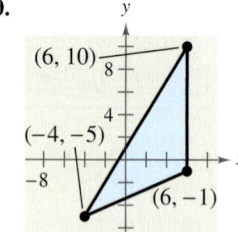

21. $(-2, 4), (2, 3), (-1, 5)$

22. $(0, -2), (-1, 4), (3, 5)$

23. $(-3, 5), (2, 6), (3, -5)$

24. $(-2, 4), (1, 5), (3, -2)$

In Exercises 25 and 26, find a value of y such that the triangle with the given vertices has an area of 4 square units.

25. $(-5, 1), (0, 2), (-2, y)$

26. $(-4, 2), (-3, 5), (-1, y)$

In Exercises 27 and 28, find a value of y such that the triangle with the given vertices has an area of 6 square units.

27. $(-2, -3), (1, -1), (-8, y)$

28. $(1, 0), (5, -3), (-3, y)$

29. *Area of a Region*   A large region of forest has been infested with gypsy moths. The region is roughly triangular, as shown in the figure. From the northernmost vertex A of the region, the distances to the other vertices are 25 miles south and 10 miles east (for vertex B), and 20 miles south and 28 miles east (for vertex C). Use a graphing utility to approximate the number of square miles in this region.

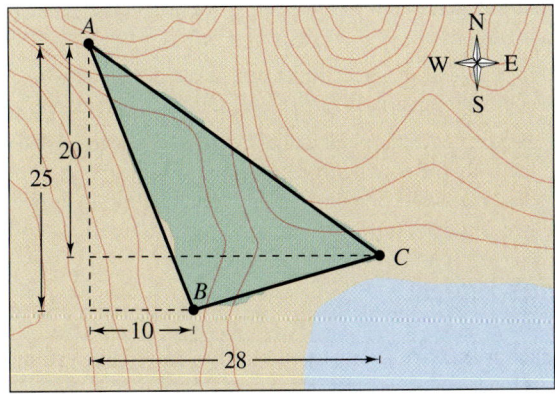

 30. *Area of a Region*   You own a triangular tract of land, as shown in the figure. To estimate the number of square feet in the tract, you start at one vertex, walk 65 feet east and 50 feet north to the second vertex, and then walk 85 feet west and 30 feet north to the third vertex. Use a graphing utility to determine how many square feet there are in the tract of land.

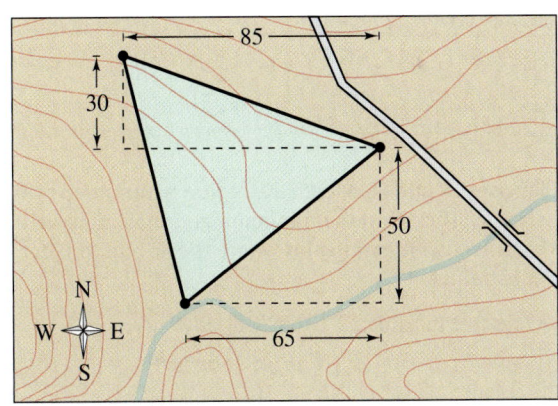

In Exercises 31–36, use a determinant to determine whether the points are collinear.

31. $(3, -1), (0, -3), (12, 5)$ **32.** $(-3, -5), (6, 1), (10, 2)$

33. $\left(2, -\frac{1}{2}\right), (-4, 4), (6, -3)$ **34.** $(0, 1), (4, -2), \left(-2, \frac{5}{2}\right)$

35. $(0, 2), (1, 2.4), (-1, 1.6)$ **36.** $(2, 3), (3, 3.5), (-1, 2)$

In Exercises 37 and 38, find y such that the points are collinear.

37. $(2, -5), (4, y), (5, -2)$ **38.** $(-6, 2), (-5, y), (-3, 5)$

In Exercises 39–44, use a determinant to find an equation of the line passing through the points.

39. $(0, 0), (5, 3)$ **40.** $(0, 0), (-2, 2)$

41. $(-4, 3), (2, 1)$ **42.** $(10, 7), (-2, -7)$

43. $\left(-\frac{1}{2}, 3\right), \left(\frac{5}{2}, 1\right)$ **44.** $\left(\frac{2}{3}, 4\right), (6, 12)$

In Exercises 45 and 46, find the uncoded 1×3 row matrices for the message. Then encode the message using the encoding matrix.

Message	*Encoding Matrix*

45. TROUBLE IN RIVER CITY $\begin{bmatrix} 1 & -1 & 0 \\ 1 & 0 & -1 \\ -6 & 2 & 3 \end{bmatrix}$

46. PLEASE SEND MONEY $\begin{bmatrix} 4 & 2 & 1 \\ -3 & -3 & -1 \\ 3 & 2 & 1 \end{bmatrix}$

In Exercises 47–50, write a cryptogram for the message using the matrix A.

$$A = \begin{bmatrix} 1 & 2 & 2 \\ 3 & 7 & 9 \\ -1 & -4 & -7 \end{bmatrix}.$$

47. CALL AT NOON

48. ICEBERG DEAD AHEAD

49. HAPPY BIRTHDAY

50. OPERATION OVERLOAD

In Exercises 51–54, use A^{-1} to decode the cryptogram.

51. $A = \begin{bmatrix} 1 & 2 \\ 3 & 5 \end{bmatrix}$

11   21   64   112   25   50   29   53   23   46
40   75   55   92

52. $A = \begin{bmatrix} -5 & 2 \\ -7 & 3 \end{bmatrix}$

−136   58   −173   72   −120   51   −95   38
−178   73   −70   28   −242   101   −115   47
−90   36   −115   49   −199   82

53. $A = \begin{bmatrix} 1 & -1 & 0 \\ 1 & 0 & -1 \\ -6 & 2 & 3 \end{bmatrix}$

9   −1   −9   38   −19   −19   28   −9   −19   −80   25
41   −64   21   31   9   −5   −4

54. $A = \begin{bmatrix} 3 & -4 & 2 \\ 0 & 2 & 1 \\ 4 & -5 & 3 \end{bmatrix}$

112   −140   83   19   −25   13   72   −76   61   95
−118   71   20   21   38   35   −23   36   42   −48   32

In Exercises 55 and 56, decode the cryptogram by using the inverse of the matrix A.

$$A = \begin{bmatrix} 1 & 2 & 2 \\ 3 & 7 & 9 \\ -1 & -4 & -7 \end{bmatrix}$$

55. 20   17 -15 -12 -56 -104 1 -25 -65
62   143   181

56. 13 -9 -59 61   112   106 -17 -73 -131 11
24   29   65   144   172

57. The following cryptogram was encoded with a 2×2 matrix.

8   21 -15 -10 -13 -13 5   10   5   25   5   19
-1 6   20   40 -18 -18 1   16

The last word of the message is _RON. What is the message?

Model It

58. *Data Analysis: Supreme Court*   The table shows the numbers y of U.S. Supreme Court cases waiting to be tried for the years 2000 through 2002. (Source: Office of the Clerk, Supreme Court of the United States)

Year	Number of cases, y
2000	8965
2001	9176
2002	9406

(a) Use the technique demonstrated in Exercises 67–70 in Section 6.3 to create a system of linear equations for the data. Let t represent the year, with $t = 0$ corresponding to 2000.

(b) Use Cramer's Rule to solve the system from part (a) and find the least squares regression parabola $y = at^2 + bt + c$.

 (c) Use a graphing utility to graph the parabola from part (b).

 (d) Use the graph from part (c) to estimate when the number of U.S. Supreme Court cases waiting to be tried will reach 10,000.

Synthesis

True or False?   **In Exercises 59–61, determine whether the statement is true or false. Justify your answer.**

59. In Cramer's Rule, the numerator is the determinant of the coefficient matrix.

60. You cannot use Cramer's Rule when solving a system of linear equations if the determinant of the coefficient matrix is zero.

61. In a system of linear equations, if the determinant of the coefficient matrix is zero, the system has no solution.

62. *Writing*   At this point in the text, you have learned several methods for solving systems of linear equations. Briefly describe which method(s) you find easiest to use and which method(s) you find most difficult to use.

Skills Review

In Exercises 63–66, use any method to solve the system of equations.

63. $\begin{cases} -x - 7y = -22 \\ 5x + y = -26 \end{cases}$

64. $\begin{cases} 3x + 8y = 11 \\ -2x + 12y = -16 \end{cases}$

65. $\begin{cases} -x - 3y + 5z = -14 \\ 4x + 2y - z = -1 \\ 5x - 3y + 2z = -11 \end{cases}$

66. $\begin{cases} 5x - y - z = 7 \\ -2x + 3y + z = -5 \\ 4x + 10y - 5z = -37 \end{cases}$

In Exercises 67 and 68, sketch the region determined by the constraints. Then find the minimum and maximum values of the objective function and where they occur, subject to the constraints.

67. Objective function:

$z = 6x + 4y$

Constraints:

$$x \geq 0$$
$$y \geq 0$$
$$x + 6y \leq 30$$
$$6x + y \leq 40$$

68. Objective function:

$z = 6x + 7y$

Constraints:

$$x \geq 0$$
$$y \geq 0$$
$$4x + 3y \geq 24$$
$$x + 3y \geq 15$$

7   Chapter Summary

What did you learn?

7   Review Exercises

7.1   In Exercises 1–4, determine the order of the matrix.

1. $\begin{bmatrix} -4 \\ 0 \\ 5 \end{bmatrix}$

2. $\begin{bmatrix} 3 & -1 & 0 & 6 \\ -2 & 7 & 1 & 4 \end{bmatrix}$

3. $[3]$

4. $\begin{bmatrix} 6 & 2 & -5 & 8 & 0 \end{bmatrix}$

In Exercises 5 and 6, write the augmented matrix for the system of linear equations.

5. $\begin{cases} 3x - 10y = 15 \\ 5x + 4y = 22 \end{cases}$

6. $\begin{cases} 8x - 7y + 4z = 12 \\ 3x - 5y + 2z = 20 \\ 5x + 3y - 3z = 26 \end{cases}$

In Exercises 7 and 8, write the system of linear equations represented by the augmented matrix. (Use variables x, y, z, and w, if applicable.)

7. $\begin{bmatrix} 5 & 1 & 7 & \vdots & -9 \\ 4 & 2 & 0 & \vdots & 10 \\ 9 & 4 & 2 & \vdots & 3 \end{bmatrix}$

8. $\begin{bmatrix} 13 & 16 & 7 & 3 & \vdots & 2 \\ 1 & 21 & 8 & 5 & \vdots & 12 \\ 4 & 10 & -4 & 3 & \vdots & -1 \end{bmatrix}$

In Exercises 9 and 10, write the matrix in row-echelon form. Remember that the row-echelon form of a matrix is not unique.

9. $\begin{bmatrix} 0 & 1 & 1 \\ 1 & 2 & 3 \\ 2 & 2 & 2 \end{bmatrix}$

10. $\begin{bmatrix} 4 & 8 & 16 \\ 3 & -1 & 2 \\ -2 & 10 & 12 \end{bmatrix}$

In Exercises 11–14, write the system of linear equations represented by the augmented matrix. Then use back-substitution to solve the system. (Use variables x, y, and z.)

11. $\begin{bmatrix} 1 & 2 & 3 & \vdots & 9 \\ 0 & 1 & -2 & \vdots & 2 \\ 0 & 0 & 1 & \vdots & 0 \end{bmatrix}$

12. $\begin{bmatrix} 1 & 3 & -9 & \vdots & 4 \\ 0 & 1 & -1 & \vdots & 10 \\ 0 & 0 & 1 & \vdots & -2 \end{bmatrix}$

13. $\begin{bmatrix} 1 & -5 & 4 & \vdots & 1 \\ 0 & 1 & 2 & \vdots & 3 \\ 0 & 0 & 1 & \vdots & 4 \end{bmatrix}$

14. $\begin{bmatrix} 1 & -8 & 0 & \vdots & -2 \\ 0 & 1 & -1 & \vdots & -7 \\ 0 & 0 & 1 & \vdots & 1 \end{bmatrix}$

In Exercises 15–24, use matrices and Gaussian elimination with back-substitution to solve the system of equations (if possible).

15. $\begin{cases} 5x + 4y = 2 \\ -x + y = -22 \end{cases}$

16. $\begin{cases} 2x - 5y = 2 \\ 3x - 7y = 1 \end{cases}$

17. $\begin{cases} 0.3x - 0.1y = -0.13 \\ 0.2x - 0.3y = -0.25 \end{cases}$

18. $\begin{cases} 0.2x - 0.1y = 0.07 \\ 0.4x - 0.5y = -0.01 \end{cases}$

19. $\begin{cases} 2x + 3y + z = 10 \\ 2x - 3y - 3z = 22 \\ 4x - 2y + 3z = -2 \end{cases}$

20. $\begin{cases} 2x + 3y + 3z = 3 \\ 6x + 6y + 12z = 13 \\ 12x + 9y - z = 2 \end{cases}$

21. $\begin{cases} 2x + y + 2z = 4 \\ 2x + 2y = 5 \\ 2x - y + 6z = 2 \end{cases}$

22. $\begin{cases} x + 2y + 6z = 1 \\ 2x + 5y + 15z = 4 \\ 3x + y + 3z = -6 \end{cases}$

23. $\begin{cases} 2x + y + z = 6 \\ -2y + 3z - w = 9 \\ 3x + 3y - 2z - 2w = -11 \\ x + z + 3w = 14 \end{cases}$

24. $\begin{cases} x + 2y + w = 3 \\ -3y + 3z = 0 \\ 4x + 4y + z + 2w = 0 \\ 2x + z = 3 \end{cases}$

In Exercises 25–28, use matrices and Gauss-Jordan elimination to solve the system of equations.

25. $\begin{cases} -x + y + 2z = 1 \\ 2x + 3y + z = -2 \\ 5x + 4y + 2z = 4 \end{cases}$

26. $\begin{cases} 4x + 4y + 4z = 5 \\ 4x - 2y - 8z = 1 \\ 5x + 3y + 8z = 6 \end{cases}$

27. $\begin{cases} 2x - y + 9z = -8 \\ -x - 3y + 4z = -15 \\ 5x + 2y - z = 17 \end{cases}$

28. $\begin{cases} -3x + y + 7z = -20 \\ 5x - 2y - z = 34 \\ -x + y + 4z = -8 \end{cases}$

In Exercises 29 and 30, use the matrix capabilities of a graphing utility to reduce the augmented matrix corresponding to the system of equations, and solve the system.

29. $\begin{cases} 3x - y + 5z - 2w = -44 \\ x + 6y + 4z - w = 1 \\ 5x - y + z + 3w = -15 \\ 4y - z - 8w = 58 \end{cases}$

30. $\begin{cases} 4x + 12y + 2z = 20 \\ x + 6y + 4z = 12 \\ x + 6y + z = 8 \\ -2x - 10y - 2z = -10 \end{cases}$

7.2 In Exercises 31–34, find x and y.

31. $\begin{bmatrix} -1 & x \\ y & 9 \end{bmatrix} = \begin{bmatrix} -1 & 12 \\ -7 & 9 \end{bmatrix}$

32. $\begin{bmatrix} -1 & 0 \\ x & 5 \\ -4 & y \end{bmatrix} = \begin{bmatrix} -1 & 0 \\ 8 & 5 \\ -4 & 0 \end{bmatrix}$

33. $\begin{bmatrix} x+3 & -4 & 4y \\ 0 & -3 & 2 \\ -2 & y+5 & 6x \end{bmatrix} = \begin{bmatrix} 5x-1 & -4 & 44 \\ 0 & -3 & 2 \\ -2 & 16 & 6 \end{bmatrix}$

34. $\begin{bmatrix} -9 & 4 & 2 & -5 \\ 0 & -3 & 7 & -4 \\ 6 & -1 & 1 & 0 \end{bmatrix} = \begin{bmatrix} -9 & 4 & x-10 & -5 \\ 0 & -3 & 7 & 2y \\ \frac{1}{2}x & -1 & 1 & 0 \end{bmatrix}$

In Exercises 35–38, if possible, find (a) $A + B$, (b) $A - B$, (c) $4A$, and (d) $A + 3B$.

35. $A = \begin{bmatrix} 2 & -2 \\ 3 & 5 \end{bmatrix}$, $B = \begin{bmatrix} -3 & 10 \\ 12 & 8 \end{bmatrix}$

36. $A = \begin{bmatrix} 5 & 4 \\ -7 & 2 \\ 11 & 2 \end{bmatrix}$, $B = \begin{bmatrix} 4 & 12 \\ 20 & 40 \\ 15 & 30 \end{bmatrix}$

37. $A = \begin{bmatrix} 5 & 4 \\ -7 & 2 \\ 11 & 2 \end{bmatrix}$, $B = \begin{bmatrix} 0 & 3 \\ 4 & 12 \\ 20 & 40 \end{bmatrix}$

38. $A = \begin{bmatrix} 6 & -5 & 7 \end{bmatrix}$, $B = \begin{bmatrix} -1 \\ 4 \\ 8 \end{bmatrix}$

In Exercises 39–42, perform the matrix operations. If it is not possible, explain why.

39. $\begin{bmatrix} 7 & 3 \\ -1 & 5 \end{bmatrix} + \begin{bmatrix} 10 & -20 \\ 14 & -3 \end{bmatrix}$

40. $\begin{bmatrix} -11 & 16 & 19 \\ -7 & -2 & 1 \end{bmatrix} - \begin{bmatrix} 6 & 0 \\ 8 & -4 \\ -2 & 10 \end{bmatrix}$

41. $-2\begin{bmatrix} 1 & 2 \\ 5 & -4 \\ 6 & 0 \end{bmatrix} + 8\begin{bmatrix} 7 & 1 \\ 1 & 2 \\ 1 & 4 \end{bmatrix}$

42. $-\begin{bmatrix} 8 & -1 & 8 \\ -2 & 4 & 12 \\ 0 & -6 & 0 \end{bmatrix} - 5\begin{bmatrix} -2 & 0 & -4 \\ 3 & -1 & 1 \\ 6 & 12 & -8 \end{bmatrix}$

In Exercises 43 and 44, use the matrix capabilities of a graphing utility to evaluate the expression.

43. $3\begin{bmatrix} 8 & -2 & 5 \\ 1 & 3 & -1 \end{bmatrix} + 6\begin{bmatrix} 4 & -2 & -3 \\ 2 & 7 & 6 \end{bmatrix}$

44. $-5\begin{bmatrix} 2 & 0 \\ 7 & -2 \\ 8 & 2 \end{bmatrix} + 4\begin{bmatrix} 4 & -2 \\ 6 & 11 \\ -1 & 3 \end{bmatrix}$

In Exercises 45–48, solve for X in the equation, given

$A = \begin{bmatrix} -4 & 0 \\ 1 & -5 \\ -3 & 2 \end{bmatrix}$ and $B = \begin{bmatrix} 1 & 2 \\ -2 & 1 \\ 4 & 4 \end{bmatrix}$.

45. $X = 3A - 2B$

46. $6X = 4A + 3B$

47. $3X + 2A = B$

48. $2A - 5B = 3X$

In Exercises 49–52, find AB, if possible.

49. $A = \begin{bmatrix} 2 & -2 \\ 3 & 5 \end{bmatrix}$, $B = \begin{bmatrix} -3 & 10 \\ 12 & 8 \end{bmatrix}$

50. $A = \begin{bmatrix} 5 & 4 \\ -7 & 2 \\ 11 & 2 \end{bmatrix}$, $B = \begin{bmatrix} 4 & 12 \\ 20 & 40 \\ 15 & 30 \end{bmatrix}$

51. $A = \begin{bmatrix} 5 & 4 \\ -7 & 2 \\ 11 & 2 \end{bmatrix}$, $B = \begin{bmatrix} 4 & 12 \\ 20 & 40 \end{bmatrix}$

52. $A = \begin{bmatrix} 6 & -5 & 7 \end{bmatrix}$, $B = \begin{bmatrix} -1 \\ 4 \\ 8 \end{bmatrix}$

In Exercises 53–60, perform the matrix operations. If it is not possible, explain why.

53. $\begin{bmatrix} 1 & 2 \\ 5 & -4 \\ 6 & 0 \end{bmatrix}\begin{bmatrix} 6 & -2 & 8 \\ 4 & 0 & 0 \end{bmatrix}$

54. $\begin{bmatrix} 1 & 5 & 6 \\ 2 & -4 & 0 \end{bmatrix}\begin{bmatrix} 6 & -2 & 8 \\ 4 & 0 & 0 \end{bmatrix}$

55. $\begin{bmatrix} 1 & 5 & 6 \\ 2 & -4 & 0 \end{bmatrix}\begin{bmatrix} 6 & 4 \\ -2 & 0 \\ 8 & 0 \end{bmatrix}$

56. $\begin{bmatrix} 1 & 3 & 2 \\ 0 & 2 & -4 \\ 0 & 0 & 3 \end{bmatrix}\begin{bmatrix} 4 & -3 & 2 \\ 0 & 3 & -1 \\ 0 & 0 & 2 \end{bmatrix}$

57. $\begin{bmatrix} 4 \\ 6 \end{bmatrix}\begin{bmatrix} 6 & -2 \end{bmatrix}$

58. $\begin{bmatrix} 4 & -2 & 6 \end{bmatrix} \begin{bmatrix} -2 & 1 \\ 0 & -3 \\ 2 & 0 \end{bmatrix}$

59. $\begin{bmatrix} 2 & 1 \\ 6 & 0 \end{bmatrix} \left(\begin{bmatrix} 4 & 2 \\ -3 & 1 \end{bmatrix} + \begin{bmatrix} -2 & 4 \\ 0 & 4 \end{bmatrix} \right)$

60. $-3 \begin{bmatrix} 1 & -1 \\ 4 & 2 \end{bmatrix} \left(\begin{bmatrix} 0 & 3 \\ 1 & 2 \end{bmatrix} \begin{bmatrix} 1 & 0 \\ 5 & -3 \end{bmatrix} \right)$

 In Exercises 61 and 62, use the matrix capabilities of a graphing utility to find the product.

61. $\begin{bmatrix} 4 & 1 \\ 11 & -7 \\ 12 & 3 \end{bmatrix} \begin{bmatrix} 3 & -5 & 6 \\ 2 & -2 & -2 \end{bmatrix}$

62. $\begin{bmatrix} -2 & 3 & 10 \\ 4 & -2 & 2 \end{bmatrix} \begin{bmatrix} 1 & 1 \\ -5 & 2 \\ 3 & 2 \end{bmatrix}$

63. *Manufacturing* A tire corporation has three factories, each of which manufactures two products. The number of units of product i produced at factory j in one day is represented by a_{ij} in the matrix

$$A = \begin{bmatrix} 80 & 120 & 140 \\ 40 & 100 & 80 \end{bmatrix}.$$

Find the production levels if production is decreased by 5%.

64. *Manufacturing* A corporation has four factories, each of which manufactures three types of cordless power tools. The number of units of cordless power tools produced at factory j in one day is represented by a_{ij} in the matrix

$$A = \begin{bmatrix} 80 & 70 & 90 & 40 \\ 50 & 30 & 80 & 20 \\ 90 & 60 & 100 & 50 \end{bmatrix}.$$

Find the production levels if production is increased by 20%.

65. *Manufacturing* A manufacturing company produces three kinds of computer games that are shipped to two warehouses. The number of units of game i that are shipped to warehouse j is represented by a_{ij} in the matrix

$$A = \begin{bmatrix} 8200 & 7400 \\ 6500 & 9800 \\ 5400 & 4800 \end{bmatrix}.$$

The price per unit is represented by the matrix

$$B = \begin{bmatrix} \$10.25 & \$14.50 & \$17.75 \end{bmatrix}.$$

Compute BA and interpret the result.

66. *Long-Distance Plans* The charges (in dollars per minute) of two long-distance telephone companies for in-state, state-to-state, and international calls are represented by C.

Company

	A	B	
In-state	0.07	0.095	
State-to-state	0.10	0.08	} Type of call
International	0.28	0.25	

$$C = \begin{bmatrix} 0.07 & 0.095 \\ 0.10 & 0.08 \\ 0.28 & 0.25 \end{bmatrix}$$

You plan to use 120 minutes on in-state calls, 80 minutes on state-to-state calls, and 20 minutes on international calls each month.

(a) Write a matrix T that represents the times spent on the phone for each type of call.

(b) Compute TC and interpret the result.

7.3 **In Exercises 67–70, show that B is the inverse of A.**

67. $A = \begin{bmatrix} -4 & -1 \\ 7 & 2 \end{bmatrix}$, $B = \begin{bmatrix} -2 & -1 \\ 7 & 4 \end{bmatrix}$

68. $A = \begin{bmatrix} 5 & -1 \\ 11 & -2 \end{bmatrix}$, $B = \begin{bmatrix} -2 & 1 \\ -11 & 5 \end{bmatrix}$

69. $A = \begin{bmatrix} 1 & 1 & 0 \\ 1 & 0 & 1 \\ 6 & 2 & 3 \end{bmatrix}$, $B = \begin{bmatrix} -2 & -3 & 1 \\ 3 & 3 & -1 \\ 2 & 4 & 1 \end{bmatrix}$

70. $A = \begin{bmatrix} 1 & -1 & 0 \\ -1 & 0 & -1 \\ 8 & -4 & 2 \end{bmatrix}$, $B = \begin{bmatrix} -2 & 1 & \frac{1}{2} \\ -3 & 1 & \frac{1}{2} \\ 2 & -2 & -\frac{1}{2} \end{bmatrix}$

In Exercises 71–74, find the inverse of the matrix (if it exists).

71. $\begin{bmatrix} -6 & 5 \\ -5 & 4 \end{bmatrix}$

72. $\begin{bmatrix} -3 & -5 \\ 2 & 3 \end{bmatrix}$

73. $\begin{bmatrix} -1 & -2 & -2 \\ 3 & 7 & 9 \\ 1 & 4 & 7 \end{bmatrix}$

74. $\begin{bmatrix} 0 & -2 & 1 \\ -5 & -2 & -3 \\ 7 & 3 & 4 \end{bmatrix}$

 In Exercises 75–78, use the matrix capabilities of a graphing utility to find the inverse of the matrix (if it exists).

75. $\begin{bmatrix} 2 & 0 & 3 \\ -1 & 1 & 1 \\ 2 & -2 & 1 \end{bmatrix}$

76. $\begin{bmatrix} 1 & 4 & 6 \\ 2 & -3 & 1 \\ -1 & 18 & 16 \end{bmatrix}$

77. $\begin{bmatrix} 1 & 3 & 1 & 6 \\ 4 & 4 & 2 & 6 \\ 3 & 4 & 1 & 2 \\ -1 & 2 & -1 & -2 \end{bmatrix}$

78. $\begin{bmatrix} 8 & 0 & 2 & 8 \\ 4 & -2 & 0 & -2 \\ 1 & 2 & 1 & 4 \\ -1 & 4 & 1 & 1 \end{bmatrix}$

In Exercises 79–82, use the formula below to find the inverse of the matrix, it it exists.

$$A^{-1} = \frac{1}{ad - bc}\begin{bmatrix} d & -b \\ -c & a \end{bmatrix}$$

79. $\begin{bmatrix} -7 & 2 \\ -8 & 2 \end{bmatrix}$

80. $\begin{bmatrix} 10 & 4 \\ 7 & 3 \end{bmatrix}$

81. $\begin{bmatrix} -\frac{1}{2} & 20 \\ \frac{3}{10} & -6 \end{bmatrix}$

82. $\begin{bmatrix} -\frac{3}{4} & \frac{5}{2} \\ -\frac{4}{5} & -\frac{8}{3} \end{bmatrix}$

In Exercises 83–90, use an inverse matrix to solve (if possible) the system of linear equations.

83. $\begin{cases} -x + 4y = 8 \\ 2x - 7y = -5 \end{cases}$

84. $\begin{cases} 5x - y = 13 \\ -9x + 2y = -24 \end{cases}$

85. $\begin{cases} -3x + 10y = 8 \\ 5x - 17y = -13 \end{cases}$

86. $\begin{cases} 4x - 2y = -10 \\ -19x + 9y = 47 \end{cases}$

87. $\begin{cases} 3x + 2y - z = 6 \\ x - y + 2z = -1 \\ 5x + y + z = 7 \end{cases}$

88. $\begin{cases} -x + 4y - 2z = 12 \\ 2x - 9y + 5z = -25 \\ -x + 5y - 4z = 10 \end{cases}$

89. $\begin{cases} -2x + y + 2z = -13 \\ -x - 4y + z = -11 \\ -y - z = 0 \end{cases}$

90. $\begin{cases} 3x - y + 5z = -14 \\ -x + y + 6z = 8 \\ -8x + 4y - z = 44 \end{cases}$

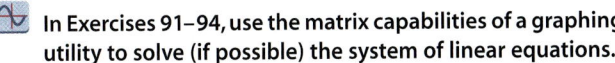 In Exercises 91–94, use the matrix capabilities of a graphing utility to solve (if possible) the system of linear equations.

91. $\begin{cases} x + 2y = -1 \\ 3x + 4y = -5 \end{cases}$

92. $\begin{cases} x + 3y = 23 \\ -6x + 2y = -18 \end{cases}$

93. $\begin{cases} -3x - 3y - 4z = 2 \\ y + z = -1 \\ 4x + 3y + 4z - -1 \end{cases}$

94. $\begin{cases} x - 3y - 2z = 8 \\ -2x + 7y + 3z = -19 \\ x - y - 3z = 3 \end{cases}$

7.4 In Exercises 95–98, find the determinant of the matrix.

95. $\begin{bmatrix} 8 & 5 \\ 2 & -4 \end{bmatrix}$

96. $\begin{bmatrix} -9 & 11 \\ 7 & -4 \end{bmatrix}$

97. $\begin{bmatrix} 50 & -30 \\ 10 & 5 \end{bmatrix}$

98. $\begin{bmatrix} 14 & -24 \\ 12 & -15 \end{bmatrix}$

In Exercises 99–102, find all (a) minors and (b) cofactors of the matrix.

99. $\begin{bmatrix} 2 & -1 \\ 7 & 4 \end{bmatrix}$

100. $\begin{bmatrix} 3 & 6 \\ 5 & -4 \end{bmatrix}$

101. $\begin{bmatrix} 3 & 2 & -1 \\ -2 & 5 & 0 \\ 1 & 8 & 6 \end{bmatrix}$

102. $\begin{bmatrix} 8 & 3 & 4 \\ 6 & 5 & -9 \\ -4 & 1 & 2 \end{bmatrix}$

In Exercises 103–106, find the determinant of the matrix. Expand by cofactors on the row or column that appears to make the computations easiest.

103. $\begin{bmatrix} -2 & 4 & 1 \\ -6 & 0 & 2 \\ 5 & 3 & 4 \end{bmatrix}$

104. $\begin{bmatrix} 4 & 7 & -1 \\ 2 & -3 & 4 \\ -5 & 1 & -1 \end{bmatrix}$

105. $\begin{bmatrix} 3 & 0 & -4 & 0 \\ 0 & 8 & 1 & 2 \\ 6 & 1 & 8 & 2 \\ 0 & 3 & -4 & 1 \end{bmatrix}$

106. $\begin{bmatrix} -5 & 6 & 0 & 0 \\ 0 & 1 & -1 & 2 \\ -3 & 4 & -5 & 1 \\ 1 & 6 & 0 & 3 \end{bmatrix}$

7.5 In Exercises 107–110, use Cramer's Rule to solve (if possible) the system of equations.

107. $\begin{cases} 5x - 2y = 6 \\ -11x + 3y = -23 \end{cases}$ **108.** $\begin{cases} 3x + 8y = -7 \\ 9x - 5y = 37 \end{cases}$

109. $\begin{cases} -2x + 3y - 5z = -11 \\ 4x - y + z = -3 \\ -x - 4y + 6z = 15 \end{cases}$ **110.** $\begin{cases} 5x - 2y + z = 15 \\ 3x - 3y - z = -7 \\ 2x - y - 7z = -3 \end{cases}$

In Exercises 111–114, use a determinant and the given vertices of a triangle to find the area of the triangle.

111.
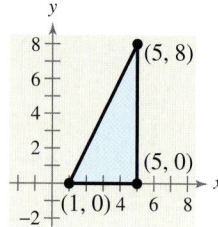

112.
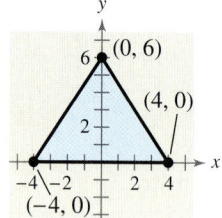

113.
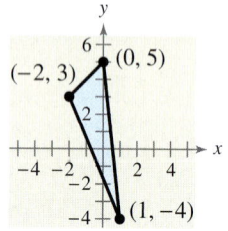

114.
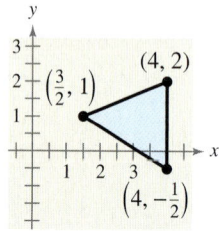

In Exercises 115 and 116, use a determinant to determine whether the points are collinear.

115. $(-1, 7), (3, -9), (-3, 15)$

116. $(0, -5), (-2, -6), (8, -1)$

In Exercises 117–120, use a determinant to find an equation of the line passing through the points.

117. $(-4, 0), (4, 4)$ **118.** $(2, 5), (6, -1)$

119. $\left(-\frac{5}{2}, 3\right), \left(\frac{7}{2}, 1\right)$ **120.** $(-0.8, 0.2), (0.7, 3.2)$

In Exercises 121 and 122, find the uncoded 1×3 row matrices for the message. Then encode the message using the encoding matrix.

	Message	Encoding Matrix

121. LOOK OUT BELOW $\begin{bmatrix} 2 & -2 & 0 \\ 3 & 0 & -3 \\ -6 & 2 & 3 \end{bmatrix}$

122. RETURN TO BASE $\begin{bmatrix} 2 & 1 & 0 \\ -6 & -6 & -2 \\ 3 & 2 & 1 \end{bmatrix}$

In Exercises 123 and 124, decode the cryptogram by using the inverse of the matrix

$$A = \begin{bmatrix} -5 & 4 & -3 \\ 10 & -7 & 6 \\ 8 & -6 & 5 \end{bmatrix}.$$

123. $-5 \quad 11 \quad -2 \quad 370 \quad -265 \quad 225 \quad -57 \quad 48 \quad -33 \ 32$
$-15 \quad 20 \quad 245 \quad -171 \quad 147$

124. $145 \quad -105 \quad 92 \quad 264 \quad -188 \quad 160 \quad 23 \quad -16 \quad 15$
$129 \quad -84 \quad 78 \quad -9 \quad 8 \quad -5 \quad 159 \quad -118 \quad 100 \quad 219$
$-152 \quad 133 \quad 370 \quad -265 \quad 225 \quad -105 \quad 84 \quad -63$

Synthesis

True or False? **In Exercises 125 and 126, determine whether the statement is true or false. Justify your answer.**

125. It is possible to find the determinant of a 4×5 matrix.

126. $\begin{vmatrix} a_{11} & a_{12} & a_{13} \\ a_{21} & a_{22} & a_{23} \\ a_{31} + c_1 & a_{32} + c_2 & a_{33} + c_3 \end{vmatrix} =$

$\begin{vmatrix} a_{11} & a_{12} & a_{13} \\ a_{21} & a_{22} & a_{23} \\ a_{31} & a_{32} & a_{33} \end{vmatrix} + \begin{vmatrix} a_{11} & a_{12} & a_{13} \\ a_{21} & a_{22} & a_{23} \\ c_1 & c_2 & c_3 \end{vmatrix}$

127. Under what conditions does a matrix have an inverse?

128. *Writing* What is meant by the cofactor of an entry of a matrix? How are cofactors used to find the determinant of the matrix?

129. Three people were asked to solve a system of equations using an augmented matrix. Each person reduced the matrix to row-echelon form. The reduced matrices were

$$\begin{bmatrix} 1 & 2 & \vdots & 3 \\ 0 & 1 & \vdots & 1 \end{bmatrix},$$

$$\begin{bmatrix} 1 & 0 & \vdots & 1 \\ 0 & 1 & \vdots & 1 \end{bmatrix},$$

and

$$\begin{bmatrix} 1 & 2 & \vdots & 3 \\ 0 & 0 & \vdots & 0 \end{bmatrix}.$$

Can all three be right? Explain.

130. *Think About It* Describe the row-echelon form of an augmented matrix that corresponds to a system of linear equations that has a unique solution.

131. Solve the equation for λ.

$$\begin{vmatrix} 2 - \lambda & 5 \\ 3 & -8 - \lambda \end{vmatrix} = 0$$

7   Chapter Test

Take this test as you would take a test in class. When you are finished, check your work against the answers given in the back of the book.

In Exercises 1 and 2, write the matrix in reduced row-echelon form.

1. $\begin{bmatrix} 1 & -1 & 5 \\ 6 & 2 & 3 \\ 5 & 3 & -3 \end{bmatrix}$

2. $\begin{bmatrix} 1 & 0 & -1 & 2 \\ -1 & 1 & 1 & -3 \\ 1 & 1 & -1 & 1 \\ 3 & 2 & -3 & 4 \end{bmatrix}$

3. Write the augmented matrix corresponding to the system of equations and solve the system.

$$\begin{cases} 4x + 3y - 2z = 14 \\ -x - y + 2z = -5 \\ 3x + y - 4z = 8 \end{cases}$$

4. Find (a) $A - B$, (b) $3A$, (c) $3A - 2B$, and (d) AB (if possible).

$$A = \begin{bmatrix} 5 & 4 \\ -4 & -4 \end{bmatrix}, \quad B = \begin{bmatrix} 4 & -1 \\ -4 & 0 \end{bmatrix}$$

In Exercises 5 and 6, find the inverse of the matrix (if it exists).

5. $\begin{bmatrix} -6 & 4 \\ 10 & -5 \end{bmatrix}$

6. $\begin{bmatrix} -2 & 4 & -6 \\ 2 & 1 & 0 \\ 4 & -2 & 5 \end{bmatrix}$

7. Use the result of Exercise 5 to solve the system.

$$\begin{cases} -6x + 4y = 10 \\ 10x - 5y = 20 \end{cases}$$

In Exercises 8–10, evaluate the determinant of the matrix.

8. $\begin{bmatrix} -9 & 4 \\ 13 & 16 \end{bmatrix}$

9. $\begin{bmatrix} \frac{5}{2} & \frac{13}{4} \\ -8 & \frac{6}{5} \end{bmatrix}$

10. $\begin{bmatrix} 6 & -7 & 2 \\ 3 & -2 & 0 \\ 1 & 5 & 1 \end{bmatrix}$

In Exercises 11 and 12, use Cramer's Rule to solve (if possible) the system of equations.

11. $\begin{cases} 7x + 6y = 9 \\ -2x - 11y = -49 \end{cases}$

12. $\begin{cases} 6x - y + 2z = -4 \\ -2x + 3y - z = 10 \\ 4x - 4y + z = -18 \end{cases}$

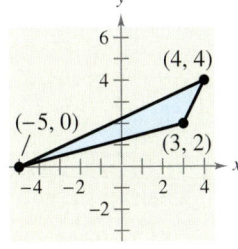

FIGURE FOR 13

13. Use a determinant to find the area of the triangle in the figure.

14. Find the uncoded 1×3 row matrices for the message KNOCK ON WOOD. Then encode the message using the matrix A below.

$$A = \begin{bmatrix} 1 & -1 & 0 \\ 1 & 0 & -1 \\ 6 & -2 & -3 \end{bmatrix}$$

15. One hundred liters of a 50% solution is obtained by mixing a 60% solution with a 20% solution. How many liters of each solution must be used to obtain the desired mixture?

Proofs in Mathematics

Proofs without words are pictures or diagrams that give a visual understanding of why a theorem or statement is true. They can also provide a starting point for writing a formal proof. The following proof shows that a 2×2 determinant is the area of a parallelogram.

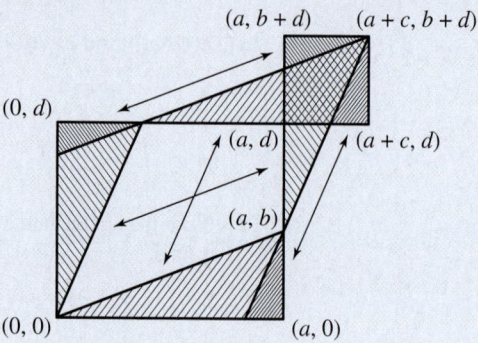

$$\begin{vmatrix} a & b \\ c & d \end{vmatrix} = ad - bc = \|\square\| - \|\square\| = \|\square\|$$

The following is a color-coded version of the proof along with a brief explanation of why this proof works.

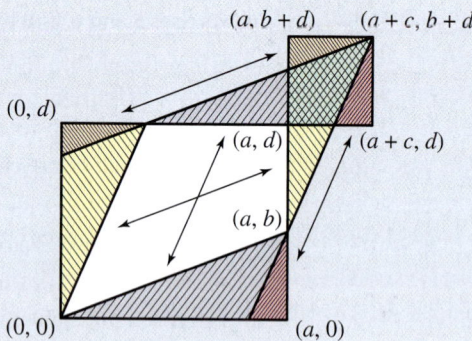

$$\begin{vmatrix} a & b \\ c & d \end{vmatrix} = ad - bc = \|\square\| - \|\square\| = \|\square\|$$

Area of $\square$ = Area of orange $\triangle$ + Area of yellow $\triangle$ + Area of blue $\triangle$ + Area of pink $\triangle$ + Area of white quadrilateral

Area of $\square$ = Area of orange $\triangle$ + Area of pink $\triangle$ + Area of green quadrilateral

Area of $\square$ = Area of white quadrilateral + Area of blue $\triangle$ + Area of yellow $\triangle$ − Area of green quadrilateral
    = Area of $\square$ − Area of $\square$

From "Proof Without Words" by Solomon W. Golomb, *Mathematics Magazine*, March 1985. Vol. 58, No. 2, pg. 107. Reprinted with permission.

This collection of thought-provoking and challenging exercises further explores and expands upon concepts learned in this chapter.

1. The columns of matrix T show the coordinates of the vertices of a triangle. Matrix A is a transformation matrix.

$$A = \begin{bmatrix} 0 & -1 \\ 1 & 0 \end{bmatrix} \qquad T = \begin{bmatrix} 1 & 2 & 3 \\ 1 & 4 & 2 \end{bmatrix}$$

(a) Find AT and AAT. Then sketch the original triangle and the two transformed triangles. What transformation does A represent?

(b) Given the triangle determined by AAT, describe the transformation process that produces the triangle determined by AT and then the triangle determined by T.

2. The matrices show the number of people (in thousands) who lived in each region of the United States in 2000 and the number of people (in thousands) projected to live in each region in 2015. The regional populations are separated into three age categories. (Source: U.S. Census Bureau)

2000

	0–17	18–64	65 +
Northeast	13,049	33,175	7,372
Midwest	16,646	39,486	8,263
South	25,569	62,235	12,437
Mountain	4,935	11,210	2,031
Pacific	12,098	28,036	4,893

2015

	0–17	18–64	65 +
Northeast	12,589	34,081	8,165
Midwest	15,886	41,038	10,101
South	25,916	68,998	17,470
Mountain	5,226	12,626	3,270
Pacific	14,906	33,296	6,565

(a) The total population in 2000 was 281,435,000 and the projected total population in 2015 is 310,133,000. Rewrite the matrices to give the information as percents of the total population.

(b) Write a matrix that gives the projected change in the percent of the population in each region and age group from 2000 to 2015.

(c) Based on the result of part (b), which region(s) and age group(s) are projected to show relative growth from 2000 to 2015?

3. Determine whether the matrix is idempotent. A square matrix is **idempotent** if $A^2 = A$.

(a) $\begin{bmatrix} 1 & 0 \\ 0 & 0 \end{bmatrix}$ (b) $\begin{bmatrix} 0 & 1 \\ 1 & 0 \end{bmatrix}$

(c) $\begin{bmatrix} 2 & 3 \\ -1 & -2 \end{bmatrix}$ (d) $\begin{bmatrix} 2 & 3 \\ 1 & 2 \end{bmatrix}$

4. Let $A = \begin{bmatrix} 1 & 2 \\ -2 & 1 \end{bmatrix}$.

(a) Show that $A^2 - 2A + 5I = O$, where I is the identity matrix of order 2.

(b) Show that $A^{-1} = \frac{1}{5}(2I - A)$.

(c) Show in general that for any square matrix satisfying

$$A^2 - 2A + 5I = O$$

the inverse of A is given by

$$A^{-1} = \frac{1}{5}(2I - A).$$

5. Two competing companies offer cable television to a city with 100,000 households. Gold Cable Company has 25,000 subscribers and Galaxy Cable Company has 30,000 subscribers. (The other 45,000 households do not subscribe.) The percent changes in cable subscriptions each year are shown in the matrix below.

Percent Changes

	From Gold	From Galaxy	From Non-subscriber
To Gold	0.70	0.15	0.15
To Galaxy	0.20	0.80	0.15
To Nonsubscriber	0.10	0.05	0.70

Percent Changes

(a) Find the number of subscribers each company will have in 1 year using matrix multiplication. Explain how you obtained your answer.

(b) Find the number of subscribers each company will have in 2 years using matrix multiplication. Explain how you obtained your answer.

(c) Find the number of subscribers each company will have in 3 years using matrix multiplication. Explain how you obtained your answer.

(d) What is happening to the number of subscribers to each company? What is happening to the number of nonsubscribers?

6. Find x such that the matrix is equal to its own inverse.

$$A = \begin{bmatrix} 3 & x \\ -2 & -3 \end{bmatrix}$$

7. Find x such that the matrix is singular.

$$A = \begin{bmatrix} 4 & x \\ -2 & -3 \end{bmatrix}$$

8. Find an example of a singular 2×2 matrix satisfying $A^2 = A$.

591

9. Verify the following equation.

$$\begin{vmatrix} 1 & 1 & 1 \\ a & b & c \\ a^2 & b^2 & c^2 \end{vmatrix} = (a - b)(b - c)(c - a)$$

10. Verify the following equation.

$$\begin{vmatrix} 1 & 1 & 1 \\ a & b & c \\ a^3 & b^3 & c^3 \end{vmatrix} = (a - b)(b - c)(c - a)(a + b + c)$$

11. Verify the following equation.

$$\begin{vmatrix} x & 0 & c \\ -1 & x & b \\ 0 & -1 & a \end{vmatrix} = ax^2 + bx + c$$

12. Use the equation given in Exercise 11 as a model to find a determinant that is equal to $ax^3 + bx^2 + cx + d$.

13. The atomic masses of three compounds are shown in the table. Use a linear system and Cramer's Rule to find the atomic masses of sulfur (S), nitrogen (N), and fluorine (F).

Compound	Formula	Atomic mass
Tetrasulphur tetranitride	S_4N_4	184
Sulfur hexafluoride	SF_6	146
Dinitrogen tetrafluoride	N_2F_4	104

14. A walkway lighting package includes a transformer, a certain length of wire, and a certain number of lights on the wire. The price of each lighting package depends on the length of wire and the number of lights on the wire. Use the following information to find the cost of a transformer, the cost per foot of wire, and the cost of a light. Assume that the cost of each item is the same in each lighting package.

- A package that contains a transformer, 25 feet of wire, and 5 lights costs $20.

- A package that contains a transformer, 50 feet of wire, and 15 lights costs $35.

- A package that contains a transformer, 100 feet of wire, and 20 lights costs $50.

15. The **transpose** of a matrix, denoted A^T, is formed by writing its columns as rows. Find the transpose of each matrix and verify that $(AB)^T = B^T A^T$.

$$A = \begin{bmatrix} -1 & 1 & -2 \\ 2 & 0 & 1 \end{bmatrix}, \quad B = \begin{bmatrix} -3 & 0 \\ 1 & 2 \\ 1 & -1 \end{bmatrix}$$

16. Use the inverse of matrix A to decode the cryptogram.

$$A = \begin{bmatrix} 1 & -2 & 2 \\ 1 & 1 & -3 \\ 1 & -1 & 4 \end{bmatrix}$$

23	13	-34	31	-34	63	25	-17	61
24	14	-37	41	-17	-8	20	-29	40
38	-56	116	13	-11	1	22	-3	-6
41	-53	85	28	-32	16			

17. A code breaker intercepted the encoded message below.

45	-35	38	-30	18	-18	35	-30	81	-60
42	-28	75	-55	2	-2	22	-21	15	
-10									

Let

$$A^{-1} = \begin{bmatrix} w & x \\ y & z \end{bmatrix}.$$

(a) You know that $[45 \quad -35]A^{-1} = [10 \quad 15]$ and that $[38 \quad -30]A^{-1} = [8 \quad 14]$, where A^{-1} is the inverse of the encoding matrix A. Write and solve two systems of equations to find w, x, y, and z.

(b) Decode the message.

 18. Let

$$A = \begin{bmatrix} 6 & 4 & 1 \\ 0 & 2 & 3 \\ 1 & 1 & 2 \end{bmatrix}.$$

Use a graphing utility to find A^{-1}. Compare $|A^{-1}|$ with $|A|$. Make a conjecture about the determinant of the inverse of a matrix.

19. Let A be an $n \times n$ matrix each of whose rows adds up to zero. Find $|A|$.

 20. Consider matrices of the form

$$A = \begin{bmatrix} 0 & a_{12} & a_{13} & a_{14} & \cdots & a_{1n} \\ 0 & 0 & a_{23} & a_{24} & \cdots & a_{2n} \\ 0 & 0 & 0 & a_{34} & \cdots & a_{3n} \\ \vdots & \vdots & \vdots & \vdots & \cdots & \vdots \\ 0 & 0 & 0 & 0 & \cdots & a_{(n-1)n} \\ 0 & 0 & 0 & 0 & \cdots & 0 \end{bmatrix}$$

(a) Write a 2×2 matrix and a 3×3 matrix in the form of A.

(b) Use a graphing utility to raise each of the matrices to higher powers. Describe the result.

(c) Use the result of part (b) to make a conjecture about powers of A if A is a 4×4 matrix. Use a graphing utility to test your conjecture.

(d) Use the results of parts (b) and (c) to make a conjecture about powers of A if A is an $n \times n$ matrix.

Sequences, Series, and Probability

8

Poker has become a popular card game in recent years. You can use the probability theory developed in this chapter to calculate the likelihood of getting different poker hands.

Jeff Greenberg/PhotoEdit, Inc.

SELECTED APPLICATIONS

Sequences, series, and probability have many real-life applications. The applications listed below represent a small sample of the applications in this chapter.

8.1 Sequences and Series

Scott Olson/Getty Images

Sequences

In mathematics, the word *sequence* is used in much the same way as in ordinary English. Saying that a collection is listed in *sequence* means that it is ordered so that it has a first member, a second member, a third member, and so on.

Mathematically, you can think of a sequence as a *function* whose domain is the set of positive integers.

$$f(1) = a_1, \ f(2) = a_2, \ f(3) = a_3, \ f(4) = a_4, \ldots, \ f(n) = a_n, \ldots$$

Rather than using function notation, however, sequences are usually written using subscript notation, as indicated in the following definition.

Definition of Sequence

An **infinite sequence** is a function whose domain is the set of positive integers. The function values

$$a_1, a_2, a_3, a_4, \ldots, a_n, \ldots$$

are the **terms** of the sequence. If the domain of the function consists of the first n positive integers only, the sequence is a **finite sequence**.

On occasion it is convenient to begin subscripting a sequence with 0 instead of 1 so that the terms of the sequence become $a_0, a_1, a_2, a_3, \ldots$.

Example 1 **Writing the Terms of a Sequence**

Write the first four terms of the sequences given by

a. $a_n = 3n - 2$ **b.** $a_n = 3 + (-1)^n$.

Solution

a. The first four terms of the sequence given by $a_n = 3n - 2$ are

$$a_1 = 3(1) - 2 = 1 \qquad \text{1st term}$$
$$a_2 = 3(2) - 2 = 4 \qquad \text{2nd term}$$
$$a_3 = 3(3) - 2 = 7 \qquad \text{3rd term}$$
$$a_4 = 3(4) - 2 = 10. \qquad \text{4th term}$$

b. The first four terms of the sequence given by $a_n = 3 + (-1)^n$ are

$$a_1 = 3 + (-1)^1 = 3 - 1 = 2 \qquad \text{1st term}$$
$$a_2 = 3 + (-1)^2 = 3 + 1 = 4 \qquad \text{2nd term}$$
$$a_3 = 3 + (-1)^3 = 3 - 1 = 2 \qquad \text{3rd term}$$
$$a_4 = 3 + (-1)^4 = 3 + 1 = 4. \qquad \text{4th term}$$

✓**CHECKPOINT** Now try Exercise 1.

Exploration

Write out the first five terms of the sequence whose nth term is

$$a_n = \frac{(-1)^{n+1}}{2n - 1}.$$

Are they the same as the first five terms of the sequence in Example 2? If not, how do they differ?

Example 2   **A Sequence Whose Terms Alternate in Sign**

Write the first five terms of the sequence given by $a_n = \dfrac{(-1)^n}{2n - 1}$.

Solution

The first five terms of the sequence are as follows.

$$a_1 = \frac{(-1)^1}{2(1) - 1} = \frac{-1}{2 - 1} = -1 \qquad \text{1st term}$$

$$a_2 = \frac{(-1)^2}{2(2) - 1} = \frac{1}{4 - 1} = \frac{1}{3} \qquad \text{2nd term}$$

$$a_3 = \frac{(-1)^3}{2(3) - 1} = \frac{-1}{6 - 1} = -\frac{1}{5} \qquad \text{3rd term}$$

$$a_4 = \frac{(-1)^4}{2(4) - 1} = \frac{1}{8 - 1} = \frac{1}{7} \qquad \text{4th term}$$

$$a_5 = \frac{(-1)^5}{2(5) - 1} = \frac{-1}{10 - 1} = -\frac{1}{9} \qquad \text{5th term}$$

✔**CHECKPOINT**   Now try Exercise 17.

Simply listing the first few terms is not sufficient to define a unique sequence—the nth term *must be given*. To see this, consider the following sequences, both of which have the same first three terms.

$$\frac{1}{2}, \frac{1}{4}, \frac{1}{8}, \frac{1}{16}, \dots, \frac{1}{2^n}, \dots$$

$$\frac{1}{2}, \frac{1}{4}, \frac{1}{8}, \frac{1}{15}, \dots, \frac{6}{(n + 1)(n^2 - n + 6)}, \dots$$

Example 3   **Finding the nth Term of a Sequence**

Write an expression for the apparent nth term (a_n) of each sequence.

a. $1, 3, 5, 7, \dots$ **b.** $2, -5, 10, -17, \dots$

Solution

a. n: 1 2 3 4 . . . n
   Terms: 1 3 5 7 . . . a_n
   Apparent pattern: Each term is 1 less than twice n, which implies that

$$a_n = 2n - 1.$$

b. n: 1   2   3   4 . . . n
   Terms: 2 -5 10 -17 . . . a_n
   Apparent pattern: The terms have alternating signs with those in the even positions being negative. Each term is 1 more than the square of n, which implies that

$$a_n = (-1)^{n+1}(n^2 + 1)$$

✔**CHECKPOINT**   Now try Exercise 37.

Technology

To graph a sequence using a graphing utility, set the mode to *sequence* and *dot* and enter the sequence. The graph of the sequence in Example 3(a) is shown below. You can use the *trace* feature or *value* feature to identify the terms.

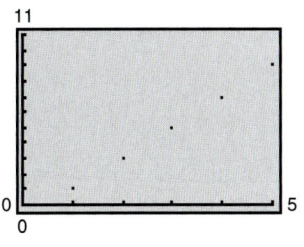

Some sequences are defined **recursively.** To define a sequence recursively, you need to be given one or more of the first few terms. All other terms of the sequence are then defined using previous terms. A well-known example is the Fibonacci sequence shown in Example 4.

Example 4   The Fibonacci Sequence: A Recursive Sequence

The Fibonacci sequence is defined recursively, as follows.

$$a_0 = 1, \ a_1 = 1, \ a_k = a_{k-2} + a_{k-1}, \text{ where } k \geq 2$$

Write the first six terms of this sequence.

Solution

$a_0 = 1$	0th term is given.
$a_1 = 1$	1st term is given.
$a_2 = a_{2-2} + a_{2-1} = a_0 + a_1 = 1 + 1 = 2$	Use recursion formula.
$a_3 = a_{3-2} + a_{3-1} = a_1 + a_2 = 1 + 2 = 3$	Use recursion formula.
$a_4 = a_{4-2} + a_{4-1} = a_2 + a_3 = 2 + 3 = 5$	Use recursion formula.
$a_5 = a_{5-2} + a_{5-1} = a_3 + a_4 = 3 + 5 = 8$	Use recursion formula.

✓**CHECKPOINT**   Now try Exercise 51.

Factorial Notation

Some very important sequences in mathematics involve terms that are defined with special types of products called **factorials.**

Definition of Factorial

If n is a positive integer, **n factorial** is defined as

$$n! = 1 \cdot 2 \cdot 3 \cdot 4 \cdots (n-1) \cdot n.$$

As a special case, zero factorial is defined as $0! = 1$.

Here are some values of $n!$ for the first several nonnegative integers. Notice that $0!$ is 1 by definition.

$0! = 1$

$1! = 1$

$2! = 1 \cdot 2 = 2$

$3! = 1 \cdot 2 \cdot 3 = 6$

$4! = 1 \cdot 2 \cdot 3 \cdot 4 = 24$

$5! = 1 \cdot 2 \cdot 3 \cdot 4 \cdot 5 = 120$

The value of n does not have to be very large before the value of $n!$ becomes extremely large. For instance, $10! = 3,628,800$.

Factorials follow the same conventions for order of operations as do exponents. For instance,

$$2n! = 2(n!) = 2(1 \cdot 2 \cdot 3 \cdot 4 \cdots n)$$

whereas $(2n)! = 1 \cdot 2 \cdot 3 \cdot 4 \cdots 2n$.

Example 5   Writing the Terms of a Sequence Involving Factorials

Write the first five terms of the sequence given by

$$a_n = \frac{2^n}{n!}.$$

Begin with $n = 0$. Then graph the terms on a set of coordinate axes.

Solution

$$a_0 = \frac{2^0}{0!} = \frac{1}{1} = 1 \qquad \text{0th term}$$

$$a_1 = \frac{2^1}{1!} = \frac{2}{1} = 2 \qquad \text{1st term}$$

$$a_2 = \frac{2^2}{2!} = \frac{4}{2} = 2 \qquad \text{2nd term}$$

$$a_3 = \frac{2^3}{3!} = \frac{8}{6} = \frac{4}{3} \qquad \text{3rd term}$$

$$a_4 = \frac{2^4}{4!} = \frac{16}{24} = \frac{2}{3} \qquad \text{4th term}$$

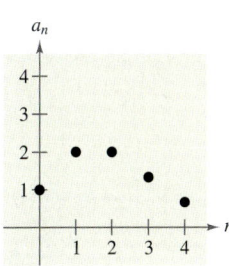

FIGURE **8.1**

Figure 8.1 shows the first five terms of the sequence.

✓**CHECKPOINT**   Now try Exercise 59.

When working with fractions involving factorials, you will often find that the fractions can be reduced to simplify the computations.

Example 6   Evaluating Factorial Expressions

Evaluate each factorial expression.

a. $\dfrac{8!}{2! \cdot 6!}$ **b.** $\dfrac{2! \cdot 6!}{3! \cdot 5!}$ **c.** $\dfrac{n!}{(n-1)!}$

Solution

a. $\dfrac{8!}{2! \cdot 6!} = \dfrac{1 \cdot 2 \cdot 3 \cdot 4 \cdot 5 \cdot 6 \cdot 7 \cdot 8}{1 \cdot 2 \cdot 1 \cdot 2 \cdot 3 \cdot 4 \cdot 5 \cdot 6} = \dfrac{7 \cdot 8}{2} = 28$

b. $\dfrac{2! \cdot 6!}{3! \cdot 5!} = \dfrac{1 \cdot 2 \cdot 1 \cdot 2 \cdot 3 \cdot 4 \cdot 5 \cdot 6}{1 \cdot 2 \cdot 3 \cdot 1 \cdot 2 \cdot 3 \cdot 4 \cdot 5} = \dfrac{6}{3} = 2$

c. $\dfrac{n!}{(n-1)!} = \dfrac{1 \cdot 2 \cdot 3 \cdots (n-1) \cdot n}{1 \cdot 2 \cdot 3 \cdots (n-1)} = n$

✓**CHECKPOINT**   Now try Exercise 69.

STUDY TIP

Note in Example 6(a) that you can simplify the computation as follows.

$$\frac{8!}{2! \cdot 6!} = \frac{8 \cdot 7 \cdot 6!}{2! \cdot 6!}$$

$$= \frac{8 \cdot 7}{2 \cdot 1} = 28$$

Technology

Most graphing utilities are able to sum the first n terms of a sequence. Check your user's guide for a *sum sequence* feature or a *series* feature.

Summation Notation

There is a convenient notation for the sum of the terms of a finite sequence. It is called **summation notation** or **sigma notation** because it involves the use of the uppercase Greek letter sigma, written as Σ.

> ### Definition of Summation Notation
>
> The sum of the first n terms of a sequence is represented by
>
> $$\sum_{i=1}^{n} a_i = a_1 + a_2 + a_3 + a_4 + \cdots + a_n$$
>
> where i is called the **index of summation,** n is the **upper limit of summation,** and 1 is the **lower limit of summation.**

STUDY TIP

Summation notation is an instruction to add the terms of a sequence. From the definition at the right, the upper limit of summation tells you where to end the sum. Summation notation helps you generate the appropriate terms of the sequence prior to finding the actual sum, which may be unclear.

Example 7   Summation Notation for Sums

Find each sum.

a. $\displaystyle\sum_{i=1}^{5} 3i$ **b.** $\displaystyle\sum_{k=3}^{6} (1 + k^2)$ **c.** $\displaystyle\sum_{i=0}^{8} \frac{1}{i!}$

Solution

a. $\displaystyle\sum_{i=1}^{5} 3i = 3(1) + 3(2) + 3(3) + 3(4) + 3(5)$

$\qquad\qquad = 3(1 + 2 + 3 + 4 + 5)$

$\qquad\qquad = 3(15)$

$\qquad\qquad = 45$

b. $\displaystyle\sum_{k=3}^{6} (1 + k^2) = (1 + 3^2) + (1 + 4^2) + (1 + 5^2) + (1 + 6^2)$

$\qquad\qquad\qquad = 10 + 17 + 26 + 37$

$\qquad\qquad\qquad = 90$

c. $\displaystyle\sum_{i=0}^{8} \frac{1}{i!} = \frac{1}{0!} + \frac{1}{1!} + \frac{1}{2!} + \frac{1}{3!} + \frac{1}{4!} + \frac{1}{5!} + \frac{1}{6!} + \frac{1}{7!} + \frac{1}{8!}$

$\qquad\quad = 1 + 1 + \frac{1}{2} + \frac{1}{6} + \frac{1}{24} + \frac{1}{120} + \frac{1}{720} + \frac{1}{5040} + \frac{1}{40{,}320}$

$\qquad\quad \approx 2.71828$

For this summation, note that the sum is very close to the irrational number $e \approx 2.718281828$. It can be shown that as more terms of the sequence whose nth term is $1/n!$ are added, the sum becomes closer and closer to e.

✓**CHECKPOINT**   Now try Exercise 73.

In Example 7, note that the lower limit of a summation does not have to be 1. Also note that the index of summation does not have to be the letter i. For instance, in part (b), the letter k is the index of summation.

Variations in the upper and lower limits of summation can produce quite different-looking summation notations for *the same sum*. For example, the following two sums have the same terms.

$$\sum_{i=1}^{3} 3(2^i) = 3(2^1 + 2^2 + 2^3)$$

$$\sum_{i=0}^{2} 3(2^{i+1}) = 3(2^1 + 2^2 + 2^3)$$

Properties of Sums

1. $\sum_{i=1}^{n} c = cn,$ c is a constant.   2. $\sum_{i=1}^{n} ca_i = c \sum_{i=1}^{n} a_i,$ c is a constant.

3. $\sum_{i=1}^{n} (a_i + b_i) = \sum_{i=1}^{n} a_i + \sum_{i=1}^{n} b_i$ 4. $\sum_{i=1}^{n} (a_i - b_i) = \sum_{i=1}^{n} a_i - \sum_{i=1}^{n} b_i$

For proofs of these properties, see Proofs in Mathematics on page 674.

Series

Many applications involve the sum of the terms of a finite or infinite sequence. Such a sum is called a **series.**

Definition of Series

Consider the infinite sequence $a_1, a_2, a_3, \ldots, a_i, \ldots$.

1. The sum of the first n terms of the sequence is called a **finite series** or the **nth partial sum** of the sequence and is denoted by

$$a_1 + a_2 + a_3 + \cdots + a_n = \sum_{i=1}^{n} a_i.$$

2. The sum of all the terms of the infinite sequence is called an **infinite series** and is denoted by

$$a_1 + a_2 + a_3 + \cdots + a_i + \cdots = \sum_{i=1}^{\infty} a_i.$$

Example 8   **Finding the Sum of a Series**

For the series $\sum_{i=1}^{\infty} \dfrac{3}{10^i}$, find (a) the third partial sum and (b) the sum.

Solution

a. The third partial sum is

$$\sum_{i=1}^{3} \frac{3}{10^i} = \frac{3}{10^1} + \frac{3}{10^2} + \frac{3}{10^3} = 0.3 + 0.03 + 0.003 = 0.333.$$

b. The sum of the series is

$$\sum_{i=1}^{\infty} \frac{3}{10^i} = \frac{3}{10^1} + \frac{3}{10^2} + \frac{3}{10^3} + \frac{3}{10^4} + \frac{3}{10^5} + \cdots$$

$$= 0.3 + 0.03 + 0.003 + 0.0003 + 0.00003 + \cdots$$

$$= 0.33333\ldots = \frac{1}{3}.$$

✔CHECKPOINT   Now try Exercise 99.

Application

Sequences have many applications in business and science. One such application is illustrated in Example 9.

Example 9 Population of the United States

For the years 1980 to 2003, the resident population of the United States can be approximated by the model

$$a_n = 226.9 + 2.05n + 0.035n^2, \qquad n = 0, 1, \ldots, 23$$

where a_n is the population (in millions) and n represents the year, with $n = 0$ corresponding to 1980. Find the last five terms of this finite sequence, which represent the U.S. population for the years 1999 to 2003. (Source: U.S. Census Bureau)

Solution

The last five terms of this finite sequence are as follows.

$a_{19} = 226.9 + 2.05(19) + 0.035(19)^2 \approx 278.5$ 1999 population

$a_{20} = 226.9 + 2.05(20) + 0.035(20)^2 = 281.9$ 2000 population

$a_{21} = 226.9 + 2.05(21) + 0.035(21)^2 \approx 285.4$ 2001 population

$a_{22} = 226.9 + 2.05(22) + 0.035(22)^2 \approx 288.9$ 2002 population

$a_{23} = 226.9 + 2.05(23) + 0.035(23)^2 \approx 292.6$ 2003 population

✔CHECKPOINT   Now try Exercise 111.

Exploration

A $3 \times 3 \times 3$ cube is created using 27 unit cubes (a unit cube has a length, width, and height of 1 unit) and only the faces of each cube that are visible are painted blue (see Figure 8.2). Complete the table below to determine how many unit cubes of the $3 \times 3 \times 3$ cube have 0 blue faces, 1 blue face, 2 blue faces, and 3 blue faces. Do the same for a $4 \times 4 \times 4$ cube, a $5 \times 5 \times 5$ cube, and a $6 \times 6 \times 6$ cube and add your results to the table below. What type of pattern do you observe in the table? Write a formula you could use to determine the column values for an $n \times n \times n$ cube.

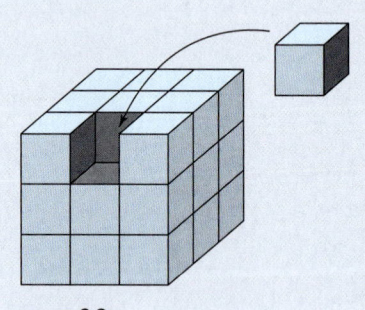

FIGURE 8.2

Number of blue cube faces	0	1	2	3
$3 \times 3 \times 3$				

8.1 | **Exercises**   The *HM mathSpace®* CD-ROM and *Eduspace®* for this text contain step-by-step solutions to all odd-numbered exercises. They also provide Tutorial Exercises for additional help.

VOCABULARY CHECK: Fill in the blanks.

1. An _____ _____ is a function whose domain is the set of positive integers.

2. The function values $a_1, a_2, a_3, a_4, \ldots$ are called the _____ of a sequence.

3. A sequence is a _____ sequence if the domain of the function consists of the first n positive integers.

4. If you are given one or more of the first few terms of a sequence, and all other terms of the sequence are defined using previous terms, then the sequence is said to be defined _____.

5. If n is a positive integer, n _____ is defined as $n! = 1 \cdot 2 \cdot 3 \cdot 4 \cdots (n-1) \cdot n$.

6. The notation used to represent the sum of the terms of a finite sequence is _____ _____ or sigma notation.

7. For the sum $\displaystyle\sum_{i=1}^{n} a_i$, i is called the _____ of summation, n is the _____ limit of summation, and 1 is the _____ limit of summation.

8. The sum of the terms of a finite or infinite sequence is called a _____.

9. The _____ _____ _____ of a sequence is the sum of the first n terms of the sequence.

PREREQUISITE SKILLS REVIEW: Practice and review algebra skills needed for this section at **www.Eduspace.com.**

In Exercises 1–22, write the first five terms of the sequence. 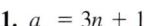 (Assume that n begins with 1.)

1. $a_n = 3n + 1$

2. $a_n = 5n - 3$

3. $a_n = 2^n$

4. $a_n = \left(\frac{1}{2}\right)^n$

5. $a_n = (-2)^n$

6. $a_n = \left(-\frac{1}{2}\right)^n$

7. $a_n = \dfrac{n+2}{n}$

8. $a_n = \dfrac{n}{n+2}$

9. $a_n = \dfrac{6n}{3n^2 - 1}$

10. $a_n = \dfrac{3n^2 - n + 4}{2n^2 + 1}$

11. $a_n = \dfrac{1 + (-1)^n}{n}$

12. $a_n = 1 + (-1)^n$

13. $a_n = 2 - \dfrac{1}{3^n}$

14. $a_n = \dfrac{2^n}{3^n}$

15. $a_n = \dfrac{1}{n^{3/2}}$

16. $a_n = \dfrac{10}{n^{2/3}}$

17. $a_n = \dfrac{(-1)^n}{n^2}$

18. $a_n = (-1)^n \left(\dfrac{n}{n+1}\right)$

19. $a_n = \frac{2}{3}$

20. $a_n = 0.3$

21. $a_n = n(n-1)(n-2)$

22. $a_n = n(n^2 - 6)$

In Exercises 23–26, find the indicated term of the sequence.

23. $a_n = (-1)^n(3n - 2)$

$a_{25} = \rule{1.5cm}{0.3pt}$

24. $a_n = (-1)^{n-1}[n(n-1)]$

$a_{16} = \rule{1.5cm}{0.3pt}$

25. $a_n = \dfrac{4n}{2n^2 - 3}$

$a_{11} = \rule{1.5cm}{0.3pt}$

26. $a_n = \dfrac{4n^2 - n + 3}{n(n-1)(n+2)}$

$a_{13} = \rule{1.5cm}{0.3pt}$

In Exercises 27–32, use a graphing utility to graph the first 10 terms of the sequence. (Assume that n begins with 1.)

27. $a_n = \dfrac{3}{4}n$

28. $a_n = 2 - \dfrac{4}{n}$

29. $a_n = 16(-0.5)^{n-1}$

30. $a_n = 8(0.75)^{n-1}$

31. $a_n = \dfrac{2n}{n+1}$

32. $a_n = \dfrac{n^2}{n^2 + 2}$

In Exercises 33–36, match the sequence with the graph of its first 10 terms. [The graphs are labeled (a), (b), (c), and (d).]

(a)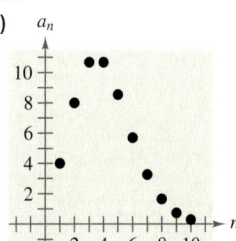

(b)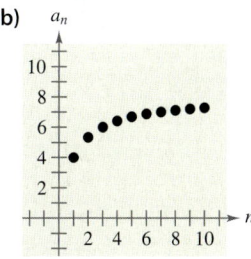

(c)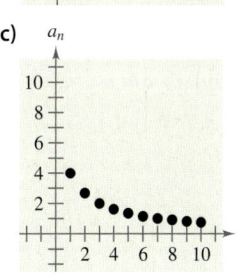

(d)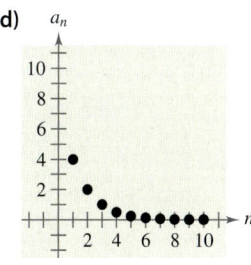

33. $a_n = \dfrac{8}{n+1}$

34. $a_n = \dfrac{8n}{n+1}$

35. $a_n = 4(0.5)^{n-1}$

36. $a_n = \dfrac{4^n}{n!}$

In Exercises 37–50, write an expression for the apparent nth term of the sequence. (Assume that n begins with 1.)

37. $1, 4, 7, 10, 13, \ldots$

38. $3, 7, 11, 15, 19, \ldots$

39. $0, 3, 8, 15, 24, \ldots$

40. $2, -4, 6, -8, 10, \ldots$

41. $\frac{-2}{3}, \frac{3}{4}, \frac{-4}{5}, \frac{5}{6}, \frac{-6}{7}, \ldots$

42. $\frac{1}{2}, \frac{-1}{4}, \frac{1}{8}, \frac{-1}{16}, \ldots$

43. $\frac{2}{1}, \frac{3}{3}, \frac{4}{5}, \frac{5}{7}, \frac{6}{9}, \ldots$

44. $\frac{1}{3}, \frac{2}{9}, \frac{4}{27}, \frac{8}{81}, \ldots$

45. $1, \frac{1}{4}, \frac{1}{9}, \frac{1}{16}, \frac{1}{25}, \ldots$

46. $1, \frac{1}{2}, \frac{1}{6}, \frac{1}{24}, \frac{1}{120}, \ldots$

47. $1, -1, 1, -1, 1, \ldots$

48. $1, 2, \frac{2^2}{2}, \frac{2^3}{6}, \frac{2^4}{24}, \frac{2^5}{120}, \ldots$

49. $1 + \frac{1}{1}, 1 + \frac{1}{2}, 1 + \frac{1}{3}, 1 + \frac{1}{4}, 1 + \frac{1}{5}, \ldots$

50. $1 + \frac{1}{2}, 1 + \frac{3}{4}, 1 + \frac{7}{8}, 1 + \frac{15}{16}, 1 + \frac{31}{32}, \ldots$

In Exercises 51–54, write the first five terms of the sequence defined recursively.

51. $a_1 = 28, \quad a_{k+1} = a_k - 4$

52. $a_1 = 15, \quad a_{k+1} = a_k + 3$

53. $a_1 = 3, \quad a_{k+1} = 2(a_k - 1)$

54. $a_1 = 32, \quad a_{k+1} = \frac{1}{2}a_k$

In Exercises 55–58, write the first five terms of the sequence defined recursively. Use the pattern to write the nth term of the sequence as a function of n. (Assume that n begins with 1.)

55. $a_1 = 6, \quad a_{k+1} = a_k + 2$

56. $a_1 = 25, \quad a_{k+1} = a_k - 5$

57. $a_1 = 81, \quad a_{k+1} = \frac{1}{3}a_k$

58. $a_1 = 14, \quad a_{k+1} = (-2)a_k$

In Exercises 59–64, write the first five terms of the sequence. (Assume that n begins with 0.)

59. $a_n = \dfrac{3^n}{n!}$

60. $a_n = \dfrac{n!}{n}$

61. $a_n = \dfrac{1}{(n + 1)!}$

62. $a_n = \dfrac{n^2}{(n + 1)!}$

63. $a_n = \dfrac{(-1)^{2n}}{(2n)!}$

64. $a_n = \dfrac{(-1)^{2n+1}}{(2n + 1)!}$

In Exercises 65–72, simplify the factorial expression.

65. $\dfrac{4!}{6!}$

66. $\dfrac{5!}{8!}$

67. $\dfrac{10!}{8!}$

68. $\dfrac{25!}{23!}$

69. $\dfrac{(n + 1)!}{n!}$

70. $\dfrac{(n + 2)!}{n!}$

71. $\dfrac{(2n - 1)!}{(2n + 1)!}$

72. $\dfrac{(3n + 1)!}{(3n)!}$

In Exercises 73–84, find the sum.

73. $\sum\limits_{i=1}^{5} (2i + 1)$

74. $\sum\limits_{i=1}^{6} (3i - 1)$

75. $\sum\limits_{k=1}^{4} 10$

76. $\sum\limits_{k=1}^{5} 5$

77. $\sum\limits_{i=0}^{4} i^2$

78. $\sum\limits_{i=0}^{5} 2i^2$

79. $\sum\limits_{k=0}^{3} \dfrac{1}{k^2 + 1}$

80. $\sum\limits_{j=3}^{5} \dfrac{1}{j^2 - 3}$

81. $\sum\limits_{k=2}^{5} (k + 1)^2(k - 3)$

82. $\sum\limits_{i=1}^{4} [(i - 1)^2 + (i + 1)^3]$

83. $\sum\limits_{i=1}^{4} 2^i$

84. $\sum\limits_{j=0}^{4} (-2)^j$

 In Exercises 85–88, use a calculator to find the sum.

85. $\sum\limits_{j=1}^{6} (24 - 3j)$

86. $\sum\limits_{j=1}^{10} \dfrac{3}{j + 1}$

87. $\sum\limits_{k=0}^{4} \dfrac{(-1)^k}{k + 1}$

88. $\sum\limits_{k=0}^{4} \dfrac{(-1)^k}{k!}$

In Exercises 89–98, use sigma notation to write the sum.

89. $\dfrac{1}{3(1)} + \dfrac{1}{3(2)} + \dfrac{1}{3(3)} + \cdots + \dfrac{1}{3(9)}$

90. $\dfrac{5}{1 + 1} + \dfrac{5}{1 + 2} + \dfrac{5}{1 + 3} + \cdots + \dfrac{5}{1 + 15}$

91. $\left[2\left(\frac{1}{8}\right) + 3\right] + \left[2\left(\frac{2}{8}\right) + 3\right] + \cdots + \left[2\left(\frac{8}{8}\right) + 3\right]$

92. $\left[1 - \left(\frac{1}{6}\right)^2\right] + \left[1 - \left(\frac{2}{6}\right)^2\right] + \cdots + \left[1 - \left(\frac{6}{6}\right)^2\right]$

93. $3 - 9 + 27 - 81 + 243 - 729$

94. $1 - \frac{1}{2} + \frac{1}{4} - \frac{1}{8} + \cdots - \frac{1}{128}$

95. $\dfrac{1}{1^2} - \dfrac{1}{2^2} + \dfrac{1}{3^2} - \dfrac{1}{4^2} + \cdots - \dfrac{1}{20^2}$

96. $\dfrac{1}{1 \cdot 3} + \dfrac{1}{2 \cdot 4} + \dfrac{1}{3 \cdot 5} + \cdots + \dfrac{1}{10 \cdot 12}$

97. $\frac{1}{4} + \frac{3}{8} + \frac{7}{16} + \frac{15}{32} + \frac{31}{64}$

98. $\frac{1}{2} + \frac{2}{4} + \frac{6}{8} + \frac{24}{16} + \frac{120}{32} + \frac{720}{64}$

In Exercises 99–102, find the indicated partial sum of the series.

99. $\sum\limits_{i=1}^{\infty} 5\left(\frac{1}{2}\right)^i$

Fourth partial sum

100. $\sum\limits_{i=1}^{\infty} 2\left(\frac{1}{3}\right)^i$

Fifth partial sum

101. $\sum\limits_{n=1}^{\infty} 4\left(-\frac{1}{2}\right)^n$

Third partial sum

102. $\sum\limits_{n=1}^{\infty} 8\left(-\frac{1}{4}\right)^n$

Fourth partial sum

In Exercises 103–106, find the sum of the infinite series.

103. $\displaystyle\sum_{i=1}^{\infty} 6\left(\tfrac{1}{10}\right)^i$

104. $\displaystyle\sum_{k=1}^{\infty} \left(\tfrac{1}{10}\right)^k$

105. $\displaystyle\sum_{k=1}^{\infty} 7\left(\tfrac{1}{10}\right)^k$

106. $\displaystyle\sum_{i=1}^{\infty} 2\left(\tfrac{1}{10}\right)^i$

107. **Compound Interest**   A deposit of $5000 is made in an account that earns 8% interest compounded quarterly. The balance in the account after n quarters is given by

$$A_n = 5000\left(1 + \frac{0.08}{4}\right)^n, \quad n = 1, 2, 3, \ldots .$$

   (a) Write the first eight terms of this sequence.

   (b) Find the balance in this account after 10 years by finding the 40th term of the sequence.

108. **Compound Interest**   A deposit of $100 is made each month in an account that earns 12% interest compounded monthly. The balance in the account after n months is given by

$$A_n = 100(101)[(1.01)^n - 1], \quad n = 1, 2, 3, \ldots .$$

   (a) Write the first six terms of this sequence.

   (b) Find the balance in this account after 5 years by finding the 60th term of the sequence.

   (c) Find the balance in this account after 20 years by finding the 240th term of the sequence.

Model It

109. **Data Analysis: Number of Stores**   The table shows the numbers a_n of Best Buy stores for the years 1998 to 2003.   (Source: Best Buy Company, Inc.)

Year	Number of stores, a_n
1998	311
1999	357
2000	419
2001	481
2002	548
2003	608

Model It (continued)

   (a) Use the *regression* feature of a graphing utility to find a linear sequence that models the data. Let n represent the year, with $n = 8$ corresponding to 1998.

   (b) Use the *regression* feature of a graphing utility to find a quadratic sequence that models the data.

   (c) Evaluate the sequences from parts (a) and (b) for $n = 8, 9, \ldots, 13$. Compare these values with those shown in the table. Which model is a better fit for the data? Explain.

   (d) Which model do you think would better predict the number of Best Buy stores in the future? Use the model you chose to predict the number of Best Buy stores in 2008.

---

110. **Medicine**   The numbers a_n (in thousands) of AIDS cases reported from 1995 to 2003 can be approximated by the model

$$a_n = 0.0457n^3 - 0.352n^2 - 9.05n + 121.4,$$
$$n = 5, 6, \ldots, 13$$

   where n is the year, with $n = 5$ corresponding to 1995. (Source: U.S. Centers for Disease Control and Prevention)

   (a) Find the terms of this finite sequence. Use the *statistical plotting* feature of a graphing utility to construct a bar graph that represents the sequence.

   (b) What does the graph in part (a) say about reported cases of AIDS?

111. **Federal Debt**   From 1990 to 2003, the federal debt of the United States rose from just over $3 trillion to almost $7 trillion. The federal debt a_n (in billions of dollars) from 1990 to 2003 is approximated by the model

$$a_n = 2.7698n^3 - 61.372n^2 + 600.00n + 3102.9,$$
$$n = 0, 1, \ldots, 13$$

   where n is the year, with $n = 0$ corresponding to 1990. (Source: U.S. Office of Management and Budget)

   (a) Find the terms of this finite sequence. Use the *statistical plotting* feature of a graphing utility to construct a bar graph that represents the sequence.

   (b) What does the pattern in the bar graph in part (a) say about the future of the federal debt?

112. Revenue The revenues a_n (in millions of dollars) for Amazon.com for the years 1996 through 2003 are shown in the figure. The revenues can be approximated by the model

$$a_n = 46.609n^2 - 119.84n - 1125.8, \quad n = 6, 7, \ldots, 13$$

where n is the year, with $n = 6$ corresponding to 1996. Use this model to approximate the total revenue from 1996 through 2003. Compare this sum with the result of adding the revenues shown in the figure. (Source: Amazon.com)

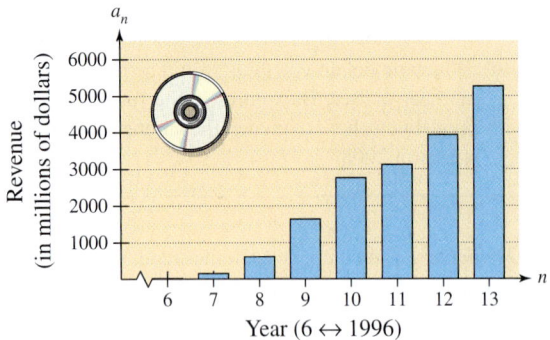

Year (6 ↔ 1996)

Synthesis

True or False? In Exercises 113 and 114, determine whether the statement is true or false. Justify your answer.

113. $\displaystyle\sum_{i=1}^{4} (i^2 + 2i) = \sum_{i=1}^{4} i^2 + 2\sum_{i=1}^{4} i$ **114.** $\displaystyle\sum_{j=1}^{4} 2^j = \sum_{j=3}^{6} 2^{j-2}$

Fibonacci Sequence In Exercises 115 and 116, use the Fibonacci sequence. (See Example 4.)

115. Write the first 12 terms of the Fibonacci sequence a_n and the first 10 terms of the sequence given by

$$b_n = \frac{a_{n+1}}{a_n}, \quad n \geq 1.$$

116. Using the definition for b_n in Exercise 115, show that b_n can be defined recursively by

$$b_n = 1 + \frac{1}{b_{n-1}}.$$

Arithmetic Mean In Exercises 117–120, use the following definition of the arithmetic mean $\bar{x}$ of a set of n measurements $x_1, x_2, x_3, \ldots, x_n$.

$$\bar{x} = \frac{1}{n} \sum_{i=1}^{n} x_i$$

117. Find the arithmetic mean of the six checking account balances $327.15, $785.69, $433.04, $265.38, $604.12, and $590.30. Use the statistical capabilities of a graphing utility to verify your result.

118. Find the arithmetic mean of the following prices per gallon for regular unleaded gasoline at five gasoline stations in a city: $1.899, $1.959, $1.919, $1.939, and $1.999. Use the statistical capabilities of a graphing utility to verify your result.

119. Proof Prove that $\displaystyle\sum_{i=1}^{n} (x_i - \bar{x}) = 0$.

120. Proof Prove that $\displaystyle\sum_{i=1}^{n} (x_i - \bar{x})^2 = \sum_{i=1}^{n} x_i^2 - \frac{1}{n}\left(\sum_{i=1}^{n} x_i\right)^2$.

In Exercises 121–124, find the first five terms of the sequence.

121. $a_n = \dfrac{x^n}{n!}$ **122.** $a_n = \dfrac{(-1)^n x^{2n+1}}{2n+1}$

123. $a_n = \dfrac{(-1)^n x^{2n}}{(2n)!}$ **124.** $a_n = \dfrac{(-1)^n x^{2n+1}}{(2n+1)!}$

Skills Review

In Exercises 125–128, determine whether the function has an inverse function. If it does, find its inverse function.

125. $f(x) = 4x - 3$ **126.** $g(x) = \dfrac{3}{x}$

127. $h(x) = \sqrt{5x + 1}$ **128.** $f(x) = (x - 1)^2$

In Exercises 129–132, find (a) $A - B$, (b) $4B - 3A$, (c) AB, and (d) BA.

129. $A = \begin{bmatrix} 6 & 5 \\ 4 & 3 \end{bmatrix}$, $B = \begin{bmatrix} -2 & 4 \\ 6 & -3 \end{bmatrix}$

130. $A = \begin{bmatrix} 10 & 7 \\ -4 & 6 \end{bmatrix}$, $B = \begin{bmatrix} 0 & -12 \\ 8 & 11 \end{bmatrix}$

131. $A = \begin{bmatrix} -2 & -3 & 6 \\ 4 & 5 & 7 \\ 1 & 7 & 4 \end{bmatrix}$, $B = \begin{bmatrix} 1 & 4 & 2 \\ 0 & 1 & 6 \\ 0 & 3 & 1 \end{bmatrix}$

132. $A = \begin{bmatrix} -1 & 4 & 0 \\ 5 & 1 & 2 \\ 0 & -1 & 3 \end{bmatrix}$, $B = \begin{bmatrix} 0 & 4 & 0 \\ 3 & 1 & -2 \\ -1 & 0 & 2 \end{bmatrix}$

In Exercises 133–136, find the determinant of the matrix.

133. $A = \begin{bmatrix} 3 & 5 \\ -1 & 7 \end{bmatrix}$ **134.** $A = \begin{bmatrix} -2 & 8 \\ 12 & 15 \end{bmatrix}$

135. $A = \begin{bmatrix} 3 & 4 & 5 \\ 0 & 7 & 3 \\ 4 & 9 & -1 \end{bmatrix}$

136. $A = \begin{bmatrix} 16 & 11 & 10 & 2 \\ 9 & 8 & 3 & 7 \\ -2 & -1 & 12 & 3 \\ -4 & 6 & 2 & 1 \end{bmatrix}$

8.2   Arithmetic Sequences and Partial Sums

© mediacolor's Alamy

Arithmetic Sequences

A sequence whose consecutive terms have a common difference is called an **arithmetic sequence.**

Definition of Arithmetic Sequence

A sequence is **arithmetic** if the differences between consecutive terms are the same. So, the sequence

$$a_1, a_2, a_3, a_4, \ldots, a_n, \ldots$$

is arithmetic if there is a number d such that

$$a_2 - a_1 = a_3 - a_2 = a_4 - a_3 = \cdots = d.$$

The number d is the **common difference** of the arithmetic sequence.

Example 1   Examples of Arithmetic Sequences

a. The sequence whose nth term is $4n + 3$ is arithmetic. For this sequence, the common difference between consecutive terms is 4.

$$7, 11, 15, 19, \ldots, 4n + 3, \ldots \qquad \text{Begin with } n = 1.$$

$$11 - 7 = 4$$

b. The sequence whose nth term is $7 - 5n$ is arithmetic. For this sequence, the common difference between consecutive terms is -5.

$$2, -3, -8, -13, \ldots, 7 - 5n, \ldots \qquad \text{Begin with } n = 1.$$

$$-3 - 2 = -5$$

c. The sequence whose nth term is $\frac{1}{4}(n + 3)$ is arithmetic. For this sequence, the common difference between consecutive terms is $\frac{1}{4}$.

$$1, \frac{5}{4}, \frac{3}{2}, \frac{7}{4}, \ldots, \frac{n + 3}{4}, \ldots \qquad \text{Begin with } n = 1.$$

$$\tfrac{5}{4} - 1 = \tfrac{1}{4}$$

✓**CHECKPOINT**   Now try Exercise 1.

The sequence $1, 4, 9, 16, \ldots$, whose nth term is n^2, is *not* arithmetic. The difference between the first two terms is

$$a_2 - a_1 = 4 - 1 = 3$$

but the difference between the second and third terms is

$$a_3 - a_2 = 9 - 4 = 5.$$

In Example 1, notice that each of the arithmetic sequences has an nth term that is of the form $dn + c$, where the common difference of the sequence is d. An arithmetic sequence may be thought of as a linear function whose domain is the set of natural numbers.

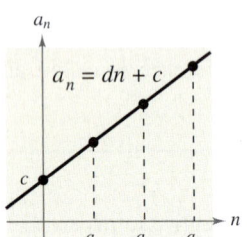

FIGURE 8.3

The nth Term of an Arithmetic Sequence

The nth term of an arithmetic sequence has the form

$$a_n = dn + c \qquad \text{Linear form}$$

where d is the common difference between consecutive terms of the sequence and $c = a_1 - d$. A graphical representation of this definition is shown in Figure 8.3. Substituting $a_1 - d$ for c in $a_n = dn + c$ yields an alternative *recursion* form for the nth term of an arithmetic sequence.

$$a_n = a_1 + (n - 1)\,d \qquad \text{Alternative form}$$

Example 2   Finding the nth Term of an Arithmetic Sequence

Find a formula for the nth term of the arithmetic sequence whose common difference is 3 and whose first term is 2.

Solution

Because the sequence is arithmetic, you know that the formula for the nth term is of the form $a_n = dn + c$. Moreover, because the common difference is $d = 3$, the formula must have the form

$$a_n = 3n + c. \qquad \text{Substitute 3 for } d.$$

Because $a_1 = 2$, it follows that

$$c = a_1 - d$$
$$= 2 - 3 \qquad \text{Substitute 2 for } a_1 \text{ and 3 for } d.$$
$$= -1.$$

So, the formula for the nth term is

$$a_n = 3n - 1.$$

The sequence therefore has the following form.

$$2, 5, 8, 11, 14, \dots, 3n - 1, \dots$$

✔**CHECKPOINT**   Now try Exercise 21.

Another way to find a formula for the nth term of the sequence in Example 2 is to begin by writing the terms of the sequence.

a_1	a_2	a_3	a_4	a_5	a_6	a_7	
2	$2 + 3$	$5 + 3$	$8 + 3$	$11 + 3$	$14 + 3$	$17 + 3$	$\cdots$
2	5	8	11	14	17	20	$\cdots$

From these terms, you can reason that the nth term is of the form

$$a_n = dn + c = 3n - 1.$$

Example 3   Writing the Terms of an Arithmetic Sequence

The fourth term of an arithmetic sequence is 20, and the 13th term is 65. Write the first 11 terms of this sequence.

Solution

You know that $a_4 = 20$ and $a_{13} = 65$. So, you must add the common difference d nine times to the fourth term to obtain the 13th term. Therefore, the fourth and 13th terms of the sequence are related by

$$a_{13} = a_4 + 9d. \qquad \text{a_4 and a_{13} are nine terms apart.}$$

Using $a_4 = 20$ and $a_{13} = 65$, you can conclude that $d = 5$, which implies that the sequence is as follows.

a_1	a_2	a_3	a_4	a_5	a_6	a_7	a_8	a_9	a_{10}	a_{11} ...
5	10	15	20	25	30	35	40	45	50	55 ...

✓**CHECKPOINT**   Now try Exercise 37.

If you know the nth term of an arithmetic sequence *and* you know the common difference of the sequence, you can find the $(n + 1)$th term by using the *recursion formula*

$$a_{n+1} = a_n + d. \qquad \text{Recursion formula}$$

With this formula, you can find any term of an arithmetic sequence, *provided* that you know the preceding term. For instance, if you know the first term, you can find the second term. Then, knowing the second term, you can find the third term, and so on.

Example 4   Using a Recursion Formula

Find the ninth term of the arithmetic sequence that begins with 2 and 9.

Solution

For this sequence, the common difference is $d = 9 - 2 = 7$. There are two ways to find the ninth term. One way is simply to write out the first nine terms (by repeatedly adding 7).

$$2, 9, 16, 23, 30, 37, 44, 51, 58$$

Another way to find the ninth term is to first find a formula for the nth term. Because the first term is 2, it follows that

$$c = a_1 - d = 2 - 7 = -5.$$

Therefore, a formula for the nth term is

$$a_n = 7n - 5$$

which implies that the ninth term is

$$a_9 = 7(9) - 5 = 58.$$

✓**CHECKPOINT**   Now try Exercise 45.

The Sum of a Finite Arithmetic Sequence

There is a simple formula for the *sum* of a finite arithmetic sequence.

STUDY TIP

Note that this formula works only for *arithmetic* sequences.

> ### The Sum of a Finite Arithmetic Sequence
>
> The sum of a finite arithmetic sequence with n terms is
>
> $$S_n = \frac{n}{2}(a_1 + a_n).$$

For a proof of the sum of a finite arithmetic sequence, see Proofs in Mathematics on page 675.

Example 5 Finding the Sum of a Finite Arithmetic Sequence

Find the sum: $1 + 3 + 5 + 7 + 9 + 11 + 13 + 15 + 17 + 19$.

Solution

To begin, notice that the sequence is arithmetic (with a common difference of 2). Moreover, the sequence has 10 terms. So, the sum of the sequence is

$$S_n = \frac{n}{2}(a_1 + a_n) \qquad \text{Formula for the sum of an arithmetic sequence}$$

$$= \frac{10}{2}(1 + 19) \qquad \text{Substitute 10 for } n, \text{ 1 for } a_1, \text{ and 19 for } a_n.$$

$$= 5(20) = 100. \qquad \text{Simplify.}$$

 **CHECKPOINT** Now try Exercise 63.

Example 6 Finding the Sum of a Finite Arithmetic Sequence

Find the sum of the integers (a) from 1 to 100 and (b) from 1 to N.

Solution

a. The integers from 1 to 100 form an arithmetic sequence that has 100 terms. So, you can use the formula for the sum of an arithmetic sequence, as follows.

$$S_n = 1 + 2 + 3 + 4 + 5 + 6 + \cdots + 99 + 100$$

$$= \frac{n}{2}(a_1 + a_n) \qquad \text{Formula for sum of an arithmetic sequence}$$

$$= \frac{100}{2}(1 + 100) \qquad \text{Substitute 100 for } n, \text{ 1 for } a_1, \text{ 100 for } a_n.$$

$$= 50(101) = 5050 \qquad \text{Simplify.}$$

b. $S_n = 1 + 2 + 3 + 4 + \cdots + N$

$$= \frac{n}{2}(a_1 + a_n) \qquad \text{Formula for sum of an arithmetic sequence}$$

$$= \frac{N}{2}(1 + N) \qquad \text{Substitute } N \text{ for } n, \text{ 1 for } a_1, \text{ and } N \text{ for } a_n.$$

 **CHECKPOINT** Now try Exercise 65.

The Granger Collection

Historical Note

A teacher of Carl Friedrich Gauss (1777–1855) asked him to add all the integers from 1 to 100. When Gauss returned with the correct answer after only a few moments, the teacher could only look at him in astounded silence. This is what Gauss did:

$$S_n = 1 + 2 + 3 + \cdots + 100$$
$$S_n = 100 + 99 + 98 + \cdots + 1$$
$$\overline{2S_n = 101 + 101 + 101 + \cdots + 101}$$

$$S_n = \frac{100 \times 101}{2} = 5050$$

The sum of the first n terms of an infinite sequence is the *nth partial sum*. The nth partial sum can be found by using the formula for the sum of a finite arithmetic sequence.

Example 7   Finding a Partial Sum of an Arithmetic Sequence

Find the 150th partial sum of the arithmetic sequence

5, 16, 27, 38, 49, ...

Solution

For this arithmetic sequence, $a_1 = 5$ and $d = 16 - 5 = 11$. So,

$$c = a_1 - d = 5 - 11 = -6$$

and the nth term is $a_n = 11n - 6$. Therefore, $a_{150} = 11(150) - 6 = 1644$, and the sum of the first 150 terms is

$$S_{150} = \frac{n}{2}(a_1 + a_{150}) \qquad \text{\textcolor{red}{\textit{n}th partial sum formula}}$$

$$= \frac{150}{2}(5 + 1644) \qquad \text{\textcolor{red}{Substitute 150 for \textit{n}, 5 for a_1, and 1644 for a_{150}.}}$$

$$= 75(1649) \qquad \text{\textcolor{red}{Simplify.}}$$

$$= 123{,}675. \qquad \text{\textcolor{red}{\textit{n}th partial sum}}$$

✔**CHECKPOINT**   Now try Exercise 69.

Applications

Example 8   Prize Money

In a golf tournament, the 16 golfers with the lowest scores win cash prizes. First place receives a cash prize of \$1000, second place receives \$950, third place receives \$900, and so on. What is the total amount of prize money?

Solution

The cash prizes awarded form an arithmetic sequence in which the common difference is $d = -50$. Because

$$c = a_1 - d = 1000 - (-50) = 1050$$

you can determine that the formula for the nth term of the sequence is $a_n = -50n + 1050$. So, the 16th term of the sequence is $a_{16} = -50(16) + 1050 = 250$, and the total amount of prize money is

$$S_{16} = 1000 + 950 + 900 + \cdots + 250$$

$$S_{16} = \frac{n}{2}(a_1 + a_{16}) \qquad \text{\textcolor{red}{\textit{n}th partial sum formula}}$$

$$= \frac{16}{2}(1000 + 250) \qquad \text{\textcolor{red}{Substitute 16 for \textit{n}, 1000 for a_1, and 250 for a_{16}.}}$$

$$= 8(1250) = \$10{,}000. \qquad \text{\textcolor{red}{Simplify.}}$$

✔**CHECKPOINT**   Now try Exercise 89.

Example 9 **Total Sales**

A small business sells \$10,000 worth of skin care products during its first year. The owner of the business has set a goal of increasing annual sales by \$7500 each year for 9 years. Assuming that this goal is met, find the total sales during the first 10 years this business is in operation.

Solution

The annual sales form an arithmetic sequence in which $a_1 = 10{,}000$ and $d = 7500$. So,

$$c = a_1 - d$$

$$= 10{,}000 - 7500$$

$$= 2500$$

and the nth term of the sequence is

$$a_n = 7500n + 2500.$$

This implies that the 10th term of the sequence is

$$a_{10} = 7500(10) + 2500$$

$$= 77{,}500. \qquad \text{See Figure 8.4.}$$

The sum of the first 10 terms of the sequence is

$$S_{10} = \frac{n}{2}(a_1 + a_{10}) \qquad \textcolor{red}{n\text{th partial sum formula}}$$

$$= \frac{10}{2}(10{,}000 + 77{,}500) \qquad \textcolor{red}{\text{Substitute 10 for } n, \text{10,000 for } a_1, \text{ and 77,500 for } a_{10}.}$$

$$= 5(87{,}500) \qquad \textcolor{red}{\text{Simplify.}}$$

$$= 437{,}500. \qquad \textcolor{red}{\text{Simplify.}}$$

So, the total sales for the first 10 years will be \$437,500.

✔CHECKPOINT Now try Exercise 91.

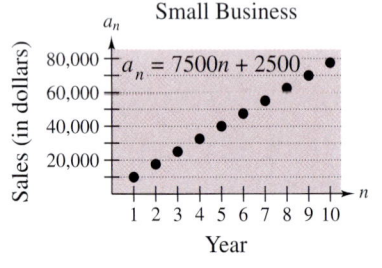

Small Business

$a_n = 7500n + 2500$

FIGURE **8.4**

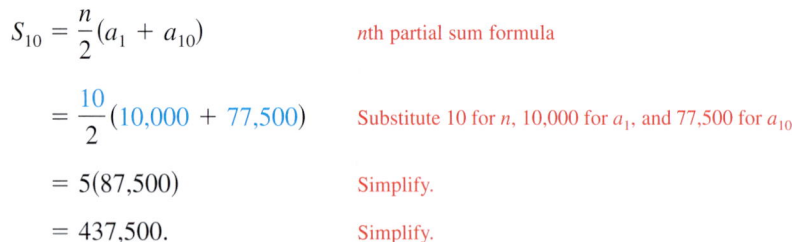

*W*RITING ABOUT *M*ATHEMATICS

Numerical Relationships Decide whether it is possible to fill in the blanks in each of the sequences such that the resulting sequence is arithmetic. If so, find a recursion formula for the sequence.

a. -7, ___, ___, ___, ___, ___, 11

b. 17, ___, ___, ___, ___, ___, ___, ___, 71

c. $2, 6$, ___, ___, 162

d. $4, 7.5$, ___, ___, ___, ___, ___, ___, ___, 39

e. $8, 12$, ___, ___, ___, 60.75

8.2 | Exercises

VOCABULARY CHECK: Fill in the blanks.

1. A sequence is called an _____ sequence if the differences between two consecutive terms are the same. This difference is called the _____ difference.

2. The nth term of an arithmetic sequence has the form _____.

3. The formula $S_n = \dfrac{n}{2}(a_1 + a_n)$ can be used to find the sum of the first n terms of an arithmetic sequence, called the _____ of a _____ _____ _____.

PREREQUISITE SKILLS REVIEW: Practice and review algebra skills needed for this section at **www.Eduspace.com**.

In Exercises 1–10, determine whether the sequence is arithmetic. If so, find the common difference.

1. $10, 8, 6, 4, 2, \ldots$
2. $4, 7, 10, 13, 16, \ldots$
3. $1, 2, 4, 8, 16, \ldots$
4. $80, 40, 20, 10, 5, \ldots$
5. $\frac{9}{4}, 2, \frac{7}{4}, \frac{3}{2}, \frac{5}{4}, \ldots$
6. $3, \frac{5}{2}, 2, \frac{3}{2}, 1, \ldots$
7. $\frac{1}{3}, \frac{2}{3}, 1, \frac{4}{3}, \frac{5}{6}, \ldots$
8. $5.3, 5.7, 6.1, 6.5, 6.9, \ldots$
9. $\ln 1, \ln 2, \ln 3, \ln 4, \ln 5, \ldots$
10. $1^2, 2^2, 3^2, 4^2, 5^2, \ldots$

In Exercises 11–18, write the first five terms of the sequence. Determine whether the sequence is arithmetic. If so, find the common difference. (Assume that n begins with 1.)

11. $a_n = 5 + 3n$
12. $a_n = 100 - 3n$
13. $a_n = 3 - 4(n - 2)$
14. $a_n = 1 + (n - 1)4$
15. $a_n = (-1)^n$
16. $a_n = 2^{n-1}$
17. $a_n = \dfrac{(-1)^n 3}{n}$
18. $a_n = (2^n)n$

In Exercises 19–30, find a formula for a_n for the arithmetic sequence.

19. $a_1 = 1, d = 3$
20. $a_1 = 15, d = 4$
21. $a_1 = 100, d = -8$
22. $a_1 = 0, d = -\frac{2}{3}$
23. $a_1 = x, d = 2x$
24. $a_1 = -y, d = 5y$
25. $4, \frac{3}{2}, -1, -\frac{7}{2}, \ldots$
26. $10, 5, 0, -5, -10, \ldots$
27. $a_1 = 5, a_4 = 15$
28. $a_1 = -4, a_5 = 16$
29. $a_3 = 94, a_6 = 85$
30. $a_5 = 190, a_{10} = 115$

In Exercises 31–38, write the first five terms of the arithmetic sequence.

31. $a_1 = 5, d = 6$
32. $a_1 = 5, d = -\frac{3}{4}$
33. $a_1 = -2.6, d = -0.4$
34. $a_1 = 16.5, d = 0.25$
35. $a_1 = 2, a_{12} = 46$
36. $a_4 = 16, a_{10} = 46$
37. $a_8 = 26, a_{12} = 42$
38. $a_3 = 19, a_{15} = -1.7$

In Exercises 39–44, write the first five terms of the arithmetic sequence. Find the common difference and write the nth term of the sequence as a function of n.

39. $a_1 = 15, \quad a_{k+1} = a_k + 4$
40. $a_1 = 6, \quad a_{k+1} = a_k + 5$
41. $a_1 = 200, \quad a_{k+1} = a_k - 10$
42. $a_1 = 72, \quad a_{k+1} = a_k - 6$
43. $a_1 = \frac{5}{8}, \quad a_{k+1} = a_k - \frac{1}{8}$
44. $a_1 = 0.375, \quad a_{k+1} = a_k + 0.25$

In Exercises 45–48, the first two terms of the arithmetic sequence are given. Find the missing term.

45. $a_1 = 5, a_2 = 11, a_{10} = $ ▢
46. $a_1 = 3, a_2 = 13, a_9 = $ ▢
47. $a_1 = 4.2, a_2 = 6.6, a_7 = $ ▢
48. $a_1 = -0.7, a_2 = -13.8, a_8 = $ ▢

In Exercises 49–52, match the arithmetic sequence with its graph. [The graphs are labeled (a), (b), (c), and (d).]

(a)
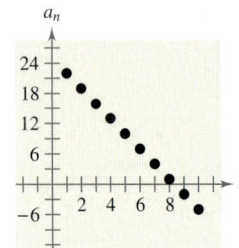

(b)
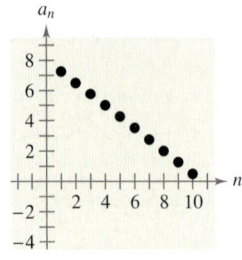

(c)
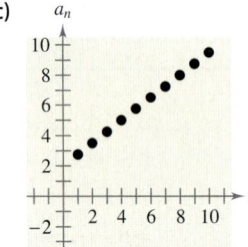

(d)
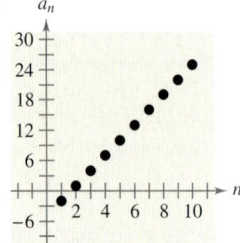

49. $a_n = -\frac{3}{4}n + 8$ **50.** $a_n = 3n - 5$

51. $a_n = 2 + \frac{3}{4}n$ **52.** $a_n = 25 - 3n$

 In Exercises 53–56, use a graphing utility to graph the first 10 terms of the sequence. (Assume that n begins with 1.)

53. $a_n = 15 - \frac{3}{2}n$ **54.** $a_n = -5 + 2n$

55. $a_n = 0.2n + 3$ **56.** $a_n = -0.3n + 8$

In Exercises 57–64, find the indicated nth partial sum of the arithmetic sequence.

57. $8, 20, 32, 44, \ldots, \quad n = 10$

58. $2, 8, 14, 20, \ldots, \quad n = 25$

59. $4.2, 3.7, 3.2, 2.7, \ldots, \quad n = 12$

60. $0.5, 0.9, 1.3, 1.7, \ldots, \quad n = 10$

61. $40, 37, 34, 31, \ldots, \quad n = 10$

62. $75, 70, 65, 60, \ldots, \quad n = 25$

63. $a_1 = 100, \ a_{25} = 220, \quad n = 25$

64. $a_1 = 15, \ a_{100} = 307, \quad n = 100$

65. Find the sum of the first 100 positive odd integers.

66. Find the sum of the integers from -10 to 50.

In Exercises 67–74, find the partial sum.

67. $\displaystyle\sum_{n=1}^{50} n$ **68.** $\displaystyle\sum_{n=1}^{100} 2n$

69. $\displaystyle\sum_{n=10}^{100} 6n$ **70.** $\displaystyle\sum_{n=51}^{100} 7n$

71. $\displaystyle\sum_{n=11}^{30} n - \sum_{n=1}^{10} n$ **72.** $\displaystyle\sum_{n=51}^{100} n - \sum_{n=1}^{50} n$

73. $\displaystyle\sum_{n=1}^{400} (2n - 1)$ **74.** $\displaystyle\sum_{n=1}^{250} (1000 - n)$

 In Exercises 75–80, use a graphing utility to find the partial sum.

75. $\displaystyle\sum_{n=1}^{20} (2n + 5)$ **76.** $\displaystyle\sum_{n=0}^{50} (1000 - 5n)$

77. $\displaystyle\sum_{n=1}^{100} \frac{n + 4}{2}$ **78.** $\displaystyle\sum_{n=0}^{100} \frac{8 - 3n}{16}$

79. $\displaystyle\sum_{i=1}^{60} \left(250 - \frac{8}{3}i\right)$ **80.** $\displaystyle\sum_{j=1}^{200} (4.5 + 0.025j)$

Job Offer   **In Exercises 81 and 82, consider a job offer with the given starting salary and the given annual raise.**

(a) Determine the salary during the sixth year of employment.

(b) Determine the total compensation from the company through six full years of employment.

	Starting Salary	*Annual Raise*
81.	$32,500	$1500
82.	$36,800	$1750

83. *Seating Capacity*   Determine the seating capacity of an auditorium with 30 rows of seats if there are 20 seats in the first row, 24 seats in the second row, 28 seats in the third row, and so on.

84. *Seating Capacity*   Determine the seating capacity of an auditorium with 36 rows of seats if there are 15 seats in the first row, 18 seats in the second row, 21 seats in the third row, and so on.

85. *Brick Pattern*   A brick patio has the approximate shape of a trapezoid (see figure). The patio has 18 rows of bricks. The first row has 14 bricks and the 18th row has 31 bricks. How many bricks are in the patio?

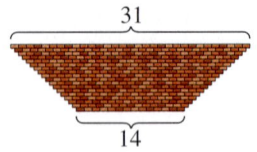

FIGURE FOR 85   FIGURE FOR 86

86. *Brick Pattern*   A triangular brick wall is made by cutting some bricks in half to use in the first column of every other row. The wall has 28 rows. The top row is one-half brick wide and the bottom row is 14 bricks wide. How many bricks are used in the finished wall?

87. Falling Object   An object with negligible air resistance is dropped from a plane. During the first second of fall, the object falls 4.9 meters; during the second second, it falls 14.7 meters; during the third second, it falls 24.5 meters; during the fourth second, it falls 34.3 meters. If this arithmetic pattern continues, how many meters will the object fall in 10 seconds?

88. Falling Object   An object with negligible air resistance is dropped from the top of the Sears Tower in Chicago at a height of 1454 feet. During the first second of fall, the object falls 16 feet; during the second second, it falls 48 feet; during the third second, it falls 80 feet; during the fourth second, it falls 112 feet. If this arithmetic pattern continues, how many feet will the object fall in 7 seconds?

89. Prize Money   A county fair is holding a baked goods competition in which the top eight bakers receive cash prizes. First places receives a cash prize of $200, second place receives $175, third place receives $150, and so on.

(a) Write a sequence a_n that represents the cash prize awarded in terms of the place n in which the baked good places.

(b) Find the total amount of prize money awarded at the competition.

90. Prize Money   A city bowling league is holding a tournament in which the top 12 bowlers with the highest three-game totals are awarded cash prizes. First place will win $1200, second place $1100, third place $1000, and so on.

(a) Write a sequence a_n that represents the cash prize awarded in terms of the place n in which the bowler finishes.

(b) Find the total amount of prize money awarded at the tournament.

91. Total Profit   A small snowplowing company makes a profit of $8000 during its first year. The owner of the company sets a goal of increasing profit by $1500 each year for 5 years. Assuming that this goal is met, find the total profit during the first 6 years of this business. What kinds of economic factors could prevent the company from meeting its profit goal? Are there any other factors that could prevent the company from meeting its goal? Explain.

92. Total Sales   An entrepreneur sells $15,000 worth of sports memorabilia during one year and sets a goal of increasing annual sales by $5000 each year for 9 years. Assuming that this goal is met, find the total sales during the first 10 years of this business. What kinds of economic factors could prevent the business from meeting its goals?

93. Borrowing Money   You borrowed $2000 from a friend to purchase a new laptop computer and have agreed to pay back the loan with monthly payments of $200 plus 1% interest on the unpaid balance.

(a) Find the first six monthly payments you will make, and the unpaid balance after each month.

(b) Find the total amount of interest paid over the term of the loan.

94. Borrowing Money   You borrowed $5000 from your parents to purchase a used car. The arrangements of the loan are such that you will make payments of $250 per month plus 1% interest on the unpaid balance.

(a) Find the first year's monthly payments you will make, and the unpaid balance after each month.

(b) Find the total amount of interest paid over the term of the loan.

Model It

95. Data Analysis: Personal Income   The table shows the per capita personal income a_n in the United States from 1993 to 2003. (Source: U.S. Bureau of Economic Analysis)

Year	Per capita personal income, a_n
1993	$21,356
1994	$22,176
1995	$23,078
1996	$24,176
1997	$25,334
1998	$26,880
1999	$27,933
2000	$29,848
2001	$30,534
2002	$30,913
2003	$31,633

(a) Find an arithmetic sequence that models the data. Let n represent the year, with $n = 3$ corresponding to 1993.

 (b) Use the *regression* feature of a graphing utility to find a linear model for the data. How does this model compare with the arithmetic sequence you found in part (a)?

 (c) Use a graphing utility to graph the terms of the finite sequence you found in part (a).

(d) Use the sequence from part (a) to estimate the per capita personal income in 2004 and 2005.

(e) Use your school's library, the Internet, or some other reference source to find the actual per capita personal income in 2004 and 2005, and compare these values with the estimates from part (d).

96. *Data Analysis: Revenue* The table shows the annual revenue a_n (in millions of dollars) for Nextel Communications, Inc. from 1997 to 2003. (Source: Nextel Communications, Inc.)

Year	Revenue, a_n
1997	739
1998	1847
1999	3326
2000	5714
2001	7689
2002	8721
2003	10,820

(a) Construct a bar graph showing the annual revenue from 1997 to 2003.

 (b) Use the *linear regression* feature of a graphing utility to find an arithmetic sequence that approximates the annual revenue from 1997 to 2003.

(c) Use summation notation to represent the *total* revenue from 1997 to 2003. Find the total revenue.

 (d) Use the sequence from part (b) to estimate the annual revenue in 2008.

Synthesis

True or False? In Exercises 97 and 98, determine whether the statement is true or false. Justify your answer.

97. Given an arithmetic sequence for which only the first two terms are known, it is possible to find the nth term.

98. If the only known information about a finite arithmetic sequence is its first term and its last term, then it is possible to find the sum of the sequence.

99. *Writing* In your own words, explain what makes a sequence arithmetic.

100. *Writing* Explain how to use the first two terms of an arithmetic sequence to find the nth term.

101. *Exploration*

(a) Graph the first 10 terms of the arithmetic sequence $a_n = 2 + 3n$.

(b) Graph the equation of the line $y = 3x + 2$.

(c) Discuss any differences between the graph of

$a_n = 2 + 3n$

and the graph of

$y = 3x + 2$.

(d) Compare the slope of the line in part (b) with the common difference of the sequence in part (a). What can you conclude about the slope of a line and the common difference of an arithmetic sequence?

102. *Pattern Recognition*

(a) Compute the following sums of positive odd integers.

$1 + 3 = $ ▢

$1 + 3 + 5 = $ ▢

$1 + 3 + 5 + 7 = $ ▢

$1 + 3 + 5 + 7 + 9 = $ ▢

$1 + 3 + 5 + 7 + 9 + 11 = $ ▢

(b) Use the sums in part (a) to make a conjecture about the sums of positive odd integers. Check your conjecture for the sum

$1 + 3 + 5 + 7 + 9 + 11 + 13 = $ ▢.

(c) Verify your conjecture algebraically.

103. *Think About It* The sum of the first 20 terms of an arithmetic sequence with a common difference of 3 is 650. Find the first term.

104. *Think About It* The sum of the first n terms of an arithmetic sequence with first term a_1 and common difference d is S_n. Determine the sum if each term is increased by 5. Explain.

Skills Review

In Exercises 105–108, find the slope and y-intercept (if possible) of the equation of the line. Sketch the line.

105. $2x - 4y = 3$

106. $9x + y = -8$

107. $x - 7 = 0$

108. $y + 11 = 0$

In Exercises 109 and 110, use Gauss-Jordan elimination to solve the system of equations.

109. $\begin{cases} 2x - y + 7z = -10 \\ 3x + 2y - 4z = 17 \\ 6x - 5y + z = -20 \end{cases}$

110. $\begin{cases} -x + 4y + 10z = 4 \\ 5x - 3y + z = 31 \\ 8x + 2y - 3z = -5 \end{cases}$

111. Make a Decision To work an extended application analyzing the median sales price of existing one-family homes in the United States from 1987 to 2003, visit this text's website at *college.hmco.com*. (Data Source: National Association of Realtors)

8.3 Geometric Sequences and Series

What you should learn

- Recognize, write, and find the *n*th terms of geometric sequences.
- Find *n*th partial sums of geometric sequences.
- Find the sum of an infinite geometric series.
- Use geometric sequences to model and solve real-life problems.

Why you should learn it

Geometric sequences can be used to model and solve real-life problems. For instance, in Exercise 99 on page 622, you will use a geometric sequence to model the population of China.

© Bob Krist/Corbis

Geometric Sequences

In Section 8.2, you learned that a sequence whose consecutive terms have a common *difference* is an arithmetic sequence. In this section, you will study another important type of sequence called a **geometric sequence.** Consecutive terms of a geometric sequence have a common *ratio*.

> ### Definition of Geometric Sequence
>
> A sequence is **geometric** if the ratios of consecutive terms are the same. So, the sequence $a_1, a_2, a_3, a_4, \ldots, a_n \ldots$ is geometric if there is a number r such that
>
> $$\frac{a_2}{a_1} = r, \quad \frac{a_3}{a_2} = r, \quad \frac{a_4}{a_3} = r, \quad r \neq 0$$
>
> and so the number r is the **common ratio** of the sequence.

Example 1   **Examples of Geometric Sequences**

a. The sequence whose *n*th term is 2^n is geometric. For this sequence, the common ratio of consecutive terms is 2.

$$2, 4, 8, 16, \ldots, 2^n, \ldots \qquad \text{Begin with } n = 1.$$

$$\frac{4}{2} = 2$$

b. The sequence whose *n*th term is $4(3^n)$ is geometric. For this sequence, the common ratio of consecutive terms is 3.

$$12, 36, 108, 324, \ldots, 4(3^n), \ldots \qquad \text{Begin with } n = 1.$$

$$\frac{36}{12} = 3$$

c. The sequence whose *n*th term is $\left(-\frac{1}{3}\right)^n$ is geometric. For this sequence, the common ratio of consecutive terms is $-\frac{1}{3}$.

$$-\frac{1}{3}, \frac{1}{9}, -\frac{1}{27}, \frac{1}{81}, \ldots, \left(-\frac{1}{3}\right)^n, \ldots \qquad \text{Begin with } n = 1.$$

$$\frac{1/9}{-1/3} = -\frac{1}{3}$$

✓**CHECKPOINT**   Now try Exercise 1.

The sequence $1, 4, 9, 16, \ldots$, whose *n*th term is n^2, is *not* geometric. The ratio of the second term to the first term is

$$\frac{a_2}{a_1} = \frac{4}{1} = 4$$

but the ratio of the third term to the second term is $\dfrac{a_3}{a_2} = \dfrac{9}{4}$.

In Example 1, notice that each of the geometric sequences has an nth term that is of the form ar^n, where the common ratio of the sequence is r. A geometric sequence may be thought of as an exponential function whose domain is the set of natural numbers.

The nth Term of a Geometric Sequence

The nth term of a geometric sequence has the form

$$a_n = a_1 r^{n-1}$$

where r is the common ratio of consecutive terms of the sequence. So, every geometric sequence can be written in the following form.

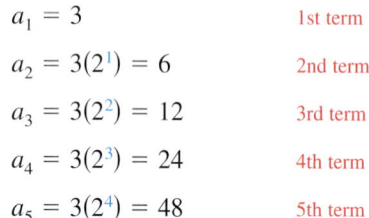

If you know the nth term of a geometric sequence, you can find the $(n+1)$th term by multiplying by r. That is, $a_{n+1} = ra_n$.

Example 2   Finding the Terms of a Geometric Sequence

Write the first five terms of the geometric sequence whose first term is $a_1 = 3$ and whose common ratio is $r = 2$. Then graph the terms on a set of coordinate axes.

Solution

Starting with 3, repeatedly multiply by 2 to obtain the following.

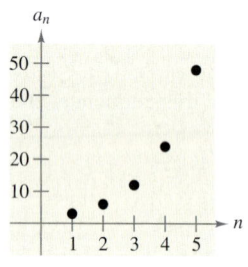

FIGURE **8.5**

$a_1 = 3$	1st term
$a_2 = 3(2^1) = 6$	2nd term
$a_3 = 3(2^2) = 12$	3rd term
$a_4 = 3(2^3) = 24$	4th term
$a_5 = 3(2^4) = 48$	5th term

Figure 8.5 shows the first five terms of this geometric sequence.

✔**CHECKPOINT**   Now try Exercise 11.

Example 3   Finding a Term of a Geometric Sequence

Find the 15th term of the geometric sequence whose first term is 20 and whose common ratio is 1.05.

Solution

$$a_{15} = a_1 r^{n-1} \qquad \text{Formula for geometric sequence}$$
$$= 20(1.05)^{15-1} \qquad \text{Substitute 20 for } a_1, \text{ 1.05 for } r, \text{ and 15 for } n.$$
$$\approx 39.599 \qquad \text{Use a calculator.}$$

✔**CHECKPOINT**   Now try Exercise 27.

Example 4   **Finding a Term of a Geometric Sequence**

Find the 12th term of the geometric sequence

$$5, 15, 45, \ldots .$$

Solution

The common ratio of this sequence is

$$r = \frac{15}{5} = 3.$$

Because the first term is $a_1 = 5$, you can determine the 12th term ($n = 12$) to be

$a_n = a_1 r^{n-1}$	Formula for geometric sequence
$a_{12} = 5(3)^{12-1}$	Substitute 5 for a_1, 3 for r, and 12 for n.
$= 5(177{,}147)$	Use a calculator.
$= 885{,}735.$	Simplify.

✔CHECKPOINT   Now try Exercise 35.

If you know any two terms of a geometric sequence, you can use that information to find a formula for the nth term of the sequence.

STUDY TIP

Remember that r is the common ratio of consecutive terms of a geometric sequence. So, in Example 5,

$$a_{10} = a_1 r^9$$
$$= a_1 \cdot r \cdot r \cdot r \cdot r^6$$
$$= a_1 \cdot \frac{a_2}{a_1} \cdot \frac{a_3}{a_2} \cdot \frac{a_4}{a_3} \cdot r^6$$
$$= a_4 r^6.$$

Example 5   **Finding a Term of a Geometric Sequence**

The fourth term of a geometric sequence is 125, and the 10th term is 125/64. Find the 14th term. (Assume that the terms of the sequence are positive.)

Solution

The 10th term is related to the fourth term by the equation

$a_{10} = a_4 r^6.$	Multiply 4th term by r^{10-4}.

Because $a_{10} = 125/64$ and $a_4 = 125$, you can solve for r as follows.

$\dfrac{125}{64} = 125 r^6$	Substitute $\frac{125}{64}$ for a_{10} and 125 for a_4.
$\dfrac{1}{64} = r^6$	Divide each side by 125.
$\dfrac{1}{2} = r$	Take the sixth root of each side.

You can obtain the 14th term by multiplying the 10th term by r^4.

$a_{14} = a_{10} r^4$	Multiply the 10th term by r^{14-10}.
$= \dfrac{125}{64}\left(\dfrac{1}{2}\right)^4$	Substitute $\frac{125}{64}$ for a_{10} and $\frac{1}{2}$ for r.
$= \dfrac{125}{1024}$	Simplify.

✔CHECKPOINT   Now try Exercise 41.

The Sum of a Finite Geometric Sequence

The formula for the sum of a *finite* geometric sequence is as follows.

The Sum of a Finite Geometric Sequence

The sum of the finite geometric sequence

$$a_1, \; a_1 r, \; a_1 r^2, \; a_1 r^3, \; a_1 r^4, \ldots, a_1 r^{n-1}$$

with common ratio $r \neq 1$ is given by $S_n = \displaystyle\sum_{i=1}^{n} a_1 \, r^{i-1} = a_1 \left(\dfrac{1 - r^n}{1 - r} \right).$

For a proof of the sum of a finite geometric sequence, see Proofs in Mathematics on page 675.

Example 6   **Finding the Sum of a Finite Geometric Sequence**

Find the sum $\displaystyle\sum_{i=1}^{12} 4(0.3)^{i-1}$.

Solution

By writing out a few terms, you have

$$\sum_{i=1}^{12} 4(0.3)^{i-1} = 4(0.3)^0 + 4(0.3)^1 + 4(0.3)^2 + \cdots + 4(0.3)^{11}.$$

Now, because $a_1 = 4$, $r = 0.3$, and $n = 12$, you can apply the formula for the sum of a finite geometric sequence to obtain

$$S_n = a_1 \left(\frac{1 - r^n}{1 - r} \right) \qquad \text{\color{red}{Formula for the sum of a sequence}}$$

$$\sum_{i=1}^{12} 4(0.3)^{i-1} = 4 \left[\frac{1 - (0.3)^{12}}{1 - 0.3} \right] \qquad \text{\color{red}{Substitute 4 for } a_1, \text{ 0.3 for } r, \text{ and 12 for } n.}$$

$$\approx 5.714. \qquad \text{\color{red}{Use a calculator.}}$$

✓**CHECKPOINT**   Now try Exercise 57.

When using the formula for the sum of a finite geometric sequence, be careful to check that the sum is of the form

$$\sum_{i=1}^{n} a_1 \, r^{i-1}. \qquad \text{\color{red}{Exponent for } r \text{ is } i - 1.}$$

If the sum is not of this form, you must adjust the formula. For instance, if the sum in Example 6 were $\displaystyle\sum_{i=1}^{12} 4(0.3)^i$, then you would evaluate the sum as follows.

$$\sum_{i=1}^{12} 4(0.3)^i = 4(0.3) + 4(0.3)^2 + 4(0.3)^3 + \cdots + 4(0.3)^{12}$$

$$= 4(0.3) + [4(0.3)](0.3) + [4(0.3)](0.3)^2 + \cdots + [4(0.3)](0.3)^{11}$$

$$= 4(0.3) \left[\frac{1 - (0.3)^{12}}{1 - 0.3} \right] \approx 1.714. \qquad \text{\color{red}{$a_1 = 4(0.3)$, $r = 0.3$, $n = 12$}}$$

Geometric Series

The summation of the terms of an infinite geometric *sequence* is called an **infinite geometric series** or simply a **geometric series.**

The formula for the sum of a *finite* geometric *sequence* can, depending on the value of r, be extended to produce a formula for the sum of an *infinite* geometric *series*. Specifically, if the common ratio r has the property that $|r| < 1$, it can be shown that r^n becomes arbitrarily close to zero as n increases without bound. Consequently,

$$a_1\left(\frac{1 - r^n}{1 - r}\right) \longrightarrow a_1\left(\frac{1 - 0}{1 - r}\right) \quad \text{as} \quad n \longrightarrow \infty.$$

This result is summarized as follows.

The Sum of an Infinite Geometric Series

If $|r| < 1$, the infinite geometric series

$$a_1 + a_1 r + a_1 r^2 + a_1 r^3 + \cdots + a_1 r^{n-1} + \cdots$$

has the sum

$$S = \sum_{i=0}^{\infty} a_1 r^i = \frac{a_1}{1 - r}.$$

Note that if $|r| \geq 1$, the series does not have a sum.

Example 7   **Finding the Sum of an Infinite Geometric Series**

Find each sum.

a. $\displaystyle\sum_{n=1}^{\infty} 4(0.6)^{n-1}$

b. $3 + 0.3 + 0.03 + 0.003 + \cdots$

Solution

a. $\displaystyle\sum_{n=1}^{\infty} 4(0.6)^{n-1} = 4 + 4(0.6) + 4(0.6)^2 + 4(0.6)^3 + \cdots + 4(0.6)^{n-1} + \cdots$

$$= \frac{4}{1 - 0.6} \qquad \frac{a_1}{1 - r}$$

$$= 10$$

b. $3 + 0.3 + 0.03 + 0.003 + \cdots = 3 + 3(0.1) + 3(0.1)^2 + 3(0.1)^3 + \cdots$

$$= \frac{3}{1 - 0.1} \qquad \frac{a_1}{1 - r}$$

$$= \frac{10}{3}$$

$$\approx 3.33$$

✓CHECKPOINT   Now try Exercise 79.

Application

Example 8 **Increasing Annuity**

A deposit of $50 is made on the first day of each month in a savings account that pays 6% compounded monthly. What is the balance at the end of 2 years? (This type of savings plan is called an **increasing annuity**.)

Solution

The first deposit will gain interest for 24 months, and its balance will be

$$A_{24} = 50\left(1 + \frac{0.06}{12}\right)^{24}$$

$$= 50(1.005)^{24}.$$

The second deposit will gain interest for 23 months, and its balance will be

$$A_{23} = 50\left(1 + \frac{0.06}{12}\right)^{23}$$

$$= 50(1.005)^{23}.$$

The last deposit will gain interest for only 1 month, and its balance will be

$$A_1 = 50\left(1 + \frac{0.06}{12}\right)^{1}$$

$$= 50(1.005).$$

The total balance in the annuity will be the sum of the balances of the 24 deposits. Using the formula for the sum of a finite geometric sequence, with $A_1 = 50(1.005)$ and $r = 1.005$, you have

$$S_{24} = 50(1.005)\left[\frac{1 - (1.005)^{24}}{1 - 1.005}\right] \qquad \text{Substitute } 50(1.005) \text{ for } A_1, 1.005 \text{ for } r, \text{ and } 24 \text{ for } n.$$

$$= \$1277.96. \qquad \text{Simplify.}$$

✔**CHECKPOINT** Now try Exercise 107.

*W*RITING ABOUT *M*ATHEMATICS

An Experiment You will need a piece of string or yarn, a pair of scissors, and a tape measure. Measure out any length of string at least 5 feet long. Double over the string and cut it in half. Take one of the resulting halves, double it over, and cut it in half. Continue this process until you are no longer able to cut a length of string in half. How many cuts were you able to make? Construct a sequence of the resulting string lengths after each cut, starting with the original length of the string. Find a formula for the nth term of this sequence. How many cuts could you theoretically make? Discuss why you were not able to make that many cuts.

8.3 | Exercises

VOCABULARY CHECK: Fill in the blanks.

1. A sequence is called a _____ sequence if the ratios between consecutive terms are the same. This ratio is called the _____ ratio.

2. The nth term of a geometric sequence has the form _____.

3. The formula for the sum of a finite geometric sequence is given by _____.

4. The sum of the terms of an infinite geometric sequence is called a _____ _____.

5. The formula for the sum of an infinite geometric series is given by _____.

PREREQUISITE SKILLS REVIEW: Practice and review algebra skills needed for this section at **www.Eduspace.com.**

In Exercises 1–10, determine whether the sequence is geometric. If so, find the common ratio.

1. $5, 15, 45, 135, \ldots$
2. $3, 12, 48, 192, \ldots$
3. $3, 12, 21, 30, \ldots$
4. $36, 27, 18, 9, \ldots$
5. $1, -\frac{1}{2}, \frac{1}{4}, -\frac{1}{8}, \ldots$
6. $5, 1, 0.2, 0.04, \ldots$
7. $\frac{1}{8}, \frac{1}{4}, \frac{1}{2}, 1, \ldots$
8. $9, -6, 4, -\frac{8}{3}, \ldots$
9. $1, \frac{1}{2}, \frac{1}{3}, \frac{1}{4}, \ldots$
10. $\frac{1}{5}, \frac{2}{7}, \frac{3}{9}, \frac{4}{11}, \ldots$

In Exercises 11–20, write the first five terms of the geometric sequence.

11. $a_1 = 2, r = 3$
12. $a_1 = 6, r = 2$
13. $a_1 = 1, r = \frac{1}{2}$
14. $a_1 = 1, r = \frac{1}{3}$
15. $a_1 = 5, r = -\frac{1}{10}$
16. $a_1 = 6, r = -\frac{1}{4}$
17. $a_1 = 1, r = e$
18. $a_1 = 3, r = \sqrt{5}$
19. $a_1 = 2, r = \dfrac{x}{4}$
20. $a_1 = 5, r = 2x$

In Exercises 21–26, write the first five terms of the geometric sequence. Determine the common ratio and write the nth term of the sequence as a function of n.

21. $a_1 = 64, \ a_{k+1} = \frac{1}{2}a_k$
22. $a_1 = 81, \ a_{k+1} = \frac{1}{3}a_k$
23. $a_1 = 7, \ a_{k+1} = 2a_k$
24. $a_1 = 5, \ a_{k+1} = -2a_k$
25. $a_1 = 6, \ a_{k+1} = -\frac{3}{2}a_k$
26. $a_1 = 48, \ a_{k+1} = -\frac{1}{2}a_k$

In Exercises 27–34, write an expression for the nth term of the geometric sequence. Then find the indicated term.

27. $a_1 = 4, r = \frac{1}{2}, n = 10$
28. $a_1 = 5, r = \frac{3}{2}, n = 8$
29. $a_1 = 6, r = -\frac{1}{3}, n = 12$
30. $a_1 = 64, r = -\frac{1}{4}, n = 10$
31. $a_1 = 100, r = e^x, n = 9$
32. $a_1 = 1, r = \sqrt{3}, n = 8$
33. $a_1 = 500, r = 1.02, n = 40$
34. $a_1 = 1000, r = 1.005, n = 60$

In Exercises 35–42, find the indicated nth term of the geometric sequence.

35. 9th term: $7, 21, 63, \ldots$
36. 7th term: $3, 36, 432, \ldots$
37. 10th term: $5, 30, 180, \ldots$
38. 22nd term: $4, 8, 16, \ldots$
39. 3rd term: $a_1 = 16, a_4 = \frac{27}{4}$
40. 1st term: $a_2 = 3, a_5 = \frac{3}{64}$
41. 6th term: $a_4 = -18, a_7 = \frac{2}{3}$
42. 7th term: $a_3 = \frac{16}{3}, a_5 = \frac{64}{27}$

In Exercises 43–46, match the geometric sequence with its graph. [The graphs are labeled (a), (b), (c), and (d).]

(a)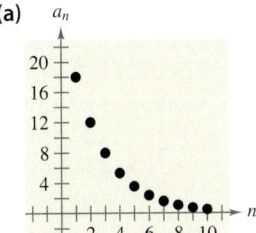

(b)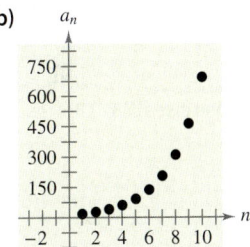

(c)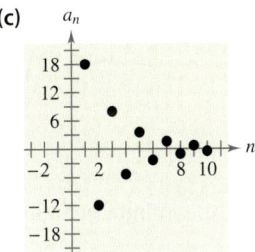

(d)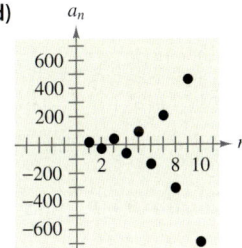

43. $a_n = 18\left(\frac{2}{3}\right)^{n-1}$
44. $a_n = 18\left(-\frac{2}{3}\right)^{n-1}$
45. $a_n = 18\left(\frac{3}{2}\right)^{n-1}$
46. $a_n = 18\left(-\frac{3}{2}\right)^{n-1}$

In Exercises 47–52, use a graphing utility to graph the first 10 terms of the sequence.

47. $a_n = 12(-0.75)^{n-1}$ **48.** $a_n = 10(1.5)^{n-1}$

49. $a_n = 12(-0.4)^{n-1}$ **50.** $a_n = 20(-1.25)^{n-1}$

51. $a_n = 2(1.3)^{n-1}$ **52.** $a_n = 10(1.2)^{n-1}$

In Exercises 53–72, find the sum of the finite geometric sequence.

53. $\displaystyle\sum_{n=1}^{9} 2^{n-1}$ **54.** $\displaystyle\sum_{n=1}^{10} \left(\tfrac{5}{2}\right)^{n-1}$

55. $\displaystyle\sum_{n=1}^{9} (-2)^{n-1}$ **56.** $\displaystyle\sum_{n=1}^{8} 5\left(-\tfrac{3}{2}\right)^{n-1}$

57. $\displaystyle\sum_{i=1}^{7} 64\left(-\tfrac{1}{2}\right)^{i-1}$ **58.** $\displaystyle\sum_{i=1}^{10} 2\left(\tfrac{1}{4}\right)^{i-1}$

59. $\displaystyle\sum_{i=1}^{6} 32\left(\tfrac{1}{4}\right)^{i-1}$ **60.** $\displaystyle\sum_{i=1}^{12} 16\left(\tfrac{1}{2}\right)^{i-1}$

61. $\displaystyle\sum_{n=0}^{20} 3\left(\tfrac{3}{2}\right)^{n}$ **62.** $\displaystyle\sum_{n=0}^{40} 5\left(\tfrac{3}{5}\right)^{n}$

63. $\displaystyle\sum_{n=0}^{15} 2\left(\tfrac{4}{3}\right)^{n}$ **64.** $\displaystyle\sum_{n=0}^{20} 10\left(\tfrac{1}{5}\right)^{n}$

65. $\displaystyle\sum_{n=0}^{5} 300(1.06)^{n}$ **66.** $\displaystyle\sum_{n=0}^{6} 500(1.04)^{n}$

67. $\displaystyle\sum_{n=0}^{40} 2\left(-\tfrac{1}{4}\right)^{n}$ **68.** $\displaystyle\sum_{n=0}^{50} 10\left(\tfrac{2}{3}\right)^{n-1}$

69. $\displaystyle\sum_{i=1}^{10} 8\left(-\tfrac{1}{4}\right)^{i-1}$ **70.** $\displaystyle\sum_{i=0}^{25} 8\left(-\tfrac{1}{2}\right)^{i}$

71. $\displaystyle\sum_{i=1}^{10} 5\left(-\tfrac{1}{3}\right)^{i-1}$ **72.** $\displaystyle\sum_{i=1}^{100} 15\left(\tfrac{2}{3}\right)^{i-1}$

In Exercises 73–78, use summation notation to write the sum.

73. $5 + 15 + 45 + \cdots + 3645$

74. $7 + 14 + 28 + \cdots + 896$

75. $2 - \dfrac{1}{2} + \dfrac{1}{8} - \cdots + \dfrac{1}{2048}$

76. $15 - 3 + \dfrac{3}{5} - \cdots - \dfrac{3}{625}$

77. $0.1 + 0.4 + 1.6 + \cdots + 102.4$

78. $32 + 24 + 18 + \cdots + 10.125$

In Exercises 79–92, find the sum of the infinite geometric series.

79. $\displaystyle\sum_{n=0}^{\infty} \left(\tfrac{1}{2}\right)^{n}$ **80.** $\displaystyle\sum_{n=0}^{\infty} 2\left(\tfrac{2}{3}\right)^{n}$

81. $\displaystyle\sum_{n=0}^{\infty} \left(-\tfrac{1}{2}\right)^{n}$ **82.** $\displaystyle\sum_{n=0}^{\infty} 2\left(-\tfrac{2}{3}\right)^{n}$

83. $\displaystyle\sum_{n=0}^{\infty} 4\left(\tfrac{1}{4}\right)^{n}$ **84.** $\displaystyle\sum_{n=0}^{\infty} \left(\tfrac{1}{10}\right)^{n}$

85. $\displaystyle\sum_{n=0}^{\infty} (0.4)^{n}$ **86.** $\displaystyle\sum_{n=0}^{\infty} 4(0.2)^{n}$

87. $\displaystyle\sum_{n=0}^{\infty} -3(0.9)^{n}$ **88.** $\displaystyle\sum_{n=0}^{\infty} -10(0.2)^{n}$

89. $8 + 6 + \dfrac{9}{2} + \dfrac{27}{8} + \cdots$ **90.** $9 + 6 + 4 + \dfrac{8}{3} + \cdots$

91. $\dfrac{1}{9} - \dfrac{1}{3} + 1 - 3 + \cdots$ **92.** $-\dfrac{125}{36} + \dfrac{25}{6} - 5 + 6 - \cdots$

In Exercises 93–96, find the rational number representation of the repeating decimal.

93. $0.\overline{36}$ **94.** $0.\overline{297}$

95. $0.3\overline{18}$ **96.** $1.3\overline{8}$

Graphical Reasoning In Exercises 97 and 98, use a graphing utility to graph the function. Identify the horizontal asymptote of the graph and determine its relationship to the sum.

97. $f(x) = 6\left[\dfrac{1 - (0.5)^{x}}{1 - (0.5)}\right]$, $\displaystyle\sum_{n=0}^{\infty} 6\left(\tfrac{1}{2}\right)^{n}$

98. $f(x) = 2\left[\dfrac{1 - (0.8)^{x}}{1 - (0.8)}\right]$, $\displaystyle\sum_{n=0}^{\infty} 2\left(\tfrac{4}{5}\right)^{n}$

Model It

99. *Data Analysis: Population* The table shows the population a_n of China (in millions) from 1998 through 2004. (Source: U.S. Census Bureau)

Year	Population, a_n
1998	1250.4
1999	1260.1
2000	1268.9
2001	1276.9
2002	1284.3
2003	1291.5
2004	1298.8

(a) Use the *exponential regression* feature of a graphing utility to find a geometric sequence that models the data. Let n represent the year, with $n = 8$ corresponding to 1998.

(b) Use the sequence from part (a) to describe the rate at which the population of China is growing.

Model It (continued)

(c) Use the sequence from part (a) to predict the population of China in 2010. The U.S. Census Bureau predicts the population of China will be 1374.6 million in 2010. How does this value compare with your prediction?

(d) Use the sequence from part (a) to determine when the population of China will reach 1.32 billion.

---

100. Compound Interest A principal of $1000 is invested at 6% interest. Find the amount after 10 years if the interest is compounded (a) annually, (b) semiannually, (c) quarterly, (d) monthly, and (e) daily.

101. Compound Interest A principal of $2500 is invested at 2% interest. Find the amount after 20 years if the interest is compounded (a) annually, (b) semiannually, (c) quarterly, (d) monthly, and (e) daily.

102. Depreciation A tool and die company buys a machine for $135,000 and it depreciates at a rate of 30% per year. (In other words, at the end of each year the depreciated value is 70% of what it was at the beginning of the year.) Find the depreciated value of the machine after 5 full years.

103. Annuities A deposit of $100 is made at the beginning of each month in an account that pays 6%, compounded monthly. The balance A in the account at the end of 5 years is

$$A = 100\left(1 + \frac{0.06}{12}\right)^1 + \cdots + 100\left(1 + \frac{0.06}{12}\right)^{60}.$$

Find A.

104. Annuities A deposit of $50 is made at the beginning of each month in an account that pays 8%, compounded monthly. The balance A in the account at the end of 5 years is

$$A = 50\left(1 + \frac{0.08}{12}\right)^1 + \cdots + 50\left(1 + \frac{0.08}{12}\right)^{60}.$$

Find A.

105. Annuities A deposit of P dollars is made at the beginning of each month in an account earning an annual interest rate r, compounded monthly. The balance A after t years is

$$A = P\left(1 + \frac{r}{12}\right) + P\left(1 + \frac{r}{12}\right)^2 + \cdots + P\left(1 + \frac{r}{12}\right)^{12t}.$$

Show that the balance is

$$A = P\left[\left(1 + \frac{r}{12}\right)^{12t} - 1\right]\left(1 + \frac{12}{r}\right).$$

106. Annuities A deposit of P dollars is made at the beginning of each month in an account earning an annual interest rate r, compounded continuously. The balance A after t years is $A = Pe^{r/12} + Pe^{2r/12} + \cdots + Pe^{12tr/12}$. Show that the balance is

$$A = \frac{Pe^{r/12}(e^{rt} - 1)}{e^{r/12} - 1}.$$

Annuities In Exercises 107–110, consider making monthly deposits of P dollars in a savings account earning an annual interest rate r. Use the results of Exercises 105 and 106 to find the balance A after t years if the interest is compounded (a) monthly and (b) continuously.

107. $P = \$50$, $r = 7\%$, $t = 20$ years
108. $P = \$75$, $r = 3\%$, $t = 25$ years
109. $P = \$100$, $r = 4\%$, $t = 40$ years
110. $P = \$20$, $r = 6\%$, $t = 50$ years

111. Annuities Consider an initial deposit of P dollars in an account earning an annual interest rate r, compounded monthly. At the end of each month, a withdrawal of W dollars will occur and the account will be depleted in t years. The amount of the initial deposit required is

$$P = W\left(1 + \frac{r}{12}\right)^{-1} + W\left(1 + \frac{r}{12}\right)^{-2} + \cdots + W\left(1 + \frac{r}{12}\right)^{-12t}.$$

Show that the initial deposit is

$$P = W\left(\frac{12}{r}\right)\left[1 - \left(1 + \frac{r}{12}\right)^{-12t}\right].$$

112. Annuities Determine the amount required in a retirement account for an individual who retires at age 65 and wants an income of $2000 from the account each month for 20 years. Use the result of Exercise 111 and assume that the account earns 9% compounded monthly.

Multiplier Effect In Exercises 113–116, use the following information. A tax rebate has been given to property owners by the state government with the anticipation that each property owner spends approximately $p\%$ of the rebate, and in turn each recipient of this amount spends $p\%$ of what they receive, and so on. Economists refer to this exchange of money and its circulation within the economy as the "multiplier effect." The multiplier effect operates on the idea that the expenditures of one individual become the income of another individual. For the given tax rebate, find the total amount put back into the state's economy, if this effect continues without end.

Tax rebate	$p\%$
113. $400	75%
114. $250	80%
115. $600	72.5%
116. $450	77.5%

117. *Geometry* The sides of a square are 16 inches in length. A new square is formed by connecting the midpoints of the sides of the original square, and two of the resulting triangles are shaded (see figure). If this process is repeated five more times, determine the total area of the shaded region.

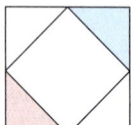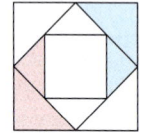

118. *Sales* The annual sales a_n (in millions of dollars) for Urban Outfitters for 1994 through 2003 can be approximated by the model

$$a_n = 54.6e^{0.172n}, \qquad n = 4, 5, \ldots, 13$$

where n represents the year, with $n = 4$ corresponding to 1994. Use this model and the formula for the sum of a finite geometric sequence to approximate the total sales earned during this 10-year period. (Source: Urban Outfitters Inc.)

119. *Salary* An investment firm has a job opening with a salary of $30,000 for the first year. Suppose that during the next 39 years, there is a 5% raise each year. Find the total compensation over the 40-year period.

120. *Distance* A ball is dropped from a height of 16 feet. Each time it drops h feet, it rebounds $0.81h$ feet.

(a) Find the total vertical distance traveled by the ball.

(b) The ball takes the following times (in seconds) for each fall.

$$s_1 = -16t^2 + 16, \qquad s_1 = 0 \text{ if } t = 1$$
$$s_2 = -16t^2 + 16(0.81), \qquad s_2 = 0 \text{ if } t = 0.9$$
$$s_3 = -16t^2 + 16(0.81)^2, \qquad s_3 = 0 \text{ if } t = (0.9)^2$$
$$s_4 = -16t^2 + 16(0.81)^3, \qquad s_4 = 0 \text{ if } t = (0.9)^3$$
$$\vdots \qquad\qquad \vdots$$
$$s_n = -16t^2 + 16(0.81)^{n-1}, \qquad s_n = 0 \text{ if } t = (0.9)^{n-1}$$

Beginning with s_2, the ball takes the same amount of time to bounce up as it does to fall, and so the total time elapsed before it comes to rest is

$$t = 1 + 2\sum_{n=1}^{\infty}(0.9)^n.$$

Find this total time.

Synthesis

True or False? **In Exercises 121 and 122, determine whether the statement is true or false. Justify your answer.**

121. A sequence is geometric if the ratios of consecutive differences of consecutive terms are the same.

122. You can find the nth term of a geometric sequence by multiplying its common ratio by the first term of the sequence raised to the $(n - 1)$th power.

123. *Writing* Write a brief paragraph explaining why the terms of a geometric sequence decrease in magnitude when $-1 < r < 1$.

124. Find two different geometric series with sums of 4.

Skills Review

In Exercises 125–128, evaluate the function for $f(x) = 3x + 1$ **and** $g(x) = x^2 - 1$.

125. $g(x + 1)$

126. $f(x + 1)$

127. $f(g(x + 1))$

128. $g(f(x + 1))$

In Exercises 129–132, completely factor the expression.

129. $9x^3 - 64x$

130. $x^2 + 4x - 63$

131. $6x^2 - 13x - 5$

132. $16x^2 - 4x^4$

In Exercises 133–138, perform the indicated operation(s) and simplify.

133. $\dfrac{3}{x + 3} \cdot \dfrac{x(x + 3)}{x - 3}$

134. $\dfrac{x - 2}{x + 7} \cdot \dfrac{2x(x + 7)}{6x(x - 2)}$

135. $\dfrac{x}{3} \div \dfrac{3x}{6x + 3}$

136. $\dfrac{x - 5}{x - 3} \div \dfrac{10 - 2x}{2(3 - x)}$

137. $5 + \dfrac{7}{x + 2} + \dfrac{2}{x - 2}$

138. $8 - \dfrac{x - 1}{x + 4} - \dfrac{4}{x - 1} - \dfrac{x + 4}{(x - 1)(x + 4)}$

139. **Make a Decision** To work an extended application analyzing the amounts spent on research and development in the United States from 1980 to 2003, visit this text's website at *college.hmco.com*. (*Data Source: U.S. National Science Foundation*)

Mathematical Induction

What you should learn

- Use mathematical induction to prove statements involving a positive integer n.
- Recognize patterns and write the nth term of a sequence.
- Find the sums of powers of integers.
- Find finite differences of sequences.

Why you should learn it

Finite differences can be used to determine what type of model can be used to represent a sequence. For instance, in Exercise 61 on page 634, you will use finite differences to find a model that represents the number of individual income tax returns filed in the United States from 1998 to 2003.

Mario Tama/Getty Images

Introduction

In this section, you will study a form of mathematical proof called **mathematical induction.** It is important that you see clearly the logical need for it, so take a closer look at the problem discussed in Example 5 in Section 8.2.

$$S_1 = 1 = 1^2$$
$$S_2 = 1 + 3 = 2^2$$
$$S_3 = 1 + 3 + 5 = 3^2$$
$$S_4 = 1 + 3 + 5 + 7 = 4^2$$
$$S_5 = 1 + 3 + 5 + 7 + 9 = 5^2$$

Judging from the pattern formed by these first five sums, it appears that the sum of the first n odd integers is

$$S_n = 1 + 3 + 5 + 7 + 9 + \cdots + (2n - 1) = n^2.$$

Although this particular formula *is* valid, it is important for you to see that recognizing a pattern and then simply *jumping to the conclusion* that the pattern must be true for all values of n is *not* a logically valid method of proof. There are many examples in which a pattern appears to be developing for small values of n and then at some point the pattern fails. One of the most famous cases of this was the conjecture by the French mathematician Pierre de Fermat (1601–1665), who speculated that all numbers of the form

$$F_n = 2^{2^n} + 1, \quad n = 0, 1, 2, \ldots$$

are prime. For $n = 0, 1, 2, 3,$ and 4, the conjecture is true.

$$F_0 = 3$$
$$F_1 = 5$$
$$F_2 = 17$$
$$F_3 = 257$$
$$F_4 = 65,537$$

The size of the next Fermat number ($F_5 = 4,294,967,297$) is so great that it was difficult for Fermat to determine whether it was prime or not. However, another well-known mathematician, Leonhard Euler (1707–1783), later found the factorization

$$F_5 = 4,294,967,297$$
$$= 641(6,700,417)$$

which proved that F_5 is not prime and therefore Fermat's conjecture was false.

Just because a rule, pattern, or formula seems to work for several values of n, you cannot simply decide that it is valid for all values of n without going through a *legitimate proof.* Mathematical induction is one method of proof.

Example 1   **Finding Binomial Coefficients**

Find each binomial coefficient.

a. $_8C_2$ **b.** $\begin{pmatrix} 10 \\ 3 \end{pmatrix}$ **c.** $_7C_0$ **d.** $\begin{pmatrix} 8 \\ 8 \end{pmatrix}$

Solution

a. $_8C_2 = \dfrac{8!}{6! \cdot 2!} = \dfrac{(8 \cdot 7) \cdot 6!}{6! \cdot 2!} = \dfrac{8 \cdot 7}{2 \cdot 1} = 28$

b. $\begin{pmatrix} 10 \\ 3 \end{pmatrix} = \dfrac{10!}{7! \cdot 3!} = \dfrac{(10 \cdot 9 \cdot 8) \cdot 7!}{7! \cdot 3!} = \dfrac{10 \cdot 9 \cdot 8}{3 \cdot 2 \cdot 1} = 120$

c. $_7C_0 = \dfrac{7!}{7! \cdot 0!} = 1$ **d.** $\begin{pmatrix} 8 \\ 8 \end{pmatrix} = \dfrac{8!}{0! \cdot 8!} = 1$

✓**CHECKPOINT**   Now try Exercise 1.

When $r \neq 0$ and $r \neq n$, as in parts (a) and (b) above, there is a simple pattern for evaluating binomial coefficients that works because there will always be factorial terms that divide out from the expression.

$$\underbrace{_8C_2 = \dfrac{\overbrace{8 \cdot 7}^{\text{2 factors}}}{8 \cdot 7}}_{\text{2 factors}} \quad \text{and} \quad \underbrace{\begin{pmatrix} 10 \\ 3 \end{pmatrix} = \dfrac{\overbrace{10 \cdot 9 \cdot 8}^{\text{3 factors}}}{3 \cdot 2 \cdot 1}}_{\text{3 factors}}$$

Example 2   **Finding Binomial Coefficients**

Find each binomial coefficient.

a. $_7C_3$ **b.** $\begin{pmatrix} 7 \\ 4 \end{pmatrix}$ **c.** $_{12}C_1$ **d.** $\begin{pmatrix} 12 \\ 11 \end{pmatrix}$

Solution

a. $_7C_3 = \dfrac{7 \cdot 6 \cdot 5}{3 \cdot 2 \cdot 1} = 35$

b. $\begin{pmatrix} 7 \\ 4 \end{pmatrix} = \dfrac{7 \cdot 6 \cdot 5 \cdot 4}{4 \cdot 3 \cdot 2 \cdot 1} = 35$

c. $_{12}C_1 = \dfrac{12}{1} = 12$

d. $\begin{pmatrix} 12 \\ 11 \end{pmatrix} = \dfrac{12!}{1! \cdot 11!} = \dfrac{(12) \cdot 11!}{1! \cdot 11!} = \dfrac{12}{1} = 12$

✓**CHECKPOINT**   Now try Exercise 7.

It is not a coincidence that the results in parts (a) and (b) of Example 2 are the same and that the results in parts (c) and (d) are the same. In general, it is true that

$$_nC_r = {_nC_{n-r}}.$$

This shows the symmetric property of binomial coefficients that was identified earlier.

Pascal's Triangle

There is a convenient way to remember the pattern for binomial coefficients. By arranging the coefficients in a triangular pattern, you obtain the following array, which is called **Pascal's Triangle.** This triangle is named after the famous French mathematician Blaise Pascal (1623–1662).

$$
\begin{array}{ccccccccccccccc}
 & & & & & & & 1 & & & & & & & \\
 & & & & & & 1 & & 1 & & & & & & \\
 & & & & & 1 & & 2 & & 1 & & & & & \\
 & & & & 1 & & 3 & & 3 & & 1 & & & & \\
 & & & 1 & & 4 & & 6 & & 4 & & 1 & & & \quad 4 + 6 = 10\\
 & & 1 & & 5 & & 10 & & 10 & & 5 & & 1 & & \\
 & 1 & & 6 & & 15 & & 20 & & 15 & & 6 & & 1 & \\
1 & & 7 & & 21 & & 35 & & 35 & & 21 & & 7 & & 1 \quad 15 + 6 = 21
\end{array}
$$

The first and last numbers in each row of Pascal's Triangle are 1. Every other number in each row is formed by adding the two numbers immediately above the number. Pascal noticed that numbers in this triangle are precisely the same numbers that are the coefficients of binomial expansions, as follows.

$$(x + y)^0 = 1 \qquad\qquad\qquad\qquad \text{0th row}$$

$$(x + y)^1 = 1x + 1y \qquad\qquad\qquad \text{1st row}$$

$$(x + y)^2 = 1x^2 + 2xy + 1y^2 \qquad\qquad \text{2nd row}$$

$$(x + y)^3 = 1x^3 + 3x^2y + 3xy^2 + 1y^3 \qquad \text{3rd row}$$

$$(x + y)^4 = 1x^4 + 4x^3y + 6x^2y^2 + 4xy^3 + 1y^4 \qquad \vdots$$

$$(x + y)^5 = 1x^5 + 5x^4y + 10x^3y^2 + 10x^2y^3 + 5xy^4 + 1y^5$$

$$(x + y)^6 = 1x^6 + 6x^5y + 15x^4y^2 + 20x^3y^3 + 15x^2y^4 + 6xy^5 + 1y^6$$

$$(x + y)^7 = 1x^7 + 7x^6y + 21x^5y^2 + 35x^4y^3 + 35x^3y^4 + 21x^2y^5 + 7xy^6 + 1y^7$$

The top row in Pascal's Triangle is called the *zeroth row* because it corresponds to the binomial expansion $(x + y)^0 = 1$. Similarly, the next row is called the *first row* because it corresponds to the binomial expansion $(x + y)^1 = 1(x) + 1(y)$. In general, the *nth row* in Pascal's Triangle gives the coefficients of $(x + y)^n$.

Example 3   Using Pascal's Triangle

Use the seventh row of Pascal's Triangle to find the binomial coefficients.

$$_8C_0,\ _8C_1,\ _8C_2,\ _8C_3,\ _8C_4,\ _8C_5,\ _8C_6,\ _8C_7,\ _8C_8$$

Solution

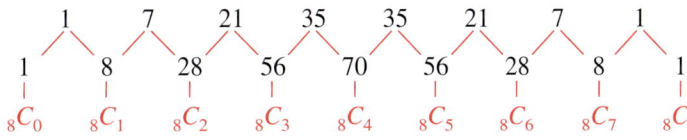

$$
\begin{array}{ccccccccc}
 & 1 & 7 & 21 & 35 & 35 & 21 & 7 & 1 \\
1 & 8 & 28 & 56 & 70 & 56 & 28 & 8 & 1 \\
_8C_0 & _8C_1 & _8C_2 & _8C_3 & _8C_4 & _8C_5 & _8C_6 & _8C_7 & _8C_8
\end{array}
$$

✓CHECKPOINT   Now try Exercise 11.

Exploration

Complete the table and describe the result.

n	r	$_nC_r$	$_nC_{n-r}$
9	5		
7	1		
12	4		
6	0		
10	7		

What characteristic of Pascal's Triangle is illustrated by this table?

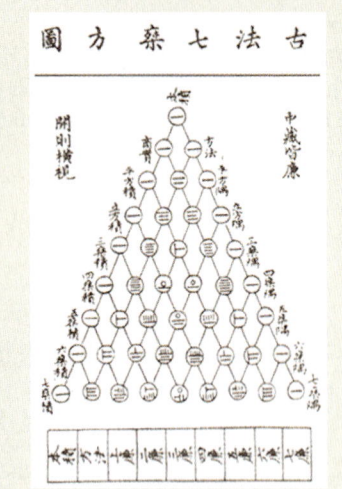

Binomial Expansions

As mentioned at the beginning of this section, when you write out the coefficients for a binomial that is raised to a power, you are **expanding a binomial.** The formulas for binomial coefficients give you an easy way to expand binomials, as demonstrated in the next four examples.

Example 4 **Expanding a Binomial**

Write the expansion for the expression

$(x + 1)^3$.

Solution

The binomial coefficients from the third row of Pascal's Triangle are

$1, 3, 3, 1.$

So, the expansion is as follows.

$$(x + 1)^3 = (1)x^3 + (3)x^2(1) + (3)x(1^2) + (1)(1^3)$$
$$= x^3 + 3x^2 + 3x + 1$$

✔CHECKPOINT Now try Exercise 15.

To expand binomials representing *differences* rather than sums, you alternate signs. Here are two examples.

$$(x - 1)^3 = x^3 - 3x^2 + 3x - 1$$
$$(x - 1)^4 = x^4 - 4x^3 + 6x^2 - 4x + 1$$

Example 5 **Expanding a Binomial**

Write the expansion for each expression.

a. $(2x - 3)^4$ **b.** $(x - 2y)^4$

Solution

The binomial coefficients from the fourth row of Pascal's Triangle are

$1, 4, 6, 4, 1.$

Therefore, the expansions are as follows.

a. $(2x - 3)^4 = (1)(2x)^4 - (4)(2x)^3(3) + (6)(2x)^2(3^2) - (4)(2x)(3^3) + (1)(3^4)$
$$= 16x^4 - 96x^3 + 216x^2 - 216x + 81$$

b. $(x - 2y)^4 = (1)x^4 - (4)x^3(2y) + (6)x^2(2y)^2 - (4)x(2y)^3 + (1)(2y)^4$
$$= x^4 - 8x^3y + 24x^2y^2 - 32xy^3 + 16y^4$$

✔CHECKPOINT Now try Exercise 19.

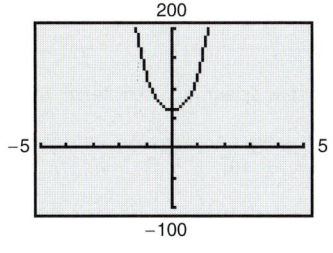
Example 6   **Expanding a Binomial**

Write the expansion for $(x^2 + 4)^3$.

Solution

Use the third row of Pascal's Triangle, as follows.

$$(x^2 + 4)^3 = (1)(x^2)^3 + (3)(x^2)^2(4) + (3)x^2(4^2) + (1)(4^3)$$

$$= x^6 + 12x^4 + 48x^2 + 64$$

✓**CHECKPOINT**   Now try Exercise 29.

Sometimes you will need to find a specific term in a binomial expansion. Instead of writing out the entire expansion, you can use the fact that, from the Binomial Theorem, the $(r + 1)$th term is $_nC_r\, x^{n-r}\, y^r$.

Example 7   **Finding a Term in a Binomial Expansion**

a. Find the sixth term of $(a + 2b)^8$.

b. Find the coefficient of the term a^6b^5 in the expansion of $(3a - 2b)^{11}$.

Solution

a. Remember that the formula is for the $(r + 1)$th term, so r is one less than the number of the term you are looking for. So, to find the sixth term in this binomial expansion, use $r = 5$, $n = 8$, $x = a$, and $y = 2b$, as shown.

$$_8C_5\, a^{8-5}(2b)^5 = 56 \cdot a^3 \cdot (2b)^5 = 56(2^5)a^3b^5 = 1792a^3b^5.$$

b. In this case, $n = 11$, $r = 5$, $x = 3a$, and $y = -2b$. Substitute these values to obtain

$$_nC_r\, x^{n-r}\, y^r = {}_{11}C_5(3a)^6(-2b)^5$$

$$= (462)(729a^6)(-32b^5)$$

$$= -10{,}777{,}536a^6b^5.$$

So, the coefficient is $-10{,}777{,}536$.

✓**CHECKPOINT**   Now try Exercise 41.

_W_RITING ABOUT _M_ATHEMATICS

Error Analysis   You are a math instructor and receive the following solutions from one of your students on a quiz. Find the error(s) in each solution. Discuss ways that your student could avoid the error(s) in the future.

a. Find the second term in the expansion of $(2x - 3y)^5$.

$$5(2x)^4(3y)^2 = 720x^4y^2$$

b. Find the fourth term in the expansion of $\left(\frac{1}{2}x + 7y\right)^6$.

$$_6C_4\left(\tfrac{1}{2}x\right)^2(7y)^4 = 9003.75x^2y^4$$

8.5 | Exercises

VOCABULARY CHECK: Fill in the blanks.

1. The coefficients of a binomial expansion are called _____ _____.

2. To find binomial coefficients, you can use the _____ _____ or _____ _____.

3. The notation used to denote a binomial coefficient is _____ or _____.

4. When you write out the coefficients for a binomial that is raised to a power, you are _____ a _____.

PREREQUISITE SKILLS REVIEW: Practice and review algebra skills needed for this section at **www.Eduspace.com.**

In Exercises 1–10, calculate the binomial coefficient.

1. $_5C_3$

2. $_8C_6$

3. $_{12}C_0$

4. $_{20}C_{20}$

5. $_{20}C_{15}$

6. $_{12}C_5$

7. $\binom{10}{4}$

8. $\binom{10}{6}$

9. $\binom{100}{98}$

10. $\binom{100}{2}$

In Exercises 11–14, evaluate using Pascal's Triangle.

11. $\binom{8}{5}$

12. $\binom{8}{7}$

13. $_7C_4$

14. $_6C_3$

In Exercises 15–34, use the Binomial Theorem to expand and simplify the expression.

15. $(x + 1)^4$

16. $(x + 1)^6$

17. $(a + 6)^4$

18. $(a + 5)^5$

19. $(y - 4)^3$

20. $(y - 2)^5$

21. $(x + y)^5$

22. $(c + d)^3$

23. $(r + 3s)^6$

24. $(x + 2y)^4$

25. $(3a - 4b)^5$

26. $(2x - 5y)^5$

27. $(2x + y)^3$

28. $(7a + b)^3$

29. $(x^2 + y^2)^4$

30. $(x^2 + y^2)^6$

31. $\left(\frac{1}{x} + y\right)^5$

32. $\left(\frac{1}{x} + 2y\right)^6$

33. $2(x - 3)^4 + 5(x - 3)^2$

34. $3(x + 1)^5 - 4(x + 1)^3$

In Exercises 35–38, expand the binomial by using Pascal's Triangle to determine the coefficients.

35. $(2t - s)^5$

36. $(3 - 2z)^4$

37. $(x + 2y)^5$

38. $(2v + 3)^6$

In Exercises 39–46, find the specified nth term in the expansion of the binomial.

39. $(x + y)^{10}$, $n = 4$

40. $(x - y)^6$, $n = 7$

41. $(x - 6y)^5$, $n = 3$

42. $(x - 10z)^7$, $n = 4$

43. $(4x + 3y)^9$, $n = 8$

44. $(5a + 6b)^5$, $n = 5$

45. $(10x - 3y)^{12}$, $n = 9$

46. $(7x + 2y)^{15}$, $n = 7$

In Exercises 47–54, find the coefficient a of the term in the expansion of the binomial.

Binomial	Term
47. $(x + 3)^{12}$	ax^5
48. $(x^2 + 3)^{12}$	ax^8
49. $(x - 2y)^{10}$	ax^8y^2
50. $(4x - y)^{10}$	ax^2y^8
51. $(3x - 2y)^9$	ax^4y^5
52. $(2x - 3y)^8$	ax^6y^2
53. $(x^2 + y)^{10}$	ax^8y^6
54. $(z^2 - t)^{10}$	az^4t^8

In Exercises 55–58, use the Binomial Theorem to expand and simplify the expression.

55. $\left(\sqrt{x} + 3\right)^4$

56. $\left(2\sqrt{t} - 1\right)^3$

57. $(x^{2/3} - y^{1/3})^3$

58. $(u^{3/5} + 2)^5$

∫ **In Exercises 59–62, expand the expression in the difference quotient and simplify.**

$$\frac{f(x + h) - f(x)}{h} \qquad \textcolor{red}{\text{Difference quotient}}$$

59. $f(x) = x^3$

60. $f(x) = x^4$

61. $f(x) = \sqrt{x}$

62. $f(x) = \dfrac{1}{x}$

In Exercises 63–68, use the Binomial Theorem to expand the complex number. Simplify your result.

63. $(1 + i)^4$

64. $(2 - i)^5$

65. $(2 - 3i)^6$

66. $\left(5 + \sqrt{-9}\right)^3$

67. $\left(-\dfrac{1}{2} + \dfrac{\sqrt{3}}{2}i\right)^3$

68. $\left(5 - \sqrt{3}i\right)^4$

Approximation In Exercises 69–72, use the Binomial Theorem to approximate the quantity accurate to three decimal places. For example, in Exercise 69, use the expansion

$$(1.02)^8 = (1 + 0.02)^8 = 1 + 8(0.02) + 28(0.02)^2 + \cdots.$$

69. $(1.02)^8$

70. $(2.005)^{10}$

71. $(2.99)^{12}$

72. $(1.98)^9$

Graphical Reasoning In Exercises 73 and 74, use a graphing utility to graph f and g in the same viewing window. What is the relationship between the two graphs? Use the Binomial Theorem to write the polynomial function g in standard form.

73. $f(x) = x^3 - 4x$, $g(x) = f(x + 4)$

74. $f(x) = -x^4 + 4x^2 - 1$, $g(x) = f(x - 3)$

Probability In Exercises 75–78, consider n independent trials of an experiment in which each trial has two possible outcomes: "success" or "failure." The probability of a success on each trial is p, and the probability of a failure is $q = 1 - p$. In this context, the term $_nC_k\, p^k q^{n-k}$ in the expansion of $(p + q)^n$ gives the probability of k successes in the n trials of the experiment.

75. A fair coin is tossed seven times. To find the probability of obtaining four heads, evaluate the term

$$_7C_4\left(\tfrac{1}{2}\right)^4\left(\tfrac{1}{2}\right)^3$$

in the expansion of $\left(\tfrac{1}{2} + \tfrac{1}{2}\right)^7$.

76. The probability of a baseball player getting a hit during any given time at bat is $\tfrac{1}{4}$. To find the probability that the player gets three hits during the next 10 times at bat, evaluate the term

$$_{10}C_3\left(\tfrac{1}{4}\right)^3\left(\tfrac{3}{4}\right)^7$$

in the expansion of $\left(\tfrac{1}{4} + \tfrac{3}{4}\right)^{10}$.

77. The probability of a sales representative making a sale with any one customer is $\tfrac{1}{3}$. The sales representative makes eight contacts a day. To find the probability of making four sales, evaluate the term

$$_8C_4\left(\tfrac{1}{3}\right)^4\left(\tfrac{2}{3}\right)^4$$

in the expansion of $\left(\tfrac{1}{3} + \tfrac{2}{3}\right)^8$.

78. To find the probability that the sales representative in Exercise 77 makes four sales if the probability of a sale with any one customer is $\tfrac{1}{2}$, evaluate the term

$$_8C_4\left(\tfrac{1}{2}\right)^4\left(\tfrac{1}{2}\right)^4$$

in the expansion of $\left(\tfrac{1}{2} + \tfrac{1}{2}\right)^8$.

Model It

79. *Data Analysis: Water Consumption* The table shows the per capita consumption of bottled water $f(t)$ (in gallons) in the United States from 1990 through 2003. (Source: Economic Research Service, U.S. Department of Agriculture)

Year	Consumption, $f(t)$
1990	8.0
1991	8.0
1992	9.7
1993	10.3
1994	11.3
1995	12.1
1996	13.0
1997	13.9
1998	15.0
1999	16.4
2000	17.4
2001	18.8
2002	20.7
2003	22.0

(a) Use the *regression* feature of a graphing utility to find a cubic model for the data. Let t represent the year, with $t = 0$ corresponding to 1990.

(b) Use a graphing utility to plot the data and the model in the same viewing window.

(c) You want to adjust the model so that $t = 0$ corresponds to 2000 rather than 1990. To do this, you shift the graph of f 10 units *to the left* to obtain $g(t) = f(t + 10)$. Write $g(t)$ in standard form.

(d) Use a graphing utility to graph g in the same viewing window as f.

(e) Use both models to estimate the per capita consumption of bottled water in 2008. Do you obtain the same answer?

(f) Describe the overall trend in the data. What factors do you think may have contributed to the increase in the per capita consumption of bottled water?

80. *Child Support* The amounts $f(t)$ (in billions of dollars) of child support collected in the United States from 1990 to 2002 can be approximated by the model

$$f(t) = 0.031t^2 + 0.82t + 6.1, \quad 0 \le t \le 12$$

where t represents the year, with $t = 0$ corresponding to 1990 (see figure). (Source: U.S. Department of Health and Human Services)

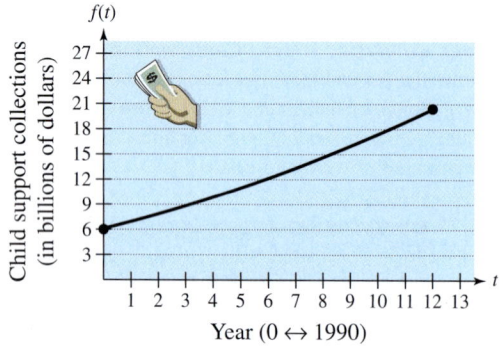

Year $(0 \leftrightarrow 1990)$

(a) You want to adjust the model so that $t = 0$ corresponds to 2000 rather than 1990. To do this, you shift the graph of f 10 units *to the left* and obtain $g(t) = f(t + 10)$. Write $g(t)$ in standard form.

 (b) Use a graphing utility to graph f and g in the same viewing window.

 (c) Use the graphs to estimate when the child support collections will exceed \$30 billion.

Synthesis

True or False? In Exercises 81–83, determine whether the statement is true or false. Justify your answer.

81. The Binomial Theorem could be used to produce each row of Pascal's Triangle.

82. A binomial that represents a difference cannot always be accurately expanded using the Binomial Theorem.

83. The x^{10}-term and the x^{14}-term of the expansion of $(x^2 + 3)^{12}$ have identical coefficients.

84. *Writing* In your own words, explain how to form the rows of Pascal's Triangle.

85. Form rows 8–10 of Pascal's Triangle.

86. *Think About It* How many terms are in the expansion of $(x + y)^n$?

87. *Think About It* How do the expansions of $(x + y)^n$ and $(x - y)^n$ differ?

 88. *Graphical Reasoning* Which two functions have identical graphs, and why? Use a graphing utility to graph the functions in the given order and in the same viewing window. Compare the graphs.

(a) $f(x) = (1 - x)^3$

(b) $g(x) = 1 - x^3$

(c) $h(x) = 1 + 3x + 3x^2 + x^3$

(d) $k(x) = 1 - 3x + 3x^2 - x^3$

(e) $p(x) = 1 + 3x - 3x^2 + x^3$

Proof In Exercises 89–92, prove the property for all integers r and n where $0 \le r \le n$.

89. $_nC_r = {_nC_{n-r}}$

90. $_nC_0 - {_nC_1} + {_nC_2} - \cdots \pm {_nC_n} = 0$

91. $_{n+1}C_r = {_nC_r} + {_nC_{r-1}}$

92. The sum of the numbers in the nth row of Pascal's Triangle is 2^n.

Skills Review

In Exercises 93–96, the graph of $y = g(x)$ is shown. Graph f and use the graph to write an equation for the graph of g.

93. $f(x) = x^2$

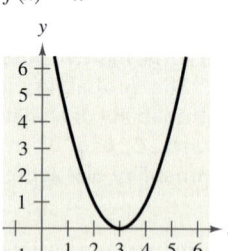

94. $f(x) = x^2$

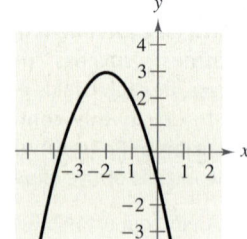

95. $f(x) = \sqrt{x}$

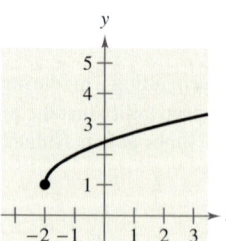

96. $f(x) = \sqrt{x}$

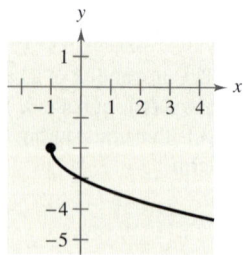

In Exercises 97 and 98, find the inverse of the matrix.

97. $\begin{bmatrix} -6 & 5 \\ -5 & 4 \end{bmatrix}$

98. $\begin{bmatrix} 1.2 & -2.3 \\ -2 & 4 \end{bmatrix}$

8.6 Counting Principles

Simple Counting Problems

This section and Section 8.7 present a brief introduction to some of the basic counting principles and their application to probability. In Section 8.7, you will see that much of probability has to do with counting the number of ways an event can occur. The following two examples describe simple counting problems.

Example 1   Selecting Pairs of Numbers at Random

Eight pieces of paper are numbered from 1 to 8 and placed in a box. One piece of paper is drawn from the box, its number is written down, and the piece of paper is *replaced in the box*. Then, a second piece of paper is drawn from the box, and its number is written down. Finally, the two numbers are added together. How many different ways can a sum of 12 be obtained?

Solution

To solve this problem, count the different ways that a sum of 12 can be obtained using two numbers from 1 to 8.

First number	4	5	6	7	8
Second number	8	7	6	5	4

From this list, you can see that a sum of 12 can occur in five different ways.

✓**CHECKPOINT**   Now try Exercise 5.

Example 2   Selecting Pairs of Numbers at Random

Eight pieces of paper are numbered from 1 to 8 and placed in a box. Two pieces of paper are drawn from the box *at the same time*, and the numbers on the pieces of paper are written down and totaled. How many different ways can a sum of 12 be obtained?

Solution

To solve this problem, count the different ways that a sum of 12 can be obtained *using two different numbers* from 1 to 8.

First number	4	5	7	8
Second number	8	7	5	4

So, a sum of 12 can be obtained in four different ways.

✓**CHECKPOINT**   Now try Exercise 7.

The difference between the counting problems in Examples 1 and 2 can be described by saying that the random selection in Example 1 occurs **with replacement,** whereas the random selection in Example 2 occurs **without replacement,** which eliminates the possibility of choosing two 6's.

The Fundamental Counting Principle

Examples 1 and 2 describe simple counting problems in which you can *list* each possible way that an event can occur. When it is possible, this is always the best way to solve a counting problem. However, some events can occur in so many different ways that it is not feasible to write out the entire list. In such cases, you must rely on formulas and counting principles. The most important of these is the **Fundamental Counting Principle.**

Fundamental Counting Principle

Let E_1 and E_2 be two events. The first event E_1 can occur in m_1 different ways. After E_1 has occurred, E_2 can occur in m_2 different ways. The number of ways that the two events can occur is $m_1 \cdot m_2$.

The Fundamental Counting Principle can be extended to three or more events. For instance, the number of ways that three events E_1, E_2, and E_3 can occur is $m_1 \cdot m_2 \cdot m_3$.

Example 3   **Using the Fundamental Counting Principle**

How many different pairs of letters from the English alphabet are possible?

Solution

There are two events in this situation. The first event is the choice of the first letter, and the second event is the choice of the second letter. Because the English alphabet contains 26 letters, it follows that the number of two-letter pairs is $26 \cdot 26 = 676$.

✔CHECKPOINT   Now try Exercise 13.

Example 4   **Using the Fundamental Counting Principle**

Telephone numbers in the United States currently have 10 digits. The first three are the *area code* and the next seven are the *local telephone number*. How many different telephone numbers are possible within each area code? (Note that at this time, a local telephone number cannot begin with 0 or 1.)

Solution

Because the first digit of a local telephone number cannot be 0 or 1, there are only eight choices for the first digit. For each of the other six digits, there are 10 choices.

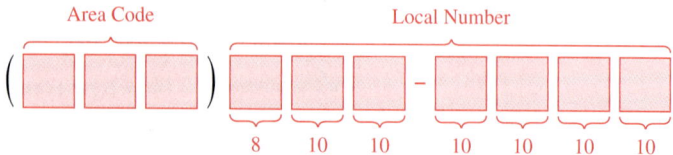

So, the number of local telephone numbers that are possible *within* each area code is $8 \cdot 10 \cdot 10 \cdot 10 \cdot 10 \cdot 10 \cdot 10 = 8{,}000{,}000$.

✔CHECKPOINT   Now try Exercise 19.

Permutations

One important application of the Fundamental Counting Principle is in determining the number of ways that *n* elements can be arranged (in order). An ordering of *n* elements is called a **permutation** of the elements.

> ## Definition of Permutation
>
> A **permutation** of *n* different elements is an ordering of the elements such that one element is first, one is second, one is third, and so on.

Example 5   **Finding the Number of Permutations of *n* Elements**

How many permutations are possible for the letters A, B, C, D, E, and F?

Solution

Consider the following reasoning.

First position: Any of the *six* letters

Second position: Any of the remaining *five* letters

Third position: Any of the remaining *four* letters

Fourth position: Any of the remaining *three* letters

Fifth position: Any of the remaining *two* letters

Sixth position: The *one* remaining letter

So, the numbers of choices for the six positions are as follows.

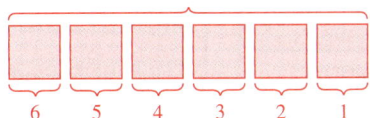

Permutations of six letters

The total number of permutations of the six letters is

$$6! = 6 \cdot 5 \cdot 4 \cdot 3 \cdot 2 \cdot 1$$

$$= 720.$$

✓**CHECKPOINT**   Now try Exercise 39.

> ## Number of Permutations of *n* Elements
>
> The number of permutations of *n* elements is
>
> $$n \cdot (n - 1) \cdots 4 \cdot 3 \cdot 2 \cdot 1 = n!.$$
>
> In other words, there are *n*! different ways that *n* elements can be ordered.

Eleven thoroughbred racehorses hold the title of Triple Crown winner for winning the Kentucky Derby, the Preakness, and the Belmont Stakes in the same year. Forty-nine horses have won two out of the three races.

Example 6   **Counting Horse Race Finishes**

Eight horses are running in a race. In how many different ways can these horses come in first, second, and third? (Assume that there are no ties.)

Solution

Here are the different possibilities.

Win (first position): *Eight* choices

Place (second position): *Seven* choices

Show (third position): *Six* choices

Using the Fundamental Counting Principle, multiply these three numbers together to obtain the following.

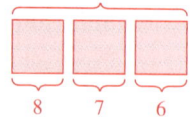

Different orders of horses

8   7   6

So, there are $8 \cdot 7 \cdot 6 = 336$ different orders.

✓**CHECKPOINT**   Now try Exercise 43.

It is useful, on occasion, to order a *subset* of a collection of elements rather than the entire collection. For example, you might want to choose and order r elements out of a collection of n elements. Such an ordering is called a **permutation of n elements taken r at a time.**

Technology

Most graphing calculators are programmed to evaluate $_nP_r$. Consult the user's guide for your calculator and then evaluate $_8P_5$. You should get an answer of 6720.

Permutations of n Elements Taken r at a Time

The number of permutations of n elements taken r at a time is

$$_nP_r = \frac{n!}{(n-r)!}$$

$$= n(n-1)(n-2) \cdots (n-r+1).$$

Using this formula, you can rework Example 6 to find that the number of permutations of eight horses taken three at a time is

$$_8P_3 = \frac{8!}{(8-3)!}$$

$$= \frac{8!}{5!}$$

$$= \frac{8 \cdot 7 \cdot 6 \cdot 5!}{5!}$$

$$= 336$$

which is the same answer obtained in the example.

Remember that for permutations, order is important. So, if you are looking at the possible permutations of the letters A, B, C, and D taken three at a time, the permutations (A, B, D) and (B, A, D) are counted as different because the *order* of the elements is different.

Suppose, however, that you are asked to find the possible permutations of the letters A, A, B, and C. The total number of permutations of the four letters would be $_4P_4 = 4!$. However, not all of these arrangements would be *distinguishable* because there are two A's in the list. To find the number of distinguishable permutations, you can use the following formula.

Distinguishable Permutations

Suppose a set of n objects has n_1 of one kind of object, n_2 of a second kind, n_3 of a third kind, and so on, with

$$n = n_1 + n_2 + n_3 + \cdots + n_k.$$

Then the number of **distinguishable permutations** of the n objects is

$$\frac{n!}{n_1! \cdot n_2! \cdot n_3! \cdots \cdots n_k!}.$$

Example 7   **Distinguishable Permutations**

In how many distinguishable ways can the letters in BANANA be written?

Solution

This word has six letters, of which three are A's, two are N's, and one is a B. So, the number of distinguishable ways the letters can be written is

$$\frac{n!}{n_1! \cdot n_2! \cdot n_3!} = \frac{6!}{3! \cdot 2! \cdot 1!}$$

$$= \frac{6 \cdot 5 \cdot 4 \cdot 3!}{3! \cdot 2!}$$

$$= 60.$$

The 60 different distinguishable permutations are as follows.

AAABNN	AAANBN	AAANNB	AABANN	AABNAN	AABNNA
AANABN	AANANB	AANBAN	AANBNA	AANNAB	AANNBA
ABAANN	ABANAN	ABANNA	ABNAAN	ABNANA	ABNNAA
ANAABN	ANAANB	ANABAN	ANABNA	ANANAB	ANANBA
ANBAAN	ANBANA	ANBNAA	ANNAAB	ANNABA	ANNBAA
BAAANN	BAANAN	BAANNA	BANAAN	BANANA	BANNAA
BNAAAN	BNAANA	BNANAA	BNNAAA	NAAABN	NAAANB
NAABAN	NAABNA	NAANAB	NAANBA	NABAAN	NABANA
NABNAA	NANAAB	NANABA	NANBAA	NBAAAN	NBAANA
NBANAA	NBNAAA	NNAAAB	NNAABA	NNABAA	NNBAAA

✓CHECKPOINT   Now try Exercise 45.

Combinations

When you count the number of possible permutations of a set of elements, *order* is important. As a final topic in this section, you will look at a method of selecting subsets of a larger set in which order is *not* important. Such subsets are called **combinations of *n* elements taken *r* at a time.** For instance, the combinations

$$\{A, B, C\} \quad \text{and} \quad \{B, A, C\}$$

are equivalent because both sets contain the same three elements, and the order in which the elements are listed is not important. So, you would count only one of the two sets. A common example of how a combination occurs is a card game in which the player is free to reorder the cards after they have been dealt.

Example 8 Combinations of *n* Elements Taken *r* at a Time

In how many different ways can three letters be chosen from the letters A, B, C, D, and E? (The order of the three letters is not important.)

Solution

The following subsets represent the different combinations of three letters that can be chosen from the five letters.

$\{A, B, C\}$	$\{A, B, D\}$
$\{A, B, E\}$	$\{A, C, D\}$
$\{A, C, E\}$	$\{A, D, E\}$
$\{B, C, D\}$	$\{B, C, E\}$
$\{B, D, E\}$	$\{C, D, E\}$

From this list, you can conclude that there are 10 different ways that three letters can be chosen from five letters.

✔**CHECKPOINT** Now try Exercise 55.

Combinations of *n* Elements Taken *r* at a Time

The number of combinations of *n* elements taken *r* at a time is

$$_nC_r = \frac{n!}{(n-r)!r!} \quad \text{which is equivalent to} \quad _nC_r = \frac{_nP_r}{r!}.$$

Note that the formula for $_nC_r$ is the same one given for binomial coefficients. To see how this formula is used, solve the counting problem in Example 8. In that problem, you are asked to find the number of combinations of five elements taken three at a time. So, $n = 5$, $r = 3$, and the number of combinations is

$$_5C_3 = \frac{5!}{2!3!} = \frac{5 \cdot 4 \cdot \overset{2}{\cancel{3!}}}{2 \cdot 1 \cdot \cancel{3!}} = 10$$

which is the same answer obtained in Example 8.

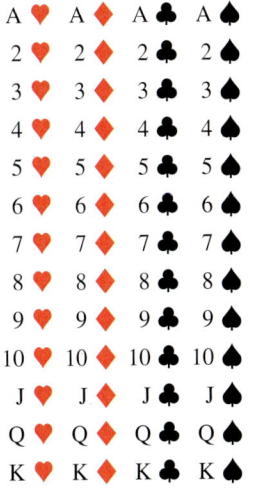

A ♥	A ♦	A ♣	A ♠
2 ♥	2 ♦	2 ♣	2 ♠
3 ♥	3 ♦	3 ♣	3 ♠
4 ♥	4 ♦	4 ♣	4 ♠
5 ♥	5 ♦	5 ♣	5 ♠
6 ♥	6 ♦	6 ♣	6 ♠
7 ♥	7 ♦	7 ♣	7 ♠
8 ♥	8 ♦	8 ♣	8 ♠
9 ♥	9 ♦	9 ♣	9 ♠
10 ♥	10 ♦	10 ♣	10 ♠
J ♥	J ♦	J ♣	J ♠
Q ♥	Q ♦	Q ♣	Q ♠
K ♥	K ♦	K ♣	K ♠

FIGURE 8.7   *Standard deck of playing cards*

Example 9   Counting Card Hands

A standard poker hand consists of five cards dealt from a deck of 52 (see Figure 8.7). How many different poker hands are possible? (After the cards are dealt, the player may reorder them, and so order is not important.)

Solution

You can find the number of different poker hands by using the formula for the number of combinations of 52 elements taken five at a time, as follows.

$$_{52}C_5 = \frac{52!}{(52-5)!5!}$$

$$= \frac{52!}{47!5!}$$

$$= \frac{52 \cdot 51 \cdot 50 \cdot 49 \cdot 48 \cdot \cancel{47!}}{5 \cdot 4 \cdot 3 \cdot 2 \cdot 1 \cdot \cancel{47!}}$$

$$= 2,598,960$$

✓CHECKPOINT   Now try Exercise 63.

Example 10   Forming a Team

You are forming a 12-member swim team from 10 girls and 15 boys. The team must consist of five girls and seven boys. How many different 12-member teams are possible?

Solution

There are $_{10}C_5$ ways of choosing five girls. The are $_{15}C_7$ ways of choosing seven boys. By the Fundamental Counting Principal, there are $_{10}C_5 \cdot {}_{15}C_7$ ways of choosing five girls and seven boys.

$$_{10}C_5 \cdot {}_{15}C_7 = \frac{10!}{5! \cdot 5!} \cdot \frac{15!}{8! \cdot 7!}$$

$$= 252 \cdot 6435$$

$$= 1,621,620$$

So, there are 1,621,620 12-member swim teams possible.

✓CHECKPOINT   Now try Exercise 65.

When solving problems involving counting principles, you need to be able to distinguish among the various counting principles in order to determine which is necessary to solve the problem correctly. To do this, ask yourself the following questions.

1. Is the order of the elements important? *Permutation*

2. Are the chosen elements a subset of a larger set in which order is not important? *Combination*

3. Does the problem involve two or more separate events? *Fundamental Counting Principle*

8.6 | Exercises

VOCABULARY CHECK: Fill in the blanks.

1. The _____ _____ _____ states that if there are m_1 ways for one event to occur and m_2 ways for a second event to occur, there are $m_1 \cdot m_2$ ways for both events to occur.

2. An ordering of n elements is called a _____ of the elements.

3. The number of permutations of n elements taken r at a time is given by the formula _____.

4. The number of _____ _____ of n objects is given by $\dfrac{n!}{n_1! n_2! n_3! \cdots n_k!}$.

5. When selecting subsets of a larger set in which order is not important, you are finding the number of _____ of n elements taken r at a time.

PREREQUISITE SKILLS REVIEW: Practice and review algebra skills needed for this section at **www.Eduspace.com.**

Random Selection   **In Exercises 1–8, determine the number of ways a computer can randomly generate one or more such integers from 1 through 12.**

1. An odd integer
2. An even integer
3. A prime integer
4. An integer that is greater than 9
5. An integer that is divisible by 4
6. An integer that is divisible by 3
7. Two *distinct* integers whose sum is 9
8. Two *distinct* integers whose sum is 8

9. *Entertainment Systems*   A customer can choose one of three amplifiers, one of two compact disc players, and one of five speaker models for an entertainment system. Determine the number of possible system configurations.

10. *Job Applicants*   A college needs two additional faculty members: a chemist and a statistician. In how many ways can these positions be filled if there are five applicants for the chemistry position and three applicants for the statistics position?

11. *Course Schedule*   A college student is preparing a course schedule for the next semester. The student may select one of two mathematics courses, one of three science courses, and one of five courses from the social sciences and humanities. How many schedules are possible?

12. *Aircraft Boarding*   Eight people are boarding an aircraft. Two have tickets for first class and board before those in the economy class. In how many ways can the eight people board the aircraft?

13. *True-False Exam*   In how many ways can a six-question true-false exam be answered? (Assume that no questions are omitted.)

14. *True-False Exam*   In how many ways can a 12-question true-false exam be answered? (Assume that no questions are omitted.)

15. *License Plate Numbers*   In the state of Pennsylvania, each standard automobile license plate number consists of three letters followed by a four-digit number. How many distinct license plate numbers can be formed in Pennsylvania?

16. *License Plate Numbers*   In a certain state, each automobile license plate number consists of two letters followed by a four-digit number. To avoid confusion between "O" and "zero" and between "I" and "one," the letters "O" and "I" are not used. How many distinct license plate numbers can be formed in this state?

17. *Three-Digit Numbers*   How many three-digit numbers can be formed under each condition?

   (a) The leading digit cannot be zero.

   (b) The leading digit cannot be zero and no repetition of digits is allowed.

   (c) The leading digit cannot be zero and the number must be a multiple of 5.

   (d) The number is at least 400.

18. *Four-Digit Numbers*   How many four-digit numbers can be formed under each condition?

   (a) The leading digit cannot be zero.

   (b) The leading digit cannot be zero and no repetition of digits is allowed.

   (c) The leading digit cannot be zero and the number must be less than 5000.

   (d) The leading digit cannot be zero and the number must be even.

19. *Combination Lock*   A combination lock will open when the right choice of three numbers (from 1 to 40, inclusive) is selected. How many different lock combinations are possible?

20. *Combination Lock* A combination lock will open when the right choice of three numbers (from 1 to 50, inclusive) is selected. How many different lock combinations are possible?

21. *Concert Seats* Four couples have reserved seats in a row for a concert. In how many different ways can they be seated if

(a) there are no seating restrictions?

(b) the two members of each couple wish to sit together?

22. *Single File* In how many orders can four girls and four boys walk through a doorway single file if

(a) there are no restrictions?

(b) the girls walk through before the boys?

In Exercises 23–28, evaluate $_nP_r$.

23. $_4P_4$

24. $_5P_5$

25. $_8P_3$

26. $_{20}P_2$

27. $_5P_4$

28. $_7P_4$

In Exercises 29 and 30, solve for *n*.

29. $14 \cdot {}_nP_3 = {}_{n+2}P_4$

30. $_nP_5 = 18 \cdot {}_{n-2}P_4$

 In Exercises 31–36, evaluate using a graphing utility.

31. $_{20}P_5$

32. $_{100}P_5$

33. $_{100}P_3$

34. $_{10}P_8$

35. $_{20}C_5$

36. $_{10}C_7$

37. *Posing for a Photograph* In how many ways can five children posing for a photograph line up in a row?

38. *Riding in a Car* In how many ways can six people sit in a six-passenger car?

39. *Choosing Officers* From a pool of 12 candidates, the offices of president, vice-president, secretary, and treasurer will be filled. In how many different ways can the offices be filled?

40. *Assembly Line Production* There are four processes involved in assembling a product, and these processes can be performed in any order. The management wants to test each order to determine which is the least time-consuming. How many different orders will have to be tested?

In Exercises 41–44, find the number of distinguishable permutations of the group of letters.

41. A, A, G, E, E, E, M

42. B, B, B, T, T, T, T, T

43. A, L, G, E, B, R, A

44. M, I, S, S, I, S, S, I, P, P, I

45. Write all permutations of the letters A, B, C, and D.

46. Write all permutations of the letters A, B, C, and D if the letters B and C must remain between the letters A and D.

47. *Batting Order* A baseball coach is creating a nine-player batting order by selecting from a team of 15 players. How many different batting orders are possible?

48. *Athletics* Six sprinters have qualified for the finals in the 100-meter dash at the NCAA national track meet. In how many ways can the sprinters come in first, second, and third? (Assume there are no ties.)

49. *Jury Selection* From a group of 40 people, a jury of 12 people is to be selected. In how many different ways can the jury be selected?

50. *Committee Members* As of January 2005, the U.S. Senate Committee on Indian Affairs had 14 members. Assuming party affiliation was not a factor in selection, how many different committees were possible from the 100 U.S. senators?

51. Write all possible selections of two letters that can be formed from the letters A, B, C, D, E, and F. (The order of the two letters is not important.)

52. *Forming an Experimental Group* In order to conduct an experiment, five students are randomly selected from a class of 20. How many different groups of five students are possible?

53. *Lottery Choices* In the Massachusetts Mass Cash game, a player chooses five distinct numbers from 1 to 35. In how many ways can a player select the five numbers?

54. *Lottery Choices* In the Louisiana Lotto game, a player chooses six distinct numbers from 1 to 40. In how many ways can a player select the six numbers?

55. *Defective Units* A shipment of 10 microwave ovens contains three defective units. In how many ways can a vending company purchase four of these units and receive (a) all good units, (b) two good units, and (c) at least two good units?

56. *Interpersonal Relationships* The complexity of interpersonal relationships increases dramatically as the size of a group increases. Determine the numbers of different two-person relationships in groups of people of sizes (a) 3, (b) 8, (c) 12, and (d) 20.

57. *Poker Hand* You are dealt five cards from an ordinary deck of 52 playing cards. In how many ways can you get (a) a full house and (b) a five-card combination containing two jacks and three aces? (A full house consists of three of one kind and two of another. For example, A-A-A-5-5 and K-K-K-10-10 are full houses.)

58. *Job Applicants* A toy manufacturer interviews eight people for four openings in the research and development department of the company. Three of the eight people are women. If all eight are qualified, in how many ways can the employer fill the four positions if (a) the selection is random and (b) exactly two selections are women?

59. *Forming a Committee* A six-member research committee at a local college is to be formed having one administrator, three faculty members, and two students. There are seven administrators, 12 faculty members, and 20 students in contention for the committee. How many six-member committees are possible?

60. *Law Enforcement* A police department uses computer imaging to create digital photographs of alleged perpetrators from eyewitness accounts. One software package contains 195 hairlines, 99 sets of eyes and eyebrows, 89 noses, 105 mouths, and 74 chins and cheek structures.

(a) Find the possible number of different faces that the software could create.

(b) A eyewitness can clearly recall the hairline and eyes and eyebrows of a suspect. How many different faces can be produced with this information?

Geometry **In Exercises 61–64, find the number of diagonals of the polygon. (A line segment connecting any two nonadjacent vertices is called a *diagonal* of the polygon.)**

61. Pentagon

62. Hexagon

63. Octagon

64. Decagon (10 sides)

Model It

65. *Lottery* Powerball is a lottery game that is operated by the Multi-State Lottery Association and is played in 27 states, Washington D.C., and the U.S. Virgin Islands. The game is played by drawing five white balls out of a drum of 55 white balls (numbered 1–55) and one red powerball out of a drum of 42 red balls (numbered 1–42). The jackpot is won by matching all five white balls in any order and the red powerball.

(a) Find the possible number of winning Powerball numbers.

(b) Find the possible number of winning Powerball numbers if the jackpot is won by matching all five white balls in order and the red power ball.

(c) Compare the results of part (a) with a state lottery in which a jackpot is won by matching six balls from a drum of 55 balls.

---

66. *Permutations or Combinations?* Decide whether each scenario should be counted using permutations or combinations. Explain your reasoning.

(a) Number of ways 10 people can line up in a row for concert tickets

(b) Number of different arrangements of three types of flowers from an array of 20 types

(c) Number of three-digit pin numbers for a debit card

(d) Number of two-scoop ice cream cones created from 31 different flavors

Synthesis

True or False? **In Exercises 67 and 68, determine whether the statement is true or false. Justify your answer.**

67. The number of letter pairs that can be formed in any order from any of the first 13 letters in the alphabet (A–M) is an example of a permutation.

68. The number of permutations of n elements can be determined by using the Fundamental Counting Principle.

69. What is the relationship between $_nC_r$ and $_nC_{n-r}$?

70. Without calculating the numbers, determine which of the following is greater. Explain.

(a) The number of combinations of 10 elements taken six at a time

(b) The number of permutations of 10 elements taken six at a time

Proof **In Exercises 71–74, prove the identity.**

71. $_nP_{n-1} = _nP_n$

72. $_nC_n = _nC_0$

73. $_nC_{n-1} = _nC_1$

74. $_nC_r = \dfrac{_nP_r}{r!}$

 75. *Think About It* Can your calculator evaluate $_{100}P_{80}$? If not, explain why.

76. *Writing* Explain in words the meaning of $_nP_r$.

Skills Review

In Exercises 77–80, evaluate the function at each specified value of the independent variable and simplify.

77. $f(x) = 3x^2 + 8$

(a) $f(3)$ (b) $f(0)$ (c) $f(-5)$

78. $g(x) = \sqrt{x-3} + 2$

(a) $g(3)$ (b) $g(7)$ (c) $g(x+1)$

79. $f(x) = -|x-5| + 6$

(a) $f(-5)$ (b) $f(-1)$ (c) $f(11)$

80. $f(x) = \begin{cases} x^2 - 2x + 5, & x \le -4 \\ -x^2 - 2, & x > -4 \end{cases}$

(a) $f(-4)$ (b) $f(-1)$ (c) $f(-20)$

In Exercises 81–84, solve the equation. Round your answer to two decimal places, if necessary.

81. $\sqrt{x-3} = x - 6$

82. $\dfrac{4}{t} + \dfrac{3}{2t} = 1$

83. $\log_2(x-3) = 5$

84. $e^{x/3} = 16$

8.7   Probability

What you should learn

- Find the probabilities of events.
- Find the probabilities of mutually exclusive events.
- Find the probabilities of independent events.
- Find the probability of the complement of an event.

Why you should learn it

Probability applies to many games of chance. For instance, in Exercise 55, on page 664, you will calculate probabilities that relate to the game of roulette.

Hank de Lespinasse/The Image Bank

The Probability of an Event

Any happening for which the result is uncertain is called an **experiment.** The possible results of the experiment are **outcomes,** the set of all possible outcomes of the experiment is the **sample space** of the experiment, and any subcollection of a sample space is an **event.**

For instance, when a six-sided die is tossed, the sample space can be represented by the numbers 1 through 6. For this experiment, each of the outcomes is *equally likely*.

To describe sample spaces in such a way that each outcome is equally likely, you must sometimes distinguish between or among various outcomes in ways that appear artificial. Example 1 illustrates such a situation.

Example 1   **Finding a Sample Space**

Find the sample space for each of the following.

a. One coin is tossed.

b. Two coins are tossed.

c. Three coins are tossed.

Solution

a. Because the coin will land either heads up (denoted by H) or tails up (denoted by T), the sample space is

$$S = \{H, T\}.$$

b. Because either coin can land heads up or tails up, the possible outcomes are as follows.

$HH =$ heads up on both coins

$HT =$ heads up on first coin and tails up on second coin

$TH =$ tails up on first coin and heads up on second coin

$TT =$ tails up on both coins

So, the sample space is

$$S = \{HH, HT, TH, TT\}.$$

Note that this list distinguishes between the two cases HT and TH, even though these two outcomes appear to be similar.

c. Following the notation of part (b), the sample space is

$$S = \{HHH, HHT, HTH, HTT, THH, THT, TTH, TTT\}.$$

Note that this list distinguishes among the cases HHT, HTH, and THH, and among the cases HTT, THT, and TTH.

✓CHECKPOINT   Now try Exercise 1.

To calculate the probability of an event, count the number of outcomes in the event and in the sample space. The *number of outcomes* in event E is denoted by $n(E)$, and the number of outcomes in the sample space S is denoted by $n(S)$. The probability that event E will occur is given by $n(E)/n(S)$.

The Probability of an Event

If an event E has $n(E)$ equally likely outcomes and its sample space S has $n(S)$ equally likely outcomes, the **probability** of event E is

$$P(E) = \frac{n(E)}{n(S)}.$$

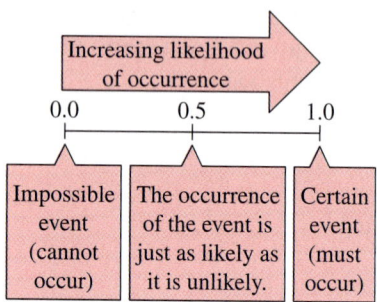

FIGURE 8.8

Because the number of outcomes in an event must be less than or equal to the number of outcomes in the sample space, the probability of an event must be a number between 0 and 1. That is,

$$0 \le P(E) \le 1$$

as indicated in Figure 8.8. If $P(E) = 0$, event E *cannot occur*, and E is called an **impossible event**. If $P(E) = 1$, event E *must occur*, and E is called a **certain event.**

Example 2 **Finding the Probability of an Event**

a. Two coins are tossed. What is the probability that both land heads up?

b. A card is drawn from a standard deck of playing cards. What is the probability that it is an ace?

Solution

a. Following the procedure in Example 1(b), let

$$E = \{HH\}$$

and

$$S = \{HH, HT, TH, TT\}.$$

The probability of getting two heads is

$$P(E) = \frac{n(E)}{n(S)} = \frac{1}{4}.$$

b. Because there are 52 cards in a standard deck of playing cards and there are four aces (one in each suit), the probability of drawing an ace is

$$P(E) = \frac{n(E)}{n(S)}$$

$$= \frac{4}{52}$$

$$= \frac{1}{13}.$$

✔CHECKPOINT   Now try Exercise 11.

FIGURE **8.9**

Example 3 **Finding the Probability of an Event**

Two six-sided dice are tossed. What is the probability that the total of the two dice is 7? (See Figure 8.9.)

Solution

Because there are six possible outcomes on each die, you can use the Fundamental Counting Principle to conclude that there are 6 · 6 or 36 different outcomes when two dice are tossed. To find the probability of rolling a total of 7, you must first count the number of ways in which this can occur.

First die	Second die
1	6
2	5
3	4
4	3
5	2
6	1

So, a total of 7 can be rolled in six ways, which means that the probability of rolling a 7 is

$$P(E) = \frac{n(E)}{n(S)} = \frac{6}{36} = \frac{1}{6}.$$

✓ CHECKPOINT   Now try Exercise 15.

STUDY TIP

You could have written out each sample space in Examples 2 and 3 and simply counted the outcomes in the desired events. For larger sample spaces, however, you should use the counting principles discussed in Section 8.6.

Example 4 **Finding the Probability of an Event**

Twelve-sided dice, as shown in Figure 8.10, can be constructed (in the shape of regular dodecahedrons) such that each of the numbers from 1 to 6 appears twice on each die. Prove that these dice can be used in any game requiring ordinary six-sided dice without changing the probabilities of different outcomes.

Solution

For an ordinary six-sided die, each of the numbers 1, 2, 3, 4, 5, and 6 occurs only once, so the probability of any particular number coming up is

$$P(E) = \frac{n(E)}{n(S)} = \frac{1}{6}.$$

For one of the 12-sided dice, each number occurs twice, so the probability of any particular number coming up is

$$P(E) = \frac{n(E)}{n(S)} = \frac{2}{12} = \frac{1}{6}.$$

✓ CHECKPOINT   Now try Exercise 17.

FIGURE **8.10**

| Example 5 | The Probability of Winning a Lottery | |

In the Arizona state lottery, a player chooses six different numbers from 1 to 41. If these six numbers match the six numbers drawn (in any order) by the lottery commission, the player wins (or shares) the top prize. What is the probability of winning the top prize if the player buys one ticket?

Solution

To find the number of elements in the sample space, use the formula for the number of combinations of 41 elements taken six at a time.

$$n(S) = {}_{41}C_6$$

$$= \frac{41 \cdot 40 \cdot 39 \cdot 38 \cdot 37 \cdot 36}{6 \cdot 5 \cdot 4 \cdot 3 \cdot 2 \cdot 1}$$

$$= 4{,}496{,}388$$

If a person buys only one ticket, the probability of winning is

$$P(E) = \frac{n(E)}{n(S)} = \frac{1}{4{,}496{,}388}.$$

 **CHECKPOINT**   Now try Exercise 21.

| Example 6 | Random Selection | |

The numbers of colleges and universities in various regions of the United States in 2003 are shown in Figure 8.11. One institution is selected at random. What is the probability that the institution is in one of the three southern regions? (Source: National Center for Education Statistics)

Solution

From the figure, the total number of colleges and universities is 4163. Because there are $700 + 284 + 386 = 1370$ colleges and universities in the three southern regions, the probability that the institution is in one of these regions is

$$P(E) = \frac{n(E)}{n(S)} = \frac{1370}{4163} \approx 0.329.$$

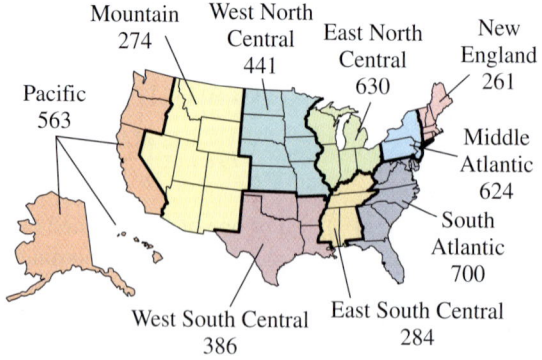

FIGURE **8.11**

 **CHECKPOINT**   Now try Exercise 33.

Mutually Exclusive Events

Two events A and B (from the same sample space) are **mutually exclusive** if A and B have no outcomes in common. In the terminology of sets, the intersection of A and B is the empty set, which is written as

$$P(A \cap B) = 0.$$

For instance, if two dice are tossed, the event A of rolling a total of 6 and the event B of rolling a total of 9 are mutually exclusive. To find the probability that one or the other of two mutually exclusive events will occur, you can *add* their individual probabilities.

Probability of the Union of Two Events

If A and B are events in the same sample space, the probability of A or B occurring is given by

$$P(A \cup B) = P(A) + P(B) - P(A \cap B).$$

If A and B are mutually exclusive, then

$$P(A \cup B) = P(A) + P(B).$$

Example 7   **The Probability of a Union of Events**

One card is selected from a standard deck of 52 playing cards. What is the probability that the card is either a heart or a face card?

Solution

Because the deck has 13 hearts, the probability of selecting a heart (event A) is

$$P(A) = \frac{13}{52}.$$

Similarly, because the deck has 12 face cards, the probability of selecting a face card (event B) is

$$P(B) = \frac{12}{52}.$$

Because three of the cards are hearts *and* face cards (see Figure 8.12), it follows that

$$P(A \cap B) = \frac{3}{52}.$$

Finally, applying the formula for the probability of the union of two events, you can conclude that the probability of selecting a heart or a face card is

$$P(A \cup B) = P(A) + P(B) - P(A \cap B)$$

$$= \frac{13}{52} + \frac{12}{52} - \frac{3}{52} = \frac{22}{52} \approx 0.423.$$

✓CHECKPOINT   Now try Exercise 45.

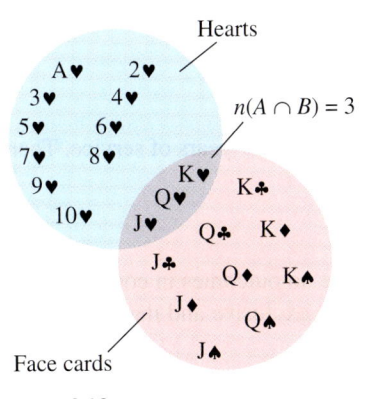

$n(A \cap B) = 3$

Hearts

Face cards

FIGURE **8.12**

The Complement of an Event

The **complement of an event** A is the collection of all outcomes in the sample space that are *not* in A. The complement of event A is denoted by A'. Because $P(A \text{ or } A') = 1$ and because A and A' are mutually exclusive, it follows that $P(A) + P(A') = 1$. So, the probability of A' is

$$P(A') = 1 - P(A).$$

For instance, if the probability of *winning* a certain game is

$$P(A) = \frac{1}{4}$$

the probability of *losing* the game is

$$P(A') = 1 - \frac{1}{4}$$
$$= \frac{3}{4}.$$

Probability of a Complement

Let A be an event and let A' be its complement. If the probability of A is $P(A)$, the probability of the complement is

$$P(A') = 1 - P(A).$$

Example 11 Finding the Probability of a Complement

A manufacturer has determined that a machine averages one faulty unit for every 1000 it produces. What is the probability that an order of 200 units will have one or more faulty units?

Solution

To solve this problem as stated, you would need to find the probabilities of having exactly one faulty unit, exactly two faulty units, exactly three faulty units, and so on. However, using complements, you can simply find the probability that all units are perfect and then subtract this value from 1. Because the probability that any given unit is perfect is 999/1000, the probability that all 200 units are perfect is

$$P(A) = \left(\frac{999}{1000}\right)^{200}$$
$$\approx 0.819.$$

So, the probability that at least one unit is faulty is

$$P(A') = 1 - P(A)$$
$$\approx 1 - 0.819.$$
$$= 0.181$$

✓CHECKPOINT Now try Exercise 51.

8.7 | Exercises

VOCABULARY CHECK:

In Exercises 1–7, fill in the blanks.

1. An _____ is an event whose result is uncertain, and the possible results of the event are called _____.

2. The set of all possible outcomes of an experiment is called the _____ _____.

3. To determine the _____ of an event, you can use the formula $P(E) = \dfrac{n(E)}{n(S)}$, where $n(E)$ is the number of outcomes in the event and $n(S)$ is the number of outcomes in the sample space.

4. If $P(E) = 0$, then E is an _____ event, and if $P(E) = 1$, then E is a _____ event.

5. If two events from the same sample space have no outcomes in common, then the two events are _____ _____.

6. If the occurrence of one event has no effect on the occurrence of a second event, then the events are _____.

7. The _____ of an event A is the collection of all outcomes in the sample space that are not in A.

8. Match the probability formula with the correct probability name.

   (a) Probability of the union of two events   (i) $P(A \cup B) = P(A) + P(B)$

   (b) Probability of mutually exclusive events   (ii) $P(A') = 1 - P(A)$

   (c) Probability of independent events   (iii) $P(A \cup B) = P(A) + P(B) - P(A \cap B)$

   (d) Probability of a complement   (iv) $P(A \text{ and } B) = P(A) \cdot P(B)$

PREREQUISITE SKILLS REVIEW: Practice and review algebra skills needed for this section at **www.Eduspace.com.**

In Exercises 1–6, determine the sample space for the experiment.

1. A coin and a six-sided die are tossed.

2. A six-sided die is tossed twice and the sum of the points is recorded.

3. A taste tester has to rank three varieties of yogurt, A, B, and C, according to preference.

4. Two marbles are selected from a bag containing two red marbles, two blue marbles, and one yellow marble. The color of each marble is recorded.

5. Two county supervisors are selected from five supervisors, A, B, C, D, and E, to study a recycling plan.

6. A sales representative makes presentations about a product in three homes per day. In each home, there may be a sale (denote by S) or there may be no sale (denote by F).

Tossing a Coin **In Exercises 7–10, find the probability for the experiment of tossing a coin three times. Use the sample space** $S = \{HHH, HHT, HTH, HTT, THH, THT, TTH, TTT\}$.

7. The probability of getting exactly one tail

8. The probability of getting a head on the first toss

9. The probability of getting at least one head

10. The probability of getting at least two heads

Drawing a Card **In Exercises 11–14, find the probability for the experiment of selecting one card from a standard deck of 52 playing cards.**

11. The card is a face card.

12. The card is not a face card.

13. The card is a red face card.

14. The card is a 6 or lower. (Aces are low.)

Tossing a Die **In Exercises 15–20, find the probability for the experiment of tossing a six-sided die twice.**

15. The sum is 4.   16. The sum is at least 7.

17. The sum is less than 11.   18. The sum is 2, 3, or 12.

19. The sum is odd and no more than 7.

20. The sum is odd or prime.

Drawing Marbles **In Exercises 21–24, find the probability for the experiment of drawing two marbles (without replacement) from a bag containing one green, two yellow, and three red marbles.**

21. Both marbles are red.

22. Both marbles are yellow.

23. Neither marble is yellow.

24. The marbles are of different colors.

In Exercises 25–28, you are given the probability that an event will happen. Find the probability that the event *will not* happen.

25. $P(E) = 0.7$ **26.** $P(E) = 0.36$

27. $P(E) = \dfrac{1}{4}$

28. $P(E) = \dfrac{2}{3}$

In Exercises 29–32, you are given the probability that an event *will not* happen. Find the probability that the event *will* happen.

29. $P(E') = 0.14$

30. $P(E') = 0.92$

31. $P(E') = \dfrac{17}{35}$

32. $P(E') = \dfrac{61}{100}$

33. *Data Analysis* A study of the effectiveness of a flu vaccine was conducted with a sample of 500 people. Some participants in the study were given no vaccine, some were given one injection, and some were given two injections. The results of the study are listed in the table.

	No vaccine	One injection	Two injections	Total
Flu	7	2	13	22
No flu	149	52	277	478
Total	156	54	290	500

A person is selected at random from the sample. Find the specified probability.

(a) The person had two injections.

(b) The person did not get the flu.

(c) The person got the flu and had one injection.

34. *Data Analysis* One hundred college students were interviewed to determine their political party affiliations and whether they favored a balanced-budget amendment to the Constitution. The results of the study are listed in the table, where D represents Democrat and R represents Republican.

	Favor	Not Favor	Unsure	Total
D	23	25	7	55
R	32	9	4	45
Total	55	34	11	100

A person is selected at random from the sample. Find the probability that the described person is selected.

(a) A person who doesn't favor the amendment

(b) A Republican

(c) A Democrat who favors the amendment

35. *Graphical Reasoning* The figure shows the results of a recent survey in which 1011 adults were asked to grade U.S. public schools. (Source: Phi Delta Kappa/Gallup Poll)

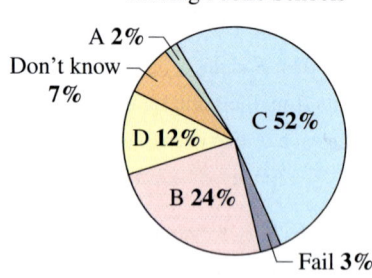

Grading Public Schools

A 2%
Don't know 7%
D 12%
C 52%
B 24%
Fail 3%

(a) Estimate the number of adults who gave U.S. public schools a B.

(b) An adult is selected at random. What is the probabilty that the adult will give the U.S. public schools an A?

(c) An adult is selected at random. What is the probabilty the adult will give the U.S. public schools a C or a D?

36. *Graphical Reasoning* The figure shows the results of a survey in which auto racing fans listed their favorite type of racing. (Source: ESPN Sports Poll/TNS Sports)

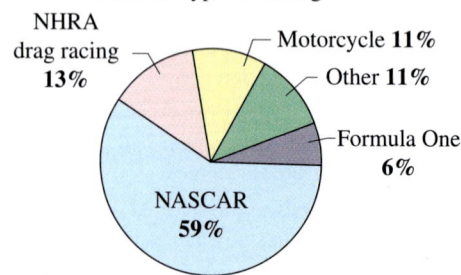

Favorite Type of Racing

NHRA drag racing 13%
Motorcycle 11%
Other 11%
Formula One 6%
NASCAR 59%

(a) What is the probability that an auto racing fan selected at random lists NASCAR racing as his or her favorite type of racing?

(b) What is the probability that an auto racing fan selected at random lists Formula One or motorcycle racing as his or her favorite type of racing?

(c) What is the probability that an auto racing fan selected at random does *not* list NHRA drag racing as his or her favorite type of racing?

37. *Alumni Association* A college sends a survey to selected members of the class of 2006. Of the 1254 people who graduated that year, 672 are women, of whom 124 went on to graduate school. Of the 582 male graduates, 198 went on to graduate school. An alumni member is selected at random. What are the probabilities that the person is (a) female, (b) male, and (c) female and did not attend graduate school?

38. *Education* In a high school graduating class of 202 students, 95 are on the honor roll. Of these, 71 are going on to college, and of the other 107 students, 53 are going on to college. A student is selected at random from the class. What are the probabilities that the person chosen is (a) going to college, (b) not going to college, and (c) on the honor roll, but not going to college?

39. *Winning an Election* Taylor, Moore, and Jenkins are candidates for public office. It is estimated that Moore and Jenkins have about the same probability of winning, and Taylor is believed to be twice as likely to win as either of the others. Find the probability of each candidate winning the election.

40. *Winning an Election* Three people have been nominated for president of a class. From a poll, it is estimated that the first candidate has a 37% chance of winning and the second candidate has a 44% chance of winning. What is the probability that the third candidate will win?

In Exercises 41–52, the sample spaces are large and you should use the counting principles discussed in Section 8.6.

41. *Preparing for a Test* A class is given a list of 20 study problems, from which 10 will be part of an upcoming exam. A student knows how to solve 15 of the problems. Find the probabilities that the student will be able to answer (a) all 10 questions on the exam, (b) exactly eight questions on the exam, and (c) at least nine questions on the exam.

42. *Payroll Mix-Up* Five paychecks and envelopes are addressed to five different people. The paychecks are randomly inserted into the envelopes. What are the probabilities that (a) exactly one paycheck will be inserted in the correct envelope and (b) at least one paycheck will be inserted in the correct envelope?

43. *Game Show* On a game show, you are given five digits to arrange in the proper order to form the price of a car. If you are correct, you win the car. What is the probability of winning, given the following conditions?

(a) You guess the position of each digit.

(b) You know the first digit and guess the positions of the other digits.

44. *Card Game* The deck of a card game is made up of 108 cards. Twenty-five each are red, yellow, blue, and green, and eight are wild cards. Each player is randomly dealt a seven-card hand.

(a) What is the probability that a hand will contain exactly two wild cards?

(b) What is the probability that a hand will contain two wild cards, two red cards, and three blue cards?

45. *Drawing a Card* One card is selected at random from an ordinary deck of 52 playing cards. Find the probabilities that (a) the card is an even-numbered card, (b) the card is a heart or a diamond, and (c) the card is a nine or a face card.

46. *Poker Hand* Five cards are drawn from an ordinary deck of 52 playing cards. What is the probability that the hand drawn is a full house? (A full house is a hand that consists of two of one kind and three of another kind.)

47. *Defective Units* A shipment of 12 microwave ovens contains three defective units. A vending company has ordered four of these units, and because each is identically packaged, the selection will be random. What are the probabilities that (a) all four units are good, (b) exactly two units are good, and (c) at least two units are good?

48. *Random Number Generator* Two integers from 1 through 40 are chosen by a random number generator. What are the probabilities that (a) the numbers are both even, (b) one number is even and one is odd, (c) both numbers are less than 30, and (d) the same number is chosen twice?

49. *Flexible Work Hours* In a survey, people were asked if they would prefer to work flexible hours—even if it meant slower career advancement—so they could spend more time with their families. The results of the survey are shown in the figure. Three people from the survey were chosen at random. What is the probability that all three people would prefer flexible work hours?

Flexible Work Hours

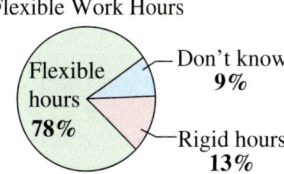

50. *Consumer Awareness* Suppose that the methods used by shoppers to pay for merchandise are as shown in the circle graph. Two shoppers are chosen at random. What is the probability that both shoppers paid for their purchases only in cash?

How Shoppers Pay for Merchandise

Mostly credit — 7%
Half cash, half credit — 30%
Only credit — 4%
Mostly cash — 27%
Only cash — 32%

51. *Backup System* A space vehicle has an independent backup system for one of its communication networks. The probability that either system will function satisfactorily during a flight is 0.985. What are the probabilities that during a given flight (a) both systems function satisfactorily, (b) at least one system functions satisfactorily, and (c) both systems fail?

52. *Backup Vehicle* A fire company keeps two rescue vehicles. Because of the demand on the vehicles and the chance of mechanical failure, the probability that a specific vehicle is available when needed is 90%. The availability of one vehicle is *independent* of the availability of the other. Find the probabilities that (a) both vehicles are available at a given time, (b) neither vehicle is available at a given time, and (c) at least one vehicle is available at a given time.

53. *A Boy or a Girl?* Assume that the probability of the birth of a child of a particular sex is 50%. In a family with four children, what are the probabilities that (a) all the children are boys, (b) all the children are the same sex, and (c) there is at least one boy?

54. *Geometry* You and a friend agree to meet at your favorite fast-food restaurant between 5:00 and 6:00 P.M. The one who arrives first will wait 15 minutes for the other, and then will leave (see figure). What is the probability that the two of you will actually meet, assuming that your arrival times are random within the hour?

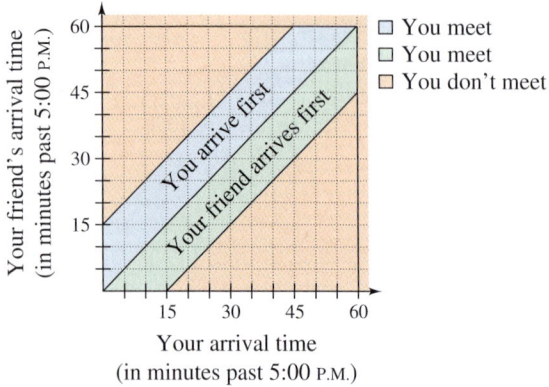

□ You meet
□ You meet
□ You don't meet

Your friend's arrival time (in minutes past 5:00 P.M.)

You arrive first
Your friend arrives first

Your arrival time (in minutes past 5:00 P.M.)

Model It

55. *Roulette* American roulette is a game in which a wheel turns on a spindle and is divided into 38 pockets. Thirty-six of the pockets are numbered 1–36, of which half are red and half are black. Two of the pockets are green and are numbered 0 and 00 (see figure). The dealer spins the wheel and a small ball in opposite directions. As the ball slows to a stop, it has an equal probability of landing in any of the numbered pockets.

(a) Find the probability of landing in the number 00 pocket.

(b) Find the probability of landing in a red pocket.

(c) Find the probability of landing in a green pocket or a black pocket.

(d) Find the probability of landing in the number 14 pocket on two consecutive spins.

(e) Find the probability of landing in a red pocket on three consecutive spins.

(f) European roulette does not contain the 00 pocket. Repeat parts (a)–(e) for European roulette. How do the probabilities for European roulette compare with the probabilities for American roulette?

56. *Estimating π* A coin of diameter d is dropped onto a paper that contains a grid of squares d units on a side (see figure).

(a) Find the probability that the coin covers a vertex of one of the squares on the grid.

(b) Perform the experiment 100 times and use the results to approximate π.

Synthesis

True or False?   **In Exercises 57 and 58, determine whether the statement is true or false. Justify your answer.**

57. If A and B are independent events with nonzero probabilities, then A can occur when B occurs.

58. Rolling a number less than 3 on a normal six-sided die has a probability of $\frac{1}{3}$. The complement of this event is to roll a number greater than 3, and its probability is $\frac{1}{2}$.

59. *Pattern Recognition and Exploration* Consider a group of n people.

(a) Explain why the following pattern gives the probabilities that the n people have distinct birthdays.

$$n = 2: \quad \frac{365}{365} \cdot \frac{364}{365} = \frac{365 \cdot 364}{365^2}$$

$$n = 3: \quad \frac{365}{365} \cdot \frac{364}{365} \cdot \frac{363}{365} = \frac{365 \cdot 364 \cdot 363}{365^3}$$

(b) Use the pattern in part (a) to write an expression for the probability that $n = 4$ people have distinct birthdays.

(c) Let P_n be the probability that the n people have distinct birthdays. Verify that this probability can be obtained recursively by

$$P_1 = 1 \quad \text{and} \quad P_n = \frac{365 - (n - 1)}{365} P_{n-1}.$$

(d) Explain why $Q_n = 1 - P_n$ gives the probability that at least two people in a group of n people have the same birthday.

(e) Use the results of parts (c) and (d) to complete the table.

n	10	15	20	23	30	40	50
P_n							
Q_n							

(f) How many people must be in a group so that the probability of at least two of them having the same birthday is greater than $\frac{1}{2}$? Explain.

60. *Think About It* A weather forecast indicates that the probability of rain is 40%. What does this mean?

Skills Review

In Exercises 61–70, find all real solutions of the equation.

61. $6x^2 + 8 = 0$

62. $4x^2 + 6x - 12 = 0$

63. $x^3 - x^2 - 3x = 0$

64. $x^5 + x^3 - 2x = 0$

65. $\dfrac{12}{x} = -3$

66. $\dfrac{32}{x} = 2x$

67. $\dfrac{2}{x - 5} = 4$

68. $\dfrac{3}{2x + 3} - 4 = \dfrac{-1}{2x + 3}$

69. $\dfrac{3}{x - 2} + \dfrac{x}{x + 2} = 1$

70. $\dfrac{2}{x} - \dfrac{5}{x - 2} = \dfrac{-13}{x^2 - 2x}$

In Exercises 71–74, sketch the graph of the solution set of the system of inequalities.

71. $\begin{cases} y \geq -3 \\ x \geq -1 \\ -x - y \geq -8 \end{cases}$

72. $\begin{cases} x \leq 3 \\ y \leq 6 \\ 5x + 2y \geq 10 \end{cases}$

73. $\begin{cases} x^2 + y \geq -2 \\ y \geq x - 4 \end{cases}$

74. $\begin{cases} x^2 + y^2 \leq 4 \\ x + y \geq -2 \end{cases}$

8 Chapter Summary

What did you learn?

Section 8.1
	Review Exercises
☐ Use sequence notation to write the terms of sequences *(p. 594)*.	1–8
☐ Use factorial notation *(p. 596)*.	9–12
☐ Use summation notation to write sums *(p. 598)*.	13–20
☐ Find the sums of infinite series *(p. 599)*.	21–24
☐ Use sequences and series to model and solve real-life problems *(p. 600)*.	25, 26

Section 8.2
☐ Recognize, write, and find the nth terms of arithmetic sequences *(p. 605)*.	27–40
☐ Find nth partial sums of arithmetic sequences *(p. 608)*.	41–46
☐ Use arithmetic sequences to model and solve real-life problems *(p. 609)*.	47, 48

Section 8.3
☐ Recognize, write, and find the nth terms of geometric sequences *(p. 615)*.	49–60
☐ Find nth partial sums of geometric sequences *(p. 618)*.	61–70
☐ Find sums of infinite geometric series *(p. 619)*.	71–76
☐ Use geometric sequences to model and solve real-life problems *(p. 620)*.	77, 78

Section 8.4
☐ Use mathematical induction to prove statements involving a positive integer n *(p. 625)*.	79–82
☐ Recognize patterns and write the nth term of a sequence *(p. 629)*.	83–86
☐ Find the sums of powers of integers *(p. 631)*.	87–90
☐ Find finite differences of sequences *(p. 632)*.	91–94

Section 8.5
☐ Use the Binomial Theorem to calculate binomial coefficients *(p. 635)*.	95–98
☐ Use Pascal's Triangle to calculate binomial coefficients *(p. 637)*.	99–102
☐ Use binomial coefficients to write binomial expansions *(p. 638)*.	103–108

Section 8.6
☐ Solve simple counting problems *(p. 643)*.	109, 110
☐ Use the Fundamental Counting Principle to solve counting problems *(p. 644)*.	111, 112
☐ Use permutations to solve counting problems *(p. 645)*.	113, 114
☐ Use combinations to solve counting problems *(p. 648)*.	115, 116

Section 8.7
☐ Find the probabilities of events *(p. 653)*.	117, 118
☐ Find the probabilities of mutually exclusive events *(p. 657)*.	119, 120
☐ Find the probabilities of independent events *(p. 659)*.	121, 122
☐ Find the probability of the complement of an event *(p. 660)*.	123, 124

8 Review Exercises

8.1 In Exercises 1–4, write the first five terms of the sequence. (Assume that n begins with 1.)

1. $a_n = 2 + \dfrac{6}{n}$

2. $a_n = \dfrac{(-1)^n 5n}{2n - 1}$

3. $a_n = \dfrac{72}{n!}$

4. $a_n = n(n - 1)$

In Exercises 5–8, write an expression for the apparent nth term of the sequence. (Assume that n begins with 1.)

5. $-2, 2, -2, 2, -2, \ldots$

6. $-1, 2, 7, 14, 23, \ldots$

7. $4, 2, \frac{4}{3}, 1, \frac{4}{5}, \ldots$

8. $1, -\frac{1}{2}, \frac{1}{3}, -\frac{1}{4}, \frac{1}{5}, \ldots$

In Exercises 9–12, simplify the factorial expression.

9. $5!$

10. $3! \cdot 2!$

11. $\dfrac{3! \cdot 5!}{6!}$

12. $\dfrac{7! \cdot 6!}{6! \cdot 8!}$

In Exercises 13–18, find the sum.

13. $\displaystyle\sum_{i=1}^{6} 5$

14. $\displaystyle\sum_{k=2}^{5} 4k$

15. $\displaystyle\sum_{j=1}^{4} \dfrac{6}{j^2}$

16. $\displaystyle\sum_{i=1}^{8} \dfrac{i}{i + 1}$

17. $\displaystyle\sum_{k=1}^{10} 2k^3$

18. $\displaystyle\sum_{j=0}^{4} (j^2 + 1)$

In Exercises 19 and 20, use sigma notation to write the sum.

19. $\dfrac{1}{2(1)} + \dfrac{1}{2(2)} + \dfrac{1}{2(3)} + \cdots + \dfrac{1}{2(20)}$

20. $\dfrac{1}{2} + \dfrac{2}{3} + \dfrac{3}{4} + \cdots + \dfrac{9}{10}$

In Exercises 21–24, find the sum of the infinite series.

21. $\displaystyle\sum_{i=1}^{\infty} \dfrac{5}{10^i}$

22. $\displaystyle\sum_{i=1}^{\infty} \dfrac{3}{10^i}$

23. $\displaystyle\sum_{k=1}^{\infty} \dfrac{2}{100^k}$

24. $\displaystyle\sum_{k=2}^{\infty} \dfrac{9}{10^k}$

25. *Compound Interest* A deposit of \$10,000 is made in an account that earns 8% interest compounded monthly. The balance in the account after n months is given by

$$A_n = 10{,}000\left(1 + \dfrac{0.08}{12}\right)^n, \quad n = 1, 2, 3, \ldots$$

(a) Write the first 10 terms of this sequence.

(b) Find the balance in this account after 10 years by finding the 120th term of the sequence.

 26. *Education* The enrollment a_n (in thousands) in Head Start programs in the United States from 1994 to 2002 can be approximated by the model

$$a_n = 1.07n^2 + 6.1n + 693, \quad n = 4, 5, \ldots, 12$$

where n is the year, with $n = 4$ corresponding to 1994. Find the terms of this finite sequence. Use a graphing utility to construct a bar graph that represents the sequence. (Source: U.S. Administration for Children and Families)

8.2 In Exercises 27–30, determine whether the sequence is arithmetic. If so, find the common difference.

27. $5, 3, 1, -1, -3, \ldots$

28. $0, 1, 3, 6, 10, \ldots$

29. $\frac{1}{2}, 1, \frac{3}{2}, 2, \frac{5}{2}, \ldots$

30. $\frac{9}{9}, \frac{8}{9}, \frac{7}{9}, \frac{6}{9}, \frac{5}{9}, \ldots$

In Exercises 31–34, write the first five terms of the arithmetic sequence.

31. $a_1 = 4, d = 3$

32. $a_1 = 6, d = -2$

33. $a_1 = 25, a_{k+1} = a_k + 3$

34. $a_1 = 4.2, a_{k+1} = a_k + 0.4$

In Exercises 35–40, find a formula for a_n for the arithmetic sequence.

35. $a_1 = 7, d = 12$

36. $a_1 = 25, d = -3$

37. $a_1 = y, d = 3y$

38. $a_1 = -2x, d = x$

39. $a_2 = 93, a_6 = 65$

40. $a_7 = 8, a_{13} = 6$

In Exercises 41–44, find the partial sum.

41. $\displaystyle\sum_{j=1}^{10} (2j - 3)$

42. $\displaystyle\sum_{j=1}^{8} (20 - 3j)$

43. $\displaystyle\sum_{k=1}^{11} \left(\tfrac{2}{3}k + 4\right)$

44. $\displaystyle\sum_{k=1}^{25} \left(\dfrac{3k + 1}{4}\right)$

45. Find the sum of the first 100 positive multiples of 5.

46. Find the sum of the integers from 20 to 80 (inclusive).

47. *Job Offer* The starting salary for an accountant is $34,000 with a guaranteed salary increase of $2250 per year. Determine (a) the salary during the fifth year and (b) the total compensation through 5 full years of employment.

48. *Baling Hay* In the first two trips baling hay around a large field, a farmer obtains 123 bales and 112 bales, respectively. Because each round gets shorter, the farmer estimates that the same pattern will continue. Estimate the total number of bales made if the farmer takes another six trips around the field.

8.3 In Exercises 49–52, determine whether the sequence is geometric. If so, find the common ratio.

49. 5, 10, 20, 40, . . .

50. 54, −18, 6, −2, . . .

51. $\frac{1}{3}, -\frac{2}{3}, \frac{4}{3}, -\frac{8}{3}, \ldots$

52. $\frac{1}{4}, \frac{2}{5}, \frac{3}{6}, \frac{4}{7}, \ldots$

In Exercises 53–56, write the first five terms of the geometric sequence.

53. $a_1 = 4, \ r = -\frac{1}{4}$

54. $a_1 = 2, \ r = 2$

55. $a_1 = 9, \ a_3 = 4$

56. $a_1 = 2, \ a_3 = 12$

In Exercises 57–60, write an expression for the nth term of the geometric sequence. Then find the 20th term of the sequence.

57. $a_1 = 16, a_2 = -8$

58. $a_3 = 6, a_4 = 1$

59. $a_1 = 100, r = 1.05$

60. $a_1 = 5, r = 0.2$

In Exercises 61–66, find the sum of the finite geometric sequence.

61. $\displaystyle\sum_{i=1}^{7} 2^{i-1}$

62. $\displaystyle\sum_{i=1}^{5} 3^{i-1}$

63. $\displaystyle\sum_{i=1}^{4} \left(\frac{1}{2}\right)^{i}$

64. $\displaystyle\sum_{i=1}^{6} \left(\frac{1}{3}\right)^{i-1}$

65. $\displaystyle\sum_{i=1}^{5} (2)^{i-1}$

66. $\displaystyle\sum_{i=1}^{4} 6(3)^{i}$

 In Exercises 67–70, use a graphing utility to find the sum of the finite geometric sequence.

67. $\displaystyle\sum_{i=1}^{10} 10\left(\frac{3}{5}\right)^{i-1}$

68. $\displaystyle\sum_{i=1}^{15} 20(0.2)^{i-1}$

69. $\displaystyle\sum_{i=1}^{25} 100(1.06)^{i-1}$

70. $\displaystyle\sum_{i=1}^{20} 8\left(\frac{6}{5}\right)^{i-1}$

In Exercises 71–76, find the sum of the infinite geometric series.

71. $\displaystyle\sum_{i=1}^{\infty} \left(\frac{7}{8}\right)^{i-1}$

72. $\displaystyle\sum_{i=1}^{\infty} \left(\frac{1}{3}\right)^{i-1}$

73. $\displaystyle\sum_{i=1}^{\infty} (0.1)^{i-1}$

74. $\displaystyle\sum_{i=1}^{\infty} (0.5)^{i-1}$

75. $\displaystyle\sum_{k=1}^{\infty} 4\left(\frac{2}{3}\right)^{k-1}$

76. $\displaystyle\sum_{k=1}^{\infty} 1.3\left(\frac{1}{10}\right)^{k-1}$

77. *Depreciation* A paper manufacturer buys a machine for $120,000. During the next 5 years, it will depreciate at a rate of 30% per year. (That is, at the end of each year the depreciated value will be 70% of what it was at the beginning of the year.)

(a) Find the formula for the nth term of a geometric sequence that gives the value of the machine t full years after it was purchased.

(b) Find the depreciated value of the machine after 5 full years.

78. *Annuity* You deposit $200 in an account at the beginning of each month for 10 years. The account pays 6% compounded monthly. What will your balance be at the end of 10 years? What would the balance be if the interest were compounded continuously?

8.4 In Exercises 79–82, use mathematical induction to prove the formula for every positive integer n.

79. $3 + 5 + 7 + \cdots + (2n + 1) = n(n + 2)$

80. $1 + \frac{3}{2} + 2 + \frac{5}{2} + \cdots + \frac{1}{2}(n + 1) = \frac{n}{4}(n + 3)$

81. $\displaystyle\sum_{i=0}^{n-1} ar^{i} = \frac{a(1 - r^{n})}{1 - r}$

82. $\displaystyle\sum_{k=0}^{n-1} (a + kd) = \frac{n}{2}[2a + (n - 1)d]$

In Exercises 83–86, find a formula for the sum of the first n terms of the sequence.

83. 9, 13, 17, 21, . . .

84. 68, 60, 52, 44, . . .

85. $1, \frac{3}{5}, \frac{9}{25}, \frac{27}{125}, \ldots$

86. $12, -1, \frac{1}{12}, -\frac{1}{144}, \ldots$

In Exercises 87–90, find the sum using the formulas for the sums of powers of integers.

87. $\displaystyle\sum_{n=1}^{30} n$

88. $\displaystyle\sum_{n=1}^{10} n^{2}$

89. $\displaystyle\sum_{n=1}^{7} (n^{4} - n)$

90. $\displaystyle\sum_{n=1}^{6} (n^{5} - n^{2})$

In Exercises 91–94, write the first five terms of the sequence beginning with the given term. Then calculate the first and second differences of the sequence. State whether the sequence has a linear model, a quadratic model, or neither.

91. $a_1 = 5$
$a_n = a_{n-1} + 5$

92. $a_1 = -3$
$a_n = a_{n-1} - 2n$

93. $a_1 = 16$
$a_n = a_{n-1} - 1$

94. $a_0 = 0$
$a_n = n - a_{n-1}$

8.5 In Exercises 95–98, use the Binomial Theorem to calculate the binomial coefficient.

95. $_6C_4$

96. $_{10}C_7$

97. $_8C_5$

98. $_{12}C_3$

In Exercises 99–102, use Pascal's Triangle to calculate the binomial coefficient.

99. $\binom{7}{3}$

100. $\binom{9}{4}$

101. $\binom{8}{6}$

102. $\binom{5}{3}$

In Exercises 103–108, use the Binomial Theorem to expand and simplify the expression. (Remember that $i = \sqrt{-1}$.)

103. $(x + 4)^4$

104. $(x - 3)^6$

105. $(a - 3b)^5$

106. $(3x + y^2)^7$

107. $(5 + 2i)^4$

108. $(4 - 5i)^3$

8.6 **109.** *Numbers in a Hat* Slips of paper numbered 1 through 14 are placed in a hat. In how many ways can you draw two numbers with replacement that total 12?

110. *Home Theater Systems* A customer in an electronics store can choose one of six speaker systems, one of five DVD players, and one of six plasma televisions to design a home theater system. How many systems can be designed?

111. *Telephone Numbers* The same three-digit prefix is used for all of the telephone numbers in a small town. How many different telephone numbers are possible by changing only the last four digits?

112. *Course Schedule* A college student is preparing a course schedule for the next semester. The student may select one of three mathematics courses, one of four science courses, and one of six history courses. How many schedules are possible?

113. *Bike Race* There are 10 bicyclists entered in a race. In how many different ways could the top three places be decided?

114. *Jury Selection* A group of potential jurors has been narrowed down to 32 people. In how many ways can a jury of 12 people be selected?

115. *Apparel* You have eight different suits to choose from to take on a trip. How many combinations of three suits could you take on your trip?

116. *Menu Choices* A local sub shop offers five different breads, seven different meats, three different cheeses, and six different vegetables. Find the total number of combinations of sandwiches possible.

8.7 **117.** *Apparel* A man has five pairs of socks, of which no two pairs are the same color. He randomly selects two socks from a drawer. What is the probability that he gets a matched pair?

118. *Bookshelf Order* A child returns a five-volume set of books to a bookshelf. The child is not able to read, and so cannot distinguish one volume from another. What is the probability that the books are shelved in the correct order?

119. *Students by Class* At a particular university, the numbers of students in the four classes are broken down by percents, as shown in the table.

Class	Percent
Freshmen	31
Sophomores	26
Juniors	25
Seniors	18

A single student is picked randomly by lottery for a cash scholarship. What is the probability that the scholarship winner is

(a) a junior or senior?

(b) a freshman, sophomore, or junior?

120. *Data Analysis* A sample of college students, faculty, and administration were asked whether they favored a proposed increase in the annual activity fee to enhance student life on campus. The results of the study are listed in the table.

	Students	Faculty	Admin.	Total
Favor	237	37	18	292
Oppose	163	38	7	208
Total	400	75	25	500

A person is selected at random from the sample. Find each specified probability.

(a) The person is not in favor of the proposal.

(b) The person is a student.

(c) The person is a faculty member and is in favor of the proposal.

121. *Tossing a Die* A six-sided die is tossed three times. What is the probability of getting a 6 on each roll?

122. *Tossing a Die* A six-sided die is tossed six times. What is the probability that each side appears exactly once?

123. *Drawing a Card* You randomly select a card from a 52-card deck. What is the probability that the card is *not* a club?

124. *Tossing a Coin* Find the probability of obtaining at least one tail when a coin is tossed five times.

Synthesis

True or False? **In Exercises 125–129, determine whether the statement is true or false. Justify your answer.**

125. $\dfrac{(n+2)!}{n!} = (n+2)(n+1)$

126. $\displaystyle\sum_{i=1}^{5}(i^3+2i) = \sum_{i=1}^{5}i^3 + \sum_{i=1}^{5}2i$

127. $\displaystyle\sum_{k=1}^{8}3k = 3\sum_{k=1}^{8}k$

128. $\displaystyle\sum_{j=1}^{6}2^j = \sum_{j=3}^{8}2^{j-2}$

129. The value of $_nP_r$ is always greater than the value of $_nC_r$.

130. *Think About It* An infinite sequence is a function. What is the domain of the function?

131. *Think About It* How do the two sequences differ?

(a) $a_n = \dfrac{(-1)^n}{n}$

(b) $a_n = \dfrac{(-1)^{n+1}}{n}$

132. *Graphical Reasoning* The graphs of two sequences are shown below. Identify each sequence as arithmetic or geometric. Explain your reasoning.

(a)

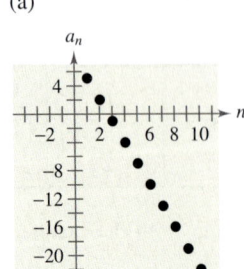

(b)

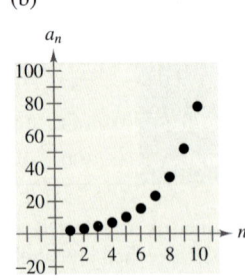

133. *Writing* Explain what is meant by a recursion formula.

134. *Writing* Explain why the terms of a geometric sequence decrease when $0 < r < 1$.

Graphical Reasoning **In Exercises 135–138, match the sequence or sum of a sequence with its graph without doing any calculations. Explain your reasoning. [The graphs are labeled (a), (b), (c), and (d).]**

(a)

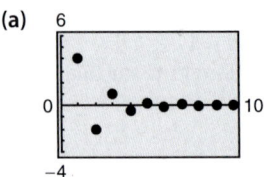

(b)

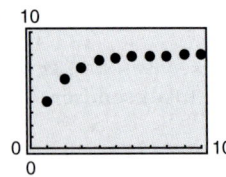

(c)

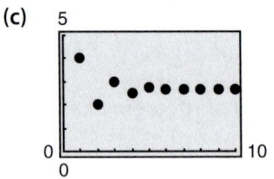

(d)

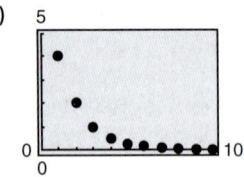

135. $a_n = 4\left(\tfrac{1}{2}\right)^{n-1}$

136. $a_n = 4\left(-\tfrac{1}{2}\right)^{n-1}$

137. $a_n = \displaystyle\sum_{k=1}^{n}4\left(\tfrac{1}{2}\right)^{k-1}$

138. $a_n = \displaystyle\sum_{k=1}^{n}4\left(-\tfrac{1}{2}\right)^{k-1}$

139. *Population Growth* Consider an idealized population with the characteristic that each member of the population produces one offspring at the end of every time period. If each member has a life span of three time periods and the population begins with 10 newborn members, then the following table shows the population during the first five time periods.

Age Bracket	Time Period				
	1	2	3	4	5
0–1	10	10	20	40	70
1–2		10	10	20	40
2–3			10	10	20
Total	10	20	40	70	130

The sequence for the total population has the property that

$$S_n = S_{n-1} + S_{n-2} + S_{n-3}, \quad n > 3.$$

Find the total population during the next five time periods.

140. The probability of an event must be a real number in what interval? Is the interval open or closed?

Take this test as you would take a test in class. When you are finished, check your work against the answers given in the back of the book.

1. Write the first five terms of the sequence $a_n = \dfrac{(-1)^n}{3n + 2}$. (Assume that n begins with 1.)

2. Write an expression for the nth term of the sequence.

$$\frac{3}{1!}, \frac{4}{2!}, \frac{5}{3!}, \frac{6}{4!}, \frac{7}{5!}, \cdots$$

3. Find the next three terms of the series. Then find the fifth partial sum of the series.

$$6 + 17 + 28 + 39 + \cdots$$

4. The fifth term of an arithmetic sequence is 5.4, and the 12th term is 11.0. Find the nth term.

5. Write the first five terms of the sequence $a_n = 5(2)^{n-1}$. (Assume that n begins with 1.)

In Exercises 6–8, find the sum.

6. $\displaystyle\sum_{i=1}^{50} (2i^2 + 5)$.

7. $\displaystyle\sum_{n=1}^{7} (8n - 5)$

8. $\displaystyle\sum_{i=1}^{\infty} 4\left(\tfrac{1}{2}\right)^i$.

9. Use mathematical induction to prove the formula.

$$5 + 10 + 15 + \cdots + 5n = \frac{5n(n + 1)}{2}$$

10. Use the Binomial Theorem to expand the expression $(x + 2y)^4$.

11. Find the coefficient of the term $a^3 b^5$ in the expansion of $(2a - 3b)^8$.

In Exercises 12 and 13, evaluate each expression.

12. (a) $_9P_2$ (b) $_{70}P_3$

13. (a) $_{11}C_4$ (b) $_{66}C_4$

14. How many distinct license plates can be issued consisting of one letter followed by a three-digit number?

15. Eight people are going for a ride in a boat that seats eight people. The owner of the boat will drive, and only three of the remaining people are willing to ride in the two bow seats. How many seating arrangements are possible?

16. You attend a karaoke night and hope to hear your favorite song. The karaoke song book has 300 different songs (your favorite song is among the 300 songs). Assuming that the singers are equally likely to pick any song and no song is repeated, what is the probability that your favorite song is one of the 20 that you hear that night?

17. You are with seven of your friends at a party. Names of all of the 60 guests are placed in a hat and drawn randomly to award eight door prizes. Each guest is limited to one prize. What is the probability that you and your friends win all eight of the prizes?

18. The weather report calls for a 75% chance of snow. According to this report, what is the probability that it will *not* snow?

8 Cumulative Test for Chapters 6–8

Take this test to review the material from earlier chapters. When you are finished, check your work against the answers given in the back of the book.

In Exercises 1–4, solve the system by the specified method.

1. Substitution

$$\begin{cases} y = 3 - x^2 \\ 2(y - 2) = x - 1 \end{cases}$$

2. Elimination

$$\begin{cases} x + 3y = -1 \\ 2x + 4y = 0 \end{cases}$$

3. Elimination

$$\begin{cases} -2x + 4y - z = 3 \\ x - 2y + 2z = -6 \\ x - 3y - z = 1 \end{cases}$$

4. Gauss-Jordan Elimination

$$\begin{cases} x + 3y - 2z = -7 \\ -2x + y - z = -5 \\ 4x + y + z = 3 \end{cases}$$

In Exercises 5 and 6, sketch the graph of the solution set of the system of inequalities.

5. $\begin{cases} 2x + y \geq -3 \\ x - 3y \leq 2 \end{cases}$

6. $\begin{cases} x - y > 6 \\ 5x + 2y < 10 \end{cases}$

7. Sketch the region determined by the constraints. Then find the minimum and maximum values, and where they occur, of the objective function $z = 3x + 2y$, subject to the indicated constraints.

$$x + 4y \leq 20$$
$$2x + y \leq 12$$
$$x \geq 0$$
$$y \geq 0$$

8. A custom-blend bird seed is to be mixed from seed mixtures costing $0.75 per pound and $1.25 per pound. How many pounds of each seed mixture are used to make 200 pounds of custom-blend bird seed costing $0.95 per pound?

9. Find the equation of the parabola $y = ax^2 + bx + c$ passing through the points $(0, 4)$, $(3, 1)$, and $(6, 4)$.

$$\begin{cases} -x + 2y - z = 9 \\ 2x - y + 2z = -9 \\ 3x + 3y - 4z = 7 \end{cases}$$

SYSTEM FOR 10 AND 11

In Exercises 10 and 11, use the system of equations at the left.

10. Write the augmented matrix corresponding to the system of equations.

11. Solve the system using the matrix found in Exercise 10 and Gauss-Jordan elimination.

In Exercises 12–15, use the following matrices to find each of the following, if possible.

$$A = \begin{bmatrix} 4 & 0 \\ -1 & 2 \end{bmatrix}, \quad B = \begin{bmatrix} -1 & 3 \\ 1 & 0 \end{bmatrix}$$

12. $A + B$

13. $-2B$

14. $A - 2B$

15. AB

$$\begin{bmatrix} 8 & 0 & -5 \\ 1 & 3 & -1 \\ -2 & 6 & 4 \end{bmatrix}$$

MATRIX FOR 16

16. Find the determinant of the matrix at the left.

17. Find the inverse of the matrix (if it exists): $\begin{bmatrix} 1 & 2 & -1 \\ 3 & 7 & -10 \\ -5 & -7 & -15 \end{bmatrix}$.

	Gym shoes	Jogging shoes	Walking shoes
Age group 14 – 17	0.09	0.09	0.03
Age group 18 – 24	0.06	0.10	0.05
Age group 25 – 34	0.12	0.25	0.12

MATRIX FOR **18**

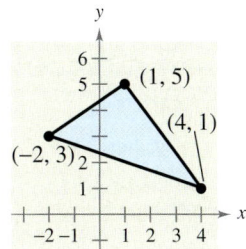

FIGURE FOR **21**

18. The percents (by age group) of the total amounts spent on three types of footwear in a recent year are shown in the matrix. The total amounts (in millions) spent by each age group on the three types of footwear were $442.20 (14–17 age group), $466.57(18–24 age group), and $1088.09 (25–34 age group). How many dollars worth of gym shoes, jogging shoes, and walking shoes were sold that year? (Source: National Sporting Goods Association)

In Exercises 19 and 20, use Cramer's Rule to solve the system of equations.

19. $\begin{cases} 8x - 3y = -52 \\ 3x + 5y = 5 \end{cases}$

20. $\begin{cases} 5x + 4y + 3z = 7 \\ -3x - 8y + 7z = -9 \\ 7x - 5y - 6z = -53 \end{cases}$

21. Find the area of the triangle shown in the figure.

22. Write the first five terms of the sequence $a_n = \dfrac{(-1)^{n+1}}{2n+3}$ (assume that n begins with 1).

23. Write an expression for the nth term of the sequence.

$$\frac{2!}{4}, \frac{3!}{5}, \frac{4!}{6}, \frac{5!}{7}, \frac{6!}{8}, \cdots$$

24. Find the sum of the first 20 terms of the arithmetic sequence 8, 12, 16, 20, ...

25. The sixth term of an arithmetic sequence is 20.6, and the ninth term is 30.2.

    (a) Find the 20th term.

    (b) Find the nth term.

26. Write the first five terms of the sequence $a_n = 3(2)^{n-1}$ (assume that n begins with 1).

27. Find the sum: $\displaystyle\sum_{i=0}^{\infty} 1.3\left(\tfrac{1}{10}\right)^{i-1}$.

28. Use mathematical induction to prove the formula

$$3 + 7 + 11 + 15 + \cdots + (4n - 1) = n(2n + 1).$$

29. Use the Binomial Theorem to expand and simplify $(z - 3)^4$.

In Exercises 30–33, evaluate the expression.

30. $_7P_3$ **31.** $_{25}P_2$ **32.** $\dbinom{8}{4}$ **33.** $_{10}C_3$

In Exercises 34 and 35, find the number of distinguishable permutations of the group of letters.

34. B, A, S, K, E, T, B, A, L, L   **35.** A, N, T, A, R, C, T, I, C, A

36. A personnel manager at a department store has 10 applicants to fill three different sales positions. In how many ways can this be done, assuming that all the applicants are qualified for any of the three positions?

37. On a game show, the digits 3, 4, and 5 must be arranged in the proper order to form the price of an appliance. If the digits are arranged correctly, the contestant wins the appliance. What is the probability of winning if the contestant knows that the price is at least $400?

Proofs in Mathematics

Properties of Sums *(p. 599)*

1. $\displaystyle\sum_{i=1}^{n} c = cn,$ c is a constant.

2. $\displaystyle\sum_{i=1}^{n} ca_i = c\sum_{i=1}^{n} a_i,$ c is a constant.

3. $\displaystyle\sum_{i=1}^{n} (a_i + b_i) = \sum_{i=1}^{n} a_i + \sum_{i=1}^{n} b_i$

4. $\displaystyle\sum_{i=1}^{n} (a_i - b_i) = \sum_{i=1}^{n} a_i - \sum_{i=1}^{n} b_i$

Infinite Series

The study of infinite series was considered a novelty in the fourteenth century. Logician Richard Suiseth, whose nickname was Calculator, solved this problem.

If throughout the first half of a given time interval a variation continues at a certain intensity; throughout the next quarter of the interval at double the intensity; throughout the following eighth at triple the intensity and so ad infinitum; The average intensity for the whole interval will be the intensity of the variation during the second subinterval (or double the intensity).

This is the same as saying that the sum of the infinite series

$$\frac{1}{2} + \frac{2}{4} + \frac{3}{8} + \cdots + \frac{n}{2^n} + \cdots$$

is 2.

Proof

Each of these properties follows directly from the properties of real numbers.

1. $\displaystyle\sum_{i=1}^{n} c = c + c + c + \cdots + c = cn$ *n* terms

The Distributive Property is used in the proof of Property 2.

2. $\displaystyle\sum_{i=1}^{n} ca_i = ca_1 + ca_2 + ca_3 + \cdots + ca_n$

$$= c(a_1 + a_2 + a_3 + \cdots + a_n) = c\sum_{i=1}^{n} a_i$$

The proof of Property 3 uses the Commutative and Associative Properties of Addition.

3.

$$\sum_{i=1}^{n} (a_i + b_i) = (a_1 + b_1) + (a_2 + b_2) + (a_3 + b_3) + \cdots + (a_n + b_n)$$

$$= (a_1 + a_2 + a_3 + \cdots + a_n) + (b_1 + b_2 + b_3 + \cdots + b_n)$$

$$= \sum_{i=1}^{n} a_i + \sum_{i=1}^{n} b_i$$

The proof of Property 4 uses the Commutative and Associative Properties of Addition and the Distributive Property.

4.

$$\sum_{i=1}^{n} (a_i - b_i) = (a_1 - b_1) + (a_2 - b_2) + (a_3 - b_3) + \cdots + (a_n - b_n)$$

$$= (a_1 + a_2 + a_3 + \cdots + a_n) + (-b_1 - b_2 - b_3 - \cdots - b_n)$$

$$= (a_1 + a_2 + a_3 + \cdots + a_n) - (b_1 + b_2 + b_3 + \cdots + b_n)$$

$$= \sum_{i=1}^{n} a_i - \sum_{i=1}^{n} b_i$$

The Sum of a Finite Arithmetic Sequence   *(p. 608)*

The sum of a finite arithmetic sequence with n terms is

$$S_n = \frac{n}{2}(a_1 + a_n).$$

Proof

Begin by generating the terms of the arithmetic sequence in two ways. In the first way, repeatedly add d to the first term to obtain

$$S_n = a_1 + a_2 + a_3 + \cdots + a_{n-2} + a_{n-1} + a_n$$
$$= a_1 + [a_1 + d] + [a_1 + 2d] + \cdots + [a_1 + (n-1)d].$$

In the second way, repeatedly subtract d from the nth term to obtain

$$S_n = a_n + a_{n-1} + a_{n-2} + \cdots + a_3 + a_2 + a_1$$
$$= a_n + [a_n - d] + [a_n - 2d] + \cdots + [a_n - (n-1)d].$$

If you add these two versions of S_n, the multiples of d subtract out and you obtain

$$2S_n = (a_1 + a_n) + (a_1 + a_n) + (a_1 + a_n) + \cdots + (a_1 + a_n) \quad \text{n terms}$$
$$2S_n = n(a_1 + a_n)$$
$$S_n = \frac{n}{2}(a_1 + a_n).$$

The Sum of a Finite Geometric Sequence   *(p. 618)*

The sum of the finite geometric sequence

$$a_1, \ a_1r, \ a_1r^2, \ a_1r^3, \ a_1r^4, \ \ldots, a_1r^{n-1}$$

with common ratio $r \neq 1$ is given by $S_n = \displaystyle\sum_{i=1}^{n} a_1 r^{i-1} = a_1\left(\frac{1 - r^n}{1 - r}\right).$

Proof

$$S_n = a_1 + a_1r + a_1r^2 + \cdots + a_1r^{n-2} + a_1r^{n-1}$$
$$rS_n = a_1r + a_1r^2 + a_1r^3 + \cdots + a_1r^{n-1} + a_1r^n \qquad \text{Multiply by } r.$$

Subtracting the second equation from the first yields

$$S_n - rS_n = a_1 - a_1r^n.$$

So, $S_n(1 - r) = a_1(1 - r^n)$, and, because $r \neq 1$, you have $S_n = a_1\left(\dfrac{1 - r^n}{1 - r}\right).$

The Binomial Theorem *(p. 635)*

In the expansion of $(x + y)^n$

$$(x + y)^n = x^n + nx^{n-1}y + \cdots + {}_nC_r\, x^{n-r}y^r + \cdots + nxy^{n-1} + y^n$$

the coefficient of $x^{n-r}y^r$ is

$${}_nC_r = \frac{n!}{(n-r)!\,r!}.$$

Proof

The Binomial Theorem can be proved quite nicely using mathematical induction. The steps are straightforward but look a little messy, so only an outline of the proof is presented.

1. If $n = 1$, you have $(x + y)^1 = x^1 + y^1 = {}_1C_0 x + {}_1C_1 y$, and the formula is valid.

2. Assuming that the formula is true for $n = k$, the coefficient of $x^{k-r}y^r$ is

$${}_kC_r = \frac{k!}{(k-r)!\,r!} = \frac{k(k-1)(k-2)\cdots(k-r+1)}{r!}.$$

To show that the formula is true for $n = k + 1$, look at the coefficient of $x^{k+1-r}y^r$ in the expansion of

$$(x + y)^{k+1} = (x + y)^k(x + y).$$

From the right-hand side, you can determine that the term involving $x^{k+1-r}y^r$ is the sum of two products.

$$\left({}_kC_r x^{k-r}y^r \right)(x) + \left({}_kC_{r-1} x^{k+1-r}y^{r-1} \right)(y)$$

$$= \left[\frac{k!}{(k-r)!\,r!} + \frac{k!}{(k+1-r)!\,(r-1)!} \right] x^{k+1-r}y^r$$

$$= \left[\frac{(k+1-r)k!}{(k+1-r)!\,r!} + \frac{k!\,r}{(k+1-r)!\,r!} \right] x^{k+1-r}y^r$$

$$= \left[\frac{k!(k+1-r+r)}{(k+1-r)!\,r!} \right] x^{k+1-r}y^r$$

$$= \left[\frac{(k+1)!}{(k+1-r)!\,r!} \right] x^{k+1-r}y^r$$

$$= {}_{k+1}C_r x^{k+1-r}y^r$$

So, by mathematical induction, the Binomial Theorem is valid for all positive integers n.

This collection of thought-provoking and challenging exercises further explores and expands upon concepts learned in this chapter.

 1. Let $x_0 = 1$ and consider the sequence x_n given by

$$x_n = \frac{1}{2} x_{n-1} + \frac{1}{x_{n-1}}, \quad n = 1, 2, \ldots$$

Use a graphing utility to compute the first 10 terms of the sequence and make a conjecture about the value of x_n as n approaches infinity.

2. Consider the sequence

$$a_n = \frac{n + 1}{n^2 + 1}.$$

 (a) Use a graphing utility to graph the first 10 terms of the sequence.

 (b) Use the graph from part (a) to estimate the value of a_n as n approaches infinity.

(c) Complete the table.

n	1	10	100	1000	10,000
a_n					

(d) Use the table from part (c) to determine (if possible) the value of a_n as n approaches infinity.

3. Consider the sequence

$$a_n = 3 + (-1)^n.$$

 (a) Use a graphing utility to graph the first 10 terms of the sequence.

(b) Use the graph from part (a) to describe the behavior of the graph of the sequence.

(c) Complete the table.

n	1	10	101	1000	10,001
a_n					

(d) Use the table from part (c) to determine (if possible) the value of a_n as n approaches infinity.

4. The following operations are performed on each term of an arithmetic sequence. Determine if the resulting sequence is arithmetic, and if so, state the common difference.

(a) A constant C is added to each term.

(b) Each term is multiplied by a nonzero constant C.

(c) Each term is squared.

5. The following sequence of perfect squares is not arithmetic.

1, 4, 9, 16, 25, 36, 49, 64, 81, . . .

However, you can form a related sequence that is arithmetic by finding the differences of consecutive terms.

(a) Write the first eight terms of the related arithmetic sequence described above. What is the nth term of this sequence?

(b) Describe how you can find an arithmetic sequence that is related to the following sequence of perfect cubes.

1, 8, 27, 64, 125, 216, 343, 512, 729, . . .

(c) Write the first seven terms of the related sequence in part (b) and find the nth term of the sequence.

(d) Describe how you can find the arithmetic sequence that is related to the following sequence of perfect fourth powers.

1, 16, 81, 256, 625, 1296, 2401, 4096, 6561, . . .

(e) Write the first six terms of the related sequence in part (d) and find the nth term of the sequence.

6. Can the Greek hero Achilles, running at 20 feet per second, ever catch a tortoise, starting 20 feet ahead of Achilles and running at 10 feet per second? The Greek mathematician Zeno said no. When Achilles runs 20 feet, the tortoise will be 10 feet ahead. Then, when Achilles runs 10 feet, the tortoise will be 5 feet ahead. Achilles will keep cutting the distance in half but will never catch the tortoise. The table shows Zeno's reasoning. From the table you can see that both the distances and the times required to achieve them form infinite geometric series. Using the table, show that both series have finite sums. What do these sums represent?

Distance (in feet)	Time (in seconds)
20	1
10	0.5
5	0.25
2.5	0.125
1.25	0.0625
0.625	0.03125

7. Recall that a *fractal* is a geometric figure that consists of a pattern that is repeated infinitely on a smaller and smaller scale. A well-known fractal is called the *Sierpinski Triangle*. In the first stage, the midpoints of the three sides are used to create the vertices of a new triangle, which is then removed, leaving three triangles. The first three stages are shown on the next page. Note that each remaining triangle is similar to the original triangle. Assume that the length of each side of the original triangle is one unit.

Write a formula that describes the side length of the triangles that will be generated in the nth stage. Write a formula for the area of the triangles that will be generated in the nth stage.

FIGURE FOR 7

8. You can define a sequence using a piecewise formula. The following is an example of a piecewise-defined sequence.

$$a_1 = 7, \quad a_n = \begin{cases} \dfrac{a_{n-1}}{2}, & \text{if } a_{n-1} \text{ is even} \\ 3a_{n-1} + 1, & \text{if } a_{n-1} \text{ is odd} \end{cases}$$

(a) Write the first 10 terms of the sequence.

(b) Choose three different values for a_1 (other than $a_1 = 7$). For each value of a_1, find the first 10 terms of the sequence. What conclusions can you make about the behavior of this sequence?

9. The numbers 1, 5, 12, 22, 35, 51, . . . are called pentagonal numbers because they represent the numbers of dots used to make pentagons, as shown below. Use mathematical induction to prove that the nth pentagonal number P_n is given by

$$P_n = \frac{n(3n-1)}{2}.$$

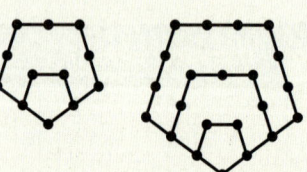

10. What conclusion can be drawn from the following information about the sequence of statements P_n?

(a) P_3 is true and P_k implies P_{k+1}.

(b) $P_1, P_2, P_3, \ldots, P_{50}$ are all true.

(c) $P_1, P_2,$ and P_3 are all true, but the truth of P_k does not imply that P_{k+1} is true.

(d) P_2 is true and P_{2k} implies P_{2k+2}.

11. Let $f_1, f_2, \ldots, f_n, \ldots$ be the Fibonacci sequence.

(a) Use mathematical induction to prove that

$$f_1 + f_2 + \cdots + f_n = f_{n+2} - 1.$$

(b) Find the sum of the first 20 terms of the Fibonacci sequence.

12. The odds in favor of an event occurring are the ratio of the probability that the event will occur to the probability that the event will not occur. The reciprocal of this ratio represents the odds against the event occurring.

(a) Six marbles in a bag are red. The odds against choosing a red marble are 4 to 1. How many marbles are in the bag?

(b) A bag contains three blue marbles and seven yellow marbles. What are the odds in favor of choosing a blue marble? What are the odds against choosing a blue marble?

(c) Write a formula for converting the odds in favor of an event to the probability of the event.

(d) Write a formula for converting the probability of an event to the odds in favor of the event.

13. You are taking a test that contains only multiple choice questions (there are five choices for each question). You are on the last question and you know that the answer is not B or D, but you are not sure about answers A, C, and E. What is the probability that you will get the right answer if you take a guess?

14. A dart is thrown at the circular target shown below. The dart is equally likely to hit any point inside the target. What is the probability that it hits the region outside the triangle?

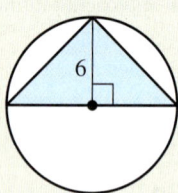

15. An event A has n possible outcomes, which have the values $x_1, x_2, \ldots, x_n$. The probabilities of the n outcomes occurring are $p_1, p_2, \ldots, p_n$. The **expected value** V of an event A is the sum of the products of the outcomes' probabilities and their values, $V = p_1 x_1 + p_2 x_2 + \cdots + p_n x_n$.

(a) To win California's Super Lotto Plus game, you must match five different numbers chosen from the numbers 1 to 47, plus one Mega number chosen from the numbers 1 to 27. You purchase a ticket for $1. If the jackpot for the next drawing is $12,000,000, what is the expected value for the ticket?

(b) You are playing a dice game in which you need to score 60 points to win. On each turn, you roll two six-sided dice. Your score for the turn is 0 if the dice do not show the same number, and the product of the numbers on the dice if they do show the same number. What is the expected value for each turn? How many turns will it take on average to score 60 points?

Answers to Odd-Numbered Exercises and Tests

Chapter P

Section P.1   (page 9)

Vocabulary Check   *(page 9)*

1. rational   **2.** irrational   **3.** absolute value
4. composite   **5.** prime   **6.** variables; constants
7. terms   **8.** coefficient   **9.** Zero-Factor Property

1. (a) $5, 1, 2$ (b) $0, 5, 1, 2$
   (c) $-9, 5, 0, 1, -4, 2, -11$
   (d) $-\frac{7}{2}, \frac{2}{3}, -9, 5, 0, 1, -4, 2, -11$ (e) $\sqrt{2}$
3. (a) 1 (b) 1 (c) $-13, 1, -6$
   (d) $2.01, -13, 1, -6, 0.666\ldots$
   (e) $0.010110111\ldots$
5. (a) $\frac{6}{3}, 8$ (b) $\frac{6}{3}, 8$ (c) $\frac{6}{3}, -1, 8, -22$
   (d) $-\frac{1}{3}, \frac{6}{3}, -7.5, -1, 8, -22$ (e) $-\pi, \frac{1}{2}\sqrt{2}$
7. 0.625 **9.** $0.\overline{123}$ **11.** $-1 < 2.5$
13.

$-4 > -8$

15.

$\frac{3}{2} < 7$

17.

$\frac{5}{6} > \frac{2}{3}$

19. (a) $x \le 5$ denotes the set of all real numbers less than or equal to 5.
   (b)   (c) Unbounded

21. (a) $x < 0$ denotes the set of all real numbers less than 0.
   (b)   (c) Unbounded

23. (a) $[4, \infty)$ denotes the set of all real numbers greater than or equal to 4.
   (b)   (c) Unbounded

25. (a) $-2 < x < 2$ denotes the set of all real numbers greater than -2 and less than 2.
   (b)   (c) Bounded

27. (a) $-1 \le x < 0$ denotes the set of all real numbers greater than or equal to -1 and less than 0.
   (b)   (c) Bounded

29. (a) $[-2, 5)$ denotes the set of all real numbers greater than or equal to -2 and less than 5.
   (b)   (c) Bounded

31. $-2 < x \le 4$ **33.** $y \ge 0$ **35.** $10 \le t \le 22$
37. $W > 65$ **39.** 10 **41.** 5 **43.** -1 **45.** -1
47. -1 **49.** $|-3| > -|-3|$ **51.** $-5 = -|5|$
53. $-|-2| = -|2|$ **55.** 51 **57.** $\frac{5}{2}$ **59.** $\frac{128}{75}$
61. $|\$113{,}356 - \$112{,}700| = \$656 > \$500$
   $0.05(\$112{,}700) = \$5635$
   Because the actual expenses differ from the budget by more than \$500, there is failure to meet the "budget variance test."
63. $|\$37{,}335 - \$37{,}640| = \$305 < \$500$
   $0.05(\$37{,}640) = \$1882$
   Because the difference between the actual expenses and the budget is less than \$500 and less than 5% of the budgeted amount, there is compliance with the "budget variance test."
65. (a)

Year	Expenditures (in billions)	Surplus or deficit (in billions)
1960	$92.2	$0.3 (s)
1970	$195.6	$2.8 (d)
1980	$590.9	$73.8 (d)
1990	$1253.2	$221.2 (d)
2000	$1788.8	$236.4 (s)

(b)

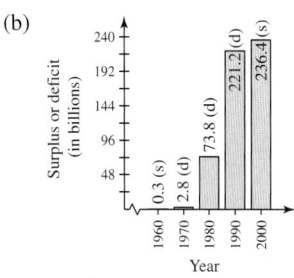

67. $|x - 5| \le 3$ **69.** $|y| \ge 6$
71. $|326 - 351| = 25$ miles
73. $7x$ and 4 are the terms; 7 is the coefficient.
75. $\sqrt{3}x^2, -8x,$ and -11 are the terms; $\sqrt{3}$ and -8 are the coefficients.
77. $4x^3, x/2,$ and -5 are the terms; 4 and $\frac{1}{2}$ are the coefficients.
79. (a) -10 (b) -6 **81.** (a) 14 (b) 2
83. (a) Division by 0 is undefined.   (b) 0
85. Commutative Property of Addition
87. Multiplicative Inverse Property
89. Distributive Property
91. Multiplicative Identity Property

CHAPTER P

93. Associative Property of Addition

95. Distributive Property

97. $\dfrac{1}{2}$ **99.** $\dfrac{3}{8}$ **101.** 48 **103.** $\dfrac{5x}{12}$

105. (a)

n	1	0.5	0.01	0.0001	0.000001
$5/n$	5	10	500	50,000	5,000,000

(b) The value of $5/n$ approaches infinity as n approaches 0.

107. False. If $a < b$, then $\dfrac{1}{a} > \dfrac{1}{b}$, where $a \neq b \neq 0$.

109. (a) No. If one variable is negative and the other is positive, the expressions are unequal.

(b) $|u + v| \leq |u| + |v|$

The expressions are equal when u and v have the same sign. If u and v differ in sign, $|u + v|$ is less than $|u| + |v|$.

111. The only even prime number is 2, because its only factors are itself and 1.

113. (a) Negative (b) Negative

115. Yes. $|a| = -a$ if $a < 0$.

Section P.2 *(page 21)*

> ### Vocabulary Check *(page 21)*
>
> **1.** exponent; base **2.** scientific notation
> **3.** square root **4.** principle nth root
> **5.** index; radicand **6.** simplest form
> **7.** conjugates **8.** rationalizing
> **9.** power; index

1. $8 \times 8 \times 8 \times 8 \times 8$ **3.** 4.9^6 **5.** (a) 27 (b) 81

7. (a) 1 (b) -9 **9.** (a) $\dfrac{243}{64}$ (b) -1

11. (a) $\dfrac{5}{6}$ (b) 4 **13.** -1600 **15.** 2.125

17. -24 **19.** 6 **21.** -54 **23.** 1

25. (a) $-125z^3$ (b) $5x^6$ **27.** $24y^2$ (b) $3x^2$

29. (a) $\dfrac{7}{x}$ (b) $\dfrac{4}{3}(x + y)^2$ **31.** (a) 1 (b) $\dfrac{1}{4x^4}$

33. (a) $-2x^3$ (b) $\dfrac{10}{x}$ **35.** (a) 3^{3n} (b) $\dfrac{b^5}{a^5}$

37. 5.73×10^7 square miles

39. 8.99×10^{-5} gram per cubic centimeter

41. 4,568,000,000 ounces

43. 0.0000000000000000016022 coulomb

45. (a) 50,000 (b) 200,000

47. (a) 954.448 (b) 3.077×10^{10}

49. (a) 67,082.039 (b) 39.791

51. (a) 3 (b) $\dfrac{3}{2}$ **53.** (a) $\dfrac{1}{8}$ (b) $\dfrac{27}{8}$

55. (a) -4 (b) 2 **57.** (a) 7.550 (b) -7.225

59. (a) -0.011 (b) 0.005 **61.** (a) 4 (b) $2\sqrt[5]{3x}$

63. (a) $2\sqrt{2}$ (b) $3\sqrt[3]{2}$ **65.** (a) $6x\sqrt{2x}$ (b) $\dfrac{18\sqrt{z}}{z^2}$

67. (a) $2x\sqrt[3]{2x^2}$ (b) $\dfrac{5|x|\sqrt{3}}{y^2}$

69. (a) $34\sqrt{2}$ (b) $22\sqrt{2}$ **71.** (a) $2\sqrt{x}$ (b) $4\sqrt{y}$

73. (a) $13\sqrt{x + 1}$ (b) $18\sqrt{5x}$

75. $\sqrt{5} + \sqrt{3} > \sqrt{5 + 3}$ **77.** $5 > \sqrt{3^2 + 2^2}$

79. $\dfrac{\sqrt{3}}{3}$ **81.** $\dfrac{5 + \sqrt{3}}{11}$ **83.** $\dfrac{2}{\sqrt{2}}$

85. $\dfrac{2}{3(\sqrt{5} - \sqrt{3})}$ **87.** $9^{1/2}$ **89.** $\sqrt[5]{32}$

91. $(-216)^{1/3}$ **93.** $81^{3/4}$ **95.** $\dfrac{2}{x}$ **97.** $\dfrac{1}{x^3}$, $x > 0$

99. (a) $\sqrt{3}$ (b) $\sqrt[3]{(x + 1)^2}$

101. (a) $2\sqrt[4]{2}$ (b) $\sqrt[8]{2x}$ **103.** $\dfrac{\pi}{2} \approx 1.57$ seconds

105. (a)

h	0	1	2	3	4	5	6
t	0	2.93	5.48	7.67	9.53	11.08	12.32

h	7	8	9	10	11	12
t	13.29	14.00	14.50	14.80	14.93	14.96

(b) $t \to 8.64\sqrt{3} \approx 14.96$

107. True. When dividing variables, you subtract exponents.

109. $a^0 = 1, a \neq 0$, using the property $\dfrac{a^m}{a^n} = a^{m-n}$:

$\dfrac{a^m}{a^m} = a^{m-m} = a^0 = 1$.

111. When any positive integer is squared, the units digit is 0, 1, 4, 5, 6, or 9. Therefore, $\sqrt{5233}$ is not an integer.

Section P.3 *(page 29)*

> ### Vocabulary Check *(page 29)*
>
> **1.** n; a_n; a_0 **2.** descending
> **3.** monomial; binomial; trinomial **4.** like terms
> **5.** First terms; Outer terms; Inner terms; Last terms
> **6.** c **7.** a **8.** b

1. d **2.** e **3.** b **4.** a **5.** f **6.** c

7. $-2x^3 + 4x^2 - 3x + 20$ (Answers will vary.)

9. $-15x^4 + 1$ (Answers will vary.)

11. (a) $-\frac{1}{2}x^5 + 14x$

(b) Degree: 5; Leading coefficient: $-\frac{1}{2}$

(c) Binomial

13. (a) $-3x^4 + 2x^2 - 5$

(b) Degree: 4; Leading coefficient: -3

(c) Trinomial

15. (a) $x^5 - 1$

(b) Degree: 5; Leading coefficient: 1

(c) Binomial

17. (a) 3

(b) Degree: 0; Leading coefficient: 3

(c) Monomial

19. (a) $-4x^5 + 6x^4 + 1$

(b) Degree: 5; Leading coefficient: -4

(c) Trinomial

21. (a) $4x^3y$

(b) Degree: 4; Leading coefficient: 4

(c) Monomial

23. Polynomial: $-3x^3 + 2x + 8$

25. Not a polynomial because it includes a term with a negative exponent

27. Polynomial: $-y^4 + y^3 + y^2$ **29.** $-2x - 10$

31. $3x^3 - 2x + 2$ **33.** $8.3x^3 + 29.7x^2 + 11$

35. $12z + 8$ **37.** $3x^3 - 6x^2 + 3x$

39. $-15z^2 + 5z$ **41.** $-4x^4 + 4x$ **43.** $7.5x^3 + 9x$

45. $-\frac{1}{2}x^2 - 12x$ **47.** $4x^3 - 2x^2 + 4$

49. $5x^2 - 4x + 11$ **51.** $-30x^3 + 57x^2 + 25x - 12$

53. $x^4 - x^3 + 5x^2 - 9x - 36$ **55.** $x^2 + 7x + 12$

57. $6x^2 - 7x - 5$ **59.** $x^4 + x^2 + 1$

61. $x^2 - 100$ **63.** $x^2 - 4y^2$ **65.** $4x^2 + 12x + 9$

67. $4x^2 - 20xy + 25y^2$ **69.** $x^3 + 3x^2 + 3x + 1$

71. $8x^3 - 12x^2y + 6xy^2 - y^3$ **73.** $16x^6 - 24x^3 + 9$

75. $m^2 - n^2 - 6m + 9$

77. $x^2 + 2xy + y^2 - 6x - 6y + 9$ **79.** $4r^4 - 25$

81. $\frac{1}{4}x^2 - 3x + 9$ **83.** $\frac{1}{9}x^2 - 4$

85. $1.44x^2 + 7.2x + 9$ **87.** $2.25x^2 - 16$

89. $2x^2 + 2x$ **91.** $u^4 - 16$ **93.** $x - y$

95. $x^2 - 2\sqrt{5}x + 5$

97. (a) $P = 22x - 25,000$ (b) $85,000$

99. (a) $500r^2 + 1000r + 500$

(b)

r	$2\frac{1}{2}\%$	3%	4%
$500(1 + r)^2$	\$525.31	\$530.45	\$540.80

r	$4\frac{1}{2}\%$	5%
$500(1 + r)^2$	\$546.01	\$551.25

(c) The amount increases with increasing r.

101. (a) $V = 4x^3 - 88x^2 + 468x$

(b)

x (cm)	1	2	3
V (cm³)	384	616	720

103. (a) $3x^2 + 8x$ (b) $30x^2$

(c) $x^2 + \frac{7}{2}x$ (d) $\frac{3}{2}x^2 + 14x + 30$

105. $44x + 308$

107. (a) Estimates will vary.

(b) The difference in safe load decreases in magnitude.

109. $(x + 1)(x + 4) = x(x + 4) + 1(x + 4)$

Distributive Property

111. False. $(4x^2 + 1)(3x + 1) = 12x^3 + 4x^2 + 3x + 1$

113. $m + n$

115. The student omitted the middle term when squaring the binomial. $(x - 3)^2 = x^2 - 6x + 9 \neq x^2 + 9$

117. No. $(x^2 + 1) + (-x^2 + 3) = 4$, which is not a second-degree polynomial. (Examples will vary.)

119. $(3 + 4)^2 = 49 \neq 25 = 3^2 + 4^2$.

If either x or y is zero, then $(x + y)^2 = x^2 + y^2$.

Section P.4   *(page 38)*

Vocabulary Check   *(page 38)*

1. factoring   2. irreducible

3. completely factored   4. factoring by grouping

5. (a) ii   (b) iii   (c) i

1. 30   **3.** $6x^2y$ **5.** $3(x + 2)$ **7.** $2x(x^2 - 3)$

9. $(x - 1)(x + 6)$ **11.** $(x + 3)(x - 1)$

13. $\frac{1}{2}(x + 8)$ **15.** $\frac{1}{2}x(x^2 + 4x - 10)$

17. $\frac{2}{3}(x - 6)(x - 3)$ **19.** $(x + 9)(x - 9)$

21. $2(4y - 3)(4y + 3)$ **23.** $\left(4x + \frac{1}{3}\right)\left(4x - \frac{1}{3}\right)$

25. $(x + 1)(x - 3)$ **27.** $(3u + 2v)(3u - 2v)$

29. $(x - 2)^2$ **31.** $(2t + 1)^2$ **33.** $(5y - 1)^2$

35. $(3u + 4v)^2$ **37.** $\left(x - \frac{2}{3}\right)^2$

39. $(x - 2)(x^2 + 2x + 4)$ **41.** $(y + 4)(y^2 - 4y + 16)$

43. $(2t - 1)(4t^2 + 2t + 1)$

45. $(u + 3v)(u^2 - 3uv + 9v^2)$ **47.** $-5(x^2 - 5)$

49. $-2(t^3 - 2t - 3)$ **51.** $(x + 2)(x - 1)$

53. $(s - 3)(s - 2)$ **55.** $-(y + 5)(y - 4)$

57. $(x - 20)(x - 10)$ **59.** $(3x - 2)(x - 1)$

61. $(5x + 1)(x + 5)$ **63.** $-(3z - 2)(3z + 1)$

65. $(x - 1)(x^2 + 2)$ **67.** $(2x - 1)(x^2 - 3)$

69. $(3 + x)(2 - x^3)$ **71.** $(3x^2 - 1)(2x + 1)$

73. $(x + 2)(3x + 4)$ **75.** $(2x - 1)(3x + 2)$

77. $(3x - 1)(5x - 2)$ **79.** $6(x + 3)(x - 3)$

81. $x^2(x - 4)$ **83.** $(x - 1)^2$ **85.** $(1 - 2x)^2$

87. $-2x(x + 1)(x - 2)$ **89.** $(9x + 1)(x + 1)$

91. $\frac{1}{81}(x + 36)(x - 18)$ **93.** $(3x + 1)(x^2 + 5)$

95. $x(x - 4)(x^2 + 1)$ **97.** $\frac{1}{4}(x^2 + 3)(x + 12)$

99. $(t + 6)(t - 8)$ **101.** $(x + 2)(x + 4)(x - 2)(x - 4)$

103. $5(x + 2)(x^2 - 2x + 4)$ **105.** $(3 - 4x)(23 - 60x)$

107. $5(1 - x)^2(3x + 2)(4x + 3)$

109. $(x - 2)^2(x + 1)^3(7x - 5)$

111. $3(x^2 + 1)^4(x^4 - x^2 + 1)^4(3x + 2)^2(33x^6 + 20x^5 + 3)$

113. b   **114.** c   **115.** a   **116.** d

CHAPTER P

117.

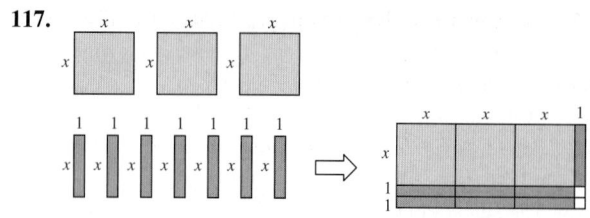

119.

121. $4\pi(r + 1)$ **123.** $4(6 - x)(6 + x)$

125. $-14, 14, -2, 2$ **127.** $-11, 11, -4, 4, -1, 1$

129. Two possible answers: $2, -12$

131. Two possible answers: $-2, -4$

133. A 3 was not factored out of the second binomial.

135. $kx(Q - x)$ **137.** True. $a^2 - b^2 = (a + b)(a - b)$

139. $(x^n + y^n)(x^n - y^n)$

141. $x^{3n} - y^{2n}$ is completely factored.

143. Answers will vary. Sample answer: $x^2 - 3$

Section P.5   *(page 47)*

Vocabulary Check   *(page 47)*

1. domain   **2.** rational expression   **3.** complex

4. smaller   **5.** equivalent   **6.** difference quotient

---

1. All real numbers   **3.** All nonnegative real numbers

5. All real numbers x such that $x \neq 2$

7. All real numbers x such that $x \geq -1$

9. $3x$, $x \neq 0$ **11.** $\dfrac{3x}{2}$, $x \neq 0$ **13.** $\dfrac{3y}{y + 1}$, $x \neq 0$

15. $\dfrac{-4y}{5}$, $y \neq \dfrac{1}{2}$ **17.** $-\dfrac{1}{2}$, $x \neq 5$

19. $y - 4$, $y \neq -4$ **21.** $\dfrac{x(x + 3)}{x - 2}$, $x \neq -2$

23. $\dfrac{y - 4}{y + 6}$, $y \neq 3$ **25.** $\dfrac{-(x^2 + 1)}{(x + 2)}$, $x \neq 2$ **27.** $z - 2$

29.

x	0	1	2	3	4	5	6
$\dfrac{x^2 - 2x - 3}{x - 3}$	1	2	3	Undef.	5	6	7
$x + 1$	1	2	3	4	5	6	7

The expressions are equivalent except at $x = 3$.

31. The expression cannot be simplified.   **33.** $\dfrac{\pi}{4}$, $r \neq 0$

35. $\dfrac{1}{5(x - 2)}$, $x \neq 1$ **37.** $\dfrac{r + 1}{r}$, $r \neq 1$

39. $\dfrac{t - 3}{(t + 3)(t - 2)}$, $t \neq -2$ **41.** $\dfrac{(x + 6)(x + 1)}{x^2}$, $x \neq 6$

43. $\dfrac{x + 5}{x - 1}$ **45.** $\dfrac{6x + 13}{x + 3}$ **47.** $-\dfrac{2}{x - 2}$

49. $-\dfrac{x^2 + 3}{(x + 1)(x - 2)(x - 3)}$ **51.** $\dfrac{2 - x}{x^2 + 1}$, $x \neq 0$

53. The error was incorrect subtraction in the numerator.

55. $\dfrac{1}{2}$, $x \neq 2$ **57.** $x(x + 1)$, $x \neq -1, 0$

59. $\dfrac{2x - 1}{2x}$, $x > 0$ **61.** $\dfrac{x^7 - 2}{x^2}$ **63.** $\dfrac{-1}{(x^2 + 1)^5}$

65. $\dfrac{2x^3 - 2x^2 - 5}{(x - 1)^{1/2}}$ **67.** $\dfrac{3x - 1}{3}$, $x \neq 0$

69. $\dfrac{-1}{x(x + h)}$, $h \neq 0$ **71.** $\dfrac{-1}{(x - 4)(x + h - 4)}$, $h \neq 0$

73. $\dfrac{1}{\sqrt{x + 2} + \sqrt{x}}$ **75.** $\dfrac{1}{\sqrt{x + h + 1} + \sqrt{x + 1}}$, $h \neq 0$

77. $\dfrac{x}{2(2x + 1)}$, $x \neq 0$

79. (a) $\dfrac{1}{16}$ minute   (b) $\dfrac{x}{16}$ minute(s)   (c) $\dfrac{60}{16} = \dfrac{15}{4}$ minutes

81. (a) 9.09%   (b) $\dfrac{288(MN - P)}{N(MN + 12P)}$; 9.09%

83. (a)

t	0	2	4	6	8	10
T	75	55.9	48.3	45	43.3	42.3

t	12	14	16	18	20	22
T	41.7	41.3	41.1	40.9	40.7	40.6

(b) The model is approaching a T-value of 40.

85. False. In order for the simplified expression to be equivalent to the original expression, the domain of the simplified expression needs to be restricted. If n is even, $x \neq -1, 1$. If n is odd, $x \neq 1$.

87. Completely factor each polynomial in the numerator and in the denominator. Then conclude that there are no common factors.

Section P.6   *(page 56)*

Vocabulary Check   *(page 56)*

1. numerator   **2.** reciprocal

---

1. Change all signs when distributing the minus sign.
$2x - (3y + 4) = 2x - 3y - 4$

3. Change all signs when distributing the minus sign.

$$\frac{4}{16x - (2x + 1)} = \frac{4}{14x - 1}$$

5. z occurs twice as a factor.

$(5z)(6z) = 30z^2$

7. The fraction as a whole is multiplied by a, not the numerator and denominator separately.

$$a\left(\frac{x}{y}\right) = \frac{ax}{y}$$

9. $\sqrt{x + 9}$ cannot be simplified.

11. Divide out common factors, not common terms.

$\dfrac{2x^2 + 1}{5x}$ cannot be simplified.

13. To get rid of negative exponents:

$$\frac{1}{a^{-1} + b^{-1}} = \frac{1}{a^{-1} + b^{-1}} \cdot \frac{ab}{ab} = \frac{ab}{b + a}.$$

15. Factor within grouping symbols before applying exponent to each factor.

$(x^2 + 5x)^{1/2} = [x(x + 5)]^{1/2} = x^{1/2}(x + 5)^{1/2}$

17. To add fractions, first find a common denominator.

$$\frac{3}{x} + \frac{4}{y} = \frac{3y + 4x}{xy}$$

19. $3x + 2$ **21.** $2x^2 + x + 15$ **23.** $\frac{1}{3}$ **25.** 2

27. $\dfrac{1}{2x^2}$ **29.** $\dfrac{25}{9}, \dfrac{49}{16}$ **31.** 1, 2   **33.** $1 - 5x$

35. $1 - 7x$ **37.** $3x - 1$ **39.** $3x^2(2x - 1)^{-3}$

41. $\dfrac{4}{3}x^{-1} + 4x^{-4} - 7x(2x)^{-1/3}$ **43.** $\dfrac{16}{x} - 5 - x$

45. $4x^{8/3} - 7x^{5/3} + \dfrac{1}{x^{1/3}}$ **47.** $\dfrac{3}{x^{1/2}} - 5x^{3/2} - x^{7/2}$

49. $\dfrac{-7x^2 - 4x + 9}{(x^2 - 3)^3(x + 1)^4}$ **51.** $\dfrac{27x^2 - 24x + 2}{(6x + 1)^4}$

53. $\dfrac{-1}{(x + 3)^{2/3}(x + 2)^{7/4}}$ **55.** $\dfrac{4x - 3}{(3x - 1)^{4/3}}$ **57.** $\dfrac{x}{x^2 + 4}$

59. $\dfrac{(3x - 2)^{1/2}(15x^2 - 4x + 45)}{2(x^2 + 5)^{1/2}}$

61. (a)

x	0.5	1.0	1.5	2.0
t	1.70	1.72	1.78	1.89

x	2.5	3.0	3.5	4.0
t	2.02	2.18	2.36	2.57

(b) $x = 0.5$ mile

(c) $\dfrac{3x\sqrt{x^2 - 8x + 20} + (x - 4)\sqrt{x^2 + 4}}{6\sqrt{x^2 + 4}\sqrt{x^2 - 8x + 20}}$

63. True. $x^{-1} + y^{-2} = \dfrac{1}{x} + \dfrac{1}{y^2} = \dfrac{y^2 + x}{xy^2}$

65. True. $\dfrac{1}{\sqrt{x} + 4} = \dfrac{1}{\sqrt{x} + 4} \cdot \dfrac{\sqrt{x} - 4}{\sqrt{x} - 4} = \dfrac{\sqrt{x} - 4}{x - 16}$

67. Add exponents when multiplying powers with like bases.

$x^n \cdot x^{3n} = x^{4n}$

69. When a binomial is squared, there is also a middle term.

$(x^n + y^n)^2 = x^{2n} + 2x^n y^n + y^{2n} \neq x^{2n} + y^{2n}$

71. The two answers are equivalent and can be obtained by factoring.

$\frac{1}{10}(2x - 1)^{5/2} + \frac{1}{6}(2x - 1)^{3/2}$

$= \frac{1}{60}(2x - 1)^{3/2}[6(2x - 1) + 10]$

$= \frac{1}{60}(2x - 1)^{3/2}(12x + 4)$

$= \frac{4}{60}(2x - 1)^{3/2}(3x + 1)$

$= \frac{1}{15}(2x - 1)^{3/2}(3x + 1)$

(a) $\frac{2}{5}(2x - 3)^{3/2}(x + 1)$ (b) $\frac{8}{15}(4 + x)^{3/2}(x - 1)$

Section P.7   *(page 64)*

Vocabulary Check   *(page 64)*

1. (a) v   (b) vi   (c) i   (d) iv   (e) iii   (f) ii

2. Cartesian   **3.** Distance Formula

4. Midpoint Formula

1. A: $(2, 6)$, B: $(-6, -2)$, C: $(4, -4)$, D: $(-3, 2)$

3.   **5.**

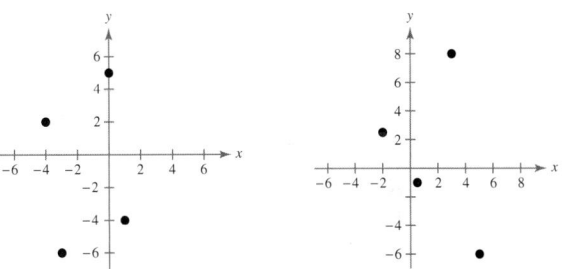

7. $(-3, 4)$ **9.** $(-5, -5)$ **11.** Quadrant IV

13. Quadrant II   **15.** Quadrant III or IV

17. Quadrant III   **19.** Quadrant I or III

21.   **23.** 8   **25.** 5

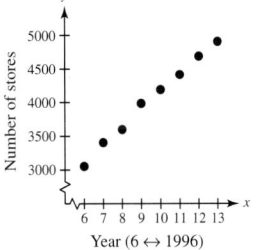

Year (6 ↔ 1996)

27. (a) 4, 3, 5   (b) $4^2 + 3^2 = 5^2$

29. (a) 10, 3, $\sqrt{109}$ (b) $10^2 + 3^2 = \left(\sqrt{109}\right)^2$

31. (a)

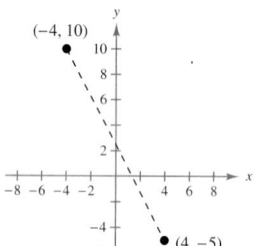

(b) 10
(c) $(5, 4)$

33. (a)

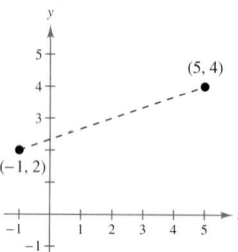

(b) 17
(c) $\left(0, \frac{5}{2}\right)$

35. (a)

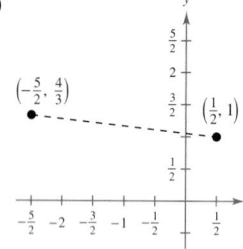

(b) $2\sqrt{10}$
(c) $(2, 3)$

37. (a)

(b) $\dfrac{\sqrt{82}}{3}$
(c) $\left(-1, \frac{7}{6}\right)$

39. (a)

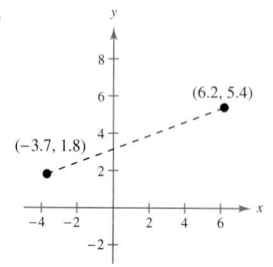

(b) $\sqrt{110.97}$
(c) $(1.25, 3.6)$

41. $\left(\sqrt{5}\right)^2 + \left(\sqrt{45}\right)^2 = \left(\sqrt{50}\right)^2$

43. $(2x_m - x_1, 2y_m - y_1)$

45. $\left(\dfrac{3x_1 + x_2}{4}, \dfrac{3y_1 + y_2}{4}\right), \left(\dfrac{x_1 + x_2}{2}, \dfrac{y_1 + y_2}{2}\right),$
$\left(\dfrac{x_1 + 3x_2}{4}, \dfrac{y_1 + 3y_2}{4}\right)$

47. $2\sqrt{505} \approx 45$ yards   **49.** \$3803.5 million

51. $(0, 1), (4, 2), (1, 4)$

53. $(-3, 6), (2, 10), (2, 4), (-3, 4)$

55. \$3.31 per pound; 2001   **57.** $\approx 250\%$

59.

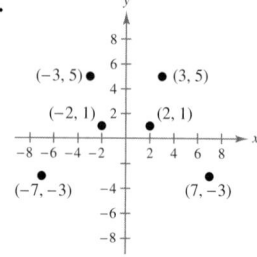

(a) The point is reflected through the y-axis.
(b) The point is reflected through the x-axis.
(c) The point is reflected through the origin.

61. (a) 1990s   (b) 11.8%; 21.2%   (c) \$6.24
(d) Answers will vary.

63. (2000, \$20,223 million); (1998, \$19,384.5 million);
(2002, \$21,061.5 million)

65. (a)

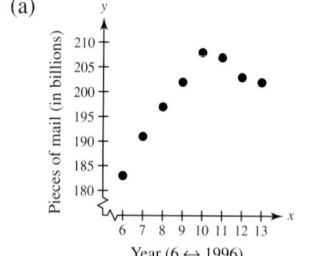

(b) 2002

(c) Answers will vary. Sample answer: Technology now enables us to transport information in many ways other than by mail. The Internet is one example.

67. (a)

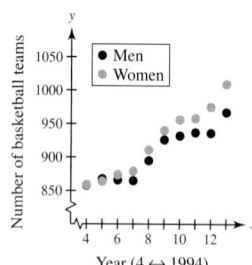

(b) 1994
(c) 2003; 42 teams

69. False. The Midpoint Formula would be used 15 times.

71. Point on x-axis: $y = 0$; Point on y-axis: $x = 0$

73. b   **74.** c   **75.** d   **76.** a

77. Use the Midpoint Formula to prove that the diagonals of the parallelogram bisect each other.
$$\left(\frac{b + a}{2}, \frac{c + 0}{2}\right) = \left(\frac{a + b}{2}, \frac{c}{2}\right)$$
$$\left(\frac{a + b + 0}{2}, \frac{c + 0}{2}\right) = \left(\frac{a + b}{2}, \frac{c}{2}\right)$$

Review Exercises *(page 70)*

1. (a) 11 (b) 11 (c) 11, -14
(d) 11, -14, $-\frac{8}{9}$, $\frac{5}{2}$, 0.4 (e) $\sqrt{6}$
3. (a) $0.8\overline{3}$ (b) 0.875

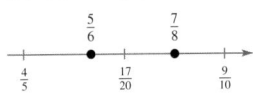

$\frac{5}{6} < \frac{7}{8}$

5. The set consists of all real numbers less than or equal to 7.

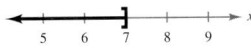

7. 155 **9.** $|x - 7| \geq 4$ **11.** $|y + 30| < 5$
13. (a) -7 (b) -19 **15.** (a) -1 (b) -3
17. Associative Property of Addition
19. Additive Identity Property **21.** -11 **23.** $\frac{1}{12}$
25. -144 **27.** (a) $192x^{11}$ (b) $\frac{y^5}{2}$, $y \neq 0$
29. (a) $\frac{3u^5}{v^4}$ (b) $\frac{1}{m^2}$ **31.** 2.1343×10^9
33. 483,700,000 **35.** (a) 9 (b) 343
37. (a) 216 (b) 32 **39.** (a) $2\sqrt{2}$ (b) $26\sqrt{2}$
41. Radicals cannot be combined by addition or subtraction unless the index and the radicand are the same.
43. $2 + \sqrt{3}$ **45.** $\frac{3}{\sqrt{7} - 1}$ **47.** 64 **49.** $6x^{9/10}$
51. $-11x^2 + 3$; Degree: 2; Leading coefficient: -11
53. $-12x^2 - 4$; Degree: 2; Leading coefficient: -12
55. $-3x^2 - 7x + 1$ **57.** $2x^3 - 10x^2 + 12x$
59. $15x^2 - 27x - 6$ **61.** $4x^2 - 12x + 9$ **63.** 41
65. (a) Explanations will vary.

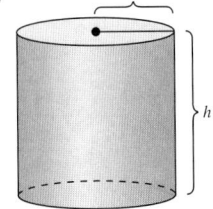

 (b) $168\pi \approx 527.79$ square inches
67. $x(x + 1)(x - 1)$ **69.** $(5x + 7)(5x - 7)$
71. $(x - 4)(x^2 + 4x + 16)$ **73.** $(x + 10)(2x + 1)$
75. $(x - 1)(x^2 + 2)$
77. All real numbers x except $x = -6$
79. $\frac{x - 8}{15}$, $x \neq -8$ **81.** $\frac{1}{x^2}$, $x \neq \pm 2$
83. $\frac{3x}{(x - 1)(x^2 + x + 1)}$ **85.** $\frac{3ax^2}{(a^2 - x)(a - x)}$
87. $\frac{-1}{2x(x + h)}$, $h \neq 0$
89. The multiplication in parentheses comes first.
$10(4 \cdot 7) = 10(28) = 280$

91. Multiply exponents when raising a power to a power.
$(3^4)^4 = 3^{16}$
93. Add the numbers in parentheses before squaring.
$(5 + 8)^2 = 13^2 \neq 5^2 + 8^2$
95. $16x^2 - 9x + 20$ **97.** $-5x^2 + 2x + 15$
99. $x^2 + 5x + \frac{7}{x}$ **101.** $\frac{2(x + 1)}{(x + 2)^2}$
103. 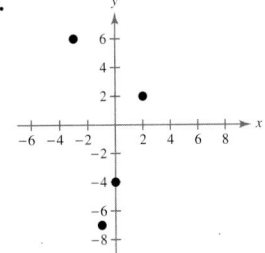 **105.** Quadrant IV

107. (a) 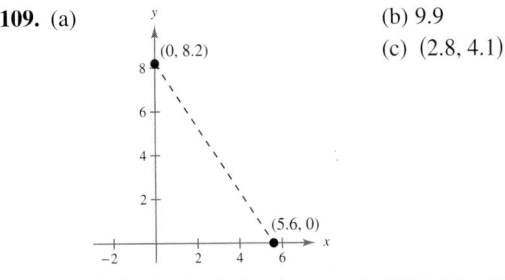 (b) 5
(c) $\left(-1, \frac{13}{2}\right)$

109. (a)
(b) 9.9
(c) (2.8, 4.1)

111. (2, 5), (4, 5), (2, 0), (4, 0) **113.** \$656.45 million
115. False. There is also a cross-product term when a binomial sum is squared.
$(x + a)^2 = x^2 + 2ax + a^2$
117. No. When $x = b/a$, the expression is undefined.

Chapter Test *(page 73)*

1. $-\frac{10}{3} > -|-4|$ **2.** 9.15
3. Additive Identity Property
4. (a) -18 (b) $\frac{4}{9}$ (c) $-\frac{27}{125}$ (d) $\frac{8}{729}$
5. (a) 25 (b) $\frac{3\sqrt{6}}{2}$ (c) 1.8×10^5 (d) 2.7×10^{13}
6. (a) $12z^8$ (b) $(u - 2)^{-7}$ (c) $\frac{3x^2}{y^2}$
7. (a) $15z\sqrt{2z}$ (b) $4x^{14/15}$ (c) $\frac{2\sqrt[3]{2v}}{v^2}$
8. $4x^4 - 3x^2 + x - 5$; Degree: 4; Leading coefficient: 4

9. $2x^2 - 3x - 5$ **10.** $x^2 - 5$ **11.** $8, x \neq 3$

12. $\dfrac{x-1}{2x}$, $x \neq \pm 1$

13. (a) $x^2(2x+1)(x-2)$ (b) $(x-2)(x+2)^2$

14. (a) $4\sqrt[3]{4}$ (b) $-3(1+\sqrt{3})$

15. All real numbers x except $x = 4$

16. $\dfrac{4}{y+4}$, $y \neq 2$ **17.** $545

18.

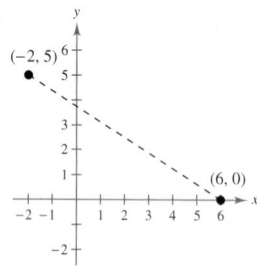

Midpoint: $\left(2, \frac{5}{2}\right)$; Distance: $\sqrt{89}$

19. $\frac{5}{6}\sqrt{3}\,x^2$

Problem Solving   *(page 75)*

1. (a) Men's: 1,150,347 cubic millimeters;
            696,910 cubic millimeters
       Women's: 696,910 cubic millimeters;
            448,921 cubic millimeters
   (b) Men's: 1.04×10^{-5} kilograms per cubic millimeter;
            6.31×10^{-6} kilograms per cubic millimeter
       Women's: 8.91×10^{-6} kilograms per cubic millimeter;
            5.74×10^{-6} kilograms per cubic millimeter
   (c) No. Iron has a greater density than cork.

3. 1.62 ounces   **5.** Answers will vary.

7. $r \approx 0.28$ **9.** 9.57 square feet

11. $y_1(0) = 0,\ y_2(0) = 2$

$y_2 = \dfrac{x(2 - 3x^2)}{\sqrt{1 - x^2}}$

13. (a) $(2, -1), (3, 0)$ (b) $\left(-\frac{4}{3}, -2\right), \left(-\frac{2}{3}, -1\right)$

Chapter 1

Section 1.1   *(page 86)*

1. (a) Yes   (b) Yes   **3.** (a) No   (b) Yes

5.

x	-1	0	1	2	$\frac{5}{2}$
y	7	5	3	1	0
(x, y)	$(-1, 7)$	$(0, 5)$	$(1, 3)$	$(2, 1)$	$\left(\frac{5}{2}, 0\right)$

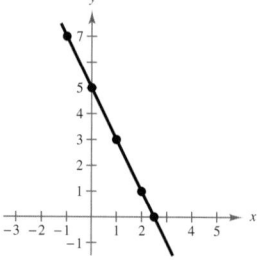

7.

x	-1	0	1	2	3
y	4	0	-2	-2	0
(x, y)	$(-1, 4)$	$(0, 0)$	$(1, -2)$	$(2, -2)$	$(3, 0)$

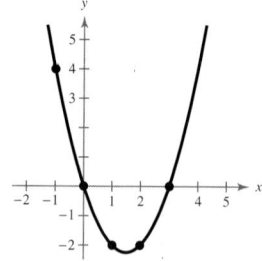

9. x-intercepts: $(\pm 2, 0)$
   y-intercept: $(0, 16)$

11. x-intercepts: $(0, 0), (2, 0)$
   y-intercept: $(0, 0)$

13.

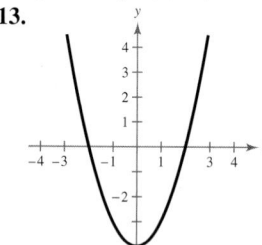

15.

17. y-axis symmetry   **19.** Origin symmetry

21. Origin symmetry   **23.** x-axis symmetry

25.

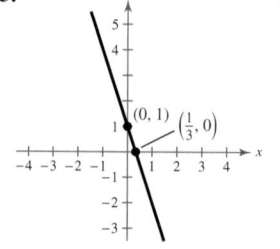

27.

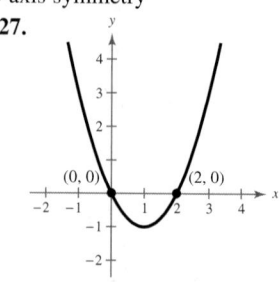

29.
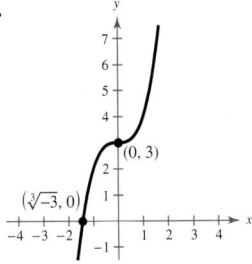

31.
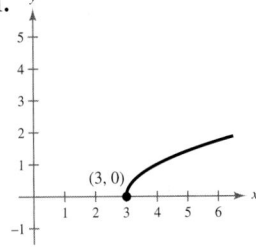

33.
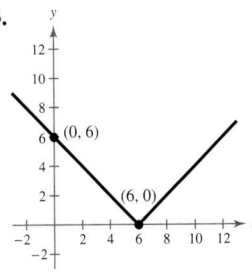

35.
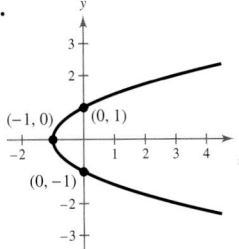

37.
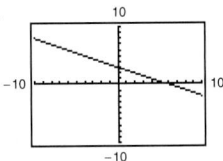

Intercepts: $(6, 0)$, $(0, 3)$

39.
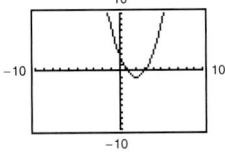

Intercepts: $(3, 0)$, $(1, 0)$, $(0, 3)$

41.
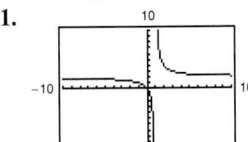

Intercept: $(0, 0)$

43.
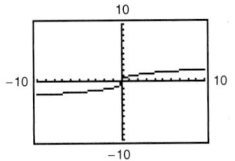

Intercept: $(0, 0)$

45.
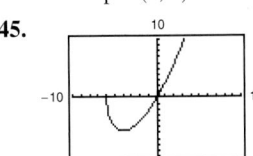

Intercepts: $(0, 0)$, $(-6, 0)$

47.
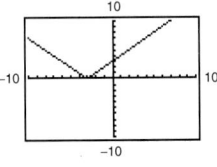

Intercepts: $(-3, 0)$, $(0, 3)$

49. $x^2 + y^2 = 16$ **51.** $(x - 2)^2 + (y + 1)^2 = 16$
53. $(x + 1)^2 + (y - 2)^2 = 5$
55. $(x - 3)^2 + (y - 4)^2 = 25$
57. Center: $(0, 0)$; Radius: 5   **59.** Center: $(1, -3)$; Radius: 3

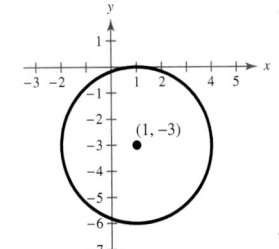

61. Center: $\left(\frac{1}{2}, \frac{1}{2}\right)$; Radius: $\frac{3}{2}$ **63.**

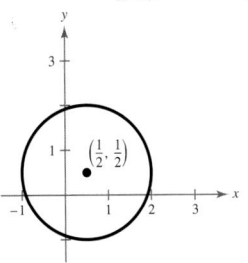

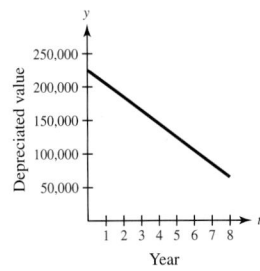

65. (a)
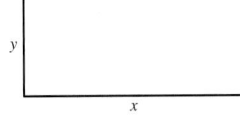
(b) Answers will vary.

(c)
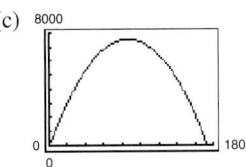
(d) $x = 86\frac{2}{3}$, $y = 86\frac{2}{3}$

(e) A regulation NFL playing field is 120 yards long and $53\frac{1}{3}$ yards wide. The actual area is 6400 square yards.

67. (a) and (b)
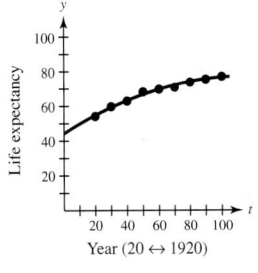

(c) 66.0 years   (d) 2005: 77.0 years; 2010: 77.1 years
(e) Answers will vary.

69. False. A graph is symmetric with respect to the x-axis if, whenever (x, y) is on the graph $(x, -y)$ is also on the graph.

71. The viewing window is incorrect. Change the viewing window. Answers will vary.

73. $9x^5$, $4x^3$, -7 **75.** $2\sqrt{2x}$ **77.** $\sqrt[3]{|t|}$

Section 1.2   *(page 94)*

Vocabulary Check   *(page 94)*

1. equation   **2.** solve   **3.** identities; conditional
4. $ax + b = 0$ **5.** extraneous

1. (a) No   (b) No   (c) Yes   (d) No
3. (a) Yes   (b) Yes   (c) No   (d) No
5. (a) Yes   (b) No   (c) No   (d) No
7. (a) No   (b) No   (c) No   (d) No
9. (a) Yes   (b) No   (c) Yes   (d) No

CHAPTER 1

11. Identity   **13.** Conditional equation   **15.** Identity
17. Identity   **19.** Conditional equation
21. Original equation
Subtract 32 from each side.
Simplify.
Divide each side by 4.
Simplify.
23. 4   **25.** -9 **27.** 5   **29.** 9   **31.** No solution
33. -4 **35.** $-\frac{6}{5}$ **37.** 9   **39.** $\frac{7}{3}$ **41.** 13
43. No solution. The x-terms sum to zero.   **45.** 10   **47.** 4
49. 3   **51.** 0   **53.** No solution. The variable is divided out.
55. No solution. The solution is extraneous.   **57.** 5
59. No solution. The solution is extraneous.
61. 0   **63.** All real numbers

65.
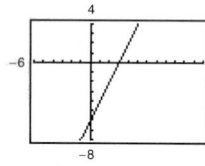
$x = 3$

67.
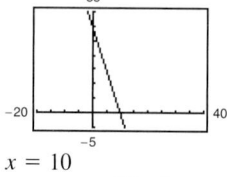
$x = 10$

69.
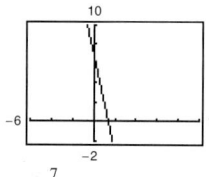
$x = \frac{7}{5}$

71. x-intercept: $\left(\frac{12}{5}, 0\right)$
y-intercept: $(0, 12)$

73. x-intercept: $\left(-\frac{1}{2}, 0\right)$ **75.** x-intercept: $(5, 0)$
y-intercept: $(0, -3)$ y-intercept: $\left(0, \frac{10}{3}\right)$
77. x-intercept: $(-20, 0)$ **79.** x-intercept: $(1.6, 0)$
y-intercept: $\left(0, \frac{8}{3}\right)$ y-intercept: $(0, -0.3)$
81. $\dfrac{1}{3 - a}$, $a \neq 3$ **83.** $\dfrac{5}{4 + a}$, $a \neq -4$
85. $\dfrac{18}{36 + a}$, $a \neq -36$ **87.** $\dfrac{-17}{10 - 2a}$, $a \neq 5$
89. 138.889   **91.** 19.993   **93.** $h = 10$ feet
95. (a) 61.2 inches
(b) Yes. The estimated height of a male with a 19-inch femur is 69.4 inches.
(c)

Height, x	Female femur length	Male femur length
60	15.48	14.79
70	19.80	19.28
80	24.12	23.77
90	28.44	28.26
100	32.76	32.75
110	37.08	37.24

100 inches

(d) $x \approx 100.59$; There would not be a problem because it is not likely for either a male or a female to be 100 inches tall (which is 8 feet 4 inches tall).

97. (a)
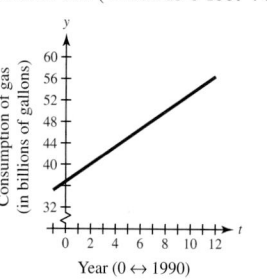
y-intercept: $(0, 36.8)$

(b) y-intercept: $(0, 36.8)$ (c) 2007; answers will vary.
99. 23,437.5 miles
101. False. $x(3 - x) = 10$
$3x - x^2 = 10$
The equation cannot be written in the form $ax + b = 0$.
103. Equivalent equations have the same solution set, and one is derived from the other by steps for generating equivalent equations.
$2x = 5, \ 2x + 3 = 8$
105. (a)

x	-1	0	1	2	3	4
$3.2x - 5.8$	-9	-5.8	-2.6	0.6	3.8	7

(b) $1 < x < 2$. The expression changes from negative to positive in this interval.
(c)

x	1.5	1.6	1.7	1.8	1.9	2
$3.2x - 5.8$	-1	-0.68	-0.36	-0.04	0.28	0.6

(d) $1.8 < x < 1.9$. To improve accuracy, evaluate the expression in this interval and determine where the sign changes.
107. $\dfrac{x - 4}{2x - 1}$, $x \neq -9$

109.
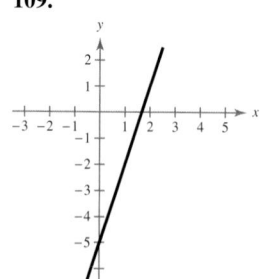

111.
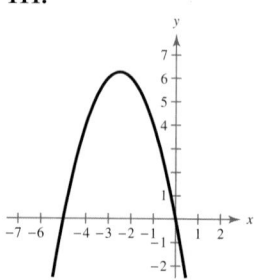

Section 1.3 *(page 105)*

Vocabulary Check *(page 105)*

1. mathematical modeling
2. verbal model; algebraic equation   **3.** $A = \pi r^2$
4. $P = 2l + 2w$ **5.** $V = s^3$ **6.** $V = \pi r^2 h$
7. $A = P\left(1 + \dfrac{r}{n}\right)^{12t}$ **8.** $I = Prt$

---

1. A number increased by 4
3. A number divided by 5
5. A number decreased by 4 is divided by 5.
7. Negative 3 is multiplied by a number increased by 2.
9. 12 is multiplied by a number and that product is multiplied by the number decreased by 5.
11. $n + (n + 1) = 2n + 1$
13. $(2n - 1)(2n + 1) = 4n^2 - 1$
15. $50t$ **17.** $0.20x$ **19.** $6x$ **21.** $25x + 1200$
23. $0.30L$ **25.** $N = p(500)$ **27.** $4x + 8x = 12x$
29. $262, 263$ **31.** $37, 185$ **33.** $-5, -4$
35. 13.5 **37.** 27% **39.** 2400 **41.** $\$22,316.98$
43. 17.2% increase   **45.** 93% increase   **47.** $\$1450.00$
49. (a)

(b) $l = 1.5w$; $p = 5w$
(c) 7.5 meters $\times$ 5 meters

51. 97   **53.** 3 hours
55. 3 hours at 58 miles per hour; 2 hours and 45 minutes at 52 miles per hour
57. (a) 3.8 hours, 3.2 hours   (b) 1.1 hours
(c) 25.6 miles
59. $66\frac{2}{3}$ kilometers per hour   **61.** 1.28 seconds
63. 1044 feet   **65.** 4.36 feet   **67.** $\$16,666.67$
69. Minivans: $\$450,000$; SUVs: $\$150,000$
71. ≈ 32.1 gallons
73. 50 pounds of each kind   **75.** 8064 units
77. $\dfrac{2A}{b}$ **79.** $\dfrac{S}{1 + R}$ **81.** $\dfrac{3V}{4\pi a^2}$
83. $\dfrac{2(h - v_0 t)}{t^2}$ **85.** $\dfrac{CC_2}{C_2 - C}$ **87.** $\dfrac{L - a + d}{d}$
89. $x = 6$ feet from the 50-pound child
91. $\sqrt[3]{\dfrac{4.47}{\pi}} \approx 1.12$ inches   **93.** 18°C   **95.** 122°F
97. False. The expression should be $\dfrac{z^3 - 8}{z^2 - 9}$.
99. (a) Negative; answers will vary.
(b) Positive; answers will vary.
101. $\dfrac{x^2}{5}$, $x \neq 0$ **103.** $\dfrac{1}{x - 5}$ **105.** $\dfrac{10\sqrt{3}}{21}$

---

107. $-\left(\sqrt{6} - \sqrt{11}\right)$

Section 1.4 *(page 120)*

Vocabulary Check *(page 120)*

1. quadratic equation
2. factoring; extracting square roots; completing the square; Quadratic Formula
3. discriminant
4. position; $-16t^2 + v_0 t + s_0$; initial velocity; initial height
5. Pythagorean Theorem

---

1. $2x^2 + 8x - 3 = 0$ **3.** $x^2 - 6x + 6 = 0$
5. $3x^2 - 90x - 10 = 0$ **7.** $0, -\frac{1}{2}$ **9.** $4, -2$
11. -5 **13.** $3, -\frac{1}{2}$ **15.** $2, -6$ **17.** $-\frac{20}{3}, -4$
19. $-a$ **21.** ± 7 **23.** $\pm \sqrt{11}$ **25.** $\pm 3\sqrt{3}$
27. $8, 16$ **29.** $-2 \pm \sqrt{14}$ **31.** $\dfrac{1 \pm 3\sqrt{2}}{2}$
33. 2 **35.** $4, -8$ **37.** $\sqrt{11} - 6, -\sqrt{11} - 6$
39. $1 \pm \dfrac{\sqrt{6}}{3}$ **41.** $2 \pm 2\sqrt{3}$ **43.** $\dfrac{-5 \pm \sqrt{89}}{4}$
45. $\dfrac{1}{(x + 1)^2 + 4}$ **47.** $\dfrac{4}{(x + 2)^2 - 7}$
49. $\dfrac{1}{\sqrt{9 - (x - 3)^2}}$
51. (a)

(b) and (c) $x = -1, -5$
(d) The answers are the same.
53. (a)

(b) and (c) $x = 3, 1$
(d) The answers are the same.
55. (a)

(b) and (c) $x = -\frac{1}{2}, \frac{3}{2}$
(d) The answers are the same.
57. (a)

(b) and (c) $x = 1, -4$
(d) The answers are the same.

59. No real solution   **61.** Two real solutions

63. No real solution   **65.** Two real solutions

67. $\frac{1}{2}, -1$ **69.** $\frac{1}{4}, -\frac{3}{4}$ **71.** $1 \pm \sqrt{3}$

73. $-7 \pm \sqrt{5}$ **75.** $-4 \pm 2\sqrt{5}$ **77.** $\frac{2}{3} \pm \frac{\sqrt{7}}{3}$

79. $-\frac{4}{3}$ **81.** $-\frac{1}{2} \pm \sqrt{2}$ **83.** $\frac{2}{7}$ **85.** $2 \pm \frac{\sqrt{6}}{2}$

87. $6 \pm \sqrt{11}$ **89.** $-\frac{3}{8} \pm \frac{\sqrt{265}}{8}$ **91.** $0.976, -0.643$

93. $1.355, -14.071$ **95.** $1.687, -0.488$

97. $-0.290, -2.200$ **99.** $1 \pm \sqrt{2}$ **101.** $6, -12$

103. $\frac{1}{2} \pm \sqrt{3}$ **105.** $-\frac{1}{2}$ **107.** $\frac{3}{4} \pm \frac{\sqrt{97}}{4}$

109. (a)

(b) $w(w + 14) = 1632$
(c) $w = 34$ feet
$l = 48$ feet

111. 6 inches × 6 inches × 2 inches

113. 19.098 feet; 9.5 trips

115. (a) $20\sqrt{5} \approx 44.72$ seconds   (b) 6.2 miles

117. (a) $s = -16t^2 + 146\frac{2}{3}t + 6.25$

(b) $s(3) = 302.25$ feet; $s(4) = 336.92$ feet;
$s(5) = 339.58$ feet;
During the interval $3 \le t \le 5$, the baseball's speed decreased due to gravity.

(c) Assuming the ball is not caught and drops to the ground, the baseball is in the air about 9.209 seconds.

119. (a)

t	7	8	9	10	11	12	13
P	4.51	4.81	5.09	5.35	5.60	5.83	6.04

The average admission price reached or surpassed $5 in 1999.

(b) Answers will vary.   (c) $6.87. Answers will vary.

121. $\frac{5\sqrt{2}}{2} \approx 3.54$ centimeters

123. ≈ 550 miles per hour and 600 miles per hour

125. 50,000 units   **127.** 258 units   **129.** 653 units

131. (a)

t	5	6	7	8
M	396.78	420.54	447.98	479.08

t	9	10	11	12
M	513.86	552.30	594.42	640.20

The total money in circulation reached $600 billion during 2001.

(b) Answers will vary.

(c) $991.98 billion; Answers will vary.

133. 354.5 miles, 433.5 miles

135. False. $b^2 - 4ac < 0$, so the quadratic equation has no real solution.

137. Yes. The student should have subtracted $15x$ from both sides to make the right side of the equation equal to zero. Factoring out an x shows that there are two solutions, $x = 0$ and $x = 6$.

139. (a) and (b) $x = -5, -\frac{10}{3}$

(c) The method used in part (a) reduces the number of algebraic steps.

141. Answers will vary. Sample answer: $x^2 - 3x - 18 = 0$

143. Answers will vary. Sample answer: $x^2 - 22x + 112 = 0$

145. Answers will vary. Sample answer: $x^2 - 2x - 1 = 0$

147. Associative Property of Multiplication

149. Additive Inverse Property   **151.** $x^2 - 3x - 18$

153. $x^3 + 3x^2 - 2x + 8$ **155.** $x^2(x - 3)(x^2 + 3x + 9)$

157. $(x + 5)(x^2 - 2)$ **159.** Answers will vary.

Section 1.5   (page 131)

Vocabulary Check   *(page 131)*

1. (a) iii   (b) i   (c) ii   **2.** $\sqrt{-1}; -1$

3. complex numbers; $a + bi$ **4.** principal square

5. complex conjugates

1. $a = -10, b = 6$ **3.** $a = 6, b = 5$ **5.** $4 + 3i$

7. $2 - 3\sqrt{3}i$ **9.** $5\sqrt{3}i$ **11.** 8

13. $-1 - 6i$ **15.** $0.3i$ **17.** $11 - i$ **19.** 4

21. $3 - 3\sqrt{2}i$ **23.** $-14 + 20i$ **25.** $\frac{1}{6} + \frac{7}{6}i$

27. $5 + i$ **29.** $12 + 30i$ **31.** 24   **33.** $-9 + 40i$

35. -10 **37.** $6 - 3i, 45$ **39.** $-1 + \sqrt{5}i, 6$

41. $-2\sqrt{5}i, 20$ **43.** $\sqrt{8}, 8$ **45.** $-5i$

47. $\frac{8}{41} + \frac{10}{41}i$ **49.** $\frac{4}{5} + \frac{3}{5}i$ **51.** $-5 - 6i$

53. $-\frac{120}{1681} - \frac{27}{1681}i$ **55.** $-\frac{1}{2} - \frac{5}{2}i$ **57.** $\frac{62}{949} + \frac{297}{949}i$

59. $-2\sqrt{3}$ **61.** -10

63. $\left(21 + 5\sqrt{2}\right) + \left(7\sqrt{5} - 3\sqrt{10}\right)i$ **65.** $1 \pm i$

67. $-2 \pm \frac{1}{2}i$ **69.** $-\frac{5}{2}, -\frac{3}{2}$ **71.** $2 \pm \sqrt{2}i$

73. $\frac{5}{7} \pm \frac{5\sqrt{15}}{7}$ **75.** $-1 + 6i$

77. $-5i$ **79.** $-375\sqrt{3}i$ **81.** i

83. (a) $z_1 = 9 + 16i, z_2 = 20 - 10i$

(b) $z = \frac{11,240}{877} + \frac{4630}{877}i$

85. (a) 16   (b) 16   (c) 16   (d) 16

87. False. If the complex number is real, the number equals its conjugate.

89. False.
$i^{44} + i^{150} - i^{74} - i^{109} + i^{61} = 1 - 1 + 1 - i + i = 1$

91. Proof   **93.** $-x^2 - 3x + 12$ **95.** $3x^2 + \frac{23}{2}x - 2$

97. -31 **99.** $\frac{27}{2}$ **101.** $a = \frac{\sqrt{3V\pi b}}{2\pi b}$ **103.** 1 liter

Section 1.6   *(page 140)*

1. $0, \pm\dfrac{3\sqrt{2}}{2}$ **3.** $\pm 3, \pm 3i$ **5.** $-6, 3 \pm 3\sqrt{3}i$

7. $-3, 0$ **9.** $3, 1, -1$ **11.** $\pm 1, \dfrac{1}{2} \pm \dfrac{\sqrt{3}}{2}i$

13. $\pm\sqrt{3}, \pm 1$ **15.** $\pm\frac{1}{2}, \pm 4$

17. $1, -2, 1 \pm \sqrt{3}i, -\dfrac{1}{2} \pm \dfrac{\sqrt{3}i}{2}$ **19.** $-\dfrac{1}{5}, -\dfrac{1}{3}$

21. $\frac{1}{4}$ **23.** $1, -\dfrac{125}{8}$

25. (a)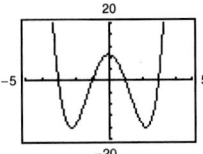

(b) $(0, 0), (3, 0), (-1, 0)$
(c) $x = 0, 3, -1$
(d) The x-intercepts and the solutions are the same.

27. (a)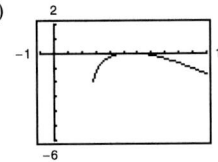

(b) $(\pm 3, 0), (\pm 1, 0)$
(c) $x = \pm 3, \pm 1$
(d) The x-intercepts and the solutions are the same.

29. 50   **31.** 26   **33.** -16 **35.** $2, -5$ **37.** 0

39. 9   **41.** $\dfrac{101}{4}$ **43.** 14   **45.** 9   **47.** $-3 \pm 16\sqrt{2}$

49. $\pm\sqrt{14}$ **51.** 1

53. (a)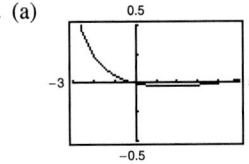

(b) $(5, 0), (6, 0)$
(c) $x = 5, 6$
(d) The x-intercepts and the solutions are the same.

55. (a)

(b) $(0, 0), (4, 0)$
(c) $x = 0, 4$
(d) The x-intercepts and the solutions are the same.

57. $2, -\dfrac{3}{2}$ **59.** $\dfrac{-3 \pm \sqrt{21}}{6}$ **61.** $4, -5$

63. $\dfrac{1 \pm \sqrt{31}}{3}$ **65.** $3, -2$ **67.** $\sqrt{3}, -3$

69. $3, \dfrac{-1 - \sqrt{17}}{2}$

71. (a)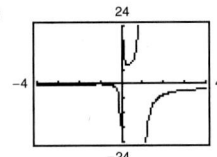

(b) $(-1, 0)$ (c) $x = -1$
(d) The x-intercept and the solution are the same.

73. (a)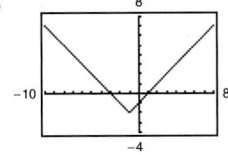

(b) $(1, 0), (-3, 0)$
(c) $x = 1, -3$
(d) The x-intercepts and the solutions are the same.

75. ± 1.038 **77.** 16.756   **79.** $x^2 - 3x - 10 = 0$
81. $21x^2 + 31x - 42 = 0$ **83.** $x^3 - 4x^2 - 3x + 12 = 0$
85. $x^4 - 1 = 0$ **87.** 34 students
89. 191.5 miles per hour   **91.** 4%
93. (a) 1999   (b) During 2005; answers will vary
95. (a)

x	T
5	162.56
10	192.31
15	212.68
20	228.20
25	240.62
30	250.83
35	259.38
40	266.60

(b) ≈ 15 pounds per square inch
(c) $x \approx 14.81$
(d) Answers will vary.

97. 500 units   **99.** 90 feet

101. (a)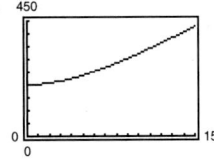

The height $h \approx 11.4$ when $S = 350$.

(b)

h	8	9	10	11	12	13
S	284.3	302.6	321.9	341.8	362.5	383.6

The height h is between 11 and 12 inches when $S = 350$.
(c) $h = 11.4$ when $S = 350$.
(d) Solving graphically or numerically yields an approximate solution. An exact solution is obtained algebraically.

103. $\dfrac{21 + \sqrt{585}}{2} \approx 22.6$ hours

$\dfrac{27 + \sqrt{585}}{2} \approx 25.6$ hours

CHAPTER 1

105. $g = \dfrac{\mu s v^2}{R}$ **107.** False. See Example 7 on page 137.

109. $6, -4$ **111.** ± 15 **113.** $a = 9, b = 9$

115. $a = 4, b = 24$ **117.** $\dfrac{25}{6x}$ **119.** $\dfrac{-3z^2 - 2z + 4}{z(z + 2)}$

121. 11

Section 1.7 *(page 150)*

1. $-1 \le x \le 5$. Bounded **3.** $x > 11$. Unbounded
5. $x < -2$. Unbounded
7. b **8.** f **9.** d **10.** c **11.** e **12.** a
13. (a) Yes (b) No (c) Yes (d) No
15. (a) Yes (b) No (c) No (d) Yes
17. (a) Yes (b) Yes (c) Yes (d) No
19. $x < 3$ **21.** $x < \frac{3}{2}$

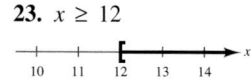 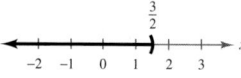

23. $x \ge 12$ **25.** $x > 2$

27. $x \ge \frac{2}{7}$ **29.** $x < 5$

31. $x \ge 4$ **33.** $x \ge 2$

35. $x \ge -4$ **37.** $-1 < x < 3$

39. $-\frac{9}{2} < x < \frac{15}{2}$ **41.** $-\frac{3}{4} < x < -\frac{1}{4}$

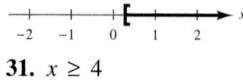 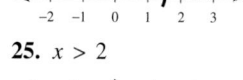

43. $10.5 \le x \le 13.5$ **45.** $-6 < x < 6$

47. $x < -2, x > 2$ **49.** No solution

51. $14 \le x \le 26$ **53.** $x \le -\frac{3}{2}, x \ge 3$

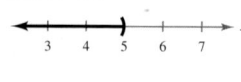

55. $x \le -5, x \ge 11$ **57.** $4 < x < 5$

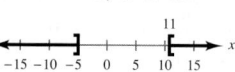 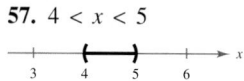

59. $x \le -\frac{29}{2}, x \ge -\frac{11}{2}$

61. **63.**

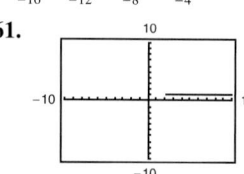 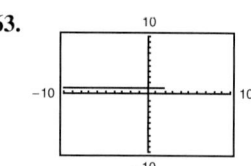

$x > 2$ $x \le 2$

65. **67.**

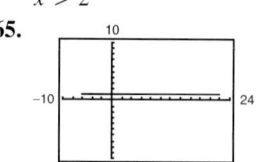 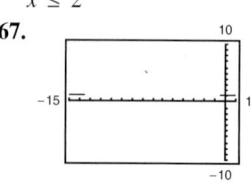

$-6 \le x \le 22$ $x \le -\frac{27}{2}, x \ge -\frac{1}{2}$

69. **71.**

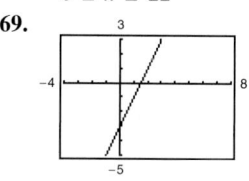 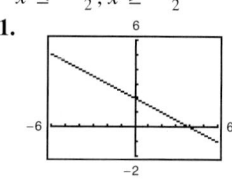

(a) $x \ge 2$ (a) $-2 \le x \le 4$
(b) $x \le \frac{3}{2}$ (b) $x \le 4$

73. (a) $1 \le x \le 5$

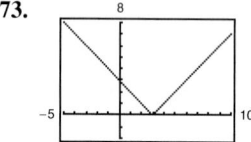 (b) $x \le -1, x \ge 7$

75. $[5, \infty)$ **77.** $[-3, \infty)$ **79.** $\left(-\infty, \frac{7}{2}\right]$
81. All real numbers within eight units of 10 **83.** $|x| \le 3$
85. $|x - 7| \ge 3$ **87.** $|x - 12| < 10$
89. $|x + 3| > 4$ **91.** $x > 6$ **93.** $r > 3.125\%$
95. $x \ge 36$ **97.** $134 \le x \le 234$
99. (a) (b) $x \ge 129$

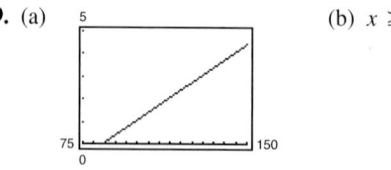

101. (a) $1 \le t \le 10$ (b) $t > 16$
103. 106.864 square inches $\le$ area ≤ 109.464 square inches
105. Might be undercharged or overcharged by \$0.19.
107. $13.7 < t < 17.5$ **109.** $20 \le h \le 80$

111. False. c has to be greater than zero.　**113.** b

115. $5\sqrt{5}; \left(-\frac{3}{2}, 7\right)$　**117.** $2\sqrt{65}; (-1, -1)$　**119.** 10

121. $\frac{1}{7}, -\frac{1}{2}$　**123.** $(-3, 10)$　**125.** Answers will vary.

Section 1.8　(page 161)

> ### Vocabulary Check　(page 161)
>
> **1.** critical; test intervals　**2.** zeros; undefined values
> **3.** $P = R - C$

1. (a) No　(b) Yes　(c) Yes　(d) No

3. (a) Yes　(b) No　(c) No　(d) Yes

5. $2, -\frac{3}{2}$　**7.** $\frac{7}{2}, 5$

9. $[-3, 3]$

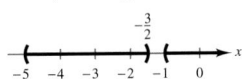

11. $(-7, 3)$

13. $(-\infty, -5] \cup [1, \infty)$

15. $(-3, 2)$

17. $(-3, 1)$

19. $\left(-\infty, -4 - \sqrt{21}\right] \cup \left[-4 + \sqrt{21}, \infty\right)$

21. $(-1, 1) \cup (3, \infty)$

23. $[-3, 2] \cup [3, \infty)$

25. $x = \frac{1}{2}$

27. $(-\infty, 0) \cup \left(0, \frac{3}{2}\right)$

29. $[-2, 0] \cup [2, \infty)$　**31.** $[-2, \infty)$

33.

35.

(a) $x \le -1, \ x \ge 3$
(b) $0 \le x \le 2$

(a) $-2 \le x \le 0,$
$\quad 2 \le x < \infty$
(b) $x \le 4$

37. $(-\infty, -1) \cup (0, 1)$

39. $(-\infty, -1) \cup (4, \infty)$

41. $(5, 15)$

43. $\left(-5, -\frac{3}{2}\right) \cup (-1, \infty)$

45. $\left(-\frac{3}{4}, 3\right) \cup [6, \infty)$

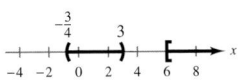

47. $(-3, -2] \cup [0, 3)$

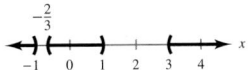

49. $(-\infty, -1) \cup \left(-\frac{2}{3}, 1\right) \cup (3, \infty)$

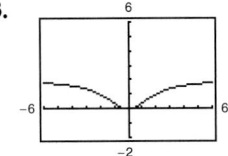

51.

53.

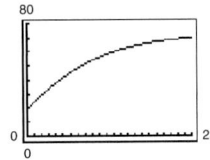

(a) $0 \le x < 2$

(a) $|x| \ge 2$

(b) $2 < x \le 4$

(b) $-\infty < x < \infty$

55. $[-2, 2]$　**57.** $(-\infty, 3] \cup [4, \infty)$

59. $(-5, 0] \cup (7, \infty)$　**61.** $(-3.51, 3.51)$

63. $(-0.13, 25.13)$　**65.** $(2.26, 2.39)$

67. (a) $t = 10$ seconds　(b) 4 seconds $< t < 6$ seconds

69. 13.8 meters $\le L \le 36.2$ meters

71. $40,000 \le x \le 50,000; 50.00 \le p \le 55.00$

73. (a)

(b)

t	24	26	28	30	32	34
C	70.5	71.6	72.9	74.6	76.8	79.6

2011

(c) $t \approx 31$

(d)

t	36	37	38	39
C	83.2	85.4	87.8	90.5

t	40	41	42	43
C	93.5	96.8	100.4	104.4

2016 to 2021

(e) $37 \le t \le 41$　(f) Answers will vary.

75. $R_1 \ge 2$ ohms

77. True. The test intervals are $(-\infty, -3), (-3, 1), (1, 4),$ and $(4, \infty)$.

79. $(-\infty, -4] \cup [4, \infty)$　**81.** $\left(-\infty, -2\sqrt{30}\right] \cup \left[2\sqrt{30}, \infty\right)$

83. (a) If $a > 0$ and $c \le 0, b$ can be any real number. If $a > 0$ and $c > 0, b < -2\sqrt{ac}$ or $b > 2\sqrt{ac}$.

(b) 0

85. $(2x + 5)^2$　**87.** $(x + 3)(x + 2)(x - 2)$　**89.** $2x^2 + x$

Review Exercises *(page 165)*

1.

x	-2	-1	0	1	2
y	-11	-8	-5	-2	1

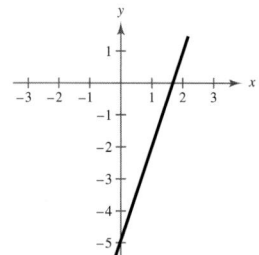

3.

x	-1	0	1	2	3	4
y	4	0	-2	-2	0	4

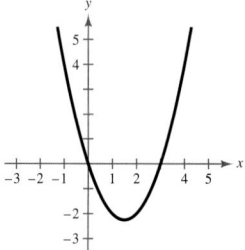

5. 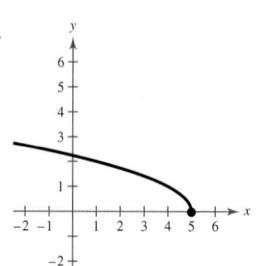 **7.**

9.

11. x-intercepts: $(1, 0)$, $(5, 0)$
y-intercept: $(0, 5)$

13. No symmetry

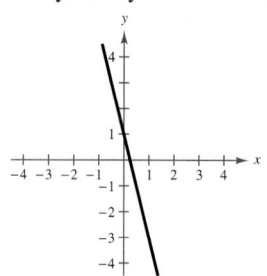

15. y-axis symmetry

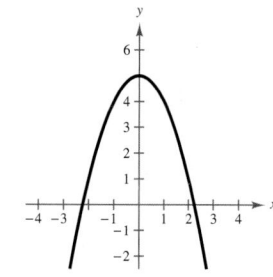

17. No symmetry

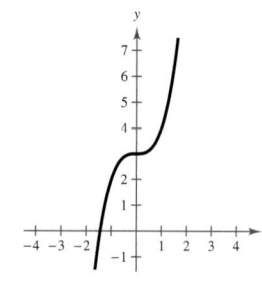

19. No symmetry

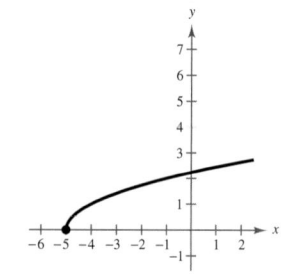

21. Center: $(0, 0)$;
Radius: 3

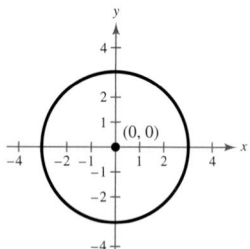

23. Center: $(-2, 0)$;
Radius: 4

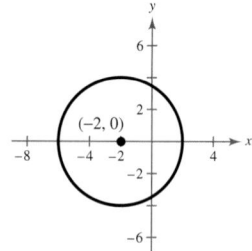

25. Center: $\left(\frac{1}{2}, -1\right)$;
Radius: 6

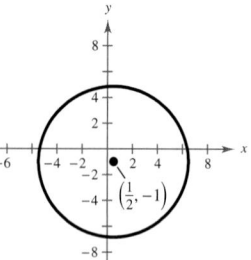

27. $(x - 2)^2 + (y + 3)^2 = 13$

29. (a)

x	0	4	8	12	16	20
F	0	5	10	15	20	25

(b) 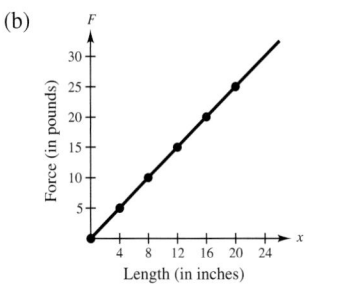   (c) 12.5 pounds

31. Identity   **33.** Identity   **35.** 20   **37.** $-\frac{1}{2}$

39. -30 **41.** 9

43. x-intercept: $\left(\frac{1}{3}, 0\right)$ **45.** x-intercept: $(4, 0)$
y-intercept: $(0, -1)$ y-intercept: $(0, -8)$

47. x-intercept: $\left(\frac{4}{3}, 0\right)$ **49.** x-intercept: $(2, 0)$
y-intercept: $\left(0, \frac{2}{3}\right)$ y-intercept: $\left(0, -\frac{5}{19}\right)$

51. $h = 10$ inches

53. September: \$325,000; October: \$364,000

55. 24 feet   **57.** Nine   **59.** $\frac{20}{7}$ liters ≈ 2.857 liters

61. $h = \dfrac{3V}{\pi r^2}$ **63.** $\frac{2}{3}$ hour   **65.** $-\frac{5}{2}, 3$ **67.** $\pm\sqrt{2}$

69. $-4 \pm 3\sqrt{2}$ **71.** $6 \pm \sqrt{6}$ **73.** $-\frac{5}{4} \pm \dfrac{\sqrt{241}}{4}$

75. (a) $x = 0, 20$
(b) ![graph]   (c) $x = 10$

77. $6 + 2i$ **79.** $-1 + 3i$ **81.** $3 + 7i$

83. $40 + 65i$ **85.** $-4 - 46i$ **87.** $\frac{23}{17} + \frac{10}{17}i$

89. $\frac{21}{13} - \frac{1}{13}i$ **91.** $\pm\dfrac{\sqrt{3}}{3}i$ **93.** $1 \pm 3i$ **95.** $0, \dfrac{12}{5}$

97. $\pm\sqrt{2}, \pm\sqrt{3}$ **99.** 5   **101.** No solution

103. $-124, 126$ **105.** $-2 \pm \dfrac{\sqrt{95}}{5}, -4$ **107.** $\pm\sqrt{10}$

109. $\frac{4}{3}$ **111.** $-5, 15$ **113.** $1, 3$ **115.** 143,203 units

117. $-7 < x \le 2$. Bounded   **119.** $x \le -10$. Unbounded

121. $(-\infty, 12]$ **123.** $\left[\frac{32}{15}, \infty\right)$ **125.** $\left(-\frac{2}{3}, 17\right]$

127. $[-4, 4]$ **129.** $(-\infty, -1) \cup (7, \infty)$

131. 353.44 square centimeters $\le$ area $\le$ 392.04 square
centimeters

133. $(-3, 9)$ **135.** $\left(-\frac{4}{3}, \frac{1}{2}\right)$ **137.** $[-4, 0] \cup [4, \infty)$

139. $(-\infty, -5) \cup (-2, \infty)$ **141.** $[-5, -1) \cup (1, \infty)$

143. $[-4, -3] \cup (0, \infty)$ **145.** 4.9%

147. False. $\sqrt{-18}\sqrt{-2} = \left(3\sqrt{2}i\right)\left(\sqrt{2}i\right) = 6i^2 = -6$
and $\sqrt{(-18)(-2)} = \sqrt{36} = 6$

149. Some solutions to certain types of equations may be
extraneous solutions, which do not satisfy the original
equations. So, checking is crucial.

Chapter Test   *(page 169)*

1. No symmetry

2. y-axis symmetry

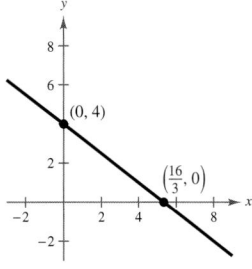

3. No symmetry

4. Origin symmetry

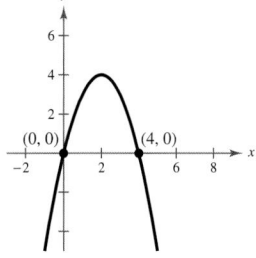
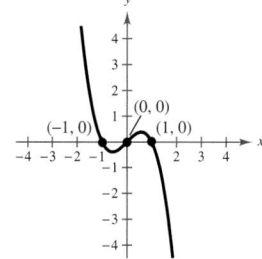

5. No symmetry

6. x-axis symmetry

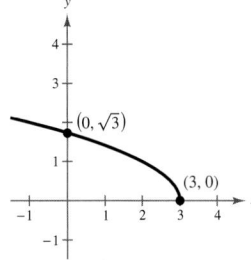
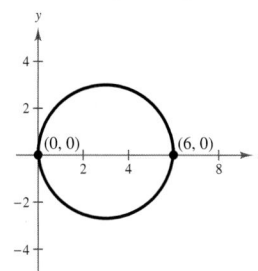

7. $\frac{128}{11}$ **8.** $-4, 5$ **9.** No solution

10. $\pm\sqrt{2}, \pm\sqrt{3}i$ **11.** 4   **12.** $-2, \frac{8}{3}$

13. $-\frac{11}{2} \le x < 3$

14. $x < -6$ or $0 < x < 4$

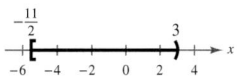
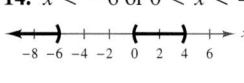

15. $x < -4$ or $x > \frac{3}{2}$

16. $x \le 10$ or $x \ge 20$

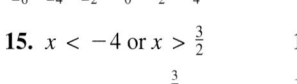
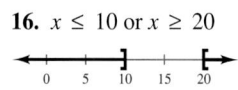

17. (a) $-3 + 5i$ (b) 7   **18.** $2 - i$

19. (a)

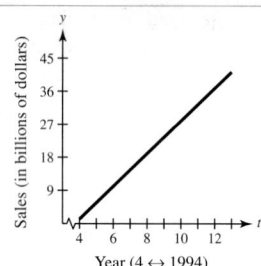

(b) \$63.5 billion dollars

(c) Answers will vary.

20. $r = 4.774$ inches

21. $93\frac{3}{4}$ kilometers per hour   **22.** $a = 80, b = 20$

Problem Solving   *(page 171)*

1.

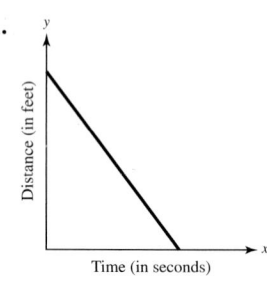

3. (a) Answers will vary. Sample answer:

$A = \pi ab$

$b = 20 - a$, since $a + b = 20$

$A = \pi a(20 - a)$

(b)

a	4	7	10	13	16
A	64π	91π	100π	91π	64π

(c) $\dfrac{10\pi + 10\sqrt{\pi(\pi-3)}}{\pi} \approx 12.12$ or

$\dfrac{10\pi - 10\sqrt{\pi(\pi-3)}}{\pi} \approx 7.88$

(d)

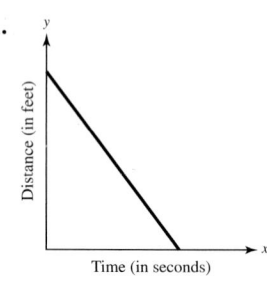

(e) $(0, 0), (20, 0)$

They represent the minimum and maximum values of a.

(f) $100\pi; a = 10; b = 10$

5. (a) ≈ 60.6 seconds   (b) ≈ 146.2 seconds

(c) The speed at which water drains decreases as the amount of water in the bathtub decreases.

7. (a) Answers will vary. Sample answers: 5, 12, 13; 8, 15, 17.

(b) Yes; yes; yes

(c) The product of the three numbers in a Pythagorean Triple is divisible by 60.

9. $x_1 + x_2 = -\dfrac{b}{a}; x_1 \cdot x_2 = \dfrac{c}{a}$

11. (a) $\frac{1}{2} - \frac{1}{2}i$ (b) $\frac{3}{10} + \frac{1}{10}i$ (c) $-\frac{1}{34} - \frac{2}{17}i$

13. (a) Yes   (b) No   (c) Yes

15. $(-\infty, -2) \cup (-1, 2) \cup (2, \infty)$

Chapter 2

Section 2.1   *(page 183)*

1. (a) L_2 (b) L_3 (c) L_1

3.

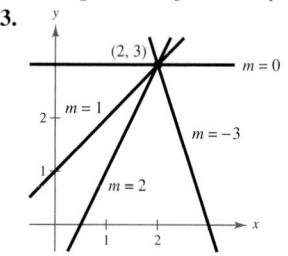

5. $\frac{3}{2}$ **7.** -4

9. $m = 5$;
y-intercept: $(0, 3)$

11. $m = -\frac{1}{2}$;
y-intercept: $(0, 4)$

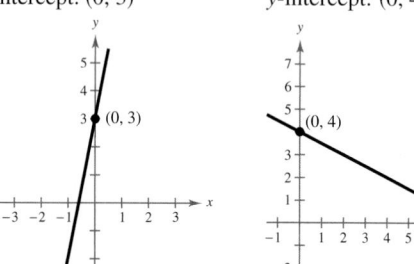

13. m is undefined.
There is no y-intercept.

15. $m = -\frac{7}{6}$;
y-intercept: $(0, 5)$

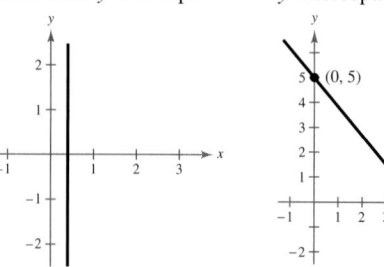

17. $m = 0$;
y-intercept: $(0, 3)$

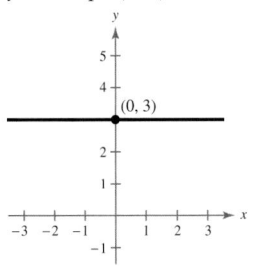

19. m is undefined.
There is no y-intercept.

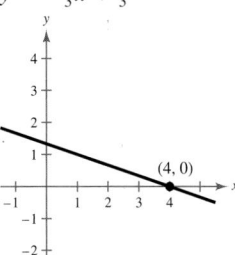

43. $y = -\frac{1}{3}x + \frac{4}{3}$

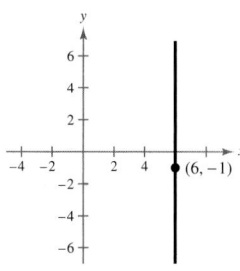

45. $x = 6$

21.

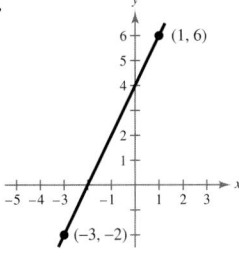

$m = 2$

23.

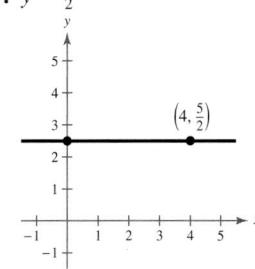

m is undefined.

47. $y = \frac{5}{2}$

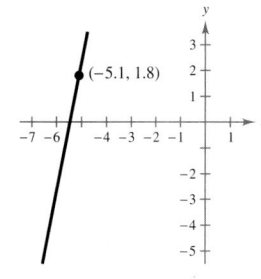

49. $y = 5x + 27.3$

25.

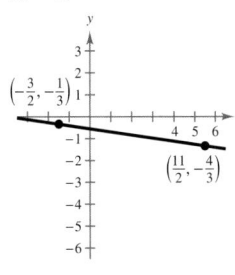

$m = -\frac{1}{7}$

27.

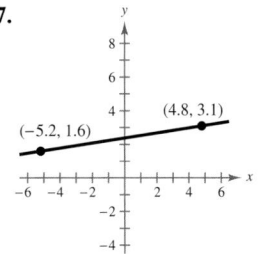

$m = 0.15$

51. $y = -\frac{3}{5}x + 2$

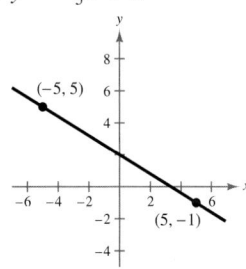

53. $x = -8$

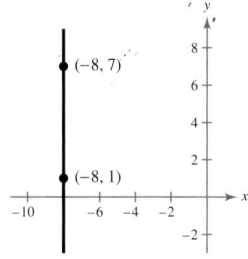

29. Answers will vary. Sample answer:
$(0, 1), (3, 1), (-1, 1)$

31. Answers will vary. Sample answer:
$(6, -5), (7, -4), (8, -3)$

33. Answers will vary. Sample answer:
$(-8, 0), (-8, 2), (-8, 3)$

35. Answers will vary. Sample answer:
$(-4, 6), (-3, 8), (-2, 10)$

37. Answers will vary. Sample answer:
$(9, -1), (11, 0), (13, 1)$

39. $y = 3x - 2$

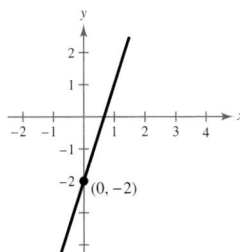

41. $y = -2x$

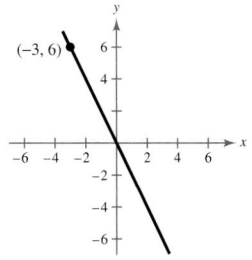

55. $y = -\frac{1}{2}x + \frac{3}{2}$

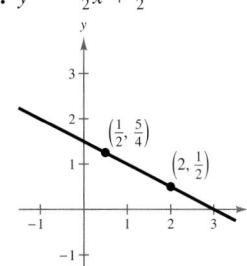

57. $y = -\frac{6}{5}x - \frac{18}{25}$

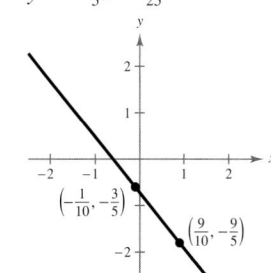

59. $y = 0.4x + 0.2$

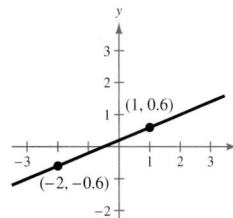

61. $y = -1$

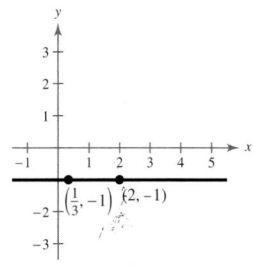

CHAPTER 2

63. $x = \frac{7}{3}$

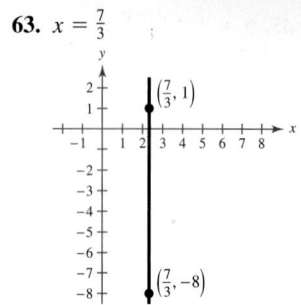

65. Perpendicular   **67.** Parallel

69. (a) $y = 2x - 3$ (b) $y = -\frac{1}{2}x + 2$

71. (a) $y = -\frac{3}{4}x + \frac{3}{8}$ (b) $y = \frac{4}{3}x + \frac{127}{72}$

73. (a) $y = 0$ (b) $x = -1$

75. (a) $x = 2$ (b) $y = 5$

77. (a) $y = x + 4.3$ (b) $y = -x + 9.3$

79. $3x + 2y - 6 = 0$ **81.** $12x + 3y + 2 = 0$

83. $x + y - 3 = 0$

85. Line (b) is perpendicular to line (c).

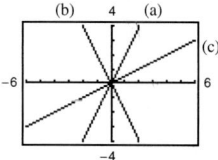

87. Line (a) is parallel to line (b).

Line (c) is perpendicular to line (a) and line (b).

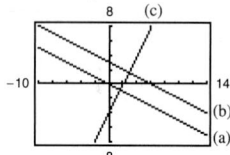

89. $3x - 2y - 1 = 0$ **91.** $80x + 12y + 139 = 0$

93. (a) Sales increasing 135 units per year

(b) No change in sales

(c) Sales decreasing 40 units per year

95. (a) Salary increased greatest from 1990 to 1992; Least from 1992 to 1994

(b) Slope of line from 1990 to 2002 is about 2351.83

(c) Salary increased an average of $2351.83 over the 12 years between 1990 and 2002

97. 12 feet   **99.** $V(t) = 3165 - 125t$

101. b; The slope is -20, which represents the decrease in the amount of the loan each week. The y-intercept is $(0, 200)$ which represents the original amount of the loan.

102. c; The slope is 2, which represents the hourly wage per unit produced. The y-intercept is $(0, 8.50)$ which represents the initial hourly wage.

103. a; The slope is 0.32, which represents the increase in travel cost for each mile driven. The y-intercept is $(0, 30)$ which represents the amount per day for food.

104. d; The slope is -100, which represents the decrease in the value of the word processor each year. The y-intercept is $(0, 750)$ which represents the initial purchase price of the computer.

105. $y = 0.4825t - 2.2325$; $y(18) \approx \$6.45$; $y(20) \approx \$7.42$

107. $V = -175t + 875$

109. (a) $y(t) = 179.5t + 40{,}571$

(b) $y(8) = 42{,}007$; $y(10) = 42{,}366$ (c) $m = 179.5$

111. $S = 0.85L$

113. (a) $C = 16.75t + 36{,}500$ (b) $R = 27t$

(c) $P = 10.25t - 36{,}500$ (d) $t \approx 3561$ hours

115. (a)

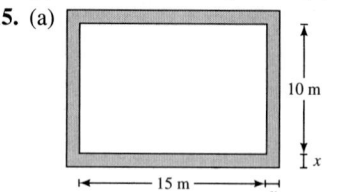

(b) $y = 8x + 50$

(c)

(d) $m = 8$, 8 meters

117. $C = 0.38x + 120$

119. (a) and (b)

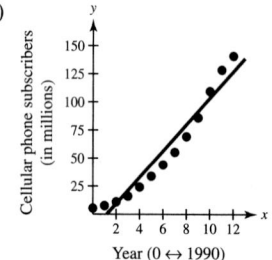

(c) Answers will vary. Sample answer:

$y = 11.72x - 14.1$

(d) Answers will vary. Sample answer: The y-intercept indicates that initially there were -14.1 million subscribers which doesn't make sense in the context of this problem. Each year, the number of cellular phone subscribers increases by 11.72 million.

(e) The model is accurate.

(f) Answers will vary. Sample answer: 196.9 million

121. False. The slope with the greatest magnitude corresponds to the steepest line.

123. Find the distance between each two points and use the Pythagorean Theorem.

125. No. The slope cannot be determined without knowing the scale on the y-axis. The slopes could be the same.

127. V-intercept: initial cost; Slope: annual depreciation

129. d   **130.** c   **131.** a   **132.** b

133. -1 **135.** $\frac{7}{2}, 7$ **137.** No solution

139. Answers will vary.

Section 2.2 *(page 197)*

Vocabulary Check *(page 197)*

1. domain; range; function
2. verbally; numerically; graphically; algebraically
3. independent; dependent   **4.** piecewise-defined
5. implied domain   **6.** difference quotient

---

1. Yes   **3.** No

5. Yes, each input value has exactly one output value.

7. No, the input values of 7 and 10 each have two different output values.

9. (a) Function
  (b) Not a function, because the element 1 in A corresponds to two elements, -2 and 1, in B.
  (c) Function
  (d) Not a function, because not every element in A is matched with an element in B.

11. Each is a function. For each year there corresponds one and only one circulation.

13. Not a function   **15.** Function   **17.** Function

19. Not a function   **21.** Function   **23.** Not a function

25. (a) -1 (b) -9 (c) $2x - 5$

27. (a) 36π (b) $\frac{9}{2}\pi$ (c) $\frac{32}{3}\pi r^3$

29. (a) 1   (b) 2.5   (c) $3 - 2|x|$

31. (a) $-\dfrac{1}{9}$ (b) Undefined   (c) $\dfrac{1}{y^2 + 6y}$

33. (a) 1   (b) -1 (c) $\dfrac{|x - 1|}{x - 1}$

35. (a) -1 (b) 2   (c) 6

37. (a) -7 (b) 4   (c) 9

39.

x	-2	-1	0	1	2
$f(x)$	1	-2	-3	-2	1

41.

t	-5	-4	-3	-2	-1
$h(t)$	1	$\frac{1}{2}$	0	$\frac{1}{2}$	1

43.

x	-2	-1	0	1	2
$f(x)$	5	$\frac{9}{2}$	4	1	0

45. 5   **47.** $\frac{4}{3}$ **49.** ± 3 **51.** $0, \pm 1$ **53.** $2, -1$

55. 3, 0   **57.** All real numbers

59. All real numbers t except $t = 0$

61. All real numbers y such that $y \geq 0$

63. All real numbers x such that $-1 \leq x \leq 1$

65. All real numbers x except $x = 0, -2$

67. All real numbers s such that $s \geq 1$ except $s = 4$

69. All real numbers x such that $x > 0$

71. $\{(-2, 4), (-1, 1), (0, 0), (1, 1), (2, 4)\}$

73. $\{(-2, 4), (-1, 3), (0, 2), (1, 3), (2, 4)\}$

75. $g(x) = cx^2$; $c = -2$ **76.** $f(x) = cx$; $c = \frac{1}{4}$

77. $r(x) = \dfrac{c}{x}$; $c = 32$ **78.** $h(x) = c\sqrt{|x|}$; $c = 3$

79. $3 + h, h \neq 0$ **81.** $3x^2 + 3xh + h^2 + 3, h \neq 0$

83. $-\dfrac{x + 3}{9x^2}, x \neq 3$ **85.** $\dfrac{\sqrt{5x} - 5}{x - 5}$ **87.** $A = \dfrac{P^2}{16}$

89. (a) The maximum volume is 1024 cubic centimeters.
  (b)

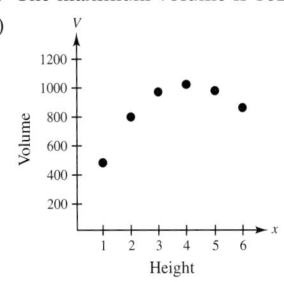

    Yes, V is a function of x.
  (c) $V = x(24 - 2x)^2, 0 < x < 12$

91. $A = \dfrac{x^2}{2(x - 2)}, x > 2$

93. Yes, the ball will be at a height of 6 feet.

95. 1990: \$27,300
1991: \$28,052
1992: \$29,168
1993: \$30,648
1994: \$32,492
1995: \$34,700
1996: \$37,272
1997: \$40,208
1998: \$41,300
1999: \$43,800
2000: \$46,300
2001: \$48,800
2002: \$51,300

97. (a) $C = 12.30x + 98,000$
  (b) $R = 17.98x$
  (c) $P = 5.68x - 98,000$

99. (a) $R = \dfrac{240n - n^2}{20}, n \geq 80$

  (b)

n	90	100	110	120	130	140	150
$R(n)$	\$675	\$700	\$715	\$720	\$715	\$700	\$675

The revenue is maximum when 120 people take the trip.

101. (a)

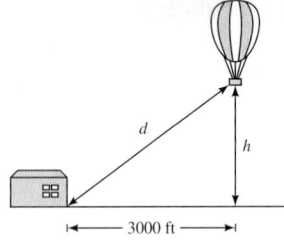

(b) $h = \sqrt{d^2 - 3000^2}$, $d \geq 3000$

103. False. The range is $[-1, \infty)$.

105. The domain is the set of inputs of the function, and the range is the set of outputs.

107. (a) Yes. The amount you pay in sales tax will increase as the price of the item purchased increases.

(b) No. The length of time that you study will not necessarily determine how well you do on an exam.

109. $\frac{15}{8}$ **111.** $-\frac{1}{5}$ **113.** $2x - 3y - 11 = 0$

115. $10x + 9y + 15 = 0$

Section 2.3   *(page 210)*

Vocabulary Check   *(page 210)*

1. ordered pairs   **2.** Vertical Line Test

3. zeros   **4.** decreasing

5. maximum   **6.** average rate of change; secant

7. odd   **8.** even

---

1. Domain: $(-\infty, -1], [1, \infty)$ **3.** Domain: $[-4, 4]$
Range: $[0, \infty)$ Range: $[0, 4]$

5. (a) 0   (b) -1 (c) 0   (d) -2

7. (a) -3 (b) 0   (c) 1   (d) -3 **9.** Function

11. Not a function   **13.** Function   **15.** $-\frac{5}{2}, 6$

17. 0   **19.** $0, \pm\sqrt{2}$ **21.** $\pm\frac{1}{2}, 6$ **23.** $\frac{1}{2}$

25.

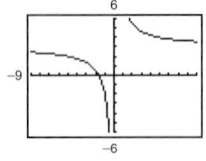

$-\frac{5}{3}$

27.

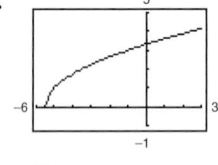

$-\frac{11}{2}$

29.

$\frac{1}{3}$

31. Increasing on $(-\infty, \infty)$

33. Increasing on $(-\infty, 0)$ and $(2, \infty)$
Decreasing on $(0, 2)$

35. Increasing on $(-\infty, 0)$ and $(2, \infty)$; Constant on $(0, 2)$

37. Increasing on $(1, \infty)$; Decreasing on $(-\infty, -1)$
Constant on $(-1, 1)$

39.

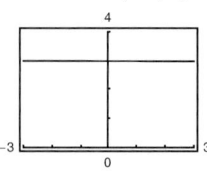

Constant on $(-\infty, \infty)$

41.

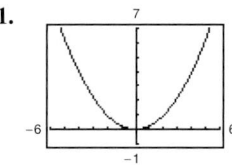

Decreasing on $(-\infty, 0)$
Increasing on $(0, \infty)$

43.

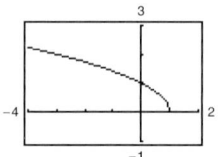

Increasing on $(-\infty, 0)$
Decreasing on $(0, \infty)$

45.

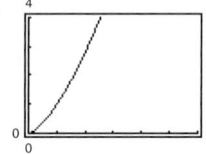

Decreasing on $(-\infty, 1)$

47.

Increasing on $(0, \infty)$

49.

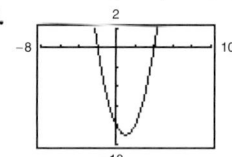

Relative minimum:
$(1, -9)$

51.

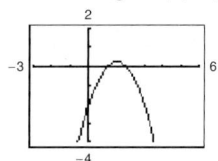

Relative maximum:
$(1.5, 0.25)$

53.

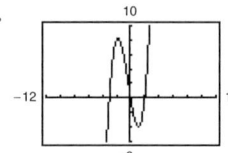

Relative maximum: $(-1.79, 8.21)$
Relative minimum: $(1.12, -4.06)$

55.

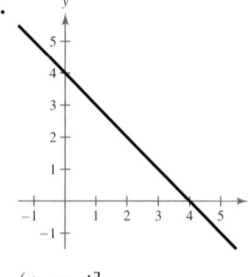

$(-\infty, 4]$

57.

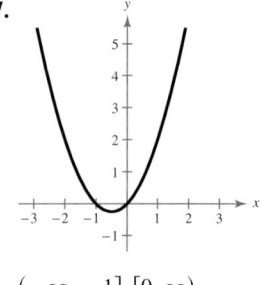

$(-\infty, -1], [0, \infty)$

59. 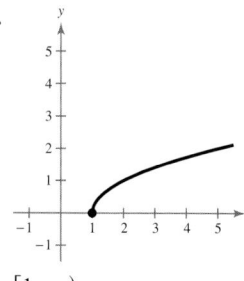 **61.**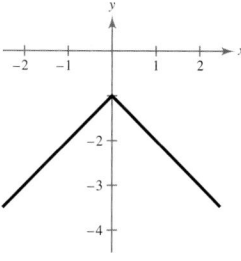

$[1, \infty)$ $f(x) < 0$ for all x

63. The average rate of change from $x_1 = 0$ to $x_2 = 3$ is -2.

65. The average rate of change from $x_1 = 1$ to $x_2 = 5$ is 18.

67. The average rate of change from $x_1 = 1$ to $x_2 = 3$ is 0.

69. The average rate of change from $x_1 = 3$ to $x_2 = 11$ is $-\frac{1}{4}$.

71. Even; y-axis symmetry   **73.** Odd; origin symmetry

75. Neither even nor odd; no symmetry

77. $h = -x^2 + 4x - 3$ **79.** $h = 2x - x^2$

81. $L = \frac{1}{2}y^2$ **83.** $L = 4 - y^2$

85. (a) 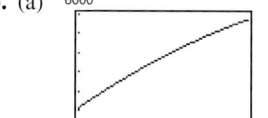   (b) 30 watts

87. (a) Ten thousands   (b) Ten millions   (c) Percents

89. (a)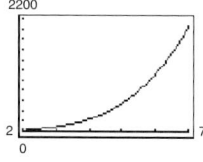

(b) The average rate of change from 2002 to 2007 is 408.56. The estimated revenue is increasing each year at a fast pace.

91. (a) $s = -16t^2 + 64t + 6$

(b)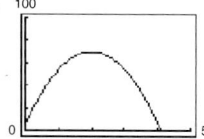

(c) Average rate of change $= 16$

(d) The slope of the secant line is positive.

(e) Secant line: $16t + 6$

(f)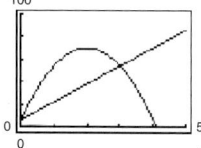

93. (a) $s = -16t^2 + 120t$

(b)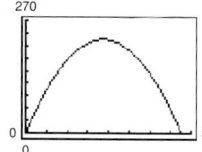

(c) Average rate of change $= -8$

(d) The slope of the secant line is negative.

(e) Secant line: $-8t + 240$

(f)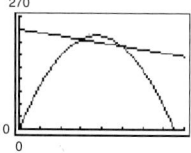

95. (a) $s = -16t^2 + 120$

(b)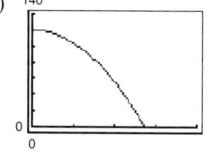

(c) Average rate of change $= -32$

(d) The slope of the secant line is negative.

(e) Secant line: $-32t + 120$

(f)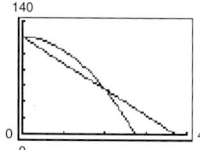

97. False. The function $f(x) = \sqrt{x^2 + 1}$ has a domain of all real numbers.

99. (a) Even. The graph is a reflection in the x-axis.

(b) Even. The graph is a reflection in the y-axis.

(c) Even. The graph is a vertical translation of f.

(d) Neither. The graph is a horizontal translation of f.

101. (a) $\left(\frac{3}{2}, 4\right)$ (b) $\left(\frac{3}{2}, -4\right)$

103. (a) $(-4, 9)$ (b) $(-4, -9)$

105. (a) 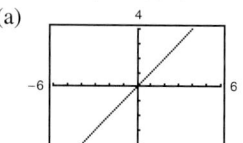   (b)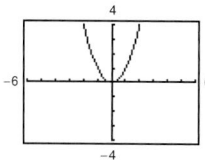

(c) 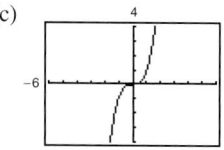   (d)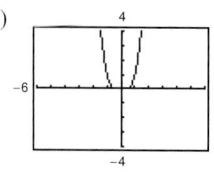

CHAPTER 2

(e) 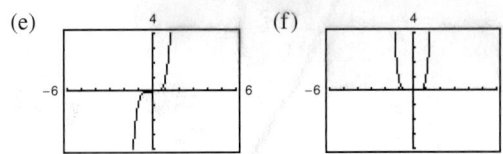 (f)

All the graphs pass through the origin. The graphs of the odd powers of x are symmetric with respect to the origin, and the graphs of the even powers are symmetric with respect to the y-axis. As the powers increase, the graphs become flatter in the interval $-1 < x < 1$.

107. 0, 10   **109.** 0, ± 1

111. (a) 37   (b) -28 (c) $5x - 43$

113. (a) -9 (b) $2\sqrt{7} - 9$
(c) The given value is not in the domain of the function.

115. $h + 4$, $h \neq 0$

Section 2.4   (page 220)

1. (a) $f(x) = -2x + 6$ **3.** (a) $f(x) = -3x + 11$
(b)   (b)

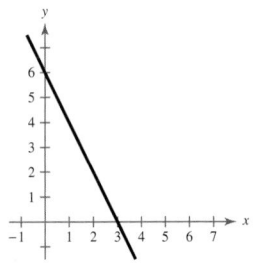

5. (a) $f(x) = -1$ **7.** (a) $f(x) = \frac{6}{7}x - \frac{45}{7}$
(b)   (b)

9. 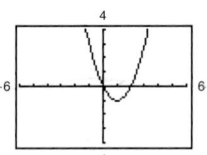   **11.**

13. 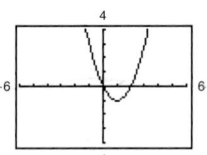   **15.**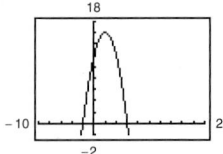

17. 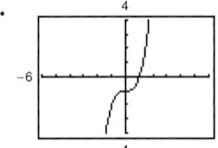   **19.**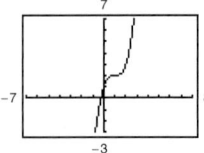

21. 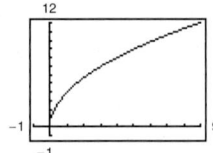   **23.**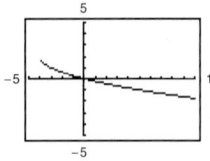

25. 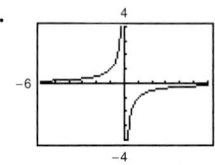   **27.**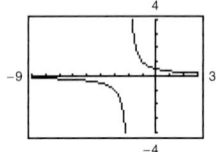

29. (a) 2   (b) 2   (c) -4 (d) 3
31. (a) 1   (b) 3   (c) 7   (d) -19
33. (a) 6   (b) -11 (c) 6   (d) -22
35. (a) -10 (b) -4 (c) -1 (d) 41

37.   **39.**

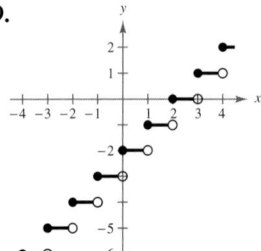

41.   **43.**

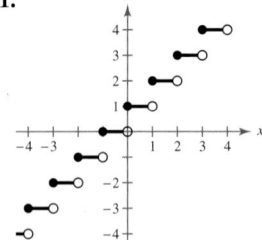   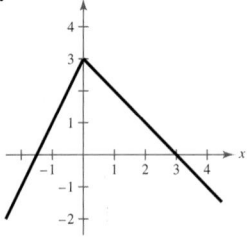

45.
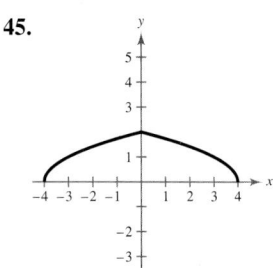

47.
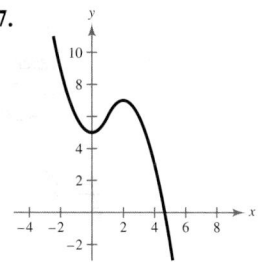

49.
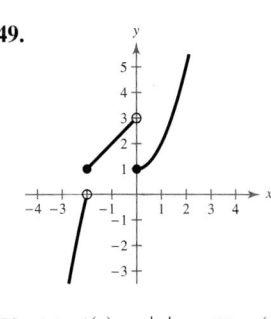

51. (a)
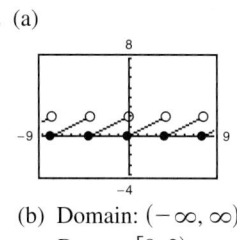

(b) Domain: $(-\infty, \infty)$;
Range: $[0, 2)$

(c) Sawtooth pattern

53. (a) $f(x) = |x|$ (b) $g(x) = |x + 2| - 1$

55. (a) $f(x) = x^3$ (b) $g(x) = (x - 1)^3 - 2$

57. (a) $f(x) = 2$ (b) $g(x) = 2$

59. (a) $f(x) = x$ (b) $g(x) = x - 2$

61. (a)
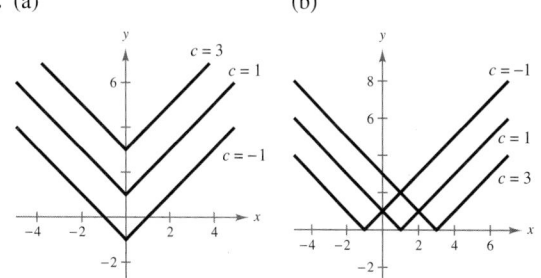
Cost (in dollars) vs Time (in minutes)

(b) $5.64

63. (a)
Cost of overnight delivery (in dollars) vs Weight (in pounds)

(b) $50.25

65. (a) $W(30) = 360; W(40) = 480;$
$W(45) = 570; W(50) = 660$

(b) $W(h) = \begin{cases} 12h, & 0 < h \le 45 \\ 18(h - 45) + 540, & h > 45 \end{cases}$

67. (a) $f(x) = \begin{cases} 0.505x^2 - 1.47x + 6.3, & 1 \le x \le 6 \\ -1.97x + 26.3, & 6 < x \le 12 \end{cases}$

Answers will vary. Sample answer: The domain is determined by inspection of a graph of the data with the two models.

(b)
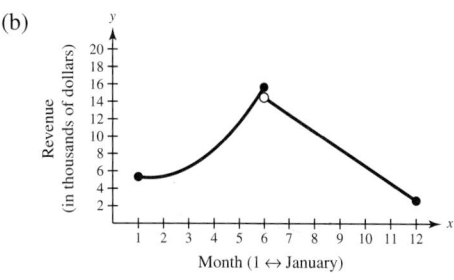
Revenue (in thousands of dollars) vs Month (1 ↔ January)

(c) $f(5) = 11.575, f(11) = 4.63$; These values represent the revenue for the months of May and November, respectively.

(d) These values are quite close to the actual data values.

69. False. A piecewise-defined function is a function that is defined by two or more equations over a specified domain. That domain may or may not include x- and y-intercepts.

71. $f(x) = \begin{cases} -\frac{4}{3}x + 6, & 0 \le x \le 3 \\ -\frac{2}{5}x + \frac{16}{5}, & 3 < x \le 8 \end{cases}$

73. $x \le 1$ **75.** Neither

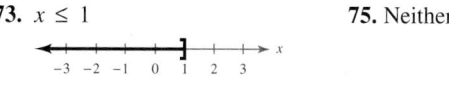

Section 2.5   *(page 228)*

Vocabulary Check   *(page 228)*

1. rigid   **2.** $-f(x); f(-x)$ **3.** nonrigid
4. horizontal shrink; horizontal stretch
5. vertical stretch; vertical shrink
6. (a) iv   (b) ii   (c) iii   (d) i

1. (a)   (b)

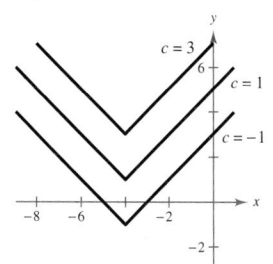

(c)

3. (a)

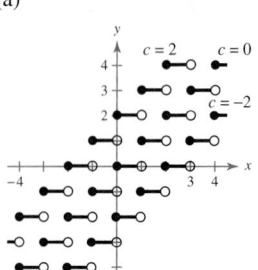

(b)

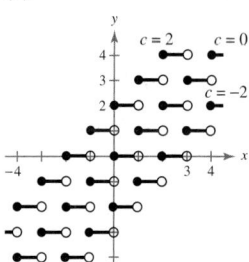

(c)

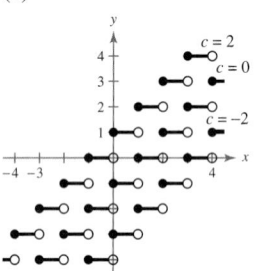

(g)

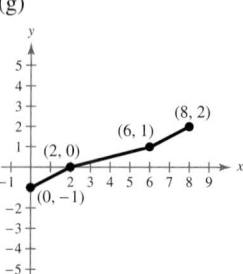

7. (a)

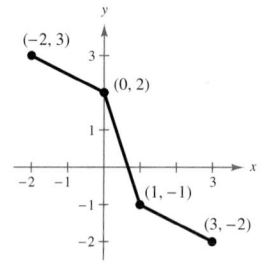

(b)

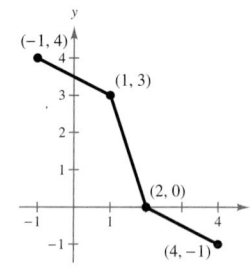

5. (a)

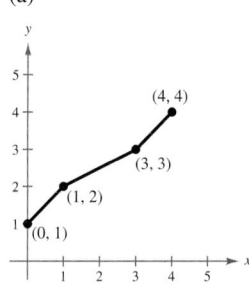

(b)

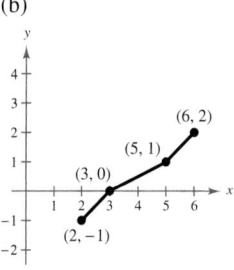

(c)

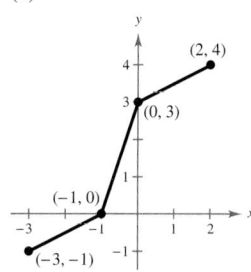

(d)

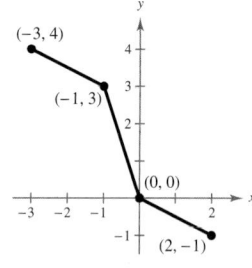

(c)

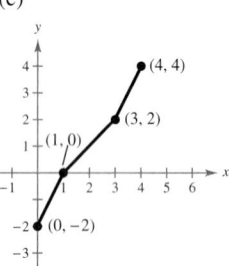

(d)

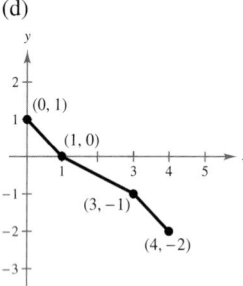

(e)

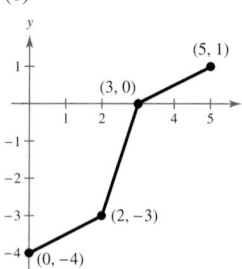

(f)

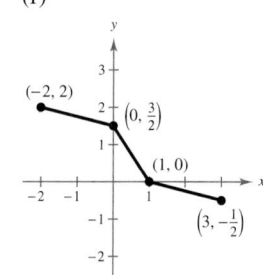

(e)

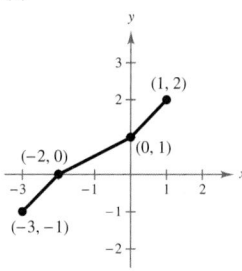

(f)

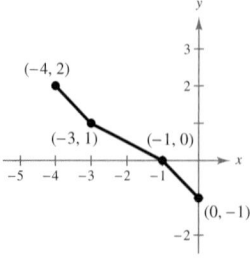

(g)

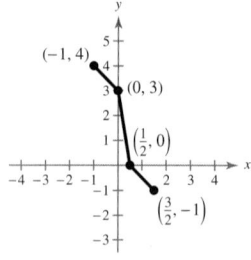

9. (a) $y = x^2 - 1$ (b) $y = 1 - (x + 1)^2$
  (c) $y = -(x - 2)^2 + 6$ (d) $y = (x - 5)^2 - 3$
11. (a) $y = |x| + 5$ (b) $y = -|x + 3|$
  (c) $y = |x - 2| - 4$ (d) $y = -|x - 6| - 1$
13. Horizontal shift of $y = x^3$; $y = (x - 2)^3$
15. Reflection in the x-axis of $y = x^2$; $y = -x^2$
17. Reflection in the x-axis and vertical shift of $y = \sqrt{x}$;
  $y = 1 - \sqrt{x}$
19. (a) $f(x) = x^2$
  (b) Reflection in the x-axis, and vertical shift 12 units
    upward, of $f(x) = x^2$
  (c)   (d) $g(x) = 12 - f(x)$

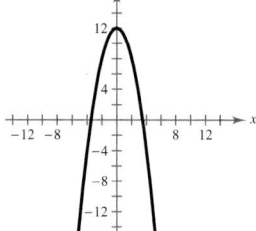

21. (a) $f(x) = x^3$
  (b) Vertical shift seven units upward, of $f(x) = x^3$
  (c)   (d) $g(x) = f(x) + 7$

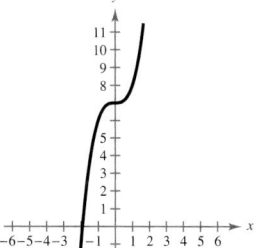

23. (a) $f(x) = x^2$
  (b) Vertical shrink of two-thirds, and vertical shift four
    units upward, of $f(x) = x^2$
  (c)   (d) $g(x) = \frac{2}{3}f(x) + 4$

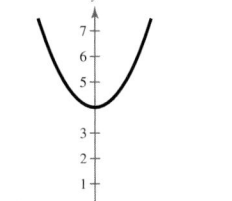

25. (a) $f(x) = x^2$
  (b) Reflection in the x-axis, horizontal shift five units to the
    left, and vertical shift two units upward, of $f(x) = x^2$

(c)   (d) $g(x) = 2 - f(x + 5)$

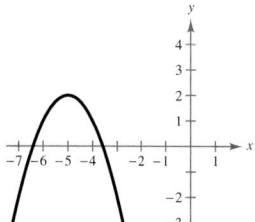

27. (a) $f(x) = \sqrt{x}$
  (b) Horizontal shrink of $\frac{1}{3}$, of $f(x) = \sqrt{x}$
  (c)   (d) $g(x) = f(3x)$

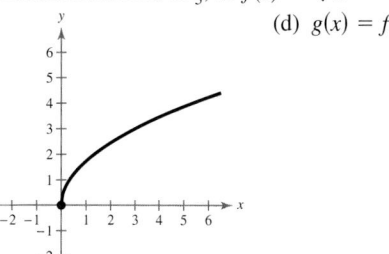

29. (a) $f(x) = x^3$
  (b) Vertical shift two units upward, and horizontal shift one
    unit to the right, of $f(x) = x^3$
  (c)   (d) $g(x) = f(x - 1) + 2$

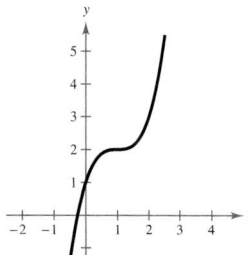

31. (a) $f(x) = |x|$
  (b) Reflection in the x-axis, and vertical shift two units
    downward, of $f(x) = |x|$
  (c)   (d) $g(x) = -f(x) - 2$

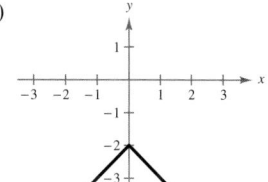

33. (a) $f(x) = |x|$
  (b) Reflection in the x-axis, horizontal shift four units to the
    left, and vertical shift eight units upward, of $f(x) = |x|$

CHAPTER 2

(c) (d) $g(x) = -f(x + 4) + 8$

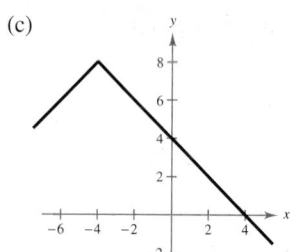

(c) (d) $g(x) = f\left(\frac{1}{2}x\right) - 4$

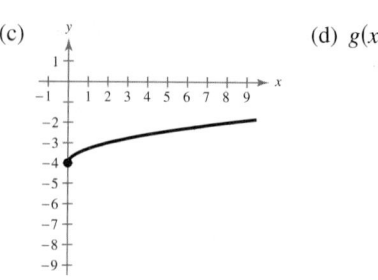

35. (a) $f(x) = [\![x]\!]$

(b) Reflection in the x-axis, and vertical shift three units upward, of $f(x) = [\![x]\!]$

(c) (d) $g(x) = 3 - f(x)$

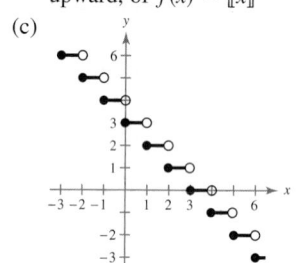

37. (a) $f(x) = \sqrt{x}$

(b) Horizontal shift of nine units to the right, of $f(x) = \sqrt{x}$

(c) (d) $g(x) = f(x - 9)$

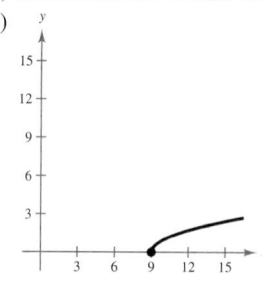

39. (a) $f(x) = \sqrt{x}$

(b) Reflection in the y-axis, horizontal shift of seven units to the right, and vertical shift two units downward, of $f(x) = \sqrt{x}$

(c) (d) $g(x) = f(7 - x) - 2$

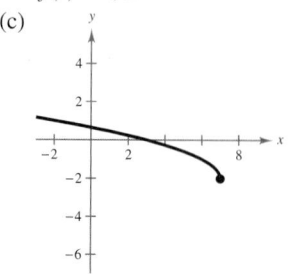

41. (a) $f(x) = \sqrt{x}$

(b) Horizontal stretch, and vertical shift four units downward, of $f(x) = \sqrt{x}$

43. $f(x) = (x - 2)^2 - 8$ **45.** $f(x) = (x - 13)^3$

47. $f(x) = -|x| - 10$ **49.** $f(x) = -\sqrt{-x + 6}$

51. (a) $y = -3x^2$ (b) $y = 4x^2 + 3$

53. (a) $y = -\frac{1}{2}|x|$ (b) $y = 3|x| - 3$

55. Vertical stretch of $y = x^3$; $y = 2x^3$

57. Reflection in the x-axis and vertical shrink of $y = x^2$; $y = -\frac{1}{2}x^2$

59. Reflection in the y-axis and vertical shrink of $y = \sqrt{x}$; $y = \frac{1}{2}\sqrt{-x}$

61. $y = -(x - 2)^3 + 2$ **63.** $y = -\sqrt{x} - 3$

65. (a) (b)

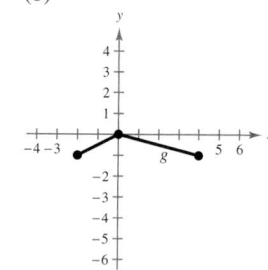

(c) (d)

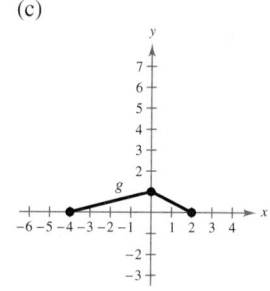
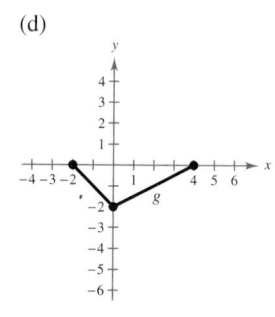

(e) (f)

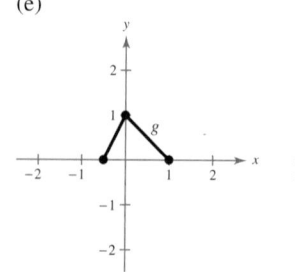
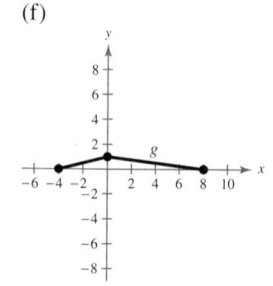

67. (a) Horizontal stretch of 0.035 and a vertical shift of 20.6 units upward.

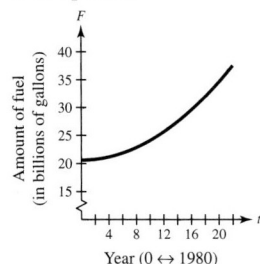

Year (0 ↔ 1980)

(b) 0.77-billion-gallon increase in fuel usage by trucks each year

(c) $f(t) = 20.6 + 0.035(t + 10)^2$. The graph is shifted 10 units to the left.

(d) 52.1 billion gallons. Yes.

69. True. $|-x| = |x|$

71. (a) $g(t) = \frac{3}{4}f(t)$ (b) $g(t) = f(t) + 10{,}000$

    (c) $g(t) = f(t - 2)$

73. $(-2, 0), (-1, 1), (0, 2)$ **75.** $\dfrac{4}{x(1 - x)}$

77. $\dfrac{3x - 2}{x(x - 1)}$ **79.** $\dfrac{(x - 4)\sqrt{x^2 - 4}}{x^2 - 4}$

81. $5(x - 3), x \neq -3$

83. (a) 38   (b) $\frac{57}{4}$ (c) $x^2 - 12x + 38$

85. All real numbers x except $x = 1$

87. All real numbers x such that $-9 \leq x \leq 9$

Section 2.6   *(page 238)*

Vocabulary Check   *(page 238)*

1. addition; subtraction; multiplication; division

2. composition   **3.** $g(x)$ **4.** inner; outer

1. 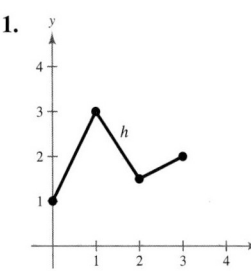   **3.**

5. (a) $2x$ (b) 4   (c) $x^2 - 4$

    (d) $\dfrac{x + 2}{x - 2}$; all real numbers x except $x = 2$

7. (a) $x^2 + 4x - 5$ (b) $x^2 - 4x + 5$ (c) $4x^3 - 5x^2$

    (d) $\dfrac{x^2}{4x - 5}$; all real numbers x except $x = \dfrac{5}{4}$

9. (a) $x^2 + 6 + \sqrt{1 - x}$ (b) $x^2 + 6 - \sqrt{1 - x}$

    (c) $(x^2 + 6)\sqrt{1 - x}$

    (d) $\dfrac{(x^2 + 6)\sqrt{1 - x}}{1 - x}$; all real numbers x such that $x < 1$

11. (a) $\dfrac{x + 1}{x^2}$ (b) $\dfrac{x - 1}{x^2}$ (c) $\dfrac{1}{x^3}$

    (d) x; all real numbers x except $x = 0$

13. 3   **15.** 5   **17.** $9t^2 - 3t + 5$ **19.** 74

21. 26   **23.** $\frac{3}{5}$

25. 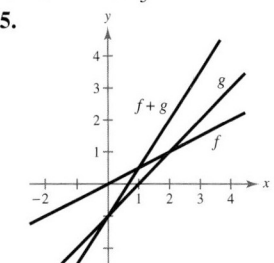   **27.**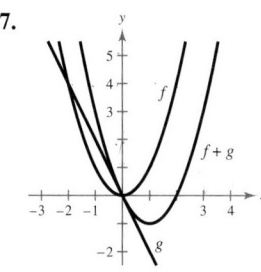

29.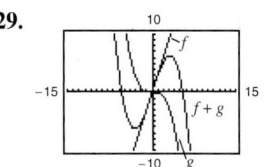

    $f(x), g(x)$

31. (a) $(x - 1)^2$ (b) $x^2 - 1$ (c) x^4

33. (a) x (b) x (c) $\sqrt[3]{\sqrt[3]{x - 1} - 1}$

35. (a) $\sqrt{x^2 + 4}$ (b) $x + 4$

    Domains of f and $g \circ f$: $x \geq -4$

    Domains of g and $f \circ g$: all real numbers

37. (a) $x + 1$ (b) $\sqrt{x^2 + 1}$

    Domains of f and $g \circ f$: all real numbers

    Domains of g and $f \circ g$: all real numbers x such that $x \geq 0$

39. (a) $|x + 6|$ (b) $|x| + 6$

    Domains of f, g, $f \circ g$, and $g \circ f$: all real numbers

41. (a) $\dfrac{1}{x + 3}$ (b) $\dfrac{1}{x} + 3$

    Domains of f and $g \circ f$: all real numbers x except $x = 0$

    Domains of g: all real numbers

    Domains of $f \circ g$: all real numbers x except $x = -3$

43. (a) 3   (b) 0   **45.** (a) 0   (b) 4

47. $f(x) = x^2, g(x) = 2x + 1$

49. $f(x) = \sqrt[3]{x}, \ g(x) = x^2 - 4$

51. $f(x) = \dfrac{1}{x}, \ g(x) = x + 2$

53. $f(x) = \dfrac{x + 3}{4 + x}, \ g(x) = -x^2$

55. $T = \frac{3}{4}x + \frac{1}{15}x^2$

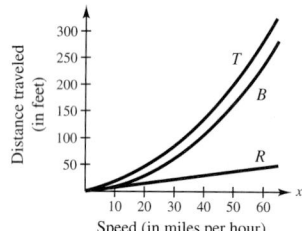

57. (a) $c(t) = \dfrac{p(t) + b(t) - d(t)}{p(t)} \times 100$

(b) $c(5)$ is the population change in the year 2005.

59. (a) $(A + N)(t) = 5.31t^2 - 102.0t + 1338$

$(A + N)(4) = 1014.96$

$(A + N)(8) = 861.84$

$(A + N)(12) = 878.64$

(b) $(A - N)(t) = 1.41t^2 - 17.6t + 132$

$(A - N)(4) = 84.16$

$(A - N)(8) = 81.44$

$(A - N)(12) = 123.84$

61. (a) $y_1 = 10.20t + 92.7$

$y_2 = 3.357t^2 - 26.46t + 379.5$

$y_3 = -0.465t^2 + 9.71t + 7.4$

(b) $y_1 + y_2 + y_3 = 2.892t^2 - 6.55t + 479.6$; this amount represents the amount spent on health care in the United States.

(c)

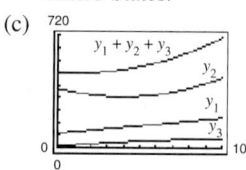

(d) In 2008, \$1298.708 billion is estimated to be spent on health services and supplies, and in 2010, \$1505.4 billion is estimated.

63. (a) $r(x) = \dfrac{x}{2}$ (b) $A(r) = \pi r^2$

(c) $(A \circ r)(x) = \pi\left(\dfrac{x}{2}\right)^2$; $(A \circ r)(x)$ represents the area of the circular base of the tank on the square foundation with side length x.

65. (a) $N(T(t)) = 30(3t^2 + 2t + 20)$ This represents the number of bacteria in the food as a function of time.

(b) $t = 2.846$ hours

67. $g(f(x))$ represents 3 percent of an amount over \$500,000.

69. False. $(f \circ g)(x) = 6x + 1$ and $(g \circ f)(x) = 6x + 6$

71. Answers will vary.   **73.** 3   **75.** $\dfrac{-4}{x(x + h)}$

77. $3x - y - 10 = 0$ **79.** $3x + 2y - 22 = 0$

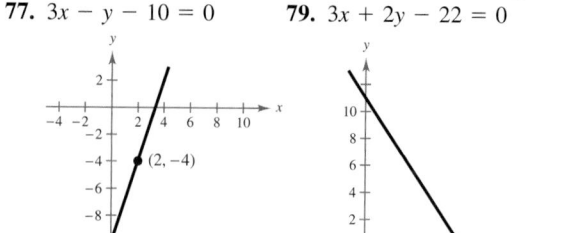

Section 2.7   *(page 248)*

1. $f^{-1}(x) = \frac{1}{6}x$ **3.** $f^{-1}(x) = x - 9$

5. $f^{-1}(x) = \dfrac{x - 1}{3}$ **7.** $f^{-1}(x) = x^3$

9. c   **10.** b   **11.** a   **12.** d

13. (a) $f(g(x)) = f\left(\dfrac{x}{2}\right) = 2\left(\dfrac{x}{2}\right) = x$

$g(f(x)) = g(2x) = \dfrac{(2x)}{2} = x$

(b)

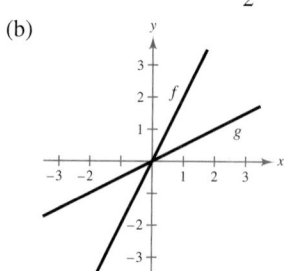

15. (a) $f(g(x)) = f\left(\dfrac{x - 1}{7}\right) = 7\left(\dfrac{x - 1}{7}\right) + 1 = x$

$g(f(x)) = g(7x + 1) = \dfrac{(7x + 1) - 1}{7} = x$

(b)

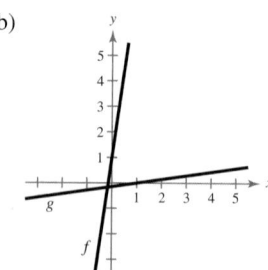

17. (a) $f(g(x)) = f\left(\sqrt[3]{8x}\right) = \dfrac{\left(\sqrt[3]{8x}\right)^3}{8} = x$

$g(f(x)) = g\left(\dfrac{x^3}{8}\right) = \sqrt[3]{8\left(\dfrac{x^3}{8}\right)} = x$

(b)

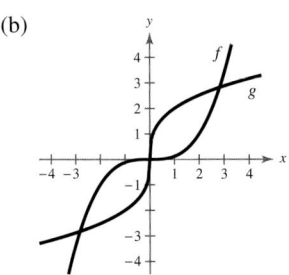

19. (a) $f(g(x)) = f(x^2 + 4)$, $x \geq 0$
$$= \sqrt{(x^2 + 4) - 4} = x$$
$g(f(x)) = g(\sqrt{x - 4})$
$$= (\sqrt{x - 4})^2 + 4 = x$$

(b)

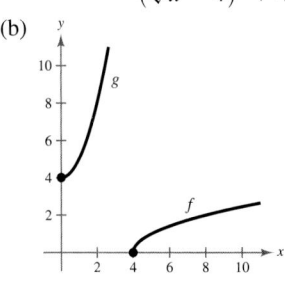

21. (a) $f(g(x)) = f(\sqrt{9 - x})$, $x \leq 9$
$$= 9 - (\sqrt{9 - x})^2 = x$$
$g(f(x)) = g(9 - x^2)$, $x \geq 0$
$$= \sqrt{9 - (9 - x^2)} = x$$

(b)

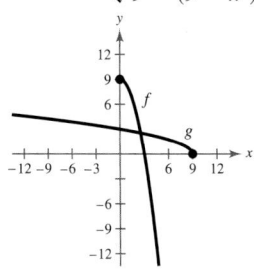

23. (a) $f(g(x)) = f\left(-\dfrac{5x + 1}{x - 1}\right) = \dfrac{-\left(\dfrac{5x + 1}{x - 1}\right) - 1}{-\left(\dfrac{5x + 1}{x - 1}\right) + 5}$

$$= \dfrac{-5x - 1 - x + 1}{-5x - 1 + 5x - 5} = x$$

$g(f(x)) = g\left(\dfrac{x - 1}{x + 5}\right) = \dfrac{-5\left(\dfrac{x - 1}{x + 5}\right) - 1}{\dfrac{x - 1}{x + 5} - 1}$

$$= \dfrac{-5x + 5 - x - 5}{x - 1 - x - 5} = x$$

(b)

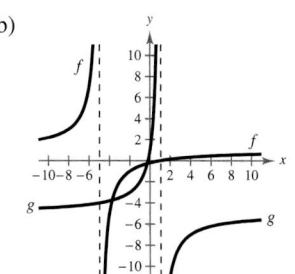

25. No   27.

x	-2	0	2	4	6	8
$f^{-1}(x)$	-2	-1	0	1	2	3

29. Yes   31. No

33.

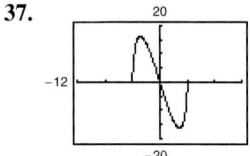

The function has an inverse.

35.

The function does not have an inverse.

37.

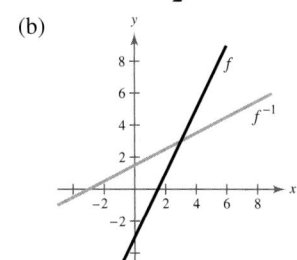

The function does not have an inverse.

39. (a) $f^{-1}(x) = \dfrac{x + 3}{2}$

(b)

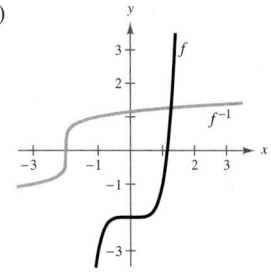

(c) The graph of f^{-1} is the reflection of the graph of f in the line $y = x$.

(d) The domains and ranges of f and f^{-1} are all real numbers.

41. (a) $f^{-1}(x) = \sqrt[5]{x + 2}$

(b)

(c) The graph of f^{-1} is the reflection of the graph of f in the line $y = x$.

(d) The domains and ranges of f and f^{-1} are all real numbers.

43. (a) $f^{-1}(x) = x^2,\ x \geq 0$

(b)

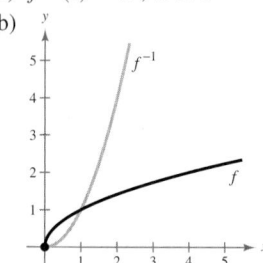

(c) The graph of f^{-1} is the reflection of the graph of f in the line $y = x$.

(d) The domains and ranges of f and f^{-1} are all real numbers x such that $x \geq 0$.

45. (a) $f^{-1}(x) = \sqrt{4 - x^2},\ 0 \leq x \leq 2$

(b)

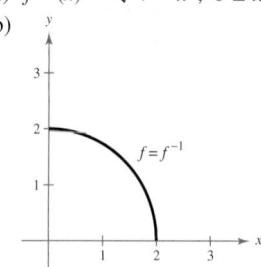

(c) The graph of f^{-1} is the same as the graph of f.

(d) The domains and ranges of f and f^{-1} are all real numbers x such that $0 \leq x \leq 2$.

47. (a) $f^{-1}(x) = \dfrac{4}{x}$

(b)

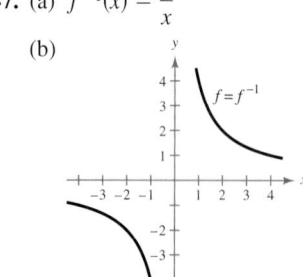

(c) The graph of f^{-1} is the same as the graph of f.

(d) The domains and ranges of f and f^{-1} are all real numbers x except $x = 0$.

49. (a) $f^{-1}(x) = \dfrac{2x + 1}{x - 1}$

(b)

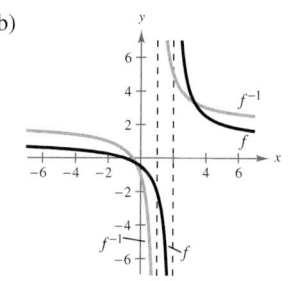

(c) The graph of f^{-1} is the reflection of the graph of f in the line $y = x$.

(d) The domain of f and the range of f^{-1} are all real numbers x except $x = 2$. The domain of f^{-1} and the range of f are all real numbers x except $x = 1$.

51. (a) $f^{-1}(x) = x^3 + 1$

(b)

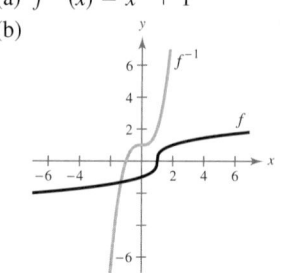

(c) The graph of f^{-1} is the reflection of the graph of f in the line $y = x$.

(d) The domains and ranges of f and f^{-1} are all real numbers.

53. (a) $f^{-1}(x) = \dfrac{5x - 4}{6 - 4x}$

(b)

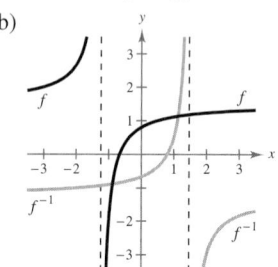

(c) The graph of f^{-1} is the reflection of the graph of f in the line $y = x$.

(d) The domain of f and the range of f^{-1} are all real numbers x except $x = -\frac{5}{4}$. The domain of f^{-1} and the range of f are all real numbers x except $x = \frac{3}{2}$.

55. No inverse   **57.** $g^{-1}(x) = 8x$ **59.** No inverse

61. $f^{-1}(x) = \sqrt{x} - 3$ **63.** No inverse

65. No inverse   **67.** $f^{-1}(x) = \dfrac{x^2 - 3}{2},\ x \geq 0$

69. 32   **71.** 600   **73.** $2\sqrt[3]{x + 3}$

75. $\dfrac{x + 1}{2}$ **77.** $\dfrac{x + 1}{2}$

79. (a) $f^{-1}(108{,}209) = 11$

(b) f^{-1} represents the year for a given number of households in the United States.

(c) $y = 1578.68t + 90{,}183.63$

(d) $f^{-1} = \dfrac{t - 90{,}183.63}{1578.68}$ (e) $f^{-1}(117{,}022) = 17$

(f) $f^{-1}(108{,}209) = 11.418$; the results are similar.

81. (a) Yes

(b) f^{-1} yields the year for a given number of miles traveled by motor vehicles.

(c) $f^{-1}(2632) = 8$

(d) No. $f(t)$ would not pass the Horizontal Line Test.

83. (a) $y = \sqrt{\dfrac{x - 245.50}{0.03}}$, $245.5 < x < 545.5$

$x = $ degrees Fahrenheit; $y = \%$ load

(b) 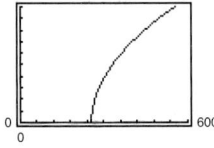   (c) $0 < x < 92.11$

85. False. $f(x) = x^2$ has no inverse.

87. Answers will vary.

89.

x	1	3	4	6
y	1	2	6	7

x	1	2	6	7
$f^{-1}(x)$	1	3	4	6

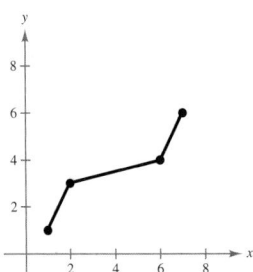

91.

x	-2	-1	3	4
y	6	0	-2	-3

x	-3	-2	0	6
$f^{-1}(x)$	4	3	-1	-2

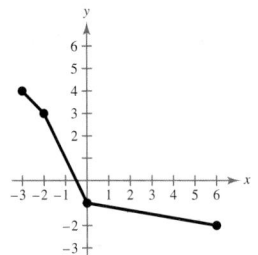

93. $k = \frac{1}{4}$ **95.** ± 8 **97.** $\frac{3}{2}$ **99.** $3 \pm \sqrt{5}$

101. $5, -\frac{10}{3}$ **103.** $16, 18$

Review Exercises   *(page 253)*

1. (a) L_2 (b) L_3 (c) L_1 (d) L_4

3. slope: -2 **5.** slope: 0
   y-intercept: -7 y-intercept: 6

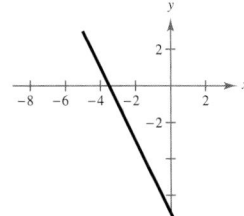 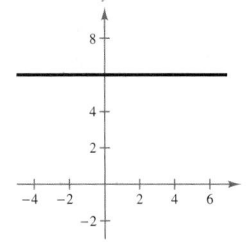

7. slope: 3 **9.** slope: $-\frac{5}{2}$
   y-intercept: 13 y-intercept: -1

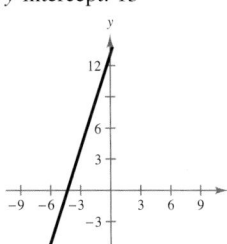 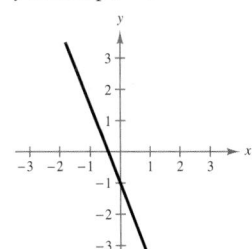

11. $t = \frac{7}{3}$ **13.** $(6, 0), (10, 1), (-2, -2)$

15.   **17.**

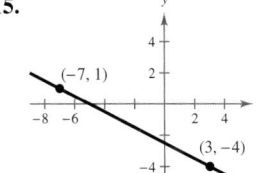 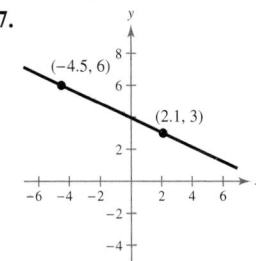

$m = -\frac{1}{2}$ $m = -\frac{5}{11}$

19. $y = \frac{3}{2}x - 5$

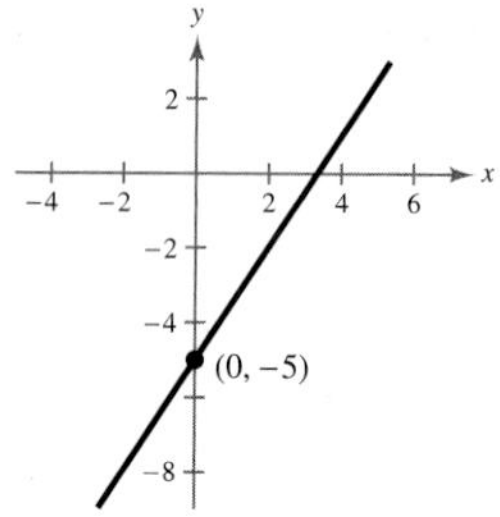

21. $y = -\frac{1}{2}x + 2$

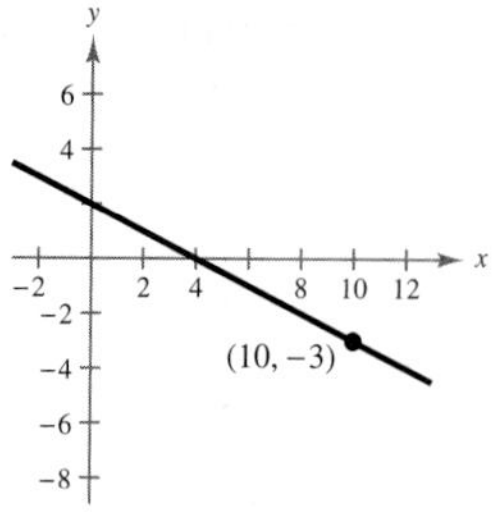

23. $x = 0$ **25.** $y = -\frac{4}{3}x + \frac{8}{3}$

27. (a) $y = \frac{5}{4}x - \frac{23}{4}$ (b) $y = -\frac{4}{5}x + \frac{2}{5}$

29. $V = 850t + 7400,\ 6 \le t \le 11$ **31.** \$210,000

33. (a) Not a function, because 20 in the domain corresponds to two values in the range and because 10 in A is not matched with any element in B.

(b) A function, because each input value has exactly one output value

(c) A function, because each input value has exactly one output value

(d) Not a function, because 30 in A is not matched with any element in B

35. No **37.** Yes

39. (a) 5 (b) 17 (c) $t^4 + 1$ (d) $t^2 + 2t + 2$

41. (a) -3 (b) -1 (c) 2 (d) 6

43. All real numbers x such that $-5 \le x \le 5$

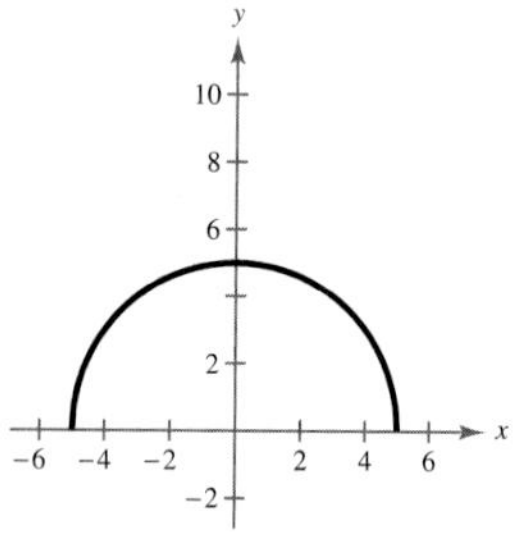

45. All real numbers s except $s = 3$

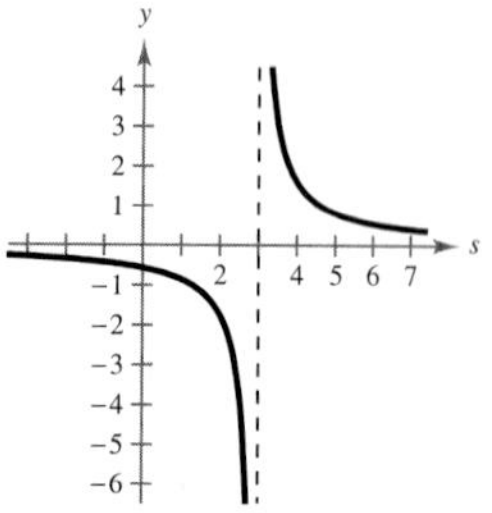

47. All real numbers x except $x = 3, -2$

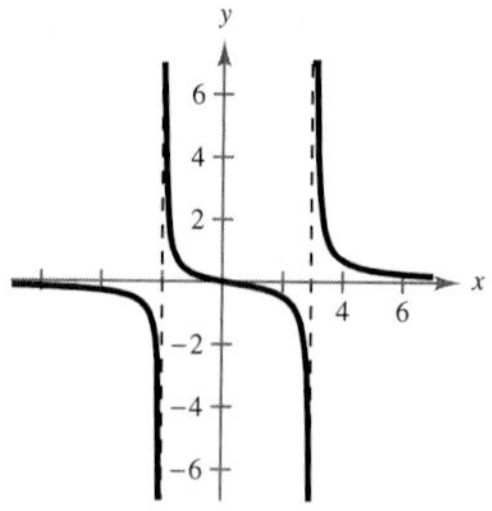

49. (a) 16 feet per second (b) 1.5 seconds

(c) -16 feet per second

51. (a) $A = x(12 - x)$ (b) $0 < x < 12$

53. $4x + 2h + 3,\ h \neq 0$ **55.** Function

57. Not a function **59.** $\frac{7}{3}, 3$ **61.** $-\frac{3}{8}$

63. Increasing on $(0, \infty)$

Decreasing on $(-\infty, -1)$

Constant on $(-1, 0)$

65.

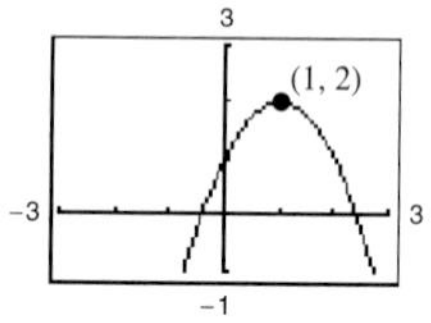

67.

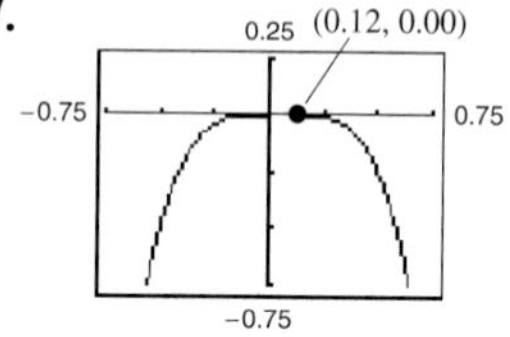

69. 4 **71.** $\dfrac{1 - \sqrt{2}}{2}$ **73.** Neither even nor odd

75. Odd

77. $f(x) = -3x$

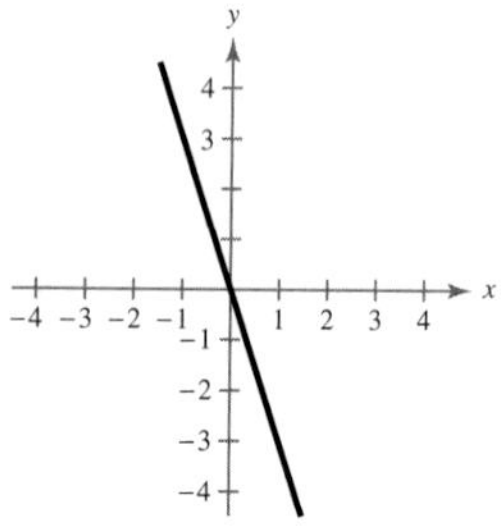

79. $f(x) = \frac{5}{3}x + \frac{10}{3}$

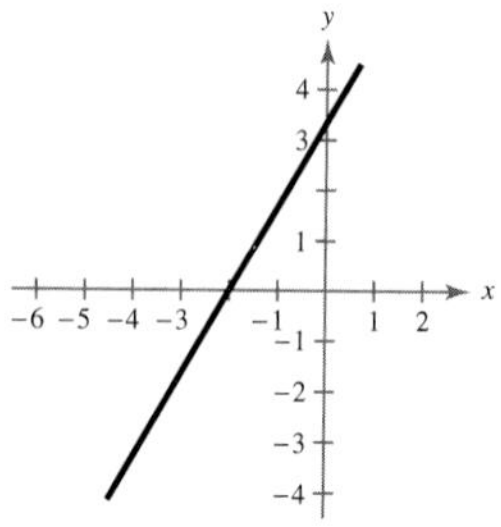

81.

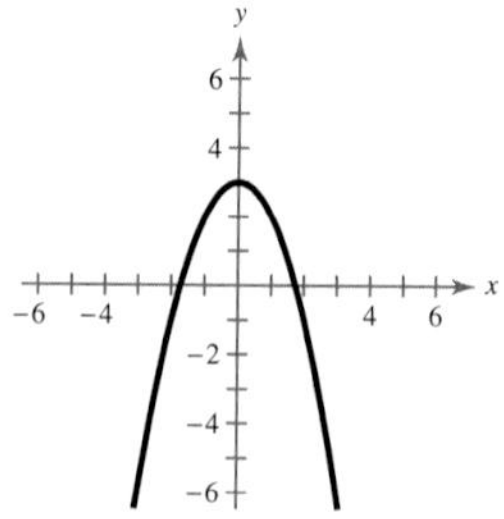

83.

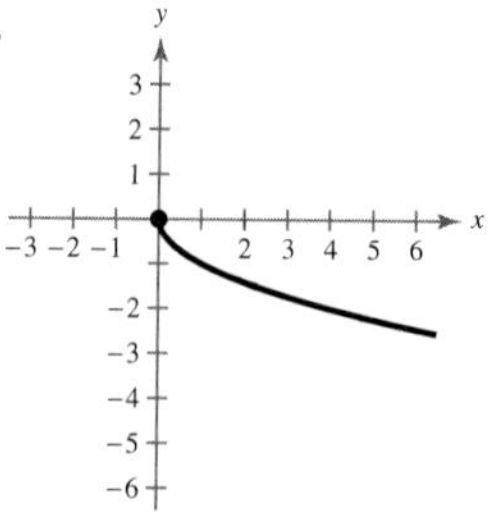

85.

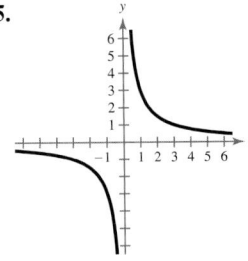

87.

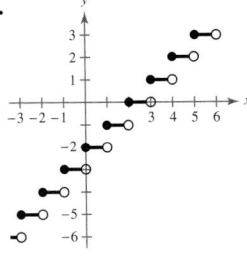

89.

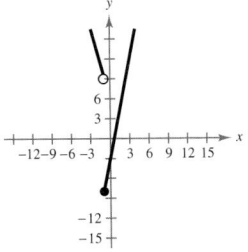

91. $y = x^3$

93. (a) $f(x) = x^2$
  (b) Vertical shift of nine units downward
  (c)   (d) $h(x) = f(x) - 9$

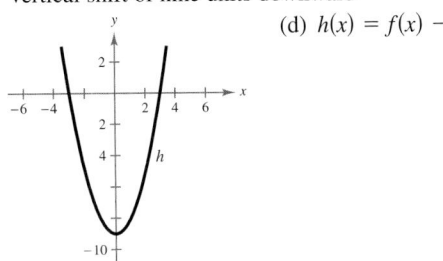

95. (a) $f(x) = \sqrt{x}$
  (b) Horizontal shift of seven units to the right
  (c)   (d) $h(x) = f(x - 7)$

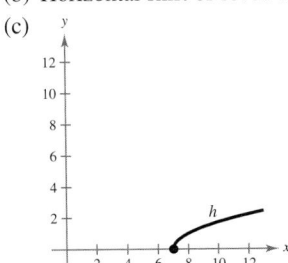

97. (a) $f(x) = x^2$
  (b) Reflection in the x-axis, horizontal shift of three units to the left, and vertical shift of one unit upward
  (c)   (d) $h(x) = -f(x + 3) + 1$

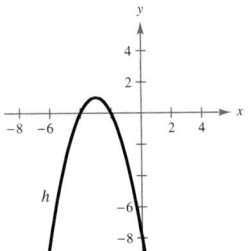

99. (a) $f(x) = [\![x]\!]$
  (b) Reflection in the x-axis and vertical shift of six units upward
  (c)   (d) $h(x) = -f(x) + 6$

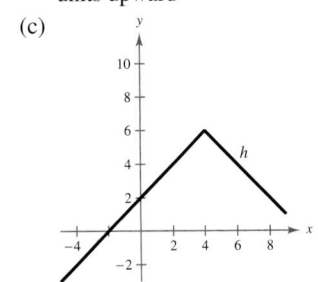

101. (a) $f(x) = |x|$
  (b) Reflections in the x-axis and the y-axis, horizontal shift of four units to the right, and vertical shift of six units upward
  (c)

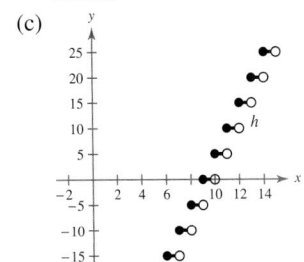

  (d) $h(x) = -f(-x + 4) + 6$

103. (a) $f(x) = [\![x]\!]$
  (b) Horizontal shift of nine units to the right and vertical stretch
  (c)   (d) $h(x) = 5f(x - 9)$

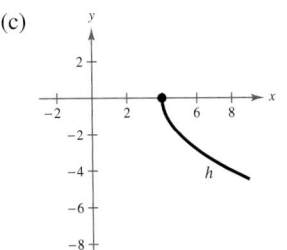

105. (a) $f(x) = \sqrt{x}$
  (b) Reflection in the x-axis, vertical stretch, and horizontal shift of four units to the right
  (c)   (d) $h(x) = -2f(x - 4)$

107. (a) $x^2 + 2x + 2$ (b) $x^2 - 2x + 4$
(c) $2x^3 - x^2 + 6x - 3$
(d) $\dfrac{x^2 + 3}{2x - 1}$; all real numbers x except $x = \dfrac{1}{2}$

109. (a) $x - \frac{8}{3}$ (b) $x - 8$
Domains of $f, g, f \circ g$, and $g \circ f$: all real numbers

111. $f(x) = x^3$, $g(x) = 6x - 5$

113. (a) $(v + d)(t) = -36.04t^2 + 804.6t - 1112$
(b)
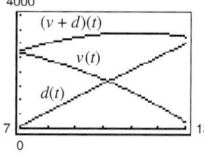
(c) $(v + d)(10) = 3330$

115. $f^{-1}(x) = x + 7$
$f(f^{-1}(x)) = x + 7 - 7 = x$
$f^{-1}(f(x)) = x - 7 + 7 = x$

117. The function has an inverse.

119.
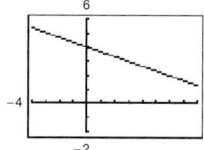
121.
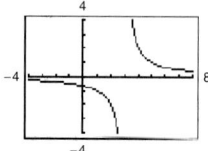

The function has an
inverse.

The function has an
inverse.

123. (a) $f^{-1}(x) = 2x + 6$
(b)
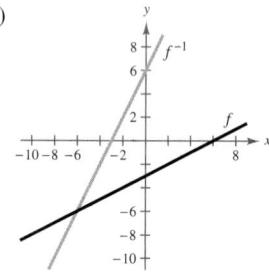
(c) The graphs are reflections of each other in the line $y = x$.
(d) Both f and f^{-1} have domains and ranges that are all real numbers.

125. (a) $f^{-1}(x) = x^2 - 1$, $x \geq 0$
(b)
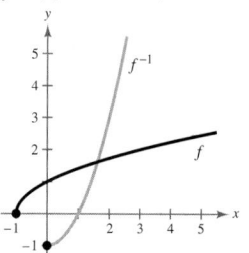
(c) The graphs are reflections of each other in the line $y = x$.

(d) The graph of f has a domain of all real numbers x such that $x \geq -1$ and a range of $[0. \infty)$. The graph of f^{-1} has a domain of all real numbers x such that $x \geq 0$ and a range of $[-1, \infty)$.

127. $x \geq 4$; $f^{-1}(x) = \sqrt{\dfrac{x}{2} + 4}$

129. False. The graph is reflected in the x-axis, shifted nine units to the left, and then shifted 13 units downward.
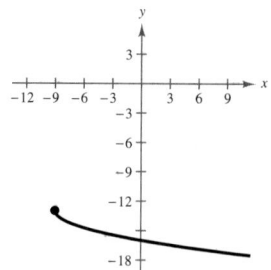

131. A function from a set A to a set B is a relation that assigns to each element x in the set A exactly one element y in the set B.

133. Answers will vary. Sample answer:
Common function: $f(x) = x^3$
The domain and range of the function are the set of all real numbers.
The function is odd.
The graph is increasing on the interval $(-\infty, \infty)$.
The graph is symmetric with respect to the origin.
The graph has an intercept at $(0, 0)$.
$f(x) = x^3 + 1$
The domain and range of the function are the set of all real numbers.
The function is neither odd nor even.
The graph is increasing on the interval $(-\infty, \infty)$.
The graph is not symmetric.
The graph has an intercept at $(0, 1)$, and $(-1, 0)$.

Chapter Test   *(page 257)*

1. $2x + y - 1 = 0$ **2.** $17x + 10y - 59 = 0$
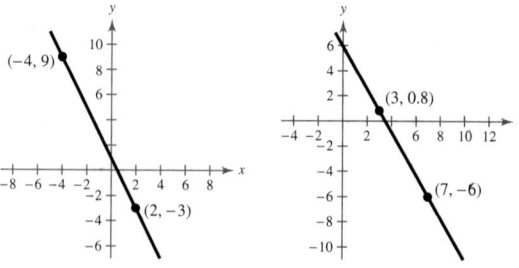

3. (a) $4x - 7y + 44 = 0$ (b) $7x + 4y - 53 = 0$
4. (a) -9 (b) 1 (c) $|x - 4| - 15$
5. (a) $-\dfrac{1}{8}$ (b) $-\dfrac{1}{28}$ (c) $\dfrac{\sqrt{x}}{x^2 - 18x}$

6. All real numbers x such that $-10 \le x \le 10$

7. All real numbers

8. (a)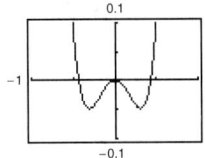

(b) Increasing on $(-0.31, 0)$, $(0.31, \infty)$
Decreasing on $(-\infty, -0.31)$, $(0, 0.31)$

(c) Even

9. (a)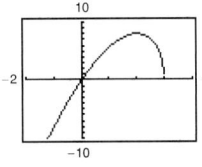

(b) Increasing on $(-\infty, 2)$
Decreasing on $(2, 3)$

(c) Neither even nor odd

10. (a)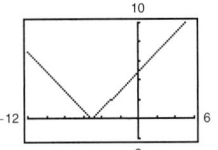

(b) Increasing on $(-5, \infty)$
Decreasing on $(-\infty, -5)$

(c) Neither even nor odd

11.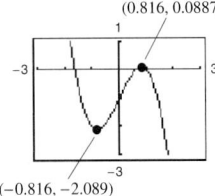

12. Average rate of change $= -3$

13.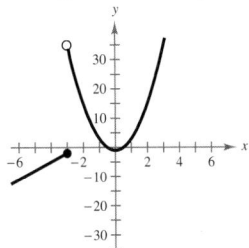

14. (a) $f(x) = [\![x]\!]$

(b) Reflection in the x-axis of $y = [\![x]\!]$

(c)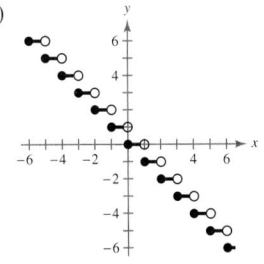

15. (a) $f(x) = \sqrt{x}$

(b) Reflection in the x-axis, horizontal shift, and vertical shift of $y = \sqrt{x}$

(c)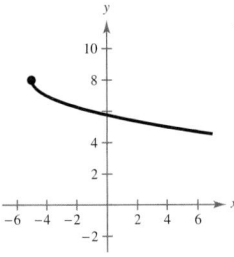

16. (a) $f(x) = |x|$

(b) Vertical shrink, horizontal shift, and vertical shift of $y = |x|$

(c)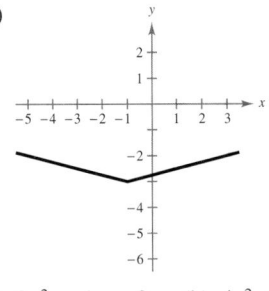

17. (a) $2x^2 - 4x - 2$ (b) $4x^2 + 4x - 12$

(c) $-3x^4 - 12x^3 + 22x^2 + 28x - 35$

(d) $\dfrac{3x^2 - 7}{-x^2 - 4x + 5}$, $x \ne 1, -5$

(e) $3x^4 + 24x^3 + 18x^2 - 120x + 68$

(f) $-9x^4 + 30x^2 - 16$

18. (a) $\dfrac{1 + 2x^{3/2}}{x}$, $x > 0$ (b) $\dfrac{1 - 2x^{3/2}}{x}$, $x > 0$

(c) $\dfrac{2\sqrt{x}}{x}$, $x > 0$ (d) $\dfrac{1}{2x^{3/2}}$, $x > 0$

(e) $\dfrac{\sqrt{x}}{2x}$, $x > 0$ (f) $\dfrac{2\sqrt{x}}{x}$, $x > 0$

19. $f^{-1}(x) = \sqrt[3]{x} - 8$ **20.** No inverse

21. $f^{-1}(x) = \left(\frac{8}{3}x\right)^{2/3}$, $x \ge 0$ **22.** \$153

Cumulative Test for Chapters P–2
(page 258)

1. $\dfrac{4x^3}{15y^5}$, $x \ne 0$ **2.** $2x^2 y \sqrt{6y}$ **3.** $5x - 6$

4. $x^3 - x^2 - 5x + 6$ **5.** $\dfrac{s - 1}{(s + 1)(s + 3)}$

6. $(x + 3)(7 - x)$ **7.** $x(x + 1)(1 - 6x)$

8. $2(3 - 2x)(9 + 6x + 4x^2)$

9. $4x^2 + 12x$ **10.** $\frac{3}{2}x^2 + 8x + \frac{5}{2}$

11.

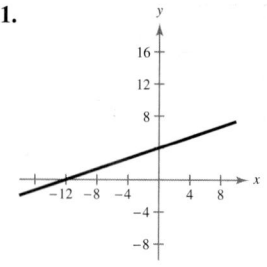

12.

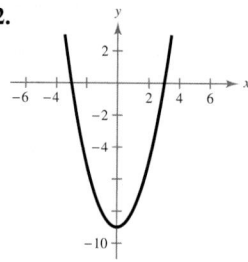

13.

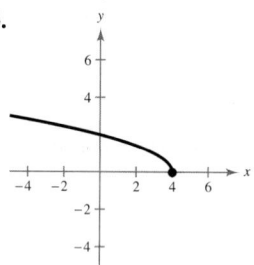

14. $x = -\frac{9}{5}$

15. $x = \frac{27}{5}$ **16.** $x = \frac{23}{6}$ **17.** $1, 3$ **18.** $2 \pm \sqrt{10}$

19. ± 4 **20.** $\dfrac{-5 \pm \sqrt{97}}{6}$ **21.** $-\dfrac{3}{2} \pm \dfrac{\sqrt{69}}{6}$

22. ± 8 **23.** $0, -12, \pm 2i$ **24.** $0, 3$ **25.** ± 8

26. 6 **27.** $-5, 9$ **28.** No solution

29. (a) Not a solution   (b) Not a solution   (c) Solution
     (d) Solution

30. (a) Not a solution   (b) Solution   (c) Not a solution
     (d) Not a solution

31. $-7 \le x \le 5$

32. $x < -\frac{3}{2}, \ x > -\frac{1}{4}$

33. $x \le -\frac{7}{5}, \ x \ge -1$

34. $x < \dfrac{1 - \sqrt{17}}{2}, \ x > \dfrac{1 + \sqrt{17}}{2}$

35. $2x - y + 2 = 0$

36. For some values of x there correspond two values of y.

37. (a) $\dfrac{3}{2}$ (b) Division by 0 is undefined.   (c) $\dfrac{s + 2}{s}$

38. Neither   **39.** Neither   **40.** Even

41. (a) Vertical shrink by $\frac{1}{2}$
     (b) Vertical shift of two units upward
     (c) Horizontal shift of two units to the left

42. (a) $5x - 2$ (b) $-3x - 4$ (c) $4x^2 - 11x - 3$
     (d) $\dfrac{x - 3}{4x + 1}$; Domain: all real numbers x except $x = -\dfrac{1}{4}$

43. (a) $\sqrt{x - 1} + x^2 + 1$ (b) $\sqrt{x - 1} - x^2 - 1$
     (c) $x^2 \sqrt{x - 1} + \sqrt{x - 1}$
     (d) $\dfrac{\sqrt{x - 1}}{x^2 + 1}$; Domain: all real numbers x such that $x \ge 1$

44. (a) $2x + 12$ (b) $\sqrt{2x^2 + 6}$
     Domain of $f \circ g$: all real numbers x such that $x \ge -6$
     Domain of $g \circ f$: all real numbers

45. (a) $|x| - 2$ (b) $|x - 2|$
     Domain of $f \circ g$ and $g \circ f$: all real numbers

46. $h^{-1}(x) = \frac{1}{5}(x + 2)$ **47.** $n = 9$

48. (a) $R(n) = n[8 - 0.05(n - 80)], \quad n \ge 80$
     (b)

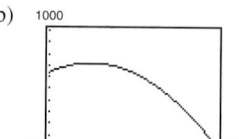

49. 32

Problem Solving   *(page 261)*

1. (a) $W_1 = 2000 + 0.07S$ (b) $W_2 = 2300 + 0.05S$
     (c)

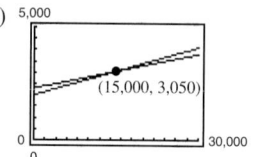

     Both jobs pay the same monthly salary if sales equal
     \$15,000.
     (d) No. Job 1 would pay \$3400 and job 2 would pay \$3300.

3. (a) The function will be even.
     (b) The function will be odd.
     (c) The function will be neither even nor odd.

5. $f(x) = a_{2n}x^{2n} + a_{2n-2}x^{2n-2} + \cdots + a_2 x^2 + a_0$
     $f(-x) = a_{2n}(-x)^{2n} + a_{2n-2}(-x)^{2n-2}$
             $+ \cdots + a_2(-x)^2 + a_0$
             $= f(x)$

7. (a) $81\frac{2}{3}$ hours   (b) $25\frac{5}{7}$ miles per hour
     (c) $y = \dfrac{-180}{7}x + 3400$
     Domain: $0 \le x \le \dfrac{1190}{9}$
     Range: $0 \le y \le 3400$

(d)

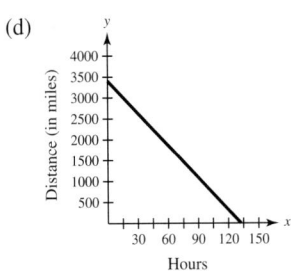

Hours

9. (a) $(f \circ g)(x) = 4x + 24$ (b) $(f \circ g)^{-1}(x) = \frac{1}{4}x - 6$

(c) $f^{-1}(x) = \frac{1}{4}x;\ g^{-1}(x) = x - 6$

(d) $(g^{-1} \circ f^{-1})(x) = \frac{1}{4}x - 6$

(e) $(f \circ g)(x) = 8x^3 + 1;\ (f \circ g)^{-1}(x) = \frac{1}{2}\sqrt[3]{x - 1};$

    $f^{-1}(x) = \sqrt[3]{x - 1};\ g^{-1}(x) = \frac{1}{2}x;$

    $(g^{-1} \circ f^{-1})(x) = \frac{1}{2}\sqrt[3]{x - 1}$

(f) Answers will vary.   (g) $(f \circ g)^{-1}(x) = (g^{-1} \circ f^{-1})(x)$

11. (a)   (b)

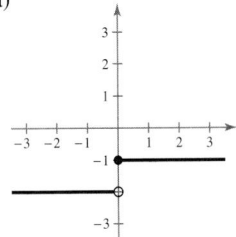 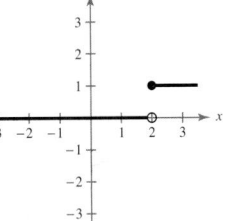

(c)   (d)

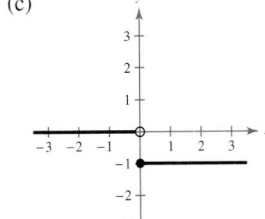 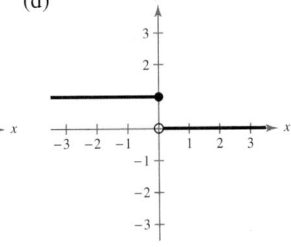

(e)   (f)

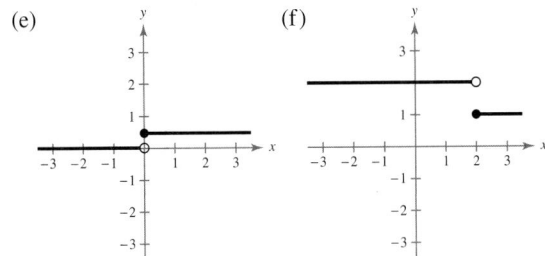

13. Proof

15. (a)

x	-4	-2	0	4
$f(f^{-1}(x))$	-4	-2	0	4

(b)

x	-3	-2	0	1
$(f + f^{-1})(x)$	5	1	-3	-5

(c)

x	-3	-2	0	1
$(f \circ f^{-1})(x)$	4	0	2	6

(d)

x	-4	-3	0	4
$|f^{-1}(x)|$	2	1	1	3

Chapter 3

Section 3.1   *(page 270)*

Vocabulary Check   *(page 270)*

1. nonnegative integer; real   **2.** quadratic; parabola

3. axis   **4.** positive; minimum

5. negative; maximum

1. g   **2.** c   **3.** b   **4.** h

5. f   **6.** a   **7.** e   **8.** d

9. (a)   (b)

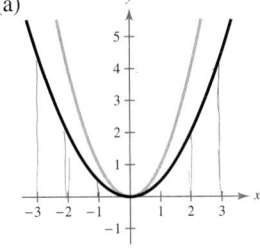 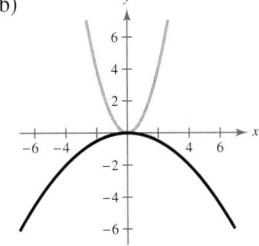

Vertical shrink

Vertical shrink and reflection in the x-axis

(c)   (d)

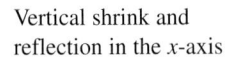 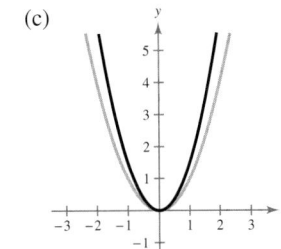 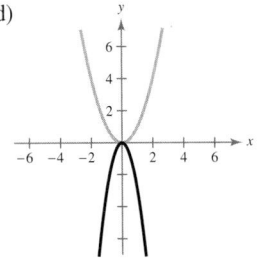

Vertical stretch

Vertical stretch and reflection in the x-axis

11. (a)   (b)

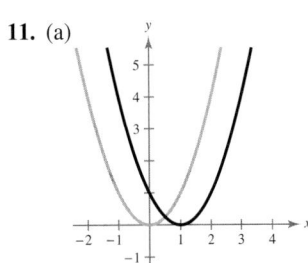 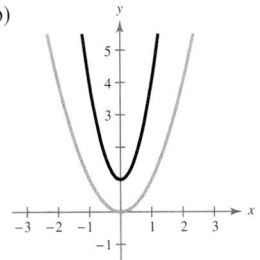

Horizontal translation

Horizontal shrink and vertical translation

(c)

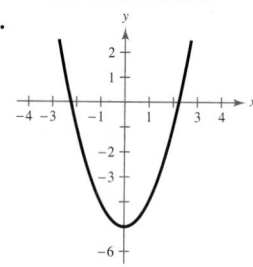

Horizontal stretch and vertical translation

(d)

Horizontal translation

13.

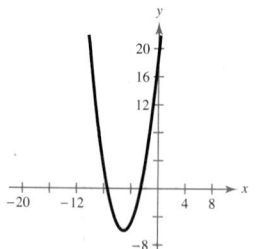

Vertex: $(0, -5)$
Axis of symmetry: y-axis
x-intercepts: $(\pm\sqrt{5}, 0)$

15.

Vertex: $(0, -4)$
Axis of symmetry: y-axis
x-intercepts: $(\pm 2\sqrt{2}, 0)$

17.

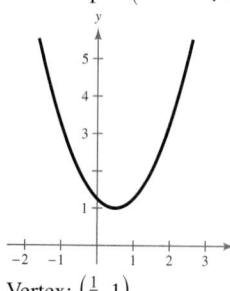

Vertex: $(-5, -6)$
Axis of symmetry: $x = -5$
x-intercepts: $(-5 \pm \sqrt{6}, 0)$

19.

Vertex: $(4, 0)$
Axis of symmetry: $x = 4$
x-intercept: $(4, 0)$

21.

Vertex: $\left(\frac{1}{2}, 1\right)$
Axis of symmetry: $x = \frac{1}{2}$
No x-intercept

23.

Vertex: $(1, 6)$
Axis of symmetry: $x = 1$
x-intercepts: $\left(1 \pm \sqrt{6}, 0\right)$

25.

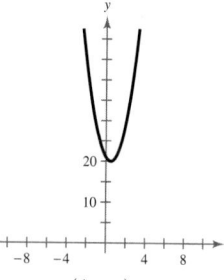

Vertex: $\left(\frac{1}{2}, 20\right)$
Axis of symmetry: $x = \frac{1}{2}$
No x-intercept

27.

Vertex: $(4, -16)$
Axis of symmetry: $x = 4$
x-intercepts: $(-4, 0), (12, 0)$

29.

Vertex: $(-1, 4)$
Axis of symmetry: $x = -1$
x-intercepts: $(1, 0), (-3, 0)$

31.

Vertex: $(-4, -5)$
Axis of symmetry: $x = -4$
x-intercepts: $\left(-4 \pm \sqrt{5}, 0\right)$

33.

Vertex: $(4, -1)$
Axis of symmetry: $x = 4$
x-intercepts: $\left(4 \pm \frac{1}{2}\sqrt{2}, 0\right)$

35.

Vertex: $(-2, -3)$
Axis of symmetry: $x = -2$
x-intercepts: $\left(-2 \pm \sqrt{6}, 0\right)$

37. $y = (x - 1)^2$ **39.** $y = -(x + 1)^2 + 4$
41. $y = -2(x + 2)^2 + 2$ **43.** $f(x) = (x + 2)^2 + 5$
45. $f(x) = -\frac{1}{2}(x - 3)^2 + 4$ **47.** $f(x) = \frac{3}{4}(x - 5)^2 + 12$
49. $f(x) = -\frac{24}{49}\left(x + \frac{1}{4}\right)^2 + \frac{3}{2}$ **51.** $f(x) = -\frac{16}{3}\left(x + \frac{5}{2}\right)^2$
53. $(\pm 4, 0)$ **55.** $(5, 0), (-1, 0)$
57.

$(0, 0), (4, 0)$

59.

$(3, 0), (6, 0)$

61.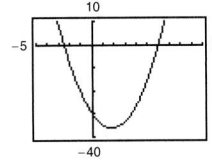

$\left(-\frac{5}{2}, 0\right), (6, 0)$

63.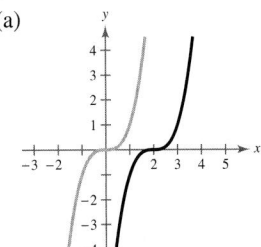

$(7, 0), (-1, 0)$

65. $f(x) = x^2 - 2x - 3$
$g(x) = -x^2 + 2x + 3$

67. $f(x) = x^2 - 10x$
$g(x) = -x^2 + 10x$

69. $f(x) = 2x^2 + 7x + 3$
$g(x) = -2x^2 - 7x - 3$

71. $55, 55$ **73.** $12, 6$

75. (a) $A = \dfrac{8x(50 - x)}{3}$

(b)

x	5	10	15	20	25	30
A	600	1067	1400	1600	1667	1600

$x = 25$ feet, $y = 33\frac{1}{3}$ feet

(c)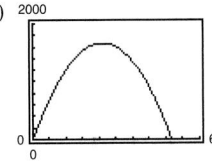

$x = 25$ feet, $y = 33\frac{1}{3}$ feet

(d) $A = -\frac{8}{3}(x - 25)^2 + \frac{5000}{3}$ (e) They are identical.

77. 16 feet **79.** 20 fixtures **81.** 350,000 units

83. (a) $\$14,000,000; \$14,375,000; \$13,500,000$
(b) 24; $\$14,400$

85. (a)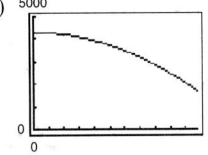

(b) 4299; answers will vary.
(c) 8879; 24

87. (a)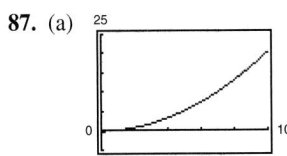

(b) 69.6 miles per hour

89. True. The equation has no real solutions, so the graph has no x-intercepts.

91. $f(x) = a\left(x + \dfrac{b}{2a}\right)^2 + \dfrac{4ac - b^2}{4a}$

93. Yes. A graph of a quadratic equation whose vertex is on the x-axis has only one x-intercept.

95. $y = -\frac{1}{3}x + \frac{5}{3}$ **97.** $y = \frac{5}{4}x + 3$ **99.** 27

101. $-\frac{1408}{49}$ **103.** 109 **105.** Answers will vary.

Section 3.2 *(page 284)*

Vocabulary Check *(page 284)*

1. continuous 2. Leading Coefficient Test
3. $n; n - 1$ 4. (a) solution; (b) $(x - a)$; (c) x-intercept
5. touches; crosses 6. standard
7. Intermediate Value

1. c **2.** g **3.** h **4.** f
5. a **6.** e **7.** d **8.** b

9. (a) (b)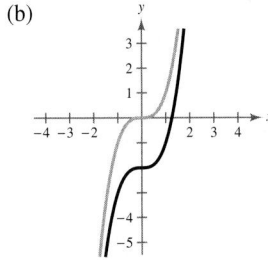

(c) 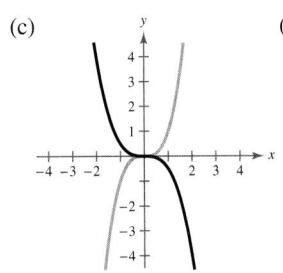 (d)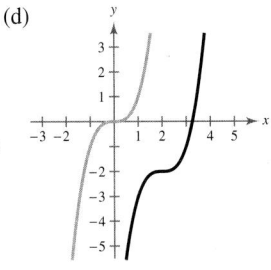

11. (a) 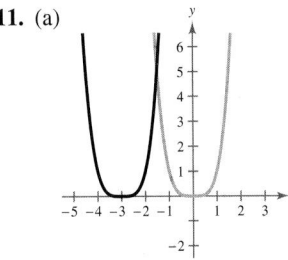 (b)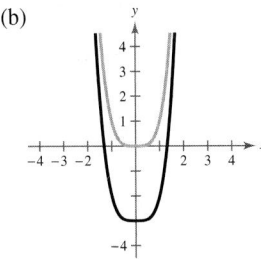

(c) 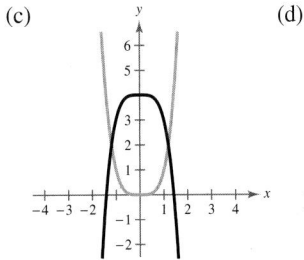 (d)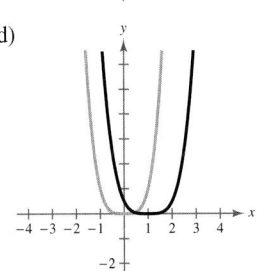

CHAPTER 3

(e)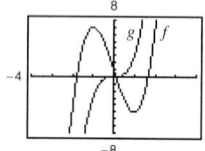

(f)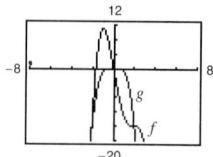

13. Falls to the left, rises to the right
15. Falls to the left, falls to the right
17. Rises to the left, falls to the right
19. Rises to the left, falls to the right
21. Falls to the left, falls to the right

23.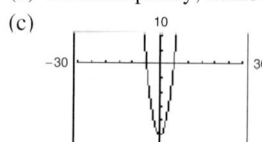

25.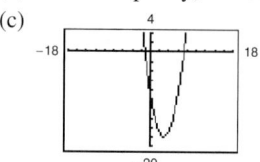

27. (a) ± 5
(b) odd multiplicity; number of turning points: 1
(c)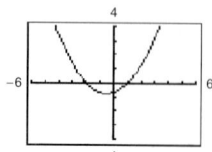

29. (a) 3
(b) even multiplicity; number of turning points: 1
(c)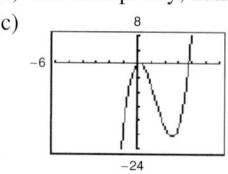

31. (a) $-2, 1$
(b) odd multiplicity; number of turning points: 1
(c)

33. (a) $0, 2 \pm \sqrt{3}$
(b) odd multiplicity; number of turning points: 2
(c)

35. (a) $0, 2$
(b) 0, odd multiplicity; 2, even multiplicity; number of turning points: 2
(c)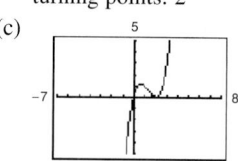

37. (a) $0, \pm\sqrt{3}$
(b) 0, odd multiplicity; $\pm\sqrt{3}$, even multiplicity; number of turning points: 4
(c)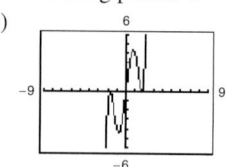

39. (a) No real zeros
(b) number of turning points: 1
(c)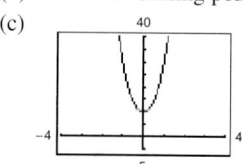

41. (a) $\pm 2, -3$
(b) odd multiplicity; number of turning points: 2
(c)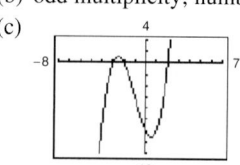

43. (a)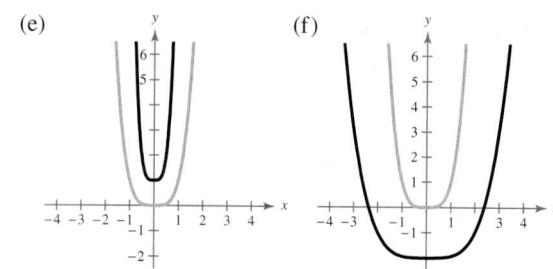

(b) x-intercepts: $(0, 0), \left(\frac{5}{2}, 0\right)$ (c) $x = 0, \frac{5}{2}$
(d) The answers in part (c) match the x-intercepts.

45. (a)

(b) x-intercepts: $(0, 0), (\pm 1, 0), (\pm 2, 0)$
(c) $x = 0, 1, -1, 2, -2$
(d) The answers in part (c) match the x-intercepts.

47. $f(x) = x^2 - 10x$ **49.** $f(x) = x^2 + 4x - 12$

51. $f(x) = x^3 + 5x^2 + 6x$

53. $f(x) = x^4 - 4x^3 - 9x^2 + 36x$

55. $f(x) = x^2 - 2x - 2$ **57.** $f(x) = x^2 + 4x + 4$

59. $f(x) = x^3 + 2x^2 - 3x$ **61.** $f(x) = x^3 - 3x$

63. $f(x) = x^4 + x^3 - 15x^2 + 23x - 10$

65. $f(x) = x^5 + 16x^4 + 96x^3 + 256x^2 + 256x$

67. (a) Falls to the left, rises to the right

     (b) $0, \pm 3$ (c) Answers will vary.

     (d)

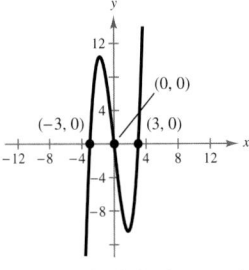

69. (a) Rises to the left, rises to the right

     (b) No zeros   (c) Answers will vary.

     (d)

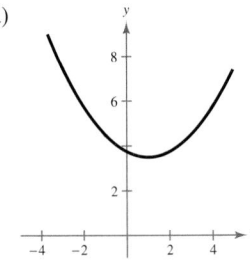

71. (a) Falls to the left, rises to the right

     (b) $0, 3$ (c) Answers will vary.

     (d)

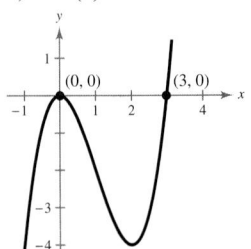

73. (a) Falls to the left, rises to the right

     (b) $0, 2, 3$ (c) Answers will vary.

     (d)

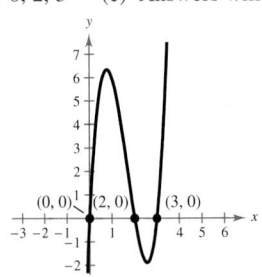

75. (a) Rises to the left, falls to the right

     (b) $-5, 0$ (c) Answers will vary.

     (d)

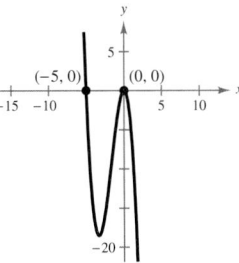

77. (a) Falls to the left, rises to the right

     (b) $0, 4$ (c) Answers will vary.

     (d)

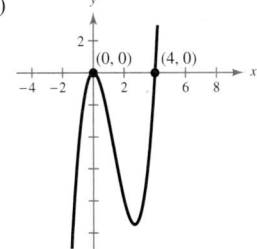

79. (a) Falls to the left, falls to the right

     (b) ± 2 (c) Answers will vary.

     (d)

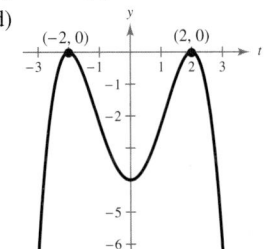

81.

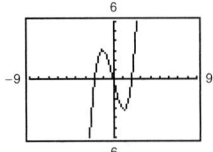

Zeros: $0, \pm 2$,
odd multiplicity

83.

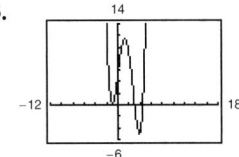

Zeros: -1,
even multiplicity;
$3, \frac{9}{2}$, odd multiplicity

85. $[-1, 0], [1, 2], [2, 3]; \approx -0.879, 1.347, 2.532$

87. $[-2, -1], [0, 1]; \approx -1.585, 0.779$

89. (a) $V = l \times w \times h$

$\qquad = (36 - 2x)(36 - 2x)x$

$\qquad = x(36 - 2x)^2$

     (b) Domain: $0 < x < 18$

     (c)

x	1	2	3	4	5	6	7
V	1156	2048	2700	3136	3380	3456	3388

6 inches $\times$ 24 inches $\times$ 24 inches

(d)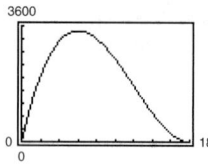

$x = 6$

91. (a) $A = -2x^2 + 12x$

(b) $V = -384x^2 + 2304x$

(c) 0 inches $< x < 6$ inches

(d)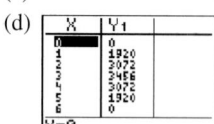

When $x = 3$, the volume is maximum at $V = 3456$; dimensions of gutter are 3 inches × 6 inches × 3 inches

(e)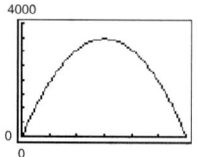

The maximum value is the same.

(f) No. Answers will vary.

93.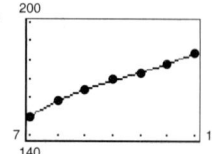

The model is a good fit.

95. Region 1: 259,370

Region 2: 223,470

Answers will vary.

97. (a)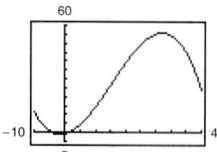

(b) $t \approx 15$

(c) Vertex: $(15.22, 2.54)$

(d) The results are approximately equal.

99. False. A fifth-degree polynomial can have at most four turning points.

101. True. The degree of the function is odd and its leading coefficient is negative, so the graph rises to the left and falls to the right.

103.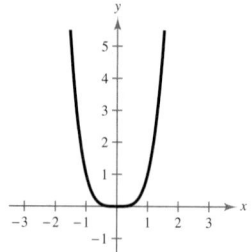

(a) Vertical shift of two units; Even

(b) Horizontal shift of two units; Neither even nor odd

(c) Reflection in the y-axis; Even

(d) Reflection in the x-axis; Even

(e) Horizontal stretch; Even

(f) Vertical shrink; Even

(g) $g(x) = x^3$; Neither odd nor even

(h) $g(x) = x^{16}$; Even

105. $(5x - 8)(x + 3)$ **107.** $x^2(4x + 5)(x - 3)$

109. $-\frac{7}{2}, 4$ **111.** $-\frac{5}{4}, \frac{1}{3}$ **113.** $1 \pm \sqrt{22}$

115. $\dfrac{-5 \pm \sqrt{185}}{4}$

117. Horizontal translation four units to the left of $y = x^2$

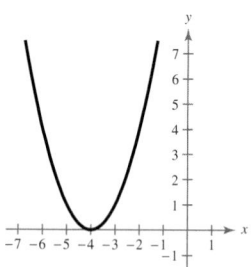

119. Horizontal translation one unit left and vertical translation five units down of $y = \sqrt{x}$

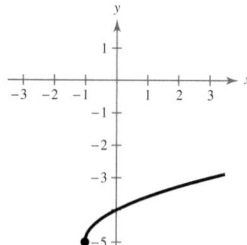

121. Vertical stretch by a factor of 2 and vertical translation nine units up of $y = [\![x]\!]$

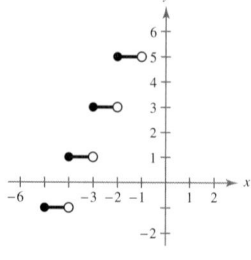

Section 3.3   *(page 295)*

1. Answers will vary.
3.

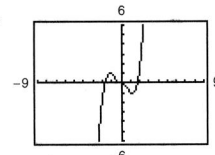

5. $2x + 4$

7. $x^2 - 3x + 1$ **9.** $x^3 + 3x^2 - 1$ **11.** $7 - \dfrac{11}{x + 2}$

13. $3x + 5 - \dfrac{2x - 3}{2x^2 + 1}$ **15.** $x^2 + 2x + 4 + \dfrac{2x - 11}{x^2 - 2x + 3}$

17. $x + 3 + \dfrac{6x^2 - 8x + 3}{(x - 1)^3}$ **19.** $3x^2 - 2x + 5$

21. $4x^2 - 9$ **23.** $-x^2 + 10x - 25$

25. $5x^2 + 14x + 56 + \dfrac{232}{x - 4}$

27. $10x^3 + 10x^2 + 60x + 360 + \dfrac{1360}{x - 6}$

29. $x^2 - 8x + 64$

31. $-3x^3 - 6x^2 - 12x - 24 - \dfrac{48}{x - 2}$

33. $-x^3 - 6x^2 - 36x - 36 - \dfrac{216}{x - 6}$

35. $4x^2 + 14x - 30$
37. $f(x) = (x - 4)(x^2 + 3x - 2) + 3,\ \ f(4) = 3$
39. $f(x) = \left(x + \tfrac{2}{3}\right)(15x^3 - 6x + 4) + \tfrac{34}{3},\ \ f\!\left(-\tfrac{2}{3}\right) = \tfrac{34}{3}$
41. $f(x) = \left(x - \sqrt{2}\right)\!\left[x^2 + \left(3 + \sqrt{2}\right)x + 3\sqrt{2}\right] - 8,$
$f\!\left(\sqrt{2}\right) = -8$
43. $f(x) = \left(x - 1 + \sqrt{3}\right)\!\left[-4x^2 + \left(2 + 4\sqrt{3}\right)x + \left(2 + 2\sqrt{3}\right)\right],$
$f\!\left(1 - \sqrt{3}\right) = 0$
45. (a) 1   (b) 4   (c) 4   (d) 1954
47. (a) 97   (b) $-\tfrac{5}{3}$ (c) 17   (d) -199
49. $(x - 2)(x + 3)(x - 1)$; Zeros: $2, -3, 1$
51. $(2x - 1)(x - 5)(x - 2)$; Zeros: $\tfrac{1}{2}, 5, 2$
53. $\left(x + \sqrt{3}\right)\!\left(x - \sqrt{3}\right)(x + 2)$; Zeros: $-\sqrt{3}, \sqrt{3}, -2$
55. $(x - 1)\!\left(x - 1 - \sqrt{3}\right)\!\left(x - 1 + \sqrt{3}\right)$;
Zeros: $1,\ 1 + \sqrt{3},\ 1 - \sqrt{3}$
57. (a) Answers will vary.   (b) $2x - 1$
(c) $f(x) = (2x - 1)(x + 2)(x - 1)$ (d) $\tfrac{1}{2}, -2, 1$
(e)

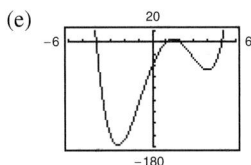

59. (a) Answers will vary.   (b) $(x - 1),\ (x - 2)$
(c) $f(x) = (x - 1)(x - 2)(x - 5)(x + 4)$
(d) $1, 2, 5, -4$
(e)

61. (a) Answers will vary.   (b) $x + 7$
(c) $f(x) = (x + 7)(2x + 1)(3x - 2)$
(d) $-7, -\tfrac{1}{2}, \tfrac{2}{3}$
(e)

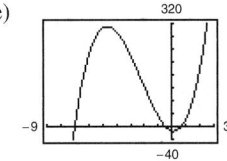

63. (a) Answers will vary.   (b) $\left(x - \sqrt{5}\right)$
(c) $f(x) = \left(x - \sqrt{5}\right)\!\left(x + \sqrt{5}\right)(2x - 1)$ (d) $\pm\sqrt{5}, \tfrac{1}{2}$
(e)

65. (a) Zeros are 2 and $\approx \pm 2.236$.
(b) $x = 2$ (c) $f(x) = (x - 2)\!\left(x - \sqrt{5}\right)\!\left(x + \sqrt{5}\right)$
67. (a) Zeros are -2, ≈ 0.268, and ≈ 3.732.
(b) $x = -2$
(c) $h(t) = (t + 2)\!\left[t - \left(2 + \sqrt{3}\right)\right]\!\left[t - \left(2 - \sqrt{3}\right)\right]$
69. $2x^2 - x - 1,\ \ x \neq \tfrac{3}{2}$ **71.** $x^2 + 3x,\ \ x \neq -2, -1$
73. (a) and (b)

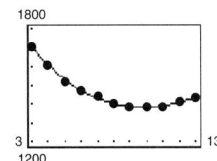

$M = -0.242x^3 + 12.43x^2 - 173.4x + 2118$

(c)

t	3	4	5	6	7	8
$M(t)$	1703	1608	1531	1473	1430	1402

t	9	10	11	12	13
$M(t)$	1388	1385	1392	1409	1433

Answers will vary.
(d) 1614 thousand. No, because the model will approach negative infinity quickly.

75. False. $-\frac{4}{7}$ is a zero of f.

77. True. The degree of the numerator is greater than the degree of the denominator.

79. $x^{2n} + 6x^n + 9$ **81.** The remainder is 0.

83. $c = -210$ **85.** 0; $x + 3$ is a factor of f.

87. $\pm\dfrac{5}{3}$ **89.** $-\dfrac{7}{5}, 2$ **91.** $\dfrac{-3 \pm \sqrt{3}}{2}$

93. $f(x) = x^3 - 7x^2 + 12x$

95. $f(x) = x^3 + x^2 - 7x - 3$

Section 3.4   (page 308)

Vocabulary Check   (page 308)

1. Fundamental Theorem of Algebra

2. Linear Factorization Theorem   **3.** Rational Zero

4. conjugate   **5.** irreducible over the reals

6. Descartes' Rule of Signs   **7.** lower; upper

1. $0, 6$ **3.** $2, -4$ **5.** $-6, \pm i$ **7.** $\pm 1, \pm 3$

9. $\pm 1, \pm 3, \pm 5, \pm 9, \pm 15, \pm 45, \pm\frac{1}{2}, \pm\frac{3}{2}, \pm\frac{5}{2}, \pm\frac{9}{2}, \pm\frac{15}{2}, \pm\frac{45}{2}$

11. $1, 2, 3$ **13.** $1, -1, 4$ **15.** $-1, -10$ **17.** $\frac{1}{2}, -1$

19. $-2, 3, \pm\frac{2}{3}$ **21.** $-1, 2$ **23.** $-6, \frac{1}{2}, 1$

25. (a) $\pm 1, \pm 2, \pm 4$

(b)

(c) $-2, -1, 2$

27. (a) $\pm 1, \pm 3, \pm\frac{1}{2}, \pm\frac{3}{2}, \pm\frac{1}{4}, \pm\frac{3}{4}$

(b)

(c) $-\frac{1}{4}, 1, 3$

29. (a) $\pm 1, \pm 2, \pm 4, \pm 8, \pm\frac{1}{2}$

(b)

(c) $-\frac{1}{2}, 1, 2, 4$

31. (a) $\pm 1, \pm 3, \pm\frac{1}{2}, \pm\frac{3}{2}, \pm\frac{1}{4}, \pm\frac{3}{4}, \pm\frac{1}{8}, \pm\frac{3}{8}, \pm\frac{1}{16}, \pm\frac{3}{16}, \pm\frac{1}{32}, \pm\frac{3}{32}$

(b)

(c) $1, \frac{3}{4}, -\frac{1}{8}$

33. (a) $\pm 1, \approx\pm 1.414$

(b) $f(x) = (x + 1)(x - 1)\left(x + \sqrt{2}\right)\left(x - \sqrt{2}\right)$

35. (a) $0, 3, 4, \approx\pm 1.414$

(b) $h(x) = x(x - 3)(x - 4)\left(x + \sqrt{2}\right)\left(x - \sqrt{2}\right)$

37. $x^3 - x^2 + 25x - 25$ **39.** $x^3 + 4x^2 - 31x - 174$

41. $3x^4 - 17x^3 + 25x^2 + 23x - 22$

43. (a) $(x^2 + 9)(x^2 - 3)$

(b) $(x^2 + 9)\left(x + \sqrt{3}\right)\left(x - \sqrt{3}\right)$

(c) $(x + 3i)(x - 3i)\left(x + \sqrt{3}\right)\left(x - \sqrt{3}\right)$

45. (a) $(x^2 - 2x - 2)(x^2 - 2x + 3)$

(b) $\left(x - 1 + \sqrt{3}\right)\left(x - 1 - \sqrt{3}\right)(x^2 - 2x + 3)$

(c) $\left(x - 1 + \sqrt{3}\right)\left(x - 1 - \sqrt{3}\right)\left(x - 1 + \sqrt{2}i\right)$
    $\left(x - 1 - \sqrt{2}i\right)$

47. $-\frac{3}{2}, \pm 5i$ **49.** $\pm 2i, 1, -\frac{1}{2}$ **51.** $-3 \pm i, \frac{1}{4}$

53. $2, -3 \pm \sqrt{2}i, 1$ **55.** $\pm 5i$; $(x + 5i)(x - 5i)$

57. $2 \pm \sqrt{3}$; $\left(x - 2 - \sqrt{3}\right)\left(x - 2 + \sqrt{3}\right)$

59. $\pm 3, \pm 3i$; $(x + 3)(x - 3)(x + 3i)(x - 3i)$

61. $1 \pm i$; $(z - 1 + i)(z - 1 - i)$

63. $2, 2 \pm i$; $(x - 2)(x - 2 + i)(x - 2 - i)$

65. $-2, 1 \pm \sqrt{2}i$; $(x + 2)\left(x - 1 + \sqrt{2}i\right)\left(x - 1 - \sqrt{2}i\right)$

67. $-\frac{1}{5}, 1 \pm \sqrt{5}i$; $(5x + 1)\left(x - 1 + \sqrt{5}i\right)\left(x - 1 - \sqrt{5}i\right)$

69. $2, \pm 2i$; $(x - 2)^2(x + 2i)(x - 2i)$

71. $\pm i, \pm 3i$; $(x + i)(x - i)(x + 3i)(x - 3i)$

73. $-10, -7 \pm 5i$ **75.** $-\frac{3}{4}, 1 \pm \frac{1}{2}i$

77. $-2, -\frac{1}{2}, \pm i$ **79.** No real zeros

81. No real zeros   **83.** One positive zero

85. One or three positive zeros

87. Answers will vary.   **89.** Answers will vary.

91. $1, -\frac{1}{2}$ **93.** $-\frac{3}{4}$ **95.** $\pm 2, \pm\frac{3}{2}$ **97.** $\pm 1, \frac{1}{4}$

99. d   **100.** a   **101.** b   **102.** c

103. (a)

(b) $V = x(9 - 2x)(15 - 2x)$

Domain: $0 < x < \frac{9}{2}$

(c)

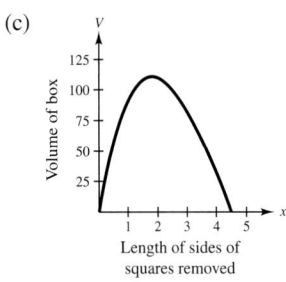

1.82 centimeters × 5.36 centimeters × 11.36 centimeters

(d) $\frac{1}{2}, \frac{7}{2}, 8$; 8 is not in the domain of V.

105. $x \approx 38.4$, or $384,000$

107. (a) $V = x^3 + 9x^2 + 26x + 24 = 120$

(b) 4 feet by 5 feet by 6 feet

109. $x \approx 40$, or 4000 units

111. No. Setting $p = 9,000,000$ and solving the resulting equation yields imaginary roots.

113. False. The most complex zeros it can have is two, and the Linear Factorization Theorem guarantees that there are three linear factors, so one zero must be real.

115. r_1, r_2, r_3 **117.** $5 + r_1, 5 + r_2, 5 + r_3$

119. The zeros cannot be determined.

121. (a) $0 < k < 4$ (b) $k = 4$ (c) $k < 0$ (d) $k > 4$

123. Answers will vary. There are infinitely many possible functions for f. Sample equation and graph:

$$f(x) = -2x^3 + 3x^2 + 11x - 6$$

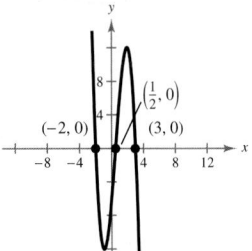

125. Answers will vary.

127. (a) $x^2 + b$ (b) $x^2 - 2ax + a^2 + b^2$

129. $-11 + 9i$ **131.** $20 + 40i$

133.

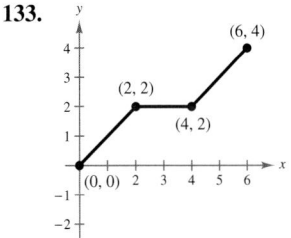

135.

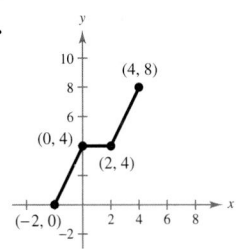

137.

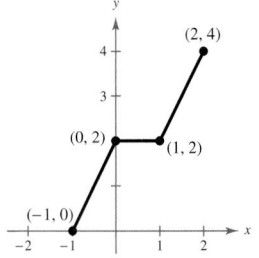

Section 3.5   *(page 319)*

Vocabulary Check   *(page 319)*

1. variation; regression   **2.** sum of square differences

3. correlation coefficient   **4.** directly proportional

5. constant of variation   **6.** directly proportional

7. inverse   **8.** combined   **9.** jointly proportional

1.

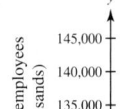
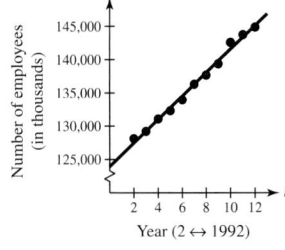

The model is a good fit for the actual data.

3.

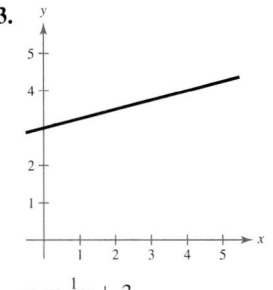

$y = \frac{1}{4}x + 3$

5.

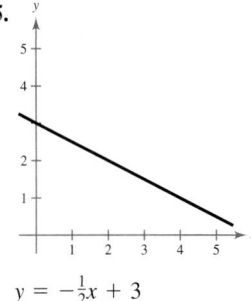

$y = -\frac{1}{2}x + 3$

7. (a) and (b)

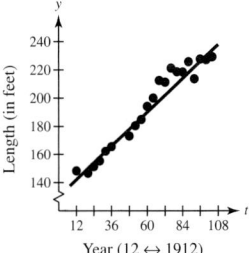

(c) $y = 1.03t + 130.27$ (d) The models are similar.

(e) Part (b): 238 feet; Part (c): 241.51 feet

(f) Answers will vary.

CHAPTER 3

9. (a)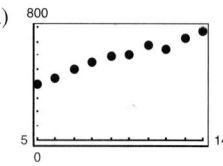

(b) $S = 38.4t + 224$

(c)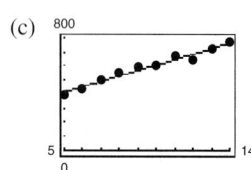

The model is a good fit.

(d) 2005: \$800 million; 2007: \$876.8 million

(e) Each year the annual gross ticket sales for Broadway shows in New York City increase by \$38.4 million.

11. Inversely

13.

x	2	4	6	8	10
$y = kx^2$	4	16	36	64	100

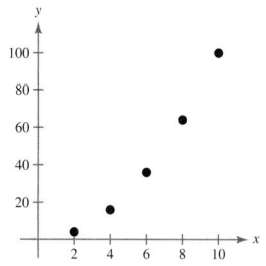

15.

x	2	4	6	8	10
$y = kx^2$	2	8	18	32	50

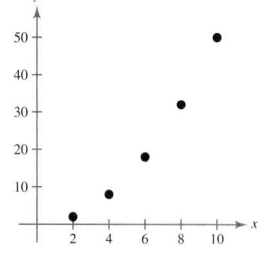

17.

x	2	4	6	8	10
$y = k/x^2$	$\frac{1}{2}$	$\frac{1}{8}$	$\frac{1}{18}$	$\frac{1}{32}$	$\frac{1}{50}$

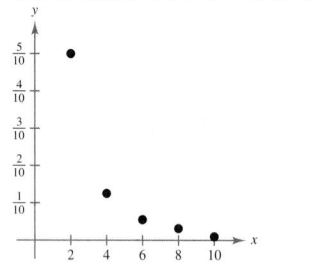

19.

x	2	4	6	8	10
$y = k/x^2$	$\frac{5}{2}$	$\frac{5}{8}$	$\frac{5}{18}$	$\frac{5}{32}$	$\frac{1}{10}$

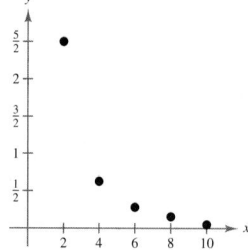

21. $y = \dfrac{5}{x}$ **23.** $y = -\dfrac{7}{10}x$ **25.** $y = \dfrac{12}{5}x$

27. $y = 205x$ **29.** $I = 0.035P$

31. Model: $y = \frac{33}{13}x$; 25.4 centimeters, 50.8 centimeters

33. $y = 0.0368x$; \$7360

35. (a) 0.05 meter   (b) $176\frac{2}{3}$ newtons

37. 39.47 pounds   **39.** $A = kr^2$ **41.** $y = \dfrac{k}{x^2}$

43. $F - \dfrac{kg}{r^2}$ **45.** $P = \dfrac{k}{V}$ **47.** $F = \dfrac{km_1m_2}{r^2}$

49. The area of a triangle is jointly proportional to its base and height.

51. The volume of a sphere varies directly as the cube of its radius.

53. Average speed is directly proportional to the distance and inversely proportional to the time.

55. $A = \pi r^2$ **57.** $y = \dfrac{28}{x}$ **59.** $F = 14rs^3$

61. $z = \dfrac{2x^2}{3y}$ **63.** ≈ 0.61 mile per hour   **65.** 506 feet

67. 1470 joules   **69.** The velocity is increased by one-third.

71. (a)

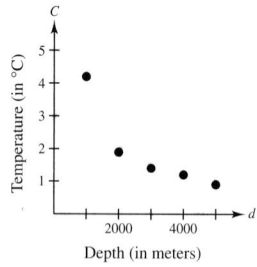

(b) Yes. $k_1 = 4200$, $k_2 = 3800$, $k_3 = 4200$, $k_4 = 4800$, $k_5 = 4500$

(c) $C = \dfrac{4300}{d}$

(d) 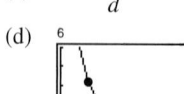   (e) ≈ 1433 meters

73. (a) 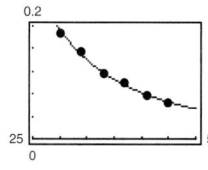   (b) 0.2857 microwatt per square centimeter

75. False. y will increase if k is positive and y will decrease if k is negative.

77. True. The closer the value of $|r|$ is to 1, the better the fit.

79. The accuracy is questionable when based on such limited data.

81. $x > 5$ **83.** $-4 < x < 5$

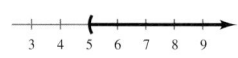

85. (a) $-\frac{5}{3}$ (b) $-\frac{7}{3}$ (c) 21   **87.** Answers will vary.

Review Exercises   *(page 326)*

1. (a)

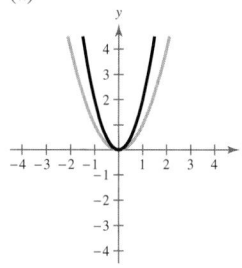

Vertical stretch

(b)

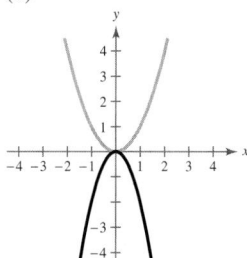

Vertical stretch and reflection in the x-axis

(c)

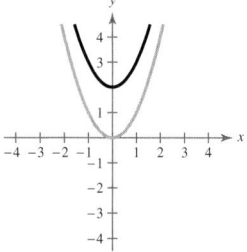

Vertical translation

(d)

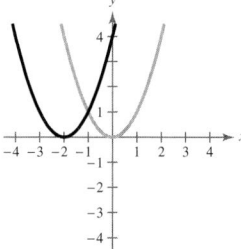

Horizontal translation

3. $g(x) = (x - 1)^2 - 1$

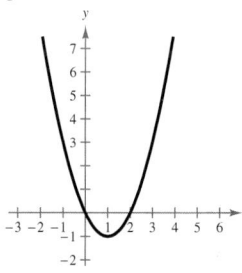

Vertex: $(1, -1)$
Axis of symmetry: $x = 1$
x-intercepts: $(0, 0)$, $(2, 0)$

5. $f(x) = (x + 4)^2 - 6$

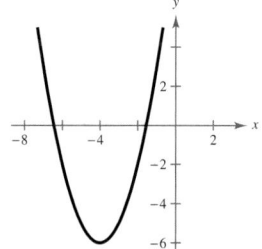

Vertex: $(-4, -6)$
Axis of symmetry: $x = -4$
x-intercepts: $\left(-4 \pm \sqrt{6}, 0\right)$

7. $f(t) = -2(t - 1)^2 + 3$

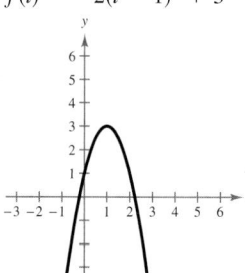

Vertex: $(1, 3)$
Axis of symmetry: $t = 1$
t-intercepts: $\left(1 \pm \dfrac{\sqrt{6}}{2}, 0\right)$

9. $h(x) = 4\left(x + \frac{1}{2}\right)^2 + 12$

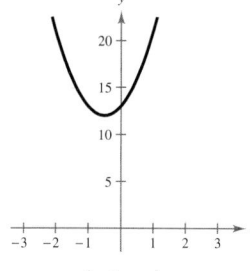

Vertex: $\left(-\frac{1}{2}, 12\right)$
Axis of symmetry: $x = -\frac{1}{2}$
No x-intercept

11. $h(x) = \left(x + \frac{5}{2}\right)^2 - \frac{41}{4}$

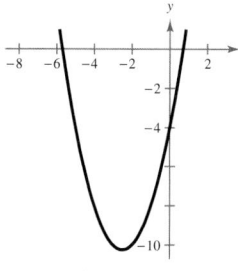

Vertex: $\left(-\frac{5}{2}, -\frac{41}{4}\right)$
Axis of symmetry: $x = -\frac{5}{2}$
x-intercepts: $\left(\dfrac{\pm\sqrt{41} - 5}{2}, 0\right)$

13. $f(x) = \frac{1}{3}\left(x + \frac{5}{2}\right)^2 - \frac{41}{12}$

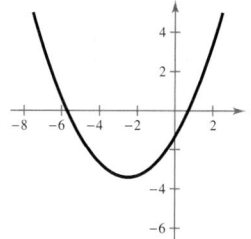

Vertex: $\left(-\frac{5}{2}, -\frac{41}{12}\right)$
Axis of symmetry: $x = -\frac{5}{2}$
x-intercepts: $\left(\dfrac{\pm\sqrt{41} - 5}{2}, 0\right)$

15. $f(x) = -\frac{1}{2}(x - 4)^2 + 1$ **17.** $f(x) = (x - 1)^2 - 4$

19. (a) $A = x\left(\dfrac{8 - x}{2}\right)$ (b) $0 < x < 8$

(c)

x	1	2	3	4	5	6
A	$\frac{7}{2}$	6	$\frac{15}{2}$	8	$\frac{15}{2}$	6

(d)

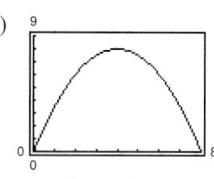

$x = 4, y = 2$

(e) $A = -\frac{1}{2}(x - 4)^2 + 8$; $x = 4, y = 2$

21. (a) $12,000; $13,750; $15,000

(b) Maximum revenue at $40; $16,000. Any price greater or less than $40 per unit will not yield as much revenue.

23. 1091 units

25.

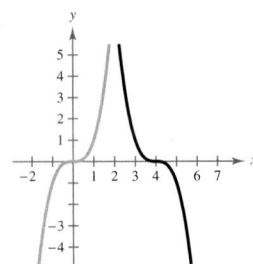

27.

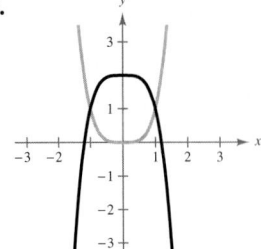

29.

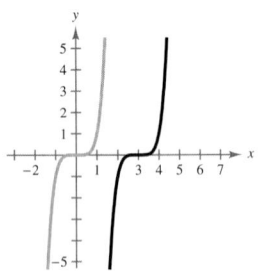

31. Falls to the left, falls to the right

33. Rises to the left, rises to the right

35. $-7, \frac{3}{2}$, odd multiplicity; turning point: 1

37. $0, \pm\sqrt{3}$, odd multiplicity; turning points: 2

39. 0, even multiplicity; $\frac{5}{3}$, odd multiplicity; turning points: 2

41. (a) Rises to the left, falls to the right   (b) -1

(c) Answers will vary.

(d)

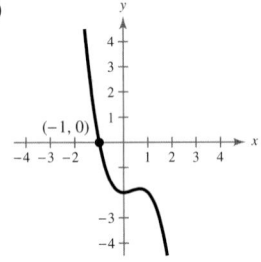

43. (a) Rises to the right, rises to the left   (b) $-3, 0, 1$

(c) Answers will vary.

(d)

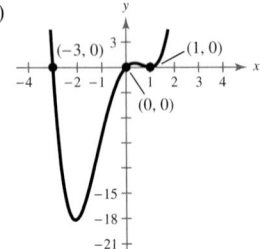

45. $[-1, 0]; \approx -0.900$

47. $[-1, 0], [1, 2]; \approx -0.200, \approx 1.772$

49. $8x + 5 + \dfrac{2}{3x - 2}$ **51.** $5x + 2$

53. $x^2 - 3x + 2 - \dfrac{1}{x^2 + 2}$

55. $6x^3 + 8x^2 - 11x - 4 - \dfrac{8}{x - 2}$

57. $2x^2 - 11x - 6$

59. (a) Yes   (b) Yes   (c) Yes   (d) No

61. (a) -421 (b) -9

63. (a) Answers will vary.

(b) $(x + 7), (x + 1)$

(c) $f(x) = (x + 7)(x + 1)(x - 4)$

(d) $-7, -1, 4$

(e)

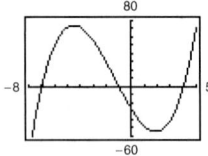

65. (a) Answers will vary.   (b) $(x + 1), (x - 4)$

(c) $f(x) = (x + 1)(x - 4)(x + 2)(x - 3)$

(d) $-2, -1, 3, 4$

(e)

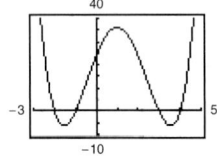

67. $0, 2$ **69.** $8, 1$ **71.** $-4, 6, \pm 2i$

73. $\pm 1, \pm 3, \pm 5, \pm 15, \pm\frac{1}{2}, \pm\frac{3}{2}, \pm\frac{5}{2}, \pm\frac{15}{2}, \pm\frac{1}{4}, \pm\frac{3}{4}, \pm\frac{5}{4}, \pm\frac{15}{4}$

75. $-1, -3, 6$ **77.** $1, 8$ **79.** $-4, 3$

81. $3x^4 - 14x^3 + 17x^2 - 42x + 24$ **83.** $4, \pm i$

85. $-3, \frac{1}{2}, 2 \pm i$ **87.** $0, 1, -5; f(x) = x(x - 1)(x + 5)$

89. $-4, 2 \pm 3i; g(x) = (x + 4)^2(x - 2 - 3i)(x - 2 + 3i)$

91. (a)

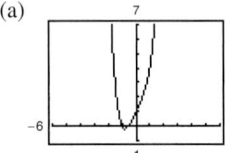

(b) Two zeros

(c) $-1, -0.54$

93. (a)

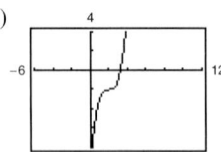

(b) One zero

(c) 3.26

95. Two or no positive real zeros, one negative real zero

97. Answers will vary.

99. (a)

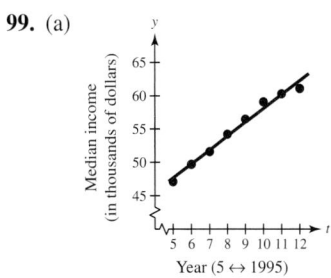

Year (5 ↔ 1995)

(b) The model is a good fit for the actual data.
101. Model: $y = \frac{8}{5}x$; 3.2 kilometers, 16 kilometers
103. A factor of 4   **105.** ≈ 2 hours, 26 minutes
107. False. A fourth-degree polynomial can have at most four zeros, and complex zeros occur in conjugate pairs.
109. Find the vertex of the quadratic function and write the function in standard form. If the leading coefficient is positive, the vertex is a minimum. If the leading coefficient is negative, the vertex is a maximum.

Chapter Test   *(page 330)*

1. (a) Reflection in the x-axis followed by a vertical translation
   (b) Horizontal translation
2. Vertex: $(-2, -1)$;
   Intercepts: $(0, 3)$, $(-3, 0)$, $(-1, 0)$
3. $y = (x - 3)^2 - 6$
4. (a) 50 feet
   (b) 5. Yes, changing the constant term results in a vertical translation of the graph and therefore changes the maximum height.
5. Rises to the left, falls to the right

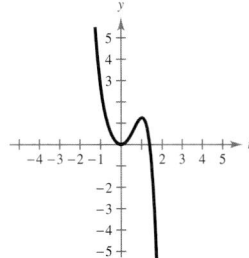

6. $3x + \dfrac{x - 1}{x^2 + 1}$ **7.** $2x^3 + 4x^2 + 3x + 6 + \dfrac{9}{x - 2}$
8. $(4x - 1)(x - \sqrt{3})(x + \sqrt{3})$;
   Solutions: $\frac{1}{4}, \pm\sqrt{3}$
9. $-2, \frac{3}{2}$ **10.** $\pm 1, -\frac{2}{3}$
11. $f(x) = x^4 - 9x^3 + 28x^2 - 30x$
12. $f(x) = x^4 - 6x^3 + 16x^2 - 24x + 16$
13. $-2, \pm\sqrt{5}i$ **14.** $-2, 4, -1 \pm \sqrt{2}i$

15. $v = 6\sqrt{s}$ **16.** $A = \dfrac{25}{6}xy$ **17.** $b = \dfrac{48}{a}$
18. $y = 476.1x + 625.5$; The model is a fairly good fit.

Problem Solving   *(page 333)*

1. (a) (i) $6, -2$ (iv) 2
       (ii) $0, -5$ (v) $1 \pm \sqrt{7}$
       (iii) $-5, 2$ (vi) $\dfrac{-3 \pm \sqrt{7}i}{2}$

(b)
(i)
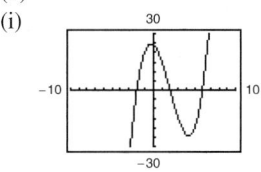
(ii)
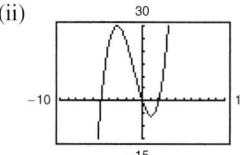

(iii)
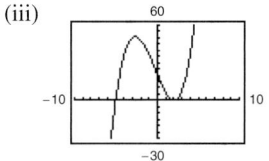
(iv)
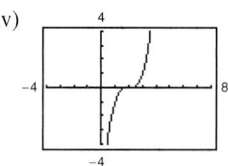

(v)
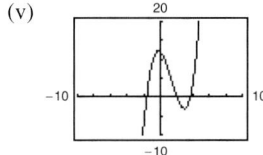
(vi)
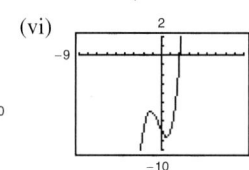

Graph (iii) touches the x-axis at $(2, 0)$, and all the other graphs pass through the x-axis at $(2, 0)$.
(c) (i) $(6, 0), (-2, 0)$ (iv) No other x-intercepts
    (ii) $(0, 0), (-5, 0)$ (v) $(-1.6, 0), (3.6, 0)$
    (iii) $(-5, 0)$ (vi) No other x-intercepts
(d) When the function has two real zeros, the results are the same. When the function has one real zero, the graph touches the x-axis at the zero. When there are no real zeros, there is no x-intercept.
3. Answers will vary.
5. 2 inches × 2 inches × 5 inches
7. False, the statement would be true if the 2 was replaced by $f(-1)$.
9. (a) $m_1 = 5$; less than   (b) $m_2 = 3$; greater than
    (c) $m_3 = 4.1$; less than   (d) $m_h = h + 4$
    (e) $m_h = 3, 5, 4.1$; The values are the same.
    (f) $m_{\tan} = 4$ since $h = 0$.
11. (a) $A(x) = \dfrac{x^2}{16} + \dfrac{1}{\pi}\left(\dfrac{100 - x}{2}\right)^2$ (b) $0 \le x \le 100$
    (c) Maximum area at $x = 0$; Minimum area at $x = 56$
    (d) Answers will vary.

CHAPTER 3

Chapter 4

Section 4.1 *(page 341)*

Vocabulary Check *(page 341)*

1. rational functions **2.** vertical asymptote

3. horizontal asymptote

---

1. (a)

x	$f(x)$	x	$f(x)$	x	$f(x)$
0.5	-2	1.5	2	5	0.25
0.9	-10	1.1	10	10	$0.\overline{1}$
0.99	-100	1.01	100	100	$0.\overline{01}$
0.999	-1000	1.001	1000	1000	$0.\overline{001}$

  (b) Vertical asymptote: $x = 1$
  Horizontal asymptote: $y = 0$
  (c) Domain: all real numbers x except $x = 1$

3. (a)

x	$f(x)$	x	$f(x)$	x	$f(x)$
0.5	-1	1.5	5.4	5	3.125
0.9	-12.79	1.1	17.29	10	$3.\overline{03}$
0.99	-147.8	1.01	152.3	100	$3.\overline{0003}$
0.999	-1498	1.001	1502	1000	3

  (b) Vertical asymptotes: $x = \pm 1$
  Horizontal asymptote: $y = 3$
  (c) Domain: all real numbers x except $x = \pm 1$

5. Domain: all real numbers x except $x = 0$
  Vertical asymptote: $x = 0$
  Horizontal asymptote: $y = 0$

7. Domain: all real numbers x except $x = 2$
  Vertical asymptote: $x = 2$
  Horizontal asymptote: $y = -1$

9. Domain: all real numbers x except $x = \pm 1$
  Vertical asymptotes: $x = \pm 1$

11. Domain: all real numbers x
  Horizontal asymptote: $y = 3$

13. d   **14.** a   **15.** c   **16.** b

17. 1   **19.** None   **21.** 6   **23.** 2

25. The domain is all real numbers x except $x = \pm 4$. There is a vertical asymptote at $x = -4$, and a horizontal asymptote at $y = 0$.

27. The domain is all real numbers x except $x = -1, 3$. There is a vertical asymptote at $x = 3$, and a horizontal asymptote at $y = 1$.

29. The domain is all real numbers x except $x = -1, \frac{1}{2}$. There is a vertical asymptote at $x = \frac{1}{2}$, and a horizontal asymptote at $y = \frac{1}{2}$.

31. The domain is all real numbers x except $x = \frac{2}{3}, 2$. There is a vertical asymptote at $x = 2$, and a horizontal asymptote at $y = 2$.

33. (a) Domain of f: all real numbers x except $x = -2$
  Domain of g: all real numbers x
  (b) $x - 2$; Vertical asymptote: none
  (c)

x	-4	-3	-2.5	-2	-1.5	-1	0
$f(x)$	-6	-5	-4.5	Undef.	-3.5	-3	-2
$g(x)$	-6	-5	-4.5	-4	-3.5	-3	-2

  (d) The functions differ only at $x - 2$, where f is undefined.

35. (a) Domain of f: all real numbers x except $x = 0, \frac{1}{2}$
  Domain of g: all real numbers x except $x = 0$
  (b) $\dfrac{1}{x}$; Vertical asymptote: $x = 0$
  (c)

x	-1	-0.5	0	0.5	2	3	4
$f(x)$	-1	-2	Undef.	Undef.	$\frac{1}{2}$	$\frac{1}{3}$	$\frac{1}{4}$
$g(x)$	-1	-2	Undef.	2	$\frac{1}{2}$	$\frac{1}{3}$	$\frac{1}{4}$

  (d) The functions differ only at $x = 0.5$, where f is undefined.

37. (a) 4   (b) Less than   (c) Greater than

39. (a) 2   (b) Greater than   (c) Less than

41. (a)

M	200	400	600	800	1000
t	0.472	0.596	0.710	0.817	0.916

M	1200	1400	1600	1800	2000
t	1.009	1.096	1.178	1.255	1.328

  (b) The model is a good fit for the experimental times.

43. (a)

  (b) $28.33 million; $170 million; $765 million
  (c) No. The function is undefined at $p = 100$.

45. (a) 333 deer, 500 deer, 800 deer   (b) 1500 deer

47. (a)

n	1	2	3	4	5	6
P	0.50	0.74	0.82	0.86	0.89	0.91

n	7	8	9	10
P	0.92	0.93	0.94	0.95

P approaches 1 as n increases.

(b) 100%

49. (a)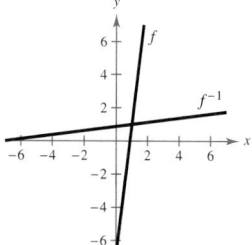

The model is a fairly good fit for the data.

(b) $1218.65 million

(c) This model has an asymptote after the year 2003, which would cause the prediction to be inaccurate.

(d) $y = 68.78t - 563$; $y = 9.329t^2 - 127.12t + 438.3$

(e) $y = \$537.5$ million; $y = \$792.6$ million. These values are well below the prediction by the given model.

(f) The quadratic model is the best fit for the data.

51. False. Polynomials do not have vertical asymptotes.

53. Sample answer: $f(x) = \dfrac{2x^2}{x^2 + 1}$

55. Sample answer: $f(x) = \dfrac{x^3}{(x + 2)(x - 1)}$

57. Answers will vary. Sample answer:

$f(x) = \dfrac{1}{x^2 + 2}; f(x) = \dfrac{1}{x - 20}$

59. $f^{-1}(x) = \dfrac{x + 7}{8}$

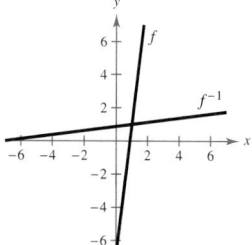

61. $x + 9 + \dfrac{42}{x - 4}$ **63.** $2x - 9 + \dfrac{34}{x + 5}$

Section 4.2   *(page 350)*

Vocabulary Check   *(page 350)*

1. slant asymptote   **2.** vertical

1. 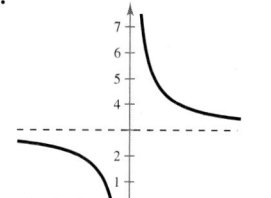   **3.**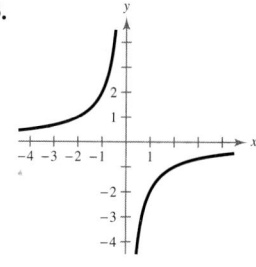

5. 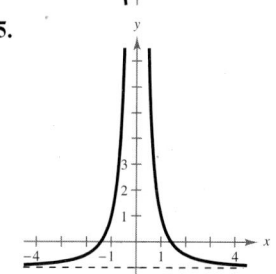   **7.**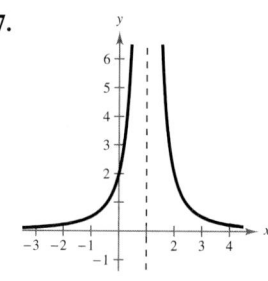

9. 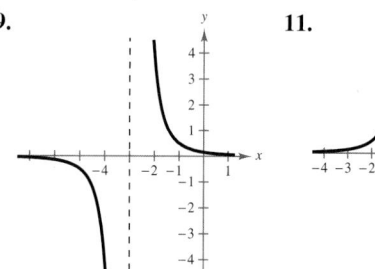   **11.**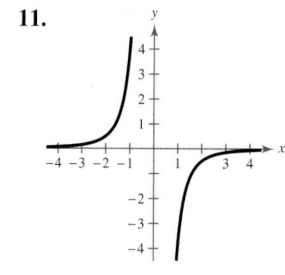

13. (a) The domain is all real numbers x except $x = -2$.
(b) y-intercept: $\left(0, \frac{1}{2}\right)$
(c) Vertical asymptote: $x = -2$
Horizontal asymptote: $y = 0$
(d)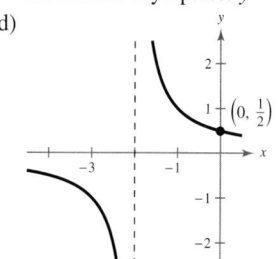

15. (a) The domain is all real numbers x except $x = -2$.
(b) y-intercept: $\left(0, \frac{1}{2}\right)$
(c) Vertical asymptote: $x = -2$
Horizontal asymptote: $y = 0$
(d)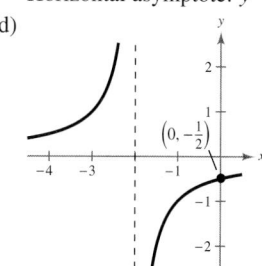

CHAPTER 4

17. (a) The domain is all real numbers x except $x = -1$.
(b) x-intercept: $\left(-\frac{5}{2}, 0\right)$
y-intercept: $(0, 5)$
(c) Vertical asymptote: $x = -1$
Horizontal asymptote: $y = 2$
(d)

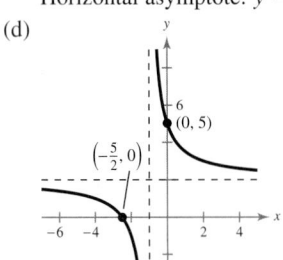

19. (a) The domain is all real numbers x except $x = -2$.
(b) x-intercept: $\left(-\frac{5}{2}, 0\right)$
y-intercept: $\left(0, \frac{5}{2}\right)$
(c) Vertical asymptote: $x = -2$
Horizontal asymptote: $y = 2$
(d)

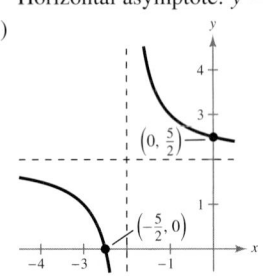

21. (a) The domain is all real numbers x.
(b) Intercept: $(0, 0)$
(c) Horizontal asymptote: $y = 1$
(d)

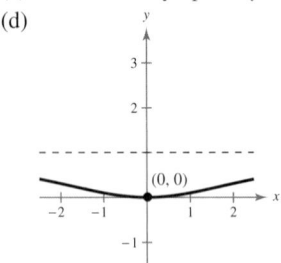

23. (a) The domain is all real numbers x except $x = \pm 3$.
(b) Intercept: $(0, 0)$
(c) Vertical asymptotes: $x = \pm 3$
Horizontal asymptote: $y = 1$
(d)

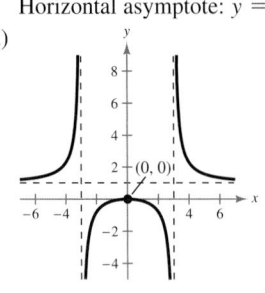

25. (a) The domain is all real numbers s.
(b) Intercept: $(0, 0)$ (c) Horizontal asymptote: $y = 0$
(d)

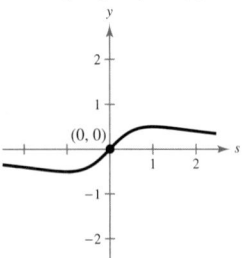

27. (a) The domain is all real numbers x except $x = 0, 4$.
(b) x-intercept: $(-1, 0)$
(c) Vertical asymptotes: $x = 0$, $x = 4$
Horizontal asymptote: $y = 0$
(d)

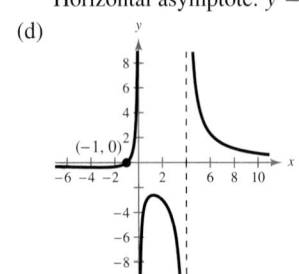

29. (a) The domain is all real numbers x except $x = -1, 2$.
(b) Intercept: $(0, 0)$
(c) Vertical asymptotes: $x = -1$, $x = 2$
Horizontal asymptote: $y = 0$
(d)

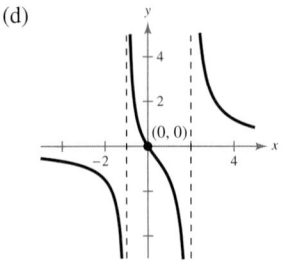

31. (a) The domain is all real numbers x except $x = \pm 2$.
(b) x-intercepts: $(1, 0)$ and $(4, 0)$
(c) Vertical asymptotes: $x = \pm 2$
Horizontal asymptote: $y = 1$
(d)

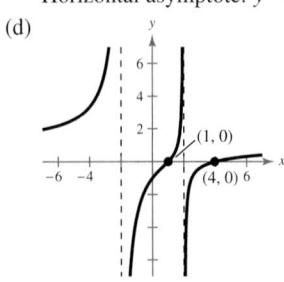

33. (a) The domain is all real numbers x except $x = -2, 7$.
   (b) Intercept: $(0, 0)$
   (c) Vertical asymptotes: $x = -2$, $x = 7$
       Horizontal asymptote: $y = 0$
   (d)

35. (a) The domain is all real numbers x except $x = \pm 1, 2$.
   (b) x-intercepts: $(3, 0)$, $\left(-\frac{1}{2}, 0\right)$; y-intercept: $\left(0, -\frac{3}{2}\right)$
   (c) Vertical asymptotes: $x = 2$, $x = \pm 1$
       Horizontal asymptote: $y = 0$
   (d)

37. (a) The domain is all real numbers x except $x = 2, -3$.
   (b) Intercept: $(0, 0)$
   (c) Vertical asymptote: $x = 2$
       Horizontal asymptote: $y = 1$
   (d)

39. (a) The domain is all real numbers x except $x = -\frac{3}{2}, 2$.
   (b) x-intercept: $\left(\frac{1}{2}, 0\right)$; y-intercept: $\left(0, \frac{1}{3}\right)$
   (c) Vertical asymptote: $x = -\frac{3}{2}$
       Horizontal asymptote: $y = 1$
   (d)

41. (a) The domain is all real numbers t except $t = -1$.
   (b) t-intercept: $(1, 0)$
       y-intercept: $(0, -1)$
   (c) Vertical asymptote: None
       Horizontal asymptote: None
   (d)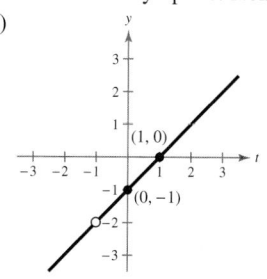

43. (a) Domain of f: all real numbers x except $x = -1$
       Domain of g: all real numbers x
   (b) $x - 1$; Vertical asymptote: none
   (c)

x	-3	-2	-1.5	-1	-0.5	0	1
$f(x)$	-4	-3	-2.5	Undef.	-1.5	-1	0
$g(x)$	-4	-3	-2.5	-2	-1.5	-1	0

   (d)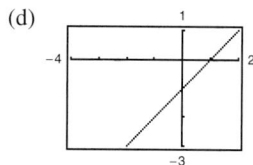

   (e) Because there are only a finite number of pixels, the graphing utility may not attempt to evaluate the function where it does not exist.

45. (a) Domain of f: all real numbers x except $x = 0, 2$
       Domain of g: all real numbers x except $x = 0$
   (b) $\dfrac{1}{x}$; Vertical asymptote: $x = 0$
   (c)

x	-0.5	0	0.5	1	1.5	2	3
$f(x)$	-2	Undef.	2	1	$\frac{2}{3}$	Undef.	$\frac{1}{3}$
$g(x)$	-2	Undef.	2	1	$\frac{2}{3}$	$\frac{1}{2}$	$\frac{1}{3}$

   (d)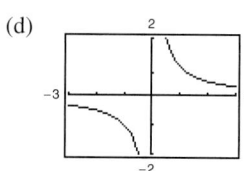

   (e) Because there are only a finite number of pixels, the graphing utility may not attempt to evaluate the function where it does not exist.

CHAPTER 4

47. (a) Domain: all real numbers x except $x = 0$
   (b) x-intercepts: $(2, 0)$, $(-2, 0)$
   (c) Vertical asymptote: $x = 0$
       Slant asymptote: $y = x$
   (d)

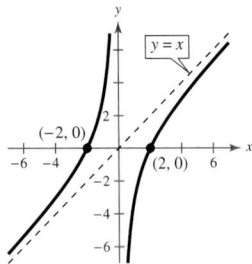

49. (a) Domain: all real numbers x except $x = 0$
   (b) No intercepts
   (c) Vertical asymptote: $x = 0$
       Slant asymptote: $y = 2x$
   (d)

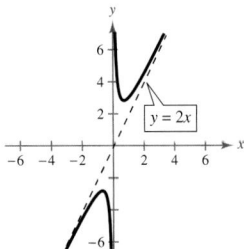

51. (a) Domain: all real numbers x except $x = 0$
   (b) No intercepts
   (c) Vertical asymptote: $x = 0$
       Slant asymptote: $y = x$
   (d)

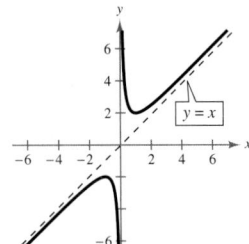

53. (a) Domain: all real numbers t except $t = -5$
   (b) Intercept: $(0, -0.2)$
   (c) Vertical asymptote: $t = -5$
       Slant asymptote: $y = -t + 5$
   (d)

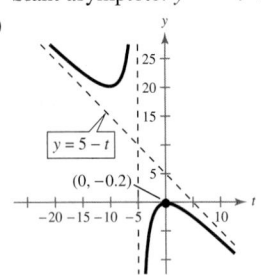

55. (a) Domain: all real numbers x except $x = \pm 1$
   (b) Intercept: $(0, 0)$
   (c) Vertical asymptotes: $x = \pm 1$
       Slant asymptote: $y = x$
   (d)

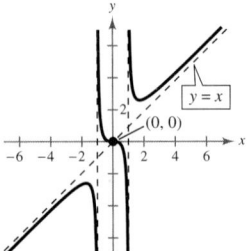

57. (a) Domain: all real numbers x except $x = 0$
   (b) No intercepts
   (c) Vertical asymptotes: $x = 0$
       Slant asymptote: $y = x + 1$
   (d)

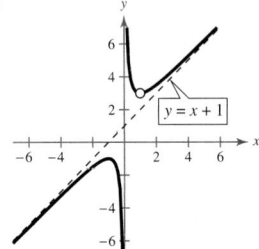

59. (a) Domain: all real numbers x except $x = 1$
   (b) y-intercept: $(0, -1)$
   (c) Vertical asymptote: $x = 1$
       Slant asymptote: $y = x$
   (d)

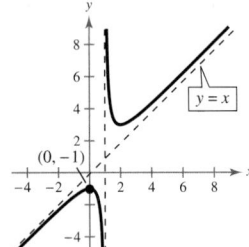

61. (a) Domain: all real numbers x except $x = -1, -2$
   (b) y-intercept: $(0, 0.5)$
       x-intercepts: $(0.5, 0)$, $(1, 0)$
   (c) Vertical asymptote: $x = -2$
       Slant asymptote: $y = 2x - 7$
   (d)

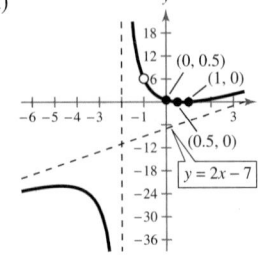

63.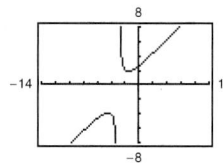

Domain: all real numbers x except $x = -3$
Vertical asymptote: $x = -3$
Slant asymptote: $y = x + 2$
$y = x + 2$

65.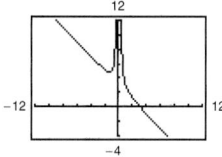

Domain: all real numbers x except $x = 0$
Vertical asymptote: $x = 0$
Slant asymptote: $y = -x + 3$
$y = -x + 3$

67. (a) $(-1, 0)$ (b) -1
69. (a) $(1, 0), (-1, 0)$ (b) ± 1
71. (a) 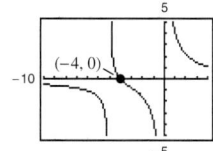 $(-4, 0)$

(b) -4
73. (a) 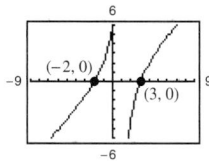 $(3, 0), (-2, 0)$

(b) $3, -2$
75. (a) Answers will vary.   (b) $[0, 950]$
(c) 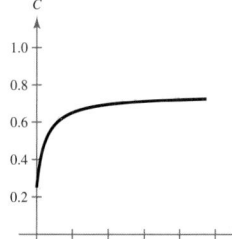   (d) Increases more slowly; 0.75

77. (a) Answers will vary.   (b) $(4, \infty)$
(c)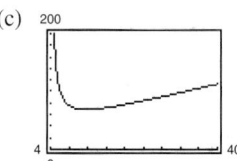

11.75 inches $\times\ 5.87$ inches

79. 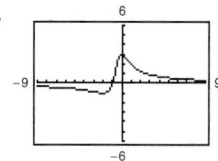   Minimum: $(-2, -1)$
Maximum: $(0, 3)$

81.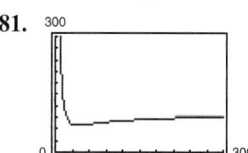

$x \approx 40.45$, or 4045 components.
83. (a) Answers will vary.
(b) Vertical asymptote: $x = 25$
Horizontal asymptote: $y = 25$
(c)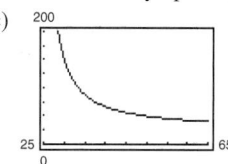

(d)

x	30	35	40	45	50	55	60
y	150	87.5	66.7	56.3	50	45.8	42.9

(e) Yes. You would expect the average speed for the round trip to be the average of the average speeds for the two parts of the trip.
(f) No. At 20 miles per hour you would use more time in one direction than is required for the round trip at an average speed of 50 miles per hour.
85. False. There are two distinct branches of the graph.
87. False. The degree of the numerator is 2 more than the degree of the denominator.
89. 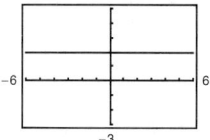   The fraction is not reduced.

91. $f(x) = \dfrac{x^2 - x - 6}{x - 2}$ **93.** $(x - 7)(x - 8)$
95. $(x - 5)(x + 2i)(x - 2i)$
97. $x \geq \frac{10}{3}$ **99.** $-3 < x < 7$

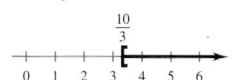   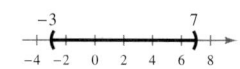

101. Answers will vary.

CHAPTER 4

Section 4.3 *(page 362)*

Vocabulary Check *(page 362)*

1. conic or conic section   **2.** circle; (h, k); r
3. parabola; directrix; focus   **4.** vertex
5. axis   **6.** ellipse; foci   **7.** major axis; center
8. minor axis   **9.** hyberbola; foci
10. transverse axis; center

1. Not shown   **2.** c   **3.** e   **4.** a   **5.** Not shown
6. h   **7.** f   **8.** b   **9.** d   **10.** g
11. Vertex: $(0, 0)$ **13.** Vertex: $(0, 0)$
      Focus: $\left(0, \frac{1}{2}\right)$ Focus: $\left(-\frac{3}{2}, 0\right)$

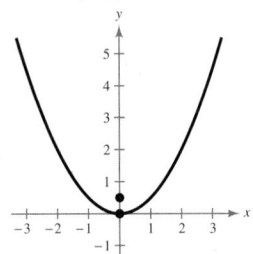
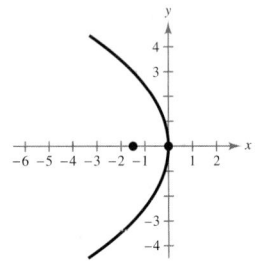

15. Vertex: $(0, 0)$
      Focus: $(0, -2)$

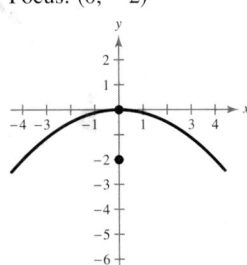

17. $y^2 = -8x$ **19.** $x^2 = -6y$ **21.** $x^2 = 4y$
23. $y^2 = -12x$ **25.** $y^2 = 9x$
27. $x^2 = \frac{3}{2}y$; Focus: $\left(0, \frac{3}{8}\right)$ **29.** $y^2 = \frac{9}{5}x$; Focus: $\left(\frac{9}{20}, 0\right)$
31. $y^2 = 6x$

33. (a)   (b) $y = \dfrac{19x^2}{51,200}$

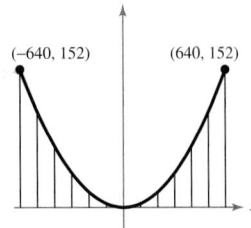

(c)

Distance, x	0	200	400	500	600
Height, y	0	$14\frac{27}{32}$	$59\frac{3}{8}$	$92\frac{99}{128}$	$133\frac{19}{32}$

35. Center: $(0, 0)$ **37.** Center: $(0, 0)$
      Vertices: $(\pm 5, 0)$ Vertices: $\left(\pm\frac{5}{3}, 0\right)$

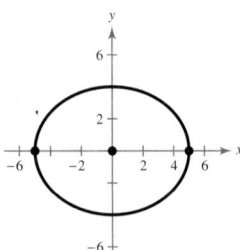
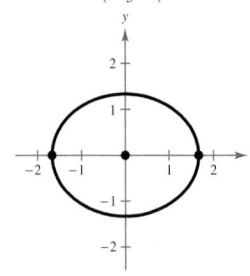

39. Center: $(0, 0)$ **41.** Center: $(0, 0)$
      Vertices: $(\pm 3, 0)$ Vertices: $(0, \pm 1)$

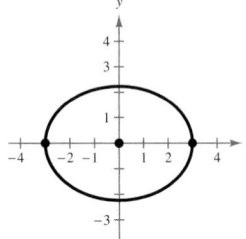
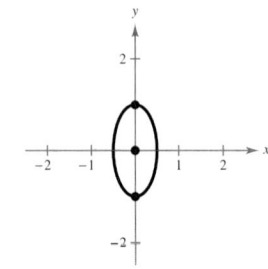

43. $\dfrac{x^2}{1} + \dfrac{y^2}{4} = 1$ **45.** $\dfrac{x^2}{4} + \dfrac{y^2}{9/4} = 1$

47. $\dfrac{x^2}{25} + \dfrac{y^2}{21} = 1$ **49.** $\dfrac{x^2}{36} + \dfrac{y^2}{11} = 1$

51. $\dfrac{21x^2}{400} + \dfrac{y^2}{25} = 1$ **53.** $\left(\pm\sqrt{5}, 0\right)$; Length of string: 6 feet

55. (a)   (b) $y = \frac{3}{4}\sqrt{400 - x^2}$

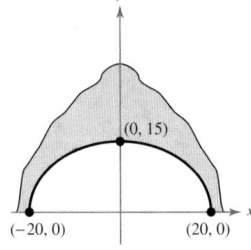

(c) Yes, with clearance of 0.52 foot.

57.   **59.**

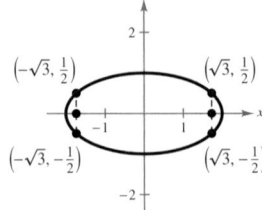
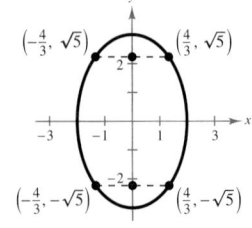

61. Center: $(0, 0)$
Vertices: $(\pm 1, 0)$

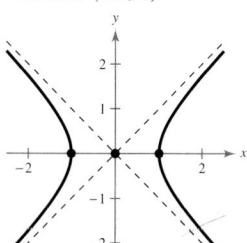

63. Center: $(0, 0)$
Vertices: $(0, \pm 1)$

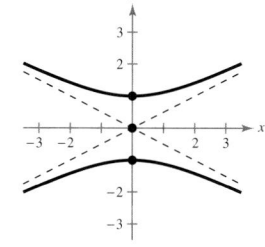

65. Center: $(0, 0)$
Vertices: $(0, \pm 5)$

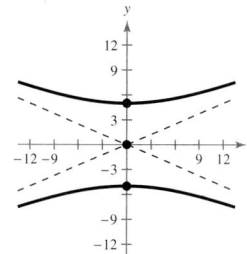

67. Center: $(0, 0)$
Vertices: $\left(0, \pm \frac{1}{2}\right)$

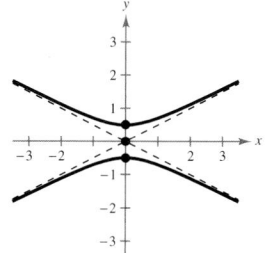

69. $\dfrac{y^2}{4} - \dfrac{x^2}{12} = 1$ **71.** $\dfrac{x^2}{1} - \dfrac{y^2}{9} = 1$

73. $\dfrac{17y^2}{1024} - \dfrac{17x^2}{64} = 1$ **75.** $\dfrac{y^2}{9} - \dfrac{x^2}{9/4} = 1$

77. (a) $\dfrac{x^2}{1} - \dfrac{y^2}{56.25} = 1$ (b) 2.405 feet **79.** 10 miles

81. False. The equation represents a hyperbola:
$$\dfrac{x^2}{144} - \dfrac{y^2}{144} = 1.$$

83. False. If the graph crossed the directrix, there would exist points nearer the directrix than the focus.

85. Circle **87.** Parabola **89.** Ellipse

91. An ellipse is a circle if $a = b$.

93.

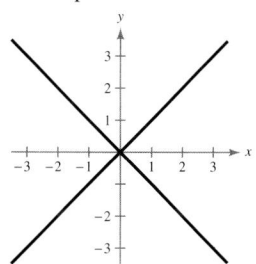

Two intersecting lines

95–97. Answers will vary.

99.

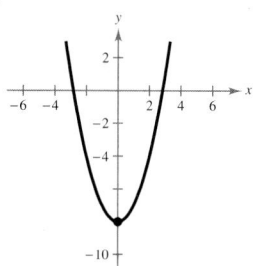

Vertex: $(0, -8)$

101.

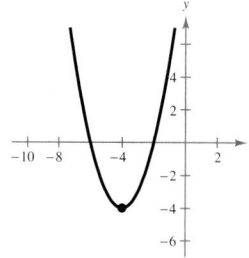

Vertex: $(-4, -4)$

103. $x^3 - 7x^2 + 17x - 15$

105. $\pm 1, \pm 2, \pm 4, \pm 5, \pm 10, \pm 20, \pm \frac{1}{2}, \pm \frac{5}{2}, \pm \frac{1}{3}, \pm \frac{2}{3},$
$\pm \frac{4}{3}, \pm \frac{5}{3}, \pm \frac{10}{3}, \pm \frac{20}{3}, \pm \frac{1}{6}, \pm \frac{5}{6}$

Section 4.4 *(page 372)*

Vocabulary Check *(page 372)*

1. g **2.** d **3.** e **4.** b
5. a **6.** f **7.** c

1. Center: $(-2, 1)$ horizontal shift two units to the left and vertical shift one unit upward

3. Center: $(1, -3)$ horizontal shift one unit to the right and vertical shift three units downward

5. Center: $(-4, -2)$ horizontal shift four units to the left and vertical shift two units downward

7. Center: $(0, 0)$ **9.** Center: $(-3, 8)$
Radius: 7 Radius: 4

11. Center: $(1, 0)$ **13.** $(x - 1)^2 + (y + 3)^2 = 1$
Radius: $\sqrt{10}$ Center: $(1, -3)$
Radius: 1

15. $\left(x + \frac{3}{2}\right)^2 + (y - 3)^2 = 1$
Center: $\left(-\frac{3}{2}, 3\right)$
Radius: 1

17. Vertex: $(1, -2)$ **19.** Vertex: $\left(5, -\frac{1}{2}\right)$
Focus: $(1, -4)$ Focus: $\left(\frac{11}{2}, -\frac{1}{2}\right)$
Directrix: $y = 0$ Directrix: $x = \frac{9}{2}$

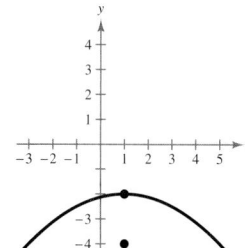

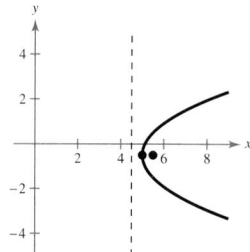

CHAPTER 4

19. (a) Domain: all real numbers x

   (b) Intercept: $(0, 0)$

   (c) Horizontal asymptote: $y = 0$

   (d)

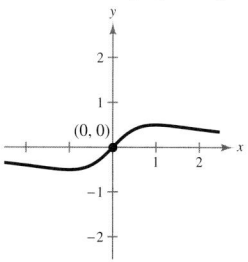

21. (a) Domain: all real numbers x

   (b) Intercept: $(0, 0)$

   (c) Horizontal asymptote: $y = -6$

   (d)

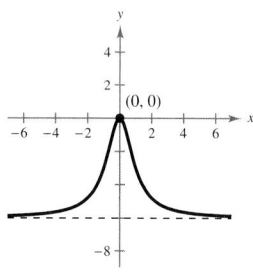

23. (a) Domain: all real numbers x except $x = 0, \frac{1}{3}$

   (b) x-intercept: $\left(\frac{3}{2}, 0\right)$

   (c) Vertical asymptote: $x = 0$

      Horizontal asymptote: $y = 2$

   (d)

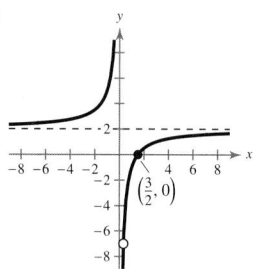

25. (a) Domain: all real numbers x

   (b) Intercept: $(0, 0)$

   (c) Slant asymptote: $y = 2x$

   (d)

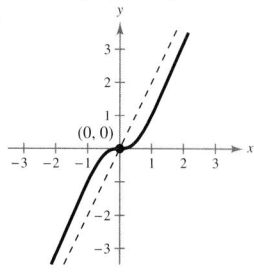

27. (a) Domain: all real numbers x except $x = -2$

   (b) y-intercept: $(0, -5)$

      x-intercept: $(2, 0)$, $(-5, 0)$

   (c) Vertical asymptote: $x = -2$

      Slant asymptote: $y = x + 1$

   (d)

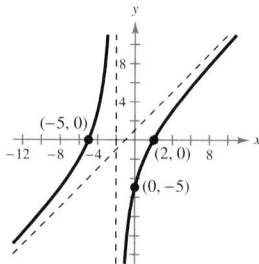

29. (a) Domain: all real numbers x except $x = \frac{4}{3}, -1$

   (b) y-intercept: $(0, -0.5)$

      x-intercepts: $\left(\frac{2}{3}, 0\right)$, $(1, 0)$

   (c) Vertical asymptote: $x = \frac{4}{3}$

      Slant asymptote: $y = x - \frac{1}{3}$

   (d)

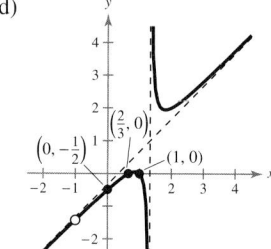

31. (a)

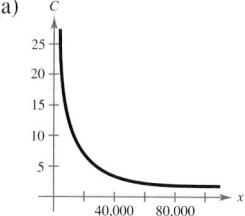

   (b) $\$100.90$, $\$10.90$, $\$1.90$

   (c) $\$0.90$ is the horizontal asymptote of the function.

33.

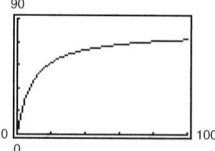

   80.3 milligrams per square decimeter per hour

35. Parabola   **37.** Hyperbola   **39.** Parabola

41. Hyperbola   **43.** $y^2 = 4x$ **45.** $y^2 = -24x$

47. $(0, 50)$ **49.** $\dfrac{x^2}{25} + \dfrac{y^2}{9} = 1$ **51.** $\dfrac{2x^2}{9} + \dfrac{y^2}{36} = 1$

53. $\dfrac{x^2}{61} + \dfrac{y^2}{25} = 1$

55. The foci should be placed 3 feet on either side of the center and have the same height as the pillars.

57. $\dfrac{y^2}{1} - \dfrac{x^2}{8} = 1$ **59.** $\dfrac{x^2}{1} - \dfrac{y^2}{4} = 1$

61. $(x - 3)^2 = -2y$

Parabola

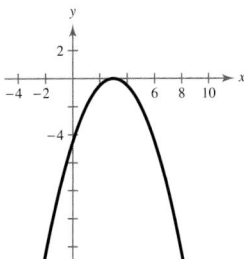

Shifted three units to the right from the origin

63. $(1, 2)$

Degenerate conic (a point)

Shifted one unit to the right and two units upward from the origin.

65. $\dfrac{(x + 5)^2}{9} + (y - 1)^2 = 1$

Ellipse

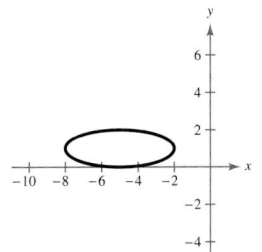

Shifted five units to the left and one unit upward from the origin.

67. $(x - 4)^2 - \dfrac{(y - 4)^2}{9} = 1$

Hyperbola

Shifted four units to the right and four units upward from the origin.

69. $(x + 6)^2 = -9(y - 4)$ **71.** $(x - 4)^2 = -8(y - 2)$

73. $\dfrac{(x - 5)^2}{25} + \dfrac{(y - 3)^2}{9} = 1$ **75.** $\dfrac{(x - 2)^2}{25} + \dfrac{y^2}{21} = 1$

77. $\dfrac{x^2}{36} - \dfrac{(y - 7)^2}{9} = 1$ **79.** $\dfrac{(x + 2)^2}{64} - \dfrac{(y - 3)^2}{36} = 1$

81. $\dfrac{5(x - 4)^2}{16} - \dfrac{5y^2}{64} = 1$ **83.** $8\sqrt{6}$ meters

85. (a) Circle
(b) $(x - 100)^2 + y^2 = 62{,}500$

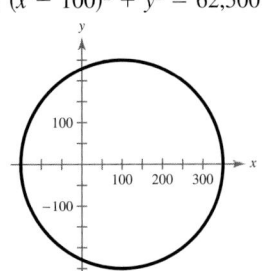

(c) Approximately 180.28 meters

87. True. See Exercise 63.

Chapter Test *(page 379)*

1. Domain: all real numbers x except $x = 4$
Vertical asymptote: $x = 4$
Horizontal asymptote: $y = 0$

2. Domain: all real numbers x
Vertical asymptote: none
Horizontal asymptote: $y = -1$

3. Domain: all real numbers x except $x = 2$
Vertical asymptote: $x = 2$
Slant asymptote: $y = x + 4$

4. x-intercepts: $(-2, 0), (2, 0)$
No y-intercept
Vertical asymptote: $x = 0$
Horizontal asymptote: $y = -1$

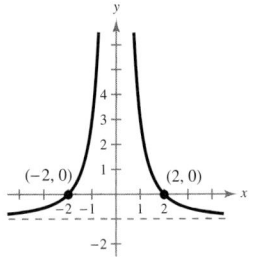

5. No x-intercepts
y-intercept: $(0, -2)$
Vertical asymptote: $x = 1$
Slant asymptote: $y = x + 1$

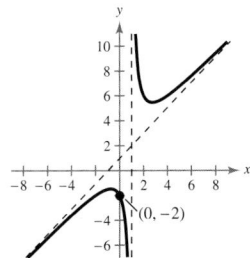

6. x-intercept: $(-1, 0)$
y-intercept: $\left(0, -\dfrac{1}{12}\right)$
Vertical asymptotes: $x = 3, x = -4$
Horizontal asymptote: $y = 0$

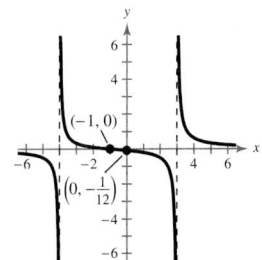

7. x-intercept: $(-1.5, 0)$
y-intercept: $(0, 0.75)$
Vertical asymptote: $x = -4$
Horizontal asymptote: $y = 2$

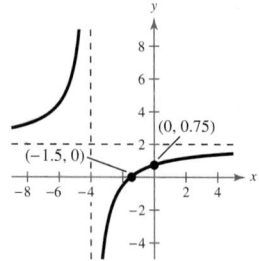

8. x-intercept: $(-3, 0)$
y-intercept: $(0, 1.5)$
No asymptotes

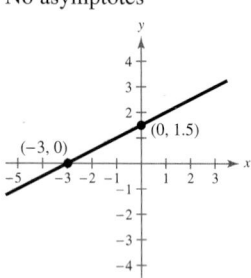

9. x-intercepts: $(-0.5, 0), (2, 0)$
y-intercept: $(0, -2)$
Vertical asymptote: $x = -1$
Slant asymptote: $y = 2x - 5$

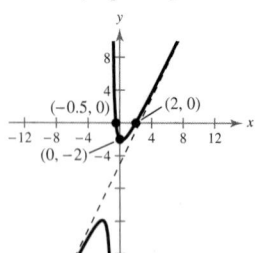

10. 6.24 inches $\times$ 12.49 inches

11. (a) Answers will vary.   (b) $A = \dfrac{x^2}{2(x-2)}$, $x > 2$

(c)

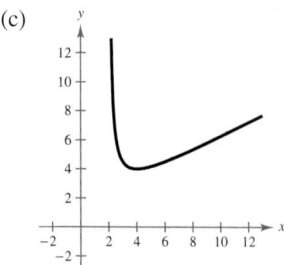

$A = 4$

12.

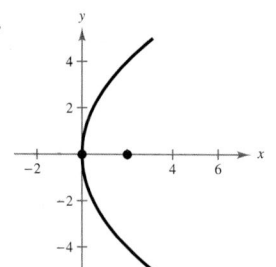

Vertex: $(0, 0)$
Focus: $(2, 0)$

13.

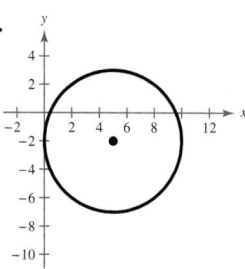

Center: $(5, -2)$

14.

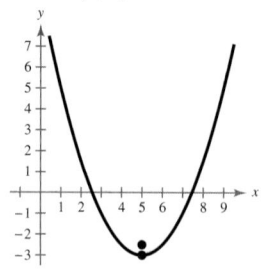

Vertex: $(5, -3)$
Focus: $\left(5, -\frac{5}{2}\right)$

15.

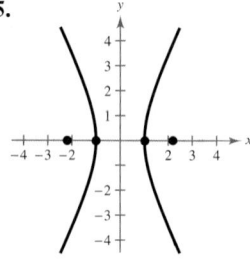

Vertices: $(\pm 1, 0)$
Foci: $\left(\pm \sqrt{5}, 0\right)$

16.

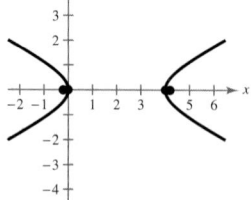

Vertices: $(0, 0), (4, 0)$
Foci: $\left(2 \pm \sqrt{5}, 0\right)$

17.

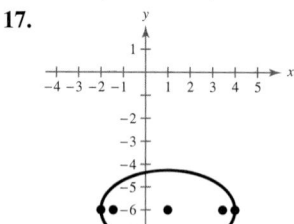

Center: $(1, -6)$
Vertices: $(4, -6), (-2, -6)$
Foci: $\left(1 \pm \sqrt{6}, -6\right)$

18. $\dfrac{(x-4)^2}{16} + \dfrac{(y-2)^2}{4} = 1$ **19.** $\dfrac{y^2}{9} - \dfrac{x^2}{4} = 1$

20. 34 meters

21. Smallest distance: $\approx 363{,}292$ kilometers
Greatest distance: $\approx 405{,}508$ kilometers

Problem Solving   *(page 381)*

1. (i) d   (ii) b   (iii) a   (iv) c

3. (a) $y_1 = 0.031x^2 - 1.57x + 21.0$

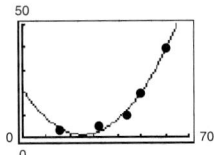

(b) $y_2 = \dfrac{-134.82}{x - 59.93}$

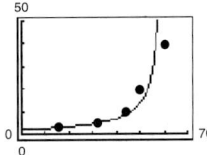

(c) The models are a good fit for the original data.

(d) $y_1 = 1.125$; $y_2 = 3.861$

The rational model is the better fit for the original data.

(e) The reciprocal model should not be used to predict the near point for a person who is 70 years old because a negative value is obtained. The quadratic model is a better fit.

5. Answers will vary.   **7.** $y^2 = 6x$; ≈ 2.04 inches

9. (a) $\dfrac{x_1}{2p}$

(b) $x^2 = 4y$, $x^2 = 8y$, $x^2 = 12y$, $x^2 = 16y$

(c) $x - y - 1 = 0$ $x - y - 2 = 0$
$x + y + 1 = 0$ $x + y + 2 = 0$

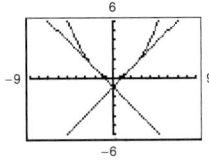

$x - y - 3 = 0$ $x - y - 4 = 0$
$x + y + 3 = 0$ $x + y + 4 = 0$

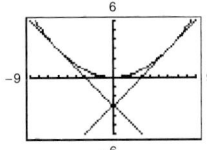 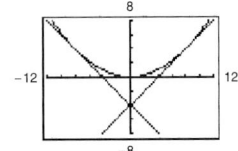

11. Proof

Chapter 5

Section 5.1   *(page 392)*

1. 946.852   **3.** 0.006   **5.** 1767.767

7. d   **8.** c   **9.** a   **10.** b

11.

x	-2	-1	0	1	2
$f(x)$	4	2	1	0.5	0.25

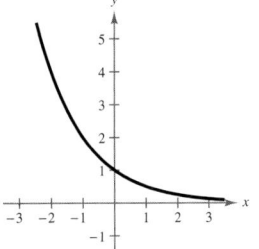

13.

x	-2	-1	0	1	2
$f(x)$	36	6	1	0.167	0.028

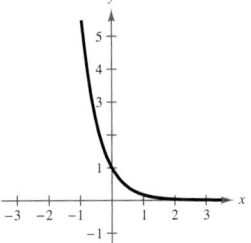

15.

x	-2	-1	0	1	2
$f(x)$	0.125	0.25	0.5	1	2

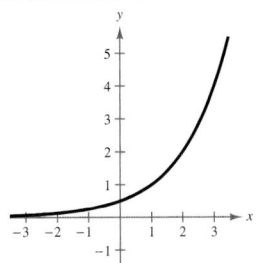

17. Shift the graph of f four units to the right.

19. Shift the graph of f five units upward.

21. Reflect the graph of f in the x-axis and y-axis and shift six units to the right.

23. 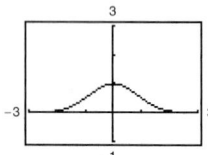　　**25.**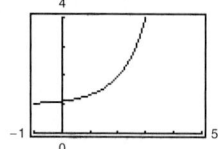

27. 0.472　　**29.** 3.857×10^{-22}　　**31.** 7166.647

33.

x	-2	-1	0	1	2
$f(x)$	0.135	0.368	1	2.718	7.389

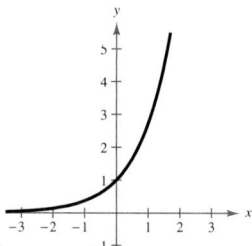

35.

x	-8	-7	-6	-5	-4
$f(x)$	0.055	0.149	0.406	1.104	3

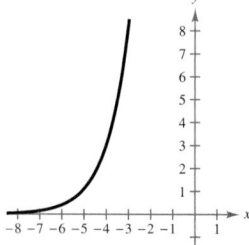

37.

x	-2	-1	0	1	2
$f(x)$	4.037	4.100	4.271	4.736	6

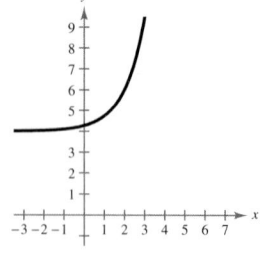

39. 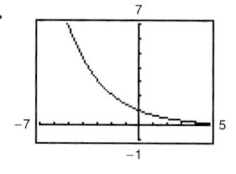　　**41.**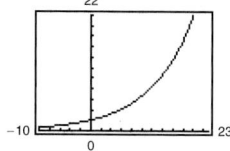

43.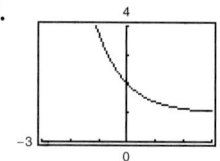

45. $x = 2$　　**47.** $x = -3$　　**49.** $x = \frac{1}{3}$　　**51.** $x = 3, -1$

53.

n	1	2	4
A	\$3200.21	\$3205.09	\$3207.57

n	12	365	Continuous
A	\$3209.23	\$3210.06	\$3210.06

55.

n	1	2	4
A	\$4515.28	\$4535.05	\$4545.11

n	12	365	Continuous
A	\$4551.89	\$4555.18	\$4555.30

57.

t	10	20	30
A	\$17,901.90	\$26,706.49	\$39,841.40

t	40	50
A	\$59,436.39	\$88,668.67

59.

t	10	20	30
A	\$22,986.49	\$44,031.56	\$84,344.25

t	40	50
A	\$161,564.86	\$309,484.08

61. \$222,822.57　　**63.** \$35.45

65. (a) $V(1) = 10,000.298$　　(b) $V(1.5) = 100,004.47$
(c) $V(2) = 1,000,059.6$

67. (a) 25 grams　　(b) 16.21 grams
(c)

69. (a)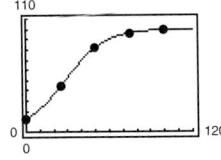

(b)

x	0	25	50	75	100
Model	12.5	44.5	81.82	96.19	99.3
Actual	12	44	81	96	99

(c) 63.14%   (d) 38 masses

71. True. As $x \to -\infty$, $f(x) \to -2$ but never reaches -2.

73. $f(x) = h(x)$ **75.** $f(x) = g(x) = h(x)$

77.   (a) $x < 0$ (b) $x > 0$

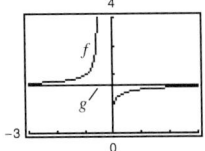

79.

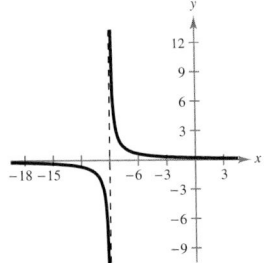

As $x \to \infty$, $f(x) \to g(x)$.
As $x \to -\infty$, $f(x) \to g(x)$.

81. $y = \pm\sqrt{25 - x^2}$

83.

85. Answers will vary.

Section 5.2   *(page 402)*

1. $4^3 = 64$ **3.** $7^{-2} = \frac{1}{49}$ **5.** $32^{2/5} = 4$
7. $36^{1/2} = 6$ **9.** $\log_5 125 = 3$ **11.** $\log_{81} 3 = \frac{1}{4}$
13. $\log_6 \frac{1}{36} = -2$ **15.** $\log_7 1 = 0$ **17.** 4   **19.** 0
21. 2   **23.** -0.0972 **25.** 1.097   **27.** 4   **29.** 1

31.   Domain: $(0, \infty)$
x-intercept: $(1, 0)$
Vertical asymptote: $x = 0$

33. 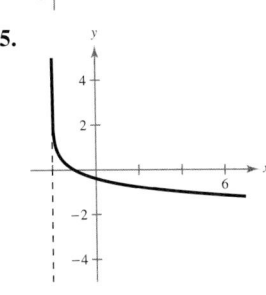   Domain: $(0, \infty)$
x-intercept: $(9, 0)$
Vertical asymptote: $x = 0$

35.   Domain: $(-2, \infty)$
x-intercept: $(-1, 0)$
Vertical asymptote: $x = -2$

37.   Domain: $(0, \infty)$
x-intercept: $(5, 0)$
Vertical asymptote: $x = 0$

39. c   **40.** f   **41.** d   **42.** e   **43.** b   **44.** a
45. $e^{-0.693\cdots} = \frac{1}{2}$ **47.** $e^{1.386\cdots} = 4$
49. $e^{5.521\cdots} = 250$ **51.** $e^0 = 1$
53. $\ln 20.0855\ldots = 3$ **55.** $\ln 1.6487\ldots = \frac{1}{2}$
57. $\ln 0.6065\ldots = -0.5$ **59.** $\ln 4 = x$ **61.** 2.913
63. -0.575 **65.** 3   **67.** $-\frac{2}{3}$

CHAPTER 5

69.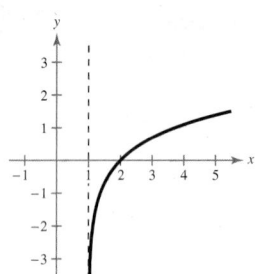

Domain: $(1, \infty)$
x-intercept: $(2, 0)$
Vertical asymptote: $x = 1$

71.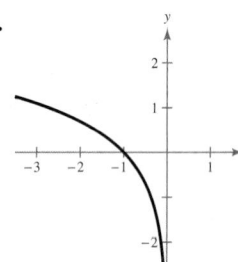

Domain: $(-\infty, 0)$
x-intercept: $(-1, 0)$
Vertical asymptote: $x = 0$

73. 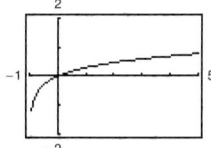   **75.**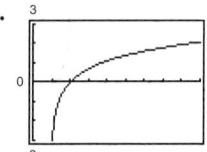

77.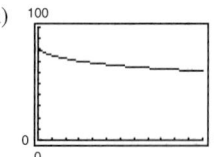

79. $x = 3$ **81.** $x = 7$ **83.** $x = 4$ **85.** $x = -5, 5$
87. (a) 30 years; 20 years   (b) \$396,234; \$301,123.20
(c) \$246,234; \$151,123.20
(d) $x = 1000$; The monthly payment must be greater than \$1000.
89. (a)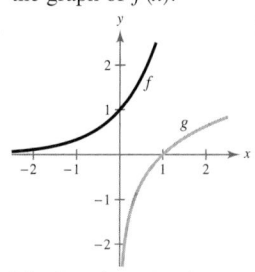
(b) 80
(c) 68.1
(d) 62.3

91. False. Reflecting $g(x)$ about the line $y = x$ will determine the graph of $f(x)$.
93. 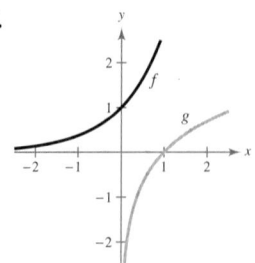   **95.**

The functions f and g
are inverses.

The functions f and g
are inverses.

97. (a)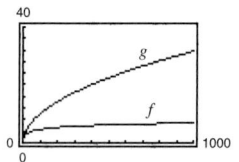

$g(x)$; The natural log function grows at a slower rate than the square root function.

(b)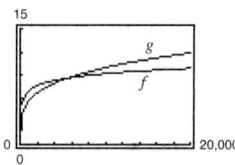

$g(x)$; The natural log function grows at a slower rate than the fourth root function.
99. (a) False   (b) True   (c) True   (d) False
101. (a)

(b) Increasing: $(1, \infty)$
Decreasing: $(0, 1)$

(c) Relative minimum: $(1, 0)$
103. 15   **105.** 4300   **107.** 1028

Section 5.3   (page 409)

Vocabulary Check   (page 409)

1. change-of-base   **2.** $\dfrac{\log x}{\log a} = \dfrac{\ln x}{\ln a}$

3. c   **4.** a   **5.** b

---

1. (a) $\dfrac{\log x}{\log 5}$ (b) $\dfrac{\ln x}{\ln 5}$ **3.** (a) $\dfrac{\log x}{\log \frac{1}{5}}$ (b) $\dfrac{\ln x}{\ln \frac{1}{5}}$

5. (a) $\dfrac{\log \frac{3}{10}}{\log x}$ (b) $\dfrac{\ln \frac{3}{10}}{\ln x}$

7. (a) $\dfrac{\log x}{\log 2.6}$ (b) $\dfrac{\ln x}{\ln 2.6}$

9. 1.771   **11.** -2.000 **13.** -0.417 **15.** 2.633
17. $\frac{3}{2}$ **19.** $-3 - \log_5 2$ **21.** $6 + \ln 5$ **23.** 2
25. $\frac{3}{4}$ **27.** 2.4   **29.** -9 is not in the domain of $\log_3 x$.
31. 4.5   **33.** $-\frac{1}{2}$ **35.** 7   **37.** 2
39. $\log_4 5 + \log_4 x$ **41.** $4 \log_8 x$ **43.** $1 - \log_5 x$
45. $\frac{1}{2} \ln z$ **47.** $\ln x + \ln y + 2 \ln z$
49. $\ln z + 2 \ln(z - 1)$ **51.** $\frac{1}{2} \log_2(a - 1) - 2 \log_2 3$
53. $\frac{1}{3} \ln x - \frac{1}{3} \ln y$ **55.** $4 \ln x + \frac{1}{2} \ln y - 5 \ln z$
57. $2 \log_5 x - 2 \log_5 y - 3 \log_5 z$
59. $\dfrac{3}{4} \ln x + \dfrac{1}{4}\ln(x^2 + 3)$ **61.** $\ln 3x$ **63.** $\log_4 \dfrac{z}{y}$

65. $\log_2(x + 4)^2$ **67.** $\log_3 \sqrt[4]{5x}$ **69.** $\ln \dfrac{x}{(x + 1)^3}$

71. $\log \dfrac{xz^3}{y^2}$ **73.** $\ln \dfrac{x}{(x^2 - 4)^4}$ **75.** $\ln \sqrt[3]{\dfrac{x(x + 3)^2}{x^2 - 1}}$

77. $\log_8 \dfrac{\sqrt[3]{y(y + 4)^2}}{y - 1}$

79. $\log_2 \frac{32}{4} = \log_2 32 - \log_2 4$; Property 2

81. $\beta = 10(\log I + 12)$; 60 dB   **83.** ≈ 3

85. $y = 256.24 - 20.8 \ln x$

87. False. $\ln 1 = 0$ **89.** False. $\ln(x - 2) \neq \ln x - \ln 2$

91. False. $u = v^2$ **93.** Answers will vary.

95. $f(x) = \dfrac{\log x}{\log 2} = \dfrac{\ln x}{\ln 2}$ **97.** $f(x) = \dfrac{\log x}{\log \frac{1}{2}} = \dfrac{\ln x}{\ln \frac{1}{2}}$

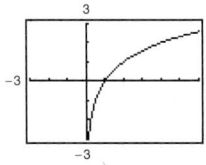   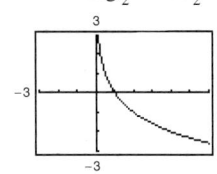

99. $f(x) = \dfrac{\log x}{\log 11.8} = \dfrac{\ln x}{\ln 11.8}$

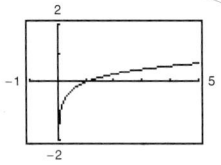

101. $f(x) = h(x)$; Property 2

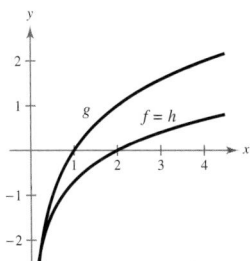

103. $\dfrac{3x^4}{2y^3}$, $x \neq 0$ **105.** 1, $x \neq 0$, $y \neq 0$

107. $-1, \dfrac{1}{3}$ **109.** $\dfrac{-1 \pm \sqrt{97}}{6}$

Section 5.4   *(page 419)*

Vocabulary Check   *(page 419)*

1. solve

2. (a) $x = y$ (b) $x = y$ (c) x (d) x

3. extraneous

1. (a) Yes   (b) No

3. (a) No   (b) Yes   (c) Yes, approximate

5. (a) Yes, approximate   (b) No   (c) Yes

7. (a) No   (b) Yes   (c) Yes, approximate

9. 2   **11.** -5 **13.** 2   **15.** $\ln 2 \approx 0.693$

17. $e^{-1} \approx 0.368$ **19.** 64   **21.** $(3, 8)$ **23.** $(9, 2)$

25. $2, -1$ **27.** $\approx 1.618, \approx -0.618$ **29.** $\dfrac{\ln 5}{\ln 3} \approx 1.465$

31. $\ln 5 \approx 1.609$ **33.** $\ln 28 \approx 3.332$

35. $\dfrac{\ln 80}{2 \ln 3} \approx 1.994$ **37.** 2   **39.** 4

41. $3 - \dfrac{\ln 565}{\ln 2} \approx -6.142$ **43.** $\dfrac{1}{3} \log\left(\dfrac{3}{2}\right) \approx 0.059$

45. $1 + \dfrac{\ln 7}{\ln 5} \approx 2.209$ **47.** $\dfrac{\ln 12}{3} \approx 0.828$

49. $-\ln \dfrac{3}{5} \approx 0.511$ **51.** 0   **53.** $\dfrac{\ln \frac{8}{3}}{3 \ln 2} + \dfrac{1}{3} \approx 0.805$

55. $\ln 5 \approx 1.609$ **57.** $\ln 4 \approx 1.386$

59. $2 \ln 75 \approx 8.635$ **61.** $\dfrac{1}{2} \ln 1498 \approx 3.656$

63. $\dfrac{\ln 4}{365 \ln\left(1 + \frac{0.065}{365}\right)} \approx 21.330$

65. $\dfrac{\ln 2}{12 \ln\left(1 + \frac{0.10}{12}\right)} \approx 6.960$

67. 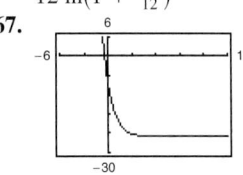   **69.**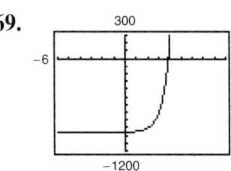

-0.427 3.847

71. 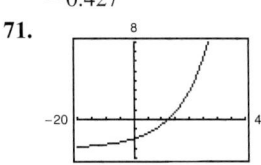   **73.**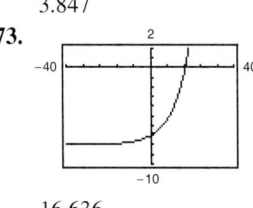

12.207 16.636

75. $e^{-3} \approx 0.050$ **77.** $\dfrac{e^{2.4}}{2} \approx 5.512$ **79.** $1,000,000$

81. $\dfrac{e^{10/3}}{5} \approx 5.606$ **83.** $e^2 - 2 \approx 5.389$

85. $e^{-2/3} \approx 0.513$ **87.** $2(3^{11/6}) \approx 14.988$

89. No solution   **91.** $1 + \sqrt{1 + e} \approx 2.928$

93. No solution   **95.** 7   **97.** $\dfrac{-1 + \sqrt{17}}{2} \approx 1.562$

99. 2   **101.** $\dfrac{725 + 125\sqrt{33}}{8} \approx 180.384$

103.

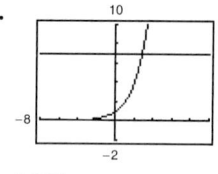

2.807

105.

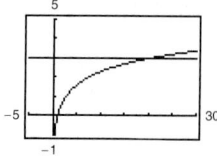

20.086

107. (a) 8.2 years   (b) 12.9 years

109. (a) 1426 units   (b) 1498 units

111. (a)

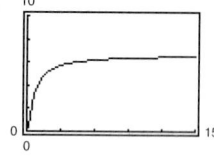

(b) $V = 6.7$; The yield will approach 6.7 million cubic feet per acre.

(c) 29.3 years

113. 2001

115. (a) $y = 100$ and $y = 0$; The range falls between 0% and 100%.

(b) Males: 69.71 inches   Females: 64.51 inches

117. (a)

x	0.2	0.4	0.6	0.8	1.0
y	162.6	78.5	52.5	40.5	33.9

(b)

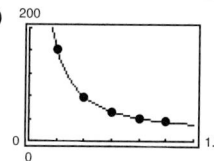

The model appears to fit the data well.

(c) 1.2 meters

(d) No. According to the model, when the number of g's is less than 23, x is between 2.276 meters and 4.404 meters, which isn't realistic in most vehicles.

119. $\log_b uv = \log_b u + \log_b v$
True by Property 1 in Section 5.3.

121. $\log_b(u - v) = \log_b u - \log_b v$
False.
$1.95 \approx \log(100 - 10) \neq \log 100 - \log 10 = 1$

123. Yes. See Exercise 93.

125. Yes. Time to double: $t = \dfrac{\ln 2}{r}$;

Time to quadruple: $t = \dfrac{\ln 4}{r} = 2\left(\dfrac{\ln 2}{r}\right)$

127. $4|x|y^2\sqrt{3y}$ **129.** $5\sqrt[3]{3}$

131.

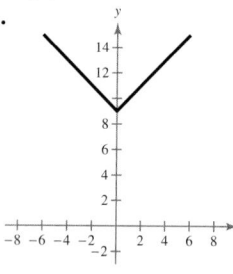

133.

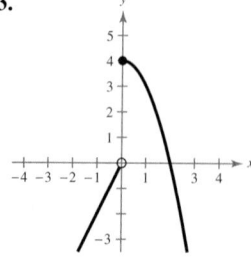

135. 1.226   **137.** -5.595

Section 5.5   (page 430)

1. c   **2.** e   **3.** b   **4.** a   **5.** d   **6.** f

	Initial Investment	Annual % Rate	Time to Double	Amount After 10 years
7.	$1000	3.5%	19.8 yr	$1419.07
9.	$750	8.9438%	7.75 yr	$1834.36
11.	$500	11.0%	6.3 yr	$1505.00
13.	$6376.28	4.5%	15.4 yr	$10,000.00

15. $112,087.09

17. (a) 6.642 years   (b) 6.330 years
(c) 6.302 years   (d) 6.301 years

19.

r	2%	4%	6%	8%	10%	12%
t	54.93	27.47	18.31	13.73	10.99	9.16

21.

r	2%	4%	6%	8%	10%	12%
t	55.48	28.01	18.85	14.27	11.53	9.69

23.

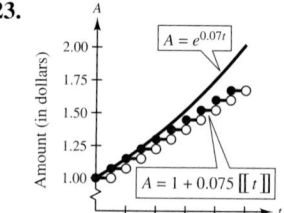

Continuous compounding

	Half-life (years)	Initial Quantity	Amount After 1000 Years
25.	1599	10 g	6.48 g
27.	5715	2.26 g	2 g
29.	24,100	2.16 g	2.1 g

31. $y = e^{0.7675x}$ **33.** $y = 5e^{-0.4024x}$

35. (a) Decreasing due to the negative exponent.
(b) 2000: population of 2430 thousand
2003: population of 2408.95 thousand
(c) 2018

37. $k = 0.2988$; $\approx 5,309,734$ hits   **39.** 3.15 hours

41. (a) $\approx 12,180$ years old   (b) ≈ 4797 years old

43. (a) $V = -6394t + 30,788$ (b) $V = 30,788e^{-0.268t}$

(c)

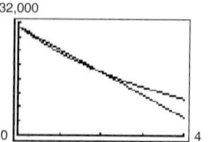

The exponential model depreciates faster.

(d)

t	1	3
$V = -6394t + 30,788$	24,394	11,606
$V = 30,788e^{-0.268t}$	23,550	13,779

(e) Answers will vary.

45. (a) $S(t) = 100(1 - e^{-0.1625t})$

(b) 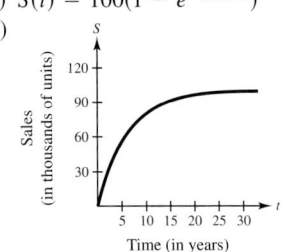   (c) 55,625

47. (a) 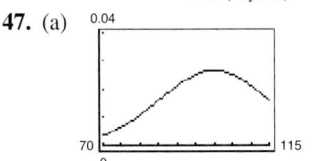   (b) 100

49. (a) 203 animals   (b) 13 years

(c) 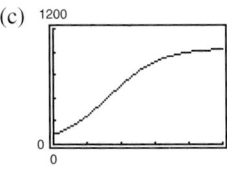   Horizontal asymptotes: $y = 0$, $y = 1000$. The population size will approach 1000 as time increases.

51. (a) $10^{7.9} \approx 79,432,823$ (b) $10^{8.3} \approx 199,526,231$

(c) $10^{4.2} \approx 15,849$

53. (a) 20 decibels   (b) 70 decibels

(c) 40 decibels   (d) 120 decibels

55. 95%   **57.** 4.64   **59.** 1.58×10^{-6} moles per liter

61. $10^{5.1}$ **63.** 3:00 A.M.

65. (a) 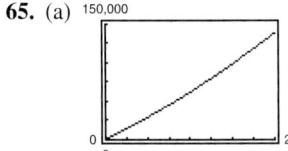   (b) ≈ 21 years; Yes

67. False. The domain can be the set of real numbers for a logistic growth function.

69. False. The graph of $f(x)$ is the graph of $g(x)$ shifted upward five units.

71. (a) Logarithmic   (b) Logistic   (c) Exponential

(d) Linear   (e) None of the above   (f) Exponential

73. (a) 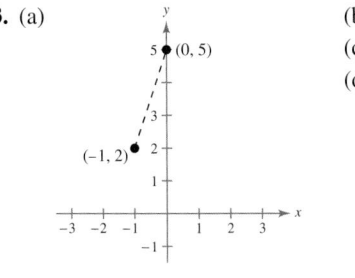   (b) $\sqrt{10}$

(c) $\left(-\frac{1}{2}, \frac{7}{2}\right)$

(d) 3

75. (a) 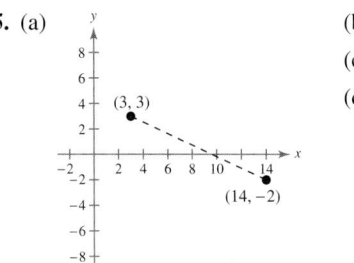   (b) $\sqrt{146}$

(c) $\left(\frac{17}{2}, \frac{1}{2}\right)$

(d) $-\frac{5}{11}$

77. (a) 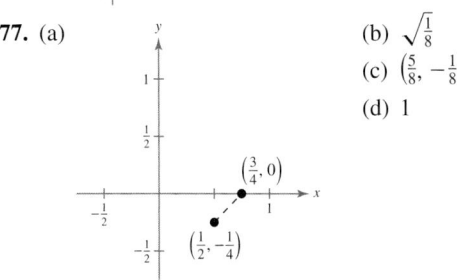   (b) $\sqrt{\frac{1}{8}}$

(c) $\left(\frac{5}{8}, -\frac{1}{8}\right)$

(d) 1

79.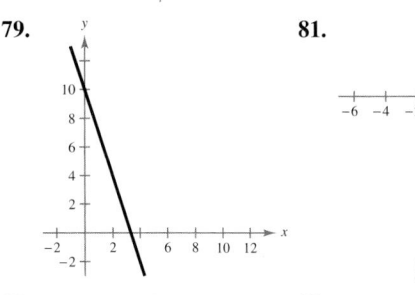

81.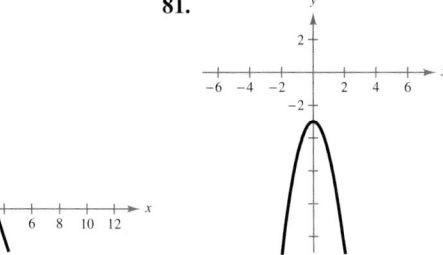

83.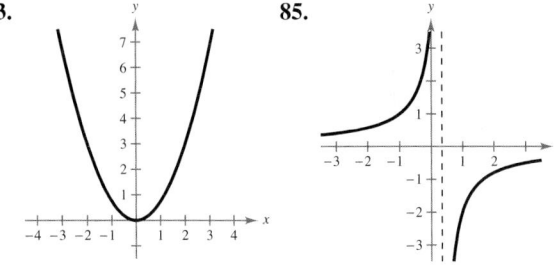

85.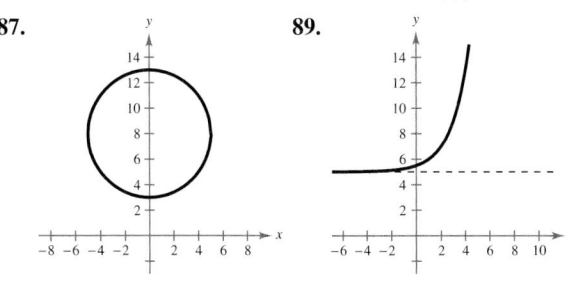

87.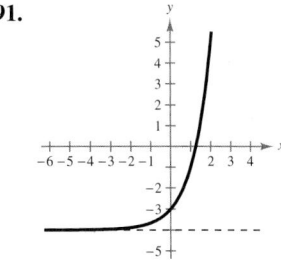

89.

91.

93. Answers will vary.

CHAPTER 5

Review Exercises   *(page 437)*

1. 76.699   **3.** 0.337   **5.** 1456.529

7. c   **8.** d   **9.** a   **10.** b

11. Shift the graph of *f* one unit to the right.

13. Reflect *f* in the *x*-axis and shift two units to the left.

15.

x	−1	0	1	2	3
f(x)	8	5	4.25	4.063	4.016

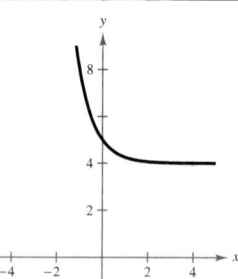

17.

x	−2	−1	0	1	2
f(x)	−0.377	−1	−2.65	−7.023	−18.61

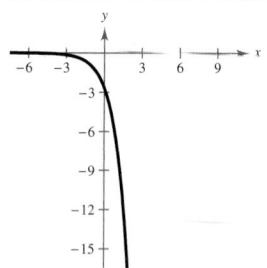

19.

x	−1	0	1	2	3
f(x)	4.008	4.04	4.2	5	9

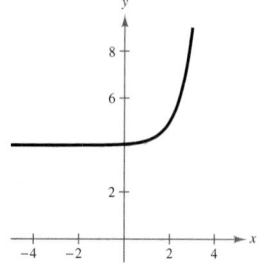

21.

x	−2	−1	0	1	2
f(x)	3.25	3.5	4	5	7

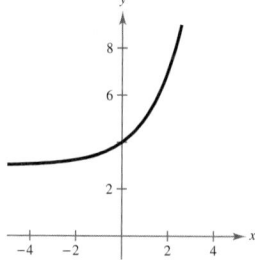

23. $x = -4$ **25.** $x = \frac{22}{5}$ **27.** 2980.958   **29.** 0.183

31.

x	−2	−1	0	1	2
h(x)	2.72	1.65	1	0.61	0.37

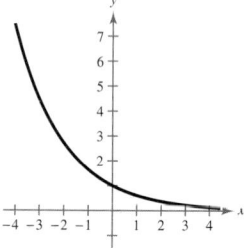

33.

x	−3	−2	−1	0	1
f(x)	0.37	1	2.72	7.39	20.09

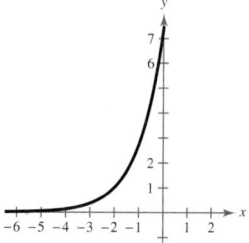

35.

n	1	2	4	12
A	$6569.98	$6635.43	$6669.46	$6692.64

n	365	Continuous
A	$6704.00	$6704.39

37. (a) 0.154   (b) 0.487   (c) 0.811

39. (a) $1,069,047.14   (b) 7.9 years

41. $\log_4 64 = 3$ **43.** $\ln 2.2255 \ldots = 0.8$

45. 3   **47.** −3   **49.** $x = 7$ **51.** $x = -5$

53. Domain: $(0, \infty)$
x-intercept: $(1, 0)$
Vertical asymptote: $x = 0$

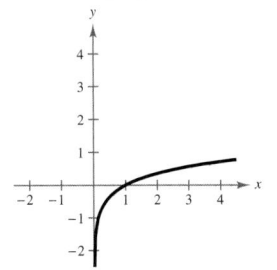

55. Domain: $(0, \infty)$
x-intercept: $(3, 0)$
Vertical asymptote: $x = 0$

57. Domain: $(-5, \infty)$
x-intercept: $(9995, 0)$
Vertical asymptote: $x = -5$

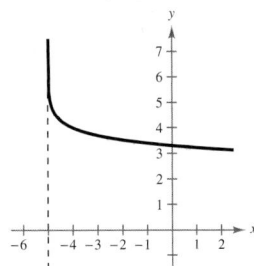

59. 3.118 **61.** -12 **63.** 2.034

65. Domain: $(0, \infty)$
x-intercept: $(e^{-3}, 0)$
Vertical asymptote: $x = 0$

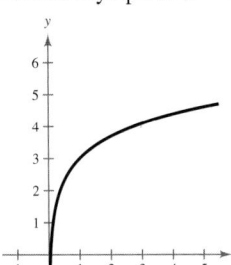

67. Domain: $(-\infty, 0), (0, \infty)$
x-intercept: $(\pm 1, 0)$
Vertical asymptote: $x = 0$

69. 53.4 inches **71.** 1.585 **73.** -2.322
75. $\log 2 + 2\log 3 \approx 1.255$ **77.** $2 \ln 2 + \ln 5 \approx 2.996$
79. $1 + 2\log_5 x$ **81.** $1 + \log_3 2 - \frac{1}{3}\log_3 x$
83. $2 \ln x + 2 \ln y + \ln z$ **85.** $\ln(x + 3) - \ln x - \ln y$
87. $\log_2 5x$ **89.** $\ln \dfrac{x}{\sqrt[4]{y}}$ **91.** $\log_8 y^7 \sqrt[3]{x + 4}$

93. $\ln \dfrac{\sqrt{2x - 1}}{(x + 1)^2}$

95. (a) $0 \le h < 18,000$

(b)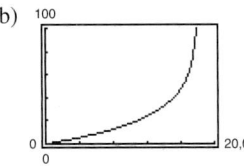

Vertical asymptote:
$h = 18,000$

(c) The plane is climbing at a slower rate, so the time required increases.
(d) 5.46 minutes
97. 3 **99.** $\ln 3 \approx 1.099$ **101.** 16
103. $e^4 \approx 54.598$ **105.** $\ln 12 \approx 2.485$ **107.** $x = 1, 3$
109. $\dfrac{\ln 22}{\ln 2} \approx 4.459$ **111.** $\dfrac{\ln 17}{\ln 5} \approx 1.760$
113. $\ln 2 \approx 0.693, \ln 5 \approx 1.609$
115. 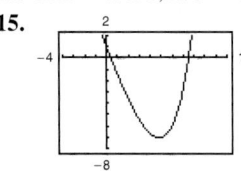 **117.**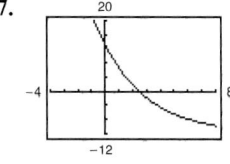

7.480; 0.392   2.447
119. $\frac{1}{3}e^{8.2} \approx 1213.650$ **121.** $\frac{1}{4}e^{7.5} \approx 452.011$
123. $3e^2 \approx 22.167$ **125.** $e^4 - 1 \approx 53.598$
127. No solution **129.** 0.900
131. 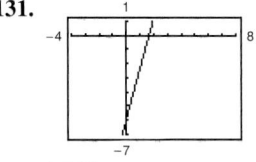 **133.**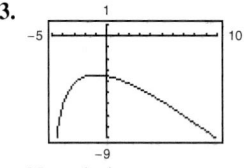

1.643   No solution
135. 15.2 years **137.** e **138.** b **139.** f
140. d **141.** a **142.** c **143.** $y = 2e^{0.1014x}$
145. 2008 **147.** (a) 13.8629% (b) \$11,486.98
149. (a) 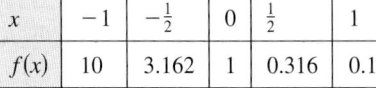 (b) 71

151. $10^{-3.5}$ watt per square centimeter
153. True by the inverse properties
155. b and d are negative.
a and c are positive.
Answers will vary.

Chapter Test *(page 441)*

1. 1123.690 **2.** 687.291 **3.** 0.497 **4.** 22.198
5.

x	-1	$-\frac{1}{2}$	0	$\frac{1}{2}$	1
$f(x)$	10	3.162	1	0.316	0.1

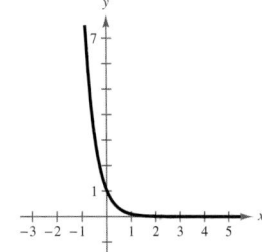

6.

x	−1	0	1	2	3
f(x)	−0.005	−0.028	−0.167	−1	−6

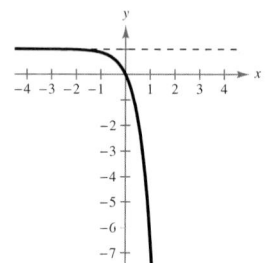

7.

x	−1	−$\frac{1}{2}$	0	$\frac{1}{2}$	1
f(x)	0.865	0.632	0	−1.718	−6.389

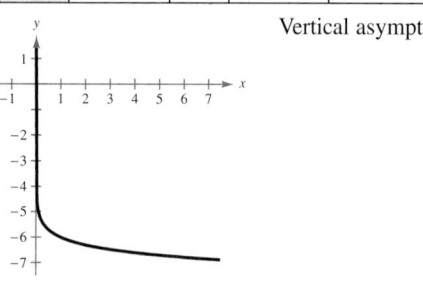

8. (a) −0.89 (b) 9.2

9.

x	$\frac{1}{2}$	1	$\frac{3}{2}$	2	4
f(x)	−5.699	−6	−6.176	−6.301	−6.602

Vertical asymptote: $x = 0$

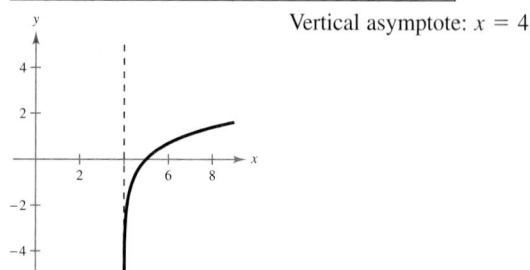

10.

x	5	7	9	11	13
f(x)	0	1.099	1.609	1.946	2.197

Vertical asymptote: $x = 4$

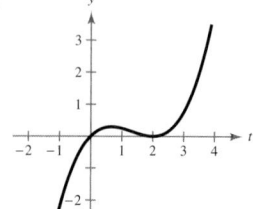

11.

x	−5	−3	−1	0	1
f(x)	1	2.099	2.609	2.792	2.946

Vertical asymptote: $x = -6$

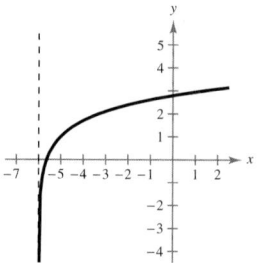

12. 1.945 **13.** 0.115 **14.** 1.328

15. $\log_2 3 + 4 \log_2 |a|$ **16.** $\ln 5 + \frac{1}{2}\ln x - \ln 6$

17. $(\log 7 + 2 \log x) - (\log y + 3 \log z)$

18. $\log_3 13y$ **19.** $\ln \dfrac{x^4}{y^4}$ **20.** $\ln\dfrac{x^2(x - 5)}{y^3}$

21. $x = -2$ **22.** $x = \dfrac{\ln 44}{-5} \approx -0.757$

23. $\dfrac{\ln 197}{4} \approx 1.321$ **24.** $e^{1/2} \approx 1.649$

25. $e^{-11/4} \approx 0.0639$ **26.** $\frac{800}{501} \approx 1.597$

27. $y = 2745e^{0.1570x}$ **28.** 55%

29. (a)

x	$\frac{1}{4}$	1	2	4	5	6
H	58.720	75.332	86.828	103.43	110.59	117.38

(b) 103 centimeters; 103.43 centimeters

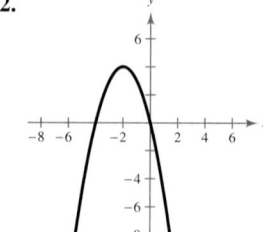

Cumulative Test for Chapters 3–5 *(page 442)*

1. $y = -\frac{3}{4}(x + 8)^2 + 5$

2.

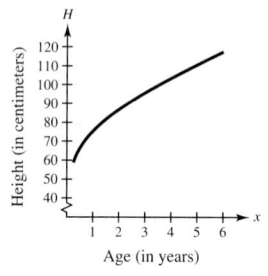

3.

4.

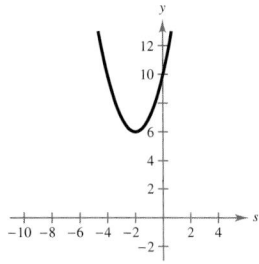

5. $-2, \pm 2i$ **6.** $-7, 0, 3$ **7.** $3x - 2 - \dfrac{3x - 2}{2x^2 + 1}$

8. $2x^3 - x^2 + 2x - 10 + \dfrac{25}{x + 2}$ **9.** 1.20

10. $x^4 + 3x^3 - 11x^2 + 9x + 70$

11. Domain: all real numbers x except $x = 3$

Vertical asymptote: $x = 3$

Horizontal asymptote: $y = 2$

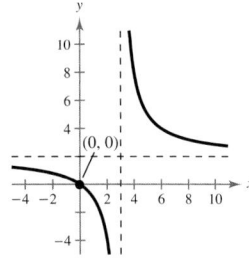

12. Domain: all real numbers x except $x = 5$

Vertical asymptote: $x = 5$

Slant asymptote: $y = 4x + 20$

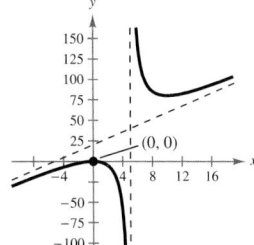

13. Intercept: $(0, 0)$

Vertical asymptotes: $x = \pm 3$

Horizontal asymptote: $y = 0$

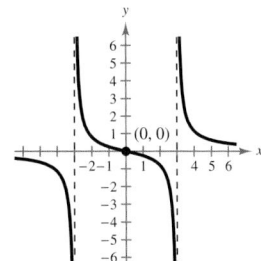

14. y-intercept: $(0, -1)$

x-intercept: $(1, 0)$

Horizontal asymptote: $y = 1$

Vertical asymptote: $x = -1$

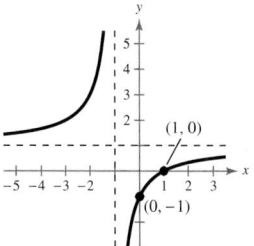

15. y-intercepts: $(0, 6)$

x-intercepts: $(-2, 0), (-3, 0)$

Slant asymptote: $y = x + 4$

Vertical asymptote: $x = -1$

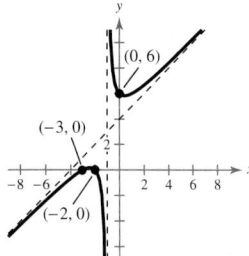

16. 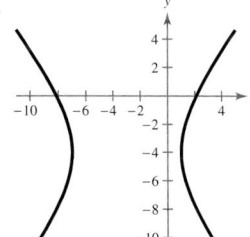 **17.**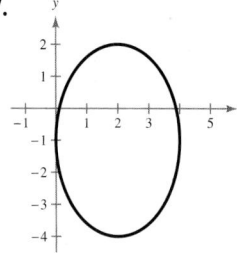

18. $(x - 3)^2 = \dfrac{3}{2}(y + 2)$ **19.** $\dfrac{(y - 2)^2}{\frac{4}{5}} - \dfrac{x^2}{\frac{16}{5}} = 1$

20. Reflect f in the x-axis and y-axis, and shift three units to the right.

21. Reflect f in the x-axis, and shift four units upward.

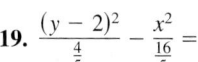

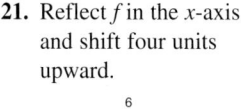

22. 1.991 **23.** -0.067 **24.** 1.717 **25.** 0.281

26. 0.302 **27.** -1.733 **28.** -4.087

29. $\ln(x + 4) + \ln(x - 4) - 4 \ln x, \; x > 4$

30. $\ln \dfrac{x^2}{\sqrt{x + 5}}, \; x > 0$ **31.** $\dfrac{\ln 12}{2} \approx 1.242$

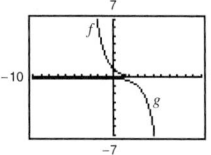

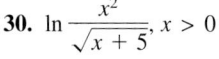

CHAPTER 5

32. $\dfrac{\ln 9}{\ln 4} + 5 \approx 6.585$ **33.** $\ln 3 \approx 1.099$ or $3 \ln 2 \approx 2.079$

34. $\dfrac{64}{5} = 12.8$ **35.** $\frac{1}{2}e^8 \approx 1490.479$

36. $e^6 - 2 \approx 401.429$

37.

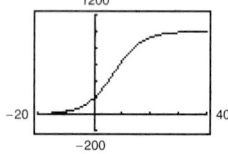

Horizontal asymptotes: $y = 0$, $y = 1000$

38. $2000

39. (a) and (c)

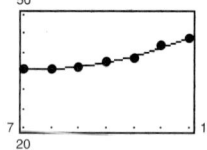

The model is a good fit for the data.

(b) $y = 0.274x^2 - 4.08x + 50.6$

(d) 65.9 Yes, this is a reasonable answer.

40. $16,302.05   **41.** 6.3 hours   **42.** 2006

43. (a) 300   (b) 570   (c) 8.959

Problem Solving   *(page 445)*

1.

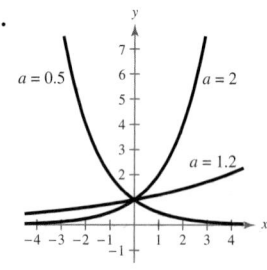

$y = 0.5^x$ and $y = 1.2^x$

$0 \le a \le 1.44$

3. As $x \to \infty$, the graph of e^x increases at a greater rate than the graph of x^n.

5. Answers will vary.

7. (a)

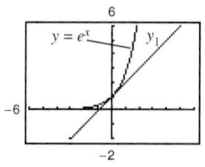

(b)

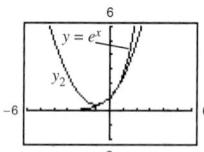

(c)

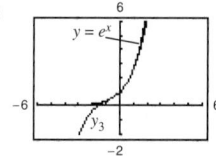

9.

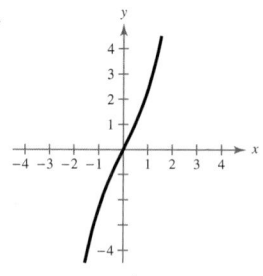

$f^{-1}(x) = \ln\left(\dfrac{x + \sqrt{x^2 + 4}}{2}\right)$

11. c   **13.** $t = \dfrac{\ln c_1 - \ln c_2}{\left(\dfrac{1}{k_2} - \dfrac{1}{k_1}\right)\ln \dfrac{1}{2}}$

15. (a) $y_1 = 252{,}606(1.0310)^t$

(b) $y_2 = 400.88t^2 - 1464.6t + 291{,}782$

(c)

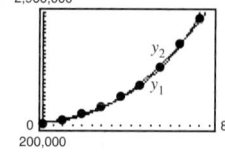

(d) The exponential model is a better fit. No, because the model is rapidly approaching infinity.

17. $1, e^2$

19. $y_4 = (x - 1) - \frac{1}{2}(x - 1)^2 + \frac{1}{3}(x - 1)^3 - \frac{1}{4}(x - 1)^4$

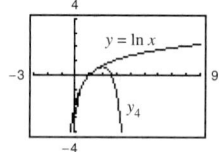

The pattern implies that

$\ln x = (x - 1) - \frac{1}{2}(x - 1)^2 + \frac{1}{3}(x - 1)^3 - \cdots .$

21.

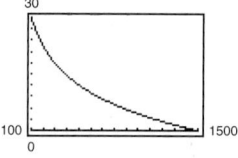

17.7 cubic feet per minute

23. (a)

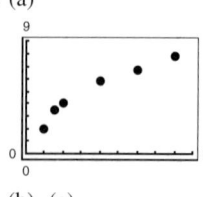

(b)–(e)

Answers will vary.

25. (a)

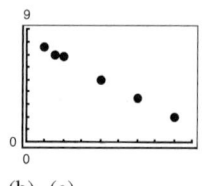

(b)–(e)

Answers will vary.

Chapter 6

Section 6.1 *(page 455)*

Vocabulary Check *(page 455)*

1. system of equations   2. solution
3. solving   4. substitution
5. point of intersection   6. break-even

---

1. (a) No   (b) No   (c) No   (d) Yes
3. (a) No   (b) Yes   (c) No   (d) No
5. $(2, 2)$ **7.** $(2, 6), (-1, 3)$
9. $(0, -5), (4, 3)$ **11.** $(0, 0), (2, -4)$
13. $(0, 1), (1, -1), (3, 1)$ **15.** $(5, 5)$ **17.** $\left(\frac{1}{2}, 3\right)$
19. $(1, 1)$ **21.** $\left(\frac{20}{3}, \frac{40}{3}\right)$ **23.** No solution
25. $(-2, 4), (0, 0)$ **27.** No solution   **29.** $(4, 3)$
31. $\left(\frac{5}{2}, \frac{3}{2}\right)$ **33.** $(2, 2), (4, 0)$ **35.** $(1, 4), (4, 7)$
37. $\left(4, -\frac{1}{2}\right)$ **39.** No solution   **41.** $(4, 3), (-4, 3)$
43. 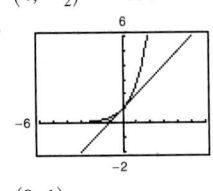   **45.**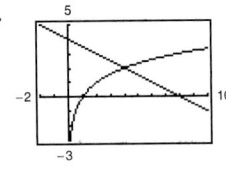
  $(0, 1)$ $(4, 2)$
47.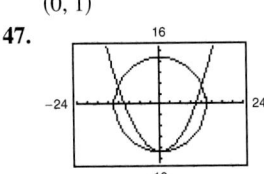
  $(0, -13), (\pm 12, 5)$
49. $(1, 2)$ **51.** $(-2, 0), \left(\frac{29}{10}, \frac{21}{10}\right)$ **53.** No solution
55. $(0.287, 1.751)$ **57.** $(-1, 0), (0, 1), (1, 0)$
59. $\left(\frac{1}{2}, 2\right), \left(-4, -\frac{1}{4}\right)$ **61.** 192 units
63. (a) 781 units   (b) 3708 units
65. (a) 8 weeks
  (b)

	1	2	3	4
$360 - 24x$	336	312	228	264
$24 + 18x$	42	60	78	96

	5	6	7	8
$360 - 24x$	240	216	192	168
$24 + 18x$	114	132	150	168

67. More than $11,666.67

69. (a) $\begin{cases} x + \quad\;\; y = 25{,}000 \\ 0.06x + 0.085y = \quad 2{,}000 \end{cases}$
  (b)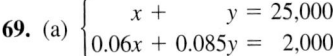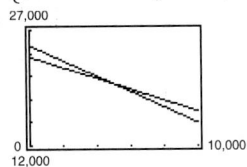
  Decreases; Interest is fixed.
  (c) $5000
71. (a) Solar: $0.1429t^2 - 4.46t + 96.8$
    Wind: $16.371t - 102.7$
  (b)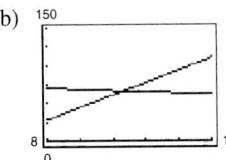
  (c) Point of intersection: $(10.3, 66.01)$. Consumption of solar and wind energy are equal at this point in time in the year 2000.
  (d) $t = 10.3, 135.47$
  (e) The results are the same, but due to the given parameters, $t = 135.47$ is not of significance.
  (f) Answers will vary.
73. 6 meters $\times$ 9 meters   **75.** 9 inches $\times$ 12 inches
77. 8 kilometers $\times$ 12 kilometers
79. False. To solve a system of equations by substitution, you can solve for either variable in one of the two equations and then back-substitute.
81. 1. *Solve* one of the equations for one variable in terms of the other.
  2. *Substitute* the expression found in Step 1 into the other equation to obtain an equation in one variable.
  3. *Solve* the equation obtained in Step 2.
  4. *Back-substitute* the value obtained in Step 3 into the expression obtained in Step 1 to find the value of the other variable.
  5. *Check* that the solution satisfies *each* of the original equations.
83. (a) $y = 2x$ (b) $y = 0$ (c) $y = x - 2$
85. $2x + 7y - 45 = 0$ **87.** $y - 3 = 0$
89. $30x - 17y - 18 = 0$
91. Domain: All real numbers x except $x = 6$
    Horizontal asymptote: $y = 0$
    Vertical asymptote: $x = 6$
93. Domain: All real numbers x except $x = \pm 4$
    Horizontal asymptote: $y = 1$
    Vertical asymptotes: $x = \pm 4$

CHAPTER 6

Section 6.2 *(page 467)*

1. $(2, 1)$

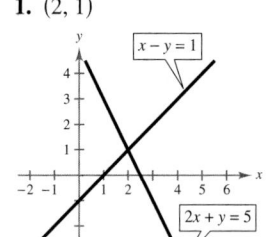

3. $(1, -1)$

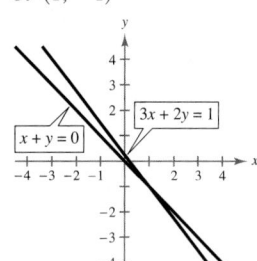

5. No solution

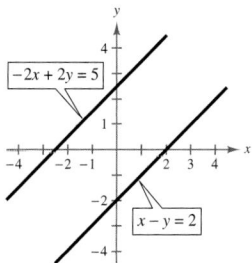

7. $\left(a, \frac{3}{2}a - \frac{5}{2}\right)$

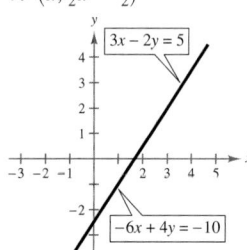

9. $\left(\frac{1}{3}, -\frac{2}{3}\right)$

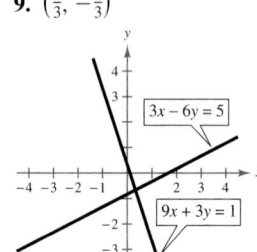

11. $\left(\frac{5}{2}, \frac{3}{4}\right)$ **13.** $(3, 4)$ **15.** $(4, -1)$ **17.** $\left(\frac{12}{7}, \frac{18}{7}\right)$
19. No solution **21.** $\left(\frac{18}{5}, \frac{3}{5}\right)$
23. Infinitely many solutions: $\left(a, -\frac{1}{2} + \frac{5}{6}a\right)$
25. $\left(\frac{90}{31}, -\frac{67}{31}\right)$ **27.** $\left(-\frac{6}{35}, \frac{43}{35}\right)$ **29.** $(5, -2)$
31. b; one solution; consistent
32. a; infinitely many solutions; consistent
33. c; one solution; consistent
34. d; no solutions; inconsistent
35. $(4, 1)$ **37.** $(2, -1)$ **39.** $(6, -3)$ **41.** $\left(\frac{43}{6}, \frac{25}{6}\right)$
43. 550 miles per hour, 50 miles per hour
45. $(80, 10)$ **47.** $(2,000,000, 100)$
49. Cheeseburger: 310 calories; fries: 230 calories

51. (a) $\begin{cases} x + y = 10 \\ 0.2x + 0.5y = 3 \end{cases}$

(b)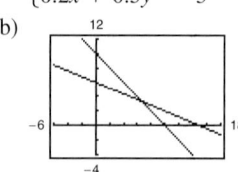

(c) 20% solution: $6\frac{2}{3}$ liters
50% solution: $3\frac{1}{3}$ liters

Decreases
53. \$6000 **55.** 400 adult, 1035 student
57. $y = 0.97x + 2.1$ **59.** $y = 0.32x + 4.1$
61. $y = -2x + 4$
63. (a) $y = 14x + 19$ (b) 41.4 bushels per acre
65. False. Two lines that coincide have infinitely many points of intersection.
67. No. Two lines will intersect only once or will coincide, and if they coincide the system will have infinitely many solutions.
69. $(39,600, 398)$. It is necessary to change the scale on the axes to see the point of intersection.
71. $k = -4$
73. $x \le -\frac{22}{3}$ **75.** $x \le \frac{19}{16}$

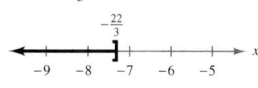 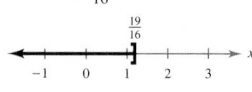

77. $-2 < x < 18$ **79.** $-5 < x < \frac{7}{2}$

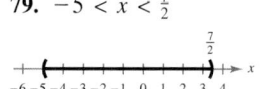

81. $\ln 6x$ **83.** $\log_9 \dfrac{12}{x}$ **85.** No solution

87. Answers will vary.

Section 6.3 *(page 479)*

1. (a) No (b) No (c) No (d) Yes
3. (a) No (b) No (c) Yes (d) No
5. $(1, -2, 4)$ **7.** $(3, 10, 2)$ **9.** $\left(\frac{1}{2}, -2, 2\right)$
11. $\begin{cases} x - 2y + 3z = 5 \\ \quad\quad y - 2z = 9 \\ 2x \quad\quad - 3z = 0 \end{cases}$

First step in putting the system in row-echelon form
13. $(1, 2, 3)$ **15.** $(-4, 8, 5)$ **17.** $(5, -2, 0)$
19. No solution **21.** $\left(-\frac{1}{2}, 1, \frac{3}{2}\right)$
23. $(-3a + 10, 5a - 7, a)$ **25.** $(-a + 3, a + 1, a)$
27. $(2a, 21a - 2, 8a)$ **29.** $\left(-\frac{3}{2}a + \frac{1}{2}, -\frac{2}{3}a + 1, a\right)$

31. $(1, 1, 1, 1)$ **33.** No solution   **35.** $(0, 0, 0)$
37. $(9a, -35a, 67a)$ **39.** $s = -16t^2 + 144$
41. $s = -16t^2 - 32t + 500$
43. $y = \frac{1}{2}x^2 - 2x$ **45.** $y = x^2 - 6x + 8$

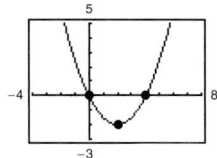

47. $x^2 + y^2 - 4x = 0$ **49.** $x^2 + y^2 + 6x - 8y = 0$

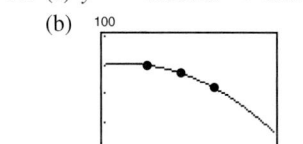

51. 6 touchdowns, 6 extra-point kicks, and 1 field goal
53. \$300,000 at 8%
\$400,000 at 9%
\$75,000 at 10%
55. $250,000 - \frac{1}{2}s$ in certificates of deposit
$125,000 + \frac{1}{2}s$ in municipal bonds
$125,000 - s$ in blue-chip stocks
s in growth stocks
57. Brand X = 4 lb   **59.** Vanilla = 2 lb
Brand Y = 9 lb   Hazelnut = 4 lb
Brand Z = 9 lb   French Roast = 4 lb
61. Television = 30 ads
Radio = 10 ads
Newspaper = 20 ads
63. (a) Not possible
(b) No gallons of 10%, 6 gallons of 15%, 6 gallons of 25%
(c) 4 gallons of 10%, No gallons of 15%, 8 gallons of 25%
65. $I_1 = 1, I_2 = 2, I_3 = 1$
67. $y = x^2 - x$ **69.** $y = -\frac{5}{24}x^2 - \frac{3}{10}x + \frac{41}{6}$
71. (a) $y = -0.0075x^2 + 1.3x + 20$
(b)

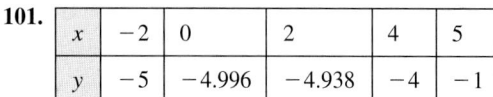

(c)

x	100	120	140
y	75	68	55

The values are the same.
(d) 24.25%   (e) 156 females
73. Touchdowns = 8; Field goals = 2;
Two-point conversions = 1; Extra-point kicks = 5

75. $x = 5$ **77.** $x = \pm\sqrt{2}/2$ or $x = 0$
$y = 5$ $y = \frac{1}{2}$ $y = 0$
$\lambda = -5$ $\lambda = 1$ $\lambda = 0$
79. False. Equation 2 does not have a leading coefficient of 1.
81. No. Answers will vary.
83. $\begin{cases} 3x + y - z = 9 \\ x + 2y - z = 0 \\ -x + y + 3z = 1 \end{cases}$ $\begin{cases} x + y + z = 5 \\ x - 2z = 0 \\ 2y + z = 0 \end{cases}$
85. $\begin{cases} x + 2y - 4z = -5 \\ -x - 4y + 8z = 13 \\ x + 6y + 4z = 7 \end{cases}$ $\begin{cases} x + 2y + 4z = 9 \\ y + 2z = 3 \\ x - 4z = -4 \end{cases}$
87. 6.375   **89.** 80,000   **91.** $11 + i$
93. $22 + 3i$ **95.** $\frac{7}{2} + \frac{7}{2}i$
97. (a) $-4, 0, 3$
(b)

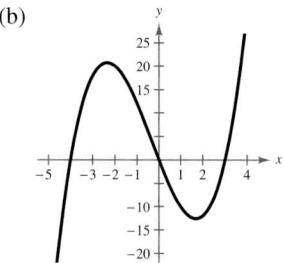

99. (a) $-4, -\frac{3}{2}, 3$
(b)

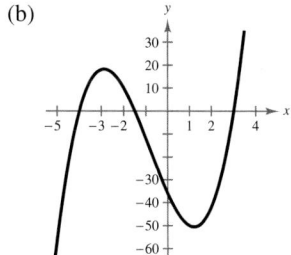

101.

x	-2	0	2	4	5
y	-5	-4.996	-4.938	-4	-1

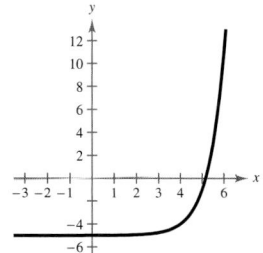

CHAPTER 6

103.

x	-2	-1	0	1	2
y	5.793	4.671	4	3.598	3.358

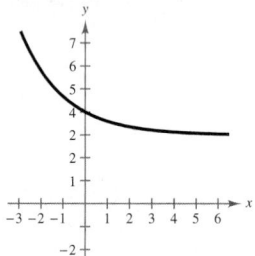

105. $(40, 40)$ **107.** Answers will vary.

Section 6.4   *(page 491)*

1. b   **2.** c   **3.** d   **4.** a

5. $\dfrac{A}{x} + \dfrac{B}{x - 14}$ **7.** $\dfrac{A}{x} + \dfrac{B}{x^2} + \dfrac{C}{x - 10}$

9. $\dfrac{A}{x - 5} + \dfrac{B}{(x - 5)^2} + \dfrac{C}{(x - 5)^3}$ **11.** $\dfrac{A}{x} + \dfrac{Bx + C}{x^2 + 10}$

13. $\dfrac{A}{x} + \dfrac{Bx + C}{x^2 + 1} + \dfrac{Dx + E}{(x^2 + 1)^2}$ **15.** $\dfrac{1}{2}\left(\dfrac{1}{x - 1} - \dfrac{1}{x + 1}\right)$

17. $\dfrac{1}{x} - \dfrac{1}{x + 1}$ **19.** $\dfrac{1}{x} - \dfrac{2}{2x + 1}$

21. $\dfrac{1}{x - 1} - \dfrac{1}{x + 2}$ **23.** $-\dfrac{3}{x} - \dfrac{1}{x + 2} + \dfrac{5}{x - 2}$

25. $\dfrac{3}{x} - \dfrac{1}{x^2} + \dfrac{1}{x + 1}$ **27.** $\dfrac{3}{x - 3} + \dfrac{9}{(x - 3)^2}$

29. $-\dfrac{1}{x} + \dfrac{2x}{x^2 + 1}$ **31.** $-\dfrac{1}{x - 1} + \dfrac{x + 2}{x^2 - 2}$

33. $\dfrac{1}{6}\left(\dfrac{2}{x^2 + 2} - \dfrac{1}{x + 2} + \dfrac{1}{x - 2}\right)$

35. $\dfrac{1}{8}\left(\dfrac{1}{2x + 1} + \dfrac{1}{2x - 1} - \dfrac{4x}{4x^2 + 1}\right)$

37. $\dfrac{1}{x + 1} + \dfrac{2}{x^2 - 2x + 3}$ **39.** $1 - \dfrac{2x + 1}{x^2 + x + 1}$

41. $2x - 7 + \dfrac{17}{x + 2} + \dfrac{1}{x + 1}$

43. $x + 3 + \dfrac{6}{x - 1} + \dfrac{4}{(x - 1)^2} + \dfrac{1}{(x - 1)^3}$

45. $\dfrac{3}{2x - 1} - \dfrac{2}{x + 1}$ **47.** $\dfrac{2}{x} - \dfrac{1}{x^2} - \dfrac{2}{x + 1}$

49. $\dfrac{1}{x^2 + 2} + \dfrac{x}{(x^2 + 2)^2}$ **51.** $2x + \dfrac{1}{2}\left(\dfrac{3}{x - 4} - \dfrac{1}{x + 2}\right)$

53. (a) $\dfrac{3}{x} - \dfrac{2}{x - 4}$

$y = \dfrac{x - 12}{x(x - 4)}$

$y = \dfrac{3}{x}, \; y = -\dfrac{2}{x - 4}$

(b)

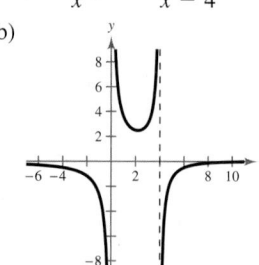 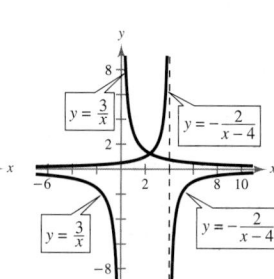

(c) The vertical asymptotes are the same.

55. (a) $\dfrac{3}{x - 3} + \dfrac{5}{x + 3}$

$y = \dfrac{2(4x - 3)}{x^2 - 9}$

$y = \dfrac{3}{x - 3}, \; y = \dfrac{5}{x + 3}$

(b)

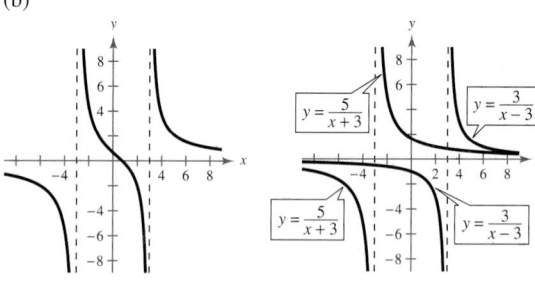

(c) The vertical asymptotes are the same.

57. (a) $\dfrac{2000}{7 - 4x} - \dfrac{2000}{11 - 7x}, \quad 0 < x \le 1$

(b) $\text{Ymax} = \left|\dfrac{2000}{7 - 4x}\right|$

$\text{Ymin} = \left|\dfrac{-2000}{11 - 7x}\right|$

(c)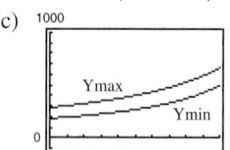

(d) Maximum: $400°\text{F}$
Minimum: $266.7°\text{F}$

59. False. The partial fraction decomposition is

$\dfrac{A}{x + 10} + \dfrac{B}{x - 10} + \dfrac{C}{(x - 10)^2}$.

61. $\dfrac{1}{2a}\left(\dfrac{1}{a + x} + \dfrac{1}{a - x}\right)$ **63.** $\dfrac{1}{a}\left(\dfrac{1}{y} + \dfrac{1}{a - y}\right)$

65.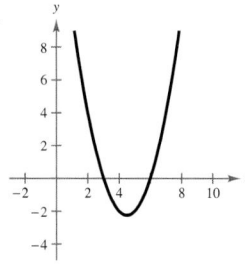

67.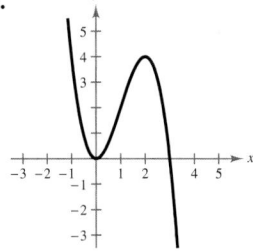

69.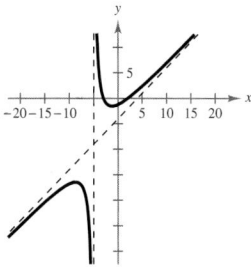

Section 6.5 *(page 500)*

Vocabulary Check *(page 500)*

1. solution   **2.** graph   **3.** linear
4. solution   **5.** consumer surplus

1.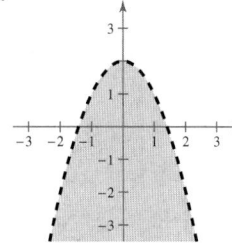

3.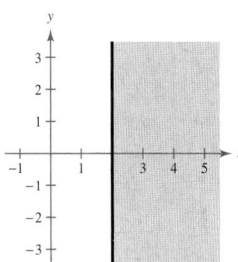

5.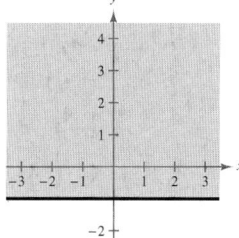

7.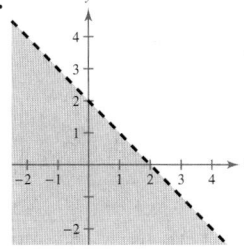

9.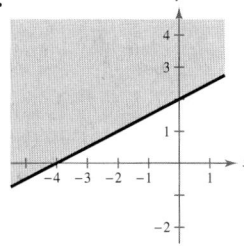

11.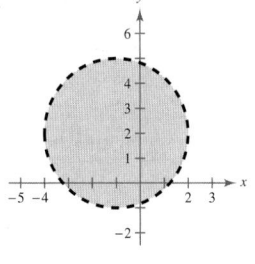

13.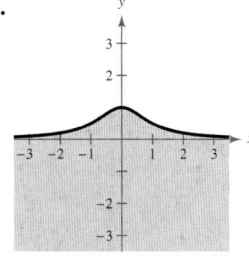

15.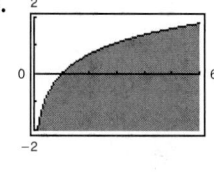

17.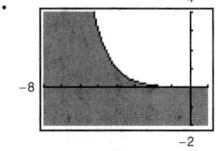

19.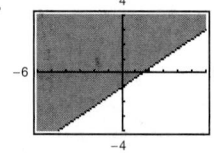

21.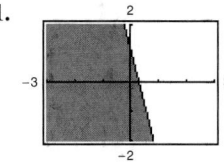

23.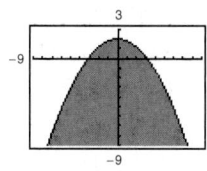

25.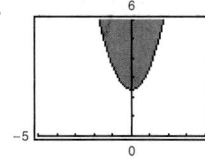

27. $y \le \frac{1}{2}x + 2$

29. $y \ge -\frac{2}{3}x + 2$
31. (a) No   (b) No   (c) Yes   (d) Yes
33. (a) Yes   (b) No   (c) Yes   (d) Yes
35.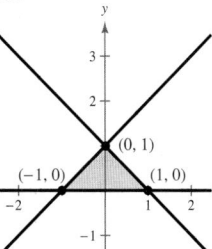

37.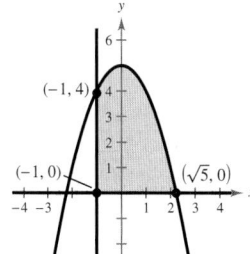

39.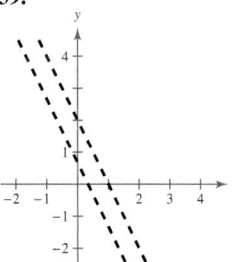

No solution

41.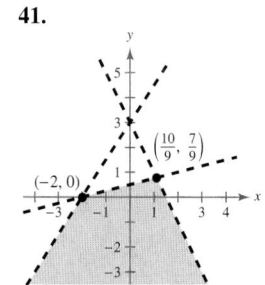

CHAPTER 6

43.

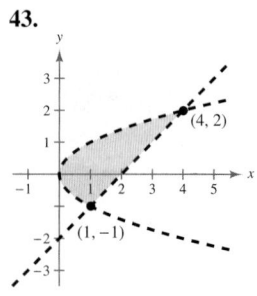

45.

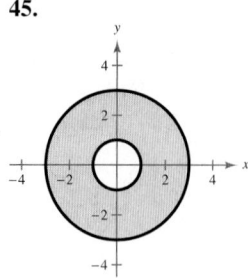

47.

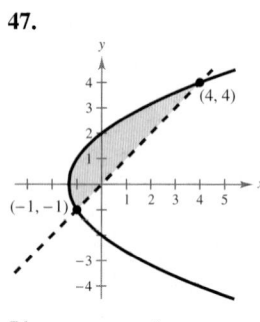

49.

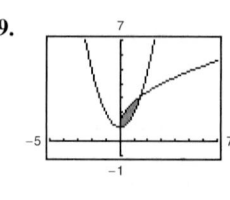

51.

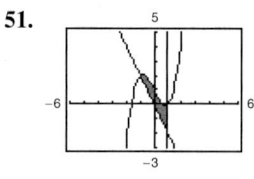

53.

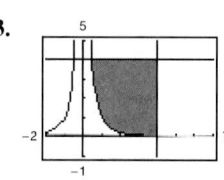

55. $\begin{cases} y \le 4 - x \\ x \ge 0 \\ y \ge 0 \end{cases}$ **57.** $\begin{cases} y \ge 4 - x \\ y \ge 2 - \frac{1}{4}x \\ x \ge 0, \ y \ge 0 \end{cases}$

59. $\begin{cases} x^2 + y^2 \le 16 \\ x \ge 0 \\ y \ge 0 \end{cases}$ **61.** $\begin{cases} 2 \le x \le 5 \\ 1 \le y \le 7 \end{cases}$ **63.** $\begin{cases} y \le \frac{3}{2}x \\ y \le -x + 5 \\ y \ge 0 \end{cases}$

65. (a)

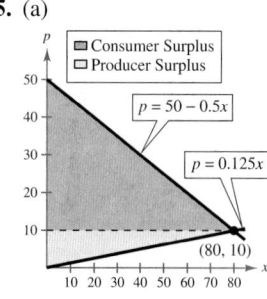

(b) Consumer surplus: $1600
Producer surplus: $400

67. (a)

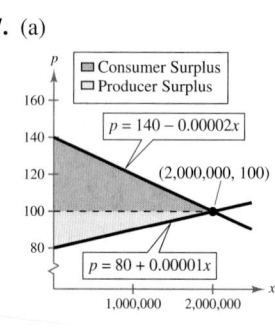

(b) Consumer surplus:
$40,000,000
Producer surplus:
$20,000,000

69. $\begin{cases} x + \frac{3}{2}y \le 12 \\ \frac{4}{3}x + \frac{3}{2}y \le 15 \\ x \ge 0 \\ y \ge 0 \end{cases}$

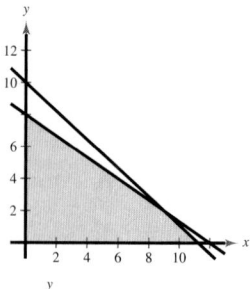

71. $\begin{cases} x + y \le 20{,}000 \\ y \ge 2x \\ x \ge 5{,}000 \\ y \ge 5{,}000 \end{cases}$

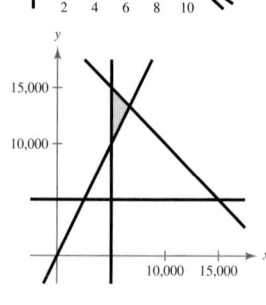

73. $\begin{cases} 55x + 70y \le 7500 \\ x \ge 50 \\ y \ge 40 \end{cases}$

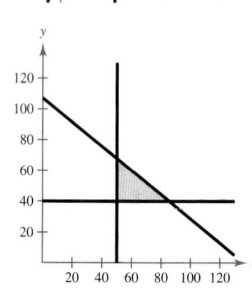

75. (a) $\begin{cases} 20x + 10y \ge 300 \\ 15x + 10y \ge 150 \\ 10x + 20y \ge 200 \\ x \ge 0 \\ y \ge 0 \end{cases}$ (b)

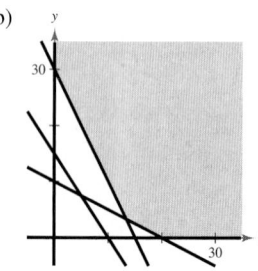

(c) Answers will vary.

77. (a) $y = 19.17t - 46.61$

(b)

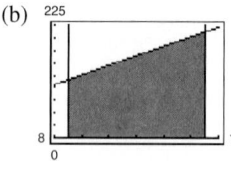

(c) Total retail sales $= \dfrac{h}{2}(a + b) = \$821.3$ billion

79. True. The figure is a rectangle with a length of 9 units and a width of 11 units.

81. The graph is a half-line on the real number line; on the rectangular coordinate system, the graph is a half-plane.

83. (a) $\begin{cases} \pi y^2 - \pi x^2 \geq 10 \\ \quad\quad\quad y > x \\ \quad\quad\quad x > 0 \end{cases}$ (b)

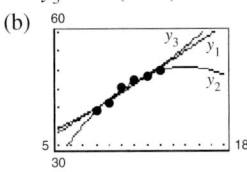

(c) The line is an asymptote to the boundary. The larger the circles, the closer the radii can be and the constraint will still be satisfied.

85. d   **86.** b   **87.** c   **88.** a

89. $5x + 3y - 8 = 0$ **91.** $28x + 17y + 13 = 0$

93. $x + y + 1.8 = 0$

95. (a) $y_1 = 2.17t + 22.5$
$y_2 = -0.241t^2 + 7.23t - 3.4$
$y_3 = 27(1.05^t)$

(b)

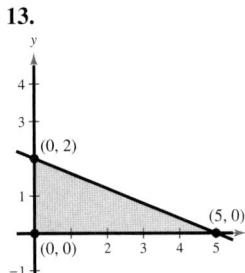

(c) The quadratic model is the best fit for the data.

(d) $48.66

Section 6.6   (page 510)

1. Minimum at $(0, 0)$: 0
Maximum at $(5, 0)$: 20

3. Minimum at $(0, 0)$: 0
Maximum at $(0, 5)$: 40

5. Minimum at $(0, 0)$: 0
Maximum at $(3, 4)$: 17

7. Minimum at $(0, 0)$: 0
Maximum at $(4, 0)$: 20

9. Minimum at $(0, 0)$: 0
Maximum at $(60, 20)$: 740

11. Minimum at $(0, 0)$: 0
Maximum at any point on the line segment connecting $(60, 20)$ and $(30, 45)$: 2100

13.

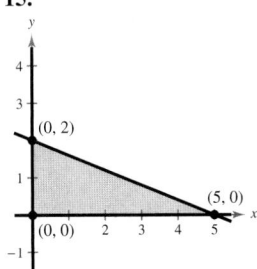

Minimum at $(0, 0)$: 0
Maximum at $(5, 0)$: 30

15.

Minimum at $(0, 0)$: 0
Maximum at $(0, 2)$: 48

17.

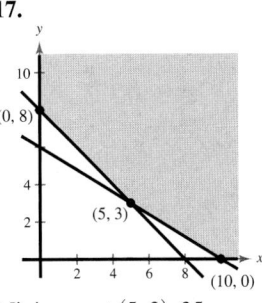

Minimum at $(5, 3)$: 35
No maximum

19.

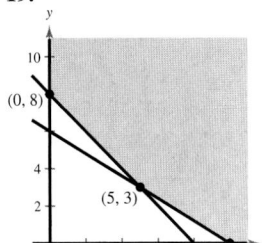

Minimum at $(10, 0)$: 20
No maximum

21.

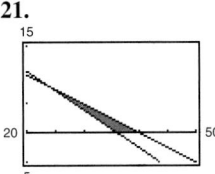

Minimum at $(24, 8)$: 104
Maximum at $(40, 0)$: 160

23.

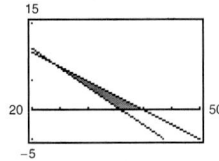

Minimum at $(36, 0)$: 36
Maximum at $(24, 8)$: 56

25. Maximum at $(3, 6)$: 12

27. Maximum at $(0, 10)$: 10

29. Maximum at $(0, 5)$: 25

31. Maximum at $\left(\frac{22}{3}, \frac{19}{6}\right)$: $\frac{271}{6}$

33.

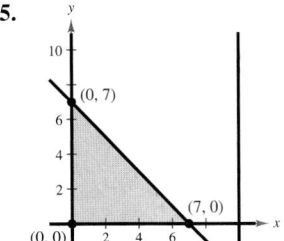

The maximum, 5, occurs at any point on the line segment connecting $(2, 0)$ and $\left(\frac{20}{19}, \frac{45}{19}\right)$.

35.

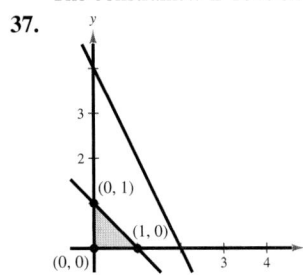

The constraint $x \leq 10$ is extraneous. Maximum at $(0, 7)$: 14

37.

The constraint $2x + y \leq 4$ is extraneous.
Maximum at $(0, 1)$: 4

39. 750 units of model A
1000 units of model B
Optimal profit: $83,750

41. 216 units of $300 model
0 units of $250 model
Optimal profit: $8640

43. Three bags of brand X
Six bags of brand Y
Optimal cost: $195

45. 0 tax returns
12 audits
Optimal revenue: $30,000

47. $62,500 to type A
$187,500 to type B
Optimal return: $23,750

49. True. The objective function has a maximum value at any point on the line segment connecting the two vertices.

51. (a) $t \geq 9$ (b) $\frac{3}{4} \leq t \leq 9$

53. $z = x + 5y$ **55.** $z = 4x + y$ **57.** $\dfrac{9}{2(x+3)}, \; x \neq 0$

59. $\dfrac{x^2 + 2x - 13}{x(x-2)}, \; x \neq \pm 3$ **61.** $\ln 3 \approx 1.099$

63. $4 \ln 38 \approx 14.550$ **65.** $\frac{1}{3}e^{12/7} \approx 1.851$

67. $(-4, 3, -7)$

Review Exercises   *(page 515)*

1. $(1, 1)$ **3.** $(0.25, 0.625)$ **5.** $(5, 4)$

7. $(0, 0), (2, 8), (-2, 8)$ **9.** $(4, -2)$

11. $(1.41, -0.66), (-1.41, 10.66)$

13.

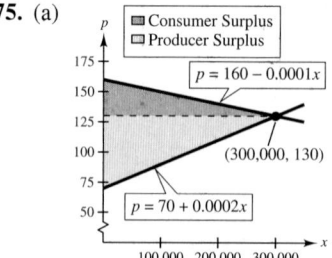

$(0, -2)$

15. 3847 units   **17.** 96 meters × 144 meters   **19.** $\left(\frac{5}{2}, 3\right)$

21. $(-0.5, 0.8)$ **23.** $(0, 0)$ **25.** $\left(\frac{8}{5}a + \frac{14}{5}, a\right)$

27. d, one solution, consistent

28. c, infinite solutions, consistent

29. b, no solution, inconsistent

30. a, one solution, consistent

31. $\left(\dfrac{500,000}{7}, \dfrac{159}{7}\right)$ **33.** $(2, -4, -5)$

35. $\left(\dfrac{24}{5}, \dfrac{22}{5}, -\dfrac{8}{5}\right)$ **37.** $(3a + 4, 2a + 5, a)$

39. $(a - 4, a - 3, a)$ **41.** $y = 2x^2 + x - 5$

43. $x^2 + y^2 - 4x + 4y - 1 = 0$

45. (a) $y = 3x^2 - 14.3x + 117.6$

(b)

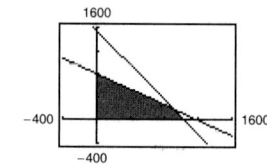

(c) 195.2; yes.

The model is a good fit

47. $16,000 at 7%
$13,000 at 9%
$1%

49. $\dfrac{A}{x} + \dfrac{B}{x+20}$ **51.** $\dfrac{A}{x} + \dfrac{B}{x^2} + \dfrac{C}{x-5}$

53. $\dfrac{3}{x+2} - \dfrac{4}{x+4}$ **55.** $1 - \dfrac{25}{8(x+5)} + \dfrac{9}{8(x-3)}$

57. $\dfrac{1}{2}\left(\dfrac{3}{x-1} - \dfrac{x-3}{x^2+1}\right)$ **59.** $\dfrac{3x}{x^2+1} + \dfrac{x}{(x^2+1)^2}$

61.

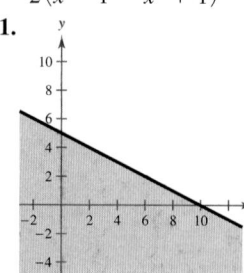

63.

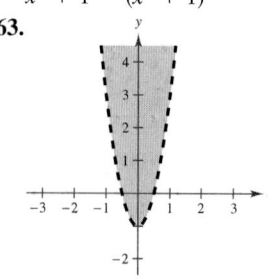

65.

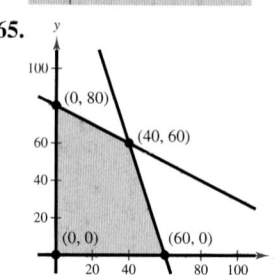

67.

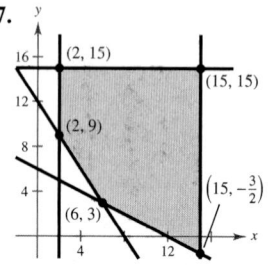

69.

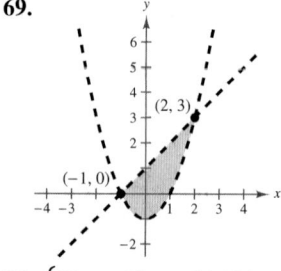

71.

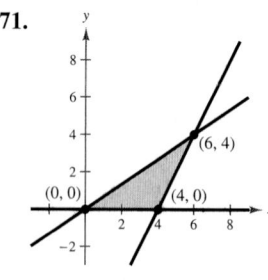

73. $\begin{cases} 20x + 30y \leq 24{,}000 \\ 12x + 8y \leq 12{,}400 \\ x \geq 0 \\ y \geq 0 \end{cases}$

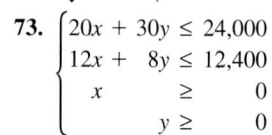

75. (a)

(b) Consumer surplus: $4,500,000
Producer surplus: $9,000,000

77.

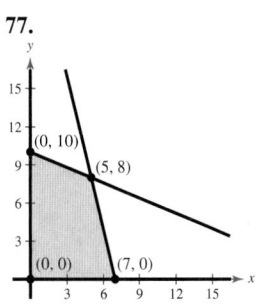

Minimum at $(0, 0)$: 0
Maximum at $(5, 8)$: 47

79.

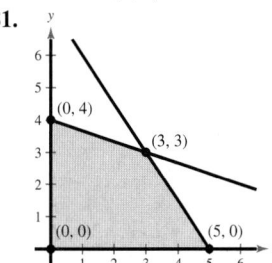

Minimum at $(15, 0)$: 26.25
No maximum

81.

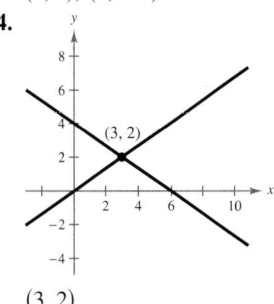

Minimum at $(0, 0)$: 0
Maximum at $(3, 3)$: 48

83. 72 haircuts
0 permanents
Optimal revenue: $1800

85. Three bags of brand X
Two bags of brand Y
Optimal cost: $105

87. False. To represent a region covered by an isosceles trapezoid, the last two inequality signs should be $\leq$.

89. $\begin{cases} x + y = 2 \\ x - y = -14 \end{cases}$

91. $\begin{cases} 3x + y = 7 \\ -6x + 3y = 1 \end{cases}$

93. $\begin{cases} x + y + z = 6 \\ x + y - z = 0 \\ x - y - z = 2 \end{cases}$

95. $\begin{cases} 2x + 2y - 3z = 7 \\ x - 2y + z = 4 \\ -x + 4y - z = -1 \end{cases}$

97. An inconsistent system of linear equations has no solution.
99. Answers will vary.

Chapter Test   *(page 519)*

1. $(-3, 4)$ **2.** $(0, -1), (1, 0), (2, 1)$
3. $(8, 4), (2, -2)$

4.

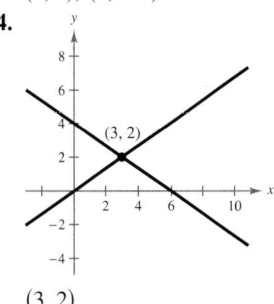

$(3, 2)$

5.

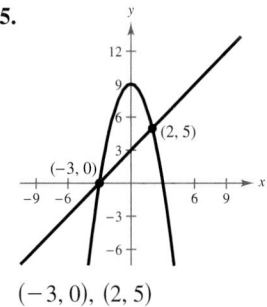

$(-3, 0), (2, 5)$

6.

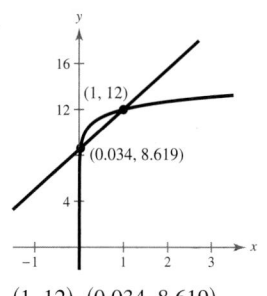

$(1, 12), (0.034, 8.619)$

7. $(1, 5)$ **8.** $(2, -1)$ **9.** $(2, -3, 1)$ **10.** No solution
11. $-\dfrac{1}{x + 1} + \dfrac{3}{x - 2}$ **12.** $\dfrac{2}{x^2} + \dfrac{3}{2 - x}$
13. $-\dfrac{5}{x} + \dfrac{3}{x + 1} + \dfrac{3}{x - 1}$ **14.** $-\dfrac{2}{x} + \dfrac{3x}{x^2 + 2}$

15.

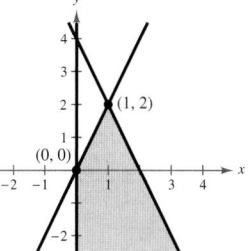

16.

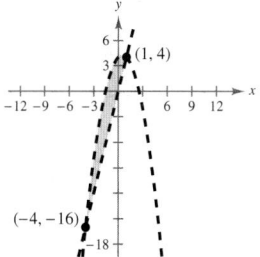

17.

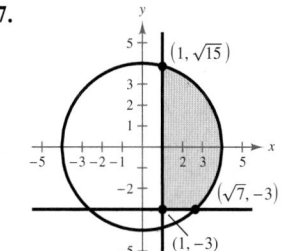

18. Maximum at $(12, 0)$: 240
Minimum at $(0, 0)$: 0

19. 8%: $20,000
8.5%: $30,000

20. $y = -\frac{1}{2}x^2 + x + 6$

21. 0 units of model I, 5300 units of model II
Optimal profit: $212,000

Problem Solving   *(page 521)*

1.

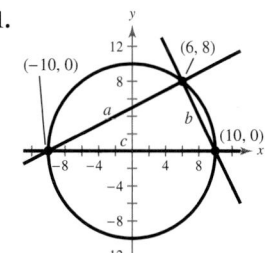

$a = 8\sqrt{5}, b = 4\sqrt{5}, c = 20$
$\left(8\sqrt{5}\right)^2 + \left(4\sqrt{5}\right)^2 = 20^2$
Therefore, the triangle is a right triangle.

CHAPTER 6

3. $ad \neq bc$ **5.** (a) One   (b) Two   (c) Four
7. 10.1 feet deep; $\approx$ 252.7 feet long   **9.** \$12.00

11. (a) $(3, -4)$ (b) $\left(\dfrac{2}{-a + 5}, \dfrac{1}{4a - 1}, \dfrac{1}{a}\right)$

13. (a) $\left(\dfrac{-5a + 16}{6}, \dfrac{5a - 16}{6}, a\right)$

(b) $\left(\dfrac{-11a + 36}{14}, \dfrac{13a - 40}{14}, a\right)$

(c) $(-a + 3, a - 3, a)$ (d) Infinitely many

15. $\begin{cases} a + \quad\ t \leq 32 \\ 0.15a \qquad \geq 1.9 \\ 193a + 772t \geq 11{,}000 \end{cases}$

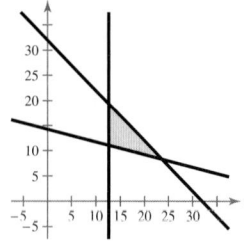

17. (a) $\begin{cases} x + y \leq 200 \\ x \qquad \geq 35 \\ 0 < y \leq 130 \end{cases}$ (b)

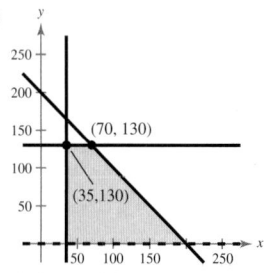

(c) No, because the total cholesterol is greater than 200 milligrams per deciliter.

(d) LDL: 140 milligrams per deciliter
HDL: 50 milligrams per deciliter
Total: 190 milligrams per deciliter

(e) $(50, 120)$; $\frac{170}{50} = 3.4 < 4$; answers will vary.

Chapter 7

Section 7.1   *(page 534)*

Vocabulary Check   *(page 534)*

1. matrix   **2.** square   **3.** main diagonal
4. row; column   **5.** augmented   **6.** coefficient
7. row-equivalent   **8.** reduced row-echelon form
9. Gauss-Jordan elimination

1. 1×2 **3.** 3×1 **5.** 2×2

7. $\begin{bmatrix} 4 & -3 & \vdots & -5 \\ -1 & 3 & \vdots & 12 \end{bmatrix}$ **9.** $\begin{bmatrix} 1 & 10 & -2 & \vdots & 2 \\ 5 & -3 & 4 & \vdots & 0 \\ 2 & 1 & 0 & \vdots & 6 \end{bmatrix}$

11. $\begin{bmatrix} 7 & -5 & 1 & \vdots & 13 \\ 19 & 0 & -8 & \vdots & 10 \end{bmatrix}$ **13.** $\begin{cases} x + 2y = 7 \\ 2x - 3y = 4 \end{cases}$

15. $\begin{cases} 2x \qquad + 5z = -12 \\ \quad y - 2z = \quad 7 \\ \qquad\qquad\quad 2 \end{cases}$

17. $\begin{cases} 9x + 12y + 3z \qquad = \quad 0 \\ -2x + 18y + 5z + 2w = \quad 10 \\ x + 7y - 8z \qquad = -4 \\ 3x \qquad\qquad + 2z \ = -10 \end{cases}$

19. $\begin{bmatrix} 1 & 4 & 3 \\ 0 & 2 & -1 \end{bmatrix}$

21. $\begin{bmatrix} 1 & 1 & 4 & -1 \\ 0 & 5 & -2 & 6 \\ 0 & 3 & 20 & 4 \end{bmatrix} \begin{bmatrix} 1 & 1 & 4 & -1 \\ 0 & 1 & -\frac{2}{5} & \frac{6}{5} \\ 0 & 3 & 20 & 4 \end{bmatrix}$

23. Add 5 times Row 2 to Row 1.

25. Interchange Row 1 and Row 2.
Add 4 times new Row 1 to Row 3.

27. (a) $\begin{bmatrix} 1 & 2 & 3 \\ 0 & -5 & -10 \\ 3 & 1 & -1 \end{bmatrix}$ (b) $\begin{bmatrix} 1 & 2 & 3 \\ 0 & -5 & -10 \\ 0 & -5 & -10 \end{bmatrix}$

(c) $\begin{bmatrix} 1 & 2 & 3 \\ 0 & -5 & -10 \\ 0 & 0 & 0 \end{bmatrix}$ (d) $\begin{bmatrix} 1 & 2 & 3 \\ 0 & 1 & 2 \\ 0 & 0 & 0 \end{bmatrix}$

(e) $\begin{bmatrix} 1 & 0 & -1 \\ 0 & 1 & 2 \\ 0 & 0 & 0 \end{bmatrix}$

The matrix is in reduced row-echelon form.

29. Reduced row-echelon form

31. Not in row-echelon form

33. $\begin{bmatrix} 1 & 1 & 0 & 5 \\ 0 & 1 & 2 & 0 \\ 0 & 0 & 1 & -1 \end{bmatrix}$ **35.** $\begin{bmatrix} 1 & -1 & -1 & 1 \\ 0 & 1 & 6 & 3 \\ 0 & 0 & 0 & 0 \end{bmatrix}$

37. $\begin{bmatrix} 1 & 0 & 0 \\ 0 & 1 & 0 \\ 0 & 0 & 1 \end{bmatrix}$ **39.** $\begin{bmatrix} 1 & 2 & 0 & 0 \\ 0 & 0 & 1 & 0 \\ 0 & 0 & 0 & 1 \\ 0 & 0 & 0 & 0 \end{bmatrix}$

41. $\begin{bmatrix} 1 & 0 & 3 & 16 \\ 0 & 1 & 2 & 12 \end{bmatrix}$ **43.** $\begin{cases} x - 2y = \quad 4 \\ \qquad y = -3 \end{cases}$
$(-2, -3)$

45. $\begin{cases} x - y + 2z = \quad 4 \\ \quad y - \ z = \quad 2 \\ \qquad\quad z = -2 \end{cases}$

$(8, 0, -2)$

47. $(3, -4)$ **49.** $(-4, -10, 4)$ **51.** $(3, 2)$
53. $(-5, 6)$ **55.** $(-1, -4)$ **57.** Inconsistent
59. $(4, -3, 2)$ **61.** $(7, -3, 4)$ **63.** $(-4, -3, 6)$
65. $(2a + 1, 3a + 2, a)$
67. $(4 + 5b + 4a, 2 - 3b - 3a, b, a)$ **69.** Inconsistent
71. $(0, 2 - 4a, a)$ **73.** $(1, 0, 4, -2)$
75. $(-2a, a, a, 0)$ **77.** Yes; $(-1, 1, -3)$ **79.** No

81. $\begin{bmatrix} 1 & 3 & \frac{3}{2} & \vdots & 4 \\ 0 & 1 & \frac{7}{4} & \vdots & -\frac{3}{2} \\ 0 & 0 & 1 & \vdots & 2 \end{bmatrix}, \begin{bmatrix} 1 & 3 & 1 & \vdots & 3 \\ 0 & 1 & 2 & \vdots & -1 \\ 0 & 0 & 1 & \vdots & 2 \end{bmatrix}$

83. $\dfrac{4x^2}{(x+1)^2(x-1)} = \dfrac{1}{x-1} + \dfrac{3}{x+1} - \dfrac{2}{(x+1)^2}$

85. \$150,000 at 7%

\$750,000 at 8%

\$600,000 at 10%

87. $y = x^2 + 2x + 5$

89. (a) $y = -0.004x^2 + 0.367x + 5$

(b)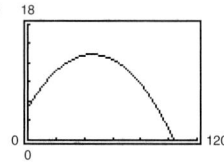

(c) 13 feet, 104 feet   (d) 13.418 feet, 103.793 feet

(e) The results are similar.

91. (a) $x_1 = s,\ x_2 = t,\ x_3 = 600 - s,\ x_4 = s - t,$

$x_5 = 500 - t,\ x_6 = s,\ x_7 = t$

(b) $x_1 = 0,\ x_2 = 0,\ x_3 = 600,\ x_4 = 0,\ x_5 = 500,$

$x_6 = 0,\ x_7 = 0$

(c) $x_1 = 0,\ x_2 = -500,\ x_3 = 600,\ x_4 = 500,$

$x_5 = 1000,\ x_6 = 0,\ x_7 = -500$

93. False. It is a 2×4 matrix.

95. False. Gaussian elimination reduces a matrix until a row-echelon form is obtained; Gauss-Jordan elimination reduces a matrix until a reduced row-echelon form is obtained.

97. (a) There exists a row with all zeros except for the entry in the last column.

(b) There are fewer rows with nonzero entries than there are variables and no rows as in (a).

99. They are the same.

101.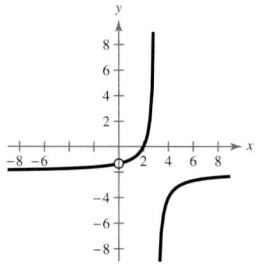

103.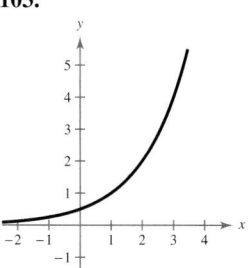

105.

Section 7.2 *(page 549)*

Vocabulary Check *(page 549)*

1. equal   **2.** scalars   **3.** zero; O **4.** identity

5. (a) iii   (b) iv   (c) i   (d) v   (e) ii

6. (a) ii   (b) iv   (c) i   (d) iii

1. $x = -4,\ y = 22$ **3.** $x = 2,\ y = 3$

5. (a) $\begin{bmatrix} 3 & -2 \\ 1 & 7 \end{bmatrix}$ (b) $\begin{bmatrix} -1 & 0 \\ 3 & -9 \end{bmatrix}$ (c) $\begin{bmatrix} 3 & -3 \\ 6 & -3 \end{bmatrix}$

(d) $\begin{bmatrix} -1 & -1 \\ 8 & -19 \end{bmatrix}$

7. (a) $\begin{bmatrix} 7 & 3 \\ 1 & 9 \\ -2 & 15 \end{bmatrix}$ (b) $\begin{bmatrix} 5 & -5 \\ 3 & -1 \\ -4 & -5 \end{bmatrix}$ (c) $\begin{bmatrix} 18 & -3 \\ 6 & 12 \\ -9 & 15 \end{bmatrix}$

(d) $\begin{bmatrix} 16 & -11 \\ 8 & 2 \\ -11 & -5 \end{bmatrix}$

9. (a) $\begin{bmatrix} 3 & 3 & -2 & 1 & 1 \\ -2 & 5 & 7 & -6 & -8 \end{bmatrix}$

(b) $\begin{bmatrix} 1 & 1 & 0 & -1 & 1 \\ 4 & -3 & -11 & 6 & 6 \end{bmatrix}$

(c) $\begin{bmatrix} 6 & 6 & -3 & 0 & 3 \\ 3 & 3 & -6 & 0 & -3 \end{bmatrix}$

(d) $\begin{bmatrix} 4 & 4 & -1 & -2 & 3 \\ 9 & -5 & -24 & 12 & 11 \end{bmatrix}$

11. (a), (b), and (d) not possible

(c) $\begin{bmatrix} 18 & 0 & 9 \\ -3 & -12 & 0 \end{bmatrix}$

13. $\begin{bmatrix} -8 & -7 \\ 15 & -1 \end{bmatrix}$ **15.** $\begin{bmatrix} -24 & -4 & 12 \\ -12 & 32 & 12 \end{bmatrix}$

17. $\begin{bmatrix} 10 & 8 \\ -59 & 9 \end{bmatrix}$ **19.** $\begin{bmatrix} -17.143 & 2.143 \\ 11.571 & 10.286 \end{bmatrix}$

21. $\begin{bmatrix} -1.581 & -3.739 \\ -4.252 & -13.249 \\ 9.713 & -0.362 \end{bmatrix}$ **23.** $\begin{bmatrix} -6 & -9 \\ -1 & 0 \\ 17 & -10 \end{bmatrix}$

25. $\begin{bmatrix} 3 & 3 \\ -\frac{1}{2} & 0 \\ -\frac{13}{2} & \frac{11}{2} \end{bmatrix}$ **27.** Not possible

29. $\begin{bmatrix} 3 & -4 \\ 10 & 16 \\ 26 & 46 \end{bmatrix}$ **31.** $\begin{bmatrix} 3 & 0 & 0 \\ 0 & -4 & 0 \\ 0 & 0 & -10 \end{bmatrix}$

Order: 3×2 Order: 3×3

CHAPTER 7

33. $\begin{bmatrix} 0 & 0 & 0 \\ 0 & 0 & 0 \\ 0 & 0 & 0 \end{bmatrix}$ **35.** $\begin{bmatrix} 41 & 7 & 7 \\ 42 & 5 & 25 \\ -10 & -25 & 45 \end{bmatrix}$

Order: 3×3

37. $\begin{bmatrix} 151 & 25 & 48 \\ 516 & 279 & 387 \\ 47 & -20 & 87 \end{bmatrix}$ **39.** Not possible

41. (a) $\begin{bmatrix} 0 & 15 \\ 6 & 12 \end{bmatrix}$ (b) $\begin{bmatrix} -2 & 2 \\ 31 & 14 \end{bmatrix}$ (c) $\begin{bmatrix} 9 & 6 \\ 12 & 12 \end{bmatrix}$

43. (a) $\begin{bmatrix} 0 & -10 \\ 10 & 0 \end{bmatrix}$ (b) $\begin{bmatrix} 0 & -10 \\ 10 & 0 \end{bmatrix}$ (c) $\begin{bmatrix} 8 & -6 \\ 6 & 8 \end{bmatrix}$

45. (a) $\begin{bmatrix} 7 & 7 & 14 \\ 8 & 8 & 16 \\ -1 & -1 & -2 \end{bmatrix}$ (b) $[13]$ (c) Not possible

47. $\begin{bmatrix} 5 & 8 \\ -4 & -16 \end{bmatrix}$ **49.** $\begin{bmatrix} -4 & 10 \\ 3 & 14 \end{bmatrix}$

51. (a) $\begin{bmatrix} -1 & 1 \\ -2 & 1 \end{bmatrix}\begin{bmatrix} x_1 \\ x_2 \end{bmatrix} = \begin{bmatrix} 4 \\ 0 \end{bmatrix}$ (b) $\begin{bmatrix} 4 \\ 8 \end{bmatrix}$

53. (a) $\begin{bmatrix} -2 & -3 \\ 6 & 1 \end{bmatrix}\begin{bmatrix} x_1 \\ x_2 \end{bmatrix} = \begin{bmatrix} -4 \\ -36 \end{bmatrix}$ (b) $\begin{bmatrix} -7 \\ 6 \end{bmatrix}$

55. (a) $\begin{bmatrix} 1 & -2 & 3 \\ -1 & 3 & -1 \\ 2 & -5 & 5 \end{bmatrix}\begin{bmatrix} x_1 \\ x_2 \\ x_3 \end{bmatrix} = \begin{bmatrix} 9 \\ -6 \\ 17 \end{bmatrix}$ (b) $\begin{bmatrix} 1 \\ -1 \\ 2 \end{bmatrix}$

57. (a) $\begin{bmatrix} 1 & -5 & 2 \\ -3 & 1 & -1 \\ 0 & -2 & 5 \end{bmatrix}\begin{bmatrix} x_1 \\ x_2 \\ x_3 \end{bmatrix} = \begin{bmatrix} -20 \\ 8 \\ -16 \end{bmatrix}$ (b) $\begin{bmatrix} -1 \\ 3 \\ -2 \end{bmatrix}$

59. $\begin{bmatrix} 84 & 60 & 30 \\ 42 & 120 & 84 \end{bmatrix}$

61. (a) $A = \begin{bmatrix} 125 & 100 & 75 \\ 100 & 175 & 125 \end{bmatrix}$

The entries represent the numbers of bushels of each crop that are shipped to each outlet.

(b) $B = [\$3.50 \quad \$6.00]$

The entries represent the profits per bushel of each crop.

(c) $BA = [\$1037.50 \quad \$1400 \quad \$1012.50]$

The entries represent the profits from both crops at each of the three outlets.

63. $\begin{bmatrix} \$15,770 & \$18,300 \\ \$26,500 & \$29,250 \\ \$21,260 & \$24,150 \end{bmatrix}$

The entries represent the wholesale and retail values of the inventories at the three outlets.

65. $P^3 = \begin{bmatrix} 0.300 & 0.175 & 0.175 \\ 0.308 & 0.433 & 0.217 \\ 0.392 & 0.392 & 0.608 \end{bmatrix}$

$P^4 = \begin{bmatrix} 0.250 & 0.188 & 0.188 \\ 0.315 & 0.377 & 0.248 \\ 0.435 & 0.435 & 0.565 \end{bmatrix}$

$P^5 = \begin{bmatrix} 0.225 & 0.194 & 0.194 \\ 0.314 & 0.345 & 0.267 \\ 0.461 & 0.461 & 0.539 \end{bmatrix}$

$P^6 = \begin{bmatrix} 0.213 & 0.197 & 0.197 \\ 0.311 & 0.326 & 0.280 \\ 0.477 & 0.477 & 0.523 \end{bmatrix}$

$P^7 = \begin{bmatrix} 0.206 & 0.198 & 0.198 \\ 0.308 & 0.316 & 0.288 \\ 0.486 & 0.486 & 0.514 \end{bmatrix}$

$P^8 = \begin{bmatrix} 0.203 & 0.199 & 0.199 \\ 0.305 & 0.309 & 0.292 \\ 0.492 & 0.492 & 0.508 \end{bmatrix}$

Approaches the matrix

$\begin{bmatrix} 0.2 & 0.2 & 0.2 \\ 0.3 & 0.3 & 0.3 \\ 0.5 & 0.5 & 0.5 \end{bmatrix}$

67. (a) $\begin{array}{cc} \text{Sales \$} & \text{Profit} \\ \begin{bmatrix} 447 & 115 \\ 624.5 & 161 \\ 731.2 & 188 \end{bmatrix} \end{array}$ (b) \$464

The entries represent the total sales and profits for each type of milk.

69. (a) $[2 \quad 0.5 \quad 3]$

(b) 120 lb 150 lb

$[473.5 \quad 588.5]$

The entries represent the total calories burned.

71. True. The sum of two matrices of different orders is undefined.

73. Not possible **75.** Not possible **77.** 2×2

79. 2×3 **81.** $AC = BC = \begin{bmatrix} 2 & 3 \\ 2 & 3 \end{bmatrix}$

83. AB is a diagonal matrix whose entries are the products of the corresponding entries of A and B.

85. $-8, \dfrac{4}{3}$ **87.** $0, \dfrac{-5 \pm \sqrt{37}}{4}$ **89.** $4, \pm\dfrac{\sqrt{15}}{3}i$

91. $\left(7, -\frac{1}{2}\right)$ **93.** $(3, -1)$

Section 7.3 *(page 560)*

Vocabulary Check *(page 560)*

1. square **2.** inverse

3. nonsingular; singular **4.** $A^{-1}B$

1–9. $AB = I$ and $BA = I$

11. $\begin{bmatrix} \frac{1}{2} & 0 \\ 0 & \frac{1}{3} \end{bmatrix}$ **13.** $\begin{bmatrix} -3 & 2 \\ -2 & 1 \end{bmatrix}$ **15.** $\begin{bmatrix} 1 & -1 \\ 2 & -1 \end{bmatrix}$

17. Does not exist   **19.** Does not exist

21. $\begin{bmatrix} 1 & 1 & -1 \\ -3 & 2 & -1 \\ 3 & -3 & 2 \end{bmatrix}$ **23.** $\begin{bmatrix} 1 & 0 & 0 \\ -\frac{3}{4} & \frac{1}{4} & 0 \\ \frac{7}{20} & -\frac{1}{4} & \frac{1}{5} \end{bmatrix}$

25. $\begin{bmatrix} -\frac{1}{8} & 0 & 0 & 0 \\ 0 & 1 & 0 & 0 \\ 0 & 0 & \frac{1}{4} & 0 \\ 0 & 0 & 0 & -\frac{1}{5} \end{bmatrix}$ **27.** $\begin{bmatrix} -175 & 37 & -13 \\ 95 & -20 & 7 \\ 14 & -3 & 1 \end{bmatrix}$

29. $\begin{bmatrix} -1.5 & 1.5 & 1 \\ 4.5 & -3.5 & -3 \\ -1 & 1 & 1 \end{bmatrix}$ **31.** $\begin{bmatrix} -12 & -5 & -9 \\ -4 & -2 & -4 \\ -8 & -4 & -6 \end{bmatrix}$

33. $\begin{bmatrix} 0 & -1.\overline{81} & 0.\overline{90} \\ -10 & 5 & 5 \\ 10 & -2.\overline{72} & -3.\overline{63} \end{bmatrix}$ **35.** Does not exist

37. $\begin{bmatrix} 1 & 0 & 1 & 0 \\ 0 & 1 & 0 & 1 \\ 2 & 0 & 1 & 0 \\ 0 & 1 & 0 & 2 \end{bmatrix}$ **39.** $\begin{bmatrix} \frac{3}{19} & \frac{2}{19} \\ -\frac{2}{19} & \frac{5}{19} \end{bmatrix}$

41. Does not exist   **43.** $\begin{bmatrix} \frac{16}{59} & \frac{15}{59} \\ -\frac{4}{59} & \frac{70}{59} \end{bmatrix}$

45. $(5, 0)$ **47.** $(-8, -6)$ **49.** $(3, 8, -11)$

51. $(2, 1, 0, 0)$ **53.** $(2, -2)$ **55.** No solution

57. $(-4, -8)$ **59.** $(-1, 3, 2)$

61. $\left(\frac{5}{16}a + \frac{13}{16}, \frac{19}{16}a + \frac{11}{16}, a \right)$ **63.** $(-7, 3, -2)$

65. $(5, 0, -2, 3)$

67. $7000 in AAA-rated bonds
$1000 in A-rated bonds
$2000 in B-rated bonds

69. $9000 in AAA-rated bonds
$1000 in A-rated bonds
$2000 in B-rated bonds

71. (a) $I_1 = -3$ amperes   (b) $I_1 = 2$ amperes
$I_2 = 8$ amperes $I_2 = 3$ amperes
$I_3 = 5$ amperes $I_3 = 5$ amperes

73. True. If B is the inverse of A, then $AB = I = BA$.

75. Answers will vary.

77. $x \geq -5$ or $x \leq -9$

79. $\dfrac{2 \ln 315}{\ln 3} \approx 10.472$ **81.** $2^{6.5} \approx 90.510$

83. Answers will vary.

Section 7.4   (page 568)

1. 5   **3.** 5   **5.** 27   **7.** 0   **9.** 6   **11.** -9

13. 72   **15.** $\frac{11}{6}$ **17.** -0.002 **19.** -4.842 **21.** 0

23. (a) $M_{11} = -5, M_{12} = 2, M_{21} = 4, M_{22} = 3$
(b) $C_{11} = -5, C_{12} = -2, C_{21} = -4, C_{22} = 3$

25. (a) $M_{11} = -4, M_{12} = -2, M_{21} = 1, M_{22} = 3$
(b) $C_{11} = -4, C_{12} = 2, C_{21} = -1, C_{22} = 3$

27. (a) $M_{11} = 3, M_{12} = -4, M_{13} = 1, M_{21} = 2, M_{22} = 2,$
$M_{23} = -4, M_{31} = -4, M_{32} = 10, M_{33} = 8$
(b) $C_{11} = 3, C_{12} = 4, C_{13} = 1, C_{21} = -2, C_{22} = 2,$
$C_{23} = 4, C_{31} = -4, C_{32} = -10, C_{33} = 8$

29. (a) $M_{11} = 30, M_{12} = 12, M_{13} = 11, M_{21} = -36,$
$M_{22} = 26, M_{23} = 7, M_{31} = -4, M_{32} = -42, M_{33} = 12$
(b) $C_{11} = 30, C_{12} = -12, C_{13} = 11, C_{21} = 36, C_{22} = 26,$
$C_{23} = -7, C_{31} = -4, C_{32} = 42, C_{33} = 12$

31. (a) -75 (b) -75 **33.** (a) 96   (b) 96

35. (a) 170   (b) 170   **37.** 0   **39.** 0   **41.** -9

43. -58 **45.** -30 **47.** -168 **49.** 0

51. 412   **53.** -126 **55.** 0   **57.** -336 **59.** 410

61. (a) -3 (b) -2 (c) $\begin{bmatrix} -2 & 0 \\ 0 & -3 \end{bmatrix}$ (d) 6

63. (a) -8 (b) 0   (c) $\begin{bmatrix} -4 & 4 \\ 1 & -1 \end{bmatrix}$ (d) 0

65. (a) -21 (b) -19 (c) $\begin{bmatrix} 7 & 1 & 4 \\ -8 & 9 & -3 \\ 7 & -3 & 9 \end{bmatrix}$ (d) 399

67. (a) 2   (b) -6 (c) $\begin{bmatrix} 1 & 4 & 3 \\ -1 & 0 & 3 \\ 0 & 2 & 0 \end{bmatrix}$ (d) -12

69–73. Answers will vary.   **75.** $-1, 4$ **77.** $-1, -4$

79. $8uv - 1$ **81.** e^{5x} **83.** $1 - \ln x$

85. True. If an entire row is zero, then each cofactor in the expansion is multiplied by zero.

87. Answers will vary.

89. A square matrix is a square array of numbers. The determinant of a square matrix is a real number.

91. (a) Columns 2 and 3 of A were interchanged.
$|A| = -115 = -|B|$
(b) Rows 1 and 3 of A were interchanged.
$|A| = -40 = -|B|$

93. (a) Multiply Row 1 by 5.
(b) Multiply Column 2 by 4 and Column 3 by 3.

95. All real numbers x

97. All real numbers x such that $-4 \le x \le 4$

99. All real numbers t such that $t > 1$

101.

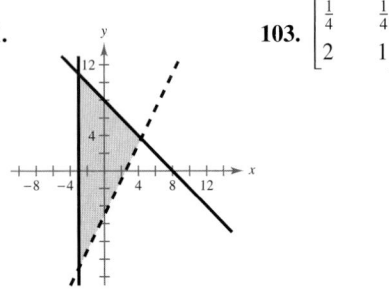

103. $\begin{bmatrix} \frac{1}{4} & \frac{1}{4} \\ 2 & 1 \end{bmatrix}$

105. Does not exist

Section 7.5   *(page 580)*

Vocabulary Check   *(page 580)*

1. Cramer's Rule   **2.** collinear

3. $A = \pm\dfrac{1}{2}\begin{vmatrix} x_1 & y_1 & 1 \\ x_2 & y_2 & 1 \\ x_3 & y_3 & 1 \end{vmatrix}$ **4.** cryptogram

5. uncoded; coded

1. $(2, -2)$ **3.** Not possible   **5.** $\left(\frac{32}{7}, \frac{30}{7}\right)$

7. $(-1, 3, 2)$ **9.** $(-2, 1, -1)$ **11.** $\left(0, -\frac{1}{2}, \frac{1}{2}\right)$

13. $(1, 2, 1)$ **15.** 7 **17.** 14 **19.** $\frac{33}{8}$ **21.** $\frac{5}{2}$

23. 28 **25.** $y = \frac{16}{5}$ or $y = 0$

27. $y = -3$ or $y = -11$ **29.** 250 square miles

31. Collinear   **33.** Not collinear   **35.** Collinear

37. $y = -3$ **39.** $3x - 5y = 0$ **41.** $x + 3y - 5 = 0$

43. $2x + 3y - 8 = 0$

45. Uncoded: $[20\ 18\ 15], [21\ 2\ 12], [5\ 0\ 9], [14\ 0\ 18],$
       $[9\ 22\ 5], [18\ 0\ 3], [9\ 20\ 25]$

   Encoded: $-52\ 10\ 27\ -49\ 3\ 34\ -49\ 13\ 27$
       $-94\ 22\ 54\ 1\ 1\ -7\ 0\ -12\ 9$
       $-121\ 41\ 55$

47. $-6\ -35\ -69\ 11\ 20\ 17\ 6\ -16\ -58\ 46\ 79\ 67$

49. $-5\ -41\ -87\ 91\ 207\ 257\ 11\ -5\ -41\ 40\ 80$
     $84\ 76\ 177\ 227$

51. HAPPY NEW YEAR

53. CLASS IS CANCELED

55. SEND PLANES   **57.** MEET ME TONIGHT RON

59. False. The denominator is the determinant of the coefficient matrix.

61. False. If the determinant of the coefficient matrix is zero, the system has either no solution or infinitely many solutions.

63. $(-6, 4)$ **65.** $(-1, 0, -3)$

67.

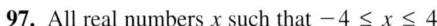

Minimum at $(0, 0)$: 0

Maximum at $(6, 4)$: 52

Review Exercises   *(page 584)*

1. 3×1 **3.** 1×1 **5.** $\begin{bmatrix} 3 & -10 & \vdots & 15 \\ 5 & 4 & \vdots & 22 \end{bmatrix}$

7. $\begin{cases} 5x + y + 7z = -9 \\ 4x + 2y = 10 \\ 9x + 4y + 2z = 3 \end{cases}$ **9.** $\begin{bmatrix} 1 & 2 & 3 \\ 0 & 1 & 1 \\ 0 & 0 & 1 \end{bmatrix}$

11. $\begin{cases} x + 2y + 3z = 9 \\ y - 2z = 2 \\ z = 0 \end{cases}$ **13.** $\begin{cases} x - 5y + 4z = 1 \\ y + 2z = 3 \\ z = 4 \end{cases}$

   $(5, 2, 0)$ $(-40, -5, 4)$

15. $(10, -12)$ **17.** $\left(-\frac{1}{5}, \frac{7}{10}\right)$ **19.** $(5, 2, -6)$

21. $\left(-2a + \frac{3}{2}, 2a + 1, a\right)$ **23.** $(1, 0, 4, 3)$

25. $(2, -3, 3)$ **27.** $(2, 3, -1)$ **29.** $(2, 6, -10, -3)$

31. $x = 12, y = -7$ **33.** $x = 1, y = 11$

35. (a) $\begin{bmatrix} -1 & 8 \\ 15 & 13 \end{bmatrix}$ (b) $\begin{bmatrix} 5 & -12 \\ -9 & -3 \end{bmatrix}$

   (c) $\begin{bmatrix} 8 & -8 \\ 12 & 20 \end{bmatrix}$ (d) $\begin{bmatrix} -7 & 28 \\ 39 & 29 \end{bmatrix}$

37. (a) $\begin{bmatrix} 5 & 7 \\ -3 & 14 \\ 31 & 42 \end{bmatrix}$ (b) $\begin{bmatrix} 5 & 1 \\ -11 & -10 \\ -9 & -38 \end{bmatrix}$

   (c) $\begin{bmatrix} 20 & 16 \\ -28 & 8 \\ 44 & 8 \end{bmatrix}$ (d) $\begin{bmatrix} 5 & 13 \\ 5 & 38 \\ 71 & 122 \end{bmatrix}$

39. $\begin{bmatrix} 17 & -17 \\ 13 & 2 \end{bmatrix}$ **41.** $\begin{bmatrix} 54 & 4 \\ -2 & 24 \\ -4 & 32 \end{bmatrix}$

43. $\begin{bmatrix} 48 & -18 & -3 \\ 15 & 51 & 33 \end{bmatrix}$ **45.** $\begin{bmatrix} -14 & -4 \\ 7 & -17 \\ -17 & -2 \end{bmatrix}$

47. $\begin{bmatrix} 3 & \frac{2}{3} \\ -\frac{4}{3} & \frac{11}{3} \\ \frac{10}{3} & 0 \end{bmatrix}$ **49.** $\begin{bmatrix} -30 & 4 \\ 51 & 70 \end{bmatrix}$

51. $\begin{bmatrix} 100 & 220 \\ 12 & -4 \\ 84 & 212 \end{bmatrix}$ **53.** $\begin{bmatrix} 14 & -2 & 8 \\ 14 & -10 & 40 \\ 36 & -12 & 48 \end{bmatrix}$

55. $\begin{bmatrix} 44 & 4 \\ 20 & 8 \end{bmatrix}$ **57.** $\begin{bmatrix} 24 & -8 \\ 36 & -12 \end{bmatrix}$ **59.** $\begin{bmatrix} 1 & 17 \\ 12 & 36 \end{bmatrix}$

61. $\begin{bmatrix} 14 & -22 & 22 \\ 19 & -41 & 80 \\ 42 & -66 & 66 \end{bmatrix}$ **63.** $\begin{bmatrix} 76 & 114 & 133 \\ 38 & 95 & 76 \end{bmatrix}$

65. $[\$274{,}150 \quad \$303{,}150]$

The merchandise shipped to warehouse 1 is worth $274,150 and the merchandise shipped to warehouse 2 is worth $303,150.

67–69. $AB = I$ and $BA = I$

71. $\begin{bmatrix} 4 & -5 \\ 5 & -6 \end{bmatrix}$ **73.** $\begin{bmatrix} 13 & 6 & -4 \\ -12 & -5 & 3 \\ 5 & 2 & -1 \end{bmatrix}$

75. $\begin{bmatrix} \frac{1}{2} & -1 & -\frac{1}{2} \\ \frac{1}{2} & -\frac{2}{3} & -\frac{5}{6} \\ 0 & \frac{2}{3} & \frac{1}{3} \end{bmatrix}$ **77.** $\begin{bmatrix} -3 & 6 & -5.5 & 3.5 \\ 1 & -2 & 2 & -1 \\ 7 & -15 & 14.5 & -9.5 \\ -1 & 2.5 & -2.5 & 1.5 \end{bmatrix}$

79. $\begin{bmatrix} 1 & -1 \\ 4 & -\frac{7}{2} \end{bmatrix}$ **81.** $\begin{bmatrix} 2 & \frac{20}{3} \\ \frac{1}{10} & \frac{1}{6} \end{bmatrix}$ **83.** $(36, 11)$

85. $(-6, -1)$ **87.** $(2, -1, -2)$ **89.** $(6, 1, -1)$
91. $(-3, 1)$ **93.** $(1, 1, -2)$ **95.** -42 **97.** 550
99. (a) $M_{11} = 4, M_{12} = 7, M_{21} = -1, M_{22} = 2$
  (b) $C_{11} = 4, C_{12} = -7, C_{21} = 1, C_{22} = 2$
101. (a) $M_{11} = 30, M_{12} = -12, M_{13} = -21,$
    $M_{21} = 20, M_{22} = 19, M_{23} = 22, M_{31} = 5,$
    $M_{32} = -2, M_{33} = 19$
  (b) $C_{11} = 30, C_{12} = 12, C_{13} = -21,$
    $C_{21} = -20, C_{22} = 19, C_{23} = -22,$
    $C_{31} = 5, C_{32} = 2, C_{33} = 19$
103. 130 **105.** 279 **107.** $(4, 7)$ **109.** $(-1, 4, 5)$
111. 16 **113.** 10 **115.** Collinear
117. $x - 2y + 4 = 0$ **119.** $2x + 6y - 13 = 0$
121. Uncoded: $[12 \quad 15 \quad 15], [11 \quad 0 \quad 15], [21 \quad 20 \quad 0],$
    $[2 \quad 5 \quad 12], [15 \quad 23 \quad 0]$
  Encoded: $-21 \ 6 \ 0 \ -68 \ 8 \ 45 \ 102 \ -42 \ -60 \ -53$
    $20 \ 21 \ 99 \ -30 \ -69$
123. SEE YOU FRIDAY
125. False. The matrix must be square.
127. The matrix must be square and its determinant nonzero.
129. No. The first two matrices describe a system of equations with one solution. The third matrix describes a system with infinitely many solutions.
131. $\lambda = \pm 2\sqrt{10} - 3$

Chapter Test *(page 589)*

1. $\begin{bmatrix} 1 & 0 & 0 \\ 0 & 1 & 0 \\ 0 & 0 & 1 \end{bmatrix}$

2. $\begin{bmatrix} 1 & 0 & -1 & 2 \\ 0 & 1 & 0 & -1 \\ 0 & 0 & 0 & 0 \\ 0 & 0 & 0 & 0 \end{bmatrix}$

3. $\begin{bmatrix} 4 & 3 & -2 & \vdots & 14 \\ -1 & -1 & 2 & \vdots & -5 \\ 3 & 1 & -4 & \vdots & 8 \end{bmatrix}, \left(1, 3, -\frac{1}{2}\right)$

4. (a) $\begin{bmatrix} 1 & 5 \\ 0 & -4 \end{bmatrix}$

5. $\begin{bmatrix} \frac{1}{2} & \frac{2}{5} \\ 1 & \frac{3}{5} \end{bmatrix}$

  (b) $\begin{bmatrix} 15 & 12 \\ -12 & -12 \end{bmatrix}$

  (c) $\begin{bmatrix} 7 & 14 \\ -4 & -12 \end{bmatrix}$

  (d) $\begin{bmatrix} 4 & -5 \\ 0 & 4 \end{bmatrix}$

6. $\begin{bmatrix} -\frac{5}{2} & 4 & -3 \\ 5 & -7 & 6 \\ 4 & -6 & 5 \end{bmatrix}$

7. $(13, 22)$ **8.** -196 **9.** 29 **10.** 43
11. $(-3, 5)$ **12.** $(-2, 4, 6)$ **13.** 7
14. Uncoded: $[11 \ 14 \ 15], [3 \ 11 \ 0], [15 \ 14 \ 0], [23 \ 15 \ 15],$
    $[4 \ 0 \ 0]$
  Encoded: $115 \ -41 \ -59 \ 14 \ -3 \ -11 \ 29 \ -15$
    $-14 \ 128 \ -53 \ -60 \ 4 \ -4 \ 0$
15. 75 liters of 60% solution
  25 liters of 20% solution

Problem Solving *(page 591)*

1. (a) $AT = \begin{bmatrix} -1 & -4 & -2 \\ 1 & 2 & 3 \end{bmatrix}$
    $AAT = \begin{bmatrix} -1 & -2 & -3 \\ -1 & -4 & -2 \end{bmatrix}$

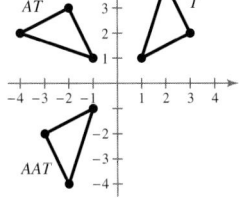

  A represents a counterclockwise rotation.
  (b) AAT is rotated clockwise 90° to obtain AT. AT is then rotated clockwise 90° to obtain T.

CHAPTER 7

3. (a) Yes   (b) No   (c) No   (d) No
5. (a) Gold Cable Company: 28,750 subscribers
     Galaxy Cable Company: 35,750 subscribers
     Nonsubscribers: 35,500
     Answers will vary.
  (b) Gold Cable Company: 30,813 subscribers
     Galaxy Cable Company: 39,675 subscribers
     Nonsubscribers: 29,513
     Answers will vary.
  (c) Gold Cable Company: 31,947 subscribers
     Galaxy Cable Company: 42,329 subscribers
     Nonsubscribers: 25,724
     Answers will vary.
  (d) Cable companies are increasing the number of sub-
     scribers, while the nonsubscribers are decreasing.
7. $x = 6$ **9–11.** Answers will vary.
13. Sulfur: 32 atomic mass units
    Nitrogen: 14 atomic mass units
    Fluorine: 19 atomic mass units
15. $A^T = \begin{bmatrix} -1 & 2 \\ 1 & 0 \\ -2 & 1 \end{bmatrix}$ $B^T = \begin{bmatrix} -3 & 1 & 1 \\ 0 & 2 & -1 \end{bmatrix}$
    $(AB)^T = \begin{bmatrix} 2 & -5 \\ 4 & -1 \end{bmatrix} = B^T A^T$
17. (a) $A^{-1} = \begin{bmatrix} 1 & -2 \\ 1 & -3 \end{bmatrix}$
  (b) JOHN RETURN TO BASE
19. $|A| = 0$

Chapter 8

Section 8.1   *(page 601)*

Vocabulary Check   *(page 601)*

1. infinite sequence   **2.** terms   **3.** finite
4. recursively   **5.** factorial
6. summation notation   **7.** index; upper; lower
8. series   **9.** nth partial sum

---

1. 4, 7, 10, 13, 16   **3.** 2, 4, 8, 16, 32
5. $-2, 4, -8, 16, -32$ **7.** $3, 2, \frac{5}{3}, \frac{3}{2}, \frac{7}{5}$
9. $3, \frac{12}{11}, \frac{9}{13}, \frac{24}{47}, \frac{15}{37}$ **11.** $0, 1, 0, \frac{1}{2}, 0$ **13.** $\frac{5}{3}, \frac{17}{9}, \frac{53}{27}, \frac{161}{81}, \frac{485}{243}$
15. $1, \frac{1}{2^{3/2}}, \frac{1}{3^{3/2}}, \frac{1}{8}, \frac{1}{5^{3/2}}$ **17.** $-1, \frac{1}{4}, -\frac{1}{9}, \frac{1}{16}, -\frac{1}{25}$
19. $\frac{2}{3}, \frac{2}{3}, \frac{2}{3}, \frac{2}{3}, \frac{2}{3}$ **21.** 0, 0, 6, 24, 60   **23.** -73 **25.** $\frac{44}{239}$

27. 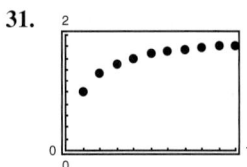   **29.**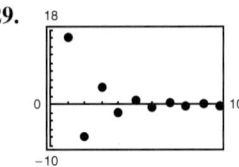

31.

33. c   **34.** b   **35.** d   **36.** a   **37.** $a_n = 3n - 2$
39. $a_n = n^2 - 1$ **41.** $a_n = \dfrac{(-1)^n(n+1)}{n+2}$
43. $a_n = \dfrac{n+1}{2n-1}$ **45.** $a_n = \dfrac{1}{n^2}$ **47.** $a_n = (-1)^{n+1}$
49. $a_n = 1 + \dfrac{1}{n}$ **51.** 28, 24, 20, 16, 12
53. 3, 4, 6, 10, 18   **55.** 6, 8, 10, 12, 14
                   $a_n = 2n + 4$
57. 81, 27, 9, 3, 1   **59.** $1, 3, \dfrac{9}{2}, \dfrac{9}{2}, \dfrac{27}{8}$
     $a_n = \dfrac{243}{3^n}$
61. $1, \dfrac{1}{2}, \dfrac{1}{6}, \dfrac{1}{24}, \dfrac{1}{120}$ **63.** $1, \dfrac{1}{2}, \dfrac{1}{24}, \dfrac{1}{720}, \dfrac{1}{40,320}$
65. $\frac{1}{30}$ **67.** 90   **69.** $n + 1$
71. $\dfrac{1}{2n(2n+1)}$ **73.** 35   **75.** 40   **77.** 30
79. $\frac{9}{5}$ **81.** 88   **83.** 30   **85.** 81   **87.** $\frac{47}{60}$
89. $\displaystyle\sum_{i=1}^{9} \dfrac{1}{3i}$ **91.** $\displaystyle\sum_{i=1}^{8}\left[2\left(\dfrac{i}{8}\right) + 3\right]$ **93.** $\displaystyle\sum_{i=1}^{6}(-1)^{i+1}3i$
95. $\displaystyle\sum_{i=1}^{20} \dfrac{(-1)^{i+1}}{i^2}$ **97.** $\displaystyle\sum_{i=1}^{5} \dfrac{2^i - 1}{2^{i+1}}$ **99.** $\dfrac{75}{16}$ **101.** $-\dfrac{3}{2}$
103. $\frac{2}{3}$ **105.** $\frac{7}{9}$
107. (a) $A_1 = \$5100.00$, $A_2 = \$5202.00$, $A_3 = \$5306.04$,
      $A_4 = \$5412.16$, $A_5 = \$5520.40$, $A_6 = \$5630.81$,
      $A_7 = \$5743.43$, $A_8 = \$5858.30$
  (b) $A_{40} = \$11,040.20$
109. (a) $b_n = 60.57n - 182$
  (b) $c_n = 1.61n^2 + 26.8n - 9.5$
  (c)

n	8	9	10	11	12	13
a_n	311	357	419	481	548	608
b_n	303	363	424	484	545	605
c_n	308	362	420	480	544	611

    The quadratic model is a better fit.
  (d) The quadratic model; 995

111. (a) $a_0 = \$3102.9$, $a_1 = \$3644.3$, $a_2 = \$4079.6$,
$a_3 = \$4425.3$, $a_4 = \$4698.2$, $a_5 = \$4914.8$,
$a_6 = \$5091.8$, $a_7 = \$5245.7$, $a_8 = \$5393.2$,
$a_9 = \$5550.9$, $a_{10} = \$5735.5$, $a_{11} = \$5963.5$,
$a_{12} = \$6251.5$, $a_{13} = \$6616.3$

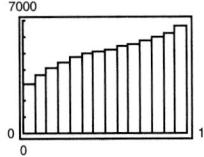

(b) The federal debt is increasing.
113. True by the Properties of Sums
115. $1, 1, 2, 3, 5, 8, 13, 21, 34, 55, 89, 144$
$1, 2, \frac{3}{2}, \frac{5}{3}, \frac{8}{5}, \frac{13}{8}, \frac{21}{13}, \frac{34}{21}, \frac{55}{34}, \frac{89}{55}$
117. $\$500.95$ **119.** Answers will vary.
121. $x, \dfrac{x^2}{2}, \dfrac{x^3}{6}, \dfrac{x^4}{24}, \dfrac{x^5}{120}$

123. $-\dfrac{x^2}{2}, \dfrac{x^4}{24}, -\dfrac{x^6}{720}, \dfrac{x^8}{40{,}320}, -\dfrac{x^{10}}{3{,}628{,}800}$

125. $f^{-1}(x) = \dfrac{x+3}{4}$ **127.** $h^{-1}(x) = \dfrac{x^2-1}{5}, x \geq 0$

129. (a) $\begin{bmatrix} 8 & 1 \\ -2 & 6 \end{bmatrix}$ (b) $\begin{bmatrix} -26 & 1 \\ 12 & -21 \end{bmatrix}$

(c) $\begin{bmatrix} 18 & 9 \\ 10 & 7 \end{bmatrix}$ (d) $\begin{bmatrix} 4 & 2 \\ 24 & 21 \end{bmatrix}$

131. (a) $\begin{bmatrix} -3 & -7 & 4 \\ 4 & 4 & 1 \\ 1 & 4 & 3 \end{bmatrix}$ (b) $\begin{bmatrix} 10 & 25 & -10 \\ -12 & -11 & 3 \\ -3 & -9 & -8 \end{bmatrix}$

(c) $\begin{bmatrix} -2 & 7 & -16 \\ 4 & 42 & 45 \\ 1 & 23 & 48 \end{bmatrix}$ (d) $\begin{bmatrix} 16 & 31 & 42 \\ 10 & 47 & 31 \\ 13 & 22 & 25 \end{bmatrix}$

133. 26 **135.** -194

Section 8.2   *(page 611)*

Vocabulary Check   *(page 611)*

1. arithmetic; common   **2.** $a_n = dn + c$
3. sum of a finite arithmetic sequence

---

1. Arithmetic sequence, $d = -2$
3. Not an arithmetic sequence
5. Arithmetic sequence, $d = -\frac{1}{4}$
7. Not an arithmetic sequence
9. Not an arithmetic sequence
11. $8, 11, 14, 17, 20$
Arithmetic sequence, $d = 3$
13. $7, 3, -1, -5, -9$
Arithmetic sequence, $d = -4$

15. $-1, 1, -1, 1, -1$
Not an arithmetic sequence
17. $-3, \frac{3}{2}, -1, \frac{3}{4}, -\frac{3}{5}$
Not an arithmetic sequence
19. $a_n = 3n - 2$ **21.** $a_n = -8n + 108$
23. $a_n = 2xn - x$ **25.** $a_n = -\frac{5}{2}n + \frac{13}{2}$
27. $a_n = \frac{10}{3}n + \frac{5}{3}$ **29.** $a_n = -3n + 103$
31. $5, 11, 17, 23, 29$ **33.** $-2.6, -3.0, -3.4, -3.8, -4.2$
35. $2, 6, 10, 14, 18$ **37.** $-2, 2, 6, 10, 14$
39. $15, 19, 23, 27, 31$; $d = 4$; $a_n = 4n + 11$
41. $200, 190, 180, 170, 160$; $d = -10$; $a_n = -10n + 210$
43. $\frac{5}{8}, \frac{1}{2}, \frac{3}{8}, \frac{1}{4}, \frac{1}{8}$; $d = -\frac{1}{8}$; $a_n = -\frac{1}{8}n + \frac{3}{4}$
45. 59 **47.** 18.6 **49.** b   **50.** d   **51.** c   **52.** a
53.

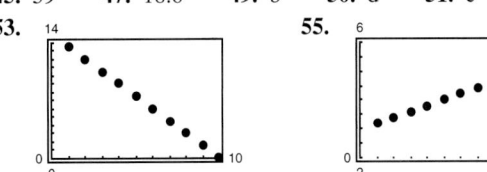

55.

57. 620 **59.** 17.4 **61.** 265 **63.** 4000
65. $10{,}000$ **67.** 1275 **69.** $30{,}030$ **71.** 355
73. $160{,}000$ **75.** 520 **77.** 2725 **79.** $10{,}120$
81. (a) $\$40{,}000$ (b) $\$217{,}500$ **83.** 2340 seats
85. 405 bricks   **87.** 490 meters
89. (a) $a_n = -25n + 225$ (b) $\$900$
91. $\$70{,}500$; answers will vary.
93. (a)

Month	1	2	3	4	5	6
Monthly payment	$220	$218	$216	$214	$212	$210
Unpaid balance	$1800	$1600	$1400	$1200	$1000	$800

(b) $\$110$
95. (a) $a_n = 1098n + 17{,}588$
(b) $a_n = 1114.9n + 17{,}795$; the models are similar.
(c)

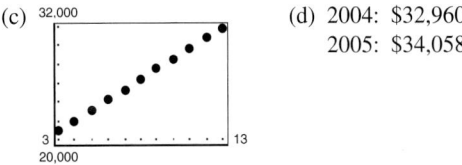

(d) 2004: $\$32{,}960$
2005: $\$34{,}058$

(e) Answers will vary.
97. True. Given a_1 and a_2, $d = a_2 - a_1$ and
$a_n = a_1 + (n-1)d$.
99. Answers will vary.

CHAPTER 8

101. (a) 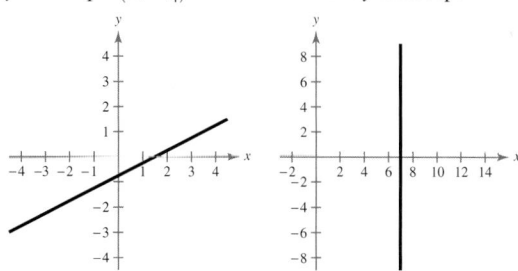 (b)

(c) The graph of $y = 3x + 2$ contains all points on the line. The graph of $a_n = 2 + 3n$ contains only points at the positive integers.

(d) The slope of the line and the common difference of the arithmetic sequence are equal.

103. 4

105. Slope: $\frac{1}{2}$;
y-intercept: $\left(0, -\frac{3}{4}\right)$

107. Slope: undefined;
No y-intercept

109. $x = 1, y = 5, z = -1$ **111.** Answers will vary.

Section 8.3 *(page 621)*

Vocabulary Check *(page 621)*

1. geometric; common **2.** $a_n = a_1 r^{n-1}$

3. $S_n = a_1\left(\dfrac{1 - r^n}{1 - r}\right)$ **4.** geometric series

5. $S = \dfrac{a_1}{1 - r}$

1. Geometric sequence, $r = 3$
3. Not a geometric sequence
5. Geometric sequence, $r = -\frac{1}{2}$
7. Geometric sequence, $r = 2$
9. Not a geometric sequence
11. 2, 6, 18, 54, 162 **13.** $1, \frac{1}{2}, \frac{1}{4}, \frac{1}{8}, \frac{1}{16}$
15. $5, -\frac{1}{2}, \frac{1}{20}, -\frac{1}{200}, \frac{1}{2000}$ **17.** $1, e, e^2, e^3, e^4$
19. $2, \dfrac{x}{2}, \dfrac{x^2}{8}, \dfrac{x^3}{32}, \dfrac{x^4}{128}$
21. 64, 32, 16, 8, 4; $r = \frac{1}{2}$; $a_n = 128\left(\frac{1}{2}\right)^n$

23. 7, 14, 28, 56, 112; $r = 2$; $a_n = \frac{7}{2}(2)^n$
25. $6, -9, \frac{27}{2}, -\frac{81}{4}, \frac{243}{8}$; $r = -\frac{3}{2}$; $a_n = -4\left(-\frac{3}{2}\right)^n$
27. $a_n = 4\left(\dfrac{1}{2}\right)^{n-1}$; $\dfrac{1}{128}$ **29.** $a_n = 6\left(-\dfrac{1}{3}\right)^{n-1}$; $-\dfrac{2}{3^{10}}$
31. $a_n = 100e^{x(n-1)}$; $100e^{8x}$
33. $a_n = 500(1.02)^{n-1}$; ≈ 1082.372 **35.** 45,927
37. 50,388,480 **39.** $a_3 = 9$ **41.** $a_6 = -2$
43. a **44.** c **45.** b **46.** d
47. **49.**
51.
53. 511 **55.** 171 **57.** 43 **59.** $\frac{1365}{32}$
61. 29,921.311 **63.** 592.647 **65.** 2092.596
67. $\frac{8}{5}$ **69.** 6.400 **71.** 3.750
73. $\displaystyle\sum_{n=1}^{7} 5(3)^{n-1}$ **75.** $\displaystyle\sum_{n=1}^{7} 2\left(-\frac{1}{4}\right)^{n-1}$ **77.** $\displaystyle\sum_{n=1}^{6} 0.1(4)^{n-1}$
79. 2 **81.** $\frac{2}{3}$ **83.** $\frac{16}{3}$ **85.** $\frac{5}{3}$ **87.** -30
89. 32 **91.** Undefined **93.** $\frac{4}{11}$ **95.** $\frac{7}{22}$
97.

Horizontal asymptote: $y = 12$
Corresponds to the sum of the series
99. (a) $a_n = 1190.88(1.006)^n$
(b) The population is growing at a rate of 0.6% per year.
(c) 1342.2 million. This value is close to the prediction.
(d) 2007
101. (a) \$3714.87 (b) \$3722.16 (c) \$3725.85
(d) \$3728.32 (e) \$3729.52
103. \$7011.89 **105.** Answers will vary.
107. (a) \$26,198.27 (b) \$26,263.88
109. (a) \$118,590.12 (b) \$118,788.73
111. Answers will vary. **113.** \$1600
115. $\approx$ \$2181.82 **117.** 126 square inches
119. \$3,623,993.23
121. False. A sequence is geometric if the ratios of consecutive terms are the same.
123. Given a real number r between -1 and 1, as the exponent n increases, r^n approaches zero.

125. $x^2 + 2x$ **127.** $3x^2 + 6x + 1$

129. $x(3x + 8)(3x - 8)$ **131.** $(3x + 1)(2x - 5)$

133. $\dfrac{3x}{x - 3},\ x \neq -3$ **135.** $\dfrac{2x + 1}{3},\ x \neq 0, -\dfrac{1}{2}$

137. $\dfrac{5x^2 + 9x - 30}{(x + 2)(x - 2)}$ **139.** Answers will vary.

Section 8.4   *(page 633)*

Vocabulary Check   *(page 633)*

1. mathematical induction   **2.** first

3. arithmetic   **4.** second

---

1. $\dfrac{5}{(k + 1)(k + 2)}$ **3.** $\dfrac{(k + 1)^2(k + 2)^2}{4}$

5–33. Answers will vary.   **35.** $S_n = n(2n - 1)$

37. $S_n = 10 - 10\left(\dfrac{9}{10}\right)^n$ **39.** $S_n = \dfrac{n}{2(n + 1)}$

41. 120   **43.** 91   **45.** 979   **47.** 70   **49.** -3402

51. 0, 3, 6, 9, 12, 15

First differences: 3, 3, 3, 3, 3

Second differences: 0, 0, 0, 0

Linear

53. 3, 1, -2, -6, -11, -17

First differences: -2, -3, -4, -5, -6

Second differences: -1, -1, -1, -1

Quadratic

55. 2, 4, 16, 256, 65,536, 4,294,967,296

First differences: 2, 12, 240, 65,280, 4,294,901,760

Second differences: 10, 228, 65,040, 4,294,836,480

Neither

57. $a_n = n^2 - n + 3$ **59.** $a_n = \dfrac{1}{2}n^2 + n - 3$

61. (a) 2.2, 2.4, 2.2, 2.3, 0.9

(b) A linear model can be used.

$a_n = 2.2n + 102.7$

(c) $a_n = 2.08n + 103.9$

(d) Part b: $a_n = 142.3$; Part c: $a_n = 141.34$

These are very similar.

63. True. P_7 may be false.

65. True. If the second differences are all zero, then the first differences are all the same and the sequence is arithmetic.

67. $4x^4 - 4x^2 + 1$ **69.** $-64x^3 + 240x^2 - 300x + 125$

71. (a) Domain: all real numbers x except $x = -3$

(b) Intercept: $(0, 0)$

(c) Vertical asymptote: $x = -3$

Horizontal asymptote: $y = 1$

(d)

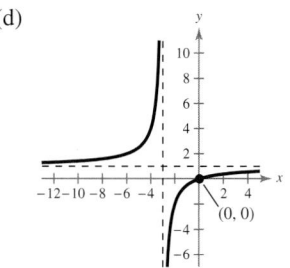

73. (a) Domain: all real numbers t except $t = 0$

(b) t-intercept: $(7, 0)$

(c) Vertical asymptote: $t = 0$

Horizontal asymptote: $y = 1$

(d)

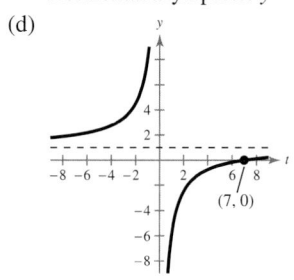

Section 8.5   *(page 640)*

Vocabulary Check   *(page 640)*

1. binomial coefficients

2. Binomial Theorem; Pascal's Triangle

3. $\dbinom{n}{r};\ {}_nC_r$ **4.** expanding a binomial

---

1. 10   **3.** 1   **5.** 15,504   **7.** 210   **9.** 4950

11. 56   **13.** 35   **15.** $x^4 + 4x^3 + 6x^2 + 4x + 1$

17. $a^4 + 24a^3 + 216a^2 + 864a + 1296$

19. $y^3 - 12y^2 + 48y - 64$

21. $x^5 + 5x^4y + 10x^3y^2 + 10x^2y^3 + 5xy^4 + y^5$

23. $r^6 + 18r^5s + 135r^4s^2 + 540r^3s^3 + 1215r^2s^4$
    $+ 1458rs^5 + 729s^6$

25. $243a^5 - 1620a^4b + 4320a^3b^2 - 5760a^2b^3$
    $+ 3840ab^4 - 1024b^5$

27. $8x^3 + 12x^2y + 6xy^2 + y^3$

29. $x^8 + 4x^6y^2 + 6x^4y^4 + 4x^2y^6 + y^8$

31. $\dfrac{1}{x^5} + \dfrac{5y}{x^4} + \dfrac{10y^2}{x^3} + \dfrac{10y^3}{x^2} + \dfrac{5y^4}{x} + y^5$

33. $2x^4 - 24x^3 + 113x^2 - 246x + 207$

35. $32t^5 - 80t^4s + 80t^3s^2 - 40t^2s^3 + 10ts^4 - s^5$

37. $x^5 + 10x^4y + 40x^3y^2 + 80x^2y^3 + 80xy^4 + 32y^5$

39. $120x^7y^3$ **41.** $360x^3y^2$ **43.** $1{,}259{,}712x^2y^7$

45. $32{,}476{,}950{,}000x^4y^8$ **47.** $1{,}732{,}104$

49. 180 **51.** $-326{,}592$ **53.** 210

55. $x^2 + 12x^{3/2} + 54x + 108x^{1/2} + 81$

57. $x^2 - 3x^{4/3}y^{1/3} + 3x^{2/3}y^{2/3} - y$

59. $3x^2 + 3xh + h^2, h \neq 0$ **61.** $\dfrac{1}{\sqrt{x+h} + \sqrt{x}}, h \neq 0$

63. -4 **65.** $2035 + 828i$ **67.** 1

69. 1.172 **71.** $510{,}568.785$

73.

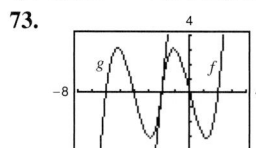

g is shifted four units to the left of f.
$g(x) = x^3 + 12x^2 + 44x + 48$

75. 0.273 **77.** 0.171

79. (a) $f(t) = 0.0025t^3 - 0.015t^2 + 0.88t + 7.7$

(b)

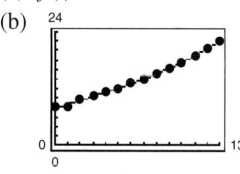

(c) $g(t) = 0.0025t^3 + 0.06t^2 + 1.33t + 17.5$

(d)

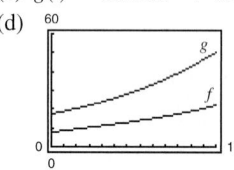

(e) $f(t)$: 33.26 gallons; $g(t)$: 33.26 gallons; yes

(f) The trend is for the per capita consumption of bottled water to increase. This may be due to the increasing concern with contaminants in tap water.

81. True. The coefficients from the Binomial Theorem can be used to find the numbers in Pascal's Triangle.

83. False. The coefficient of the x^{10}-term is $1{,}732{,}104$ and the coefficient of the x^{14}-term is $192{,}456$.

85.
```
      1   8   28   56   70   56   28    8    1
    1   9   36   84  126  126   84   36    9    1
  1  10  45  120  210  252  210  120   45   10    1
```

87. The signs of the terms in the expansion of $(x - y)^n$ alternate between positive and negative.

89–91. Answers will vary.

93.

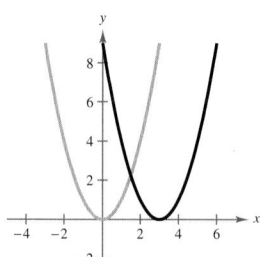

$g(x) = (x - 3)^2$

95.

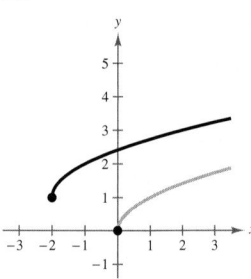

$g(x) = \sqrt{x + 2} + 1$

97. $\begin{bmatrix} 4 & -5 \\ 5 & -6 \end{bmatrix}$

Section 8.6   *(page 650)*

Vocabulary Check   *(page 650)*

1. Fundamental Counting Principle   **2.** permutation

3. $_nP_r = \dfrac{n!}{(n-r)!}$ **4.** distinguishable permutations

5. combinations

---

1. 6   **3.** 5   **5.** 3   **7.** 8   **9.** 30   **11.** 30

13. 64   **15.** 175,760,000

17. (a) 900   (b) 648   (c) 180   (d) 600

19. 64,000   **21.** (a) 40,320   (b) 384   **23.** 24

25. 336   **27.** 120   **29.** $n = 5$ or $n = 6$

31. 1,860,480   **33.** 970,200   **35.** 15,504   **37.** 120

39. 11,880   **41.** 420   **43.** 2520

45. ABCD, ABDC, ACBD, ACDB, ADBC, ADCB, BACD, BADC, CABD, CADB, DABC, DACB, BCAD, BDAC, CBAD, CDAB, DBAC, DCAB, BCDA, BDCA, CBDA, CDBA, DBCA, DCBA

47. 1,816,214,400   **49.** 5,586,853,480

51. AB, AC, AD, AE, AF, BC, BD, BE, BF, CD, CE, CF, DE, DF, EF

53. 324,632   **55.** (a) 35   (b) 63   (c) 203

57. (a) 3744   (b) 24   **59.** 292,600

61. 5   **63.** 20

65. (a) 146,107,962

(b) If the jackpot is won, there is only one winning number.

(c) There are 28,989,675 possible winning numbers in the state lottery, which is considerably less than the possible number of winning Powerball numbers.

67. False. It is an example of a combination.

69. They are equal.

71–73. Proof

75. No. For some calculators the number is too great.

77. (a) 35   (b) 8   (c) 83

79. (a) -4 (b) 0   (c) 0   **81.** 8.30   **83.** 35

Section 8.7 *(page 661)*

1. $\{(H, 1), (H, 2), (H, 3), (H, 4), (H, 5), (H, 6),$
$(T, 1), (T, 2), (T, 3), (T, 4), (T, 5), (T, 6)\}$

3. $\{ABC, ACB, BAC, BCA, CAB, CBA\}$

5. $\{AB, AC, AD, AE, BC, BD, BE, CD, CE, DE\}$

7. $\frac{3}{8}$ **9.** $\frac{7}{8}$ **11.** $\frac{3}{13}$ **13.** $\frac{3}{26}$ **15.** $\frac{1}{12}$ **17.** $\frac{11}{12}$

19. $\frac{1}{3}$ **21.** $\frac{1}{5}$ **23.** $\frac{2}{5}$ **25.** 0.3 **27.** $\frac{3}{4}$ **29.** 0.86

31. $\frac{18}{35}$ **33.** (a) 58% (b) 95.6% (c) 0.4%

35. (a) 243 (b) $\frac{1}{50}$ (c) $\frac{16}{25}$

37. (a) $\frac{112}{209}$ (b) $\frac{97}{209}$ (c) $\frac{274}{627}$

39. $P(\{\text{Taylor wins}\}) = \frac{1}{2}$
$P(\{\text{Moore wins}\}) = P(\{\text{Jenkins wins}\}) = \frac{1}{4}$

41. (a) $\frac{21}{1292}$ (b) $\frac{225}{646}$ (c) $\frac{49}{323}$

43. (a) $\frac{1}{120}$ (b) $\frac{1}{24}$ **45.** (a) $\frac{5}{13}$ (b) $\frac{1}{2}$ (c) $\frac{4}{13}$

47. (a) $\frac{14}{55}$ (b) $\frac{12}{55}$ (c) $\frac{54}{55}$ **49.** 0.4746

51. (a) 0.9702 (b) 0.9998 (c) 0.0002

53. (a) $\frac{1}{16}$ (b) $\frac{1}{8}$ (c) $\frac{15}{16}$

55. (a) $\frac{1}{38}$ (b) $\frac{9}{19}$ (c) $\frac{10}{19}$ (d) $\frac{1}{1444}$ (e) $\frac{729}{6859}$

(f) The probabilities are slightly better in European roulette.

57. True. Two events are independent if the occurrence of one has no effect on the occurrence of the other.

59. (a) As you consider successive people with distinct birthdays, the probabilities must decrease to take into account the birth dates already used. Because the birth dates of people are independent events, multiply the respective probabilities of distinct birthdays.

(b) $\frac{365}{365} \cdot \frac{364}{365} \cdot \frac{363}{365} \cdot \frac{362}{365}$ (c) Answers will vary.

(d) Q_n is the probability that the birthdays are *not* distinct, which is equivalent to at least two people having the same birthday.

(e)

n	10	15	20	23	30	40	50
P_n	0.88	0.75	0.59	0.49	0.29	0.11	0.03
Q_n	0.12	0.25	0.41	0.51	0.71	0.89	0.97

(f) 23

61. No real solution **63.** $0, \dfrac{1 \pm \sqrt{13}}{2}$ **65.** -4

67. $\frac{11}{2}$ **69.** -10

71.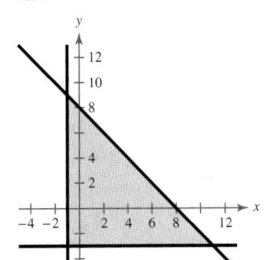

73.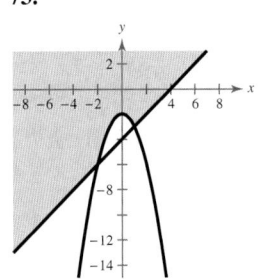

Review Exercises *(page 667)*

1. $8, 5, 4, \frac{7}{2}, \frac{16}{5}$ **3.** $72, 36, 12, 3, \frac{3}{5}$ **5.** $a_n = 2(-1)^n$

7. $a_n = \dfrac{4}{n}$ **9.** 120 **11.** 1 **13.** 30 **15.** $\dfrac{205}{24}$

17. 6050 **19.** $\displaystyle\sum_{k=1}^{20} \frac{1}{2k}$ **21.** $\dfrac{5}{9}$ **23.** $\dfrac{2}{99}$

25. (a) $A_1 = \$10,067, A_2 = \$10,134, A_3 = \$10,201,$
$A_4 = \$10,269, A_5 = \$10,338, A_6 = \$10,407,$
$A_7 = \$10,476, A_8 = \$10,546, A_9 = \$10,616,$
$A_{10} = \$10,687$
(b) $A_{120} = \$22,196.40$

27. Arithmetic sequence, $d = -2$

29. Arithmetic sequence, $d = \frac{1}{2}$ **31.** 4, 7, 10, 13, 16

33. 25, 28, 31, 34, 37 **35.** $a_n = 12n - 5$

37. $a_n = 3ny - 2y$ **39.** $a_n = -7n + 107$

41. 80 **43.** 88 **45.** 25,250

47. (a) \$43,000 (b) \$192,500

49. Geometric sequence, $r = 2$

51. Geometric sequence, $r = -2$ **53.** $4, -1, \frac{1}{4}, -\frac{1}{16}, \frac{1}{64}$

55. $9, 6, 4, \frac{8}{3}, \frac{16}{9}$ or $9, -6, 4, -\frac{8}{3}, \frac{16}{9}$

57. $a_n = 16\left(-\frac{1}{2}\right)^{n-1}$; $\approx -3.052 \times 10^{-5}$

59. $a_n = 100(1.05)^{n-1}$; ≈ 252.695

61. 127 **63.** $\frac{15}{16}$ **65.** 31 **67.** 24.85

69. 5486.45 **71.** 8 **73.** $\frac{10}{9}$ **75.** 12

77. (a) $a_t = 120,000(0.7)^t$ (b) \$20,168.40

79–81. Answers will vary. **83.** $S_n = n(2n + 7)$

85. $S_n = \frac{5}{2}\left[1 - \left(\frac{3}{5}\right)^n\right]$ **87.** 465 **89.** 4648

91. 5, 10, 15, 20, 25
First differences: 5, 5, 5, 5
Second differences: 0, 0, 0
Linear

93. 16, 15, 14, 13, 12
First differences: $-1, -1, -1, -1$
Second differences: 0, 0, 0
Linear

95. 15 **97.** 56 **99.** 35 **101.** 28

103. $x^4 + 16x^3 + 96x^2 + 256x + 256$

105. $a^5 - 15a^4b + 90a^3b^2 - 270a^2b^3 + 405ab^4 - 243b^5$

107. $41 + 840i$ **109.** 11 **111.** 10,000 **113.** 720

115. 56 **117.** $\frac{1}{9}$ **119.** (a) 43% (b) 82%

CHAPTER 8

121. $\frac{1}{216}$ **123.** $\frac{3}{4}$

125. True. $\dfrac{(n + 2)!}{n!} = \dfrac{(n + 2)(n + 1)n!}{n!} = (n + 2)(n + 1)$

127. True by Properties of Sums

129. False. When r equals 0 or 1, then the results are the same.

131. In the sequence in part (a), the odd-numbered terms are negative, whereas in the sequence in part (b), the even-numbered terms are negative.

133. Each term of the sequence is defined in terms of preceding terms.

135. d   **136.** a   **137.** b   **138.** c

139. 240, 440, 810, 1490, 2740

Chapter Test   (page 671)

1. $-\dfrac{1}{5}, \dfrac{1}{8}, -\dfrac{1}{11}, \dfrac{1}{14}, -\dfrac{1}{17}$ **2.** $a_n = \dfrac{n + 2}{n!}$

3. 50, 61, 72; 140   **4.** $a_n = 0.8n + 1.4$

5. 5, 10, 20, 40, 80   **6.** 86,100   **7.** 189

8. 4   **9.** Answers will vary.

10. $x^4 + 8x^3y + 24x^2y^2 + 32xy^3 + 16y^4$ **11.** $-108{,}864$

12. (a) 72   (b) 328,440   **13.** (a) 330   (b) 720,720

14. 26,000   **15.** 720   **16.** $\frac{1}{15}$ **17.** 3.908×10^{-10}

18. 25%

Cumulative Test for Chapters 6–8   (page 672)

1. $(1, 2), \left(-\frac{3}{2}, \frac{3}{4}\right)$ **2.** $(2, -1)$

3. $(4, 2, -3)$ **4.** $(1, -2, 1)$

5.

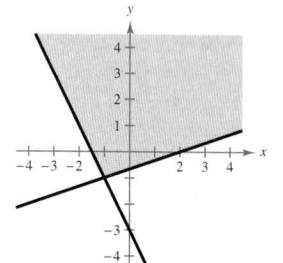

6.

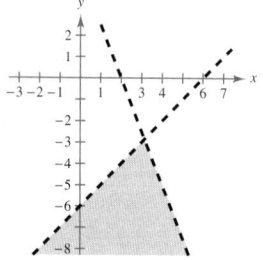

7.

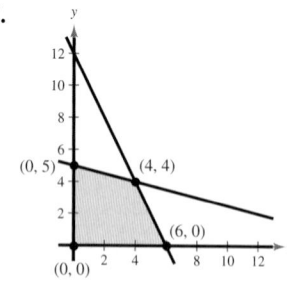

Maximum at $(4, 4)$: $z = 20$

Minimum at $(0, 0)$: $z = 0$

8. $0.75 mixture: 120 pounds; $1.25 mixture: 80 pounds

9. $y = \frac{1}{3}x^2 - 2x + 4$

10. $\begin{bmatrix} -1 & 2 & -1 & \vdots & 9 \\ 2 & -1 & 2 & \vdots & -9 \\ 3 & 3 & -4 & \vdots & 7 \end{bmatrix}$ **11.** $(-2, 3, -1)$

12. $\begin{bmatrix} 3 & 3 \\ 0 & 2 \end{bmatrix}$ **13.** $\begin{bmatrix} 2 & -6 \\ -2 & 0 \end{bmatrix}$ **14.** $\begin{bmatrix} 6 & -6 \\ -3 & 2 \end{bmatrix}$

15. $\begin{bmatrix} -4 & 12 \\ 3 & -3 \end{bmatrix}$ **16.** 84   **17.** $\begin{bmatrix} -175 & 37 & -13 \\ 95 & -20 & 7 \\ 14 & -3 & 1 \end{bmatrix}$

18. Gym shoes: $198.36 million
Jogging shoes: $358.48 million
Walking shoes: $167.17 million

19. $(-5, 4)$ **20.** $(-3, 4, 2)$ **21.** 9

22. $\dfrac{1}{5}, -\dfrac{1}{7}, \dfrac{1}{9}, -\dfrac{1}{11}, \dfrac{1}{13}$ **23.** $a_n = \dfrac{(n + 1)!}{n + 3}$

24. 920   **25.** (a) 65.4   (b) $a_n = 3.2n + 1.4$

26. 3, 6, 12, 24, 48   **27.** $\frac{13}{9}$ **28.** Answers will vary.

29. $z^4 - 12z^3 + 54z^2 - 108z + 81$ **30.** 210   **31.** 600

32. 70   **33.** 120   **34.** 453,600   **35.** 151,200

36. 720   **37.** $\frac{1}{4}$

Problem Solving   (page 677)

1. 1, 1.5, 1.41$\overline{6}$, 1.414215686,
1.414213562, 1.414213562, . . .
x_n approaches $\sqrt{2}$.

3. (a)

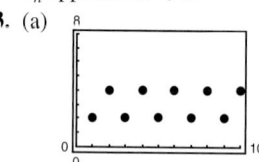

(b) If n is odd, $a_n = 2$, and if n is even, $a_n = 4$.

(c)

n	1	10	101	1000	10,001
a_n	2	4	2	4	2

(d) It is not possible to find the value of a_n as n approaches infinity.

5. (a) 3, 5, 7, 9, 11, 13, 15, 17; $a_n = 2n + 1$

(b) To obtain the arithmetic sequence, find the differences of consecutive terms of the sequence of perfect cubes. Then find the differences of consecutive terms of this sequence.

(c) 12, 18, 24, 30, 36, 42, 48; $a_n = 6n + 6$

(d) To obtain the arithmetic sequence, find the third sequence obtained by taking differences of consecutive terms in consecutive sequences.

(e) 60, 84, 108, 132, 156, 180; $a_n = 24n + 36$

7. $s_n = \left(\dfrac{1}{2}\right)^{n-1}$

$a_n = \dfrac{\sqrt{3}}{4} s_n^{\,2}$

9–11. (a) Answers will vary.   (b) 17,710

13. $\frac{1}{3}$ **15.** (a) $-$0.71   (b) 2.53, 24 turns

Index

FORMULAS FROM GEOMETRY

Triangle:

$h = a \sin \theta$

$\text{Area} = \dfrac{1}{2}bh$

$c^2 = a^2 + b^2 - 2ab \cos \theta$ (Law of Cosines)

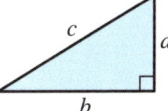

Right Triangle:

Pythagorean Theorem
$c^2 = a^2 + b^2$

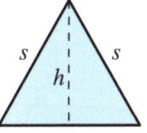

Equilateral Triangle:

$h = \dfrac{\sqrt{3}s}{2}$

$\text{Area} = \dfrac{\sqrt{3}s^2}{4}$

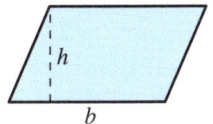

Parallelogram:

$\text{Area} = bh$

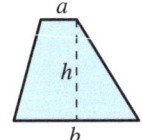

Trapezoid:

$\text{Area} = \dfrac{h}{2}(a + b)$

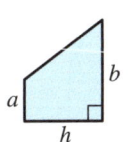

Circle:

$\text{Area} = \pi r^2$

$\text{Circumference} = 2\pi r$

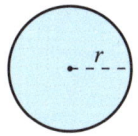

Sector of Circle:

$\text{Area} = \dfrac{\theta r^2}{2}$

$s = r\theta$

θ in radians

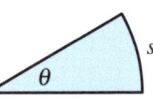

Circular Ring:

$\text{Area} = \pi(R^2 - r^2)$

$\quad\quad\quad = 2\pi pw$

p = average radius,

w = width of ring

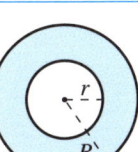
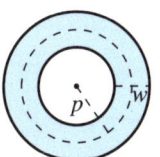

Sector of Circular Ring:

$\text{Area} = \theta pw$

p = average radius,

w = width of ring,

θ in radians

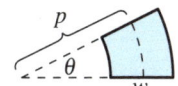

Ellipse:

$\text{Area} = \pi ab$

$\text{Circumference} \approx 2\pi\sqrt{\dfrac{a^2 + b^2}{2}}$

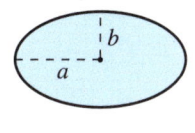

Cone:

$\text{Volume} = \dfrac{Ah}{3}$

A = area of base

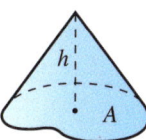

Right Circular Cone:

$\text{Volume} = \dfrac{\pi r^2 h}{3}$

$\text{Lateral Surface Area} = \pi r \sqrt{r^2 + h^2}$

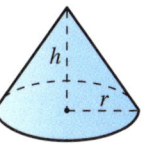

Frustum of Right Circular Cone:

$\text{Volume} = \dfrac{\pi(r^2 + rR + R^2)h}{3}$

$\text{Lateral Surface Area} = \pi s(R + r)$

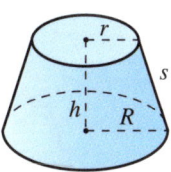

Right Circular Cylinder:

$\text{Volume} = \pi r^2 h$

$\text{Lateral Surface Area} = 2\pi rh$

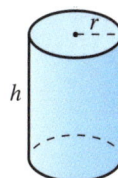

Sphere:

$\text{Volume} = \dfrac{4}{3}\pi r^3$

$\text{Surface Area} = 4\pi r^2$

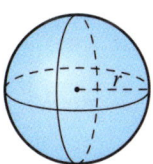

Wedge:

$A = B \sec \theta$

A = area of upper face,

B = area of base

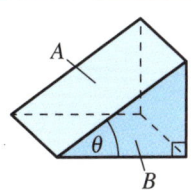

ALGEBRA

Factors and Zeros of Polynomials:

Given the polynomial $p(x) = a_n x^n + a_{n-1}x^{n-1} + \cdots + a_1 x + a_0$. If $p(b) = 0$, then b is a *zero* of the polynomial and a solution of the equation $p(x) = 0$. Furthermore, $(x - b)$ is a *factor* of the polynomial.

Fundamental Theorem of Algebra:

An *n*th degree polynomial has n (not necessarily distinct) zeros.

Quadratic Formula:

If $p(x) = ax^2 + bx + c, a \neq 0$ and $b^2 - 4ac \geq 0$, then the real zeros of p are $x = \left(-b \pm \sqrt{b^2 - 4ac}\right)/2a$.

Example

If $p(x) = x^2 + 3x - 1$, then $p(x) = 0$ if
$$x = \frac{-3 \pm \sqrt{13}}{2}.$$

Special Factors:

$x^2 - a^2 = (x - a)(x + a)$

$x^3 - a^3 = (x - a)(x^2 + ax + a^2)$

$x^3 + a^3 = (x + a)(x^2 - ax + a^2)$

$x^4 - a^4 = (x - a)(x + a)(x^2 + a^2)$

$x^4 + a^4 = \left(x^2 + \sqrt{2}ax + a^2\right)\left(x^2 - \sqrt{2}ax + a^2\right)$

$x^n - a^n = (x - a)(x^{n-1} + ax^{n-2} + \cdots + a^{n-1})$, for n odd

$x^n + a^n = (x + a)(x^{n-1} - ax^{n-2} + \cdots + a^{n-1})$, for n odd

$x^{2n} - a^{2n} = (x^n - a^n)(x^n + a^n)$

Examples

$x^2 - 9 = (x - 3)(x + 3)$

$x^3 - 8 = (x - 2)(x^2 + 2x + 4)$

$x^3 + 4 = \left(x + \sqrt[3]{4}\right)\left(x^2 - \sqrt[3]{4}x + \sqrt[3]{16}\right)$

$x^4 - 4 = \left(x - \sqrt{2}\right)\left(x + \sqrt{2}\right)(x^2 + 2)$

$x^4 + 4 = (x^2 + 2x + 2)(x^2 - 2x + 2)$

$x^5 - 1 = (x - 1)(x^4 + x^3 + x^2 + x + 1)$

$x^7 + 1 = (x + 1)(x^6 - x^5 + x^4 - x^3 + x^2 - x + 1)$

$x^6 - 1 = (x^3 - 1)(x^3 + 1)$

Binomial Theorem:

$(x + a)^2 = x^2 + 2ax + a^2$

$(x - a)^2 = x^2 - 2ax + a^2$

$(x + a)^3 = x^3 + 3ax^2 + 3a^2 x + a^3$

$(x - a)^3 = x^3 - 3ax^2 + 3a^2 x - a^3$

$(x + a)^4 = x^4 + 4ax^3 + 6a^2 x^2 + 4a^3 + a^4$

$(x - a)^4 = x^4 - 4ax^3 + 6a^2 x^2 - 4a^3 x + a^4$

$(x + a)^n = x^n + nax^{n-1} + \dfrac{n(n - 1)}{2!}a^2 x^{n-2} + \cdots + na^{n-1}x + a^n$

$(x - a)^n = x^n - nax^{n-1} + \dfrac{n(n - 1)}{2!}a^2 x^{n-2} - \cdots \pm na^{n-1}x \mp a^n$

Examples

$(x + 3)^2 = x^2 + 6x + 9$

$(x^2 - 5)^2 = x^4 - 10x^2 + 25$

$(x + 2)^3 = x^3 + 6x^2 + 12x + 8$

$(x - 1)^3 = x^3 - 3x^2 + 3x - 1$

$\left(x + \sqrt{2}\right)^4 = x^4 + 4\sqrt{2}x^3 + 12x^2 + 8\sqrt{2}x + 4$

$(x - 4)^4 = x^4 - 16x^3 + 96x^2 - 256x + 256$

$(x + 1)^5 = x^5 + 5x^4 + 10x^3 + 10x^2 + 5x + 1$

$(x - 1)^6 = x^6 - 6x^5 + 15x^4 - 20x^3 + 15x^2 - 6x + 1$

Rational Zero Test:

If $p(x) = a_n x^n + a_{n-1}x^{n-1} + \cdots + a_1 x + a_0$ has integer coefficients, then every *rational* zero of $p(x) = 0$ is of the form $x = r/s$, where r is a factor of a_0 and s is a factor of a_n.

Example

If $p(x) = 2x^4 - 7x^3 + 5x^2 - 7x + 3$, then the only possible *rational* zeros are $x = \pm 1, \pm\frac{1}{2}, \pm 3$, and $\pm\frac{3}{2}$. By testing, you find the two rational zeros to be $\frac{1}{2}$ and 3.

Factoring by Grouping:

$acx^3 + adx^2 + bcx + bd = ax^2(cx + d) + b(cx + d)$
$$= (ax^2 + b)(cx + d)$$

Example

$3x^3 - 2x^2 - 6x + 4 = x^2(3x - 2) - 2(3x - 2)$
$$= (x^2 - 2)(3x - 2)$$